Howard Lidz Architect
115 S. 22nd Street
Philadelphia, PA 19103

ITEM LOCATION

U.C.I. Division	Page
1.1 Overhead	1
2.1 Site Clearing and Exploration	16
2.2 Building Demolition	20
2.3 Earthwork	26
2.4 Caissons & Piling	35
2.5 Site Drainage & Utilities	39
3.1 Formwork & Expansion Joints	59
3.2 Reinforcing Steel	69
3.3 Cast in Place Concrete	72
4.1 Mortar & Accessories	82
4.2 Brick Masonry	84
5.1 Structural Metal	98
5.2 Metal Joists & Decking	104
6.1 Rough Carpentry	117
6.2 Finish Carpentry	124
7.1 Waterproofing	135
7.2 Insulation	137
7.3 Shingles	141
8.1 Metal Doors	158
8.2 Wood & Plastic Doors	161
8.3 Special Doors	167
8.4 Entrances & Storefronts	171
8.5 Metal Windows	172
9.1 Lath & Plaster	185
9.2 Drywall	190
9.3 Tile & Terrazzo	193

U.C.I. Division	Page
1.5 Contractor, Equipment	9
2.6 Roads and Walks	47
2.7 Site Improvements	51
2.8 Lawns & Planting	56
2.9 Heavy Construction	58
3.4 Precast Concrete	80
3.5 Cementitious Decks	81
4.3 Block & Tile Masonry	89
4.4 Stonework	95
5.4 Misc. & Ornamental Metal	106
5.8 Expansion Control & Fasteners	112
6.3 Laminated Construction	132
6.4 Architectural Woodwork	133
7.4 Roofing & Siding	143
7.6 Sheet Metal Work	150
7.8 Roof Accessories	156
8.6 Wood Windows	174
8.7 Finish Hardware & Specialties	176
8.8 Glass & Glazing	182
8.9 Window/Curtain Walls	184
9.5 Acoustical Treatment	196
9.6 Flooring	198
9.8 Painting & Wall Covering	203
10.1 Specialties	207
11.1 Architectural Equipment	221
12.1 Furnishings	234
13.1 Special Construction	239
14.1 Conveying Systems	250

U.C.I. Division	Page
15.1 Pipe & Fittings	253
15.2 Plumbing Fixtures	266
15.3 Plumbing Appliances	272
15.4 Fire Extinguishing Systems	273
16.0 Raceways	298
16.1 Conductors & Grounding	301
16.2 Boxes & Wiring Devices	303
16.3 Starters, Boards & Switches	305
16.4 Transformers & Bus Duct	309

U.C.I. Division	Page
15.5 Heating	276
15.6 HVAC Piping Specialties	286
15.7 Air Conditioning & Ventilating	288
16.5 Power Systems & Capacitors	311
16.6 Lighting	312
16.7 Lighting Utilities	315
16.8 Special Systems	316
17.1 Square Foot, Cubic Foot and Percentage of Total Costs	320
18.1 Listed Alphabetically	329
19.1 Listed Alphabetically by State	333

1. GENERAL REQUIREMENTS
2. SITE WORK
3. CONCRETE
4. MASONRY
5. METALS
6. WOOD & PLASTICS
7. MOISTURE-THERMAL CONTROL
8. DOORS, WINDOWS & GLASS
9. FINISHES
10. SPECIALTIES
11. EQUIPMENT
12. FURNISHINGS
13. SPECIAL CONSTRUCTION
14. CONVEYING SYSTEMS
15. MECHANICAL
16. ELECTRICAL
17. SQUARE FOOT
18. REPAIR & REMODELING
19. CITY COST INDEXES

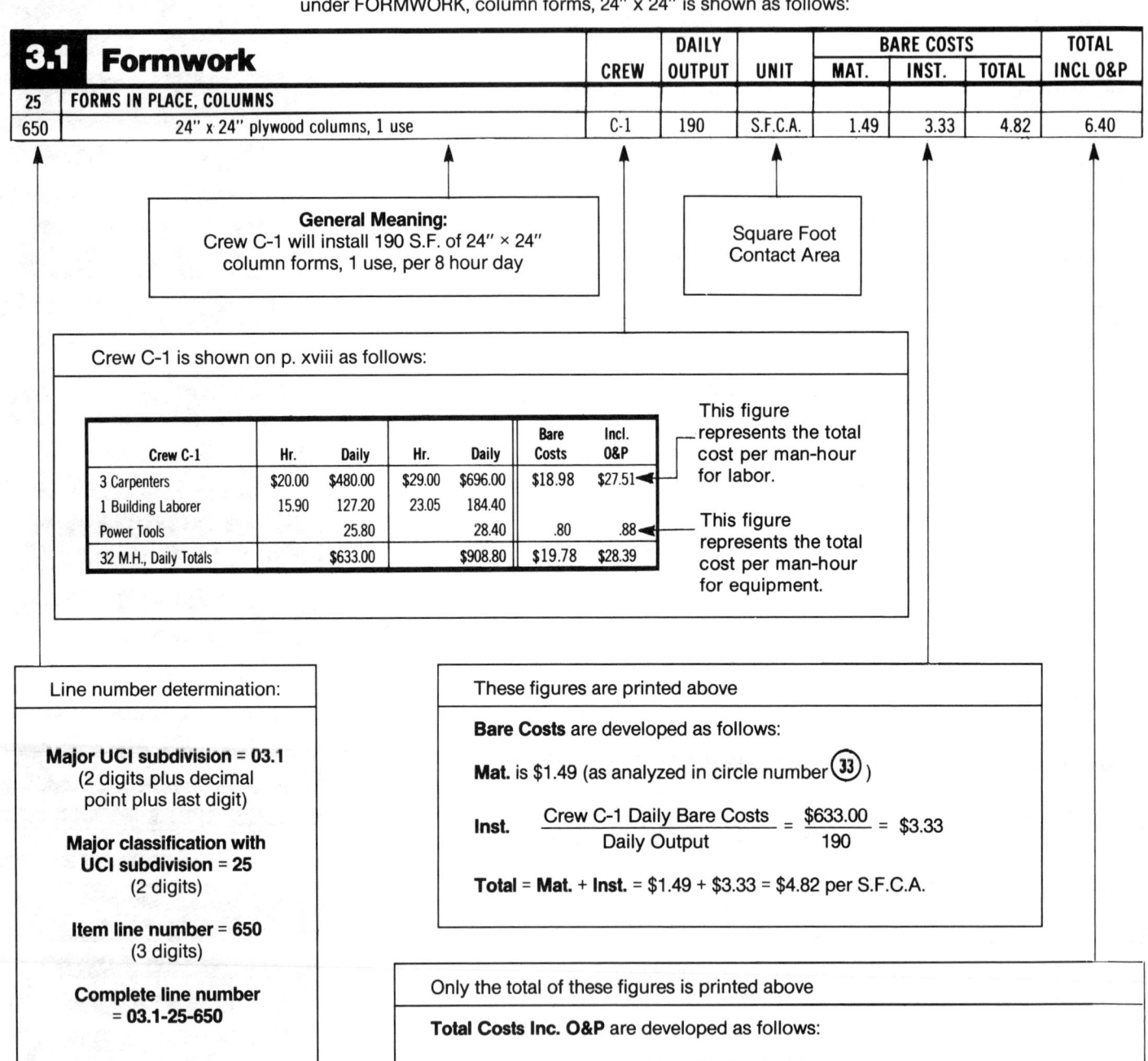

FOREWORD

OBJECT OF THE BOOK

The object of this book is to provide annually updated average unit prices on a wide variety of building construction items, compiled to assist construction industry professionals in the preparation of engineering estimates. Presently there are about 20,000 unit prices included in the book.

PRICE VARIATIONS

The prices quoted are usually averaged for the major cities in the United States, although in some cases they are obtained from a definite locality. Data accumulated from actual job costs in 1985 and material dealer's 1986 quotations have been combined with negotiated 1986 labor rates to revise previous figures. Hundreds of contractors, subcontractors and manufacturers have furnished cost information on their products.

Variations in wage rates, labor efficiency, union restrictions and material prices will result in local fluctuations. The relative material, labor and total construction cost index listed by major division for 162 key U.S. and Canadian cities is shown in Division 19. Indexes for Canadian cities reflect Canadian dollars. Shown in Division 19 is the historic trend of construction costs from Jan. 1, 1986 back to 1942.

Local, regional or national shortages of construction material can severely influence material costs as well as cause considerable job delays with a corresponding increase in indirect job costs.

The specific cost figures when totaled and multiplied by the appropriate city cost index should give a total cost for an entire building that will be close to the low bid and inside the bid range. That is, total cost derived from prices in this book will seldom be below the low bidder or above the average of all bidders.

Sales tax is not included in material prices. See ⓢ, page 348 for amount of sales tax by states.

The square foot and cubic foot cost range of 59 building types are shown in Division 17. See pages 413 and 414 for a more detailed description.

No guaranty or warranty is made by R.S. Means Co., Inc. or the editors as to the correctness or sufficiency of the information contained in this book. R.S. Means Co., Inc. or the editors assume no responsibility or liability in connnection with use of this book.

FACTORS AFFECTING COSTS

Quality The prices in this book for materials and workmanship are in line with U.S. Government specifications and represent good sound construction.

Overtime No allowance has been made for overtime.

Productivity The daily output shown in the book is based on an eight hour day in daylight hours.

Size of Job The book is aimed primarily at industrial and commercial buildings costing $500,000 and up or large multi-family or custom single-family housing projects. The costs are also for new construction of complete buildings rather than repairs or minor alterations. Material prices given are usually large quantity purchases, but quite a few less than carload lot prices are included. WITH REASONABLE EXERCISE OF JUDGMENT THE FIGURES CAN BE USED FOR ANY BUILDING WORK, but do not apply to civil engineering structures such as bridges, dams, highways or the like.

Repair and Remodeling Costs and information are listed in Division 18.

Location Material prices are for metropolitan areas. Beyond a 20 mile radius of large cities, extra trucking or other transportation charges will increase the material costs slightly. This may be offset by lower wage rates, but these two factors should be given consideration in preparing an estimate, especially if the job site is remote. Highly specialized subcontract items may require high travel and per diem expenses for mechanics.

Other Factors which affect costs are weather, season of year, contractor management, local union restrictions, building code requirements, availability of adequate energy, skilled labor force and building materials as well as general business conditions which influence the "in-place" costs of all building items. Substitute materials and construction methods may have to be employed that may increase installed cost and/or life cycle costs above those anticipated. These factors, difficult to evaluate, are not dependent on the section of the country where the job is located, and leave a wider margin of error that is unavoidable and increasingly important.

MATERIAL COSTS

Costs of material in line items are developed by contacting manufacturers, dealers, distributors, and contractors throughout the United States. These costs are adjusted to reflect the 30 city average, January, 1986, quoted prices.

The material and equipment costs shown in this book are average bare costs. When these costs are part of the "Total Incl. O & P", they have been increased by 10% to provide for handling costs incurred by the installing contractor.

If more current costs of material are available at a particular location, adjustments can be made to the line items to reflect any difference in cost. Costs can be adjusted for a particular city by applying the City Cost Index from Division 19.

Many of these material costs are developed under ⓞⓞ numbers in the back of the book. These circle reference numbers are indexed to the left of the line items. By reference to the circle numbers, the estimator can determine what materials have been considered in a particular item.

CONSTRUCTION EQUIPMENT COSTS

The power equipment description and cost is included in each crew tabulation that utilizes power equipment. The daily equipment costs are based on dividing the weekly bare rental rate by 5 (number of working days per week), plus the hourly operating cost times 8 (hours per day). Division 1.5 contains a listing of power equipment and accessories normally encountered in building construction projects. The cost figures listed in the last column to the right in Division 1.5, under the heading "Crew Equipment Cost", is the daily equipment bare cost included in the crew tabulations. To derive the daily equipment cost, including the subcontractor's overhead and profit, 10% is added to the bare costs derived above. In addition, the total equipment cost for each crew is divided by the total daily man-hours of the crew to determine the equipment cost per man-hour for the crew. This hourly cost is listed both as bare and with O & P to simplify estimating based solely on man-hours.

LABOR COSTS

Labor rates used are averaged from 30 major U.S. cities and rounded to the nearest 5¢. Figures used are the latest available from Trade Union Agreements negotiated by the unions and the employers in the various major cities. The labor rates used are those in effect on January 1, 1986. Labor costs should be factored as the new wage agreements for 1986 are negotiated. There are several reliable national indexes available which will give precise indications as to the percentage change in labor rates as the year advances. ("Means Construction Cost Indexes" is one service available.)

Labor rates are listed two ways: **Base rate** which includes fringe benefits but does not include insurance, taxes or contractors' overhead and profit; **Billing rate** which includes fringe benefits, insurance, taxes and contractors' overhead and profit. Both rates for the various trades are listed on the inside back cover. Detailed breakdown of how the billing rate was figured is shown on page

344. Labor costs can be factored to correspond with local rates (see Division 19 for labor factor for 162 key cities listed by major construction division). Relative worker productivity must also be considered.

To calculate man-hours for the various line items, divide the total daily crew hours by the daily output listed for the line item.

On large jobs built outside the metropolitan area or in smaller cities, skilled labor is generally imported and wages tend toward larger city averages regardless of local rates.

It may be noticed that in some cases labor unit costs over the last several years have sometimes held steady or even decreased in spite of the increase in hourly wages. Where this has happened, it is a result of new mechanical equipment being used in the field or of new methods of handling materials. Factory prefabrication of components and many other innovations have reduced both helpers and skilled mechanics time. Severe competition between subcontractors may tend to keep bid prices from rising as rapidly as the labor rates.

HOW THE BOOK IS ARRANGED

NUMBERING All cost information has been divided into the 16 Uniform Construction Index divisions plus a S. F. and C. F. cost division, a Repair and Remodeling division, and a City Cost Index division. These divisions are patterned after the Uniform System adopted by the American Institute of Architects, Associated General Contractors of America, Inc. and The Construction Specifications Institute, Inc. This system is widely used by most segments of the building construction industry. These major divisions are easily located in this book by using the flip tabs on each page. The major divisions are then divided into their major UCI subdivisions shown in boldface type at the top of each page along with the major UCI subdivision number. The individual line items making up the UCI subdivision are then listed alphabetically. Sub-items and other sizes are indented below each descriptive item. An outline of the major components and their numbering system is shown on the first page of the book.

Each unit price has been assigned a three-segment code, the last three numbers of which are listed in the left column of each page. A graphic explanation of the numbering system is shown on page ii of the Foreword.

DESCRIPTIONS Line items are listed with their unit price information to the right. Sub-items and additional sizes are indented beneath appropriate line items. Often the first line or two of the bold face item will have descriptive matter that will pertain to all line items following the bold face listing.

CREW To the right of the item described is the typical crew needed to install the item. Typical crews are tabulated on p. viii to xxiv of the Foreword. When an installation is done by one trade and requires no power equipment, the appropriate trade is listed directly in the crew column. For instance "2 Carp" would indicate that the installation is done with 2 Carpenters. If the installing crew was C-2 (shown on p. xviii of the Foreword) this crew would be made up of 1 Carpenter Foreman, 4 Carpenters, 1 Building Laborer plus an allowance for power tools. The crew costs are listed both with bare labor rates, and with billing rates as shown on p. 344 and the inside back cover. For each type of cost (bare and including Sub's O & P), the total cost per eight-hour day and the composite hourly cost for the crew are listed.

Local labor rates may be substituted to arrive at precise local daily crew costs. Abbreviations for the individual trades are shown on the inside back cover.

EQUIPMENT The power equipment required for each crew is included in the particular crew cost. Refer to the earlier paragraph, entitled "Construction Equipment", for details on the power equipment required by each crew.

DAILY OUTPUT To the right of the crew column is the Daily Output column. This is the number of units that a given crew will install in an 8-hour day. In other words, the total daily crew cost divided by the Daily Output will produce the unit installation costs. To calculate the unit equipment cost, divide the daily equipment costs by the Daily Output.

To calculate unit man-hours, divide the total daily crew hours by the Daily Output.

UNIT To the right of the Daily Output column is the Unit column which describes the unit upon which the price, production and crew are based, as S.F. (square foot), etc. See p. 415 to 417 for a list of abbreviations.

CONSTRUCTION COSTS The next 3 columns contain costs that do not include the Subcontractor's Overhead and Profit. The first column under the BARE COST heading is MAT. which is the unit material cost for the item (without Subs O & P). The next column is the INST. figures. This column is the installation cost which includes labor and equipment. As mentioned above this cost is calculated by dividing the daily crew or labor cost (using bare costs in both cases) by the Daily Output. The last column under the BARE COST heading is TOTAL which is the arithmetical sum of the two previous columns. The last column is the TOTAL INCL. O & P. This column is the sum of the bare material cost plus 10%, which is then added to the installation costs including Subs O & P. In this case the crew costs and the labor costs used would include the Subs O & P. When the costs are divided by the Daily Output, a unit installation cost including Subs O & P is developed. A sample calculation is shown on p. ii of the Foreword. In a few cases, an M, L or E will be found next to the prices in this column. This indicates that the particular cost figures include only Material (M), Labor (L) or Equipment (E).

SPECIAL On a few pages the column headings are EQUIP. and LABOR. The EQUIP. column includes equipment rental and operating costs while the LABOR column represents the labor to operate the equipment. These costs are bare costs and do not include the Sub's O & P.

GENERAL ARRANGEMENT In the margin and preceding the item there are frequently found large numbers in circles. These refer to breakdowns in the back of the book. The breakdowns show how the editors arrived at the tabulated figures, and are often of value in listing materials for purchase. There are numerous notes on economy in construction in these sections.

Division 17 contains Square Foot and Cubic Foot costs plus percentages of total contract for Plumbing, Heating, Ventilating and Air Conditioning, Electrical and the sum of Mechanical and Electrical. The results shown are in ascending order of cost and percentage. These figures include contractor's overhead and profit but do not include architectural fees or land cost.

Repair and Remodeling costs and information are shown in Division 18.

Division 19 shows relative construction costs listed by division for material, labor and total costs for 162 major U.S. and Canadian cities.

CROSS-REFERENCES Numerous cross-references are included to aid in the rapid location of items that could logically be listed under 2 or more subdivisions. A cross-reference referring to, say, Division 7.4 would mean that the entire division (Roofing & Siding) was the intended cross-reference. A cross reference of 7.4-15 would indicate Built-Up Roofing and a cross-reference of 7.4-15-900 would indicate line 900 of Division 7.4-15, or, in this case, the bond charge for built-up roofs.

ROUNDING OF COSTS In general, all unit prices in excess of $5.00 have been rounded to make them easier to use and still maintain adequate precision of the results. The rounding rules generally followed in this book are as follows:

Price from $.01 to $5.00 rounded to the nearest 1¢
from $5.01 to $20.00 rounded to the nearest 5¢
from $20.01 to $100.00 rounded to the nearest $1
from $100.01 to $1,000.00 rounded to the nearest $5
from $1,000.01 to $10,000.00 rounded to the nearest $25
from $10,000.01 to $50,000.00 rounded to the nearest $100
over $50,000.01 rounded to the nearest $500

OVERHEAD, PROFIT, CONTINGENCIES

GENERAL Prices given in this book are listed two ways: (1) BARE COSTS to which an allowance for Contractor overhead and profit must be added and (2) TOTAL INCL. O & P containing the original Subcontractor's overhead and profit (in other words, the Contractor doing the work), thus it is necessary to add a General Contractor mark-up percentage for all subcontracted items unless the owner is receiving multiple prime contracts. The General Contractor's normal mark-up ranges from 0 to 15%. The mark-up depends on economic conditions plus the supervision and trouble-shooting expected by the General Contractor. For purposes of this book it is best to add an allowance of 10% to the figures including O & P. In using the bare cost figures, experience shows that an addition of 25% for General Contractor's mark-up plus the Subcontractor's (if any) overhead and profit.

OVERHEAD AND PROFIT Breakdowns on pages 342 to 350 give details on overhead and indicate that it will run 20% to 22% of the bare cost for General Contractors. For an **engineering** estimate a profit margin of 3% to 5% seems reasonable, so an allowance of 25% should be added to all bare costs (costs that do not include Sub's O & P) in the book. A General Contractor should be able to tell very closely from his records what his own overhead will be. He will then add whatever profit and contingency percentage circumstances dictate. Subcontractor's overhead and profit allowances are detailed on p. 344. These are general figures and deviations for individual companies are certainly to be expected.

SUBCONTRACTORS Usually a considerable portion of all large jobs is subcontracted. In fact the percentage done by Subcontractors is constantly increasing and may run over 90%. Since the workmen employed by these companies do nothing else but install their particular product, they soon become experts in that line. The result is, installation by these firms is accomplished so efficiently that the total in-place cost, even with the subcontractor's overhead and profit, is no more and often less than if the principal contractor had handled the installation himself. There is, moreover, the big advantage of having the work done right. Companies who sell specialties are anxious that their product perform well and consequently the installation will be the best possible.

CONTINGENCIES The allowance for contingencies generally is to provide for unforeseen construction difficulties. On alterations or repair jobs, 20% is none too much. If drawings are final and only field contingencies are being considered, 2% or 3% is probably sufficient and often nothing need be added. As far as the contract is concerned, future changes in plans will be covered by extras. The contractor should consider inflationary price trends and possible material shortages during the course of the job. This escalation factor is dependent upon both economic conditions and the anticipated time between the estimate and actual construction. If drawings are not complete or approved or a budget cost is wanted, it is wise to add 5% to 10%. Contingencies, then, are a matter of judgment. Additional allowances are shown on page 1 for contingencies, page 2 for job conditions and page 329 for factors to convert prices for Repair and Remodeling jobs.

ESTIMATING PRECISION

In making an engineering estimate ignore the cents column. Just give the total per unit cost to the nearest dollar. The cents will average up in a column of figures. An engineering estimate of $257,323.37 is inappropriate: $257,325 is certainly more sensible and $257,000 is better and just as likely to be right.

If you follow this simple instruction, the time saved is tremendous with an added important advantage. Using round figures your mind is left free to exercise judgment and common sense rather than being overcome and befuddled by a mass of computations.

When you have finished, roughly check the big items for location of decimal point. That is important. A large error can creep in if you write down $300 when it should be $3,000. Also check the list to be sure you have not omitted any large item. A common error is to overlook, let us say, heating or to forget finished flooring or painting or otherwise commit a gross omission. No amount of accuracy in prices can compensate for such an oversight.

It is important to keep the bare costs and costs that already include the Sub's O & P separate since different mark-ups will have to be applied to each category. Organize your estimating procedures to minimize confusion and simplify checking to insure against omissions and/or duplications.

ESTIMATING GUIDELINES

The following suggestions are made to enable the estimator to perform unit price estimating in a logical, easy to check and thorough manner.

1. Use pre-printed or columnar forms for orderly sequence of dimensions and locations and for recording telephone quotations.
2. Use only the front side of each paper or form except for certain pre-printed summary forms.
3. Be consistent in listing dimensions: for example, length x width x height. This helps in re-checking to insure that, say, the total length of partitions is appropriate for the building area.
4. Use printed (rather than measured) dimensions where given.
5. Add up multiple printed dimensions for a single entry where possible.
6. Measure all other dimensions carefully.
7. Use each set of dimensions to calculate multiple related quantities.
8. Convert foot and inch measurements to decimal feet when listing. Memorize decimal equivalents to .01 parts of a foot (1/8" equals approximately .01').
9. Do not "round off" quantities until the final summary.
10. Mark drawings with different colors as items are taken off.
11. Keep similar items together; different items separate.
12. Identify location and drawing numbers on your estimate to aid in future checking for completeness.
13. Measure or list everything on the drawings or mentioned in the specifications.
14. It may be necessary to list items not called for to make the job complete.
15. Be alert for: notes on plans such as N.T.S. (not to scale); changes in scale throughout the drawings; reduced size drawings; discrepancies between the specifications and the drawings.

16. Develop a consistent pattern of performing an estimate, for example:
    a. Start the quantity take-off at the lower floor and move to the next higher floor.
    b. Proceed from the main section of the building to the wings.
    c. Proceed from south to north or vice versa, clockwise or counterclockwise.
    d. Take off floor plan quantities first, elevations next, then detail drawings.
17. List all gross dimensions that can be either used again for different quantities, or used as a rough check of other quantities for verification (exterior perimeter, gross floor area, individual floor areas, etc.).
18. Utilize design symmetry or repetition (repetitive floors, repetitive wings, symmetrical design around a center line, similar room layouts, etc.).

    Note: extreme caution is needed here to not omit or duplicate an area.
19. Do not convert units until the final total is obtained. For instance, when estimating concrete work, keep all units to the nearest cubic foot, then summarize and convert to cubic yards.
20. When figuring alternates, it is best to total all items involved in the basic system, then total all items involved in the alternates. Thus you work with positive numbers in all cases. When adds and deducts are used, whether to add or subtract a portion of an item, is often confusing, especially on a complicated or involved alternate.

THE COMPANY AND THE EDITORS

R.S. Means Co., Inc., is a 44-year old firm actively engaged in construction cost publishing and consulting throughout North America. The company helps you cut down on the time and energy you use when organizing and utilizing cost data. A thoroughly experienced and highly qualified staff of over ninety professionals at R. S. Means works daily at collecting, analyzing and disseminating reliable cost information for your needs.

Each editor contributing to this publication is a full-time Means employee. All these staff members have years of practical contractor experience and thorough engineering training prior to joining the firm. Each Means construction expert contributes to the maintenance of a complete, continually-updated construction cost data system.

Constant flow of new products means that even the most sophisticated construction professional finds little time to examine and evaluate all the construction cost possibilities when they are so many and diverse. R. S. Means fills the gap between theoretical and practical knowledge in the field of construction engineering. The Means organization is always prepared to assist you and help in the solution of construction problems through the services of its four major divisions: Construction and Cost Data Publishing, Computer Software Services, Consulting, and Educational Seminars.

CREWS

Crew No.	Bare Costs		Incl. Subs O & P		Cost Per Man-hour	
Crew A-1	Hr.	Daily	Hr.	Daily	Bare Costs	Incl. O&P
1 Building Laborer	$15.90	$127.20	$23.05	$184.40	$15.90	$23.05
1 Gas Eng. Power Tool		36.60		40.25	4.57	5.03
8 M.H., Daily Totals		$163.80		$224.65	$20.47	$28.08
Crew A-2	Hr.	Daily	Hr.	Daily	Bare Costs	Incl. O&P
2 Building Laborers	$15.90	$254.40	$23.05	$368.80	$16.05	$23.18
1 Truck Driver (light)	16.35	130.80	23.45	187.60		
1 Light Truck, 1.5 Ton		57.40		63.15	2.39	2.63
24 M.H., Daily Totals		$442.60		$619.55	$18.44	$25.81
Crew A-3	Hr.	Daily	Hr.	Daily	Bare Costs	Incl. O&P
1 Truck Driver (heavy)	$16.60	$132.80	$23.80	$190.40	$16.60	$23.80
1 Dump Truck, 12 ton		220.20		242.20	27.52	30.27
8 M.H., Daily Totals		$353.00		$432.60	$44.12	$54.07
Crew A-4	Hr.	Daily	Hr.	Daily	Bare Costs	Incl. O&P
2 Carpenters	$20.00	$320.00	$29.00	$464.00	$19.75	$28.48
1 Painter, Ordinary	19.25	154.00	27.45	219.60		
24 M.H., Daily Totals		$474.00		$683.60	$19.75	$28.48
Crew A-5	Hr.	Daily	Hr.	Daily	Bare Costs	Incl. O&P
2 Building Laborers	$15.90	$254.40	$23.05	$368.80	$15.95	$23.09
.25 Truck Driver (light)	16.35	32.70	23.45	46.90		
.25 Light Truck, 1.5 Ton		14.35		15.80	.79	.87
18 M.H., Daily Totals		$301.45		$431.50	$16.74	$23.96
Crew A-6	Hr.	Daily	Hr.	Daily	Bare Costs	Incl. O&P
1 Chief Of Party	$19.45	$155.60	$28.20	$225.60	$18.47	$26.80
1 Instrument Man	17.50	140.00	25.40	203.20		
16 M.H., Daily Totals		$295.60		$428.80	$18.47	$26.80
Crew A-7	Hr.	Daily	Hr.	Daily	Bare Costs	Incl. O&P
1 Chief Of Party	$19.45	$155.60	$28.20	$225.60	$17.50	$25.46
1 Instrument Man	17.50	140.00	25.40	203.20		
1 Rodman/Chainman	15.55	124.40	22.80	182.40		
24 M.H., Daily Totals		$420.00		$611.20	$17.50	$25.46
Crew A-8	Hr.	Daily	Hr.	Daily	Bare Costs	Incl. O&P
1 Chief Of Party	$19.45	$155.60	$28.20	$225.60	$17.01	$24.80
1 Instrument Man	17.50	140.00	25.40	203.20		
2 Rodmen/Chainmen	15.55	248.80	22.80	364.80		
32 M.H., Daily Totals		$544.40		$793.60	$17.01	$24.80
Crew B-1	Hr.	Daily	Hr.	Daily	Bare Costs	Incl. O&P
1 Labor Foreman (outside)	$17.90	$143.20	$25.95	$207.60	$16.56	$24.01
2 Building Laborers	15.90	254.40	23.05	368.80		
24 M.H., Daily Totals		$397.60		$576.40	$16.56	$24.01
Crew B-2	Hr.	Daily	Hr.	Daily	Bare Costs	Incl. O&P
1 Labor Foreman (outside)	$17.90	$143.20	$25.95	$207.60	$16.30	$23.63
4 Building Laborers	15.90	508.80	23.05	737.60		
40 M.H., Daily Totals		$652.00		$945.20	$16.30	$23.63

Crew No.	Bare Costs		Incl. Subs O & P		Cost Per Man-hour	
Crew B-3	Hr.	Daily	Hr.	Daily	Bare Costs	Incl. O&P
1 Labor Foreman (outside)	$17.90	$143.20	$25.95	$207.60	$17.25	$24.91
2 Building Laborers	15.90	254.40	23.05	368.80		
1 Equip. Oper. (med.)	20.60	164.80	29.85	238.80		
2 Truck Drivers (heavy)	16.60	265.60	23.80	380.80		
F.E. Loader, T.M., 2.5 C.Y.		600.00		660.00		
2 Dump Trucks, 16 Ton		559.20		615.10	24.15	26.56
48 M.H., Daily Totals		$1987.20		$2471.10	$41.40	$51.47
Crew B-4	Hr.	Daily	Hr.	Daily	Bare Costs	Incl. O&P
1 Labor Foreman (outside)	$17.90	$143.20	$25.95	$207.60	$16.35	$23.65
4 Building Laborers	15.90	508.80	23.05	737.60		
1 Truck Driver (heavy)	16.60	132.80	23.80	190.40		
1 Lowbed Tractor Trailer		224.30		246.75		
1 Platform Trailer		96.00		105.60	6.67	7.34
48 M.H., Daily Totals		$1105.10		$1487.95	$23.02	$30.99
Crew B-5	Hr.	Daily	Hr.	Daily	Bare Costs	Incl. O&P
1 Labor Foreman (outside)	$17.90	$143.20	$25.95	$207.60	$18.06	$26.18
4 Building Laborers	15.90	508.80	23.05	737.60		
2 Equip. Oper. (med.)	20.60	329.60	29.85	477.60		
1 Mechanic	21.80	174.40	31.60	252.80		
1 Air Compr., 250 C.F.M.		111.40		122.55		
Air Tools & Accessories		25.80		28.40		
2-50 Ft. Air Hoses, 1.5" Dia.		13.20		14.50		
F.E. Loader, T.M., 2.5 C.Y.		600.00		660.00	11.72	12.89
64 M.H., Daily Totals		$1906.40		$2501.05	$29.78	$39.07
Crew B-6	Hr.	Daily	Hr.	Daily	Bare Costs	Incl. O&P
2 Building Laborers	$15.90	$254.40	$23.05	$368.80	$17.08	$24.76
1 Equip. Oper. (light)	19.45	155.60	28.20	225.60		
1 Backhoe Loader, 48 H.P.		156.00		171.60	6.50	7.15
24 M.H., Daily Totals		$566.00		$766.00	$23.58	$31.91
Crew B-7	Hr.	Daily	Hr.	Daily	Bare Costs	Incl. O&P
1 Labor Foreman (outside)	$17.90	$143.20	$25.95	$207.60	$17.01	$24.66
4 Building Laborers	15.90	508.80	23.05	737.60		
1 Equip. Oper. (med.)	20.60	164.80	29.85	238.80		
1 Chipping Machine		137.00		150.70		
F.E. Loader, T.M., 2.5 C.Y.		600.00		660.00	15.35	16.88
48 M.H., Daily Totals		$1553.80		$1994.70	$32.36	$41.54
Crew B-8	Hr.	Daily	Hr.	Daily	Bare Costs	Incl. O&P
1 Labor Foreman (outside)	$17.90	$143.20	$25.95	$207.60	$17.70	$25.59
2 Building Laborers	15.90	254.40	23.05	368.80		
2 Equip. Oper. (med.)	20.60	329.60	29.85	477.60		
1 Equip. Oper. Oiler	17.50	140.00	25.40	203.20		
2 Truck Drivers (heavy)	16.60	265.60	23.80	380.80		
1 Hyd. Crane, 25 Ton		384.80		423.30		
F.E. Loader, T.M., 2.5 C.Y.		600.00		660.00		
2 Dump Trucks, 16 Ton		559.20		615.10	24.12	26.53
64 M.H., Daily Totals		$2676.80		$3336.40	$41.82	$52.12
Crew B-9	Hr.	Daily	Hr.	Daily	Bare Costs	Incl. O&P
1 Labor Foreman (outside)	$17.90	$143.20	$25.95	$207.60	$16.30	$23.63
4 Building Laborers	15.90	508.80	23.05	737.60		
1 Air Compr., 250 C.F.M.		111.40		122.55		
Air Tools & Accessories		25.80		28.40		
2-50 Ft. Air Hoses, 1.5" Dia.		13.20		14.50	3.76	4.13
40 M.H., Daily Totals		$802.40		$1110.65	$20.06	$27.76

CREWS

Crew No.	Bare Costs		Incl. Subs O & P		Cost Per Man-hour	
Crew B-10	Hr.	Daily	Hr.	Daily	Bare Costs	Incl. O&P
1 Equip. Oper. (med.)	$20.60	$164.80	$29.85	$238.80	$19.03	$27.58
.5 Building Laborer	15.90	63.60	23.05	92.20		
12 M.H., Daily Totals		$228.40		$331.00	$19.03	$27.58
Crew B-10A	Hr.	Daily	Hr.	Daily	Bare Costs	Incl. O&P
1 Equip. Oper. (med.)	$20.60	$164.80	$29.85	$238.80	$19.03	$27.58
.5 Building Laborer	15.90	63.60	23.05	92.20		
1 Roll. Compact., 2K Lbs.		75.60		83.15	6.30	6.92
12 M.H., Daily Totals		$304.00		$414.15	$25.33	$34.50
Crew B-10B	Hr.	Daily	Hr.	Daily	Bare Costs	Incl. O&P
1 Equip. Oper. (med.)	$20.60	$164.80	$29.85	$238.80	$19.03	$27.58
.5 Building Laborer	15.90	63.60	23.05	92.20		
1 Dozer, 200 H.P.		634.80		698.30	52.90	58.19
12 M.H., Daily Totals		$863.20		$1029.30	$71.93	$85.77
Crew B-10C	Hr.	Daily	Hr.	Daily	Bare Costs	Incl. O&P
1 Equip. Oper. (med.)	$20.60	$164.80	$29.85	$238.80	$19.03	$27.58
.5 Building Laborer	15.90	63.60	23.05	92.20		
1 Dozer, 200 H.P.		634.80		698.30		
1 Vibratory Roller, Towed		72.85		80.15	58.97	64.87
12 M.H., Daily Totals		$936.05		$1109.45	$78.00	$92.45
Crew B-10D	Hr.	Daily	Hr.	Daily	Bare Costs	Incl. O&P
1 Equip. Oper. (med.)	$20.60	$164.80	$29.85	$238.80	$19.03	$27.58
.5 Building Laborer	15.90	63.60	23.05	92.20		
1 Dozer, 200 H.P		634.80		698.30		
1 Sheepsft. Roller, Towed		81.10		89.20	59.65	65.62
12 M.H., Daily Totals		$944.30		$1118.50	$78.68	$93.20
Crew B-10E	Hr.	Daily	Hr.	Daily	Bare Costs	Incl. O&P
1 Equip. Oper. (med.)	$20.60	$164.80	$29.85	$238.80	$19.03	$27.58
.5 Building Laborer	15.90	63.60	23.05	92.20		
1 Tandem Roller, 5 Ton		85.00		93.50	7.08	7.79
12 M.H., Daily Totals		$313.40		$424.50	$26.11	$35.37
Crew B-10F	Hr.	Daily	Hr.	Daily	Bare Costs	Incl. O&P
1 Equip. Oper. (med.)	$20.60	$164.80	$29.85	$238.80	$19.03	$27.58
.5 Building Laborer	15.90	63.60	23.05	92.20		
1 Tandem Roller, 10 Ton		175.20		192.70	14.60	16.05
12 M.H., Daily Totals		$403.60		$523.70	$33.63	$43.63
Crew B-10G	Hr.	Daily	Hr.	Daily	Bare Costs	Incl. O&P
1 Equip. Oper. (med.)	$20.60	$164.80	$29.85	$238.80	$19.03	$27.58
.5 Building Laborer	15.90	63.60	23.05	92.20		
1 Sheepsft. Roll., 130 H.P.		290.60		319.65	24.21	26.63
12 M.H., Daily Totals		$519.00		$650.65	$43.24	$54.21
Crew B-10H	Hr.	Daily	Hr.	Daily	Bare Costs	Incl. O&P
1 Equip. Oper. (med.)	$20.60	$164.80	$29.85	$238.80	$19.03	$27.58
.5 Building Laborer	15.90	63.60	23.05	92.20		
1 Diaphr. Water Pump, 2"		19.30		21.25		
1-20 Ft. Suction Hose, 2"		3.40		3.75		
2-50 Ft. Disch. Hoses, 2"		7.20		7.90	2.49	2.74
12 M.H., Daily Totals		$258.30		$363.90	$21.52	$30.32

Crew No.	Bare Costs		Incl. Subs O & P		Cost Per Man-hour	
Crew B-10I	Hr.	Daily	Hr.	Daily	Bare Costs	Incl. O&P
1 Equip. Oper. (med.)	$20.60	$164.80	$29.85	$238.80	$19.03	$27.58
.5 Building Laborer	15.90	63.60	23.05	92.20		
1 Diaphr. Water Pump, 4"		40.20		44.20		
1-20 Ft. Suction Hose, 4"		7.20		7.90		
2-50 Ft. Disch. Hoses, 4"		15.20		16.70	5.21	5.73
12 M.H., Daily Totals		$291.00		$399.80	$24.24	$33.31
Crew B-10J	Hr.	Daily	Hr.	Daily	Bare Costs	Incl. O&P
1 Equip. Oper. (med.)	$20.60	$164.80	$29.85	$238.80	$19.03	$27.58
.5 Building Laborer	15.90	63.60	23.05	92.20		
1 Centr. Water Pump, 3"		24.10		26.50		
1-20 Ft. Suction Hose, 3"		4.60		5.05		
2-50 Ft. Disch. Hoses, 3"		10.00		11.00	3.22	3.54
12 M.H., Daily Totals		$267.10		$373.55	$22.25	$31.12
Crew B-10K	Hr.	Daily	Hr.	Daily	Bare Costs	Incl. O&P
1 Equip. Oper. (med.)	$20.60	$164.80	$29.85	$238.80	$19.03	$27.58
.5 Building Laborer	15.90	63.60	23.05	92.20		
1 Centr. Water Pump, 6"		61.60		67.75		
1-20 Ft. Suction Hose, 6"		14.40		15.85		
2-50 Ft. Disch. Hoses, 6"		26.00		28.60	8.50	9.35
12 M.H., Daily Totals		$330.40		$443.20	$27.53	$36.93
Crew B-10L	Hr.	Daily	Hr.	Daily	Bare Costs	Incl. O&P
1 Equip. Oper. (med.)	$20.60	$164.80	$29.85	$238.80	$19.03	$27.58
.5 Building Laborer	15.90	63.60	23.05	92.20		
1 Dozer, 75 H.P.		206.00		226.60	17.16	18.88
12 M.H., Daily Totals		$434.40		$557.60	$36.19	$46.46
Crew B-10M	Hr.	Daily	Hr.	Daily	Bare Costs	Incl. O&P
1 Equip. Oper. (med.)	$20.60	$164.80	$29.85	$238.80	$19.03	$27.58
.5 Building Laborer	15.90	63.60	23.05	92.20		
1 Dozer, 300 H.P.		931.00		1024.10	77.58	85.34
12 M.H., Daily Totals		$1159.40		$1355.10	$96.61	$112.92
Crew B-10N	Hr.	Daily	Hr.	Daily	Bare Costs	Incl. O&P
1 Equip. Oper. (med.)	$20.60	$164.80	$29.85	$238.80	$19.03	$27.58
.5 Building Laborer	15.90	63.60	23.05	92.20		
F.E. Loader, T.M., 1.5 C.Y		276.40		304.05	23.03	25.33
12 M.H., Daily Totals		$504.80		$635.05	$42.06	$52.91
Crew B-10O	Hr.	Daily	Hr.	Daily	Bare Costs	Incl. O&P
1 Equip. Oper. (med.)	$20.60	$164.80	$29.85	$238.80	$19.03	$27.58
.5 Building Laborer	15.90	63.60	23.05	92.20		
F.E. Loader, T.M., 2.25 C.Y.		421.20		463.30	35.10	38.60
12 M.H., Daily Totals		$649.60		$794.30	$54.13	$66.18
Crew B-10P	Hr.	Daily	Hr.	Daily	Bare Costs	Incl. O&P
1 Equip. Oper. (med.)	$20.60	$164.80	$29.85	$238.80	$19.03	$27.58
.5 Building Laborer	15.90	63.60	23.05	92.20		
F.E. Loader, T.M., 2.5 C.Y		600.00		660.00	50.00	55.00
12 M.H., Daily Totals		$828.40		$991.00	$69.03	$82.58
Crew B-10Q	Hr.	Daily	Hr.	Daily	Bare Costs	Incl. O&P
1 Equip. Oper. (med.)	$20.60	$164.80	$29.85	$238.80	$19.03	$27.58
.5 Building Laborer	15.90	63.60	23.05	92.20		
F.E. Loader, T.M., 5 C.Y.		1010.00		1111.00	84.16	92.58
12 M.H., Daily Totals		$1238.40		$1442.00	$103.19	$120.16

CREWS

Crew No.	Bare Costs		Incl. Subs O & P		Cost Per Man-hour	
Crew B-10R	Hr.	Daily	Hr.	Daily	Bare Costs	Incl. O&P
1 Equip. Oper. (med.)	$20.60	$164.80	$29.85	$238.80	$19.03	$27.58
.5 Building Laborer	15.90	63.60	23.05	92.20		
F.E. Loader, W.M., 1 C.Y.		202.40		222.65	16.86	18.55
12 M.H., Daily Totals		$430.80		$553.65	$35.89	$46.13
Crew B-10S	Hr.	Daily	Hr.	Daily	Bare Costs	Incl. O&P
1 Equip. Oper. (med.)	$20.60	$164.80	$29.85	$238.80	$19.03	$27.58
.5 Building Laborer	15.90	63.60	23.05	92.20		
F.E. Loader, W.M., 1.5 C.Y.		271.00		298.10	22.58	24.84
12 M.H., Daily Totals		$499.40		$629.10	$41.61	$52.42
Crew B-10T	Hr.	Daily	Hr.	Daily	Bare Costs	Incl. O&P
1 Equip. Oper. (med.)	$20.60	$164.80	$29.85	$238.80	$19.03	$27.58
.5 Building Laborer	15.90	63.60	23.05	92.20		
F.E. Loader, W.M., 2.5 C.Y.		443.20		487.50	36.93	40.62
12 M.H., Daily Totals		$671.60		$818.50	$55.96	$68.20
Crew B-10U	Hr.	Daily	Hr.	Daily	Bare Costs	Incl. O&P
1 Equip. Oper. (med.)	$20.60	$164.80	$29.85	$238.80	$19.03	$27.58
.5 Building Laborer	15.90	63.60	23.05	92.20		
F.E. Loader, W.M., 5.5 C.Y.		845.00		929.50	70.41	77.45
12 M.H., Daily Totals		$1073.40		$1260.50	$89.44	$105.03
Crew B-10W	Hr.	Daily	Hr.	Daily	Bare Costs	Incl. O&P
1 Equip. Oper. (med.)	$20.60	$164.80	$29.85	$238.80	$19.03	$27.58
.5 Building Laborer	15.90	63.60	23.05	92.20		
1 Dozer, 105 H.P.		346.00		380.60	28.83	31.71
12 M.H., Daily Totals		$574.40		$711.60	$47.86	$59.29
Crew B-10X	Hr.	Daily	Hr.	Daily	Bare Costs	Incl. O&P
1 Equip. Oper. (med.)	$20.60	$164.80	$29.85	$238.80	$19.03	$27.58
.5 Building Laborer	15.90	63.60	23.05	92.20		
1 Dozer, 410 H.P.		1157.00		1272.70	96.41	106.05
12 M.H., Daily Totals		$1385.40		$1603.70	$115.44	$133.63
Crew B-10Y	Hr.	Daily	Hr.	Daily	Bare Costs	Incl. O&P
1 Equip. Oper. (med.)	$20.60	$164.80	$29.85	$238.80	$19.03	$27.58
.5 Building Laborer	15.90	63.60	23.05	92.20		
1 Vibratory Drum Roller		281.40		309.55	23.45	25.79
12 M.H., Daily Totals		$509.80		$640.55	$42.48	$53.37
Crew B-11	Hr.	Daily	Hr.	Daily	Bare Costs	Incl. O&P
1 Equipment Oper. (med.)	$20.60	$164.80	$29.85	$238.80	$18.25	$26.45
1 Building Laborer	15.90	127.20	23.05	184.40		
16 M.H., Daily Totals		$292.00		$423.20	$18.25	$26.45
Crew B-11A	Hr.	Daily	Hr.	Daily	Bare Costs	Incl. O&P
1 Equipment Oper. (med.)	$20.60	$164.80	$29.85	$238.80	$18.25	$26.45
1 Building Laborer	15.90	127.20	23.05	184.40		
1 Dozer, 200 H.P.		634.80		698.30	39.67	43.64
16 M.H., Daily Totals		$926.80		$1121.50	$57.92	$70.09
Crew B-11B	Hr.	Daily	Hr.	Daily	Bare Costs	Incl. O&P
1 Equipment Oper. (med.)	$20.60	$164.80	$29.85	$238.80	$18.25	$26.45
1 Building Laborer	15.90	127.20	23.05	184.40		
1 Dozer, 200 H.P.		634.80		698.30		
1 Air Powered Tamper		13.10		14.40		
1 Air Compr. 365 C.F.M.		158.00		173.80		
2-50 Ft. Air Hoses, 1.5" Dia.		13.20		14.50	51.19	56.31
16 M.H., Daily Totals		$1111.10		$1324.20	$69.44	$82.76

Crew No.	Bare Costs		Incl. Subs O & P		Cost Per Man-hour	
Crew B-11C	Hr.	Daily	Hr.	Daily	Bare Costs	Incl. O&P
1 Equipment Oper. (med.)	$20.60	$164.80	$29.85	$238.80	$18.25	$26.45
1 Building Laborer	15.90	127.20	23.05	184.40		
1 Backhoe Loader, 48 H.P.		156.00		171.60	9.75	10.72
16 M.H., Daily Totals		$448.00		$594.80	$28.00	$37.17
Crew B-11K	Hr.	Daily	Hr.	Daily	Bare Costs	Incl. O&P
1 Equipment Oper. (med.)	$20.60	$164.80	$29.85	$238.80	$18.25	$26.45
1 Building Laborer	15.90	127.20	23.05	184.40		
1 Trencher, 8' D., 16" W.		420.60		462.65	26.28	28.91
16 M.H., Daily Totals		$712.60		$885.85	$44.53	$55.36
Crew B-11L	Hr.	Daily	Hr.	Daily	Bare Costs	Incl. O&P
1 Equipment Oper. (med.)	$20.60	$164.80	$29.85	$238.80	$18.25	$26.45
1 Building Laborer	15.90	127.20	23.05	184.40		
1 Grader, 30,000 Lbs.		422.40		464.65	26.40	29.04
16 M.H., Daily Totals		$714.40		$887.85	$44.65	$55.49
Crew B-11M	Hr.	Daily	Hr.	Daily	Bare Costs	Incl. O&P
1 Equipment Oper. (med.)	$20.60	$164.80	$29.85	$238.80	$18.25	$26.45
1 Building Laborer	15.90	127.20	23.05	184.40		
1 Backhoe Loader, 80 H.P.		234.20		257.60	14.63	16.10
16 M.H., Daily Totals		$526.20		$680.80	$32.88	$42.55
Crew B-12	Hr.	Daily	Hr.	Daily	Bare Costs	Incl. O&P
1 Equip. Oper. (crane)	$21.05	$168.40	$30.50	$244.00	$19.27	$27.95
1 Equip. Oper. Oiler	17.50	140.00	25.40	203.20		
16 M.H., Daily Totals		$308.40		$447.20	$19.27	$27.95
Crew B-12A	Hr.	Daily	Hr.	Daily	Bare Costs	Incl. O&P
1 Equip. Oper. (crane)	$21.05	$168.40	$30.50	$244.00	$19.27	$27.95
1 Equip. Oper. Oiler	17.50	140.00	25.40	203.20		
1 Hyd. Excavator, 1 C.Y.		443.80		488.20	27.73	30.51
16 M.H., Daily Totals		$752.20		$935.40	$47.00	$58.46
Crew B-12B	Hr.	Daily	Hr.	Daily	Bare Costs	Incl. O&P
1 Equip. Oper. (crane)	$21.05	$168.40	$30.50	$244.00	$19.27	$27.95
1 Equip. Oper. Oiler	17.50	140.00	25.40	203.20		
1 Hyd. Excavator, 1.5 C.Y.		579.60		637.55	36.22	39.84
16 M.H., Daily Totals		$888.00		$1084.75	$55.49	$67.79
Crew B-12C	Hr.	Daily	Hr.	Daily	Bare Costs	Incl. O&P
1 Equip. Oper. (crane)	$21.05	$168.40	$30.50	$244.00	$19.27	$27.95
1 Equip. Oper. Oiler	17.50	140.00	25.40	203.20		
1 Hyd. Excavator, 2 C.Y.		801.00		881.10	50.06	55.06
16 M.H., Daily Totals		$1109.40		$1328.30	$69.33	$83.01
Crew B-12D	Hr.	Daily	Hr.	Daily	Bare Costs	Incl. O&P
1 Equip. Oper. (crane)	$21.05	$168.40	$30.50	$244.00	$19.27	$27.95
1 Equip. Oper. Oiler	17.50	140.00	25.40	203.20		
1 Hyd. Excavator, 3.5 C.Y.		1412.00		1553.20	88.25	97.07
16 M.H., Daily Totals		$1720.40		$2000.40	$107.52	$125.02
Crew B-12E	Hr.	Daily	Hr.	Daily	Bare Costs	Incl. O&P
1 Equip. Oper. (crane)	$21.05	$168.40	$30.50	$244.00	$19.27	$27.95
1 Equip. Oper. Oiler	17.50	140.00	25.40	203.20		
1 Hyd. Excavator, .5 C.Y.		262.00		288.20	16.37	18.01
16 M.H., Daily Totals		$570.40		$735.40	$35.64	$45.96

CREWS

Crew B-12F	Hr.	Daily	Hr.	Daily	Bare Costs	Incl. O&P
1 Equip. Oper. (crane)	$21.05	$168.40	$30.50	$244.00	$19.27	$27.95
1 Equip. Oper. Oiler	17.50	140.00	25.40	203.20		
1 Hyd. Excavator, .75 C.Y.		340.20		374.20	21.26	23.38
16 M.H., Daily Totals		$648.60		$821.40	$40.53	$51.33

Crew B-12G	Hr.	Daily	Hr.	Daily	Bare Costs	Incl. O&P
1 Equip. Oper. (crane)	$21.05	$168.40	$30.50	$244.00	$19.27	$27.95
1 Equip. Oper. Oiler	17.50	140.00	25.40	203.20		
1 Power Shovel, .5 C.Y.		278.40		306.25		
1 Clamshell Bucket, .5 C.Y.		29.65		32.60	19.25	21.17
16 M.H., Daily Totals		$616.45		$786.05	$38.52	$49.12

Crew B-12H	Hr.	Daily	Hr.	Daily	Bare Costs	Incl. O&P
1 Equip. Oper. (crane)	$21.05	$168.40	$30.50	$244.00	$19.27	$27.95
1 Equip. Oper. Oiler	17.50	140.00	25.40	203.20		
1 Power Shovel, 1 C.Y.		387.60		426.35		
1 Clamshell Bucket, 1 C.Y.		40.15		44.15	26.73	29.40
16 M.H., Daily Totals		$736.15		$917.70	$46.00	$57.35

Crew B-12I	Hr.	Daily	Hr.	Daily	Bare Costs	Incl. O&P
1 Equip. Oper. (crane)	$21.05	$168.40	$30.50	$244.00	$19.27	$27.95
1 Equip. Oper. Oiler	17.50	140.00	25.40	203.20		
1 Power Shovel, .75 C.Y.		346.20		380.80		
1 Dragline Bucket, .75 C.Y.		21.45		23.60	22.97	25.27
16 M.H., Daily Totals		$676.05		$851.60	$42.24	$53.22

Crew B-12J	Hr.	Daily	Hr.	Daily	Bare Costs	Incl. O&P
1 Equip. Oper. (crane)	$21.05	$168.40	$30.50	$244.00	$19.27	$27.95
1 Equip. Oper. Oiler	17.50	140.00	25.40	203.20		
1 Gradall, 3 Ton, .5 C.Y.		440.40		484.45	27.52	30.27
16 M.H., Daily Totals		$748.80		$931.65	$46.79	$58.22

Crew B-12K	Hr.	Daily	Hr.	Daily	Bare Costs	Incl. O&P
1 Equip. Oper. (crane)	$21.05	$168.40	$30.50	$244.00	$19.27	$27.95
1 Equip. Oper. Oiler	17.50	140.00	25.40	203.20		
1 Gradall, 3 Ton, 1 C.Y.		629.00		691.90	39.31	43.24
16 M.H., Daily Totals		$937.40		$1139.10	$58.58	$71.19

Crew B-12L	Hr.	Daily	Hr.	Daily	Bare Costs	Incl. O&P
1 Equip. Oper. (crane)	$21.05	$168.40	$30.50	$244.00	$19.27	$27.95
1 Equip. Oper. Oiler	17.50	140.00	25.40	203.20		
1 Power Shovel, .5 C.Y.		278.40		306.25		
1 F.E. Attachment, .5 C.Y.		42.05		46.25	20.02	22.03
16 M.H., Daily Totals		$628.85		$799.70	$39.29	$49.98

Crew B-12M	Hr.	Daily	Hr.	Daily	Bare Costs	Incl. O&P
1 Equip. Oper. (crane)	$21.05	$168.40	$30.50	$244.00	$19.27	$27.95
1 Equip. Oper. Oiler	17.50	140.00	25.40	203.20		
1 Power Shovel, .75		346.20		380.80		
1 F.E. Attachment, .75 C.Y.		78.80		86.70	26.56	29.21
16 M.H., Daily Totals		$733.40		$914.70	$45.83	$57.16

Crew B-12N	Hr.	Daily	Hr.	Daily	Bare Costs	Incl. O&P
1 Equip. Oper. (crane)	$21.05	$168.40	$30.50	$244.00	$19.27	$27.95
1 Equip. Oper. Oiler	17.50	140.00	25.40	203.20		
1 Power Shovel, 1 C.Y.		387.60		426.35		
1 F.E. Attachment, 1 C.Y.		110.55		121.60	31.13	34.24
16 M.H., Daily Totals		$806.55		$995.15	$50.40	$62.19

Crew B-12O	Hr.	Daily	Hr.	Daily	Bare Costs	Incl. O&P
1 Equip. Oper. (crane)	$21.05	$168.40	$30.50	$244.00	$19.27	$27.95
1 Equip. Oper. Oiler	17.50	140.00	25.40	203.20		
1 Power Shovel, 1.5 C.Y.		483.40		531.75		
1 F.E. Attachment, 1.5 C.Y.		121.55		133.70	37.80	41.59
16 M.H., Daily Totals		$913.35		$1112.65	$57.07	$69.54

Crew B-12P	Hr.	Daily	Hr.	Daily	Bare Costs	Incl. O&P
1 Equip. Oper. (crane)	$21.05	$168.40	$30.50	$244.00	$19.27	$27.95
1 Equip. Oper. Oiler	17.50	140.00	25.40	203.20		
1 Crawler Crane, 40 Ton		483.40		531.75		
1 Dragline Bucket, 1.5 C.Y.		31.65		34.80	32.19	35.40
16 M.H., Daily Totals		$823.45		$1013.75	$51.46	$63.35

Crew B-12Q	Hr.	Daily	Hr.	Daily	Bare Costs	Incl. O&P
1 Equip. Oper. (crane)	$21.05	$168.40	$30.50	$244.00	$19.27	$27.95
1 Equip. Oper. Oiler	17.50	140.00	25.40	203.20		
1 Hyd. Excavator, 5/8 C.Y.		274.20		301.60	17.13	18.85
16 M.H., Daily Totals		$582.60		$748.80	$36.40	$46.80

Crew B-12R	Hr.	Daily	Hr.	Daily	Bare Costs	Incl. O&P
1 Equip. Oper. (crane)	$21.05	$168.40	$30.50	$244.00	$19.27	$27.95
1 Equip. Oper. Oiler	17.50	140.00	25.40	203.20		
1 Hyd. Excavator, 1.5 C.Y.		579.60		637.55	36.22	39.84
16 M.H., Daily Totals		$888.00		$1084.75	$55.49	$67.79

Crew B-12S	Hr.	Daily	Hr.	Daily	Bare Costs	Incl. O&P
1 Equip. Oper. (crane)	$21.05	$168.40	$30.50	$244.00	$19.27	$27.95
1 Equip. Oper. Oiler	17.50	140.00	25.40	203.20		
1 Hyd. Excavator, 2.5 C.Y.		1166.00		1282.60	72.87	80.16
16 M.H., Daily Totals		$1474.40		$1729.80	$92.14	$108.11

Crew B-12T	Hr.	Daily	Hr.	Daily	Bare Costs	Incl. O&P
1 Equip. Oper. (crane)	$21.05	$168.40	$30.50	$244.00	$19.27	$27.95
1 Equip. Oper. Oiler	17.50	140.00	25.40	203.20		
1 Crawler Crane, 75 Ton		740.00		814.00		
1 F.E. Attachment, 3 C.Y.		228.60		251.45	60.53	66.59
16 M.H., Daily Totals		$1277.00		$1512.65	$79.80	$94.54

Crew B-12V	Hr.	Daily	Hr.	Daily	Bare Costs	Incl. O&P
1 Equip. Oper. (crane)	$21.05	$168.40	$30.50	$244.00	$19.27	$27.95
1 Equip. Oper. Oiler	17.50	140.00	25.40	203.20		
1 Crawler Crane, 75 Ton		740.00		814.00		
1 Dragline Bucket, 3 C.Y.		59.10		65.00	49.94	54.93
16 M.H., Daily Totals		$1107.50		$1326.20	$69.21	$82.88

Crew B-13	Hr.	Daily	Hr.	Daily	Bare Costs	Incl. O&P
1 Labor Foreman (outside)	$17.90	$143.20	$25.95	$207.60	$17.15	$24.86
4 Building Laborers	15.90	508.80	23.05	737.60		
1 Equip. Oper. (crane)	21.05	168.40	30.50	244.00		
1 Equip. Oper. Oiler	17.50	140.00	25.40	203.20		
1 Hyd. Crane, 25 Ton		384.80		423.30	6.87	7.55
56 M.H., Daily Totals		$1345.20		$1815.70	$24.02	$32.41

Crew B-14	Hr.	Daily	Hr.	Daily	Bare Costs	Incl. O&P
1 Labor Foreman (outside)	$17.90	$143.20	$25.95	$207.60	$16.82	$24.39
4 Building Laborers	15.90	508.80	23.05	737.60		
1 Equip. Oper. (light)	19.45	155.60	28.20	225.60		
1 Backhoe Loader, 48 H.P.		156.00		171.60	3.25	3.57
48 M.H., Daily Totals		$963.60		$1342.40	$20.07	$27.96

CREWS

Crew No.	Bare Costs		Incl. Subs O & P		Cost Per Man-hour	
Crew B-15	Hr.	Daily	Hr.	Daily	Bare Costs	Incl. O&P
1 Equipment Oper. (med)	$20.60	$164.80	$29.85	$238.80	$17.64	$25.42
.5 Building Laborer	15.90	63.60	23.05	92.20		
2 Truck Drivers (heavy)	16.60	265.60	23.80	380.80		
2 Dump Trucks, 16 Ton		559.20		615.10		
1 Dozer, 200 H.P.		634.80		698.30	42.64	46.90
28 M.H., Daily Totals		$1688.00		$2025.20	$60.28	$72.32
Crew B-16	Hr.	Daily	Hr.	Daily	Bare Costs	Incl. O&P
1 Labor Foreman (outside)	$17.90	$143.20	$25.95	$207.60	$16.57	$23.96
2 Building Laborers	15.90	254.40	23.05	368.80		
1 Truck Driver (heavy)	16.60	132.80	23.80	190.40		
1 Dump Truck, 16 Ton		279.60		307.55	8.73	9.61
32 M.H., Daily Totals		$810.00		$1074.35	$25.30	$33.57
Crew B-17	Hr.	Daily	Hr.	Daily	Bare Costs	Incl. O&P
2 Building Laborers	$15.90	$254.40	$23.05	$368.80	$16.96	$24.52
1 Equip. Oper. (light)	19.45	155.60	28.20	225.60		
1 Truck Driver (heavy)	16.60	132.80	23.80	190.40		
1 Backhoe Loader, 48 H.P.		156.00		171.60		
1 Dump Truck, 12 Ton		220.20		242.20	11.75	12.93
32 M.H., Daily Totals		$919.00		$1198.60	$28.71	$37.45
Crew B-18	Hr.	Daily	Hr.	Daily	Bare Costs	Incl. O&P
1 Labor Foreman (outside)	$17.90	$143.20	$25.95	$207.60	$16.56	$24.01
2 Building Laborers	15.90	254.40	23.05	368.80		
1 Vibrating Compactor		39.85		43.85	1.66	1.82
24 M.H., Daily Totals		$437.45		$620.25	$18.22	$25.83
Crew B-19	Hr.	Daily	Hr.	Daily	Bare Costs	Incl. O&P
1 Pile Driver Foreman	$22.10	$176.80	$34.65	$277.20	$20.26	$30.88
4 Pile Drivers	20.10	643.20	31.50	1008.00		
2 Equip. Oper. (crane)	21.05	336.80	30.50	488.00		
1 Equip. Oper. Oiler	17.50	140.00	25.40	203.20		
1 Crane, 40 Ton & Access.		483.40		531.75		
60 L.F. Leads, 15K Ft. Lbs.		48.00		52.80		
1 Hammer, 15K Ft. Lbs.		200.60		220.65		
1 Air Compr., 600 C.F.M.		242.80		267.10		
2-50 Ft. Air Hoses, 3" Dia.		19.10		21.00	15.52	17.08
64 M.H., Daily Totals		$2290.70		$3069.70	$35.78	$47.96
Crew B-20	Hr.	Daily	Hr.	Daily	Bare Costs	Incl. O&P
1 Plumber Foreman (out)	$24.55	$196.40	$35.50	$284.00	$21.00	$30.38
1 Plumber	22.55	180.40	32.60	260.80		
1 Building Laborer	15.90	127.20	23.05	184.40		
24 M.H., Daily Totals		$504.00		$729.20	$21.00	$30.38
Crew B-21	Hr.	Daily	Hr.	Daily	Bare Costs	Incl. O&P
1 Plumber Foreman (out)	$24.55	$196.40	$35.50	$284.00	$21.00	$30.40
1 Plumber	22.55	180.40	32.60	260.80		
1 Building Laborer	15.90	127.20	23.05	184.40		
.5 Equip. Oper. (crane)	21.05	84.20	30.50	122.00		
.5 S.P. Crane, 5 Ton		90.10		99.10	3.21	3.53
28 M.H., Daily Totals		$678.30		$950.30	$24.21	$33.93

Crew No.	Bare Costs		Incl. Subs O & P		Cost Per Man-hour	
Crew B-22	Hr.	Daily	Hr.	Daily	Bare Costs	Incl. O&P
1 Plumber Foreman (out)	$24.55	$196.40	$35.50	$284.00	$21.01	$30.40
1 Plumber	22.55	180.40	32.60	260.80		
1 Building Laborer	15.90	127.20	23.05	184.40		
.75 Equip. Oper. (crane)	21.05	126.30	30.50	183.00		
.75 S.P. Crane, 5 Ton		135.15		148.65	4.50	4.95
30 M.H., Daily Totals		$765.45		$1060.85	$25.51	$35.35
Crew B-23	Hr.	Daily	Hr.	Daily	Bare Costs	Incl. O&P
1 Labor Foreman (outside)	$17.90	$143.20	$25.95	$207.60	$16.30	$23.63
4 Building Laborers	15.90	508.80	23.05	737.60		
Truck & Drill Rig		146.60		161.25		
Appropriate Fixtures		67.20		73.90	5.34	5.87
40 M.H., Daily Totals		$865.80		$1180.35	$21.64	$29.50
Crew B-24	Hr.	Daily	Hr.	Daily	Bare Costs	Incl. O&P
1 Cement Finisher	$19.20	$153.60	$27.00	$216.00	$18.36	$26.35
1 Building Laborer	15.90	127.20	23.05	184.40		
1 Carpenter	20.00	160.00	29.00	232.00		
24 M.H., Daily Totals		$440.80		$632.40	$18.36	$26.35
Crew B-25	Hr.	Daily	Hr.	Daily	Bare Costs	Incl. O&P
1 Labor Foreman (outside)	$17.90	$143.20	$25.95	$207.60	$17.04	$24.70
7 Building Laborers	15.90	890.40	23.05	1290.80		
2 Equip. Oper. (med.)	20.60	329.60	29.85	477.60		
1 Paving Machine		544.40		598.85		
1 Tandem Roller, 10 Ton		175.20		192.70	8.99	9.89
80 M.H., Daily Totals		$2082.80		$2767.55	$26.03	$34.59
Crew B-26	Hr.	Daily	Hr.	Daily	Bare Costs	Incl. O&P
1 Labor Foreman (outside)	$17.90	$143.20	$25.95	$207.60	$17.76	$25.87
6 Building Laborers	15.90	763.20	23.05	1106.40		
2 Equip. Oper. (med.)	20.60	329.60	29.85	477.60		
1 Rodman (reinf.)	21.75	174.00	33.65	269.20		
1 Cement Finisher	19.20	153.60	27.00	216.00		
1 Grader, 30,000 Lbs.		422.40		464.65		
1 Paving Mach. & Equip.		1125.00		1237.50	17.58	19.34
88 M.H., Daily Totals		$3111.00		$3978.95	$35.34	$45.21
Crew B-27	Hr.	Daily	Hr.	Daily	Bare Costs	Incl. O&P
1 Labor Foreman (outside)	$17.90	$143.20	$25.95	$207.60	$16.40	$23.77
3 Building Laborers	15.90	381.60	23.05	553.20		
1 Berm Machine		81.70		89.85	2.55	2.80
32 M.H., Daily Totals		$606.50		$850.65	$18.95	$26.57
Crew B-28	Hr.	Daily	Hr.	Daily	Bare Costs	Incl. O&P
2 Carpenters	$20.00	$320.00	$29.00	$464.00	$18.63	$27.01
1 Building Laborer	15.90	127.20	23.05	184.40		
24 M.H., Daily Totals		$447.20		$648.40	$18.63	$27.01
Crew B-29	Hr.	Daily	Hr.	Daily	Bare Costs	Incl. O&P
1 Labor Foreman (outside)	$17.90	$143.20	$25.95	$207.60	$17.15	$24.86
4 Building Laborers	15.90	508.80	23.05	737.60		
1 Equip. Oper. (crane)	21.05	168.40	30.50	244.00		
1 Equip. Oper. Oiler	17.50	140.00	25.40	203.20		
1 Gradall, 3 Ton, 1/2 C.Y.		440.40		484.45	7.86	8.65
56 M.H., Daily Totals		$1400.80		$1876.85	$25.01	$33.51

CREWS

Crew No.	Bare Costs		Incl. Subs O & P		Cost Per Man-hour	
Crew B-30	Hr.	Daily	Hr.	Daily	Bare Costs	Incl. O&P
1 Equip. Oper. (med.)	$20.60	$164.80	$29.85	$238.80	$17.93	$25.81
2 Truck Drivers (heavy)	16.60	265.60	23.80	380.80		
1 Hyd. Excavator, 1.5 C.Y.		579.60		637.55		
2 Dump Trucks, 16 Ton		559.20		615.10	47.45	52.19
24 M.H., Daily Totals		$1569.20		$1872.25	$65.38	$78.00
Crew B-31	Hr.	Daily	Hr.	Daily	Bare Costs	Incl. O&P
1 Labor Foreman (outside)	$17.90	$143.20	$25.95	$207.60	$17.12	$24.82
3 Building Laborers	15.90	381.60	23.05	553.20		
1 Carpenter	20.00	160.00	29.00	232.00		
1 Air Compr., 250 C.F.M.		111.40		122.55		
1 Sheeting Driver		5.05		5.55		
2-50 Ft Air Hoses, 1.5" Dia.		13.20		14.50	3.24	3.56
40 M.H., Daily Totals		$814.45		$1135.40	$20.36	$28.38
Crew B-32	Hr.	Daily	Hr.	Daily	Bare Costs	Incl. O&P
2 Equip. Oper. (med.)	$20.60	$329.60	$29.85	$477.60	$20.60	$29.85
1 Grader, 30,000 lbs.		422.40		464.65		
1 Tandem Roller, 10 Ton		175.20		192.70	37.35	41.08
16 M.H., Daily Totals		$927.20		$1134.95	$57.95	$70.93
Crew B-33	Hr.	Daily	Hr.	Daily	Bare Costs	Incl. O&P
1 Equip. Oper. (med.)	$20.60	$164.80	$29.85	$238.80	$19.25	$27.90
.5 Building Laborer	15.90	63.60	23.05	92.20		
.25 Equip. Oper. (med.)	20.60	41.20	29.85	59.70		
14 M.H., Daily Totals		$269.60		$390.70	$19.25	$27.90
Crew B-33A	Hr.	Daily	Hr.	Daily	Bare Costs	Incl. O&P
1 Equip. Oper. (med.)	$20.60	$164.80	$29.85	$238.80	$19.25	$27.90
.5 Building Laborer	15.90	63.60	23.05	92.20		
.25 Equip. Oper. (med.)	20.60	41.20	29.85	59.70		
1 Scraper, Towed, 7 C.Y.		89.85		98.85		
1 Dozer, 300 H.P.		931.00		1024.10		
.25 Dozer, 300 H.P.		232.75		256.05	89.54	98.50
14 M.H., Daily Totals		$1523.20		$1769.70	$108.79	$126.40
Crew B-33B	Hr.	Daily	Hr.	Daily	Bare Costs	Incl. O&P
1 Equip. Oper. (med.)	$20.60	$164.80	$29.85	$238.80	$19.25	$27.90
.5 Building Laborer	15.90	63.60	23.05	92.20		
.25 Equip. Oper. (med.)	20.60	41.20	29.85	59.70		
1 Scraper, Towed, 10 C.Y.		120.30		132.35		
1 Dozer, 300 H.P.		931.00		1024.10		
.25 Dozer, 300 H.P.		232.75		256.05	91.71	100.89
14 M.H., Daily Totals		$1553.65		$1803.20	$110.96	$128.79
Crew B-33C	Hr.	Daily	Hr.	Daily	Bare Costs	Incl. O&P
1 Equip. Oper. (med.)	$20.60	$164.80	$29.85	$238.80	$19.25	$27.90
.5 Building Laborer	15.90	63.60	23.05	92.20		
.25 Equip. Oper. (med.)	20.60	41.20	29.85	59.70		
1 Scraper, Towed, 12 C.Y.		120.30		132.35		
1 Dozer, 300 H.P.		931.00		1024.10		
.25 Dozer, 300 H.P.		232.75		256.05	91.71	100.89
14 M.H., Daily Totals		$1553.65		$1803.20	$110.96	$128.79
Crew B-33D	Hr.	Daily	Hr.	Daily	Bare Costs	Incl. O&P
1 Equip. Oper. (med.)	$20.60	$164.80	$29.85	$238.80	$19.25	$27.90
.5 Building Laborer	15.90	63.60	23.05	92.20		
.25 Equip. Oper. (med.)	20.60	41.20	29.85	59.70		
1 S.P. Scraper, 14 C.Y.		964.00		1060.40		
.25 Dozer, 300 H.P.		232.75		256.05	85.48	94.03
14 M.H., Daily Totals		$1466.35		$1707.15	$104.73	$121.93
Crew B-33E	Hr.	Daily	Hr.	Daily	Bare Costs	Incl. O&P
1 Equip. Oper. (med.)	$20.60	$164.80	$29.85	$238.80	$19.25	$27.90
.5 Building Laborer	15.90	63.60	23.05	92.20		
.25 Equip. Oper. (med.)	20.60	41.20	29.85	59.70		
1 S.P. Scraper, 24 C.Y.		1115.00		1226.50		
.25 Dozer, 300 H.P.		232.75		256.05	96.26	105.89
14 M.H., Daily Totals		$1617.35		$1873.25	$115.51	$133.79
Crew B-33F	Hr.	Daily	Hr.	Daily	Bare Costs	Incl. O&P
1 Equip. Oper. (med.)	$20.60	$164.80	$29.85	$238.80	$19.25	$27.90
.5 Building Laborer	15.90	63.60	23.05	92.20		
.25 Equip. Oper. (med.)	20.60	41.20	29.85	59.70		
1 Elev. Scraper, 11 C.Y.		632.00		695.20		
.25 Dozer, 300 H.P.		232.75		256.05	61.76	67.94
14 M.H., Daily Totals		$1134.35		$1341.95	$81.01	$95.84
Crew B-33G	Hr.	Daily	Hr.	Daily	Bare Costs	Incl. O&P
1 Equip. Oper. (med.)	$20.60	$164.80	$29.85	$238.80	$19.25	$27.90
.5 Building Laborer	15.90	63.60	23.05	92.20		
.25 Equip. Oper. (med.)	20.60	41.20	29.85	59.70		
1 Elev. Scraper, 20 C.Y.		948.00		1042.80		
.25 Dozer, 300 H.P.		232.75		256.05	84.33	92.77
14 M.H., Daily Totals		$1450.35		$1689.55	$103.58	$120.67
Crew B-34	Hr.	Daily	Hr.	Daily	Bare Costs	Incl. O&P
1 Truck Driver (heavy)	$16.60	$132.80	$23.80	$190.40	$16.60	$23.80
8 M.H., Daily Totals		$132.80		$190.40	$16.60	$23.80
Crew B-34A	Hr.	Daily	Hr.	Daily	Bare Costs	Incl. O&P
1 Truck Driver (heavy)	$16.60	$132.80	$23.80	$190.40	$16.60	$23.80
1 Dump Truck, 12 Ton		220.20		242.20	27.52	30.27
8 M.H., Daily Totals		$353.00		$432.60	$44.12	$54.07
Crew B-34B	Hr.	Daily	Hr.	Daily	Bare Costs	Incl. O&P
1 Truck Driver (heavy)	$16.60	$132.80	$23.80	$190.40	$16.60	$23.80
1 Dump Truck, 16 Ton		279.60		307.55	34.95	38.44
8 M.H., Daily Totals		$412.40		$497.95	$51.55	$62.24
Crew B-34C	Hr.	Daily	Hr.	Daily	Bare Costs	Incl. O&P
1 Truck Driver (heavy)	$16.60	$132.80	$23.80	$190.40	$16.60	$23.80
1 Truck Tractor, 40 Ton		300.40		330.45		
1 Dump Trailer, 16.5 C.Y.		80.15		88.15	47.56	52.32
8 M.H., Daily Totals		$513.35		$609.00	$64.16	$76.12
Crew B-34D	Hr.	Daily	Hr.	Daily	Bare Costs	Incl. O&P
1 Truck Driver (heavy)	$16.60	$132.80	$23.80	$190.40	$16.60	$23.80
1 Truck Tractor, 40 Ton		300.40		330.45		
1 Dump Trailer, 20 C.Y.		95.20		104.70	49.45	54.39
8 M.H., Daily Totals		$528.40		$625.55	$66.05	$78.19

CREWS

Crew No.	Bare Costs		Incl. Subs O & P		Cost Per Man-hour	
Crew B-34E	Hr.	Daily	Hr.	Daily	Bare Costs	Incl. O&P
1 Truck Driver (heavy)	$16.60	$132.80	$23.80	$190.40	$16.60	$23.80
1 Truck, Off Highway, 25 Ton		648.00		712.80	81.00	89.10
8 M.H., Daily Totals		$780.80		$903.20	$97.60	$112.90
Crew B-35	Hr.	Daily	Hr.	Daily	Bare Costs	Incl. O&P
1 Plumber Foreman (ins)	$23.05	$184.40	$33.35	$266.80	$20.79	$30.09
1 Plumber	22.55	180.40	32.60	260.80		
1 Welder (plumber)	22.55	180.40	32.60	260.80		
1 Plumber Apprentice	18.04	144.32	26.10	208.80		
1 Equip. Oper. (crane)	21.05	168.40	30.50	244.00		
1 Equip. Oper. Oiler	17.50	140.00	25.40	203.20		
1 Electric Welding Mach.		18.60		20.45		
1 Hyd. Excavator, .75 C.Y.		340.20		374.20	7.47	8.22
48 M.H., Daily Totals		$1356.72		$1839.05	$28.26	$38.31
Crew B-36	Hr.	Daily	Hr.	Daily	Bare Costs	Incl. O&P
1 Labor Foreman (outside)	$17.90	$143.20	$25.95	$207.60	$17.57	$25.47
2 Building Laborers	15.90	254.40	23.05	368.80		
1 Equip. Oper. (med.)	20.60	164.80	29.85	238.80		
1 Dozer, 200 H.P.		634.80		698.30		
1 Aggregate Spreader		51.20		56.30	21.43	23.58
32 M.H., Daily Totals		$1248.40		$1569.80	$39.00	$49.05
Crew B-37	Hr.	Daily	Hr.	Daily	Bare Costs	Incl. O&P
1 Labor Foreman (outside)	$17.90	$143.20	$25.95	$207.60	$16.82	$24.39
4 Building Laborers	15.90	508.80	23.05	737.60		
1 Equip. Oper. (light)	19.45	155.60	28.20	225.60		
1 Tandem Roller, 5 Ton		85.00		93.50	1.77	1.94
48 M.H., Daily Totals		$892.60		$1264.30	$18.59	$26.33
Crew B-38	Hr.	Daily	Hr.	Daily	Bare Costs	Incl. O&P
2 Building Laborers	$15.90	$254.40	$23.05	$368.80	$17.08	$24.76
1 Equip. Oper. (light)	19.45	155.60	28.20	225.60		
1 Backhoe Loader, 48 H.P.		156.00		171.60		
1 Demol. Hammer, Hyd.		212.45		233.70	15.35	16.88
24 M.H., Daily Totals		$778.45		$999.70	$32.43	$41.64
Crew B-39	Hr.	Daily	Hr.	Daily	Bare Costs	Incl. O&P
1 Labor Foreman (outside)	$17.90	$143.20	$25.95	$207.60	$16.82	$24.39
4 Building Laborers	15.90	508.80	23.05	737.60		
1 Equipment Oper. (light)	19.45	155.60	28.20	225.60		
1 Air Compr., 250 C.F.M.		111.40		122.55		
Air Tools & Accessories		25.80		28.40		
2-50 Ft. Air Hoses, 1.5" Dia		13.20		14.50	3.13	3.44
48 M.H., Daily Totals		$958.00		$1336.25	$19.95	$27.83
Crew B-40	Hr.	Daily	Hr.	Daily	Bare Costs	Incl. O&P
1 Pile Driver Foreman	$22.10	$176.80	$34.65	$277.20	$20.26	$30.88
4 Pile Drivers	20.10	643.20	31.50	1008.00		
2 Equip. Oper. (crane)	21.05	336.80	30.50	488.00		
1 Equip. Oper. Oiler	17.50	140.00	25.40	203.20		
1 Crane, 40 Ton		483.40		531.75		
Vibratory Hammer & Gen.		993.00		1092.30	23.06	25.37
64 M.H., Daily Totals		$2773.20		$3600.45	$43.32	$56.25

Crew No.	Bare Costs		Incl. Subs O & P		Cost Per Man-hour	
Crew B-41	Hr.	Daily	Hr.	Daily	Bare Costs	Incl. O&P
1 Labor Foreman (outside)	$17.90	$143.20	$25.95	$207.60	$16.57	$24.02
4 Building Laborers	15.90	508.80	23.05	737.60		
.25 Equip. Oper. (crane)	21.05	42.10	30.50	61.00		
.25 Equip. Oper. Oiler	17.50	35.00	25.40	50.80		
.25 Crawler Crane, 40 Ton		120.85		132.95	2.74	3.02
44 M.H., Daily Totals		$849.95		$1189.95	$19.31	$27.04
Crew B-42	Hr.	Daily	Hr.	Daily	Bare Costs	Incl. O&P
1 Labor Foreman (outside)	$17.90	$143.20	$25.95	$207.60	$17.71	$26.01
4 Building Laborers	15.90	508.80	23.05	737.60		
1 Equip. Oper. (crane)	21.05	168.40	30.50	244.00		
1 Equip. Oper. Oiler	17.50	140.00	25.40	203.20		
1 Welder	21.70	173.60	34.10	272.80		
1 Hyd. Crane, 25 Ton		384.80		423.30		
1 Gas Welding Machine		56.80		62.50		
1 Horz. Boring Csg. Mch.		482.60		530.85	14.44	15.88
64 M.H., Daily Totals		$2058.20		$2681.85	$32.15	$41.89
Crew B-43	Hr.	Daily	Hr.	Daily	Bare Costs	Incl. O&P
1 Labor Foreman (outside)	$17.90	$143.20	$25.95	$207.60	$17.35	$25.16
3 Building Laborers	15.90	381.60	23.05	553.20		
1 Equip. Oper. (crane)	21.05	168.40	30.50	244.00		
1 Equip. Oper. Oiler	17.50	140.00	25.40	203.20		
1 Drill Rig & Augers		546.25		600.85	11.38	12.51
48 M.H., Daily Totals		$1379.45		$1808.85	$28.73	$37.67
Crew B-44	Hr.	Daily	Hr.	Daily	Bare Costs	Incl. O&P
1 Pile Driver Foreman	$22.10	$176.80	$34.65	$277.20	$20.06	$30.58
4 Pile Drivers	20.10	643.20	31.50	1008.00		
2 Equip. Oper. (crane)	21.05	336.80	30.50	488.00		
1 Building Laborer	15.90	127.20	23.05	184.40		
1 Crane, 40 Ton, & Access.		845.95		930.55		
45 L.F. Leads, 15K Ft. Lbs.		36.00		39.60	13.78	15.15
64 M.H., Daily Totals		$2165.95		$2927.75	$33.84	$45.73
Crew B-45	Hr.	Daily	Hr.	Daily	Bare Costs	Incl. O&P
1 Equip. Oper. (med.)	$20.60	$164.80	$29.85	$238.80	$18.60	$26.82
1 Truck Driver (heavy)	16.60	132.80	23.80	190.40		
1 Dist. Tank Truck, 3K Gal.		243.20		267.50	15.20	16.71
16 M.H., Daily Totals		$540.80		$696.70	$33.80	$43.53
Crew B-46	Hr.	Daily	Hr.	Daily	Bare Costs	Incl. O&P
1 Pile Driver Foreman	$22.10	$176.80	$34.65	$277.20	$18.33	$27.80
2 Pile Drivers	20.10	321.60	31.50	504.00		
3 Building Laborers	15.90	381.60	23.05	553.20		
1 Chain Saw, 36" Long		28.20		31.00	.58	.64
48 M.H., Daily Totals		$908.20		$1365.40	$18.91	$28.44
Crew B-47	Hr.	Daily	Hr.	Daily	Bare Costs	Incl. O&P
1 Blast Foreman (outside)	$17.90	$143.20	$25.95	$207.60	$17.75	$25.73
1 Driller	15.90	127.20	23.05	184.40		
1 Equip. Oper. (light)	19.45	155.60	28.20	225.60		
1 Crawler Type Drill, 4"		172.20		189.40		
1 Air Compr., 600 C.F.M.		242.80		267.10		
2-50 Ft. Air Hoses, 3" Dia		19.10		21.00	18.08	19.89
24 M.H., Daily Totals		$860.10		$1095.10	$35.83	$45.62

CREWS

Crew No.	Bare Costs		Incl. Subs O & P		Cost Per Man-hour	
Crew B-48	Hr.	Daily	Hr.	Daily	Bare Costs	Incl. O&P
1 Labor Foreman (outside)	$17.90	$143.20	$25.95	$207.60	$17.65	$25.60
3 Building Laborers	15.90	381.60	23.05	553.20		
1 Equip. Oper. (crane)	21.05	168.40	30.50	244.00		
1 Equip. Oper. Oiler	17.50	140.00	25.40	203.20		
1 Equip. Oper. (light)	19.45	155.60	28.20	225.60		
1 Centr. Water Pump, 6"		61.60		67.75		
1-20 Ft. Suction Hose, 6"		14.40		15.85		
1-50 Ft. Disch. Hose, 6"		13.00		14.30		
1 Drill Rig & Augers		546.25		600.85	11.34	12.47
56 M.H., Daily Totals		$1624.05		$2132.35	$28.99	$38.07
Crew B-49	Hr.	Daily	Hr.	Daily	Bare Costs	Incl. O&P
1 Labor Foreman (outside)	$17.90	$143.20	$25.95	$207.60	$18.39	$27.10
3 Building Laborers	15.90	381.60	23.05	553.20		
2 Equip. Oper. (crane)	21.05	336.80	30.50	488.00		
2 Equip. Oper. Oilers	17.50	280.00	25.40	406.40		
1 Equip. Oper. (light)	19.45	155.60	28.20	225.60		
2 Pile Drivers	20.10	321.60	31.50	504.00		
1 Hyd. Crane, 25 Ton		384.80		423.30		
1 Centr. Water Pump, 6"		61.60		67.75		
1-20 Ft. Suction Hose, 6"		14.40		15.85		
1-50 Ft. Disch. Hose, 6"		13.00		14.30		
1 Drill Rig & Augers		546.25		600.85	11.59	12.75
88 M.H., Daily Totals		$2638.85		$3506.85	$29.98	$39.85
Crew B-50	Hr.	Daily	Hr.	Daily	Bare Costs	Incl. O&P
2 Pile Driver Foremen	$22.10	$353.60	$34.65	$554.40	$19.43	$29.56
6 Pile Drivers	20.10	964.80	31.50	1512.00		
2 Equip. Oper. (crane)	21.05	336.80	30.50	488.00		
1 Equip. Oper. Oiler	17.50	140.00	25.40	203.20		
3 Building Laborers	15.90	381.60	23.05	553.20		
1 Crane, 40 Ton		483.40		531.75		
60 L.F. Leads, 15K Ft. Lbs.		48.00		52.80		
1 Hammer, 15K Ft. Lbs.		200.60		220.65		
1 Air Compr., 600 C.F.M.		242.80		267.10		
2-50 Ft. Air Hoses, 3" Dia		19.10		21.00		
1 Chain Saw, 36" Long		28.20		31.00	9.12	10.03
112 M.H., Daily Totals		$3198.90		$4435.10	$28.55	$39.59
Crew B-51	Hr.	Daily	Hr.	Daily	Bare Costs	Incl. O&P
1 Labor Foreman (outside)	$17.90	$143.20	$25.95	$207.60	$16.35	$23.65
4 Building Laborers	15.90	508.80	23.05	737.60		
1 Truck Driver (heavy)	16.60	132.80	23.80	190.40		
1 Light Truck, 1.5 Ton		57.40		63.15	1.19	1.31
48 M.H., Daily Totals		$842.20		$1198.75	$17.54	$24.96
Crew B-52	Hr.	Daily	Hr.	Daily	Bare Costs	Incl. O&P
1 Carpenter Foreman	$22.00	$176.00	$31.90	$255.20	$18.58	$26.97
1 Carpenter	20.00	160.00	29.00	232.00		
3 Building Laborers	15.90	381.60	23.05	553.20		
1 Cement Finisher	19.20	153.60	27.00	216.00		
.5 Rodman (reinf.)	21.75	87.00	33.65	134.60		
.5 Equip. Oper. (med.)	20.60	82.40	29.85	119.40		
.5 F.E. Ldr., T.M., 2.5 C.Y.		300.00		330.00	5.35	5.89
56 M.H., Daily Totals		$1340.60		$1840.40	$23.93	$32.86
Crew B-53	Hr.	Daily	Hr.	Daily	Bare Costs	Incl. O&P
1 Equip. Oper. (light)	$19.45	$155.60	$28.20	$225.60	$19.45	$28.20
1 Trencher, Chain, 12 H.P.		135.40		148.95	16.92	18.61
8 M.H., Daily Totals		$291.00		$374.55	$36.37	$46.81

Crew No.	Bare Costs		Incl. Subs O & P		Cost Per Man-hour	
Crew B-54	Hr.	Daily	Hr.	Daily	Bare Costs	Incl. O&P
1 Equip. Oper. (light)	$19.45	$155.60	$28.20	$225.60	$19.45	$28.20
1 Trencher, Chain, 40 H.P.		199.80		219.80	24.97	27.47
8 M.H., Daily Totals		$355.40		$445.40	$44.42	$55.67
Crew B-55	Hr.	Daily	Hr.	Daily	Bare Costs	Incl. O&P
1 Building Laborer	$15.90	$127.20	$23.05	$184.40	$16.12	$23.25
1 Truck Driver (light)	16.35	130.80	23.45	187.60		
1 Flatbed Truck w/Auger		146.60		161.25		
1 Truck, 3 Ton		67.20		73.90	13.36	14.69
16 M.H., Daily Totals		$471.80		$607.15	$29.48	$37.94
Crew B-56	Hr.	Daily	Hr.	Daily	Bare Costs	Incl. O&P
1 Building Laborer	$15.90	$127.20	$23.05	$184.40	$17.67	$25.62
1 Equip. Oper. (light)	19.45	155.60	28.20	225.60		
1 Crawler Type Drill, 4"		172.20		189.40		
1 Air Compr., 600 C.F.M.		242.80		267.10		
1-50 Ft. Air Hose, 3" dia		9.55		10.50	26.53	29.18
16 M.H., Daily Totals		$707.35		$877.00	$44.20	$54.80
Crew B-57	Hr.	Daily	Hr.	Daily	Bare Costs	Incl. O&P
1 Labor Foreman (outside)	$17.90	$143.20	$25.95	$207.60	$17.95	$26.02
2 Building Laborers	15.90	254.40	23.05	368.80		
1 Equip. Oper. (crane)	21.05	168.40	30.50	244.00		
1 Equip. Oper. (light)	19.45	155.60	28.20	225.60		
1 Equip. Oper. Oiler	17.50	140.00	25.40	203.20		
1 Power Shovel, 1 C.Y.		387.60		426.35		
1 Clamshell Bucket, 1 C.Y.		40.15		44.15		
1 Centr. Water Pump, 6"		61.60		67.75		
1-20 Ft. Suction Hose, 6"		14.40		15.85		
20-50 Ft. Disch. Hoses, 6"		260.00		286.00	15.91	17.50
48 M.H., Daily Totals		$1625.35		$2089.30	$33.86	$43.52
Crew B-58	Hr.	Daily	Hr.	Daily	Bare Costs	Incl. O&P
2 Building Laborers	$15.90	$254.40	$23.05	$368.80	$17.08	$24.76
1 Equip. Oper. (light)	19.45	155.60	28.20	225.60		
1 Backhoe Loader, 48 H.P.		156.00		171.60		
1 Small Helicopter		1080.00		1188.00	51.50	56.65
24 M.H., Daily Totals		$1646.00		$1954.00	$68.58	$81.41
Crew B-59	Hr.	Daily	Hr.	Daily	Bare Costs	Incl. O&P
1 Truck Driver (heavy)	$16.60	$132.80	$23.80	$190.40	$16.60	$23.80
1 Wtr. Tank Trk., 5K Gal.		224.30		246.75		
Appropriate Fixtures		214.20		235.60	54.81	60.29
8 M.H., Daily Totals		$571.30		$672.75	$71.41	$84.09
Crew B-60	Hr.	Daily	Hr.	Daily	Bare Costs	Incl. O&P
1 Labor Foreman (outside)	$17.90	$143.20	$25.95	$207.60	$18.16	$26.33
2 Building Laborers	15.90	254.40	23.05	368.80		
1 Equip. Oper. (crane)	21.05	168.40	30.50	244.00		
2 Equip. Oper. (light)	19.45	311.20	28.20	451.20		
1 Equip. Oper. Oiler	17.50	140.00	25.40	203.20		
1 Crawler Crane, 40 Ton		483.40		531.75		
45 L.F. Leads, 15K Ft. Lbs.		36.00		39.60		
1 Backhoe Loader, 48 H.P.		156.00		171.60	12.06	13.26
56 M.H., Daily Totals		$1692.60		$2217.75	$30.22	$39.59

CREWS

Crew No.		Bare Costs		Incl. Subs O & P		Cost Per Man-hour	
Crew B-61	Hr.	Daily	Hr.	Daily	Bare Costs	Incl. O&P
1 Labor Foreman (outside)	$17.90	$143.20	$25.95	$207.60	$17.01	$24.66
3 Building Laborers	15.90	381.60	23.05	553.20		
1 Equip. Oper. (light)	19.45	155.60	28.20	225.60		
1 Cement Mixer, 2 C.Y.		201.80		222.00		
1 Air Compr., 160 C.F.M.		70.25		77.30	6.80	7.48
40 M.H., Daily Totals		$952.45		$1285.70	$23.81	$32.14
Crew B-62	Hr.	Daily	Hr.	Daily	Bare Costs	Incl. O&P
2 Building Laborers	$15.90	$254.40	$23.05	$368.80	$17.08	$24.76
1 Equip. Oper. (light)	19.45	155.60	28.20	225.60		
1 Loader, Skid Steer		96.00		105.60	4.00	4.40
24 M.H., Daily Totals		$506.00		$700.00	$21.08	$29.16
Crew B-63	Hr.	Daily	Hr.	Daily	Bare Costs	Incl. O&P
4 Building Laborers	$15.90	$508.80	$23.05	$737.60	$16.61	$24.08
1 Equip. Oper. (light)	19.45	155.60	28.20	225.60		
1 Loader, Skid Steer		96.00		105.60	2.40	2.64
40 M.H., Daily Totals		$760.40		$1068.80	$19.01	$26.72
Crew B-64	Hr.	Daily	Hr.	Daily	Bare Costs	Incl. O&P
1 Building Laborer	$15.90	$127.20	$23.05	$184.40	$16.12	$23.25
1 Truck Driver (light)	16.35	130.80	23.45	187.60		
1 Power Mulcher (small)		104.05		114.45		
1 Light Truck, 1.5 Ton		57.40		63.15	10.09	11.10
16 M.H., Daily Totals		$419.45		$549.60	$26.21	$34.35
Crew B-65	Hr.	Daily	Hr.	Daily	Bare Costs	Incl. O&P
1 Building Laborer	$15.90	$127.20	$23.05	$184.40	$16.12	$23.25
1 Truck Driver (light)	16.35	130.80	23.45	187.60		
1 Power Mulcher (large)		162.00		178.20		
1 Light Truck, 1.5 Ton		57.40		63.15	13.71	15.08
16 M.H., Daily Totals		$477.40		$613.35	$29.83	$38.33
Crew B-66	Hr.	Daily	Hr.	Daily	Bare Costs	Incl. O&P
1 Equip. Oper. (light)	$19.45	$155.60	$28.20	$225.60	$19.45	$28.20
1 Backhoe Ldr. w/Attchmt.		117.60		129.35	14.70	16.16
8 M.H., Daily Totals		$273.20		$354.95	$34.15	$44.36
Crew B-67	Hr.	Daily	Hr.	Daily	Bare Costs	Incl. O&P
1 Millwright	$20.75	$166.00	$29.35	$234.80	$20.10	$28.77
1 Equip. Oper. (light)	19.45	155.60	28.20	225.60		
1 Forklift		169.40		186.35	10.58	11.64
16 M.H., Daily Totals		$491.00		$646.75	$30.68	$40.41
Crew B-68	Hr.	Daily	Hr.	Daily	Bare Costs	Incl. O&P
2 Millwrights	$20.75	$332.00	$29.35	$469.60	$20.31	$28.96
1 Equip. Oper. (light)	19.45	155.60	28.20	225.60		
1 Forklift		169.40		186.35	7.05	7.76
24 M.H., Daily Totals		$657.00		$881.55	$27.36	$36.72
Crew B-69	Hr.	Daily	Hr.	Daily	Bare Costs	Incl. O&P
1 Labor Foreman (outside)	$17.90	$143.20	$25.95	$207.60	$17.35	$25.16
3 Highway Laborers	15.90	381.60	23.05	553.20		
1 Equip Oper. (crane)	21.05	168.40	30.50	244.00		
1 Equip Oper. Oiler	17.50	140.00	25.40	203.20		
1 Truck Crane, 80 Ton		844.00		928.40	17.58	19.34
48 M.H., Daily Totals		$1677.20		$2136.40	$34.93	$44.50

Crew No.		Bare Costs		Incl. Subs O & P		Cost Per Man-hour	
Crew B-70	Hr.	Daily	Hr.	Daily	Bare Costs	Incl. O&P
1 Labor Foreman (outside)	$17.90	$143.20	$25.95	$207.60	$18.20	$26.37
3 Highway Laborers	15.90	381.60	23.05	553.20		
3 Equip. Oper. (med.)	20.60	494.40	29.85	716.40		
1 Motor Grader, 30,000 Lb.		422.40		464.65		
1 Grader Attach., Ripper		39.15		43.05		
1 Road Sweeper, S.P.		147.20		161.90		
1 F.E. Loader, 1-3/4 C.Y.		271.00		298.10	15.70	17.28
56 M.H., Daily Totals		$1898.95		$2444.90	$33.90	$43.65
Crew B-71	Hr.	Daily	Hr.	Daily	Bare Costs	Incl. O&P
1 Labor Foreman (outside)	$17.90	$143.20	$25.95	$207.60	$18.20	$26.37
3 Highway Laborers	15.90	381.60	23.05	553.20		
3 Equip. Oper. (med.)	20.60	494.40	29.85	716.40		
1 Pvmt. Profiler, 450 H.P.		1708.00		1878.80		
1 Road Sweeper, S.P.		147.20		161.90		
1 F.E. Loader, 1-3/4 C.Y.		271.00		298.10	37.96	41.76
56 M.H., Daily Totals		$3145.40		$3816.00	$56.16	$68.13
Crew B-72	Hr.	Daily	Hr.	Daily	Bare Costs	Incl. O&P
1 Labor Foreman (outside)	$17.90	$143.20	$25.95	$207.60	$18.50	$26.81
3 Highway Laborers	15.90	381.60	23.05	553.20		
4 Equip. Oper. (med.)	20.60	659.20	29.85	955.20		
1 Mix Paver 165 H.P.		986.00		1084.60		
1 Roller, Pneu. Tire, 12 T.		158.60		174.45		
1 Pvmt. Profiler, 450 H.P.		1708.00		1878.80		
1 Hammermill, 250 H.P.		750.80		825.90		
1 Windrow Loader		449.40		494.35	63.32	69.65
64 M.H., Daily Totals		$5236.80		$6174.10	$81.82	$96.46
Crew B-73	Hr.	Daily	Hr.	Daily	Bare Costs	Incl. O&P
1 Labor Foreman (outside)	$17.90	$143.20	$25.95	$207.60	$19.08	$27.66
2 Highway Laborers	15.90	254.40	23.05	368.80		
5 Equip. Oper. (med.)	20.60	824.00	29.85	1194.00		
1 Road Mixer, 310 H.P.		1364.00		1500.40		
1 Roller, Tandem, 12 Ton		175.20		192.70		
1 Hammermill, 250 H.P.		750.80		825.90		
1 Motor Grader, 30,000 Lb.		422.40		464.65		
.5 F.E. Loader, 1-3/4 C.Y.		135.50		149.05		
.5 Water Truck, 5000 Gal.		112.15		123.35		
Appropriate Fixtures		107.10		117.80	47.92	52.71
64 M.H., Daily Totals		$4288.75		$5144.25	$67.00	$80.37
Crew B-74	Hr.	Daily	Hr.	Daily	Bare Costs	Incl. O&P
1 Labor Foreman (outside)	$17.90	$143.20	$25.95	$207.60	$18.67	$27.00
1 Highway Laborer	15.90	127.20	23.05	184.40		
4 Equip. Oper. (med.)	20.60	659.20	29.85	955.20		
2 Truck Drivers (heavy)	16.60	265.60	23.80	380.80		
1 Motor Grader, 30,000 Lb.		422.40		464.65		
1 Grader Attach., Ripper		39.15		43.05		
2 Stabilizers, 310 H.P.		1554.00		1709.40		
1 Flatbed Truck, 3 Ton		67.20		73.90		
1 Chem. Spreader, Towed		78.40		86.25		
1 Vibr. Roller, 29,000 Lb.		345.50		380.05		
1 Water Truck, 5000 Gal.		214.20		235.60		
Appropriate Fixtures		224.30		246.75	46.01	50.61
64 M.H., Daily Totals		$4140.35		$4967.65	$64.68	$77.61

CREWS

Crew No.	Bare Costs		Incl. Subs O & P		Cost Per Man-hour	
Crew B-75	Hr.	Daily	Hr.	Daily	Bare Costs	Incl. O&P
1 Labor Foreman (outside)	$17.90	$143.20	$25.95	$207.60	$18.97	$27.45
1 Highway Laborer	15.90	127.20	23.05	184.40		
4 Equip. Oper. (med.)	20.60	659.20	29.85	955.20		
1 Truck Driver (heavy)	16.60	132.80	23.80	190.40		
1 Motor Grader, 30,000 Lb.		422.40		464.65		
1 Grader Attach., Ripper		39.15		43.05		
2 Stabilizers, 310 H.P.		1554.00		1709.40		
1 Dist. Truck, 3000 Gal.		243.20		267.50		
1 Vibr. Roller, 29,000 Lb.		345.50		380.05	46.50	51.15
56 M.H., Daily Totals		$3666.65		$4402.25	$65.47	$78.60
Crew B-76	Hr.	Daily	Hr.	Daily	Bare Costs	Incl. O&P
1 Dock Builder Foreman	$22.10	$176.80	$34.65	$277.20	$20.24	$30.95
5 Dock Builders	20.10	804.00	31.50	1260.00		
2 Equip. Oper. (crane)	21.05	336.80	30.50	488.00		
1 Equip. Oper. Oiler	17.50	140.00	25.40	203.20		
1 Crawler Crane, 60 Ton		640.00		704.00		
1 Barge, 400 Ton		360.95		397.05		
1 Hammer, 15K. Ft. Lbs.		200.60		220.65		
60 L.F. Leads, 15K. Ft. Lbs.		48.00		52.80		
1 Air Compr., 600 C.F.M.		242.80		267.10		
2-50 Ft. Air Hoses, 3" Dia.		19.10		21.00	20.99	23.09
72 M.H., Daily Totals		$2969.05		$3891.00	$41.23	$54.04
Crew B-77	Hr.	Daily	Hr.	Daily	Bare Costs	Incl. O&P
1 Labor Foreman (outside)	$17.90	$143.20	$25.95	$207.60	$16.40	$23.77
3 Highway Laborers	15.90	381.60	23.05	553.20		
1 Crack Cleaner, 25 H.P.		63.40		69.75		
1 Crack Filler, Trailer Mtd.		105.60		116.15		
1 Flatbed Truck, 3 Ton		67.20		73.90	7.38	8.11
32 M.H., Daily Totals		$761.00		$1020.60	$23.78	$31.88
Crew B-78	Hr.	Daily	Hr.	Daily	Bare Costs	Incl. O&P
1 Labor Foreman (outside)	$17.90	$143.20	$25.95	$207.60	$16.30	$23.63
4 Highway Laborers	15.90	508.80	23.05	737.60		
1 Paint Striper, S.P.		158.40		174.25		
1 Flatbed Truck, 3 Ton		67.20		73.90		
1 Pickup Truck, 3/4 Ton		63.60		69.95	7.23	7.95
40 M.H., Daily Totals		$941.20		$1263.30	$23.53	$31.58
Crew B-79	Hr.	Daily	Hr.	Daily	Bare Costs	Incl. O&P
1 Labor Foreman (outside)	$17.90	$143.20	$25.95	$207.60	$16.40	$23.77
3 Highway Laborers	15.90	381.60	23.05	553.20		
1 Thermo. Striper, T.M.		198.00		217.80		
1 Flatbed Truck, 3 Ton		67.20		73.90		
2 Pickup Trucks, 3/4 Ton		127.20		139.90	12.26	13.48
32 M.H., Daily Totals		$917.20		$1192.40	$28.66	$37.25
Crew B-80	Hr.	Daily	Hr.	Daily	Bare Costs	Incl. O&P
1 Labor Foreman (outside)	$17.90	$143.20	$25.95	$207.60	$17.75	$25.73
1 Highway Laborer	15.90	127.20	23.05	184.40		
1 Equip. Oper. (light)	19.45	155.60	28.20	225.60		
1 Flatbed Truck, 3 Ton		67.20		73.90		
1 Post Driver, T.M.		192.00		211.20	10.80	11.87
24 M.H., Daily Totals		$685.20		$902.70	$28.55	$37.60

Crew No.	Bare Costs		Incl. Subs O & P		Cost Per Man-hour	
Crew B-81	Hr.	Daily	Hr.	Daily	Bare Costs	Incl. O&P
1 Equip. Oper. (med.)	$20.60	$164.80	$29.85	$238.80	$18.60	$26.82
1 Truck Driver (heavy)	16.60	132.80	23.80	190.40		
1 Hydromulcher, T.M.		136.20		149.80		
1 Tractor Truck, 4x2		224.30		246.75	22.53	24.78
16 M.H., Daily Totals		$658.10		$825.75	$41.13	$51.60
Crew B-82	Hr.	Daily	Hr.	Daily	Bare Costs	Incl. O&P
1 Highway Laborer	$15.90	$127.20	$23.05	$184.40	$17.67	$25.62
1 Equip. Oper. (light)	19.45	155.60	28.20	225.60		
1 Horiz. Borer, 6 H.P.		33.00		36.30	2.06	2.26
16 M.H., Daily Totals		$315.80		$446.30	$19.73	$27.88
Crew B-83	Hr.	Daily	Hr.	Daily	Bare Costs	Incl. O&P
1 Tugboat Captain	$20.60	$164.80	$29.85	$238.80	$18.25	$26.45
1 Tugboat Hand	15.90	127.20	23.05	184.40		
1 Tugboat, 250 H.P.		396.00		435.60	24.75	27.22
16 M.H., Daily Totals		$688.00		$858.80	$43.00	$53.67
Crew B-84	Hr.	Daily	Hr.	Daily	Bare Costs	Incl. O&P
1 Equip. Oper. (med.)	$20.60	$164.80	$29.85	$238.80	$20.60	$29.85
1 Rotary Mower/Tractor		146.20		160.80	18.27	20.10
8 M.H., Daily Totals		$311.00		$399.60	$38.87	$49.95
Crew B-85	Hr.	Daily	Hr.	Daily	Bare Costs	Incl. O&P
1 Highway Laborer	$15.90	$127.20	$23.05	$184.40	$17.70	$25.56
1 Equip. Oper. (med.)	20.60	164.80	29.85	238.80		
1 Truck Driver (heavy)	16.60	132.80	23.80	190.40		
1 Aerial Lift Truck		282.40		310.65		
1 Brush Chipper, 130 H.P.		137.00		150.70		
1 Pruning Saw, Rotary		16.30		17.95	18.15	19.97
24 M.H., Daily Totals		$860.50		$1092.90	$35.85	$45.53
Crew B-86	Hr.	Daily	Hr.	Daily	Bare Costs	Incl. O&P
1 Equip. Oper. (med.)	$20.60	$164.80	$29.85	$238.80	$20.60	$29.85
1 Stump Chipper, S.P.		160.00		176.00	20.00	22.00
8 M.H., Daily Totals		$324.80		$414.80	$40.60	$51.85
Crew B-87	Hr.	Daily	Hr.	Daily	Bare Costs	Incl. O&P
1 Common Laborer	$15.90	$127.20	$23.05	$184.40	$19.66	$28.49
4 Equip. Oper. (med.)	20.60	659.20	29.85	955.20		
2 Feller Bunchers, 50 H.P.		527.60		580.35		
1 Log Chipper, 22" Tree		1554.00		1709.40		
1 Dozer, 105 H.P.		206.00		226.60		
1 Chainsaw, Gas, 36" Long		28.20		31.00	57.89	63.68
40 M.H., Daily Totals		$3102.20		$3686.95	$77.55	$92.17
Crew B-88	Hr.	Daily	Hr.	Daily	Bare Costs	Incl. O&P
1 Common Laborer	$15.90	$127.20	$23.05	$184.40	$19.92	$28.87
6 Equip. Oper. (med.)	20.60	988.80	29.85	1432.80		
2 Feller Bunchers, 50 H.P.		527.60		580.35		
1 Log Chipper, 22" Tree		1554.00		1709.40		
2 Log Skidders, 50 H.P.		472.00		519.20		
1 Dozer, 105 H.P.		206.00		226.60		
1 Chainsaw, Gas, 36" Long		28.20		31.00	49.78	54.75
56 M.H., Daily Totals		$3903.80		$4683.75	$69.70	$83.62

CREWS

Crew No.	Bare Costs		Incl. Subs O & P		Cost Per Man-hour	
Crew B-89	Hr.	Daily	Hr.	Daily	Bare Costs	Incl. O&P
1 Equip. Oper. (light)	$19.45	$155.60	$28.20	$225.60	$19.45	$28.20
1 Stake Body, 3 Ton		648.00		712.80		
1 Air Compressor		51.40		56.55		
1 Water Tank, 65 Gal.		18.45		20.30		
1 Generator, 10 K.W.		54.60		60.05		
1 Concrete Saw		36.60		40.25	101.13	111.24
8 M.H., Daily Totals		$964.65		$1115.55	$120.58	$139.44
Crew B-90	Hr.	Daily	Hr.	Daily	Bare Costs	Incl. O&P
1 Labor Foreman (outside)	$17.90	$143.20	$25.95	$207.60	$17.21	$24.88
3 Highway Laborers	15.90	381.60	23.05	553.20		
2 Equip. Oper. (light)	19.45	311.20	28.20	451.20		
2 Truck Drivers (heavy)	16.60	265.60	23.80	380.80		
1 Asphalt Emulsion Paver		1364.00		1500.40		
1 Dist. Truck, 2000 Gal.		212.80		234.10	24.63	27.10
64 M.H., Daily Totals		$2678.40		$3327.30	$41.84	$51.98
Crew B-91	Hr.	Daily	Hr.	Daily	Bare Costs	Incl. O&P
1 Labor Foreman (outside)	$17.90	$143.20	$25.95	$207.60	$18.58	$26.90
2 Highway Laborers	15.90	254.40	23.05	368.80		
4 Equip. Oper. (med.)	20.60	659.20	29.85	955.20		
1 Truck Driver (heavy)	16.60	132.80	23.80	190.40		
1 Dist. Truck, 3000 Gal.		243.20		267.50		
1 Aggreg. Spreader, S.P.		367.60		404.35		
1 Roller, Pneu. Tire, 12 Ton		158.60		174.45		
1 Roller, Steel, 10 Ton		175.20		192.70	14.75	16.23
64 M.H., Daily Totals		$2134.20		$2761.00	$33.33	$43.13
Crew B-92	Hr.	Daily	Hr.	Daily	Bare Costs	Incl. O&P
1 Labor Foreman (outside)	$17.90	$143.20	$25.95	$207.60	$16.40	$23.77
3 Highway Laborers	15.90	381.60	23.05	553.20		
1 Crack Cleaner, 25 H.P.		63.40		69.75		
1 Air Compressor		51.40		56.55		
1 Tar Kettle, T.M.		33.40		36.75		
1 Flatbed Truck, 3 Ton		67.20		73.90	6.73	7.40
32 M.H., Daily Totals		$740.20		$997.75	$23.13	$31.17
Crew B-93	Hr.	Daily	Hr.	Daily	Bare Costs	Incl. O&P
1 Equip. Oper. (med.)	$20.60	$164.80	$29.85	$238.80	$20.60	$29.85
1 Feller Buncher, 50 H.P.		263.80		290.20	32.97	36.27
8 M.H., Daily Totals		$428.60		$529.00	$53.57	$66.12
Crew C-1	Hr.	Daily	Hr.	Daily	Bare Costs	Incl. O&P
3 Carpenters	$20.00	$480.00	$29.00	$696.00	$18.98	$27.51
1 Building Laborer	15.90	127.20	23.05	184.40		
Power Tools		25.80		28.40	.80	.88
32 M.H., Daily Totals		$633.00		$908.80	$19.78	$28.39
Crew C-2	Hr.	Daily	Hr.	Daily	Bare Costs	Incl. O&P
1 Carpenter Foreman (out)	$22.00	$176.00	$31.90	$255.20	$19.65	$28.49
4 Carpenters	20.00	640.00	29.00	928.00		
1 Building Laborer	15.90	127.20	23.05	184.40		
Power Tools		34.40		37.85	.72	.78
48 M.H., Daily Totals		$977.60		$1405.45	$20.37	$29.27

Crew No.	Bare Costs		Incl. Subs O & P		Cost Per Man-hour	
Crew C-3	Hr.	Daily	Hr.	Daily	Bare Costs	Incl. O&P
1 Rodman Foreman	$23.75	$190.00	$36.70	$293.60	$20.25	$30.70
4 Rodmen (reinf.)	21.75	696.00	33.65	1076.80		
1 Equip. Oper. (light)	19.45	155.60	28.20	225.60		
2 Building Laborers	15.90	254.40	23.05	368.80		
Stressing Equipment		33.60		36.95		
Grouting Equipment		53.65		59.00	1.36	1.49
64 M.H., Daily Totals		$1383.25		$2060.75	$21.61	$32.19
Crew C-4	Hr.	Daily	Hr.	Daily	Bare Costs	Incl. O&P
1 Rodman Foreman	$23.75	$190.00	$36.70	$293.60	$22.25	$34.41
3 Rodmen (reinf.)	21.75	522.00	33.65	807.60		
Stressing Equipment		33.60		36.95	1.05	1.15
32 M.H., Daily Totals		$745.60		$1138.15	$23.30	$35.56
Crew C-5	Hr.	Daily	Hr.	Daily	Bare Costs	Incl. O&P
1 Rodman Foreman	$23.75	$190.00	$36.70	$293.60	$21.32	$32.45
4 Rodmen (reinf.)	21.75	696.00	33.65	1076.80		
1 Equip. Oper. (crane)	21.05	168.40	30.50	244.00		
1 Equip. Oper. Oiler	17.50	140.00	25.40	203.20		
1 Hyd. Crane, 25 Ton		384.80		423.30	6.87	7.55
56 M.H., Daily Totals		$1579.20		$2240.90	$28.19	$40.00
Crew C-6	Hr.	Daily	Hr.	Daily	Bare Costs	Incl. O&P
1 Labor Foreman (outside)	$17.90	$143.20	$25.95	$207.60	$16.78	$24.19
4 Building Laborers	15.90	508.80	23.05	737.60		
1 Cement Finisher	19.20	153.60	27.00	216.00		
2 Gas Engine Vibrators		54.30		59.75	1.13	1.24
48 M.H., Daily Totals		$859.90		$1220.95	$17.91	$25.43
Crew C-7	Hr.	Daily	Hr.	Daily	Bare Costs	Incl. O&P
1 Labor Foreman (outside)	$17.90	$143.20	$25.95	$207.60	$17.15	$24.75
5 Building Laborers	15.90	636.00	23.05	922.00		
1 Cement Finisher	19.20	153.60	27.00	216.00		
1 Equip. Oper. (med.)	20.60	164.80	29.85	238.80		
2 Gas Engine Vibrators		54.30		59.75		
1 Concrete Bucket, 1 C.Y.		15.55		17.10		
1 Hyd. Crane, 55 Ton		527.60		580.35	9.33	10.26
64 M.H., Daily Totals		$1695.05		$2241.60	$26.48	$35.01
Crew C-8	Hr.	Daily	Hr.	Daily	Bare Costs	Incl. O&P
1 Labor Foreman (outside)	$17.90	$143.20	$25.95	$207.60	$17.80	$25.56
3 Building Laborers	15.90	381.60	23.05	553.20		
2 Cement Finishers	19.20	307.20	27.00	432.00		
1 Equip. Oper. (med.)	20.60	164.80	29.85	238.80		
1 Concrete Pump (small)		511.20		562.30	9.12	10.04
56 M.H., Daily Totals		$1508.00		$1993.90	$26.92	$35.60
Crew C-9	Hr.	Daily	Hr.	Daily	Bare Costs	Incl. O&P
1 Cement Finisher	$19.20	$153.60	$27.00	$216.00	$19.20	$27.00
1 Gas Finishing Mach.		28.40		31.25	3.55	3.90
8 M.H., Daily Totals		$182.00		$247.25	$22.75	$30.90
Crew C-10	Hr.	Daily	Hr.	Daily	Bare Costs	Incl. O&P
1 Building Laborer	$15.90	$127.20	$23.05	$184.40	$18.10	$25.68
2 Cement Finishers	19.20	307.20	27.00	432.00		
2 Gas Finishing Mach.		56.80		62.50	2.36	2.60
24 M.H., Daily Totals		$491.20		$678.90	$20.46	$28.28

CREWS

Crew No.	Bare Costs		Incl. Subs O & P		Cost Per Man-hour	
Crew C-11	Hr.	Daily	Hr.	Daily	Bare Costs	Incl. O&P
1 Struc. Steel Foreman	$23.70	$189.60	$37.25	$298.00	$21.38	$33.08
6 Struc. Steel Workers	21.70	1041.60	34.10	1636.80		
1 Equip. Oper. (crane)	21.05	168.40	30.50	244.00		
1 Equip. Oper. Oiler	17.50	140.00	25.40	203.20		
1 Truck Crane, 150 Ton		1030.00		1133.00	14.30	15.73
72 M.H., Daily Totals		$2569.60		$3515.00	$35.68	$48.81
Crew C-12	Hr.	Daily	Hr.	Daily	Bare Costs	Incl. O&P
1 Carpenter Foreman (out)	$22.00	$176.00	$31.90	$255.20	$19.82	$28.74
3 Carpenters	20.00	480.00	29.00	696.00		
1 Building Laborer	15.90	127.20	23.05	184.40		
1 Equip. Oper. (crane)	21.05	168.40	30.50	244.00		
1 Hyd. Crane, 12 Ton		235.00		258.50	4.89	5.38
48 M.H., Daily Totals		$1186.60		$1638.10	$24.71	$34.12
Crew C-13	Hr.	Daily	Hr.	Daily	Bare Costs	Incl. O&P
1 Struc. Steel Worker	$21.70	$173.60	$34.10	$272.80	$21.13	$32.40
1 Welder	21.70	173.60	34.10	272.80		
1 Carpenter	20.00	160.00	29.00	232.00		
1 Gas Welding Machine		56.80		62.50	2.36	2.60
24 M.H., Daily Totals		$564.00		$840.10	$23.49	$35.00
Crew C-14	Hr.	Daily	Hr.	Daily	Bare Costs	Incl. O&P
1 Carpenter Foreman (out)	$22.00	$176.00	$31.90	$255.20	$19.41	$28.53
5 Carpenters	20.00	800.00	29.00	1160.00		
4 Building Laborers	15.90	508.80	23.05	737.60		
4 Rodmen (reinf.)	21.75	696.00	33.65	1076.80		
2 Cement Finishers	19.20	307.20	27.00	432.00		
1 Equip. Oper. (crane)	21.05	168.40	30.50	244.00		
1 Equip. Oper. Oiler	17.50	140.00	25.40	203.20		
1 Crane, 80 Ton, & Tools		844.00		928.40		
Power Tools		25.80		28.40		
2 Gas Finishing Mach.		56.80		62.50	6.43	7.07
144 M.H., Daily Totals		$3723.00		$5128.10	$25.84	$35.60
Crew C-15	Hr.	Daily	Hr.	Daily	Bare Costs	Incl. O&P
1 Carpenter Foreman (out)	$22.00	$176.00	$31.90	$255.20	$18.87	$27.41
2 Carpenters	20.00	320.00	29.00	464.00		
3 Building Laborers	15.90	381.60	23.05	553.20		
2 Cement Finishers	19.20	307.20	27.00	432.00		
1 Rodman (reinf.)	21.75	174.00	33.65	269.20		
Power Tools		17.20		18.90		
1 Gas Finishing Mach.		28.40		31.25	.63	.69
72 M.H., Daily Totals		$1404.40		$2023.75	$19.50	$28.10
Crew C-16	Hr.	Daily	Hr.	Daily	Bare Costs	Incl. O&P
1 Labor Foreman (outside)	$17.90	$143.20	$25.95	$207.60	$18.67	$27.36
3 Building Laborers	15.90	381.60	23.05	553.20		
2 Cement Finishers	19.20	307.20	27.00	432.00		
1 Equip. Oper. (med.)	20.60	164.80	29.85	238.80		
2 Rodmen (reinf.)	21.75	348.00	33.65	538.40		
1 Concrete Pump (small)		511.20		562.30	7.10	7.80
72 M.H., Daily Totals		$1856.00		$2532.30	$25.77	$35.16
Crew C-17	Hr.	Daily	Hr.	Daily	Bare Costs	Incl. O&P
2 Skilled Worker Foremen	$22.50	$360.00	$32.85	$525.60	$20.90	$30.49
8 Skilled Workers	20.50	1312.00	29.90	1913.60		
80 M.H., Daily Totals		$1672.00		$2439.20	$20.90	$30.49

Crew No.	Bare Costs		Incl. Subs O & P		Cost Per Man-hour	
Crew C-17A	Hr.	Daily	Hr.	Daily	Bare Costs	Incl. O&P
2 Skilled Worker Foremen	$22.50	$360.00	$32.85	$525.60	$20.90	$30.49
8 Skilled Workers	20.50	1312.00	29.90	1913.60		
.125 Crane, 80 Ton, & Tools		109.70		120.70		
		1.10		1.25		
		3.70		4.05	1.43	1.57
80 M.H., Daily Totals		$1786.50		$2565.20	$22.33	$32.06
Crew C-17B	Hr.	Daily	Hr.	Daily	Bare Costs	Incl. O&P
2 Skilled Worker Foremen	$22.50	$360.00	$32.85	$525.60	$20.90	$30.49
8 Skilled Workers	20.50	1312.00	29.90	1913.60		
.25 Crane, 80 Ton		211.00		232.10		
Appropriate Fixtures		2.15		2.35		
		7.10		7.80	2.75	3.02
80 M.H., Daily Totals		$1892.25		$2681.45	$23.65	$33.51
Crew C-17C	Hr.	Daily	Hr.	Daily	Bare Costs	Incl. O&P
2 Skilled Worker Foremen	$22.50	$360.00	$32.85	$525.60	$20.90	$30.49
8 Skilled Workers	20.50	1312.00	29.90	1913.60		
.375 Crane, 80 Ton		320.70		352.80		
Appropriate Fixtures		3.25		3.60		
		10.80		11.85	4.18	4.60
80 M.H., Daily Totals		$2006.75		$2807.45	$25.08	$35.09
Crew C-17D	Hr.	Daily	Hr.	Daily	Bare Costs	Incl. O&P
2 Skilled Worker Foremen	$22.50	$360.00	$32.85	$525.60	$20.90	$30.49
8 Skilled Workers	20.50	1312.00	29.90	1913.60		
.5 Crane, 80 Ton		422.00		464.20		
Appropriate Fixtures		4.30		4.75		
		14.20		15.60	5.50	6.05
80 M.H., Daily Totals		$2112.50		$2923.75	$26.40	$36.54
Crew C-17E	Hr.	Daily	Hr.	Daily	Bare Costs	Incl. O&P
2 Skilled Worker Foremen	$22.50	$360.00	$32.85	$525.60	$20.90	$30.49
8 Skilled Workers	20.50	1312.00	29.90	1913.60		
1 Hyd. Jack with Rods		47.65		52.40	.59	.65
80 M.H., Daily Totals		$1719.65		$2491.60	$21.49	$31.14
Crew C-18	Hr.	Daily	Hr.	Daily	Bare Costs	Incl. O&P
.125 Labor Foreman (out)	$17.90	$17.90	$25.95	$25.95	$16.12	$23.37
1 Building Laborer	15.90	127.20	23.05	184.40		
1 Concrete Cart, 10 C.F.		36.20		39.80	4.02	4.42
9 M.H., Daily Totals		$181.30		$250.15	$20.14	$27.79
Crew C-19	Hr.	Daily	Hr.	Daily	Bare Costs	Incl. O&P
.125 Labor Foreman (out)	$17.90	$17.90	$25.95	$25.95	$16.12	$23.37
1 Building Laborer	15.90	127.20	23.05	184.40		
1 Concrete Cart, 18 C.F.		57.70		63.45	6.41	7.05
9 M.H., Daily Totals		$202.80		$273.80	$22.53	$30.42
Crew C-20	Hr.	Daily	Hr.	Daily	Bare Costs	Incl. O&P
1 Labor Foreman (outside)	$17.90	$143.20	$25.95	$207.60	$17.15	$24.75
5 Building Laborers	15.90	636.00	23.05	922.00		
1 Cement Finisher	19.20	153.60	27.00	216.00		
1 Equip. Oper. (med.)	20.60	164.80	29.85	238.80		
2 Gas Engine Vibrators		54.30		59.75		
1 Concrete Pump (small)		511.20		562.30	8.83	9.71
64 M.H., Daily Totals		$1663.10		$2206.45	$25.98	$34.46

CREWS

Crew No.	Bare Costs		Incl. Subs O & P		Cost Per Man-hour	
Crew C-21	Hr.	Daily	Hr.	Daily	Bare Costs	Incl. O&P
1 Labor Foreman (outside)	$17.90	$143.20	$25.95	$207.60	$17.15	$24.75
5 Building Laborers	15.90	636.00	23.05	922.00		
1 Cement Finisher	19.20	153.60	27.00	216.00		
1 Equip. Oper. (med.)	20.60	164.80	29.85	238.80		
2 Gas Engine Vibrators		54.30		59.75		
1 Concrete Conveyer		119.10		131.00	2.70	2.98
64 M.H., Daily Totals		$1271.00		$1775.15	$19.85	$27.73
Crew C-22	Hr.	Daily	Hr.	Daily	Bare Costs	Incl. O&P
1 Rodman Foreman	$23.75	$190.00	$36.70	$293.60	$22.01	$33.95
4 Rodmen (reinf.)	21.75	696.00	33.65	1076.80		
.125 Equip. Oper. (crane)	21.05	21.05	30.50	30.50		
.125 Equip. Oper. Oiler	17.50	17.50	25.40	25.40		
.125 Hyd. Crane, 25 Ton		50.00		55.05	1.19	1.31
42 M.H., Daily Totals		$974.55		$1481.35	$23.20	$35.26
Crew C-23	Hr.	Daily	Hr.	Daily	Bare Costs	Incl. O&P
2 Skilled Worker Foremen	$22.50	$360.00	$32.85	$525.60	$20.65	$30.10
6 Skilled Workers	20.50	984.00	29.90	1435.20		
1 Equip. Oper. (crane)	21.05	168.40	30.50	244.00		
1 Equip. Oper. Oiler	17.50	140.00	25.40	203.20		
1 Crane, 90 Ton		790.00		869.00	9.87	10.86
80 M.H., Daily Totals		$2442.40		$3277.00	$30.52	$40.96
Crew C-24	Hr.	Daily	Hr.	Daily	Bare Costs	Incl. O&P
2 Skilled Worker Foremen	$22.50	$360.00	$32.85	$525.60	$20.65	$30.10
6 Skilled Workers	20.50	984.00	29.90	1435.20		
1 Equip. Oper. (crane)	21.05	168.40	30.50	244.00		
1 Equip. Oper. Oiler	17.50	140.00	25.40	203.20		
1 Truck Crane, 150 Ton		1030.00		1133.00	12.87	14.16
80 M.H., Daily Totals		$2682.40		$3541.00	$33.52	$44.26
Crew D-1	Hr.	Daily	Hr.	Daily	Bare Costs	Incl. O&P
1 Bricklayer	$20.50	$164.00	$29.20	$233.60	$18.25	$26.00
1 Bricklayer Helper	16.00	128.00	22.80	182.40		
16 M.H., Daily Totals		$292.00		$416.00	$18.25	$26.00
Crew D-2	Hr.	Daily	Hr.	Daily	Bare Costs	Incl. O&P
3 Bricklayers	$20.50	$492.00	$29.20	$700.80	$18.81	$26.85
2 Bricklayer Helpers	16.00	256.00	22.80	364.80		
.5 Carpenter	20.00	80.00	29.00	116.00		
44 M.H., Daily Totals		$828.00		$1181.60	$18.81	$26.85
Crew D-3	Hr.	Daily	Hr.	Daily	Bare Costs	Incl. O&P
3 Bricklayers	$20.50	$492.00	$29.20	$700.80	$18.76	$26.75
2 Bricklayer Helpers	16.00	256.00	22.80	364.80		
.25 Carpenter	20.00	40.00	29.00	58.00		
42 M.H., Daily Totals		$788.00		$1123.60	$18.76	$26.75
Crew D-4	Hr.	Daily	Hr.	Daily	Bare Costs	Incl. O&P
1 Bricklayer	$20.50	$164.00	$29.20	$233.60	$17.98	$25.75
2 Bricklayer Helpers	16.00	256.00	22.80	364.80		
1 Equip. Oper. (light)	19.45	155.60	28.20	225.60		
Power Equipment		111.40		122.55		
Appropriate Fixtures		35.00		38.50		
		8.40		9.25	4.83	5.32
32 M.H., Daily Totals		$730.40		$994.30	$22.81	$31.07
Crew D-5	Hr.	Daily	Hr.	Daily	Bare Costs	Incl. O&P
1 Bricklayer	$20.50	$164.00	$29.20	$233.60	$20.50	$29.20
1 Power Tool		28.30		31.15	3.53	3.89
8 M.H., Daily Totals		$192.30		$264.75	$24.03	$33.09
Crew D-6	Hr.	Daily	Hr.	Daily	Bare Costs	Incl. O&P
3 Bricklayers	$20.50	$492.00	$29.20	$700.80	$18.32	$26.12
3 Bricklayer Helpers	16.00	384.00	22.80	547.20		
.25 Carpenter	20.00	40.00	29.00	58.00		
50 M.H., Daily Totals		$916.00		$1306.00	$18.32	$26.12
Crew D-7	Hr.	Daily	Hr.	Daily	Bare Costs	Incl. O&P
1 Tile Layer	$19.75	$158.00	$27.70	$221.60	$17.67	$24.77
1 Tile Layer Helper	15.60	124.80	21.85	174.80		
16 M.H., Daily Totals		$282.80		$396.40	$17.67	$24.77
Crew D-8	Hr.	Daily	Hr.	Daily	Bare Costs	Incl. O&P
3 Bricklayers	$20.50	$492.00	$29.20	$700.80	$18.70	$26.64
2 Bricklayer Helpers	16.00	256.00	22.80	364.80		
40 M.H., Daily Totals		$748.00		$1065.60	$18.70	$26.64
Crew D-9	Hr.	Daily	Hr.	Daily	Bare Costs	Incl. O&P
3 Bricklayers	$20.50	$492.00	$29.20	$700.80	$18.25	$26.00
3 Bricklayer Helpers	16.00	384.00	22.80	547.20		
48 M.H., Daily Totals		$876.00		$1248.00	$18.25	$26.00
Crew D-10	Hr.	Daily	Hr.	Daily	Bare Costs	Incl. O&P
1 Bricklayer Foreman	$22.50	$180.00	$32.05	$256.40	$19.21	$27.47
1 Bricklayer	20.50	164.00	29.20	233.60		
2 Bricklayer Helpers	16.00	256.00	22.80	364.80		
1 Equip. Oper. (crane)	21.05	168.40	30.50	244.00		
1 Truck Crane, 12.5 Ton		224.20		246.60	5.60	6.16
40 M.H., Daily Totals		$992.60		$1345.40	$24.81	$33.63
Crew D-11	Hr.	Daily	Hr.	Daily	Bare Costs	Incl. O&P
1 Bricklayer Foreman	$22.50	$180.00	$32.05	$256.40	$19.66	$28.01
1 Bricklayer	20.50	164.00	29.20	233.60		
1 Bricklayer Helper	16.00	128.00	22.80	182.40		
24 M.H., Daily Totals		$472.00		$672.40	$19.66	$28.01
Crew D-12	Hr.	Daily	Hr.	Daily	Bare Costs	Incl. O&P
1 Bricklayer Foreman	$22.50	$180.00	$32.05	$256.40	$18.75	$26.71
1 Bricklayer	20.50	164.00	29.20	233.60		
2 Bricklayer Helpers	16.00	256.00	22.80	364.80		
32 M.H., Daily Totals		$600.00		$854.80	$18.75	$26.71
Crew D-13	Hr.	Daily	Hr.	Daily	Bare Costs	Incl. O&P
1 Bricklayer Foreman	$22.50	$180.00	$32.05	$256.40	$19.34	$27.72
1 Bricklayer	20.50	164.00	29.20	233.60		
2 Bricklayer Helpers	16.00	256.00	22.80	364.80		
1 Carpenter	20.00	160.00	29.00	232.00		
1 Equip. Oper. (crane)	21.05	168.40	30.50	244.00		
1 Truck Crane, 12.5 Ton		224.20		246.60	4.67	5.13
48 M.H., Daily Totals		$1152.60		$1577.40	$24.01	$32.85
Crew E-1	Hr.	Daily	Hr.	Daily	Bare Costs	Incl. O&P
1 Welder Foreman	$23.70	$189.60	$37.25	$298.00	$21.61	$33.18
1 Welder	21.70	173.60	34.10	272.80		
1 Equip. Oper. (light)	19.45	155.60	28.20	225.60		
1 Gas Welding Machine		56.80		62.50	2.36	2.60
24 M.H., Daily Totals		$575.60		$858.90	$23.97	$35.78

CREWS

Crew No.	Bare Costs		Incl. Subs O & P		Cost Per Man-hour	
Crew E-2	Hr.	Daily	Hr.	Daily	Bare Costs	Incl. O&P
1 Struc. Steel Foreman	$23.70	$189.60	$37.25	$298.00	$21.29	$32.79
4 Struc. Steel Workers	21.70	694.40	34.10	1091.20		
1 Equip. Oper. (crane)	21.05	168.40	30.50	244.00		
1 Equip. Oper. Oiler	17.50	140.00	25.40	203.20		
1 Crane, 90 Ton		790.00		869.00	14.10	15.51
56 M.H., Daily Totals		$1982.40		$2705.40	$35.39	$48.30
Crew E-3	Hr.	Daily	Hr.	Daily	Bare Costs	Incl. O&P
1 Struc. Steel Foreman	$23.70	$189.60	$37.25	$298.00	$22.36	$35.15
1 Struc. Steel Worker	21.70	173.60	34.10	272.80		
1 Welder	21.70	173.60	34.10	272.80		
1 Gas Welding Machine		56.80		62.50		
1 Torch, Gas & Air		49.60		54.55	4.43	4.87
24 M.H., Daily Totals		$643.20		$960.65	$26.79	$40.02
Crew E-4	Hr.	Daily	Hr.	Daily	Bare Costs	Incl. O&P
1 Struc. Steel Foreman	$23.70	$189.60	$37.25	$298.00	$22.20	$34.88
3 Struc. Steel Workers	21.70	520.80	34.10	818.40		
1 Gas Welding Machine		56.80		62.50	1.77	1.95
32 M.H., Daily Totals		$767.20		$1178.90	$23.97	$36.83
Crew E-5	Hr.	Daily	Hr.	Daily	Bare Costs	Incl. O&P
2 Struc. Steel Foremen	$23.70	$379.20	$37.25	$596.00	$21.61	$33.50
5 Struc. Steel Workers	21.70	868.00	34.10	1364.00		
1 Equip. Oper. (crane)	21.05	168.40	30.50	244.00		
1 Welder	21.70	173.60	34.10	272.80		
1 Equip. Oper. Oiler	17.50	140.00	25.40	203.20		
1 Crane, 90 Ton		790.00		869.00		
1 Gas Welding Machine		56.80		62.50		
1 Torch, Gas & Air		49.60		54.55	11.20	12.32
80 M.H., Daily Totals		$2625.60		$3666.05	$32.81	$45.82
Crew E-6	Hr.	Daily	Hr.	Daily	Bare Costs	Incl. O&P
3 Struc. Steel Foreman	$23.70	$568.80	$37.25	$894.00	$21.63	$33.55
9 Struc. Steel Workers	21.70	1562.40	34.10	2455.20		
1 Equip. Oper. (crane)	21.05	168.40	30.50	244.00		
1 Welder	21.70	173.60	34.10	272.80		
1 Equip. Oper. Oiler	17.50	140.00	25.40	203.20		
1 Equip. Oper. (light)	19.45	155.60	28.20	225.60		
1 Crane, 90 Ton		790.00		869.00		
1 Gas Welding Machine		56.80		62.50		
1 Torch, Gas & air		49.60		54.55		
1 Air Compr., 160 C.F.M.		70.25		77.30		
2 Impact Wrenches		33.00		36.30	7.80	8.59
128 M.H., Daily Totals		$3768.45		$5394.45	$29.43	$42.14
Crew E-7	Hr.	Daily	Hr.	Daily	Bare Costs	Incl. O&P
1 Struc. Steel Foreman	$23.70	$189.60	$37.25	$298.00	$21.61	$33.50
4 Struc. Steel Workers	21.70	694.40	34.10	1091.20		
1 Welder Foreman	23.70	189.60	37.25	298.00		
2 Welders	21.70	347.20	34.10	545.60		
1 Equip. Oper. (crane)	21.05	168.40	30.50	244.00		
1 Equip. Oper. Oiler	17.50	140.00	25.40	203.20		
1 Crane, 90 Ton		790.00		869.00		
2 Gas Welding Machines		113.60		124.95	11.29	12.42
80 M.H., Daily Totals		$2632.80		$3673.95	$32.90	$45.92

Crew No.	Bare Costs		Incl. Subs O & P		Cost Per Man-hour	
Crew E-8	Hr.	Daily	Hr.	Daily	Bare Costs	Incl. O&P
1 Struc. Steel Foreman	$23.70	$189.60	$37.25	$298.00	$21.46	$33.18
4 Struc. Steel Workers	21.70	694.40	34.10	1091.20		
1 Welder Foreman	23.70	189.60	37.25	298.00		
4 Welders	21.70	694.40	34.10	1091.20		
1 Equip. Oper. (crane)	21.05	168.40	30.50	244.00		
1 Equip. Oper. Oiler	17.50	140.00	25.40	203.20		
1 Equip. Oper. (light)	19.45	155.60	28.20	225.60		
1 Crane, 90 Ton		790.00		869.00		
4 Gas Welding Machines		227.20		249.90	9.78	10.75
1 4 M.H., Daily Totals		$3249.20		$4570.10	$31.24	$43.93
Crew E-9	Hr.	Daily	Hr.	Daily	Bare Costs	Incl. O&P
2 Struc. Steel Foremen	$23.70	$379.20	$37.25	$596.00	$21.63	$33.55
5 Struc. Steel Workers	21.70	868.00	34.10	1364.00		
1 Welder Foreman	23.70	189.60	37.25	298.00		
5 Welders	21.70	868.00	34.10	1364.00		
1 Equip. Oper. (crane)	21.05	168.40	30.50	244.00		
1 Equip. Oper. Oiler	17.50	140.00	25.40	203.20		
1 Equip. Oper. (light)	19.45	155.60	28.20	225.60		
1 Crane, 90 Ton		790.00		869.00		
5 Gas Welding Machines		284.00		312.40		
1 Torch, Gas & Air		49.60		54.55	8.77	9.65
128 M.H., Daily Totals		$3892.40		$5530.75	$30.40	$43.20
Crew E-10	Hr.	Daily	Hr.	Daily	Bare Costs	Incl. O&P
1 Welder Foreman	$23.70	$189.60	$37.25	$298.00	$22.70	$35.67
1 Welder	21.70	173.60	34.10	272.80		
4 Gas Welding Machines		227.20		249.90		
1 Truck, 3 Ton		67.20		73.90	18.40	20.23
16 M.H., Daily Totals		$657.60		$894.60	$41.10	$55.90
Crew E-11	Hr.	Daily	Hr.	Daily	Bare Costs	Incl. O&P
2 Painters, Struc. Steel	$20.00	$320.00	$32.35	$517.60	$18.83	$28.98
1 Building Laborer	15.90	127.20	23.05	184.40		
1 Equip. Oper. (light)	19.45	155.60	28.20	225.60		
Gas Engine Equipment		111.40		122.55		
Appropriate Fixtures		35.00		38.50		
		8.40		9.25	4.83	5.32
32 M.H., Daily Totals		$757.60		$1097.90	$23.66	$34.30
Crew E-12	Hr.	Daily	Hr.	Daily	Bare Costs	Incl. O&P
1 Welder Foreman	$23.70	$189.60	$37.25	$298.00	$21.57	$32.72
1 Equip. Oper. (light)	19.45	155.60	28.20	225.60		
1 Gas Welding Machine		56.80		62.50	3.55	3.90
16 M.H., Daily Totals		$402.00		$586.10	$25.12	$36.62
Crew E-13	Hr.	Daily	Hr.	Daily	Bare Costs	Incl. O&P
1 Welder Foreman	$23.70	$189.60	$37.25	$298.00	$22.28	$34.23
.5 Equip. Oper. (light)	19.45	77.80	28.20	112.80		
1 Gas Welding Machine		56.80		62.50	4.73	5.20
12 M.H., Daily Totals		$324.20		$473.30	$27.01	$39.43
Crew E-14	Hr.	Daily	Hr.	Daily	Bare Costs	Incl. O&P
1 Welder Foreman	$23.70	$189.60	$37.25	$298.00	$23.70	$37.25
1 Gas Welding Machine		56.80		62.50	7.10	7.81
8 M.H., Daily Totals		$246.40		$360.50	$30.80	$45.06

CREWS

Crew No.	Bare Costs		Incl. Subs O & P		Cost Per Man-hour	
	Hr.	Daily	Hr.	Daily	Bare Costs	Incl. O&P
Crew E-15						
2 Painters, Struc. Steel	$20.00	$320.00	$32.35	$517.60	$20.00	$32.35
1 Paint Sprayer, 17 C.F.M.		27.00		29.70	1.68	1.85
16 M.H., Daily Totals		$347.00		$547.30	$21.68	$34.20
Crew F-1	Hr.	Daily	Hr.	Daily	Bare Costs	Incl. O&P
1 Carpenter	$20.00	$160.00	$29.00	$232.00	$20.00	$29.00
Power Tools		8.60		9.45	1.07	1.18
8 M.H., Daily Totals		$168.60		$241.45	$21.07	$30.18
Crew F-2	Hr.	Daily	Hr.	Daily	Bare Costs	Incl. O&P
2 Carpenters	$20.00	$320.00	$29.00	$464.00	$20.00	$29.00
Power Tools		17.20		18.90	1.07	1.18
16 M.H., Daily Totals		$337.20		$482.90	$21.07	$30.18
Crew F-3	Hr.	Daily	Hr.	Daily	Bare Costs	Incl. O&P
4 Carpenters	$20.00	$640.00	$29.00	$928.00	$20.21	$29.30
1 Equip. Oper. (crane)	21.05	168.40	30.50	244.00		
1 Hyd. Crane, 12 Ton		235.00		258.50		
Power Tools		17.20		18.90	6.30	6.93
40 M.H., Daily Totals		$1060.60		$1449.40	$26.51	$36.23
Crew F-4	Hr.	Daily	Hr.	Daily	Bare Costs	Incl. O&P
4 Carpenters	$20.00	$640.00	$29.00	$928.00	$19.75	$28.65
1 Equip. Oper. (crane)	21.05	168.40	30.50	244.00		
1 Equip. Oper. Oiler	17.50	140.00	25.40	203.20		
1 Hyd. Crane, 55 Ton		527.60		580.35		
Power Tools		17.20		18.90	11.35	12.48
48 M.H., Daily Totals		$1493.20		$1974.45	$31.10	$41.13
Crew F-5	Hr.	Daily	Hr.	Daily	Bare Costs	Incl. O&P
1 Carpenter Foreman	$22.00	$176.00	$31.90	$255.20	$20.50	$29.72
3 Carpenters	20.00	480.00	29.00	696.00		
Power Tools		17.20		18.90	.53	.59
32 M.H., Daily Totals		$673.20		$970.10	$21.03	$30.31
Crew F-6	Hr.	Daily	Hr.	Daily	Bare Costs	Incl. O&P
2 Carpenters	$20.00	$320.00	$29.00	$464.00	$18.57	$26.92
2 Building Laborers	15.90	254.40	23.05	368.80		
1 Equip. Oper. (crane)	21.05	168.40	30.50	244.00		
1 Hyd. Crane, 12 Ton		235.00		258.50		
Power Tools		17.20		18.90	6.30	6.93
40 M.H., Daily Totals		$995.00		$1354.20	$24.87	$33.85
Crew F-7	Hr.	Daily	Hr.	Daily	Bare Costs	Incl. O&P
2 Carpenters	$20.00	$320.00	$29.00	$464.00	$17.95	$26.02
2 Building Laborers	15.90	254.40	23.05	368.80		
Power Tools		17.20		18.90	.53	.59
32 M.H., Daily Totals		$591.60		$851.70	$18.48	$26.61
Crew G-1	Hr.	Daily	Hr.	Daily	Bare Costs	Incl. O&P
1 Roofer Foreman	$20.80	$166.40	$31.80	$254.40	$17.64	$26.98
4 Roofers, Composition	18.80	601.60	28.75	920.00		
2 Roofer Helpers	13.75	220.00	21.05	336.80		
Application Equipment		101.35		111.50	1.80	1.99
56 M.H., Daily Totals		$1089.35		$1622.70	$19.44	$28.97

Crew No.	Bare Costs		Incl. Subs O & P		Cost Per Man-hour	
Crew G-2	Hr.	Daily	Hr.	Daily	Bare Costs	Incl. O&P
1 Plasterer	$19.90	$159.20	$28.35	$226.80	$17.43	$24.96
1 Plasterer Helper	16.50	132.00	23.50	188.00		
1 Building Laborer	15.90	127.20	23.05	184.40		
Grouting Equipment		107.25		118.00	4.46	4.91
24 M.H., Daily Totals		$525.65		$717.20	$21.89	$29.87
Crew G-3	Hr.	Daily	Hr.	Daily	Bare Costs	Incl. O&P
2 Sheet Metal Workers	$22.70	$363.20	$33.15	$530.40	$19.30	$28.10
2 Building Laborers	15.90	254.40	23.05	368.80		
Power Tools		27.00		29.70	.84	.92
32 M.H., Daily Totals		$644.60		$928.90	$20.14	$29.02
Crew G-4	Hr.	Daily	Hr.	Daily	Bare Costs	Incl. O&P
1 Labor Foreman (outside)	$17.90	$143.20	$25.95	$207.60	$16.56	$24.01
2 Building Laborers	15.90	254.40	23.05	368.80		
1 Light Truck, 1.5 Ton		57.40		63.15		
1 Air Compr., 160 C.F.M.		70.25		77.30	5.31	5.85
24 M.H., Daily Totals		$525.25		$716.85	$21.87	$29.86
Crew G-5	Hr.	Daily	Hr.	Daily	Bare Costs	Incl. O&P
1 Roofer Foreman	$20.80	$166.40	$31.80	$254.40	$17.18	$26.28
2 Roofers, Composition	18.80	300.80	28.75	460.00		
2 Roofer Helpers	13.75	220.00	21.05	336.80		
Application Equipment		101.35		111.50	2.53	2.78
40 M.H., Daily Totals		$788.55		$1162.70	$19.71	$29.06
Crew H-1	Hr.	Daily	Hr.	Daily	Bare Costs	Incl. O&P
2 Glaziers	$20.15	$322.40	$28.75	$460.00	$20.92	$31.42
2 Struc. Steel Workers	21.70	347.20	34.10	545.60		
32 M.H., Daily Totals		$669.60		$1005.60	$20.92	$31.42
Crew H-2	Hr.	Daily	Hr.	Daily	Bare Costs	Incl. O&P
2 Glaziers	$20.15	$322.40	$28.75	$460.00	$18.73	$26.85
1 Building Laborer	15.90	127.20	23.05	184.40		
24 M.H., Daily Totals		$449.60		$644.40	$18.73	$26.85
Crew J-1	Hr.	Daily	Hr.	Daily	Bare Costs	Incl. O&P
3 Plasterers	$19.90	$477.60	$28.35	$680.40	$18.54	$26.41
2 Plasterer Helpers	16.50	264.00	23.50	376.00		
1 Mixing Machine, 6 C.F.		37.80		41.60	.94	1.04
40 M.H., Daily Totals		$779.40		$1098.00	$19.48	$27.45
Crew J-2	Hr.	Daily	Hr.	Daily	Bare Costs	Incl. O&P
3 Plasterers	$19.90	$477.60	$28.35	$680.40	$18.80	$26.73
2 Plasterer Helpers	16.50	264.00	23.50	376.00		
1 Lather	20.10	160.80	28.35	226.80		
1 Mixing Machine, 6 C.F.		37.80		41.60	.78	.86
48 M.H., Daily Totals		$940.20		$1324.80	$19.58	$27.59
Crew J-3	Hr.	Daily	Hr.	Daily	Bare Costs	Incl. O&P
1 Terrazzo Worker	$19.90	$159.20	$27.90	$223.20	$17.92	$25.12
1 Terrazzo Helper	15.95	127.60	22.35	178.80		
1 Mixing Mach. & Grinder		31.20		34.30		
Appropriate Fixtures		87.20		95.90	7.40	8.13
16 M.H., Daily Totals		$405.20		$532.20	$25.32	$33.25
Crew J-4	Hr.	Daily	Hr.	Daily	Bare Costs	Incl. O&P
1 Tile Layer	$19.75	$158.00	$27.70	$221.60	$17.67	$24.77
1 Tile Layer Helper	15.60	124.80	21.85	174.80		
16 M.H., Daily Totals		$282.80		$396.40	$17.67	$24.77

CREWS

Crew No.	Bare Costs		Incl. Subs O & P		Cost Per Man-hour	
Crew K-1	Hr.	Daily	Hr.	Daily	Bare Costs	Incl. O&P
1 Carpenter	$20.00	$160.00	$29.00	$232.00	$18.17	$26.22
1 Truck Driver (light)	16.35	130.80	23.45	187.60		
1 Truck w/Power Equip.		176.75		194.40	11.04	12.15
16 M.H., Daily Totals		$467.55		$614.00	$29.21	$38.37
Crew K-2	Hr.	Daily	Hr.	Daily	Bare Costs	Incl. O&P
1 Struc. Steel Foreman	$23.70	$189.60	$37.25	$298.00	$20.58	$31.60
1 Struc. Steel Worker	21.70	173.60	34.10	272.80		
1 Truck Driver (light)	16.35	130.80	23.45	187.60		
1 Truck w/Power Equip.		176.75		194.40	7.36	8.10
24 M.H., Daily Totals		$670.75		$952.80	$27.94	$39.70
Crew L-1	Hr.	Daily	Hr.	Daily	Bare Costs	Incl. O&P
1 Electrician	$22.40	$179.20	$32.20	$257.60	$22.47	$32.40
1 Plumber	22.55	180.40	32.60	260.80		
16 M.H., Daily Totals		$359.60		$518.40	$22.47	$32.40
Crew L-2	Hr.	Daily	Hr.	Daily	Bare Costs	Incl. O&P
1 Carpenter	$20.00	$160.00	$29.00	$232.00	$17.77	$25.90
1 Helper	15.55	124.40	22.80	182.40		
16 M.H., Daily Totals		$284.40		$414.40	$17.77	$25.90
Crew L-3	Hr.	Daily	Hr.	Daily	Bare Costs	Incl. O&P
1 Carpenter	$20.00	$160.00	$29.00	$232.00	$21.27	$30.83
.5 Electrician	22.40	89.60	32.20	128.80		
.5 Sheet Metal Worker	22.70	90.80	33.15	132.60		
16 M.H., Daily Totals		$340.40		$493.40	$21.27	$30.83
Crew L-4	Hr.	Daily	Hr.	Daily	Bare Costs	Incl. O&P
2 Skilled Workers	$20.50	$328.00	$29.90	$478.40	$18.85	$27.53
1 Helper	15.55	124.40	22.80	182.40		
24 M.H., Daily Totals		$452.40		$660.80	$18.85	$27.53
Crew L-5	Hr.	Daily	Hr.	Daily	Bare Costs	Incl. O&P
1 Struc. Steel Foreman	$23.70	$189.60	$37.25	$298.00	$21.89	$34.03
5 Struc. Steel Workers	21.70	868.00	34.10	1364.00		
1 Equip. Oper. (crane)	21.05	168.40	30.50	244.00		
1 Hyd. Crane, 25 Ton		384.80		423.30	6.87	7.55
56 M.H., Daily Totals		$1610.80		$2329.30	$28.76	$41.58
Crew L-6	Hr.	Daily	Hr.	Daily	Bare Costs	Incl. O&P
1 Plumber	$22.55	$180.40	$32.60	$260.80	$22.50	$32.46
.5 Electrician	22.40	89.60	32.20	128.80		
12 M.H., Daily Totals		$270.00		$389.60	$22.50	$32.46
Crew L-7	Hr.	Daily	Hr.	Daily	Bare Costs	Incl. O&P
2 Carpenters	$20.00	$320.00	$29.00	$464.00	$19.17	$27.75
1 Building Laborer	15.90	127.20	23.05	184.40		
.5 Electrician	22.40	89.60	32.20	128.80		
28 M.H., Daily Totals		$536.80		$777.20	$19.17	$27.75
Crew L-8	Hr.	Daily	Hr.	Daily	Bare Costs	Incl. O&P
2 Carpenters	$20.00	$320.00	$29.00	$464.00	$20.51	$29.72
.5 Plumber	22.55	90.20	32.60	130.40		
20 M.H., Daily Totals		$410.20		$594.40	$20.51	$29.72
Crew L-9	Hr.	Daily	Hr.	Daily	Bare Costs	Incl. O&P
1 Labor Foreman (inside)	$16.40	$131.20	$23.75	$190.00	$18.02	$26.67
2 Building Laborers	15.90	254.40	23.05	368.80		
1 Struc. Steel Worker	21.70	173.60	34.10	272.80		
.5 Electrician	22.40	89.60	32.20	128.80		
36 M.H., Daily Totals		$648.80		$960.40	$18.02	$26.67
Crew M-1	Hr.	Daily	Hr.	Daily	Bare Costs	Incl. O&P
3 Elevator Constructors	$22.65	$543.60	$32.90	$789.60	$21.51	$31.26
1 Elevator Apprentice	18.12	144.96	26.35	210.80		
Hand Tools		61.00		67.10	1.90	2.09
32 M.H., Daily Totals		$749.56		$1067.50	$23.41	$33.35
Crew M-2	Hr.	Daily	Hr.	Daily	Bare Costs	Incl. O&P
2 Millwrights	$20.75	$332.00	$29.35	$469.60	$20.75	$29.35
Power Tools		17.20		18.90	1.07	1.18
16 M.H., Daily Totals		$349.20		$488.50	$21.82	$30.53
Crew Q-1	Hr.	Daily	Hr.	Daily	Bare Costs	Incl. O&P
1 Plumber	$22.55	$180.40	$32.60	$260.80	$20.29	$29.35
1 Plumber Apprentice	18.04	144.32	26.10	208.80		
16 M.H., Daily Totals		$324.72		$469.60	$20.29	$29.35
Crew Q-2	Hr.	Daily	Hr.	Daily	Bare Costs	Incl. O&P
2 Plumbers	$22.55	$360.80	$32.60	$521.60	$21.04	$30.43
1 Plumber Apprentice	18.04	144.32	26.10	208.80		
24 M.H., Daily Totals		$505.12		$730.40	$21.04	$30.43
Crew Q-3	Hr.	Daily	Hr.	Daily	Bare Costs	Incl. O&P
1 Plumber Foreman (ins)	$23.05	$184.40	$33.35	$266.80	$21.54	$31.16
2 Plumbers	22.55	360.80	32.60	521.60		
1 Plumber Apprentice	18.04	144.32	26.10	208.80		
32 M.H., Daily Totals		$689.52		$997.20	$21.54	$31.16
Crew Q-4	Hr.	Daily	Hr.	Daily	Bare Costs	Incl. O&P
1 Plumber Foreman (ins)	$23.05	$184.40	$33.35	$266.80	$21.54	$31.16
1 Plumber	22.55	180.40	32.60	260.80		
1 Welder (plumber)	22.55	180.40	32.60	260.80		
1 Plumber Apprentice	18.04	144.32	26.10	208.80		
1 Electric Welding Mach.		18.60		20.45	.58	.63
32 M.H., Daily Totals		$708.12		$1017.65	$22.12	$31.79
Crew Q-5	Hr.	Daily	Hr.	Daily	Bare Costs	Incl. O&P
1 Steamfitter	$22.75	$182.00	$32.90	$263.20	$20.47	$29.60
1 Steamfitter Apprentice	18.20	145.60	26.30	210.40		
16 M.H., Daily Totals		$327.60		$473.60	$20.47	$29.60
Crew Q-6	Hr.	Daily	Hr.	Daily	Bare Costs	Incl. O&P
2 Steamfitters	$22.75	$364.00	$32.90	$526.40	$21.23	$30.70
1 Steamfitter Apprentice	18.20	145.60	26.30	210.40		
24 M.H., Daily Totals		$509.60		$736.80	$21.23	$30.70
Crew Q-7	Hr.	Daily	Hr.	Daily	Bare Costs	Incl. O&P
1 Steamfitter Foreman (ins)	$23.25	$186.00	$33.60	$268.80	$21.73	$31.42
2 Steamfitters	22.75	364.00	32.90	526.40		
1 Steamfitter Apprentice	18.20	145.60	26.30	210.40		
32 M.H., Daily Totals		$695.60		$1005.60	$21.73	$31.42

CREWS

Crew No.	Bare Costs		Incl. Subs O & P		Cost Per Man-hour	
Crew Q-8	Hr.	Daily	Hr.	Daily	Bare Costs	Incl. O&P
1 Steamfitter Foreman (ins)	$23.25	$186.00	$33.60	$268.80	$21.73	$31.42
1 Steamfitter	22.75	182.00	32.90	263.20		
1 Welder (steamfitter)	22.75	182.00	32.90	263.20		
1 Steamfitter Apprentice	18.20	145.60	26.30	210.40		
1 Electric Welding Mach.		18.60		20.45	.58	.63
32 M.H., Daily Totals		$714.20		$1026.05	$22.31	$32.05
Crew Q-9	Hr.	Daily	Hr.	Daily	Bare Costs	Incl. O&P
1 Sheet Metal Worker	$22.70	$181.60	$33.15	$265.20	$20.43	$29.85
1 Sheet Metal Apprentice	18.16	145.28	26.55	212.40		
16 M.H., Daily Totals		$326.88		$477.60	$20.43	$29.85
Crew Q-10	Hr.	Daily	Hr.	Daily	Bare Costs	Incl. O&P
2 Sheet Metal Workers	$22.70	$363.20	$33.15	$530.40	$21.18	$30.95
1 Sheet Metal Apprentice	18.16	145.28	26.55	212.40		
24 M.H., Daily Totals		$508.48		$742.80	$21.18	$30.95
Crew Q-11	Hr.	Daily	Hr.	Daily	Bare Costs	Incl. O&P
1 Sheet Metal Foreman (ins)	$23.20	$185.60	$33.90	$271.20	$21.69	$31.68
2 Sheet Metal Workers	22.70	363.20	33.15	530.40		
1 Sheet Metal Apprentice	18.16	145.28	26.55	212.40		
32 M.H., Daily Totals		$694.08		$1014.00	$21.69	$31.68
Crew Q-12	Hr.	Daily	Hr.	Daily	Bare Costs	Incl. O&P
1 Sprinkler Installer	$23.25	$186.00	$33.80	$270.40	$20.92	$30.42
1 Sprinkler Apprentice	18.60	148.80	27.05	216.40		
16 M.H., Daily Totals		$334.80		$486.80	$20.92	$30.42
Crew Q-13	Hr.	Daily	Hr.	Daily	Bare Costs	Incl. O&P
1 Sprinkler Foreman (ins)	$23.75	$190.00	$34.50	$276.00	$22.21	$32.28
2 Sprinkler Installers	23.25	372.00	33.80	540.80		
1 Sprinkler Apprentice	18.60	148.80	27.05	216.40		
32 M.H., Daily Totals		$710.80		$1033.20	$22.21	$32.28
Crew Q-14	Hr.	Daily	Hr.	Daily	Bare Costs	Incl. O&P
1 Asbestos Worker	$22.75	$182.00	$33.55	$268.40	$20.47	$30.20
1 Asbestos Apprentice	18.20	145.60	26.85	214.80		
16 M.H., Daily Totals		$327.60		$483.20	$20.47	$30.20
Crew Q-15	Hr.	Daily	Hr.	Daily	Bare Costs	Incl. O&P
1 Plumber	$22.55	$180.40	$32.60	$260.80	$20.29	$29.35
1 Plumber Apprentice	18.04	144.32	26.10	208.80		
1 Electric Welding Mach.		18.60		20.45	1.16	1.27
16 M.H., Daily Totals		$343.32		$490.05	$21.45	$30.62
Crew Q-16	Hr.	Daily	Hr.	Daily	Bare Costs	Incl. O&P
2 Plumbers	$22.55	$360.80	$32.60	$521.60	$21.04	$30.43
1 Plumber Apprentice	18.04	144.32	26.10	208.80		
1 Electric Welding Mach.		18.60		20.45	.77	.85
24 M.H., Daily Totals		$523.72		$750.85	$21.81	$31.28
Crew Q-17	Hr.	Daily	Hr.	Daily	Bare Costs	Incl. O&P
1 Steamfitter	$22.75	$182.00	$32.90	$263.20	$20.47	$29.60
1 Steamfitter Apprentice	18.20	145.60	26.30	210.40		
1 Electric Welding Mach.		18.60		20.45	1.16	1.27
16 M.H., Daily Totals		$346.20		$494.05	$21.63	$30.87
Crew Q-18	Hr.	Daily	Hr.	Daily	Bare Costs	Incl. O&P
2 Steamfitters	$22.75	$364.00	$32.90	$526.40	$21.23	$30.70
1 Steamfitter Apprentice	18.20	145.60	26.30	210.40		
1 Electric Welding Mach.		18.60		20.45	.77	.85
24 M.H., Daily Totals		$528.20		$757.25	$22.00	$31.55
Crew Q-19	Hr.	Daily	Hr.	Daily	Bare Costs	Incl. O&P
1 Steamfitter	$22.75	$182.00	$32.90	$263.20	$21.11	$30.46
1 Steamfitter Apprentice	18.20	145.60	26.30	210.40		
1 Electrician	22.40	179.20	32.20	257.60		
24 M.H., Daily Totals		$506.80		$731.20	$21.11	$30.46
Crew Q-20	Hr.	Daily	Hr.	Daily	Bare Costs	Incl. O&P
1 Sheet Metal Worker	$22.70	$181.60	$33.15	$265.20	$20.82	$30.32
1 Sheet Metal Apprentice	18.16	145.28	26.55	212.40		
.5 Electrician	22.40	89.60	32.20	128.80		
20 M.H., Daily Totals		$416.48		$606.40	$20.82	$30.32
Crew Q-21	Hr.	Daily	Hr.	Daily	Bare Costs	Incl. O&P
2 Steamfitters	$22.75	$364.00	$32.90	$526.40	$21.52	$31.07
1 Steamfitter Apprentice	18.20	145.60	26.30	210.40		
1 Electrician	22.40	179.20	32.20	257.60		
32 M.H., Daily Totals		$688.80		$994.40	$21.52	$31.07
Crew R-1	Hr.	Daily	Hr.	Daily	Bare Costs	Incl. O&P
1 Electrician Foreman	$22.90	$183.20	$32.95	$263.60	$20.20	$29.19
3 Electricians	22.40	537.60	32.20	772.80		
2 Helpers	15.55	248.80	22.80	364.80		
48 M.H., Daily Totals		$969.60		$1401.20	$20.20	$29.19
Crew R-2	Hr.	Daily	Hr.	Daily	Bare Costs	Incl. O&P
1 Electrician Foreman	$22.90	$183.20	$32.95	$263.60	$20.32	$29.37
3 Electricians	22.40	537.60	32.20	772.80		
2 Helpers	15.55	248.80	22.80	364.80		
1 Equip. Oper. (crane)	21.05	168.40	30.50	244.00		
1 S.P. Crane, 5 Ton		180.20		198.20	3.21	3.53
56 M.H., Daily Totals		$1318.20		$1843.40	$23.53	$32.90
Crew R-3	Hr.	Daily	Hr.	Daily	Bare Costs	Incl. O&P
1 Electrician Foreman	$22.90	$183.20	$32.95	$263.60	$22.33	$32.18
1 Electrician	22.40	179.20	32.20	257.60		
.5 Equip. Oper. (crane)	21.05	84.20	30.50	122.00		
.5 S.P. Crane, 5 Ton		90.10		99.10	4.50	4.95
20 M.H., Daily Totals		$536.70		$742.30	$26.83	$37.11
Crew R-4	Hr.	Daily	Hr.	Daily	Bare Costs	Incl. O&P
1 Struc. Steel Foreman	$23.70	$189.60	$37.25	$298.00	$22.24	$34.35
3 Struc. Steel Workers	21.70	520.80	34.10	818.40		
1 Electrician	22.40	179.20	32.20	257.60		
1 Gas Welding Machine		56.80		62.50	1.42	1.56
40 M.H., Daily Totals		$946.40		$1436.50	$23.66	$35.91

1.1 Overhead

		DAILY		BARE COSTS			TOTAL
	CREW	OUTPUT	UNIT	MAT.	INST.	TOTAL	INCL O&P
02-001 **ARCHITECTURAL FEES** New construction, minimum			Project				4.90%
005 Maximum							16%
010 (10) For alteration work, to $500,000, add to fee			↓				50%
015 Over $500,000, add to fee							25%
04-001 **CLEANING UP** After job completion, allow			Job Cost				.30%
003 Rubbish removal, see division 2.1-43							
005 Cleanup of floor area, continuous, per day	A-5	12	M.S.F.	1.49	25	26.49	38
010 Final	"	11.50	"	1.58	26	27.58	39
06-001 **CONSTRUCTION COST INDEX** (Div. 19) for 162 major U.S. and							
002 Canadian cities, total cost, min. (Greensboro NC)			%				81.00
005 Average							100%
010 Maximum (Anchorage, AK)			↓				134.10
08-001 (60) **CONSTRUCTION ECONOMIES**							
09-001 **CONSTRUCTION MANAGEMENT FEES** $1,000,000 job, minimum			Project				4.50%
005 Maximum							7.50%
030 $5,000,000 job, minimum							2.50%
035 Maximum			↓				4%
10-001 (9) **CONSTRUCTION TIME** Requirements							
11-001 **CONTINGENCIES** Allowance to add at conceptual stage			Project				15%
005 Schematic stage							10%
010 Preliminary working drawing stage							7%
015 Final working drawing stage			↓				2%
12-001 (13)(17) **CONTRACTOR EQUIPMENT** See division 1.5							
14-001 **CREWS** For building construction, see p. viii-xxiv							
010							
15-001 **ENGINEERING FEES** Educational planning consultant, minimum			Project				.50%
010 Maximum			"				2.50%
020 (11) Electrical, minimum			Contract				4.10%
030 Maximum							10.10%
040 Elevator & conveying systems, minimum							2.50%
050 Maximum							5%
060 Food service & kitchen equipment, minimum							8%
070 Maximum							12%
080 Landscaping & site development, minimum							2.50%
090 Maximum							6%
100 Mechanical (plumbing & HVAC), minimum							4.10%
110 Maximum			↓				10.10%
120 Structural, minimum			Project				1%
130 Maximum			"				2.50%
16-001 **HISTORICAL COST INDEXES** (Div. 19) Back to 1942							
18-001 **INSURANCE** Builders risk, standard, minimum			Job Cost				.10%
005 Maximum							.50%
020 (2) All-risk type, minimum							.12%
025 Maximum			↓				.68%
040 Contractor's equipment floater, minimum			Value				.50%
045 Maximum			"				2.50%
060 Public liability, average			Job Cost				.82%
061							
080 (7) Workers' compensation & employer's liability, average							
085 by trade, carpentry, general			Payroll		10.10%		
090 Clerical					.57%		
095 Concrete					8.77%		
100 Electrical					4%		
105 Excavation					6.87%		
110 Glazing			↓		7.90%		
115 Insulation					7.35%		

1.1 Overhead

		CREW	DAILY OUTPUT	UNIT	BARE COSTS MAT.	BARE COSTS INST.	BARE COSTS TOTAL	TOTAL INCL O&P
120	Lathing			Payroll		6.30%		
125	Masonry					7.60%		
130	Painting & decorating					7.70%		
135	Pile driving					17%		
140	Plastering					7.70%		
145	Plumbing					4.80%		
150	Roofing					18.20%		
155	Sheet metal work (HVAC)					6.30%		
160	Steel erection, structural					19.30%		
165	Tile work, interior ceramic					5.40%		
170	Waterproofing, brush or hand caulking					4.43%		
180	Wrecking					20.70%		
200	Range of 36 trades in 50 states, excl. wrecking, minimum					1.04%		
210	Average					9.30%		
220	Maximum					54.13%		
19-001	**JOB CONDITIONS** Modifications to total							
002	project cost summaries							
010	Economic conditions, favorable, deduct			Project				2%
020	Unfavorable, add							5%
030	Hoisting conditions, favorable, deduct							1%
040	Unfavorable, add							5%
050	General Contractor management, experienced, deduct							2%
060	Inexperienced, add							10%
070	Labor availability, surplus, deduct							1%
080	Shortage, add							10%
090	Material storage area, available, deduct							1%
100	Not available, add							2%
110	Subcontractor availability, surplus, deduct							5%
120	Shortage, add							12%
130	Work space, available, deduct							1%
140	Not available, add							4%
20-001	**LABOR INDEX** (Div. 19) For 162 major U.S. and Canadian cities							
002	Minimum (Roanoke, VA)			%		66.90%		
005	Average					100%		
010	Maximum (Anchorage, AK)					141.30%		
22-001	**MAIN OFFICE EXPENSE** Average for General Contractors							
002	As a percentage of their annual volume							
005	Annual volume under 1 million dollars			% Vol.			13.60%	
010	④ Up to 2.5 million dollars						8%	
015	Up to 4.0 million dollars						6.80%	
020	Up to 7.0 million dollars						5.60%	
025	Up to 10 million dollars						5.10%	
030	Over 10 million dollars						3.90%	
040	Field personnel clerk, average			Week		195		285L
050	Field engineer, minimum					455		665L
055	Average					605		885L
060	Maximum					685		995L
065	General purpose laborer, average					630		920L
070	Project Manager, minimum					865		1,250L
075	Average					965		1,400L
080	Maximum					1,100		1,600L
100	Superintendent, minimum					820		1,200L
105	Average					910		1,325L
110	Maximum					1,025		1,500L
120	Timekeeper, average					495		725L
24-001	**MARK-UP** For General Contractors for change							
010	of scope of job as bid							
020	Extra work, by subcontractors, add			%				10%
025	By General Contractor, add			"				15%

1.1 Overhead

			DAILY		BARE COSTS			TOTAL
		CREW	OUTPUT	UNIT	MAT.	INST.	TOTAL	INCL O&P
24	040 Omitted work, by subcontractors, deduct			%				5%
	045 By General Contractor, deduct							7.50%
	060 Overtime work, by subcontractor, add							15%
	065 By General Contractor, add							10%
	100 (6) Subcontractors, on their own labor, minimum					40%		
	110 Maximum			↓		60.80%		
26-001	MATERIAL INDEX (Div. 19) For 162 major U.S. and Canadian cities							
	002 Minimum (Greensboro, NC)			%	93.80%			
	004 Average				100%			
	006 Maximum (Anchorage, AK)			↓	125.30%			
27-001	MODELS Cardboard & paper, 1 building, minimum			Ea.	315		315	345M
	005 Maximum				775		775	855M
	010 2 buildings, minimum				445		445	490M
	015 Maximum			↓	1,055		1,055	1,150M
	020 Plexiglass and metal, basic layout			S.F.Flr.	.03		.03	.03M
	021 Including equipment and personnel			"	.10		.10	.11M
	030 Site plan layout, minimum			Ea.	510		510	560M
	035 Maximum			"	1,230		1,230	1,350M
28-001	(4) OVERHEAD As percent of direct costs, minimum			%				5%
	005 Average							12%
	010 (6) Maximum			↓				22%
30-001	OVERHEAD & PROFIT Allowance to add to items in this							
	002 book that do not include Subs O&P, average			%				25%
	010 (4) Allowance to add to items in this book that							
	011 do include Subs O&P, minimum			%				5%
	015 Average							10%
	020 Maximum							15%
	030 Typical, by size of project, under $100,000							30%
	035 $500,000 project							25%
	040 $2,000,000 project							20%
	045 Over $10,000,000 project			↓				15%
32-001	(3) OVERTIME For early completion of projects or where							
	002 labor shortages exist, add to usual labor, up to			Costs		118%		
34-001	(1) PERFORMANCE BOND For buildings, minimum			Job Cost				.39%
	010 Maximum							1.20%
36-001	PERMITS Rule of thumb, most cities, minimum							.50%
	010 Maximum			↓				2%
38-001	PHOTOGRAPHS 8" x 10", 4 shots, 2 prints ea., std. mounting			Set	75		75	83M
	010 Hinged linen mounts				87		87	96M
	020 8" x 10", 4 shots, 2 prints each, in color				156		156	170M
	030 For I.D. slugs, add to all above				2.40		2.40	2.64M
	050 Aerial photos, initial fly-over, 6 shots, 1 print ea., 8" x 10"				604		604	665M
	055 11" x 14" prints				662		662	730M
	060 16" x 20" prints				772		772	850M
	070 For full color prints, add				40%			
	075 Add for traffic control area			↓	210		210	230M
	090 For over 30 miles from airport, add per			Mile	3.62		3.62	3.98M
	100 Vertical photography, 4 to 6 shots with							
	101 different scales, 1 print each			Set	840		840	925M
	150 Time lapse equipment, camera and projector, buy				3,255		3,255	3,575M
	155 Rent per month			↓	420		420	460M
	170 Cameraman and film, including processing, B.&W.			Day	462		462	510M
	172 Color			"	525		525	580M
40-001	RENDERINGS Water color, matted, 20" x 30", eye level,							
	002 1 building, minimum			Ea.	315		315	345M
	005 Average				520		520	570M
	010 Maximum				625		625	690M
	100 5 buildings, minimum			↓	520		520	570M
	110 Maximum				935		935	1,025M

1.1 Overhead

				CREW	DAILY OUTPUT	UNIT	BARE COSTS MAT.	BARE COSTS INST.	BARE COSTS TOTAL	TOTAL INCL O&P
40	200		Aerial perspective, 1 building, minimum			Ea.	320		320	350M
	210		Maximum				625		625	690M
	300		5 buildings, minimum				625		625	690M
	310		Maximum			↓	1,350		1,350	1,475M
	42-001		**SAFETY NETS** No supports, stock sizes, nylon, 4" mesh			S.F.	1.05		1.05	1.15M
	010		Polypropylene, 6" mesh				1.14		1.14	1.25M
	020		Small mesh debris nets, 3/4" mesh, stock sizes				.55		.55	.60M
	022		Combined 4" mesh and 3/4" mesh, stock sizes				1.55		1.55	1.70M
	030		Monthly rental, 4" mesh, stock sizes, 1st month				.20		.20	.22M
	032		2nd month rental				.10		.10	.11M
	034		Subsequent months rental			↓	.09		.09	.09M
	040		Minimum rental is three months							
	43-001		**SCAFFOLDING, STEEL TUBULAR** 1 use per month, no plank							
	009		Building exterior, 1 to 5 stories	3 Carp	16.80	C.S.F.	12.50	29	41.50	55
	020		6 to 12 stories	4 Carp	15		11.20	43	54.20	74
	031	(14)	13 to 20 stories	5 Carp	16.75		9.60	48	57.60	80
	046		Building interior walls, (area) up to 16' high	3 Carp	22.70		11.60	21	32.60	43
	056		16' to 40' high		18.70		11.95	26	37.95	50
	080		Building interior floor area, up to 30' high	↓	85	↓	3.90	5.65	9.55	12.45
	090		Over 30' high	4 Carp	96.50		4.40	6.65	11.05	14.45
	600		Scaffold steel tubular, suspended slab form supports to 8'-2" high							
	610		1 use per month	4 Carp	31	C.S.F.	11.10	21	32.10	42
	615		2 uses per month	"	43	"	5.53	14.90	20.43	28
	650		Steel tubular, suspended slab form supports to 14'-8" high							
	660		1 use per month	4 Carp	16	C.S.F.	17.35	40	57.35	77
	665		2 uses per month	"	22	"	8.65	29	37.65	52
	44-001		**SCAFFOLDING SPECIALTIES**							
	005									
	120		Sidewalk bridge, heavy duty steel posts & beams, including							
	121		parapet protection & waterproofing							
	122		8' to 10' wide, 2 posts	3 Carp	15	L.F.	10.44	32	42.44	58
	123		3 posts	"	10	"	15.65	48	63.65	87
	150		Sidewalk bridge using tubular steel							
	151		scaffold frames, including planking	3 Carp	45	L.F.	3.50	10.65	14.15	19.30
	160		For 2 uses per month, deduct from all above				50%			
	170		For 1 use every 2 months, add to all above				100%			
	190		Catwalks, 32" wide, no guardrails, 6' span, buy			Ea.	104		104	115M
	200		10' span, buy			"	170		170	185M
	280		Hand winch-operated masons							
	281		scaffolding, no plank moving required							
	290		98' long, 10'-6" high, buy			Ea.	4,700		4,700	5,175M
	300		Rent per month				280		280	310M
	310		28'-6" high, buy				8,220		8,220	9,050M
	320		Rent per month				430		430	475M
	340		196' long, 28'-6" high, buy				15,400		15,400	16,900M
	350		Rent per month				835		835	920M
	360		64'-6" high, buy				26,150		26,150	28,800M
	370		Rent per month			↓	1,671		1,671	1,850M
	372		Putlog, standard, 8' span, with hangers, buy				73		73	80M
	373		Rent per month				10.50		10.50	11.55M
	375		12' span, buy			Ea.	103		103	115M
	376		Trussed type, 14' span, buy			"	167		167	185M
	377		Rent per month				16		16	17.60M
	379		20' span, buy			Ea.	190		190	210M
	380		Rolling ladders with handrails, 30" wide, buy, 2 step				140		140	155M
	400		7 step				315		315	345M
	405		10 step			↓	410		410	450M
	406									
	410		Rolling towers, buy, 5' wide, 7' long, 9' high			Ea.	850		850	935M
	420		For 5' high added sections, add			"	165		165	180M

1.1 Overhead

				Crew	Daily Output	Unit	Bare Costs Mat.	Bare Costs Inst.	Bare Costs Total	Total Incl O&P
44	430		Complete incl. wheels, railings, etc.							
	440		up to 20' high, rent per month			Ea.	94		94	105M
	650		Stair unit, interior, for scaffolding, buy			↓	265		265	290M
	655		Rent per month				23.60		23.60	26M
46	001	**SCHEDULING**	Critical path, as % of architectural fee, minimum			%				2%
	010		Maximum			"				4%
	030		Computer-update, micro, no plots, minimum			Ea.				150
	040		Including plots, maximum			"				2,000
	060		Rule of thumb, CPM scheduling, small job			Job Cost				.10%
	065		Large job							.05%
	070		Cost control, small job							.15%
	075		Large job			↓				.04%
48	001	**SMALL TOOLS**	As % of contractor's work, minimum			Total				.50%
	010		Maximum			"				1%
50	001	**SURVEYING**	Conventional, topographical, minimum	A-7	3.30	Acre	11.55	125	136.55	200
	010		Maximum	A-8	.60		34.65	905	939.65	1,350
	030		Lot location and lines, minimum	A-7	2		18.95	210	228.95	325
	032		Average	"	1.25		34.65	335	369.65	525
	040		Maximum	A-8	1	↓	56.70	545	601.70	855
	060		Monuments, 3' long	A-7	10	Ea.	7.90	42	49.90	70
	080		Property lines, perimeter, cleared land	"	1,000	L.F.	.02	.42	.44	.63
	090		Wooded land	A-8	875	"	.02	.62	.64	.93
	110		Crew for building layout, 2 man crew	A-6	1	Day		295	295	430L
	120		3 man crew	A-7	1			420	420	610L
	130		4 man crew	A-8	1	↓		545	545	795L
	150		Aerial surveying, including ground control, minimum fee, 10 acres			Total				3,780
	151		100 acres							4,090
	155		From existing photography, deduct			↓				630
	160		2' contours, 10 acres			Acre				315
	165		20 acres							210
	180		50 acres							68
	185		100 acres							58
	200		1000 acres							13
	205		10,000 acres			↓				8.40
	215		For 1' contours and							
	216		dense urban areas, add to above			Acre				40%
	300		Inertial guidance system for							
	301		locating coordinates, rent per day			Ea.				3,000
52	001	**SWING STAGING**	For masonry, 5' wide to 7' long, hand operated							
	002		cable type, with 150' cables, buy			Ea.	1,690		1,690	1,850M
	003		Rent per week			"	9.70		9.70	10.65M
	060		Lightweight (not for masons) 20' long for 200' height,							
	061		manual type, buy			Ea.	3,120		3,120	3,425M
	062		Rent per month				345		345	380M
	070		Powered, electric or air, to 300' high, buy				10,000		10,000	11,000M
	071		Rent per month				865		865	950M
	080		Over 300' per 100 L.F., add per month			↓	50		50	55M
	090									
	100		Single bosun's chair or work basket 3' x 3.5', electric, buy			Ea.	6,100		6,100	6,700M
	101		Rent per month			"	420		420	460M
54	001	**TARPAULINS**	Cotton duck, 10 oz. to 13.13 oz. per S.Y., minimum			S.F.	.25		.25	.27M
	005		Maximum				.42		.42	.46M
	010		Polyvinyl coated nylon, 14 oz. to 18 oz., minimum				.34		.34	.37M
	015		Maximum				.42		.42	.46M
	020		Reinforced polyethylene 3 mils thick, white				.04		.04	.04M
	030		4 mils thick, white, clear or black				.05		.05	.05M
	040		5.5 mils thick, clear				.06		.06	.06M
	050		White, fire retardant				.11		.11	.12M
	060		7.5 mils, oil resistant, fire retardant				.12		.12	.13M
	070		8.5 mils, black			↓	.15		.15	.16M

1.1 Overhead

			CREW	DAILY OUTPUT	UNIT	BARE COSTS MAT.	BARE COSTS INST.	BARE COSTS TOTAL	TOTAL INCL O&P
54	071	Woven polyethylene, 6 mils thick			S.F.	.26		.26	.28M
	074	Mylar polyester, non-reinforced, 7 mils thick			"	.82		.82	.90M
	56-001	**TAXES** Sales tax, State, County & City, average			%	4.28%			
	005	Maximum				8%			
	020	Social Security, on first $40,000 of wages					7.05%		
	021								
	030	Unemployment, MA, combined Federal and State, minimum			%		2.30%		
	035	Average					5.50%		
	040	Maximum					6.50%		
	041								
	58-001	**TEMPORARY CONSTRUCTION** See also Division 1.5-15							
	002	Air supported structures see also Division 13.1-05							
	028	Barricades, see also Division 1.5-15-160							
	029								
	030	Barricades, 5' high, 3 rail @ 2" x 8", fixed	2 Carp	30	L.F.	8.58	10.65	19.23	25
	035	Movable	"	20	"	9.20	16	25.20	33
	050	Stock units, 6' high, 8' wide, plain, buy			Ea.	343		343	375M
	055	With reflective tape, buy			"	440		440	485M
	060	Break-a-way 3" PVC pipe barricade							
	061	with 3 ea. 1' x 4' reflectorized panels, buy			Ea.	242		242	265M
	150	Plywood with steel legs, 32" wide				45		45	50M
	170	Telescoping Christmas tree, 9' high, 5 flags, buy				90		90	99M
	190	Traffic cones, PVC, 18" high				5.60		5.60	6.15M
	195	28" high				11.19		11.19	12.30M
	210	Catwalks, no handrails, 3 joists 2" x 4"	2 Carp	55	L.F.	2.83	5.80	8.63	11.55
	215	3 joists, 3" x 6"	"	40		5.30	8	13.30	17.40
	230	Fencing, chain link, 5' high	2 Sswk	100		4.64	3.47	8.11	10.55
	235	6' high		75		7.36	4.63	11.99	15.35
	250	Rented chain link, 6' high, to 500'		100		1.82	3.47	5.29	7.45
	255	Over 1000' (up to 12 mo.)		110		1.51	3.16	4.67	6.60
	270	Plywood, painted, 2" x 4" frame, 4' high	A-4	135		3.28	3.51	6.79	8.65
	275	4" x 4" frame, 8' high	"	110		5.65	4.31	9.96	12.45
	290	Wire mesh on 4" x 4" posts, 4' high	2 Carp	100		3.70	3.20	6.90	8.70
	295	8' high		80		5.56	4	9.56	11.90
	310	Guardrail, wooden, 3' high, 1" x 6" on 2" x 4" posts		200		1.28	1.60	2.88	3.73
	315	2" x 6" on 4" x 4" posts		165		2.11	1.94	4.05	5.15
	320	Portable metal with base pads, buy				10.80		10.80	11.90M
	321	Typical installation, assume 10 reuses	2 Carp	600		1.26	.53	1.79	2.16
	330	Heat, incl. fuel and operation, per week, 12 hrs. per day	1 Skwk	8.75	CSF Flr	14.16	18.75	32.91	43
	335	24 hrs. per day	"	4.50		18.80	36	54.80	74
	350	Lighting, incl. service lamps, wiring & outlets, minimum	1 Elec	34		1.74	5.25	6.99	9.50
	355	Maximum	"	17		4	10.55	14.55	19.55
	370	Power for temporary lighting only, per month, minimum/month						.85	.88
	375	Maximum/month						2.21	2.55
	390	Power for job duration incl. elevator, etc., minimum						42	49
	395	Maximum						86	100
	420	Office trailer, furnished, no hookups, 20' x 8', buy	2 Skwk	1	Ea.	3,300	330	3,630	4,100
	425	Rent per month						115	125M
	427	32' x 8', buy	2 Skwk	.70		5,800	470	6,270	7,075
	428	Rent per month						170	185M
	430	50' x 10', buy	2 Skwk	.60		10,000	545	10,545	11,800
	440	Rent per month						315	345M
	443	50' x 12', buy	2 Skwk	.50		11,100	655	11,755	13,200
	444	Rent per month						340	375M
	446	For air conditioning, rent per month, add				32		32	35M
	447	For delivery, add per			Mile	1.37		1.37	1.50M
	450	Portable buildings, prefab, on skids, economy, 8' x 8'	2 Carp	265	S.F.	51	1.21	52.21	58
	460	Deluxe, 8' x 12'		150		67	2.13	69.13	77
	470	Ramp, 3/4" plywood on 2" x 6" joists, 16" O.C.		300		1.28	1.07	2.35	2.95
	475	On 2" x 10" joists, 16" O.C.		275		1.90	1.16	3.06	3.78

1.1 Overhead

			DAILY		BARE COSTS			TOTAL
		CREW	OUTPUT	UNIT	MAT.	INST.	TOTAL	INCL O&P
58	480 Roads, gravel fill, no surfacing, 4" gravel thickness	B-14	715	S.Y.	.60	1.35	1.95	2.54
	490 8" gravel thickness	"	615	"	1.14	1.57	2.71	3.44
	520 Sidewalks, 2" x 12" planks, 2 uses	1 Carp	350	S.F.	.36	.46	.82	1.06
	522 Exterior plywood, 2 uses, 1/2" thick		750		.19	.21	.40	.52
	523 5/8" thick		650		.25	.25	.50	.63
	525 3/4" thick		600		.34	.27	.61	.76
	526 Signs, hi-intensity reflectorized, no posts, buy				8.40		8.40	9.25M
	527 Stair tread protection, 2" x 12" planks, 1 use	1 Carp	75	Tread	.94	2.13	3.07	4.13
	529 Exterior plywood, 1/2" thick, 1 use		65		.43	2.46	2.89	4.04
	530 3/4" thick, 1 use		60		.78	2.67	3.45	4.72
	550 Storage vans, trailer mounted, 16' x 8', buy	2 Skwk	1.80	Ea.	2,200	180	2,380	2,675
	552 Rent per month				75		75	83M
	555 28' x 10', buy	2 Skwk	1.40		2,600	235	2,835	3,200
	557 Rent per month				80		80	88M
	560 Surveyor stakes, hardwood, 1" x 1" x 48" long			C	25		25	28M
	562 2" x 2" x 18" long				25		25	28M
	564 2" x 2" x 24" long				29		29	32M
	566 2" x 4" x 24" long				40		40	44M
	580 Toilet, portable, see Division 1.5-15-641							
60-001	**TESTING** For concrete building costing $1,000,000, minimum			Project				5,000
	002 Maximum							50,000
	005 Steel building, minimum							5,000
	007 Maximum							10,000
	010 For building costing, $10,000,000, minimum							35,000
	015 Maximum							50,000
	020 Asphalt testing, compressive strength Marshall stability, set of 3			Ea.				100
	022 Density, set of 3							39
	025 Extraction, individual tests on sample							80
	030 Penetration							25
	035 Mix design, 5 specimens							640
	036 Additional specimen							80
	040 Specific gravity							31
	042 Swell test							62
	045 Water effect and cohesion, set of 6							154
	047 Water effect and plastic flow							62
	060 Concrete testing, aggregates, abrasion							62
	065 Absorption							26
	080 Petrographic analysis							450
	090 Specific gravity							25
	100 Sieve analysis, washed							67
	105 Unwashed							36
	120 Sulfate soundness							68
	130 Weight per cubic foot							16
	150 Cement, physical tests							255
	160 Chemical tests							309
	180 Compressive strength, cylinders, delivered to lab							9
	190 Picked up by lab, minimum							8
	195 Average							10
	200 Maximum							19
	220 Compressive strength, cores (not incl. drilling)							26
	225 Core drilling, 4" diameter (plus technician)			Inch				15
	226 Technician for core drilling			Hr.				33
	230 Patching core holes			Ea.				21
	240 Drying shrinkage at 28 days							245
	250 Flexural test beams							32
	260 Mix design, one batch mix							155
	265 Added trial batches							104
	280 Modulus of elasticity							124
	290 Tensile test, cylinders							52

1.1 Overhead

			CREW	DAILY OUTPUT	UNIT	BARE COSTS MAT.	BARE COSTS INST.	BARE COSTS TOTAL	TOTAL INCL O&P
60	300	Water-Cement ratio curve, 3 batches			Ea.				460
	310	4 batches							615
	330	Masonry testing, absorption, per 5 brick							51
	335	Chemical resistance, per 2 brick							51
	340	Compressive strength, per 5 brick							41
	342	Efflorescence, per 5 brick							41
	344	Imperviousness, per 5 brick							41
	347	Modulus of rupture, per 5 brick							36
	350	Moisture, block only							21
	355	Mortar, compressive strength, set of 3							21
	410	Reinforcing steel, bend test							19
	420	Tensile test, up to #8 bar							26
	422	#9 to #11 bar							31
	424	#14 bar and larger							61
	440	Soil testing, Atterberg limits, liquid and plastic limits							30
	451	Hydrometer analysis and specific gravity							50
	460	Sieve analysis, washed							41
	470	Unwashed							41
	471	Consolidation test							410
	472	Density of undisturbed sample							24
	475	Moisture content							7.50
	480	Permeability, variable or constant head							67
	490	Proctor compaction, 4" standard mold							113
	495	6" modified mold							124
	510	Shear tests, triaxial, minimum							154
	515	Maximum							288
	530	Direct shear, minimum			↓				98
	535	Maximum							268
	540	Subsurface exploration, mobilization			Mile				4.25
	541	Difficult access for rig, add			Hr.				83
	542	Auger borings, drill rig, incl. samples			L.F.				10.50
	543	Hand			↓				16
	545	Drill and sample every 5', split spoon							14
	546	Extra samples			Ea.				17
	555	Technician for inspection, per day, earthwork							150
	565	Bolting							165
	575	Roofing							150
	579	Welding			↓				165
	582	Non-destructive testing, dye penetrant			Day				265
	584	Magnetic particle							265
	586	Radiography							400
	588	Ultrasonic			↓				290
	600	Welding certification, minimum			Ea.				68
	610	Maximum			"				210
	62-001	**WATCHMAN** Service, monthly basis, uniformed man, minimum			Hr.				5.95
	010	Maximum							11
	020	Man and command dog (man dog), minimum							8.40
	030	Maximum			↓				12.50
	050	Sentry dog, leased, with job patrol (yard dog), 1 dog			Week				155
	060	2 dogs			"				220
	080	Purchase, trained sentry dog, minimum			Ea.				830
	090	Maximum			↓				1,550
	64-001	**WEATHER STATION** Remote recording, minimum				3,190		3,190	3,500M
	010	Maximum			↓	5,790		5,790	6,375M
	68-001	**WINTER PROTECTION** Reinforced plastic on wood	2 Clab	750	S.F.	.23	.34	.57	.74
	010	framing to close openings							
	020	Tarpaulins hung over scaffolding, 8 uses, not incl. scaffolding		1,500		.11	.17	.28	.37
	030	Prefab, fiberglass panels, steel frame, 8 uses	↓	1,200	↓	.44	.21	.65	.79

1.5 Contractor Equipment

		UNIT	HOURLY OPER. COST	RENT PER DAY	RENT PER WEEK	RENT PER MONTH	CREWS EQUIPMENT COST
05-001 002	**EARTHWORK EQUIPMENT RENTAL** Without operators unless noted by*						
004	⑬ Aggregate Spreader, push type 8' to 12' wide	Ea.	.65	79	230	690	51.20
005	Augers for truck or trailer mounting, vertical drilling						
006	4" to 36" diam., 54 H.P., gas, 10' spindle travel	Ea.	6.70	155	465	1,475	146.60
007	14' spindle travel		7	185	550	1,650	166
008	Auger, horizontal boring machine, 12" to 36" diameter, 45 H.P.		6.15	595	1,775	5,375	404.20
009	12" to 48" diameter, 65 H.P.		8.45	625	2,075	6,100	482.60
010	⑰ Backhoe, diesel hydraulic, crawler mounted, 1/2 C.Y. capacity		6.50	495*	1,050	3,425	262
012	5/8 C.Y. capacity		7.40	520*	1,075	3,525	274.20
014	3/4 C.Y. capacity		10.65	590*	1,275	5,000	340.20
015	1 C.Y. capacity		13.60	920*	1,675	4,900	443.80
020	1-1/2 C.Y. capacity		17.45	990*	2,200	6,375	579.60
030	2 C.Y. capacity		27	1,200*	2,925	8,825	801
032	2-1/2 C.Y. capacity		42	1,475*	4,150	11,800	1,166
034	3-1/2 C.Y. capacity		49	1,775*	5,100	15,200	1,412
035	Gradall type, truck mounted, 3 ton @ 15' radius, 5/8 C.Y.		17.55	870*	1,500	4,550	440.40
037	1 C.Y. capacity		23	1,050*	2,225	6,625	629
040	Backhoe-loader, wheel type, 40 to 45 H.P., 5/8 C.Y. capacity		4.70	335*	400	1,250	117.60
045	45 H.P. to 60 H.P., 3/4 C.Y. capacity		5.75	405*	550	1,700	156
046	80 H.P., 1-1/4 C.Y. capacity		8.65	465*	825	2,325	234.20
050	Brush chipper, gas engine, 6" cutter head, 35 H.P.		3.13	125	395	1,150	104.05
055	12" cutter head, 130 H.P.		6.25	165	435	1,275	137
060	15" cutter head, 165 H.P.		8	200	490	1,500	162
075	Bucket, clamshell, general purpose, 3/8 C.Y.		.42	35	105	305	24.35
080	1/2 C.Y.		.58	40	125	380	29.65
085	3/4 C.Y.		.65	52	150	460	35.20
090	1 C.Y.		.77	54	170	500	40.15
095	1-1/2 C.Y.		1.10	70	210	625	50.80
100	2 C.Y.		1.20	85	280	855	65.60
101	⑰ Bucket, dragline, medium duty, 1/2 C.Y.		.37	26	67	200	16.35
102	3/4 C.Y.		.43	30	90	280	21.45
103	1 C.Y.		.50	36	110	320	26
104	1-1/2 C.Y.		.58	44	135	385	31.65
105	2 C.Y.		.70	58	175	520	40.60
107	3 C.Y.		1.01	90	255	750	59.10
120	Compactor, roller, 2 drum, 2000 lb., operator walking		2.45	93	280	835	75.60
125	Rammer compactor, gas, 1000 lb. blow		1.02	52	150	450	38.15
130	Vibratory plate, gas, 13" plate, 1000 lb. blow		.54	41	120	355	28.30
135	24" plate, 5000 lb. blow		.73	50	170	495	39.85
137	Curb builder, 14 H.P., gas, single screw		2.85	88	265	800	75.80
139	Double screw	↓	2.59	100	305	925	81.70
175	Extractor, piling, see lines 250 to 275						
180							
186	Grader, self-propelled, 25,000 lb.	Ea.	11.85	405	1,225	4,325	339.80
191	30,000 lb.		16.55	485	1,450	5,100	422.40
192	40,000 lb.		22	635	1,900	6,300	556
193	55,000 lb.		28.10	950	2,850	9,100	794.80
195	Hammer, pavement demo., hyd., gas, self-prop., 1000 to 1250 lb.		6.18	265	815	2,450	212.45
200	1300 to 1500 lb.		6.65	290	865	2,600	226.20
205	⑬ Pile driving hammer, steam or air, 4150 ft.-lb. @ 225 BPM		3.83	130	390	1,150	108.65
210	8750 ft.-lb. @ 145 BPM		5.14	200	605	1,650	162.10
215	15,000 ft.-lb. @ 60 BPM		5.95	285	765	1,975	200.60
220	24,450 ft.-lb. @ 111 BPM	↓	7.25	365	1,125	2,900	283
225	Leads, 15,000 ft.-lb. hammers	L.F.				12	.80
230	24,450 ft.-lb. hammers and heavier	"				13	.85
235	Diesel type hammer, 22,400 ft.-lb.	Ea.	8.40	435	1,275	3,025	322.20
240	41,300 ft.-lb.		12.55	615	2,125	4,950	525.40
245	141,000 ft.-lb.		21	1,050	3,150	9,550	798
250	Vib. elec. hammer/ext., 200 KW diesel generator, 34 H.P.	↓	13.45	580	1,750	5,250	457.60

1.5 Contractor Equipment

		UNIT	HOURLY OPER. COST	RENT PER DAY	RENT PER WEEK	RENT PER MONTH	CREWS EQUIPMENT COST
05 255	80 H.P.	Ea.	18.20	1,050	3,150	9,500	775.60
260	150 H.P.		21	1,375	4,125	10,200	993
270	Extractor, steam or air, 700 ft.-lb.		.94	155	435	1,125	94.50
275	1000 ft.-lb.		1.41	235	630	1,500	137.30
300	Roller, tandem, gas, 3 to 5 ton		3.25	105	295	900	85
305	Diesel, 8 to 12 ton		6.65	185	610	1,575	175.20
310	Towed type, vibratory, gas 12.5 H.P., 2 ton		2.23	94	275	820	72.85
315	Sheepsfoot, double 60" x 60"		2.39	105	310	925	81.10
320	Pneumatic tire diesel roller, 12 ton		8.45	145	455	1,375	158.60
325	21 to 25 ton		13.15	325	960	2,925	297.20
330	Sheepsfoot roller, self-propelled, 4 wheel, 130 H.P.		10.70	360	1,025	3,125	290.60
332	300 H.P.		16	445	1,325	4,025	393
335	Vibratory steel drum & pneumatic tire, diesel, 18,000 lb.		12.05	310	925	2,775	281.40
340	29,000 lb.		15.06	370	1,125	3,350	345.50
345	Scrapers, towed type, 7 to 9 C.Y. capacity		2.73	115	340	1,025	89.85
350	12 to 17 C.Y. capacity		4.54	140	420	1,250	120.30
355	Self-propelled, 4 x 4 drive, 2 engine, 14 C.Y. capacity		38	1,375*	3,300	10,250	964
360	1 engine, 24 C.Y. capacity		45	1,525*	3,775	11,200	1,115
365	Self-loading, 11 C.Y. capacity		34	860*	1,800	5,350	632
370	22 C.Y. capacity		51	1,200*	2,700	8,200	948
385	Shovels, see Cranes division 1.5-20						
386	Shovel front attachment, mechanical, 1/2 C.Y.	Ea.	.63	60	185	535	42.05
387	3/4 C.Y.		2.60	96	290	855	78.80
388	1 C.Y.		2.82	150	440	1,300	110.55
389	1-1/2 C.Y.		3.07	170	485	1,475	121.55
391	3 C.Y.		5.70	295	915	2,725	228.60
411	Tractor, crawler, with bulldozer, torque converter, diesel 75 H.P.		7	445*	750	2,250	206
415	105 H.P.		10.75	585*	1,300	4,000	346
420	140 H.P.		14.30	690*	1,625	4,850	439.40
426	200 H.P.		19.35	890*	2,400	7,850	634.80
431	300 H.P.		27	1,175*	3,575	10,200	931
436	410 H.P.		39	1,400*	4,225	13,000	1,157
438	700 H.P.		69	2,200*	7,100	21,200	1,972
440	Loader, crawler, torque converter, diesel, 1-1/2 C.Y.,80 H.P.		10.55	515*	960	2,850	276.40
445	1-1/2 to 1-3/4 C.Y., 95 H.P.		12.50	550*	1,075	3,300	315
451	1-3/4 to 2-1/4 C.Y., 130 H.P.		16.40	665*	1,450	4,450	421.20
453	2-1/2 to 3-1/4 C.Y., 190 H.P.		25	835*	2,000	6,000	600
456	4-1/2 to 5 C.Y., 275 H.P.		35	1,250*	3,650	11,100	1,010
461	Tractor loader, wheel, torque conv 4 x 4, 1 to 1-1/4 C.Y., 65 H.P.		8.30	425*	680	2,175	202.40
462	1-1/2 to 1-3/4 C.Y., 80 H.P.		10.75	495*	925	2,775	271
465	1-3/4 to 2 C.Y., 100 H.P.		12.60	550*	1,125	3,350	325.80
471	2-1/2 to 3-1/2 C.Y., 130 H.P.		17.90	655*	1,500	4,525	443.20
473	3 to 4-1/2 C.Y., 170 H.P.		22	760*	1,825	4,925	541
476	5-1/4 to 5-3/4 C.Y., 270 H.P.		35	1,050*	2,825	9,325	845
481	7 to 8 C.Y., 375 H.P.		59	1,475*	4,100	10,700	1,292
487	12-1/2 C.Y., 690 H.P.		96	2,650*	8,075	24,400	2,383
488	Wheeled, skid steer, 10 C.F., 30 H.P. gas		3.50	90	340	840	96
489	1 C.Y., 78 H.P., diesel		4.85	155	510	1,475	140.80
490	Trencher, chain, boom type, gas, operator walking, 12 H.P.		8.05	100	355	1,075	135.40
491	Operator riding, 40 H.P.		12.85	170	485	1,400	199.80
500	Wheel type, gas, 4' deep, 12" wide		11.25	450*	750	2,250	240
510	Diesel, 6' deep, 20" wide		15	625*	1,475	4,500	415
515	Ladder type, gas, 5' deep, 8" wide		8.55	495*	970	2,850	262.40
520	Diesel, 8' deep, 16" wide		10.70	665*	1,675	5,250	420.60
525	Truck, dump, tandem, 12 ton payload		13.15	350*	575	1,725	220.20
530	Three axle dump, 16 ton payload		16.20	435*	750	2,275	279.60
535	Dump trailer only, rear dump, 16-1/2 C.Y.		4.02	80	240	715	80.15
540	20 C.Y.		4.90	92	280	845	95.20
545	Flatbed, single axle, 1-1/2 ton rating		3.80	46	135	415	57.40
550	3 ton rating		4.40	58	160	505	67.20

1.5 Contractor Equipment

		UNIT	HOURLY OPER. COST	RENT PER DAY	RENT PER WEEK	RENT PER MONTH	CREWS EQUIPMENT COST
05 555	Off highway rear dump, 25 ton capacity	Ea.	26	840*	2,200	6,750	648
560	35 ton capacity	"	35	1,150*	3,150	9,475	910
10-001	**CONCRETE EQUIPMENT RENTAL**						
010	without operators unless noted by *						
020	Bucket, concrete lightweight, 1/2 C.Y.	Ea.	.12	22	55	140	11.95
030	1 C.Y.		.12	30	73	220	15.55
040	1-1/2 C.Y.		.19	30	89	260	19.30
050	2 C.Y.		.19	36	110	330	23.50
060	Cart, concrete, operator walking, 10 C.F.		.90	49	145	400	36.20
070	Operator riding, 18 C.F.		1.59	70	225	630	57.70
080	Conveyer for concrete, portable, gas, 16" wide, 26' long		1.44	78	215	620	54.50
090	46' long		1.67	140	440	1,300	101.35
100	56' long		1.89	175	520	1,500	119.10
110	Core drill, electric, 2-1/2 H.P., 1" to 8" bit diameter		.59	46	115	370	27.70
120	Finisher, concrete floor, gas, riding trowel, 48" diameter		1.65	93	260	630	65.20
130	Gas, manual, 3 blade, 36" trowel		.42	31	93	275	21.95
140	4 blade, 48" trowel		.55	39	120	325	28.40
150	Float, hand-operated (Bull float) 48" wide		.12	10	28	48	6.55
160	Grinder, concrete and terrazzo, electric, floor		.90	40	120	355	31.20
170	Wall grinder		.30	18	45	190	11.40
180	Mixer, powered, mortar and concrete, gas, 6 C.F., 18 H.P.		2.10	35	105	305	37.80
190	10 C.F., 25 H.P.		3.05	45	135	370	51.40
200	16 C.F.		3.15	93	310	915	87.20
210	Concrete, stationary, tilt drum, 2 C.Y.		5.85	255	775	2,325	201.80
212	Pump, concrete, truck mounted, 4" line, 80' boom		18.90	790*	1,800	5,275	511.20
214 (46)	5" line, 110' boom		29	845*	2,000	6,000	632
216	Mud jack, 47 C.F. per hr.		1.14	28	83	250	25.70
218	225 C.F. per hr.		3.03	145	415	1,225	107.25
260	Saw, concrete, manual, gas, 18 H.P.		1.70	38	115	340	36.60
265	Self-propelled, gas, 30 H.P.		2.72	56	170	505	55.75
270	Vibrators, concrete, electric, 60 cycle, 2 H.P.		.24	22	66	200	15.10
280	3 H.P.		.31	30	90	245	20.50
290	Gas engine, 3 H.P.		.69	22	68	200	19.10
300	5 H.P.	↓	.92	34	99	275	27.15
310	Concrete transit mixer, hydraulic drive						
311							
312	6 x 4, 250 H.P., 8 C.Y., rear discharge	Ea.	18.10	610	2,000	5,700	544.80
320	Front discharge		18.40	630	2,075	6,100	562.20
330	6 x 6, 285 H.P., 12 C.Y., rear discharge		21	730	2,375	6,900	643
340	Front discharge	↓	22	765	2,450	7,250	666
15-001	**GENERAL EQUIPMENT RENTAL**						
010							
015	Aerial lift, scissor type, to 15' high, 1000 lb. capacity	Ea.	.16	79	255	765	52.30
016	To 25' high, 2000 lb. capacity		.19	120	375	1,125	76.50
017	Telescoping boom to 40' high, 750 lb. capacity		.22	245	720	2,325	145.75
018	2000 lb. capacity		.25	310	1,025	3,100	207
019	To 60' high, 750 lb. capacity	↓	.28	555	1,275	3,400	257.25
019							
020	Air compressor, portable, gas engine, 60 C.F.M.	Ea.	3.95	35	99	305	51.40
030	160 C.F.M.		4.78	53	160	480	70.25
040	Diesel engine, rotary screw, 250 C.F.M.		7.05	89	275	800	111.40
050	365 C.F.M.		10.75	120	360	1,075	158
060	600 C.F.M.		17.85	165	500	1,500	242.80
070	750 C.F.M.		22	215	630	1,875	302
080	For silenced models, small sizes, add		3%	5%	5%		
090	Large sizes, add	↓	4%	7%	7%		
092	Air tools and accessories						
093	Breaker, pavement, 60 lb.	Ea.	.44	17	47	140	12.90
094	80 lb.		.58	18	53	155	15.25
095	Drills, hand (jackhammer) 65 lb.	↓	.60	17	57	175	16.20

1.5 Contractor Equipment

			UNIT	HOURLY OPER. COST	RENT PER DAY	RENT PER WEEK	RENT PER MONTH	CREWS EQUIPMENT COST
15	096	Wagon, swing boom, 4" drifter	Ea.	7.40	190	565	1,650	172.20
	097	5" drifter		7.70	215	830	2,400	227.60
	098	Dust control per drill		.14	11	26	77	6.30
	099	Hammer, chipping, 12 lb.		.16	15	40	115	9.30
	100	Hose, air with couplings, 50' long, 3/4" diameter		.10	4	13	27	3.40
	110	1" diameter		.10	6	15	36	3.80
	120	1-1/2" diameter		.10	11	29	84	6.60
	130	2" diameter		.11	13	40	120	8.90
	140	2-1/2" diameter		.12	14	42	125	9.35
	141	3" diameter		.12	14	43	130	9.55
	145	Drill, steel, 7/8" x 2'			4	11	17	2.20
	146	7/8" x 6'			6	12	28	2.40
	152	Moil points		.67	2.80	9	13	7.15
	153	Sheeting driver for 60 lb. breaker		.08	9	20	55	4.65
	154	For 125 lb. breaker		.08	9	22	66	5.05
	155	Spade		.13	5	12	36	3.45
	156	Tamper, single, 35 lb.		.44	16	48	145	13.10
	157	Triple, 140 lb.		1.46	31	99	295	31.50
	158	Wrenches, impact, air powered up to 3/4" bolt		.08	20	60	175	12.65
	159	Up to 1-1/4" bolt		.09	27	79	225	16.50
	160	Barricades, barrels, reflectorized 1 to 50 barrels			.50	1.48	4.48	.30
	161	100 to 200 barrels			.43	1.28	3.78	.25
	162	Barrels with flashers, 1 to 50 barrels			.80	2.38	7.10	.45
	163	100 to 200 barrels			.73	2.15	6.45	.40
	164	Barrels with steady burn type C lights			.87	2.44	7.90	.50
	165	Illuminated board, trailer mounted, with generator		1.10	81	310	785	70.80
	167	Portable, stock, with flashers, 1 to 6 units			.56	1.64	4.80	.30
	168	25 to 50 units			.50	1.36	4.10	.25
	170	Carts, brick, hand powered, 1000 lb. capacity		1	14.45	42	125	16.40
	180	Gas engine, 1500 lb., 7-1/2 ft. lift		1.22	71	210	625	51.75
	183	Distributor, asphalt, trailer mtd, 2000 gal., 38 H.P. diesel		4.85	265	870	2,600	212.80
	184	3000 gal., 38 H.P. diesel		5.40	305	1,000	2,675	243.20
	185	Drill, rotary hammer, electric, 1-1/2" diameter		.13	33	97	225	20.45
	186	Carbide bit for above			11	31	46	6.20
	187	Emulsion sprayer, 65 gal., 5 H.P. engine		1.03	18	51	175	18.45
	188	200 gal., 5 H.P. engine		1.53	35	105	345	33.25
	190	Fencing, see division 1.1-58 and 2.7-09						
	192	Floodlight, mercury, vapor or quartz, on tripod						
	193	1000 watt	Ea.	.05	20	60	180	12.40
	194	2000 watt		.08	35	100	300	20.65
	202	Forklift, wheeled, for brick, 18', 3000 lb.		6.15	145	470	1,375	143.20
	204	28', 4000 lb.		7.55	185	545	1,650	169.40
	210	Generator, electric, gas engine, 1.5 KW to 3 KW		.52	23	70	205	18.15
	220	5 KW		.89	40	145	360	36.10
	230	10 KW		1.95	60	195	585	54.60
	240	25 KW		3.78	91	325	845	95.25
	250	Diesel engine, 20 KW		3.30	61	200	600	66.40
	260	50 KW		3.90	135	345	1,025	100.20
	270	100 KW		8.23	215	550	1,575	175.85
	280	250 KW		16.80	300	910	2,725	316.40
	285	Hammer, hydraulic, for mounting on boom, to 500 ft.-lb.		2.36	185	560	1,675	130.90
	286	500 to 1200 ft.-lb.		2.65	270	805	2,400	182.20
	290	Heaters, space, oil or electric, 50 MBH		1.03	15	45	135	17.25
	300	100 MBH		1.29	22	63	180	22.90
	310	300 MBH		1.77	33.68	100	300	34.15
	315	500 MBH		2.24	49	145	430	46.90
	320	Hose, water, suction with coupling, 20' long, 2" diameter			7	17	46	3.40
	321	3" diameter			9	23	66	4.60
	322	4" diameter			12	36	110	7.20
	323	6" diameter			28	72	185	14.40

1.5 Contractor Equipment

		UNIT	HOURLY OPER. COST	RENT PER DAY	RENT PER WEEK	RENT PER MONTH	CREWS EQUIPMENT COST
324	8" diameter	Ea.		38	90	275	18
325	Discharge hose with coupling, 50' long, 2" diameter			7	18	47	3.60
326	3" diameter			10	25	64	5
327	4" diameter			13	38	100	7.60
328	6" diameter			25	65	165	13
329	8" diameter			42	91	190	18.20
330	Ladders, extension type, 16' to 36' long			10	29	70	5.80
340	40' to 60' long			20	60	175	12
341	Level, laser type, for pipe laying, self leveling			145	435	815	87
343	Manual leveling			130	385	665	77
344	Rotary beacon with rod and sensor			145	435	815	87
346	Builders level with tripod and rod			15	51	125	10.20
350	Light towers, towable, with generator, 1000 watt		1.30	80	205	620	51.40
360	2000 watt		1.94	115	345	1,025	84.50
370	Mixer, powered, plaster and mortar, 6 C.F., 7 H.P.		.95	35	105	300	28.60
380	10 C.F., 9 H.P.		1.24	45	135	370	36.90
390	Paint sprayers complete, 8 CFM			32	87	260	17.40
400	17 CFM			40	135	340	27
402	Pavers, bituminous, rubber tires, 8' wide, 52 H.P., gas		11.80	390	1,300	3,800	354.40
403	8' wide, 64 H.P., diesel		10.95	405	1,475	4,325	382.60
405	Crawler, 10' wide, 78 H.P., gas		19.30	670	1,950	5,850	544.40
406	10' wide, 87 H.P., diesel		15.40	785	2,600	7,650	643.20
407	Concrete paver, 12' to 24' wide, 250 H.P.		31	1,325	4,100	12,600	1,068
408	Placer-spreader-trimmer, 24' wide, 300 H.P.		35	1,400	4,225	12,700	1,125
410	Pump, centrifugal gas pump, 1-1/2"-4 MGPH		.42	16	48	135	12.95
420	2"-8 MGPH		.46	22	64	190	16.50
430	3"-15 MGPH		1.01	27	80	225	24.10
440	6"-90 MGPH		1.70	80	240	700	61.60
450	Submersible electric pump, 1-1/4"-55 GPM		.14	18	50	150	11.10
460	1-1/2"-83 GPM		.22	22	67	185	15.15
470	2"-120 GPM		.34	31	85	240	19.70
480	3"-300 GPM		.60	41	115	350	27.80
490	4"-560 GPM		1.60	51	150	425	42.80
500	6"-1590 GPM		2.10	135	400	1,200	96.80
510	Diaphragm pump, gas, single, 1-1/2" diameter		.45	14.35	41	125	11.80
520	2" diameter		.59	24	73	205	19.30
530	3" diameter		.93	30	100	250	27.45
540	Double, 4" diameter		1.15	50	155	450	40.20
550	Trash pump, self-priming, gas, 2" diameter		.44	28	84	250	20.30
560	Diesel, 4" diameter		1.25	60	190	580	48
565	Diesel, 6" diameter		2.65	89	280	815	77.20
566	Rollers, see division 1.5-05						
570	Salamanders, L.P. gas fired, 100,000 B.T.U.	Ea.	.52	14	35	110	11.15
572	Sandblaster, portable, open top, 3 C.F. capacity		1.25	44	125	330	35
573	6 C.F. capacity		2.55	52	150	440	50.40
574	Accessories for above			13.65	42	125	8.40
580	Saw, chain, gas engine, 18" long		.50	26	93	260	22.60
590	36" long		.65	31	115	325	28.20
595	60" long		1.10	38	145	400	37.80
600	Masonry, table mounted, 14" diameter, 5 H.P.		.79	34	110	240	28.30
610	Circular, hand held, electric, 7" diameter		.15	14	37	110	8.60
620	12" diameter		.16	25	75	195	16.30
630	Steam cleaner, 100 gallons per hour		.29	34	105	305	23.30
631	200 gallons per hour		.75	44	130	375	32
635	Torch, cutting, acetylene-oxygen, 150' hose						
636	Hourly operating cost includes tips and gas	Ea.	5.20	14	40	140	49.60
641	Toilet, portable chemical				23	73	4.60
642	Recycle flush type				28	88	5.60
643	Toilet, fresh water flush, garden hose,					135	9
644	Hoisted, non-flush, for high rise					72	4.80

1.5 Contractor Equipment

		UNIT	HOURLY OPER. COST	RENT PER DAY	RENT PER WEEK	RENT PER MONTH	CREWS EQUIPMENT COST
15 645	Toilet trailers, minimum	Ea.				63	4.20
646	Maximum	"				98	6.50
647	Trailer, office, see division 1.1-58-420						
648							
650	Trailers, platform, flush deck, 2 axle, 25 ton capacity	Ea.	1.05	89	290	1,025	66.40
660	40 ton capacity		1.25	130	430	1,500	96
670	3 axle, 50 ton capacity		2.35	155	535	1,850	125.80
680	75 ton capacity	↓	2.98	225	785	2,750	180.85
685	Trailer, storage, see division 1.1-58-550						
686							
690	Water tank, engine driven discharge, 5000 gallons	Ea.	9.20	240	703	2,100	214.20
700	10,000 gallons		11	345	1,025	3,025	293
702	Transit with tripod	↓		26	77	230	15.40
703							
705	Trench box, 8,000 lbs. 8' x 16'	Ea.	1.32	185	410	1,175	92.55
707	12,000 lbs., 10' x 20'		1.63	265	625	1,750	138.05
710	Truck, pickup, 3/4 ton, 2 wheel drive		4.95	38	120	380	63.60
720	4 wheel drive		5.10	49	180	530	76.80
730	Tractor, 4 x 2, 30 ton capacity, 195 H.P.		9.16	215	755	2,575	224.30
741	250 H.P.		13.55	240	835	2,950	275.40
750	6 x 2, 40 ton capacity, 240 H.P.		12.55	295	1,000	3,550	300.40
760	6 x 4, 45 ton capacity, 240 H.P.		14.20	300	1,050	3,100	323.60
770	Welder, electric, 200 amp		.54	17	51	150	14.50
780	300 amp		.70	21	65	180	18.60
790	Gas engine, 200 amp		3.12	35	100	300	44.95
800	300 amp		4.35	37	110	320	56.80
810	Wheelbarrow, any size			9	24	56	4.80
820	Wrecking ball, 4000 lb.	↓		24	73	220	14.60
20-001	**LIFTING & HOISTING EQUIPMENT RENTAL**						
010	without operators unless noted by*						
020	Crane, climbing, 106' jib, 6000 lb. capacity, 410 FPM	Ea.	12.65			7,750	617.85
030	101' jib, 10,250 lb. capacity, 270 FPM	"	16.65			10,000	799.85
040	Tower, static, 130' high, 106' jib,						
050	6200 lb. capacity at 400 FPM	Ea.	12.65			8,300	654.50
065	Crawler, cable, 1/2 C.Y., 15 tons at 12' radius		12.30	660*	900	2,700	278.40
070	3/4 C.Y., 20 tons at 12' radius		12.65	810*	1,225	3,800	346.20
080	1 C.Y., 25 tons at 12' radius		13.45	895*	1,400	4,375	387.60
090	1-1/2 C.Y., 40 tons at 12' radius		16.05	935*	1,775	5,625	483.40
100	2 C.Y., 50 tons at 12' radius		20	1,225*	2,400	6,275	640
110	3 C.Y., 75 tons at 12' radius		25	1,375*	2,700	7,100	740
120	100 ton capacity, standard boom		33	1,075	2,925	7,950	849
130	165 ton capacity, standard boom		47	1,250	4,100	12,200	1,196
140	200 ton capacity, 150' boom		53	1,750	5,425	16,200	1,509
150	450' boom		54	1,950	6,100	18,400	1,652
160	Truck mounted, cable operated, 6 x 4, 20 tons at 10' radius		9.85	780*	1,125	3,375	303.80
170	25 tons at 10' radius		12	940*	1,475	4,375	391
180	8 x 4, 30 tons at 10' radius		14.70	1,000*	1,600	4,850	437.60
190	40 tons at 12' radius		14.85	1,050*	1,825	5,500	483.80
200	8 x 4, 60 tons at 15' radius		19.50	1,100*	2,000	6,075	556
205	82 tons at 15' radius		21	1,200*	2,240	6,775	616
210	90 tons at 15' radius		25	1,425*	2,950	8,875	790
220	115 tons at 15' radius		29	1,575*	3,350	10,100	902
230	150 tons at 18' radius		35	1,700*	3,750	11,200	1,030
235	165 tons at 18' radius		36	1,750*	3,925	11,700	1,073
240	Truck mounted, hydraulic, 12 ton capacity		8.75	695*	825	2,500	235
250	25 ton capacity		13.10	880*	1,400	4,375	384.80
255	33 ton capacity		13.60	925*	1,475	4,625	403.80
260	55 ton capacity		18.45	1,075*	1,900	5,900	527.60
270	80 ton capacity	↓	23	1,425*	3,300	10,700	844
275							

1.5 Contractor Equipment

		UNIT	HOURLY OPER. COST	RENT PER DAY	RENT PER WEEK	RENT PER MONTH	CREWS EQUIPMENT COST
20 280	Self-propelled, 4 x 4, with telescoping boom, 5 ton	Ea.	5.40	550*	685	1,950	180.20
290	12-1/2 ton capacity		9.40	450*	745	2,125	224.20
300	15 ton capacity		9.70	495*	900	2,800	257.60
310	25 ton capacity		11.90	790*	1,475	4,325	390.20
320	Derricks, guy, 20 ton capacity, 60' boom, 75' mast		3.85	140	420	1,250	114.80
330	100' boom, 115' mast		4.37	180	525	1,550	139.95
340	Stiffleg, 20 ton capacity, 70' boom, 37' mast		3.20	185	550	1,650	135.60
350	100' boom, 47' mast		3.73	225	675	2,050	164.85
355	Helicopter, small, lift to 1250 lbs. maximum		190	1,925*			1,520
357							
360	Hoists, chain type, overhead, manual, 3/4 ton	Ea.		10	29	85	5.80
390	10 ton			19	61	165	12.20
400	Hoist and tower, 4000 lb. capacity, portable, 40' high		3.05			1,075	96.05
410	For each added 10' section, add					53	3.50
420	Hoist and single tubular tower, 5000 lb., 100' high		3.74			1,675	141.60
430	For each added 6'-6" section, add					45	3
440	Hoist and double tubular tower, 5000 lb., 100' high		3.20			2,025	160.60
450	For each added 6'-6" section, add					88	5.85
455	Hoist and tower, mast type, 6000 lb., 100' high		5.10			1,600	147.45
457	For each added 10' section, add					60	4
460 (13)	Hoist and tower, personnel, electric, 2000 lb., 100' @ 125 FPM		3.80			2,375	188.70
470	3000 lb., 100' @ 200 FPM		4.27			2,600	207.50
480	3000 lb., 150' @ 300 FPM		4.70			2,700	217.60
490	4000 lb., 100' @ 300 FPM		4.98			3,700	286.50
500	6000 lb., 100' @ 275 FPM		5.90			3,800	300.50
510	For added heights up to 500', add	L.F.				6.10	.40
520	Jacks, hydraulic, 20 ton	Ea.		11	32	80	6.40
550	100 ton	"		16	56	160	11.20
600 (12)	Jacks, hydraulic, climbing with 50' jackrods						
601	and control consoles, minimum 3 mo. rental						
610	30 ton capacity	Ea.				715	47.65
615	For each added 10' jackrod section, add					17.65	1.15
630	50 ton capacity					1,325	88.30
635	For each added 10' jackrod section, add					25	1.65
650	125 ton capacity					3,750	250
655	For each added 10' jackrod section, add					185	12.30
660	Cable jack, 10 ton capacity with 200' cable					650	43.30
665	For each added 50' of cable, add					38	2.50

		UNIT	HOURLY OPER. COST	RENT 1ST MO.	RENT 2ND MO.	RENT 3RD MO.
25-001	**WELLPOINT EQUIPMENT RENTAL** See also division 2.3-55					
002						
010 (23)	Combination jetting & wellpoint pump, 60 H.P. diesel	Ea.	2.55	2,150	1,520	1,230
020	High pressure gas jet pump, 200 H.P., 300 psi	"	7	1,600	1,440	1,130
030	Discharge pipe, 8" diameter	L.F.		3.62	1.50	1.16
035	12" diameter			4.73	2.34	1.78
040	Header pipe, flows up to 150 G.P.M., 4" diameter			1.95	1.10	.83
050	400 G.P.M., 6" diameter			3	1.21	.89
060	800 G.P.M., 8" diameter			4.74	2.34	1.55
070	1500 G.P.M., 10" diameter			4.90	2.50	2.22
080	2500 G.P.M., 12" diameter			5.45	4.33	2.40
090	4500 G.P.M., 16" diameter			9.48	5.90	4.12
095	For quick coupling aluminum and plastic pipe, add			10.25	10.25	10.25
110	Wellpoint, 25' long, with fittings & riser pipe, 1-1/2" or 2" diameter	Ea.		26	18.40	14.50
120	Wellpoint pump, diesel powered, 4" diameter, 20 H.P.		2.50	1,800	1,225	782
130	6" diameter, 30 H.P.		3.65	2,050	1,350	1,236
140	8" suction, 40 H.P.		3.97	2,425	1,450	1,300
150	10" suction, 75 H.P.		4.75	2,675	1,650	1,490
160	12" suction, 100 H.P.		5.90	3,600	2,425	2,210
170	12" suction, 175 H.P.		6.20	4,275	2,400	2,190

2.1 Exploration & Clearing

		CREW	DAILY OUTPUT	UNIT	BARE COSTS EQUIP.	BARE COSTS LABOR	BARE COSTS TOTAL	TOTAL INCL O&P
05-001	**BORINGS** Initial field stake out and determination of elevations	A-6	1	Day		295	295	430L
002								
010	Drawings showing boring details			Total		165	165	240L
020	Report and recommendations from P.E.					365	365	530L
030	Mobilization and demobilization, minimum	B-55	4.80	↓	45	54	99	125
035	For over 100 miles, per added mile		450	Mile	.48	.57	1.05	1.35
060	Auger holes in earth, no samples, 2-1/2" diameter		78.60	L.F.	2.72	3.28	6	7.70
065	4" diameter		67.50		3.17	3.82	6.99	9
080	Cased borings in earth, with samples, 2-1/2" diameter		55.50		3.85	4.65	8.50	10.95
085	4" diameter	↓	32.60		6.55	7.90	14.45	18.65
100	Drilling in rock, "BX" core, no sampling	B-56	34.90		12.15	8.10	20.25	25
105	With casing & sampling		31.70		13.40	8.90	22.30	28
120	"NX" core, no sampling		25.92		16.40	10.90	27.30	34
125	With casing & sampling	↓	25	↓	17	11.30	28.30	35
140	Drill rig and crew with light duty rig	B-55	1	Day	215	260	475	605
145	With heavy duty rig	B-56	1	"	425	285	710	875
10-001	**CLEAR AND GRUB** Light, trees to 6" diam., cut & chip	B-7	1	Acre	735	815	1,550	2,000
015	Grub stumps and remove	B-30	2		570	215	785	935
020	Medium, trees to 10" diam., cut & chip	B-7	.80		920	1,025	1,945	2,500
025	Grub stumps and remove	B-30	1.50		760	285	1,045	1,250
030	Heavy, trees to 16" diam., cut & chip	B-7	.70		1,050	1,175	2,225	2,850
035	Grub stumps and remove	B-30	1.20		950	360	1,310	1,550
040	If burning is allowed, reduce cut & chip			↓				33%
040								
100	Stump removal on site by hydraulic backhoe, 1-1/2 C.Y.							
105	8" to 12" diameter	B-30	33	Ea.	35	13.05	48.05	57
110	14" to 18" diameter		25		46	17.20	63.20	75
115	19" to 24" diameter	↓	16	↓	71	27	98	115
200	Trees, on site using chain saws and chipper,							
205	not incl. stumps, up to 6" diameter	B-7	47	Ea.	15.70	17.40	33.10	42
210	7" to 12" diameter		30		25	27	52	66
215	13" to 18" diameter		23		32	36	68	87
220	19" to 24" diameter	↓	17		43	48	91	115
230	For machine load, 2 mile haul to dump, add			↓		30	30	38L
15-001	**CLEARING** Brush with brush saw & rake	1 Clab	565	S.Y.		.23	.23	.33L
010	By hand	"	280			.45	.45	.66L
030	With dozer, ball and chain, light clearing	B-11A	3,675		.17	.08	.25	.31
040	Medium clearing	"	3,110	↓	.20	.09	.29	.36

		CREW	DAILY OUTPUT	UNIT	BARE COSTS MAT.	BARE COSTS INST.	BARE COSTS TOTAL	TOTAL INCL O&P
20-001	**CORE DRILLING** Reinf. concrete slab up to 6" thick, 1" diameter core	B-89	75	Ea.		12.85	12.85	14.85
015	Each added inch thick, add		600			1.61	1.61	1.86
030	3" diameter core		60			16.10	16.10	18.60
035	Each added inch thick, add		400			2.41	2.41	2.79
050	4" diameter core		45			21	21	25
055	Each added inch thick, add		300			3.22	3.22	3.72
070	6" diameter core		30			32	32	37
075	Each added inch thick, add		200			4.82	4.82	5.60
090	8" diameter core		24			40	40	46
095	Each added inch thick, add		150			6.45	6.45	7.45
110	10" diameter core		18			54	54	62
115	Each added inch thick, add		110			8.75	8.75	10.15
130	12" diameter core		15			64	64	74
135	Each added inch thick, add		90			10.70	10.70	12.40
150	14" diameter core		13			74	74	86
155	Each added inch thick, add		80			12.05	12.05	13.95
170	18" diameter core		10.50			92	92	105
175	Each added inch thick, add	↓	60			16.10	16.10	18.60
176	For horizontal holes, add to above			↓				30%
177	Prestressed hollow core plank, 6" thick							

For expanded coverage of these items see *Means' Site Work Cost Data 1986*

2.1 Exploration & Clearing

		CREW	DAILY OUTPUT	UNIT	BARE COSTS MAT.	BARE COSTS INST.	BARE COSTS TOTAL	TOTAL INCL O&P
20	178	1" diameter core	B-89	80	Ea.		12.05	12.05	13.95
	179	Each added inch thick, add		660			1.46	1.46	1.69
	180	3" diameter core		65			14.85	14.85	17.15
	181	Each added inch thick, add		440			2.19	2.19	2.54
	182	4" diameter core		50			19.30	19.30	22
	183	Each added inch thick, add		330			2.92	2.92	3.38
	184	6" diameter core		35			28	28	32
	185	Each added inch thick, add		220			4.38	4.38	5.05
	186	8" diameter core		27			36	36	41
	187	Each added inch thick, add		167			5.80	5.80	6.70
	188	10" diameter core		20			48	48	56
	189	Each added inch thick, add		120			8.05	8.05	9.30
	190	12" diameter core		17			57	57	66
	191	Each added inch thick, add		100			9.65	9.65	11.15
	195	Minimum charge for above, 3" diameter core		9			105	105	125
	200	4" diameter core		8.50			115	115	130
	205	6" diameter core		7.75			125	125	145
	210	8" diameter core		7			140	140	160
	215	10" diameter core		6			160	160	185
	220	12" diameter core		5			195	195	225
	225	14" diameter core		4.33			225	225	260
	230	18" diameter core	↓	4	↓		240	240	280

		CREW	DAILY OUTPUT	UNIT	BARE COSTS EQUIP.	BARE COSTS LABOR	BARE COSTS TOTAL	TOTAL INCL O&P
25-001	**DEMOLITION** Large urban projects, including disposal, steel	B-8	21,500	C.F.	.07	.05	.12	.16
010	Concrete		15,300		.10	.07	.17	.22
020	Masonry		20,100		.08	.06	.14	.17
030	Mixture of types, average	↓	20,100		.08	.06	.14	.17
050	Small bldgs. or single bldgs., no salvage included, steel	B-3	14,800		.08	.06	.14	.17
060	Concrete		11,300		.10	.07	.17	.22
070	Masonry		14,800		.08	.06	.14	.17
080	Wood		14,800	↓	.08	.06	.14	.17
200	Single family, one story house, wood, minimum		1.30	Ea.	890	635	1,525	1,900
210	Maximum		.78		1,475	1,050	2,525	3,175
250	Two family, two story house, wood, minimum		.93		1,250	890	2,140	2,650
255	Maximum		.57		2,025	1,450	3,475	4,325
270	Three family, three story house, wood, minimum		.77		1,500	1,075	2,575	3,200
275	Maximum	↓	.39	↓	2,975	2,125	5,100	6,325
27-001	**DEMOLITION, FOOTINGS & FOUNDATIONS**							
010	Floors, concrete slab on grade							
020	4" thick, plain concrete	B-9	500	S.F.	.30	1.30	1.60	2.22
030	Reinforced wire mesh		470		.32	1.39	1.71	2.36
035	Rod, reinforced		400		.38	1.63	2.01	2.78
050	6" thick, plain concrete		375		.40	1.74	2.14	2.96
060	Reinforced, wire mesh		340		.44	1.92	2.36	3.27
065	Rod, reinforced	↓	300	↓	.50	2.17	2.67	3.70
100	Footings, concrete, 1' thick, 2' wide	B-5	300	L.F.	2.50	3.85	6.35	8.35
110	1'-6" thick, 2' wide		250		3	4.62	7.62	10
120	3' wide		200		3.75	5.80	9.55	12.50
140	2' thick, 3' wide	↓	175		4.29	6.60	10.89	14.30
150	Average reinforcing, add							10%
160	Heavy reinforcing, add			↓				20%
300	Walls, block, 4" thick	B-9	1,010	S.F.	.15	.65	.80	1.10
310	6" thick		965		.16	.68	.84	1.15
320	8" thick		920		.16	.71	.87	1.21
330	12" thick	↓	880		.17	.74	.91	1.26
340	For horizontal reinforcing, add							10%
350	For vertical reinforcing, add							20%
400	Concrete, plain concrete, 6" thick	B-9	160		.94	4.08	5.02	6.95
410	8" thick	"	140	↓	1.07	4.66	5.73	7.95

For expanded coverage of these items see *Means' Site Work Cost Data 1986*

2.1 Exploration & Clearing

			DAILY		BARE COSTS			TOTAL
		CREW	OUTPUT	UNIT	EQUIP.	LABOR	TOTAL	INCL O&P
27	420	10" thick	B-9	120	S.F.	1.25	5.45	6.70	9.25
	430	12" thick	"	100		1.50	6.50	8	11.10
	440	For average reinforcing, add							10%
	450	For heavy reinforcing, add							20%
	900	For congested sites or small quantities, add up to							200%
	905								
	910	Add for disposal, on site	B-11A	232	C.Y.	2.74	1.26	4	4.83
	920	To five miles	B-34D	76	"	5.20	1.75	6.95	8.25
28-001		**DEMOLITION, FRAMING, HEAVY**							
	010	Concrete, average reinforcing, beams, 8" x 10"	B-9	120	L.F.	1.25	5.45	6.70	9.25
	020	10" x 12"		110		1.37	5.95	7.32	10.10
	030	12" x 14"		90		1.67	7.25	8.92	12.35
	050	Columns, 8" x 8"		120		1.25	5.45	6.70	9.25
	060	10" x 10"		120		1.25	5.45	6.70	9.25
	070	12" x 12"		110		1.37	5.95	7.32	10.10
	080	14" x 14"		100		1.50	6.50	8	11.10
	100	Girders, 14" x 16"		85		1.77	7.65	9.42	13.05
	110	16" x 18"		80		1.88	8.15	10.03	13.90
	120	Slabs, elevated, 6" thick		600	S.F.	.25	1.09	1.34	1.85
	125	8" thick		450		.33	1.45	1.78	2.47
	130	10" thick		360		.42	1.81	2.23	3.09
	140	Add for heavy reinforcement							20%
	200	Steel framing beams, 4" x 6"	B-13	500	L.F.	.77	1.92	2.69	3.63
	210	4" x 8"		400		.96	2.40	3.36	4.54
	215	8" x 12"		250		1.54	3.84	5.38	7.25
	230	Columns, 6" x 6"		400		.96	2.40	3.36	4.54
	235	8" x 8"		350		1.10	2.74	3.84	5.20
	240	10" x 10"		320		1.20	3	4.20	5.65
	250	Girders, 10" x 12"		225		1.71	4.27	5.98	8.05
	255	10" x 14"		200		1.92	4.80	6.72	9.10
	260	10" x 16"		165		2.33	5.80	8.13	11
	265	10" x 24"		125		3.08	7.70	10.78	14.55
	300	Wood framing beams, 6" x 8"	B-2	275			2.37	2.37	3.44L
	310	6" x 10"		220			2.96	2.96	4.30L
	320	6" x 12"		185			3.52	3.52	5.10L
	330	8" x 12"		140			4.66	4.66	6.75L
	340	10" x 12"		110			5.95	5.95	8.60L
30-001		**DISPOSAL** Urban buildings with salvage value allowed,							
	002	including loading and 5 mile haul to dump							
	010	Steel frame	B-3	430	C.Y.	2.70	1.93	4.63	5.75
	020	Concrete frame		365		3.18	2.27	5.45	6.75
	030	Masonry construction		445		2.60	1.86	4.46	5.55
	040	Wood frame		247		4.69	3.35	8.04	10
32-001		**DUMP CHARGES** Typical urban city, fees only							
	010	Building construction materials			C.Y.				3.85M
	020	Demolition lumber, trees, brush							6.60M
	030	Rubbish only							3.85M
	100	Reclamation station, usual charge			Ton				16.50M
35-001		**EXPLOSIVE DEMOLITION** large projects, not including disposal,							
	002	based on building volume, steel building	C-17	16,900	C.F.		.10	.10	.14L
	010	Concrete building		16,900			.10	.10	.14L
	020	Masonry building		16,900			.10	.10	.14L
	040	Disposal of material, minimum	B-3	445	C.Y.	2.60	1.86	4.46	5.55
	050	Maximum	"	365	"	3.18	2.27	5.45	6.75
37-001		**GUTTING** Complete building interior, residential, incl. disposal, minimum	B-16	400	S.F.Flr.	.70	1.33	2.03	2.69
	002	Maximum		360		.78	1.47	2.25	2.98
	010	Commercial building, minimum		350		.80	1.52	2.32	3.07
	020	Maximum		250		1.12	2.12	3.24	4.30

For expanded coverage of these items see *Means' Site Work Cost Data 1986*

2.1 Exploration & Clearing

		CREW	DAILY OUTPUT	UNIT	BARE COSTS MAT.	BARE COSTS INST.	BARE COSTS TOTAL	TOTAL INCL O&P
40-001	**MOVING BUILDINGS** One day move, up to 24' wide, not incl. new fndtns.							
002	Reset on new foundations, patch & hook-up, average move			Total				5,000
010	Wood or steel frame bldg., based on ground floor area	B-4	185	S.F.		5.95	5.95	8.05
020	Masonry bldg., based on ground floor area	"	137			8.05	8.05	10.85
040	For 24' to 42' wide, add			↓		1.28	1.28	1.40L
045	For each additional day on road, add	B-4	1	Day		1,100	1,100	1,500
100	Construct new basement, move building, 1 day							
101	move, patch & hook-up, based on ground floor area	B-3	130	S.F.	4.95	15.30	20.25	24
43-001	**RUBBISH HANDLING** The following are to be added to the							
010	demolition prices above							
030	Chute, circular, prefabricated steel, 18" diameter	B-1	40	L.F.	8.50	9.95	18.45	24
035	30" diameter	"	30	"	16.50	13.25	29.75	37
050	Dumpster, (debris box container), 5 C.Y., rent per week incl. pickup			Ea.				85
055	10 C.Y. capacity							110
060	30 C.Y. capacity							150
065	40 C.Y. capacity			↓				185
100	Dust partition, 6 mil polyethylene, 4' x 8' panels, 1" x 3" frame	1 Carp	800	S.F.	.13	.20	.33	.43
110	2" x 4" frame	"	1,000	"	.25	.16	.41	.51
150	Load, haul, dump into chute, 50' haul	B-6	1,100	C.F.		.51	.51	.70
155	100' haul		900			.63	.63	.85
160	Over 100' haul, add per 100 L.F.	↓	2,500			.23	.23	.31
180	In elevators, per 10 floors, add		1,000	↓		.57	.57	.77
300	Loading & trucking, including 2 mile haul, chute loaded	B-16	32	C.Y.		25	25	34
310	Hand loaded	"	40			20	20	27
320	Machine loaded	B-6	50			11.30	11.30	15.30
330	Wheeled and ramp dump loaded	B-1	32			12.45	12.45	18L
500	Overhaul, per mile, up to 8 C.Y. truck	B-34B	1,165			.35	.35	.43
510	Over 8 C.Y. truck	"	1,550			.27	.27	.32
44-001	**SAW CUTTING** Incl. cost of blade, asphalt, first 3 inches of depth	B-89	2,400	L.F.	.22	.40	.62	.71
010	Each additional inch of depth		3,200		.05	.30	.35	.40
050	Concrete slabs, mesh reinforcing, per inch of depth		2,250		.22	.43	.65	.74
055	Rod reinforcing, per inch of depth		1,850		.32	.52	.84	.96
100	Concrete walls, plain, per inch of depth		500		.22	1.93	2.15	2.47
110	Rod reinforcing, per inch of depth		400		.32	2.41	2.73	3.14
200	Masonry walls, brick, per inch of depth		400		.22	2.41	2.63	3.03
210	Block, per inch of depth	↓	500		.22	1.93	2.15	2.47
600	Wood sheathing, to 1" thick, on walls	1 Carp	200			.80	.80	1.16L
610	On roof	"	250	↓		.64	.64	.93L

		CREW	DAILY OUTPUT	UNIT	BARE COSTS EQUIP	BARE COSTS LABOR	BARE COSTS TOTAL	TOTAL
45-001	**SITE DEMOLITION** Abandon catch basin or manhole	B-6	7	Ea.	22	59	81	110
002	Remove existing catch basin or manhole	↓	4		39	105	144	190
003	Catch basin or manhole frames & covers stored		13		12	32	44	59
004	Remove and reset	↓	7	↓	22	59	81	110
060	Fencing, barbed wire, 3 strand	2 Clab	430	L.F.		.59	.59	.86L
065	5 strand		280			.91	.91	1.32L
070	Chain link, remove only		310			.82	.82	1.19L
075	Remove and reset		50			5.10	5.10	7.35L
080	Guard rail, remove only		85	↓		2.99	2.99	4.34L
085	Remove and reset		35			7.25	7.25	10.55L
090	Hydrants, fire, remove only	2 Plum	4.70	Ea.		77	77	110L
095	Remove and reset	"	1.40	"		260	260	375L
100	Masonry walls, block or tile, solid	B-5	1,800	C.F.	.42	.64	1.06	1.39
110	Cavity		2,200		.34	.53	.87	1.14
120	Brick, solid		900		.83	1.28	2.11	2.78
130	With block		1,130		.66	1.02	1.68	2.21
140	Stone, with mortar		900		.83	1.28	2.11	2.78
150	Dry set	↓	1,500	↓	.50	.77	1.27	1.67
170	Pavement, with power equipment, bituminous roads	B-38	690	S.Y.	.53	.59	1.12	1.45
180	Bituminous driveways	"	680	"	.54	.60	1.14	1.47

For expanded coverage of these items see *Means' Site Work Cost Data 1986*

2.1 Exploration & Clearing

			CREW	DAILY OUTPUT	UNIT	BARE COSTS EQUIP.	BARE COSTS LABOR	BARE COSTS TOTAL	TOTAL INCL O&P
45	190	Concrete to 6" thick, mesh reinforced	B-38	255	S.Y.	1.44	1.61	3.05	3.92
	200	Rod reinforced		200	"	1.84	2.05	3.89	5
	210	Concrete 7" to 24" thick, plain		13.10	C.Y.	28	31	59	76
	220	Reinforced	↓	9.50	"	39	43	82	105
	230	With hand held air equipment, bituminous	B-39	1,900	S.F.	.08	.43	.51	.70
	232	Concrete to 6" thick, no reinforcing		1,200		.13	.67	.80	1.11
	234	Mesh reinforced		830		.18	.97	1.15	1.61
	236	Rod reinforced	↓	765	↓	.20	1.06	1.26	1.75
	240	Curbs, concrete, plain	B-6	325	L.F.	.48	1.26	1.74	2.36
	250	Reinforced		220		.71	1.86	2.57	3.48
	260	Granite curbs		355		.44	1.15	1.59	2.16
	270	Bituminous curbs		830		.19	.49	.68	.92
	290	Pipe, concrete, not incl. excavation, 12" diameter		175		.89	2.34	3.23	4.38
	293	15" diameter		150		1.04	2.73	3.77	5.10
	296	24" diameter		120		1.30	3.42	4.72	6.40
	300	36" diameter		90		1.73	4.56	6.29	8.50
	320	Steel, welded connections, 4" diameter		160		.98	2.56	3.54	4.79
	330	10" diameter	↓	80	↓	1.95	5.15	7.10	9.60
	350	Railroad track, ties and track	B-14	110	↓	1.42	7.35	8.77	12.20
	360	Ballast		500	C.Y.	.31	1.62	1.93	2.68
	370	Remove and re-install ties & track with new bolts & spikes		50	L.F.	3.12	16.15	19.27	27
	380	Turnouts with new bolts & spikes	↓	1	Ea.	155	810	965	1,350
	400	Sidewalks, bituminous, 2-1/2" thick	B-6	325	S.Y.	.48	1.26	1.74	2.36
	405	Brick, set in mortar		185		.84	2.22	3.06	4.14
	410	Concrete, plain		160		.98	2.56	3.54	4.79
	420	Mesh reinforced	↓	150	↓	1.04	2.73	3.77	5.10
	500	Slab on grade, plain	B-5	45	C.Y.	16.70	26	42.70	56
	510	Mesh reinforcing		33		23	35	58	76
	520	Rod reinforcing	↓	25		30	46	76	100
	550	For congested sites or small quantities, add up to							200%
	555	For disposal on site, add	B-11A	232		2.74	1.26	4	4.83
	560	To 5 miles, add	B-34D	76		5.20	1.75	6.95	8.25
	50-001	TEST PITS Hand digging, light soil	1 Clab	4.50			28	28	41L
	010	Heavy soil	"	2.50	↓		51	51	74L
	60-001	TORCH CUTTING Steel, 1" thick plate	A-1	95	L.F.	.39	1.34	1.73	2.36
	010	1" diameter bar	"	210	Ea.	.17	.61	.78	1.07
	050	Oxygen lance cutting reinforced concrete walls							
	055	12" to 16" thick walls	1 Clab	10	L.F.		12.70	12.70	18.40L
	060	24" thick walls	"	6	"		21	21	31L

2.2 Building Demolition

		CREW	DAILY OUTPUT	UNIT	BARE COSTS EQUIP.	BARE COSTS LABOR	BARE COSTS TOTAL	TOTAL INCL O&P
02-001	ASBESTOS CONTROL METHODS							
010	Removal, including cleanup							
020	Beams, W10 x 19	B-2	152	L.F.		4.29	4.29	6.20L
030	W12 x 22		135			4.83	4.83	7L
040	W14 x 26		115			5.65	5.65	8.20L
050	W16 x 31		102			6.40	6.40	9.25L
060	W18 x 40		91			7.15	7.15	10.40L
070	W24 x 55		72			9.05	9.05	13.15L
080	W30 x 108		54			12.05	12.05	17.50L
090	W36 x 150		46	↓		14.15	14.15	21L
100	Ceilings, fluted		370	S.F.		1.76	1.76	2.55L
110	Smooth	↓	400	"		1.63	1.63	2.36L

For expanded coverage of these items see *Means' Site Work Cost Data 1986*

2.2 Building Demolition

		CREW	DAILY OUTPUT	UNIT	BARE COSTS MAT.	BARE COSTS INST.	BARE COSTS TOTAL	TOTAL INCL O&P
02 120	Wire lath, complete	B-2	500	S.F.		1.30	1.30	1.89L
200	Columns, W6 x 12		177	L.F.		3.68	3.68	5.35L
210	W8 x 31		103			6.35	6.35	9.20L
220	W10 x 49		82			7.95	7.95	11.55L
230	W12 x 40	↓	88	↓		7.40	7.40	10.75L
300	Encapsulation, with sealants							
310	Ceilings and walls, minimum	B-2	13,333	S.F.	.20	.05	.25	.29
320	Maximum		6,666		.25	.10	.35	.42
330	Columns and beams, minimum		8,000		.20	.08	.28	.34
340	Maximum	↓	3,636	↓	.25	.18	.43	.53
350	Pipes, 12" diameter or less, minimum		3,076	L.F.	.20	.21	.41	.53
360	Maximum	↓	1,538	"	.25	.42	.67	.89
400	Dust protection, polyethylene sheeting, 6 mil thick, 4" duct tape	1 Carp	800	S.F.	.13	.20	.33	.43
06-001	**CEILING DEMOLITION**							
020	Drywall, on wood frame	1 Clab	400	S.F.		.32	.32	.46L
022	On metal frame		380			.33	.33	.48L
024	On suspension system, incl. system		360			.35	.35	.51L
100	Plaster, lime & horse hair, on wood lath, incl. lath		350			.36	.36	.53L
102	On metal lath		285			.45	.45	.65L
110	Gypsum, on gypsum lath		360			.35	.35	.51L
112	On metal lath		250			.51	.51	.74L
150	Tile, wood fiber, 12" x 12", glued		450			.28	.28	.41L
154	Stapled		500			.25	.25	.37L
158	On suspension system, incl. system	↓	380	↓		.33	.33	.48L
159								
200	Wood, tongue & groove, 1" x 4"	1 Clab	500	S.F.		.25	.25	.37L
204	1" x 8"		550			.23	.23	.33L
240	Plywood or wood fiberboard, 4' x 8' sheets	↓	600	↓		.21	.21	.31L

		CREW	DAILY OUTPUT	UNIT	BARE COSTS EQUIP.	BARE COSTS LABOR	BARE COSTS TOTAL	TOTAL INCL O&P
08-001	**CUTOUT DEMOLITION** Concrete, elevated slab, light reinf., under 6 C.F.	B-9	65	C.F.	2.31	10.05	12.36	17.10
005	Light reinforcing, over 6 C.F.		75	"	2.01	8.70	10.71	14.80
020	Slab on grade to 6" thick, not reinforced under 8 S.F.		85	S.F.	1.77	7.65	9.42	13.05
025	Not reinforced, over 8 S.F.		175	"	.86	3.73	4.59	6.35
060	Walls, not reinforced, under 6 C.F.		60	C.F.	2.51	10.85	13.36	18.50
065	Not reinforced, over 6 C.F.		65		2.31	10.05	12.36	17.10
100	Concrete, elevated slab, bar reinforced, under 6 C.F.		45		3.34	14.50	17.84	25
105	Bar reinforced, over 6 C.F.	↓	50	↓	3.01	13.05	16.06	22
120	Slab on grade to 6" thick, bar reinforced, under 8 S.F.		75	S.F.	2.01	8.70	10.71	14.80
125	Bar reinforced, over 8 S.F.		105	"	1.43	6.20	7.63	10.60
140	Walls, bar reinforced, under 6 C.F.		50	C.F.	3.01	13.05	16.06	22
145	Bar reinforced, over 6 C.F.	↓	55	"	2.73	11.85	14.58	20
200	Brick, to 4 S.F. opening, not including toothing							
204	4" thick	B-9	30	Ea.	5	22	27	37
206	8" thick		18		8.35	36	44.35	62
208	12" thick		10		15.05	65	80.05	110
240	Concrete block, to 4 S.F. opening, 2" thick		35		4.30	18.65	22.95	32
242	4" thick		30		5	22	27	37
244	8" thick		27		5.55	24	29.55	41
246	12" thick		24		6.25	27	33.25	46
260	Gypsum block, to 4 S.F. opening, 2" thick		80		1.88	8.15	10.03	13.90
262	4" thick		70		2.15	9.30	11.45	15.85
264	8" thick		55		2.73	11.85	14.58	20
280	Terra cotta, to 4 S.F. opening, 4" thick		70		2.15	9.30	11.45	15.85
284	8" thick		65		2.31	10.05	12.36	17.10
288	12" thick	↓	50	↓	3.01	13.05	16.06	22
300	Toothing masonry cutouts, brick, soft old mortar	1 Brhe	40	V.L.F.		3.20	3.20	4.56L
310	Hard mortar	"	30	"		4.27	4.27	6.05L

For expanded coverage of these items see *Means' Site Work Cost Data 1986*

2.2 Building Demolition

			DAILY		BARE COSTS			TOTAL
		CREW	OUTPUT	UNIT	EQUIP.	LABOR	TOTAL	INCL O&P
08 320	Block, soft old mortar	1 Brhe	70	V.L.F.		1.83	1.83	2.60L
340	Hard mortar	"	50	"		2.56	2.56	3.64L
600	Walls, interior, not including re-framing,							
601	openings to 5 S.F.							
610	Drywall to 5/8" thick	A-1	24	Ea.	1.53	5.30	6.83	9.35
620	Paneling to 3/4" thick		20		1.83	6.35	8.18	11.25
630	Plaster, on gypsum lath		20		1.83	6.35	8.18	11.25
634	On wire lath	↓	14	↓	2.61	9.10	11.71	16.05
700	Wood frame, not including re-framing, openings to 5 S.F.							
720	Floors, sheathing & flooring to 2" thick	A-1	5	Ea.	7.30	25	32.30	45
731	Roofs, sheathing to 1" thick, not including roofing		6		6.10	21	27.10	37
741	Walls, sheathing to 1" thick, not including siding	↓	7	↓	5.25	18.15	23.40	32
32-001	**DOOR DEMOLITION**							
002								
020	Doors, exterior, 1-3/4" thick, single, 3' x 7' high	1 Clab	16	Ea.		7.95	7.95	11.50L
022	Double, 6' x 7' high		12			10.60	10.60	15.35L
050	Interior, 1-3/8" thick, single, 3' x 7' high		20			6.35	6.35	9.20L
052	Double, 6' x 7' high		16			7.95	7.95	11.50L
070	Bi-folding, 3' x 6'-8" high		20			6.35	6.35	9.20L
072	6' x 6'-8" high		18			7.05	7.05	10.25L
090	Bi-passing, 3' x 6'-8" high	↓	16			7.95	7.95	11.50L
094	6' x 6'-8" high		14			9.10	9.10	13.15L
150	Remove and reset, minimum	1 Carp	8			20	20	29L
152	Maximum	"	6			27	27	39L
200	Frames, including trim, metal	A-1	8		4.58	15.90	20.48	28
220	Wood	"	14		2.61	9.10	11.71	16.05
300	Special doors, counter doors	F-2	6		2.87	53	55.87	80
310	Double acting		10		1.72	32	33.72	48
320	Floor door (trap type)		8		2.15	40	42.15	60
330	Glass, sliding, incl. frames		12		1.43	27	28.43	40
340	Overhead, commercial, 12' x 12' high		4		4.30	80	84.30	120
344	20' x 16' high		3		5.75	105	110.75	160
350	Residential, 9' x 7' high		8		2.15	40	42.15	60
354	16' x 7' high		7		2.46	46	48.46	69
360	Remove & reset, minimum		4		4.30	80	84.30	120
362	Maximum		2.50		6.90	130	136.90	195
370	Roll-up grille		5		3.44	64	67.44	97
380	Revolving door		2		8.60	160	168.60	240
390	Swing door	↓	3	↓	5.75	105	110.75	160
36-001	**FLOORING DEMOLITION**							
020	Brick with mortar	A-1	300	S.F.	.12	.42	.54	.75
040	Carpet, bonded, including surface scraping	1 Clab	1,000			.13	.13	.18L
044	Scrim applied		4,000			.03	.03	.05L
048	Tackless	↓	4,500			.03	.03	.04L
060	Composition	A-1	200		.18	.64	.82	1.12
080	Resilient, sheet goods (linoleum)	1 Clab	700			.18	.18	.26L
082	For gym floors		450			.28	.28	.41L
090	Tile, 12" x 12"	↓	500	↓		.25	.25	.37L
100								
200	Tile, ceramic, thin set	A-1	400	S.F.	.09	.32	.41	.56
202	Mud set		350		.10	.36	.46	.64
220	Marble, slate, thin set		400		.09	.32	.41	.56
222	Mud set		350		.10	.36	.46	.64
260	Terrazzo, thin set		250		.15	.51	.66	.90
262	Mud set		225		.16	.57	.73	1
264	Cast in place	↓	175	↓	.21	.73	.94	1.28
265								
300	Wood, block, on end	A-1	400	S.F.	.09	.32	.41	.56
320	Parquet	"	450	"	.08	.28	.36	.50

For expanded coverage of these items see *Means' Site Work Cost Data 1986*

	2.2	Building Demolition	CREW	DAILY OUTPUT	UNIT	BARE COSTS			TOTAL INCL O&P
						EQUIP.	LABOR	TOTAL	
36	340	Strip flooring, interior, 2-1/4" x 25/32" thick	A-1	325	S.F.	.11	.39	.50	.69
	350	Exterior, porch flooring, 1" x 4"		220		.17	.58	.75	1.02
	380	Subfloor, tongue & groove, 1" x 6"		325		.11	.39	.50	.69
	382	1" x 8"		430		.09	.30	.39	.52
	384	1" x 10"		520		.07	.24	.31	.43
	400	Plywood, nailed		600		.06	.21	.27	.37
	410	Glued & nailed	↓	400	↓	.09	.32	.41	.56
	44-001	**FRAMING DEMOLITION**							
	002								
	340	Fascia boards, 1" x 6"	1 Clab	500	L.F.		.25	.25	.37L
	344	1" x 8"		450			.28	.28	.41L
	348	1" x 10"		400			.32	.32	.46L
	380	Headers over openings, 2 @ 2" x 6"		110			1.16	1.16	1.67L
	384	2 @ 2" x 8"		100			1.27	1.27	1.84L
	388	2 @ 2" x 10"		90			1.41	1.41	2.05L
	424	Joists, 2" x 8"		470			.27	.27	.39L
	428	2" x 12"		440			.29	.29	.42L
	540	Posts, 4" x 4"		400			.32	.32	.46L
	544	6" x 6"		200			.64	.64	.92L
	548	8" x 8"		150			.85	.85	1.23L
	550	10" x 10"		120			1.06	1.06	1.54L
	580	Rafters, ordinary, 2" x 6"		425			.30	.30	.43L
	584	2" x 8"		360	↓		.35	.35	.51L
	620	Stairs & stringers, minimum		20	Riser		6.35	6.35	9.20L
	624	Maximum		13	"		9.80	9.80	14.15L
	660	Studs, 2" x 4"		1,000	L.F.		.13	.13	.18L
	664	2" x 6"	↓	800	"		.16	.16	.23L
	48-001	**HVAC DEMOLITION**							
	002								
	010	Air conditioner, split unit, 3 ton	Q-5	2	Ea.		165	165	235L
	015	Package unit, 3 ton	Q-6	3			170	170	245L
	030	Boiler, electric	Q-19	2			255	255	365L
	034	Gas or oil, steel, under 150 MBH	Q-6	3			170	170	245L
	038	Over 150 MBH	"	2	↓		255	255	370L
	100	Ductwork, 4" high, 8" wide	1 Clab	200	L.F.		.64	.64	.92L
	110	6" high, 8" wide		165			.77	.77	1.12L
	120	10" high, 12" wide	↓	125	↓		1.02	1.02	1.47L
	300	Mechanical equipment, light items	Q-5	.90	Ton		365	365	525L
	360	Heavy items	"	1.10	"		300	300	430L
	370	Deduct for salvage (when applicable), minimum			Job				50
	375	Maximum			"				200
	52-001	**MASONRY DEMOLITION**							
	002								
	100	Chimney, 16" x 16", soft old mortar	A-1	24	V.L.F.	1.53	5.30	6.83	9.35
	102	Hard mortar		18		2.03	7.05	9.08	12.50
	108	20" x 20", soft old mortar		12		3.05	10.60	13.65	18.70
	110	Hard mortar		10		3.66	12.70	16.36	22
	114	20" x 32", soft old mortar		10		3.66	12.70	16.36	22
	116	Hard mortar		8		4.58	15.90	20.48	28
	120	48" x 48", soft old mortar		5		7.30	25	32.30	45
	122	Hard mortar		4		9.15	32	41.15	56
	200	Columns, 8" x 8", soft old mortar		48		.76	2.65	3.41	4.68
	202	Hard mortar		40		.92	3.18	4.10	5.60
	206	16" x 16", soft old mortar		16		2.29	7.95	10.24	14.05
	210	Hard mortar		14		2.61	9.10	11.71	16.05
	214	24" x 24", soft old mortar		8		4.58	15.90	20.48	28
	216	Hard mortar		6		6.10	21	27.10	37
	220	36" x 36", soft old mortar		4		9.15	32	41.15	56
	222	Hard mortar	↓	3	↓	12.20	42	54.20	75

For expanded coverage of these items see *Means' Site Work Cost Data 1986*

2.2 Building Demolition		CREW	DAILY OUTPUT	UNIT	BARE COSTS			TOTAL INCL O&P
					EQUIP.	LABOR	TOTAL	
52 300	Copings, precast or masonry, to 8" wide							
302	Soft old mortar	A-1	180	L.F.	.20	.71	.91	1.25
304	Hard mortar	"	160	"	.23	.80	1.03	1.40
310	To 12" wide							
312	Soft old mortar	A-1	160	L.F.	.23	.80	1.03	1.40
314	Hard mortar	"	140	"	.26	.91	1.17	1.60
400	Fireplace, brick, 30" x 24" opening							
402	Soft old mortar	A-1	2	Ea.	18.30	64	82.30	110
404	Hard mortar	"	1.25	"	29	100	129	180
407								
410	Stone, soft old mortar	A-1	1.50	Ea.	24	85	109	150
412	Hard mortar		1	"	37	125	162	225
500	Veneers, brick, soft old mortar		140	S.F.	.26	.91	1.17	1.60
502	Hard mortar		125		.29	1.02	1.31	1.80
510	Granite & marble, 2" thick		180		.20	.71	.91	1.25
512	4" thick		170		.22	.75	.97	1.32
514	Stone, 4" thick		180		.20	.71	.91	1.25
516	8" thick		175		.21	.73	.94	1.28
540	Alternate pricing method, stone, 4" thick		60	C.F.	.61	2.12	2.73	3.74
542	8" thick		85	"	.43	1.50	1.93	2.64
56-001	**MILLWORK & TRIM DEMOLITION**							
002								
100	Cabinets, wood, base cabinets	1 Clab	40	L.F.		3.18	3.18	4.61L
102	Wall cabinets	"	40	"		3.18	3.18	4.61L
106	Remove and reset, base cabinets	1 Carp	9	Ea.		17.80	17.80	26L
107	Wall cabinets	"	10	"		16	16	23L
110	Steel, painted, base cabinets	1 Clab	30	L.F.		4.24	4.24	6.15L
112	Wall cabinets		30	"		4.24	4.24	6.15L
120	Casework, large area		160	S.F.		.80	.80	1.15L
122	Selective		100	"		1.27	1.27	1.84L
150	Counter top, minimum		100	L.F.		1.27	1.27	1.84L
151	Maximum		60			2.12	2.12	3.07L
155	Remove and reset, minimum	1 Carp	25			6.40	6.40	9.25L
156	Maximum	"	20			8	8	11.60L
200	Paneling, 4' x 8' sheets, 1/4" thick	1 Clab	500	S.F.		.25	.25	.37L
210	Boards, 1" x 4"		350			.36	.36	.53L
212	1" x 6"		375			.34	.34	.49L
214	1" x 8"		400			.32	.32	.46L
300	Trim, baseboard, to 6" wide		600	L.F.		.21	.21	.31L
304	12" wide		500			.25	.25	.37L
308	Remove and reset, minimum	1 Carp	200			.80	.80	1.16L
309	Maximum	"	150			1.07	1.07	1.55L
310	Ceiling trim	1 Clab	500			.25	.25	.37L
312	Chair rail		600			.21	.21	.31L
314	Railings with balusters		120			1.06	1.06	1.54L
316	Wainscoting		350	S.F.		.36	.36	.53L
60-001	**PLUMBING DEMOLITION**							
002								
102	Fixtures, including 10' piping							
110	Bath tubs, cast iron	1 Plum	4	Ea.		45	45	65L
112	Fiberglass		6			30	30	43L
114	Steel		5			36	36	52L
120	Lavatory, wall hung		10			18.05	18.05	26L
122	Counter top		8			23	23	33L
130	Sink, steel or cast iron, single		8			23	23	33L
132	Double		7			26	26	37L
140	Water closet, floor mounted		8			23	23	33L
142	Wall mounted		7			26	26	37L
150	Urinal, floor mounted		4			45	45	65L
152	Wall mounted		7			26	26	37L

For expanded coverage of these items see *Means' Site Work Cost Data 1986*

2.2 Building Demolition

		Description	CREW	DAILY OUTPUT	UNIT	BARE COSTS EQUIP.	BARE COSTS LABOR	BARE COSTS TOTAL	TOTAL INCL O&P
60	160	Water fountains, free standing	1 Plum	8	Ea.		23	23	33L
	162	Recessed		6	"		30	30	43L
	200	Piping, metal, to 2" diameter		200	L.F.		.90	.90	1.30L
	205	To 4" diameter	↓	150			1.20	1.20	1.74L
	210	To 8" diameter	2 Plum	100			3.61	3.61	5.20L
	215	To 16" diameter	"	60			6	6	8.70L
	216	Deduct for salvage, aluminum scrap			Ton				700
	217	Brass scrap							600
	218	Copper scrap							900
	220	Lead scrap							200
	222	Steel scrap			↓				50
	225	Water heater, 40 gal.	1 Plum	6	Ea.		30	30	43L
64-001		**ROOFING & SIDING DEMOLITION**							
	002								
	100	Deck, roof, concrete plank	B-13	1,680	S.F.	.23	.57	.80	1.08
	110	Gypsum plank		3,900		.10	.25	.35	.47
	115	Metal decking	↓	3,500		.11	.27	.38	.52
	120	Wood, boards, tongue and groove, 2" x 6"	B-9	2,400		.06	.27	.33	.46
	122	2" x 10"		2,600		.06	.25	.31	.43
	128	Standard planks, 1" x 6"		2,700		.06	.24	.30	.41
	132	1" x 8"		2,900		.05	.22	.27	.38
	134	1" x 12"	↓	3,000	↓	.05	.22	.27	.37
	200	Gutters, aluminum or wood, edge hung	1 Clab	200	L.F.		.64	.64	.92L
	210	Built-in		100	"		1.27	1.27	1.84L
	250	Roof accessories, plumbing vent flashing		14	Ea.		9.10	9.10	13.15L
	260	Adjustable metal chimney flashing	↓	9	"		14.15	14.15	20L
	300	Roofing, built-up, 5 ply	B-2	1,600	S.F.		.41	.41	.59L
	301	5 ply roof, including gravel		890			.73	.73	1.05L
	310	Gravel removal, minimum		5,000			.13	.13	.19L
	312	Maximum		2,000			.33	.33	.47L
	340	Roof insulation board		3,900			.17	.17	.24L
	400	Shingles, asphalt strip		3,500			.19	.19	.27L
	410	Slate		2,500			.26	.26	.38L
	430	Wood	↓	2,200	↓		.30	.30	.43L
	450	Skylight to 10 S.F.	1 Clab	4	Ea.		32	32	46L
	451								
	500	Siding, metal, horizontal	1 Clab	300	S.F.		.42	.42	.61L
	502	Vertical		280			.45	.45	.66L
	520	Wood, boards, vertical		280			.45	.45	.66L
	522	Clapboards, horizontal		260			.49	.49	.71L
	524	Shingles		250			.51	.51	.74L
	526	Textured plywood	↓	500	↓		.25	.25	.37L
68-001		**WALLS & PARTITIONS DEMOLITION**							
	002								
	010	Brick, 4" to 12" thick	B-9	220	C.F.	.68	2.96	3.64	5.05
	015								
	020	Concrete block, 4" thick	B-9	1,000	S.F.	.15	.65	.80	1.11
	028	8" thick	"	810		.19	.80	.99	1.37
	100	Drywall, nailed	1 Clab	1,000			.13	.13	.18L
	102	Glued & nailed		900			.14	.14	.20L
	150	Fiberboard, nailed		900			.14	.14	.20L
	152	Glued & nailed		800			.16	.16	.23L
	200	Movable walls, metal, 5' high		300			.42	.42	.61L
	202	8' high	↓	400			.32	.32	.46L
	220	Metal studs, finish 2 sides, fiberboard	B-1	520			.76	.76	1.11L
	225	Lath and plaster		260			1.53	1.53	2.22L
	230	Plasterboard (drywall)		520			.76	.76	1.11L
	235	Plywood		450			.88	.88	1.28L
	290	Plasterboard (drywall), one side	↓	2,000	↓		.20	.20	.29L
	295								

For expanded coverage of these items see *Means' Site Work Cost Data 1986*

2.2 Building Demolition

		CREW	DAILY OUTPUT	UNIT	BARE COSTS EQUIP.	BARE COSTS LABOR	BARE COSTS TOTAL	TOTAL INCL O&P
68 300	Plaster, lime & horsehair, on wood lath	1 Clab	400	S.F.		.32	.32	.46L
302	On metal lath		335			.38	.38	.55L
340	Gypsum or perlite, on gypsum lath		410			.31	.31	.45L
342	On metal lath		300			.42	.42	.61L
360	Plywood, one side	B-1	1,500			.27	.27	.38L
375	Terra cotta block and plaster, to 6" thick	"	175			2.27	2.27	3.29L
380	Toilet partitions, slate or marble	1 Clab	5	Ea.		25	25	37L
382	Hollow metal	"	8	"		15.90	15.90	23L
72-001	**WINDOW DEMOLITION**							
002								
020	Aluminum, including trim, to 12 S.F.	A-1	12	Ea.	3.05	10.60	13.65	18.70
024	To 25 S.F.		8		4.58	15.90	20.48	28
028	To 50 S.F.		4		9.15	32	41.15	56
032	Storm windows, to 12 S.F.		20		1.83	6.35	8.18	11.25
036	To 25 S.F.		16		2.29	7.95	10.24	14.05
040	To 50 S.F.		12		3.05	10.60	13.65	18.70
060	Glass, minimum	1 Clab	200	S.F.		.64	.64	.92L
062	Maximum	"	150	"		.85	.85	1.23L
100	Steel, including trim, to 12 S.F.	A-1	10	Ea.	3.66	12.70	16.36	22
102	To 25 S.F.		7		5.25	18.15	23.40	32
104	To 50 S.F.		3		12.20	42	54.20	75
200	Wood, including trim, to 12 S.F.	1 Clab	16			7.95	7.95	11.50L
202	To 25 S.F.		12			10.60	10.60	15.35L
206	To 50 S.F.		6			21	21	31L
502	Remove & reset window, minimum	1 Carp	6			27	27	39L
504	Average		4			40	40	58L
508	Maximum		2			80	80	115L

2.3 Earthwork

		CREW	DAILY OUTPUT	UNIT	BARE COSTS EQUIP.	BARE COSTS LABOR	BARE COSTS TOTAL	TOTAL INCL O&P
03-001	**BACKFILL** By hand, no compaction, light soil	1 Clab	14	C.Y.		9.10	9.10	13.15L
010	Heavy soil		12			10.60	10.60	15.35L
030	Compaction in 6" layers, hand tamp, add		20.60			6.15	6.15	8.95L
040	Roller compaction, add	B-10A	100		.76	2.28	3.04	4.14
050	Air tamp, add	B-9	190		.79	3.43	4.22	5.85
060	Vibrating plate, add	A-1	60		.61	2.12	2.73	3.74
080	Compaction in 12" layers, hand tamp, add	1 Clab	34			3.74	3.74	5.40L
090	Roller compaction, add	B-10A	150		.50	1.52	2.02	2.76
100	Air tamp, add	B-9	285		.53	2.29	2.82	3.90
110	Vibrating plate, add	A-1	90		.41	1.41	1.82	2.50
130	Dozer backfilling, bulk, up to 300' haul, no compaction	B-10B	1,200		.53	.19	.72	.86
140	Air tamped	B-11B	240		3.41	1.22	4.63	5.50
160	Compacted backfill, 6" to 12" lifts, vibrating roller	B-10C	800		.88	.29	1.17	1.39
170	Sheepsfoot roller	B-10D	750		.95	.30	1.25	1.49
190	Dozer backfilling, trench, up to 300' haul, no compaction	B-10B	900		.71	.25	.96	1.14
200	Air tamped	B-11B	235		3.49	1.24	4.73	5.65
220	Compacted backfill, 6" to 12" lifts, vibrating roller	B-10C	700		1.01	.33	1.34	1.58
230	Sheepsfoot roller	B-10D	650		1.10	.35	1.45	1.72

					BARE COSTS MAT.	BARE COSTS INST.	BARE COSTS TOTAL	
05-001	**BORROW** Buy and load at pit, haul 1-1/2 hour round trip							
002	and spread, with 180 H.P. dozer, no compaction							
010	Bank run gravel	B-15	600	C.Y.	6.50	2.81	9.31	10.55
020	Common borrow	"	600	"	5	2.81	7.81	8.90

For expanded coverage of these items see *Means' Site Work Cost Data 1986*

2.3 Earthwork

		Crew	Daily Output	Unit	Bare Costs Mat.	Bare Costs Inst.	Bare Costs Total	Total Incl O&P
05	030 Crushed stone, 1-1/2"	B-15	600	C.Y.	13.35	2.81	16.16	18.05
	032 3/4"		600		13.60	2.81	16.41	18.35
	034 1/2"		600		14	2.81	16.81	18.80
	036 3/8"		600		14.50	2.81	17.31	19.35
	040 Sand, washed, concrete		600		11.75	2.81	14.56	16.30
	050 Dead or bank sand		600		8	2.81	10.81	12.20
	060 Select structural fill		600		8.50	2.81	11.31	12.75
	070 Screened loam		600		13.50	2.81	16.31	18.25
	080 Topsoil, weed free		600		9.80	2.81	12.61	14.15
	090 For 5 mile haul, add	B-34B	335			1.23	1.23	1.49

		Crew	Daily Output	Unit	Bare Costs Equip.	Bare Costs Labor	Bare Costs Total	Total Incl O&P
08-001	COMPACTION Rolling with road roller, 5 tons	B-10E	8	Hr.	10.65	29	39.65	53
	010 10 tons	B-10F	8	"	22	29	51	65
	030 Sheepsfoot or wobbly wheel roller, 8" lifts, common fill	B-10G	360	C.Y.	.81	.63	1.44	1.81
	040 Select fill	"	430		.68	.53	1.21	1.51
	050 Terra Probe, deep sand, vibrating 30,000 C.Y., minimum	B-43	1,779		.31	.47	.78	1.02
	052 Maximum	"	1,078		.51	.77	1.28	1.68
	054 Mobilization & demobilization, minimum	B-8	.63	Total	2,450	1,800	4,250	5,300
	056 Maximum	"	.48	"	3,225	2,350	5,575	6,950
	060 Vibratory plate, 8" lifts, common fill	A-1	75	C.Y.	.49	1.70	2.19	3
	070 Select fill	"	90	"	.41	1.41	1.82	2.50
	090 Vibroflotation compacted sand cylinder, minimum	B-60	750	V.L.F.	.90	1.36	2.26	2.96
	095 Maximum		325		2.08	3.13	5.21	6.80
	110 Vibro replacement compacted stone cylinder, minimum		500		1.35	2.03	3.38	4.44
	115 Maximum		250		2.70	4.07	6.77	8.85
	130 Mobilization and demobilization, minimum		.47	Total	1,425	2,175	3,600	4,725
	140 Maximum		.14	"	4,825	7,275	12,100	15,800
10-001	DEWATERING Excavate drainage trench, 2' wide, 2' deep	B-11C	90	C.Y.	1.73	3.24	4.97	6.60
	010 2' wide, 3' deep, with backhoe loader	"	100		1.56	2.92	4.48	5.95
	020 Excavate sump pits by hand, light soil	1 Clab	7.10			17.90	17.90	26L
	030 Heavy soil	"	3.50			36	36	53L
	050 Pumping 8 hr., attended 2 hrs. per day, including 20 L.F.							
	055 of suction hose & 100 L.F. discharge hose							
	060 2" diaphragm pump used for 8 hours	B-10H	4	Day	7.50	57	64.50	91
	065 4" diaphragm pump used for 8 hours	B-10I	4		15.65	57	72.65	100
	080 8 hrs. attended, 2" diaphragm pump	B-10H	1		30	230	260	365
	090 3" centrifugal pump	B-10J	1		39	230	269	375
	100 4" diaphragm pump	B-10I	1		63	230	293	400
	110 6" centrifugal pump	B-10K	1		100	230	330	445

		Crew	Daily Output	Unit	Bare Costs Mat.	Bare Costs Inst.	Bare Costs Total	Total Incl O&P
	130 Re-lay CMP incl. excavation 3' deep, 12" diameter	B-6	115	L.F.	4.90	4.92	9.82	12.05
	140 18" diameter		100	"	7.40	5.65	13.05	15.80
	160 Sump hole construction, incl. excavation and gravel, pit		1,250	C.F.	.25	.45	.70	.89
	170 With 12" gravel collar, 12" pipe, corrugated, 16 ga.		70	L.F.	8.20	8.10	16.30	19.95
	180 15" pipe, corrugated, 16 ga.		55		10.05	10.30	20.35	25
	190 18" pipe, corrugated, 16 ga.		50		11.75	11.30	23.05	28
	200 24" pipe, corrugated, 14 ga.		40		17.75	14.15	31.90	39
	220 Wood lining, up to 4' x 4', add		300	S.F.C.A.	.78	1.89	2.67	3.41

		Crew	Daily Output	Unit	Bare Costs Equip.	Bare Costs Labor	Bare Costs Total	Total Incl O&P
12-001	DREDGING Mobilization and demobilization., add to below, minimum	B-8	.53	Total	2,925	2,125	5,050	6,300
	010 Maximum	"	.10	"	15,400	11,300	26,700	33,400
	030 Barge mounted clamshell excavation into scows,							
	031 Dumped 20 miles at sea, minimum	B-57	310	C.Y.	2.46	2.78	5.24	6.75
	040 Maximum	"	213	"	3.59	4.05	7.64	9.80
	050 Barge mounted dragline or clamshell, hopper dumped,							
	051 pumped 1000 ft. to shore dump, minimum	B-57	340	C.Y.	2.25	2.53	4.78	6.15
	060 Maximum	"	243	"	3.14	3.55	6.69	8.60

For expanded coverage of these items see *Means' Site Work Cost Data 1986*

2.3 Earthwork

		CREW	DAILY OUTPUT	UNIT	BARE COSTS EQUIP.	BARE COSTS LABOR	BARE COSTS TOTAL	TOTAL INCL O&P
12	100 Hydraulic method, pumped 1000 ft. to shore dump, minimum	B-57	460	C.Y.	1.66	1.87	3.53	4.54
	110 Maximum		310		2.46	2.78	5.24	6.75
	140 Into scows dumped 20 miles, minimum		425		1.80	2.03	3.83	4.92
	150 Maximum	↓	243		3.14	3.55	6.69	8.60
	160 For inland rivers in South, deduct			↓	30%	30%		

		CREW	DAILY OUTPUT	UNIT	BARE COSTS MAT.	BARE COSTS INST.	BARE COSTS TOTAL	TOTAL INCL O&P
15-001	DRILLING AND BLASTING only, rock, under 1500 C.Y.	B-47	75	C.Y.	1.05	11.45	12.50	15.75
010	Over 1500 C.Y.	"	85		1.05	10.10	11.15	14.05
030	Bulk excavating, can vary greatly, average							14
050	Pits, average			↓				18
070	Drilling only, 2" hole for rock bolts, average	B-47	395	L.F.		2.18	2.18	2.77
080	2-1/2" hole for pre-splitting, average		340	"		2.53	2.53	3.22
130	Deep hole method, up to 1500 C.Y.		50	C.Y.	1.05	17.20	18.25	23
140	Over 1500 C.Y.		66	"	1.05	13.05	14.10	17.75
160	Quarry operations, 2-1/2" to 3-1/2" diameter, up to 1500 C.Y.		350	L.F.		2.46	2.46	3.13
170	Over 1500 C.Y.		425	"		2.02	2.02	2.58
190	Restricted areas, up to 1500 C.Y.		13	C.Y.	1.05	66	67.05	85
200	Over 1500 C.Y.		20		1.05	43	44.05	56
220	Trenches, up to 1500 C.Y.		22		1.05	39	40.05	51
230	Over 1500 C.Y.		26		1.05	33	34.05	43
250	Pier holes, up to 1500 C.Y.		22		1.05	39	40.05	51
260	Over 1500 C.Y.		31		1.05	28	29.05	36
280	Boulders loaded only, not including hauling		85		1.05	10.10	11.15	14.05
290	Drilled, blasted and loaded, not incl. hauling	↓	30	↓	1.05	29	30.05	38
310	Jackhammer operator with compressor	B-9	27.70	Hr.		29	29	40
330	Track mounted air drill with operator and foreman	B-47	14.40	"		60	60	76
350	Blasting caps			Ea.	1.60		1.60	1.76M
370	Explosives			Lb.	1.55		1.55	1.70M
390	Blasting mats, rent, for first day			Ea.	37		37	41M
400	Per added day				21		21	23M
420	Preblast survey for 6 room house, individual lot, minimum	A-6	2.40			125	125	180L
430	Maximum	"	1.35	↓		220	220	320L
450	City block within zone of influence, minimum	A-8	25,200	S.F.		.02	.02	.03L
460	Maximum	"	15,100	"		.04	.04	.05L

		CREW	DAILY OUTPUT	UNIT	BARE COSTS EQUIP.	BARE COSTS LABOR	BARE COSTS TOTAL	TOTAL INCL O&P
16-001	EXCAVATING, BULK Medium earth piled or truck							
002	loaded, no trucks or haul included							
005	For mobilization and demobilization, see division 2.3-35							
010	For hauling, see division 2.3-30							
020	Backhoe, hydraulic, crawler mtd., 1 C.Y. cap.=45 C.Y./hr.	B-12A	360	C.Y.	1.23	.86	2.09	2.60
025	1-1/2 C.Y. cap. = 60 C.Y./hr.	B-12B	480		1.21	.64	1.85	2.26
026	2 C.Y. cap. = 75 C.Y./hr.	B-12C	600		1.34	.51	1.85	2.21
030	3-1/2 C.Y. cap. = 150 C.Y./hr.	B-12D	1,200		1.18	.26	1.44	1.67
031	Wheel mounted, 1/2 C.Y. cap. = 20 C.Y./hr.	B-12E	160		1.64	1.93	3.57	4.60
036	3/4 C.Y. cap. = 30 C.Y./hr.	B-12F	240		1.42	1.29	2.71	3.42
050	Clamshell, 1/2 C.Y. cap. = 20 C.Y./hr.	B-12G	160		1.93	1.93	3.86	4.91
055	1 C.Y. cap. = 35 C.Y./hr.	B-12H	280		1.53	1.10	2.63	3.28
070	Dozer, 50' haul, 75 H.P. = 50 C.Y./hr.	B-10L	400		.52	.57	1.09	1.39
075	300 H.P. = 150 C.Y./hr.	B-10M	1,200		.78	.19	.97	1.13
085	150' haul, 75 H.P. = 25 C.Y./hr.	B-10L	200		1.03	1.14	2.17	2.79
090	300 H.P. = 100 C.Y./hr.	B-10M	800		1.16	.29	1.45	1.69
100	Dragline, 3/4 C.Y. cap. = 35 C.Y./hr.	B-12I	280		1.31	1.10	2.41	3.04
105	1-1/2 C.Y. cap. = 65 C.Y./hr.	B-12P	520		.99	.59	1.58	1.95
120	Front end loader, track mtd., 1-1/2 C.Y. cap. = 70 C.Y./hr.	B-10N	560		.49	.41	.90	1.13
125	2-1/2 C.Y. cap. = 95 C.Y./hr.	B-10O	760		.55	.30	.85	1.05
130	3-1/2 C.Y. cap. = 130 C.Y./hr.	B-10P	1,040		.58	.22	.80	.95
135	4-1/2 C.Y. cap. = 160 C.Y./hr.	B-10Q	1,280	↓	.79	.18	.97	1.13

2.3 Earthwork

			CREW	DAILY OUTPUT	UNIT	BARE COSTS EQUIP.	LABOR	TOTAL	TOTAL INCL O&P
16	150	Wheel mounted, 3/4 C.Y. cap. = 45 C.Y./hr.	B-10R	360	C.Y.	.56	.63	1.19	1.54
	155	1-1/2 C.Y. cap. = 80 C.Y./hr.	B-10S	640		.42	.36	.78	.98
	160	2-1/4 C.Y. cap. = 100 C.Y./hr.	B-10T	800		.55	.29	.84	1.02
	165	5 C.Y. cap. = 185 C.Y./hr.	B-10U	1,480		.57	.15	.72	.85
	180	Gradall, 30 inch bucket, 1/2 C.Y. = 30 C.Y./hr.	B-12J	240		1.84	1.29	3.13	3.88
	185	48 inch bucket, 1 C.Y. = 45 C.Y./hr.	B-12K	360		1.75	.86	2.61	3.16
	240	Scrapers, towed, 10 C.Y. capacity, 1/4 push dozer, 1500' haul	B-33B	600		2.14	.45	2.59	3.01
	245	5000' haul	"	440		2.92	.61	3.53	4.10
	260	15 C.Y. capacity, 1/4 push dozer, 1500' haul	B-33C	800		1.61	.34	1.95	2.25
	265	5000' haul	"	560		2.29	.48	2.77	3.22
	300	Self-propelled scrapers, 15 C.Y. capacity, 1/4 dozer, 1500' haul	B-33D	800		1.50	.34	1.84	2.13
	305 (16)	5000' haul	"	560		2.14	.48	2.62	3.05
	315 (17)	25 C.Y. capacity, 1/4 push dozer, 1500' haul	B-33E	1,000		1.35	.27	1.62	1.87
	320	5000' haul	"	600		2.25	.45	2.70	3.12
	330	Elevating scrapers, self propelled, 10 C.Y., 1/4 dozer, 1500' haul	B-33F	600		1.44	.45	1.89	2.24
	335	5000' haul	"	440		1.97	.61	2.58	3.05
	350	20 C.Y. capacity, 1/4 push dozer, 1500' haul	B-33G	840		1.41	.32	1.73	2.01
	355	5000' haul	"	680		1.74	.40	2.14	2.48
	370	Shovel, 1/2 C.Y. capacity = 30 C.Y./hr.	B-12L	240		1.34	1.29	2.63	3.33
	375	3/4 C.Y. capacity = 50 C.Y./hr.	B-12M	400		1.06	.77	1.83	2.29
	380	1 C.Y. capacity = 60 C.Y./hr.	B-12N	480		1.04	.64	1.68	2.07
	385	1-1/2 C.Y. capacity = 90 C.Y./hr.	B-12O	720		.84	.43	1.27	1.55
	400	For soft soil or sand, deduct					15%		
	410	For heavy soil or stiff clay, add					60%		
	420	For wet excavation with clamshell or dragline, add					100%		
	425	All other equipment, add					50%		
	440	Clamshell in sheeting or cofferdam, minimum	B-12H	222		1.93	1.39	3.32	4.13
	445	Maximum	"	59		7.25	5.25	12.50	15.55
17	001	EXCAVATING, STRUCTURAL Hand, pits to 6' deep, sandy soil	1 Clab	8			15.90	15.90	23L
	010	Heavy soil or clay		4			32	32	46L
	030	Pits 6' to 12' deep, sandy soil		5			25	25	37L
	050	Heavy soil or clay		3			42	42	61L
	070	Pits 12' to 18' deep, sandy soil		4			32	32	46L
	090	Heavy soil or clay		2			64	64	92L
	110	Hand loading trucks from stock pile, sandy soil		12			10.60	10.60	15.35L
	130	Heavy soil or clay		8			15.90	15.90	23L
	150	For wet or muck hand excavation, add to above				50%	50%		
18	001	EXCAVATING, TRENCH or continuous footing, 4' wide							
	002	bottom with banks, sloped 1/2 to 1, damp sandy loam							
	003	3' deep, 3/8 C.Y. tractor backhoe = 245 L.F./day	B-11C	150	C.Y.	1.04	1.95	2.99	3.97
	004	1/2 C.Y. tractor backhoe = 325 L.F./day	B-11M	200		1.17	1.46	2.63	3.40
	005	4' deep, 3/8 C.Y. tractor backhoe = 170 L.F./day	B-11C	150		1.04	1.95	2.99	3.97
	006	1/2 C.Y. tractor backhoe = 225 L.F./day	B-11M	200		1.17	1.46	2.63	3.40
	007	5' deep, 3/8 C.Y. tractor backhoe = 125 L.F./day	B-11C	150		1.04	1.95	2.99	3.97
	008	1/2 C.Y. tractor backhoe = 165 L.F./day	B-11M	200		1.17	1.46	2.63	3.40
	009	6' deep, 1/2 C.Y. tractor backhoe = 130 L.F./day	"	200		1.17	1.46	2.63	3.40
	010	5/8 C.Y. hydraulic backhoe = 160 L.F./day	B-12Q	250		1.10	1.23	2.33	3
	020	8' deep, 1/2 C.Y. tractor backhoe = 85 L.F./day	B-11M	200		1.17	1.46	2.63	3.40
	030	1/2 C.Y. Gradall = 85 L.F./day	B-12J	200		2.20	1.54	3.74	4.66
	040	3/4 C.Y. hydraulic backhoe = 120 L.F./day	B-12F	300		1.13	1.03	2.16	2.74
	050	10' deep, 1 C.Y. hydraulic backhoe = 120 L.F./day	B-12A	400		1.11	.77	1.88	2.34
	060	1 C.Y. Gradall = 120 L.F./day	B-12K	400		1.57	.77	2.34	2.85
	070	12' deep, 1 C.Y. hydraulic backhoe = 90 L.F./day	B-12A	400		1.11	.77	1.88	2.34
	080	1-1/4 C.Y. hydraulic backhoe = 115 L.F./day	B-12R	500		1.16	.62	1.78	2.17
	090	14' deep, 1-1/4 C.Y. hydraulic backhoe = 90 L.F./day	"	500		1.16	.62	1.78	2.17
	100	1-1/2 C.Y. hydraulic backhoe = 105 L.F./day	B-12B	600		.97	.51	1.48	1.81
	110	16' deep, 2 C.Y. hydraulic backhoe = 115 L.F./day	B-12C	800		1	.39	1.39	1.66
	120	18' deep, 2-1/2 C.Y. hydraulic backhoe = 115 L.F./day	B-12S	1,000		1.17	.31	1.48	1.73
	130	20' deep, 3-1/2 C.Y. hydraulic backhoe = 135 L.F./day	B-12D	1,400		1.01	.22	1.23	1.43

For expanded coverage of these items see *Means' Site Work Cost Data 1986*

2.3 Earthwork

			CREW	DAILY OUTPUT	UNIT	BARE COSTS EQUIP.	BARE COSTS LABOR	BARE COSTS TOTAL	TOTAL INCL O&P
18	140	By hand with pick and shovel to 6' deep, light soil	1 Clab	8	C.Y.		15.90	15.90	23L
	150	Heavy soil	"	4			32	32	46L
	170	For tamping backfilled trenches, air tamp, add	A-1	100		.37	1.27	1.64	2.25
	190	Vibrating plate, add		90	↓	.41	1.41	1.82	2.50
	210	Trim sides and bottom for concrete pours, regular		600	S.F.	.06	.21	.27	.37
	230	Hardpan	↓	180	"	.20	.71	.91	1.25
	240	Pier and spread footing excavation, add to above			C.Y.	30%	30%		
19-001		**EXCAVATING, UTILITY TRENCH** Sandy clay soil							
	002								
	005	Trenching with chain trencher, 12 H.P., operator walking							
	010	4" wide trench, 12" deep	B-53	800	L.F.	.17	.19	.36	.47
	015	18" deep		750		.18	.21	.39	.50
	020	24" deep	↓	700	↓	.19	.22	.41	.54
	025								
	030	6" wide trench, 12" deep	B-53	650	L.F.	.21	.24	.45	.58
	035	18" deep		600		.23	.26	.49	.62
	040	24" deep		550		.25	.28	.53	.68
	045	36" deep		450		.30	.35	.65	.83
	060	8" wide trench, 12" deep		475		.29	.33	.62	.79
	065	18" deep		400		.34	.39	.73	.94
	070	24" deep		350		.39	.44	.83	1.07
	075	36" deep	↓	300	↓	.45	.52	.97	1.25
	100	Backfill by hand including compaction, add							
	105	4" wide trench, 12" deep	A-1	800	L.F.	.05	.16	.21	.28
	110	18" deep		530		.07	.24	.31	.42
	115	24" deep		400		.09	.32	.41	.56
	130	6" wide trench, 12" deep		540		.07	.24	.31	.42
	135	18" deep		405		.09	.31	.40	.55
	140	24" deep		270		.14	.47	.61	.83
	145	36" deep		180		.20	.71	.91	1.25
	160	8" wide trench, 12" deep		400		.09	.32	.41	.56
	165	18" deep		265		.14	.48	.62	.85
	170	24" deep		200		.18	.64	.82	1.12
	175	36" deep	↓	135	↓	.27	.94	1.21	1.66
	200	Chain trencher, 40 H.P. operator riding							
	205	6" wide trench and backfill, 12" deep	B-54	1,200	L.F.	.17	.13	.30	.37
	210	18" deep		1,000		.20	.16	.36	.45
	215	24" deep		975		.20	.16	.36	.46
	220	36" deep		900		.22	.17	.39	.49
	225	48" deep		750		.27	.21	.48	.59
	230	60" deep		650		.31	.24	.55	.69
	240	8" wide trench and backfill, 12" deep		1,000		.20	.16	.36	.45
	245	18" deep		950		.21	.16	.37	.47
	250	24" deep		900		.22	.17	.39	.49
	255	36" deep		800		.25	.19	.44	.56
	260	48" deep		650		.31	.24	.55	.69
	270	12" wide trench and backfill, 12" deep		975		.20	.16	.36	.46
	275	18" deep		860		.23	.18	.41	.52
	280	24" deep		800		.25	.19	.44	.56
	285	36" deep		725		.28	.21	.49	.61
	300	16" wide trench and backfill, 12" deep		900		.22	.17	.39	.49
	305	18" deep		750		.27	.21	.48	.59
	310	24" deep	↓	700	↓	.29	.22	.51	.64
	320	Compaction with vibratory plate, add				50%	50%		
20-001		**FILL** Spread dumped material, no compaction, by dozer	B-10B	1,000	C.Y.	.63	.23	.86	1.03
	010	By hand	1 Clab	12	"		10.60	10.60	15.35L

2.3 Earthwork

			CREW	DAILY OUTPUT	UNIT	BARE COSTS MAT.	BARE COSTS INST.	BARE COSTS TOTAL	TOTAL INCL O&P
20	050	Gravel fill, compacted, under floor slabs, 3" deep (at $4.05 ton)	B-14	10,000	S.F.	.08	.10	.18	.22
	060	6" deep		9,000		.15	.11	.26	.31
	070	9" deep		7,200		.23	.13	.36	.44
	080	12" deep		6,000		.30	.16	.46	.55
	100	Alternate pricing method, 3" deep		90	C.Y.	8	10.70	18.70	24
	110	6" deep		160		8	6	14	17.20
	120	9" deep		200		8	4.82	12.82	15.50
	130	12" deep		220		8	4.38	12.38	14.90
	150	For fill under exterior paving, see division 2.6-07							

			CREW	DAILY OUTPUT	UNIT	BARE COSTS EQUIP.	BARE COSTS LABOR	BARE COSTS TOTAL	TOTAL INCL O&P
22-001		GRADING Site excav. & fill, not incl. mobilization, demobilization or							
	002	compaction. Includes 1/4 push dozer per scraper.							
	010	Dozer 300' haul, 75 H.P., = 20 C.Y./hr.	B-10L	160	C.Y.	1.29	1.43	2.72	3.49
	020	300 H.P., 70 C.Y./hr.	B-10M	560		1.66	.41	2.07	2.42
	040	Scraper, towed, 7 C.Y. 300' haul, 55 C.Y./hr.	B-33A	440		2.85	.61	3.46	4.02
	050	1000' haul, 25 C.Y./hr.	"	200		6.25	1.35	7.60	8.85
	070	10 C.Y. 300' haul, 85 C.Y./hr.	B-33B	680		1.89	.40	2.29	2.65
	080	1000' haul, 50 C.Y./hr.	"	400		3.21	.67	3.88	4.51
	100	Self-propelled scraper, 15 C.Y., 1000' haul, 95 C.Y./hr.	B-33D	760		1.57	.35	1.92	2.25
	110	2000' haul, 70 C.Y./hr.	"	560		2.14	.48	2.62	3.05
	130	25 C.Y. 1000' haul, 200 C.Y./hr.	B-33E	1,600		.84	.17	1.01	1.17
	140	2000' haul, 160 C.Y./hr.	"	1,280		1.05	.21	1.26	1.46
	160	For dozer with ripper, 200 H.P., add, minimum	B-11A	1,980		.32	.15	.47	.57
	170	Add, maximum	"	990		.64	.29	.93	1.13
	180	300 H.P., add, minimum	B-10M	3,670		.25	.06	.31	.37
	190	Add, maximum	"	1,840		.51	.12	.63	.74
	210	Fine grade, 3 passes with motor grader	B-11L	1,600	S.Y.	.26	.18	.44	.55
	220	With grader plus rolling	B-32	1,600		.37	.21	.58	.71
	240	Hand grading, finish	1 Clab	75			1.70	1.70	2.46L
	250	Rough		130			.98	.98	1.42L
	252	Alternate pricing method, finish		675	S.F.		.19	.19	.27L
	254	Rough		1,200	"		.11	.11	.15L

			CREW	DAILY OUTPUT	UNIT	BARE COSTS MAT.	BARE COSTS INST.	BARE COSTS TOTAL	TOTAL INCL O&P
25-001		GROUTING, CHEMICAL Soil stabilization, phenolic resin,							
	002	medium gradation stone, minimum	B-61	10.50	C.Y.	105	91	196	240
	010	Average		5.90		130	160	290	360
	020	Maximum		3		140	315	455	585
28-001		GROUTING, PRESSURE Cement and sand, 1:1 mix, minimum		124	Bag	6	7.70	13.70	16.95
	010	Maximum		51	"	6.85	18.70	25.55	33
	020	Cement and sand, 1:1 mix, minimum		120	C.F.	4	7.95	11.95	15.10
	030	Maximum		51		5	18.70	23.70	31
	040	Cement grout, minimum (1 bag = 1 C.F.)		137		5.30	6.95	12.25	15.20
	050	Maximum		57		6.25	16.70	22.95	29
	070	Alternate pricing method: (Add for materials)							
	071	5 man crew and equipment	B-61	1	Day		950	950	1,275

			CREW	DAILY OUTPUT	UNIT	BARE COSTS EQUIP.	BARE COSTS LABOR	BARE COSTS TOTAL	TOTAL INCL O&P
30-001		HAULING Earth 6 C.Y. dump truck 1/4 mile round trip, 5.0 loads/hr.	B-34A	240	C.Y.	.92	.55	1.47	1.80
	003	1/2 mile round trip, 4.1 loads/hr.		197		1.12	.67	1.79	2.20
	004	1 mile round trip, 3.3 loads/hr.		160		1.38	.83	2.21	2.70
	010	2 mile round trip, 2.6 loads/hr.		125		1.76	1.06	2.82	3.46
	015	3 mile round trip, 2.1 loads/hr.		100		2.20	1.33	3.53	4.33
	020	4 mile round trip, 1.8 loads/hr.		85		2.59	1.56	4.15	5.10
	031	12 C.Y. dump truck, 1/4 mile round trip 3.7 loads/hr.	B-34B	356		.79	.37	1.16	1.40
	032	1/2 mile round trip, 3.2 loads/hr.		308		.91	.43	1.34	1.62
	033	1 mile round trip 2.7 loads/hr.		260		1.08	.51	1.59	1.92
	040	2 mile round trip, 2.2 loads/hr.		210		1.33	.63	1.96	2.37

For expanded coverage of these items see *Means' Site Work Cost Data 1986*

2.3 Earthwork

		CREW	DAILY OUTPUT	UNIT	BARE COSTS EQUIP.	BARE COSTS LABOR	BARE COSTS TOTAL	TOTAL INCL O&P
30 045	3 mile round trip, 1.9 loads/hr.	B-34B	180	C.Y.	1.55	.74	2.29	2.77
050	4 mile round trip, 1.6 loads/hr.		150		1.86	.89	2.75	3.32
054	5 mile round trip, 1 load/hr.		98		2.85	1.36	4.21	5.10
055	10 mile round trip, .75 load/hr.		49		5.70	2.71	8.41	10.15
056	20 mile round trip, .5 load/hr.	↓	32		8.75	4.15	12.90	15.55
060	16.5 C.Y. dump trailer, 1 mile round trip, 2.6 loads/hr.	B-34C	340		1.12	.39	1.51	1.79
070	2 mile round trip, 2.1 loads/hr.		275		1.38	.48	1.86	2.21
100	3 mile round trip, 1.8 loads/hr.		235		1.62	.57	2.19	2.59
110	4 mile round trip, 1.6 loads/hr.		210		1.81	.63	2.44	2.90
111	5 mile round trip, 1 load/hr.		132		2.88	1.01	3.89	4.61
112	10 mile round trip, .75 load/hr.		100		3.81	1.33	5.14	6.10
113	20 mile round trip, .5 load/hr.	↓	66		5.75	2.01	7.76	9.25
115	20 C.Y. dump trailer, 1 mile round trip, 2.5 loads/hr.	B-34D	400		.99	.33	1.32	1.56
120	2 mile round trip, 2 loads/hr.		320		1.24	.42	1.66	1.95
122	3 mile round trip, 1.7 loads/hr.		270		1.47	.49	1.96	2.32
124	4 mile round trip, 1.5 loads/hr.		240		1.65	.55	2.20	2.61
124	5 mile round trip, 1.1 load/hr.		172		2.30	.77	3.07	3.64
125	10 mile round trip, .85 load/hr.		136		2.91	.98	3.89	4.60
125	20 mile round trip, .6 load/hr.	↓	96		4.12	1.38	5.50	6.50
130	Hauling in medium traffic, add					20	20%	20%
140	Heavy traffic, add					30	30%	30%
160	Grading at dump, if required, by dozer	B-10B	1,000	↓	.63	.23	.86	1.03
180	Spotter at dump or cut, if required	1 Clab	8	Hr.		15.90	15.90	23L

					BARE COSTS MAT.	BARE COSTS INST.	BARE COSTS TOTAL	
32-001 002	**HORIZONTAL BORING** Casing only, 100' minimum, not incl. jacking pits or dewatering							
010	Roadwork, 1/2" thick wall, 24" diameter casing	B-42	10	L.F.	35	205	240	305
020	36" diameter		9.50		50	215	265	335
030 040	48" diameter	↓	9	↓	85	230	315	390
050	Railroad work, 24" diameter	B-42	7	L.F.	35	295	330	420
060	36" diameter		6.50		50	315	365	470
070	48" diameter	↓	6	↓	85	345	430	540
090	For ledge, add						125	155

					BARE COSTS EQUIP.	BARE COSTS LABOR	BARE COSTS TOTAL	
35-001	**MOBILIZATION AND DEMOBILIZATION** Dozer, 105 H.P.	B-34C	6.20	Ea.	61	21	82	98
010	300 H.P.		4.45		86	30	116	135
030	Scraper, towed type (incl. tractor), 6 C.Y. capacity		5.85		65	23	88	105
040	10 C.Y.		4.30		89	31	120	140
060	Self-propelled scraper, 15 C.Y.		3.10		125	43	168	195
070	24 C.Y.		1.90		200	70	270	320
090	Shovel, backhoe or dragline, 3/4 C.Y.		4.30		89	31	120	140
100	1-1/2 C.Y.		2.90		130	46	176	210
120	Tractor shovel or front end loader, 1 C.Y.		6.20		61	21	82	98
130	2-1/4 C.Y.	↓	4.45	↓	86	30	116	135

					BARE COSTS MAT.	BARE COSTS INST.	BARE COSTS TOTAL	
36-001	**RIP-RAP** Random, filter stone dumped from trucks							
010	Machine placed slope protection	B-12G	62	C.Y.	8	9.95	17.95	21
011	3/8 to 1/4 C.Y. pieces, $35 ton, 18" thick, grouted	B-13	80	S.Y.	25	16.80	41.80	50
020	Not grouted		53		36.50	25	61.50	74
040	Gabions, galvanized steel mesh boxes, stone filled, 6" deep		190		10.50	7.10	17.60	21
050	9" deep		163		13.25	8.25	21.50	26
060	12" deep		153		18	8.80	26.80	32
070	18" deep		102		24	13.20	37.20	44
080	36" deep	↓	60	↓	42.50	22	64.50	77

For expanded coverage of these items see *Means' Site Work Cost Data 1986*

2.3 Earthwork

		CREW	DAILY OUTPUT	UNIT	BARE COSTS MAT.	BARE COSTS INST.	BARE COSTS TOTAL	TOTAL INCL O&P
38-001	**SHEET PILING** Steel, not incl. wales, 22 psf, 15' excav., left in place	B-40	10.80	Ton	625	255	880	1,025
010	Pull & salvage		7.20		156	385	541	670
030	㉑ 20' deep excavation, 27 psf, left in place		12.90		625	215	840	965
040	Pull & salvage		8.60		156	320	476	590
060	25' deep excavation, 38 psf, left in place		19		625	145	770	875
070	Pull & salvage		12.70		156	220	376	455
090	40' deep excavation, 38 psf, left in place		21.20		625	130	755	855
100	Pull & salvage		14.20		155	195	350	425
120	15' deep excavation, 22 psf, left in place		985	S.F.	6.90	2.82	9.72	11.25
130	Pull & salvage		655		1.72	4.23	5.95	7.40
150	20' deep excavation, 27 psf, left in place		960		8.45	2.89	11.34	13.05
160	Pull & salvage		640		2.11	4.33	6.44	7.95
180	25' deep excavation, 38 psf, left in place		1,000		11.90	2.77	14.67	16.70
190	Pull & salvage		670		2.97	4.14	7.11	8.65
210	Rent sheet piling and wales, first month			Ton	150		150	165M
220	Per added month				14		14	15.40M
230	Rental piling left in place, add to rental				460		460	505M
250	Wales, connections & struts, 2/3 salvage				150		150	165M
270	High strength piling, 50,000 psi, add				50		50	55M
280	55,000 psi, add				55		55	61M
300	Tie rod, not upset, 1-1/2" to 4" diameter with turnbuckle				1,175		1,175	1,300M
310	No turnbuckle				1,015		1,015	1,125M
330	Upset, 1-3/4" to 4" diameter with turnbuckle				1,450		1,450	1,600M
340	No turnbuckle				1,275		1,275	1,400M
360	Lightweight, 18" to 28" wide, 7 ga., 9.22 psf, and							
361	9 ga., 8.6 psf, minimum			Lb.	.30		.30	.33M
370	Average				.31		.31	.34M
375	Maximum				.33		.33	.36M
390	⑳ Wood, solid sheeting, incl. wales, braces and spacers,							
391	pull & salvage, 8' deep excavation	B-31	330	S.F.	.80	2.47	3.27	4.32
400	10' deep, 50 S.F./hr. in & 150 S.F./hr. out		300		.90	2.71	3.61	4.77
410	12' deep, 45 S.F./hr. in & 135 S.F./hr. out		270		1	3.02	4.02	5.30
420	14' deep, 42 S.F./hr. in & 126 S.F./hr. out		250		1.25	3.26	4.51	5.90
430	16' deep, 40 S.F./hr. in & 120 S.F./hr. out		240		1.35	3.39	4.74	6.20
440	18' deep, 38 S.F./hr. in & 114 S.F./hr. out		230		1.35	3.54	4.89	6.40
450	20' deep, 35 S.F./hr. in & 105 S.F./hr. out		210		1.35	3.88	5.23	6.90
452	Left in place, 8' deep, 55 S.F./hr.		440		1.55	1.85	3.40	4.29
454	10' deep, 50 S.F./hr.		400		1.75	2.04	3.79	4.76
456	12' deep, 45 S.F./hr.		360		1.95	2.26	4.21	5.30
457	16' deep, 40 S.F./hr.		320		2.75	2.55	5.30	6.55
458	18' deep, 38 S.F./hr.		305		2.70	2.67	5.37	6.70
459	20' deep, 35 S.F./hr.		280		2.65	2.91	5.56	6.95
470	Alternate pricing, left in place, 8' deep		1.76	M.B.F.	390	465	855	1,075
480	Pull and salvage, 8' deep		1.32	"	195	615	810	1,075
500	For creosoting add cost of treatment to lumber							
510	See under Lumber treating, division 6.1-57							
40-001	**SOLDIER BEAMS & LAGGING** H piles with 3" wood sheeting							
002	horizontal between piles, including removal of wales & braces							
010	No hydrostatic head, 15' deep, 1 line of braces, minimum	B-50	545	S.F.	4.90	5.85	10.75	13.55
020	Maximum		495		5.40	6.45	11.85	14.90
040	15' to 22' deep with 2 lines of braces, 10" H, minimum		360		6.35	8.90	15.25	19.30
050	Maximum		330		7	9.70	16.70	21
070	23' to 35' deep with 3 lines of braces, 12" H, minimum		325		7.25	9.85	17.10	22
080	Maximum		295		8	10.85	18.85	24
100	36' to 45' deep with 4 lines of braces, 14" H, minimum		290		8.55	11.05	19.60	25
110	Maximum		265		9.40	12.05	21.45	27
130	No hydrostatic head, left in place, 15' deep, 1 line of braces, min.		635		6.90	5.05	11.95	14.55
140	Maximum		575		7.60	5.55	13.15	16.05
160	15' to 22' deep with 2 lines of braces, minimum		455		10.60	7.05	17.65	21
170	Maximum		415		11.70	7.70	19.40	24

For expanded coverage of these items see *Means Site Work Cost Data 1986*

2.3 Earthwork

		Crew	Daily Output	Unit	Bare Costs Mat.	Bare Costs Inst.	Bare Costs Total	Total Incl O&P
40 190	23' to 35' deep with 3 lines of braces, minimum	B-50	420	S.F.	12.40	7.60	20	24
200	Maximum		380		13.65	8.40	22.05	27
220	36' to 45' deep with 4 lines of braces, minimum		385		14.55	8.30	22.85	28
230	Maximum	↓	350		16.40	9.15	25.55	31
235	Lagging only, 3" thick wood between piles 8' O.C., minimum	B-46	400		1	2.27	3.27	4.51
237	Maximum		250		1.50	3.63	5.13	7.10
240	Open sheeting for trenches to 10' deep, minimum		1,736		.36	.52	.88	1.18
245	Maximum	↓	1,510		.42	.60	1.02	1.37
250	Tie-back method of sheeting, add, minimum						20%	20%
255	Maximum			↓			60%	60%
270	Tie-backs only, based on tie-backs total length, minimum	B-46	86.80	L.F.	7.50	10.45	17.95	24
275	Maximum		38.50	"	16.50	24	40.50	54
350	Tie-backs only, typical average, 25' long		2	Ea.	320	455	775	1,025
360	35' long	↓	1.58	"	410	575	985	1,325
42-001	**SHORING** Existing building, with timber, no salvage allowance	B-51	2.20	M.B.F.	390	385	775	975
100	With 35 ton screw jacks, per box and jack	"	3.60	Jack	25	235	260	360
45-001	**SLURRY TRENCH** Excavated slurry trench in wet soils							
002	backfilled with 3000 psi concrete, no reinforcing steel							
005	Minimum	C-7	333	C.F.	3.35	5.10	8.45	10.40
010	Maximum		200	"	5.65	8.50	14.15	17.40
020	Alternate pricing method, minimum		150	S.F.	7.25	11.30	18.55	23
030	Maximum	↓	120		9.35	14.15	23.50	29
050	Reinforced slurry trench, minimum	B-48	177		5.50	9.20	14.70	18.10
060	Maximum	"	69	↓	16	24	40	49
080	Dump disposal, 2 mile haul, excavation, add	B-34B	99	C.Y.		4.17	4.17	5.05
090	Bentonite castings, add	"	40	"		10.30	10.30	12.45
50-001	**TERMITE PRETREATMENT**							
002	Slab & walls, residential	1 Skwk	1,508	S.F.Flr.	.11	.11	.22	.28
010	Commercial, minimum		2,496		.05	.07	.12	.15
020	Maximum		1,645	↓	.08	.10	.18	.23
040	Insecticides for termite control, minimum		14.20	Gal.	10.50	11.55	22.05	28
050	Maximum	↓	11	"	15.75	14.90	30.65	39
52-001	**UNDERPINNING FOUNDATIONS** Including excavation,							
002	forming, reinforcing, concrete and equipment							
010	5' to 16' below grade, up to 500 C.Y.	B-52	2.30	C.Y.	136.50	585	721.50	950
020	Over 1000 C.Y.		2.50		120	535	655	870
040	16' to 25' below grade, up to 500 C.Y.		2		147	670	817	1,075
050	Over 1000 C.Y.		2.10		131.25	640	771.25	1,025
070	26' to 40' below grade, up to 500 C.Y.		1.60		162.75	840	1,002.75	1,325
080	Over 1000 C.Y.	↓	1.80		147	745	892	1,175
090	For under 50 C.Y., add				10%	40%		
100	For 50 C.Y. to 100 C.Y., add			↓	5%	20%		

		Crew	Daily Output	Unit	Bare Costs Equip.	Bare Costs Labor	Bare Costs Total	Total Incl O&P
55-001	**WELLPOINTS** For wellpoint equipment rental, see division 1.5-25							
010	Installation and removal of single stage system							
011	Labor only, .75 man-hours per L.F., minimum	1 Clab	10.70	L.F.Hdr.		11.90	11.90	17.20L
020	2.0 man-hours per L.F., maximum	"	4	"		32	32	46L
040	Pump operation, 4 @ 6 hr. shifts							
041	Per 24 hour day	4 Eqlt	1.27	Day	490		490	710L
050	Per 168 hour week, 160 hr. straight, 8 hr. double time		.18	Week	3,450		3,450	5,025L
055	Per 4.3 week month	↓	.04	Month	15,600		15,600	22,600L
060	(23) Complete installation, operation, equipment rental, fuel &							
061	removal of system with 2" wellpoints 5' O.C.							
070	100' long header, 6" diameter, first month			L.F.Hdr.	194	86	280	374
080	Thereafter, per month				152	61	213	286
100	200' long header, 8" diameter, first month				110	51	161	212
110	Thereafter, per month				74	35	109	145
130	500' long header, 8" diameter, first month				59	27	86	115
140	Thereafter, per month			↓	30	18	48	63

For expanded coverage of these items see *Means' Site Work Cost Data 1986*

2.3 Earthwork

		CREW	DAILY OUTPUT	UNIT	BARE COSTS EQUIP.	BARE COSTS LABOR	BARE COSTS TOTAL	TOTAL INCL O&P
55	160	1,000' long header, 10" diameter, first month			L.F.Hdr.	54	20	74	100
	170	Thereafter, per month			"	15	12	27	35
	190	Note: above figures include pumping 168 hrs. per week							
	191	and include the pump operator and one stand-by pump.							

		CREW	DAILY OUTPUT	UNIT	BARE COSTS MAT.	BARE COSTS INST.	BARE COSTS TOTAL	TOTAL INCL O&P
57-001	WELLS Dewatering excavations 10' to 20' deep, 2' diameter							
002	with steel casing, minimum	B-6	165	V.L.F.	1.60	3.43	5.03	6.40
005	Average	↓	98		3.15	5.80	8.95	11.30
010	Maximum		49	↓	8.25	11.55	19.80	25
030	For pumps for above, see division 1.5-15-410							
050	For other types of wells, see division 2.5-40							

2.4 Caissons & Piling

		CREW	DAILY OUTPUT	UNIT	BARE COSTS MAT.	BARE COSTS INST.	BARE COSTS TOTAL	TOTAL INCL O&P
05-001	CAISSONS Incl. excav., concrete, 50 lbs. reinf. per C.Y., but not incl.							
002	mobilization, boulder removal, disposal or pre-drilling							
010	Open style, machine drilled, to 50' deep, in stable ground, no							
011	(19) casings or ground water, 18" diam., .065 C.Y./L.F.	B-43	200	V.L.F.	7.80	6.90	14.70	17.60
020	24" diameter, .116 C.Y./L.F.		190		13.90	7.25	21.15	25
030	30" diameter, .182 C.Y./L.F.		150		22	9.20	31.20	36
040	36" diameter, .262 C.Y./L.F.		125		32	11.05	43.05	50
050	48" diameter, .465 C.Y./L.F.		100		55	13.80	68.80	79
060	60" diameter, .727 C.Y./L.F.		90		87	15.35	102.35	115
070	72" diameter, 1.05 C.Y./L.F.		80		130	17.25	147.25	165
080	84" diameter, 1.43 C.Y./L.F.	↓	75	↓	170	18.40	188.40	210
085								
100	For bell excavation and concrete, add							
102	4' bell diameter, 24" shaft, .444 C.Y.	B-43	20	Ea.	38	69	107	130
104	6' bell diameter, 30" shaft, 1.57 C.Y.		5.70		135	240	375	465
106	8' bell diameter, 36" shaft, 3.72 C.Y.		2.40		315	575	890	1,100
108	9' bell diameter, 48" shaft, 4.48 C.Y.		2		385	690	1,075	1,325
110	10' bell diameter, 60" shaft, 5.24 C.Y.		1.70		445	810	1,255	1,550
112	12' bell diameter, 72" shaft, 8.74 C.Y.		1		740	1,375	2,115	2,625
114	14' bell diameter, 84" shaft, 13.6 C.Y.	↓	.70		1,165	1,975	3,140	3,875
120	Open style, machine drilled, to 50' deep, in wet ground, pulled							
130	casing and pumping, 18" diameter, .065 C.Y./L.F.	B-48	160	V.L.F.	7.80	10.15	17.95	22
140	24" diameter, .116 C.Y./L.F.		125		13.90	13	26.90	32
150	30" diameter, .182 C.Y./L.F.		85		22	19.10	41.10	49
160	36" diameter, .262 C.Y./L.F.	↓	60		32	27	59	71
170	48" diameter, .465 C.Y./L.F.	B-49	55		55	48	103	125
180	60" diameter, .727 C.Y./L.F.		35		87	75	162	195
190	72" diameter, 1.05 C.Y./L.F.		30		130	88	218	260
200	84" diameter, 1.43 C.Y./L.F.	↓	25	↓	170	105	275	325
205								
210	For bell excavation and concrete, add							
212	4' bell diameter, 24" shaft, .444 C.Y.	B-48	19.80	Ea.	38	82	120	150
214	6' bell diameter, 30" shaft, 1.57 C.Y.	↓	5.70		135	285	420	525
216	8' bell diameter, 36" shaft, 3.72 C.Y.		2.40		315	675	990	1,225
218	9' bell diameter, 48" shaft, 4.48 C.Y.	B-49	3.30		385	800	1,185	1,475
220	10' bell diameter, 60" shaft, 5.24 C.Y.		2.80		445	940	1,385	1,750
222	12' bell diameter, 72" shaft, 8.74 C.Y.		1.60		740	1,650	2,390	3,000
224	14' bell diameter, 84" shaft, 13.6 C.Y.	↓	1	↓	1,165	2,650	3,815	4,800
230	Open style, machine drilled, to 50' deep, in soft rocks and							
240	medium hard shales, 18" diameter, .065 C.Y./L.F.	B-49	50	V.L.F.	7.80	53	60.80	79

For expanded coverage of these items see *Means' Site Work Cost Data 1986*

2.4 Caissons & Piling

			CREW	DAILY OUTPUT	UNIT	BARE COSTS MAT.	BARE COSTS INST.	BARE COSTS TOTAL	TOTAL INCL O&P
05	250	24" diameter, .116 C.Y./L.F.	B-49	30	V.L.F.	13.90	88	101.90	130
	260	30" diameter, .182 C.Y./L.F.		20		22	130	152	200
	270	36" diameter, .262 C.Y./L.F.		15		32	175	207	270
	280	48" diameter, .465 C.Y./L.F.		10		55	265	320	410
	290	60" diameter, .727 C.Y./L.F.		7		87	375	462	595
	300	72" diameter, 1.05 C.Y./L.F.		6		130	440	570	725
	310	84" diameter, 1.43 C.Y./L.F.		5		170	530	700	890
	320	For bell excavation and concrete, add							
	322	4' bell diameter, 24" shaft, .444 C.Y.	B-49	10.90	Ea.	38	240	278	365
	324	6' bell diameter, 30" shaft, 1.57 C.Y.		3.10		135	850	985	1,275
	326	8' bell diameter, 36" shaft, 3.72 C.Y.		1.30		315	2,025	2,340	3,050
	328	9' bell diameter, 48" shaft, 4.48 C.Y.		1.10		385	2,400	2,785	3,600
	330	10' bell diameter, 60" shaft, 5.24 C.Y.		.90		445	2,925	3,370	4,375
	332	12' bell diameter, 72" shaft, 8.74 C.Y.		.60		740	4,400	5,140	6,650
	334	14' bell diameter, 84" shaft, 13.6 C.Y.		.40		1,165	6,600	7,765	10,000
	360	For rock excavation, sockets, add, minimum		120	C.F.		22	22	29
	365	Average		95			28	28	37
	370	Maximum		48			55	55	73
	390	For 50' to 100' deep, add			V.L.F.	7%	7%		
	400	For 100' to 150' deep, add				25%	25%		
	410	For 150' to 200' deep, add				30%	30%		
	420	For casings left in place, add			Lb.	.50		.50	.55M
	430	For other than 50 lb. reinf. per C.Y., add or deduct			"	.51		.51	.56M
	440	For steel "I" beam cores, add	B-49	8.30	Ton	500	320	820	975
	450	Load and haul excess excavation, 2 miles	B-34B	178	C.Y.		2.32	2.32	2.80
	500	Bottom inspection	1 Skwk	1.20	Ea.		135	135	200L
10-	001	MOBILIZATION Set up & remove, air compressor, 600 C.F.M.	A-5	3.30			91	91	130
	010	1200 C.F.M.	"	2.20			135	135	195
	020	Crane, with pile leads and pile hammer, 75 ton	B-8	.50			5,350	5,350	6,675
	030	150 ton	"	.31			8,625	8,625	10,800
	050	Drill rig, complete to 36", minimum	B-4	1.38			800	800	1,075
	060	Up to 84"	"	.92			1,200	1,200	1,625
	080	Auxiliary boiler, small	A-5	1.66			180	180	260
	090	Large	"	.83			365	365	520
	110	Rule of thumb: complete pile driving set up, small	B-19	.45			5,100	5,100	6,825
	120	Large	"	.27			8,475	8,475	11,400
15-	001	PILES, CONCRETE 200 piles, 60' long							
	002	unless specified otherwise, not incl. pile caps or mobilization							
	010	Cast in place, thin wall shell pile, straight sided,							
	011	not incl. reinforcing, 8" diam., 16 ga., 5.8 lb./L.F.	B-19	700	V.L.F.	3.50	3.27	6.77	8.25
	020	10" diameter, 16 ga. corrugated, 7.3 lb./L.F.		650		4.60	3.52	8.12	9.80
	030	12" diameter, 16 ga. corrugated, 8.7 lb./L.F.		600		5.85	3.82	9.67	11.55
	040	14" diameter, 16 ga. corrugated, 10.0 lb./L.F.		550		7.05	4.16	11.21	13.35
	050	16" diameter, 16 ga. corrugated, 11.6 lb./L.F.		500		8.55	4.58	13.13	15.55
	080	Cast in place friction pile, 50' long, fluted,							
	081	tapered steel, 4000 psi concrete, no reinforcing							
	090	12" diameter, 7 ga.	B-19	600	V.L.F.	10.75	3.82	14.57	16.95
	100	14" diameter, 7 ga.		560		11.80	4.09	15.89	18.45
	110	16" diameter, 7 ga.		520		13.80	4.41	18.21	21
	120	18" diameter, 7 ga.		480		16.20	4.77	20.97	24
	130	End bearing, fluted, constant diameter,							
	132	4000 psi concrete, no reinforcing							
	134	12" diameter, 7 ga.	B-19	600	V.L.F.	11.05	3.82	14.87	17.25
	136	14" diameter, 7 ga.		560		13.85	4.09	17.94	21
	138	16" diameter, 7 ga.		520		15.90	4.41	20.31	23
	140	18" diameter, 7 ga.		480		17.35	4.77	22.12	25
	150	For reinforcing steel, add			Lb.	.45		.45	.49M
	170	For ball or pedestal end, add	B-19	11	C.Y.	62	210	272	345
	190	For lengths above 60', concrete, add	"	11	"	62	210	272	345
	200	For steel thin shell, pipe only			Lb.	.45		.45	.49M

For expanded coverage of these items see *Means' Site Work Cost Data 1986*

2.4 Caissons & Piling

		CREW	DAILY OUTPUT	UNIT	BARE COSTS MAT.	BARE COSTS INST.	BARE COSTS TOTAL	TOTAL INCL O&P
15 220	Precast, prestressed, straight cylinder, 12" diam., 2-3/8" wall	B-19	720	V.L.F.	7.15	3.18	10.33	12.15
230	14" diameter, 2-1/2" wall		680		8.80	3.37	12.17	14.20
250	16" diameter, 3" wall		640		10.70	3.58	14.28	16.55
260	18" diameter, 3" wall		600		12.35	3.82	16.17	18.70
280	20" diameter, 3-1/2" wall		560		16.05	4.09	20.14	23
290	24" diameter, 3-1/2" wall		520		20	4.41	24.41	28
310	Precast, prestressed, 40' long, 10" thick, square		700		5.60	3.27	8.87	10.55
320	12" thick, square		680		7.35	3.37	10.72	12.60
340	14" thick, square		600		11.25	3.82	15.07	17.50
350	Octagonal		640		11	3.58	14.58	16.90
370	16" thick, square		560		13.15	4.09	17.24	19.95
380	Octagonal		600		12.90	3.82	16.72	19.30
400	18" thick, square		520		15.75	4.41	20.16	23
410	Octagonal		560		15.55	4.09	19.64	23
430	20" thick, square		480		17.85	4.77	22.62	26
440	Octagonal		520		17.65	4.41	22.06	25
460	24" thick, square		440		26	5.20	31.20	36
470	Octagonal		480		25	4.77	29.77	34
475	Mobilization for 10,000 L.F. pile job, add		3,300			.69	.69	.93
480	25,000 L.F. pile job, add		8,500			.27	.27	.36
20-001	PILES, STEEL Not including mobilization or demobilization							
010	Step tapered, round, concrete filled							
011	8" tip, 60 ton capacity, 30' depth	B-19	760	V.L.F.	4.70	3.01	7.71	9.20
012	60' depth		740		5.25	3.10	8.35	9.90
013	80' depth		700		5.50	3.27	8.77	10.45
014								
015	10" tip, 90 ton capacity, 30' depth	B-19	700	V.L.F.	5.65	3.27	8.92	10.60
016	60' depth		690		5.85	3.32	9.17	10.90
017	80' depth		670		6.30	3.42	9.72	11.50
019	12" tip, 120 ton capacity, 30' depth		660		6.55	3.47	10.02	11.85
020	60' depth		630		6.80	3.64	10.44	12.35
021	80' depth		590		7.30	3.88	11.18	13.25
023								
025	"H" Sections, 8" x 8", 36 lb. per L.F.	B-19	640	V.L.F.	8.15	3.58	11.73	13.75
040	10" x 10", 42 lb. per L.F.		610		9.50	3.76	13.26	15.50
050	57 lb. per L.F.		610		12.85	3.76	16.61	19.15
070	12" x 12", 53 lb. per L.F.		590		12	3.88	15.88	18.40
080	74 lb. per L.F.		590		16.80	3.88	20.68	24
100	14" x 14", 73 lb. per L.F.		540		17	4.24	21.24	24
110	89 lb. per L.F.		540		21	4.24	25.24	29
130	14" x 14", 102 lb. per L.F.		510		21.65	4.49	26.14	30
140	117 lb. per L.F.		510		25	4.49	29.49	34
160	Splice or standard points, not in leads, 8" or 10"	1 Sswl	5	Ea.	41	35	76	100
170	12" or 14"		4		67	43	110	140
190	Heavy duty points, not in leads, 10" wide		4		72	43	115	145
210	14" wide		3.50		105	50	155	195
260	Pipe piles, 8" diameter, 29 lb. per L.F., no concrete	B-19	500	V.L.F.	8.10	4.58	12.68	15.05
270	Concrete filled		460		8.90	4.98	13.88	16.45
290	10" diameter, 34 lb. per L.F., no concrete		500		9.45	4.58	14.03	16.55
300	Concrete filled		450		10.75	5.10	15.85	18.65
320	12" diameter, 44 lb. per L.F., no concrete		475		12.25	4.82	17.07	19.95
330	Concrete filled		415		14.10	5.50	19.60	23
350	14" diameter, 46 lb. per L.F., no concrete		430		12.75	5.35	18.10	21
360	Concrete filled		355		15.30	6.45	21.75	25
380	16" diameter, 52 lb. per L.F., no concrete		385		14.45	5.95	20.40	24
390	Concrete filled		335		17.75	6.85	24.60	29
410	18" diameter, 59 lb. per L.F., no concrete		355		16.40	6.45	22.85	27
420	Concrete filled		310		21	7.40	28.40	33
440	Splices for pipe piles, not in leads 8" or 10" diam.	1 Sswl	4.67	Ea.	28	37	65	89
450	12" or 14" diameter	"	3.79	"	38	46	84	115

For expanded coverage of these items see *Means' Site Work Cost Data 1986*

2.4 Caissons & Piling

		CREW	DAILY OUTPUT	UNIT	BARE COSTS MAT.	BARE COSTS INST.	BARE COSTS TOTAL	TOTAL INCL O&P
20 460	16" diameter	1 Sswl	3.03	Ea.	46	57	103	140
480	Standard points, 8" or 10" diameter		4.61		23	38	61	84
490	12" or 14" diameter		4.05		33	43	76	105
500	16" diameter		3.37		39	52	91	125
520	Heavy duty points, 10" or 12" diameter		2.89		33	60	93	130
530	14" or 16" diameter		2.02		51	86	137	190
550	For reinforcing steel, add		1,150	Lb.	.30	.15	.45	.57
570	For thick wall sections, add			"	.27		.27	.29M
25-001	**PILES, WOOD** Untreated, friction or end bearing, not including							
005	mobilization or demobilization							
010	Up to 30' long, 12" butts, 8" points	B-19	625	V.L.F.	3.40	3.67	7.07	8.65
020	30 to 39' long, 12" butts, 8" points		700		3.40	3.27	6.67	8.15
030	40 to 49' long, 12" butts, 7" points		720		3.40	3.18	6.58	8
040	50 to 59' long, 13" butts, 7" points		800		3.55	2.86	6.41	7.75
050	60 to 69' long, 13" butts, 7" points		840		4.10	2.73	6.83	8.15
060	70 to 80' long, 13" butts, 6" points		840		4.60	2.73	7.33	8.70
080	Treated piles, 12 lb. creosote per C.F.,							
081	friction or end bearing, ASTM class B							
100	Up to 30' long, 12" butts, 8" points	B-19	625	V.L.F.	5.55	3.67	9.22	11
110	30 to 39' long, 12" butts, 8" points		700		5.55	3.27	8.82	10.50
120	40 to 49' long, 12" butts, 7" points		720		5.10	3.18	8.28	9.85
130	50 to 59' long, 13" butts, 7" points		800		6.15	2.86	9.01	10.60
140	60 to 69' long, 13" butts, 6" points		840		6.20	2.73	8.93	10.45
150	70 to 80' long, 13" butts, 6" points		840		7.40	2.73	10.13	11.80
170	Boot for pile tip, minimum	1 Pile	27	Ea.	10.30	5.95	16.25	21
180	Maximum		21		31	7.65	38.65	46
200	Point for pile tip, minimum		20		12.50	8.05	20.55	26
210	Maximum		15		42	10.70	52.70	63
230	Splice for piles over 50' long, minimum	B-46	35		31	26	57	73
240	Maximum		20		36	45	81	110
260	Concrete encasement with wire mesh and tube		331	V.L.F.	5.40	2.74	8.14	10.05
270	Mobilization for 10,000 L.F. pile job, add	B-19	3,300			.69	.69	.93
280	25,000 L.F. pile job, add	"	8,500			.27	.27	.36
30-001	**PILING SPECIAL COSTS** Concrete pile caps							
010	Concrete pile cap for pile groups, 40 to 80 ton capacity							
015	2 pile cap	C-17B	15	Ea.	145	125	270	340
020	3 pile cap		15		145	125	270	340
025	4 pile cap		11		210	170	380	475
030	5 pile cap		8		325	235	560	695
035	6 pile cap		6.70		380	280	660	820
040	7 pile cap		5.30		530	355	885	1,100
050	Cutoffs, concrete piles, plain	1 Pile	5.50			29	29	46L
060	With steel thin shell		38			4.23	4.23	6.65L
070	Steel pile or "H" piles		19			8.45	8.45	13.25L
080	Wood piles		38			4.23	4.23	6.65L

		CREW	DAILY OUTPUT	UNIT	BARE COSTS EQUIP.	BARE COSTS LABOR	BARE COSTS TOTAL	TOTAL INCL O&P
090	Pre-augering up to 30' deep, average soil, 24" diameter	B-43	180	L.F.	3.03	4.63	7.66	10.05
092	36" diameter		115		4.75	7.25	12	15.75
096	48" diameter		70		7.80	11.90	19.70	26
098	60" diameter		50		10.90	16.65	27.55	36

		CREW	DAILY OUTPUT	UNIT	BARE COSTS MAT.	BARE COSTS INST.	BARE COSTS TOTAL	TOTAL INCL O&P
100	Testing, any type piles, test load is twice the design load							
105	50 ton design load, 100 ton test	B-19	.47	Ea.	2,500	4,875	7,375	9,275
110	100 ton design load, 200 ton test		.31		3,700	7,400	11,100	14,000
115	150 ton design load, 300 ton test		.25		4,400	9,175	13,575	17,100
120	200 ton design load, 400 ton test		.22		5,000	10,400	15,400	19,500
125	400 ton design load, 800 ton test		.20		6,000	11,500	17,500	21,900

2.4 Caissons & Piling

			CREW	DAILY OUTPUT	UNIT	BARE COSTS MAT.	BARE COSTS INST.	BARE COSTS TOTAL	TOTAL INCL O&P
30	150	Wet conditions, soft damp ground, add				40%	40%		
	160	Swampy, wet ground conditions, add				40%	40%		
	170	Barge mounted driving rig, add				30%	30%		
35-001	(21)	Sheet piling, Steel, see division 2.3-38							
40-001		**PRESSURE INJECTED FOOTINGS** or Displacement Caissons, incl.							
	010	mobilization and demobilization, up to 50 miles							
	020	Uncased shafts, 30 to 80 tons capacity, 17" diam., 10' depth	B-44	88	V.L.F.	10.55	25	35.55	45
	030	25' depth		165		7	13.15	20.15	25
	040	80-150 tons capacity, 22" diameter, 10' depth		80		13.85	27	40.85	52
	050	20' depth		130		10.85	16.65	27.50	34
	070	Cased shafts, 10 to 30 ton capacity, 10-5/8" diam., 20' depth		175		7	12.40	19.40	24
	080	30' depth		240		6.25	9	15.25	19.05
	085	30 to 60 ton capacity, 12" diameter, 20' depth		160		9.65	13.55	23.20	29
	090	40' depth		230		8	9.40	17.40	22
	100	80 to 100 ton capacity, 16" diameter, 20' depth		160		12.85	13.55	26.40	32
	110	40' depth		230		11	9.40	20.40	25
	120	110 to 140 ton capacity, 17-5/8" diameter, 20' depth		160		13.80	13.55	27.35	33
	130	40' depth		230		12	9.40	21.40	26
	140	140 to 175 ton capacity, 19" diameter, 20' depth		130		15.65	16.65	32.30	40
	150	40' depth		210		13.85	10.30	24.15	29
	160	Load test, 80 ton design load, 160 ton test	B-19	.32	Ea.	3,500	7,150	10,650	13,400
	165	150 ton design load, 300 ton test	"	.25	"	4,500	9,175	13,675	17,200
	170	Over 30' long, L.F. cost tends to be lower							
	190	Maximum depth is about 90'							

2.5 Site Drainage & Utilities

			CREW	DAILY OUTPUT	UNIT	BARE COSTS MAT.	BARE COSTS INST.	BARE COSTS TOTAL	TOTAL INCL O&P
02-001		**CATCH BASINS OR MANHOLES** Including footing & excavation,							
	002	not including frame and cover							
	005	Brick, 4' inside diameter, 4' deep	D-1	1	Ea.	315	290	605	765
	010	6' deep		.70		440	415	855	1,075
	015	8' deep		.50		560	585	1,145	1,450
	020	For depths over 8', add		4	V.L.F.	73	73	146	185
	040	Concrete blocks (radial), 4' I.D., 4' deep		1.50	Ea.	200	195	395	495
	050	6' deep		1		270	290	560	715
	060	8' deep		.70		345	415	760	975
	070	For depths over 8', add		5.50	V.L.F.	52	53	105	135
	080	Concrete, cast in place, 8" thick, 4' deep	B-6	2	Ea.	280	285	565	690
	090	6' deep		1.50		370	375	745	920
	100	8' deep		1		490	565	1,055	1,300
	110	For depths over 8', add		8	V.L.F.	62	71	133	165
	111	Precast 4', I.D., 4' deep		4.10	Ea.	220	140	360	430
	112	6' deep		3		300	190	490	585
	113	8' deep		2		380	285	665	800
	114	For depths over 8', add		16	V.L.F.	52	35	87	105
	115	5' I.D., 4' deep		3	Ea.	330	190	520	620
	116	6' deep		2		450	285	735	880
	117	8' deep		1.50		570	375	945	1,150
	118	For depths over 8', add		12	V.L.F.	73	47	120	145
	119	6' I.D., 4' deep		2	Ea.	540	285	825	975
	120	6' deep		1.50		700	375	1,075	1,275
	121	8' deep		1		875	565	1,440	1,725
	122	For depths over 8', add		8	V.L.F.	115	71	186	220

For expanded coverage of these items see *Means' Site Work Cost Data 1986*

2.5 Site Drainage & Utilities

			DAILY			BARE COSTS		TOTAL
		CREW	OUTPUT	UNIT	MAT.	INST.	TOTAL	INCL O&P
02 125	Slab tops, precast, 8" thick							
130	4' diameter manhole	B-6	8	Ea.	80	71	151	185
140	5' diameter manhole		7.50		90	75	165	200
150	6' diameter manhole		7		150	81	231	275
160	Frames and covers, 24" square, 500 lb.		7.80		150	73	223	265
170	26" D shape, 600 lb.		7		165	81	246	290
180	Light traffic, 18" diameter, 100 lb.		10		60	57	117	145
190	24" diameter, 300 lb.		8.70		115	65	180	215
200	36" diameter, 900 lb.		5.80		300	98	398	460
210	Heavy traffic, 24" diameter, 400 lb.		7.80		145	73	218	260
220	36" diameter, 1150 lb.		3		405	190	595	700
230	Mass. State standard, 26" diameter, 475 lb.		7		160	81	241	285
240	30" diameter, 620 lb.		7		205	81	286	335
250	Watertight, 24" diameter, 350 lb.		7.80		240	73	313	360
260	26" diameter, 500 lb.		7		285	81	366	425
270	32" diameter, 575 lb.		6		300	94	394	460
280	3 piece cover & frame, 10" deep,							
290	1200 lbs., for heavy equipment	B-6	3	Ea.	400	190	590	695
300	Raised for paving 1-1/4" to 2" high,							
310	4 piece expansion ring							
320	20" to 26" diameter	1 Clab	3	Ea.	78	42	120	145
330	30" to 36" diameter	"	3		105	42	147	175
332	Frames and covers, existing, raised for paving 2", including							
334	row of brick, concrete collar, up to 12" wide frame	B-9	18		23	45	68	87
336	20" to 26" wide frame		11		28	73	101	130
338	30" to 36" wide frame		9		40	89	129	165
340	Inverts, single channel brick	D-1	3		44	97	141	185
350	Concrete		5		34	58	92	120
360	Triple channel, brick		2		60	145	205	275
370	Concrete		3		39	97	136	180
380	Steps, heavyweight cast iron, 7" x 9"	1 Bric	40		5.85	4.10	9.95	12.25
390	8" x 9"		40		8.75	4.10	12.85	15.45
400	Standard sizes, galvanized steel		40		8.75	4.10	12.85	15.45
410	Aluminum		40		7.80	4.10	11.90	14.40
05-001	**ELECTRIC UTILITIES** See division 16.7							
07-001	**EXCAVATION AND BACKFILL** See division 2.3-18							
010	Hand excavate and trim for bells after trench excavation							
020	8" pipe	1 Clab	155	L.F.		.82	.82	1.19L
030	18" pipe	"	130	"		.98	.98	1.42L
08-001	**GAS SERVICE AND DISTRIBUTION** Not including excavation							
005	or backfill.							
010	Polyethylene, 60 psi, coils, 1/2" diameter, SDR 7	B-20	450	L.F.	.12	1.12	1.24	1.75
015	1-1/4" diameter, SDR 10		400		.49	1.26	1.75	2.36
020	2" diameter, SDR 11		360		.80	1.40	2.20	2.91
025	3" diameter, SDR 11.5		300		1.69	1.68	3.37	4.29
030	40' joints with coupling, 3" diameter, SDR 11.5	B-21	300		1.81	2.26	4.07	5.15
035	4" diameter, SDR 11		260		3	2.61	5.61	6.95
040	6" diameter, SDR 21		240		3.35	2.83	6.18	7.65
045	8" diameter, SDR 21		200		6.35	3.39	9.74	11.75
050	Steel, schedule 40, plain end, tar coated & wrapped							
055	1" diameter	Q-4	300	L.F.	4.06	2.36	6.42	7.85
060	2" diameter		280		3.65	2.53	6.18	7.65
065	3" diameter		260		6.65	2.72	9.37	11.25
070	4" diameter	B-35	255		9.38	5.30	14.68	17.55
075	5" diameter		220		19.46	6.15	25.61	30
080	6" diameter		180		20.09	7.55	27.64	32
085	8" diameter		140		29.09	9.70	38.79	45
10-001	**FILL** In trench, crushed bank run	B-6	260	C.Y.	13.50	2.18	15.68	17.80
010	Screened 3/4" to 1/2"	"	260	"	14	2.18	16.18	18.35

For expanded coverage of these items see *Means' Site Work Cost Data 1986*

2.5 Site Drainage & Utilities

		CREW	DAILY OUTPUT	UNIT	BARE COSTS MAT.	BARE COSTS INST.	BARE COSTS TOTAL	TOTAL INCL O&P
12-001	**LINING PIPE** With cement, including bypass & cleaning, 6" to 12" pipe	C-17E	160	L.F.	5.50	10.75	16.25	22
020	16" to 20" pipe		145		6.05	11.85	17.90	24
025	24" to 36" pipe		130		7.25	13.25	20.50	27
030	48" to 72" pipe for transmission line		110		11.50	15.65	27.15	35
27-001	**PIPE, DRAINAGE AND SEWAGE** Excavation & backfill							
002	not included							
010	Asbestos Cement, class 2400, 6" diameter	B-20	330	L.F.	3.20	1.53	4.73	5.75
012	8" diameter	B-21	380		4.20	1.79	5.99	7.10
014	10" diameter		330		5.65	2.06	7.71	9.10
016	12" diameter		280		7.15	2.42	9.57	11.25
018	14" diameter		245		8.20	2.77	10.97	12.90
020	16" diameter		210		10.15	3.23	13.38	15.70
030	Class 4000, truckload lots, 10" diameter		320		5.05	2.12	7.17	8.50
032	12" diameter		265		6.40	2.56	8.96	10.65
034	14" diameter		235		8.60	2.89	11.49	13.50
036	16" diameter		200		10.95	3.39	14.34	16.80
038	18" diameter		150		12.85	4.52	17.37	20
040	24" diameter	B-22	115		20.85	6.65	27.50	32
042	30" diameter		80		29	9.55	38.55	45
044	36" diameter		65		38.70	11.80	50.50	59
090	Bituminous fiber, plain, 2" diameter	2 Plum	400		.87	.90	1.77	2.26
092	3" diameter		400		1.10	.90	2	2.51
094	4" diameter		380		1.40	.95	2.35	2.91
096	5" diameter		360		2.40	1	3.40	4.09
098	6" diameter		340		2.65	1.06	3.71	4.45
100	8" diameter		300		6.20	1.20	7.40	8.55
130	⑮ Concrete, non-reinforced, extra strength, B&S or T&G joints							
132	6" diameter	B-20	265	L.F.	2.75	1.90	4.65	5.80
134	8" diameter	B-21	300		2.95	2.26	5.21	6.40
136	10" diameter		260		3.30	2.61	5.91	7.30
138	12" diameter		198		4.05	3.43	7.48	9.25
140	15" diameter		165		4.70	4.11	8.81	10.95
142	18" diameter		130		5.90	5.20	11.10	13.80
144	21" diameter		105		7.30	6.45	13.75	17.10
146	24" diameter		90		8.90	7.55	16.45	20
160	Reinforced Concrete Culvert, class 3, no gaskets							
162	12" diameter, 106#/L.F.	B-21	190	L.F.	4.75	3.57	8.32	10.25
164	15" diameter, 150#/L.F.		155		5.80	4.38	10.18	12.50
166	18" diameter, 200#/L.F.		122		7.20	5.55	12.75	15.70
168	24" diameter, 325#/L.F.		88		11.50	7.70	19.20	23
170	30" diameter, 385#/L.F.	B-13	80		16	16.80	32.80	40
172	36" diameter, 525#/L.F.		60		23	22	45	56
174	42" diameter, 685#/L.F.		55		29	24	53	65
176	48" diameter, 865#/L.F.		50		38	27	65	78
178	60" diameter, 1070#/L.F.		35		60	38	98	120
180	72" diameter, 1880#/L.F.		30		85	45	130	155
182	84" diameter, 2500#/L.F.		24		135	56	191	225
184	96" diameter, 3000#/L.F.		20		160	67	227	265
190	With gaskets, class 3, 12" diameter	B-21	210		5.75	3.23	8.98	10.85
192	15" diameter		175		6.80	3.88	10.68	12.90
195	18" diameter		150		8.35	4.52	12.87	15.50
197	24" diameter		100		13	6.80	19.80	24
200	30" diameter	B-13	85		17.50	15.85	33.35	41
203	36" diameter		70		25	19.20	44.20	53
205	48" diameter		58		40	23	63	75
210	72" diameter		38		110	35	145	170
211	Flared ends, 6'-1" long, 12" diameter	B-21	190		30	3.57	33.57	38
212	15" diameter		155		34	4.38	38.38	44
213	6'-2" long, 18" diameter		122		40	5.55	45.55	52
214	24" diameter		88		48	7.70	55.70	64

For expanded coverage of these items see *Means' Site Work Cost Data 1986*

2.5 Site Drainage & Utilities

		CREW	DAILY OUTPUT	UNIT	BARE COSTS MAT.	BARE COSTS INST.	BARE COSTS TOTAL	TOTAL INCL O&P
27 215	30" diameter	B-13	80	L.F.	60	16.80	76.80	89
216	36" diameter	"	60		85	22	107	125
220	Bituminous coated, per diameter inch, 1 coat, add				.19		.19	.20M
225	3 coat, add				.29		.29	.31M
230	Coal tar epoxy coating, per diameter inch, 1 coat, add				.16		.16	.17M
235	3 coat, add				.24		.24	.26M
240	Vitrified plate lined, add to above, 30" to 36" diameter			S.F.C.A.	2.60		2.60	2.86M
245	42" to 54" diameter, add				2.40		2.40	2.64M
250	60" to 72" diameter, add				2.20		2.20	2.42M
255	Over 72" diameter, add				2.10		2.10	2.31M
260	Radius pipe, add to pipe prices, 12" to 60" diameter			L.F.	50%	20%		
265	Over 60" diameter, add			"	20%	20%		
270	Reinforced elliptical, 8' lengths, C507 class 3							
271								
275	14" x 23" inside, round equivalent 18" diameter	B-21	82	L.F.	30	8.25	38.25	45
280	24" x 38" inside, round equivalent 30" diameter	B-22	58		40	13.20	53.20	62
285	29" x 45" inside, round equivalent 36" diameter		52		50	14.70	64.70	75
290	38" x 60" inside, round equivalent 48" diameter		38		75	20	95	110
295	48" x 76" inside, round equivalent 60" diameter		26		110	29	139	160
300	58" x 91" inside, round equivalent 72" diameter		22		160	35	195	225
350	Corrugated Metal Pipe, galvanized or aluminum, bituminous							
351	coated with paved invert, 20' to 30' lengths							
355	8" diameter 16 ga.	B-20	330	L.F.	4.35	1.53	5.88	7
360	10" diameter 16 ga.		260		5.50	1.94	7.44	8.85
365	12" diameter 16 ga.		210		6.45	2.40	8.85	10.55
370	15" diameter 16 ga.	B-21	210		8.10	3.23	11.33	13.45
375	18" diameter 16 ga.		190		9.75	3.57	13.32	15.75
380	24" diameter 14 ga.		160		14.90	4.24	19.14	22
385	30" diameter 14 ga.		120		18.70	5.65	24.35	28
390	36" diameter 12 ga.	B-13	120		28	11.20	39.20	46
395	48" diameter 12 ga.		100		37	13.45	50.45	59
400	60" diameter 10 ga.		75		59	17.95	76.95	89
405	72" diameter 8 ga.		45		83	30	113	130
410	Corrugated Metal Pipe, galvanized or aluminum, plain							
413	8" diameter 16 ga.	B-20	355	L.F.	3.55	1.42	4.97	5.95
415	10" diameter 16 ga.		280		4.50	1.80	6.30	7.55
417	12" diameter 16 ga.		220		5.20	2.29	7.49	9.05
420	15" diameter 16 ga.	B-21	220		6.45	3.08	9.53	11.40
423	18" diameter 16 ga.		205		7.75	3.31	11.06	13.15
425	24" diameter 14 ga.		175		12.55	3.88	16.43	19.25
427	30" diameter 14 ga.		130		15.65	5.20	20.85	25
430	36" diameter 12 ga.	B-13	130		24	10.35	34.35	40
433	48" diameter 12 ga.		110		33	12.25	45.25	53
435	60" diameter 10 ga.		78		53	17.25	70.25	82
450	Corrugated Metal Pipe, end sections, 18" diameter	B-21	16	Ea.	75	42	117	140
455	30" diameter	"	12	"	150	57	207	245
465	Corrugated steel or alum. oval arch culverts, coated & paved							
470	17" x 11" 16 ga., 15" equivalent	B-22	200	L.F.	8.50	3.83	12.33	14.65
475	21" x 15" 16 ga., 18" equivalent		150		9.95	5.10	15.05	18
480	28" x 20" 14 ga., 24" equivalent		125		15.35	6.10	21.45	25
485	35" x 24" 14 ga., 30" equivalent		100		19.25	7.65	26.90	32
490	42" x 29" 12 ga., 36" equivalent	B-13	100		28	13.45	41.45	49
495	49" x 33" 12 ga., 42" equivalent		90		34	14.95	48.95	58
500	57" x 38" 12 ga., 48" equivalent		70		38	19.20	57.20	68
510	Corrugated steel or alum. oval arch culverts, plain							
513	17" x 13" 16 ga., 15" equivalent	B-22	225	L.F.	6.70	3.40	10.10	12.10
515	21" x 15" 16 ga., 18" equivalent		175		8.10	4.37	12.47	14.95
517	28" x 20" 14 ga., 24" equivalent		150		13	5.10	18.10	21
520	35" x 24" 14 ga., 30" equivalent		108		16.25	7.10	23.35	28
523	42" x 29" 12 ga., 36" equivalent	B-13	108		25	12.45	37.45	44

For expanded coverage of these items see *Means' Site Work Cost Data 1986*

2.5 Site Drainage & Utilities

		CREW	DAILY OUTPUT	UNIT	BARE COSTS MAT.	BARE COSTS INST.	BARE COSTS TOTAL	TOTAL INCL O&P
27 525	49" x 33" 12 ga., 42" equivalent	B-13	92	L.F.	30	14.60	44.60	53
527	57" x 38" 12 ga., 48" equivalent		75	"	33	17.95	50.95	61
580	End sections, 17" x 13"		22	Ea.	50	61	111	140
585	42" x 29"	↓	17	"	150	79	229	270
595	Multi-plate arch, steel	B-20	1,690	Lb.	.52	.30	.82	1
596								
605	Polyvinyl chloride pipe, schedule 40, 4" diameter	B-20	375	L.F.	3.15	1.34	4.49	5.40
615	6" diameter		350		5.56	1.44	7	8.20
625	8" diameter	↓	335		8.50	1.50	10	11.55
630	10" diameter	B-21	330		12.60	2.06	14.66	16.75
635	12" diameter		320		16.65	2.12	18.77	21
640	16" diameter	↓	190		33.90	3.57	37.47	42
650	Vitrified clay sewer pipe, premium joint, C-200, 4" diameter	B-20	265		1.65	1.90	3.55	4.57
655	4' and 5' lengths, 6" diameter	"	200		2.60	2.52	5.12	6.50
660	8" diameter	B-21	200		3.70	3.39	7.09	8.80
665	10" diameter		190		5.10	3.57	8.67	10.60
670	12" diameter		150		7.15	4.52	11.67	14.20
675	15" diameter		110		14	6.15	20.15	24
680	18" diameter		88		20	7.70	27.70	33
685	24" diameter	↓	45		44	15.05	59.05	70
690	30" diameter	B-22	31		48	25	73	87
695	36" diameter	"	20		70	38	108	130
705	3' lengths, add to above				30%	30%		
710	2' lengths, add to above				40%	60%		
715	For die cast slip joints compared to premium joints				85%	105%		
720	For plain joints compared to premium joints				75%	105%		
725	For standard strength compared to extra strength			↓	80%	90%		
29-001	**SUBDRAINAGE** Foundation underdrains							
005	Asbestos Cement, class 4000 underdrain, perforated							
010	4" diameter	B-20	390	L.F.	1.90	1.29	3.19	3.96
015	6" diameter	"	380		3	1.33	4.33	5.20
020	8" diameter	B-21	370		5	1.83	6.83	8.05
025	10" diameter		365		6.25	1.86	8.11	9.50
030	12" diameter	↓	360		8.40	1.88	10.28	11.90
035	Unperforated, 4" diameter	B-20	390		1.35	1.29	2.64	3.35
040	6" diameter	"	380		2.65	1.33	3.98	4.83
045	8" diameter	B-21	370		3.75	1.83	5.58	6.70
050	10" diameter		365		5	1.86	6.86	8.10
055	12" diameter	↓	360		6.25	1.88	8.13	9.50
060	Bituminous fiber, perforated underdrain, 3" diameter	2 Plum	400		1.10	.90	2	2.51
065	4" diameter		380		1.40	.95	2.35	2.91
070	5" diameter		360		2.40	1	3.40	4.09
075	6" diameter	↓	340	↓	2.65	1.06	3.71	4.45
080	Corrugated steel or aluminum perforated, asphalt coated							
085	6" diameter, 18 ga.	B-20	380	L.F.	2.50	1.33	3.83	4.67
090	8" diameter, 16 ga.	"	370		3.90	1.36	5.26	6.25
095	10" diameter, 16 ga.	B-21	360		4.90	1.88	6.78	8.05
100	12" diameter, 16 ga.		285		5.90	2.38	8.28	9.80
105	18" diameter, 16 ga.	↓	205	↓	8.90	3.31	12.21	14.45
110	Plain steel or aluminum, corrugated							
115	6" diameter, 18 ga.	B-20	380	L.F.	2.15	1.33	3.48	4.28
120	8" diameter, 16 ga.	"	370		3.30	1.36	4.66	5.60
125	10" diameter, 16 ga.	B-21	360		4.25	1.88	6.13	7.30
130	12" diameter, 16 ga.		285		5.25	2.38	7.63	9.10
135	18" diameter, 16 ga.	↓	205		7.25	3.31	10.56	12.60
140	Porous wall concrete underdrain, std. strength, 4" diameter	B-20	335		1.30	1.50	2.80	3.61
145	6" diameter	"	315		1.40	1.60	3	3.85
150	8" diameter	B-21	310		2.25	2.19	4.44	5.55
155	12" diameter	"	285	↓	4.80	2.38	7.18	8.60

For expanded coverage of these items see *Means' Site Work Cost Data 1986*

2.5 Site Drainage & Utilities

		CREW	DAILY OUTPUT	UNIT	BARE COSTS MAT.	BARE COSTS INST.	BARE COSTS TOTAL	TOTAL INCL O&P
29	160	15" diameter	B-21	230	L.F.	5.50	2.95	8.45	10.20
	165	18" diameter	"	165		7.20	4.11	11.31	13.70
	170	Extra strength, 6" diameter	B-20	315		1.70	1.60	3.30	4.18
	175	8" diameter	B-21	310		2.55	2.19	4.74	5.85
	180	10" diameter		285		5	2.38	7.38	8.85
	185	12" diameter		230		5.40	2.95	8.35	10.05
	190	15" diameter		200		6.10	3.39	9.49	11.45
	195	18" diameter	↓	165		8.90	4.11	13.01	15.55
	200	Vitrified clay, perforated, 2' lengths (C-211), 4" diameter	B-20	400		1.80	1.26	3.06	3.80
	205	6" diameter		315		3.05	1.60	4.65	5.65
	210	8" diameter	↓	290		5.25	1.74	6.99	8.30
	215	12" diameter	B-21	275		7.35	2.47	9.82	11.55
	220	Channel pipe, 4" diameter	B-20	430		1.30	1.17	2.47	3.13
	225	6" diameter		335		1.75	1.50	3.25	4.10
	230	8" diameter	↓	295		3.20	1.71	4.91	6
	235	12" diameter	B-21	280		6.20	2.42	8.62	10.20
	320	PVC, perforated, 4" diameter	B-20	430		.55	1.17	1.72	2.30
	330	6" diameter	"	400	↓	1.20	1.26	2.46	3.14
	30-001	**PIPING, WATER DISTRIBUTION SYSTEMS** Pipe laid in trench,							
	002	excavation and backfill not included							
	010	Asbestos cement pipe, 150 psi, C.L. lots, 4" diameter	B-20	300	L.F.	3.15	1.68	4.83	5.90
	015	6" diameter	"	290		4.65	1.74	6.39	7.65
	020	8" diameter	B-21	290		6.25	2.34	8.59	10.15
	025	10" diameter		270		9.30	2.51	11.81	13.75
	030	12" diameter		250		12.80	2.71	15.51	17.90
	035	14" diameter		200		15.90	3.39	19.29	22
	040	16" diameter		180		19	3.77	22.77	26
	045	18" diameter	↓	150		21	4.52	25.52	29
	060	For 100 psi pipe, 8" diameter, deduct				10%			
	070	16" diameter, deduct				15%			
	090	For 200 psi pipe, 8" diameter, add				15%			
	100	16" diameter, add			↓	20%			
	140	Ductile Iron pipe, class 250 water piping, 18' lengths							
	141	Mechanical joint, 4" diameter	B-20	144	L.F.	5	3.50	8.50	10.55
	142	6" diameter	"	126		5.50	4	9.50	11.85
	143	8" diameter	B-21	108		7.50	6.30	13.80	17.05
	144	10" diameter		90		10	7.55	17.55	22
	145	12" diameter		72		12.75	9.40	22.15	27
	146	14" diameter		54		16.50	12.55	29.05	36
	147	16" diameter	↓	46		19.20	14.75	33.95	42
	148	18" diameter	B-22	38		22	20	42	52
	149	24" diameter	"	36		33	21	54	66
	155	Tyton joint, 4" diameter	B-20	158		4.50	3.19	7.69	9.55
	156	6" diameter	"	138		5.10	3.65	8.75	10.90
	157	8" diameter	B-21	118		7.05	5.75	12.80	15.80
	158	10" diameter		100		9.50	6.80	16.30	19.95
	159	12" diameter		80		12	8.50	20.50	25
	160	14" diameter		60		15	11.30	26.30	32
	161	16" diameter	↓	54		17.75	12.55	30.30	37
	162	18" diameter	B-22	44		21	17.40	38.40	47
	163	20" diameter		42		23	18.25	41.25	51
	164	24" diameter	↓	40		29	19.15	48.15	58
	170	Copper tubing, type K, 20' joints, 1-1/2" diameter	B-20	360		2.55	1.40	3.95	4.83
	171	2" diameter		315		3.98	1.60	5.58	6.70
	172	2-1/2" diameter		220		5.70	2.29	7.99	9.60
	173	3" diameter	↓	200	↓	7.65	2.52	10.17	12.05
	175	Gate valves with boxes, cast iron							
	178	4" diameter	B-20	6	Ea.	250	84	334	395
	180	6" diameter	"	5		320	100	420	500
	182	8" diameter	B-21	4	↓	485	170	655	770

For expanded coverage of these items see *Means' Site Work Cost Data 1986*

2.5 Site Drainage & Utilities

		CREW	DAILY OUTPUT	UNIT	BARE COSTS MAT.	BARE COSTS INST.	BARE COSTS TOTAL	TOTAL INCL O&P
30 184	10" diameter	B-21	3.50	Ea.	750	195	945	1,100
186	12" diameter		3		900	225	1,125	1,300
188	14" diameter		2		2,125	340	2,465	2,825
190	16" diameter		2		2,850	340	3,190	3,600
195	Butterfly valves with boxes, cast iron							
197	4" diameter	B-20	6	Ea.	195	84	279	335
199	6" diameter	"	5		200	100	300	365
201	8" diameter	B-21	4		400	170	570	680
203	10" diameter		3.50		575	195	770	905
205	12" diameter		3		600	225	825	975
207	14" diameter		2		1,200	340	1,540	1,800
209	16" diameter		2		1,400	340	1,740	2,025
215	Hydrant, fire, no excavation or backfill included	2 Plum	1.70		790	210	1,000	1,175
220	Wall, bronze, non-freeze type, 3/4" diam. for 12" wall		11		127	33	160	185
225	For 18" wall		11		135	33	168	195
230	For 24" wall		11		142	33	175	205
235	Post type, non freeze, 4' depth of burr, 3/4" connection		4		120	90	210	260
236								
240	Polyvinyl chloride pipe, class 150, S.D.R.-18, 4" diameter	B-20	200	L.F.	1.80	2.52	4.32	5.65
245	6" diameter	"	180		3.84	2.80	6.64	8.30
250	8" diameter	B-21	160		4	4.24	8.24	10.35
255	10" diameter		140		4.80	4.85	9.65	12.05
260	12" diameter		100		11.65	6.80	18.45	22
265	Class 160, S.D.R.-26, 1-1/2" diameter	B-20	300		.32	1.68	2	2.78
270	2" diameter		250		.50	2.02	2.52	3.47
275	2-1/2" diameter		250		.74	2.02	2.76	3.73
280	3" diameter		200		1.10	2.52	3.62	4.86
285	4" diameter		200		1.81	2.52	4.33	5.65
290	6" diameter		180		3.84	2.80	6.64	8.30
295	8" diameter	B-21	160		4	4.24	8.24	10.35
800	Fittings, ductile iron, mechanical joint							
801	90° bend 4" diameter	B-20	37	Ea.	65	13.60	78.60	91
802	6" diameter		25		80	20	100	115
804	8" diameter		21		115	24	139	160
806	10" diameter	B-21	21		165	32	197	225
808	12" diameter		18		220	38	258	295
810	14" diameter		16		375	42	417	470
812	16" diameter		14		425	48	473	535
814	18" diameter		10		720	68	788	885
816	20" diameter		8		850	85	935	1,050
818	24" diameter		6		1,200	115	1,315	1,475
820	Wye or tee, 4" diameter	B-20	25		90	20	110	130
822	6" diameter		17		120	30	150	175
824	8" diameter		14		170	36	206	240
826	10" diameter	B-21	14		265	48	313	360
828	12" diameter		12		350	57	407	465
830	14" diameter		10		560	68	628	710
832	16" diameter		8		665	85	750	850
834	18" diameter		6		1,000	115	1,115	1,250
836	20" diameter		4		1,200	170	1,370	1,550
838	24" diameter		3		1,800	225	2,025	2,300
32-001	**SEPTIC TANKS** Not incl. excav. or piping, precast, 1,000 gallon	B-21	8	Ea.	300	85	385	450
010	2,000 gallon		5		580	135	715	830
020	5,000 gallon		1.70		2,375	400	2,775	3,175
030	15,000 gallon	B-13	1.30		8,100	1,025	9,125	10,300
040	25,000 gallon		.80		11,880	1,675	13,555	15,300
050	40,000 gallon		.60		17,890	2,250	20,140	22,700
060	Fiberglass, 1,000 gallon	B-21	6		385	115	500	580
070	1,500 gallon	"	4		490	170	660	775

For expanded coverage of these items see *Means' Site Work Cost Data 1986*

2.5 Site Drainage & Utilities

			DAILY		BARE COSTS			TOTAL
		CREW	OUTPUT	UNIT	MAT.	INST.	TOTAL	INCL O&P
32	100 Distribution boxes, concrete, 5 outlets	2 Plum	16	Ea.	30	23	53	66
	110 12 outlets	"	8		190	45	235	275
	115 Leaching field chambers, 13' x 3'-7" x 1'-4", standard	B-13	16		175	84	259	305
	120 Heavy duty, 8' x 4' x 1'-6"		14		140	96	236	285
	130 13' x 3'-9" x 1'-6"		12		220	110	330	395
	135 20' x 4' x 1'-6"	↓	5		480	270	750	890
	140 Leaching pit, precast concrete, 3' pit	B-21	8		190	85	275	330
	150 6' pit		4.70		240	145	385	465
	200 Velocity reducing pit, precast conc., 6' diameter, 3' deep	↓	4.70	↓	180	145	325	400
	201							
	220 Excavation for septic tank, 3/4 C.Y. backhoe	B-12F	145	C.Y.		4.47	4.47	5.65
	240 4' trench for disposal field, 3/4 C.Y. backhoe	"	335	L.F.		1.94	1.94	2.45
	260 Gravel fill, run of the bank	B-6	150	C.Y.	3.15	3.77	6.92	8.55
	280 Crushed stone, 3/4"	"	150	"	11	3.77	14.77	17.20
35-001	**SEWAGE PUMPING STATIONS** Prefabricated steel, concrete							
	002 or fiberglass, 200 GPM	C-17D	.17	Total	22,000	12,400	34,400	41,400
	020 1,000 GPM	"	.07	"	32,500	30,200	62,700	77,500
	021							
	050 Add for generator unit, 200 GPM, steel	C-17D	.34	Total	16,800	6,225	23,025	27,100
	060 Concrete		.51		10,500	4,150	14,650	17,300
	100 Add for generator unit, 1,000 GPM, steel		.30		17,850	7,050	24,900	29,400
	120 Concrete	↓	.38		15,000	5,550	20,550	24,200
	150 For wet well, if required, add	B-23	.50	↓	4,300	1,725	6,025	7,100
37-001	**SEWAGE TREATMENT** Not incl. fencing or external piping							
	002 Steel packaged, blown air aeration plants							
	010 1,000 GPD	C-17D	.69	Total	6,500	3,050	9,550	11,400
	020 5,000 GPD	"	.19		20,000	11,100	31,100	37,400
	030 15,000 GPD	C-14	.20		33,000	18,600	51,600	62,000
	040 30,000 GPD		.10		64,000	37,200	101,200	121,500
	050 50,000 GPD		.08		75,000	46,500	121,500	146,500
	060 100,000 GPD		.04		130,000	93,000	223,000	271,000
	070 200,000 GPD		.03		185,000	124,000	309,000	374,500
	080 500,000 GPD	↓	.01	↓	480,000	372,500	852,500	1,041,000
	100 Concrete, extended aeration, primary and secondary treatment							
	101 10,000 GPD	C-14	.10	Total	32,000	37,200	69,200	86,500
	110 30,000 GPD		.07		45,000	53,000	98,000	123,000
	120 50,000 GPD		.04		75,000	93,000	168,000	210,500
	140 100,000 GPD		.03		90,000	124,000	214,000	270,000
	150 500,000 GPD	↓	.01		420,000	372,500	792,500	975,000
	170 Municipal wastewater treatment facility							
	172 1.0 MGD			Total	1,160K	1,325K	2,485K	3,325K
	174 1.5 MGD				1,680K	2,000K	3,680K	4,900K
	176 2.0 MGD				1,995K	2,200K	4,195K	5,575K
	178 3.0 MGD				2,310K	2,625K	4,935K	6,550K
	180 5.0 MGD			↓	3,570K	3,875K	7,445K	9,925K
	200 Holding tank system, not incl. excavation or backfill							
	201 Recirculating chemical water closet	2 Plum	4	Ea.	495	90	585	675
	210 For voltage converter, add	"	16		130	23	153	175
	220 For high level alarm, add	1 Plum	7.80	↓	65	23	88	105
38-001	**UTILITY VAULTS** Precast concrete, 6" thick							
	005 5' x 10' x 6' high, I.D.	B-13	2	Ea.	1,410	675	2,085	2,450
	010 6' x 10' x 6' high, I.D.		2		1,480	675	2,155	2,525
	015 5' x 12' x 6' high, I.D.		2		1,650	675	2,325	2,725
	020 6' x 12' x 6' high, I.D.		1.80		1,740	745	2,485	2,925
	025 6' x 13' x 6' high, I.D.		1.50		2,100	895	2,995	3,525
	030 8' x 14' x 7' high, I.D.	↓	1	↓	2,400	1,350	3,750	4,450
	035 Hand hole, precast concrete, 1-1/2" thick							
	040 1'-4" x 2'-4" x 1'-3", I.D., light duty	B-1	4	Ea.	140	99	239	300
	045 4'-6" x 5'-10" x 2'-7", O.D., heavy duty	B-6	3	"	310	190	500	595

For expanded coverage of these items see *Means' Site Work Cost Data 1986*

2.5 Site Drainage & Utilities

		CREW	DAILY OUTPUT	UNIT	BARE COSTS MAT.	BARE COSTS INST.	BARE COSTS TOTAL	TOTAL INCL O&P
39-001	**WASTEWATER TREATMENT SYSTEM** Fiberglass, 1,000 gallon	B-21	1.29	Ea.	2,100	525	2,625	3,050
010	1,500 gallon	"	1.03	"	4,600	660	5,260	5,975
40-001	**WELLS** Domestic water, drilled and cased, including casing							
010	4" to 6" diameter	B-23	160	V.L.F.	7	5.40	12.40	15.10
020	8" diameter	"	127	"	8.75	6.80	15.55	18.90
040	Gravel pack well, 40' deep, incl. gravel & casing, complete							
050	24" diameter casing x 18" diameter screen	B-23	.13	Total	16,000	6,650	22,650	26,700
060	36" diameter casing x 18" diameter screen		.12	"	18,000	7,225	25,225	29,600
080	Observation wells, 1-1/4" riser pipe	↓	163	V.L.F.	8.75	5.30	14.05	16.85
090	For flush Buffalo roadway box, add	1 Skwk	16.60	Ea.	23	9.90	32.90	40
120	Test well, 2-1/2" diameter, up to 50' deep (15 to 50 GPM)	B-23	1.51	"	400	575	975	1,225
130	Over 50' deep, add	"	121.80	L.F.	11.50	7.10	18.60	22
150	Pumps, installed in wells to 100' deep, 4" submersible							
151	1/2 H.P.	B-21	1.80	Ea.	285	375	660	840
152	3/4 H.P.		1.65		330	410	740	940
160	1 H.P.		1.10		370	615	985	1,275
170	1-1/2 H.P.		1		445	680	1,125	1,450
180	2 H.P.		1		690	680	1,370	1,700
190	3 H.P.		.90		805	755	1,560	1,950
200	5 H.P.		.90		1,055	755	1,810	2,225
300	6" submersible, 25' to 150' deep, 25 H.P., 249 to 297 GPM	↓	.70	↓	3,285	970	4,255	4,975
310	25' to 500' deep, 30 H.P., 100 to 300 GPM		.50		4,110	1,350	5,460	6,425

2.6 Roads & Walks

		CREW	DAILY OUTPUT	UNIT	BARE COSTS MAT.	BARE COSTS INST.	BARE COSTS TOTAL	TOTAL INCL O&P
02-001	**ASPHALT BLOCKS** Premolded, 6"x12" x 1-1/4", w/bed & neopr. adhesive	D-1	135	S.F.	2.35	2.16	4.51	5.65
010	3" thick		130		3.40	2.25	5.65	6.95
030	Hexagonal tile, 8" wide, 1-1/4" thick		135		2.60	2.16	4.76	5.95
040	2" thick		130		3.20	2.25	5.45	6.70
050	Square, 8" x 8", 1-1/4" thick		135		2.60	2.16	4.76	5.95
060	2" thick	↓	130		3.20	2.25	5.45	6.70
090	For exposed aggregate (ground finish) add				.50		.50	.55M
091	For colors, add			↓	.30		.30	.33M
05-001	**BASE** Prepare and roll sub-base, small areas	B-32	670	S.Y.		1.38	1.38	1.69
010	Large areas	"	1,500			.62	.62	.76
07-001	**BASE COURSE** Crushed 3/4" stone @ $10.50 per ton, delivered, 3" deep	B-36	4,000		1.65	.31	1.96	2.21
010	6" deep		3,000		3.35	.42	3.77	4.21
020	9" deep		2,200		5	.57	5.57	6.20
030	12" deep	↓	1,800	↓	6.65	.69	7.34	8.20
035	Bank run gravel @ $4.50 per ton, delivered, spread to sub-grade							
037	6" deep	B-10B	3,000	S.Y.	1.37	.29	1.66	1.85
039	9" deep		2,200		2.05	.39	2.44	2.72
040	12" deep	↓	1,800	↓	2.73	.48	3.21	3.57
070	Liquid applications, emulsion	B-45	3,000	Gal.	1.05	.18	1.23	1.39
080	Prime and seal, cut back asphalt		3,000	"	1.20	.18	1.38	1.55
100	Macadam penetration, 2 gal. per S.Y., 4" thick		3,000	S.Y.	2.15	.18	2.33	2.60
110	6" thick, 3 gal. per S.Y.		2,000		3.25	.27	3.52	3.92
120	8" thick, 4 gal. per S.Y.	↓	1,500		4.35	.36	4.71	5.25
600	Stabilization fabric, polypropylene, 6 oz./S.Y.	B-6	10,000		1.35	.06	1.41	1.56
10-001	**BITUMINOUS** Paving, wearing course, @ $27.30 per ton, 1-1/2" thick	B-25	3,300		2.23	.63	2.86	3.29
010	3" thick		1,650		4.45	1.26	5.71	6.55
030	Binder course, @ $25.50 per ton, 1-1/2" thick		3,300		2.08	.63	2.71	3.13
040	2" thick		2,300		2.77	.91	3.68	4.25
050	3" thick		1,650		4.16	1.26	5.42	6.25
060	4" thick	↓	1,200	↓	5.54	1.74	7.28	8.40

For expanded coverage of these items see *Means' Site Work Cost Data 1986*

2.6 Roads & Walks

			CREW	DAILY OUTPUT	UNIT	BARE COSTS MAT.	BARE COSTS INST.	BARE COSTS TOTAL	TOTAL INCL O&P
10	080	Sand finish course, @ $28.50 per ton, 3/4" thick	B-25	5,700	S.Y.	1.16	.37	1.53	1.76
	090	1" thick	"	3,850		1.55	.54	2.09	2.42
	120	Gravel surfacing on asphalt, screened and rolled	B-36	10,000		.06	.12	.18	.22
	130	Plus sealer and binder	"	2,600		.25	.48	.73	.88
	140	Patching utility trenches, 2" thick asphalt	B-37	390		3.10	2.29	5.39	6.65
	150	3" thick asphalt	"	260		4.65	3.43	8.08	10
	300	Pavement overlay, polypropylene, 4 oz./S.Y., ideal conditions	B-63	10,000		.75	.08	.83	.93
	310	Adverse conditions	"	5,000	↓	.75	.15	.90	1.04
12-001		BITUMINOUS CONCRETE At the plant (145 lb. per C.F.)			Ton	27.30		27.30	30M
	020	All weather patching mix				28.50		28.50	31M
	030	Berm mix			↓	28.50		28.50	31M
15-001		BRICK PAVING 4" x 8" x 1-1/2", without joints (4.5 brick/S.F.)	D-1	110	S.F.	1.80	2.65	4.45	5.75
	010	Grouted, 3/8" joint (3.9 brick/S.F.)		90		1.60	3.24	4.84	6.40
	020	4" x 8" x 2-1/4", without joints (4.5 bricks/S.F.)		110		1.95	2.65	4.60	5.95
	030	Grouted, 3/8" joint (3.9 brick/S.F.)	↓	90	↓	1.70	3.24	4.94	6.50
17-001		CALCIUM CHLORIDE Delivered, 100 lb. bags, truckload lots			Ton	235		235	260M
	020	Solution, 4 lb. flake per gallon, tank truck delivery			Gal.	.48		.48	.52M
20-001		CONCRETE PAVING With mesh, not incl. base, joints or							
	002	finish, 4500 psi concrete, 6" thick	B-26	2,000	S.Y.	11.55	1.56	13.11	14.70
	010	8" thick		1,500		14.70	2.07	16.77	18.80
	020	9" thick		1,350		16.30	2.30	18.60	21
	030	10" thick		1,200		17.90	2.59	20.49	23
	040	12" thick		1,000		21	3.11	24.11	27
	050	15" thick	↓	750	↓	25.80	4.15	29.95	34
	051								
	070	Finishing, broom finish, add to above	2 Cefi	135	S.Y.		2.28	2.28	3.20L
	080	Belt dragged, add to above	"	285			1.08	1.08	1.52L
	100	Curing, with sprayed membrane	2 Clab	1,050	↓	.15	.24	.39	.52
	120	Expansion joints, see division 3.1-10							
	140	Reinforcing steel mats, #6 @ 8" O.C, one way	4 Rodm	210	S.Y.	5.60	3.31	8.91	11.30
	150	#9 @ 8" O.C., one way	"	160	"	12.70	4.35	17.05	21
	170	Coloring concrete, 5 sack mix, black, 1.8 lb. per sack, add			C.Y.	8.30		8.30	9.15M
	180	7.5 lb. per sack, add				35		35	39M
	190	Brown, red, yellow, 1.8 lb. per sack, add				9.90		9.90	10.90M
	200	9.4 lb. per sack, add				52		52	57M
	210	Green, 1.8 lb. per sack, add				29		29	32M
	220	9.4 lb. per sack, add			↓	150		150	165M
22-001		CURBS Bituminous, plain, 8" wide, 6" high, 50 L.F. per ton	B-27	1,000	L.F.	.57	.61	1.18	1.48
	010	8" wide, 8" high, 44 L.F. per ton		900		.66	.67	1.33	1.67
	015	Bituminous berm, 12" w., 3"to 6" 35 L.F./ton, before pavement		700		.82	.87	1.69	2.12
	020	12" w. 1-1/2"to 4" H.,60 L.F. per ton; laid with pavement	↓	1,050		.48	.58	1.06	1.34
	030	Concrete, 6" x 18", cast in place, straight	C-2	500		2.89	1.96	4.85	6
	040	6" x 18" radius	"	450		3.09	2.17	5.26	6.50
	045	Precast, 6" x 18", straight	B-29	350		4.70	4	8.70	10.55
	060	6" x 18" radius		325		7.10	4.31	11.41	13.60
	100	Granite, split face, straight, 5" x 16"		500		9.25	2.80	12.05	13.95
	110	6" x 18" (see also division 4.4-15)		450		12.65	3.11	15.76	18.10
	130	Radius curbing, 6" x 18", over 10' radius		260	↓	16.83	5.40	22.23	26
	140	Corners, 2' radius		80	Ea.	61	17.50	78.50	91
	160	Edging, 4-1/2" x 12", straight		300	L.F.	5.25	4.67	9.92	12.05
	180	Curb inlets, (guttermouth) straight	↓	41	Ea.	117	34	151	175
	185	Monolithic concrete curb and gutter, cast in place							
	186	with 6" high curb and 6" thick gutter							
	190	24" wide, .055 C.Y. per L.F.	C-2	375	L.F.	4.46	2.61	7.07	8.65
	195	30" wide, .066 C.Y. per L.F.	"	340	"	5.25	2.88	8.13	9.90
25-001		FILL Borrow, load, 1 mile haul, compact and shape							
	002	for embankments	B-15	1,200	C.Y.	3.50	1.41	4.91	5.55
	010	Select fill for shoulders & embankments	"	1,200	"	5.50	1.41	6.91	7.75
	020	For hauling over 1 mile, (see 2.3-30, also) add to above per C.Y.			Mile		.60		.76

For expanded coverage of these items see *Means' Site Work Cost Data 1986*

2.6 Roads & Walks

		CREW	DAILY OUTPUT	UNIT	BARE COSTS MAT.	BARE COSTS INST.	BARE COSTS TOTAL	TOTAL INCL O&P
27-001	**FINE GRADE** Area to be paved with grader, small area	B-11L	800	S.Y.		.89	.89	1.11
010	Large area	"	2,000	"		.36	.36	.44
30-001	**GUARDRAIL** Corrugated steel, galv. steel posts, 6'-3" O.C.	B-43	1,000	L.F.	8.25	1.38	9.63	10.90
020	End sections, galvanized, flared		175	Ea.	33	7.90	40.90	47
030	Wrap around end		175	"	43	7.90	50.90	58
040	Timber guardrail, 4" x 8" with 6" x 8" wood posts, treated		960	L.F.	6.20	1.44	7.64	8.70
060	Cable guardrail, 3 at 3/4" cables, steel posts	B-80	305		7.75	2.25	10	11.50
070	Wood posts		340		7.20	2.02	9.22	10.60
090	Guide rail, steel box beam, 6" x 6"		120		26	5.70	31.70	36
110	Median barrier, steel box beam, 6" x 8"		115		29	5.95	34.95	40
120	Impact barrier, barrel type	B-16	30	Ea.	180	27	207	235
140	Resilient guard fence & light shield, 6' high	B-2	130	L.F.	14	5	19	23
150	Concrete posts, individual, 6'-5", triangular	B-80	110	Ea.	17.50	6.25	23.75	27
155	Square	"	110	"	18	6.25	24.25	28
200	Precast median, 3'-6" high, single face, 1'-3" wide	B-29	380	L.F.	17.50	3.69	21.19	24
220	Double face, 2'-0" wide	"	340		20	4.12	24.12	28
32-001	**PAINTING** Lines on pavement, reflectorized white or yellow, 4" wide	B-78	20,000		.03	.05	.08	.10
020	6" wide		11,000		.04	.09	.13	.16
050	8" wide		10,000		.06	.09	.15	.19
060	12" wide		4,000		.09	.24	.33	.41
061								
062	Arrows or gore lines	B-78	2,300	S.F.	.20	.41	.61	.77
064	Temporary paint, white or yellow	"	15,000	L.F.	.03	.06	.09	.12
066	Removal	A-1	360		.36	.46	.82	1.02
068	Temporary tape, white		776		.15	.21	.36	.45
070	Yellow		679		.16	.24	.40	.51
071	Thermoplastic, white or yellow, 4" wide	B-79	15,000		.25	.06	.31	.35
073	6" wide		14,000		.37	.07	.44	.49
074	8" wide		12,000		.49	.08	.57	.64
075	12" wide		6,000		.73	.15	.88	1
076	Arrows		660	S.F.	1.10	1.39	2.49	3.02
077	Gore lines		2,500		.71	.37	1.08	1.26
078	Letters		660		1.10	1.39	2.49	3.02
079	Layout of pavement marking	A-2	25,000	L.F.		.02	.02	.02L
080	Parking stall, paint, white	B-78	860	Stall	.67	1.09	1.76	2.21
100	Street letters and numbers	"	1,600	S.F.	.21	.59	.80	1.02
35-001	**PARKING** Barriers, timber with saddles, treated type							
010	4" x 4" for cars	B-2	520	L.F.	2	1.25	3.25	4.02
020	6" x 6" for trucks		520	"	3.86	1.25	5.11	6.05
040	Folding with individual padlocks		50	Ea.	50	13.05	63.05	74
060	Flexible fixed stanchion, 2' high, 3" diameter		100	"	12.50	6.50	19	23
080	Parking lot control, see division 11.1-40							
100	Precast concrete, incl. dowels, 6" x 10" x 6'-0"	B-2	120	Ea.	18	5.45	23.45	28
110	8" x 13" x 6'-0"	"	120	"	21	5.45	26.45	31
37-001	**SEALCOATING** 2 coat tar pitch emulsion, over 10,000 S.Y.	B-36	10,000	S.Y.	.31	.12	.43	.50
010	Under 1000 S.Y.	B-1	1,050		.31	.38	.69	.89
030	Petroleum resistant, over 10,000 S.Y.	B-36	10,000		.45	.12	.57	.65
040	Under 1000 S.Y.	B-1	1,050		.45	.38	.83	1.04
060	Non-skid pavement renewal, over 10,000 S.Y.	B-36	10,000		.50	.12	.62	.71
070	under 1000 S.Y.	B-1	1,050		.50	.38	.88	1.10
080	Prepare and clean surface for above	A-2	8,545			.05	.05	.07
100	Hand seal bituminous curbing	B-1	4,420	L.F.	.19	.09	.28	.34
40-001	**SIDEWALKS** Bituminous, no base included, 2" thick	B-37	660	S.Y.	3.10	1.35	4.45	5.35
010	2-1/2" thick	"	520	"	3.88	1.72	5.60	6.70
015	Brick on 4" thick sand bed laid flat, 4.5 per S.F.	D-1	110	S.F.	1.88	2.65	4.53	5.85
020	Laid on edge, 7.2 per S.F. (see also 2.6-15)		70		2.92	4.17	7.09	9.15
025	For 4" thick concrete bed and joints, add		595		.72	.49	1.21	1.49
028	For steam cleaning, add	A-1	950		.05	.17	.22	.29
030	Concrete, 3000 psi, cast in place with 6 x 6 - #10/10 mesh,							
031	broomed finish, no base, 4" thick	B-24	600	S.F.	.83	.73	1.56	1.97

For expanded coverage of these items see *Means' Site Work Cost Data 1986*

2.6 Roads & Walks

			CREW	DAILY OUTPUT	UNIT	BARE COSTS MAT.	BARE COSTS INST.	BARE COSTS TOTAL	TOTAL INCL O&P
40	035	5" thick	B-24	545	S.F.	.99	.81	1.80	2.25
	040	6" thick	"	510		1.15	.86	2.01	2.51
	045	For bank run gravel base, 4" thick, add	B-6	2,500		.10	.23	.33	.42
	052	8" thick, add	"	1,600		.20	.35	.55	.70
	055	Exposed aggregate finish, add to above, minimum	D-1	1,875		.06	.16	.22	.29
	060	Maximum		455		.18	.64	.82	1.11
	065	Precast concrete patio blocks, 2" thick, natural		265		.56	1.10	1.66	2.19
	070	Green		265		.79	1.10	1.89	2.44
	075	Exposed local aggregate, natural		265		1.80	1.10	2.90	3.55
	080	Colors		265		2	1.10	3.10	3.77
	085	Exposed granite or limestone aggregate		265		1.80	1.10	2.90	3.55
	090	Exposed white tumblestone aggregate		265		1.85	1.10	2.95	3.60
	100	Crushed stone, 1" thick, white marble	2 Clab	1,700		.31	.15	.46	.56
	105	Bluestone	"	1,700		.07	.15	.22	.29
	110	Flagging, bluestone, irregular, 1" thick,	D-1	81		1.45	3.60	5.05	6.75
	115	Snapped random rectangular, 1" thick		92		2.10	3.17	5.27	6.85
	120	1-1/2" thick		85		2.50	3.44	5.94	7.65
	125	2" thick		83		2.85	3.52	6.37	8.15
	130	Slate, natural cleft, irregular, 3/4" thick		92		1.30	3.17	4.47	5.95
	135	Random rectangular, gauged, 1/2" thick		105		2.90	2.78	5.68	7.15
	140	Random rectangular, butt joint, gauged, 1/4" thick		150		3.10	1.95	5.05	6.20
	150	For interior setting, add				25%	25%		
	155	Granite blocks, 3-1/2" x 3-1/2" x 3-1/2"	D-1	92	S.F.	4	3.17	7.17	8.90
	160	4" to 12" long, 3" to 5" wide, 3" to 5" thick		98		3.50	2.98	6.48	8.10
	165	6" to 15" long, 3" to 6" wide, 3" to 5" thick		105		1.46	2.78	4.24	5.55
	165								
	170	Redwood, prefabricated, 4' x 4' sections	F-2	316	S.F.	2.16	1.07	3.23	3.90
	175	Redwood planks, 1" thick, on sleepers	"	240	"	1.42	1.41	2.83	3.57
42-001		SIGNS Stock. Stop signs, 24" x 24", no posts, .080" alum., reflectorized	B-80	70	Ea.	17	9.80	26.80	32
	010	High intensity		70		32	9.80	41.80	48
	030	30" x 30", reflectorized		70		27	9.80	36.80	43
	040	High intensity		70		46	9.80	55.80	64
	060	Miscellaneous directional signs, 12" x 18", reflectorized		70		7	9.80	16.80	21
	070	High intensity		70		12	9.80	21.80	26
	090	18" x 24", stock signs, reflectorized		70		18	9.80	27.80	33
	100	High intensity		70		28	9.80	37.80	44
	120	24" x 24", stock signs, reflectorized		70		18	9.80	27.80	33
	130	High intensity		70		32	9.80	41.80	48
	150	Add to above for steel posts, galvanized, 10'-0" upright, bolted		200		22	3.43	25.43	29
	160	12'-0" upright, bolted		140		25	4.89	29.89	34
	180	Highway insruct. road signs, extruded, over 20 S. F., reflectorized		350	S.F.	9	1.96	10.96	12.50
	200	High intensity		350		12	1.96	13.96	15.80
	220	Highway, suspended over road, 80 S.F. min., reflectorized		165		12	4.15	16.15	18.65
	230	High intensity		165		16	4.15	20.15	23
45-001		STEPS Including excavation, borrow and concrete base, where applicable							
	010	Bricks	D-1	23	LF Riser	18	12.70	30.70	38
	020	Railroad ties	2 Clab	25		6.65	10.20	16.85	22
	030	Bluestone treads, 12" x 2" or 12" x 1-1/2"	B-28	30		10.85	14.90	25.75	34
	050	Concrete, cast in place, see division 3.1-14-680							
	060	Precast concrete, see division 3.4-50							
47-001		TERRACES Compared to sidewalks, deduct			S.F.			10%	10%
50-001		TRAFFIC SIGNALS Mid block pedestrian crosswalk,							
	002	with pushbutton and mast arms	R-2	.30	Total	9,500	4,400	13,900	16,600
	010	Intersection, 8 signals (2 each direction), programmed	"	.15		20,000	8,800	28,800	34,300
	012	For each additional traffic phase controller, add	L-9	1.20		1,200	540	1,740	2,125
	020	Semi-actuated, detectors in side street only, add		.81		2,000	800	2,800	3,375
	030	Fully-actuated, detectors in all streets, add		.49		4,000	1,325	5,325	6,350
	040	For pedestrian pushbutton, add		.70		3,200	925	4,125	4,900
	050	Optically programmed signal only, add		1.64		1,400	395	1,795	2,125

For expanded coverage of these items see *Means' Site Work Cost Data 1986*

2.6 Roads & Walks

		CREW	DAILY OUTPUT	UNIT	BARE COSTS MAT.	INST.	TOTAL	TOTAL INCL O&P
50	060 School flashing system, programmed	L-9	.41	Signal	5,500	1,575	7,075	8,400

2.7 Site Improvements

		CREW	DAILY OUTPUT	UNIT	BARE COSTS MAT.	INST.	TOTAL	TOTAL INCL O&P
02-001	**BACKSTOPS** Baseball, prefabricated, 30' wide, 12' high & 1 overhang	B-1	2	Ea.	1,150	200	1,350	1,550
010	40' wide, 12' high & 2 overhangs		1.50		1,550	265	1,815	2,100
030	Basketball, steel, single pole		8		420	50	470	535
040	Double pole		6		630	66	696	790
060	Tennis, wire mesh with ends, pair of ends		4	Set	575	99	674	775
070	Enclosed court		1.30	Ea.	3,300	305	3,605	4,075
090	Handball or squash court, outdoor, wood	2 Carp	.50		2,175	640	2,815	3,325
100	Masonry	D-1	.30		915	975	1,890	2,400
05-001	**BENCHES** Park, precast concrete pedestals, w/backs, wood rails, 4' long	2 Clab	5		200	51	251	295
010	8' long		4		340	64	404	465
030	Fiberglass, backless, one piece, 4' long		10		275	25	300	340
040	8' long		7		575	36	611	685
050	Steel barstock pedestals with backs, 2" x 3" wood rails, 4' long		10		460	25	485	545
051	8' long		7		550	36	586	660
052	3" x 8" wood plank, 4' long		10		510	25	535	600
053	8' long		7		595	36	631	705
054	Backless, 4" x 4" wood plank, 4' square		10		495	25	520	580
055	8' long		7		475	36	511	575
060	Aluminum pedestals, with backs, aluminum slats, 8' long		8		125	32	157	185
061	15' long		5		210	51	261	305
062	Portable, aluminum slats, 8' long		8		150	32	182	210
063	15' long		5		245	51	296	345
080	Cast iron pedestals, back & arms, wood slats, 4' long		8		415	32	447	505
082	8' long		5		695	51	746	840
084	Backless, wood slats, 4' long		8		355	32	387	435
086	8' long		5		595	51	646	730
170	Steel frame, fir seat, 10' long		10		115	25	140	165
07-001	**BLEACHERS** Outdoor, portable, 3 to 5 tiers, up to 300' long, minimum	2 Sswk	120	Seat	7.50	2.89	10.39	12.80
010	Maximum, less than 15' long, prefabricated		80		31	4.34	35.34	41
020	6 to 20 tiers, minimum, up to 300' long		120		11.75	2.89	14.64	17.45
030	Max., under 15', (highly prefabricated, on wheels)		800		63	.43	63.43	70
050	Permanent grandstands, wood seat, steel frame, 24" row							
060	3 to 15 tiers, minimum	2 Sswk	60	Seat	14.50	5.80	20.30	25
070	Maximum		48		25	7.25	32.25	39
090	16 to 30 tiers, minimum		60		25	5.80	30.80	37
095	Average		55		27	6.30	33.30	40
100	Maximum		48		31	7.25	38.25	45
120	Seat backs only, 30" row, fiberglass		160		13.50	2.17	15.67	18.25
130	Steel and wood		160		16.75	2.17	18.92	22
150	Seat renewal, cover with fiberglass on wood, with backs	2 Carp	100		16.75	3.20	19.95	23
160	Plain bench, no backs	"	200		7.50	1.60	9.10	10.55
09-001	**FENCE, CHAIN LINK** Industrial 6' high plus 3 strands							
002	barbed wire, 2" line post @ 10' O.C., 1-5/8" top rail							
020	9 ga. wire, galv. steel	B-1	250	L.F.	4.75	1.59	6.34	7.55
030	Aluminized steel		250		5.95	1.59	7.54	8.85
050	6 ga. wire, galv. steel		250		7.50	1.59	9.09	10.55
060	Aluminized steel		250		8.75	1.59	10.34	11.95
080	6 ga. wire, 6' high but omit barbed wire, galv. steel		260		7.15	1.53	8.68	10.10
090	Aluminized steel		260		8.50	1.53	10.03	11.55

For expanded coverage of these items see *Means' Site Work Cost Data 1986*

2.7 Site Improvements

				Daily		Bare Costs			Total
			Crew	Output	Unit	Mat.	Inst.	Total	Incl O&P
09	110	Add for corner posts, 3" diam., galv. steel	B-1	40	Ea.	39	9.95	48.95	57
	120	Aluminized steel		40		48	9.95	57.95	67
	130	Add for braces, galv. steel		80		10.50	4.97	15.47	18.75
	135	Aluminized steel		80		13	4.97	17.97	22
	140	Gate for 6' high fence, 1-5/8" frame, 3' wide, galv. steel		35		58	11.35	69.35	80
	150	Aluminized steel		35		70	11.35	81.35	93
	200	5'-0" high fence, 9 ga., no barbed wire, 2" line post,							
	201	10' O.C., 1-5/8" top rail							
	210	Galvanized steel	B-1	315	L.F.	4.25	1.26	5.51	6.50
	220	Aluminized steel		315	"	5.20	1.26	6.46	7.55
	240	Gate, 4' wide, 5' high, 2" frame, galv. steel		60	Ea.	72	6.65	78.65	89
	250	Aluminized steel		60	"	77	6.65	83.65	94
	270	Motor operator for gates, not including gates or							
	271	electrical wiring, for swinging gate 15' wide	B-1	1	Opng.	1,250	400	1,650	1,950
	280	For swinging gate up to 30' wide (pair)		1		2,250	400	2,650	3,050
	290	For sliding gate up to 45' long (pair)		1		2,200	400	2,600	3,000
	310	Overhead slide gate, chain link, 6' high, to 18' wide		22	L.F.	39	18.05	57.05	69
	311	Cantilever type		28	"	26	14.20	40.20	49
10	001	**FENCE, CHAIN LINK, RESIDENTIAL** 11 ga. wire, 1-5/8" line							
	331	post @ 10' O.C., 1-3/8" top rail,							
	335	4' high, galvanized steel	B-1	400	L.F.	3.15	.99	4.14	4.91
	340	Aluminized steel		400	"	3.70	.99	4.69	5.50
	360	Gate, 3' wide, 1-3/8" frame, galv. steel		60	Ea.	34	6.65	40.65	47
	370	Aluminized steel		60	"	42	6.65	48.65	56
	390	3' high, galvanized steel		500	L.F.	2.60	.80	3.40	4.01
	400	Aluminized steel		500	"	3.15	.80	3.95	4.62
	420	Gate, 3' wide, 1-3/8" frame, galv. steel		70	Ea.	26	5.70	31.70	37
	430	Aluminized steel		70	"	31.50	5.70	37.20	43
	440	Tennis courts, 11 ga. wire, 1-3/4" mesh, 2-1/2" line posts,							
	445	1-5/8" top rail							
	451	10' high, galvanized steel	B-1	155	L.F.	7.25	2.57	9.82	11.70
	455	Vinyl covered wire and pipe		155		8.05	2.57	10.62	12.55
	460	12' high, galvanized steel		130		8.50	3.06	11.56	13.80
	480	Vinyl covered wire and pipe		130		9.50	3.06	12.56	14.90
	500	Corner posts for above, 3" diameter, 10' high		30	Ea.	80	13.25	93.25	105
	520	12' high		30	"	95	13.25	108.25	125
11	001	**FENCE, MISC. METAL** Chicken wire, posts @ 4', 1" mesh, 4' high		410	L.F.	.83	.97	1.80	2.32
	010	2" mesh, 6' high		350		.72	1.14	1.86	2.44
	020	Galv. steel, 12 ga., 2" x 4" mesh, posts 5' O.C., 3' high		300		1.05	1.33	2.38	3.08
	030	5' high		300		1.55	1.33	2.88	3.63
	040	14 ga., 1" x 2" mesh, 3' high		300		.93	1.33	2.26	2.94
	050	5' high		300		1.39	1.33	2.72	3.45
	100	Kennel fencing, 1-1/2" mesh, 3'-6" wide, 6'-2" high, 6' long	2 Clab	4	Ea.	190	64	254	300
	105	12' long		4		240	64	304	355
	120	Top covers, 1-1/2" mesh, 6' long		15		36	16.95	52.95	64
	125	14' long		12		65	21	86	100
	130	For kennel doors, see division 8.3-36							
	131								
	450	Security fence, prison grade, set in concrete, 12' high	B-1	25	L.F.	15.50	15.90	31.40	40
	460	16' high		20		19.60	19.90	39.50	50
	480	Snow fence on steel posts 10' O.C., 4' high		260		.75	1.53	2.28	3.04
	481								
	530	Tubular picket, steel, 6' sections, 1-9/16" posts, 4' high	B-1	170	L.F.	12.25	2.34	14.59	16.85
	540	2" posts, 5' high		140		16.75	2.84	19.59	23
	560	2" posts, 6' high		115		17.95	3.46	21.41	25
	570	Staggered picket 1-9/16" posts, 4' high		170		10.95	2.34	13.29	15.45
	580	2" posts, 5' high		140		17.75	2.84	20.59	24
	590	2" posts, 6' high		115		18.95	3.46	22.41	26
	620	Gates, 4' high, 3' wide		20	Ea.	95	19.90	114.90	135
	630	5' high, 3' wide		14	"	125	28	153	180

2.7 Site Improvements

		CREW	DAILY OUTPUT	UNIT	BARE COSTS MAT.	BARE COSTS INST.	BARE COSTS TOTAL	TOTAL INCL O&P
11 640	6' high, 3' wide	B-1	14	Ea.	137	28	165	190
650	4' wide		10	"	159	40	199	235
12-001	**FENCE, WOOD** Picket, No. 2 cedar picket, Gothic, 2 rail, 3' high		160	L.F.	3.20	2.49	5.69	7.10
005	Gate, 3'-6" wide		45	Ea.	28	8.85	36.85	44
040	3 rail, 4' high		150	L.F.	3.50	2.65	6.15	7.70
050	Gate, 3-1/2' wide		41	Ea.	34	9.70	43.70	51
060	Open rail, rustic, No. 1 cedar, 2 rail, 3' high		160	L.F.	2.50	2.49	4.99	6.35
065	Gate, 3' wide		45	Ea.	32	8.85	40.85	48
070	3 rail, 4' high		150	L.F.	3.10	2.65	5.75	7.25
090	Gate, 3' wide		47	Ea.	43	8.45	51.45	60
120	Stockade, No. 2 cedar, treated wood rails, 6' high		160	L.F.	4	2.49	6.49	8
125	Gate, 3' wide		47	Ea.	33	8.45	41.45	49
130	No. 1 cedar, 3-1/4" cedar rails, 6' high		150	L.F.	9.75	2.65	12.40	14.55
150	Gate, 3' wide		47	Ea.	80	8.45	88.45	100
152	Open rail, split, No. 1 cedar, 2 rail, 3' high		160	L.F.	2.85	2.49	5.34	6.75
154	3 rail, 4'-0" high		150		3.75	2.65	6.40	7.95
330	Board, shadow box, 1" x 6", treated pine, 6' high		160		6	2.49	8.49	10.20
340	No. 1 cedar, 6' high		150		11.40	2.65	14.05	16.40
390	Basket weave, No. 1 cedar, 6' high		160		12.75	2.49	15.24	17.65
395	Gate, 3'-6" wide		34	Ea.	80	11.70	91.70	105
400	Treated pine, 6' high		150	L.F.	6.50	2.65	9.15	11
420	Gate, 3'-6" wide		47	Ea.	38	8.45	46.45	54
15-001	**FOUNTAINS** Incl. fiberglass pools, pumps, piping and lights							
020	4' diameter pool, 18" diameter spray ring	Q-1	2	Ea.	550	160	710	840
030	6' diameter pool, 24" diameter spray ring		1.50		905	215	1,120	1,300
040	7.5' diameter pool, 48" diameter spray ring		1		1,365	325	1,690	1,975
050	Rain curtains, 3' rain bar, 2' x 4' x 1' pool		2		520	160	680	805
060	7' rain bar, 2' x 8' x 1' pool		1		1,130	325	1,455	1,725
20-001	**GOAL POSTS** Steel, football, double post	B-1	1.50	Pr.	700	265	965	1,150
010	Deluxe, single post		1.50		980	265	1,245	1,450
030	Football, convertable to soccer		1.50		1,080	265	1,345	1,575
050	Soccer, regulation		2		980	200	1,180	1,375
25-001	**PLANTER BLOCKS** Precast concrete, interlocking							
002	"V" blocks for retaining soil	D-1	205	S.F.	2.60	1.42	4.02	4.89
27-001	**PLANTERS** Concrete, sandblasted, precast, 48" diameter, 24" high	2 Clab	15	Ea.	100	16.95	116.95	165
010	Fluted, precast, 7' diameter, 36" high		10		1,700	25	1,725	1,900
030	Fiberglass, circular, 36" diameter, 24" high		15		450	16.95	466.95	520
040	60" diameter, 24" high		10		800	25	825	915
060	Square, 24" side, 36" high		15		550	16.95	566.95	630
070	48" side, 36" high		15		1,050	16.95	1,066.95	1,175
090	Planter/bench, 72" square, 36" high		5		2,600	51	2,651	2,925
100	96" square, 27" high		5		4,300	51	4,351	4,800
120	Wood, square, 48" side, 24" high		15		550	16.95	566.95	630
130	Circular, 48" diameter, 30" high		10		650	25	675	750
150	72" diameter, 30" high		10		1,050	25	1,075	1,200
160	Planter/bench, 72"		5		1,600	51	1,651	1,825
30-001	**PLAYGROUND EQUIPMENT** See also individual items							
020	Bike rack, 10' long, permanent	B-1	12	Ea.	215	33	248	285
040	Horizontal monkey ladder, 14' long		4		380	99	479	560
050	Parallel bars, 10' long		4		190	99	289	355
060	Posts, tether ball set, 2-3/8" O.D.		8		90	50	140	170
080	Poles, multiple purpose, 10'-6" long		4	Pr.	95	99	194	250
100	Ground socket for movable posts, 2-3/8" post		4		40	99	139	190
110	3-1/2" post		4		65	99	164	215
130	See-saw, steel, 2 units		6	Ea.	310	66	376	435
140	4 units		4		580	99	679	780
150	6 units		3		850	135	985	1,125
170	Shelter, fiberglass golf tee, 3 person, turnable		4.60		800	86	886	1,000
190	Slides, stainless steel bed, 12' long, 6' high		3		560	135	695	810
200	20' long, 10' high		2		925	200	1,125	1,300

For expanded coverage of these items see Means' Site Work Cost Data 1986

2.7 Site Improvements

						Bare Costs			Total
			Crew	Daily Output	Unit	Mat.	Inst.	Total	Incl O&P
30	220	Swings, 8' high, plain seats, 4 seats	B-1	2	Ea.	405	200	605	735
	230	8 seats		1.30		730	305	1,035	1,250
	250	12' high, 4 seats		2		650	200	850	1,000
	260	8 seats		1.30		1,030	305	1,335	1,575
	280	Whirlers, 8' diameter		3		950	135	1,085	1,225
	290	10' diameter		3		1,050	135	1,185	1,350
32-001		**RETAINING WALLS** Asbestos bonded steel bin, excavation							
	002	and backfill not included, 10' wide							
	010	4' high, design A, 5.5' deep	E-2	650	S.F.	14	3.05	17.05	19.55
	020	8' high, design A, 5.5' deep		615		15	3.22	18.22	21
	030	10' high, design B, 7.7' deep		580		17.50	3.42	20.92	24
	040	12' high, design B, 7.7' deep		530		18	3.74	21.74	25
	050	16' high, design B, 7.7' deep		515		20	3.85	23.85	27
	060	16' high, design C, 9.9' deep		500		23	3.96	26.96	31
	070	20' high, design C, 9.9' deep		470		24	4.22	28.22	32
	080	20' high, design D, 12.1' deep		460		25	4.31	29.31	33
	090	24' high, design D, 12.1' deep		455		26	4.36	30.36	35
	100	24' high, design E, 14.3' deep		450		28	4.41	32.41	37
	110	28' high, design E, 14.3' deep		440		29	4.51	33.51	38
	130	For plain galvanized bin type walls, deduct				12%			
	180	Gravity concrete with vertical face including							
	185	Excavation and backfill, no reinforcing							
	190	6' high, no surcharge (47)	C-17C	36	L.F.	37	56	93	120
	200	33° surcharge		32		40	63	103	130
	220	8' high, no surcharge		27		51	74	125	160
	230	33° surcharge		24		65	84	149	190
	250	10' high, no surcharge		19		75	105	180	230
	260	33° surcharge		18		94	110	204	260
	280	Reinforced concrete cantilever, incl. excavation, backfill & reinf.							
	290	6' high, 33° surcharge	C-17C	35	L.F.	34	57	91	120
	300	8' high, 33° surcharge		29		42	69	111	145
	310	10' high, 33° surcharge		20		59	100	159	205
	320	20' high, 500 lb. per L.F. surcharge		7.50		180	270	450	570
	350	Concrete cribbing, incl. excavation and backfill							
	370	12' high, open face	B-13	210	S.F.	14.30	6.40	20.70	24
	390	Closed face	"	210	"	14.85	6.40	21.25	25
	410	Concrete filled slurry trench, see division 2.3-45							
	420								
	430	Stone filled gabions, not incl. excavation,							
	431	stone @ $6.50 per ton delivered, 3' deep							
	435	Galvanized, 6' high, 33° surcharge	B-13	49	L.F.	16.50	27	43.50	55
	450	Highway surcharge		27		33	50	83	105
	460	9' high, up to 33° surcharge		24		37	56	93	115
	470	Highway surcharge		16		58	84	142	175
	490	12' high, up to 33° surcharge		14		58	96	154	195
	500	Highway surcharge		11		82	120	202	255
	595	For PVC coating, add							20%
	650	For higher walls, add components as necessary							
35-001		**RUNNING TRACK** Gravel & cinders over stone base	B-36	127	S.Y.	3.10	9.85	12.95	15.75
	010	Rubber-vermiculite asphalt pavement, 1" thick	B-25	1,350		3.25	1.54	4.79	5.65
	030	Colored rubberized asphalt	"	1,350		5.60	1.54	7.14	8.20
	040	Artificial resilient mat over asphalt	B-2	101		15	6.45	21.45	26
37-001		**SPRINKLER SYSTEM** For lawns							
	010	Golf courses with fully automatic system	C-17	.05	9 Holes	65,000	33,400	98,400	120,500
	020	24' diam. head at 15' O.C incl. piping, minimum	B-20	70	Head	13.75	7.20	20.95	26
	030	Maximum		40		32	12.60	44.60	53
	050	60' diameter head, automatic operation, minimum		28		44	18	62	74
	060	Maximum		23		120	22	142	165

2.7 Site Improvements

			CREW	DAILY OUTPUT	UNIT	BARE COSTS MAT.	BARE COSTS INST.	BARE COSTS TOTAL	TOTAL INCL O&P
37	080	Residential system, custom, 1" supply	B-20	2,619	S.F.Grnd	.20	.19	.39	.50
	090	1-1/2" supply	"	2,311	"	.18	.22	.40	.51
40-001		**STONE WALLS** Including excavation, concrete footing and							
	002	stone 3' below grade. Price is exposed face area.							
	020	Decorative random stone, to 6' high, 1'-6" thick, dry set	D-1	35	S.F.	6.25	8.35	14.60	18.75
	030	Mortar set		40		7.35	7.30	14.65	18.50
	050	Cut stone, to 6' high, 1'-6" thick, dry set		35		9.95	8.35	18.30	23
	060	Mortar set		40		10.50	7.30	17.80	22
	080	Retaining wall, random stone, 6' to 10' high, 2' thick, dry set		45		7.75	6.50	14.25	17.75
	090	Mortar set		50		9.25	5.85	15.10	18.50
	110	Cut stone, 6' to 10' high, 2' thick, dry set		45		12.70	6.50	19.20	23
	120	Mortar set	↓	50	↓	13.45	5.85	19.30	23
42-001		**TENNIS COURTS** Bituminous pvmt. incl. fill, 2-1/2" thick, single court	B-37	450	S.Y.	5.35	1.98	7.33	8.70
	020	Double court	"	825		5.35	1.08	6.43	7.40
	030	Clay courts	B-1	160		5.25	2.49	7.74	9.40
	040	Pulverized natural greenstone with 4" base, fast dry		68		7.35	5.85	13.20	16.55
	080	Rubber-acrylic base resilient pavement	↓	275		5.25	1.45	6.70	7.85
	100	Colored sealer, acrylic emulsion, 3 coats	2 Clab	540		1.58	.47	2.05	2.42
	110	2 color seal coating	"	400		1.58	.64	2.22	2.66
	120	For preparing old courts, add	1 Clab	825	↓		.15	.15	.22L
	140	Posts for nets, 3-1/2" diameter with eye bolts	B-1	3.40	Pr.	130	115	245	315
	150	With pulley & reel		3.40	"	150	115	265	335
	170	Net, 42' long, nylon thread with binder		50	Ea.	160	7.95	167.95	190
	180	All metal	↓	6.50	"	280	61	341	395
	200	Paint markings on asphalt, 2 coats	1 Pord	2.50	Court	20	62	82	110
	201								
	220	Complete court with fence, etc., bituminous, minimum	B-1	.09	Court	8,400	4,425	12,825	15,600
	230	Maximum		.07		16,380	5,675	22,055	26,300
	280	Clay courts, minimum		.09		8,610	4,425	13,035	15,900
	290	Maximum	↓	.07	↓	13,650	5,675	19,325	23,200
43-001		**PLATFORM PADDLE TENNIS COURTS** Complete with lighting, etc.							
	010	Aluminum slat deck with aluminum frame	B-1	.08	Court	18,900	4,975	23,875	28,000
	050	Aluminum slat deck & wood frame		.08		19,950	4,975	24,925	29,200
	080	Aluminum deck heater, add		1.18		2,625	335	2,960	3,375
	090	Douglas fir planking & wood frame 2" x 6" x 30'		.08		15,750	4,975	20,725	24,500
	100	Plywood deck with steel frame		.08		18,900	4,975	23,875	28,000
	110	Steel slat deck with wood frame	↓	.08	↓	19,950	4,975	24,925	29,200
47-001		**TURF, ARTIFICIAL** Not including asphalt base or drainage, but							
	002	including cushion pad, over 50,000 S.F.							
	020	1/2" pile and 5/16" cushion pad, standard	C-17	3,200	S.F.	4.75	.52	5.27	6
	030	Deluxe		2,560		5.50	.65	6.15	7
	050	1/2" pile and 5/8" cushion pad, standard		2,844		6.95	.59	7.54	8.50
	060	Deluxe	↓	2,327	↓	7.75	.72	8.47	9.55
	080	For bituminous asphalt base, 2-1/2" thick,							
	090	with 6" crushed stone sub-base, add	B-25	16,000	S.F.	.60	.13	.73	.83
50-001		**TRASH CLOSURES** Steel with pullover cover							
	002	2'-3" wide, 4'-7" high, 6'-2" long	2 Clab	5	Ea.	630	51	681	765
	010	10'-1" long		4		840	64	904	1,025
	030	Wood, 10' wide, 6' high, 10' long		1.20		680	210	890	1,050
55-001		**TRASH RECEPTACLES** Fiberglass, 2' square, 18" high		30		190	8.50	198.50	220
	010	2' square, 2'-6" high		30		250	8.50	258.50	285
	030	Circular, 2' diameter, 18" high		30		165	8.50	173.50	195
	040	2' diameter, 2'-6" high	↓	30	↓	205	8.50	213.50	240

For expanded coverage of these items see Means' Site Work Cost Data 1986

2.8 Lawns & Planting

		CREW	DAILY OUTPUT	UNIT	BARE COSTS MAT.	BARE COSTS INST.	BARE COSTS TOTAL	TOTAL INCL O&P
05-001	EDGING Redwood, untreated, 1" x 4"	F-2	500	L.F.	.63	.67	1.30	1.66
010	2" x 4"		330		.62	1.02	1.64	2.15
020	Steel edge strips, 1/4" x 5" including stakes		330		2.29	1.02	3.31	3.98
030	3/16" x 4"		330		1.70	1.02	2.72	3.33
050	Brick edging, set on edge, 3 brick per L.F.	D-1	135		1.25	2.16	3.41	4.46
060	Set flat, 1-1/2 brick per L.F.	"	370		.63	.79	1.42	1.82
07-001	EROSION CONTROL Jute mesh, 100 S.Y. per roll, 4' wide, stapled	B-1	2,500	S.Y.	.59	.16	.75	.88
010	Plastic netting, stapled, 2" x 1" mesh, 20 mil		2,500		.32	.16	.48	.58
020	Polypropylene mesh, stapled, 6.5 oz/S.Y.		2,500		1.55	.16	1.71	1.94
030	Tobacco netting, #2, stapled		2,500		.03	.16	.19	.26
10-001	GROUND COVER Plants, pachysandra, in prepared beds		10	C	10	40	50	69
020	Vinca minor		10	"	39	40	79	100
060	Stone chips, in 50 lb. bags, Georgia marble		520	Bag	2.15	.76	2.91	3.47
070	Onyx gemstone		260		8.25	1.53	9.78	11.30
080	Quartz		260		3.60	1.53	5.13	6.20
090	Pea gravel, truckload lots		28	C.Y.	13.05	14.20	27.25	35
15-001	HEDGE PLANTS Barberry, 2' to 3' high		57	Ea.	21	7	28	33
020	Privet, 18" to 24" high		80		2	4.97	6.97	9.40
040	Boxwood, 18" to 20" high		80		20	4.97	24.97	29
060	Ilex, 15" to 18" high		96		12.50	4.14	16.64	19.75
25-001	LOAM OR TOPSOIL Remove and stockpile on site							
002	200 H.P. dozer, 6" deep, 200' haul	B-10B	865	C.Y.		1	1	1.19
010	300' haul		520			1.66	1.66	1.98
015	500' haul		225			3.84	3.84	4.57
020	Alternate method: 6" deep, 200' haul		5,090	S.Y.		.17	.17	.20
025	500' haul		1,325	"		.65	.65	.78
040	Spread from pile to rough finish grade, with 1.5 C.Y. F.E. loader	B-10S	200	C.Y.		2.50	2.50	3.15
050	200' by hand	1 Clab	14			9.10	9.10	13.15L
060	Top dress by hand, 1 C.Y. for 600 S.F.	"	11.50		9.20	11.05	20.25	26
061								
070	Furnish and place, truck dumped @ $9.20 per C.Y., 4" deep	B-10S	1,300	S.Y.	1.02	.38	1.40	1.61
080	6" deep	"	820		1.55	.61	2.16	2.47
30-001	MULCH Hand spread 2" deep, wood chips @ $6.00 per C.Y., delivered	1 Clab	220		.38	.58	.96	1.26
010	Peat moss @ $10.50 per 6 C.F. compressed bale	"	220		.75	.58	1.33	1.66
35-001	PLANT BED Preparation, 18" deep, by machine	A-1	4,000	S.F.	.50	.04	.54	.61
010	By hand	2 Clab	310	"	.50	.82	1.32	1.74
40-001	PLANTING Moving shrubs on site, 12" ball	B-1	28	Ea.		14.20	14.20	21L
010	24" ball	"	22			18.05	18.05	26L
030	Moving trees on site, 36" ball	B-6	3.75			150	150	205
040	60" ball	"	1			565	565	765
45-001	SEEDING Mechanical seeding, $1.85/lb., 215 lb./acre	A-1	.31	Acre	400	530	930	1,175
010	$1.85/lb., 44 lb./M.S.Y.	"	1,550	S.Y.	.08	.11	.19	.23
030	Fine grading and seeding incl. lime, fertilizer & seed,							
031	with equipment	B-14	1,000	S.Y.	.15	.96	1.11	1.51
040	Fertilizer applied 35 lbs. per M.S.F.	A-1	200	M.S.F.	5.25	.82	6.07	6.90
041								
060	Limestone applied 50 lbs. per M.S.F.	A-1	200	M.S.F.	2	.82	2.82	3.32
080	Grass seed applied 4.5 lbs. per M.S.F.	"	200	"	8.35	.82	9.17	10.30
100	Hydraulic seeding for large areas, incl. seed and fertilizer	B-81	4,000	S.Y.	.11	.16	.27	.33
110	With wood fiber mulch added	"	4,000		.19	.16	.35	.42
130	Seed only, over 100 lbs., field seed, minimum			Lb.	.40		.40	.44M
140	Maximum				3.25		3.25	3.57M
150	Lawn seed, minimum				.70		.70	.77M
160	Maximum				3.85		3.85	4.23M
180	Aerial operations, seeding only, field seed	B-58	50	Acre	440	33	473	525
190	Lawn seed		50		400	33	433	480
210	Seed & liquid fertilizer, field seed		50		510	33	543	600
220	Lawn seed		50		470	33	503	555
50-001	SODDING In East, 1 inch deep, incl. fine grade, on level ground	B-14	1,000	S.Y.	1.47	.96	2.43	2.96
020	On slopes	"	800	"	1.57	1.20	2.77	3.41

2.8 Lawns & Planting

			Crew	Daily Output	Unit	Bare Costs Mat.	Bare Costs Inst.	Bare Costs Total	Total Incl O&P
50	120	(29) In Midwest on level ground, prepared area, over 400 S.Y.	B-14	1,000	S.Y.	.95	.96	1.91	2.39
	123	100 S.Y. area		800		1.32	1.20	2.52	3.13
	126	50 S.Y. area		750		1.90	1.28	3.18	3.88
	130	On slopes, 400 S.Y. area		720		1	1.34	2.34	2.96
	170	Polyurethane with ceramic chips for median strip, minimum		1,153	S.F.	.55	.84	1.39	1.77
	180	Maximum		875	"	.78	1.10	1.88	2.39
55-001		**STEPS & SIDEWALKS** See division 2.6-40 & 45							
60-001		**TRAVEL** To all nursery items, for 10 to 20 miles, add			All				5%
	010	30 to 50 miles, add			"				10%
65-001		**TREES AND SHRUBS** Guying trees, 2" to 4" diameter	B-6	16	Ea.	22	35	57	72
	010	6" to 10" diameter	B-3	8	"	78	250	328	395
	030	Trees in place, balled & burlapped							
	050	(30) Pines, black, 2-1/2' to 3' high	B-1	50	Ea.	14.40	7.95	22.35	27
	070	Yews, spreading, 12" to 15" spread		60		8.75	6.65	15.40	19.25
	080	Upright, 2' to 2.5' high		52		24.75	7.65	32.40	38
	100	Junipers, prostrate 15" to 18" spread		80		13.90	4.97	18.87	23
	110	Pfitzer, 2' to 2.5' spread		44		21	9.05	30.05	36
	120	Upright junipers, 4-1/2' to 5' high	B-17	55		26	16.70	42.70	50
	140	Spruce, 4' to 5' high		75		34	12.25	46.25	53
	160	Birch, 6' to 8' high, 3 stems		20		35	46	81	98
	180	Flowering crab, 6' to 8' high		20		34	46	80	97
	200	Hawthorn, 8' to 10' high		20		64	46	110	130
	220	Maple, Norway, 8' to 10' high, 1-1/2" to 1-3/4" caliper		10		37	92	129	160
	240	Oak, 2-1/2" to 3" caliper		3		100	305	405	510
	260	Willow, 6' to 8' high		25		22	37	59	72
	280	Rhododendron, 18" to 24" high	B-1	48		14	8.30	22.30	27
	290	3' to 4' high	B-17	75		37	12.25	49.25	57
	310	Japanese holly, 12" to 15" high	B-1	71		5.30	5.60	10.90	13.95
	320	18" to 24" high		47		9.55	8.45	18	23
	340	American holly, 15" to 18" high		96		13	4.14	17.14	20
	350	5' to 6' high	B-17	55		55	16.70	71.70	82
	370	Canadian hemlock, 2-1/2' to 3' high	B-1	36		24	11.05	35.05	42
	380	5' to 6' high	B-17	33		58	28	86	100
	400	Douglas fir, 3' to 4' high		75		15.90	12.25	28.15	33
	410	6' to 7' high		25		58	37	95	110
	430	Arborvitae, globe, 12" to 15" high	B-1	96		7.50	4.14	11.64	14.25
	440	Pyramidal, 4' to 5' high	B-17	33		21	28	49	59
	460	White pine, 4' to 5' high		75		32	12.25	44.25	51
	470	7' to 8' high		20		59	46	105	125
	490	Juniper, andorra, 18" to 24" spread	B-1	80		14.30	4.97	19.27	23
	500	Blue pfitzer, 2' to 2.5' spread		44		21	9.05	30.05	36
	520	Euonymus, compacta, 15" to 18" high		80		7.50	4.97	12.47	15.45
	530	4' to 5' high	B-17	50		34	18.40	52.40	61
	550	Dogwood, pink, 5' to 6' high		30		30	31	61	73
	560	1-1/2" caliper		14		85	66	151	180
	580	Dogwood, white, 4' to 5' high		75		19.10	12.25	31.35	37
	590	1-1/4" caliper		18		64	51	115	135
	610	Shrubs in place, forsythia, 2' to 3' high	B-1	60		6.35	6.65	13	16.60
	620	Weigelia, 3' to 4' high		70		17	5.70	22.70	27
	630	Deutzia, 12" to 15" high		96		6.35	4.14	10.49	13
	640	Spirea, 3' to 4' high		70		19	5.70	24.70	29
	650	Honeysuckle, 3' to 4' high		60		8.50	6.65	15.15	18.95
	660	Flowering almond, 2' to 3' high		36		7.50	11.05	18.55	24

For expanded coverage of these items see Means' Site Work Cost Data 1986

2.9 Heavy Construction

		CREW	DAILY OUTPUT	UNIT	BARE COSTS MAT.	BARE COSTS INST.	BARE COSTS TOTAL	TOTAL INCL O&P
05-001	**BRIDGES** Pedestrian, spans over streams, roadways, etc.							
002	including erection, not including foundations							
005	Precast concrete, complete in place, 8' wide, 60' span	E-2	215	S.F.Deck	19.65	9.20	28.85	34
010	100' span		185		20.80	10.70	31.50	38
015	120' span		160		23.10	12.40	35.50	42
020	150' span		145		25.20	13.65	38.85	46
030	Steel, trussed or arch spans, complete in place, 8' wide, 40' span		320		24.15	6.20	30.35	35
040	50' span		395		23.10	5	28.10	32
050	60' span		465		23.10	4.26	27.36	31
060	80' span		570		27.30	3.48	30.78	35
070	100' span		465		37.80	4.26	42.06	47
080	120' span		365		49.35	5.45	54.80	62
090	150' span		310		52.50	6.40	58.90	66
100	160' span		255		53.55	7.75	61.30	70
110	10' wide, 80' span		640		27.30	3.10	30.40	34
120	120' span		415		37.80	4.78	42.58	48
130	150' span		445		46.20	4.45	50.65	57
140	200' span	↓	205		57.75	9.65	67.40	77
160	Wood, laminated type, complete in place, 80' span	C-12	203		18.90	5.85	24.75	29
170	130' span	"	153	↓	19.95	7.75	27.70	33
10-001	**DOCKS** Floating, recreational, prefabricated aluminum or							
002	concrete over polystyrene, no pilings included	F-3	330	S.F.	19.50	3.21	22.71	26
020	Pile supported, shore constructed, bare, 3" decking		130		12	8.15	20.15	24
025	4" decking		120		13	8.85	21.85	26
040	Floating, small boat, prefab, no shore facilities, minimum		318		6	3.34	9.34	11.15
050	Maximum		196	↓	20	5.40	25.40	29
070	Per slip, minimum (180 S.F. each)		1.59	Ea.	1,100	665	1,765	2,125
080	Maximum	↓	1.40		4,000	760	4,760	5,425
15-001	**RAILROAD** Car bumpers W.D., standard	B-14	2		985	480	1,465	1,750
010	W.A., heavy duty		2		1,910	480	2,390	2,775
020	Derails hand throw (sliding)		10		495	96	591	680
030	Hand throw with standard timbers, open stand & target		5.50	↓	830	175	1,005	1,150
040	Resurface and realign existing track		200	L.F.		4.82	4.82	6.70
060	For crushed stone ballast @ $7.00 per ton, add	↓	500	"	4.70	1.93	6.63	7.85
080	Siding, yard spur, level grade,							
081	㉗ 100 lb. rail, new material on wood ties	B-14	57	L.F.	38.50	16.90	55.40	66
100	㉘ Steel ties embedded in concrete with 100# rail, fasterners & plates	"	22	"	57	44	101	125
101								
120	Switch timber, for a #8 switch, creosoted	B-14	3.70	M.F.B.M.	495	260	755	905
130	Complete set of timbers, 3.7 M.F.B.M for #8 switch		1	Total	1,825	965	2,790	3,350
140	Ties, concrete, 8'-6" long, 30" O.C.	↓	80	Ea.	72	12.05	84.05	96
141								
160	Wood, creosoted, 6" x 8" x 8'-6", C.L. lots	B-14	90	Ea.	16	10.70	26.70	33
170	L.C.L. lots		90		17.50	10.70	28.20	34
190	Heavy duty, 7" x 9" x 8'-6", C.L. lots		70		21	13.75	34.75	42
200	L.C.L. lots	↓	70	↓	22.50	13.75	36.25	44
220	Turnouts, #8, incl. 100 lb. rails, plates, bars, frog, switch points,							
230	timbers and ballast 6" below bottom of tie	B-14	.50	Ea.	12,500	1,925	14,425	16,400
240	Wheel stops (mat'l. ranges $250 to $750)	"	14	Pr.	380	69	449	515
20-001	**ROOF BOLTS** 5/8" diameter, 24" long			Ea.	2.70		2.70	2.97M
020	60" long				4.65		4.65	5.10M
040	3/4" diameter, 36" long				4.60		4.60	5.05M
060	96" long				9.50		9.50	10.45M
080	Washers for above, 1/4" thick, 4" x 4"				1		1	1.10M
100	6" x 6"			↓	1.10		1.10	1.21M

For expanded coverage of these items see Means' Site Work Cost Data 1986

3.1 Formwork

		CREW	DAILY OUTPUT	UNIT	BARE COSTS MAT.	BARE COSTS INST.	BARE COSTS TOTAL	TOTAL INCL O&P
05-001	ACCESSORIES Anchor bolts, incl. nut & washer							
002	1/2" diameter, 6" long	2 Carp	175	Ea.	.32	1.83	2.15	3
005	10" long		170		.45	1.88	2.33	3.22
010	12" long		165		.55	1.94	2.49	3.42
020	5/8" diameter, 12" long		60		1.10	5.35	6.45	8.95
025	18" long		55		1.50	5.80	7.30	10.10
030	24" long		50		1.70	6.40	8.10	11.15
035	3/4" diameter, 8" long		50		1.20	6.40	7.60	10.60
040	12" long		45		1.45	7.10	8.55	11.90
045	18" long		40		1.80	8	9.80	13.55
050	24" long		35		2	9.15	11.15	15.45
060	7/8" diameter, 12" long		40		1.80	8	9.80	13.55
065	18" long		35		2	9.15	11.15	15.45
070	24" long		30		2.90	10.65	13.55	18.65
080	1" diameter, 12" long		35		2.15	9.15	11.30	15.60
085	18" long		30		2.60	10.65	13.25	18.30
090	24" long		25		4	12.80	16.80	23
095	36" long		20		5.20	16	21.20	29
120	1-1/2" diameter, 18" long		22		7.50	14.55	22.05	29
125	24" long		16		10	20	30	40
130	36" long		12		12.80	27	39.80	53
135	Larger sizes		255	Lb.	.57	1.25	1.82	2.45
150	For galvanized bolts, add			"	60%			
155	For sleeves, add			Ea.	50%	25%		
170	Anchor bolt sleeves, plastic, for 5/8" to 7/8" bolts	1 Carp	100		2.25	1.60	3.85	4.79
175	For 1-1/2" to 2-1/4" bolts		80		9.85	2	11.85	13.75
180	For 2-1/2" to 3" bolts		60		16.60	2.67	19.27	22
190								
200	Banding iron, 3/4" x 22 ga., 14 L.F./lb. or							
205	1/2" x 14 ga., 7 L.F./lb.			Lb.	.62		.62	.68M
240	Chamfer strips, polyvinyl chloride, 1/2" wide with leg	1 Carp	535	L.F.	.33	.30	.63	.80
245	3/4" chamfer with leg		525		.40	.30	.70	.88
250	1" radius with leg		515		.37	.31	.68	.86
255	1" chamfer with leg		515		.53	.31	.84	1.03
260	1-1/2" radius		500		.88	.32	1.20	1.43
270	Wood, 1/2" wide		535		.07	.30	.37	.51
275	3/4" wide		525		.08	.30	.38	.53
280	1" wide		515		.11	.31	.42	.57
300	Column clamps, adjustable, up to 24" x 24" columns, buy			Set	31		31	34M
305	Rent per month				2.65		2.65	2.91M
315	For sizes up to 30" x 30", buy				39		39	43M
320	Rent per month				2.65		2.65	2.91M
330	For sizes up to 36" x 36", buy				45		45	50M
335	Rent per month				4.50		4.50	4.95M
350	Chain and wedge type, 36" x 36", buy				48		48	53M
355	Rent per month				4.50		4.50	4.95M
370	60" x 60", buy				75		75	83M
375	Rent per month				5		5	5.50M
400	Bull winch clamps (band iron), 36" x 36", buy				30		30	33M
405	Rent per month				2.65		2.65	2.91M
415	48" x 48", buy				30		30	33M
420	Rent per month				2.65		2.65	2.91M
430	Friction collars, 2'-6" diameter, buy				530		530	585M
435	Rent per month				58		58	64M
450	4'-0" diameter, buy				650		650	715M
455	Rent per month				70		70	77M
470	Dovetail anchor system, anchor slot, filled, galv., 24 ga.	1 Carp	425	L.F.	.18	.38	.56	.74
472	Galv., 20 ga.		400		.38	.40	.78	1
474	16 oz. copper, foam filled		375		.75	.43	1.18	1.44
476	26 ga. stainless steel, foam filled		375		.69	.43	1.12	1.38

For expanded coverage of these items see Means' *Concrete & Masonry Cost Data 1986*

3.1 Formwork

			CREW	DAILY OUTPUT	UNIT	BARE COSTS MAT.	BARE COSTS INST.	BARE COSTS TOTAL	TOTAL INCL O&P
05	480	Brick anchor, corrugated, galv., 3-1/2" long, 16 ga.	1 Bric	10.50	C	8	15.60	23.60	31
	485	3-1/2" long, 12 ga.		10.50		15.30	15.60	30.90	39
	490	Galv., not corrugated, 3-1/2" long, 16 ga.		10.50		4.90	15.60	20.50	28
	495	3-1/2" long, 12 ga.		10.50		12.15	15.60	27.75	36
	500	Cavity wall, corrugated, galv., 5" long, 16 ga.		10.50		17.65	15.60	33.25	42
	505	5" long, 12 ga.	↓	10.50		28	15.60	43.60	53
	510	Furring anchors, corrugated, galv., 1-1/2" long, 16 ga.	1 Carp	10.50		6.65	15.25	21.90	29
	515	1-1/2" long, 12 ga.	"	10.50		11.50	15.25	26.75	35
	520	Stone anchors, 3-1/2" long, galv., 1/8" x 1" wide	1 Bric	10.50		26	15.60	41.60	51
	525	1/4" x 1" wide	"	10.50	↓	38	15.60	53.60	64
	540	Form braces for footings, solid steel, adjustable			Ea.	6		6	6.60M
	545	Form spreaders for footer, adjustable				3.50		3.50	3.85M
	550	Nail stakes, 3/4" diameter, 18" long				1.60		1.60	1.76M
	555	24" long				1.75		1.75	1.92M
	560	30" long				2		2	2.20M
	565	36" long			↓	2.10		2.10	2.31M
	580	Form oil, coverage varies greatly, minimum			Gal.	3.50		3.50	3.85M
	585	Maximum			"	4.90		4.90	5.40M
	590	Form patches, 1-3/4" diameter			C	2.75		2.75	3.02M
	595	2-3/4" diameter				4.90		4.90	5.40M
	610	Hangers, floor slab forms from structural steel, coil rod				140		140	155M
	615	Double wire beam hangers				115		115	125M
	620	Fascia tie, coil type				74		74	81M
	625	Snap tie type				66		66	73M
	630	Snap tie beam hanger				85		85	94M
	635	Soffit spacer				9		9	9.90M
	645	Hangers, non-fireproof steel & precast beams, coil rod, 4 ga., 12' lg.			↓	95		95	105M
	650	Wire saddle, 4 ga.				105		105	115M
	660	Hanging wire, black annealed, 9 gauge			Cwt.	60		60	66M
	665	16 gauge			"	50		50	55M
	675	Inserts, all purpose, steel one size for 3/8" to 5/8" nuts	1 Carp	.70	C	130	230	360	475
	680	Ceiling, one size for 1/4" to 5/8" nuts		.65	"	58	245	303	420
	685	Continuous slotted 12 ga., 1-5/8" x 1-3/8", 3" long		65	Ea.	2.30	2.46	4.76	6.10
	690	6" long, 12 ga.		65		3.20	2.46	5.66	7.10
	692	12" long, 8 ga.		65		7.80	2.46	10.26	12.15
	694	24" long, 8 ga.		65		12.25	2.46	14.71	17.05
	696	36" long, 8 ga.		60		16.80	2.67	19.47	22
	698	60" long, 8 ga.		55		25	2.91	27.91	32
	705	Threaded inserts, for 1/4" bolts		65		1.35	2.46	3.81	5.05
	710	For 7/8" bolts		65	↓	1.50	2.46	3.96	5.20
	720	Wedge type for shelf angles, 1/2" x 2" bolts		60	Set	1.45	2.67	4.12	5.45
	722	5/8" x 2" bolts		60		1.60	2.67	4.27	5.60
	724	3/4" x 3" bolts		60		2	2.67	4.67	6.05
	730	Extra long, 3/4" x 3" bolts	↓	60	↓	3	2.67	5.67	7.15
	735	Horseshoe shimming washers, 1/4" thick			C	24		24	26M
	740	For galvanized inserts, add			"	30%			
	750	Reglet, standard, 26 ga., 8' lengths, galv. steel	1 Carp	120	L.F.	.55	1.33	1.88	2.54
	755	Stainless steel		120	"	1.70	1.33	3.03	3.80
	765	Sleeves, plastic type, 1 use, 9" long, 2" diameter		100	Ea.	.75	1.60	2.35	3.14
	770	4" diameter		90		1.20	1.78	2.98	3.90
	775	6" diameter		75		1.65	2.13	3.78	4.91
	780	12" dameter	↓	60	↓	3.85	2.67	6.52	8.10
	820	Wall ties for walls 8" thick, 3000 lb. capacity, snap tie			C	35		35	39M
	825	5000 lb. capacity				55		55	61M
	830	Over 10" to 12" thick, 3000 lb. capacity				39		39	43M
	835	• 5000 lb. capacity				60		60	66M
	850	Add for each inch over 12" to 18", 3000 lb. cap.				1.75		1.75	1.92M
	855	5000 lb. capacity				1.95		1.95	2.14M
	870	Add for each inch over 18" thick, 3000 lb. cap.				1.90		1.90	2.09M
	875	5000 lb. capacity			↓	2.10		2.10	2.31M

For expanded coverage of these items see Means' *Concrete & Masonry Cost Data 1986*

3.1 Formwork

		Description	CREW	DAILY OUTPUT	UNIT	BARE COSTS MAT.	BARE COSTS INST.	BARE COSTS TOTAL	TOTAL INCL O&P
05	900	For cones, add to above, 1" plastic			C	29		29	32M
	905	2" wood			"	35		35	39M
	915	Single waler snap brackets			Ea.	2.95		2.95	3.24M
	925	Smooth pencil rod, 20' length, 1/4" diameter				2		2	2.20M
	930	3/8" diameter				3.35		3.35	3.68M
	940	Rod clamps for 1/4" & 3/8" diameter rod				1.47		1.47	1.61M
	950	Threaded rods, continuous, 1/2" diameter			L.F.	.53		.53	.58M
	955	1" diameter			"	1.90		1.90	2.09M
	970	Wedges, 3000 lb. capacity			C	90		90	99M
	975	5000 lb. capacity			"	110		110	120M
10-001		EXPANSION JOINT Keyed cold joint, 24 ga., incl. stakes, 3-1/2" high	1 Carp	200	L.F.	.68	.80	1.48	1.91
	005	4-1/2" high		200		.71	.80	1.51	1.94
	015	7-1/2" high		190		1.15	.84	1.99	2.49
	030	Poured asphalt, plain, 1/2" x 1"	1 Clab	450		.20	.28	.48	.63
	035	1" x 2"		400		.83	.32	1.15	1.37
	050	Neoprene, liquid, cold applied, 1/2" x 1"		450		1.25	.28	1.53	1.78
	055	1" x 2"		400		4.85	.32	5.17	5.80
	070	Polyurethane, 1 part, 1/2" x 1"		400		1.47	.32	1.79	2.08
	075	1" x 2"		350		5.90	.36	6.26	7
	090	Rubberized asphalt, hot or cold applied, 1/2" x 1"		450		.23	.28	.51	.66
	095	1" x 2"		400		.79	.32	1.11	1.33
	110	Hot applied, fuel resistant, 1/2" x 1"		450		.25	.28	.53	.68
	115	1" x 2"		400		.84	.32	1.16	1.38
	200	Premolded, bituminous fiber, 1/2" x 6"	1 Carp	375		.42	.43	.85	1.08
	205	1" x 12"		300		1.89	.53	2.42	2.85
	225	Cork with resin binder, 1/2" x 6"		375		.74	.43	1.17	1.43
	230	1" x 12"		300		3.15	.53	3.68	4.24
	250	Neoprene sponge, closed cell, 1/2" x 6"		375		1.05	.43	1.48	1.77
	255	1" x 12"		300		4.70	.53	5.23	5.95
	275	Polyethylene foam, 1/2" x 6"		375		.84	.43	1.27	1.54
	280	1" x 12"		300		3.35	.53	3.88	4.46
	300	Polyethylene backer rod, 3/8" diameter		500		.04	.32	.36	.51
	305	3/4" diameter		495		.10	.32	.42	.58
	310	1" diameter		490		.13	.33	.46	.62
	311								
	350	Polyurethane foam, with polybutylene, 1/2" x 1/2"	1 Carp	475	L.F.	.58	.34	.92	1.13
	355	1" x 1"		450		1.16	.36	1.52	1.79
	375	Polyurethane foam, regular, closed cell, 1/2" x 6"		375		.53	.43	.96	1.20
	380	1" x 12"		300		2.10	.53	2.63	3.08
	400	Polyvinyl chloride foam, closed cell, 1/2" x 6"		375		1.31	.43	1.74	2.06
	405	1" x 12"		300		4.70	.53	5.23	5.95
	425	Rubber, gray sponge, 1/2" x 6"		375		2.31	.43	2.74	3.16
	430	1" x 12"		300		9.30	.53	9.83	11
	450	Lead wool for joints, 1 ton lots			Lb.	1.37		1.37	1.50M
	455	Retail			"	1.58		1.58	1.73M
	500	For installation in walls, add			L.F.		75%		
	525	For installation in boxouts, add			"		25%		
15-001		FORMS IN PLACE, AVERAGE For reinforced concrete bldg., 1 use	C-2	165	S.F.C.A.	1.48	5.90	7.38	10.15
	005	2 use		305		.84	3.21	4.05	5.55
	010	3 use		415		.62	2.36	2.98	4.07
	015	4 use		515		.49	1.90	2.39	3.27
20-001		FORMS IN PLACE, BEAMS AND GIRDERS							
	002	See also Elevated Slabs, division 3.1-35							
	050	(32) Beam and Girder, exterior spandrel, 12" wide, 1 use	C-2	225	S.F.C.A.	1.87	4.34	6.21	8.30
	055	2 use		275		1.09	3.55	4.64	6.30
	060	(33) 3 use		295		.82	3.31	4.13	5.65
	065	4 use		310		.69	3.15	3.84	5.30
	100	(34) Exterior spandrel, 18" wide, 1 use		250		1.56	3.91	5.47	7.35
	105	2 use		275		.93	3.55	4.48	6.15

For expanded coverage of these items see Means' *Concrete & Masonry Cost Data 1986*

3.1 Formwork

		CREW	DAILY OUTPUT	UNIT	BARE COSTS			TOTAL INCL O&P
					MAT.	INST.	TOTAL	
110	3 use	C-2	305	S.F.C.A.	.72	3.21	3.93	5.40
115	4 use		315		.61	3.10	3.71	5.15
150	Exterior spandrel, 24" wide, 1 use		265		1.52	3.69	5.21	7
155	2 use		290		.90	3.37	4.27	5.85
160	3 use		315		.69	3.10	3.79	5.20
165	4 use		325		.59	3.01	3.60	4.97
200	Interior beams, 12" wide, 1 use		300		1.54	3.26	4.80	6.40
205	2 use		340		.94	2.88	3.82	5.15
210	3 use		364		.74	2.69	3.43	4.68
215	4 use		377		.64	2.59	3.23	4.43
250	Interior beams, 24" wide, 1 use		320		1.44	3.06	4.50	6
255	2 use		365		.85	2.68	3.53	4.79
260	3 use		385		.65	2.54	3.19	4.37
265	4 use		395		.55	2.47	3.02	4.16
300	Beam and Girder, encasing steel frame, hung, 1 use		325		1.47	3.01	4.48	5.95
305	2 use		390		.89	2.51	3.40	4.58
310	3 use		415		.67	2.36	3.03	4.12
315	4 use		430		.59	2.27	2.86	3.92
350	Beam bottoms only, 24" wide, 1 use		230		1.90	4.25	6.15	8.20
355	2 use		265		1.11	3.69	4.80	6.50
360	3 use		280		.85	3.49	4.34	5.95
365	4 use		290		.72	3.37	4.09	5.65
400	Beam sides only, vertical, 36" high, 1 use		335		1.36	2.92	4.28	5.70
405	2 use		405		.79	2.41	3.20	4.34
410	3 use		430		.61	2.27	2.88	3.94
415	4 use		445		.51	2.20	2.71	3.72
450	Sloped sides, 36" high, 1 use		305		1.46	3.21	4.67	6.20
455	2 use		370		.87	2.64	3.51	4.76
460	3 use		405		.68	2.41	3.09	4.22
465	4 use		425		.58	2.30	2.88	3.94
500	Upstanding beams, 36" high, 1 use		225		2.29	4.34	6.63	8.75
505	2 use		255		1.36	3.83	5.19	7
510	3 use		275		1.06	3.55	4.61	6.30
515	4 use	↓	280	↓	.90	3.49	4.39	6
25-001	**FORMS IN PLACE, COLUMNS**							
050	Round fiberglass, 4 use per mo., rent, 12" diameter	C-1	160	L.F.	1.42	3.96	5.38	7.25
055	16" diameter		150		1.63	4.22	5.85	7.85
060	18" diameter		140		1.73	4.52	6.25	8.40
065	24" diameter		135		2.25	4.69	6.94	9.20
070	28" diameter		130		2.40	4.87	7.27	9.65
080	30" diameter		125		2.70	5.05	7.75	10.25
085	36" diameter		120		2.90	5.30	8.20	10.75
150	Round fiber tube, 1 use, 8" diameter		155		1.65	4.08	5.73	7.70
155	10" diameter		155		2.30	4.08	6.38	8.40
160	12" diameter		150		2.85	4.22	7.07	9.20
165	14" diameter		145		3.65	4.37	8.02	10.30
170	16" diameter		140		4.75	4.52	9.27	11.70
175	20" diameter		135		7.50	4.69	12.19	15
180	24" diameter		130		9.50	4.87	14.37	17.45
185	30" diameter		125		14.75	5.05	19.80	24
190	36" diameter		115		18	5.50	23.50	28
195	42" diameter		100		34	6.35	40.35	46
200	48" diameter	↓	85		45	7.45	52.45	60
220	For seamless type, add				15%			
305	Round, steel, 4 use per mo., rent, 6' long							
310	Heavy duty, 20" diameter	C-1	105		2	6.05	8.05	10.85
315	24" diameter		85		2.10	7.45	9.55	13
320	30" diameter		70		2.36	9.05	11.41	15.60
325	36" diameter	C-2	60		2.63	10.55	13.18	18.05
330	48" diameter	↓	50	↓	4	12.65	16.65	23

For expanded coverage of these items see Means' *Concrete & Masonry Cost Data 1986*

3.1 Formwork

		Description	CREW	DAILY OUTPUT	UNIT	BARE COSTS MAT.	BARE COSTS INST.	BARE COSTS TOTAL	TOTAL INCL O&P
25	335	60" diameter	C-1	45	L.F.	5.15	14.05	19.20	26
	400	Column capitals, 4 use per mo., 24" column, 4' cap diam.		12	Ea.	13.75	53	66.75	91
	405	5' cap diameter		11		13.45	58	71.45	97
	410	6' cap diameter		10		17.30	63	80.30	110
	415	7' cap diameter	↓	9		19	70	89	120
	450	For second and succeeding months, deduct			↓	50%			
	500	(32) Plywood, 8" x 8" columns, 1 use	C-1	165	S.F.C.A.	1.55	3.84	5.39	7.20
	505	2 use		195		.90	3.25	4.15	5.65
	510	(33) 3 use		210		.68	3.01	3.69	5.10
	515	4 use		215		.58	2.94	3.52	4.86
	550	12" x 12" plywood columns, 1 use		180		1.50	3.52	5.02	6.70
	555	2 use		210		.85	3.01	3.86	5.25
	560	3 use		220		.63	2.88	3.51	4.82
	565	4 use		225		.52	2.81	3.33	4.61
	600	16" x 16" plywood columns, 1 use		185		1.43	3.42	4.85	6.50
	605	2 use		215		.80	2.94	3.74	5.10
	610	3 use		230		.59	2.75	3.34	4.60
	615	4 use		235		.49	2.69	3.18	4.41
	650	24" x 24" plywood columns, 1 use		190		1.49	3.33	4.82	6.40
	655	2 use		216		.83	2.93	3.76	5.10
	660	3 use		230		.60	2.75	3.35	4.61
	665	4 use		238		.50	2.66	3.16	4.37
	700	36" x 36" plywood columns, 1 use		200		1.54	3.17	4.71	6.25
	705	2 use		230		.84	2.75	3.59	4.88
	710	3 use		245		.61	2.58	3.19	4.38
	715	4 use		250		.49	2.53	3.02	4.17
	750	Steel framed plywood, 4 use per mo., rent, 8" x 8"		290		.56	2.18	2.74	3.75
	755	10" x 10"		300		.44	2.11	2.55	3.51
	760	12" x 12"		310		.37	2.04	2.41	3.34
	765	16" x 16"		335		.35	1.89	2.24	3.10
	770	20" x 20"		350		.31	1.81	2.12	2.94
	775	24" x 24"		365		.26	1.73	1.99	2.78
	30-001	**FORMS IN PLACE, CULVERT** 5' to 8' square or rectangular, 1 use		170		1.40	3.72	5.12	6.90
	005	2 use		180		.84	3.52	4.36	5.95
	010	(33) 3 use		190		.61	3.33	3.94	5.45
	015	4 use	↓	200	↓	.56	3.17	3.73	5.15
	35-001	**FORMS IN PLACE, ELEVATED SLABS**							
	005	See also corrugated form deck, division 5.2-30-610							
	100	Flat plate to 15' high, 1 use	C-2	470	S.F.	1.27	2.08	3.35	4.39
	105	2 use		520		.80	1.88	2.68	3.58
	110	3 use		545		.58	1.79	2.37	3.22
	115	4 use		560		.48	1.75	2.23	3.04
	150	15' to 20' high ceilings, 4 use		495		.58	1.97	2.55	3.48
	160	21' to 35' high ceilings, 4 use		450		.66	2.17	2.83	3.85
	200	Flat slab with drop panels, to 15' high, 1 use		449		1.59	2.18	3.77	4.88
	205	2 use		509		.97	1.92	2.89	3.83
	210	3 use		532		.77	1.84	2.61	3.49
	215	4 use		544		.66	1.80	2.46	3.31
	225	15' to 20' high ceilings, 4 use		480		.85	2.04	2.89	3.86
	235	21' to 35' high ceilings, 4 use		435		1.03	2.25	3.28	4.36
	300	Floor slab hung from steel beams, 1 use		485		1.28	2.02	3.30	4.31
	305	2 use		535		.74	1.83	2.57	3.44
	310	3 use		550		.56	1.78	2.34	3.17
	315	4 use		565		.47	1.73	2.20	3
	350	(36) Floor slab, with 20" metal pans, 1 use		415		2.27	2.36	4.63	5.90
	355	2 use		445		1.38	2.20	3.58	4.68
	360	3 use		475		1.05	2.06	3.11	4.11
	365	4 use	↓	500	↓	.91	1.96	2.87	3.81

For expanded coverage of these items see Means' *Concrete & Masonry Cost Data 1986*

3.1 Formwork

			CREW	DAILY OUTPUT	UNIT	BARE COSTS MAT.	BARE COSTS INST.	BARE COSTS TOTAL	TOTAL INCL O&P
35	370	Floor slab with 30" metal pans, 1 use	C-2	418	S.F.	2.23	2.34	4.57	5.80
	372	2 use		455		1.39	2.15	3.54	4.62
	374	3 use		470		1.08	2.08	3.16	4.18
	376	4 use		480		.93	2.04	2.97	3.95
	400	Floor slab with 19" metal domes, 1 use (36)		405		2.27	2.41	4.68	5.95
	405	2 use		435		1.38	2.25	3.63	4.75
	410	3 use		465		1.05	2.10	3.15	4.18
	415	4 use		495		.91	1.97	2.88	3.84
	450	With 30" fiberglass domes, 1 use		405		2.23	2.41	4.64	5.90
	452	2 use		450		1.39	2.17	3.56	4.65
	453	3 use		460		1.08	2.13	3.21	4.24
	455	4 use		470		.93	2.08	3.01	4.01
	500	Box out for slab openings, over 16" deep, 1 use		190	S.F.C.A.	1.82	5.15	6.97	9.40
	505	2 use		240	"	1.13	4.07	5.20	7.10
	550	Shallow slab box outs, to 10 S.F.		42	Ea.	6.57	23	29.57	41
	555	Over 10 S.F. (use perimeter)		400	L.F.	.77	2.44	3.21	4.36
	600	Bulkhead forms for slab, with keyway, 1 use, 2 piece		500		.96	1.96	2.92	3.87
	610	3 piece (see also edge forms)		460		1.30	2.13	3.43	4.49
	650	Curb forms, wood, 6" to 12" high, on elevated slabs, 1 use	C-1	180	S.F.C.A.	1.33	3.52	4.85	6.50
	655	2 use		205		.74	3.09	3.83	5.25
	660	3 use		220		.54	2.88	3.42	4.72
	665	4 use		225		.45	2.81	3.26	4.53
	700	Edge forms to 6" high, on elevated slab, 4 use		500	L.F.	.20	1.27	1.47	2.04
	710	7" to 12" high, 4 use		350	S.F.C.A.	.40	1.81	2.21	3.04
	750	Depressed area forms to 12" high, 4 use		300	L.F.	.40	2.11	2.51	3.47
	755	12" to 24" high, 4 use		175		.50	3.62	4.12	5.75
	800	Perimeter deck and rail for elevated slabs, straight		90		5	7.05	12.05	15.60
	805	Curved		65		6.85	9.75	16.60	22
	850	Void forms, round fiber, 3" diameter		450		.32	1.41	1.73	2.37
	855	4" diameter, void		425		.34	1.49	1.83	2.51
	860	6" diameter, void		400		.61	1.58	2.19	2.94
	865	8" diameter, void		375		1	1.69	2.69	3.52
	870	10" diameter, void		350		1.70	1.81	3.51	4.47
	875	12" diameter, void		300		2.10	2.11	4.21	5.35
	880	Metal end closures, loose, minimum			C	25		25	28M
	885	Maximum			"	120		120	130M
	40-001	**FORMS IN PLACE, EQUIPMENT FOUNDATIONS** 1 use	C-2	160	S.F.C.A.	1.45	6.10	7.55	10.40
	005	2 use		190		.85	5.15	6	8.35
	010	3 use		200		.65	4.89	5.54	7.75
	015	4 use		205		.55	4.77	5.32	7.45
	45-001	**FORMS IN PLACE, FOOTINGS** Continuous wall, 1 use	C-1	375		.75	1.69	2.44	3.25
	005	2 use		440		.43	1.44	1.87	2.54
	010	3 use		470		.32	1.35	1.67	2.29
	015	4 use		485		.27	1.31	1.58	2.17
	050	Dowel supports for footings or beams, 1 use		500	L.F.	.50	1.27	1.77	2.37
	100	Integral starter wall, to 4" high, 1 use		400		.74	1.58	2.32	3.09
	150	Keyway, 4 uses, tapered wood, 2" x 4"	1 Carp	530		.07	.30	.37	.51
	155	2" x 6"		500		.09	.32	.41	.56
	200	Tapered plastic, 2" x 3"		530		.35	.30	.65	.82
	205	2" x 4"		500		.45	.32	.77	.96
	225	For keyway hung from supports, add		150		.42	1.07	1.49	2.01
	226								
	300	Pile cap, square or rectangular, 1 use (33)	C-1	290	S.F.C.A.	1.24	2.18	3.42	4.50
	305	2 use		346		.73	1.83	2.56	3.43
	310	3 use		371		.53	1.71	2.24	3.03
	315	4 use		383		.44	1.65	2.09	2.86
	400	Triangular or hexagonal caps, 1 use		225		1.41	2.81	4.22	5.60
	405	2 use		280		.82	2.26	3.08	4.15
	410	3 use		305		.62	2.08	2.70	3.66
	415	4 use		315		.53	2.01	2.54	3.47

For expanded coverage of these items see Means' *Concrete & Masonry Cost Data 1986*

3.1 Formwork

			Crew	Daily Output	Unit	Bare Costs Mat.	Bare Costs Inst.	Bare Costs Total	Total Incl O&P
45	500	Spread footings, 1 use	C-1	305	S.F.C.A.	.88	2.08	2.96	3.95
	505	2 use		371		.50	1.71	2.21	3
	510	3 use		401		.39	1.58	1.97	2.70
	515	4 use		414		.32	1.53	1.85	2.55
	600	Supports for dowels, plinths or templates, 2' x 2'		25	Ea.	2.60	25	27.60	39
	605	4' x 4' footing		22		5.65	29	34.65	48
	610	8' x 8' footing		20		11.40	32	43.40	58
	615	12' x 12' footing		17		18.50	37	55.50	74
	700	Plinths, 1 use		250	S.F.C.A.	1.42	2.53	3.95	5.20
	710	4 use		270		.41	2.34	2.75	3.82
50-001		FORMS IN PLACE, GRADE BEAM 1 use	C-2	530		1.25	1.84	3.09	4.03
	005	2 use		580		.71	1.69	2.40	3.20
	010	3 use		600		.53	1.63	2.16	2.93
	015	4 use		605		.44	1.62	2.06	2.81
52-001		FORMS IN PLACE, MAT FOUNDATION 1 use		290		1.17	3.37	4.54	6.15
	005	2 use		310		.65	3.15	3.80	5.25
	010	3 use		330		.47	2.96	3.43	4.78
	012	4 use		350		.38	2.79	3.17	4.43
55-001		FORMS IN PLACE, SLAB ON GRADE							
	100	Bulkhead forms with keyway, 1 use, 2 piece	C-1	510	L.F.	.34	1.24	1.58	2.16
	105	3 piece (see also edge forms)		400		.43	1.58	2.01	2.74
	110	4 piece		350		.62	1.81	2.43	3.28
	200	Curb forms, wood, 6" to 12" high, on grade, 1 use		215	S.F.C.A.	1.10	2.94	4.04	5.45
	205	2 use		250		.63	2.53	3.16	4.33
	210	3 use		265		.47	2.39	2.86	3.95
	215	4 use		275		.40	2.30	2.70	3.74
	300	Edge forms, to 6" high, 4 use, on grade		600	L.F.	.16	1.06	1.22	1.69
	305	7" to 12" high, 4 use, on grade		435	S.F.C.A.	.51	1.46	1.97	2.65
	350	For depressed slabs, 4 use, to 12" high		300	L.F.	.42	2.11	2.53	3.49
	355	To 24" high		175		.50	3.62	4.12	5.75
	400	For slab blockouts, 1 use to 12" high		200		.42	3.17	3.59	5
	405	To 24" high		120		.50	5.30	5.80	8.10
	500	Screed, 24 ga. metal key joint, see division 3.1-10							
	502	Wood, incl. wood stakes, 1" x 3"	C-1	900	L.F.	.26	.70	.96	1.30
	505	2" x 4"		900	"	.79	.70	1.49	1.88
	600	Trench forms in floor, 1 use		160	S.F.C.A.	1.26	3.96	5.22	7.05
	605	2 use		175		.68	3.62	4.30	5.95
	610	3 use		180		.49	3.52	4.01	5.60
	615	4 use		185		.39	3.42	3.81	5.35
60-001		FORMS IN PLACE, STAIRS (Slant length x width), 1 use	C-2	165	S.F.	2.18	5.90	8.08	10.90
	005	2 use		170		1.27	5.75	7.02	9.65
	010	3 use		180		.97	5.45	6.42	8.90
	015	4 use		190		.81	5.15	5.96	8.30
	100	Alternate pricing method (0.7 L.F./S.F.), 1 use		100	LF Riser	3.17	9.80	12.97	17.55
	105	2 use		105		1.95	9.30	11.25	15.55
	110	3 use		110		1.54	8.90	10.44	14.45
	115	4 use		115		1.33	8.50	9.83	13.70
	200	Stairs, cast on sloping ground (length x width), 1 use		220	S.F.	1.44	4.44	5.88	7.95
	210	4 use		240	"	.89	4.07	4.96	6.85
65-001		FORMS IN PLACE, WALLS							
	002								
	010	Box out for wall openings, to 16" thick, to 10 S.F.	C-2	24	Ea.	12.20	41	53.20	72
	015	Over 10 S.F. (use perimeter)	"	280	L.F.	1.15	3.49	4.64	6.30
	025	Brick shelf, 4" wide, add to wall forms, use wall area							
	026	above shelf, 1 use	C-2	240	S.F.C.A.	1.20	4.07	5.27	7.20
	030	2 use		275		.71	3.55	4.26	5.90
	035	4 use		300		.47	3.26	3.73	5.20
	050	Bulkhead forms for walls, with keyway, 1 use, 2 piece		265	L.F.	1.37	3.69	5.06	6.80
	055	3 piece		175	"	1.78	5.60	7.38	10

For expanded coverage of these items see Means' *Concrete & Masonry Cost Data 1986*

3.1 Formwork

		Crew	Daily Output	Unit	Bare Costs Mat.	Bare Costs Inst.	Bare Costs Total	Total Incl O&P
070	Buttress forms, to 8' high, 1 use	C-2	350	S.F.C.A.	1.32	2.79	4.11	5.45
075	2 use		430		.78	2.27	3.05	4.13
080	3 use		460		.60	2.13	2.73	3.72
085	4 use		480		.51	2.04	2.55	3.49
100	Corbel (haunch) forms, up to 12" wide, add to wall forms, 1 use		150	L.F.	1.38	6.50	7.88	10.90
105	2 use		170		.82	5.75	6.57	9.15
110	3 use		175		.63	5.60	6.23	8.70
115	4 use		180		.54	5.45	5.99	8.40
200	Job built plyform wall forms, to 8' high, 1 use		370	S.F.C.A.	1.27	2.64	3.91	5.20
205	2 use		435		.76	2.25	3.01	4.07
210	3 use		495		.57	1.97	2.54	3.47
215	4 use		505		.50	1.94	2.44	3.33
240	Over 8' high to 16' high, 1 use		280		1.34	3.49	4.83	6.50
245	2 use		345		.82	2.83	3.65	4.98
250	3 use		375		.65	2.61	3.26	4.46
255	4 use		395		.56	2.47	3.03	4.17
270	Over 16' high, 1 use		235		1.52	4.16	5.68	7.65
275	2 use		290		.95	3.37	4.32	5.90
280	3 use		315		.76	3.10	3.86	5.30
285	4 use		330		.67	2.96	3.63	5
300	For architectural finish, add		1,820		.30	.54	.84	1.10
330	For battered walls, 1 side battered, add			S.F.C.A.	10%	10%		
335	2 sides battered, add				15%	15%		
400	Radial wall forms, smooth curved, 1 use	C-2	245		1.47	3.99	5.46	7.35
405	2 use		300		.92	3.26	4.18	5.70
410	3 use		325		.74	3.01	3.75	5.15
415	4 use		335		.65	2.92	3.57	4.91
430	Curved, with 2' chords, 1 use		290		1.24	3.37	4.61	6.20
435	2 use		355		.95	2.75	3.70	5
440	3 use		385		.85	2.54	3.39	4.59
445	4 use		400		.81	2.44	3.25	4.40
460	Retaining wall forms, battered, to 8' high, 1 use		300		1.40	3.26	4.66	6.20
465	2 use		355		.86	2.75	3.61	4.90
470	3 use		375		.71	2.61	3.32	4.53
475	4 use		390		.63	2.51	3.14	4.30
490	Over 8' high to 16' high, 1 use		240		1.53	4.07	5.60	7.55
495	2 use		295		.92	3.31	4.23	5.80
500	3 use		305		.72	3.21	3.93	5.40
505	4 use		320		.62	3.06	3.68	5.05
520	For elevated walls, add to above					10%		
550	For gang wall forming, 192 S.F. sections, deduct			S.F.C.A.	10%	10%		
555	384 S.F. sections, deduct			"	20%	20%		
575	Liners for forms (add to wall forms), A.B.S. plastic							
580	Aged wood, 4" wide, 1 use	1 Carp	250	S.F.C.A.	3.50	.64	4.14	4.78
582	2 use		400		1.90	.40	2.30	2.67
584	4 use		750		1.12	.21	1.33	1.54
590	Fractured rope rib, 1 use		250		5.50	.64	6.14	7
600	4 use		750		1.78	.21	1.99	2.27
610	Ribbed look, 1/2" & 3/4" deep, 1 use		300		3.70	.53	4.23	4.84
620	4 use		800		1.10	.20	1.30	1.50
630	Rustic brick pattern, 1 use		250		3.35	.64	3.99	4.61
640	4 use		750		1.07	.21	1.28	1.49
650	Striated, random, 3/8" x 3/8" deep, 1 use		300		3.70	.53	4.23	4.84
660	4 use		800		1.10	.20	1.30	1.50
680	Rustication strips, A.B.S. plastic, 2 piece snap-on							
685	1" deep x 1-3/8" wide, 1 use	C-2	400	L.F.	2.50	2.44	4.94	6.25
690	2 use		600		1.40	1.63	3.03	3.88
695	4 use		800		.83	1.22	2.05	2.67
705	Wood, beveled edge, 3/4" deep, 1 use		600		.12	1.63	1.75	2.47
710	1" deep, 1 use		450		.18	2.17	2.35	3.32

For Expanded Coverage of these items see *Means' Concrete & Masonry Cost Data 1986*

3.1 Formwork

			CREW	DAILY OUTPUT	UNIT	BARE COSTS MAT.	BARE COSTS INST.	BARE COSTS TOTAL	TOTAL INCL O&P
65	720	For solid board finish, uniform, 1 use, add to wall forms	C-2	300	S.F.C.A.	.45	3.26	3.71	5.20
	730	Non-uniform finish, 1 use, add to wall forms	"	250		.40	3.91	4.31	6.05
	750	Lintel or sill forms, 1 use	1 Carp	30		1.36	5.35	6.71	9.20
	752	2 use		34		.76	4.71	5.47	7.65
	754	3 use		36		.56	4.44	5	7.05
	756	4 use		37		.46	4.32	4.78	6.75
	780	Modular prefabricated plywood, to 8' high, 1 use per month	C-2	910		.74	1.07	1.81	2.36
	782	2 use per month		930		.42	1.05	1.47	1.97
	784	3 use per month		950		.31	1.03	1.34	1.82
	786	4 use per month		970		.25	1.01	1.26	1.72
	800	To 16' high, 1 use per month		550		.91	1.78	2.69	3.56
	802	2 use per month		570		.54	1.72	2.26	3.06
	804	3 use per month		590		.41	1.66	2.07	2.83
	806	4 use per month		610		.36	1.60	1.96	2.70
	860	Pilasters, 1 use		270		1.44	3.62	5.06	6.80
	862	2 use		330		.91	2.96	3.87	5.25
	864	3 use		370		.73	2.64	3.37	4.60
	866	4 use		385		.65	2.54	3.19	4.37
	900	Steel framed plywood, to 8' high, 1 use per month		600		1.03	1.63	2.66	3.48
	902	2 use per month		640		.57	1.53	2.10	2.82
	904	3 use per month		655		.38	1.49	1.87	2.56
	906	4 use per month		665		.34	1.47	1.81	2.49
	920	Over 8' to 16' high, 1 use per month		455		1.17	2.15	3.32	4.38
	922	2 use per month		505		.65	1.94	2.59	3.50
	924	3 use per month		525		.44	1.86	2.30	3.16
	926	4 use per month		530		.38	1.84	2.22	3.07
	940	Over 16' to 20' high, 1 use per month		425		1.24	2.30	3.54	4.67
	942	2 use per month		435		.70	2.25	2.95	4
	944	3 use per month		455		.49	2.15	2.64	3.63
	946	4 use per month		465		.43	2.10	2.53	3.50
70-001		GAS STATION FORMS Curb fascia, with template,							
	005	12 ga. steel, left in place, 9" high	1 Carp	50	L.F.	4.15	3.20	7.35	9.20
	100	Sign or light bases, 18" diameter, 9" high		9	Ea.	25	17.80	42.80	53
	105	30" diameter, 13" high		8		45	20	65	78
	200	Island forms, 10' long, 9" high, 3'-6" wide	C-1	10		110	63	173	210
	205	4' wide		9		120	70	190	235
	250	20' long, 9" high, 4' wide		6		210	105	315	380
	255	5' wide		5		230	125	355	435
75-001		SCAFFOLDING See division 1.1-44							
80-001		SHORES Erect and strip, by hand, horizontal members							
	050	Aluminum joists and stringers	2 Carp	60	Ea.		5.35	5.35	7.75L
	060	Steel, adjustable beams		45			7.10	7.10	10.30L
	070	Wood joists		50			6.40	6.40	9.25L
	080	Wood stringers		30			10.65	10.65	15.45L
	100	Vertical members to 10' high		55			5.80	5.80	8.45L
	105	To 13' high		50			6.40	6.40	9.25L
	110	To 16' high		45			7.10	7.10	10.30L
	150	Reshoring		1,400	S.F.	.11	.23	.34	.45
	160	Flying truss system	C-17D	9,600	S.F.C.A.		.22	.22	.30

			UNIT	BUY	RENT 1ST MO.	RENT 3RD MO.	
	176	Horizontal, aluminum joists, 6' to 30' spans	L.F.	7.55	.35	.32	
	177	Aluminum stringers, 12' & 16' spans	"	11	.55	.50	
	181	Horizontal, steel beam, adjustable, 4' to 7' span	Ea.	80	3.25	3	
	183	6' to 10' span		105	6.50	3.45	
	192	9' to 15' span		185	7.20	6.30	
	194	12' to 20' span		225	14.20	7.75	
	197	Steel stringer, 6' to 15' span	L.F.	6	.30	.27	

For expanded coverage of these items see Means' *Concrete & Masonry Cost Data 1986*

3.1 Formwork

			CREW	DAILY OUTPUT	UNIT	BUY	RENT 1ST MO.	RENT 3RD MO.	TOTAL INCL O&P
80	300	Rent for job duration, aluminum			S.F.Flr.		.16		
	305	Steel			"		.13		
	350	Vertical, adjustable steel, 5'-7" to 9'-6" high, 10,000# capacity			Ea.	40	2.75	2.25	
	355	7'-3" to 12'-10" high, 7800# capacity				50	3.20	2.50	
	360	8'-10" to 12'-4" high, 10,000# capacity				55	3.40	2.70	
	365	8'-10" to 16'-1" high, 3800# capacity			↓	60	4.75	3.75	
	400	Frame shoring systems, aluminum, 10,000# per leg,							
	405	6' wide, 5' & 6' high			Ea.	170	14	13.50	
	410	5' to 7' post with base, jack screw & top plate			Set	35	2.50	2.25	
	501	Steel, 10,000# per leg							
	504	2' & 4' wide, 3',4', 5' & 6' high			Ea.	70	3.50	3.20	
	525	6' extension tube with adjusting collar				65	2.65	2.50	
	555	Base plate				7	.35	.30	
	560	12" adjustable leg				28	1.44	1.10	
	565	Top plate			↓	13	.70	.65	
	575	Flying truss system			S.F.C.A.	6.50	.50	.45	

			CREW	DAILY OUTPUT	UNIT	BARE COSTS			TOTAL INCL O&P
						MAT.	INST.	TOTAL	
85	001	**SLIPFORMS** Silos, minimum	C-17E	3,885	S.F.C.A.	.20	.44	.64	.86
	005	Maximum		1,095		.20	1.57	1.77	2.50
	100	(37) Buildings, minimum		3,660		.26	.47	.73	.97
	105	Maximum	↓	875	↓	.30	1.97	2.27	3.18
90	001	**WATERSTOP** Polyvinyl chloride, ribbed 3/16" thick, 4" wide	1 Carp	155	L.F.	.85	1.03	1.88	2.43
	005	3/16" thick, 6" wide		145		1.20	1.10	2.30	2.92
	050	Ribbed, PVC, with center bulb, 3/16" thick, 9" wide		135		1.70	1.19	2.89	3.59
	055	3/8" thick		130		2.50	1.23	3.73	4.53
	080	Dumbbell type, PVC, 6" wide, 3/16" thick		150		1.35	1.07	2.42	3.03
	085	3/8" thick		145		2.50	1.10	3.60	4.35
	100	9" wide, 3/8" thick, PVC, plain		130		3.50	1.23	4.73	5.65
	105	Center bulb		130		4.75	1.23	5.98	7
	125	Split PVC, 3/8" thick, 6" wide		145		2.30	1.10	3.40	4.13
	130	9" wide		130		3.40	1.23	4.63	5.50
	200	Rubber, flat dumbbell, 3/8" thick, 6" wide		145		4.15	1.10	5.25	6.15
	205	9" wide		135		6.85	1.19	8.04	9.25
	250	Flat dumbbell split, 3/8" thick, 6" wide		145		5.50	1.10	6.60	7.65
	255	9" wide		135		8	1.19	9.19	10.50
	300	Center bulb, 1/4" thick, 6" wide		145		4.15	1.10	5.25	6.15
	305	9" wide		135		8	1.19	9.19	10.50
	350	Center bulb split, 3/8" thick, 6" wide		145		6.30	1.10	7.40	8.55
	355	9" wide	↓	135	↓	9.70	1.19	10.89	12.40
	500	Waterstop fittings, rubber, flat							
	501	dumbbell or center bulb, 3/8" thick,							
	520	Field Union, 6" wide	1 Carp	50	Ea.	6.75	3.20	9.95	12.05
	525	9" wide		50		9	3.20	12.20	14.55
	550	Flat Cross, 6" wide		30		24	5.35	29.35	34
	555	9" wide		30		42	5.35	47.35	54
	600	Flat Tee, 6" wide		30		22	5.35	27.35	32
	605	9" wide		30		36	5.35	41.35	47
	650	Flat Ell, 6" wide		40		21	4	25	29
	655	9" wide		40		35	4	39	44
	700	Vertical Tee, 6" wide		25		19	6.40	25.40	30
	705	9" wide		25		32	6.40	38.40	44
	750	Vertical Ell, 6" wide		35		17.25	4.57	21.82	26
	755	9" wide	↓	35	↓	30	4.57	34.57	40

3.2 Reinforcing Steel

	UNIT	MATERIAL ONLY			
		PLAIN	GALV.	STNLSS.	PLASTIC
01-001 ACCESSORIES Materials only.					
002 See also Form Accessories, division 3.1-05					
010 Beam bolsters, (BB) standard, lower, up to 1-1/2" high ㊹	M.L.F.	198	270	700	315
011 2-1/2" to 3" high		515	575	1,525	755
020 Upper, standard, (BBU) to 1-1/2" high		500			
021 2-1/2" to 3" high		870			
030 Beam bolster with plate, (BBP) to 1-1/2" high		720			
031 2-1/2" to 3" high		1,020			
050 Slab bolsters, continuous, plain, (SB) 3/4" to 1" high		148	162	445	195
051 1" to 2" high		180	200	640	230
053 For bolsters with wire runners (SBR), add		245			
054 For bolsters with plates (SBP), add	↓	495			
070 Clip or bar ties, 16 ga., 3" long	M	6.30			
071 4" long		7.40			
072 6" long		8.40			
073 8" long	↓	9.45			
090 Flange clips, expandable flanges, 10 ga., 12" O.C., continuous,					
091 galvanized, over 500 L.F., 4" to 8"	C.L.F.		24		
092 9" to 12"			27		
093 17" to 24"	↓		33		
120 High chairs, individual, no plates (1 HC), to 3" high	M	220	300	615	320
121 5" high		310	410	870	410
122 8" high		670	860	1,900	860
123 12" high		1,400	1,600	3,600	1,725
124 15" high		2,625	2,925	7,025	2,825
125 For each added 1" up to 24" high, add		145	170	275	170
140 Individual high chairs, with plates, (HCP) to 5" high, add		750			
141 Over 5" high, add		800			
150 Bar chair, (BC) for 3/4" to 1" high		105	115	215	195
155 Joist chair, (JC) for up to 1-1/2" high	↓	160	200	315	275
170 Continuous high chairs, legs 8" O.C., (CHC) to 4" high	M.L.F.	294	345	1,100	350
172 6" high		435	580	1,300	560
174 8" high		600	790	1,375	790
176 12" high		1,165	1,600	3,500	1,600
178 15" high		1,330	1,810	3,925	1,810
180 For each added 1" up to 24" high, add		110	150	330	150
190 For continuous bottom plate, (CHCP), add		695			
194 For upper continuous high chairs, (CHCU), add		320			
196 For galvanized wire runners, add	↓	36			
200					
210 Paper tubing, 4' lengths, for #2 & #3 bar	C.L.F.	13			
212 For #6 bar	"	23			
220 Screed base, adjustable, 1/2" diameter (SCBA), 2-1/2" high	C	52	54		
222 5-1/2" high		59	67		
230 3/4" diameter, 2-1/2" high		65	86		
232 5-1/2" high		91	112		
240 Screed holder, 1/2" diameter (SCCH), for 1" I.D. pipe, 6" long		67			
242 12" long		115			
250 3/4" diameter, for 1-1/2" I.D. pipe, 6" long		100			
252 12" long		160			
270 Screw anchor for bolts (SCA), 1/2" diameter		35			
272 1" diameter		110			
274 1-1/2" diameter		183			
280 Screw eye bolts, (SAEB), 1/2" x 5" long		430			
282 1" x 9" long		1,850			
284 1-1/2" x 14" long		4,600			
290 Screw anchor bolts, (SAB), 1/2" x up to 7" long		185			
292 1" x up to 12" long		615			
300 Slab lifting inserts, (LCB-1) single, 3/4" diameter, 4" high		195			
301 6" high	↓	240			

For expanded coverage of these items see Means' Concrete & Masonry Cost Data 1986

3.2 Reinforcing Steel

			UNIT	MATERIAL ONLY			
				PLAIN	GALV.	STNLSS.	PLASTIC
01	303	7" high	C		275		
	310	1" diameter, 5" high			305		
	312	7" high			330		
	320	Double lifting inserts (LCB-2), 1" diameter, 5" high			625		
	322	7" high			670		
	330	1-1/4" diameter, 5" high	↓		690		
	350	Sleeper clips for wood sleepers, 20 ga., 2" wide	M		245		
	352	4" wide			300		
	360	Spacers, plastic, for 1" bar clearance, average					35
	362	For 2" bar clearance, average	↓				40
	380	Subgrade chairs, (SCSGC) 1/2" diameter, 3-1/2" high	C	205			
	385	12" high		595			
	390	3/4" diameter, 3-1/2" high		270			
	395	12" high		620			
	420	Subgrade stakes, (SCSGS) 1/2" diameter, 16" long		165			
	425	24" long		240			
	430	3/4" diameter, 16" long		180			
	435	28" long	↓	280			
	450	Tie wire, 16 ga. annealed steel, under 500 lbs.	Cwt.	62			
	452	2,000 to 4,000 lbs.	"	55			
	455	Tie wire holder, plastic case	Ea.	25			
	460	Aluminum case	"	29			
02-001		COATED REINFORCING Add to material					
	020	Epoxy coated	Cwt.	43			
	100	Galvanized, 3/8" diameter			19		
	110	1/2" diameter			18		
	120	5/8" diameter			16		
	130	3/4" diameter or over			15		
	140	For over 20 tons, 3/4" diameter or larger, minimum			14		
	150	Maximum	↓		16		

			CREW	DAILY OUTPUT	UNIT	BARE COSTS			TOTAL INC. O&P
						MAT.	INST.	TOTAL	
03-001		PRESTRESSING STEEL Post-tensioned in field							
	002								
	010	Grouted strand, 50' span, 100 kip	C-3	1,200	Lb.	1.41	1.15	2.56	3.27
	015	300 kip		2,700		1.14	.51	1.65	2.02
	030	100' span, grouted, 100 kip		1,700		.92	.81	1.73	2.22
	035	300 kip		3,200		.83	.43	1.26	1.56
	050	200' span, grouted, 100 kip		2,700		.85	.51	1.36	1.70
	055	300 kip		3,500		.80	.40	1.20	1.47
	080	Grouted bars, 50' span, 42 kip		2,600		.92	.53	1.45	1.80
	085	143 kip		3,200		.82	.43	1.25	1.55
	100	75' span, grouted, 42 kip		3,200		.77	.43	1.20	1.49
	105	143 kip	↓	4,200		.71	.33	1.04	1.27
	120	Ungrouted strand, 50' span, 100 kip	C-4	1,275		.89	.58	1.47	1.87
	125	200 kip		1,475		.89	.51	1.40	1.75
	140	100' span, ungrouted, 100 kip		1,500		.77	.50	1.27	1.61
	145	200 kip		1,650		.77	.45	1.22	1.54
	160	200' span, ungrouted, 100 kip		1,500		.71	.50	1.21	1.54
	165	200 kip		1,700		.71	.44	1.15	1.45
	180	Ungrouted bars, 50' span, 42 kip		1,400		.82	.53	1.35	1.71
	185	143 kip		1,700		.77	.44	1.21	1.52
	200	75' span, ungrouted, 42 kip		1,800		.77	.41	1.18	1.48
	205	143 kip		2,200		.71	.34	1.05	1.30
	220	Ungrouted single strand, 100' slab, 25 kip		1,200		.82	.62	1.44	1.85
	225	35 kip	↓	1,475	↓	.77	.51	1.28	1.62
04-001		REINFORCING IN PLACE A615 Grade 60							
	002								
	010	Beams & Girders, #3 to #7	4 Rodm	1.60	Ton	510	435	945	1,225
	015	#8 to #14	"	2.70	"	500	260	760	950

For expanded coverage of these items see Means' Concrete & Masonry Cost Data 1986

3.2 Reinforcing Steel

			CREW	DAILY OUTPUT	UNIT	BARE COSTS MAT.	BARE COSTS INST.	BARE COSTS TOTAL	TOTAL INCL O&P
04	020	Columns, #3 to #7	4 Rodm	1.50	Ton	510	465	975	1,275
	025	#8 to #14		2.30		500	305	805	1,025
	030	Spirals, hot rolled, 8" to 15" diameter (38)		2.20		1,075	315	1,390	1,675
	032	15" to 24" diameter		2.20		1,025	315	1,340	1,625
	033	24" to 36" diameter		2.20		1,000	315	1,315	1,600
	034	36" to 48" diameter		2.20		1,025	315	1,340	1,625
	036	48" to 64" diameter		2.20		1,075	315	1,390	1,675
	038	64" to 84" diameter		2.20		1,200	315	1,515	1,800
	039	84" to 96" diameter		2.20		1,150	315	1,465	1,750
	040	Elevated slabs, #4 to #7		2.90		510	240	750	930
	050	Footings, #4 to #7		2.10		500	330	830	1,050
	055	#8 to #14		3.60		490	195	685	840
	060	Slab on grade, #3 to #7		2.30		505	305	810	1,025
	065								
	070	Walls, #3 to #7	4 Rodm	3	Ton	505	230	735	915
	075	#8 to #14		4		490	175	665	810
	100	Typical in place, 10 ton lots, average		1.70		510	410	920	1,200
	110	Over 50 ton lots, average		2.30		490	305	795	1,000
	120	High strength steel, Grade 75, #14 bars only, add (39)				40		40	44M
	200	Unloading, sorting, piling, add to above	C-5	100			15.80	15.80	22
	220	Crane cost for handling, add to above, minimum		135			11.70	11.70	16.60
	221	Average		92			17.15	17.15	24
	222	Maximum		35			45	45	64
	240	Dowels, 2' long, deformed, #3 bar	2 Rodm	140	Ea.	.73	2.49	3.22	4.65
	241	#4 bar		125		.87	2.78	3.65	5.25
	242	#5 bar		110		1.05	3.16	4.21	6.05
	243	#6 bar		105		1.31	3.31	4.62	6.55
	245	Longer and heavier dowels		450	Lb.	.34	.77	1.11	1.57
	250	Smooth dowels, 12" long, 1/4" or 3/8" diameter		140	Ea.	.55	2.49	3.04	4.45
	252	5/8" diameter		125		.96	2.78	3.74	5.35
	253	3/4" diameter		110		1.16	3.16	4.32	6.15
	260								
	270	Dowel caps, crimp type, 5" long, 1/2" to 3/4" diameter	2 Rodm	800	Ea.	.06	.44	.50	.74
	272	1-1/4" diameter	"	750	"	.08	.46	.54	.81
05	001	**SPLICING REINFORCING BARS** Including holding bars in							
	002	place while splicing							
	010	Butt weld columns, #4 bars	C-5	190	Ea.	.70	8.30	9	12.55
	011	#6 bars		150		1.05	10.55	11.60	16.10
	013	#10 bars		95		1.40	16.60	18	25
	015	#14 bars		65		1.80	24	25.80	36
	028	Column splice clamps, sleeve & wedge, or end bearing							
	030	#7 or #8 bars	C-5	190	Ea.	2.30	8.30	10.60	14.30
	031	#9 or #10 bars		170		2.35	9.30	11.65	15.75
	032	#11 bars		160		3.20	9.85	13.05	17.55
	033	#14 bars		150		3.70	10.55	14.25	19
	034	#18 bars		140		5.80	11.30	17.10	22
	050	Reducer inserts for above, #14 to #18 bar				2.30		2.30	2.53M
	052	#14 to #11 bar				1.95		1.95	2.14M
	054	#11 to #10 bar				.60		.60	.66M
	055	#10 to #9 bar				.60		.60	.66M
	056	#9 to #8 bar				.60		.60	.66M
	058	#8 to #7 bar				.60		.60	.66M
	060	For bolted speed sleeve type, deduct					15%		
	070								
	080	Mechanical butt splice, sleeve type with filler metal, compression							
	081	only, all grades, columns only #11 bars	C-5	68	Ea.	9.45	23	32.45	43
	090	#14 bars		62		9.65	25	34.65	47
	092	#18 bars		62		11.85	25	36.85	49
	100	125% yield point, grade 60, columns only, #6 bars		68		13	23	36	47
	102	#7, 8 bars		68		11.90	23	34.90	46

For expanded coverage of these items see Means' *Concrete & Masonry Cost Data 1986*

3.2 Reinforcing Steel

		CREW	DAILY OUTPUT	UNIT	BARE COSTS MAT.	BARE COSTS INST.	BARE COSTS TOTAL	TOTAL INCL O&P
05	103	#9 bars	C-5	68	Ea.	11.60	23	34.60	46
	104	#10 bars		68		12.60	23	35.60	47
	105	#11 bars		68		15.70	23	38.70	50
	106	#14 bars		62		19.85	25	44.85	58
	107	#18 bars	↓	62	↓	29	25	54	68
	108								
	120	Full tension, grade 60 steel, columns,							
	122	slabs or beams, #6, 7, 8 bars	C-5	68	Ea.	11.50	23	34.50	46
	123	#9 bars		68		13.10	23	36.10	47
	124	#10 bars		68		14.35	23	37.35	49
	125	#11 bars		68		17.05	23	40.05	52
	126	#14 bars		62		23	25	48	61
	127	#18 bars	↓	62		37	25	62	77
	140	If equipment handling not required, deduct			↓		50%		
	160	Mechanical threaded type, bar threading not included,							
	170	straight bars, #10 & #11	C-5	190	Ea.	10.10	8.30	18.40	23
	175	#14 bars		170		13.60	9.30	22.90	28
	180	#18 bars		100		21.80	15.80	37.60	46
	200	#11 to #14 transition		190		13.05	8.30	21.35	26
	210	#11 to #18 & #14 to #18 transition		100		23	15.80	38.80	48
	240	Bent bars, #10 & #11		140		19.80	11.30	31.10	38
	250	#14		120		26	13.15	39.15	47
	260	#18		90		39	17.55	56.55	68
	280	#11 to #14 transition		100		26	15.80	41.80	51
	290	#11 to #18 & #14 to #18 transition	↓	90	↓	39	17.55	56.55	68
06-001		WELDED WIRE FABRIC Rolls, 6 x 6 = #10/10 (W1.4/W1.4) 21 lb.	2 Rodm	35	C.S.F.	7.65	9.95	17.60	24
	020	6 x 6 = #8/8 (W2.1/W2.1) 30 lb. per C.S.F.		31		10.50	11.25	21.75	29
	030	6 x 6 = #6/6 (W2.9/W2.9) 42 lb. per C.S.F.		29		14.80	12	26.80	35
	040	6 x 6 = #4/4 (W4/W4) 58 lb. per C.S.F.		27		19.80	12.90	32.70	42
	050	4 x 4 = #10/10 (W1.4/W1.4) 31 lb. per C.S.F.		31		11.75	11.25	23	30
	060	4 x 4 = #8/8 (W2.1/W2.1) 44 lb. per C.S.F.		29		14.05	12	26.05	34
	065	4 x 4 = #6/6 (W2.9/W2.9) 61 lb. per C.S.F.		27		20	12.90	32.90	42
	070	4 x 4 = #4/4 (W4/W4) 85 lb. per C.S.F.		25		30	13.90	43.90	55
	080	2 x 2 = #14 galv. @ 21 lb., beam & column wrap		6.50		13.40	54	67.40	98
	090	2 x 2 = #12 galv. for gunite reinforcing	↓	6.50	↓	18.50	54	72.50	105
	095	Material prices for above include 10% lap							
	098								
	100	Specially fabricated heavier gauges, in sheets	4 Rodm	50	C.S.F.		13.90	13.90	22L
	101	Material only, minimum			Ton	580		580	640M
	102	Average			↓	675		675	745M
	103	Maximum				835		835	920M

3.3 Cast in Place Concrete

		CREW	DAILY OUTPUT	UNIT	BARE COSTS MAT.	BARE COSTS INST.	BARE COSTS TOTAL	TOTAL INCL O&P
01-001	AGGREGATE Expanded shale, C.L. lots, 43.5 to 52 lb. per C.F., minimum			Ton	32		32	35M
005	Maximum			"	37		37	41M
010	Lightweight vermiculite or perlite, 4 C.F. bag, C.L. lots			Bag	5		5	5.50M
015	L.C.L. lots			"	5.65		5.65	6.20M
025	Sand & stone, loaded at pit, crushed bank gravel ㊷			Ton	4.45		4.45	4.89M
030	Fill, bank run				2.25		2.25	2.47M
035	Sand, washed, for concrete				6.55		6.55	7.20M
040	For plaster or brick				7		7	7.70M
045	Stone, 3/4" to 1-1/2"				7.75		7.75	8.50M
050	3/8" roofing stone & 1/2" pea stone			↓	8.75		8.75	9.60M

For expanded coverage of these items see Means' *Concrete & Masonry Cost Data 1986*

3.3 Cast in Place Concrete

				DAILY		BARE COSTS			TOTAL
			CREW	OUTPUT	UNIT	MAT.	INST.	TOTAL	INCL O&P
01	055	For trucking 10 miles, add to the above			Ton	1.55		1.55	1.70M
	060	30 miles, add to the above			"	2.25		2.25	2.47M
	085	Sand & stone, loaded at pit, crushed bank gravel			C.Y.	6.25		6.25	6.85M
	090	Fill, bank run				3.15		3.15	3.46M
	095	Sand, washed, for concrete				9.15		9.15	10.05M
	100	For plaster or brick				9.80		9.80	10.80M
	105	Stone, 3/4" to 1-1/2"				10.85		10.85	11.95M
	110	3/8" roofing stone & 1/2" pea stone				12.25		12.25	13.45M
	115	For trucking 10 miles, add to the above				1.55		1.55	1.70M
	120	30 miles, add to the above			↓	2.25		2.25	2.47M
	131	Quartz chips, 50 lb. bags			Cwt.	11.50		11.50	12.65M
	133	Silica chips, 50 lb. bags				6.25		6.25	6.85M
	141	White marble, 3/8" to 1/2", 50 lb. bags			↓	9.80		9.80	10.80M
	143	3/4" F.O.B. plant			Ton	21		21	23M
02-001		**ANCHOR BOLTS** See division 3.1-05							
04-001		**CEMENT** Calcium aluminate, high temperature, by truck			Bag	21.50		21.50	24M
	002	By rail				19.40		19.40	21M
	005	Masonry cement, gray, T.L. or C.L. lots, Boston				4.70		4.70	5.15M
	006	L.T.L. or L.C.L. lots, Boston				5.05		5.05	5.55M
	007	T.L., U.S. average				4.75		4.75	5.20M
	010	White masonry cement, T.L. or C.L. lots				13		13	14.30M
	012	L.T.L. or L.C.L. lots				13.75		13.75	15.10M
	020	Portland, plain or air entrained, T.L. or C.L., Boston				5.20		5.20	5.70M
	022	L.T.L. or L.C.L. lots, Boston				6.10		6.10	6.70M
	024	T.L., lots, U.S. average			↓	5.15		5.15	5.65M
	030	Portland, trucked in bulk, Boston			Cwt.	3.23		3.23	3.55M
	032	U.S. average			"	3.33		3.33	3.66M
	040	Portland, high early strength, T.L. lots, bags			Bag	6.15		6.15	6.75M
	042	L.T.L. or L.C.L. lots				6.70		6.70	7.35M
	050	White high early strength, T.L. or C.L. lots, bags				13.60		13.60	14.95M
	052	L.T.L. or L.C.L. lots				14		14	15.40M
	060	White portland cement, T.L. or C.L. lots, bags				13		13	14.30M
	062	L.T.L. or L.C.L. lots			↓	13.60		13.60	14.95M
08-001		**CONCRETE ADMIXTURES & SURFACE TREATMENTS**							
	002								
	004	Abrasives, aluminum oxide, over 20 tons			Lb.	.60		.60	.66M
	007	Under 1 ton				.65		.65	.71M
	010	Silicon carbide, black, over 20 tons				.85		.85	.93M
	012	Under 1 ton			↓	.90		.90	.99M
	020	Air entraining agent, .7 to 1.5 oz. per bag, 55 gallon lots			Gal.	4		4	4.40M
	022	5 gallon lots				5.25		5.25	5.75M
	030	Bonding agent, acrylic latex (200-250 S.F. per gallon)				19		19	21M
	032	Epoxy resin (70-80 S.F. per gallon)			↓	46		46	51M
	040	Calcium chloride, 100 lb. bags, FOB plant, truckload lots			Ton	245		245	270M
	042	Less than truckload lots			Bag	15.10		15.10	16.60M
	050	Carbon black, liquid, 2 to 8 lbs. per bag of cement			Lb.	2.25		2.25	2.47M
	060	Colors, integral, 2 to 10 lb. per bag of cement, minimum				.80		.80	.88M
	061	Average				1.30		1.30	1.43M
	062	Maximum			↓	3		3	3.30M
	070	Curing compound, 200 to 400 S.F. per gallon, 55 gal. lots			Gal.	5.45		5.45	6M
	072	5 gallon lots				6.50		6.50	7.15M
	080	Premium grade, 450 S.F. per gallon, 55 gallon lots				11.50		11.50	12.65M
	082	5 gallon lots				12		12	13.20M
	090	Dustproofing compound, 200-600 S.F./gal., 55 gallon lots				6		6	6.60M
	092	5 gallon lots			↓	7.20		7.20	7.90M
	100	Epoxy dustproof coating, colors, 300-400 S.F. per coat,							
	101	or transparent, 400-600 S.F. per coat			Gal.	45		45	50M
	110	Hardeners, metallic, 55 lb. bags, natural (grey)			Lb.	.35		.35	.38M
	120	Colors, average			"	.55		.55	.60M

For expanded coverage of these items see Means' *Concrete & Masonry Cost Data 1986*

3.3 Cast in Place Concrete

		Description	CREW	DAILY OUTPUT	UNIT	BARE COSTS MAT.	BARE COSTS INST.	BARE COSTS TOTAL	TOTAL INCL O&P
08	130	Non-metallic, 55 lb. bags, natural (grey), minimum			Lb.	.30		.30	.33M
	131	Maximum				.40		.40	.44M
	132	Non-metallic, colors, minimum				.35		.35	.38M
	134	Maximum				.45		.45	.49M
	140	Non-metallic, non-slip, 100 lb. bags, minimum				.35		.35	.38M
	142	Maximum			↓	.45		.45	.49M
	150	Solution type, 300 to 400 S.F. per gallon			Gal.	4.80		4.80	5.30M
	151								
	155	Release agent, for tilt slabs			Gal.	7.50		7.50	8.25M
	157	For forms, average				5		5	5.50M
	160	Sealer, hardener and dustproofer, clear, 450 S.F., minimum				7		7	7.70M
	162	Maximum			↓	16.50		16.50	18.15M
	170	Colors (300-400 S.F. per gallon)				15		15	16.50M
	171								
	180	Set accelerator for below freezing, 1 to 1-1/2 gal. per C.Y.			Gal.	4.20		4.20	4.62M
	190	Set retarder, 2 to 4 fl. oz. per bag of cement			"	12.80		12.80	14.10M
	200	Waterproofing, integral 1 lb. per bag of cement			Lb.	.72		.72	.79M
	210	Powdered metallic, 40 lbs. per 100 S.F., minimum			↓	.77		.77	.84M
	212	Maximum				1		1	1.10M
	220	Water reducing admixture, average			Gal.	7.75		7.75	8.50M
10-001		**CONCRETE, FIELD MIX** FOB forms 2250 psi	(45)		C.Y.	44.60		44.60	49M
002		3000 psi				47.45		47.45	52M
12-001		**CONCRETE, READY MIX** Regular weight, 2000 psi	(43)			45.55		45.55	50M
010		2500 psi				47.25		47.25	52M
015		3000 psi	(42)			48.90		48.90	54M
020		3500 psi				50.55		50.55	56M
025		3750 psi				51.50		51.50	57M
030		4000 psi				52.20		52.20	57M
035		4500 psi				54.15		54.15	60M
040		5000 psi				55.15		55.15	61M
100		For high early strength cement, add				10%			
101		For structural lightweight with regular sand, add				27%			
200		For all lightweight aggregate, add			↓	50%			
300		For integral colors, 2500 psi, 5 bag mix							
310		Red, yellow or brown, 1.8 lb. per bag, add			C.Y.	12.60		12.60	13.85M
320		9.4 lb. per bag, add				66		66	73M
340		Black, 1.8 lb. per bag, add				13.10		13.10	14.40M
350		7.5 lb. per bag, add				55		55	61M
370		Green, 1.8 lb. per bag, add				27		27	30M
380		7.5 lb. per bag, add			↓	115		115	125M
14-001		**CONCRETE IN PLACE** Including forms (4 uses), reinforcing							
005		steel, including finishing unless otherwise indicated							
010		Average for concrete framed building,	(35)						
011		including finishing		C-17B	15.75	C.Y.	98	120	218	280
013		Average for substructure only, simple design, incl. finishing	(47)	29.07		69	65	134	170
015		Average for superstructure only, including finishing		13.42	↓	105	140	245	315
020		Base, granolithic, 1" x 5" high, straight	C-10	175	L.F.	.12	2.81	2.93	4.01
022		Cove	(50)	140	"	.12	3.51	3.63	4.98
030		Beams, 5 kip per L.F., 10' span	C-17A	6.28	C.Y.	168	285	453	595
035		25' span		7.40		135	240	375	495
050		Chimney foundations, minimum	(122)	26.70		89	67	156	195
051		Maximum		19.70		102	91	193	240
070		Columns, square, 12" x 12", minimum reinforcing	(49)	4.60		180	390	570	755
072		Average reinforcing		4.10		255	435	690	910
074		Maximum reinforcing	C-17B	3.84		380	495	875	1,125
080		16" x 16", minimum reinforcing	C-17A	6.25		160	285	445	585
082		Average reinforcing	"	4.93		250	360	610	795
084		Maximum reinforcing	C-17B	4.34		420	435	855	1,075
090		24" x 24", minimum reinforcing	C-17A	9.08		140	195	335	435
092		Average reinforcing	"	6.90	↓	210	260	470	605

For expanded coverage of these items see Means' *Concrete & Masonry Cost Data 1986*

3.3 Cast in Place Concrete

			DAILY		BARE COSTS			TOTAL
		CREW	OUTPUT	UNIT	MAT.	INST.	TOTAL	INCL O&P
14 094	Maximum reinforcing	C-17A	5.65	C.Y.	345	315	660	835
100	36" x 36", minimum reinforcing	C-17B	13.39		130	140	270	345
102	Average reinforcing		9.61		200	195	395	495
104	Maximum reinforcing		7.50		325	250	575	715
120	Columns, round, tied, 16" diameter, minimum reinforcing		13.02		225	145	370	455
122	Average reinforcing		8.30		375	230	605	735
124	Maximum reinforcing		6.05		510	315	825	1,000
130	20" diameter, minimum reinforcing		17.35		210	110	320	385
132	Average reinforcing		10.43		340	180	520	630
134	Maximum reinforcing		7.47		460	255	715	865
140	24" diameter, minimum reinforcing		22.18		200	85	285	340
142	Average reinforcing		11.86		290	160	450	545
144	Maximum reinforcing		8.10		460	235	695	835
150	36" diameter, minimum reinforcing		32.40		180	58	238	280
152	Average reinforcing		16.57		245	115	360	430
154	Maximum reinforcing		11.15		335	170	505	610
170	Curbs, formed in place, 6" x 18", straight,	C-15	400	L.F.	2.90	3.51	6.41	8.25
175	Curb and gutter	"	170	"	4.65	8.25	12.90	17
190	Elevated slabs, flat slab, 125 psf Sup. Load, 20' span	C-17A	13.36	C.Y.	110	135	245	315
195	30' span	C-17B	18.25		105	105	210	260
210	Flat plate, 125 psf Sup. Load, 15' span	C-17A	10.28		115	175	290	375
215	25' span	C-17B	17.01		97	110	207	265
230	Waffle const., 30" domes, 125 psf Sup. Load, 20' span		14.10		120	135	255	320
235	30' span		17.02		113	110	223	280
250	One way joists, 30" pans, 125 psf Sup. Load, 15' span	C-17A	11.07		110	160	270	355
255	25' span		11.04		125	160	285	365
270	One way beam & slab, 125 psf Sup. Load, 15' span		7.49		125	240	365	480
275	25' span		10.15		120	175	295	385
290	Two way beam & slab, 125 psf Sup. Load, 15' span		8.22		120	215	335	445
295	25' span	C-17B	12.23		105	155	260	335
310	Elevated slabs including finish, not							
311	including forms or reinforcing							
315	Regular concrete, 4" slab	C-8	2,685	S.F.	.64	.56	1.20	1.45
320	6" slab		2,585		1.02	.58	1.60	1.89
325	2-1/2" thick floor fill		2,685		.44	.56	1	1.23
330	Lightweight, 110# per C.F., 2-1/2" thick floor fill		2,585		.56	.58	1.14	1.39
340	Cellular concrete, 1-5/8" fill, under 5000 S.F.		2,000		.25	.75	1	1.27
345	Over 10,000 S.F.		2,200		.21	.69	.90	1.14
350	Add per floor for 3 to 6 stories high		31,800			.05	.05	.06
352	For 7 to 20 stories high		21,200			.07	.07	.09
380	Footings, spread under 1 C.Y.	C-17B	31.82	C.Y.	71	59	130	160
385	Over 5 C.Y.	C-17C	70.45		68	28	96	115
390	Footings, strip, 18" x 9", plain	C-17B	34.22		61	55	116	145
395	36" x 12", reinforced		49.07		67	39	106	130
400	Foundation mat, under 10 C.Y.		32.32		113	59	172	205
405	Over 20 C.Y.		47.37		100	40	140	165
420	Grade walls, 8" thick, 8' high	C-17A	10.16		100	175	275	365
425	14' high	C-20	7.30		150	230	380	465
426	12" thick, 8' high	C-17A	13.50		120	130	250	320
427	14' high	C-20	11.60		113	145	258	315
430	15" thick, 8' high	C-17B	20.01		85	95	180	225
435	12' high	C-20	14.80		95	110	205	255
450	18' high	"	12		109	140	249	305
451								
465	Ground slab, not including finish, 4" thick	C-17C	75.28	C.Y.	61	27	88	105
470	6" thick	"	113.47	"	58	17.70	75.70	89
475	Ground slab, incl. troweled finish, not incl. forms							
476	or reinforcing, over 10,000 S.F., 4" thick slab	C-8	3,520	S.F.	.71	.43	1.14	1.35
482	6" thick slab		3,610		1.05	.42	1.47	1.71
484	8" thick slab		3,275		1.44	.46	1.90	2.19

For expanded coverage of these items see Means' *Concrete & Masonry Cost Data 1986*

3.3 Cast in Place Concrete

			CREW	DAILY OUTPUT	UNIT	BARE COSTS MAT.	BARE COSTS INST.	BARE COSTS TOTAL	TOTAL INCL O&P
14	490	12" thick slab	C-8	2,875	S.F.	2.15	.52	2.67	3.06
	495	15" thick slab	"	2,560	"	2.70	.59	3.29	3.75
	520	Lift slab in place above the foundation, incl. forms,							
	521	(48) reinforcing, concrete and columns, minimum	C-17E	745	S.F.	1.87	2.31	4.18	5.40
	525	Average		655		3	2.63	5.63	7.10
	530	Maximum	↓	430	↓	3.25	4	7.25	9.35
	550	Lightweight, ready mix, including screed finish only,							
	551	(54) not including forms or reinforcing							
3	555	(43) 1:4 for structural roof decks	C-8	80	C.Y.	69.30	18.85	88.15	100
	560	1:6 for ground slab with radiant heat		90		67	16.75	83.75	96
	565	1:3:2 with sand aggregate, roof deck		80		69	18.85	87.85	100
	570	Ground slab	↓	105		69	14.35	83.35	95
	590	Pile caps, incl. forms and reinf., square or rectangular, under 5 C.Y.	C-17C	47.26		70	42	112	135
	595	Over 10 C.Y.		76.47		69	26	95	115
	600	Triangular or hexagonal, under 5 C.Y.		47.95		68	42	110	135
	605	Over 10 C.Y.	↓	77.85		70	26	96	115
	620	Retaining walls, gravity, 4' high (see also division 2.7-32)	C-17B	19.10		75	99	174	225
	625	10' high		27.10		68	70	138	175
	630	Cantilever, level backfill loading, 8' high		16.30		89	115	204	260
	635	16' high		17		94	110	204	260
	680	Stairs, not including safety treads, free standing	C-15	120	L.F.Nose	4.55	11.70	16.25	22
	685	Cast on ground		180	"	3.40	7.80	11.20	15
	700	Stair landings, free standing		285	S.F.	1.85	4.93	6.78	9.15
	705	Cast on ground	↓	685	"	1.08	2.05	3.13	4.14
16-001		CURING With burlap, 4 uses assumed, 7.5 oz.	2 Clab	55	C.S.F.	2.20	4.63	6.83	9.10
	010	12 oz.		55		3.25	4.63	7.88	10.25
	020	With waterproof curing paper, 2 ply, reinforced		70		4.50	3.63	8.13	10.20
	030	With sprayed membrane curing compound	↓	95	↓	1.70	2.68	4.38	5.75
	040	Curing blankets, 1" to 2" thick, buy, minimum			S.F.	.60		.60	.66M
	045	Maximum				1.50		1.50	1.65M
	050	Electrically heated pads, 110 volts, 10 watts per S.F.				3		3	3.30M
	060	20 watts per S.F.			↓	4		4	4.40M
18-001		CUTTING Concrete see division 2.1-44							
20-001		DAMPPROOFING See division 7.1							
22-001		(13) EQUIPMENT For placing concrete see division 1.5-10, 1.5-20							
24-001		EXPANSION JOINT See division 3.1-10							
26-001		FINISHING FLOORS Monolithic, screed finish	1 Cefi	900	S.F.		.17	.17	.24L
	005	Darby finish	"	750			.20	.20	.29L
	010	Float finish	C-9	725			.25	.25	.34
	015	Broom finish		675			.27	.27	.37
	020	Steel trowel finish, for resilient tile		625			.29	.29	.40
	025	For finish floor	↓	550			.33	.33	.45
	040	Integral topping and finish, using 1:1:2 mix, 3/16" thick	C-10	1,000		.03	.49	.52	.71
	045	(54) 1/2" thick		950		.08	.52	.60	.80
	050	3/4" thick		850		.12	.58	.70	.93
	060	1" thick		750		.16	.65	.81	1.08
	080	Granolithic topping, laid after, 1:1:1-1/2 mix, 1/2" thick		590		.10	.83	.93	1.26
	082	3/4" thick		580		.15	.85	1	1.34
	085	(50) 1" thick		575		.21	.85	1.06	1.41
	095	2" thick		500		.43	.98	1.41	1.83
	120	Heavy duty, 1:1:2, 3/4" thick, preshrunk, gray, 20 MSF		320		.18	1.54	1.72	2.32
	130	100 MSF		380		.18	1.29	1.47	1.98
	135	For colors, .50 psf, add, minimum		1,650		.40	.30	.70	.85
	140	Maximum	↓	1,500		1.60	.33	1.93	2.21
	160	Exposed local aggregate finish, minimum	1 Cefi	625		.06	.25	.31	.41
	165	Maximum	"	465	↓	.18	.33	.51	.66

For expanded coverage of these items see *Means' Concrete & Masonry Cost Data 1986*

3.3 Cast in Place Concrete

		CREW	DAILY OUTPUT	UNIT	BARE COSTS MAT.	BARE COSTS INST.	BARE COSTS TOTAL	TOTAL INCL O&P
26	180	Floor abrasives, .25 psf, add to above, aluminum oxide	C-9	850	S.F.	.08	.21	.29	.38
	185	Silicon carbide		850		.11	.21	.32	.41
	200	Floor hardeners, metallic, light service, .50 psf, add		850		.16	.21	.37	.47
	205	Medium service, .75 psf, add		750		.24	.24	.48	.59
	210	Heavy service, 1.0 psf, add		650		.31	.28	.59	.72
	215	Extra heavy, 1.5 psf, add		575		.36	.32	.68	.83
	230	Non-metallic, add to above, light service, .50 psf add		850		.05	.21	.26	.35
	235	Medium service, .75 psf, add		750		.07	.24	.31	.41
	240	Heavy service, 1.00 psf, add		650		.09	.28	.37	.48
	245	Extra heavy, 1.50 psf, add	↓	575	↓	.13	.32	.45	.57
	260	Add for colored hardeners, metallic				50%			
	265	Non-metallic			↓	25%			
	280	Trap rock wearing surface for monolithic floors							
	281	2.0 psf, add to above	C-10	1,250	S.F.	.20	.39	.59	.76
	300	Floor coloring, dusted on, 0.5 psf per S.F., add to above, minimum	C-9	1,300		.45	.14	.59	.69
	305	Maximum	"	625	↓	1.60	.29	1.89	2.16
	310	Colors only, minimum			Lb.	.80		.80	.88M
	312	Maximum			"	3.20		3.20	3.52M
	320	Integral colors, see division 3.3-12-300							
	325								
	360	1/2" topping using 5 lb. per bag, regular colors	C-10	590	S.F.	.11	.83	.94	1.27
	365	Blue or green	"	590		.14	.83	.97	1.30
	380	Dustproofing, silicate liquids, 1 coat	1 Cefi	1,900		.02	.08	.10	.14
	385	2 coats		1,300		.03	.12	.15	.20
	400	Epoxy coating, 1 coat, clear		1,500		.25	.10	.35	.42
	405	Colors		1,500		.25	.10	.35	.42
	440	Stair finish, float		275			.56	.56	.79L
	450	Steel trowel finish		200			.77	.77	1.08L
	460	Silicon carbide finish, .25 psf	↓	150	↓	.35	1.02	1.37	1.83
28-001		**FINISHING WALLS** Break ties and patch voids	1 Cefi	540	S.F.	.01	.28	.29	.41
	005	Burlap rub with grout		450		.05	.34	.39	.54
	010	Carborundum rub, dry		270		.05	.57	.62	.86
	015	Wet rub		175		.04	.88	.92	1.28
	030	Bush hammer, green concrete		170		.01	.90	.91	1.28
	035	Cured concrete		110		.02	1.40	1.42	1.99
	060	Float finish, 1/16" thick	↓	300	↓	.01	.51	.52	.73
	065								
	070	Sandblast, light penetration	C-10	1,100	S.F.	.12	.45	.57	.75
	075	Heavy penetration	"	375	"	.35	1.31	1.66	2.20
	080	Board finish, see division 3.1-65-575 & 720							
	085	Rustication strips, see division 3.1-65-680							
30-001		**FLOOR PATCHING** See division 3.3-37							
32-001		**GROUT** Column or machine bases, non-shrink, metallic grout, 1" deep	1 Cefi	35	S.F.	3.05	4.39	7.44	9.55
	005	2" deep		25		6.10	6.15	12.25	15.35
	030	Non-shrink, non-metallic grout, 1" deep		35		2.30	4.39	6.69	8.70
	035	2" deep	↓	25	↓	4.60	6.15	10.75	13.70
34-001		**GUNITE**							
	002	Applied in 1" layers, no mesh included	C-8	1,550	S.F.	.70	.97	1.67	2.06
	010	Mesh for gunite 2 x 2, #12, to 3" thick	2 Rodm	600		.80	.58	1.38	1.78
	015	Over 3" thick	"	400		1	.87	1.87	2.44
	030	Typical in place, including mesh, 2" thick, minimum	C-16	665		1.30	2.79	4.09	5.25
	035	Maximum		395		1.90	4.70	6.60	8.50
	050	4" thick, minimum	↓	595		1.90	3.12	5.02	6.35
	055	Maximum		275		2.60	6.75	9.35	12.05
	090	Prepare old walls, no scaffolding, minimum	C-10	1,000		.45	.49	.94	1.17
	095	Maximum	"	275		1.25	1.79	3.04	3.84
	110	For high finish requirement or close tolerance, add, minimum					50%		
	115	Maximum					110%		

For expanded coverage of these items see *Means' Concrete & Masonry Cost Data 1986*

3.3 Cast in Place Concrete

		CREW	DAILY OUTPUT	UNIT	BARE COSTS MAT.	BARE COSTS INST.	BARE COSTS TOTAL	TOTAL INCL O&P
36-001 002	(51) **INSULATING CONCRETE** See division 3.3-14-550 and division 3.5-25							
37-001 002	**PATCHING CONCRETE**							
010	Floors, 1/4" thick, small areas, regular grout	1 Cefi	170	S.F.	.73	.90	1.63	2.07
015	Epoxy grout	"	100	"	2.60	1.54	4.14	5
200	Walls, including chipping, cleaning and epoxy grout							
210	Minimum	1 Cefi	65	S.F.	.12	2.36	2.48	3.46
215	Average	↓	50		.18	3.07	3.25	4.52
220	Maximum	↓	40	↓	.29	3.84	4.13	5.70

		CREW	DAILY OUTPUT	UNIT	BARE COSTS EQUIP.	BARE COSTS LABOR	BARE COSTS TOTAL	TOTAL INCL O&P
38-001 002	**PLACING CONCRETE** and vibrating, including labor & equipment							
005	(46) Beams, elevated, small beams, pumped	C-20	40	C.Y.	14.15	27	41.15	55
010	With crane and bucket	C-7	45		13.30	24	37.30	50
020	Large beams, pumped	C-20	60		9.45	18.30	27.75	37
025	With crane and bucket	C-7	65		9.20	16.90	26.10	34
040	Columns, square or round, 12" thick, pumped	C-20	45		12.55	24	36.55	49
045	With crane and bucket	C-7	40		14.95	27	41.95	56
060	18" thick, pumped	C-20	60		9.45	18.30	27.75	37
065	With crane and bucket	C-7	55		10.85	19.95	30.80	41
080	24" thick, pumped	C-20	92		6.15	11.95	18.10	24
085	With crane and bucket	C-7	70		8.55	15.70	24.25	32
100	36" thick, pumped	C-20	140		4.04	7.85	11.89	15.75
105	With crane and bucket	C-7	100		5.95	11	16.95	22
140	Elevated slabs, less than 6" thick, pumped	C-20	110		5.15	10	15.15	20
145	With crane and bucket	C-7	95		6.30	11.55	17.85	24
150	6" to 10" thick, pumped	C-20	130		4.35	8.45	12.80	16.95
155	With crane and bucket	C-7	110		5.45	10	15.45	20
160	Slabs over 10" thick, pumped	C-20	150		3.77	7.30	11.07	14.70
165	With crane and bucket	C-7	130		4.60	8.45	13.05	17.25
190	Footings, continuous, shallow, direct chute	C-6	120		.45	6.70	7.15	10.15
195	Pumped	C-20	100		5.65	11	16.65	22
200	With crane and bucket	C-7	90		6.65	12.20	18.85	25
210	Deep continuous footings, direct chute	C-6	155		.35	5.20	5.55	7.90
215	Pumped	C-20	120		4.71	9.15	13.86	18.40
220	With crane and bucket	C-7	110		5.45	10	15.45	20
240	Footings, spread, under 1 C.Y., direct chute	C-6	55		.99	14.65	15.64	22
245	Pumped	C-20	50		11.30	22	33.30	44
250	With crane and bucket	C-7	45		13.30	24	37.30	50
260	Spread footings, over 5 C.Y., direct chute	C-6	110		.49	7.30	7.79	11.10
265	Pumped	C-20	105		5.40	10.45	15.85	21
270	With crane and bucket	C-7	100		5.95	11	16.95	22
290	Foundation mats, over 20 C.Y., direct chute	C-6	350		.16	2.30	2.46	3.49
295	Pumped	C-20	325		1.74	3.38	5.12	6.80
300	With crane and bucket	C-7	300		1.99	3.66	5.65	7.45
320	Grade beams, direct chute	C-6	150		.36	5.35	5.71	8.15
325	Pumped	C-20	130		4.35	8.45	12.80	16.95
330	With crane and bucket	C-7	120		4.98	9.15	14.13	18.70
350	High rise, for more than 5 stories, pumped, add per story	C-20	2,100		.27	.52	.79	1.05
351	With crane and bucket, add per story	C-7	2,100		.28	.52	.80	1.07
370	Pile caps, under 5 C.Y., direct chute	C-6	90		.60	8.95	9.55	13.55
375	Pumped	C-20	85		6.65	12.90	19.55	26
380	With crane and bucket	C-7	80		7.45	13.70	21.15	28
385	Pile cap, 5 C.Y. to 10 C.Y., direct chute	C-6	175		.31	4.60	4.91	7
390	Pumped	C-20	160		3.53	6.85	10.38	13.80
395	With crane and bucket	C-7	150		3.98	7.30	11.28	14.95
400	Pile cap, over 10 C.Y., direct chute	C-6	215		.25	3.75	4	5.70
405	Pumped	C-20	195	↓	2.90	5.65	8.55	11.30

For expanded coverage of these items see *Means' Concrete & Masonry Cost Data 1986*

3.3 Cast in Place Concrete

		CREW	DAILY OUTPUT	UNIT	BARE COSTS EQUIP.	BARE COSTS LABOR	BARE COSTS TOTAL	TOTAL INCL O&P
38 410	With crane and bucket	C-7	185	C.Y.	3.23	5.95	9.18	12.10
430	Slab on grade, 4" thick, direct chute	C-6	110		.49	7.30	7.79	11.10
435	Pumped	C-20	120		4.71	9.15	13.86	18.40
440	With crane and bucket	C-7	110		5.45	10	15.45	20
460	Slab over 6" thick, direct chute	C-6	165		.33	4.88	5.21	7.40
465	Pumped	C-20	165		3.43	6.65	10.08	13.35
470	With crane and bucket	C-7	145		4.12	7.55	11.67	15.45
490	Walls, 8" thick, direct chute	C-6	90		.60	8.95	9.55	13.55
495	Pumped	C-20	85		6.65	12.90	19.55	26
500	With crane and bucket	C-7	80		7.45	13.70	21.15	28
505	12" thick, direct chute	C-6	100		.54	8.05	8.59	12.20
510	Pumped	C-20	95		5.95	11.55	17.50	23
520	With crane and bucket	C-7	90		6.65	12.20	18.85	25
530	15" thick, direct chute	C-6	105		.52	7.65	8.17	11.65
535	Pumped	C-20	100		5.65	11	16.65	22
540	With crane and bucket	C-7	95		6.30	11.55	17.85	24
560	Wheeled concrete dumping, add to placing costs above							
561 (46)	Walking cart, 50' haul, add	C-18	32	C.Y.	1.13	4.53	5.66	7.80
562	150' haul, add		24		1.51	6.05	7.56	10.40
570	250' haul, add		18		2.01	8.05	10.06	13.90
580	Riding cart, 50' haul, add	C-19	80		.72	1.81	2.53	3.42
581	150' haul, add		60		.96	2.42	3.38	4.56
590	250' haul, add		45		1.28	3.22	4.50	6.10

		CREW	DAILY OUTPUT	UNIT	BARE COSTS MAT.	BARE COSTS INST.	BARE COSTS TOTAL	TOTAL INCL O&P
40-001 (55)	**PRESTRESSED CONCRETE** pretensioned, see division 3.4							
002								
010 (40)	Post tensioned in place, small job, see also division 3.2-03	C-17B	8.50	C.Y.	190	225	415	525
020	Large job	"	10	"	130	190	320	410
42-001	**SAWING CONCRETE** See division 2.1-44							
48-001	**STAIR TREAD INSERTS** Cast iron, abrasive, 3" wide	1 Carp	90	L.F.	4.80	1.78	6.58	7.85
002	4" wide		80		6.60	2	8.60	10.15
004	6" wide		75		8.90	2.13	11.03	12.90
005	9" wide		70		13.20	2.29	15.49	17.85
010	12" wide		65		19.20	2.46	21.66	25
030	Cast aluminum, compared to cast iron, deduct				10%			
050	Extruded aluminum safety tread, 3" wide	1 Carp	75		3.15	2.13	5.28	6.55
055	4" wide		75		4.50	2.13	6.63	8.05
060	6" wide		75		8.10	2.13	10.23	12
065	9" wide to resurface stairs		70		10	2.29	12.29	14.30
170	Cement filled pan type, plain	1 Cefi	115	S.F.	1.50	1.34	2.84	3.53
175	Non-slip	"	100	"	2.30	1.54	3.84	4.69
52-001	**WATERPROOFING AND DAMPPROOFING** See Division 7.1							
005	Integral waterproofing, add to cost of regular concrete			C.Y.	3.60		3.60	3.96M
54-001	**WINTER PROTECTION** for heated ready mix, add, minimum				2.60		2.60	2.86M
005	Maximum				2.85		2.85	3.13M
010	Protecting concrete and temporary heat, add, minimum	2 Clab	6,000	S.F.	.06	.04	.10	.13
015	Maximum, see also division 1.1-68	"	2,000	"	.40	.13	.53	.62

For expanded coverage of these items see *Means' Concrete & Masonry Cost Data 1986*

3.4 Precast Concrete

		CREW	DAILY OUTPUT	UNIT	BARE COSTS MAT.	BARE COSTS INST.	BARE COSTS TOTAL	TOTAL INCL O&P
01-001	**BEAMS** Rectangular, to 20' spans, small beams	C-11	320	L.F.	37	8.05	45.05	52
005	Large beams		240		125	10.70	135.70	150
030	30' spans, small beams		480		60	5.35	65.35	73
035	Large beams		360		125	7.15	132.15	145
060	40' spans, small beams		640		82	4.02	86.02	96
065	Large beams	↓	480		145	5.35	150.35	165
100	Inverted tee beams, add to above, small beams				15%			
105	Large beams				.05%			
05-001	**COLUMNS** Rectangular to 12' high, small columns	C-11	120		50	21	71	84
005	Large columns		96		80	27	107	125
030	24' high, small columns		192		75	13.40	88.40	100
035	Large columns	↓	144	↓	105	17.85	122.85	140
15-001	**CRIBBING** See under Retaining Walls, division 2.7-32							
20-001	**CURBS,** roadway type, see division 2.6-22							
25-001	**JOISTS** 40 psf L.L., 6" deep for 12' spans	C-12	600	L.F.	4.60	1.98	6.58	7.80
005	8" deep for 16' spans		575		5.35	2.06	7.41	8.75
010	10" deep for 20' spans		550		5.60	2.16	7.76	9.15
015	12" deep for 24' spans	↓	525	↓	8.10	2.26	10.36	12.05
30-001	**LIFT SLAB** See division 3.3-14-520							
45-001	**SLABS** Prestressed roof and floor members, grouted, 4" deep	C-11	4,875	S.F.	2.50	.53	3.03	3.47
005	6" deep		5,850		2.60	.44	3.04	3.46
010	8" deep		7,200		2.85	.36	3.21	3.62
015	10" deep		8,800		3.10	.29	3.39	3.81
020	12" deep	↓	9,500	↓	3.50	.27	3.77	4.22
50-001	**STAIRS** concrete treads on steel stringers, 3' wide	C-12	75	Riser	42	15.80	57.80	68
002								
005	Building stairs, 3'-4" wide, 9' high, straight	C-11	7	Flight	1,145	365	1,510	1,750
010	Dogleg type with divider wall, 8' x 15'	"	5		1,510	515	2,025	2,375
030	Front entrance, 5' wide with 48" platform, 2 risers	C-12	16		230	74	304	355
035	5 risers		12		265	99	364	430
050	6' wide, 2 risers		15		250	79	329	385
055	5 risers		11		280	110	390	455
070	7' wide, 2 risers		14		320	85	405	470
075	5 risers	↓	10		375	120	495	575
120	Basement entrance stairs, steel bulkhead doors, minimum	B-51	22		400	38	438	495
125	Maximum	"	11	↓	550	77	627	715
55-001	**TEES** Prestressed quad tee, short spans, roof	C-11	7,200	S.F.	2.70	.36	3.06	3.46
005	Floor		7,200		2.80	.36	3.16	3.57
020	Double tee, floor members, 60' span		8,400		6.50	.31	6.81	7.55
025	80' span		8,000		7	.32	7.32	8.15
030	Roof members, 30' span		4,800		5.45	.54	5.99	6.75
035	50' span		6,400		6.10	.40	6.50	7.25
040	Wall members, up to 55' high	↓	3,600	↓	5.95	.71	6.66	7.50
045								
050	Single tee roof members, 40' span	C-11	3,200	S.F.	4.90	.80	5.70	6.50
055	80' span		5,120		6.20	.50	6.70	7.50
060	100' span		6,000		8.65	.43	9.08	10.10
065	120' span	↓	6,000	↓	8.80	.43	9.23	10.25
100	Double tees, floor members							
110	Lightweight, 20" x 8' wide, 45' span	C-11	20	Ea.	1,765	130	1,895	2,125
115	24" x 8' wide, 50' span		18		1,925	145	2,070	2,325
120	32" x 10' wide, 60' span		16		3,320	160	3,480	3,875
125	Standard weight, 12" x 8' wide, 20' span		22		665	115	780	890
130	16" x 8' wide, 25' span		20		895	130	1,025	1,150
135	18" x 8' wide, 30' span		20		1,175	130	1,305	1,475
140	20" x 8' wide, 45' span	↓	18	↓	1,285	145	1,430	1,600

For expanded coverage of these items see *Means' Concrete & Masonry Cost Data 1986*

3.4 Precast Concrete

			CREW	DAILY OUTPUT	UNIT	BARE COSTS MAT.	BARE COSTS INST.	BARE COSTS TOTAL	TOTAL INCL O&P
55	145	24" x 8' wide, 50' span	C-11	16	Ea.	1,630	160	1,790	2,025
	150	32" x 10' wide, 60' span	"	14	"	3,000	185	3,185	3,550
	200	Roof members							
	205	Lightweight, 20" x 8' wide, 40' span	C-11	20	Ea.	1,390	130	1,520	1,700
	210	24" x 8' wide, 50' span		18		1,870	145	2,015	2,250
	215	32" x 10' wide, 60' span		16		3,100	160	3,260	3,625
	220	Standard weight, 12" x 8' wide, 30' span		22		930	115	1,045	1,175
	225	16" x 8' wide, 30' span		20		985	130	1,115	1,250
	230	18" x 8' wide, 30' span		20		1,095	130	1,225	1,375
	235	20" x 8' wide, 40' span		18		1,175	145	1,320	1,500
	240	24" x 8' wide, 50' span		16		1,500	160	1,660	1,875
	245	32" x 10' wide, 60' span	↓	14	↓	2,620	185	2,805	3,125
60-001		TILT UP Wall panel construction, walls only, 5-1/2" thick	C-14	2515	S.F.	1.69	1.48	3.17	3.95
	010	Walls only, 7-1/2" thick		2115		2.41	1.76	4.17	5.05
	050	Walls and columns, 5-1/2" thick walls, 12" x 12" columns		1565		2.52	2.38	4.90	6.05
	055	(53) 7-1/2" thick wall, 12" x 12" columns	↓	1370	↓	3.29	2.71	6.00	7.35
	080	Columns only, site precast, minimum		200	L.F.	19.25	18.60	37.85	47
	085	Maximum	↓	105	"	27.80	35	62.80	79
65-001		WALL PANELS, High rise, 4' x 8', 4" thick, smooth gray	C-11	256	S.F.	7	10.05	17.05	21
	010	Exposed aggregate		256		7.60	10.05	17.65	22
	040	(52) 8' x 8', 4" thick, smooth gray		576		7	4.46	11.46	13.80
	050	Exposed aggregate		576		7.50	4.46	11.96	14.35
	100	20' x 10', 6" thick, smooth gray		1,800		8.90	1.43	10.33	11.75
	110	Exposed aggregate		1,800		9.70	1.43	11.13	12.60
	150	30' x 10', 6" thick, smooth gray		2,100		8.90	1.22	10.12	11.45
	160	White face	↓	2,100	↓	10.15	1.22	11.37	12.85
	200	For low rise construction, deduct					25%		

3.5 Cementitious Decks

			CREW	DAILY OUTPUT	UNIT	BARE COSTS MAT.	BARE COSTS INST.	BARE COSTS TOTAL	TOTAL INCL O&P
05-001		CONCRETE CHANNEL SLABS 2-3/4" or 3-1/2" thick, straight	C-12	1,575	S.F.	2.20	.75	2.95	3.46
	005	Chopped up		785		2.75	1.51	4.26	5.10
	020	6" thick, span to 20'		1,300		2.65	.91	3.56	4.18
	030	8" thick, span to 24'		1,100		3.10	1.08	4.18	4.90
10-001		CONCRETE PLANK Lightweight, nailable, T&G, 2" thick		1,800		2.30	.66	2.96	3.44
	005	2-3/4" thick		1,575		2.45	.75	3.20	3.74
	010	3-3/4" thick	↓	1,375		3	.86	3.86	4.49
	015	For premium ceiling finish, add				.22		.22	.24M
	020	For sloping roofs, slope over 4 in 12, add					25%		
	025	Slope over 6 in 12, add					150%		
15-001		FORMBOARD Including sub-purlins, asbestos cement, 1/4" thick	C-13	2,950	↓	1.30	.19	1.49	1.71
	005								
	010	Fiberglass, 1" thick, economy	C-13	2,700	S.F.	.65	.21	.86	1.03
	015	Painted		2,700		.80	.21	1.01	1.19
	030	Mineral fiber, unpainted, 1" thick		2,700		.75	.21	.96	1.14
	035	1-1/2" thick		2,500		1.10	.23	1.33	1.55
	050	Wood fiber, 1" thick	↓	2,700	↓	.65	.21	.86	1.03
	055								
	100	Poured gypsum, 2" thick, add to formboard above	C-8	7,000	S.F.	.55	.22	.77	.89
	110	3" thick	"	5,800	"	.80	.26	1.06	1.22
25-001		INSULATING Lightweight cellular concrete roof fill							
	002	Portland cement and foaming agent	C-8	50	C.Y.	69	30	99	115
	010	(51) Poured vermiculite or perlite, field mix,							
	011	1:6 roof fill, add formboards above	C-8	50	C.Y.	67	30	97	115

For expanded coverage of these items see Means' *Concrete & Masonry Cost Data 1986*

3.5 Cementitious Decks

			CREW	DAILY OUTPUT	UNIT	BARE COSTS MAT.	BARE COSTS INST.	BARE COSTS TOTAL	TOTAL INCL O&P
25	020	Ready mix, 1:6 mix, roof fill, 2" thick	C-8	10,000	S.F.	.41	.15	.56	.65
	025	3" thick	"	7,700	"	.65	.20	.85	.97
	040	Expanded volcanic glass rock, with binder, minimum	2 Carp	1,500	B.F.	.20	.21	.41	.53
	045	Maximum	"	1,200	"	.35	.27	.62	.77
30-001	(55)	PRESTRESSED Roof and floor members, see division 3.4							
40-001		WOOD FIBER Lightweight cement T&G planks, 1" thick	2 Carp	1,000	S.F.	.60	.32	.92	1.12
	010	2" thick		950		.74	.34	1.08	1.30
	015	2-1/2" thick		925		.88	.35	1.23	1.47
	020	3" thick		900		1.05	.36	1.41	1.67
	040	Long span plank, 2" thick		950		.95	.34	1.29	1.53
	045	3" thick		900		1.18	.36	1.54	1.81
	060	For polyurethane insulation, 1" thick, add				.60		.60	.66M
	065	2" thick, add				.80		.80	.88M
	100	Add for bulb tees, sub-purlins and grout, 6' span	E-1	5,000		.34	.12	.46	.55
	110	8' span	"	4,200		.40	.14	.54	.64
45-001		WOOD PLANK Roof decks, see division 6.1-62 & 6.3-20							

4.1 Mortar & Masonry Accessories

			CREW	DAILY OUTPUT	UNIT	BARE COSTS MAT.	BARE COSTS INST.	BARE COSTS TOTAL	TOTAL INCL O&P
06-001		ANCHOR BOLTS Hooked type with nut, 1/2" diam., 8" long	1 Bric	200	Ea.	.45	.82	1.27	1.66
	003	12" long		190		.55	.86	1.41	1.83
	004	5/8" diameter, 8" long		180		.80	.91	1.71	2.18
	005	12" long		170		1.10	.96	2.06	2.58
	006	3/4" diameter, 8" long		160		1.20	1.03	2.23	2.78
	007	12" long		150		1.45	1.09	2.54	3.15
10-001		CEMENT Gypsum 80 lb. bag, T.L. lots			Bag	8.75		8.75	9.60M
	005	L.T.L. lots				9.05		9.05	9.95M
	010	Masonry, 70 lb. bag, T.L. lots				4.75		4.75	5.20M
	015	L.T.L. lots				5		5	5.50M
	020	White masonry cement, 70 lb. bag, T.L. lots				13		13	14.30M
	025	L.T.L. lots				13.75		13.75	15.10M
15-001		COLORS 50 lb. bags (2 bags per M bricks),							
	002	range 2 to 10 lb. per bag of cement, minimum			Lb.	1.65		1.65	1.81M
	005	(56) Average				2.75		2.75	3.02M
	010	Maximum				3.90		3.90	4.29M
20-001		CONTROL JOINT Rubber, 4" and wider wall	1 Bric	600	L.F.	1.35	.27	1.62	1.87
	005	PVC, 4" wall		600		1.30	.27	1.57	1.82
	010	Rubber, 6" wall only		500		2.70	.33	3.03	3.44
	012	PVC, 6" wall		500		1.75	.33	2.08	2.39
	014	Rubber, 8" and wider wall		400		2.75	.41	3.16	3.61
	016	PVC, 8" wall		400		1.80	.41	2.21	2.56
	018	PVC, 12" wall		300		2.15	.55	2.70	3.14
	030	Boxboard, asphalt impregnated, 1/4" thick		500	S.F.	.32	.33	.65	.82
25-001		FIREPLACE ACCESSORIES Chimney screens, galv., 13" x 13" flue		8	Ea.	35	21	56	68
	005	Galv., 24" x 24" flue		5		95	33	128	150
	020	Stainless steel, 13" x 13" flue		8		185	21	206	235
	025	20" x 20" flue		5		295	33	328	370
	040	Cleanout doors and frames, cast iron, 8" x 8"		12		12	13.65	25.65	33
	045	12" x 12"		10		26	16.40	42.40	52
	050	18" x 24"		8		70	21	91	105
	055	Cast iron frame, steel door, 24" x 30"		5		160	33	193	225
	080	Damper, rotary control, steel, 30" opening		6		40	27	67	83
	085	Cast iron, 30" opening		6		45	27	72	88

For expanded coverage of these items see *Means' Concrete & Masonry Cost Data 1986*

4.1 Mortar & Masonry Accessories

			CREW	DAILY OUTPUT	UNIT	BARE COSTS MAT.	BARE COSTS INST.	BARE COSTS TOTAL	TOTAL INCL O&P
25	088	36" opening	1 Bric	6	Ea.	55	27	82	99
	090	48" opening		6		75	27	102	120
	092	60" opening		6		155	27	182	210
	095	72" opening		5		265	33	298	340
	100	84" opening		5		430	33	463	520
	105	96" opening		4		460	41	501	565
	120	Steel plate, poker control, 60" opening		8		125	21	146	165
	125	84" opening		5		240	33	273	310
	140	"Universal" type, chain operated, 32" x 20" opening		8		95	21	116	135
	145	48" x 24" opening		5		160	33	193	225
	160	Dutch Oven door and frame, cast iron, 12" x 15" opening		13		55	12.60	67.60	78
	165	Copper plated, 12" x 15" opening		13		93	12.60	105.60	120
	180	Fireplace forms with registers, 25" opening		3		300	55	355	410
	190	34" opening		2.50		340	66	406	465
	200	48" opening		2		650	82	732	830
	210	72" opening		1.50		1,025	110	1,135	1,275
	240	Squirrel and bird screens, galvanized, 8" x 8" flue		16		28	10.25	38.25	45
	245	13" x 13" flue		12		35	13.65	48.65	58
30	001	**GROUTING** Bond beams and lintels, 8" deep, pumped, not incl. block,							
	002	8" thick, 0.2 C.F. per L.F.	D-4	1,750	L.F.	.48	.42	.90	1.10
	005	10" thick, 0.25 C.F. per L.F.		1,500		.60	.49	1.09	1.32
	006	12" thick, 0.3 C.F. per L.F.		1,300		.72	.56	1.28	1.56
	020	(59) Concrete block cores, solid, 4" thick, by hand, .067 C.F./S.F.	D-3	1,200	S.F.	.15	.66	.81	1.10
	025	8" thick, pumped .258 C.F. per S.F.	D-4	850		.61	.86	1.47	1.84
	030	10" thick, .340 C.F. per S.F.		825		.81	.89	1.70	2.10
	035	12" thick, .422 C.F. per S.F.		800		1	.91	1.91	2.34
	050	Cavity walls, 2" space, pumped, .167 C.F. per S.F.		2,000		.39	.37	.76	.93
	055	3" space, shoring not incl., .250 C.F. per S.F.		1,500		.60	.49	1.09	1.32
	060	4" space, .333 C.F. per S.F.		1,250		.79	.58	1.37	1.66
	070	6" space, .500 C.F. per S.F.		950		1.20	.77	1.97	2.37
	080	Door frames, 3' x 7' opening, 2.5 C.F. per opening		60	Opng.	6	12.15	18.15	23
	085	6' x 7' opening, 3.5 C.F. per opening		45	"	8.40	16.25	24.65	31
35	001	**JOINT REINFORCING** Steel bars, placed horizontal, #3 & #4 bars	1 Bric	450	Lb.	.30	.36	.66	.85
	002	#5 & #6 bars		800		.30	.21	.51	.62
	005	Placed vertical, #3 & #4 bars		350		.30	.47	.77	1
	006	#5 & #6 bars		650		.30	.25	.55	.69
	020	(58) Wire strips, regular truss, to 6" wide, galvanized		30	C.L.F.	12.50	5.45	17.95	22
	025	12" wide		20		15.50	8.20	23.70	29
	040	Cavity truss with drip section to 6" wide		30		13	5.45	18.45	22
	045	12" wide		20		16	8.20	24.20	29
40	001	**LIME** Masons, hydrated, 50 lb. bag, T.L. lots			Bag	3.90		3.90	4.29M
	005	L.T.L. lots				4.25		4.25	4.67M
	020	Finish, double hydrated, 50 lb. bag, T.L. lots				5.40		5.40	5.95M
	025	L.T.L. lots				6.15		6.15	6.75M
45	001	**LINTELS** Steel angles, minimum	1 Bric	1,000	Lb.	.33	.16	.49	.60
	005	Maximum, see also division 5.4-50		500	"	.41	.33	.74	.92
	020	Steel angles, 3-1/2" x 3", 1/4" thick, 2'-6" long		50	Ea.	5.20	3.28	8.48	10.40
	025	4'-6" long		45		9.30	3.64	12.94	15.40
	040	4" x 3-1/2", 1/4" thick, 5'-0" long		40		11.85	4.10	15.95	18.85
	045	9'-0" long		35		21.25	4.69	25.94	30
	080	Precast concrete, 4" x 8", stock units to 5' long	D-1	175	L.F.	3.85	1.67	5.52	6.60
	085	To 12' long	D-4	190		4.50	3.84	8.34	10.20
	100	6" wide, 8" high, solid, stock units to 5' long		185		5.90	3.95	9.85	11.85
	105	To 12' long		190		6.50	3.84	10.34	12.40
	120	8" wide, 8" high, stock units to 5' long		185		7.80	3.95	11.75	13.95
	125	To 12' long		190		9.50	3.84	13.34	15.70
	140	10" wide, 8" high, stock units to 14' long		180		10.40	4.06	14.46	16.95
	145	12" wide, 8" high, stock units to 19' long		185		11.70	3.95	15.65	18.25
50	001	(57) **MORTAR** Glass block			C.F.	4.80		4.80	5.30M
	010	Gypsum cement mortar			"	4.25		4.25	4.67M

For expanded coverage of these items see *Means' Concrete & Masonry Cost Data 1986*

4.1 Mortar & Masonry Accessories	CREW	DAILY OUTPUT	UNIT	BARE COSTS MAT.	BARE COSTS INST.	BARE COSTS TOTAL	TOTAL INCL O&P
50 020 Masonry cement 1:3 mix, type N			C.F.	2.20		2.20	2.42M
025 (56) 1:1:6 mix, type M				2.25		2.25	2.47M
040 Portland cement and lime, 1:1:6 mix, type N				2.10		2.10	2.31M
045 (56) 1:1/4:3 mix, type M				2.65		2.65	2.91M
55-001 POINTING MORTAR, Expanding, 50 lb. drum, gray			Lb.	.55		.55	.60M
005 White			"	.65		.65	.71M
60-001 SAND For mortar, screened and washed, at the pit			Ton	7		7	7.70M
005 With 10 mile haul				8		8	8.80M
010 With 30 mile haul				10		10	11M
020 (42) Screened and washed, at the pit			C.Y.	9.80		9.80	10.80M
025 With 10 mile haul				11.20		11.20	12.30M
030 With 30 mile haul				14		14	15.40M
65-001 WALL PLUGS For nailing to brickwork, 26 ga., galvanized, plain	1 Bric	10.50	C	7.80	15.60	23.40	31
005 Wood filled		10.50		13	15.60	28.60	37
70-001 WALL TIES To brick veneer, galv., corrugated, 7/8" x 7", 24 gauge		10.50		3.10	15.60	18.70	26
005 16 gauge		10.50		9	15.60	24.60	32
020 Buck anchors, galv., corrugated, 8" x 2", 16 gauge		10.50		42	15.60	57.60	68
025 14 gauge		10.50		49	15.60	64.60	76
060 Cavity wall, 6" long, Z type, galvanized, 1/8" diameter		10.50		14	15.60	29.60	38
065 3/16" diameter		10.50		6.10	15.60	21.70	29
080 8" long, Z type, 3/16" diameter, galvanized		10.50		20	15.60	35.60	44
085 Copperweld		10.50		30	15.60	45.60	55
100 Rectangular type, 3/16" diameter, galv., 2" x 6"		10.50		13	15.60	28.60	37
105 2" x 8"		10.50		14.80	15.60	30.40	39
120 2" x 8" or 4" x 6", stainless		10.50		46	15.60	61.60	73
125 Copperweld		10.50		55	15.60	70.60	83
150 Stone anchors, galv., U or Z shaped, 6" long, 1/8" x 1"		10.50		42	15.60	57.60	68
155 1/4" x 1"		10.50		86	15.60	101.60	115
180 12" long, 1/8" x 1"		10.50		60	15.60	75.60	88
185 1/4" x 1"		10.50		130	15.60	145.60	165
200 For asphalt and red lead painted, add				9.85		9.85	10.85M
220 For dovetail anchor slot & accessories see division 3.1-05							
75-001 WATERPROOFING admixture, add to brickwork			M	2.60		2.60	2.86M
80-001 VENT BOX Extruded aluminum, 4" deep, 8" x 2-1/2"	1 Bric	30	Ea.	19	5.45	24.45	29
005 8" x 5"		25		29	6.55	35.55	41
010 16" x 2-1/2"		25		34	6.55	40.55	47
015 16" x 8"		22		47	7.45	54.45	62
020 24" x 5"		22		53	7.45	60.45	69
025 24" x 8"		20		64	8.20	72.20	82
040 For baked enamel finish, add				35%			
050 For cast aluminum, painted, add				60%			
100 Stainless steel ventilators, 6" x 6"	1 Bric	25		45	6.55	51.55	59
105 8" x 8"		24		48	6.85	54.85	63
110 12" x 12"		23		49	7.15	56.15	64
115 12" x 6"		24		50	6.85	56.85	65
120 Foundation vent, galv., 1-1/4" thick, 8"H, 16"L, no damper		30		6.10	5.45	11.55	14.50
125 For damper, add				.80		.80	.88M

4.2 Brick Masonry	CREW	DAILY OUTPUT	UNIT	BARE COSTS MAT.	BARE COSTS INST.	BARE COSTS TOTAL	TOTAL INCL O&P
02-001 ADOBE BRICK Unstabilized, with adobe mortar (Southwestern State)							
006 Brick, 4" x 3" x 8" @ $180 per M (6.0 per S.F.)	D-2	1.60	M	210	520	730	970
008 4" x 4" x 8" @ $250 per M (4.5 per S.F.)		1.50		290	550	840	1,100
010 4" x 4" x 14" @ $360 per M (2.5 per S.F.)		1.35		430	615	1,045	1,350

For expanded coverage of these items see *Means' Concrete & Masonry Cost Data 1986*

4.2 Brick Masonry

			Daily		Bare Costs			Total
		Crew	Output	Unit	Mat.	Inst.	Total	Incl O&P
02 012	8" x 3" x 16" @ $412 per M (3.0 per S.F.)	D-2	1.40	M	470	590	1,060	1,350
014	4" x 3" x 12" @ $250 per M (4.0 per S.F.)		1.45		295	570	865	1,150
016	6" x 3" x 12" @ $260 per M (4.0 per S.F.)		1.45		305	570	875	1,150
018	4" x 4" x 16" @ $470 per M (2.25 per S.F.)		1.35		515	615	1,130	1,450
020	8" x 4" x 16" @ $550 per M (2.25 per S.F.)		1.30		600	635	1,235	1,575
022	12" x 4" x 16" @ $1,000 per M (2.25 per S.F.)		1.25		1,075	660	1,735	2,125
026	Brick, 4" x 3" x 8" @ $180 per M (6.0 per S.F.)		266	S.F.	1.25	3.11	4.36	5.80
028	4" x 4" x 8" @ $250 per M (4.5 per S.F.)		333.01		1.30	2.49	3.79	4.98
030	4" x 4" x 14" @ $360 per M (2.5 per S.F.)		540		1.05	1.53	2.58	3.34
032	8" x 3" x 16" @ $412 per M (3.0 per S.F.)		466		1.40	1.78	3.18	4.08
034	4" x 3" x 12" @ $250 per M (4.0 per S.F.)		362		1.20	2.29	3.49	4.58
036	6" x 3" x 12" @ $260 per M (4.0 per S.F.)		362		1.20	2.29	3.49	4.58
038	4" x 4" x 16" @ $470 per M (2.25 per S.F.)		600		2.30	1.38	3.68	4.50
040	8" x 4" x 16" @ $550 per M (2.25 per S.F.)		577		2.65	1.44	4.09	4.96
042	12" x 4" x 16" @ $1,000 per M (2.25 per S.F.)		555		4.70	1.49	6.19	7.30
043								
044	Adobe, partially stabilized, add			S.F.	10%			
048	Fully stabilized, add			"	30%			
03-001	CAULKING See division 7.1-20 and 18.1-30							
04-001	(121) CHIMNEY RADIAL BRICK Industrial see division 13.1-17							
06-001	CHIMNEY Brick @ $205 per M, 16" x 16", 8" flue, scaff. not incl.	D-1	18.20	V.L.F.	9.50	16.05	25.55	33
005	16" x 20" with one 8" x 12" flue		16		11.75	18.25	30	39
010	(65) 16" x 24" with two 8" x 8" flues		14		14.15	21	35.15	45
015	20" x 20" with one 12" x 12" flue		13.70		13.95	21	34.95	46
020	20" x 24" with two 8" x 12" flues		12		18.35	24	42.35	55
025	20" x 32" with two 12" x 12" flues		10		23.40	29	52.40	67
08-001	CLEAN AND POINT Smooth brick (see also division 4.5)	1 Bric	300	S.F.	.08	.55	.63	.87
010	Rough brick	"	265	"	.10	.62	.72	.99
09-001	COLUMNS Brick @ $205 per M, 8" x 8", 9 brick, scaff. not incl.	D-1	56	V.L.F.	2.15	5.20	7.35	9.80
010	12" x 8", 13.5 brick		37		3.25	7.90	11.15	14.80
020	12" x 12", 20.3 brick		25		4.85	11.70	16.55	22
030	16" x 12", 27 brick		19		6.45	15.35	21.80	29
040	16" x 16", 36 brick		14		8.60	21	29.60	39
050	20" x 16", 45 brick		11		10.75	27	37.75	50
060	20" x 20", 56.3 brick		9		13.45	32	45.45	61
070	24" x 20", 67.5 brick		7		16.10	42	58.10	77
080	24" x 24", 81 brick		6		19.35	49	68.35	91
100	36" x 36", 182.3 brick		3		43.50	97	140.50	185
10-001	(61) COMMON BRICK Standard size, material only, minimum			M	185		185	205M
005	Average (select common)			"	205		205	225M
12-001	COPING For 12" wall, stock units, aluminum	D-1	80	L.F.	6	3.65	9.65	11.80
005	Precast concrete, stock units, 6" wide		100		4	2.92	6.92	8.55
010	10" wide		90		5	3.24	8.24	10.10
011	12" wide		85		5.15	3.44	8.59	10.55
015	14" wide		80		5.40	3.65	9.05	11.15
025								
030	Limestone for 12" wall, 4" thick	D-1	90	L.F.	8.25	3.24	11.49	13.70
035	6" thick		80		11.50	3.65	15.15	17.85
050	Marble to 4" thick, no wash, 9" wide		90		14.85	3.24	18.09	21
055	12" wide		80		19.80	3.65	23.45	27
070	Terra cotta, 9" wide		90		2.90	3.24	6.14	7.80
075	12" wide		80		4.75	3.65	8.40	10.45
14-001	CUT AND POINT See division 18.1-45							

					Colors		
					Red	Buff	Gray
18-001	FACE BRICK Prices Jan. 1986, C.L. lots, material only,						
002	not including truck delivery						

For expanded coverage of these items see *Means' Concrete & Masonry Cost Data 1986*

4.2 Brick Masonry

				CREW	DAILY OUTPUT	UNIT	COLORS RED	COLORS BUFF	COLORS GRAY	TOTAL INCL O&P
18	030	(61)	Boston, standard size, 8" x 2-2/3" x 4", minimum			M	240	260	270	
	035		Maximum				350	365	370	
	045		Econo size, 8" x 4" x 4", minimum				380	390	400	
	050		Maximum				450	455	460	
	055		Jumbo, 12" x 4" x 6", minimum				900	915	915	
	060		Maximum				1,100	1,100	1,100	
	065		Norwegian, 12" x 3-1/5" x 4", minimum				455	510	510	
	070		Maximum				540	550	565	
	085		Standard glazed, plain colors, 8" x 2-2/3" x 4", minimum				750	750	750	
	090		Maximum				880	880	880	
	100		Glazed, deep trim shades, 8" x 2-2/3" x 4", min.				830	830	830	
	105		Maximum				1,040	1,040	1,040	
	108		Utility 12" x 4" x 4"				660	660	660	
	112		8" x 8" x 4"				930	1,000	1,000	
	114		8" x 16" x 4"			↓	1,820	1,940	1,940	
	117									
	120		Chicago, standard size, 8" x 2-2/3" x 4", minimum			M	230	240	260	
	125		Maximum				300	305	335	
	126		Engineer size, 8" x 3-1/5" x 4", minimum				260	275	335	
	127		Maximum				300	385	395	
	135		King size, 10" x 2-3/4" x 4", minimum				270	270	335	
	136		Maximum				315	345	355	
	140		Standard glazed, 8" x 2-2/3" x 4", minimum				505	505	505	
	145		Maximum				570	570	570	
	146		Jumbo glazed, 12" x 4" x 6", minimum				1,125	1,145	1,145	
	148		Maximum				1,375	1,375	1,375	
	171		New York City, standard size, industrial, 8" x 2-2/3" x 4"				275	355	365	
	173		Office building, standard size, 8" x 2-2/3" x 4"				310	395	395	
	175		Standard size, single glazed, 8" x 2-2/3" x 4"				630	630	630	
	177		Double glazed, standard size, 8" x 2-2/3" x 4"				785	785	785	
	180		Jumbo, unglazed, 12" x 4" x 6"				920	945	960	
	185		Jumbo, colored glazed ceramic, 12" x 4" x 6"				1,200	1,100	1,140	
	205		Utility brick, glazed, 12" x 4" x 4"				812	812	812	
	210		8" x 8" x 4"				1,160	1,550	1,250	
	215		8" x 16" x 4"			↓	2,275	2,425	2,425	

				CREW	DAILY OUTPUT	UNIT	BARE COSTS MAT.	BARE COSTS INST.	BARE COSTS TOTAL	TOTAL INCL O&P
19-001		FIRE BRICK 9" x 2-1/2" x 4-1/2", low duty, 2000° F		D-1	.60	M	850	485	1,335	1,625
005		High duty, 3000°F		"	.60	"	1,100	485	1,585	1,900
20-001		FIRE CLAY Gray, high duty, 100 lb. bag				Bag	11		11	12.10M
005		100 lb. drum, premixed (400 brick per drum)				Drum	25		25	28M
22-001		FIREPLACE For prefabricated fireplace, see division 10.1-30								
002										
010		Brick fireplace, not incl. foundations or chimneys								
011		30" x 24" opening, plain brickwork		D-1	.40	Ea.	235	730	965	1,300
020		Fireplace box only (110 brick)			2		100	145	245	320
030		Elaborate brickwork and details			.20		440	1,450	1,890	2,575
040		For hearth, add		↓	2	↓	65	145	210	280
041										
060		Plain brickwork, incl. metal circulator		D-1	.50	Ea.	425	585	1,010	1,300
080		Face brick only, standard size, 8" x 2-2/3" x 4"		"	.30	M	225	975	1,200	1,625
24-001		FLAGGING See division 2.6-40-110								
26-001		FLASHING See division 7.6-25								
28-001		FLOORING Acidproof shales, red, 8" x 3-3/4" x 1-1/4" thick		D-7	.43	M	380	660	1,040	1,350
005		2-1/4" thick		D-1	.40		405	730	1,135	1,475
020		Acid proof clay brick, 8" x 3-3/4" x 2-1/4" thick			.40		405	730	1,135	1,475
025		9" x 4-1/2" x 3" thick		↓	.37	↓	675	790	1,465	1,875

For expanded coverage of these items see *Means' Concrete & Masonry Cost Data 1986*

4.2 Brick Masonry

			CREW	DAILY OUTPUT	UNIT	BARE COSTS MAT.	BARE COSTS INST.	BARE COSTS TOTAL	TOTAL INCL O&P
28	026	Cast ceramic, pressed, 4" x 8" x 1/2", unglazed	D-7	100	S.F.	3.60	2.83	6.43	7.90
	027	Glazed		100		4.40	2.83	7.23	8.80
	028	Hand molded flooring, 4" x 8" x 3/4", unglazed		95		4.35	2.98	7.33	8.95
	029	Glazed		95		5.30	2.98	8.28	10
	030	8" hexagonal, 3/4" thick, unglazed		85		4.85	3.33	8.18	10
	031	Glazed		85		6.80	3.33	10.13	12.15
	040	Heavy duty industrial, cement mortar bed	D-1	80		2.65	3.65	6.30	8.10
	045	Acid proof joints	"	65		3.80	4.49	8.29	10.60
	050	Pavers, 8" x 4", 1" to 1-1/4" thick, red	D-7	95		1.80	2.98	4.78	6.15
	051	Ironspot	D-1	95		2.60	3.07	5.67	7.25
	054	1-3/8" to 1-3/4" thick, red		95		1.85	3.07	4.92	6.40
	056	Ironspot		95		2.80	3.07	5.87	7.45
	058	2-1/4" thick, red		90		1.90	3.24	5.14	6.70
	059	Ironspot		90		3	3.24	6.24	7.90
	060	Sidewalk or patios, on sand bed, laid flat, no mortar, 4.5 per S.F.		110		1.75	2.65	4.40	5.70
	065	Laid on edge, 7 per S.F.		70		2.75	4.17	6.92	8.95
	080	For basket weave pattern, add			S.F.or M	15%			
	085	For herringbone pattern, add			"	20%			
	086	For acid-resistant joints, add	D-1	2,100	S.F.	.40	.14	.54	.64
	087	For epoxy joints, add		600		1	.49	1.49	1.79
	088	For Furan underlayment, add		600		1.40	.49	1.89	2.23
	089	For waxed surface, steam cleaned, add	D-5	1,000		.05	.19	.24	.32
30	001	**FLUE LINING**, Square, including mortar joints, 8" x 8"	D-1	125	V.L.F.	2.60	2.34	4.94	6.20
	010	8" x 12"		103		3.75	2.83	6.58	8.15
	020	12" x 12"		93		5	3.14	8.14	9.95
	030	12" x 18"		84		9.75	3.48	13.23	15.70
	040	18" x 18"		75		13	3.89	16.89	19.85
	050	20" x 20"		66		19.50	4.42	23.92	28
	060	24" x 24"		56		26	5.20	31.20	36
	080								
	100	Round, 18" diameter	D-1	66	V.L.F.	20	4.42	24.42	28
	110	24" diameter	"	47	"	29	6.20	35.20	41
34	001	**LINTELS** See division 4.1-45 & 5.4-50							
36	001	**PARGETING** See division 7.1-25							
40	001	**REINFORCING** See division 4.1-35							
44	001	**SAND BLAST** Building face, wet system, minimum	B-9	1,700	S.F.	.10	.47	.57	.76
	008	Average		1,100		.12	.73	.85	1.14
	090	Maximum		700		.14	1.15	1.29	1.74
	200	Dry system, minimum		3,000		.10	.27	.37	.48
	210	Average		1,750		.12	.46	.58	.77
	500	Maximum		1,500		.14	.53	.67	.89
46	001	**SAWING** Brick or block by hand, per inch depth	D-5	300	L.F.	.16	.64	.80	1.06
48	001	**SCAFFOLDING AND SWING STAGING** See division 1.1-43, 44 & 52							
	002								
	003	Scaffolding & accessories included in division 4.2 and 4.3							
50	001	**SIMULATED BRICK** Aluminum, baked on colors	1 Carp	200	S.F.	1.25	.80	2.05	2.53
	005	Fiberglass panels		200		2.75	.80	3.55	4.18
	010	Urethane pieces cemented in mastic		150		3.65	1.07	4.72	5.55
	015	Vinyl siding panels		200		1.20	.80	2	2.48
	016	Cement base, brick, incl. mastic	D-1	100		3.65	2.92	6.57	8.20
	017	Corner		50	V.L.F.	7.50	5.85	13.35	16.55
	018	Stone face, incl. mastic		100	S.F.	4.25	2.92	7.17	8.85
	019	Corner		50	V.L.F.	6.10	5.85	11.95	15.05
52	001	**STEAM CLEAN** Building, incl. scaffolding minimum	B-9	2,500	S.F.	.15	.32	.47	.61
	010	Maximum	"	1,000	"	.15	.80	.95	1.28

For expanded coverage of these items see *Means' Concrete & Masonry Cost Data 1986*

4.2 Brick Masonry

			CREW	DAILY OUTPUT	UNIT	BARE COSTS MAT.	BARE COSTS INST.	BARE COSTS TOTAL	TOTAL INCL O&P
52	030	Common face brick	B-9	1,200	S.F.	.15	.67	.82	1.09
	040	Wire cut face brick	"	900	"	.15	.89	1.04	1.40
54-001		**STEPS** With select common at $205 per M	D-1	.30	M	243	975	1,218	1,650
56-001		**VENEER** 4" thick, sel. common, 8" x 2-2/3" x 4" @ $205/M (6.75/S.F.)	D-2	1.55	M	263	535	798	1,050
	005	Standard 8" x 2-2/3" x 4", running bond, red face, $225 per M		1.50		285	550	835	1,100
	010	Buff or gray face, brick at $260 per M (6.75/S.F.)		1.50		320	550	870	1,150
	015	Full header every 6th course (7.88/S.F.)		1.45		320	570	890	1,175
	020	English, full header every 2nd course (10.13/S.F.)		1.40		320	590	910	1,200
	025	Flemish, alternate header every course (9.00/S.F.)		1.35		320	615	935	1,225
	030	Flemish, alt. header every 6th course (7.13/S.F.)		1.45		320	570	890	1,175
	032	Full headers throughout (13.50/S.F.)		1.40		320	590	910	1,200
	034	Rowlock course (13.50 per S.F.)		1.35		320	615	935	1,225
	036	Rowlock stretcher (4.50 per S.F.)		1.40		320	590	910	1,200
	038	Soldier course (6.75 per S.F.)		1.35		320	615	935	1,225
	040	Sailor course (4.50 per S.F.)		1.30		320	635	955	1,250
	045	Glazed face, 8" x 2-2/3" x 4" $750 per M, running bond		1.40		825	590	1,415	1,750
	047	Full header every 6th course (7.88 per S.F.)		1.35		825	615	1,440	1,775
	050	Jumbo 12" x 4" x 6" running bond, $900 per M (3.00 S.F.)		1.30		1,056	635	1,691	2,075
	060	Norman 12" x 2-2/3" x 4" run. bond, $395 per M (4.50 per S.F.)		1.45		482	570	1,052	1,350
	065	Norwegian 12" x 3-1/5" x 4" at $455 per M (3.75 per S.F.)		1.40		553	590	1,143	1,450
	070	Economy 8" x 4" x 4" at $380 per M (4.50 per S.F.)		1.40		464	590	1,054	1,350
	075	Engineer 8" x 3-1/5" x 4" at $265 per M (5.63 per S.F.)		1.45		332	570	902	1,175
	080	Roman 12" x 2" x 4" at $470 per M (6.00 per S.F.)		1.50		549	550	1,099	1,400
	085	SCR 12" x 2-2/3" x 6" at $555 per M (4.50 per S.F.)		1.40		668	590	1,258	1,575
	090	Utility 12" x 4" x 4" at $650 per M (3.00 per S.F.)		1.35		766	615	1,381	1,725
	092	8" x 8" x 4" at $930 per M (2.25 per S.F.)		.99		1,069	835	1,904	2,375
	094	8" x 16" x 4" at $1820 per M (1.13 per S.F.)	↓	.51		2,057	1,625	3,682	4,575
	101	For battered walls, add					30%		
	102	For corbels, add					60%		
	103	For curved walls, add					30%		
	104	For pits and trenches, deduct			↓		20%		
	105	Std., sel. common, 8" x 2-2/3" x 4" $205 per M (6.75 per S.F.)	D-2	230	S.F.	1.80	3.60	5.40	7.10
	150	Standard 8" x 2-2/3" x 4", running bond, red face, @ $225 per M		220		1.95	3.76	5.71	7.50
	155	Buff or gray face, brick at $260 per M (6.75 per S.F.)		220		2.15	3.76	5.91	7.75
	160	Full header every 6th course (7.88 per S.F.)		185		2.50	4.48	6.98	9.15
	165	English, full header every 2nd course (10.13 per S.F.)		140		3.25	5.90	9.15	12
	170	Flemish, alter. header every course (9.00 per S.F.)		150		2.90	5.50	8.40	11.05
	180	Flemish, alt. header every 6th course (7.13/S.F.)		205		2.30	4.04	6.34	8.30
	182	Full headers throughout (13.50 per S.F.)		105		4.30	7.90	12.20	16
	184	Rowlock course (13.50 per S.F.)		100		4.30	8.30	12.60	16.55
	186	Rowlock stretcher (4.50 per S.F.)		310		1.45	2.67	4.12	5.40
	188	Soldier course (6.75 per S.F.)		200		2.15	4.14	6.29	8.25
	190	Sailor course (4.50 per S.F.)		290		1.45	2.86	4.31	5.65
	195	Glazed face, brick at $750 per M, running bond		210		5.60	3.94	9.54	11.80
	197	Full header every 6th course (7.88 per S.F.)		170		6.50	4.87	11.37	14.10
	200	Jumbo 12" x 4" x 6" running bond, @ $900 per M (3.00 per S.F.)		435		3.15	1.90	5.05	6.20
	205	Norman, 12" x 2-2/3" x 4" running bond, @ $395/M (4.50/S.F.)		320		2.15	2.59	4.74	6.05
	210	Norwegian 12" x 3-1/5" x 4" at $455/M (3.75 per S.F.)		375		2.05	2.21	4.26	5.40
	220	Economy 8" x 4" x 4" $380 per M (4.50 per S.F.)		310		2.10	2.67	4.77	6.10
	230	Engineer 8" x 3-1/5" x 4" at $265 per M (5.63 per S.F.)		260		1.85	3.18	5.03	6.60
	240	Roman 12" x 2" x 4" at $470 per M (6.00 per S.F.)		250		3.30	3.31	6.61	8.35
	250	SCR 12" x 2-2/3" x 6" at $555 per M (4.50 per S.F.)		310		3	2.67	5.67	7.10
	260	Utility 12" x 4" x 4" at $650 per M (3.00 per S.F.)		450		2.30	1.84	4.14	5.15
	262	8" x 8" x 4" at $930 per M (2.25 per S.F.)		440		2.40	1.88	4.28	5.35
	264	8" x 16" x 4" at $1820 per M (1.13 per S.F.)	↓	455		2.35	1.82	4.17	5.20
	270	For cavity wall construction, add					15%		
	280	For stacked bond, add					10%		
	290	For interior veneer construction, add			↓		15%		

For expanded coverage of these items see *Means' Concrete & Masonry Cost Data 1986*

4.2	Brick Masonry	CREW	DAILY OUTPUT	UNIT	BARE COSTS MAT.	BARE COSTS INST.	BARE COSTS TOTAL	TOTAL INCL O&P
56 300	For curved walls, add			S.F.		30%		
58-001	**VENT BOX** See division 4.1-80							
60-001	**WALLS** Conversion to S.F. areas and factors							
006	for other than common bond							
010	Common, 8" x 2-2/3" x 4" at $185 per M, 4" wall, as face brick	D-2	1.45	M	239	570	809	1,075
015	4" thick as back-up	"	1.60		239	520	759	1,000
020	8" thick wall, 13.50 bricks per S.F.	D-3	1.80		236	440	676	885
025	12" thick wall, 20.25 bricks per S.F.		1.90		240	415	655	855
030	16" thick wall, 27.00 bricks per S.F.	↓	2		246	395	641	830
050	Reinf., straight hard, 8" x 2-2/3" x 4" at $205 per M, 4" wall	D-2	1.45		261.74	570	831.74	1,100
055	8" thick wall, 13.50 bricks per S.F.	D-3	1.80		271.19	440	711.19	925
060	12" thick wall, 20.25 bricks per S.F.		1.90		254.26	415	669.26	870
065	16" thick wall, 27.00 bricks per S.F.	↓	1.95	↓	240.83	405	645.83	840
079	Alternate method of figuring (S.F.):							
080	Common, 8" x 2-2/3" x 4" at $185 per M, 4" wall, as face brick	D-2	215	S.F.	1.61	3.85	5.46	7.25
085	4" thick, as back-up	"	240		1.61	3.45	5.06	6.70
090	8" thick wall, 13.50 bricks per S.F.	D-3	135		3.19	5.85	9.04	11.85
100	12" thick wall, 20.25 bricks per S.F.		95		4.86	8.30	13.16	17.15
105	16" thick wall, 27.00 bricks per S.F.	↓	75		6.64	10.50	17.14	22
120	Reinf., straight hard, 8" x 2-2/3" x 4" at $205 per M, 4" wall	D-2	205		1.98	4.04	6.02	7.95
125	8" thick wall, 13.50 bricks per S.F.	D-3	130		3.73	6.05	9.78	12.75
130	12" thick wall, 20.25 bricks per S.F.		90		5.59	8.75	14.34	18.65
135	16" thick wall, 27.00 bricks per S.F.	↓	70	↓	7.57	11.25	18.82	24
62-001	**WALL PANELS** Prefabricated, 4" thick, minimum	C-11	775	S.F.	4.50	3.32	7.82	9.50
010	Maximum	"	500		5.60	5.15	10.75	13.20
020	4" brick + 2" concrete back-up, add				2.20		2.20	2.42M
030	4" brick + 1" urethane + 3" concrete back-up, add			↓	2.80		2.80	3.08M
64-001	**WALL PLUGS** See division 4.1-65							
66-001	**WALL TIES** See division 4.1-70							
68-001	**WASHING BRICK** (Incl. in brickwork), smooth brick	1 Bric	560	S.F.	.04	.29	.33	.46
005	Rough brick	"	400		.05	.41	.46	.64
70-001	**WINDOW SILL** Bluestone, natural cleft, 12" wide, 1-1/2" thick	D-1	85		5.80	3.44	9.24	11.25
005	2" thick		75	↓	6.20	3.89	10.09	12.35
010	Cut stone, 5" x 8" plain		48	L.F.	6.50	6.10	12.60	15.80
020	Face brick on edge, brick @ $225 per M		80		1.40	3.65	5.05	6.75
040	Marble, 12" wide, 1" thick		85		9.55	3.44	12.99	15.40
060	Precast concrete, stock sections, 6" wide		70		3.50	4.17	7.67	9.80
065	10" wide		60		4.60	4.87	9.47	12
070	14" wide		50		5	5.85	10.85	13.80
090	Slate, colored, unfading, 12" wide, 1" thick		85		7.40	3.44	10.84	13.05
095	2" thick		70		11	4.17	15.17	18.05
120	Stainless steel, stock		125		7.90	2.34	10.24	12
125	Custom	↓	125	↓	11.70	2.34	14.04	16.20

4.3	Block & Tile Masonry	CREW	DAILY OUTPUT	UNIT	BARE COSTS MAT.	BARE COSTS INST.	BARE COSTS TOTAL	TOTAL INCL O&P
07-001	**CONCRETE BLOCK, BACK-UP,** incl. scaffolding							
002	Sand aggregate, tooled joint 1 side							
100	Reinforced, alternate courses, 4" thick	D-8	435	S.F.	.92	1.72	2.64	3.46
110	6" thick	"	415	"	1.10	1.80	2.90	3.78

For expanded coverage of these items see *Means' Concrete & Masonry Cost Data 1986*

4.3 Block & Tile Masonry

			CREW	DAILY OUTPUT	UNIT	BARE COSTS MAT.	BARE COSTS INST.	BARE COSTS TOTAL	TOTAL INCL O&P
07	115	8" thick	D-8	395	S.F.	1.29	1.89	3.18	4.12
	120	10" thick	"	385		1.88	1.94	3.82	4.84
	125	12" thick	D-9	365	↓	1.92	2.40	4.32	5.55
12-001		**CONCRETE BLOCK BOND BEAM** Incl. scaffolding							
	002	Not including grout or reinforcing							
	010	Regular block, 8" high, 8" thick	D-8	565	L.F.	1.10	1.32	2.42	3.10
	015	12" thick	D-9	510		1.50	1.72	3.22	4.10
	050	Lightweight, 8" high, 8" thick	D-8	575		1.30	1.30	2.60	3.28
	055	12" thick	D-9	520	↓	1.65	1.68	3.33	4.22
	100								
	200	Including grout and 2 #5 bars							
	210	Regular block, 8" high, 8" thick	D-8	300	L.F.	2.07	2.49	4.56	5.85
	215	12" thick	D-9	250		2.65	3.50	6.15	7.90
	250	Lightweight, 8" high, 8" thick	D-8	305		2.27	2.45	4.72	6
	255	12" thick	D-9	255	↓	2.80	3.44	6.24	7.95
17-001		**CONCRETE BLOCK, DECORATIVE** Incl. scaffolding							
	002	Embossed, simulated brick face, not reinforced							
	010	8" x 16" units, 4" thick	D-8	400	S.F.	1.32	1.87	3.19	4.12
	020	8" thick		340		1.60	2.20	3.80	4.89
	025	12" thick	↓	300	↓	2.36	2.49	4.85	6.15
	040	Embossed both sides							
	050	8" thick	D-8	300	S.F.	2.15	2.49	4.64	5.90
	055	12" thick	"	275	"	2.70	2.72	5.42	6.85
	100	Fluted high strength							
	110	Flutes 1 side, 8" x 16" x 4" thick	D-8	345	S.F.	1.55	2.17	3.72	4.79
	115	Flutes 2 sides, 8" x 16" x 4" thick		335		1.86	2.23	4.09	5.25
	120	8" thick	↓	300	↓	2.75	2.49	5.24	6.60
	125	For special colors, add				.22		.22	.24M
	140	Deep grooved, smooth face							
	145	8" x 16" x 4" thick	D-8	345	S.F.	1.35	2.17	3.52	4.57
	150	8" thick	"	300	"	2.07	2.49	4.56	5.85
	200	Formbloc, incl. inserts & reinforcing							
	210	8" x 16" x 8" thick	D-8	345	S.F.	2.44	2.17	4.61	5.75
	215	12" thick	"	310	"	2.70	2.41	5.11	6.40
	250	Ground face							
	260	8" x 16" x 4" thick	D-8	345	S.F.	2.37	2.17	4.54	5.70
	265	6" thick		310		2.86	2.41	5.27	6.60
	270	8" thick	↓	290		3.37	2.58	5.95	7.40
	275	12" thick	D-9	265		4.73	3.31	8.04	9.90
	290	For special colors, add, minimum				5%			
	295	Maximum			↓	15%			
	400	Slump block							
	410	4" face height x 16" x 4" thick	D-1	165	S.F.	1.37	1.77	3.14	4.03
	415	6" thick		160		1.75	1.83	3.58	4.53
	420	8" thick		155		2.24	1.88	4.12	5.15
	425	10" thick		140		3.41	2.09	5.50	6.70
	430	12" thick		130		3.60	2.25	5.85	7.15
	440	6" face height x 16" x 6" thick		155		1.39	1.88	3.27	4.21
	445	8" thick		150		1.75	1.95	3.70	4.70
	450	10" thick		130		2.62	2.25	4.87	6.10
	455	12" thick	↓	120	↓	2.74	2.43	5.17	6.50
	500	Split rib profile units, 1" deep ribs, 8 ribs							
	510	8" x 16" x 4" thick	D-8	345	S.F.	1.70	2.17	3.87	4.96
	515	6" thick		325		2.08	2.30	4.38	5.55
	520	8" thick	↓	305		2.67	2.45	5.12	6.45
	525	12" thick	D-9	275	↓	3.08	3.19	6.27	7.95
	535								
	540	For special deeper colors, 4" thick, add			S.F.	.18		.18	.19M
	545	12" thick, add			"	.36		.36	.39M

For expanded coverage of these items see *Means' Concrete & Masonry Cost Data 1986*

4.3 Block & Tile Masonry

			CREW	DAILY OUTPUT	UNIT	BARE COSTS MAT.	BARE COSTS INST.	BARE COSTS TOTAL	TOTAL INCL O&P
17	560	For white, 4" thick, add			S.F.	.74		.74	.81M
	565	6" thick, add				.96		.96	1.05M
	570	8" thick, add				1.24		1.24	1.36M
	575	12" thick, add			↓	1.46		1.46	1.60M
	600	Split face or scored split face							
	610	8" x 16" x 4" thick	D-8	350	S.F.	1.59	2.14	3.73	4.79
	615	6" thick		315		1.92	2.37	4.29	5.50
	620	8" thick	↓	295		2.23	2.54	4.77	6.05
	625	12" thick	D-9	270	↓	2.58	3.24	5.82	7.45
	635								
	640	For special deeper colors, 4" thick, add			S.F.	.21		.21	.23M
	645	6" thick, add				.30		.30	.33M
	650	8" thick, add				.36		.36	.39M
	655	12" thick, add				.45		.45	.49M
	665	For white, 4" thick, add				.57		.57	.62M
	670	6" thick, add				.97		.97	1.06M
	675	8" thick, add				1.02		1.02	1.12M
	680	12" thick, add			↓	1.31		1.31	1.44M
	700	Scored ground face, 2 to 5 scores							
	710	8" x 16" x 4" thick	D-8	345	S.F.	2.52	2.17	4.69	5.85
	715	6" thick		310		2.96	2.41	5.37	6.70
	720	8" thick	↓	290		3.53	2.58	6.11	7.55
	725	12" thick	D-9	265	↓	4.49	3.31	7.80	9.65
	800	Hexagonal face profile units, 8" x 16" units							
	810	4" thick, hollow	D-8	345	S.F.	1.44	2.17	3.61	4.67
	820	Solid		345		1.90	2.17	4.07	5.20
	830	6" thick, hollow		310	↓	1.85	2.41	4.26	5.45
	835	8" thick, hollow		290		2.64	2.58	5.22	6.60
	850	For stacked bond, add		4.20	M.S.F.		180	180	255L
	855	For high rise construction, add per story	↓	67.80	"		11.05	11.05	15.70L
	860	For scored block, per score, per face, add			Ea.	.05		.05	.05M
	865	For honed or ground face, per face, add				.13		.13	.14M
	870	For honed or ground end, per end, add				.95		.95	1.04M
	875	For bullnose block, add			↓	1.06		1.06	1.16M
	880	For special colors, add			S.F.	.15		.15	.16M
27-001		**CONCRETE BLOCK FOUNDATION WALL** Incl. scaffolding							
	005	Sand aggregate, trowel cut joints, not reinf., parged 1/2" thick							
	020	Regular, 8" x 16" x 6" thick	D-8	450	S.F.	1.15	1.66	2.81	3.63
	025	8" thick		430		1.36	1.74	3.10	3.97
	030	10" thick	↓	420		1.88	1.78	3.66	4.61
	035	12" thick	D-9	395		1.94	2.22	4.16	5.30
	050	Solid, 8" x 16" block, 6" thick	D-8	440		1.44	1.70	3.14	4.01
	055	8" thick	"	415		1.79	1.80	3.59	4.54
	060	12" thick	D-9	380	↓	2.70	2.31	5.01	6.25
32-001		**CONCRETE BLOCK, HIGH STRENGTH** Incl. scaffolding							
	005	Hollow, reinforced, 8" x 16" units							
	020	3000 psi, 4" thick	D-8	430	S.F.	.96	1.74	2.70	3.53
	025	6" thick		400		1.10	1.87	2.97	3.87
	030	8" thick	↓	375		1.48	1.99	3.47	4.47
	035	12" thick	D-9	340		2	2.58	4.58	5.85
	050	5000 psi, 4" thick	D-8	430		1.10	1.74	2.84	3.69
	055	6" thick		400		1.29	1.87	3.16	4.08
	060	8" thick	↓	375		1.74	1.99	3.73	4.76
	065	12" thick	D-9	340		2.34	2.58	4.92	6.25
	100	For 75% solid block, add			↓	35%			
	105	For 100% solid block, add				55%			
37-001		**CONCRETE BLOCK, INTERLOCKING** Incl. scaffolding							
	010	Not including grout or reinforcing							

For expanded coverage of these items see *Means' Concrete & Masonry Cost Data 1986*

4.3 Block & Tile Masonry

			CREW	DAILY OUTPUT	UNIT	BARE COSTS MAT.	BARE COSTS INST.	BARE COSTS TOTAL	TOTAL INCL O&P
37	020	8" x 16" units, 8" thick	D-1	245	S.F.	1.37	1.19	2.56	3.20
	030	12" thick		220		2.05	1.33	3.38	4.15
	035	16" thick	↓	185		2	1.58	3.58	4.45
	040	Including grout & reinforcing, 8" thick	D-4	245		2.20	2.98	5.18	6.50
	045	(70) 12" thick		220		3.50	3.32	6.82	8.35
	050	16" thick	↓	185	↓	4.40	3.95	8.35	10.20
42-001		**CONCRETE BLOCK, LEAD LINED** Incl. scaffolding							
	002								
	020	12" x 12" x 4" thick, 1/16" core	D-9	165	S.F.	13.30	5.30	18.60	22
	030	1/4" core	"	100	"	21	8.75	29.75	36
47-001		**CONCRETE BLOCK, LINTELS** Incl. scaffolding							
	010	Including grout and reinforcing							
	020	8" high, 6" thick, 1 #4 bar	D-4	300	L.F.	2.06	2.43	4.49	5.60
	025	2 #4 bars		295		2.26	2.48	4.74	5.85
	040	16" high, 1 #4 bar		275		2.93	2.66	5.59	6.85
	045	2 #4 bars		270		3.13	2.71	5.84	7.15
	100	8" high, 8" thick, 1 #4 bar		275		2.16	2.66	4.82	6
	110	2 #4 bars		270		2.35	2.71	5.06	6.25
	115	2 #5 bars		270		2.59	2.71	5.30	6.55
	120	2 #6 bars		265		2.87	2.76	5.63	6.90
	150	16" high, 1 #4 bar		250		3.40	2.92	6.32	7.70
	160	2 #3 bars		245		3.43	2.98	6.41	7.85
	165	2 #4 bars		245		3.60	2.98	6.58	8
	170	2 #5 bars	↓	240	↓	3.84	3.04	6.88	8.35
52-001		**CONCRETE BLOCK, PARTITIONS** Incl. scaffolding							
	010	Acoustical slotted block							
	020	N.R.C., .50 type a, 4" thick	D-8	315	S.F.	1.99	2.37	4.36	5.55
	021	N.R.C. .65 type r, 4" thick		315		2.88	2.37	5.25	6.55
	025	N.R.C. .50 type a, 6" thick		290		2.30	2.58	4.88	6.20
	026	N.R.C. .65 type r, 6" thick		290		3.47	2.58	6.05	7.50
	040	N.R.C. .45 type a-1, 8" thick		265		3.12	2.82	5.94	7.45
	041	N.R.C. .55 type q, 8" thick		265		4.41	2.82	7.23	8.85
	050	N.R.C. .60 type r, 8" thick		265		4.10	2.82	6.92	8.55
	060	N.R.C. .65 type rr, 8" thick		265		4.10	2.82	6.92	8.55
	070	N.R.C. .65 type rsr, 8" thick		265		5.34	2.82	8.16	9.90
	071	N.R.C. .70 type r, 12" thick	↓	245	↓	5.65	3.05	8.70	10.55
	100	Lightweight block, tooled joints, 2 sides							
	110	Not reinforced, 8" x 16" x 4" thick	D-8	435	S.F.	.96	1.72	2.68	3.51
	115	6" thick		405		1.25	1.85	3.10	4.01
	120	8" thick		380		1.49	1.97	3.46	4.44
	125	10" thick	↓	365		2.10	2.05	4.15	5.25
	130	12" thick	D-9	345		2.15	2.54	4.69	6
	200	Not reinforced, 8" x 24" x 4" thick		460		.88	1.90	2.78	3.68
	210	6" thick		440		1.18	1.99	3.17	4.13
	215	8" thick		415		1.39	2.11	3.50	4.54
	220	10" thick		385		1.68	2.28	3.96	5.10
	225	12" thick	↓	365		1.79	2.40	4.19	5.40
	280	Solid, not reinforced, 8" x 16" x 2" thick	D-8	440		.79	1.70	2.49	3.29
	290	4" thick		420		1.03	1.78	2.81	3.67
	295	6" thick		390		1.27	1.92	3.19	4.13
	300	8" thick		365		1.66	2.05	3.71	4.75
	305	10" thick	↓	350		2.51	2.14	4.65	5.80
	310	12" thick	D-9	330	↓	2.54	2.65	5.19	6.60
	599								
	600	Stud block, 8" x 16", 6" + 2", plain	D-8	410	S.F.	1.44	1.82	3.26	4.18
	605	Embossed		390		1.93	1.92	3.85	4.86
	610	8" + 2", plain		390		1.66	1.92	3.58	4.56
	615	Embossed		370		2.23	2.02	4.25	5.35
	620	10" + 2", plain		370		1.99	2.02	4.01	5.05
	625	Embossed	↓	350	↓	2.46	2.14	4.60	5.75

4.3 Block & Tile Masonry

			CREW	DAILY OUTPUT	UNIT	BARE COSTS MAT.	BARE COSTS INST.	BARE COSTS TOTAL	TOTAL INCL O&P
52	650	Stud block wall, 8" x 16" unit, 6" + 2" + 2"	D-8	350	S.F.	1.72	2.14	3.86	4.94
	655	8" + 2" + 2"		340		1.94	2.20	4.14	5.25
	670	Special jamb block, 6" + 2"		390		2.20	1.92	4.12	5.15
	675	8" + 2"		370		2.20	2.02	4.22	5.30
	680	10" + 2"		350		2.45	2.14	4.59	5.75
53-001		COLUMN BLOCK Pilaster incl. reinf., grout and scaffolding							
	016	1 piece unit, 16" x 16"	D-1	26	V.L.F.	8.30	11.25	19.55	25
	017	2 piece units, 16" x 20"		24		10.60	12.15	22.75	29
	018	20" x 20"		22		14.60	13.25	27.85	35
	019	22" x 24"		18		17.65	16.20	33.85	43
	020	20" x 32"		14		21.60	21	42.60	53
54-001		CHIMNEY BLOCK Incl. scaffolding							
	002								
	022	1 piece, with 8" x 8" flue, 16" x 16"	D-1	28	V.L.F.	6.35	10.45	16.80	22
	023	2 piece, 16" x 16"		26		6.70	11.25	17.95	23
	024	2 piece with 8" x 12" flue, 16" x 20"		24		8.96	12.15	21.11	27
	025	With 12" x 12" flue, 20" x 20"		22		12.40	13.25	25.65	33
57-001		CONCRETE BLOCK, SOLAR SCREEN Incl. scaffolding							
	020	4" thick, 6" x 6" units	D-8	180	S.F.	2.15	4.16	6.31	8.30
	030	8" x 8" units		270		2.73	2.77	5.50	6.95
	035	12" x 12" units		300		1.71	2.49	4.20	5.45
	050	8" thick, 8" x 16" units		330		2	2.27	4.27	5.45
60-001		GLASS BLOCK Incl. scaffolding							
	010	4" thick, plain, under 1000 S.F., 6" x 6" block	D-8	115	S.F.	16.25	6.50	22.75	27
	015	8" x 8" block		160		13.65	4.68	18.33	22
	020	12" x 12" block		175		12.70	4.27	16.97	20
	030	Plain, 1000 to 5000 S.F., 6" x 6" block		135		14.60	5.55	20.15	24
	035	8" x 8" block		190		11.10	3.94	15.04	17.80
	040	12" x 12" block		215		11.70	3.48	15.18	17.85
	049								
	050	Plain, over 5000 S.F., 6" x 6" block	D-8	145	S.F.	14.15	5.15	19.30	23
	055	8" x 8" block		215		10.60	3.48	14.08	16.60
	060	12" x 12" block		240		11.10	3.12	14.22	16.65
	070	For solar reflective blocks, add				65%			
	080	Plain, 4" x 8" blocks, under 1000 S.F.	D-8	145		17.20	5.15	22.35	26
	085	Over 5000 S.F.		170		13.60	4.40	18	21
	100	3-1/8" thick thinline, plain, under 1000 S.F., 6" x 6" block		115		13.95	6.50	20.45	25
	105	8" x 8" block		160		10.95	4.68	15.63	18.70
	120	Plain, over 5000 S.F., 6" x 6" block		145		13.05	5.15	18.20	22
	125	8" x 8" block		215		8	3.48	11.48	13.75
	140	For cleaning block after installation (both sides), add		1,000		.05	.75	.80	1.12
65-001		GLAZED CONCRETE BLOCK Incl. scaffolding							
	010	Not reinforced, 2" thick, solid	D-8	360	S.F.	3.56	2.08	5.64	6.90
	020	Hollow core, 8" x 16" units, 4" thick		345		3.98	2.17	6.15	7.45
	025	6" thick		330		4.22	2.27	6.49	7.85
	030	8" thick		310		5.52	2.41	7.93	9.50
	035	10" thick		295		4.91	2.54	7.45	9
	040	12" thick	D-9	280		5.30	3.13	8.43	10.30
	070	Double face, 8" x 16" units, 4" thick	D-8	310		6.28	2.41	8.69	10.35
	075	6" thick		290		7.01	2.58	9.59	11.40
	080	8" thick		270		7.35	2.77	10.12	12.05
	100	Corner, bullnose or square, 8" x 16" units, 2" thick		315	Ea.	6.32	2.37	8.69	10.35
	105	4" thick		285	"	5.52	2.62	8.14	9.80
	120	Cap and sill, 8" x 16" units, 2" thick		420	L.F.	5.52	1.78	7.30	8.60
	125	4" thick		380		6	1.97	7.97	9.40
	150	Cove base, 8" high, 2" thick		315		3.78	2.37	6.15	7.55
	155	4" thick		285		3.95	2.62	6.57	8.10

For expanded coverage of these items see *Means' Concrete & Masonry Cost Data 1986*

4.3 Block & Tile Masonry

			DAILY		BARE COSTS			TOTAL
		CREW	OUTPUT	UNIT	MAT.	INST.	TOTAL	INCL O&P
65 160	6" thick	D-8	265	L.F.	4.24	2.82	7.06	8.70
165	8" thick	"	245	"	4.45	3.05	7.50	9.25
70-001	**GROUTING** See division 4.1-30							
75-001	**INSULATION** See also division 7.2-25							
010	Inserts, styrofoam, plant installed, add to block prices							
020	8" x 16" units, 6" thick			S.F.	.63		.63	.69M
025	8" thick				.63		.63	.69M
030	10" thick				.75		.75	.82M
035	12" thick				.77		.77	.84M
050	8" x 8" units, 8" thick				.50		.50	.55M
055	12" thick			↓	.65		.65	.71M
80-001	**REINFORCING** See division 4.1-35							
83-001	**STRUCTURAL FACING TILE** Incl. scaffolding, 6T series, 5-1/3" x 12"							
002	select and clear, 2.3 pieces per S.F., 2" thick, glazed 1 side	D-8	225	S.F.	2.95	3.32	6.27	8
010	4" thick, glazed 1 side		220		3.65	3.40	7.05	8.85
015	Glazed 2 sides		195		5.80	3.84	9.64	11.85
025	6" thick, glazed 1 side		210		5.65	3.56	9.21	11.30
030	Glazed 2 sides		185		6.60	4.04	10.64	13
040	8" thick, glazed 1 side	↓	180	↓	6.25	4.16	10.41	12.80
045								
050	6T special shapes, group 1	D-8	300	Ea.	3.25	2.49	5.74	7.15
055	Group 2		275		3.75	2.72	6.47	8
060	Group 3		250		5.50	2.99	8.49	10.30
065	Group 4		225		9.75	3.32	13.07	15.45
070	Group 5		200		11.65	3.74	15.39	18.15
075	Group 6		190	↓	17.45	3.94	21.39	25
100	6T fire rated, 4" thick, 1 hr. rating		210	S.F.	6.70	3.56	10.26	12.45
105	6" thick, 2 hr. rating	↓	200		10.40	3.74	14.14	16.75
120	For stack bond with ground ends, add				8%	10%		
130	6T acoustic, 4" thick	D-8	210		6.25	3.56	9.81	11.95
140	For designer colors, add			↓	25%			
170								
200	8W series, 8" x 16", grade S.S., 1.125 pieces per S.F.							
205	2" thick, glazed 1 side	D-8	360	S.F.	3.40	2.08	5.48	6.70
210	4" thick, glazed 1 side		345		3.65	2.17	5.82	7.10
215	Glazed 2 sides		325		6.25	2.30	8.55	10.15
220	6" thick, glazed 1 side		330		5.20	2.27	7.47	8.95
225	8" thick, glazed 1 side		310	↓	6.10	2.41	8.51	10.15
250	8W special shapes, group 1		270	Ea.	4.70	2.77	7.47	9.10
255	Group 2		260		5.55	2.88	8.43	10.20
260	Group 3		250		11.25	2.99	14.24	16.65
265	Group 4		240		16.75	3.12	19.87	23
270	Group 5		230		24.50	3.25	27.75	32
275	Group 6		220	↓	34.25	3.40	37.65	43
300	8W fire rated, 4" thick, 1 hr.		345	S.F.	6.60	2.17	8.77	10.35
310	8W acoustic, 4" thick, add	↓	345		7.25	2.17	9.42	11.05
320	For designer colors, add			↓	25%			
87-001	**SURFACE BONDING** Block walls with fiberglass mortar,							
002	gray or white colors, not incl. block work	1 Bric	540	S.F.	.35	.30	.65	.82
90-001	**TERRA COTTA** Coping, split type, not glazed, 9" wide	D-1	90	L.F.	3.05	3.24	6.29	8
010	13" wide		80		5	3.65	8.65	10.70
020	Split type, glazed, 9" wide		90		3.40	3.24	6.64	8.35
025	13" wide	↓	80	↓	5.60	3.65	9.25	11.35
050	Partition or back-up blocks, scored, in C.L. lots							
060	Split furring tiles, 12" x 12", 2" thick	D-8	600	S.F.	2.05	1.25	3.30	4.03
070	Non-load bearing, 12" x 12", 3" thick		550		2.05	1.36	3.41	4.19
075	4" thick	↓	500	↓	2.70	1.50	4.20	5.10

For expanded coverage of these items see *Means' Concrete & Masonry Cost Data 1986*

4.3 Block & Tile Masonry

			CREW	DAILY OUTPUT	UNIT	BARE COSTS MAT.	BARE COSTS INST.	BARE COSTS TOTAL	TOTAL INCL O&P
90	080	6" thick	D-8	450	S.F.	3.05	1.66	4.71	5.70
	085	8" thick		400		4.15	1.87	6.02	7.25
	100	Load bearing, 12" x 12", 4" thick, in walls		500		2.25	1.50	3.75	4.61
	105	In floors		750		2.25	1	3.25	3.90
	120	6" thick, in walls		450		3.05	1.66	4.71	5.70
	125	In floors		675		3.05	1.11	4.16	4.93
	140	8" thick, in walls		400		4.15	1.87	6.02	7.25
	145	In floors		575		4.15	1.30	5.45	6.40
	160	10" thick, in walls		350		5.50	2.14	7.64	9.10
	165	In floors		500		5.50	1.50	7	8.20
	180	12" thick, in walls		300		6.15	2.49	8.64	10.30
	185	In floors		450		6.15	1.66	7.81	9.15
	200	For reinforcing with steel rods, add to above				15%	5%		
	210	For smooth tile instead of scored, add				.03		.03	.03M
	220	For L.C.L. quantities, add				10%	10%		

4.4 Stonework

			CREW	DAILY OUTPUT	UNIT	BARE COSTS MAT.	BARE COSTS INST.	BARE COSTS TOTAL	TOTAL INCL O&P
05-001		ASHLAR VENEER 4" + or - thick, random or random rectangular							
	020	Medium priced stone	D-8	130	S.F.	5.85	5.75	11.60	14.65
	030	High priced stone	"	140	"	6.86	5.35	12.21	15.15
	040	(68)							
	060	Seam face, split joints, medium price stone	D-8	120	S.F.	5.50	6.25	11.75	14.95
	070	High price stone		125		8	6	14	17.30
	100	Split or rock face, split joints, medium price stone		120		5.50	6.25	11.75	14.95
	110	High price stone		125		8	6	14	17.30
06-001		BLUESTONE Cut to size							
	002								
	050	Sills, natural cleft, 10" wide to 6' long, 1-1/2" thick	D-11	70	L.F.	5.40	6.75	12.15	15.55
	055	2" thick	"	63		6	7.50	13.50	17.25
	100	Stair treads, natural cleft, 12" wide, 6' long, 1-1/2" thick	D-10	115		6.30	8.65	14.95	18.65
	105	2" thick		105		7	9.45	16.45	21
	110	Smooth finish, 1-1/2" thick		115		10	8.65	18.65	23
	120								
	130	Thermal finish, 1-1/2" thick	D-10	115	L.F.	9.20	8.65	17.85	22
	135	2" thick	"	105	"	10.65	9.45	20.10	25
07-001		CAST CERAMIC FLOORING See division 4.2-28							
10-001		FACING PANELS Stone aggregate mounted on plywood,							
	002	see division 7.4-30							
15-001		GRANITE Cut to size							
	005	Veneer, polished face, 3/4" to 1-1/2" thick							
	015	Low price, gray, light gray, etc.	D-10	130	S.F.	12	7.65	19.65	24
	022	High price, red, black, etc.	"	130	"	29	7.65	36.65	42
	030	1-1/2" to 2-1/2" thick, veneer							
	035	Low price, gray, light gray, etc.	D-10	130	S.F.	14	7.65	21.65	26
	055	High price, red, black, etc.	"	130	"	18	7.65	25.65	30
	070	2-1/2" to 4" thick, veneer							
	075	Low price, gray, light gray, etc.	D-10	110	S.F.	22	9	31	36
	095	High price, red, black, etc.	"	110		27	9	36	42
	100	Deduct from polished for bush hammered finish, deduct				5%			
	105	Coarse rubbed finish, deduct				10%			
	110	Honed finish, deduct				5%			
	115	Thermal finish, deduct				18%			

For expanded coverage of these items see *Means' Concrete & Masonry Cost Data 1986*

4.4 Stonework

			CREW	DAILY OUTPUT	UNIT	BARE COSTS MAT.	BARE COSTS INST.	BARE COSTS TOTAL	TOTAL INCL O&P
15	245	For radius under 5', add			L.F.	100%			
	250	Steps, copings, etc., finished on more than one surface							
	255	Minimum	D-10	50	C.F.	50	19.85	69.85	82
	260	Maximum	"	50	"	75	19.85	94.85	110
	280	Pavers, 4" x 4" x 4" blocks, split face and joints							
	285	Minimum	D-11	80	S.F.	6.50	5.90	12.40	15.55
	290	Maximum	"	80	"	13	5.90	18.90	23
	320								
	350	Curbing, city street type, 6" x 18", split face,							
	351	sawn top, radius nosing, 4' to 7' lengths	D-10	75	L.F.	12	13.25	25.25	31
	360	Highway type, 5" x 16", split face,							
	361	sawn top, 4' to 7' lengths	D-10	300	L.F.	9.50	3.31	12.81	14.95
	400	Soffits, 2" thick, minimum	D-13	35	S.F.	25	33	58	73
	410	Maximum		35		50	33	83	100
	420	4" thick, minimum		35		30	33	63	78
	430	Maximum	↓	35	↓	60	33	93	110
20-001		**LIGHTWEIGHT NATURAL STONE** Lava type							
	002								
	010	Veneer, rubble face, sawed back, irregular shapes	D-10	130	S.F.	4	7.65	11.65	14.75
	020	Sawed face and back, irregular shapes	"	130	"	6	7.65	13.65	16.95
25-001		**LIMESTONE** See also Ashlar Veneer, division 4.4-05							
	002	Veneer facing panels							
	010	Sawn finish, 2" thick, to 3' x 5' panels	D-10	130	S.F.	8.40	7.65	16.05	19.60
	015	Smooth finish, 2" thick, to 3' x 5' panels		130		10	7.65	17.65	21
	030	3" thick, to 4' x 9' panels		225		11	4.41	15.41	18.10
	035	4" thick, to 5' x 11' panels	↓	275		13	3.61	16.61	19.20
	050	Texture finish, light stick, 4-1/2" thick, 5' x 12'	D-4	300		9.50	2.43	11.93	13.75
	075	5" thick, to 5' x 14' panels	D-10	275		14.50	3.61	18.11	21
	100	Medium ribbed, textured finish, 4-1/2" thick, to 5' x 12'		275		9.50	3.61	13.11	15.35
	105	5" thick, to 5' x 14' panels		275		14.50	3.61	18.11	21
	120	Deep ribbed, textured finish, 4-1/2" thick, to 5' x 10'		275		9.50	3.61	13.11	15.35
	125	5" thick, to 5' x 14' panels		275		14.50	3.61	18.11	21
	140	Sugar cube, textured finish, 4-1/2" thick, to 5' x 12'		275		9.50	3.61	13.11	15.35
	145	5" thick, to 5' x 14' panels		275	↓	14.50	3.61	18.11	21
	200	Coping, smooth finish, top & 2 sides	↓	30	C.F.	45	33	78	94
	205								
	210	Sills, lintels, jambs, smooth finish, average	D-10	20	C.F.	45	50	95	115
	215	Detailed		20	"	60	50	110	135
	230	Steps, extra hard, 14" wide, 6" rise	↓	50	L.F.	25	19.85	44.85	54
30-001		**MARBLE** Ashlar, split face, 4" + or - thick, random							
	004	lengths 1' to 4' & heights 2" to 7-1/2", average	D-8	175	S.F.	6.85	4.27	11.12	13.60
	010	Base, polished, 3/4" or 7/8" thick, polished, 6" high	D-10	65	L.F.	8.10	15.25	23.35	30
	030	Carvings or bas relief, from templates, average		80	S.F.	90	12.40	102.40	115
	035	Maximum	↓	80	"	250	12.40	262.40	290
	060	Columns, cornices, mouldings, etc.							
	065	Hand or special machine cut, average	D-10	35	C.F.	90	28	118	135
	070	Maximum	"	35	"	200	28	228	260
	100	Facing, polished finish, cut to size, 3/4" to 7/8" thick							
	105	Average	D-10	130	S.F.	13	7.65	20.65	25
	110	Maximum	"	130	"	40	7.65	47.65	54
	120								
	130	1-1/4" thick, average	D-10	125	S.F.	17	7.95	24.95	29
	135	Maximum		125		35	7.95	42.95	49
	150	2" thick, average		120		20	8.25	28.25	33
	155	Maximum	↓	120	↓	40	8.25	48.25	55
	220	Window sills, 6" x 2" thick	D-1	85	L.F.	10	3.44	13.44	15.90
	250	Flooring, polished tiles, 12" x 12" x 3/8" thick							
	251	Thin set, average	D-11	90	S.F.	5.50	5.25	10.75	13.50
	260	Maximum	"	90	"	17	5.25	22.25	26

For expanded coverage of these items see *Means' Concrete & Masonry Cost Data 1986*

4.4 Stonework

			CREW	DAILY OUTPUT	UNIT	BARE COSTS MAT.	BARE COSTS INST.	BARE COSTS TOTAL	TOTAL INCL O&P
30	270	Mortar bed, average	D-11	65	S.F.	5.50	7.25	12.75	16.40
	274	Maximum	"	65		17	7.25	24.25	29
	278	Travertine, 1-1/4" thick, average	D-10	130		13	7.65	20.65	25
	279	Maximum	"	130		17.50	7.65	25.15	30
	280	Patio blocks, non-slip, 7/8" thick	D-11	75		6.50	6.30	12.80	16.10
	290	Shower or toilet partitions, 7/8" thick partitions							
	305	3/4" or 1-1/4" thick stiles, polished 2 sides, average	D-11	75	S.F.	22	6.30	28.30	33
	320	Soffits, add to above prices			"	20%	100%		
	321	Stairs, risers, 7/8" thick x 6" high	D-10	115	L.F.	8.75	8.65	17.40	21
	336	Treads, 12" wide x 1-1/4" thick	"	115	"	11.50	8.65	20.15	24
	350	Thresholds, 3' long, 7/8" thick, 4" to 5" wide, plain	D-12	24	Ea.	8	25	33	44
	355	Beveled		24	"	9	25	34	46
	370	Window stools, polished, 7/8" thick, 5" wide		85	L.F.	8	7.05	15.05	18.85
35-001		**PATIO BLOCKS** See also division 2.6-40							
40-001		**ROUGH STONE WALL**							
	002								
	010	Random fieldstone, under 18" thick	D-12	60	C.F.	8.60	10	18.60	24
	015	Over 18" thick	"	63	"	7.75	9.50	17.25	22
45-001		**SANDSTONE OR BROWNSTONE**							
	010	Sawed face veneer, 2-1/2" thick, to 2' x 4' panels	D-10	130	S.F.	8.80	7.65	16.45	20
	015	4" thick, to 3'-6" x 8' panels		100		8.40	9.95	18.35	23
	030	Split face, random sizes		100		5.85	9.95	15.80	19.90
	035	Cut stone trim (sandstone)							
	036	Ribbon stone, 4" thick, 5' pieces	D-8	120	L.F.	68	6.25	74.25	84
	037	Cove stone, 4" thick, 5' pieces		105		58	7.10	65.10	74
	038	Cornice stone, 10" to 12" wide		90		89	8.30	97.30	110
	039	Band stone, 4" thick, 5' pieces		145		47	5.15	52.15	59
	041	Window and door trim, 3" to 4" wide		160		35	4.68	39.68	45
	042	Key stone, 18" long		60	Ea.	29	12.45	41.45	50
50-001		**SIMULATED STONE**							
	010	Insulated fiberglass panels	L-4	200	S.F.	5.50	2.26	7.76	9.35
53-001		**SLATE** Pennsylvania, blue gray to gray black; Vermont,							
	005	Unfading green, mottled green & purple, gray & purple							
	010	Virginia, blue black							
	015								
	020	Exterior paving, natural cleft, 1" thick							
	050	24" x 24", Pennsylvania	D-12	120	S.F.	4.25	5	9.25	11.80
	055	Vermont		120		6.40	5	11.40	14.15
	060	Virginia		120		7.40	5	12.40	15.25
	100	Interior flooring, natural cleft, 1/2" thick							
	130	24" x 24" Pennsylvania	D-12	120	S.F.	3	5	8	10.40
	135	Vermont		120		5.30	5	10.30	12.95
	140	Virginia		120		6.40	5	11.40	14.15
	200	Facing panels, 1-1/4" thick, to 4' x 6' panels							
	210	Natural cleft finish, Pennsylvania	D-10	180	S.F.	11	5.50	16.50	19.55
	210	Vermont		180		12.25	5.50	17.75	21
	211	Virginia		180		13.25	5.50	18.75	22
	215	Sand rubbed finish, add				1.25		1.25	1.37M
	220	Honed finish, add				2		2	2.20M
	250	Ribbon, natural cleft finish, 1" thick, to 9 S.F.	D-10	80		4.90	12.40	17.30	22
	270	1-1/2" thick		78		6.20	12.75	18.95	24
	285	2" thick		76		12.50	13.05	25.55	31
	300	Roofing, see division 7.3-35							
	350	Stair treads, sand finish, 1" thick x 12" wide							
	360	3 L.F. to 6 L.F.	D-10	120	L.F.	12.95	8.25	21.20	25
	370	Ribbon, sand finish, 1" thick x 12" wide							
	375	To 6 L.F.	D-10	120	L.F.	8.25	8.25	16.50	20

For expanded coverage of these items see *Means' Concrete & Masonry Cost Data 1986*

4.4 Stonework

		CREW	DAILY OUTPUT	UNIT	BARE COSTS MAT.	BARE COSTS INST.	BARE COSTS TOTAL	TOTAL INCL O&P
400	Stools or sills, sand finish, 1" thick, 6" wide	D-12	160	L.F.	4.50	3.75	8.25	10.30
420	10" wide		90		7.80	6.65	14.45	18.10
440	2" thick, 6" wide		140		7.90	4.29	12.19	14.80
460	10" wide	↓	90		11.50	6.65	18.15	22
480	For lengths over 3', add			↓	25%			

4 5.1 Structural Metals

		CREW	DAILY OUTPUT	UNIT	BARE COSTS MAT.	BARE COSTS INST.	BARE COSTS TOTAL	TOTAL INCL O&P
05-001	**ALUMINUM** Structural shapes, 1" to 10" members, under 1 ton	E-2	1,050	Lb.	1.80	1.89	3.69	4.56
010	Over 5 tons		1,330		1.60	1.49	3.09	3.79
030	Extrusions, over 5 tons, stock shapes		1,330		2.20	1.49	3.69	4.45
040	Custom shapes	↓	1,330	↓	2.40	1.49	3.89	4.67
10-001	**ANCHOR BOLTS** See division 3.1-05 & 4.1-06							
15-001	(77) **BOLTS** High strength, see division 5.1-50-520							
20-001	**CANOPY FRAMING** 6" and 8" members	E-4	3,000	Lb.	.56	.26	.82	1.01
26-001	**COLUMNS** Aluminum, extruded, stock units, 6" diameter	E-4	240	L.F.	7.10	3.20	10.30	12.70
010	8" diameter		170		9.60	4.51	14.11	17.50
020	10" diameter		150		12.10	5.10	17.20	21
030	12" diameter		140		22	5.50	27.50	33
040	15" diameter	↓	120	↓	28	6.40	34.40	41
041	Caps and bases, plain, 6" diameter			Set	11		11	12.10M
042	8" diameter				16		16	17.60M
043	10" diameter				23		23	25M
044	12" diameter				49		49	54M
045	15" diameter				78		78	86M
046	Caps, ornamental, minimum				75		75	83M
047	Maximum			↓	480		480	530M
050	For square columns, add to column prices above			L.F.	30%			
060	Tubular aluminum	E-4	1,500	Lb.	4.10	.51	4.61	5.30
070	Residential, flat, 8' high, plain		20	Ea.	26	38	64	88
072	Fancy		20		42	38	80	105
074	Corner type, plain		20		47	38	85	110
076	Fancy	↓	20		68	38	106	135
080	Steel, concrete filled, extra strong pipe, 3-1/2" diameter	E-2	660	L.F.	8.75	3	11.75	13.70
085	4-1/2" diameter		900		10	2.20	12.20	14
090	5-1/2" diameter		1,140		12	1.74	13.74	15.55
093	6-5/8" diameter		1,200		17	1.65	18.65	21
100	Lightweight units, 3-1/2" diameter		780		2.75	2.54	5.29	6.50
105	4" diameter		900	↓	3	2.20	5.20	6.30
110	For galvanizing, add			Lb.	.25		.25	.27M
130	For web ties, angles, etc., add per added lb.	1 Sswk	945		.65	.18	.83	1
150	Steel pipe, extra strong, no concrete, 3" to 5" O.D.	E-2	12,960		.60	.15	.75	.87
160	6" to 12" O.D.		39,100	↓	.60	.05	.65	.73
170	Steel pipe, extra strong, no concrete, 3" diameter x 12'-0"		60	Ea.	88	33	121	140
175	4" diameter x 12'-0"		58		135	34	169	195
180	6" diameter x 12'-0"		54		265	37	302	340
185	8" diameter x 14'-0"		50		325	40	365	410
190	10" diameter x 16'-0"		48		670	41	711	795
195	12" diameter x 18'-0"		45	↓	900	44	944	1,050
330	Square structural tubing, 4" to 6" square, light section		11,270	Lb.	.78	.18	.96	1.10
360	Heavy section	↓	27,600	"	.78	.07	.85	.96

5.1 Structural Metals

		Description	CREW	DAILY OUTPUT	UNIT	BARE COSTS MAT.	BARE COSTS INST.	BARE COSTS TOTAL	TOTAL INCL O&P
26	400	Concrete filled, light section, add			L.F.	.45		.45	.49M
	435	Heavy section, add			"	.45		.45	.49M
	450	Square structural tubing, 4" x 4" x 1/4" x 12'-0"	E-2	58	Ea.	120	34	154	180
	455	6" x 6" x 1/4" x 12'-0"		54		215	37	252	285
	460	8" x 8" x 3/8" x 14'-0"		50		395	40	435	490
	465	10" x 10" x 1/2" x 16'-0"		48		730	41	771	860
	510	Rectangular structural tubing, 5" to 6" wide, light section		9,500	Lb.	.89	.21	1.10	1.26
	520	Heavy section		31,200		.87	.06	.93	1.04
	530	7" to 10" wide, light section		37,000		.73	.05	.78	.88
	540	Heavy section		68,000		.71	.03	.74	.82
	550	Rectangular structural tubing, 5" x 3" x 1/4" x 12'-0"		58	Ea.	110	34	144	170
	555	6" x 4" x 1/4" x 12'-0"		54		170	37	207	235
	560	8" x 4" x 3/8" x 12'-0"		54		240	37	277	315
	565	10" x 6" x 1/2" x 14'-0"		50		390	40	430	485
	570	12" x 8" x 5/8" x 16'-0"		48		730	41	771	860
	575								
	580	Adjustable jack post, 8' maximum height, 2-3/4" diameter			Ea.	15		15	16.50M
	585	4" diameter			"	25		25	28M
	600	Prefabricated fireproof with steel jackets and one coat							
	610	shop paint, 2 to 4 hour rated, minimum	E-2	27,000	Lb.	.55	.07	.62	.71
	620	Average		35,000		.60	.06	.66	.74
	625	Maximum		43,000		.70	.05	.75	.83
	640	Mild steel, flat, 9" wide, stock units, painted, plain	E-4	160	L.F.	2.95	4.80	7.75	10.60
	645	Fancy		160		6.95	4.80	11.75	15
	650	Corner columns, painted, plain		160		4.85	4.80	9.65	12.70
	655	Fancy		160		10.50	4.80	15.30	18.90
	680	Wide flange, A36 steel, 2 tier, W 8 x 24	E-2	1,080		9.95	1.84	11.79	13.45
	685	W 8 x 31		1,080		12.05	1.84	13.89	15.75
	690	W 8 x 48		1,032		19.65	1.92	21.57	24
	695	W 8 x 67		984		24	2.01	26.01	29
	700	W 10 x 45		1,032		18.90	1.92	20.82	23
	705	W 10 x 68		984		26	2.01	28.01	31
	710	W 10 x 112		960		42	2.07	44.07	49
	715	W 12 x 50		1,032		19.40	1.92	21.32	24
	720	W 12 x 87		984		33	2.01	35.01	39
	725	W 12 x 120		960		46	2.07	48.07	53
	730	W 12 x 190		912		79	2.17	81.17	90
	735	W 14 x 74		984		29	2.01	31.01	35
	740	W 14 x 120		960		47	2.07	49.07	55
	745	W 14 x 176		912		67	2.17	69.17	77
30-001		**LIGHTGAGE FRAMING**							
	020	For steel studs see division 9.1-16 & 9.2-20							
	040	Angle framing, 4" and larger	E-4	3,000	Lb.	.58	.26	.84	1.03
	045	Less than 4" angles		1,800		.63	.43	1.06	1.35
	060	Channel framing, 8" and larger		3,500		.53	.22	.75	.92
	065	Less than 8" channels		2,000		.58	.38	.96	1.23
	100	Continuous slotted channel framing system, minimum	2 Sswk	2,400		.89	.14	1.03	1.21
	120	Maximum	"	1,600		1.58	.22	1.80	2.08
	130	Cross bracing, rods, 3/4" diameter	E-3	700		.63	.92	1.55	2.07
	131	7/8" diameter		700		.53	.92	1.45	1.96
	132	1" diameter		700		.53	.92	1.45	1.96
	133	Angle, 5" x 5" x 3/8"		2,800		.37	.23	.60	.75
	135	Hanging lintels, average		850		.63	.76	1.39	1.82
	138	Roof frames, 3'-0" square, 5' span	E-2	4,200		.53	.47	1	1.23
	140	Tie rod, not upset, 1-1/2" to 4" diameter, with turnbuckle	2 Sswk	800		.60	.43	1.03	1.34
	142	No turnbuckle		700		.53	.50	1.03	1.36
	150	Upset, 1-3/4" to 4" diameter, with turnbuckle		800		.69	.43	1.12	1.44
	152	No turnbuckle		700		.59	.50	1.09	1.43
	160	Tubular aluminum framing for window wall, minimum	(105)	600		2.75	.58	3.33	3.93
	180	Maximum		500		8	.69	8.69	9.90

5.1 Structural Metals

		CREW	DAILY OUTPUT	UNIT	BARE COSTS MAT.	BARE COSTS INST.	BARE COSTS TOTAL	TOTAL INCL O&P
35-001	**PRE-ENGINEERED STEEL BUILDINGS** For Hangars see 13.1-40							
010	Building shell above the foundations with 26 ga. colored							
011	roofing and siding, minimum	E-2	1,800	S.F.Flr.	3.15	1.10	4.25	4.97
020	Maximum	"	1,000	"	6.55	1.98	8.53	9.90
080	Accessory items: add to the basic building cost above							
100	Eave overhang, 2' wide, 26 ga., with soffit	E-2	360	L.F.	5.55	5.50	11.05	13.60
120	4' wide, without soffit		300		8.15	6.60	14.75	18
130	With soffit		250		10.75	7.95	18.70	23
150	6' wide, without soffit		250		13	7.95	20.95	25
160	With soffit		200		18	9.90	27.90	33
180	Entrance canopy, incl. frame, 4' x 4'		25	Ea.	92	79	171	210
190	4' x 8'		19	"	150	105	255	305
210	End wall roof overhang, 4' wide, without soffit		850	L.F.	6.10	2.33	8.43	9.90
220	With soffit		500	"	8.75	3.96	12.71	15.05
230	Doors, H.M. self-framing, incl. butts, lockset and trim							
235	Single leaf, 3070 (3' x 7'), economy	2 Sswk	5	Opng.	265	69	334	400
240	Deluxe		4		300	87	387	465
245	Glazed		4		285	87	372	450
250	3670 (3'-6" x 7')		4		275	87	362	440
255	4070 (4' x 7')		3		305	115	420	515
260	Double leaf, 6070 (6' x 7')		2		465	175	640	785
265	Glazed		2		515	175	690	840
270	Framing only, for openings, 3' x 7'	E-2	25		75	79	154	190
280	10' x 10'		21		145	94	239	290
300	For windows below, 2020 (2' x 2')		25		57	79	136	170
310	4030 (4' x 3')		22		64	90	154	195
330	Flashings, 26 ga., corner or eave, painted	2 Sswk	240	L.F.	1.46	1.45	2.91	3.88
340	Galvanized		240		1.34	1.45	2.79	3.75
360	Rake flashing, painted		240		1.80	1.45	3.25	4.25
370	Galvanized		240		1.62	1.45	3.07	4.05
420	Ridge flashing, 18" wide, painted		240		1.70	1.45	3.15	4.14
430	Galvanized		240		1.57	1.45	3.02	4
450	Gutter, eave type, 26 ga., painted		320		1.75	1.09	2.84	3.63
470	Galvanized		320		1.73	1.09	2.82	3.61
490	Valley type, between buildings, painted		120		3.45	2.89	6.34	8.35
500	Galvanized		120		3.20	2.89	6.09	8.05
520	Insulation, rated .6 lb density, vinyl faced							
525	1-1/2" thick, R5	2 Carp	2,300	S.F.	.22	.14	.36	.44
530	3" thick, R10		2,300		.32	.14	.46	.55
535	4" thick, R11		2,300		.42	.14	.56	.66
540	Foil faced, 1-1/2" thick, R5		2,300		.22	.14	.36	.44
545	2" thick, R6		2,300		.24	.14	.38	.47
550	3" thick, R10		2,300		.30	.14	.44	.53
555	4" thick, R13		2,300		.38	.14	.52	.62
560	Metalized polyester facing, 1-1/2" thick, R6		2,300		.24	.14	.38	.47
565	2" thick, R8		2,300		.28	.14	.42	.51
570	3" thick, R11		2,300		.42	.14	.56	.66
575	4" thick, R13		2,300		.48	.14	.62	.73
580	Vinyl, scrim foil (SD), 1-1/2" thick, R5		2,300		.26	.14	.40	.49
585	2" thick, R6		2,300		.28	.14	.42	.51
590	3" thick, R10		2,300		.36	.14	.50	.60
595	4" thick, R13		2,300		.43	.14	.57	.67
610	For heavy duty vinyl and foil, add				.08		.08	.08M
630	For .75 lb. density, add				.08		.08	.08M
650	Sash, single slide, glazed, with screens, 2020 (2'x 2')	E-1	22	Opng.	83	26	109	130
655	3030 (3' x 3')		14		110	41	151	180
670	4030 (4' x 3')		13		145	44	189	225
675	6040 (6' x 4')		12		150	48	198	235
680	Double slide sash, 3030 (3' x 3')		14		125	41	166	200
690	6040 (6' x 4')		12		210	48	258	305

5.1 Structural Metals

		CREW	DAILY OUTPUT	UNIT	BARE COSTS MAT.	BARE COSTS INST.	BARE COSTS TOTAL	TOTAL INCL O&P
710	Fixed glass, no screens, 3030 (3' x 3')	E-1	14	Opng.	80	41	121	150
720	6040 (6' x 4')		12		140	48	188	225
740	Prefinished storm sash, 3030 (3' x 3')		70		28	8.20	36.20	43
760	6040 (6' x 4')		60		40	9.60	49.60	58
780	Siding and roofing, see division 7.4							
780								
781	Skylight, fiberglass panels, to 30 S.F.	E-1	10	Ea.	150	58	208	250
782	Larger sizes, add for excess over 30 S.F.	"	300	S.F.	1.25	1.92	3.17	4.24
800	Roof vents, circular with damper, birdscreen							
801	and operator hardware, painted							
810	26 ga., 12" diameter	1 Sswk	4	Ea.	100	43	143	180
815	20" diameter		3		140	58	198	245
820	24 ga., 24" diameter		2		160	87	247	310
825	Galvanized		2		145	87	232	295
830	Continuous, 26 ga., 10' long, 9" wide	2 Sswk	4		200	87	287	355
840	12" wide	"	4		235	87	322	395
40-001	**SPACE FRAME** Steel 4' modular, 40' to 70' spans, 5.5 psf, minimum	E-2	556	S.F.	7.85	3.57	11.42	13.50
020	Maximum		365		11.25	5.45	16.70	19.80
040	5' modular, 4.5 psf minimum		585		6.75	3.39	10.14	12.05
050	Maximum		405		9.65	4.89	14.54	17.30
070	Add to above for galvanizing, 4' modular				1		1	1.10M
080	5' modular				.90		.90	.99M
45-001	**STRESSED SKIN** Roof and ceiling system, spans to 100', minimum	E-2	1,150		2.80	1.72	4.52	5.45
020	Maximum	"	760		5.60	2.61	8.21	9.70
47-001	**STRUCTURAL STEEL MEMBERS** Common WF sizes, spans 10' to 45'							
002	including bolted connections and erection							
010	W 6 x 9	E-2	600	L.F.	5.45	3.30	8.75	10.50
030	W 8 x 10		600		5.65	3.30	8.95	10.70
050	X 31		500		12.90	3.96	16.86	19.60
070	W 10 x 22		660		12.65	3	15.65	18
090	X 49		540		18.95	3.67	22.62	26
110	W 12 x 14		880		7.05	2.25	9.30	10.85
130	X 22		880		9.90	2.25	12.15	13.95
150	X 26		880		11.40	2.25	13.65	15.60
170	X 72		640		26	3.10	29.10	33
190	W 14 x 26		990		10.60	2	12.60	14.40
210	X 30		900		11.85	2.20	14.05	16.05
230	X 34		810		13.10	2.45	15.55	17.75
250	X 120		720		43	2.75	45.75	51
270	W 16 x 26		1,000		10	1.98	11.98	13.70
290	X 31		900		11.95	2.20	14.15	16.15
310	X 40		800		14.60	2.48	17.08	19.45
330	W 18 x 35	E-5	960		12.75	2.74	15.49	17.85
350	x 40		960		14.20	2.74	16.94	19.45
370	X 50		912		18.20	2.88	21.08	24
390	X 55		912		20	2.88	22.88	26
410	W 21 x 44		1,064		15.60	2.47	18.07	21
430	x 50		1,064		17.70	2.47	20.17	23
450	X 62		1,036		21	2.53	23.53	27
470	X 68		1,036		24	2.53	26.53	30
490	W 24 x 55		1,110		20	2.37	22.37	25
510	X 62		1,110		22	2.37	24.37	28
530	X 68		1,110		23	2.37	25.37	29
550	X 76		1,110		26	2.37	28.37	32
570	X 84		1,080		28	2.43	30.43	34
590	W 27 x 94		1,190		32	2.21	34.21	38
610	W 30 x 99		1,200		34	2.19	36.19	40
630	X 108		1,200		37	2.19	39.19	44
650	X 116		1,160		40	2.26	42.26	47
670	W 33 x 118		1,176		40	2.23	42.23	47

5.1 Structural Metals

		CREW	DAILY OUTPUT	UNIT	BARE COSTS MAT.	BARE COSTS INST.	BARE COSTS TOTAL	TOTAL INCL O&P
690	W33 x 130	E-5	1,134	L.F.	44	2.32	46.32	52
710	X 141		1,134		48	2.32	50.32	56
730	W 36 x 135		1,170		46	2.24	48.24	54
750	X 150		1,170		51	2.24	53.24	59
770	X 194		1,125		67	2.33	69.33	77
790	X 230		1,125		80	2.33	82.33	91
810	X 300	↓	1,035	↓	105	2.54	107.54	120
50-001	**STRUCTURAL STEEL PROJECTS** Bolted, unless mentioned otherwise							
020	Apts., nursing homes, etc., steel bearing, 1 to 2 stories	E-5	10.30	Ton	825	255	1,080	1,275
030	3 to 6 stories	"	10.10		840	260	1,100	1,275
040	7 to 15 stories	E-6	14.20		865	265	1,130	1,325
050	Over 15 stories	"	13.90		895	270	1,165	1,375
070	Offices, hospitals, etc., steel bearing, 1 to 2 stories	E-5	10.30		800	255	1,055	1,225
080	3 to 6 stories	E-6	14.40		830	260	1,090	1,300
090	7 to 15 stories		14.20		865	265	1,130	1,325
100	Over 15 stories	↓	13.90		890	270	1,160	1,375
110	For multi-story masonry wall bearing construction, add					30%		
130	Industrial bldgs., 1 story, beams & girders, steel bearing	E-5	12.90		815	205	1,020	1,175
140	Masonry bearing	"	10	↓	815	265	1,080	1,275
150	Industrial bldgs., 1 story, under 10 tons,							
151	steel from warehouse, trucked	E-2	7.50	Ton	960	265	1,225	1,425
160	1 story with roof trusses, steel bearing	E-5	10.60		955	250	1,205	1,400
170	Masonry bearing	"	8.30		955	315	1,270	1,500
190	Monumental structures, banks, stores, etc., minimum	E-6	13		835	290	1,125	1,325
200	Maximum	"	9		1,450	420	1,870	2,200
220	Churches, minimum	E-5	11.60		750	225	975	1,150
230	Maximum	"	5.20		1,075	505	1,580	1,900
280	Power stations, fossil fuels, minimum	E-6	11		750	345	1,095	1,325
290	Maximum		5.70		1,275	660	1,935	2,350
295	Nuclear fuels, non-safety steel, minimum		7		855	540	1,395	1,700
300	Maximum		5.50		1,175	685	1,860	2,275
304	Safety steel, minimum		2.50		1,275	1,500	2,775	3,550
307	Maximum	↓	1.50		1,600	2,500	4,100	5,350
310	Roof trusses, minimum	E-5	13		855	200	1,055	1,225
320	Maximum		8.30		1,400	315	1,715	1,975
321	Schools, minimum		14.50		825	180	1,005	1,150
322	Maximum	↓	8.30		1,275	315	1,590	1,850
340	Welded construction, simple commercial bldgs., 1 to 2 stories	F-7	7.60		855	345	1,200	1,425
350	7 to 15 stories	E-9	8.30		965	470	1,435	1,725
370	Welded rigid frame, 1 story, minimum	E-7	15.80		855	165	1,020	1,175
380	Maximum	"	5.50		1,125	480	1,605	1,900
400	High strength steels, add to A36 price, minimum				75		75	83M
410	Maximum			↓	150		150	165M
430	Column base plates, light	2 Sswk	2,000	Lb.	.40	.17	.57	.71
440	Heavy plates	E-2	15,000	"	.35	.13	.48	.57
460	Castellated beams, light sections, to 50#/L.F., minimum		10.70	Ton	1,050	185	1,235	1,400
470	Maximum		7		1,300	285	1,585	1,825
490	Heavy sections, over 50# per L.F., minimum	↓	11.70	↓	915	170	1,085	1,250
500	Maximum		7.80		1,200	255	1,455	1,675
520	High strength bolts in place, light reaming, 3/4" bolts, average	2 Sswk	165	Ea.	.85	2.10	2.95	4.24
530	7/8" bolts, average	"	160	"	1.25	2.17	3.42	4.78
550	Steel domes							
551								
570	Steel estimating weights per S.F.							
571								
590	Galvanizing structural steel, under 1 ton, add to above			Ton	350		350	385M
600	Over 20 ton, add to above			"	250		250	275M
610	Cold galvanizing, brush	1 Psst	1,600	S.F.	.15	.10	.25	.33
611								

5.1 Structural Metals

		CREW	DAILY OUTPUT	UNIT	BARE COSTS MAT.	BARE COSTS INST.	BARE COSTS TOTAL	TOTAL INCL O&P
612	Steel surface treatments, sand blast, on grade							
613	Brush off blast	E-11	2,250	S.F.	.10	.34	.44	.60
614	Commercial		1,352		.26	.56	.82	1.10
615	Near white		1,040		.36	.73	1.09	1.45
616	White	↓	845	↓	.52	.90	1.42	1.87
617	Wire brush, hand	1 Psst	695	S.F.	.03	.23	.26	.41
618	Power tool	"	533	"	.03	.30	.33	.52
651	Paints & protective coatings, sprayed							
652	Alkyds, primer (79)	E-15	1,830	S.F.	.05	.19	.24	.35
654	Gloss topcoats		1,830		.04	.19	.23	.34
656	Silicone alkyd		1,830		.09	.19	.28	.40
661	Epoxy, primer		1,350		.06	.26	.32	.47
663	Intermediate or topcoat		1,350		.10	.26	.36	.52
665	Enamel coat		1,350		.07	.26	.33	.48
670	Epoxy ester, primer		1,350		.04	.26	.30	.45
672	Topcoats	↓	1,350	↓	.05	.26	.31	.46
676								
681	Latex primer	E-15	1,830	S.F.	.05	.19	.24	.35
683	Topcoats		1,830		.05	.19	.24	.35
691	Universal primers, one part, phenolic, modified alkyd		1,830		.06	.19	.25	.37
694	Two part, epoxy spray		1,350		.07	.26	.33	.48
700	Zinc rich primers, self cure, spray, inorganic,		890		.14	.39	.53	.77
701	Epoxy, spray, organic	↓	890	↓	.15	.39	.54	.78
702	Above grade, simple structures, add					25%		
703	Intricate structures, add					50%		
55-001	**VIBRATION PADS**							
030	Laminated synthetic rubber impregnated cotton duck	2 Sswk	20	B.F.	18	17.35	35.35	47
060	Neoprene bearing pads, 1/2" thick		24	S.F.	14	14.45	28.45	38
070	1" thick		20		28	17.35	45.35	58
090	Fabric reinforced neoprene, 5000 psi, 1/2" thick		24		37	14.45	51.45	63
100	1" thick		20		75	17.35	92.35	110
120	Felt surfaced vinyl pads, cork and sisal, 5/8" thick		24		18	14.45	32.45	43
130	1" thick		20		30	17.35	47.35	60
150	Teflon bonded to 10 ga. carbon steel, 1/32" layer		24		28	14.45	42.45	54
160	3/32" layer		24		40	14.45	54.45	67
180	Bonded to 10 ga. stainless steel, 1/32" layer		24		35	14.45	49.45	61
190	3/32" layer	↓	24	↓	50	14.45	64.45	78
210	Circular, encased, rule of thumb			Kip	1.35		1.35	1.48M
60-001	**WELD ROD** Steel, type E6010, 1/8" diameter, less than 1000#			Lb.	.66		.66	.72M
010	1000# to 2500#				.59		.59	.64M
020	2500# to 5000#				.56		.56	.61M
040	Steel, type E6011, 1/8" diameter, less than 1000#				.67		.67	.73M
050	1000# to 2500#				.60		.60	.66M
060	2500# to 5000#				.57		.57	.62M
065	Steel, type E7018, (low hydrogen) 1/8" diam., less than 1000#				.64		.64	.70M
066	1000# to 2500#				.57		.57	.62M
067	2500# to 5000#				.54		.54	.59M
070	Steel, type E7024, (jet weld) 1/8" diam., less than 1000#				.63		.63	.69M
071	1000# to 2500#				.56		.56	.61M
072	2500# to 5000#				.53		.53	.58M
080	Deduct for 5/32" diameter, type E6010 or type E6011				.02		.02	.02M
081	Semi-automatic coils, 1/16" diameter, 3000# lots				.56		.56	.61M
155	Aluminum, type 4043, 1/8" diameter, less than 50#				3.34		3.34	3.67M
160	50# to 100#				3.07		3.07	3.37M
181	100# to 500#			↓	3		3	3.30M
185								
190	Cast iron, 1/8" diameter			Lb.	.64		.64	.70M
195								

5.1 Structural Metals

		CREW	DAILY OUTPUT	UNIT	BARE COSTS MAT.	BARE COSTS INST.	BARE COSTS TOTAL	TOTAL INCL O&P
200	Stainless steel, type 308-15, 1/8" diam., less than 499#			Lb.	3.01		3.01	3.31M
210	500# to 999#				2.76		2.76	3.03M
222	Over 1000#			↓	2.66		2.66	2.92M
65-001	**WELDING** Field. Cost per welder, no operating engineer	E-14	8	Hr.	1.65	31	32.65	47
020	With 1/2 operating engineer	E-13	8		1.65	41	42.65	61
030	With 1 operating engineer	E-12	8	↓	1.65	50	51.65	75
040	(80)							
050	With no operating engineer, minimum	E-14	13.30	Ton	1.30	18.55	19.85	29
060	Maximum	"	2.50		5.25	99	104.25	150
080	With one operating engineer per welder, minimum	E-12	13.30		1.30	30	31.30	46
090	Maximum	"	2.50		5.25	160	165.25	240
120	Continuous fillet, stick welding, incl. equipment							
130	Single pass, 1/8" thick, 0.1#/L.F.	E-14	240	L.F.	.07	1.03	1.10	1.58
140	3/16" thick, 0.2#/L.F.		120		.13	2.05	2.18	3.15
150	1/4" thick, 0.3#/L.F.		80		.20	3.08	3.28	4.73
161	5/16" thick, 0.4#/L.F.		60		.26	4.11	4.37	6.30
180	3 passes, 3/8" thick, 0.5#/L.F.		48		.33	5.15	5.48	7.85
201	4 passes, 1/2" thick, 0.7#/L.F.		34		.46	7.25	7.71	11.10
220	5 to 6 passes, 3/4" thick, 1.3#/L.F.		19		.86	12.95	13.81	19.90
240	8 to 11 passes, 1" thick, 2.4#/L.F.	↓	10	↓	1.58	25	26.58	38
250								
260	For all position welding, add, minimum			L.F.		20%		
270	Maximum					300%		
290	For semi-automatic welding, deduct, minimum					5%		
300	Maximum			↓		15%		
400	Cleaning and welding plates, bars, or rods							
401	to existing beams, columns, or trusses	E-14	12	L.F.	.33	21	21.33	30

5.2 Metal Joists & Decks

		CREW	DAILY OUTPUT	UNIT	BARE COSTS MAT.	BARE COSTS INST.	BARE COSTS TOTAL	TOTAL INCL O&P
10-001	**BULB TEE** Subpurlins, 40 psf L.L., painted, 5' span	E-1	5,900	S.F.	.17	.10	.27	.33
010	8' span		4,200		.25	.14	.39	.48
030	11' span	↓	2,700	↓	.48	.21	.69	.85
060	(78) For galvanizing, add			Lb.	.20		.20	.22M
20-001	**LIGHTGAGE JOISTS** Punched, double nailable, 10" deep, 14 ga.	E-1	1,000	L.F.	2.85	.58	3.43	3.99
070	12 gauge		1,000		4	.58	4.58	5.25
090	12" deep, 14 gauge		880		3.35	.65	4	4.66
100	(75) 12 gauge	↓	880	↓	4.65	.65	5.30	6.10
110	(79) For galvanizing, add			Lb.	.12		.12	.13M
30-001	**METAL DECKING** Steel floor panels, over 15,000 S.F.							
020	Cellular units, galvanized, 2" deep, 20-20 gauge	E-4	1,460	S.F.	2.25	.53	2.78	3.28
030	18-18 gauge		1,390		2.50	.55	3.05	3.60
040	3" deep, galvanized, 20-20 gauge		1,375		2.43	.56	2.99	3.53
050	18-20 gauge		1,350		2.65	.57	3.22	3.79
060	18-18 gauge		1,290		3	.59	3.59	4.21
070	16-18 gauge		1,230		3.20	.62	3.82	4.48
080	16-16 gauge		1,150		3.45	.67	4.12	4.82
100	4-1/2" deep, galvanized, 20-18 gauge		1,100		4.40	.70	5.10	5.90
110	18-18 gauge		1,040		4.95	.74	5.69	6.60
120	16-18 gauge		980		5.30	.78	6.08	7.05
130	16-16 gauge	↓	935		5.60	.82	6.42	7.40
150	For acoustical deck, add				.70		.70	.77M
170	For cells used for ventilation, add			↓	.25		.25	.27M

5.2 Metal Joists & Decks

		CREW	DAILY OUTPUT	UNIT	BARE COSTS MAT.	BARE COSTS INST.	BARE COSTS TOTAL	TOTAL INCL O&P
190	For multi-story or congested site, add			S.F.		50%		
200								
210	Open type, galv., 1-1/2" deep, 22 ga., under 50 square	E-4	4,500	S.F.	.68	.17	.85	1.01
240	Over 50 square		4,900		.58	.16	.74	.88
260	20 ga., under 50 square		3,865		.80	.20	1	1.19
270	Over 50 square		4,170		.68	.18	.86	1.03
290	18 ga., under 50 square		3,800		1.05	.20	1.25	1.47
300	Over 50 square		4,100		.88	.19	1.07	1.26
320	3" deep, over 50 sq., 22 gauge		3,600		.80	.21	1.01	1.21
330	20 gauge		3,400		.94	.23	1.17	1.38
340	18 gauge		3,200		1.20	.24	1.44	1.69
350	16 gauge		3,000		1.50	.26	1.76	2.04
370	4-1/2" deep, long span roof, 20 gauge		2,700		1.50	.28	1.78	2.09
380	18 gauge		2,460		1.98	.31	2.29	2.66
390	16 gauge		2,350		2.50	.33	2.83	3.25
410	6" deep, long span, 18 gauge		2,000		2.20	.38	2.58	3.01
420	16 gauge		1,930		2.50	.40	2.90	3.36
430	14 gauge		1,860		2.95	.41	3.36	3.88
450	7-1/2" deep, long span, 18 gauge		1,690		2.35	.45	2.80	3.28
460	16 gauge		1,590		2.85	.48	3.33	3.88
480	For painted instead of galvanized, deduct				.08		.08	.08M
500	For acoustical perforated, with fiberglass, add				.35		.35	.38M
520	Non-cellular composite deck, galv., 2" deep, 22 gauge	E-4	3,860		.72	.20	.92	1.10
530	20 gauge		3,600		.78	.21	.99	1.19
540	18 gauge		3,380		1.02	.23	1.25	1.47
550	16 gauge		3,200		1.20	.24	1.44	1.69
570	3" deep, galv., 22 gauge		3,200		.75	.24	.99	1.19
580	20 gauge		3,000		.88	.26	1.14	1.36
590	18 gauge		2,850		1.10	.27	1.37	1.62
600	16 gauge		2,700		1.33	.28	1.61	1.90
610	Slab form, steel 28 gauge, 9/16" deep, uncoated	E-1	4,000		.30	.14	.44	.54
620	Galvanized		4,000		.36	.14	.50	.61
630	24 gauge, 1-5/16" deep, uncoated		3,800		.50	.15	.65	.78
640	Galvanized		3,800		.54	.15	.69	.82
650	22 gauge, 1-5/16" deep, uncoated		3,700		.58	.16	.74	.87
660	Galvanized		3,700		.70	.16	.86	1
700	Sheet metal edge closure form, 12" wide with 2 bends							
710	18 gauge	E-14	360	L.F.	1.82	.68	2.50	3
720	16 gauge	"	360	"	2.08	.68	2.76	3.29
750								
800	Metal deck and trench, 2" thick, 20 ga. combination,							
801	60% cellular, 40% non-cellular, inserts and trench	R-4	1,100	S.F.	3.85	.86	4.71	5.55
40-001	**OPEN WEB JOISTS** H series, horizontal bridging,							
002	truckload lots, span up to 30', minimum	E-7	15	Ton	700	175	875	1,025
003	Average		12		725	220	945	1,100
010	Maximum		9		750	295	1,045	1,225
030	Span 30' to 50', minimum		17		600	155	755	875
040	Maximum		10		650	265	915	1,075
060	LH series, bolted cross bridging							
061	Spans to 96' minimum	E-7	16	Ton	625	165	790	915
062	Average		13		650	205	855	1,000
070	Maximum		11		675	240	915	1,075
090	DLH series, bolted cross bridging							
091	Spans to 144' (shipped in 2 pieces), minimum	E-7	16	Ton	625	165	790	915
092	Average		13		650	205	855	1,000
100	Maximum		11		700	240	940	1,100
120	For welded cross bridging, add					30%		
140	For L.T.L. lots, add				10%	15%		
160	For shop prime paint other than mfrs. standard, add				75		75	83M
165	For bottom chord extensions, add per chord			Ea.	4.75		4.75	5.20M

5.2 Metal Joists & Decks

		CREW	DAILY OUTPUT	UNIT	BARE COSTS MAT.	BARE COSTS INST.	BARE COSTS TOTAL	TOTAL INCL O&P
40	300 Joist girders, average	E-5	13	Ton	670	200	870	1,025
	400 Trusses, factory fabricated WT chords, average	"	11	"	1,050	240	1,290	1,500

5.4 Misc. & Ornamental Metals

		CREW	DAILY OUTPUT	UNIT	BARE COSTS MAT.	BARE COSTS INST.	BARE COSTS TOTAL	TOTAL INCL O&P
02-001	ALUMINUM Miscellaneous, fabricated and erected	E-4	780	Lb.	5.75	.98	6.73	7.85
04-001	AREA WALL Galvanized steel, 20 ga., 3'-2" wide, 1' deep	1 Sswk	29	Ea.	7	6	13	17.10
010	3' deep		23		22	7.55	29.55	36
030	16 ga., galv., 3'-2" wide, 1' deep		29		14.50	6	20.50	25
040	3' deep		23		41	7.55	48.55	57
060	Welded grating for above, 15 lbs., painted		45		17	3.86	20.86	25
070	Galvanized		45		20	3.86	23.86	28
090	Translucent plastic cap for above	↓	60	↓	9	2.89	11.89	14.45
06-001	BUMPER RAILS For garages, 12 ga. rail, 6" wide, with steel							
002	posts 12'-6" O.C., Minimum	E-4	190	L.F.	8.25	4.04	12.29	15.30
003	Average		165		9.85	4.65	14.50	18
010	Maximum		140		12.35	5.50	17.85	22
030	12" channel rail, minimum		160		13.40	4.80	18.20	22
040	Maximum		120	↓	21	6.40	27.40	33
08-001	CHECKERED PLATE 1/4" & 3/8", 2000 to 5000 S.F., bolted		2,900	Lb.	.44	.26	.70	.89
010	Welded		4,400	"	.42	.17	.59	.73
030	Pit or trench cover and frame, 1/4" plate, 2' to 3' wide	↓	100	S.F.	14	7.65	21.65	27
040	For galvanizing, add			Lb.	.20		.20	.22M
050	Platforms, 1/4" plate, no handrails included, rectangular	E-4	4,200		.70	.18	.88	1.05
060	Circular	"	2,500	↓	.90	.31	1.21	1.46
10-001	CONSTRUCTION CASTINGS							
002	Manhole covers and frames see division 2.5-02 & 5.4-54							
010	Column bases, cast iron, 16" x 16" x 2-1/2"	E-4	46	Ea.	55	16.70	71.70	86
020	32" x 32" x 3-3/4"	"	23		200	33	233	270
040	Cast aluminum for wood columns, 8" x 8"	1 Carp	32		25	5	30	35
050	12" x 12"	"	32	↓	40	5	45	51
060	Miscellaneous C.I. castings, light sections	E-4	3,200	Lb.	.65	.24	.89	1.08
110	Heavy sections		4,200		.50	.18	.68	.83
130	Special low volume items	↓	3,200		1.50	.24	1.74	2.02
150	For ductile iron, add			↓	100%			
12-001	CORNER GUARDS Steel angle with anchors, set on forms 1" x 1", 1.5#/L.F.	E-4	320	L.F.	2.15	2.40	4.55	6.05
010	2" x 2" angles, 3.5#/L.F.		300		2.95	2.56	5.51	7.15
020	3" x 3" angles, 6#/L.F.		275		3.90	2.79	6.69	8.60
030	4" x 4" angles, 8#/L.F.	↓	240		5.25	3.20	8.45	10.70
035	For angles drilled and anchored to masonry, add				15%	120%		
037	Drilled and anchored to concrete, add				20%	170%		
040	For galvanized angles, add				35%			
045	For stainless steel angles, add			↓	100%			
050	Cast iron wheel guards, 3'-0" high	E-4	24	Ea.	90	32	122	150
060	5'-0" high		19		180	40	220	260
080	Pipe bumper for truck doors, 8' long, 6" diameter		20		135	38	173	205
090	8" diameter	↓	20	↓	200	38	238	280
14-001	CORNER PROTECTION							
002	Acrylic and vinyl, high impact type, stainless frame	1 Sswk	45	L.F.	14.50	3.86	18.36	22
003	Shock-absorbing water filled bumpers, corner, 26" long		14	Ea.	37	12.40	49.40	60
005	41" long		14		42	12.40	54.40	66
007	Half round, 26" long		14		35	12.40	47.40	58
009	40" long	↓	14	↓	39	12.40	51.40	62

5.4 Misc. & Ornamental Metals

			CREW	DAILY OUTPUT	UNIT	BARE COSTS MAT.	BARE COSTS INST.	BARE COSTS TOTAL	TOTAL INCL O&P
14	010	Stainless steel, 16 ga., with adhesive	1 Sswk	80	L.F.	26	2.17	28.17	32
	020	12 ga. stainless, with adhesive	"	80		32	2.17	34.17	39
	030	For screw-on installation, add					100%		
	050	Vinyl adhesive type, 3-3/8" wide	1 Carp	128		3.10	1.25	4.35	5.20
16-001		CRANE RAIL Box beam bridge, no equipment included	E-4	3,400	Lb.	.68	.23	.91	1.09
	020	Running track only, 104 lb. per yard		5,600	"	.34	.14	.48	.58
18-001		CURB EDGING Steel angle with anchors, set on forms, 1" x 1", 1.5#/L.F.		350	L.F.	1.85	2.19	4.04	5.40
	010	2" x 2" angles, 3.5#/L.F.		330		2.50	2.32	4.82	6.30
	020	3" x 3" angles, 6#/L.F.		300		3.30	2.56	5.86	7.55
	030	4" x 4" angles, 8#/L.F.		275		4.45	2.79	7.24	9.20
	100	6" x 4" angles, 12 #/L.F.		250		6.10	3.07	9.17	11.45
	105	Steel channels with anchors, on forms, 3" channel, 5#/L.F.		290		3.05	2.65	5.70	7.40
	110	4" channel, 6#/L.F.		270		3.45	2.84	6.29	8.15
	120	6" channel, 9#/L.F.		255		4.60	3.01	7.61	9.70
	130	8" channel, 12#/L.F.		225		6.20	3.41	9.61	12.05
	140	10" channel, 16#/L.F.		180		8	4.26	12.26	15.35
	150	12" channel, 22#/L.F.		140		10.50	5.50	16	19.95
	200	For curved edging, add				35%	10%		
20-001		CURB FASCIA For gas stations, etc., see division 3.1-70							
	010	As edging for sitework, see division 2.8-05							
22-001		DECORATIVE COVERING Extruded aluminum, stock sections							
	002	Doors	1 Sswk	17	S.F.	43	10.20	53.20	63
	010	Walls, minimum		41		14.25	4.23	18.48	22
	020	Maximum		26		24	6.70	30.70	37
	060	Doors, extruded bronze sections		12		56	14.45	70.45	84
	070	14 ga. stainless steel		18		32	9.65	41.65	50
24-001		DOOR FRAMES Steel channels with anchors and bar stops							
	010	6" channel @ 8.2#/L.F., 3' x 7' door, weighs 200#	E-4	13	Ea.	150	59	209	255
	020	8" channel @ 11.5#/L.F., 6' x 8' door, weighs 300#		9		210	85	295	360
	030	8' x 12' door, weighs 450#		6.50		300	120	420	510
	040	10" channel @ 15.3#/L.F., 10' x 10' door, weighs 525#		6		340	130	470	570
	050	12' x 12' door, weighs 625#		5.50		385	140	525	640
	060	12" channel @ 20.7#/L.F., 12' x 12' door, weighs 825#		4.50		485	170	655	795
	070	12' x 16' door, weighs 1000#		4		625	190	815	980
	080	For frames without bar stops, light sections, deduct				15%			
	090	Heavy sections, deduct				10%			
26-001		DOOR PROTECTION Acrylic and neoprene cover for door							
	002	frame, high impact type	1 Carp	100	L.F.	3.65	1.60	5.25	6.35
28-001		EXPANSION JOINT See division 5.8-05							
30-001		FIRE ESCAPE							
	020	2' wide balcony, 1" x 1/4" bars 1-1/2" O.C.	1 Sswk	5	L.F.	25	35	60	82
	040	1st story cantilever, standard		.09	Ea.	1,100	1,925	3,025	4,250
	070	Platform & stair, 36" x 40"		.17	Flight	500	1,025	1,525	2,150
	090	For 3'-6" wide escapes, add to above				150%	150%		
32-001		FIRE ESCAPE LADDERS One story, collapsible			Ea.	500		500	550M
	010	Portable			"	125		125	140M
36-001		FLOOR GRATING, ALUMINUM							
	002	Erection, compared to steel			S.F.		90%		
	010	Pressure-locked gratings, to prices listed in back of book							
	020	For straight cuts, add			L.F.	2		2	2.20M
	030	For curved cuts, add				2.50		2.50	2.75M
	040	For straight banding, add				3		3	3.30M
	050	For curved banding, add				3.75		3.75	4.12M
	060	For aluminum checkered plate nosings, add				2.65		2.65	2.91M
	070	For straight toe plate, add				4.40		4.40	4.84M
	080	For curved toe plate, add				6		6	6.60M
	100	For cast aluminum abrasive nosings, add				3.50		3.50	3.85M
	120	Expanded aluminum, .65# per S.F., on grade	E-4	1,050	S.F.	2.70	.73	3.43	4.09

5.4 Misc. & Ornamental Metals

		CREW	DAILY OUTPUT	UNIT	BARE COSTS MAT.	BARE COSTS INST.	BARE COSTS TOTAL	TOTAL INCL O&P
140	Extruded I bars are 10% less than 3/16" bars							
160	Heavy duty, all extruded panels, 3/4" deep, 1.7 # per S.F.	E-4	1,100	S.F.	6.80	.70	7.50	8.55
170	1-1/4" deep, 2.5# per S.F.		1,000		10.10	.77	10.87	12.30
180	1-3/4" deep, 4.8# per S.F.		925		12.80	.83	13.63	15.35
190	2-1/4" deep, 6.1# per S.F.	↓	875		15.50	.88	16.38	18.40
210	For safety serrated surface, add				.55		.55	.60M
240	Close spaced aluminum grating with vinyl tread inserts	E-4	255		31	3.01	34.01	39
260	For bottom drainage pan, add	"	510	↓	10.25	1.50	11.75	13.60
38-001	**FLOOR GRATING, FIBERGLASS** Reinforced polyester, fire retardant							
010	1" x 4" grid, 1" thick	E-4	510	S.F.	9.90	1.50	11.40	13.20
020	1-1/2" thick		500		12.70	1.53	14.23	16.35
030	Fiberglass reinforced epoxy, 1" x 4" grid, 1-1/2" thick	↓	500	↓	16.55	1.53	18.08	21
40-001	**FLOOR GRATING PLANKS** Aluminum, 9" wide							
002	14 ga., 2-1/2" rib	E-4	950	L.F.	7	.81	7.81	8.95
020	Galvanized steel, 9" wide, 14 ga., 2-1/2" rib		950		4.05	.81	4.86	5.70
030	4" rib		950		4.65	.81	5.46	6.35
050	12 gauge, 2-1/2" rib		950		4.85	.81	5.66	6.60
060	3" rib		950		5.55	.81	6.36	7.35
080	Stainless steel, type 304, 16 ga., 2" rib		950		10.05	.81	10.86	12.30
090	Type 316		950	↓	13.20	.81	14.01	15.75
41-001	**FLOOR GRATING, STEEL** Labor for installing, on grade		845	S.F.		.91	.91	1.40
010	Elevated		460	"		1.67	1.67	2.56
030	(82) Platforms, to 12' high, rectangular		3,150	Lb.	.69	.24	.93	1.13
040	Circular	↓	2,300	"	.76	.33	1.09	1.35
060	Rectangular material prices are listed in back of book							
080	For straight cuts, add			L.F.	2.65		2.65	2.91M
090	For curved cuts, add				3.10		3.10	3.41M
100	For straight banding, add				2.65		2.65	2.91M
110	For curved banding, add				3.55		3.55	3.90M
120	For checkered plate nosings, add				2.65		2.65	2.91M
130	For straight toe or kick plate, add				4.45		4.45	4.89M
140	For curved toe or kick plate, add				5.25		5.25	5.75M
150	For abrasive nosings, add			↓	4.45		4.45	4.89M
151	For stair treads, see division 5.4-66							
160	For serrated safety gratings, minimum, add			S.F.	.45		.45	.49M
170	Maximum, add				.62		.62	.68M
200	Stainless steel gratings, close spaced, on grade	E-4	340	↓	68	2.26	70.26	78
205								
210	Standard grating, 3.3# per S.F.	E-4	300	S.F.	23	2.56	25.56	29
220	4.5# per S.F.		275		32	2.79	34.79	39
240	Expanded steel grating, on grade, 3.0# per S.F.		900		1.55	.85	2.40	3.01
250	3.14# per S.F.		900		1.60	.85	2.45	3.07
260	4.0# per S.F.		850		2.05	.90	2.95	3.64
265	4.27# per S.F.		850		2.20	.90	3.10	3.81
270	5.0# per S.F.		800		2.60	.96	3.56	4.33
280	6.25# per S.F.		750		2.95	1.02	3.97	4.82
290	7.0# per S.F.	↓	700		3.40	1.10	4.50	5.40
310	For flattened expanded steel grating, add				5%			
330	For installation above grade, add			↓		.50	.50	.72L
42-001	**GRATING FRAME** Aluminum, for gratings 1" to 1-1/2" deep	1 Sswk	70	L.F.	3.95	2.48	6.43	8.25
010	For each corner, add			Ea.	3.25		3.25	3.57M
44-001	**LADDER** Steel, bolted to concrete, with cage	E-4	50	V.L.F.	42	15.35	57.35	70
010	Without cage		85		18.25	9.05	27.30	34
030	Aluminum, bolted to concrete, with cage		50		53	15.35	68.35	82
040	Without cage	↓	85	↓	24	9.05	33.05	40
46-001	**LAMP POSTS** Only, 6' high, stock units, aluminum	1 Carp	16	Ea.	22	10	32	39
010	Mild steel, plain	"	16	"	16	10	26	32
48-001	**LEAD** In 20' x 8' sheets from the mill			Cwt.	31		31	34M

5.4 Misc. & Ornamental Metals

		CREW	DAILY OUTPUT	UNIT	BARE COSTS MAT.	BARE COSTS INST.	BARE COSTS TOTAL	TOTAL INCL O&P
50-001	**LINTELS** Plain steel angles, under 500 lb. see also division 4.1-45	1 Bric	500	Lb.	.41	.33	.74	.92
010	500 to 1000 lb.		600		.35	.27	.62	.77
020	1000 to 2000 lb.		600		.34	.27	.61	.76
030	2000 to 4000 lb.	↓	600		.33	.27	.60	.75
050	For built-up angles and plates, add to above				.20		.20	.22M
070	For engineering, add to above				.10		.10	.11M
090	For galvanizing, add to above, under 500 lb.				.18		.18	.19M
100	Over 2000 lb.			↓	.15		.15	.16M
54-001	**MANHOLE COVERS** And frames (see also division 2.5-02)							
020	Aluminum, plain cover, lightweight, 12" x 12" opening	1 Sswk	9	Ea.	90	19.30	109.30	130
030	18" x 18" opening		8		180	22	202	230
040	24" x 24" opening		7		320	25	345	390
060	Heavyweight, 24" x 24" opening	↓	6		360	29	389	440
100	For stainless steel covers, add			↓	65%	10%		
56-001	**PIPE SUPPORT** Framing, under 10#/L.F.	E-4	3,900	Lb.	.56	.20	.76	.92
020	10.1 to 15#/L.F.		4,300		.49	.18	.67	.81
040	15.1 to 20#/L.F.		4,800		.43	.16	.59	.72
060	Over 20#/L.F.		5,400	↓	.39	.14	.53	.65
58-001	**RAILINGS, COMMERCIAL** Aluminum pipe rail, anodized		195	L.F.	32	3.93	35.93	41
090	Aluminum balcony rail, 1-1/2" posts, with pickets		195		21	3.93	24.93	29
100	With expanded metal panels		195		34	3.93	37.93	43
120	Hammered wire glass panel inserts		195		45	3.93	48.93	56
130	Porcelain enamel panel inserts	↓	195		45	3.93	48.93	56
140	For anodizing finish all above, add				2		2	2.20M
160	Mild steel, ornamental rounded top rail	E-4	195		27	3.93	30.93	36
170	As above but flat top rail		195		24	3.93	27.93	32
180	As above but pitch down stairs	↓	175	↓	27	4.38	31.38	36
185								
190	Residential, stock units, mild steel, deluxe	E-4	315	L.F.	5.60	2.44	8.04	9.90
191	Economy		315		4	2.44	6.44	8.15
240	Steel pipe, welded, 1-1/2" round, painted		200		20	3.84	23.84	28
250	Galvanized		200		30	3.84	33.84	39
270	Steel pipe with 4" x 1/4" toe plate, painted		200		22	3.84	25.84	30
280	Galvanized	↓	200		32	3.84	35.84	41
300	For stainless steel, add to painted				300%			
320	For curved rails, add				30%	30%		
340	Wall rail, with returns, aluminum pipe	E-4	255		13	3.01	16.01	18.90
350	Steel pipe		255		8	3.01	11.01	13.40
360	Mild steel pipe		255		10	3.01	13.01	15.60
370	Stainless steel pipe	↓	255		30	3.01	33.01	38
380	For galvanizing, add to steel pipe				3		3	3.30M
400	For gypsum stud wall mounted, add	E-4	340		.90	2.26	3.16	4.46
60-001	**RAILINGS, INDUSTRIAL** Welded, 2 rail, 3'-6" high, 1-1/2" pipe		255		21	3.01	24.01	28
010	2" angle rail	↓	255		19	3.01	22.01	26
020	For 4" high kick plate, 10 gauge, add				1.65		1.65	1.81M
030	1/4" thick, add				2.25		2.25	2.47M
050	For curved rails, add			↓	30%	30%		
61-001	**RAILINGS, ORNAMENTAL** Aluminum, bronze or stainless, minimum	1 Sswk	24	L.F.	50	7.25	57.25	66
010	Maximum		9		300	19.30	319.30	360
020	Aluminum pipe rail, minimum		15		35	11.55	46.55	57
030	Maximum		8		100	22	122	145
040	Hand-forged wrought iron, minimum		12		50	14.45	64.45	78
050	Maximum		8		200	22	222	255
060	Composite metal and wood or glass, minimum		6		100	29	129	155
070	Maximum	↓	5		250	35	285	330
62-001	**ROLLING GRILLE SUPPORTS** Overhead framed	E-4	36	↓	14	21	35	48
63-001	**SOLAR SCREENS** Or resurfacing, incl. supports, minimum	E-4	510	S.F.	4	1.50	5.50	6.70
010	Maximum	"	170	"	25	4.51	29.51	34

5.4 Misc. & Ornamental Metals

			CREW	DAILY OUTPUT	UNIT	BARE COSTS MAT.	BARE COSTS INST.	BARE COSTS TOTAL	TOTAL INCL O&P
63	030	Stock, no supports or frames incl., alum. extrusions, minimum			S.F.	3.50		3.50	3.85M
	050	Maximum				12		12	13.20M
	070	Slotted aluminum sheets, minimum				1.20		1.20	1.32M
	080	Maximum			↓	3		3	3.30M
64-001		**STAIR** Steel, 3'-6" wide, grating tread, safety nosing, steel							
	002	stringers and pipe railing, stock units	E-4	45	Riser	100	17.05	117.05	135
	020	Cement fill metal pan and picket rail		45		135	17.05	152.05	175
	040	Cast iron tread and pipe rail	↓	45		125	17.05	142.05	165
	060	For isolated stairs, add					100%		
	080	Custom steel stairs, minimum	E-4	45		120	17.05	137.05	160
	081	Average		35		170	22	192	220
	090	Maximum	↓	25		210	31	241	280
	110	For 4' wide stairs, add				10%	5%		
	130	For 5' wide stairs, add			↓	20%	10%		
	150	Landing, steel pan, conventional	E-4	160	S.F.	22	4.80	26.80	32
	160	Pre-erected		255	"	15	3.01	18.01	21
	170	Pre-erected, steel pan tread, 3'-6" wide, with flat bar rail		85	Riser	100	9.05	109.05	125
	180	With picket rail		17		135	45	180	220
	181	Spiral aluminum, 5'-0" diameter, stock units		45		135	17.05	152.05	175
	182	Custom units		45		200	17.05	217.05	245
	183	Stock units, 4'-0" diameter, safety treads		50		140	15.35	155.35	180
	184	Oak treads		50		89	15.35	104.35	120
	185	5'-0" diameter, safety treads		45		145	17.05	162.05	185
	186	Oak treads		45		95	17.05	112.05	130
	187	6'-0" diameter, safety treads		40		160	19.20	179.20	205
	188	Oak treads		40		110	19.20	129.20	150
	190	Spiral, cast iron, 4'-0" diameter, ornamental, minimum		45		125	17.05	142.05	165
	192	Maximum		25		185	31	216	250
	200	Spiral, steel, industrial checkered plate, 4' diameter		45		76	17.05	93.05	110
	220	Stock units, 6'-0" diameter		40		120	19.20	139.20	160
	240	Custom units, 4' to 6' diameter, minimum		45		55	17.05	72.05	87
	250	Maximum		30	↓	225	26	251	285
	390	Inclined ladder type, 3' wide, steel, 8' vertical rise		100	V.L.F.	73	7.65	80.65	92
	400	Aluminum	↓	100	"	100	7.65	107.65	120
66-001		**STAIR TREADS** Aluminum grating, 3' long, 1" x 3/16" bars, 6" wide	1 Sswk	24	Ea.	18	7.25	25.25	31
	010	12" wide		22		36	7.90	43.90	52
	020	1-1/2" x 3/16" bars, 6" wide		24		25	7.25	32.25	39
	030	12" wide	↓	22		50	7.90	57.90	67
	040	For abrasive nosings, add				10		10	11M
	050	For narrow mesh, add			↓	17%			
	060	Treads, 12" x 3'-6", not incl. stringers. See also division 3.3-48							
	070	Cast aluminum, abrasive, 5/16" thick	1 Sswk	15	Ea.	46	11.55	57.55	69
	080	3/8" thick		15		53	11.55	64.55	76
	090	1/2" thick		15		57	11.55	68.55	81
	100	Cast bronze, abrasive, 3/8" thick		8		320	22	342	385
	110	1/2" thick		8		425	22	447	500
	120	Cast iron, abrasive, 3/8" thick		15		43	11.55	54.55	65
	130	1/2" thick	↓	15	↓	45	11.55	56.55	68
	140	Fiberglass reinforced plastic with safety nosing,							
	150	1-1/2" thick, 12" wide, 24" long	1 Sswk	22	Ea.	62	7.90	69.90	81
	160	30" long		22		65	7.90	72.90	84
	170	36" long	↓	22	↓	71	7.90	78.90	91
	200	Steel grating, see back of book for material prices for 3'							
	201	long treads, install only	1 Sswk	15	Ea.		11.55	11.55	18.20L
	210	Add for abrasive nosing, 3' long, painted				10		10	11M
	220	Galvanized				14		14	15.40M
	230	Painting 3' treads, in shop, nonstandard paint				.75		.75	.82M
	240	Added coats, standard paint				.40		.40	.44M
	250	Expanded steel, 2-1/2" deep, 9" x 3' long, 18 gauge	1 Sswk	20		12	8.70	20.70	27
	260	14 gauge	"	20	↓	15	8.70	23.70	30

5.4 Misc. & Ornamental Metals

		CREW	DAILY OUTPUT	UNIT	BARE COSTS MAT.	BARE COSTS INST.	BARE COSTS TOTAL	TOTAL INCL O&P
68-001	**TOILET PARTITION SUPPORTS** Overhead framing for ceiling							
002	hung partitions, 3' wide	E-4	16	Stall	60	48	108	140
70-001	**TRENCH COVER** Steel gratings with bar stops and angle							
002	frame, to 18" wide	1 Sswk	20	L.F.	32	8.70	40.70	49
010	Frame only (both sides of trench), 1" grating		45		5.75	3.86	9.61	12.40
015	2" grating	↓	35	↓	9.50	4.96	14.46	18.25
020	Aluminum, stock units, including frames and							
021	3/8" plain cover plate, 2" opening	E-4	205	L.F.	13.75	3.74	17.49	21
030	6" opening		185		17.85	4.15	22	26
040	10" opening		170		24	4.51	28.51	33
050	16" opening	↓	155		30	4.95	34.95	41
070	Add per inch for additional widths to 24"				1.15		1.15	1.26M
090	For custom fabrication, add				50%			
110	For 1/4" plain cover plate, deduct				15%			
150	For cover recessed for tile, 1/4" thick, deduct				15%			
160	3/8" thick, add				5%			
180	For checkered plate cover, 1/4" thick, deduct				15%			
190	3/8" thick, add				2%			
210	For slotted or round holes in cover, 1/4" thick, add				3%			
220	3/8" thick, add				4%			
230	For abrasive cover, add			↓	15%			
72-001	**WEATHERVANES** Residential types, minimum	1 Carp	8	Ea.	15	20	35	45
010	Maximum	"	2	"	550	80	630	720
74-001	**WINDOW GUARDS** Expanded metal, steel angle frame, permanent	E-4	350	S.F.	10	2.19	12.19	14.35
002	Steel bars, 1/2" x 1/2", spaced 5" O.C.	"	290	"	5.50	2.65	8.15	10.10
003	Hinge mounted, add			Opng.	16		16	17.60M
004	Removable type, add			"	11		11	12.10M
005	For galvanized guards, add			S.F.	35%			
007	For pivoted or projected type, add				110%	40%		
010	Mild steel, stock units, economy	E-4	405		2.50	1.89	4.39	5.65
020	Deluxe		405	↓	5.25	1.89	7.14	8.70
040	Woven wire, stock units, 3/8" channel frame, 3' x 5' opening		40	Opng.	65	19.20	84.20	100
050	4' x 6' opening	↓	38		105	20	125	145
080	Basket guards for above, add				100		100	110M
100	Swinging guards for above, add			↓	30		30	33M
76-001	**WIRE** Barbed wire, imported, steel, galvanized, 15-1/2 ga.			M.L.F.	28		28	31M
020	12-1/2 ga.				33		33	36M
040	Aluminum barbed wire, 12-1/2 ga.			↓	110		110	120M
050	Helical fence topping, stainless steel			C.L.F.	120		120	130M
060	Hardware cloth, galv., 1/4" mesh, 23 ga., 2' wide			C.S.F.	45		45	50M
070	3' wide				43		43	47M
090	1/2" mesh, 19 ga., 2' wide				42		42	46M
100	4' wide				40		40	44M
120	Chain link fabric, 2" mesh, steel, galvanized, 6 ga.				67		67	74M
130	9 ga.				34		34	37M
150	1" mesh, 11 ga.				115		115	125M
160	1-3/4" mesh, 11 ga., vinyl coated				46		46	51M
180	Aluminized chain link fabric, 2" mesh, 6 ga.				140		140	155M
190	9 ga.				72		72	79M
210	Welded wire fabric, 1" x 2", 14 ga.				54		54	59M
220	2" x 4", 12-1/2 ga.			↓	27		27	30M
78-001	**WIRE ROPE** 6 x 19, fiber core, 1/2" diameter			L.F.	.88		.88	.96M
010	1" diameter				2.72		2.72	2.99M
030	6 x 19, fiber core, galvanized, 1/2" diameter				1.21		1.21	1.33M
040	1" diameter			↓	3.72		3.72	4.09M

5.8 Expansion Control & Fasteners

		CREW	DAILY OUTPUT	UNIT	BARE COSTS MAT.	BARE COSTS INST.	BARE COSTS TOTAL	TOTAL INCL O&P
02-001	**BOLTS & HEX NUTS** Steel, A307							
010	1/4" diameter, 1/2" long			Ea.	.04		.04	.04M
020	1" long				.05		.05	.05M
030	2" long				.06		.06	.06M
040	3" long				.10		.10	.11M
050	4" long				.11		.11	.12M
060	3/8" diameter, 1" long				.10		.10	.11M
070	2" long				.14		.14	.15M
080	3" long				.18		.18	.19M
090	4" long				.23		.23	.25M
100	5" long				.29		.29	.31M
110	1/2" diameter, 1-1/2" long				.23		.23	.25M
120	2" long				.26		.26	.28M
130	4" long				.34		.34	.37M
140	6" long				.60		.60	.66M
150	8" long				.77		.77	.84M
160	5/8" diameter, 1-1/2" long				.42		.42	.46M
170	2" long				.44		.44	.48M
180	4" long				.64		.64	.70M
190	6" long				.71		.71	.78M
200	8" long				1.30		1.30	1.43M
210	10" long				1.46		1.46	1.60M
220	3/4" diameter, 2" long				.61		.61	.67M
230	4" long				.87		.87	.95M
240	6" long				1.10		1.10	1.21M
250	8" long				1.55		1.55	1.70M
260	10" long				1.90		1.90	2.09M
270	12" long				2.17		2.17	2.38M
280	1" diameter, 3" long				1.65		1.65	1.81M
290	6" long				2.30		2.30	2.53M
300	12" long				4.50		4.50	4.95M
310	For galvanized, add				20%			
320	For stainless, add				150%			
03-001	**DRILLING** And layout for anchors, per							
005	inch of depth, concrete or brick walls							
010	1/4" diameter	1 Carp	32	Ea.	.05	5	5.05	7.30
020	3/8" diameter		29		.05	5.50	5.55	8.05
030	1/2" diameter		26		.05	6.15	6.20	8.95
040	5/8" diameter		25		.05	6.40	6.45	9.35
050	3/4" diameter		23		.06	6.95	7.01	10.15
060	7/8" diameter		21		.07	7.60	7.67	11.10
070	1" diameter		20		.08	8	8.08	11.70
080	1-1/4" diameter		18		.09	8.90	8.99	13
090	1-1/2" diameter		16		.10	10	10.10	14.60
100	For ceiling installations add					40%		
110	Drilling & layout for drywall or plaster walls							
120	Holes, 1/4" diameter	1 Carp	150	Ea.	.05	1.07	1.12	1.60
130	3/8" diameter		140		.05	1.14	1.19	1.71
140	1/2" diameter		130		.05	1.23	1.28	1.84
150	3/4" diameter		120		.05	1.33	1.38	1.99
160	1" diameter		110		.06	1.45	1.51	2.17
170	1-1/4" diameter		100		.07	1.60	1.67	2.40
180	1-1/2" diameter		90		.08	1.78	1.86	2.66
190	For ceiling installations add					40%		
200	Minimum labor/equipment charge	1 Carp	4	Job		40	40	58L
05-001	**EXPANSION JOINT ASSEMBLIES** Stock units							
002								
020	Floor cover assemblies, 1" space, aluminum	1 Sswk	38	L.F.	9	4.57	13.57	17.10
030	Bronze or stainless	"	38	"	30	4.57	34.57	40

5.8 Expansion Control & Fasteners

		Crew	Daily Output	Unit	Bare Costs Mat.	Inst.	Total	Total Incl O&P
05 050	2" space, aluminum	1 Sswk	38	L.F.	12	4.57	16.57	20
060	Bronze or stainless		38		40	4.57	44.57	51
080	Wall and ceiling assemblies, 1" space, aluminum		38		5	4.57	9.57	12.70
090	Bronze or stainless		38		25	4.57	29.57	35
110	2" space, aluminum		38		7	4.57	11.57	14.90
120	Bronze or stainless		38		30	4.57	34.57	40
140	Floor to wall assemblies, 1" space, aluminum		38		7	4.57	11.57	14.90
150	Bronze or stainless		38		30	4.57	34.57	40
170	Gym floor angle covers, aluminum, 3" x 3" angle		46		7.50	3.77	11.27	14.20
180	3" x 4" angle		46		8.50	3.77	12.27	15.30
200	Roof closures, aluminum, 1" space, flat roof, low profile		57		14	3.05	17.05	20
210	High profile		57		15	3.05	18.05	21
230	Roof to wall, 1" space, low profile		57		6	3.05	9.05	11.40
240	High profile		57		7	3.05	10.05	12.50
12-001	**EXPANSION ANCHORS** Bolts & shields							
010	Bolt anchors for concrete, brick or stone, no layout and drilling							
020	Expansion shields, zinc, 1/4" diameter, 1" long, single	1 Carp	90	Ea.	.46	1.78	2.24	3.08
030	1-3/8" long, double		85		.53	1.88	2.41	3.31
040	3/8" diameter, 1-1/2" long, single		85		.76	1.88	2.64	3.56
050	2" long, double		80		.98	2	2.98	3.98
060	1/2" diameter, 2-1/4" long, single		80		1.23	2	3.23	4.25
070	2-1/2" long, double		75		1.30	2.13	3.43	4.52
080	5/8" diameter, 2-5/8" long, single		75		1.76	2.13	3.89	5.05
090	3" long, double		70		1.87	2.29	4.16	5.35
100	3/4" diameter, 2-3/4" long, single		70		2.55	2.29	4.84	6.10
110	4" long, double		65		3.55	2.46	6.01	7.45
120	7/8" diameter, 5-1/2" long, double		65		8.30	2.46	10.76	12.70
130	1" diameter, 6" long, double		60		9.10	2.67	11.77	13.85
150	Self drilling, steel, 1/4" diameter bolt		26		.56	6.15	6.71	9.55
160	3/8" diameter bolt		23		.87	6.95	7.82	11.05
170	1/2" diameter bolt		20		1.32	8	9.32	13.05
180	5/8" diameter bolt		18		2.25	8.90	11.15	15.35
190	3/4" diameter bolt		16		3.70	10	13.70	18.55
200	7/8" diameter bolt		14		5.55	11.45	17	23
210	Hollow wall anchors for gypsum board,							
220	plaster, tile or wall board							
230	1/8" diameter, short			Ea.	.22		.22	.24M
240	Long				.26		.26	.28M
250	3/16" diameter, short				.33		.33	.36M
260	Long				.38		.38	.41M
270	1/4" diameter, short				.47		.47	.51M
280	Long				.50		.50	.55M
300	Toggle bolts, bright steel, 1/8" diameter, 2" long	1 Carp	85		.19	1.88	2.07	2.94
310	4" long		80		.26	2	2.26	3.18
320	3/16" diameter, 3" long		80		.22	2	2.22	3.14
330	6" long		75		.34	2.13	2.47	3.47
340	1/4" diameter, 3" long		75		.26	2.13	2.39	3.38
350	6" long		70		.37	2.29	2.66	3.72
360	3/8" diameter, 3" long		70		.57	2.29	2.86	3.94
370	6" long		60		.83	2.67	3.50	4.78
380	1/2" diameter, 4" long		60		1.02	2.67	3.69	4.99
390	6" long		50		1.52	3.20	4.72	6.30
400	Nailing anchors							
410	Nylon anchor, stainless nail, 1/4" diameter, 1" long			C	10.85		10.85	11.95M
420	1-1/2" long				14.30		14.30	15.75M
430	2" long				18.60		18.60	20M
440	Zamac anchor, stainless nail, 1/4" diameter, 1" long				22		22	24M
450	1-1/2" long				25		25	28M
460	2" long				35		35	39M
500	Screw anchors for concrete, masonry,							

5.8 Expansion Control & Fasteners	CREW	DAILY OUTPUT	UNIT	BARE COSTS MAT.	BARE COSTS INST.	BARE COSTS TOTAL	TOTAL INCL O&P
12 510 stone & tile, no layout or drilling included							
520 Jute fiber, #6, #8, & #10, 1" long			Ea.	.06		.06	.06M
530 #12, 2" long				.13		.13	.14M
540 #14, 2" long				.16		.16	.17M
550 #16, 2" long				.19		.19	.20M
560 #20, 2" long				.26		.26	.28M
570 Lag screw, 1/4" diameter, short				.29		.29	.31M
580 Long				.31		.31	.34M
590 3/8" diameter, short				.47		.47	.51M
600 Long				.57		.57	.62M
610 1/2" diameter, short				.62		.62	.68M
620 Long				.92		.92	1.01M
630 3/4" diameter, short				.94		.94	1.03M
640 Long				1.41		1.41	1.55M
660 Lead, #6 & #8, 3/4" long				.10		.10	.11M
670 #10 & #12, 1-1/2" long				.13		.13	.14M
680 #16 & #18, 1-1/2" long				.18		.18	.19M
690 Plastic, #6 & #8, 3/4" long				.03		.03	.03M
700 #8 & #10, 7/8" long				.03		.03	.03M
710 #10 & #12, 1" long				.04		.04	.04M
720 #14 & #16, 1-1/2" long			▼	.05		.05	.05M
800 Wedge bolts, not including layout or drilling							
805 Carbon steel, 1/4" diameter, 1-3/4" long	1 Carp	150	Ea.	.31	1.07	1.38	1.89
810 3" long		145		.46	1.10	1.56	2.10
815 3/8" diameter, 2-1/8" long		150		.42	1.07	1.49	2.01
820 5" long		145		.78	1.10	1.88	2.46
825 1/2" diameter, 2-3/4" long		140		.83	1.14	1.97	2.57
830 7" long		130		1.45	1.23	2.68	3.38
835 5/8" diameter, 3-1/2" long		130		1.48	1.23	2.71	3.41
840 8-1/2" long		115		2.35	1.39	3.74	4.60
845 3/4" diameter, 4-1/4" long		115		2.15	1.39	3.54	4.38
850 10" long		100		4	1.60	5.60	6.70
855 1" diameter, 6" long		100		6	1.60	7.60	8.90
860 12" long		80		8.65	2	10.65	12.40
865 1-1/4" diameter, 9" long		70		11.70	2.29	13.99	16.20
870 12" long	▼	60	▼	13.20	2.67	15.87	18.40
875 For type 303 stainless steel, add				300%			
880 For type 316 stainless steel, add				350%			
15-001 (77) HIGH-STRENGTH BOLTS Structural steel, see division 5.1-50-520							
17-001 LAG SCREWS 1/4" diameter, steel, 2" long	1 Carp	140	Ea.	.15	1.14	1.29	1.82
010 3/8" diameter, 3" long		105		.27	1.52	1.79	2.51
020 1/2" diameter, 3" long		95		.45	1.68	2.13	2.94
030 5/8" diameter, 3" long	▼	85	▼	.80	1.88	2.68	3.61
20-001 MACHINE SCREWS Steel, #8, 1" long, round head			C	2.75		2.75	3.02M
011 #8, 2" long				4.70		4.70	5.15M
020 #10 x 1" long				2.70		2.70	2.97M
030 #10 x 2" long			▼	4.90		4.90	5.40M

			UNIT	HEAD	ANCHOR BOLT	STUD & NUT	
22-001 MACHINERY ANCHORS Flush mounted heads to receive studs							
002 Material only (add 3 components for total price)							
020 For 1/2" stud, minimum			Ea.	14.10	2.35	2.10	
030 Maximum				17.20	2.85	2.55	
050 For 3/4" stud, minimum				17.55	2.95	2.60	
060 Maximum				22.10	3.70	3.25	
080 For 1" stud, minimum				22.50	3.75	3.30	
090 Maximum				25.50	4.25	3.75	
110 For 1-1/4" stud, minimum				28.50	4.75	4.20	
120 Maximum			▼	31.25	5.20	4.60	

5.8 Expansion Control & Fasteners

		CREW	DAILY OUTPUT	UNIT	HEAD	ANCHOR BOLT	STUD & NUT	TOTAL INCL O&P
140	For 1-1/2" stud, minimum			Ea.			7.20	
150	Maximum			"			8	

					BARE MAT. ONLY			
					PLAIN	GALV.	ALUM.	
25-001	**NAILS** Prices of material only, copper			Lb.	3.25			
040	Stainless steel				4.40			
060	Common 3d to 20d				.39	.48	1.85	
070	30d to 60d				.38	.47	1.85	
080	Annular or spiral thread, 4d to 60d				.62	.74		
100	Finish nails, 4d to 10d				.41	.49	1.95	
120	Drywall nails				.80	.96		
140	Flooring nails, hardened steel, 2d to 10d				.70	.84		
160	Masonry nails, hardened steel, 3/4" to 3" long				.70	.84		
180	Concrete nails, hardened steel				.70	.84		
200	Roofing nails, threaded					.62	2.55	
210	Threaded, with washers					.80		
230	Compressed lead head, threaded					.91		
240	Screw-down					.96		
260	Siding nails, plain shank					.55	1.55	
270	Threaded					.60		
290	Add to prices above for cement coating				.04			
310	Zinc or tin plating				.10			
27-001	**RIVETS** 1/2" grip length							
002								
010	Aluminum rivet & mandrel, 1/8" diameter			C			3.02	
020	3/16" diameter						4.65	
030	Aluminum rivet, steel mandrel, 1/8" diameter						3.15	
040	3/16" diameter						5.10	
050	Copper rivet, steel mandrel, 1/8" diameter				3.85			
055								
060	Monel rivet, steel mandrel, 1/8" diameter			C	13.70			
070	3/16" diameter				25			
080	Stainless rivet & mandrel, 1/8" diameter				8.45			
090	3/16" diameter				18.20			
100	Stainless rivet, steel mandrel, 1/8" diameter				7.10			
110	3/16" diameter				12			
120	Steel rivet and mandrel, 1/8" diameter				2.65			
130	3/16" diameter				5.10			
140	Hand riveting tool, minimum			Ea.	20			
150	Maximum				275			
160	Power riveting tool, minimum				180			
170	Maximum				825			
30-001	**SHEET METAL SCREWS** Steel, standard, #8 x 3/4"			C	2.10	2.50		
010	#10 x 1"				3.10	3.65		
030	With washers, #14 x 1"				10	11.10		
040	#14 x 2"				14.60	16.50		
060	Self-drilling, with washers, (pinch point), #8, 3/4" long				4.70	5.70		
070	#10 x 3/4" long				5.50	6.50		
090	Stainless steel with aluminum or neoprene washers, #14 x 1"				13.30			
100	#14 x 2" long				26			

					BARE COSTS			
					MAT.	INST.	TOTAL	
32-001	**STUDS** .22 caliber stud driver, buy, minimum			Ea.	225		225	250M
010	Maximum			"	330		330	365M
030	Powder charges for above, low velocity			C	12		12	13.20M
040	Standard velocity				16		16	17.60M
060	Drive pins & studs, 1/4" & 3/8" diam., to 3" long, minimum				29		29	32M
070	Maximum				48		48	53M
080	Pneumatic stud driver for 1/8" diameter studs			Ea.	800		800	880M
090	Drive pins for above, 1/2" to 3/4" long			M	35		35	39M

5.8 Expansion Control & Fasteners

		CREW	DAILY OUTPUT	UNIT	BARE COSTS MAT.	BARE COSTS INST.	BARE COSTS TOTAL	TOTAL INCL O&P
35-001	**TIMBER CONNECTORS** Add up cost of each part for total							
002	cost of connection							
020	Bolts, machine, sq. hd. with nut & washer, 1/2" diameter, 4" long	1 Carp	140	Ea.	.69	1.14	1.83	2.42
030	7-1/2" long		130		.85	1.23	2.08	2.72
050	3/4" diameter, 7-1/2" long		130		1.50	1.23	2.73	3.43
060	15" long		95	↓	1.85	1.68	3.53	4.48
080	Drilling bolt holes in timber, 1/2" diameter		450	Inch		.36	.36	.52L
090	1" diameter		350	"		.46	.46	.66L
110	Framing anchors, 2 or 3 dimensional, 10 gauge, no nails incl.		175	Ea.	.31	.91	1.22	1.67
130	Joist and beam hangers, 18 ga. galv., for 2" x 4" joist		175		.31	.91	1.22	1.67
140	2" x 6" to 2" x 10" joist		165		.59	.97	1.56	2.05
160	16 ga. galv., 3" x 6" to 3" x 10" joist		160		.98	1	1.98	2.53
170	3" x 10" to 3" x 14" joist		160		1.40	1	2.40	2.99
180	4" x 6" to 4" x 10" joist		155		1	1.03	2.03	2.60
190	4" x 10" to 4" x 14" joist		155		1.50	1.03	2.53	3.15
200	2-2" x 6" to 2-2" x 10" joist		150		.70	1.07	1.77	2.32
210	2-2" x 10" to 2-2" x 14" joist		150		1.05	1.07	2.12	2.70
230	3/16" thick for 6" x 8" joist		145		3.50	1.10	4.60	5.45
240	6" x 10" joist		140		4	1.14	5.14	6.05
250	6" x 12" joist		135		6.95	1.19	8.14	9.35
270	1/4" thick, 6" x 14" joist		130		7	1.23	8.23	9.50
280	Joist anchors, 1/4" x 1-1/4" x 18"	↓	140		2.60	1.14	3.74	4.52
290	Plywood clips, extruded aluminum H clip, for 3/4" panels				.05		.05	.05M
300	Galvanized 18 ga. back-up clip				.04		.04	.04M
320	Post framing, 16 ga. galv. for 4" x 4" base, 2 piece	1 Carp	130		3	1.23	4.23	5.10
330	Cap		130		2	1.23	3.23	3.98
350	Rafter anchors, 18 ga. galv., 1-1/2" wide, 5-1/4" long		145		.33	1.10	1.43	1.96
360	10-3/4" long		145		.60	1.10	1.70	2.26
380	Shear plates, 2-5/8" diameter		120		.95	1.33	2.28	2.98
390	4" diameter		115		2.30	1.39	3.69	4.55
400	Sill anchors, (embedded in concrete or block), 18-5/8" long		115		.50	1.39	1.89	2.57
410	Spike grids, 4" x 4", flat or curved		120		4.35	1.33	5.68	6.70
440	Split rings, 2-1/2" diameter		120		.53	1.33	1.86	2.52
450	4" diameter		110		.98	1.45	2.43	3.19
470	Strap ties, 14 ga., 1-3/8" wide, 12" long		180		.60	.89	1.49	1.95
480	24" long		160		1	1	2	2.55
500	Toothed rings, 2-5/8" or 4" diameter		90	↓	.70	1.78	2.48	3.35
520	Truss plates, toothed, 18 gauge, for 36' span	↓	17	Truss	6	9.40	15.40	20
540	Washers, 2" x 2" x 1/8"			Ea.	.16		.16	.17M
550	3" x 3" x 3/16"			"	.40		.40	.44M
40-001	**TRACK** Railroad, bolts			Cwt.	62		62	68M
010	Joint bars				50		50	55M
020	Spikes			↓	37		37	41M
030	Tie plates				55		55	61M
45-001	**WELDED SHEAR CONNECTORS** 3/4" diameter, 3-3/16" long	E-10	1,030	Ea.	.40	.64	1.04	1.31
002	3-3/8" long		1,030		.40	.64	1.04	1.31
020	3-7/8" long		1,030		.42	.64	1.06	1.33
030	4-3/16" long		1,030		.44	.64	1.08	1.35
050	4-7/8" long		1,030		.48	.64	1.12	1.40
060	5-3/16" long		1,030		.53	.64	1.17	1.45
080	5-3/8" long		1,030		.54	.64	1.18	1.46
090	6-3/16" long		1,000		.59	.66	1.25	1.54
100	7-3/16" long		1,000		.71	.66	1.37	1.68
110	8-3/16" long		1,000		.81	.66	1.47	1.79
150	7/8" diameter, 3-11/16" long		1,030		.56	.64	1.20	1.48
160	4-3/16" long		1,030		.60	.64	1.24	1.53
170	5-3/16" long		1,030		.72	.64	1.36	1.66
180	6-3/16" long		1,000		.84	.66	1.50	1.82
190	7-3/16" long		1,000		.92	.66	1.58	1.91
200	8-3/16" long	↓	1,000	↓	1	.66	1.66	1.99

5.8 Expansion Control & Fasteners

		CREW	DAILY OUTPUT	UNIT	BARE COSTS MAT.	BARE COSTS INST.	BARE COSTS TOTAL	TOTAL INCL O&P
46-001	WELDED STUDS 1/4" diameter, 2-11/16" long	E-10	1,030	Ea.	.21	.64	.85	1.10
010	4-1/8" long		1,030		.22	.64	.86	1.11
020	3/8" diameter, 4-1/8" long		1,030		.30	.64	.94	1.20
030	6-1/8" long		1,030		.39	.64	1.03	1.30
040	1/2" diameter, 2-1/8" long		1,030		.31	.64	.95	1.21
050	3-1/8" long		1,030		.36	.64	1	1.26
060	4-1/8" long		1,030		.41	.64	1.05	1.32
070	5-5/16" long		1,030		.51	.64	1.15	1.43
080	6-1/8" long		1,000		.57	.66	1.23	1.52
090	8-1/8" long		1,000		.68	.66	1.34	1.64
100	5/8" diameter, 2-11/16" long		1,030		.44	.64	1.08	1.35
110	6-9/16" long		1,000		.69	.66	1.35	1.65
120	8-3/16" long	↓	1,000	↓	.84	.66	1.50	1.82

					BARE MAT. ONLY STEEL	BARE MAT. ONLY BRASS		
50-001	WOOD SCREWS #9, 1" long			C	2.25	9.40		
010	#9, 2" long				3.70	18.10		
020	#12, 1" long				3	15.80		
030	#12, 2" long				5.50	28		
040	#12, 3" long				8.60			
050	#14, 2" long				6.75	40		
060	#14, 3" long				10.10			
070	#14, 4" long			↓	17			

6.1 Rough Carpentry

		CREW	DAILY OUTPUT	UNIT	BARE COSTS MAT.	BARE COSTS INST.	BARE COSTS TOTAL	TOTAL INCL O&P
02-001	BLOCKING							
260	Miscellaneous, to wood construction							
262	2" x 4"	F-1	.17	M.B.F.	315	990	1,305	1,775
266	2" x 8"	"	.27	"	340	625	965	1,275
272	To steel construction							
274	2" x 4"	F-1	.14	M.B.F.	315	1,200	1,515	2,075
278	2" x 8"	"	.21	"	340	805	1,145	1,525
05-001	BRACING Let-in, with 1" x 6" boards, studs 16" O.C.	F-1	1.50	C.L.F.	17.80	110	127.80	180
020	Studs at 24" O.C.		2.30		17.80	73	90.80	125
030	Let-in, "T" shaped, 20 ga. galv. steel, studs at 16" O.C.		5.80		53	29	82	100
040	Studs at 24" O.C.		6		53	28	81	99
050	16 ga. galv. steel straps, studs at 16" O.C.		6		53	28	81	99
060	Studs at 24" O.C.		6.20	↓	53	27	80	97
07-001	BRIDGING Wood, for joists 16" O.C., 1" x 3"		1.30	C.Pr.	36	130	166	225
010	2" x 3" bridging	↓	1.30		64	130	194	255
030	Steel, galvanized, 18 ga., for 2" x 10" joists at 12" O.C.	1 Carp	1.30		86	125	211	275
040	24" O.C.		1.40		125	115	240	305
060	For 14" joists at 16" O.C.		1.30		125	125	250	315
070	24" O.C.		1.40		137	115	252	315
090	Compression type, 16" O.C., 2" x 8" joists		2		84	80	164	210
100	2" x 12" joists	↓	2	↓	90	80	170	215
10-001	DOCK BUMPERS Bolts not included, 2" x 6" to 4" x 8", average	F-1	.30	M.B.F.	380	560	940	1,225
13-001	FRAMING, BEAMS & GIRDERS							
002								
350	Single, 2" x 6"	F-2	.70	M.B.F.	320	480	800	1,050
352	2" x 8"	"	.86	"	340	390	730	935

For expanded coverage of these items see Means' *Interior Cost Data 1986*

6.1 Rough Carpentry

			CREW	DAILY OUTPUT	UNIT	BARE COSTS MAT.	BARE COSTS INST.	BARE COSTS TOTAL	TOTAL INCL O&P
13	354	2" x 10"	F-2	1	M.B.F.	365	335	700	885
	356	2" x 12"		1.10		380	305	685	855
	358	2" x 14"		1.17		400	290	690	855
	360	3" x 8"		1.10		505	305	810	995
	362	3" x 10"		1.25		505	270	775	940
	364	3" x 12"		1.35		505	250	755	915
	366	3" x 14"	↓	1.40		505	240	745	900
	368	4" x 8"	F-3	2.66		500	400	900	1,100
	370	4" x 10"		3.16		500	335	835	1,000
	372	4" x 12"		3.60		500	295	795	955
	374	4" x 14"	↓	3.96	↓	500	270	770	915
	376								
	400	Double, 2" x 6"	F-2	.40	M.B.F.	320	845	1,165	1,550
	402	2" x 8"		.50		340	675	1,015	1,350
	404	2" x 10"		.58		365	580	945	1,225
	406	2" x 12"		.65		380	520	900	1,150
	408	2" x 14"	↓	.70	↓	400	480	880	1,125
	410								
	500	Triple, 2" x 6"	F-2	.28	M.B.F.	320	1,200	1,520	2,075
	502	2" x 8"		.33		340	1,025	1,365	1,825
	504	2" x 10"		.37		365	910	1,275	1,700
	506	2" x 12"		.40		380	845	1,225	1,625
	508	2" x 14"	↓	.44	↓	400	765	1,165	1,550
16-001		**FRAMING, CEILINGS**							
002									
	640	Suspended, 2" x 3"	F-2	.50	M.B.F.	315	675	990	1,300
	645	2" x 4"		.59		315	570	885	1,175
	650	2" x 6"		.80		320	420	740	955
	655	2" x 8"	↓	.86	↓	340	390	730	935
19-001		**FRAMING, JOISTS**							
	265	Joists, 2" x 4"	F-2	.83	M.B.F.	315	405	720	930
	268	2" x 6"		1.25		320	270	590	740
	270	2" x 8"		1.46		340	230	570	705
	272	2" x 10"		1.49		365	225	590	725
	274	2" x 12"		1.75		380	195	575	695
	276	2" x 14"		1.79		400	190	590	710
	278	3" x 6"		1.39		505	245	750	905
	278	3" x 8"		1.90		505	175	680	810
	280	3" x 10"		1.95		505	175	680	805
	282	3" x 12"		1.80		505	185	690	825
	284	4" x 6"	↓	1.60		500	210	710	850
	285	4" x 8"	F-3	4.15		500	255	755	900
	286	4" x 10"	F-2	2		500	170	670	790
	288	4" x 12"	"	1.80	↓	500	185	685	820
22-001		**FRAMING, MISCELLANEOUS**							
002									
	850	Firestops, 2" x 4"	F-2	.51	M.B.F.	315	660	975	1,300
	852	2" x 6"		.60		320	560	880	1,150
	860	Nailers, treated, wood construction, 2" x 4"		.53		485	635	1,120	1,450
	862	2" x 6"		.75		490	450	940	1,175
	864	2" x 8"		.93		510	365	875	1,075
	866	Steel construction, 2" x 4"		.50		315	675	990	1,300
	868	2" x 6"		.70		320	480	800	1,050
	870	2" x 8"		.87		340	390	730	930
	876	Rough bucks, treated, for doors or windows, 2" x 6"		.40		490	845	1,335	1,750
	878	2" x 8"		.51		510	660	1,170	1,500
	880	Stair stringers, 2" x 10"		.22		365	1,525	1,890	2,600
	882	2" x 12"	↓	.26	↓	380	1,300	1,680	2,275

For expanded coverage of these items see Means' *Interior Cost Data 1986*

6.1 Rough Carpentry

			CREW	DAILY OUTPUT	UNIT	BARE COSTS MAT.	BARE COSTS INST.	BARE COSTS TOTAL	TOTAL INCL O&P
22	884	3" x 10"	F-2	.31	M.B.F.	505	1,100	1,605	2,125
	886	3" x 12"	"	.38	"	505	885	1,390	1,825
25-001		**FRAMING, COLUMNS**							
	002								
	040	4" x 4"	F-2	.52	M.B.F.	500	650	1,150	1,475
	042	4" x 6"		.55		500	615	1,115	1,425
	044	4" x 8"		.59		500	570	1,070	1,375
	046	6" x 6"		.65		500	520	1,020	1,300
	048	6" x 8"		.70		500	480	980	1,250
	050	6" x 10"	↓	.75	↓	500	450	950	1,200
28-001		**FRAMING, ROOFS**							
	606								
	607	Fascia boards, 2" x 8"	F-2	.30	M.B.F.	340	1,125	1,465	1,975
	608	2" x 10"		.30		365	1,125	1,490	2,000
	700	Rafters, to 4 in 12 pitch, 2" x 6"		1		320	335	655	835
	706	2" x 8"		1.26		340	270	610	755
	730	Hip and valley rafters, 2" x 6"		.76		320	445	765	985
	736	2" x 8"		.96		340	350	690	875
	754	Hip and valley jacks, 2" x 6"		.60		320	560	880	1,150
	760	2" x 8"	↓	.65	↓	340	520	860	1,125
	778	For slopes steeper than 4 in 12, add				30%			
	779	For dormers or complex roofs, add				50%			
	780	Rafter tie, 1" x 4", #3	F-2	.27	M.B.F.	315	1,250	1,565	2,125
	781								
	782	Ridge board, #2 or better, 1" x 6"	F-2	.30	M.B.F.	315	1,125	1,440	1,950
	784	1" x 8"		.37		330	910	1,240	1,675
	786	1" x 10"		.42		330	805	1,135	1,525
	788	2" x 6"		.50		320	675	995	1,325
	790	2" x 8"		.60		340	560	900	1,175
	792	2" x 10"		.66		365	510	875	1,125
	794	Roof cants, split, 4" x 4"		.86		620	390	1,010	1,250
	796	6" x 6"		1.80		620	185	805	950
	798	Roof curbs, untreated, 2" x 6"		.52		320	650	970	1,275
	800	2" x 12"	↓	.80	↓	380	420	800	1,025
31-001		**FRAMING, SILLS**							
	181								
	448	Ledgers, nailed, 2" x 4"	F-2	.50	M.B.F.	315	675	990	1,300
	448	2" x 6"		.60		320	560	880	1,150
	448	Bolted, not including bolts, 3" x 8"		.65		505	520	1,025	1,300
	448	3" x 12"		.70		505	480	985	1,250
	449	Mud sills, redwood, construction grade, 2" x 4"		.59		870	570	1,440	1,775
	449	2" x 6"		.78		860	430	1,290	1,575
	450	Sills, 2" x 4"		.40		315	845	1,160	1,550
	452	2" x 6"		.55		320	615	935	1,225
	454	2" x 8"		.67		340	505	845	1,100
	460	Treated, 2" x 4"		.36		490	935	1,425	1,875
	462	2" x 6"		.50		490	675	1,165	1,500
	464	2" x 8"		.60		490	560	1,050	1,350
	470	4" x 4"		.60		620	560	1,180	1,475
	472	4" x 6"		.70		620	480	1,100	1,375
	474	4" x 8"		.80		620	420	1,040	1,275
	476	4" x 10"	↓	.87	↓	620	390	1,010	1,225
34-001		**FRAMING, SLEEPERS**							
	002								
	030	On concrete, treated, 1" x 2"	F-2	.39	M.B.F.	490	865	1,355	1,775
	032	1" x 3"		.50		490	675	1,165	1,500
	034	2" x 4"		.99		490	340	830	1,025
	036	2" x 6"	↓	1.30	↓	490	260	750	910
37-001		**FRAMING, SOFFITS & CANOPIES**							
	002								

For expanded coverage of these items see Means' *Interior Cost Data 1986*

6.1 Rough Carpentry

			CREW	DAILY OUTPUT	UNIT	BARE COSTS MAT.	BARE COSTS INST.	BARE COSTS TOTAL	TOTAL INCL O&P
37	130	Canopy or soffit framing, 1" x 4"	F-2	.30	M.B.F.	315	1,125	1,440	1,950
	134	1" x 8"		.50		320	675	995	1,325
	136	2" x 4"		.41		315	820	1,135	1,525
	140	2" x 8"		.67		340	505	845	1,100
	142	3" x 4"		.50		505	675	1,180	1,525
	146	3" x 8"	↓	.60	↓	505	560	1,065	1,350
40-001		FRAMING, WALLS							
	586	Headers over openings, 2" x 6"	F-2	.36	M.B.F.	320	935	1,255	1,700
	588	2" x 8"		.45		340	750	1,090	1,450
	590	2" x 10"		.53		365	635	1,000	1,325
	592	2" x 12"		.60		380	560	940	1,225
	594	4" x 12"		.76		500	445	945	1,175
	596	6" x 12"		.84		500	400	900	1,125
	600	Plates, untreated, 2" x 3"		.43		315	785	1,100	1,475
	602	2" x 4"		.53		315	635	950	1,250
	604	2" x 6"		.75		320	450	770	995
	612	Studs, 8' high wall, 2" x 3"		.60		315	560	875	1,150
	614	2" x 4"		.92		315	365	680	870
	616	2" x 6"		1		320	335	655	835
	618	3" x 4"	↓	.80		505	420	925	1,150
	820	For 12' high wall, deduct			↓		5%		
	821								
	822	For stub wall, 6' high, add			M.B.F.		20%		
	824	3' high, add					40%		
	825	For second story & above, add					5%		
	830	Dormer & gable, add					15%		
43-001		FRAMING, HEAVY Mill timber, beams, single 6" x 10"	F-2	1.10		500	305	805	990
	010	Single 8" x 16"		1.20		510	280	790	965
	020	Built from 2" lumber, multiple 2" x 14"		.90		400	375	775	975
	021	Built from 3" lumber, multiple 3" x 6"		.70		505	480	985	1,250
	022	Multiple 3" x 8"		.80		505	420	925	1,150
	023	Multiple 3" x 10"		.90		505	375	880	1,100
	024	Multiple 3" x 12"	↓	1	↓	505	335	840	1,050
	024								
	025	Built from 4" lumber, multiple 4" x 6"	F-2	.80	M.B.F.	500	420	920	1,150
	026	Multiple 4" x 8"		.90		500	375	875	1,075
	027	Multiple 4" x 10"		1		500	335	835	1,025
	028	Multiple 4" x 12"	↓	1.10	↓	500	305	805	990
	028								
	029	Columns, structural grade, 1500f, 4" x 4"	F-2	.60	M.B.F.	500	560	1,060	1,350
	030	6" x 6"		.65		500	520	1,020	1,300
	040	8" x 8"		.70		510	480	990	1,250
	050	10" x 10"		.75		520	450	970	1,225
	060	12" x 12"		.80		520	420	940	1,175
	080	Floor planks, 2" thick, T & G, 2" x 6"		1.05		475	320	795	980
	090	2" x 10"		1.10		500	305	805	990
	110	3" thick, 3" x 6"		1.05		665	320	985	1,200
	120	3" x 10"		1.10		670	305	975	1,175
	140	Girders, structural grade, 12" x 12"		.80		520	420	940	1,175
	150	10" x 16"	↓	1	↓	520	335	855	1,050
	205	Roof planks, see division 6.1-62							
	230	Roof purlins, 4" thick, structural grade	F-2	1.05	M.B.F.	500	320	820	1,000
	250	Roof trusses, add timber connectors, division 5.8-35	"	.45	"	505	750	1,255	1,625
50-001		FURRING Wood strips, on walls, 1" x 2", on wood	1 Carp	550	L.F.	.06	.29	.35	.49
	030	On masonry		495		.06	.32	.38	.53
	040	On concrete		260		.07	.62	.69	.97
	060	1" x 3", wood strips, on walls, on wood		550		.09	.29	.38	.52
	070	On masonry		495		.10	.32	.42	.58
	080	On concrete	↓	260	↓	.11	.62	.73	1.01

For expanded coverage of these items see *Means' Interior Cost Data 1986*

6.1 Rough Carpentry

		Description	CREW	DAILY OUTPUT	UNIT	BARE COSTS MAT.	BARE COSTS INST.	BARE COSTS TOTAL	TOTAL INCL O&P
50	085	1" x 3", wood strips, on ceilings, on wood	1 Carp	350	L.F.	.09	.46	.55	.76
	090	On masonry	↓	320		.10	.50	.60	.83
	095	On concrete	↓	210	↓	.11	.76	.87	1.23
54-001		GROUNDS For casework, 1" x 2" wood strips, on wood	1 Carp	330	L.F.	.06	.48	.54	.77
	010	On masonry	↓	285		.06	.56	.62	.88
	020	On concrete	↓	250	↓	.07	.64	.71	1
	030								
	040	For plaster, 3/4" deep, on wood	1 Carp	450	L.F.	.06	.36	.42	.58
	050	On masonry	↓	225		.06	.71	.77	1.10
	060	On concrete		175		.07	.91	.98	1.40
	070	On metal lath	↓	200	↓	.07	.80	.87	1.24
55-001		INSULATION See division 7.2							
56-001		LAMINATED See division 6.3							

		Description	CREW	DAILY OUTPUT	UNIT	BARE COSTS 20 MBF	BARE COSTS 5 MBF	BARE COSTS 1 MBF	
57-001		LUMBER TREATMENT Creosoted 8 lbs. per C.F., add			M.B.F.	160	170	195	
	020	For every added 2#/C.F., add per increment				21	21	21	
	040	Fire retardant, wet				170	180	200	
	050	KDAT				190	200	220	
	070	Salt treated, water borne, .40 lb. retention				110	125	205	
	080	Oil borne, 8 lb. retention				135	145	160	
	100	Kiln dried lumber, 1" & 2" thick, soft woods				80	85	100	
	110	Hard woods				85	90	105	
	150	For small size 1" stock, add				10	10	10	
	170	For full size rough lumber, add			↓	20%	20%	20%	

		Description	CREW	DAILY OUTPUT	UNIT	BARE COSTS MAT.	BARE COSTS INST.	BARE COSTS TOTAL	TOTAL INCL O&P
58-001		PARTITIONS Wood stud with single bottom plate and							
	002	double top plate, no waste, std. & better lumber							
	018	2" x 4" studs, 8' high, studs 12" O.C.	F-2	80	L.F.	3.05	4.22	7.27	9.40
	020	16" O.C.		100		2.32	3.37	5.69	7.40
	030	24" O.C.		125		1.75	2.70	4.45	5.80
	038	10' high, studs 12" O.C.		80		3.61	4.22	7.83	10
	040	16" O.C.		100		2.75	3.37	6.12	7.85
	050	24" O.C.		125		2.07	2.70	4.77	6.15
	058	12' high, studs 12" O.C.		65		4.39	5.20	9.59	12.25
	060	16" O.C.		80		3.35	4.22	7.57	9.70
	070	24" O.C.		100		2.48	3.37	5.85	7.55
	078	2" x 6" studs, 8' high, studs 12" O.C.		70		4.42	4.82	9.24	11.75
	080	16" O.C.		90		3.34	3.75	7.09	9.05
	090	24" O.C.		115		2.52	2.93	5.45	6.95
	098	10' high, studs 12" O.C.		70		5.20	4.82	10.02	12.60
	100	16" O.C.		90		3.98	3.75	7.73	9.75
	110	24" O.C.		115		2.96	2.93	5.89	7.45
	118	12' high, studs 12" O.C.		55		6	6.15	12.15	15.40
	120	16" O.C.		70		4.56	4.82	9.38	11.90
	130	24" O.C.		90		3.34	3.75	7.09	9.05
	140	For horizontal blocking, 2" x 4", add		600		.21	.56	.77	1.04
	150	2" x 6", add		600		.32	.56	.88	1.16
	160	For openings, add	↓	250	↓		1.35	1.35	1.93
	170	Headers for above openings, material only, add			M.B.F.	365		365	400M
	180	For rough hardware, see division 6.1-68							

		Description	CREW	DAILY OUTPUT	UNIT	BARE COSTS 20 MSF	BARE COSTS 5 MSF	BARE COSTS 1 MSF	
60-001		PLYWOOD TREATMENT Fire retardant, 1/4" thick			M.S.F.	160	185	225	
	003	3/8" thick			"	160	185	225	

For expanded coverage of these items see Means' *Interior Cost Data 1986*

6.1 Rough Carpentry

			DAILY		BARE COSTS			TOTAL
		CREW	OUTPUT	UNIT	20 MSF	5 MSF	1 MSF	INCL O&P
60	005	1/2" thick			M.S.F.	160	185	225	
	007	5/8" thick				180	195	240	
	010	3/4" thick				190	220	265	
	020	For KDAT, add				40	40	40	
	050	Salt treated water borne, .25 lb., wet, 1/4" thick				80	100	145	
	053	3/8" thick				85	105	155	
	055	1/2" thick				100	120	160	
	057	5/8" thick				115	135	185	
	060	3/4" thick				125	145	200	
	080	For KDAT add				47	47	47	
	090	For .40 lb., per C.F. retention, add				32	32	32	
	100	For certification stamp, add			↓	30	30	30	

						BARE COSTS			
						MAT.	INST.	TOTAL	
	62-001	**ROOF DECKS** For laminated decks, see division 6.3							
	020	For cementitious decks, see division 3.5							
	040	Cedar planks, 3.65 B.F. per S.F., 3" thick	F-2	320	S.F.	2.96	1.05	4.01	4.77
	050	4.65 B.F. per S.F., 4" thick		250		3.77	1.35	5.12	6.10
	070	Douglas fir, 3" thick		320		2.23	1.05	3.28	3.96
	080	4" thick		250		2.84	1.35	4.19	5.05
	100	Hemlock, 3" thick		320		2.15	1.05	3.20	3.87
	110	4" thick		250		2.75	1.35	4.10	4.96
	130	Western white spruce, 3" thick		320		2.23	1.05	3.28	3.96
	140	4" thick	↓	250	↓	2.84	1.35	4.19	5.05
	66-001	**ROOF TRUSSES** For timber connector trusses, see div. 5.8-35							
	010	Fink (W) or King post type, 2'-0" O.C.							
	020	(85) Metal plate connected, 4 in 12 slope							
	021	24' to 29' span	F-3	3,000	S.F.Flr.	.91	.35	1.26	1.48
	030	30' to 43' span		3,000		.98	.35	1.33	1.56
	040	44' to 60' span	↓	3,000		1.47	.35	1.82	2.10
	060	For change in roof pitch, add				.06		.06	.06M
	070	Glued and nailed, add			↓	50%			
	68-001	**ROUGH HARDWARE** Average % of carpentry material, minimum				.50%			
	020	Maximum				1.50%			
	70-001	**SHEATHING** Plywood on roof, CDX							
	003	5/16" thick	F-2	1,600	S.F.	.26	.21	.47	.59
	005	(83) (86) 3/8" thick		1,525		.29	.22	.51	.64
	010	1/2" thick		1,400		.37	.24	.61	.75
	020	5/8" thick		1,300		.41	.26	.67	.82
	030	3/4" thick		1,200		.46	.28	.74	.91
	050	Plywood on walls with exterior CDX, 3/8" thick		1,200		.29	.28	.57	.72
	060	1/2" thick		1,125		.37	.30	.67	.84
	070	5/8" thick		1,050		.41	.32	.73	.91
	080	3/4" thick	↓	975		.46	.35	.81	1
	100	For shear wall construction, add					20%		
	120	For structural 1 exterior plywood, add				10%			
	140	With boards, on roof 1" x 6" boards, laid horizontal	F-2	725		.70	.47	1.17	1.44
	150	Laid diagonal		650		.70	.52	1.22	1.51
	170	1" x 8" boards, laid horizontal		875		.70	.39	1.09	1.32
	180	Laid diagonal	↓	725		.70	.47	1.17	1.44
	200	For steep roofs, add					40%		
	220	For dormers, hips and valleys, add				5%	50%		
	240	Boards on walls, 1" x 6" boards, laid regular	F-2	650		.70	.52	1.22	1.51
	250	Laid diagonal		585		.70	.58	1.28	1.60
	270	1" x 8" boards, laid regular		765		.70	.44	1.14	1.40
	280	Laid diagonal		650		.70	.52	1.22	1.51
	285	Gypsum, weatherproof, 1/2" thick		1,050		.22	.32	.54	.70
	290	Sealed, 4/10" thick		1,100		.20	.31	.51	.66
	300	Wood fiber, regular, no vapor barrier, 1/2" thick		1,200		.37	.28	.65	.81
	310	5/8" thick	↓	1,200	↓	.48	.28	.76	.93

For expanded coverage of these items see Means' *Interior Cost Data 1986*

6.1 Rough Carpentry

			CREW	DAILY OUTPUT	UNIT	BARE COSTS MAT.	BARE COSTS INST.	BARE COSTS TOTAL	TOTAL INCL O&P
70	330	No vapor barrier, in colors, 1/2" thick	F-2	1,200	S.F.	.49	.28	.77	.94
	340	5/8" thick		1,200		.60	.28	.88	1.06
	360	With vapor barrier one side, white, 1/2" thick		1,200		.48	.28	.76	.93
	370	Vapor barrier 2 sides		1,200		.74	.28	1.02	1.22
	380	Asphalt impregnated, 25/32" thick		1,200		.27	.28	.55	.70
	385	Intermediate, 1/2" thick	↓	1,200	↓	.23	.28	.51	.66
74-001		STRESSED SKIN PLYWOOD ROOF PANELS 3/8" group 1 top							
	002	skin, 3/8" exterior AC bottom skin							
	003	1150f stringers, 4' x 8' panels							
	010	4-1/4" deep	F-3	2,075	S.F.Roof	2.28	.51	2.79	3.21
	020	6-1/8" deep		1,725		2.40	.61	3.01	3.48
	030	8-1/8" deep		1,475		2.72	.72	3.44	3.97
	050	3/8" top skin, no bottom skin, 5-3/4" deep		1,725		1.89	.61	2.50	2.92
	060	7-3/4" deep	↓	1,475		2.15	.72	2.87	3.35
	080	For 3-1/2" factory fiberglass insulation, add				.22		.22	.24M
	100	For 1/2" thick top skin, add			↓	.10		.10	.11M
	150	Floor panels, substitute 5/8" underlayment as							
	151	top skin, add to roof panels above			S.F.Flr.	.13		.13	.14M
	200	Curved roof panels, 3/8" structural 1 top skin,							
	201	3/8" exterior AC bottom skin, laminated ribs							
	220	8' radius, 2-1/4" deep, tie rods not req'd.	F-3	1,150	S.F.Flr.	4.60	.92	5.52	6.30
	240	10' radius, 1-1/2" deep, tie rods are included		950		2.99	1.12	4.11	4.81
	260	10' radius, 3-3/8" deep, tie rods not req'd.		1,150		5.56	.92	6.48	7.40
	280	12' radius, 2" deep, tie rods are included		950		3.44	1.12	4.56	5.30
	300	12' radius, 4-1/2" deep, tie rods not req'd.	↓	1,150	↓	5.82	.92	6.74	7.65
	320	Cost of tie rods included where required							
	400	Folded plate roofs, structural 1 top skin with intermediate							
	401	rafters and end chord. Cost of tie rods included							
	420	Slope 7 in 12, 4' fold, 2" thick, 32' span	F-3	850	S.F.Flr.	2.53	1.25	3.78	4.49
	440	Slope 8-1/2 in 12, 5' fold, 4" thick, 56' span		950		5.50	1.12	6.62	7.60
	460	Slope 10 in 12, 8' fold, 4" thick, 52' span		950		3.30	1.12	4.42	5.15
	480	Lope 10 in 12, 8' fold, 4" thick, 72' span	↓	1,025	↓	5.88	1.03	6.91	7.90
	600	Box beams, structural 1 web							
	620	24" deep, 2-2" x 4" flanges, 2 webs @ 3/8"	F-3	295	L.F.	7.82	3.60	11.42	13.50
	640	24" deep, 3-2" x 4" flanges, 2 webs @ 1/2"		260		9.87	4.08	13.95	16.45
	660	48" deep, 3-2" x 6" flanges, 2 webs @ 3/4"	↓	140	↓	17.70	7.60	25.30	30
	680	48" deep, 6-2" x 6" flanges, 4 webs @ 3/8",							
	681	including 2 interior webs	F-3	115	L.F.	33.60	9.20	42.80	50
	700	For exterior AC outer webs, add				.81		.81	.89M
	720	For medium density overlaid outer webs, add			↓	1.31		1.31	1.44M
78-001		STRUCTURAL JOISTS Fabricated "I" joists with wood flanges,							
	010	Plywood webs, incl. bridging & blocking, panels 24" O.C.							
	120	15' to 24' span, 50 psf live load	F-5	2,400	S.F.	1.08	.28	1.36	1.59
	130	55 psf live load		2,250		1.12	.30	1.42	1.66
	150	55 psf live load	↓	2,400		1.34	.28	1.62	1.88
	160	Tubular steel open webs, 45 psf, 24" O.C., 40' span	F-3	6,250		1.20	.17	1.37	1.55
	170	55' span		5,150		1.16	.21	1.37	1.56
	180	70' span		9,250		1.68	.11	1.79	2
	190	85 psf live load, 26' span	↓	2,300	↓	1.51	.46	1.97	2.29
82-001		SUBFLOOR Plywood, CDX, 1/2" thick	F-2	1,500	S.F.	.37	.22	.59	.73
	010	5/8" thick		1,350		.41	.25	.66	.81
	020	(83) (86) 3/4" thick		1,250		.46	.27	.73	.89
	030	1-1/8" thick, 2-4-1 including underlayment		1,050		.77	.32	1.09	1.31
	050	With boards, 1" x 10" S4S, laid regular		1,100		.71	.31	1.02	1.22
	060	Laid diagonal		900		.71	.37	1.08	1.32
	080	1" x 8" S4S, laid regular		1,000		.70	.34	1.04	1.25
	090	Laid diagonal		850		.70	.40	1.10	1.34
	110	Wood fiber, T&G, 2' x 8' planks, 1" thick		1,000		.97	.34	1.31	1.55
	120	1-3/8" thick	↓	900	↓	1.28	.37	1.65	1.94

For expanded coverage of these items see Means' *Interior Cost Data 1986*

6.1 Rough Carpentry

		CREW	DAILY OUTPUT	UNIT	BARE COSTS MAT.	BARE COSTS INST.	BARE COSTS TOTAL	TOTAL INCL O&P
86-001	UNDERLAYMENT Plywood, underlayment grade, 3/8" thick	F-2	1,500	S.F.	.35	.22	.57	.71
010	1/2" thick		1,450		.40	.23	.63	.77
020	5/8" thick		1,400		.50	.24	.74	.89
030	3/4" thick		1,300		.57	.26	.83	1
050	Particle board, 3/8" thick		1,500		.21	.22	.43	.55
060	1/2" thick		1,450		.22	.23	.45	.58
080	5/8" thick		1,400		.25	.24	.49	.62
090	3/4" thick		1,300		.31	.26	.57	.71
110	Hardboard, underlayment grade, 4' x 4', .215" thick	↓	1,500	↓	.24	.22	.46	.59

6.2 Finish Carpentry

		CREW	DAILY OUTPUT	UNIT	BARE COSTS MAT.	BARE COSTS INST.	BARE COSTS TOTAL	TOTAL INCL O&P
03-001	BEAMS, DECORATIVE Rough sawn cedar, non-load bearing, 4" x 4"	2 Carp	180	L.F.	1.53	1.78	3.31	4.26
010	4" x 6"		170		2.29	1.88	4.17	5.25
020	4" x 8"		160		3.05	2	5.05	6.25
030	4" x 10"		150		3.89	2.13	6.02	7.35
040	4" x 12"		140		4.68	2.29	6.97	8.45
050	8" x 8"		130	↓	6.25	2.46	8.71	10.45
090	Connector plates, steel, with bolts, straight		75	Ea.	12.50	4.27	16.77	19.95
100	Tee	↓	50	"	19	6.40	25.40	30
06-001	CABINETS Corner china cabinets, stock pine,							
002	80" high, unfinished, minimum	2 Carp	6.60	Ea.	250	48	298	345
010	Maximum	"	4.40	"	390	73	463	535
030	Built-in drawer units, pine, 18" deep, 32" high, unfinished							
040	Minimum	2 Carp	53	L.F.	26	6.05	32.05	37
050	Maximum	"	40	"	40	8	48	56
070	Kitchen base cabinets, hardwood, not incl. counter tops,							
071	24" deep, 35" high, prefinished							
080	One top drawer, one door below, 12" wide	2 Carp	24.80	Ea.	92.50	12.90	105.40	120
084	18" wide		23.30		105	13.75	118.75	135
088	24" wide	↓	22.30	↓	116	14.35	130.35	150
089								
100	Four drawers, 12" wide	2 Carp	24.80	Ea.	142	12.90	154.90	175
104	18" wide		23.30		147	13.75	160.75	180
106	24" wide		22.30		168	14.35	182.35	205
120	Two top drawers, two doors below, 27" wide		22		142	14.55	156.55	175
126	36" wide		20.30		168	15.75	183.75	210
130	48" wide		18.90		184	16.95	200.95	225
150	Range or sink base, two doors below, 30" wide		21.40		126	14.95	140.95	160
154	36" wide		20.30		142	15.75	157.75	180
158	48" wide	↓	18.90		158	16.95	174.95	200
180	For sink front units, deduct				40		40	44M
200	Corner base cabinets, 36" wide, standard	2 Carp	18		147	17.80	164.80	185
210	Lazy Susan with revolving door	"	16.50	↓	184	19.40	203.40	230
390								
400	Kitchen wall cabinets, hardwood, 12" deep with two doors							
405	12" high, 30" wide	2 Carp	24.80	Ea.	71	12.90	83.90	97
410	36" wide		24		75	13.35	88.35	100
440	15" high, 30" wide		24		74	13.35	87.35	100
444	36" wide		22.70		79	14.10	93.10	105
470	24" high, 30" wide		23.30		93	13.75	106.75	120
472	36" wide		22.70		103	14.10	117.10	135
500	30" high, one door, 12" wide		22		64	14.55	78.55	91
504	18" wide	↓	20.90	↓	76	15.30	91.30	105

For expanded coverage of these items see Means' *Interior Cost Data 1986*

6.2 Finish Carpentry

		CREW	DAILY OUTPUT	UNIT	BARE COSTS MAT.	BARE COSTS INST.	BARE COSTS TOTAL	TOTAL INCL O&P
06 506	24" wide	2 Carp	20.30	Ea.	82	15.75	97.75	115
530	Two doors, 27" wide		19.80		99	16.15	115.15	130
534	36" wide		18.80		116	17	133	150
538	48" wide		18.40		142	17.40	159.40	180
600	Corner wall, 30" high, 24" wide		18		82	17.80	99.80	115
605	30" wide		17.20		91	18.60	109.60	125
610	36" wide		16.50		100	19.40	119.40	140
650	Revolving Lazy Susan		15.20		120	21	141	165
700	Broom cabinet, 84" high, 24" deep, 18" wide		10		210	32	242	275
750	Oven cabinets, 84" high, 24" deep, 27" wide		8		220	40	260	300
775	Valance board trim		396	L.F.	6	.81	6.81	7.75
776								
900	For deluxe models of all cabinets, add to above			Ea.	40%			
950	For custom built in place, add to above			"	25%	10%		
952								
955	Rule of thumb, kitchen cabinets not including							
956	appliances & counter top, minimum	2 Carp	30	L.F.	95	10.65	105.65	120
960	Maximum	"	25	"	160	12.80	172.80	195
961	For metal cabinets, see division 12.1-15							
09-001	COLUMNS For base plates, see division 5.4-10							
005	Aluminum, round colonial, 6" diameter	2 Carp	80	V.L.F.	6.95	4	10.95	13.45
010	8" diameter		62.25		8.60	5.15	13.75	16.90
020	10" diameter		55		12.95	5.80	18.75	23
025	Fir, stock units, hollow round, 6" diameter		80		9.15	4	13.15	15.85
030	8" diameter		80		11.45	4	15.45	18.40
035	10" diameter		70		14.35	4.57	18.92	22
040	Solid turned, to 8' high, 3-1/2" diameter		80		4.10	4	8.10	10.30
050	4-1/2" diameter		75		6.30	4.27	10.57	13.10
060	5-1/2" diameter		70		8.40	4.57	12.97	15.85
080	Square columns, built-up, 5" x 5"		65		5.60	4.92	10.52	13.30
090	Solid, 3-1/2" x 3-1/2"		130		4.40	2.46	6.86	8.40
160	Pine, tapered, T & G, 10' to 14' high, 12" diam., minimum		100		37.30	3.20	40.50	46
170	Maximum		65		85	4.92	89.92	100
190	12' to 16' high, 14" diameter, minimum		100		59	3.20	62.20	70
200	Maximum		65		90	4.92	94.92	105
220	12' to 20' high, 18" diameter, minimum		65		80	4.92	84.92	95
230	Maximum		50		125	6.40	131.40	145
250	20' high, 20" diameter, minimum		40		94	8	102	115
260	Maximum		35		130	9.15	139.15	155
280	For flat pilasters, deduct				33%			
300	For splitting into halves, add			Ea.	52		52	57M
400	Rough sawn cedar posts, 4" x 4"	2 Carp	250	V.L.F.	1.20	1.28	2.48	3.17
410	4" x 6"		235		2.25	1.36	3.61	4.45
420	6" x 6"		220		3.40	1.45	4.85	5.85
430	8" x 8"		200		5.10	1.60	6.70	7.95
10-001	CONVECTOR COVERS Laminated plastic on 3/4"							
002	thick particle board, 12" wide, minimum	1 Carp	16	L.F.	17.75	10	27.75	34
005	Average		14		22	11.45	33.45	41
010	Maximum		12		29	13.35	42.35	51
030	Add to above for grille, minimum		150	S.F.	1.50	1.07	2.57	3.20
040	Maximum		75	"	4	2.13	6.13	7.50
12-001	COUNTER TOP Stock, plastic lam., 25" wide with backsplash, minimum		30	L.F.	6	5.35	11.35	14.35
010	Maximum (Also see 6.4-80)		25		13.70	6.40	20.10	24
030	Custom plastic, 7/8" thick, aluminum molding, no splash		30		14	5.35	19.35	23
040	Cove splash		30		18.59	5.35	23.94	28
060	1-1/4" thick, no splash		28		16.30	5.70	22	26
070	Square splash		28		20.80	5.70	26.50	31
090	Square edge, plastic face, 7/8" thick, no splash		30		18	5.35	23.35	28
100	With splash		30		23	5.35	28.35	33

For expanded coverage of these items see Means' *Interior Cost Data 1986*

6.2 Finish Carpentry

			CREW	DAILY OUTPUT	UNIT	BARE COSTS MAT.	BARE COSTS INST.	BARE COSTS TOTAL	TOTAL INCL O&P
12	120	For stainless channel edge, 7/8" thick, add			L.F.	1.60		1.60	1.76M
	130	1-1/4" thick, add				2		2	2.20M
	150	For solid color suede finish, add			↓	1.50		1.50	1.65M
	170	For end splash, add			Ea.	11		11	12.10M
	190	For cut outs, standard, add, minimum	1 Carp	32		1	5	6	8.35
	200	Maximum		8	↓	1	20	21	30
	210	Postformed, including backsplash and front edge		30	L.F.	6.50	5.35	11.85	14.90
	211	Mitred, add		12	Ea.		13.35	13.35	19.30L
	220	Built-in place, 25" wide, plastic laminate		25	L.F.	8.25	6.40	14.65	18.35
	230	Ceramic tile mosaic	↓	25		20.25	6.40	26.65	32
	250	Marble, stock, with splash, 1/2" thick, minimum	1 Bric	17		23	9.65	32.65	39
	270	3/4" thick, maximum	"	13		75	12.60	87.60	100
	290	Maple, solid, laminated, 1-1/2" thick, no splash	1 Carp	28		28	5.70	33.70	39
	300	With square splash		28	↓	32	5.70	37.70	43
	320	Stainless steel		24	S.F.	45	6.65	51.65	59
	340	Recessed cutting block with trim, 16" x 20" x 1"		8	Ea.	34.40	20	54.40	67
	360	Table tops, plastic laminate, square edge, 7/8" thick		45	S.F.	6.50	3.56	10.06	12.30
	370	1-1/8" thick		40	"	6.95	4	10.95	13.45
	15-001	CUPOLA Stock units, 1" redwood, 2' x 2' x 2', aluminum roof		4.10	Ea.	110	39	149	180
	010	Copper roof		3.80		215	42	257	300
	030	3' x 3' base, 3' high, aluminum roof		3.70		220	43	263	305
	040	Copper roof		3.30		280	48	328	380
	060	31" x 31" base, 51" high, aluminum roof		3.70		310	43	353	405
	070	Copper roof		3.30		425	48	473	540
	090	Hexagonal, 31" wide, 37" high, copper roof		4		345	40	385	435
	100	35" wide, 43" high, copper roof	↓	3.50		465	46	511	580
	120	For deluxe stock units, add to above				30%			
	140	For custom built units, add to above				50%	50%		
	160	Fiberglass, 5'-0" square base, 9'-3" high	F-3	6		2,600	175	2,775	3,100
	170	6'-0" square base, 8'-0" high	"	5	↓	2,800	210	3,010	3,375
	17-001	(97) DOORS AND FRAMES See division 8.1 & 8.2							
	20-001	FIREPLACE MANTEL BEAMS Rough texture wood, 4" x 8"	1 Carp	36	L.F.	3.50	4.44	7.94	10.30
	010	4" x 10"		35	"	4.35	4.57	8.92	11.40
	030	Laminated hardwood, 2-1/4" x 10-1/2" wide, 6' long		5	Ea.	85	32	117	140
	040	8' long		5	"	115	32	147	175
	060	Brackets for above, rough sawn		12	Pr.	7.95	13.35	21.30	28
	070	Laminated		12	"	11.50	13.35	24.85	32
	22-001	FIREPLACE MANTELS 6" molding, 6' x 3'-6" opening, minimum		5.50	Opng.	94	29	123	145
	010	Maximum		4.50		110	36	146	175
	030	Prefabricated pine, colonial type, stock, deluxe		2		490	80	570	655
	040	Economy	↓	3	↓	190	53	243	285
	25-001	FLOORING, WOOD See division 9.6-40							
	27-001	GRILLES And panels, hardwood, sanded							
	002	2' x 4' to 4' x 8', custom designs, unfinished, minimum	1 Carp	38	S.F.	10.30	4.21	14.51	17.45
	005	Average		30		16	5.35	21.35	25
	010	Maximum		19		24.20	8.40	32.60	39
	030	As above, but prefinished, minimum		38		13.25	4.21	17.46	21
	040	Maximum	↓	19	↓	32	8.40	40.40	47
	30-001	HARDWARE Finish, see division 6.4 & 8.7							
	010	Rough, see division 5.8							
	32-001	LOUVERS Redwood, 2'-0" opening, full circle	1 Carp	16	Ea.	78	10	88	100
	010	Half circle, 2'-0" diameter		16		65	10	75	86
	020	Octagonal		16		57.50	10	67.50	78
	030	Triangular, 5/12 pitch, 5'-0" at base	↓	16	↓	66	10	76	87
	35-001	(86) LUMBER Finish							
	36-001	MILLWORK See division 6.4							

For expanded coverage of these items see *Means' Interior Cost Data 1986*

6.2 Finish Carpentry

		CREW	DAILY OUTPUT	UNIT	BARE COSTS MAT.	BARE COSTS INST.	BARE COSTS TOTAL	TOTAL INCL O&P
37-001	**MOLDINGS, BASE**							
002								
050	Base, stock pine, 9/16" x 3-1/2"	1 Carp	240	L.F.	.58	.67	1.25	1.60
055	9/16" x 4-1/2"		200		.77	.80	1.57	2.01
056	Base shoe, oak, 3/4" x 1"	↓	240	↓	.86	.67	1.53	1.91
41-001	**MOLDINGS, CASINGS**							
002								
009	Apron, stock pine, 5/8" x 2"	1 Carp	250	L.F.	.40	.64	1.04	1.37
011	5/8" x 3-1/2"		220		1.08	.73	1.81	2.24
030	Band, stock pine, 11/16" x 1-1/8"		270		.25	.59	.84	1.13
035	11/16" x 1-3/4"		250		.38	.64	1.02	1.35
070	Casing, stock pine, 11/16" x 2-1/2"		240		.38	.67	1.05	1.38
075	11/16" x 3-1/2"	↓	215	↓	.77	.74	1.51	1.93
43-001	**MOLDINGS, CEILINGS**							
002								
060	Bed, stock pine, 9/16" x 1-3/4"	1 Carp	270	L.F.	.27	.59	.86	1.16
065	9/16" x 2"		240		.40	.67	1.07	1.41
120	Cornice molding, stock pine, 9/16" x 1-3/4"		330		.28	.48	.76	1.01
130	9/16" x 2-1/4"		300		.42	.53	.95	1.23
240	Cove scotia, stock pine, 9/16" x 1-3/4"		270		.33	.59	.92	1.22
250	11/16" x 2-3/4"		255		.77	.63	1.40	1.76
260	Crown, stock pine, 9/16" x 3-5/8"		250		.75	.64	1.39	1.75
270	11/16" x 4-5/8"	↓	220	↓	1.40	.73	2.13	2.59
46-001	**MOLDINGS, EXTERIOR**							
150	Cornice, boards, pine, 1" x 2"	1 Carp	330	L.F.	.16	.48	.64	.88
170	1" x 6"		200		.54	.80	1.34	1.75
200	1" x 12"		180		1.65	.89	2.54	3.10
220	Three piece, built-up, pine, minimum		80		.65	2	2.65	3.61
230	Maximum		65		3.25	2.46	5.71	7.15
300	Trim, exterior, sterling pine, corner board, 1" x 4"		200		.32	.80	1.12	1.51
310	1" x 6"		200		.54	.80	1.34	1.75
335	Fascia, 1" x 6"		250		.54	.64	1.18	1.52
337	1" x 8"		225		.74	.71	1.45	1.84
340	Moldings, back band		250		.35	.64	.99	1.31
350	Casing		250		.60	.64	1.24	1.59
360	Crown		250		.83	.64	1.47	1.84
370	Porch rail with balusters		22		5	7.25	12.25	16.05
380	Screen		395		.09	.41	.50	.69
410	Verge board, sterling pine, 1" x 4"		200		.32	.80	1.12	1.51
420	1" x 6"		200		.54	.80	1.34	1.75
430	2" x 6"		165		.87	.97	1.84	2.36
440	2" x 8"	↓	165		1.16	.97	2.13	2.68
470	For redwood trim, add			↓	100%			
49-001	**MOLDINGS, TRIM**							
002								
020	Astragal, stock pine, 11/16" x 1-3/4"	1 Carp	255	L.F.	.45	.63	1.08	1.40
025	1-5/16" x 2-3/16"		240		.83	.67	1.50	1.88
080	Chair rail, stock pine, 5/8" x 2-1/2"		270		.60	.59	1.19	1.52
090	5/8" x 3-1/2"		240		.95	.67	1.62	2.01
100	Closet pole, stock pine, 1-1/8" diameter		200		.60	.80	1.40	1.82
110	Fir, 1-5/8" diameter		200		.65	.80	1.45	1.87
330	Half round, stock pine, 1/4" x 1/2"		270		.07	.59	.66	.94
335	1/2" x 1"	↓	255	↓	.20	.63	.83	1.13
340	Handrail, fir, single piece, stock, hardware not included							
345	1-1/2" x 1-3/4"	1 Carp	80	L.F.	.65	2	2.65	3.61
347	Pine, 1-1/2" x 1-3/4"		80		.67	2	2.67	3.64
350	1-1/2" x 2-1/2"		76		.97	2.11	3.08	4.12
360	Lattice, stock pine, 1/4" x 1-1/8"		270		.16	.59	.75	1.03
370	1/4" x 1-3/4"	↓	250	↓	.20	.64	.84	1.15

For expanded coverage of these items see *Means' Interior Cost Data 1986*

6.2 Finish Carpentry

			CREW	DAILY OUTPUT	UNIT	BARE COSTS MAT.	BARE COSTS INST.	BARE COSTS TOTAL	TOTAL INCL O&P
49	380	Miscellaneous, custom, pine or cedar, 1" x 1"	1 Carp	270	L.F.	.11	.59	.70	.98
	390	Nominal 1" x 3"		240		.31	.67	.98	1.31
	410	Birch or oak, custom, nominal 1" x 1"		240		.16	.67	.83	1.14
	420	Nominal 1" x 3"		215		.47	.74	1.21	1.60
	440	Walnut, custom, nominal 1" x 1"		215		.27	.74	1.01	1.38
	450	Nominal 1" x 3"		200		.77	.80	1.57	2.01
	470	Teak, custom, nominal 1" x 1"		215		.63	.74	1.37	1.77
	480	Nominal 1" x 3"		200		1.77	.80	2.57	3.11
	490	Quarter round, stock pine, 1/4" x 1/4"		275		.07	.58	.65	.92
	495	3/4" x 3/4"		255		.17	.63	.80	1.10
	560	Wainscot moldings, 1-1/8" x 9/16", 2 ft. high, minimum		76	S.F.	5.30	2.11	7.41	8.90
	570	Maximum		65	"	10.50	2.46	12.96	15.10
51-001		**MOLDINGS, WINDOW AND DOOR**							
002									
	280	Door moldings, stock, decorative, 1-1/8" wide, plain	1 Carp	17	Set	18.75	9.40	28.15	34
	290	Detailed	"	17	"	32	9.40	41.40	49
	300								
	310	Door trim, interior, including headers,							
	315	stops and casings, 2 sides, pine, 2-1/2" wide	1 Carp	5.90	Opng.	19.25	27	46.25	60
	317	4-1/2" wide		5.30	"	25	30	55	71
	320	Glass beads, stock pine, 1/4" x 11/16"		285	L.F.	.14	.56	.70	.97
	325	3/8" x 1/2"		275		.17	.58	.75	1.03
	327	3/8" x 7/8"		270		.20	.59	.79	1.08
	330								
	485	Parting bead, stock pine, 3/8" x 3/4"	1 Carp	275	L.F.	.15	.58	.73	1.01
	487	1/2" x 3/4"		255		.21	.63	.84	1.14
	500	Stool caps, stock pine, 11/16" x 3-1/2"		200		.70	.80	1.50	1.93
	510	1-1/16" x 3-1/4"		150		1.79	1.07	2.86	3.51
	530	Threshold, oak, 3' long, inside, 5/8" x 3-5/8"		32	Ea.	4.15	5	9.15	11.80
	540	Outside, 1-1/2" x 7-5/8"		16	"	15.80	10	25.80	32
	590	Window trim sets, including casings, header, stops,							
	591	stool and apron, 2-1/2" wide, minimum	1 Carp	13	Opng.	9.50	12.30	21.80	28
	595	Average		10		12.50	16	28.50	37
	600	Maximum		6		15	27	42	55
53-001		**PANELING, BOARDS**							
002									
	640	Wood board paneling, 3/4" thick, knotty pine	F-2	300	S.F.	.75	1.12	1.87	2.43
	650	Rough sawn cedar		300		1.45	1.12	2.57	3.20
	670	Redwood, clear, 1" x 4" boards		300		2.65	1.12	3.77	4.52
	690	Aromatic cedar, closet lining, boards		275		1.60	1.23	2.83	3.52
56-001		**PANELING, HARDBOARD**							
	002	Not incl. furring or trim, hardboard, tempered, 1/8" thick	F-2	500	S.F.	.22	.67	.89	1.21
	010	1/4" thick		500		.38	.67	1.05	1.38
	030	Tempered pegboard, 1/8" thick		500		.23	.67	.90	1.22
	040	1/4" thick		500		.37	.67	1.04	1.37
	060	Untempered hardboard, natural finish, 1/8" thick		500		.20	.67	.87	1.19
	070	1/4" thick		500		.27	.67	.94	1.26
	090	Untempered pegboard, 1/8" thick		500		.21	.67	.88	1.20
	100	1/4" thick		500		.35	.67	1.02	1.35
	120	Plastic faced hardboard, 1/8" thick		500		.39	.67	1.06	1.39
	130	1/4" thick		500		.54	.67	1.21	1.56
	150	Plastic faced pegboard, 1/8" thick		500		.38	.67	1.05	1.38
	160	1/4" thick		500		.47	.67	1.14	1.48
	180	Wood grained, plain or grooved, 1/4" thick, minimum		500		.34	.67	1.01	1.34
	190	Maximum		425		.64	.79	1.43	1.84
	210	Moldings for hardboard, wood or aluminum, minimum		500	L.F.	.25	.67	.92	1.24
	220	Maximum		425	"	.65	.79	1.44	1.85
58-001		**PANELING, PLYWOOD**							
002									

For expanded coverage of these items see *Means' Interior Cost Data 1986*

6.2 Finish Carpentry

			CREW	DAILY OUTPUT	UNIT	BARE COSTS MAT.	BARE COSTS INST.	BARE COSTS TOTAL	TOTAL INCL O&P
58	240	(76) Plywood, prefinished, 1/4" thick, 4' x 8' sheets							
	241	with vertical grooves. Birch faced, minimum	F-2	500	S.F.	.50	.67	1.17	1.52
	242	Average		420		.80	.80	1.60	2.03
	243	Maximum		350		1.15	.96	2.11	2.64
	260	Mahogany, African		400		1.30	.84	2.14	2.64
	270	Philippine (Lauan)		500		.34	.67	1.01	1.34
	290	Oak or Cherry, minimum		500		1.20	.67	1.87	2.29
	300	Maximum		400		2.25	.84	3.09	3.68
	320	Rosewood		320		8	1.05	9.05	10.30
	340	Teak		400		2.20	.84	3.04	3.63
	360	Chestnut		375		3.30	.90	4.20	4.92
	380	Pecan		400		1.30	.84	2.14	2.64
	390	Walnut, minimum		500		1.85	.67	2.52	3
	395	Maximum		400		3.50	.84	4.34	5.05
	400	Plywood, prefinished, 3/4" thick, stock grades, minimum		320		.80	1.05	1.85	2.39
	410	Maximum		224		3.75	1.51	5.26	6.30
	430	Architectural grade, minimum		224		2.75	1.51	4.26	5.20
	440	Maximum		160		4	2.11	6.11	7.40
	460	Plywood, unfin. "A" face, birch, V.C., 1/2" thick, natural		450		1.30	.75	2.05	2.50
	470	Select		450		1.42	.75	2.17	2.64
	490	Veneer core, 3/4" thick, natural		320		1.32	1.05	2.37	2.96
	500	Select		320		1.50	1.05	2.55	3.16
	520	Lumber core, 3/4" thick, natural		320		1.85	1.05	2.90	3.54
	525								
	550	Plywood, unfinished, knotty pine, 1/4" thick, A2 grade	F-2	450	S.F.	1.30	.75	2.05	2.50
	560	A3 grade		450		1.40	.75	2.15	2.61
	580	3/4" thick, veneer core, A2 grade		320		1.70	1.05	2.75	3.38
	590	A3 grade		320		1.40	1.05	2.45	3.05
	610	Aromatic cedar, 1/4" thick, plywood		400		1.25	.84	2.09	2.58
	620	1/4" thick, particle board		400		.63	.84	1.47	1.90
59-001		PARTITIONS See divisions 9.1-30, 9.2-31 and 10.1-45 to 62							
61-001		RAILING Custom design, architectural grade, hardwood, minimum	1 Carp	38	L.F.	11	4.21	15.21	18.20
	010	Maximum		30		25	5.35	30.35	35
	030	Stock interior railing with spindles 6" O.C., 4' long		40		23	4	27	31
	040	8' long		48		21	3.33	24.33	28
64-001		SHELVING Pine, clear grade, no edge band, 1" x 8"	F-1	115		.65	1.47	2.12	2.81
	010	1" x 10"		110		.82	1.53	2.35	3.10
	020	1" x 12"		105		1.04	1.61	2.65	3.44
	030								
	040	For lumber edge band, by hand, add			L.F.	1.25		1.25	1.37M
	042	By machine, add				.77		.77	.84M
	060	Plywood, 3/4" thick with lumber edge, 12" wide	F-1	75		.98	2.25	3.23	4.30
	070	24" wide		70		1.84	2.41	4.25	5.45
	090	Bookcase, pine, clear grade, 8" shelves, 12" O.C.		70	S.F.Face	3.60	2.41	6.01	7.40
	100	12" wide shelves		65	"	4.25	2.59	6.84	8.40
	120	Adjustable closet rod and shelf, 12" wide, 3' long		20	Ea.	7.25	8.45	15.70	20
	130	8' long		15	"	15.50	11.25	26.75	33
	150	Prefinished shelves with supports, stock, 8" wide		75	L.F.	5.10	2.25	7.35	8.85
	160	10" wide		70	"	5.70	2.41	8.11	9.70
	180	Custom, high quality dadoed pine shelving units, minimum			S.F.Face			23	28
	190	Maximum			"			31.50	40
65-001		SHUTTERS, EXTERIOR Aluminum, louvered, 1'-4" wide, 3'-0" long	1 Carp	10	Pr.	24	16	40	50
	040	6'-8" long		9		40	17.80	57.80	70
	100	Pine, louvered, primed, each 1'-2" wide, 3'-3" long		10		27	16	43	53
	110	4'-7" long		10		36	16	52	63
	150	Each 1'-6" wide, 3'-3" long		10		31	16	47	57
	160	4'-7" long		10		43	16	59	70
	162	Hemlock, louvered, 1'-2" wide, 5'-7" long		10		45	16	61	73
	163	Each 1'-4" wide, 2'-2" long		10		26	16	42	52

For expanded coverage of these items see *Means' Interior Cost Data 1986*

6.2 Finish Carpentry

			CREW	DAILY OUTPUT	UNIT	BARE COSTS MAT.	BARE COSTS INST.	BARE COSTS TOTAL	TOTAL INCL O&P
65	167	4'-3" long	1 Carp	10	Pr.	31	16	47	57
	169	5'-11" long		10		44	16	60	72
	170	Door blinds, 6'-9" long, each 1'-3" wide		9		53	17.80	70.80	84
	171	1'-6" wide		9		54	17.80	71.80	85
	172	Hemlock, solid raised panel, each 1'-4" wide, 3'-3" long		10		44	16	60	72
	174	4'-3" long		10		55	16	71	84
	177	5'-11" long	↓	10	↓	76	16	92	105
	179								
	180	Door blinds, 6'-9" long, each 1'-3" wide	1 Carp	9	Pr.	85	17.80	102.80	120
	190	1'-6" wide		9		86	17.80	103.80	120
	250	Polystyrene, solid raised panel, each 1'-4" wide, 3'-3" long		10		34	16	50	61
	270	4'-7" long		10		46	16	62	74
	350	For brick walls, add			↓	5.10		5.10	5.60M
	400								
	450	Polystyrene, louvered, each 1'-2" wide, 3'-3" long	1 Carp	10	Pr.	29	16	45	55
	475	5'-3" long		10		41	16	57	68
	600	Vinyl, louvered, each 1'-2" x 4'-7" long		10		35	16	51	62
	620	Each 1'-4" x 6'-8" long	↓	9	↓	60	17.80	77.80	92
	67-001	**SIDING, BOARDS**							
	002								
	320	Wood, cedar bevel, short lengths, A grade, 1/2" x 6"	1 Carp	250	S.F.	1.30	.64	1.94	2.36
	330	1/2" x 8"		275		1.16	.58	1.74	2.12
	350	3/4" x 10", clear grade, 3' to 16'		300		1.34	.53	1.87	2.25
	360	"B" grade		300		1.32	.53	1.85	2.22
	380	Cedar, rough sawn, 1" x 4", B & Btr., natural		240		1.56	.67	2.23	2.68
	390	Stained		240		1.61	.67	2.28	2.74
	410	1" x 12", board & batten, #3 & Btr., natural		260		1	.62	1.62	1.99
	420	Stained		260		1.05	.62	1.67	2.05
	440	1" x 8" channel siding, #3 & Btr., natural		250		1.10	.64	1.74	2.14
	450	Stained		250		1.15	.64	1.79	2.19
	470	Redwood, clear, beveled, vertical grain, 1/2" x 4"		200		1.65	.80	2.45	2.97
	480	1/2" x 8"		250		1.32	.64	1.96	2.38
	500	3/4" x 10"		300		1.60	.53	2.13	2.53
	520	Channel siding, 1" x 10", clear		285		1.05	.56	1.61	1.97
	525	Redwood, T&G boards, clear, 1" x 4"	F-2	300		2.40	1.12	3.52	4.25
	527	1" x 8"	"	375		2.05	.90	2.95	3.54
	540	White pine, rough sawn, 1" x 8", natural	1 Carp	275		.49	.58	1.07	1.38
	550	Stained	"	275	↓	.57	.58	1.15	1.47
	70-001	**SIDING, SHEETS**							
	002	Siding, hardboard, 7/16" thick, prime painted, lap,							
	003	plain or grooved finish	F-2	750	S.F.	.40	.45	.85	1.08
	010	Board finish, 7/16" thick, lap or grooved, primed		750		.62	.45	1.07	1.33
	020	Stained		750		.67	.45	1.12	1.38
	070	Particle board, overlaid, 3/8" thick		750		.60	.45	1.05	1.30
	090	Plywood, medium density overlaid, 3/8" thick		750		.57	.45	1.02	1.27
	100	1/2" thick		700		.73	.48	1.21	1.49
	110	3/4" thick		650		.91	.52	1.43	1.74
	160	Texture 1-11, cedar, 5/8" thick, natural		675		1.30	.50	1.80	2.15
	170	Factory stained		675		1.35	.50	1.85	2.20
	190	Texture 1-11, fir, 5/8" thick, natural		675		.56	.50	1.06	1.33
	200	Factory stained		675		.61	.50	1.11	1.39
	205	Texture 1-11, S.Y.P., 5/8" thick, natural		675		.52	.50	1.02	1.29
	210	Factory stained		675		.57	.50	1.07	1.34
	220	Rough sawn cedar, 3/8" thick, natural		675		.96	.50	1.46	1.77
	230	Factory stained		675		1.01	.50	1.51	1.83
	250	Rough sawn fir, 3/8" thick, natural		675		.38	.50	.88	1.13
	260	Factory stained		675		.43	.50	.93	1.19
	280	Redwood, textured siding, 5/8" thick		675		1.12	.50	1.62	1.95
	300	Polyvinyl chloride coated, 3/8" thick	↓	750	↓	.76	.45	1.21	1.48

For expanded coverage of these items see *Means' Interior Cost Data 1986*

6.2 Finish Carpentry

		CREW	DAILY OUTPUT	UNIT	BARE COSTS MAT.	BARE COSTS INST.	BARE COSTS TOTAL	TOTAL INCL O&P
73-001	**SOFFITS** Wood fiber, no vapor barrier, 15/32" thick	F-2	525	S.F.	.40	.64	1.04	1.36
010	5/8" thick		525		.50	.64	1.14	1.47
030	As above, 5/8" thick, with factory finish		525		.58	.64	1.22	1.56
050	Hardboard, 3/8" thick, slotted		525		.62	.64	1.26	1.60
100	Exterior AC plywood, 1/4" thick		420		.38	.80	1.18	1.57
110	1/2" thick	↓	420	↓	.54	.80	1.34	1.74
115	For aluminum soffit, see division 7.6-54							
76-001	**STAIR PARTS** Balusters, turned, 30" high, pine, minimum	1 Carp	28	Ea.	3.73	5.70	9.43	12.40
010	Maximum		26		5	6.15	11.15	14.40
030	30" high birch balusters, minimum		28		4.75	5.70	10.45	13.50
040	Maximum		26		6	6.15	12.15	15.50
060	42" high, pine balusters, minimum		27		4.40	5.95	10.35	13.45
070	Maximum		25		5.20	6.40	11.60	15
090	42" high birch balusters, minimum		27		6.25	5.95	12.20	15.45
100	Maximum		25	↓	6.60	6.40	13	16.55
105	Baluster, stock pine, 1-1/16" x 1-1/16"		240	L.F.	.46	.67	1.13	1.47
110	1-5/8" x 1-5/8"		220	"	.93	.73	1.66	2.08
120	Newels, 3-1/4" wide, starting, minimum		7	Ea.	31	23	54	67
130	Maximum		6		150	27	177	205
150	Landing, minimum		5		45	32	77	96
160	Maximum		4	↓	145	40	185	215
180	Railings, oak, built-up, minimum		60	L.F.	3.75	2.67	6.42	8
190	Maximum		55		8	2.91	10.91	13
210	Add for sub rail	↓	110	↓	1.60	1.45	3.05	3.87
211								
230	Risers, Beech, 3/4" x 7-1/2" high	1 Carp	64	L.F.	3.75	2.50	6.25	7.75
240	Fir, 3/4" x 7-1/2" high		64		.97	2.50	3.47	4.69
260	Oak, 3/4" x 7-1/2" high		64		3	2.50	5.50	6.90
280	Pine, 3/4" x 7-1/2" high		66		.97	2.42	3.39	4.58
285	Skirt board, pine, 1" x 10"		55		1.26	2.91	4.17	5.60
290	1" x 12"		52	↓	1.54	3.08	4.62	6.15
300	Treads, oak, 1-1/16" x 9-1/2" wide, 3' long		18	Ea.	14.15	8.90	23.05	28
310	4' long		17		19.80	9.40	29.20	35
330	1-1/16" x 11-1/2" wide, 3' long		18		17	8.90	25.90	32
340	6' long	↓	14		36	11.45	47.45	56
360	Beech treads, add			↓	40%			
380	For mitered return nosings, add			L.F.	1.83		1.83	2.01M
79-001	**STAIRS, PREFABRICATED**							
011	Box stairs, 3' wide, oaktards, no handrails, 2' high	2 Carp	5	Flight	110	64	174	215
020	4 ft. high		4		185	80	265	320
030	6 ft. high		3.50		290	91	381	450
040	8 ft. high		3		360	105	465	550
060	With pine treads for carpet, 2 ft. high		5		67	64	131	165
070	4 ft. high		4		115	80	195	240
080	6 ft. high		3.50		180	91	271	330
090	8 ft. high	↓	3	↓	210	105	315	385
110	For 4' wide stairs, add				10%			
150	Prefabricated stair rail with balusters, 5 risers	2 Carp	15	Ea.	105	21	126	145
160								
170	Basement stairs, prefabricated, soft wood,							
171	open risers, 3' wide, 8' high	2 Carp	4	Flight	120	80	200	250
190	Open stairs, prefabricated prefinished poplar, metal stringers,							
191	treads 3'-6" wide, no railings							
200	3 ft. high	2 Carp	5	Flight	295	64	359	415
210	4 ft. high		4		370	80	450	525
220	6 ft. high		3.50		640	91	731	835
230	8 ft. high	↓	3	↓	940	105	1,045	1,200
250	For prefab. 3 piece wood railings & balusters, add for							
260	3 ft. high stairs	2 Carp	15	Ea.	100	21	121	140

For expanded coverage of these items see *Means' Interior Cost Data 1986*

6.2 Finish Carpentry

			CREW	DAILY OUTPUT	UNIT	BARE COSTS MAT.	BARE COSTS INST.	BARE COSTS TOTAL	TOTAL INCL O&P
79	270	4 ft. high stairs	2 Carp	14	Ea.	122	23	145	165
	280	6 ft. high stairs	↓	13		185	25	210	240
	290	8 ft. high stairs		12		250	27	277	315
	310	For 3'-6" x 3'-6" platform, add	↓	4	↓	85	80	165	210
	330	Curved stairways, 3'-3" wide, prefabricated, oak, unfinished,							
	331	incl. curved balustrade system, open one side							
	340	9' high	2 Carp	.70	Flight	4,300	455	4,755	5,400
	350	10' high		.70		4,700	455	5,155	5,825
	370	Open two sides, 9' high		.50		6,900	640	7,540	8,525
	380	10' high		.50		7,500	640	8,140	9,175
	400	Residential, wood, oak treads, prefabricated		1.50		630	215	845	1,000
	420	Built-in place	↓	.44	↓	750	725	1,475	1,875
	440	Spiral, oak, 4'-6" diameter, unfinished, prefabricated,							
	450	incl. railing, 9' high	2 Carp	1.50	Flight	2,900	215	3,115	3,500
88-001		VANITIES							
	002								
	800	Vanity bases, 2 doors, 30" high, 21" deep, 24" wide	2 Carp	11	Ea.	102	29	131	155
	805	30" wide		9.80		110	33	143	170
	810	36" wide		8		147	40	187	220
	815	48" wide	↓	6.60		190	48	238	280
	900	For deluxe models of all vanities, add to above			↓	40%			
	950	For custom built in place, add to above				25%	10%		

6.3 Laminated Construction

			CREW	DAILY OUTPUT	UNIT	BARE COSTS MAT.	BARE COSTS INST.	BARE COSTS TOTAL	TOTAL INCL O&P
10-001		LAMINATED FRAMING Not including decking							
	002								
	020	Straight roof beams, 20' clear span, 8' O.C.	F-3	2,560	S.F.Flr.	2.05	.41	2.46	2.82
	030	16' O.C.	F-3	3,200	S.F.Flr.	1.38	.33	1.71	1.97
	050	40' clear span, beams 8' O.C.		3,200		3.90	.33	4.23	4.74
	060	Beams 16' O.C.	↓	3,840		2.95	.28	3.23	3.62
	080	60' clear span, 8' O.C.	F-4	2,880		6.15	.52	6.67	7.45
	090	Beams 16' O.C.	"	3,840		4.66	.39	5.05	5.65
	110	Tudor arches, 30' to 40' clear span, frames 8' O.C.	F-3	1,680		5.35	.63	5.98	6.75
	120	Frames 16' O.C.	"	2,240		3.65	.47	4.12	4.66
	140	50' to 60' clear span, frames 8' O.C.	F-4	2,200		5.30	.68	5.98	6.75
	150	Frames 16' O.C.		2,640		4.30	.57	4.87	5.50
	170	Radial arches, 60' clear span, 8' O.C.		1,920		5.50	.78	6.28	7.10
	180	16' O.C.		2,880		3.60	.52	4.12	4.65
	200	100' clear span, frames @ 8' O.C.		1,600		4.20	.93	5.13	5.85
	210	16' O.C.		2,400		4.30	.62	4.92	5.55
	230	120' clear span, 8' O.C.		1,440		6.45	1.04	7.49	8.45
	240	Frames 16' O.C.	↓	1,920		5.85	.78	6.63	7.45
	260	Bowstring trusses, 20' O.C., 40' clear span	F-3	2,400		2.50	.44	2.94	3.35
	270	60' clear span	F-4	3,600		2.35	.41	2.76	3.13
	280	100' clear span		4,000		3.70	.37	4.07	4.56
	290	120' clear span	↓	3,600		4	.41	4.41	4.95
	310	For premium appearance, add to S.F. prices				5%			
	330	For industrial type, deduct				15%			
	350	For stain and varnish, add				5%			
	390	For 3/4" laminations, add to straight				25%			
	410	Add to curved			↓	15%			
	430	Alternate pricing method: (use nominal footage of							
	431	components). Straight beams, camber less than 6"	F-3	3.50	M.B.F.	1,500	305	1,805	2,075

For expanded coverage of these items see *Means' Interior Cost Data 1986*

6.3 Laminated Construction

		CREW	DAILY OUTPUT	UNIT	BARE COSTS MAT.	BARE COSTS INST.	BARE COSTS TOTAL	TOTAL INCL O&P
440	Columns, including hardware	F-3	2	M.B.F.	1,670	530	2,200	2,550
460	Curved members, radius over 32 ft.		2.50		1,725	425	2,150	2,475
470	Radius 10 ft. to 31 ft.	↓	3		1,890	355	2,245	2,550
490	For complicated shapes, add maximum			↓	100%			
500								
510	For pressure treating, add to straight			M.B.F.	35%			
520	Add to curved			"	45%			
20-001	LAMINATED ROOF DECK Pine or hemlock, 3" thick	F-2	425	S.F.	5.20	.79	5.99	6.85
010	4" thick		325		6.30	1.04	7.34	8.40
030	Cedar, 3" thick		425		6.25	.79	7.04	8
040	4" thick		325		7.45	1.04	8.49	9.70
060	Fir, 3" thick		425		5.20	.79	5.99	6.85
070	4" thick	↓	325	↓	6.26	1.04	7.30	8.35

6.4 Architectural Woodwork

		CREW	DAILY OUTPUT	UNIT	BARE COSTS MAT.	BARE COSTS INST.	BARE COSTS TOTAL	TOTAL INCL O&P
05-001	**CASEWORK, FRAMES**							
005	Base cabinets, counter storage, 36" high one bay							
010	18" wide	1 Carp	2.70	Ea.	46.15	59	105.15	135
040	Two bay, 36" wide		2.20		83.50	73	156.50	195
110	Three bay, 54" wide		1.50		118	105	223	285
280	Book cases, one bay, 7' high, 18" wide		2.40		61.60	67	128.60	165
350	Two bay, 36" wide		1.60		109	100	209	265
410	Three bay, 54" wide		1.20		187	135	322	400
510	Coat racks, one bay, 7' high, 24" wide		4.50		64	36	100	120
530	Two bay, 48" wide		2.75		102	58	160	195
580	Three bay, 72" wide		2.10		143	76	219	270
610	Wall mounted cabinet, one bay, 24" high, 18" wide		3.60		23	44	67	90
680	Two bay, 36" wide		2.20		42	73	115	150
740	Three bay, 54" wide		1.70		61	94	155	205
840	30" high, one bay, 18" wide		3.60		26.50	44	70.50	94
900	Two bay, 36" wide		2.15		47.50	74	121.50	160
940	Three bay, 54" wide		1.60		69.50	100	169.50	220
980	Wardrobe, 7' high, single, 24" wide		2.70		72.50	59	131.50	165
988	Partition & adjustable shelves, 48" wide		1.70		105	94	199	250
995	Partition, adjustable shelves & drawers, 48" wide	↓	1.40	↓	165	115	280	345
08-001	**CABINET DOORS**							
200	Glass panel, hardwood frame							
220	12" wide, 18" high	1 Carp	34	Ea.	4.50	4.71	9.21	11.75
260	30" high		32		6.35	5	11.35	14.25
445	18" wide, 18" high		32		5.45	5	10.45	13.25
455	30" high	↓	29	↓	7.25	5.50	12.75	15.95
480								
500	Hardwood, raised panel							
510	12" wide, 18" high	1 Carp	16	Ea.	8.70	10	18.70	24
520	30" high		15		13.90	10.65	24.55	31
550	18" wide, 18" high		15		13	10.65	23.65	30
560	30" high	↓	14	↓	20.80	11.45	32.25	39
590								
600	Plastic laminate on particle board							
610	12" wide, 18" high	1 Carp	25	Ea.	2.50	6.40	8.90	12
614	30" high		23		4.70	6.95	11.65	15.25
650	18" wide, 18" high		24		4.25	6.65	10.90	14.35
660	30" high	↓	22	↓	7.05	7.25	14.30	18.30

For expanded coverage of these items see *Means' Interior Cost Data 1986*

6.4 Architectural Woodwork

			CREW	DAILY OUTPUT	UNIT	BARE COSTS MAT.	BARE COSTS INST.	BARE COSTS TOTAL	TOTAL INCL O&P
08	680								
	700	Plywood, with edge band							
	701	12" wide, 18" high	1 Carp	27	Ea.	4.85	5.95	10.80	13.90
	712	30" high		25		8	6.40	14.40	18.05
	765	18" wide, 18" high		26		7.25	6.15	13.40	16.90
	775	30" high	↓	24	↓	12	6.65	18.65	23
12-001		**DRAWERS**							
	010	Solid hardwood front							
	100	4" high, 12" wide	1 Carp	17	Ea.	12	9.40	21.40	27
	120	18" wide	"	16	"	15.20	10	25.20	31
	280	Plastic laminate on particle board front							
	300	4" high, 12" wide	1 Carp	17	Ea.	10.70	9.40	20.10	25
	320	18" wide	"	16	"	13.25	10	23.25	29
	540	Plywood, flush panel front							
	600	4" high, 12" wide	1 Carp	17	Ea.	11.15	9.40	20.55	26
	620	18" wide	"	16	"	14	10	24	30
15-001		**CABINET HARDWARE**							
	100	Catches, minimum	1 Carp	235	Ea.	.34	.68	1.02	1.36
	104	Maximum	"	80	"	1.75	2	3.75	4.82
	200	Door/drawer pulls, handles							
	220	Handles and pulls, projecting, metal, minimum	1 Carp	160	Ea.	.68	1	1.68	2.20
	224	Maximum		68		5.40	2.35	7.75	9.35
	230	Wood, minimum		160		.36	1	1.36	1.85
	234	Maximum		68		1.90	2.35	4.25	5.50
	260	Flush, metal, minimum		160		1.35	1	2.35	2.93
	264	Maximum		68	↓	9.25	2.35	11.60	13.60
	300	Drawer tracks/glides, minimum		48	Pr.	5.30	3.33	8.63	10.65
	304	Maximum		24		14.50	6.65	21.15	26
	400	Cabinet hinges, minimum		160	↓	.90	1	1.90	2.44
	404	Maximum	↓	68	Ea.	5.10	2.35	7.45	9
80-001		**TOPS** Counter and table, plastic laminate							
	100	Counter top, 24" wide, 1-1/2" thick edging							
	150	Plastic laminate edging, minimum	1 Carp	25	L.F.	2.50	6.40	8.90	12
	154	Maximum		22		6.50	7.25	13.75	17.70
	250	Hardwood edging, minimum		20		3.15	8	11.15	15.05
	254	Maximum		16		7	10	17	22
	260	Backsplash, add to above, minimum		36		.75	4.44	5.19	7.25
	264	Maximum	↓	34		1.50	4.71	6.21	8.45
	270	For metal cove, add			↓	.55		.55	.60M
	290	Postformed backsplash, add to above							
	292	Minimum	1 Carp	21	L.F.	.85	7.60	8.45	12
	296	Maximum		19	"	1.70	8.40	10.10	14.05
	350	Well openings (for typewriters etc.)		2.50	Ea.		64	64	93L
	390	Cutouts for sinks, lavatories	↓	12	"		13.35	13.35	19.30L
	450								
	500	Table tops, 1-1/2" thick, plastic laminate edge							
	520	24" x 24", minimum	1 Carp	15	Ea.	7.70	10.65	18.35	24
	522	Maximum		14		18.20	11.45	29.65	37
	550	30" x 30" minimum		10		12	16	28	36
	552	Maximum		9		23	17.80	40.80	51
	570	30" x 72", minimum		6		29.75	27	56.75	71
	572	Maximum	↓	5.50	↓	69	29	98	120
	585								
	600	Table tops, 1-1/2" thick, hardwood edging							
	620	24" x 24", minimum	1 Carp	13.50	Ea.	11.25	11.85	23.10	30
	622	Maximum		12.50		19.70	12.80	32.50	40
	680	30" x 30", minimum		8.80		16.50	18.20	34.70	45
	682	Maximum		7.80		25.25	21	46.25	58
	710	30" x 72", minimum		5.20		36.50	31	67.50	85
	712	Maximum	↓	5	↓	72.60	32	104.60	125

For expanded coverage of these items see *Means' Interior Cost Data 1986*

7.1 Waterproofing

		CREW	DAILY OUTPUT	UNIT	BARE COSTS MAT.	BARE COSTS INST.	BARE COSTS TOTAL	TOTAL INCL O&P
05-001	**BENTONITE** Panels, 4' x 4', for walls, 3/16" thick	1 Rofc	625	S.F.	.65	.24	.89	1.08
010	Under slabs, 5/8" thick	"	900	"	.85	.17	1.02	1.19
030	Granular bentonite, 50 lb. bags (.625 C.F.)			Bag	9.50		9.50	10.45M
040	3/8" thick, troweled on	1 Rofc	475	S.F.	.69	.32	1.01	1.24
050	Drain board, expanded polystyrene, binder encapsulated, 1" thick		600		.49	.25	.74	.92
051	2" thick		600		.98	.25	1.23	1.46
052	3" thick		600		1.43	.25	1.68	1.96
053	4" thick		600		1.90	.25	2.15	2.47
060	Filter fabric, minimum				.08		.08	.08M
061	Maximum				.10		.10	.11M
070	Vapor retarder, 4 mil polyethelene				.05		.05	.05M
10-001	**BITUMINOUS ASPHALT COATING** For foundation							
002								
003	Brushed on, below grade, 1 coat	1 Rofc	665	S.F.	.08	.23	.31	.43
010	2 coat		500		.15	.30	.45	.63
030	Sprayed on, below grade, 1 coat, 25.6 S.F./gal.		830		.12	.18	.30	.41
040	2 coat, 20.5 S.F./gal.		500		.15	.30	.45	.63
050	Asphalt coating, with fibers			Gal.	15.60		15.60	17.15M
060	Troweled on, asphalt with fibers, 1/16" thick	1 Rofc	500	S.F.	.50	.30	.80	1.01
070	1/8" thick		400		1	.38	1.38	1.68
100	1/2" thick		350		4.30	.43	4.73	5.40
300	Asphalt roof coating			Gal.	9.75		9.75	10.70M
320	Asphalt base, fibered aluminum coating				10		10	11M
330	Asphalt primer, 5 gallon				9		9	9.90M
340	Glass fibered roof & patching cement, 5 gallon				12.50		12.50	13.75M
345	Reinforcing glass membrane, 450 S.F./roll			Ea.	72		72	79M
347								
350	Neoprene roof coating, 5 gal., 2 gal./sq.			Gal.	24		24	26M
370	Roof patch & flashing cement, 5 gallon			"	35		35	39M
400	Protective board, asphalt coated, in mastic, 1/4" thick	1 Rofc	450	S.F.	.38	.33	.71	.93
500								
600	Roof resaturant, glass fibered, 3 gal./sq.			Gal.	11		11	12.10M
620	Mineral rubber, 3 gal./sq.			"	10		10	11M
15-001	**BUILDING PAPER** Aluminum and kraft laminated, foil 1 side	1 Carp	19	Sq.	3.20	8.40	11.60	15.70
010	Foil 2 sides		19		6.40	8.40	14.80	19.25
030	Asphalt, two ply, #30, for subfloors		37		6.60	4.32	10.92	13.55
040	Asphalt felt sheathing paper, #15		37		3	4.32	7.32	9.55
060	Polyethylene vapor barrier, standard, .002" thick		37		1.50	4.32	5.82	7.90
070	.004" thick		37		1.80	4.32	6.12	8.25
090	.006" thick		37		2.70	4.32	7.02	9.25
100	.008" thick		37		3.60	4.32	7.92	10.25
120	.010" thick		37		4.27	4.32	8.59	10.95
130	Clear reinforced, fire retardant, .008" thick		37		7.55	4.32	11.87	14.55
135	Cross laminated type, .003" thick		37		6.40	4.32	10.72	13.30
140	.004" thick		37		7.75	4.32	12.07	14.80
150	Red rosin paper, 5 sq. rolls, 4 lbs. per square		37		1.75	4.32	6.07	8.20
160	5 lbs. per square		37		2	4.32	6.32	8.45
180	Reinf. waterproof, .002" polyethylene backing, 1 side		37		6.25	4.32	10.57	13.15
190	2 sides		37		7.50	4.32	11.82	14.50
210	Roof deck vapor barrier, class 1 metal decks	1 Rofc	37		5.30	4.06	9.36	12.05
220	For all other decks	"	37		3.25	4.06	7.31	9.80
240	Waterproofed kraft with sisal or fiberglass fibers, minimum	1 Carp	37		4.80	4.32	9.12	11.55
250	Maximum	"	37		12	4.32	16.32	19.45
20-001	**CAULKING AND SEALANTS**							
002	Acoustical sealant, elastomeric			Gal.	19		19	21M
003	Backer rod, polyethylene, 1/4" diameter	1 Bric	4.60	C.L.F.	3.65	36	39.65	55
005	1/2" diameter		4.60		6.25	36	42.25	58
007	3/4" diameter		4.60		9.35	36	45.35	61
009	1" diameter		4.60		18.35	36	54.35	71

7.1 Waterproofing		CREW	DAILY OUTPUT	UNIT	BARE COSTS			TOTAL INCL O&P
					MAT.	INST.	TOTAL	
20 010	Caulking compound, oil base, bulk							
020	Brilliant white color			Gal.	10.75		10.75	11.80M
030	Aluminum pigment and other colors			"	13.80		13.80	15.20M
050	Bulk, in place, 1/4" x 1/2", 154 L.F./gal.	1 Bric	260	L.F.	.07	.63	.70	.98
060	1/2" x 1/2", 77 L.F./gal.		250		.14	.66	.80	1.09
080	3/4" x 3/4", 34 L.F./gal.		230		.32	.71	1.03	1.37
090	3/4" x 1", 26 L.F./gal.		200		.41	.82	1.23	1.62
100	1" x 1", 19 L.F./gal.	↓	180	↓	.57	.91	1.48	1.92
110	Acrylic based, bulk			Gal.	15.25		15.25	16.75M
120	Cartridges				22		22	24M
140	Butyl based, bulk				12.20		12.20	13.40M
150	Cartridges			↓	17		17	18.70M
170	Bulk, in place 1/4" x 1/2", 154 L.F./gal.	1 Bric	230	L.F.	.08	.71	.79	1.10
180	1/2" x 1/2", 77 L.F./gal.	"	180	"	.16	.91	1.07	1.47
185	Hypalon, bulk			Gal.	34		34	37M
190	Cartridges				45		45	50M
200	Latex based, bulk				13.85		13.85	15.25M
210	Cartridges			↓	23		23	25M
230	Polysulfide compounds, 1 component, bulk				47		47	52M
260	1 or 2 component, in place, 1/4" x 1/4", 308 L.F./gal.	1 Bric	145	L.F.	.15	1.13	1.28	1.78
270	1/2" x 1/4", 154 L.F./gal.		135		.31	1.21	1.52	2.07
290	3/4" x 3/8", 68 L.F./gal.		130		.69	1.26	1.95	2.56
300	1" x 1/2", 38 L.F./gal.	↓	130	↓	1.24	1.26	2.50	3.16
320	Polyurethane, 1 or 2 component, bulk			Gal.	35		35	39M
350	1 or 2 component, in place, 1/4" x 1/4", 308 L.F./gal.	1 Bric	150	L.F.	.11	1.09	1.20	1.68
360	1/2" x 1/4", 154 L.F./gal.		145		.23	1.13	1.36	1.86
380	3/4" x 3/8", 68 L.F./gal.		130		.51	1.26	1.77	2.36
390	1" x 1/2", 38 L.F./gal.	↓	110	↓	.92	1.49	2.41	3.13
410	Silicone rubber, bulk			Gal.	72		72	79M
420	Cartridges			"	75		75	83M
440	Neoprene gaskets, closed cell, adhesive, 1/8" x 2"	1 Bric	240	L.F.	.71	.68	1.39	1.75
450	1/8" x 6"		215		1.35	.76	2.11	2.57
470	1/4" x 6"		200		1.65	.82	2.47	2.98
480	1/2" x 12"	↓	165		5.15	.99	6.14	7.10
500	Polyvinyl chloride gaskets, closed cell, adhesive, standard							
510	1/4" x 2"	1 Bric	230	L.F.	.42	.71	1.13	1.48
530	1/2" x 6"	"	185	"	1.10	.89	1.99	2.47
550	Resin epoxy coating, 2 component, heavy duty			Gal.	30		30	33M
580	Tapes, sealant, P.V.C. foam adhesive, 1/16" x 1/4"			C.L.F.	14.60		14.60	16.05M
590	1/16" x 1/2"				24		24	26M
595	1/16" x 1"				42		42	46M
600	1/8" x 1/2"			↓	28		28	31M
620	Urethane foam, 1 component, handy pack, 0.24 C.F.			Ea.	6.95		6.95	7.65M
630	11.5 C.F. pack			C.F.	11.55		11.55	12.70M
24-001	CEMENTITIOUS WATERPROOFING One coat cement base							
002	1/8" application, sprayed on	G-2	3,000	S.F.	.93	.18	1.11	1.26
25-001	CEMENT PARGING 2 coats, 1/2" thick, regular P.C.	D-1	864		.11	.34	.45	.60
010	(89) Waterproofed portland cement	"	864	↓	.12	.34	.46	.61
30-001	CONTROL JOINTS See division 4.1-20							
35-001	EXPANSION JOINTS See division 3.1-10; 5.8-05 & 7.6-15							
40-001	ELASTOMERIC WATERPROOFING EPDM, plain, 1/32" thick	2 Rofc	580	S.F.	.41	.52	.93	1.24
010	1/16" thick		570		.53	.53	1.06	1.39
030	Nylon reinforced sheets, 1/32" thick		580		.48	.52	1	1.32
040	1/16" thick	↓	570		.58	.53	1.11	1.45
060	Vulcanizing splicing tape for above, 2" wide			C.L.F.	64		64	70M
070	4" wide			"	100		100	110M
090	Adhesive bonding, 60 S.F. per gal.			Gal.	14.50		14.50	15.95M
100	Splicing, 75 S.F. per gal.			"	18.95		18.95	21M

7.1 Waterproofing

			CREW	DAILY OUTPUT	UNIT	BARE COSTS MAT.	BARE COSTS INST.	BARE COSTS TOTAL	TOTAL INCL O&P
40	120	Neoprene sheets, plain, 1/32" thick	2 Rofc	580	S.F.	.58	.52	1.10	1.43
	130	1/16" thick		570		1.10	.53	1.63	2.02
	150	Nylon reinforced, 1/32" thick		580		.61	.52	1.13	1.46
	160	1/16" thick		570		1.13	.53	1.66	2.05
	180	1/8" thick	↓	500	↓	2.15	.60	2.75	3.29
	190	Adhesive, splicing, 150 S.F. per gal. per coat			Gal.	14.75		14.75	16.20M
	210	Fiberglass reinforced, fluid applied, 1/8" thick	2 Rofc	500	S.F.	.91	.60	1.51	1.92
	220	Polyethylene and rubberized asphalt sheets, 1/8" thick		550		.61	.55	1.16	1.51
	240	Polyvinyl chloride sheets, plain, 10 mils thick		580		.13	.52	.65	.94
	250	20 mils thick		570		.20	.53	.73	1.03
	270	30 mils thick		560		.26	.54	.80	1.11
	280	Asbestos back PVC, 45 mils thick	↓	550	↓	.62	.55	1.17	1.52
	300	Adhesives, trowel grade, 40-100 S.F. per gal.			Gal.	8.90		8.90	9.80M
	310	Brush grade, 100-250 S.F. per gal.			"	8.40		8.40	9.25M
	330	Tar extended polyurethane, fluid applied, 50 mils thick	2 Rofc	665	S.F.	1	.45	1.45	1.79
	360	Vinyl plastic, sprayed on, 25 to 40 mils thick	"	475	"	.82	.63	1.45	1.87
45-001		FLASHING See division 7.6-25							
50-001		MEMBRANE WATERPROOFING On slabs, 1 ply, felt	G-1	3,000	S.F.	.13	.36	.49	.68
	010	Fabric		2,100		.16	.52	.68	.95
	030	2 ply, felt		2,500		.27	.44	.71	.95
	040	Fabric		1,650		.33	.66	.99	1.35
	060	3 ply, felt		2,100		.41	.52	.93	1.22
	070	Fabric	↓	1,550		.50	.70	1.20	1.60
	090	For installation on walls, add					15%		
	100	For 1/4" backer board, add	2 Rofc	3,500		.39	.09	.48	.56
	105	For protector board, 3/8" thick, add		3,500		.72	.09	.81	.92
	106	1/2" thick, add		3,500	↓	.88	.09	.97	1.10
	107	Fiberglass fabric, black, 20/10 mesh		116	Sq.	11.15	2.59	13.74	16.25
	108	White, 20/10 mesh		116	"	14.55	2.59	17.14	19.95
	110	Fluid neoprene, 4 coats, 50 mil		200	S.F.	2.47	1.50	3.97	5
	120	90 mil		120		4.85	2.51	7.36	9.15
	130	Fluid elastomeric copolymer compound, 32 mils thick	↓	4,400	↓	2.08	.07	2.15	2.39
55-001		METALLIC COATING Iron compound							
	005	Chip concrete, surface preparation only	1 Cefi	125	S.F.	.11	1.23	1.34	1.85
	010	Complete, including chipping, on floors, 5/8" thick		96		.72	1.60	2.32	3.04
	020	1" thick		103		1.22	1.49	2.71	3.44
	040	On walls, 5/8" thick		35		.72	4.39	5.11	6.95
	050	1" thick	↓	32	↓	1.22	4.80	6.02	8.10
65-001		PITCH Coal tar			Ton	500		500	550M
	040	Tar roof cement, 5 gal. lots			Gal.	10.50		10.50	11.55M
66-001		RUBBER COATING Water base liquid, roller applied	2 Rofc	7,000	S.F.	.49	.04	.53	.60
70-001		SILICONE OR STEARATE Sprayed on masonry, 1 coat	1 Rofc	4,000	S.F.	.25	.04	.29	.33
	010	2 coats	"	2,000	"	.42	.08	.50	.58

7.2 Insulation

			CREW	DAILY OUTPUT	UNIT	BARE COSTS MAT.	BARE COSTS INST.	BARE COSTS TOTAL	TOTAL INCL O&P
02-001		BLOWN-IN INSULATION Ceilings, with open access							
	002	Cellulose, 3-1/2" thick, R-11	G-4	3,000	S.F.	.12	.18	.30	.37
	003	5-3/16" thick, R-19		1,900		.18	.28	.46	.58
	005	6-1/2" thick, R-22		1,500		.23	.35	.58	.73
	010	8-11/16" thick, R-26		1,300		.29	.40	.69	.87
	012	10-7/8" thick, R-38	↓	900	↓	.38	.58	.96	1.21

7.2 Insulation

		CREW	DAILY OUTPUT	UNIT	BARE COSTS MAT.	BARE COSTS INST.	BARE COSTS TOTAL	TOTAL INCL O&P
02 100	Fiberglass, 5" thick, R-11	G-4	1,900	S.F.	.12	.28	.40	.51
105	6" thick, R-13		1,500		.15	.35	.50	.64
110	8-1/2" thick, R-19		1,100		.19	.48	.67	.86
120	10" thick, R-22		900		.25	.58	.83	1.07
130	12" thick, R-26	↓	750	↓	.32	.70	1.02	1.31
131								
200	Mineral wool, 4" thick, R-11	G-4	2,200	S.F.	.17	.24	.41	.51
205	6" thick, R-13		1,500		.27	.35	.62	.77
210	9" thick, R-19	↓	1,000	↓	.46	.53	.99	1.22
249								
250	Wall installation, incl. drilling & patching from outside, two 1" diam.							
251	holes @ 16" O.C., top & mid-point of wall							
270	For masonry	G-4	415	S.F.	.01	1.27	1.28	1.74
280	For wood siding		840		.01	.63	.64	.86
290	For stucco/plaster	↓	665	↓	.01	.79	.80	1.09
05-001	**BREATHER VENTS** See division 7.2-77							
10-001	(51) **CEMENTITIOUS INSULATION** See division 3.5-25							
15-001	**CLIPS** To fasten insulation (.5 per S.F.), galv., 1" or 2" insulation			C	13		13	14.30M
010	6" insulation				21		21	23M
030	Screws with plates, galv., for 1" or 2" insulation				23		23	25M
050	For 6" insulation			↓	135		135	150M
20-001	**HEATING INSULATION** See division 15.5-65							
25-001	**MASONRY INSULATION** Vermiculite or perlite, poured							
010	In cores of concrete block, 4" thick wall	D-1	4,800	S.F.	.40	.06	.46	.53
020	6" thick wall		3,000		.60	.10	.70	.80
030	8" thick wall		2,400		.85	.12	.97	1.11
040	10" thick wall		1,850		1.05	.16	1.21	1.38
050	12" thick wall	↓	1,200	↓	1.25	.24	1.49	1.72
055	For sand fill, deduct from above				70%			
060	Poured cavity wall, vermiculite or perlite, water repellant	D-1	250	C.F.	1.33	1.17	2.50	3.13
070	Foamed in place, urethane in 2-5/8" cavity	G-2	1,035	S.F.	.15	.51	.66	.86
080	For each 1" added thickness, add	"	2,372		.05	.22	.27	.36
30-001	**PERIMETER INSULATION** Asphalt impregnated cork, 1/2" thick, R1.12	1 Carp	690		1	.23	1.23	1.44
010	1" thick, R2.24		680		2.25	.24	2.49	2.82
060	Polystyrene, molded bead board, 1" thick, R4		680		.15	.24	.39	.51
070	2" thick, R8		675	↓	.30	.24	.54	.67
35-001	**POURED INSULATION** Cellulose fiber, R3.8 per inch		200	C.F.	.41	.80	1.21	1.61
004	Ceramic type (perlite), R3.2 per inch		200		1.39	.80	2.19	2.69
008	Fiberglass wool, R4 per inch		200		.27	.80	1.07	1.46
010	Mineral wool, R3 per inch		200		.73	.80	1.53	1.96
030	Polystyrene, R4 per inch		200		1.35	.80	2.15	2.64
040	Vermiculite or perlite, R2.7 per inch		200		1.39	.80	2.19	2.69
070	Wood fiber, R3.85 per inch	↓	200	↓	.44	.80	1.24	1.64
40-001	**REFLECTIVE** Aluminum foil on 40 lb. kraft, foil 1 side, R9	1 Carp	19	C.S.F.	2.65	8.40	11.05	15.10
010	Multilayered with air spaces, 2 ply, R14		19		13.10	8.40	21.50	27
050	3 ply, R17		15		16.70	10.65	27.35	34
060	5 ply, R22	↓	15	↓	25	10.65	35.65	43
45-001	**REFRIGERATION INSULATION** See division 13.1-57							
47-001	**ROOF DECK FORM BOARD** See division 3.5-15							
50-001	**ROOF DECK INSULATION**							
003	Fiberboard, mineral, 1" thick, R2.78	1 Rofc	800	S.F.	.25	.19	.44	.56
008	1-1/2" thick, R4		800		.38	.19	.57	.71
010	2" thick, R5.26	↓	800	↓	.52	.19	.71	.86

7.2 Insulation

			CREW	DAILY OUTPUT	UNIT	BARE COSTS MAT.	BARE COSTS INST.	BARE COSTS TOTAL	TOTAL INCL O&P
50	030	Fiberglass, in 3' x 4' or 4' x 8' sheets							
	040	15/16" thick, R3.3	1 Rofc	1,000	S.F.	.38	.15	.53	.65
	046	1-1/16" thick, R3.8		1,000		.43	.15	.58	.70
	060	1-5/16" thick, R5.3		1,000		.55	.15	.70	.84
	065	1-5/8" thick, R5.7		1,000		.65	.15	.80	.95
	070	1-7/8" thick, R7.7		1,000		.68	.15	.83	.98
	080	2-1/4" thick, R8	↓	800	↓	.69	.19	.88	1.05
	090	Fiberglass and urethane composite, 3' x 4' sheets							
	100	1-11/16" thick, R11.1	1 Rofc	1,000	S.F.	.59	.15	.74	.88
	120	2" thick, R14.3		800		.69	.19	.88	1.05
	130	2-5/8" thick, R18.2	↓	800	↓	.86	.19	1.05	1.23
	150	Foamglass, 2' x 4' sheets, rectangular							
	151	1-1/2" thick R 3.95	1 Rofc	800	S.F.	1.50	.19	1.69	1.94
	152	2" thick R 5.26		800		1.90	.19	2.09	2.38
	153	3" thick R7.89		700		2.25	.21	2.46	2.80
	154	4" thick R 10.53		700	↓	4.10	.21	4.31	4.84
	160	Tapered 1/16", 1/8" or 1/4" per foot	↓	600	B.F.	1.02	.25	1.27	1.51
	165	Perlite, 2' x 4' sheets							
	165	3/4" thick, R2.08	1 Rofc	800	S.F.	.21	.19	.40	.52
	166	1" thick, R2.78		800		.28	.19	.47	.60
	167	1-1/2" thick, R4.17		800		.42	.19	.61	.75
	168	2" thick, R5.26	↓	700	↓	.56	.21	.77	.94
	170	Perlite/urethane composite							
	171	1-1/4" thick, R5.88	1 Rofc	1,000	S.F.	.63	.15	.78	.92
	172	1-1/2" thick, R7.2		1,000		.66	.15	.81	.96
	173	1-3/4" thick, R10		1,000		.68	.15	.83	.98
	174	2" thick, R12.5		800		.71	.19	.90	1.07
	175	2-1/2" thick, R14.3		750		.82	.20	1.02	1.21
	176	3" thick, R20	↓	700	↓	.96	.21	1.17	1.38
	180	Phenolic foam, 2' x 4' sheets							
	181	1-3/16" thick, R10	1 Rofc	1,000	S.F.	.69	.15	.84	.99
	182	1-3/8" thick, R11.1		1,000		.80	.15	.95	1.11
	183	1-3/4" thick, R14.3		1,000		1.02	.15	1.17	1.35
	184	2" thick, R16.7		800		1.15	.19	1.34	1.55
	185	2-1/2" thick, R20		800		1.45	.19	1.64	1.88
	186	3" thick, R25	↓	800	↓	1.75	.19	1.94	2.21
	190	Polystyrene							
	191	Extruded, 2.3#/C.F., 1" thick, R5.26	1 Rofc	1,500	S.F.	.29	.10	.39	.47
	192	2" thick, R10		1,250		.57	.12	.69	.81
	193	3" thick, R15		1,000		.86	.15	1.01	1.18
	201	Expanded bead board, 1" thick, R3.57		1,500		.15	.10	.25	.32
	210	2" thick, R7.14	↓	1,250	↓	.30	.12	.42	.51
	220	Urethane, felt both sides							
	221	1" thick, R6.7	1 Rofc	1,000	S.F.	.45	.15	.60	.73
	222	1-1/2" thick, R11.11		1,000		.56	.15	.71	.85
	223	2" thick, R14.3		800		.61	.19	.80	.96
	224	2-1/2" thick, R20		800		.75	.19	.94	1.11
	225	3" thick, R25	↓	800	↓	.88	.19	1.07	1.26
	230	Urethane and gypsum board composite							
	231	1-5/8" thick, R7.7	1 Rofc	1,000	S.F.	.75	.15	.90	1.06
	232	2" thick, R10		800		.97	.19	1.16	1.35
	233	2-1/2" thick, R14.3		800		1.08	.19	1.27	1.48
	234	3" thick, R18.2	↓	800	↓	1.13	.19	1.32	1.53
	282								
60-001		**SANDWICH PANELS** See division 7.4-43							
65-001		**SHEATHING** See division 6.1-70							
70-001		**SPRAYED** Fibrous or cementitious, finished surface, wall, 1" thick, R3.7	G-2	2,050	S.F.	.09	.26	.35	.45
	010	Attic, 5.2" thick, R19	"	1,550	"	.25	.34	.59	.74

7.2 Insulation

			Crew	Daily Output	Unit	Bare Costs Mat.	Bare Costs Inst.	Bare Costs Total	Total Incl O&P
70	030	Foam type on roofs, incl. preparation							
	060	Urethane, 3 lb./C.F., 1" thick, R7.7	G-2	770	S.F.	.05	.68	.73	.99
	070	2" thick, R15.4	"	475	"	.15	1.11	1.26	1.67
75-001		**VAPOR BARRIER** See division 7.1-15							
77-001		**VENTS, ONE-WAY** For insulated decks, 1 per M.S.F., plastic, minimum	1 Rofc	40	Ea.	4.75	3.76	8.51	11
	010	Maximum		20		52	7.50	59.50	69
	030	Aluminum		30		12.50	5	17.50	21
	050	Copper	↓	28		14.70	5.35	20.05	24
	080	Fiber board baffles, 12" wide for 16" O.C. rafter spacing	1 Carp	90		.43	1.78	2.21	3.05
	090	For 24" O.C. rafter spacing	"	110	↓	.64	1.45	2.09	2.81
80-001		**WALL INSULATION, RIGID**							
	004	Fiberglass, 1.5#/C.F., unfaced, 1" thick, R4.1	1 Carp	1,000	S.F.	.18	.16	.34	.43
	006	1-1/2" thick, R6.2		1,000		.35	.16	.51	.62
	008	2" thick, R8.3		1,000		.46	.16	.62	.74
	012	3" thick, R12.4		800		.70	.20	.90	1.06
	037	3#/C.F., unfaced, 1" thick, R4.3		1,000		.50	.16	.66	.78
	039	1-1/2" thick, R6.5		1,000		.75	.16	.91	1.06
	040	2" thick, R8.7		890		1.01	.18	1.19	1.37
	042	2-1/2" thick, R10.9		800		1.26	.20	1.46	1.68
	044	3" thick, R13		800		1.50	.20	1.70	1.94
	052	Foil faced, 1" thick, R4.3		1,000		.94	.16	1.10	1.27
	054	1-1/2" thick, R6.5		1,000		1.18	.16	1.34	1.53
	056	2" thick, R8.7		890		1.43	.18	1.61	1.83
	058	2-1/2" thick, R10.9		800		1.68	.20	1.88	2.14
	060	3" thick, R13		800		2.73	.20	2.93	3.29
	067	6#/C.F., unfaced, 1" thick, R4.3		1,000		.91	.16	1.07	1.23
	069	1-1/2" thick, R6.5		890		1.37	.18	1.55	1.77
	070	2" thick, R8.7		800		1.82	.20	2.02	2.29
	072	2-1/2" thick, R10.9		800		2.28	.20	2.48	2.80
	074	3" thick, R13		730		2.75	.22	2.97	3.34
	082	Foil faced, 1" thick, R4.3		1,000		1.32	.16	1.48	1.68
	084	1-1/2" thick, R6.5		890		1.76	.18	1.94	2.20
	085	2" thick, R8.7		800		2.21	.20	2.41	2.72
	088	2-1/2" thick, R10.9		800		2.63	.20	2.83	3.18
	090	3" thick, R13	↓	730	↓	3.10	.22	3.32	3.73
	100								
	150	Foamglass, 1-1/2" thick, R2.64	1 Carp	800	S.F.	1.42	.20	1.62	1.85
	155	2" thick, R5.26		730		1.98	.22	2.20	2.50
	170	Perlite, 1" thick, R2.77		800		.28	.20	.48	.60
	175	2" thick, R5.55		730		.56	.22	.78	.93
	190	Polystyrene, extruded blue, 2.2#/C.F., 3/4" thick, R4		800		.36	.20	.56	.69
	194	1-1/2" thick, R8.1		730		.59	.22	.81	.97
	196	2" thick, R10.8		730		.78	.22	1	1.18
	210	Molded bead board, white, 1" thick, R3.85		800		.15	.20	.35	.45
	212	1-1/2" thick, R5.6		730		.23	.22	.45	.57
	214	2" thick, R7.7		730		.31	.22	.53	.66
	235	Sheathing, insulating foil faced fiberboard, 3/8" thick	↓	670	↓	.20	.24	.44	.57
	245								
	251	Urethane, no paper backing, 1/2" thick, R2.9	1 Carp	800	S.F.	.22	.20	.42	.53
	252	1" thick, R5.8		800		.42	.20	.62	.75
	254	1-1/2" thick, R8.7		730		.63	.22	.85	1.01
	256	2" thick, R11.7		730		.84	.22	1.06	1.24
	271	Fire resistant, 1/2" thick, R2.9		800		.27	.20	.47	.59
	272	1" thick, R5.8		800		.57	.20	.77	.92
	274	1-1/2" thick, R8.7		730		.78	.22	1	1.18
	276	2" thick, R11.7	↓	730	↓	1.04	.22	1.26	1.46
85-001		**WALL OR CEILING INSULATION, NON-RIGID**							
	004	Fiberglass, kraft faced, batts or blankets							

7.2 Insulation

		CREW	DAILY OUTPUT	UNIT	BARE COSTS MAT.	BARE COSTS INST.	BARE COSTS TOTAL	TOTAL INCL O&P
006	3-1/2" thick, R11, 11" wide	1 Carp	1,150	S.F.	.20	.14	.34	.42
008	15" wide		1,600		.20	.10	.30	.36
010	23" wide		1,600		.22	.10	.32	.39
014	6" thick, R19, 11" wide		1,000		.32	.16	.48	.58
016	15" wide		1,350		.32	.12	.44	.52
018	23" wide		1,600		.32	.10	.42	.50
020	9" thick, R30, 15" wide		1,150		.50	.14	.64	.75
022	23" wide		1,350		.50	.12	.62	.72
024	12" thick, R38, 15" wide		1,000		.67	.16	.83	.97
026	23" wide		1,350		.67	.12	.79	.91
035								
040	Fiberglass, foil faced, batts or blankets							
042	3-1/2" thick, R11, 15" wide	1 Carp	1,600	S.F.	.22	.10	.32	.39
044	23" wide		1,600		.22	.10	.32	.39
046	6" thick, R19, 15" wide		1,350		.34	.12	.46	.55
048	23" wide		1,600		.34	.10	.44	.52
050	9" thick, R30, 15" wide		1,150		.53	.14	.67	.78
055	23" wide		1,350		.53	.12	.65	.75
070								
080	Fiberglass, unfaced, batts or blankets							
082	3-1/2" thick, R11, 15" wide	1 Carp	1,350	S.F.	.18	.12	.30	.37
083	23" wide		1,600		.18	.10	.28	.34
086	6" thick, R19, 15" wide		1,150		.30	.14	.44	.53
088	23" wide		1,350		.30	.12	.42	.50
090	9" thick, R30, 15" wide		1,000		.48	.16	.64	.76
092	23" wide		1,150		.48	.14	.62	.73
094	12" thick, R38, 15" wide		1,000		.63	.16	.79	.92
096	23" wide		1,150		.63	.14	.77	.89
130	Mineral fiber batts, kraft faced							
132	3-1/2" thick, R13	1 Carp	1,600	S.F.	.50	.10	.60	.69
134	6" thick, R19		1,600		.60	.10	.70	.80
138	10" thick, R30		1,350		.95	.12	1.07	1.22
190	For foil backing 2 sides, add				.06		.06	.06M

7.3 Shingles

		CREW	DAILY OUTPUT	UNIT	BARE COSTS MAT.	BARE COSTS INST.	BARE COSTS TOTAL	TOTAL INCL O&P
05-001	ALUMINUM Shingles, mill finish, .020" thick	1 Carp	2.50	Sq.	115	64	179	220
010	.030" thick	"	2.50		146	64	210	255
030	For colors, anodized finish, add				32		32	35M
040	For bonderized finish, add				64		64	70M
060	Ridge cap, .020" thick	1 Carp	170	L.F.	3.98	.94	4.92	5.75
070	.030" thick		170		5.70	.94	6.64	7.65
090	Valley section for above, .020" thick		170		1.46	.94	2.40	2.97
100	.030" thick		170		1.57	.94	2.51	3.09
120	For 1" factory applied polystyrene insulation, add			Sq.	22.50		22.50	25M
150	Shakes, corrugated, .019" thick, fluropon coated, 46 lb. per sq.	1 Carp	2.30		183	70	253	300
07-001	ALUMINUM Tiles, .019" thick, mission tile		2.50		226	64	290	340
020	Spanish tiles		3		180	53	233	275
10-001	ASBESTOS Mineral fiber strip shingles, 14" x 30", 325 lb. per square		4		105	40	145	175
010	12" x 24", 167 lb. per square		3.50		85	46	131	160
011	Starters, 8" x 30", 255 lb. per 100 L.F.		3		62	53	115	145
012	Hip & ridge shingles, 4-3/4" x 14", 380 lbs. per 100 L.F.		1	C.L.F.	300	160	460	560
020	Shakes, 9.35" x 16", 500 lb. per square (siding)		2.20	Sq.	155	73	228	275
030	Hip & ridge shingles, 5-3/8" x 14"		1	C.L.F.	300	160	460	560

7.3 Shingles

			CREW	DAILY OUTPUT	UNIT	BARE COSTS MAT.	BARE COSTS INST.	BARE COSTS TOTAL	TOTAL INCL O&P
10	040	Hexagonal shape, 16" x 16"	1 Carp	3	Sq.	115	53	168	205
	050	Square, 16" x 16"	"	3	"	105	53	158	195
15-001		**ASPHALT SHINGLES**							
	010	Standard strip shingles							
	015	Inorganic, class A, 210-235 lb./square, 3 bundles/square	1 Rofc	5.50	Sq.	33	27	60	78
	020	Organic, class C, 235-240 lb./square, 3 bundles/square	"	5	"	33	30	63	82
	025	Standard, laminated multi-layered shingles							
	030	Class A, 240-260 lb./square, 3 bundles/square	1 Rofc	4.50	Sq.	54	33	87	110
	035	Class C, 260-300 lb./square, 4 bundles/square	"	4	"	50	38	88	115
	040	Premium, laminated multi-layered shingles							
	045	Class A, 260-300 lb./square, 4 bundles/square	1 Rofc	3.50	Sq.	84	43	127	160
	050	Class C, 300-385 lb./square, 5 bundles/square	"	3	"	80	50	130	165
	070	Hip and ridge roll	↓	400	L.F.	.55	.38	.93	1.18
	090	Ridge shingles	↓	330	"	.65	.46	1.11	1.41
	100	For steep roofs, add			S.F.		50%		
20-001		**CLAY TILE** 8-1/4" x 11" exposure, colors, 730 lb. per sq.							
	020	Lanai tile or Classic tile	1 Rots	1.65	Sq.	295	92	387	465
	030	Americana, most colors		1.65		295	92	387	465
	035	Green, gray or brown		1.65		295	92	387	465
	040	Blue		1.65		785	92	877	1,000
	060	Spanish tile, 900 lb. per sq., red		1.80		250	84	334	405
	080	Buff, green, gray, brown		1.80		430	84	514	600
	090	Blue		1.80		805	84	889	1,025
	110	Mission tile, 1220 lb. per sq., machine scored finish, red		1.15		490	130	620	740
	170	French tile, 935 lb. per sq., smooth finish, red		1.35		450	110	560	665
	175	Blue or green		1.35		920	110	1,030	1,175
	180	Norman tile, 1600 lb. per sq.		1		1,050	150	1,200	1,375
	220	Williamsburg tile, 950 lb. per sq., aged cedar		1.35		305	110	415	505
	225	Gray or green	↓	1.35	↓	305	110	415	505
25-001		**CONCRETE TILE** Including installation of accessories							
	002	Corrugated, 13" x 16-1/2", 90 per sq., 950 lb. per sq.							
	005	Earthtone colors, nailed to wood deck	1 Rots	1.35	Sq.	85	110	195	265
	015	Custom blues		1.35		270	110	380	470
	020	Custom greens	↓	1.35	↓	105	110	215	285
	030								
	050	Shakes, 13" x 16-1/2", 90 per sq., 950 lb. per sq.							
	060	All colors, nailed to wood deck	1 Rots	1.50	Sq.	75	100	175	235
	150	Accessory pieces, ridge & hip, 10" x 16-1/2", 8 lbs. each			Ea.	1.21		1.21	1.33M
	170	Rake, 6-1/2" x 16-3/4", 9 lbs. each				1.22		1.22	1.34M
	180	Mansard hip, 10" x 16-1/2", 9.2 lbs. each				1.58		1.58	1.73M
	190	Hip starter, 10" x 16-1/2", 10.5 lbs. each				3.12		3.12	3.43M
	200	3 or 4 way apex, 10" each side, 11.5 lbs. each			↓	5.70		5.70	6.25M
30-001		**PORCELAIN ENAMEL** 22 ga., 10" x 10", 225 lb. per sq., minimum	1 Rots	1.30	Sq.	435	115	550	655
	010	Maximum	"	1	"	540	150	690	825
35-001		**SLATE** Including felt underlay & nails, Buckingham, Virginia, black							
	010	3/16" thick	1 Rots	1.75	Sq.	600	87	687	795
	020	1/4" thick	"	1.75	"	600	87	687	795
	021								
	090	Pennsylvania black, Bangor, #1 clear	1 Rots	1.75	Sq.	600	87	687	795
	120	Vermont, unfading colors, green, mottled green		1.75		590	87	677	780
	130	Semi-weathering green & gray		1.75		600	87	687	795
	140	Purple		1.75		640	87	727	835
	150	Black or gray		1.75		640	87	727	835
	160	Red		1.75		1,500	87	1,587	1,775
	170	Variegated purple	↓	1.75	↓	770	87	857	980
40-001		**STEEL** Shingles, galvanized, 26 gauge	1 Rots	2.20	Sq.	42	69	111	150
	020	24 gauge	"	2.20	"	44	69	113	155

7.3 Shingles

			CREW	DAILY OUTPUT	UNIT	BARE COSTS MAT.	BARE COSTS INST.	BARE COSTS TOTAL	TOTAL INCL O&P
40	030	For colored galvanized shingles, add			Sq.	36		36	40M
	050	For 1" factory applied polystyrene insulation, add				24		24	26M
42-001		STEEL TILE Ceramic chip coated, 170 lb. per sq., minimum	1 Rots	2.50		158	61	219	265
	010	Maximum		2		209	76	285	345
	020	Mineral surfaced zinc coated steel, 26 ga., 150 lb./sq.		3		116	51	167	205
	030	Ridge cap		200	L.F.	1.38	.76	2.14	2.68
45-001		WOOD 16" No. 1 red cedar shingles, 5X, 5" exposure, on roof	1 Carp	2.50	Sq.	98	64	162	200
	020	7-1/2" exposure, on walls		2.05		65	78	143	185
	030	18" No. 1 red cedar perfections, 5-1/2" exposure, on roof		2.75		100	58	158	195
	050	7-1/2" exposure, on walls		2.25		67	71	138	175
	060	Resquared, and rebutted, 5-1/2" exposure, on roof		3		80	53	133	165
	090	7-1/2" exposure, on walls		2.45		53	65	118	155
	100	Add to above for fire retardant shingles, 16" long				85		85	94M
	105	18" long				86		86	95M
	110	Hand-split red cedar shakes, on walls, 24" long, 10" exposure	1 Carp	2.50		77	64	141	175
	120	18" long, 8-1/2" exposure	"	2		63	80	143	185
	170	Add to above for fire retardant shakes, 24" long				93		93	100M
	180	18" long				150		150	165M
	200	White cedar shingles, 16" long, extras, 5" exposure, on roof	1 Carp	2.40		90	67	157	195
	210	7-1/2" exposure, on walls		2		60	80	140	180
	230	For #15 organic felt underlayment on roof, 1 layer, add		64		2.98	2.50	5.48	6.90
	240	2 layers, add		32		5.96	5	10.96	13.80
	250	For plastic-coated steel foil underlayment, class B roofs, add		64		45	2.50	47.50	53
	260	For steep roofs, add to above					50%		
	270	Panelized systems, No.1 cedar shingles on 5/16" CDX plywood							
	280	On walls, 8' strips, 7" or 14" exposure	F-2	700	S.F.	2.10	.48	2.58	3
	290	Matching flush corners	"	400	L.F.	2	.84	2.84	3.41
	350	On roofs, 8' strips, 7" or 14" exposure	1 Carp	3	Sq.	220	53	273	320
	360	Matching lap corners		200	L.F.	1.05	.80	1.85	2.31
	370	Matching rake corners		200		1.35	.80	2.15	2.64
	380	Matching valley sheets		200		5.70	.80	6.50	7.45

7.4 Roofing & Siding

			CREW	DAILY OUTPUT	UNIT	BARE COSTS MAT.	BARE COSTS INST.	BARE COSTS TOTAL	TOTAL INCL O&P
03-001		ALUMINUM ROOFING Corrugated or ribbed, .0175" thick, natural	G-3	1,000	S.F.	.40	.64	1.04	1.37
	030	Painted		1,000		.45	.64	1.09	1.42
	040	Corrugated, .0215" thick, on steel frame, natural finish		1,000		.55	.64	1.19	1.53
	060	Painted		1,000		.60	.64	1.24	1.59
	070	Corrugated, on steel frame, natural, .024" thick		1,000		1.05	.64	1.69	2.08
	090	.032" thick		1,000		1.42	.64	2.06	2.49
	100	Painted, .024" thick		1,000		1.11	.64	1.75	2.15
	120	.032" thick		1,000		1.48	.64	2.12	2.56
	130	V-Beam, on steel frame construction, .032" thick, natural		1,000		1.56	.64	2.20	2.64
	150	Painted		1,000		1.68	.64	2.32	2.78
	160	.040" thick, natural		1,000		1.95	.64	2.59	3.07
	180	Painted		1,000		2.17	.64	2.81	3.32
	190	.050" thick, natural		1,000		2.29	.64	2.93	3.45
	210	Painted		1,000		2.40	.64	3.04	3.57
	220	For roofing on wood frame, deduct		4,600		.05	.14	.19	.26
	240	Ridge cap, .032" thick, natural		800	L.F.	3.62	.81	4.43	5.15
06-001		ALUMINUM SIDING .019" thick, on steel construction, natural		775	S.F.	.55	.83	1.38	1.80
	010	Painted		775		.60	.83	1.43	1.86
	040	Farm type, .021" thick on steel frame, natural		775		.65	.83	1.48	1.91
	060	Painted		775		.71	.83	1.54	1.98

7.4 Roofing & Siding

			CREW	DAILY OUTPUT	UNIT	BARE COSTS MAT.	BARE COSTS INST.	BARE COSTS TOTAL	TOTAL INCL O&P
06	070	Industrial type, corrugated, .024" thick, on steel, natural	G-3	775	S.F.	1.11	.83	1.94	2.42
	090	Painted		775		1.18	.83	2.01	2.50
	100	.032" thick, natural		775		1.43	.83	2.26	2.77
	120	Painted		775		1.47	.83	2.30	2.82
	130	V-Beam, on steel frame, .032" thick, natural		775		1.55	.83	2.38	2.90
	150	Painted		775		1.64	.83	2.47	3
	160	.040" thick, natural		775		1.97	.83	2.80	3.37
	180	Painted		775		2.10	.83	2.93	3.51
	190	.050" thick, natural		775		2.35	.83	3.18	3.78
	210	Painted		775		2.45	.83	3.28	3.89
	220	Ribbed, 4" profile, on steel frame, .032" thick, natural		775		1.45	.83	2.28	2.79
	240	Painted		775		1.58	.83	2.41	2.94
	250	.040" thick, natural		775		1.95	.83	2.78	3.34
	270	Painted		775		2.13	.83	2.96	3.54
	275	.050" thick, natural		775		2.44	.83	3.27	3.88
	276	Painted		775		2.59	.83	3.42	4.05
	280	For 8" profile instead of 4" profile, natural, deduct				.16		.16	.17M
	300	Painted, deduct				.11		.11	.12M
	310	Perforated, corrugated, 14% openings, .024" thick, natural	G-3	775		1.25	.83	2.08	2.57
	315	Painted		775		1.30	.83	2.13	2.63
	330	For siding on wood frame, deduct from above		2,800		.02	.23	.25	.35
	335								
	340	Screw fasteners, aluminum (see also division 5.8-30)			M	118		118	130M
	360	Stitch screws, 3/4" long			"	87		87	96M
	363	Flashing, sidewall, .032" thick	G-3	800	L.F.	2.90	.81	3.71	4.35
	365	End wall, .040" thick		800		2.95	.81	3.76	4.41
	367	Closure strips, corrugated, .032" thick		800		.25	.81	1.06	1.44
	368	Ribbed, 4" or 8", .032" thick		800		.22	.81	1.03	1.40
	369	V-beam, .040" thick		800		.30	.81	1.11	1.49
	370								
	380	Horizontal, colored clapboard, 8" or 10" wide, plain	1 Carp	255	S.F.	1.15	.63	1.78	2.17
	390	Insulated		255		1.35	.63	1.98	2.39
	400	Vertical board & batten, colored, non-insulated		255		1.20	.63	1.83	2.23
	420	For simulated wood design, add				.03		.03	.03M
	430	Corners for above, outside	1 Carp	255	V.L.F.	1.60	.63	2.23	2.67
	450	Inside corners	"	255	"	.75	.63	1.38	1.73
	460	Sandwich panels, 1" insulation, single story	G-3	395	S.F.	4.80	1.63	6.43	7.65
	490	Multi-story	"	345		6.30	1.87	8.17	9.60
	510	For baked enamel finish 1 side, add				.35		.35	.38M
	520	See also Metal Facing Panels, division 7.4-42							
12-001		ASPHALT Coated felt, #30, 2 sq. per roll, not mopped	1 Rofc	58	Sq.	10.40	2.59	12.99	15.40
	020	#15, 4 sq. per roll, plain or perforated, not mopped		58		9.50	2.59	12.09	14.40
	030	Roll roofing, smooth, #55		15		11.70	10.05	21.75	28
	050	#90		15		14.50	10.05	24.55	31
	052	Mineralized		15		15.90	10.05	25.95	33
	054	D.C. (Double coverage), 19" selvage edge		10		30	15.05	45.05	56
	058	Adhesive (lap cement)			Gal.	3.70		3.70	4.07M
	060	Steep, flat or dead level asphalt, 10 ton lots, bulk			Ton	265		265	290M
	080	Packaged			"	325		325	360M
13-001		ASPHALT PANELS Corrugated, 1/8" thick, smooth surface	1 Rofc	335	S.F.	1.29	.45	1.74	2.11
	010	Granulated surface		335	"	1.60	.45	2.05	2.45
	020	Ridge pieces, 1/8" thick, smooth		400	L.F.	4.33	.38	4.71	5.35
	030	Granulated		400	"	5.05	.38	5.43	6.15
	040	Closure strips, 2" thick			Ea.	1.04		1.04	1.14M
	050	PVC skylight sheets, 28" x 96"	1 Rofc	12.50	"	16.75	12.05	28.80	37
15-001		BUILT-UP ROOFING							
	002								
	012	Asphalt flood coat with gravel/slag surfacing, not including							
	014	Insulation, flashing or wood nailers							

7.4 Roofing & Siding

		CREW	DAILY OUTPUT	UNIT	BARE COSTS MAT.	BARE COSTS INST.	BARE COSTS TOTAL	TOTAL INCL O&P
15 020	Asbestos base sheet, 3 plies #15 asbestos felt, mopped	G-1	22	Sq.	42	50	92	120
035	On nailable decks		21		56	52	108	140
050	4 plies #15 asbestos felt, mopped		20		39	54	93	125
055	On nailable decks	↓	19	↓	48	57	105	140
060								
070	Coated glass base sheet, 2 plies glass (type IV), mopped	G-1	22	Sq.	42	50	92	120
085	3 plies glass, mopped		20		51	54	105	135
095	On nailable decks		19		56	57	113	145
100	3 plies glass fiber felt (type IV), mopped		22		40	50	90	120
105	On nailable decks		21		43	52	95	125
110	4 plies glass fiber felt (type IV), mopped		20		49	54	103	135
115	On nailable decks		19		53	57	110	145
120	Organic base sheet, 3 plies #15 organic felt, mopped		20		42	54	96	125
125	On nailable decks		19		46	57	103	135
130	4 plies #15 organic felt, mopped	↓	22	↓	39	50	89	115
200	Asphalt flood coat, smooth surface							
220	Asbestos base sheet & 3 plies #15 asbestos felt, mopped	G-1	24	Sq.	41	45	86	115
240	On nailable decks		23		52	47	99	130
260	4 plies #15 asbestos felt, mopped		24		37	45	82	110
270	On nailable decks	↓	23	↓	46	47	93	120
290	Coated glass fiber base sheet, mopped, and 2 plies of							
291	glass fiber felt (type IV)	G-1	25	Sq.	40	44	84	110
310	On nailable decks	"	24	"	44	45	89	115
315								
320	3 plies, mopped	G-1	23	Sq.	49	47	96	125
330	On nailable decks		22		52	50	102	130
350	3 plies glass fiber felt (type IV), mopped		25		38	44	82	105
360	On nailable decks		24		41	45	86	115
380	4 plies glass fiber felt (type IV), mopped		23		47	47	94	120
390	On nailable decks		22		51	50	101	130
400	Organic base sheet & 3 plies #15 organic felt, mopped		24		40	45	85	110
420	On nailable decks		23		43	47	90	120
430	4 plies #15 organic felt, mopped	↓	22	↓	37	50	87	115
440								
450	Coal tar pitch with gravel/slag surfacing							
460	4 plies #15 asbestos felt, mopped	G-1	21	Sq.	82	52	134	165
480	3 plies glass fiber felt (type IV), mopped	"	19	"	70	57	127	160
500	Coated glass fiber base sheet, and 2 plies of							
501	glass fiber felt, type IV, mopped	G-1	19	Sq.	78	57	135	170
530	On nailable decks		18		78	61	139	175
540	On wood decks		18		78	61	139	175
560	4 plies glass fiber felt (type IV), mopped		21		90	52	142	175
580	On nailable decks		20		90	54	144	180
590	On wood decks		20		90	54	144	180
600	4 plies #15 organic felt, mopped	↓	21	↓	79	52	131	165
660	Asphalt mineral surface, roll roofing							
670	1 ply #15 organic felt, 2 plies mineral surfaced							
680	selvage, edge roofing, lap 19", nailed & mopped	G-1	27	Sq.	38	40	78	100
700	3 plies glass fiber felt (type IV), 1 ply mineral surfaced							
710	selvage, edge roofing, lapped 18", mopped	G-1	25	Sq.	58	44	102	130
740	Coated glass fiber base sheet, 2 plies of glass fiber							
750	felt (type IV), 1 ply mineral surfaced selvage							
760	edge roofing, lapped 18", mopped	G-1	25	Sq.	62	44	106	135
770	On nailable decks	"	24	"	67	45	112	140
780	3 plies glass fiber felt (type III), 1 ply mineral surfaced							
790	selvage, edge roofing, lapped 18", mopped	G-1	25	Sq.	56	44	100	125
791	Cold applied, 3-ply system (components listed below)	G-5	50	"		15.75	15.75	23
792	Spunbond poly. fabric, 1.35 oz/S.F., 36" wide, 10.8 Sq./roll			Ea.	60		60	66M
793	49" wide, 14.6 Sq./roll			↓	81		81	89M
794	2.10 oz./S.Y., 36" wide, 10.8 Sq./roll			↓	94		94	105M

7.4 Roofing & Siding

			CREW	DAILY OUTPUT	UNIT	BARE COSTS MAT.	BARE COSTS INST.	BARE COSTS TOTAL	TOTAL INCL O&P
15	795	49" wide, 14.6 Sq./roll			Ea.	129		129	140M
	796	Base & finish coat, 3 gal./Sq., 5 gal./can			Gal.	5		5	5.50M
	797	Coating, ceramic granules, 1/2 Sq./bag			Ea.	9.40		9.40	10.35M
	798	Aluminum, 2 gal./Sq.			Gal.	9		9	9.90M
	799	Emulsion, fibered or non-fibered, 4 gal./Sq.			"	4		4	4.40M
	900	Bond charge, asphalt felt roofs, 10 year bond			Sq.	11.50		11.50	12.65M
18-001		CANTS 4" x 4" treated timber, cut diagonally	1 Rofc	325	L.F.	1.10	.46	1.56	1.92
	010	Foamglass		325		.47	.46	.93	1.22
	030	Mineral or fiber, trapezoidal, 1"x 4" x 48"		325		.24	.46	.70	.97
	040	1-1/2" x 5-5/8" x 48"	↓	325	↓	.30	.46	.76	1.04
21-001		CONCRETE See division 3.3, 3.4, & 3.5							
24-001		CURTAIN WALL See division 8.9							
27-001		ELASTOMERIC ROOFING See also single-ply membranes, division 7.4-52							
	010	For Elastomeric waterproofing, see division 7.1-40							
	030	Hypalon neoprene, fluid applied, 20 mil thick, not-reinforced	G-1	1,135	S.F.	1.35	.96	2.31	2.91
	060	Non-woven polyester, reinforced		960		1.36	1.13	2.49	3.19
	070	5 coat neoprene deck, 60 mil thick, under 10,000 S.F.		325		3.25	3.35	6.60	8.55
	090	Over 10,000 S.F.		625		3.15	1.74	4.89	6.05
	100	Neoprene membrane, 1/16" thick		1,375		1.25	.79	2.04	2.56
	130	Vinyl plastic traffic deck, sprayed, 2 to 4 mils thick		625		1.12	1.74	2.86	3.83
	150	Vinyl and neoprene membrane traffic deck	↓	1,550	↓	1	.70	1.70	2.15
	160	Polyurethane spray-on with 20 mil white hypalon coating applied							
	170	1" thick, R7, minimum	G-2	875	S.F.	1.35	.60	1.95	2.30
	180	Maximim		805		1.95	.65	2.60	3.04
	190	2" thick, R14, minimum		685		1.98	.77	2.75	3.22
	200	Maximum		575		2.95	.91	3.86	4.49
	210	3" thick, R21, minimum		500		2.39	1.05	3.44	4.06
	220	Maximum	↓	440	↓	3.60	1.19	4.79	5.60
30-001		EPOXY PANELS Exposed aggregate in epoxy matrix mounted							
	020	on plywood or wood fiber panels							
	030	Small size aggregate	F-3	445	S.F.	4.25	2.38	6.63	7.95
	050	Large size aggregate	"	445		5.10	2.38	7.48	8.85
	060	For fire rated plywood or wood fiber, add				.68		.68	.74M
	080	For asbestos board panels, add				1.05		1.05	1.15M
	090	For 1-1/2" polyurethane insulating core, add				1.05		1.05	1.15M
	110	Solid polyester panels, up to 5' x 14', 3/4" thick	F-3	335	↓	12.10	3.17	15.27	17.65
	130	Polystyrene panels, aggregate face, 2' x 2' x 2-1/4", R8.4							
	140	Straight panels, minimum	D-1	325	S.F.	4.15	.90	5.05	5.85
	150	Maximum		600		4.85	.49	5.34	6.05
	160	Corners, angles, returns, etc., minimum		225		5.65	1.30	6.95	8.05
	170	Maximum	↓	450		7.20	.65	7.85	8.85
	180	For wood, steel or painted masonry substrate, add			↓	.50		.50	.55M
33-001		FELT Asphalt asbestos, #15, no mopping	1 Rofc	58	Sq.	2.98	2.59	5.57	7.25
	020	#43		58		5.55	2.59	8.14	10.05
	030	Base sheet, #45		58		5.60	2.59	8.19	10.15
	040	#50		58		6.80	2.59	9.39	11.45
	050	Cap, mineral surfaced	↓	58	↓	30	2.59	32.59	37
	051								
	060	Flashing membrane, #65	1 Rofc	16	Sq.	7.80	9.40	17.20	23
	080	Coal tar asbestos, #15, no mopping		58		10.40	2.59	12.99	15.40
	090	Asphalt felt, #15, 4 sq. per roll, no mopping		58		3	2.59	5.59	7.25
	110	#30, 2 sq. per roll		58		6	2.59	8.59	10.55
	120	Double coated, #30		58		6.45	2.59	9.04	11.05
	140	#40		58		7.80	2.59	10.39	12.55
	150	Tarred felt, organic, #15, 4 sq. rolls		58		4.75	2.59	7.34	9.20
	155	#30, 2 sq. roll		58		10.15	2.59	12.74	15.15
	170	Add for mopping above felts, per ply, asphalt, 20 lbs. per sq.		28		2.07	5.35	7.42	10.50
	180	Coal tar mopping, 30 lbs. per sq.	↓	27	↓	6.55	5.55	12.10	15.75

7.4 Roofing & Siding

			CREW	DAILY OUTPUT	UNIT	BARE COSTS MAT.	BARE COSTS INST.	BARE COSTS TOTAL	TOTAL INCL O&P
33	190	Flood coat, with asphalt (60 lbs. per sq.)	2 Rofc	16.30	Sq.	6.20	18.45	24.65	35
	200	With coal tar (75 lbs. per sq.)	"	15.30	"	16.25	19.65	35.90	48
36-001		**FIBERGLASS** Corrugated panels, roofing, 6 oz. per S.F.	G-3	1,000	S.F.	.90	.64	1.54	1.92
	010	8 oz. per S.F.		1,000		1.45	.64	2.09	2.52
	030	Corrugated siding, 4 oz. per S.F.		880		.75	.73	1.48	1.88
	040	5 oz. per S.F.		880		.85	.73	1.58	1.99
	060	8 oz. siding, textured		880		1.35	.73	2.08	2.54
	070	Fire retardant		880		1.40	.73	2.13	2.60
	090	Flat panels, 6 oz. per S.F., clear or colors		880		.96	.73	1.69	2.11
	110	Fire retardant, class A		880		1	.73	1.73	2.16
	130	8 oz. per S.F., clear or colors	↓	880	↓	1.50	.73	2.23	2.71
	150								
	170	Sandwich panels, fiberglass, 1-9/16" thick, to 250 S.F.	G-3	180	S.F.	15	3.58	18.58	22
	180	250 S.F. and up		240		10	2.69	12.69	14.85
	190	As above, but 2-3/4" thick, to 250 S.F.		265		15	2.43	17.43	20
	200	250 S.F. and up	↓	300	↓	10	2.15	12.15	14.10
40-001		**INTEGRATED SIDING** Fabric reinforced synthetic exterior							
	002	finish, on 1" polystyrene insulation board							
	010	Minimum	1 Plas	135	S.F.	1.25	1.18	2.43	3.06
	020	Maximum		90		2	1.77	3.77	4.72
	030	For insulation, 2" polystyrene, add	↓	725		.36	.22	.58	.71
	040	For heavy duty protective fabric, add			↓	1.26		1.26	1.38M
42-001		**METAL FACING PANELS** Field assembled, insulated sandwich							
	010	wall panels, over 5000 S.F.							
	030	16 ga. aluminum exterior, 1-1/2" fiberglass and							
	040	18 ga. galvanized steel interior	G-3	195	S.F.	4.98	3.31	8.29	10.25
	060	18 ga. galvanized steel both sides		195		4	3.31	7.31	9.15
	070	20 ga. stainless steel exterior face		195		5.87	3.31	9.18	11.20
	090	20 ga. protected metal or baked enamel exterior		195		3.60	3.31	6.91	8.70
	100	16 ga. porcelain on aluminum exterior	↓	195	↓	4.65	3.31	7.96	9.90
	120	Factory made sandwich wall panels, acrylic coated, 10 mil alum.							
	130	face on 5/16" plywood core, with foil back	G-3	375	S.F.	2.04	1.72	3.76	4.72
	140	28 ga. porcelain enamel face, insulated cores, hardboard							
	150	stabilizer and galvanized back, 1" thick	G-3	375	S.F.	7.75	1.72	9.47	11
	160	2" thick	"	375	"	9.65	1.72	11.37	13.10
	170								
	180	As above, but hardboard back, 1" thick	G-3	375	S.F.	5.68	1.72	7.40	8.75
	190	2" thick		375		6.60	1.72	8.32	9.75
	210	As above, but porcelain enamel both sides, 1" thick		375		8.70	1.72	10.42	12.05
	220	2" thick		375		9.60	1.72	11.32	13.05
	240	Porcelain enamel both sides, 2' or 4' wide		375		7.40	1.72	9.12	10.60
	250	2" thick, fire rated, 3' wide	↓	375	↓	15.80	1.72	17.52	19.85
	270	Porcelain enamel on steel, 2 faces, laminated to hardboard core							
	280	1/4" thick	G-3	375	S.F.	4.05	1.72	5.77	6.95
	300	Interior, galvanized steel		375		3.48	1.72	5.20	6.30
	310	Porcelain enamel 2 sides, 1/4" thick, mineral fiber core		375		4.75	1.72	6.47	7.70
	330	Porcelain enameled steel on 3/8" plywood		375		3.50	1.72	5.22	6.35
	340	Copper faced (.010" thick), aluminum back, 3/8" thick		375		5.40	1.72	7.12	8.40
	360	Veneer panels, 3/4" thick		375		5.70	1.72	7.42	8.75
	370	Structural panels, 3" thick	↓	330	↓	6.10	1.95	8.05	9.50
43-001		**MINERAL FIBER CEMENT PANELS** Including panels, fasteners,							
	010	accessories, trim & sealant							
	013	Architectural, textured finish, 1/8" thick, minimum	G-3	500	S.F.	4.62	1.29	5.91	6.95
	014	Maximum		500		5.70	1.29	6.99	8.15
	015	1/4" thick, minimum		500		5.95	1.29	7.24	8.40
	020	Maximum		500		7	1.29	8.29	9.55
	025	3/8" thick, minimum		500		7.75	1.29	9.04	10.40
	030	Maximum		500		9.20	1.29	10.49	12
	035	5/8" thick, minimum	↓	300		10	2.15	12.15	14.10
	040	Maximum	↓	300	↓	12.90	2.15	15.05	17.30

7.4 Roofing & Siding		CREW	DAILY OUTPUT	UNIT	BARE COSTS MAT.	BARE COSTS INST.	BARE COSTS TOTAL	TOTAL INCL O&P
43 100	Corrugated, 3/8" thick, as roofing, on steel frame	G-3	750	S.F.	7.55	.86	8.41	9.55
110	On wood frame		895		7.55	.72	8.27	9.35
120	3/8" thick, as siding, one story steel frame		600		7.30	1.07	8.37	9.60
130	One story wood frame		765		7.30	.84	8.14	9.25
200	Flat sheets, 1/8" thick		1,200		3.50	.54	4.04	4.62
210	1/4" thick		1,020		5.20	.63	5.83	6.65
220	3/8" thick		885		6.75	.73	7.48	8.45
230	5/8" thick		795		9.45	.81	10.26	11.55
300	Glasweld, mineral enamel coating, 1/8" thick		600		3.63	1.07	4.70	5.55
310	1/4" thick	↓	322	↓	5.15	2	7.15	8.55
400	Sandwich panel, Glasweld face and back							
410	1" thick, perlite core	G-3	322	S.F.	6.65	2	8.65	10.20
420	Polyurethane core		322		6.65	2	8.65	10.20
450	2" thick, perlite core		322		7.30	2	9.30	10.90
460	Polyurethane core	↓	322	↓	7.25	2	9.25	10.85
45-001	**PROTECTED METAL** Roofing, metallic adhesive,							
010	incombustible. For L.C.L. add 15% to total							
030	Box rib, colored, 24 gauge	G-3	495	S.F.	3.52	1.30	4.82	5.75
040	22 gauge		495		4.15	1.30	5.45	6.45
060	Corrugated, colored, 24 gauge		495		3.14	1.30	4.44	5.35
070	22 gauge		495		3.46	1.30	4.76	5.70
090	Deep rib, 4" deep, colored, 22 gauge		495		4.21	1.30	5.51	6.50
100	Siding		420		4.21	1.53	5.74	6.85
120	Siding, box rib, colored, 24 gauge		420		3.41	1.53	4.94	5.95
130	22 gauge		420		3.57	1.53	5.10	6.15
150	Corrugated, colored, 24 gauge		420		3.41	1.53	4.94	5.95
160	22 gauge	↓	420	↓	3.57	1.53	5.10	6.15
48-001	**SHEET METAL ROOFING** See division 7.6							
51-001	**SHINGLES** See division 7.3							
52-001	**SINGLE-PLY MEMBRANE**							
002								
020	Chlorinated polyethylene(CPE), 40 mils, 0.31 P.S.F.							
030	Partially adhered with mechanical fasteners	G-5	5,100	S.F.	1.47	.15	1.62	1.84
080	Chlorosulfonated polyethylene-hypalon (CSPE), 35 mils, 0.25 PSF							
090	Fully adhered with neoprene latex	G-5	3,500	S.F.	1.28	.23	1.51	1.74
100	45 mils, 0.29 P.S.F.							
110	Loose-laid & ballasted with stone (10 P.S.F.)	G-5	7,000	S.F.	1.44	.11	1.55	1.75
120	Partially adhered with fastening strips		5,100		1.61	.15	1.76	2
130	Plates with adhesive attachment	↓	5,100	↓	1.75	.15	1.90	2.15
350	Ethylene propylene diene monomer (EPDM), 45 mils, 0.28 P.S.F.							
360	Loose-laid & ballasted with stone (10 P.S.F.)	G-5	7,000	S.F.	.58	.11	.69	.80
370	Partially adhered with batten strips		5,100		2	.15	2.15	2.43
380	Fully adhered with adhesive	↓	3,500	↓	.80	.23	1.03	1.21
400	55 mils, 0.40 P.S.F.							
410	Loose-laid & ballasted with stone (10 P.S.F.)	G-5	7,000	S.F.	.64	.11	.75	.87
420	Partially adhered to plates @ 4' O.C. with adhesive		5,100		2.05	.15	2.20	2.48
430	Fully adhered with adhesive	↓	3,500	↓	.85	.23	1.08	1.27
450	60 mils, 0.35 P.S.F.							
460	Loose-laid & ballasted with stone (10 P.S.F.)	G-5	7,000	S.F.	.68	.11	.79	.91
470	Partially adhered with bar anchors		5,100		2.09	.15	2.24	2.53
480	Fully adhered with adhesive	↓	3,500	↓	.90	.23	1.13	1.32
485	Vulcanizing tape for membrane, 2" x 50' roll			Ea.	34		34	37M
490	Batten strips, 10' sections				2.70		2.70	2.97M
491	Vulcanizing tape for batten strips, 4" x 50' roll			↓	76		76	84M
493	Plate anchors			M	75		75	83M
497	Adhesive for fully adhered systems, 60 S.F./gal.			Gal.	16		16	17.60M
498								

7.4 Roofing & Siding

		Crew	Daily Output	Unit	Bare Costs Mat.	Bare Costs Inst.	Bare Costs Total	Total Incl O&P
52 500	Modified bitumen							
530	120 mils, 0.92 P.S.F., fully adhered with solvent	G-5	2,800	S.F.	.80	.28	1.08	1.30
540	150 mils, 0.82 P.S.F.							
550	Loose-laid & ballasted with gravel (4 P.S.F.)	G-5	3,200	S.F.	.80	.25	1.05	1.24
560	Partially adhered with torch welding		2,500		.70	.32	1.02	1.24
570	Fully adhered with torch welding		2,000		.70	.39	1.09	1.35
580	Hot asphalt attachment	↓	2,000	↓	.75	.39	1.14	1.41
581								
600	160 mils, 0.78 to 1.2 P.S.F., with asphalt emulsion coating							
610	Loose-laid & ballasted with stone/gravel (10 P.S.F.)	G-5	3,200	S.F.	.85	.25	1.10	1.30
620	Partially adhered with torch welding		2,500		.75	.32	1.07	1.29
630	Fully adhered with torch welding		2,000		.75	.39	1.14	1.41
640	Hot asphalt attachment	↓	2,000	↓	.80	.39	1.19	1.46
641								
680	Neoprene, 60 mils, 0.45 P.S.F.							
700	Partially adhered with mechanical fasteners	G-5	5,100	S.F.	1.75	.15	1.90	2.15
710	Fully adhered with contact adhesive	"	3,500		1.47	.23	1.70	1.95
711	Uncured neoprene, 60 mils, for flashing	1 Rofc	600	↓	1.04	.25	1.29	1.53
750	Polyisobutylene (PIB), 100 mils, 0.57 P.S.F.							
760	Loose-laid & ballasted with stone/gravel (10 P.S.F.)	G-5	7,000	S.F.	1.40	.11	1.51	1.71
770	Partially adhered with adhesive		5,100		1.30	.15	1.45	1.66
780	Hot asphalt attachment		5,100		1.35	.15	1.50	1.71
790	Fully adhered with contact cement	↓	3,500	↓	1.40	.23	1.63	1.87
791								
820	Polyvinyl chloride (P.V.C.)							
825	45 mils, 0.30 P.S.F.							
830	Loose-laid & ballasted with stone/gravel (10 P.S.F.)	G-5	7,000	S.F.	.60	.11	.71	.83
835	Partially adhered with mechanical fasteners	"	5,100	"	.90	.15	1.05	1.22
840	48 mils, 0.33 to 0.38 P.S.F.							
845	Loose-laid & ballasted with stone/gravel (10 P.S.F.)	G-5	7,000	S.F.	.70	.11	.81	.94
850	Partially adhered with mechanical & solvent weld		5,100		1.06	.15	1.21	1.39
855	Fully adhered with cold emulsion		3,500		1.18	.23	1.41	1.63
860	Hot asphalt attachment		3,500		1.35	.23	1.58	1.82
865	60 mils, 0.40#, partially adhered with PVC coated strips	↓	5,100		1.47	.15	1.62	1.84
868	Uncured neoprene, 60 mils, for flashing	1 Rofc	600		1.02	.25	1.27	1.51
869	Separator sheet	G-5	4,000	↓	.03	.20	.23	.32
870	Reinforced PVC, 48 mils, 0.33 P.S.F.							
875	Loose-laid & ballasted with stone/gravel (12 P.S.F.)	G-5	7,000	S.F.	1.75	.11	1.86	2.09
880	Partially adhered with mechanical fasteners		5,100		2.20	.15	2.35	2.65
885	Fully adhered with adhesive	↓	3,500	↓	2.42	.23	2.65	2.99
54-001 (93)	STEEL DECKING See division 5.2-30							
57-001	STEEL ROOFING Galv., corrugated or ribbed, on steel frame, 29 ga.	G-3	1,100	S.F.	.58	.59	1.17	1.48
010	26 gauge		1,050		.78	.61	1.39	1.74
030 (93)	24 gauge		1,000		1.53	.64	2.17	2.61
040	22 gauge		950		1.28	.68	1.96	2.39
060	Colored, corrugated or ribbed, on steel frame, 26 gauge		1,050		.95	.61	1.56	1.93
070	24 gauge		1,000		1.53	.64	2.17	2.61
090	Factory insulated, 26 gauge with 1" polystyrene, galvanized		800		1.60	.81	2.41	2.92
100	Colored		800	↓	2.09	.81	2.90	3.46
120	Ridge roll, galvanized, 10" wide		800	L.F.	.98	.81	1.79	2.24
121	20" wide	↓	750	"	1.40	.86	2.26	2.78
60-001	STEEL SIDING Beveled, vinyl coated, 8" wide	1 Carp	265	S.F.	.58	.60	1.18	1.51
005	10" wide	"	275		.52	.58	1.10	1.42
008 (74)	Galv., corrugated or ribbed, on steel frame, 29 gauge	G-3	800		.51	.81	1.32	1.72
010	26 gauge		795		.69	.81	1.50	1.93
030	24 gauge		790		.95	.82	1.77	2.22
040	22 gauge		785		1.11	.82	1.93	2.40
060	20 gauge		770		1.27	.84	2.11	2.60
070	Colored, corrugated or ribbed, on steel frame, 10 yr. finish, 29 ga.	↓	800	↓	.60	.81	1.41	1.82

7.4 Roofing & Siding

			CREW	DAILY OUTPUT	UNIT	BARE COSTS MAT.	BARE COSTS INST.	BARE COSTS TOTAL	TOTAL INCL O&P
60	090	26 gauge	G-3	795	S.F.	.85	.81	1.66	2.10
	100	24 gauge		790		1.30	.82	2.12	2.61
	120	Factory sandwich panel, 26 ga., 1" insulation, galvanized		380		1.48	1.70	3.18	4.07
	130	Colored 1 side		380		2	1.70	3.70	4.64
	150	Galvanized 2 sides		380		1.70	1.70	3.40	4.31
	160	Colored 2 sides		380		2.27	1.70	3.97	4.94
	180	Acrylic paint face, regular paint liner	↓	380		1.95	1.70	3.65	4.59
	190	For 2" thick polystyrene, add				.68		.68	.74M
	200	22 ga., galv., 2" insulation, baked enamel exterior	G-3	360		6	1.79	7.79	9.20
	210	P.V.F. exterior finish	"	360	↓	6.45	1.79	8.24	9.70
63-001	(121)	THIN SHELL Roofs, see division 6.1-74							
66-001		VAPOR BARRIER See division 7.1-15							
69-001		VINYL SIDING Solid PVC panels, 8" to 10" wide, plain	1 Carp	255	S.F.	.65	.63	1.28	1.62
	010	Insulated		255		.90	.63	1.53	1.90
	020	Soffit and fascia		205	↓	1.05	.78	1.83	2.29
	030	Window and door trim moldings		185	L.F.	.32	.86	1.18	1.61
	050	Corner posts, outside corner		205		.90	.78	1.68	2.12
	060	Inside corner		205		.49	.78	1.27	1.67
	080	Corrugated vinyl sheets, .090" thick		235	S.F.	2.70	.68	3.38	3.96
	090	.120" thick		225		3.60	.71	4.31	4.99
	110	Flat sheets with asbestos fibers, colored, 1/16" thick		235		1.48	.68	2.16	2.61
	120	1/8" thick		225		2.18	.71	2.89	3.43
	140	Insulated sandwich panels with 1/16" skin, 1" thick		180		3.40	.89	4.29	5.05
	150	1-1/2" thick	↓	190		5.50	.84	6.34	7.25
72-001		WALKWAY For built-up roofs, asphalt impregnated, 3' x 6' x 1/2" thick	1 Rofc	400		1.25	.38	1.63	1.95
	010	3' x 3' x 3/4" thick, hot application	"	400		1.71	.38	2.09	2.46
	030	Concrete patio blocks, 2" thick, natural	1 Clab	115		.89	1.11	2	2.58
	040	Colors	"	115	↓	.93	1.11	2.04	2.63
75-001	(105)	WINDOW WALL See division 8.9-80							
78-001		WOOD ROOF DECK See division 6.1-62							
81-001		WOOD SIDING See division 6.2-67 & 70							

7.6 Sheet Metal Work

			CREW	DAILY OUTPUT	UNIT	BARE COSTS MAT.	BARE COSTS INST.	BARE COSTS TOTAL	TOTAL INCL O&P
05-001		COPPER ROOFING Batten seam, over 10 squares, 16 oz., 130 lb. per sq.	1 Shee	1.10	Sq.	175	165	340	435
	020	18 oz., 145 lb. per sq.		1		190	180	370	475
	030	20 oz., 160 lb. per sq.		1		210	180	390	495
	040	Standing seam, over 10 squares, 16 oz., 125 lb. per sq.		1.30		170	140	310	390
	060	18 oz., 140 lb. per sq.		1.20		185	150	335	425
	070	20 oz., 150 lb. per sq.		1.10		205	165	370	465
	090	Flat seam, over 10 squares, 16 oz., 115 lb. per sq.		1.20		160	150	310	395
	100	20 oz., 145 lb. per sq.	↓	1.10		195	165	360	455
	120	For abnormal conditions or small areas, add				25%	100%		
	130	For lead-coated copper, add			↓	18%			
10-001		DOWNSPOUTS Aluminum 2" x 3", .020" thick, embossed	1 Shee	190	L.F.	.50	.96	1.46	1.95
	010	Enameled		190		.55	.96	1.51	2
	030	Enameled, .024" thick, 2" x 3"		180		.71	1.01	1.72	2.25
	040	3" x 4"		140		1.20	1.30	2.50	3.21
	060	Round, corrugated aluminum, 3" diameter, .020" thick		190		.95	.96	1.91	2.44
	070	4" diameter, .025" thick	↓	140	↓	1.50	1.30	2.80	3.54

7.6 Sheet Metal Work

						BARE COSTS			TOTAL
			CREW	DAILY OUTPUT	UNIT	MAT.	INST.	TOTAL	INCL O&P
10	090	Wire strainer, round, 2" diameter	1 Shee	155	Ea.	.78	1.17	1.95	2.57
	100	4" diameter		155		1	1.17	2.17	2.81
	120	Rectangular, perforated, 2" x 3"		145		1.35	1.25	2.60	3.31
	130	3" x 4"		145	↓	2.20	1.25	3.45	4.25
	150	Copper, round, 16 oz., stock, 2" diameter		190	L.F.	2.55	.96	3.51	4.20
	160	3" diameter		190		3.25	.96	4.21	4.97
	180	4" diameter		145		4.25	1.25	5.50	6.50
	190	5" diameter		130		4.50	1.40	5.90	7
	210	Rectangular, corrugated copper, stock, 2" x 3"		190		3.40	.96	4.36	5.15
	220	3" x 4"		145		4.55	1.25	5.80	6.85
	240	Rectangular, plain copper, stock, 2" x 3"		190		3.55	.96	4.51	5.30
	250	3" x 4"		145	↓	4.30	1.25	5.55	6.55
	270	Wire strainers, rectangular, 2" x 3"		145	Ea.	1.75	1.25	3	3.75
	280	3" x 4"		145		2.75	1.25	4	4.85
	300	Round, 2" diameter		145		1.65	1.25	2.90	3.64
	310	3" diameter		145		2.40	1.25	3.65	4.47
	330	4" diameter		145		3.70	1.25	4.95	5.90
	340	5" diameter		115	↓	5.15	1.58	6.73	7.95
	360	Lead-coated copper, round, stock, 2" diameter		190	L.F.	4.40	.96	5.36	6.25
	370	3" diameter		190		5.40	.96	6.36	7.35
	390	4" diameter		145		6.55	1.25	7.80	9.05
	400	5" diameter, corrugated		130		10	1.40	11.40	13.05
	420	6" diameter, corrugated		105		12.95	1.73	14.68	16.75
	430	Rectangular, corrugated, stock, 2" x 3"		190		5.90	.96	6.86	7.90
	450	Plain, stock, 2" x 3"		190		6.95	.96	7.91	9.05
	460	3" x 4"		145		8.70	1.25	9.95	11.40
	480	Steel, galvanized, round, corrugated, 2" or 3" diam., 28 gauge		190		.45	.96	1.41	1.89
	490	4" diameter, 28 gauge		145		.57	1.25	1.82	2.46
	510	5" diameter, 28 gauge		130		.71	1.40	2.11	2.82
	520	26 gauge		130		.90	1.40	2.30	3.03
	540	6" diameter, 28 gauge		105		1.03	1.73	2.76	3.66
	550	26 gauge		105		1.17	1.73	2.90	3.81
	570	Rectangular, corrugated, 28 gauge, 2" x 3"		190		.45	.96	1.41	1.89
	580	3" x 4"		145		1.27	1.25	2.52	3.23
	600	Rectangular, plain, 28 gauge, galvanized, 2" x 3"		190		.42	.96	1.38	1.86
	610	3" x 4"		145		1.27	1.25	2.52	3.23
	630	Epoxy painted, 24 gauge, corrugated, 2" x 3"		190		.87	.96	1.83	2.35
	640	3" x 4"		145	↓	1.12	1.25	2.37	3.06
	660	Wire strainers, rectangular, 2" x 3"		145	Ea.	1.35	1.25	2.60	3.31
	670	3" x 4"		145		2.20	1.25	3.45	4.25
	690	Round strainers, 2" or 3" diameter		145		.91	1.25	2.16	2.83
	700	4" diameter		145		1	1.25	2.25	2.93
	720	5" diameter		145		1.05	1.25	2.30	2.98
	730	6" diameter		115		1.26	1.58	2.84	3.69
	750	Steel pipe, black, extra heavy, 4" diameter		20	L.F.	11.25	9.10	20.35	26
	760	6" diameter		18		24	10.10	34.10	41
	780	Stainless steel tubing, schedule 5, 2" x 3" or 3" diameter		190		8.15	.96	9.11	10.35
	790	3" x 4" or 4" diameter		145		10.90	1.25	12.15	13.80
	810	4" x 5" or 5" diameter	↓	135	↓	16.70	1.35	18.05	20
	815								
	820	Vinyl, rectangular, 2" x 3"	1 Shee	210	L.F.	.80	.86	1.66	2.14
	830	Round, 2-1/2"	"	220		.80	.83	1.63	2.09
12-001		DRIP EDGE Aluminum, .016" thick, 5" girth, mill finish	1 Carp	400		.14	.40	.54	.73
	010	White finish		400		.28	.40	.68	.89
	020	8" girth		400		.21	.40	.61	.81
	030	28" girth		100		1.13	1.60	2.73	3.56
	040	Galvanized, 5" girth		400		.15	.40	.55	.74
	050	8" girth	↓	400	↓	.22	.40	.62	.82
13-001		ELBOWS Aluminum, 2" x 3", embossed	1 Shee	100	Ea.	1.70	1.82	3.52	4.52
	010	Enameled	"	100	"	1.70	1.82	3.52	4.52

7.6 Sheet Metal Work

			CREW	DAILY OUTPUT	UNIT	BARE COSTS MAT.	BARE COSTS INST.	BARE COSTS TOTAL	TOTAL INCL O&P
13	020	3" x 4", .025" thick, embossed	1 Shee	100	Ea.	2.05	1.82	3.87	4.91
	030	Enameled		100		2.05	1.82	3.87	4.91
	040	Round corrugated, 3", embossed, .020" thick		100		1.70	1.82	3.52	4.52
	050	4", .025" thick		100		1.85	1.82	3.67	4.69
	060	Copper, 16 oz. round, corrugated, 2" diameter		100		6	1.82	7.82	9.25
	070	3" diameter		100		7.40	1.82	9.22	10.80
	080	4" diameter		100		9.80	1.82	11.62	13.45
	090	5" diameter		100		11.95	1.82	13.77	15.80
	100	2" x 3" corrugated		100		7.80	1.82	9.62	11.25
	110	3" x 4" corrugated		100		10.45	1.82	12.27	14.15
	130	Vinyl, 2-1/2" diameter, 45° or 75°		100		1.85	1.82	3.67	4.69
	140	Tee Y junction	↓	75	↓	10.65	2.42	13.07	15.25
15-001		EXPANSION JOINT Butyl, 1/16" thick, 29" wide	1 Rofc	165	L.F.	3.95	.91	4.86	5.75
	030	Butyl or neoprene center with foam insulation, metal flanges							
	040	Aluminum, .032" thick for openings to 2-1/2"	1 Rofc	165	L.F.	7.40	.91	8.31	9.55
	060	For joint openings to 3-1/2"		165		5.95	.91	6.86	7.95
	070	Copper, 16 oz. for openings to 2-1/2"		165		9	.91	9.91	11.30
	090	For joint openings to 3-1/2"		165		11.10	.91	12.01	13.60
	100	Galvanized steel, 26 ga. for openings to 2-1/2"		165		4.60	.91	5.51	6.45
	120	For joint openings to 3-1/2"		165		3.60	.91	4.51	5.35
	130	Lead-coated copper, 16 oz. for openings to 2-1/2"		165		9.55	.91	10.46	11.90
	150	For joint openings to 3-1/2"		165		11.35	.91	12.26	13.90
	160	Stainless steel, .018", for openings to 2-1/2"		165		6.50	.91	7.41	8.55
	180	For joint openings to 3-1/2"		165		10.90	.91	11.81	13.40
	190	Neoprene, double-seal type with thick center, 4-1/2" wide	↓	125	↓	5.75	1.20	6.95	8.15
	191								
	195	Polyethylene bellows, with galv. steel flat flanges	1 Rofc	100	L.F.	3.14	1.50	4.64	5.75
	196	With galvanized angle flanges	"	100		3.34	1.50	4.84	5.95
	200	Roof joint with extruded aluminum cover, 2"	1 Shee	115	↓	24	1.58	25.58	29
	205								
	210	Roof joint, plastic curbs, foam center, standard	1 Rofc	100	L.F.	7.80	1.50	9.30	10.90
	220	Large		100	"	9.60	1.50	11.10	12.85
	230	Transitions, regular, minimum		10	Ea.	31	15.05	46.05	57
	235	Maximum		4		135	38	173	205
	240	Large, minimum		9		39	16.70	55.70	68
	245	Maximum	↓	3	↓	135	50	185	225
	250	Roof to wall joint with extruded aluminum cover	1 Shee	115	L.F.	17.30	1.58	18.88	21
	260	See also division 3.1-10, 5.8-05							
	270	Wall joint, closed cell foam on PVC cover, 9" wide	1 Rofc	125	L.F.	2.80	1.20	4	4.92
	280	12" wide	"	115	"	3.18	1.31	4.49	5.50
20-001		FASCIA Aluminum, reverse board and batten,							
	010	.032" thick, colored, no furring included	1 Shee	145	S.F.	2.25	1.25	3.50	4.30
	030	Steel, galv. and enameled, stock, no furring, long panels		145		1.95	1.25	3.20	3.97
	060	Short panels		115		3.05	1.58	4.63	5.65
25-001		FLASHING Aluminum, mill finish, .013" thick		145		.27	1.25	1.52	2.13
	003	.016" thick		145		.31	1.25	1.56	2.17
	006	.019" thick		145		.63	1.25	1.88	2.52
	010	.032" thick		145		.77	1.25	2.02	2.68
	020	.040" thick		145		1.30	1.25	2.55	3.26
	030	.050" thick	↓	145		1.60	1.25	2.85	3.59
	040	Painted finish, add				.17		.17	.18M
	050	Fabric-backed 2 sides, .004" thick	1 Shee	330		.37	.55	.92	1.21
	070	.016" thick		330		.91	.55	1.46	1.80
	075	Mastic-backed, self adhesive		460		1.95	.39	2.34	2.72
	080	Mastic-coated 2 sides, .004" thick		330		.44	.55	.99	1.29
	100	.005" thick		330		.55	.55	1.10	1.41
	110	.016" thick	↓	330	↓	.97	.55	1.52	1.87
	130	Asphalt flashing cement, 5 gallon			Gal.	13.40		13.40	14.75M
	160	Copper, 16 oz., sheets, under 6000 lbs.	1 Shee	115	S.F.	1.11	1.58	2.69	3.53
	170	Over 6000 lbs.	"	155	"	1.11	1.17	2.28	2.93

7.6 Sheet Metal Work

		CREW	DAILY OUTPUT	UNIT	BARE COSTS MAT.	BARE COSTS INST.	BARE COSTS TOTAL	TOTAL INCL O&P
190	20 oz. sheets, under 6000 lbs.	1 Shee	110	S.F.	1.40	1.65	3.05	3.95
200	Over 6000 lbs.		145		1.40	1.25	2.65	3.37
220	24 oz. sheets, under 6000 lbs.		105		1.67	1.73	3.40	4.36
230	Over 6000 lbs.		135		1.67	1.35	3.02	3.80
250	32 oz. sheets, under 6000 lbs.		100		2.22	1.82	4.04	5.10
260	Over 6000 lbs.		130		2.22	1.40	3.62	4.48
280	Copper, paperbacked 1 side, 2 oz.		330		.75	.55	1.30	1.63
290	3 oz.		330		1.10	.55	1.65	2.01
310	Paperbacked 2 sides, copper, 2 oz.		330		.95	.55	1.50	1.85
315	3 oz.		330		1.15	.55	1.70	2.07
320	5 oz.		330		1.95	.55	2.50	2.95
325	7 oz.		330		2.15	.55	2.70	3.17
340	Mastic-backed 2 sides, copper, 2 oz.		330		.95	.55	1.50	1.85
350	3 oz.		330		1.26	.55	1.81	2.19
370	5 oz.		330		2.10	.55	2.65	3.11
380	Fabric-backed 2 sides, copper, 2 oz.		330		1.10	.55	1.65	2.01
400	3 oz.		330		1.37	.55	1.92	2.31
410	5 oz.		330		2.15	.55	2.70	3.17
430	Copper-clad stainless steel, .015" thick, under 500 lbs.		115		2.10	1.58	3.68	4.62
440	Over 2000 lbs.		155		2.05	1.17	3.22	3.97
460	.018" thick, under 500 lbs.		100		2.70	1.82	4.52	5.60
470	Over 2000 lbs.	▼	145	▼	2.40	1.25	3.65	4.47
490	Fabric, asphalt-saturated cotton, specification grade	1 Rofc	35	S.Y.	1.21	4.30	5.51	7.90
500	Utility grade		35		.74	4.30	5.04	7.40
520	Open-mesh fabric, saturated, 40 oz. per S.Y.		35		1.55	4.30	5.85	8.30
530	Close-mesh fabric, saturated, 17 oz. per S.Y.		35		1.55	4.30	5.85	8.30
550	Fiberglass, resin-coated		35		1.42	4.30	5.72	8.15
560	Asphalt-coated, 40 oz. per S.Y.		35	▼	2.12	4.30	6.42	8.90
580	Lead, 2.5 lb. per S.F., up to 12" wide		135	S.F.	2.20	1.11	3.31	4.12
590	Over 12" wide	▼	135		2.95	1.11	4.06	4.95
610	Lead-coated copper, fabric-backed, 2 oz.	1 Shee	330		.92	.55	1.47	1.82
620	5 oz.		330		1.35	.55	1.90	2.29
640	Mastic-backed 2 sides, 2 oz.		330		.59	.55	1.14	1.45
650	5 oz.		330		1.02	.55	1.57	1.93
670	Paperbacked 1 side, 2 oz.		330		.57	.55	1.12	1.43
680	3 oz.		330		.83	.55	1.38	1.72
700	Paperbacked 2 sides, 2 oz.		330		.64	.55	1.19	1.51
710	5 oz.	▼	330		1.28	.55	1.83	2.21
730	Polyvinyl chloride, black, .010" thick	1 Rofc	285		.47	.53	1	1.32
740	.020" thick		285		.85	.53	1.38	1.74
760	.030" thick		285		1.28	.53	1.81	2.22
770	.056" thick		285		2.38	.53	2.91	3.43
790	Black or white for exposed roofs, .060" thick		285		2.50	.53	3.03	3.56
800	Asbestos-backed for parking decks, .045" thick		285		.80	.53	1.33	1.69
805	PVC (19 mils) coated galv. steel (24 mils), 4' x 8' sheets	▼	240	▼	1.20	.63	1.83	2.28
806	PVC tape, 5" x 45 mils, for joint covers, 100 L.F./roll			Ea.	80		80	88M
810	Rubber, butyl, 1/32" thick	1 Rofc	285	S.F.	.61	.53	1.14	1.48
820	1/16" thick		285		.90	.53	1.43	1.80
830	Neoprene, cured, 1/16" thick		285		1.42	.53	1.95	2.37
840	1/8" thick	▼	285		2.90	.53	3.43	4
850	Shower pan, bituminous membrane, 7 oz.	1 Shee	155		1.40	1.17	2.57	3.25
855	3 ply copper and fabric, 3 oz.		155		1.70	1.17	2.87	3.58
860	7 oz.		155		3.65	1.17	4.82	5.75
865	Copper, 16 oz.		100		2.75	1.82	4.57	5.70
870	Lead on copper and fabric, 5 oz.		155		1.40	1.17	2.57	3.25
880	7 oz.		155		2.75	1.17	3.92	4.74
885	Polyvinyl chloride, .030" thick	▼	160	▼	.45	1.14	1.59	2.15
887								
890	Stainless steel sheets, 32 ga., .010" thick	1 Shee	155	S.F.	1.70	1.17	2.87	3.58
900	28 ga., .015" thick	"	155	"	2.10	1.17	3.27	4.02

7.6 Sheet Metal Work

			CREW	DAILY OUTPUT	UNIT	BARE COSTS MAT.	BARE COSTS INST.	BARE COSTS TOTAL	TOTAL INCL O&P
25	910	26 ga., .018" thick	1 Shee	155	S.F.	2.60	1.17	3.77	4.57
	920	24 ga., .025" thick	"	155		3.20	1.17	4.37	5.25
	929	For mechanically keyed flashing, add				50%			
	930	Stainless steel, paperbacked 2 sides, .005" thick	1 Shee	330		1.05	.55	1.60	1.96
	940	Terne coated stainless steel, .015" thick, 28 ga.		155		2.10	1.17	3.27	4.02
	950	.018" thick, 26 ga.		155		2.15	1.17	3.32	4.08
	960	Zinc and copper alloy, .020" thick		155		1.05	1.17	2.22	2.87
	970	.027" thick		155		1.35	1.17	2.52	3.20
	980	.032" thick		155		1.60	1.17	2.77	3.47
	990	.040" thick		155		2	1.17	3.17	3.91
30-001		GRAVEL STOP Aluminum, .050" thick, 4" height, mill finish		145	L.F.	3.65	1.25	4.90	5.85
	008	Duranodic finish		145		4.30	1.25	5.55	6.55
	010	Painted		145		3.80	1.25	5.05	6
	030	6" face height, .050" thick, mill finish		135		4.20	1.35	5.55	6.60
	035	Duranodic finish		135		5	1.35	6.35	7.45
	040	Painted		135		4.30	1.35	5.65	6.70
	060	8" face height, .050" thick, mill finish		125		5.10	1.45	6.55	7.75
	065	Duranodic finish		125		5.95	1.45	7.40	8.65
	070	Painted		125		5.45	1.45	6.90	8.10
	090	12" face height, 2 piece, mill finish		100		8.55	1.82	10.37	12.05
	095	Duranodic finish		100		9.75	1.82	11.57	13.40
	100	Painted		100		10.30	1.82	12.12	14
	120	Copper, 16 oz., 3" face height		145		3.55	1.25	4.80	5.75
	130	6" face height		135		4.60	1.35	5.95	7.05
	135	Galv. steel, 24 ga., 4" leg, plain, with continuous cleat, 4" face		145		1.80	1.25	3.05	3.81
	136	6" face height		145		1.95	1.25	3.20	3.97
	150	Polyvinyl chloride, 6" face height		135		2.88	1.35	4.23	5.15
	160	9" face height		125		3.42	1.45	4.87	5.90
	180	Stainless steel, 24 ga., 6" face height		135		5.75	1.35	7.10	8.30
	190	12" face height		100		11	1.82	12.82	14.75
	210	20 ga., 6" face height		135		6.50	1.35	7.85	9.10
	220	12" face height		100		12.90	1.82	14.72	16.85
33-001		GUTTERS Aluminum, stock units, 5" box, .027" thick, plain		120		.95	1.51	2.46	3.26
	010	Enameled		120		.90	1.51	2.41	3.20
	030	5" box type, .032" thick, plain		120		1.15	1.51	2.66	3.48
	040	Enameled		120		1.10	1.51	2.61	3.42
	060	5" x 6" combination fascia & gutter, .032" thick, enameled		60		2.50	3.03	5.53	7.15
	070	Copper, half round, 16 oz., stock units, 4" wide		120		3.75	1.51	5.26	6.35
	090	5" wide		120		4	1.51	5.51	6.60
	100	6" wide		115		4.90	1.58	6.48	7.70
	120	K type copper gutter, stock, 4" wide		120		4.75	1.51	6.26	7.45
	130	5" wide		120		5.70	1.51	7.21	8.50
	150	Lead coated copper, half round, stock, 4" wide		120		3.40	1.51	4.91	5.95
	160	6" wide		115		4.10	1.58	5.68	6.80
	180	K type lead coated copper, stock, 4" wide		120		3.45	1.51	4.96	6
	190	5" wide		120		4.25	1.51	5.76	6.90
	210	Stainless steel, half round or box, stock, 4" wide		120		4.25	1.51	5.76	6.90
	220	5" wide		120		4.50	1.51	6.01	7.15
	240	Steel, galv., half round or box, 28 ga., 5" wide, plain		120		.65	1.51	2.16	2.93
	250	Enameled		120		.70	1.51	2.21	2.98
	270	26 ga. galvanized steel, stock, 5" wide		120		.78	1.51	2.29	3.07
	280	6" wide		120		.85	1.51	2.36	3.15
	300	Vinyl, O.G., 4" wide	1 Carp	110		1.15	1.45	2.60	3.37
	310	5" wide		110		1.30	1.45	2.75	3.54
	320	4" half round, stock units		110		.90	1.45	2.35	3.10
	325	Joint connectors			Ea.	1.99		1.99	2.18M
	330	Wood, clear treated cedar, fir or hemlock, 3" x 4"	1 Carp	100	L.F.	3.65	1.60	5.25	6.35
	340	4" x 5"		100		4.75	1.60	6.35	7.55
36-001		GUTTER GUARD 6" wide strip, aluminum mesh		500		.30	.32	.62	.79
	010	Vinyl mesh		500		.15	.32	.47	.63

7.6 Sheet Metal Work

		CREW	DAILY OUTPUT	UNIT	BARE COSTS MAT.	BARE COSTS INST.	BARE COSTS TOTAL	TOTAL INCL O&P
39-001	**LEAD ROOFING** 3 lb. per S.F., batten seam	1 Shee	1.20	Sq.	440	150	590	705
010	Flat seam	"	1.30	"	395	140	535	640
42-001	**LOUVERS** Aluminum with screen, residential, 8" x 8"	1 Carp	38	Ea.	3.50	4.21	7.71	9.95
010	12" x 12"		38		4.97	4.21	9.18	11.55
020	12" x 18"		35		7.55	4.57	12.12	14.95
025	14" x 24"		30		10	5.35	15.35	18.75
030	18" x 24"		27		12	5.95	17.95	22
050	30" x 24"		24		24	6.65	30.65	36
070	Triangle, adjustable, small		20		12.25	8	20.25	25
080	Large	↓	15	↓	12.90	10.65	23.55	30
120	Extruded aluminum, see division 15.7-79							
130								
210	Midget, aluminum, 3/4" deep, 1" diameter	1 Carp	85	Ea.	.43	1.88	2.31	3.20
215	3" diameter		60		1.14	2.67	3.81	5.10
220	4" diameter		50		1.43	3.20	4.63	6.20
225	6" diameter	↓	30	↓	2.01	5.35	7.36	9.95
230	Ridge vent strip, mill finish	1 Shee	155	L.F.	1.51	1.17	2.68	3.37
231								
233	Soffit vent, continuous, 3" wide, aluminum, mill finish	1 Carp	200	L.F.	.21	.80	1.01	1.39
234	Baked enamel finish		200	"	.27	.80	1.07	1.46
240	Under eaves vent, aluminum, mill finish, 16" x 4"		75	Ea.	.85	2.13	2.98	4.03
250	16" x 8"		75		1.09	2.13	3.22	4.29
700	Vinyl wall louvers, 1-1/2" deep, 8" x 8"		38		2.25	4.21	6.46	8.60
702	12" x 12"		38		2.75	4.21	6.96	9.15
708	12" x 18"		35		3.50	4.57	8.07	10.45
720	14" x 24"	↓	30	↓	5	5.35	10.35	13.25
45-001	**MANSARD** Colored aluminum, with battens, .032" thick, custom							
050	Concave or convex surfaces	1 Shee	55	S.F.	31	3.30	34.30	39
052	Projected or flared surfaces		75		10.75	2.42	13.17	15.35
054	Straight surfaces		115		6.25	1.58	7.83	9.20
060	Stock units, .032" aluminum		115	↓	2.50	1.58	4.08	5.05
080	For framing, to 5' high, add		115	L.F.	2.05	1.58	3.63	4.56
090	Soffits, to 1' wide	↓	125	S.F.	1.15	1.45	2.60	3.39
48-001	**MONEL ROOFING** Batten seam, over 10 squares, .018" thick	1 Shee	1.20	Sq.	470	150	620	740
010	.021" thick		1.15		535	160	695	820
030	Standing seam, .018" thick		1.35		455	135	590	695
040	.021" thick		1.30		525	140	665	780
060	Flat seam, .018" thick		1.30		450	140	590	700
070	.021" thick	↓	1.20	↓	515	150	665	790
51-001	**REGLET** Aluminum, .025" thick, in concrete parapet	1 Carp	225	L.F.	1.25	.71	1.96	2.41
010	Copper, 10 oz.		225		1.15	.71	1.86	2.30
030	16 oz.		225		1.65	.71	2.36	2.85
040	Galvanized steel, 24 gauge		225		.42	.71	1.13	1.49
060	Stainless steel, .020" thick		225		1.70	.71	2.41	2.90
070	Zinc and copper alloy, 20 oz.	↓	225		1.35	.71	2.06	2.52
090	Counter flashing for above, 12" wide, .032" aluminum	1 Shee	150		1.45	1.21	2.66	3.36
100	Copper, 10 oz.		150		1.35	1.21	2.56	3.25
120	16 oz.		150		1.60	1.21	2.81	3.53
130	Galvanized steel, .020" thick		150		.58	1.21	1.79	2.41
150	Stainless steel, .020" thick		150		1.95	1.21	3.16	3.91
160	Zinc and copper alloy, 20 oz.	↓	150	↓	2.50	1.21	3.71	4.52
54-001	**SOFFIT** Aluminum, residential, stock units, .020" thick	1 Carp	210	S.F.	.70	.76	1.46	1.87
010	Baked enamel on steel, 16 or 18 gauge		105		5.05	1.52	6.57	7.75
030	Polyvinyl chloride, white, solid		230		.76	.70	1.46	1.84
040	Perforated	↓	230		.79	.70	1.49	1.88
050	For colors, add			↓	.05		.05	.05M
57-001	**STAINLESS STEEL ROOFING** Type 304, batten seam, 28 gauge	1 Shee	1.20	Sq.	265	150	415	515
010	26 gauge	"	1.15	"	345	160	505	610

7.6 Sheet Metal Work

		Description	CREW	DAILY OUTPUT	UNIT	BARE COSTS MAT.	BARE COSTS INST.	BARE COSTS TOTAL	TOTAL INCL O&P
57	020	For standing seam construction, deduct			Sq.	2%			
	050	For flat seam construction, deduct				3%			
	080	For lead or terne coated stainless, 28 gauge, add				70		70	77M
	090	For 26 gauge, add				79		79	87M
60-001		TERMITE Shields, zinc, 10" wide, .012" thick	1 Carp	350	L.F.	1	.46	1.46	1.76
	010	.020" thick	"	350	"	.70	.46	1.16	1.43
63-001		ZINC Copper alloy roofing, batten seam, .020" thick	1 Shee	1.20	Sq.	360	150	510	615
	010	.027" thick		1.15		445	160	605	720
	030	.032" thick		1.10		525	165	690	820
	040	.040" thick		1.05		670	175	845	990
	060	For standing seam construction, deduct				2%			
	070	For flat seam construction, deduct				3%			

7.8 Roof Accessories

		Description	CREW	DAILY OUTPUT	UNIT	BARE COSTS MAT.	BARE COSTS INST.	BARE COSTS TOTAL	TOTAL INCL O&P
10-001		CEILING HATCHES 2'-6" x 2'-6", single leaf, steel frame & cover	G-3	11	Ea.	268	59	327	380
	010	Aluminum cover		11		288	59	347	400
	030	2'-6" x 3'-0", single leaf, steel frame & steel cover		11		288	59	347	400
	040	Aluminum cover		11		326	59	385	445
15-001		ROOF DRAINS See division 15.1-16							
20-001		ROOF HATCHES With curb, 1" fiberglass insulation, 2'-6" x 3'-0"							
	050	Aluminum curb and cover	G-3	10	Ea.	361	64	425	490
	052	Galvanized steel		10		340	64	404	465
	054	Plain steel, primed		10		304	64	368	425
	060	2'-6" x 4'-6", aluminum curb & cover		9		517	72	589	670
	080	Galvanized steel		9		489	72	561	640
	090	Plain steel, primed		9		430	72	502	575
	120	2'-6" x 8'-0", aluminum curb and cover		6.60		897	98	995	1,125
	140	Galvanized steel		6.60		850	98	948	1,075
	150	Plain steel, primed		6.60		742	98	840	955
	180	For plexiglass panels, add to above				200		200	220M
	200	For galv. curb and alum. cover, deduct from aluminum				14		14	15.40M
30-001		ROOF VENTS Mushroom for built-up roofs, aluminum	1 Rofc	30		17.10	5	22.10	26
	010	PVC, 6" high	"	30		20.90	5	25.90	31
40-001		SKYLIGHT Plastic roof domes, flush or curb mounted, ten or							
	010	more units, curb not included, "L" frames							
	030	Nominal size under 10 S.F., double	G-3	130	S.F.	17	4.96	21.96	26
	040	Single		160		13	4.03	17.03	20
	060	10 S.F. to 20 S.F., double		315		14	2.05	16.05	18.35
	070	Single		395		10	1.63	11.63	13.35
	090	20 S.F. to 30 S.F., double		395		12	1.63	13.63	15.55
	100	Single		465		9	1.39	10.39	11.90
	120	30 S.F. to 65 S.F., double		465		10	1.39	11.39	13
	130	Single		610		8	1.06	9.06	10.30
	150	For insulated 4" curbs, double, add				15%			
	160	Single, add				30%			
	180	For integral insulated 9" curbs, double, add				30%			
	190	Single, add				45%			
	210	Ceiling plastic domes compared with single roof domes				95%	100%		
	211								
	212	Ventilating insulated plexiglass dome with							
	213	curb mounting, 36" x 36"	G-3	12	Ea.	332	54	386	445
	215	52" x 52"		12		445	54	499	565
	216	28" x 52"		10		368	64	432	500

7.8 Roof Accessories

			CREW	DAILY OUTPUT	UNIT	BARE COSTS MAT.	BARE COSTS INST.	BARE COSTS TOTAL	TOTAL INCL O&P
40	217	36" x 52"	G-3	10	Ea.	390	64	454	520
	218	For electric opening system, add			"	200		200	220M
	220	Field fabricated, factory type, aluminum and wire glass	G-3	120	S.F.	10.15	5.35	15.50	18.90
	230	Insulated safety glass with aluminum frame		160		69	4.03	73.03	82
	240	Sandwich panels, fiberglass, for walls, 1-9/16" thick, to 250 S.F.		200		12	3.22	15.22	17.85
	250	250 S.F. and up		265		9.75	2.43	12.18	14.25
	270	As above, but for roofs, 2-3/4" thick, to 250 S.F.		295		18	2.19	20.19	23
	280	250 S.F. and up		330		14.50	1.95	16.45	18.75
	300	Prefabricated glass block with metal frame, minimum		265		49	2.43	51.43	57
	310	Maximum	↓	160		61	4.03	65.03	73
	320	With precast concrete structural frame, minimum	C-11	1,500		43	1.71	44.71	50
	330	Maximum	"	1,000	↓	55	2.57	57.57	64
50-001		**SKYROOFS** Translucent panels, 2-3/4" thick, under 5000 S.F.	G-3	395	S.F.Hor.	14.30	1.63	15.93	18.10
	010	Over 5000 S.F.		465		12.70	1.39	14.09	15.95
	030	Continuous vaulted, semi-circular, to 8' wide, double glazed		145		28	4.45	32.45	37
	040	Single glazed		160		18	4.03	22.03	26
	060	To 20' wide, single glazed		175		24	3.68	27.68	32
	070	Over 20' wide, single glazed		200		25	3.22	28.22	32
	090	Motorized opening type, single glazed, 1/3 opening		145		30	4.45	34.45	39
	100	Full opening	↓	130	↓	37	4.96	41.96	48
	120	Pyramid type units, self-supporting, to 30' clear opening,							
	130	square or circular, single glazed, minimum	G-3	200	S.F.Hor.	22	3.22	25.22	29
	131	Average		165		25	3.91	28.91	33
	140	Maximum		130		27	4.96	31.96	37
	150	Grid type, 4' to 10' modules, single glass glazed, minimum		200		21	3.22	24.22	28
	155	Maximum		128		28	5.05	33.05	38
	160	Preformed acrylic, minimum		300		14.20	2.15	16.35	18.70
	165	Maximum	↓	175	↓	26	3.68	29.68	34
	180	Dome type units, self-supporting, to 100' clear opening, circular,							
	190	rise to span ratio = 0.20							
	192	Minimum	G-3	197	S.F.Hor.	15	3.27	18.27	21
	195	Maximum		113		37	5.70	42.70	49
	210	Rise to span ratio = 0.33, minimum		169		25	3.81	28.81	33
	215	Maximum		101		40	6.40	46.40	53
	220	Rise to span ratio = 0.50, minimum		148		38	4.36	42.36	48
	225	Maximum		87		44	7.40	51.40	59
	240	Ridge units, continuous, to 8' wide, double		130		89	4.96	93.96	105
	250	Single		200		58	3.22	61.22	68
	270	Ridge and furrow units, over 4' O.C., double, minimum		200		18.05	3.22	21.27	25
	275	Maximum		120		35	5.35	40.35	46
	280	Single, minimum		214		17.70	3.01	20.71	24
	285	Maximum		153	↓	27	4.21	31.21	36
	300	Rolling roof, translucent panels, flat roof, residential, minimum		253	S.F.	15.75	2.55	18.30	21
	303	Maximum		160		25	4.03	29.03	33
	310	Lean-to skyroof, long span, double, minimum		197		18.90	3.27	22.17	26
	315	Maximum		101		35	6.40	41.40	48
	330	Single, minimum		321		13.20	2.01	15.21	17.40
	335	Maximum	↓	160	↓	23	4.03	27.03	31
55-001		**SMOKE HATCHES** Unlabeled, not including hand winch operator							
	010								
	020	For 3'-0" long, add to roof hatches from division 7.8 20			Ea.	25%	5%		
	030	For 8'-0" long, add to roof hatches from division 7.8-20			"	10%	5%		
60-001		**SMOKE VENTS** Metal cover, heavy duty, low profile, 4' x 4'							
	010	Aluminum	G-3	13	Ea.	1,015	50	1,065	1,200
	020	Galvanized steel	"	13	"	910	50	960	1,075
	025								
	030	4' x 8' aluminum	G-3	8	Ea.	1,310	81	1,391	1,550
	040	Galvanized steel	"	8	↓	1,215	81	1,296	1,450
	050	Sloped cover style, deduct			↓	10%			

7.8 Roof Accessories

		CREW	DAILY OUTPUT	UNIT	BARE COSTS MAT.	BARE COSTS INST.	BARE COSTS TOTAL	TOTAL INCL O&P
70-001	SNOW GUARDS Adjust., not incl. guard pipes, galv., 12" x 6" plate	G-3	130	Ea.	32	4.96	36.96	42
020	24" x 12" plate		130		40	4.96	44.96	51
040	Bronze, 12" x 6" plate		130		57	4.96	61.96	70
050	24" x 12" plate	↓	130		105	4.96	109.96	125
060	Polycarbonate, 3-1/4" wide x 5-1/2" high	1 Shee	200	↓	4.05	.91	4.96	5.80
80-001	SOLAR ROOF DRAINS Plastic mushroom type	1 Rofc	30	Ea.	370	5	375	415
010	Vents	"	30	"	52	5	57	65
90-001	VENTILATORS See division 15.7-96							

8.1 Metal Doors

		CREW	DAILY OUTPUT	UNIT	BARE COSTS MAT.	BARE COSTS INST.	BARE COSTS TOTAL	TOTAL INCL O&P
10-001	STEEL FRAMES, KNOCK DOWN 18 ga., up to 5-3/4" deep							
002	6'-8" high, 3'-0" wide, single	F-2	16	Ea.	52	21	73	87
004	6'-0" wide, double		14		59	24	83	99
010	7'-0" high, 3'-0" wide, single		16		55	21	76	91
014	6'-0" wide, double		14		64	24	88	105
280	18 ga. drywall, up to 4-7/8" deep, 7'-0" high, 3'-0" wide, single		16		60	21	81	96
284	6'-0" wide, double		14		69	24	93	110
360	16 ga., up to 5-3/4" deep, 7'-0" high, 4'-0" wide, single		15		70	22	92	110
364	8'-0" wide, double		12		85	28	113	135
370	8'-0" high, 4'-0" wide, single		15		76	22	98	115
374	8'-0" wide, double		12		91	28	119	140
400	6-3/4" deep, 7'-0" high, 4'-0" wide, single		15		75	22	97	115
404	8'-0" wide, double		12		90	28	118	140
410	8'-0" high, 4'-0" wide, single		15		81	22	103	120
414	8'-0" wide, double		12		96	28	124	145
440	8-3/4" deep, 7'-0" high, 4'-0" wide, single		15		84	22	106	125
444	8'-0" wide, double		12		96	28	124	145
450	8'-0" high, 4'-0" wide, single		15		90	22	112	130
454	8'-0" wide, double		12		103	28	131	155
480	16 ga. drywall, up to 3-7/8" deep, 7'-0" high, 3'-0" wide, single		16		70	21	91	105
484	6'-0" wide, double	↓	14		80	24	104	120
490	For welded frames, add			↓	15		15	16.50M
520								
540	16 ga. "B" label, up to 5-3/4" deep, 7'-0" high, 4'-0" wide, single	F-2	15	Ea.	76	22	98	115
544	8'-0" wide, double		12		86	28	114	135
580	6-3/4" deep, 7'-0" high, 4'-0" wide, single		15		82	22	104	120
584	8'-0" wide, double		12		94	28	122	145
620	8-3/4" deep, 7'-0" high, 4'-0" wide, single		15		87	22	109	130
624	8'-0" wide, double	↓	12	↓	105	28	133	155
630	For "A" label use same price as "B" label							
640	For baked enamel finish, add				40%	90%		
650	For galvanizing, add				10%			
660	For porcelain enamel finish, add				100%	150%		
790	Transom lite frames, fixed, add	F-2	155	S.F.	8	2.18	10.18	11.90
800	Movable, add	"	130	"	9.50	2.59	12.09	14.15
21-001	COMMERCIAL STEEL DOORS Flush, full panel, hollow core, 1-3/8" thick							
002	20 ga., 2'-0" x 6'-8"	F-2	20	Ea.	108	16.85	124.85	145
004	2'-6" x 6'-8"		18		114	18.75	132.75	150
006	3'-0" x 6'-8"		17		120	19.85	139.85	160
010	3'-0" x 7'-0"	↓	17		130	19.85	149.85	170
012	For vision lite, add			↓	50		50	55M

For expanded coverage of these items see *Means' Interior Cost Data 1986*

8.1 Metal Doors

		CREW	DAILY OUTPUT	UNIT	BARE COSTS MAT.	BARE COSTS INST.	BARE COSTS TOTAL	TOTAL INCL O&P
21 014	For narrow lite, add			Ea.	60		60	66M
016	For bottom louver, add			"	115		115	125M
023	For baked enamel finish, add				40%	90%		
026	For galvanizing, add				30%			
029	For porcelain enamel finish, add				100%	150%		
030								
032	Half glass, 20 ga., 2'-0" x 6'-8"	F-2	20	Ea.	145	16.85	161.85	185
034	2'-6" x 6'-8"		18		150	18.75	168.75	190
036	3'-0" x 6'-8"		17		160	19.85	179.85	205
040	3'-0" x 7'-0"		17		165	19.85	184.85	210
102	Hollow core, 1-3/4" thick, full panel, 20 ga., 2'-6" x 6'-8"		18		120	18.75	138.75	160
104	3'-0" x 6'-8"		17		125	19.85	144.85	165
106	3'-0" x 7'-0"		17		135	19.85	154.85	175
108	4'-0" x 7'-0"		15		175	22	197	225
110	4'-0" x 8'-0"		13		220	26	246	280
112	18 ga., 2'-6" x 6'-8"		17		140	19.85	159.85	180
114	3'-0" x 6'-8"		16		150	21	171	195
116	3'-0" x 7'-0"		16		160	21	181	205
118	4'-0" x 7'-0"		14		195	24	219	250
120	4'-0" x 8'-0"		14		225	24	249	280
122	Half glass, 20 ga., 2'-6" x 6'-8"		20		155	16.85	171.85	195
124	3'-0" x 6'-8"		18		165	18.75	183.75	210
126	3'-0" x 7'-0"		18		175	18.75	193.75	220
128	4'-0" x 7'-0"		16		215	21	236	265
130	4'-0" x 8'-0"		13		245	26	271	305
132	18 ga., 2'-6" x 6'-8"		18		185	18.75	203.75	230
134	3'-0" x 6'-8"		17		175	19.85	194.85	220
136	3'-0" x 7'-0"		17		220	19.85	239.85	270
138	4'-0" x 7'-0"		15		250	22	272	305
140	4'-0" x 8'-0"		14		280	24	304	340
172	Composite, 1-3/4" thick, full panel, 18 ga., 3'-0" x 6'-8"		15		165	22	187	215
174	2'-6" x 7'-0"		16		155	21	176	200
176	3'-0" x 7'-0"		15		175	22	197	225
180	4'-0" x 8'-0"		13		255	26	281	320
182	Half glass, 18 ga., 3'-0" x 6'-8"		16		220	21	241	270
184	2'-6" x 7'-0"		17		195	19.85	214.85	245
186	3'-0" x 7'-0"		16		235	21	256	290
190	4'-0" x 8'-0"	↓	14	↓	305	24	329	370
23-001	**FIRE DOOR** Steel, flush, "B" label, 90 minute							
002	Full panel, 20 ga., 2'-0" x 6'-8"	F-2	20	Ea.	145	16.85	161.85	185
004	2'-6" x 6'-8"		18		150	18.75	168.75	190
006	3'-0" x 6'-8"		17		165	19.85	184.85	210
008	3'-0" x 7'-0"	↓	17	↓	170	19.85	189.85	215
012								
014	18 ga., 3'-0" x 6'-8"	F-2	16	Ea.	195	21	216	245
016	2'-6" x 7'-0"		17		185	19.85	204.85	230
018	3'-0" x 7'-0"		16		200	21	221	250
020	4'-0" x 7'-0"	↓	15	↓	235	22	257	290
022	For "A" label, 3 hour, 18 ga., use same price as "B" label							
024	For vision lite, add			Ea.	75		75	83M
052	Flush, "B" label 90 min., composite, 20 ga., 2'-0" x 6'-8"	F-2	18		160	18.75	178.75	205
054	2'-6" x 6'-8"		17		170	19.85	189.85	215
056	3'-0" x 6'-8"		16		180	21	201	230
058	3'-0" x 7'-0"		16		195	21	216	245
064	Flush, "A" label 3 hour, composite, 18 ga., 3'-0" x 6'-8"		15		215	22	237	270
066	2'-6" x 7'-0"		16		205	21	226	255
068	3'-0" x 7'-0"		15		235	22	257	290
070	4'-0" x 7'-0"		14		275	24	299	335
24-001	**RESIDENTIAL DOOR** Steel, 24 ga., embossed, full panel, 2'-8" x 6'-8"		16		120	21	141	160
004	3'-0" x 6'-8"	↓	15	↓	130	22	152	175

For expanded coverage of these items see *Means' Interior Cost Data 1986*

8.1 Metal Doors

		CREW	DAILY OUTPUT	UNIT	BARE COSTS MAT.	BARE COSTS INST.	BARE COSTS TOTAL	TOTAL INCL O&P
24 006	3'-0" x 7'-0"	F-2	15	Ea.	140	22	162	185
022	Half glass, 2'-8" x 6'-8"		17		165	19.85	184.85	210
024	3'-0" x 6'-8"		16		175	21	196	225
026	3'-0" x 7'-0"		16		185	21	206	235
072	Raised plastic face, full panel, 2'-8" x 6'-8"		16		130	21	151	175
074	3'-0" x 6'-8"		15		140	22	162	185
076	3'-0" x 7'-0"		15		160	22	182	210
082	Half glass, 2'-8" x 6'-8"		17		165	19.85	184.85	210
084	3'-0" x 6'-8"		16		175	21	196	225
086	3'-0" x 7'-0"		16		195	21	216	245
132	Flush face, full panel, 2'-6" x 6'-8"		16		100	21	121	140
134	3'-0" x 6'-8"		15		115	22	137	160
136	3'-0" x 7'-0"		15		130	22	152	175
142	Half glass, 2'-8" x 6'-8"		17		135	19.85	154.85	175
144	3'-0" x 6'-8"		16		145	21	166	190
146	3'-0" x 7'-0"		16		155	21	176	200
230	Interior, residential, closet, bi-fold, 6'-8" x 2'-0" wide		16		38	21	59	72
233	3'-0" wide		16		45	21	66	80
236	4'-0" wide		15		60	22	82	98
240	5'-0" wide		14		70	24	94	110
242	6'-0" wide		13		78	26	104	125
30-001	**ALUMINUM FRAMES** Entrance, 3' x 7' opening, clear finish	2 Sswk	7	Opng.	130	50	180	220
010	Bronze finish		7		150	50	200	245
050	6' x 7' opening, clear finish		6		160	58	218	265
052	Bronze finish		6		190	58	248	300
100	With 3' high transoms, 3' x 10' opening, clear finish		6.50		200	53	253	305
105	Bronze finish		6.50		235	53	288	340
110	Black finish		6.50		255	53	308	365
150	With 3' high transoms, 6' x 10' opening, clear finish		5.50		220	63	283	340
155	Bronze finish		5.50		260	63	323	385
160	Black finish		5.50		280	63	343	405
40-001 002	**ALUMINUM DOORS & FRAMES** Entrance, narrow stile, including hardware & closer, clear finish, not incl. glass, 3' x 7' opening	2 Sswk	2	Ea.	530	175	705	855
010	3' x 10' opening, 3' high transom		1.80		600	195	795	965
020	3'-6" x 10' opening, 3' high transom		1.80		630	195	825	995
030	6' x 7' opening		1.30	Pr.	830	265	1,095	1,325
040	6' x 10' opening, 3' high transom		1.10	"	1,040	315	1,355	1,650
100	Add to above for wide stile doors			Leaf	30%			
110	Full vision doors, with 1/2" glass, add				55%			
120	Non-standard size, add				35%			
130	Light bronze finish, add				15%			
140	Dark bronze finish, add				18%			
150	Black finish, add				27%			
160	Concealed panic device, add				330		330	365M
170	Electric striker release, add			Opng.	235		235	260M
180	Floor check, add	F-2	3	Leaf	215	110	325	395
190	Concealed closer, add	"	1.30	"	135	260	395	520
85-001 002	**STORM DOORS & FRAMES** Aluminum, residential, combination storm and screen							
040	Anodized, 6'-8" x 2'-6" wide	F-2	15	Ea.	90	22	112	130
042	2'-8" wide		14		95	24	119	140
044	3'-0" wide		14		100	24	124	145
050	For 7'-0" door, add				5%			
100	Mill finish, 6'-8" x 2'-6" wide	F-2	15		75	22	97	115
102	2'-8" wide		14		85	24	109	130
104	3'-0" wide		14		90	24	114	135
110	For 7'-0" door, add				5%			
150	White painted, 6'-8" x 2'-6" wide	F-2	15	Ea.	90	22	112	130
152	2'-8" wide	"	14	"	95	24	119	140

For expanded coverage of these items see *Means' Interior Cost Data 1986*

8.1 Metal Doors

			CREW	DAILY OUTPUT	UNIT	BARE COSTS MAT.	BARE COSTS INST.	BARE COSTS TOTAL	TOTAL INCL O&P
85	154	3'-0" wide	F-2	14	Ea.	100	24	124	145
	160	For 7'-0" door, add			"	5%			
	200	Wood door & screen, see division 8.2-32							
	202								

8.2 Wood & Plastic Doors

		CREW	DAILY OUTPUT	UNIT	BARE COSTS MAT.	BARE COSTS INST.	BARE COSTS TOTAL	TOTAL INCL O&P
12-001	**WOOD FRAMES**							
040	Exterior frame, incl. ext. trim, pine, 5/4 x 4-9/16" deep	F-2	375	L.F.	2.48	.90	3.38	4.02
042	5-3/16" deep		375		2.76	.90	3.66	4.32
044	6-9/16" deep		375		3.14	.90	4.04	4.74
060	Oak, 5/4 x 4-9/16" deep		350		3.10	.96	4.06	4.79
062	5-3/16" deep		350		3.38	.96	4.34	5.10
064	6-9/16" deep		350		3.95	.96	4.91	5.70
080	Walnut, 5/4 x 4-9/16" deep		350		4.57	.96	5.53	6.40
082	5-3/16" deep		350		5.25	.96	6.21	7.15
084	6-9/16" deep		350		5.85	.96	6.81	7.80
100	Sills, 8/4 x 8" deep, oak, no horns		100		5.05	3.37	8.42	10.40
102	2" horns		100		5.35	3.37	8.72	10.70
104	3" horns		100		6.25	3.37	9.62	11.70
110	8/4 x 10" deep, oak, no horns		90		6.45	3.75	10.20	12.45
112	2" horns		90		6.90	3.75	10.65	12.95
114	3" horns		90		7.55	3.75	11.30	13.65
200	Exterior, colonial, frame & trim, 3' opng., in-swing, minimum	F-2	22	Ea.	160	15.35	175.35	200
202	Maximum		20		375	16.85	391.85	435
210	5'-4" opening, in-swing, minimum		17		270	19.85	289.85	325
212	Maximum		15		560	22	582	650
214	Out-swing, minimum		17		290	19.85	309.85	345
216	Maximum		15		575	22	597	665
240	6'-0" opening, in-swing, minimum		16		280	21	301	340
242	Maximum		10		590	34	624	695
246	Out-swing, minimum		16		295	21	316	355
248	Maximum		10		605	34	639	715
260	For two sidelights, add, minimum		30	Opng.	135	11.25	146.25	165
262	Maximum		20	"	245	16.85	261.85	295
270	Custom birch frame, 3'-0" opening		16	Ea.	95	21	116	135
275	6'-0" opening		16	"	125	21	146	170
300	Interior frame, pine, 11/16" x 3-5/8" deep		375	L.F.	1.02	.90	1.92	2.41
302	4-9/16" deep		375		1.28	.90	2.18	2.70
304	5-3/16" deep		375		1.53	.90	2.43	2.97
320	Oak, 11/16" x 3-5/8" deep		350		1.25	.96	2.21	2.75
322	4-9/16" deep		350		1.49	.96	2.45	3.02
324	5-3/16" deep		350		1.79	.96	2.75	3.35
340	Walnut, 11/16" x 3-5/8" deep		350		1.39	.96	2.35	2.91
342	4-9/16" deep		350		1.82	.96	2.78	3.38
344	5-3/16" deep		350		2.34	.96	3.30	3.95
360	Pocket door frame		16	Ea.	58	21	79	94
380	Threshold, oak, 5/8" x 3-5/8" deep		200	L.F.	1.09	1.69	2.78	3.61
382	4-5/8" deep		190		1.45	1.77	3.22	4.14
384	5-5/8" deep		180		1.85	1.87	3.72	4.72
402	For casing, see division 6.2-41							
21-001	**WOOD DOOR, ARCHITECTURAL** Flush, interior, 7 ply, hollow core,							
002	Lauan face, 2'-0" x 6'-8"	F-2	17	Ea.	63	19.85	82.85	98

For expanded coverage of these items see *Means' Interior Cost Data 1986*

8.2 Wood & Plastic Doors

			CREW	DAILY OUTPUT	UNIT	BARE COSTS MAT.	BARE COSTS INST.	BARE COSTS TOTAL	TOTAL INCL O&P
21	004	2'-6" x 6'-8"	F-2	17	Ea.	65	19.85	84.85	100
	008	3'-0" x 6'-8"		17		68	19.85	87.85	105
	010	4'-0" x 6'-8"		16		81	21	102	120
	012	Birch face, 2'-0" x 6'-8"		17		65	19.85	84.85	100
	014	2'-6" x 6'-8"		17		67	19.85	86.85	100
	018	3'-0" x 6'-8"		17		70	19.85	89.85	105
	020	4'-0" x 6'-8"		16		84	21	105	125
	022	Oak face, 2'-0" x 6'-8"		17		82	19.85	101.85	120
	024	2'-6" x 6'-8"		17		85	19.85	104.85	120
	028	3'-0" x 6'-8"		17		87	19.85	106.85	125
	030	4'-0" x 6'-8"		16		100	21	121	140
	032	Walnut face, 2'-0" x 6'-8"		17		120	19.85	139.85	160
	034	2'-6" x 6'-8"		17		120	19.85	139.85	160
	038	3'-0" x 6'-8"		17		125	19.85	144.85	165
	040	4'-0" x 6'-8"	▼	16		150	21	171	195
	042	For 7'-0" high, add				4		4	4.40M
	044	For 8'-0" high, add				10		10	11M
	046	For 8'-0" high walnut, add				16		16	17.60M
	048	For prefinishing, clear, add				20		20	22M
	050	For prefinishing, stain, add				24		24	26M
	132	M.D. overlay on hardboard, 2'-0" x 6'-8"	F-2	17		68	19.85	87.85	105
	134	2'-6" x 6'-8"		17		70	19.85	89.85	105
	138	3'-0" x 6'-8"		17		73	19.85	92.85	110
	140	4'-0" x 6'-8"	▼	16		85	21	106	125
	142	For 7'-0" high, add				4		4	4.40M
	144	For 8'-0" high, add				10		10	11M
	172	H.P. plastic laminate, 2'-0" x 6'-8"	F-2	16		105	21	126	145
	174	2'-6" x 6'-8"		16		115	21	136	155
	178	3'-0" x 6'-8"		15		120	22	142	165
	180	4'-0" x 6'-8"	▼	14		135	24	159	185
	182	For 7'-0" high, add				10		10	11M
	184	For 8'-0" high, add				25		25	28M
	202	5 ply particle core, lauan face, 2'-6" x 6'-8"	F-2	15		65	22	87	105
	204	3'-0" x 6'-8"		14		75	24	99	115
	208	3'-0" x 7'-0"		13		82	26	108	125
	210	4'-0" x 7'-0"		12		93	28	121	145
	212	Birch face, 2'-6" x 6'-8"		15		69	22	91	110
	214	3'-0" x 6'-8"		14		79	24	103	120
	218	3'-0" x 7'-0"		13		86	26	112	130
	220	4'-0" x 7'-0"		12		97	28	125	145
	222	Oak face, 2'-6" x 6'-8"		15		85	22	107	125
	224	3'-0" x 6'-8"		14		95	24	119	140
	228	3'-0" x 7'-0"		13		105	26	131	155
	230	4'-0" x 7'-0"		12		125	28	153	180
	232	Walnut face, 2'-0" x 6'-8"		15		125	22	147	170
	234	2'-6" x 6'-8"		14		145	24	169	195
	238	3'-0" x 6'-8"		13		155	26	181	210
	240	4'-0" x 6'-8"	▼	12		185	28	213	245
	244	For 8'-0" high, add				8		8	8.80M
	246	For 8'-0" high walnut, add				14		14	15.40M
	248	For solid wood core, add				30		30	33M
	272	For prefinishing, clear, add				27		27	30M
	274	For prefinishing, stain, add				30		30	33M
	332	M.D. overlay on hardboard, 2'-6" x 6'-8"	F-2	14		68	24	92	110
	334	3'-0" x 6'-8"		13		78	26	104	125
	338	3'-0" x 7'-0"		12		85	28	113	135
	340	4'-0" x 7'-0"	▼	10		95	34	129	155
	344	For 8'-0" height, add				10		10	11M
	346	For solid wood core, add				30		30	33M
	372	H.P. plastic laminate, 2'-6" x 6'-8"	F-2	13	▼	125	26	151	175

For expanded coverage of these items see *Means' Interior Cost Data 1986*

8.2 Wood & Plastic Doors

			CREW	DAILY OUTPUT	UNIT	BARE COSTS MAT.	BARE COSTS INST.	BARE COSTS TOTAL	TOTAL INCL O&P
21	374	3'-0" x 6'-8"	F-2	12	Ea.	130	28	158	185
	378	3'-0" x 7'-0"	↓	11		140	31	171	200
	380	4'-0" x 7'-0"	↓	8		200	42	242	280
	384	For 8'-0" height, add				36		36	40M
	386	For solid wood core, add				30		30	33M
	400	Exterior, flush, solid wood core, birch, 1-3/4" x 7'-0" x 2'-6"	F-2	15		95	22	117	135
	402	2'-8" wide		15		98	22	120	140
	404	3'-0" wide		14		105	24	129	150
	410	Oak faced 1-3/4" x 7'-0" x 2'-6" wide		15		110	22	132	155
	412	2'-8" wide		15		115	22	137	160
	414	3'-0" wide		14		120	24	144	165
	420	Walnut faced 1-3/4" x 7'-0" x 2'-6" wide		15		185	22	207	235
	422	2'-8" wide		15		195	22	217	245
	424	3'-0" wide	↓	14		220	24	244	275
	430	For 6'-8" high door, deduct from 7'-0" door			↓	10%			
22-001		**WOOD DOORS, DECORATOR**							
	400	Hand carved door, mahogany, simple design							
	402	1-3/4" x 7'-0" x 3'-0" wide	F-2	14	Ea.	310	24	334	375
	404	3'-6" wide		13		405	26	431	485
	420	Rosewood, 1-3/4" x 7'-0" x 3'-0" wide		14		405	24	429	480
	422	3'-6" wide	↓	13		550	26	576	640
	428	For 6'-8" high door, deduct from 7'-0" door				10%			
	432	For detailed design, add				50%			
	434	For hand carved back, add				20%			
	436	For ornate mahogany door, 2-1/4" thick, add				20%			
	438	For ornate rosewood door, 2-1/4" thick, add				20%			
	440	For custom finish, add				80		80	88M
	460	Side panel, mahogany, simple design, 7'-0" x 1'-0" wide	F-2	21		63	16.05	79.05	92
	462	1'-2" wide		20		68	16.85	84.85	99
	464	1'-4" wide		19		74	17.75	91.75	105
	480	Rosewood, simple design 7'-0" x 1'-0" wide		21		105	16.05	121.05	140
	482	1'-2" wide		20		115	16.85	131.85	150
	484	1'-4" wide	↓	19		145	17.75	162.75	185
	490	For detailed design, add				50%			
	492	For hand carved back, add				20%			
	652	Interior cafe doors, 2'-6" opening, stock, panel pine	F-2	16		76	21	97	115
	654	3'-0" opening		16		84	21	105	125
	656	2'-6" opening		16		65	21	86	100
	800	3'-0" opening	↓	16	↓	75	21	96	115
	880	Pre-hung doors, see division 8.2-50							
23-001		**WOOD FIRE DOORS** Mineral core, 3 ply stile, "B" label,							
	004	1 hour, birch face, 2'-6" x 6'-8"	F-2	14	Ea.	92	24	116	135
	008	3'-0" x 6'-8"		13		105	26	131	155
	009	3'-0" x 7'-0"		12		120	28	148	170
	010	4'-0" x 7'-0"	↓	12	↓	140	28	168	195
	011								
	014	Oak face, 2'-6" x 6'-8"	F-2	14	Ea.	110	24	134	155
	018	3'-0" x 6'-8"		13		125	26	151	175
	019	3'-0" x 7'-0"		12		140	28	168	195
	020	4'-0" x 7'-0"		12		170	28	198	225
	024	Walnut face, 2'-6" x 6'-8"		14		170	24	194	220
	028	3'-0" x 6'-8"		13		185	26	211	240
	029	3'-0" x 7'-0"		12		195	28	223	255
	030	4'-0" x 7'-0"		12		225	28	253	290
	044	M.D. overlay on hardboard, 2'-6" x 6'-8"		15		88	22	110	130
	048	3'-0" x 6'-8"		14		105	24	129	150
	049	3'-0" x 7'-0"		13		120	26	146	170
	050	4'-0" x 7'-0"	↓	12	↓	140	28	168	195

For expanded coverage of these items see *Means' Interior Cost Data 1986*

8.2 Wood & Plastic Doors

			CREW	DAILY OUTPUT	UNIT	BARE COSTS MAT.	BARE COSTS INST.	BARE COSTS TOTAL	TOTAL INCL O&P
23	054	H.P. plastic laminate, 2'-6" x 6'-8"	F-2	13	Ea.	190	26	216	245
	058	3'-0" x 6'-8"		12		210	28	238	270
	059	3'-0" x 7'-0"		11		220	31	251	285
	060	4'-0" x 7'-0"		10		250	34	284	325
	074	90 minutes, birch face, 2'-6" x 6'-8"		14		115	24	139	160
	078	3'-0" x 6'-8"		13		125	26	151	175
	079	3'-0" x 7'-0"		12		135	28	163	190
	080	4'-0" x 7'-0"		12		160	28	188	215
	084	Oak face, 2'-6" x 6'-8"		14		130	24	154	175
	088	3'-0" x 6'-8"		13		145	26	171	195
	089	3'-0" x 7'-0"		12		155	28	183	210
	090	4'-0" x 7'-0"		12		185	28	213	245
	094	Walnut face, 2'-6" x 6'-8"		14		190	24	214	245
	098	3'-0" x 6'-8"		13		200	26	226	255
	099	3'-0" x 7'-0"		12		220	28	248	280
	100	4'-0" x 7'-0"		12		250	28	278	315
	114	M.D. overlay on hardboard, 2'-6" x 6'-8"		15		110	22	132	155
	118	3'-0" x 6'-8"		14		125	24	149	170
	119	3'-0" x 7'-0"		13		135	26	161	185
	120	4'-0" x 7'-0"		12		160	28	188	215
	124	For 8'-0" height, add				16		16	17.60M
	126	For 8'-0" height walnut, add				22		22	24M
	134	H.P. plastic laminate, 2'-6" x 6'-8"	F-2	13		215	26	241	275
	138	3'-0" x 6'-8"		12		230	28	258	295
	139	3'-0" x 7'-0"		11		240	31	271	310
	140	4'-0" x 7'-0"		10		270	34	304	345
	220	Custom architectural "A" label, flush, 1-3/4" thick, birch,							
	221	Solid core							
	222	(97) 2'-6" x 7'-0"	F-2	15	Ea.	325	22	347	390
	226	3'-0" x 7'-0"		14		350	24	374	420
	230	4'-0" x 7'-0"		13		430	26	456	510
	242	4'-0" x 8'-0"		11		475	31	506	565
	246	For 6'-8" high door, deduct from 7'-0" door				10%			
	248	For oak veneer, add				50%			
	250	For walnut veneer, add				75%			
25-001		**WOOD DOORS, PANELED** Interior, six panel, hollow core, 1-3/4" thick							
	004	Molded hardboard, 2'-0" x 6'-8"	F-2	17	Ea.	35	19.85	54.85	67
	006	2'-6" x 6'-8"		17		37	19.85	56.85	69
	008	3'-0" x 6'-8"		17		40	19.85	59.85	72
	014	Embossed print, molded hardboard, 2'-0" x 6'-8"		17		53	19.85	72.85	87
	016	2'-6" x 6'-8"		17		58	19.85	77.85	92
	018	3'-0" x 6'-8"		17		63	19.85	82.85	98
	054	Six panel, solid, 1-3/8" thick, pine, 2'-0" x 6'-8"		15		115	22	137	160
	056	2'-6" x 6'-8"		14		125	24	149	170
	058	3'-0" x 6'-8"		13		145	26	171	195
	102	Two panel, bored rail, solid, 1-3/8" thick, pine, 1'-6" x 6'-8"		16		135	21	156	180
	104	2'-0" x 6'-8"		15		195	22	217	245
	106	2'-6" x 6'-8"		14		235	24	259	295
	134	Two panel, solid, 1-3/8" thick, fir, 2'-0" x 6'-8"		15		125	22	147	170
	136	2'-6" x 6'-8"		14		130	24	154	175
	138	3'-0" x 6'-8"		13		140	26	166	190
	174	Five panel, solid, 1-3/8" thick, fir, 2'-0" x 6'-8"		15		135	22	157	180
	176	2'-6" x 6'-8"		14		140	24	164	190
	178	3'-0" x 6'-8"		13		145	26	171	195
32-001		**WOOD DOORS, RESIDENTIAL**							
	020	Exterior, combination storm & screen, pine							
	022	Cross buck, 6'-9" x 2'-6" wide	F-2	11	Ea.	135	31	166	190
	026	2'-8" wide	"	10	"	135	34	169	195

For expanded coverage of these items see *Means' Interior Cost Data 1986*

8.2 Wood & Plastic Doors

			CREW	DAILY OUTPUT	UNIT	BARE COSTS MAT.	BARE COSTS INST.	BARE COSTS TOTAL	TOTAL INCL O&P
32	028	3'-0" wide	F-2	9	Ea.	140	37	177	210
	030	7'-1" x 3'-0" wide		9		145	37	182	215
	040	Full lite, 6'-9" x 2'-6" wide		11		125	31	156	180
	042	2'-8" wide		10		125	34	159	185
	044	3'-0" wide		9		130	37	167	195
	050	7'-1" x 3'-0" wide		9		135	37	172	200
	070	Dutch door, pine, 1-3/4" x 6'-8" x 2'-8" wide, minimum		12		185	28	213	245
	072	Maximum		10		225	34	259	295
	080	3'-0" wide, minimum		12		195	28	223	255
	082	Maximum		10		240	34	274	310
	100	(92) Entrance door, colonial, 1-3/4" x 6'-8" x 2'-8" wide		16		155	21	176	200
	102	6 panel pine, 3'-0" wide		15		170	22	192	220
	110	8 panel pine, 2'-8" wide		16		195	21	216	245
	112	3'-0" wide		15		220	22	242	275
	120	For tempered safety glass lites, add				16		16	17.60M
	122								
	130	(92) Flush, birch, solid core, 1-3/4" x 6'-8" x 2'-8" wide	F-2	16	Ea.	105	21	126	145
	132	3'-0" wide		15		110	22	132	155
	134	7'-0" x 2'-8" wide		16		115	21	136	155
	136	3'-0" wide		15		120	22	142	165
	138	For tempered safety glass lites, add				35		35	39M
	270	Interior, closet, bi-folding, with hardware, no frame or trim incl.							
	272	Flush, birch, 6'-6" or 6'-8" x 2'-6" wide	F-2	13	Ea.	46	26	72	88
	274	3'-0" wide		13		49	26	75	91
	276	4'-0" wide		12		77	28	105	125
	278	5'-0" wide		11		82	31	113	135
	280	6'-0" wide		10		89	34	123	145
	300	Raised panel pine, 6'-6" or 6'-8" x 2'-6" wide		13		120	26	146	170
	302	3'-0" wide		13		130	26	156	180
	304	4'-0" wide		12		210	28	238	270
	306	5'-0" wide		11		240	31	271	310
	308	6'-0" wide		10		260	34	294	335
	320	Louvered, pine, 6'-6" or 6'-8" x 2'-6" wide		13		82	26	108	125
	322	3'-0" wide		13		88	26	114	135
	324	4'-0" wide		12		150	28	178	205
	326	5'-0" wide		11		160	31	191	220
	328	6'-0" wide		10		170	34	204	235
	440	Bi-passing closet, incl. hardware, no frame or trim incl.,							
	442	Flush, lauan, 6'-8" x 4'-0" wide	F-2	12	Opng.	62	28	90	110
	444	5'-0" wide		11		72	31	103	125
	446	6'-0" wide		10		81	34	115	135
	460	Flush, birch, 6'-8" x 4'-0" wide		12		79	28	107	125
	462	5'-0" wide		11		92	31	123	145
	464	6'-0" wide		10		105	34	139	165
	480	Louvered, pine, 6'-8" x 4'-0" wide		12		110	28	138	160
	482	5'-0" wide		11		125	31	156	180
	484	6'-0" wide		10		140	34	174	200
	500	Paneled, pine, 6'-8" x 4'-0" wide		12		215	28	243	275
	502	5'-0" wide		11		235	31	266	300
	504	6'-0" wide		10		260	34	294	335
	505								
	610	Folding accordion, closet, not including frame							
	612	Vinyl, 2 layer, stock (see also division 10.1-45)	F-2	400	S.F.	2.50	.84	3.34	3.96
	614	Woven mahogany and vinyl, stock		400		3.20	.84	4.04	4.73
	616	Wood slats with vinyl overlay, stock		400		5.50	.84	6.34	7.25
	618	Economy vinyl, stock		400		1.15	.84	1.99	2.47
	620	Rigid PVC		400		1.60	.84	2.44	2.97
	622	For custom folding, add to above			Ea.	25%			
	623								
	740	Passage doors, flush, no frame included							

For expanded coverage of these items see *Means' Interior Cost Data 1986*

8.2 Wood & Plastic Doors

			CREW	DAILY OUTPUT	UNIT	BARE COSTS MAT.	BARE COSTS INST.	BARE COSTS TOTAL	TOTAL INCL O&P
32	742	Lauan, hollow core, 1-3/8" x 6'-8" x 1'-6" wide	F-2	19	Ea.	16	17.75	33.75	43
	744	2'-0" wide		18		16	18.75	34.75	44
	746	2'-6" wide		18		18	18.75	36.75	47
	748	2'-8" wide		18		19	18.75	37.75	48
	750	3'-0" wide		17		20	19.85	39.85	50
	770	Birch, hollow core, 1-3/8" x 6'-8" x 1'-6" wide		19		20	17.75	37.75	47
	772	2'-0" wide		18		23	18.75	41.75	52
	774	2'-6" wide		18		26	18.75	44.75	55
	776	2'-8" wide		18		27	18.75	45.75	57
	778	3'-0" wide		17		29	19.85	48.85	60
	800	Pine louvered, 1-3/8" x 6'-8" x 1'-6" wide		19		45	17.75	62.75	75
	802	2'-0" wide		18		55	18.75	73.75	87
	804	2'-6" wide		18		72	18.75	90.75	105
	806	2'-8" wide		18		74	18.75	92.75	110
	808	3'-0" wide		17		78	19.85	97.85	115
	830	Pine paneled, 1-3/8" x 6'-8" x 1'-6" wide		19		65	17.75	82.75	97
	832	2'-0" wide		18		85	18.75	103.75	120
	834	2'-6" wide		18		90	18.75	108.75	125
	836	2'-8" wide		18		94	18.75	112.75	130
	838	3'-0" wide	F-2	17	Ea.	105	19.85	124.85	145
	855	(98) For over 20 doors, deduct				15%			
8	50-001	**PRE-HUNG DOORS**							
	030	Exterior, wood, combination storm & screen, 6'-9" x 2'-6" wide	F-2	15	Ea.	145	22	167	190
	032	2'-8" wide		15		145	22	167	190
	034	3'-0" wide		15		150	22	172	195
	036	For 7'-0" high door, add				7%			
	037	For aluminum storm doors, see division 8.1-85							
	160	Entrance door, flush, birch, solid core							
	162	4-5/8" solid jamb, 1-3/4" x 6'-8" x 2'-8" wide	F-2	16	Ea.	160	21	181	205
	164	3'-0" wide	"	16		165	21	186	210
	168	For 7'-0" high door, add				10%			
	200	Entrance door, colonial, 6 panel pine							
	202	4-5/8" solid jamb, 1-3/4" x 6'-8" x 2'-8" wide	F-2	16	Ea.	265	21	286	320
	204	3'-0" wide	"	16		275	21	296	335
	206	For 7'-0" high door, add				10%			
	220	For 5-5/8" solid jamb, add				10		10	11M
	230	French door, 6'-8" x 6'-0" wide, 1/2" insul. glass and grille	F-2	7	Pr.	605	48	653	735
	400	Interior, passage door, 4-5/8" solid jamb							
	440	Lauan, flush, solid core, 1-3/8" x 6'-8" x 2'-6" wide	F-2	20	Ea.	65	16.85	81.85	96
	442	2'-8" wide		20		68	16.85	84.85	99
	444	3'-0" wide		19		74	17.75	91.75	105
	460	Hollow core, 1-3/8" x 6'-8" x 2'-6" wide		20		49	16.85	65.85	78
	462	2'-8" wide		20		50	16.85	66.85	79
	464	3'-0" wide		19		53	17.75	70.75	84
	470	For 7'-0" high door, add				10%			
	500	Birch, flush, solid core, 1-3/8" x 6'-8" x 2'-6" wide	F-2	20		130	16.85	146.85	165
	502	2'-8" wide		20		110	16.85	126.85	145
	504	3'-0" wide		19		135	17.75	152.75	175
	520	Hollow core, 1-3/8" x 6'-8" x 2'-6" wide		20		63	16.85	79.85	93
	522	2'-8" wide		20		65	16.85	81.85	96
	524	3'-0" wide		19		67	17.75	84.75	99
	528	For 7'-0" high door, add				10%			
	550	Pine louvered, 1-3/8" x 6'-8" x 2'-6" wide	F-2	20		140	16.85	156.85	180
	552	2'-8" wide		20		145	16.85	161.85	185
	554	3'-0" wide		19		150	17.75	167.75	190
	600	Paneled, 1-3/8" x 6'-8" x 2'-6" wide		20		155	16.85	171.85	195
	602	2'-8" wide		20		155	16.85	171.85	195
	604	3'-0" wide		19		160	17.75	177.75	200
	620	(98)							

For expanded coverage of these items see *Means' Interior Cost Data 1986*

8.2 Wood & Plastic Doors

		CREW	DAILY OUTPUT	UNIT	BARE COSTS MAT.	BARE COSTS INST.	BARE COSTS TOTAL	TOTAL INCL O&P
650	For 5-5/8" solid jamb, add			Ea.	5.25		5.25	5.75M
652	For split jamb, deduct			"	5.25		5.25	5.75M

8.3 Special Doors

		CREW	DAILY OUTPUT	UNIT	BARE COSTS MAT.	BARE COSTS INST.	BARE COSTS TOTAL	TOTAL INCL O&P
02-001	**ACCESS PANELS** Metal, see division 9.5-15							
03-001	**ACOUSTICAL** Incl. framed seals, 3' x 7', wood, 27 STC rating	F-2	1.50	Ea.	375	225	600	735
010	Steel, 40 STC rating		1.50		1,050	225	1,275	1,475
020	45 STC rating		1.50		1,200	225	1,425	1,650
030	48 STC rating		1.50		1,400	225	1,625	1,850
040	53 STC rating	↓	1.50	↓	1,500	225	1,725	1,975
04-001	**BULKHEAD CELLAR DOORS** Steel, not incl. sides, minimum	1 Carp	5.50	Ea.	160	29	189	220
010	Maximum		5.10		185	31	216	250
050	With sides and foundation plates, minimum		4.70		170	34	204	235
060	Maximum	↓	4.30	↓	210	37	247	285
06-001	**COLD STORAGE** Single, galvanized steel							
030	Horizontal sliding, 5' x 7', manual operation, 2" thick	F-2	2	Ea.	2,125	170	2,295	2,575
040	4" thick		2		2,400	170	2,570	2,875
050	6" thick		2		2,700	170	2,870	3,200
080	Power operation, 2" thick		1.90		4,200	175	4,375	4,875
090	4" thick		1.90		4,675	175	4,850	5,400
100	6" thick		1.90		4,775	175	4,950	5,500
130	9' x 10', manual operation, 2" insulation		1.70		3,125	200	3,325	3,725
140	4" insulation		1.70		3,800	200	4,000	4,475
150	6" insulation		1.70		4,000	200	4,200	4,675
180	Power operation, 2" insulation		1.60		5,300	210	5,510	6,125
190	4" insulation		1.60		5,925	210	6,135	6,825
200	6" insulation	↓	1.70	↓	6,125	200	6,325	7,025
230	For stainless steel face, add			S.F.	20%			
300	Hinged, lightweight, 3' x 6'-6", galvanized 1 face, 2" thick	2 Carp	2	Ea.	965	160	1,125	1,300
305	4" thick		1.90		975	170	1,145	1,325
330	Aluminum doors, 3' x 6'-6", 4" thick		1.90		1,200	170	1,370	1,575
335	6" thick		1.40		1,300	230	1,530	1,750
360	Stainless steel, 3' x 6'-6", 4" thick		1.90		1,600	170	1,770	2,000
365	6" thick		1.40		2,450	230	2,680	3,025
390	Galvanized 2 face, 3' x 6'-6", 4" thick		1.90		900	170	1,070	1,225
395	6" thick	↓	1.40	↓	940	230	1,170	1,375
500	Bi-parting, electric operated							
501	6' x 8' opening, galvanized faces, for cooler	2 Carp	.80	Opng.	5,125	400	5,525	6,225
505	For freezer		.80		6,050	400	6,450	7,225
530	For door buck framing and door protection, add		2.50		500	130	630	735
600	Galvanized batten door, galvanized hinges, 4' x 7'		2		1,100	160	1,260	1,450
605	6' x 8'		1.80		1,550	180	1,730	1,975
650	Fire door, 3 hr., 6' x 8', single slide		.80		6,700	400	7,100	7,950
655	Double, bi-parting		.70		7,800	455	8,255	9,250
09-001	**COUNTER DOORS** 4' high roll-up, 6' long, galv. steel or aluminum		2		600	160	760	890
030	Galvanized steel, UL label		1.80		760	180	940	1,100
060	Stainless steel, 4' high roll-up, 6' long		2		905	160	1,065	1,225
070	10' long		1.80		1,150	180	1,330	1,525
12-001	**DARKROOM DOORS** Revolving, standard, 2 way, 30" diameter		3.50		765	91	856	975
005	3 way, 43" diameter		3		1,175	105	1,280	1,450
100	4 way, 68" diameter	↓	2	↓	1,950	160	2,110	2,375
101								

For expanded coverage of these items see *Means' Interior Cost Data 1986*

8.3 Special Doors

				Daily		Bare Costs			Total
			Crew	Output	Unit	Mat.	Inst.	Total	Incl O&P
12	200	Hinged safety, 2 way, 30" diameter	2 Carp	3.20	Opng.	970	100	1,070	1,200
	250	3 way, 43" diameter		2.90		1,500	110	1,610	1,800
	300	Pop out safety, 2 way, 30" diameter		3.10		1,350	105	1,455	1,625
	400	3 way, 43" diameter	↓	2.80	↓	1,650	115	1,765	1,975
	930	For complete dark rooms, see division 13.1-25							
	13-001	**DOUBLE ACTING** With vision panel, incl. frame, closer & hardware							
	100	.063" aluminum, 7'-2" high, 3'-4" wide	2 Carp	4.20	Pr.	550	76	626	715
	105	6'-4" wide	"	4	"	975	80	1,055	1,200
	200	Solid core wood, 1-3/4" thick, metal frame, stainless steel							
	201	base plate, 7' high opening, 4' wide	2 Carp	4	Pr.	635	80	715	815
	205	7' wide	"	3.80	"	1,175	84	1,259	1,425
	14-001	**FLEXIBLE TRANSPARENT STRIP ENTRANCE**							
	002	See division 13.1-01							
	15-001	**FLOOR, COMMERCIAL** Aluminum tile, steel frame, single leaf, 2'x2' opng.	2 Sswk	3.50	Opng.	310	99	409	495
	005	3'-6" x 3'-6" opening		3.50		550	99	649	760
	050	Double leaf, 4' x 4' opening		3		780	115	895	1,050
	055	5' x 5' opening		3		1,125	115	1,240	1,425
	18-001	**FLOOR, INDUSTRIAL** Steel 300 psf L.L., single leaf, 2' x 2' opening		6		340	58	398	465
	005	3' x 3' opening		5.50		490	63	553	640
	030	Double leaf, 4' x 4' opening		5		720	69	789	900
	035	5' x 5' opening		4.50		960	77	1,037	1,175
	100	Aluminum, 300 psf L.L., single leaf, 2' x 2' opening		6		390	58	448	520
	105	3' x 3' opening		5.50		550	63	613	705
	150	Double leaf, 4' x 4' opening		5		840	69	909	1,025
	155	5' x 5' opening		4.50		1,250	77	1,327	1,500
	200	Aluminum, 150 psf L.L., single leaf, 2' x 2' opening		6		275	58	333	395
	205	3' x 3' opening		5.50		415	63	478	555
	250	Double leaf, 4' x 4' opening		5		685	69	754	865
	255	5' x 5' opening	↓	4.50		1,050	77	1,127	1,275
	21-001	**GLASS, SLIDING** Vinyl clad, 1" insulated glass, 6'-0" x 6'-10" high	2 Carp	4		640	80	720	820
	010	8'-0" x 6'-10" high		4		755	80	835	945
	050	3 leafs, 9'-0" x 6'-10" high		3		940	105	1,045	1,200
	060	12'-0" x 6'-10" high		3		1,175	105	1,280	1,450
	24-001	**GLASS, SWING** Tempered, 1/2" thick, incl. hardware, 3' x 7' opening	2 Glaz	2		1,600	160	1,760	2,000
	010	6' x 7' opening	"	1.40	↓	3,200	230	3,430	3,850
	27-001	**HANGAR DOOR** Bi-fold, overhead, 20 PSF wind load, incl. electric operator							
	010	12' high	2 Sswk	240	S.F.	10	1.45	11.45	13.25
	020	16' high		230		10.50	1.51	12.01	13.90
	030	20' high	↓	220	↓	12	1.58	13.58	15.70
	30-001	**JALOUSIE** 1-3/4" thick, fir, with screens, plexiglass	2 Carp	3.20	Opng.	270	100	370	440
	010	With tempered glass		3.20		280	100	380	455
	33-001	**KALAMEIN** Interior, flush type, 3' x 7'	↓	4.30	↓	230	74	304	360
	36-001	**KENNEL** 2 way, swinging type, 17" x 19" opening	2 Carp	11	Opng.	65	29	94	115
	010	17" x 24" opening	"	11	"	75	29	104	125
	39-001	**OVERHEAD, COMMERCIAL** Frames not included							
	100	Stock, sectional, heavy duty, wood, 1-3/4" thick, 8' x 8' high	2 Carp	2	Ea.	395	160	555	665
	110	10' x 10' high		1.80		550	180	730	865
	120	12' x 12' high		1.50		780	215	995	1,175
	130	Chain hoist, 14' x 14' high		1.30		1,300	245	1,545	1,775
	140	12' x 16' high		1		1,375	320	1,695	1,975
	150	20' x 8' high		.80		1,125	400	1,525	1,825
	160	20' x 16' high		.60		2,450	535	2,985	3,475
	180	Center mullion openings, 8' high		4		290	80	370	435
	190	20' high	↓	2		550	160	710	835
	210	For medium duty custom doors, deduct				5%	5%		
	215	For medium duty stock doors, deduct				20%	5%		
	230	Fiberglass and aluminum, heavy duty, sectional, 12' x 12' high	2 Carp	1.50		870	215	1,085	1,275
	245	Chain hoist, 20' x 20' high	"	.50	↓	2,200	640	2,840	3,350

For expanded coverage of these items see *Means' Interior Cost Data 1986*

8.3 Special Doors

			CREW	DAILY OUTPUT	UNIT	BARE COSTS MAT.	BARE COSTS INST.	BARE COSTS TOTAL	TOTAL INCL O&P
39	260	Steel, 24 ga. sectional, manual, 8' x 8' high	2 Carp	2	Ea.	335	160	495	600
	265	10' x 10' high		1.80		495	180	675	800
	270	12' x 12' high		1.50		695	215	910	1,075
	280	Chain hoist, 20' x 14' high	↓	.70	↓	1,675	455	2,130	2,500
	285	For 1-1/4" rigid insulation and 26 ga. galv.							
	286	back panel, add			S.F.	2		2	2.20M
	290	For electric trolley operator, to 14' x 14', add	1 Carp	2	Ea.	500	80	580	665
	295	Over 14' x 14', add	"	1	"	600	160	760	890
42-001		RESIDENTIAL GARAGE DOORS Including hardware, no frame							
	002								
	005	Hinged, wood, custom, double door, 9' x 7'	2 Carp	3	Ea.	210	105	315	385
	007	16' x 7'		2		275	160	435	535
	020	Overhead, sectional, incl. hardware, fiberglass, 9' x 7', standard		8		295	40	335	380
	022	Deluxe		8		325	40	365	415
	030	16' x 7', standard		6		485	53	538	610
	032	Deluxe		6		540	53	593	670
	050	Hardboard, 9' x 7', standard		8		195	40	235	270
	052	Deluxe		8		265	40	305	350
	060	16' x 7', standard		6		285	53	338	390
	062	Deluxe		6		530	53	583	660
	070	Metal, 9' x 7', standard		8		240	40	280	320
	072	Deluxe		8		375	40	415	470
	080	16' x 7', standard		6		345	53	398	455
	082	Deluxe		6		710	53	763	860
	090	Wood, 9' x 7', standard		8		210	40	250	290
	092	Deluxe		8		620	40	660	740
	100	16' x 7', standard		6		410	53	463	530
	102	Deluxe	↓	6		785	53	838	940
	180	Door hardware only, sectional	1 Carp	4		50	40	90	115
	182	One side only	"	7		30	23	53	66
	300	Swing-up, including hardware, fiberglass, 9' x 7', standard	2 Carp	8		195	40	235	270
	302	Deluxe		8		295	40	335	380
	310	16' x 7' high		6		340	53	393	450
	312	Deluxe		6		400	53	453	515
	320	Hardboard, 9' x 7', standard		8		215	40	255	295
	322	Deluxe		8		285	40	325	370
	330	16' x 7', standard		6		335	53	388	445
	332	Deluxe		6		435	53	488	555
	340	Metal, 9' x 7', standard		8		215	40	255	295
	342	Deluxe		8		345	40	385	435
	350	16' x 7', standard		6		370	53	423	485
	352	Deluxe		6		620	53	673	760
	360	Wood, 9' x 7', standard		8		160	40	200	235
	362	Deluxe		8		285	40	325	370
	370	16' x 7', standard		6		310	53	363	420
	372	Deluxe	↓	6		450	53	503	570
	390	Door hardware only, swing up	1 Carp	4		50	40	90	115
	392	One side only		7		30	23	53	66
	400	For electric operator, economy, add		8		180	20	200	225
	410	Deluxe, including remote control	↓	8	↓	260	20	280	315
	450	For electronic control, 1 transmitter, add to operator			Total	50		50	55M
	460	2 transmitters, add to operator			"	85		85	94M
45-001		ROLLING SERVICE DOORS Steel, manual, 20 ga., 8' x 8' high, standard	2 Sswk	1.60	Ea.	620	215	835	1,025
	010	10' x 10' high		1.40		770	250	1,020	1,225
	020	20' x 10' high		1		1,650	345	1,995	2,350
	030	12' x 12' high		1.20		965	290	1,255	1,525
	040	20' x 12' high		.90		1,675	385	2,060	2,450
	050	14' x 14' high		.80		1,300	435	1,735	2,100
	060	20' x 16' high		.60		2,250	580	2,830	3,375
	070	10' x 20' high	↓	.50	↓	1,425	695	2,120	2,650

For expanded coverage of these items see *Means' Interior Cost Data 1986*

8.3 Special Doors

			CREW	DAILY OUTPUT	UNIT	BARE COSTS MAT.	BARE COSTS INST.	BARE COSTS TOTAL	TOTAL INCL O&P
45	100	12' x 12', crank operated, crank on door side	2 Sswk	.80	Ea.	1,250	435	1,685	2,050
	110	Crank thru wall	"	.70		1,300	495	1,795	2,200
	130	For vision panel, add			↓	50		50	55M
	140	For 22 ga., deduct			S.F.	.25		.25	.27M
	160	3' x 7' pass door within rolling steel door, new construction			Ea.	400		400	440M
	170	Existing construction	2 Sswk	2		500	175	675	825
	200	Class A fire doors, manual, 20 ga., 8' x 8' high		1.40		1,200	250	1,450	1,700
	210	10' x 10' high		1.10		1,600	315	1,915	2,250
	220	20' x 10' high		.80		2,800	435	3,235	3,750
	230	12' x 12' high		1		2,000	345	2,345	2,750
	240	20' x 12' high		.80		3,200	435	3,635	4,200
	250	14' x 14' high		.60		2,600	580	3,180	3,775
	260	20' x 16' high		.50		3,500	695	4,195	4,950
	270	10' x 20' high	↓	.40	↓	2,750	870	3,620	4,400
	300	For 18 ga. doors, add			S.F.	.75		.75	.82M
	330	For enamel finish, add			"	.52		.52	.57M
	360	For safety edge bottom bar, pneumatic, add			L.F.	10		10	11M
	370	Electric, add				14		14	15.40M
	400	For weatherstripping, extruded rubber, jambs, add				7		7	7.70M
	410	Hood, add				9		9	9.90M
	420	Sill, add			↓	3.50		3.50	3.85M
	450	Motor operators, to 14' x 14' opening	2 Sswk	5	Ea.	400	69	469	550
	460	Over 14' x 14', jack shaft type	"	5		500	69	569	660
	470	For fire door, fusible link, add			↓	100		100	110M
51-001		ROLL UP GRILLE Aluminum, manual operated, mill finish	2 Sswk	82	S.F.	14.20	4.23	18.43	22
010		Bronze anodized		82	"	17.80	4.23	22.03	26
040		Steel, manual operated, 10' x 10' high		1	Opng.	900	345	1,245	1,525
050		15' x 8' high	↓	.80		1,250	435	1,685	2,050
100		For safety edge bottom bar, add				220		220	240M
110		For motor operation, add	2 Sswk	5	↓	670	69	739	845
52-001		SECURITY GATES See division 10.1-72							
54-001		SHOCK ABSORBING Rigid, no frame, insulated, 1-13/16" thick, 5' x 7'	2 Sswk	1.90	Opng.	1,450	185	1,635	1,875
010		8' x 8'		1.80		1,875	195	2,070	2,375
050		Flexible, frame not incl., 5' x 7' opening, economy		2		1,250	175	1,425	1,650
060		Deluxe		1.90		1,800	185	1,985	2,275
100		8' x 8' opening, economy		2		1,900	175	2,075	2,375
110		Deluxe	↓	1.90	↓	2,200	185	2,385	2,700
57-001		SLIDING Steel, up to 50' x 18', electric, standard duty, minimum	L-5	360	S.F.	12.20	4.47	16.67	19.90
010		Maximum		340		21	4.74	25.74	30
050		Heavy duty, minimum		297		15.50	5.40	20.90	25
060		Maximum	↓	277	↓	46	5.80	51.80	59
60-001		SWING Tubular steel, 7' high, single, 3'-4" opening	2 Sswk	2.50	Ea.	285	140	425	530
010		Double, 6'-0" opening	"	2	Pr.	505	175	680	830
63-001		TELESCOPING STEEL DOORS Baked enamel finish,							
100		Overhead, .03" thick, electric operated, 10' x 10'	E-3	.80	Ea.	4,450	805	5,255	6,100
200		20' x 10'		.60		5,800	1,075	6,875	7,975
300		20' x 16'	↓	.40	↓	6,700	1,600	8,300	9,775
66-001	(99)	TIN CLAD 3 ply, 6' x 7', double sliding, manual	2 Carp	1	Opng.	1,940	320	2,260	2,600
100		For electric operator, add	1 Elec	2	"	1,400	90	1,490	1,675
69-001		VAULT FRONT Door and frame, 32" x 78", clear opening							
010		1 hour test, weighs 750 lbs.	2 Sswk	1.50	Opng.	1,350	230	1,580	1,850
020		2 hour test, 32" door, weighs 950 lbs.		1.30		1,875	265	2,140	2,475
025		40" door, weighs 1130 lbs.		1		2,100	345	2,445	2,850
030		4 hour test, 32" door, weighs 1025 lbs.		1.20		2,125	290	2,415	2,800
035		40" door, weighs 1140 lbs.	↓	.90		2,325	385	2,710	3,175
050		For stainless steel front, including frame, add to above			↓	1,625		1,625	1,800M
055		Back, add				1,625		1,625	1,800M
060		For time lock, two movement, add	1 Elec	2	Ea.	1,000	90	1,090	1,225
065		Three movement, add	"	2	"	1,550	90	1,640	1,825

For expanded coverage of these items see *Means' Interior Cost Data 1986*

8.3 Special Doors

		CREW	DAILY OUTPUT	UNIT	BARE COSTS MAT.	BARE COSTS INST.	BARE COSTS TOTAL	TOTAL INCL O&P
080	Day gate, painted, wire mesh, 32" wide	2 Sswk	1.50	Ea.	900	230	1,130	1,350
085	40" wide		1.40		1,100	250	1,350	1,600
090	Aluminum, 32" wide		1.50		1,325	230	1,555	1,825
095	40" wide		1.40		1,550	250	1,800	2,100
205	Security vault door, 3-1/2" thick, class 5R, minimum		.40	Opng.	15,500	870	16,370	18,400
210	Maximum		.30		24,000	1,150	25,150	28,200
215	7" thick, class 9R, minimum		.40		21,500	870	22,370	25,000
220	Maximum		.30		31,000	1,150	32,150	35,900
225	For Western states, add				5%			
230								
78-001	VERTICAL LIFT Doors, motor operator, incl. frame, 16' x 16' high	E-3	.50	Ea.	11,500	1,275	12,775	14,600
010	32' x 24' high	E-2	.75	"	21,200	2,650	23,850	26,900

8.4 Entrances & Storefronts

		CREW	DAILY OUTPUT	UNIT	BARE COSTS MAT.	BARE COSTS INST.	BARE COSTS TOTAL	TOTAL INCL O&P
05-001	ALUMINUM & GLASS DOORS See division 8.1-40							
10-001	BALANCED DOORS Incl. hdwre & frame, alum. & glass, 3' x 7', economy	2 Sswk	.90	Ea.	2,225	385	2,610	3,050
015	Premium		.70		3,750	495	4,245	4,900
050	Stainless steel and glass, 3' x 7', economy		.90		3,525	385	3,910	4,475
060	Premium		.70		6,000	495	6,495	7,375
20-001	REVOLVING DOORS 6'-6" to 7'-0" diameter							
002	6'-10" to 7' high, stock units, minimum	4 Sswk	.75	Opng.	15,000	925	15,925	18,000
005	Average		.60		19,000	1,150	20,150	22,700
010	Maximum		.45		23,000	1,550	24,550	27,700
100	Stainless steel		.30		28,000	2,325	30,325	34,400
110	Solid bronze		.15		35,000	4,625	39,625	45,800
150	For automatic controls, add	2 Elec	2		3,000	180	3,180	3,550
25-001	SLIDING ENTRANCE 12' x 7'-6" opening, 5' x 7' door, two way traffic,							
002	mat activated, panic pushout, incl. operator & hardware,							
003	not incl. glass or glazing	2 Glaz	.70	Opng.	7,000	460	7,460	8,350
30-001	SLIDING PANEL Mall fronts, aluminum & glass, 15' x 9' high	2 Glaz	1.30	Opng.	1,700	250	1,950	2,225
010	24' x 9' high		.70		2,700	460	3,160	3,625
020	48' x 9' high, with fixed panels		.90		5,200	360	5,560	6,225
050	For bronze finish, add				15%			
35-001	STAINLESS STEEL and glass entrance unit, narrow stiles							
002	3' x 7' opening, including hardware, minimum	2 Sswk	1.60	Opng.	1,350	215	1,565	1,825
005	Average		1.40		2,500	250	2,750	3,150
010	Maximum		1.20		3,450	290	3,740	4,250
100	For solid bronze entrance units, statuary finish, add				60%			
110	Without statuary finish, add				45%			
40-001	STOREFRONT SYSTEMS Aluminum frame, clear 3/8" plate glass,							
002	incl. 3' x 7' door with hardware (400 sq. ft. max. wall)							
050	Wall height to 12' high, commercial grade	2 Glaz	150	S.F.	11.55	2.15	13.70	15.75
060	Institutional grade		130		13.50	2.48	15.98	18.40
070	Monumental grade		115		18.20	2.80	21	24
090								
100	6' x 7' door with hardware, commerical grade	2 Glaz	135	S.F.	14.50	2.39	16.89	19.35
110	Institutional grade		115		17.50	2.80	20.30	23
120	Monumental grade		100		24	3.22	27.22	31
150	For bronze anodized finish, add				15%			
160	For black anodized finish, add				20%			
170	For stainless steel framing, add to monumental				75%			

For expanded coverage of these items see *Means' Interior Cost Data 1986*

8.4 Entrances & Storefronts

			CREW	DAILY OUTPUT	UNIT	BARE COSTS MAT.	BARE COSTS INST.	BARE COSTS TOTAL	TOTAL INCL O&P
40	200	For no 3' x 7' door and hardware, deduct			S.F.	3.50		3.50	3.85M
	250	For no 6' x 7' door and hardware, deduct			"	6.50		6.50	7.15M
	60-001	**SWING DOORS** Aluminum entrance, 6' x 7', incl. hardware & operator	2 Sswk	.70	Opng.	4,500	495	4,995	5,725
	002	For anodized finish, add			"	380		380	420M

8.5 Metal Windows

			CREW	DAILY OUTPUT	UNIT	BARE COSTS MAT.	BARE COSTS INST.	BARE COSTS TOTAL	TOTAL INCL O&P
	10-001	**ALUMINUM SASH** Stock, grade 2, glazing & trim not included, casement	2 Sswk	200	S.F.	10.65	1.74	12.39	14.45
	005	Double hung		200		6.35	1.74	8.09	9.70
	010	Fixed casement		200		4.95	1.74	6.69	8.15
	015	Picture window		200		5.60	1.74	7.34	8.90
	020	Projected window		200		12.60	1.74	14.34	16.60
	025	Single hung		200		5.90	1.74	7.64	9.20
	030	Sliding		200		8.10	1.74	9.84	11.65
	100	Mullions for above, tubular		240	L.F.	1.55	1.45	3	3.98
	200	Custom aluminum sash, grade 3, glazing not included, minimum		200	S.F.	15.70	1.74	17.44	20
	210	Maximum		85	"	20.65	4.08	24.73	29
	20-001	**ALUMINUM WINDOWS** Including frame and glazing, grade 2							
	002	See also division 5.1-35-650 thru 760							
	100	Stock units, casement, 3'-1" x 3'-2" opening	2 Sswk	10	Ea.	165	35	200	235
	105	Add for storms				30		30	33M
	160	Projected, with screen, 3'-1" x 3'-2" opening	2 Sswk	10		115	35	150	180
	170	Add for storms				32		32	35M
	200	4'-5" x 5'-3" opening	2 Sswk	8		160	43	203	245
	210	Add for storms				46		46	51M
	250	Enamel finish windows, 3'-1" x 3'-2"	2 Sswk	10		105	35	140	170
	260	4'-5" x 5'-3"		8		180	43	223	265
	300	Single hung, 2' x 3' opening, enameled, standard glazed		10		78	35	113	140
	310	Insulating glass		10		125	35	160	190
	330	2'-8" x 6'-8" opening, standard glazed		8		165	43	208	250
	340	Insulating glass		8		205	43	248	295
	370	3'-4" x 5'-0" opening, standard glazed		9		105	39	144	175
	380	Insulating glass		9		140	39	179	215
	382	Folding type, 3' x 4' opening, standard glass		10		120	35	155	185
	384	Insulating glass		10		135	35	170	205
	386	4' x 5' opening, standard glass		10		175	35	210	245
	388	Insulating glass		10		200	35	235	275
	389	Awning type, 3' x 3' opening standard glass		14		165	25	190	220
	390	Insulating glass		14		185	25	210	240
	391	3' x 4' opening, standard glass		10		190	35	225	265
	392	Insulating glass		168		210	2.07	212.07	235
	393	3' x 5'-4" opening, standard glass		10		230	35	265	310
	394	Insulating glass		10		275	35	310	355
	395	4' x 5'-4" opening, standard glass		9		255	39	294	340
	396	Insulating glass		9		290	39	329	380
	400	Sliding aluminum, 3' x 2' opening, standard glazed		10		53	35	88	115
	410	Insulating glass		10		88	35	123	150
	430	5' x 3' opening, standard glazed		9		89	39	128	160
	440	Insulating glass		9		145	39	184	220
	460	8' x 4' opening, standard glazed		6		128	58	186	230
	470	Insulating glass		6		255	58	313	370
	500	9' x 5' opening, standard glazed		4		200	87	287	355
	510	Insulating glass		4		335	87	422	505
	550	Sliding, with thermal barrier and screen, 6' x 4', 2 track		8		275	43	318	370
	570	4 track		8		340	43	383	440

8.5 Metal Windows

		CREW	DAILY OUTPUT	UNIT	BARE COSTS MAT.	BARE COSTS INST.	BARE COSTS TOTAL	TOTAL INCL O&P
600	For above units with bronze finish, add			Ea.	8%			
620	For installation in concrete openings, add					80%		
30-001	JALOUSIES Aluminum incl. glazing & screens, stock, 1'-7" x 3'-2"	2 Sswk	10		63	35	98	125
010	2'-3" x 4'-0"		10		115	35	150	180
020	3'-1" x 2'-0"		10		110	35	145	175
030	3'-1" x 5'-3"		10		150	35	185	220
100	Mullions for above, 2'-0" long		80		6.50	4.34	10.84	13.95
110	5'-3" long		80		11.55	4.34	15.89	19.50
40-001	LOUVERS See division 4.1-80, 6.2-32 & 7.6-42							
50-001	SCREENS For metal sash, aluminum or bronze mesh, flat screen	2 Sswk	1,200	S.F.	2.30	.29	2.59	2.98
050	Wicket screen, inside window		1,000		3.45	.35	3.80	4.34
080	Security screen, aluminum frame with stainless steel cloth		1,200		12.60	.29	12.89	14.30
090	Steel grate, painted, on steel frame		1,600		6.85	.22	7.07	7.90
100	For solar louvers, add		160		13.15	2.17	15.32	17.85
400	See also division 5.4-63 & 5.4-74							
60-001	STEEL SASH Custom units, glazing and trim not included,							
010	Casement, 100% vented	2 Sswk	200	S.F.	17.15	1.74	18.89	22
020	50% vented		200		13.15	1.74	14.89	17.20
030	Fixed		200		8.90	1.74	10.64	12.50
100	Projected, commercial, 40% vented		200		17.10	1.74	18.84	22
110	Intermediate, 50% vented		200		18.90	1.74	20.64	24
150	Industrial, horizontally pivoted		200		14.35	1.74	16.09	18.50
160	Fixed		200		12.60	1.74	14.34	16.60
200	Industrial security sash, 50% vented		200		22.55	1.74	24.29	28
210	Fixed		200		18.30	1.74	20.04	23
250	Picture window		200		7.75	1.74	9.49	11.25
300	Double hung		200		21	1.74	22.74	26
500	Mullions for above, open interior face		240	L.F.	3.90	1.45	5.35	6.55
510	With interior cover		240	"	7.75	1.45	9.20	10.80
70-001	STEEL WINDOWS Stock, including frame, trim and insulating glass							
002	See also division 5.1-35-650							
100	Stock units, double hung, 2'-8" x 4'-6" opening	2 Sswk	12	Ea.	125	29	154	185
110	2'-4" x 3'-9" opening		12		105	29	134	160
150	Commercial projected, 3'-9" x 5'-5" opening		10		225	35	260	300
160	6'-9" x 4'-1" opening		7		265	50	315	370
200	Intermediate projected, 2'-9" x 4'-1" opening		12		115	29	144	170
210	4'-1" x 5'-5" opening		10		175	35	210	245
85-001	STORM WINDOWS Aluminum, residential							
030	Basement, mill finish, incl. fiberglass screen							
032	1'-10" x 1'-0" high	F-2	30	Ea.	11.40	11.25	22.65	29
034	2'-9" x 1'-6" high	"	30	"	17	11.25	28.25	35
160	Double-hung, combination, storm & screen							
200	Average quality, anodized, 2'-0" x 3'-5" high	F-2	30	Ea.	35.90	11.25	47.15	56
202	2'-6" x 5'-0" high		28		40	12.05	52.05	61
204	4'-0" x 6'-0" high		25		49	13.50	62.50	73
240	White painted, 2'-0" x 3'-5" high		30		34	11.25	45.25	54
244	4'-0" x 6'-0" high		25		47	13.50	60.50	71
260	Mill finish, 2'-0" x 3'-5" high		30		30	11.25	41.25	49
262	2'-6" x 5'-0" high		28		35	12.05	47.05	56
264	4'-0" x 6-8" high		25		43	13.50	56.50	67

For expanded coverage of these items see Means' *Interior Cost Data 1986*

8.6 Wood Windows

		CREW	DAILY OUTPUT	UNIT	BARE COSTS MAT.	BARE COSTS INST.	BARE COSTS TOTAL	TOTAL INCL O&P
10-001	**AWNING WINDOW** Including frame, screen, and exterior trim							
002								
010	Average quality, builders model, 34" x 22", standard glazed	1 Carp	10	Ea.	70	16	86	100
020	Insulating glass		10		90	16	106	120
030	40" x 28", standard glazed		9		85	17.80	102.80	120
040	Insulating glass		9		115	17.80	132.80	150
050	48" x 36", standard glazed		8		105	20	125	145
060	Insulating glass		8		145	20	165	190
200	Metal clad, deluxe, insulating glass, 34" x 22"		10		180	16	196	220
210	40" x 22"		10		185	16	201	225
220	36" x 28"		9		195	17.80	212.80	240
230	40" x 28"		9		205	17.80	222.80	250
240	48" x 28"		8		230	20	250	280
250	60" x 36"	↓	8	↓	270	20	290	325
20-001	**BOW-BAY WINDOW** Including frame, screen and exterior trim,							
002	end panels operable							
100	Awning type, builders model, 8'-0" x 5'-0" high, standard glazed	2 Carp	10	Ea.	510	32	542	605
105	Insulating glass		10		590	32	622	695
110	12'-0" x 6'-0" high, standard glazed		6		880	53	933	1,050
120	Insulating glass, 6 panels		6		965	53	1,018	1,150
160	Metal clad, deluxe, insul. glass, 6'-0" x 4'-0" high, 3 panels		10		970	32	1,002	1,125
164	9'-0" x 4'-0" high, 4 panels		8		1,300	40	1,340	1,500
168	10'-0" x 5'-0" high, 5 panels		7		1,700	46	1,746	1,925
172	12'-0" x 6'-0" high, 6 panels		6		2,100	53	2,153	2,375
200	Casement type, bldrs. model, 8'-0" x 5'-0" high, standard glazed		10		450	32	482	540
205	Insulating glass		10		600	32	632	705
210	12'-0" x 6'-0" high, 6 panels, standard glazed		6		700	53	753	845
220	Insulating glass		6		930	53	983	1,100
260	Metal clad, deluxe, insul. glass, 8'-0" x 5'-0" high, 4 panels		10		1,150	32	1,182	1,300
264	10'-0" x 5'-0" high, 5 panels		8		1,475	40	1,515	1,675
268	10'-0" x 6'-0" high, 5 panels		7		1,620	46	1,666	1,850
272	12'-0" x 6'-0" high, 6 panels		6		1,980	53	2,033	2,250
300	Double hung type, bldrs. model, 8'-0" x 4'-0" high, std. glazed		10		650	32	682	760
305	Insulating glass		10		720	32	752	840
310	9'-0" x 5'-0" high, 4 panels, standard glazed		6		700	53	753	845
320	Insulating glass		6		810	53	863	970
360	Metal clad, deluxe, insul. glass, 7'-0" x 4'-0" high, 3 panels		10		1,125	32	1,157	1,275
364	8'-0" x 4'-0" high, 4 panels		8		1,150	40	1,190	1,325
368	8'-0" x 5'-0" high, 4 panels		7		1,250	46	1,296	1,450
372	9'-0" x 5'-0" high, 4 panels	↓	6	↓	1,375	53	1,428	1,600
25-001	**WEATHERSTRIPPING** See division 8.7-70							
31-001	**CASEMENT WINDOW** Including frame, screen, and exterior trim							
002								
010	Average quality, bldrs. model, 2'-0" x 3'-0" high, standard glazed	1 Carp	10	Ea.	96	16	112	130
015	Insulating glass		10		120	16	136	155
020	2'-0" x 4'-6" high, standard glazed		9		125	17.80	142.80	165
025	Insulating glass		9		160	17.80	177.80	200
030	2'-0" x 6'-0" high, standard glazed		8		140	20	160	185
035	Insulating glass		8		190	20	210	240
200	Metal clad, deluxe, insulating glass, 2'-0" x 3'-0" high		10		185	16	201	225
204	2'-0" x 4'-0" high		9		225	17.80	242.80	275
208	2'-0" x 5'-0" high		8		250	20	270	305
212	2'-0" x 6'-0" high	↓	8	↓	300	20	320	360
220	For multiple leaf units, deduct for stationary sash							
221	2' high			Ea.	17		17	18.70M
230	4'-6" high				20		20	22M
240	6' high				24		24	26M
300	For installation, add per leaf			↓		15%		

8.6 Wood Windows		CREW	DAILY OUTPUT	UNIT	BARE COSTS			TOTAL INCL O&P
					MAT.	INST.	TOTAL	
40-001	**DOUBLE HUNG** Including frame, screen, and exterior trim							
002								
010	Average quality, bldrs. model, 2'-0" x 3'-0" high, standard glazed	1 Carp	10	Ea.	68	16	84	98
015	Insulating glass		10		100	16	116	135
020	3'-0" x 4'-0" high, standard glazed		9		91	17.80	108.80	125
025	Insulating glass		9		145	17.80	162.80	185
030	4'-0" x 4'-6" high, standard glazed		8		110	20	130	150
035	Insulating glass		8		160	20	180	205
200	Metal clad, deluxe, insulating glass, 2'-6" x 3'-0" high		10		190	16	206	230
210	3'-0" x 3'-6" high		10		220	16	236	265
220	3'-0" x 4'-0" high		9		235	17.80	252.80	285
230	3'-0" x 4'-6" high		9		252	17.80	269.80	305
240	3'-0" x 5'-0" high		8		280	20	300	335
250	3'-6" x 6'-0" high		8		315	20	335	375
45-001	**WOOD SCREENS** Over 3 S.F., 3/4" frames	2 Carp	375	S.F.	1.90	.85	2.75	3.33
50-001	**PICTURE WINDOW** Including frame, screen, and exterior trim							
002								
010	Average quality, bldrs. model, 4'-0" x 4'-0" high, standard glazed	2 Carp	12	Ea.	180	27	207	235
015	Insulating glass		12		210	27	237	270
020	4'-0" x 4'-6" high, standard glazed		11		195	29	224	255
025	Insulating glass		11		225	29	254	290
030	5'-0" x 4'-0" high, standard glazed		11		215	29	244	280
035	Insulating glass		11		253	29	282	320
040	6'-0" x 4'-6" high, standard glazed		10		235	32	267	305
045	Insulating glass		10		280	32	312	355
200	Metal clad, deluxe, insulating glass, 4'-0" x 4'-0" high		12		280	27	307	345
210	4'-6" x 6'-6" high		11		415	29	444	500
220	5'-6" x 6'-6" high		10		625	32	657	735
230	6'-6" x 6'-6" high		10		740	32	772	860
61-001	**SLIDING WINDOW** Including frame, screen, and exterior trim							
002								
010	Average quality, bldrs. model, 3'-0" x 2'-0" high, standard glazed	1 Carp	10	Ea.	75	16	91	105
012	Insulating glass		10		100	16	116	135
020	4'-0" x 3'-6" high, standard glazed		9		100	17.80	117.80	135
022	Insulating glass		9		145	17.80	162.80	185
030	6'-0" x 5'-0" high, standard glazed		8		170	20	190	215
032	Insulating glass		8		235	20	255	285
200	Metal clad, deluxe, insulating glass, 3'-0" x 3'-0" high		10		260	16	276	310
205	4'-0" x 3'-6" high		9		320	17.80	337.80	380
210	5'-0" x 4'-0" high		9		385	17.80	402.80	450
215	6'-0" x 5'-0" high		8		545	20	565	630
65-001	**WINDOW GRILLE OR MUNTIN** Snap-in type							
002	Colonial or diamond pattern							
200	Wood, awning window, glass size 28" x 16" high	1 Carp	30	Ea.	10.60	5.35	15.95	19.40
206	44" x 24" high		28		14.50	5.70	20.20	24
210	Casement, glass size, 20" x 36" high		30		22	5.35	27.35	32
218	20" x 56" high		28		48	5.70	53.70	61
220	Double hung, glass size, 16" x 24" high		24	Set	23	6.65	29.65	35
228	32" x 32" high		22	"	43	7.25	50.25	58
250	Picture, glass size, 48" x 48" high		30	Ea.	44	5.35	49.35	56
258	60" x 68" high		28	"	72	5.70	77.70	87
260	Sliding, glass size, 14" x 36" high		24	Set	26	6.65	32.65	38
268	36" x 36" high		22	"	44	7.25	51.25	59
70-001	**WOOD SASH** Including glazing but not including trim							
002								
005	Custom, 5'-0" x 4'-0", 1" dbl. glazed, 3/16" thick lites	2 Carp	3.20	Ea.	264	100	364	435
010	1/4" thick lites		5		305	64	369	430
020	1" thick, triple glazed		5		360	64	424	490
025								

For expanded coverage of these items see *Means' Interior Cost Data 1986*

8.6 Wood Windows

			CREW	DAILY OUTPUT	UNIT	BARE COSTS MAT.	BARE COSTS INST.	BARE COSTS TOTAL	TOTAL INCL O&P
70	030	7'-0" x 4'-6" high, 1" double glazed, 3/16" thick lites	2 Carp	4.30	Ea.	360	74	434	505
	040	1/4" thick lites		4.30		410	74	484	560
	050	1" thick, triple glazed	↓	4.30	↓	456	74	530	610
	055								
	060	8'-6" x 5'-0" high, 1" double glazed, 3/16" thick lites	2 Carp	3.50	Ea.	480	91	571	660
	070	1/4" thick lites		3.50		530	91	621	715
	080	1" thick, triple glazed	↓	3.50	↓	580	91	671	770
	090	Window frames only, based on perimeter length			L.F.	1.40		1.40	1.54M
	120	Window sill, stock, per lineal foot				4.30		4.30	4.73M
	125	Casing, stock			↓	1.55		1.55	1.70M
	300	Replacement sash, double hung, double glazing, window to 12 S.F.	1 Carp	64	S.F.	9.90	2.50	12.40	14.50
	310	12 S.F. to 20 S.F.		94		9.10	1.70	10.80	12.50
	320	20 S.F. and over	↓	106		8.50	1.51	10.01	11.55
	380	Triple glazing for above, add			↓	1.92		1.92	2.11M
	700	Sash, single lite, 2'-0" x 2'-0" high	1 Carp	20	Ea.	42	8	50	58
	705	2'-6" x 2'-0" high		19		48	8.40	56.40	65
	710	2'-6" x 2'-6" high		18		50	8.90	58.90	68
	715	3'-0" x 2'-0" high	↓	17		54	9.40	63.40	73
80-001		**WOOD SCREENS** Over 3 S.F., 3/4" frames	2 Carp	375	S.F.	2.05	.85	2.90	3.49
	010	1-1/8" frames	"	375	"	2.25	.85	3.10	3.71

8.7 Finish Hardware & Specialties

			CREW	DAILY OUTPUT	UNIT	BARE COSTS MAT.	BARE COSTS INST.	BARE COSTS TOTAL	TOTAL INCL O&P
01-001		**AVERAGE** Percentage for hardware, total job cost, minimum							.75%
	005	Maximum							3.50%
	050	Total hardware for building, average distribution				85%	15%		
	100	Door hardware, apartment, interior			Door	76		76	84M
	150	Hospital bedroom, minimum				95		95	105M
	200	Maximum			↓	450		450	495M
	210	Pocket door			Ea.	79		79	87M
	225	School, single exterior, incl. lever, not incl. panic device			Door	265		265	290M
	250	Single interior, regular use, no lever included				120		120	130M
	255	Including handicap lever				195		195	215M
	260	Heavy use, incl. lever and closer				220		220	240M
	285	Stairway, single interior			↓	145		145	160M
	310	Double exterior, with panic device			Pr.	630		630	695M
	330								
	360	Toilet, public, single interior			Door	91		91	100M
02-001		**ASTRAGALS** One piece overlapping							
	040	Cadmium plated steel, flat, 3/16" x 2"	1 Carp	90	L.F.	3.45	1.78	5.23	6.35
	060	Prime coated steel, flat, 1/8" x 3"		90		3.20	1.78	4.98	6.10
	080	Stainless steel, flat, 3/32" x 1-5/8"		90		5.55	1.78	7.33	8.70
	100	Aluminum, flat, 1/8" x 2"		90		1.60	1.78	3.38	4.34
	120	Nail on, "T" extrusion		120		.85	1.33	2.18	2.87
	130	Vinyl bulb insert		105		1.90	1.52	3.42	4.30
	160	Screw on, "T" extrusion		90		1.35	1.78	3.13	4.06
	170	Vinyl insert		75		1.90	2.13	4.03	5.20
	200	"L" extrusion, neoprene bulbs		75		1.35	2.13	3.48	4.58
	210	Neoprene sponge insert		75		3.70	2.13	5.83	7.15
	220	Magnetic		75		5.50	2.13	7.63	9.15
	240	Spring hinged security seal, with cam		75		4.10	2.13	6.23	7.60
	260	Spring loaded locking bolt, vinyl insert		45		4.55	3.56	8.11	10.15
	280	Neoprene sponge strip, "Z" shaped, aluminum		60		4.20	2.67	6.87	8.50
	290	Solid neoprene strip, nail on aluminum strip	↓	90	↓	1.85	1.78	3.63	4.61

For expanded coverage of these items see *Means' Interior Cost Data 1986*

8.7 Finish Hardware & Specialties

		CREW	DAILY OUTPUT	UNIT	BARE COSTS MAT.	BARE COSTS INST.	BARE COSTS TOTAL	TOTAL INCL O&P
300	One piece stile protection							
302	Neoprene fabric loop, nail on aluminum strips	1 Carp	60	L.F.	3.40	2.67	6.07	7.60
311	Flush mounted aluminum extrusion, 1/2" x 1-1/4"		60		1.90	2.67	4.57	5.95
314	3/4" x 1-3/8"		60		2.45	2.67	5.12	6.55
316	1-1/8" x 1-3/4"		60		3.70	2.67	6.37	7.95
330	Mortise, 9/16" x 3/4"		60		1.90	2.67	4.57	5.95
332	13/16" x 1-3/8"		60		2.65	2.67	5.32	6.80
360	Spring bronze strip, nail on type		105		1.60	1.52	3.12	3.97
362	Screw on, with retainer		75		1.85	2.13	3.98	5.15
380	Flexible stainless steel housing, pile insert, 1/2" door		105		3.75	1.52	5.27	6.35
382	3/4" door		105		4.20	1.52	5.72	6.85
400	Extruded aluminum retainer, flush mount, pile insert		105		1.20	1.52	2.72	3.53
408	Mortise, felt insert		90		3.15	1.78	4.93	6.05
416	Mortise with spring, pile insert		90		1.85	1.78	3.63	4.61
440	Rigid vinyl retainer, mortise, pile insert		105		1.35	1.52	2.87	3.69
460	Wool pile filler strip, aluminum backing	↓	105	↓	1	1.52	2.52	3.31
500	Two piece overlapping astragal, extruded aluminum retainer							
501	Pile insert	1 Carp	60	L.F.	1.75	2.67	4.42	5.80
502	Vinyl bulb insert		60		1.30	2.67	3.97	5.30
504	Vinyl flap insert		60		1.85	2.67	4.52	5.90
506	Solid neoprene flap insert		60		1.85	2.67	4.52	5.90
508	Hypalon rubber flap insert		60		1.85	2.67	4.52	5.90
509	Snap on cover, pile insert		60		2.15	2.67	4.82	6.25
540	Magnetic aluminum, surface mounted		60		6.20	2.67	8.87	10.70
550	Interlocking aluminum, 5/8" x 1" neoprene bulb insert		45		1.75	3.56	5.31	7.10
560	Adjustable aluminum, 9/16" x 21/32", pile insert	↓	45		5.75	3.56	9.31	11.50
579	For vinyl bulb, add				.44		.44	.48M
580	Magnetic, adjustable, 9/16" x 21/32"	1 Carp	45	↓	6.20	3.56	9.76	11.95
600	Two piece stile protection							
601	Cloth backed rubber loop, 1" gap, nail on aluminum strips	1 Carp	45	L.F.	2.40	3.56	5.96	7.80
604	Screw on aluminum strips		45		2.65	3.56	6.21	8.05
610	1-1/2" gap, screw on aluminum extrusion		45		3.35	3.56	6.91	8.85
624	Vinyl fabric loop, slotted aluminum extrusion, 1" gap		45		1.15	3.56	4.71	6.40
630	1-1/4" gap	↓	45	↓	1.35	3.56	4.91	6.65
05-001	AUTOMATIC OPENERS Swing doors, single	2 Skwk	.80	Ea.	1,700	410	2,110	2,475
010	Single operating pair		.50	Pr.	3,050	655	3,705	4,300
040	For double simultaneous doors, one way, add		1.20		250	275	525	675
050	Two way, add		.90	↓	320	365	685	885
100	Sliding doors, 3' wide, including track & hanger, single		.60	Opng.	3,000	545	3,545	4,100
130	Bi-parting		.50		3,600	655	4,255	4,925
145	Activating carpet, single door, one way, add		2.20		550	150	700	820
155	Two way, add	↓	1.30	↓	800	250	1,050	1,250
175	Handicap opener, button operating	2 Carp	8	Ea.	900	40	940	1,050
10-001	AUTOMATIC OPERATORS Industrial, sliding doors, to 6' wide	2 Skwk	.60	Opng.	3,500	545	4,045	4,650
020	To 12' wide	"	.40	"	4,200	820	5,020	5,825
040	Over 12' wide, add per L.F. of excess			L.F.	470		470	515M
100	Swing doors, to 5' wide	2 Skwk	.80	Ea.	2,050	410	2,460	2,850
115	Add for controls, wall pushbutton, 3 button		4		125	82	207	255
120	Ceiling pull cord	↓	4.30	↓	90	76	166	210
11-001	BOLTS, FLUSH Standard, concealed	1 Carp	7	Ea.	10.80	23	33.80	45
080	Automatic fire exit	"	5		220	32	252	290
160	For electric release, add	1 Elec	3	↓	68	60	128	160
12-001	BUMPER PLATES 1-1/2" x 3/4" U channel	2 Carp	80	L.F.	2.15	4	6.15	8.15
100	Tear drop, spring-steel, 4" high		15	Ea.	29.50	21	50.50	63
110	8" high		15		48	21	69	84
120	10" high	↓	15	↓	60	21	81	97
13-001	DETECTION SYSTEMS See division 16.8-15							

For expanded coverage of these items see *Means' Interior Cost Data 1986*

8.7 Finish Hardware & Specialties

		CREW	DAILY OUTPUT	UNIT	BARE COSTS MAT.	BARE COSTS INST.	BARE COSTS TOTAL	TOTAL INCL O&P
15-001	**DOOR CLOSER** Rack and pinion	1 Carp	6.50	Ea.	48	25	73	88
002	Adjustable backcheck, 3 way mount, all sizes, regular arm		6		56	27	83	100
004	Hold open arm		6		62	27	89	105
010	Fusible link		6.50		53	25	78	94
020	Non sized, regular arm		6		64	27	91	110
024	Hold open arm		6		71	27	98	115
040	4 way mount, non sized, regular arm		6		82	27	109	130
044	Hold open arm	↓	6	↓	88	27	115	135
195								
200	Backcheck and adjustable power, hinge face mount							
201	All sizes, regular arm	1 Carp	6.50	Ea.	54	25	79	95
204	Hold open arm		6.50		59	25	84	100
240	Top jamb mount, all sizes, regular arm		6		54	27	81	98
244	Hold open arm		6		59	27	86	105
280	Top face mount, all sizes, regular arm		6.50		54	25	79	95
284	Hold open arm		6.50		59	25	84	100
400	Backcheck, overhead concealed, all sizes, regular arm		5.50		89	29	118	140
404	Concealed arm		5		99	32	131	155
440	Compact overhead, concealed, all sizes, regular arm		5.50		115	29	144	170
444	Concealed arm		5		125	32	157	185
480	Concealed in door, all sizes, regular arm		5.50		62	29	91	110
484	Concealed arm		5		52	32	84	105
490	Floor concealed, all sizes, single acting		2.20		64	73	137	175
494	Double acting	↓	2.20		79	73	152	190
500	For cast aluminum cylinder, deduct				11		11	12.10M
504	For delayed action, add				11		11	12.10M
508	For fusible link arm, add				8.50		8.50	9.35M
512	For shock absorbing arm, add				15.75		15.75	17.30M
516	For spring power adjustment, add				15.75		15.75	17.30M
600	Closer-holder, hinge face mount, all sizes, exposed arm	1 Carp	6.50		61	25	86	105
700	Electronic closer-holder, hinge facemount, concealed arm		5		105	32	137	160
740	With built-in detector		5		320	32	352	400
20-001	**DOORSTOPS** Holder and bumper, floor or wall		24		16.50	6.65	23.15	28
130	Wall bumper		24		3.20	6.65	9.85	13.20
160	Floor bumper, 1" high		24		2.60	6.65	9.25	12.50
19?	Plunger type, door mounted		24		12.75	6.65	19.40	24
25-001	**FLOOR CHECKS** For over 3' wide doors, single acting		2.50		320	64	384	445
050	Double acting	↓	2.50	↓	425	64	489	560
33-001	**HINGES** Full mortise, average frequency, steel base, 4-1/2"x4-1/2", USP			Pr.	17.65		17.65	19.40M
010	5" x 5", USP				30		30	33M
020	6" x 6", USP				59		59	65M
040	Brass base, 4-1/2" x 4-1/2", US10				39		39	43M
050	5" x 5", US10				49		49	54M
060	6" x 6", US10				86		86	95M
080	Stainless steel base, 4-1/2" x 4-1/2", US32			↓	60		60	66M
090	For non removable pin, add			Ea.	2.75		2.75	3.02M
091	For floating pin, driven tips, add				3.38		3.38	3.71M
093	For hospital type tip on pin, add				9.65		9.65	10.60M
094	For steeple type tip on pin, add			↓	8.85		8.85	9.75M
100	Full mortise, high frequency, steel base, 4-1/2" x 4-1/2", USP			Pr.	45		45	50M
110	5" x 5", USP				55		55	61M
120	6" x 6", USP				99		99	110M
140	Brass base, 4-1/2" x 4-1/2", US10				71		71	78M
150	5" x 5", US10				85		85	94M
160	6" x 6", US10				125		125	140M
180	Stainless steel base, 4-1/2" x 4-1/2", US32				105		105	115M
193	For hospital type tip on pin, add			Ea.	11.45		11.45	12.60M
200	Full mortise low frequency, steel base, 4-1/2" x 4-1/2", USP			Pr.	10		10	11M
210	5" x 5", USP			↓	21.50		21.50	24M
220	6" x 6", USP				36		36	40M

For expanded coverage of these items see *Means' Interior Cost Data 1986*

8.7 Finish Hardware & Specialties

		CREW	DAILY OUTPUT	UNIT	BARE COSTS MAT.	BARE COSTS INST.	BARE COSTS TOTAL	TOTAL INCL O&P
240	Brass base, 4-1/2" x 4-1/2", US10			Pr.	31		31	34M
250	5" x 5", US10				45		45	50M
280	Stainless steel base, 4-1/2" x 4-1/2", US32			▼	47		47	52M
34-001	**SPECIAL HINGES** Paumelle, high frequency							
002	Steel base, 6" x 4-1/2", US10			Pr.	70		70	77M
010	Bronze base, 5" x 4-1/2", US10				97		97	105M
020	Paumelle, average frequency, steel base, 4-1/2" x 3-1/2", US10				46		46	51M
040	Olive knuckle, low frequency, bronze base, 6" x 4-1/2", US10			▼	90		90	99M
100	Electric hinge with concealed conductor, average frequency							
101	Steel base, 4-1/2" x 4-1/2", US26D			Pr.	78		78	86M
110	Bronze base, 4-1/2" x 4-1/2", US26D			"	90		90	99M
120	Electric hinge with concealed conductor, high frequency							
121	Steel base, 4-1/2" x 4-1/2", US26D			Pr.	92		92	100M
160	Double weight, 800 lb., steel base, removable pin, 5" x 6", USP				100		100	110M
170	Steel base-welded pin, 5" x 6", USP				94		94	105M
180	Triple weight, 2000 lb., steel base, welded pin, 5" x 6", USP			▼	125		125	140M
181								
200	Pivot reinf., high frequency, steel base, 7-3/4" door plate, USP			Pr.	130		130	145M
220	Bronze base, 7-3/4" door plate, US10			"	180		180	200M
300	Swing clear, full mortise, full or half surface, high frequency,							
301	Steel base, 5" high, USP			Pr.	115		115	125M
320	Swing clear, full mortise, average frequency							
321	Steel base, 4-1/2" high, USP			Pr.	89		89	98M
400	Wide throw, average frequency, steel base, 4-1/2" x 6", USP				50		50	55M
420	High frequency, steel base, 4-1/2" x 6", USP			▼	130		130	145M
460	Spring hinge, single acting, 6" flange, steel			Ea.	23		23	25M
470	Brass				42		42	46M
490	Double acting, 6" flange, steel				47		47	52M
495	Brass			▼	75		75	83M
800	Continuous hinge, steel	2 Carp	64	L.F.	2.65	5	7.65	10.15
35-001	**KICK PLATE** 6" high, for 3' door, aluminum	1 Carp	15	Ea.	12.75	10.65	23.40	29
050	Bronze	"	15	"	33.50	10.65	44.15	52
40-001	**LOCKSET** Standard duty, cylindrical, passage doors							
002	Non-keyed, passage	1 Carp	12	Ea.	21	13.35	34.35	42
010	Privacy		12		29	13.35	42.35	51
040	Keyed, single cylinder function		10		47	16	63	75
042	Hotel		8		60	20	80	95
100	Heavy duty with sectional trim, non-keyed, passages		12		81	13.35	94.35	110
110	Privacy		12		100	13.35	113.35	130
140	Keyed, single cylinder function		10		115	16	131	150
142	Hotel		8		145	20	165	190
160	Communicating	▼	10		130	16	146	165
169	For re-core cylinder, add				18.75		18.75	21M
170	Residential, interior door, minimum	1 Carp	16		8	10	18	23
172	Maximum		8		23	20	43	54
180	Exterior, minimum		14		15	11.45	26.45	33
182	Maximum	▼	8	▼	80	20	100	115
41-001	**DEADLOCKS** Mortise, heavy duty, outside key	1 Carp	9	Ea.	83	17.80	100.80	115
002	Double cylinder		9		91	17.80	108.80	125
010	Medium duty, outside key		10		62	16	78	91
011	Double cylinder		10		73	16	89	105
100	Tubular, standard duty, outside key		10		31	16	47	57
101	Double cylinder		10		41	16	57	68
120	Night latch, outside key	▼	10	▼	41	16	57	68
42-001	**ENTRANCE LOCKS** Cylinder, grip handle, deadlocking latch	1 Carp	9	Ea.	80	17.80	97.80	115
002	Deadbolt	"	8	"	93	20	113	130

For expanded coverage of these items see *Means' Interior Cost Data 1986*

8.7 Finish Hardware & Specialties

			CREW	DAILY OUTPUT	UNIT	BARE COSTS MAT.	INST.	TOTAL	TOTAL INCL O&P
42	010	Push and pull plate, dead bolt	1 Carp	8	Ea.	90	20	110	130
	090	For handicapped lever, add			"	60		60	66M
	43-001	**MORTISE LOCKSET** Commerical, wrought knobs and full escutcheon trim							
	002	Non-keyed, passage, minimum	1 Carp	9	Ea.	80	17.80	97.80	115
	003	Maximum		8		150	20	170	195
	004	Privacy, minimum		9		90	17.80	107.80	125
	005	Maximum		8		175	20	195	220
	010	Keyed, office/entrance/apartment, minimum		8		110	20	130	150
	011	Maximum		7		210	23	233	265
	012	Single cylinder, typical, minimum		8		110	20	130	150
	013	Maximum		7		190	23	213	240
	020	Hotel, minimum		7		120	23	143	165
	021	Maximum		6		220	27	247	280
	030	Communication, double cylinder, minimum		8		125	20	145	165
	031	Maximum		7		175	23	198	225
	100	Wrought knobs and sectional trim, non-keyed, passage, minimum		10		60	16	76	89
	101	Maximum		9		130	17.80	147.80	170
	104	Privacy, minimum		10		80	16	96	110
	105	Maximum		9		145	17.80	162.80	185
	110	Keyed, entrance,office/apartment, minimum		9		100	17.80	117.80	135
	111	Maximum		8		175	20	195	220
	112	Single cylinder, typical, minimum		9		100	17.80	117.80	135
8	113	Maximum		8		165	20	185	210
	195								
	200	Cast knobs and full escutcheon trim							
	201	Non-keyed, passage, minimum	1 Carp	9	Ea.	155	17.80	172.80	195
	202	Maximum		8		250	20	270	305
	204	Privacy, minimum		9		165	17.80	182.80	205
	205	Maximum		8		270	20	290	325
	212	Keyed, single cylinder, typical, minimum		8		185	20	205	230
	213	Maximum		7		295	23	318	360
	220	Hotel, minimum		7		190	23	213	240
	221	Maximum		6		330	27	357	400
	300	Cast knob and sectional trim, non-keyed, passage, minimum		10		115	16	131	150
	301	Maximum		10		225	16	241	270
	304	Privacy, minimum		10		125	16	141	160
	305	Maximum		10		235	16	251	280
	310	Keyed, office/entrance/apartment, minimum		9		140	17.80	157.80	180
	311	Maximum		9		260	17.80	277.80	310
	312	Single cylinder, typical, minimum		9		150	17.80	167.80	190
	313	Maximum		9		295	17.80	312.80	350
	319	For re-core cylinder, add				18		18	19.80M
	320								
	400	Keyless, pushbutton type, with deadbolt, standard	1 Carp	9	Ea.	71	17.80	88.80	105
	410	Heavy duty	"	9		95	17.80	112.80	130
	415	Card type, 1 time zone, minimum				250		250	275M
	420	Maximum				715		715	785M
	425	3 time zones, minimum				600		600	660M
	430	Maximum				1,450		1,450	1,600M
	435	System with printer, and control console, 3 zones			Total	7,625		7,625	8,400M
	440	6 zones			"	10,000		10,000	11,000M
	445	For each door, minimum, add			Ea.	1,100		1,100	1,200M
	450	Maximum, add			"	1,600		1,600	1,750M
	45-001	**PANIC DEVICE** For rim locks, single door, exit only	1 Carp	6	Ea.	200	27	227	260
	002	Outside key and pull		5		245	32	277	315
	020	Bar and vertical rod, exit only		5		320	32	352	400
	021	Outside key and pull		4		370	40	410	465
	040	Bar and concealed rod		4		350	40	390	445
	050								

For expanded coverage of these items see *Means' Interior Cost Data 1986*

8.7 Finish Hardware & Specialties

		CREW	DAILY OUTPUT	UNIT	BARE COSTS MAT.	BARE COSTS INST.	BARE COSTS TOTAL	TOTAL INCL O&P
45 060	Touch bar, exit only	1 Carp	6	Ea.	270	27	297	335
061	Outside key and pull		5		325	32	357	405
070	Touch bar and vertical rod, exit only		5		385	32	417	470
071	Outside key and pull		4		445	40	485	545
100	Mortise, bar, exit only		4		250	40	290	335
160	Touch bar, exit only		4		350	40	390	445
200	Narrow stile, rim mounted, bar, exit only		6		275	27	302	340
201	Outside key and pull		5		320	32	352	400
220	Bar and vertical rod, exit only		5		325	32	357	405
221	Outside key and pull		4		370	40	410	465
240	Bar and concealed rod, exit only		3		350	53	403	460
300	Mortise, bar, exit only		4		260	40	300	345
360	Touch bar, exit only	↓	4	↓	370	40	410	465
50-001	**PUSH-PULL** Push plate, pull plate, aluminum	1 Carp	12	Ea.	30	13.35	43.35	52
050	Bronze		12		46	13.35	59.35	70
150	Pull handle and push bar, aluminum		11		78	14.55	92.55	105
200	Bronze		10		100	16	116	135
300	Push plate both sides, aluminum		14		6.50	11.45	17.95	24
350	Bronze		13		15	12.30	27.30	34
400	Door pull, designer style, cast aluminum, minimum		12		45	13.35	58.35	69
500	Maximum		8		190	20	210	240
600	Cast bronze, minimum		12		47	13.35	60.35	71
700	Maximum		8		220	20	240	270
800	Walnut, minimum		12		40	13.35	53.35	63
900	Maximum		8		220	20	240	270
60-001	**THRESHOLD** 3' long door saddles, aluminum, minimum		20		20.50	8	28.50	34
010	Maximum		12		62	13.35	75.35	88
050	Bronze, minimum		20		36	8	44	51
060	Maximum		12		105	13.35	118.35	135
070	Rubber, 1/2" thick, 5-1/2" wide		20		21	8	29	35
080	2-3/4" wide		20	↓	11.45	8	19.45	24
70-001	**WEATHERSTRIPPING** Window, double hung, 3' x 5', zinc		7.20	Opng.	13	22	35	47
010	Bronze		7.20		25	22	47	60
050	As above but heavy duty, zinc		4.60		16.10	35	51.10	68
060	Bronze		4.60		30	35	65	83
100	Doors, wood frame, interlocking, for 3' x 7' door, zinc		3		18.20	53	71.20	97
110	Bronze		3		30	53	83	110
130	6' x 7' opening, zinc		2		22	80	102	140
140	Bronze	↓	2	↓	36	80	116	155
170	Wood frame, spring type, bronze							
180	3' x 7' door	1 Carp	7.60	Opng.	8.85	21	29.85	40
190	6' x 7' door	"	7	"	12.50	23	35.50	47
220	Metal frame, spring type, bronze							
230	3' x 7' door	1 Carp	3	Opng.	35	53	88	115
240	6' x 7' door	"	2.50		45	64	109	140
250	For stainless steel, spring type, add			↓	100%			
251								
270	Metal frame, extruded sections, 3' x 7' door, aluminum	1 Carp	2	Opng.	25	80	105	145
280	Bronze		2		66	80	146	190
310	6' x 7' door, aluminum		1.20		29	135	164	225
320	Bronze	↓	1.20	↓	66	135	201	265
350	Threshold weatherstripping							
360								
365	Door sweep, flush mounted, aluminum	1 Carp	25	Ea.	10.90	6.40	17.30	21
370	Vinyl		25		3.60	6.40	10	13.25
500	Garage door bottom weatherstrip, 12' aluminum, clear		14		7.30	11.45	18.75	25
501	Bronze		14		10.65	11.45	22.10	28
505	Bottom protection, 12' aluminum, clear		14		23	11.45	34.45	42
510	Bronze	↓	14	↓	31	11.45	42.45	51

For expanded coverage of these items see *Means' Interior Cost Data 1986*

8.8 Glass & Glazing

		CREW	DAILY OUTPUT	UNIT	BARE COSTS MAT.	BARE COSTS INST.	BARE COSTS TOTAL	TOTAL INCL O&P
03-001	ACOUSTICAL GLASS UNITS 1 lite at 3/8", 1 lite at 3/16", for 1" thick	2 Glaz	100	S.F.	14	3.22	17.22	20
010	For 4" thick		80		19	4.03	23.03	27
04-001	BEVELED GLASS With design patterns, 1/4" thick, 1/2" bevel, minimum		150		40	2.15	42.15	47
005	Average		125		86	2.58	88.58	98
010	Maximum		100		120	3.22	123.22	135
06-001	CURTAIN WALL See division 8.9							
09-001	FACETED Color tinted glass, 3/4" thick, minimum	2 Glaz	95	S.F.	19.20	3.39	22.59	26
010	Maximum	"	75		31.95	4.30	36.25	41
12-001	FULL VISION Window system with 3/4" glass mullions, 10' high	H-2	130		13.35	3.46	16.81	19.65
010	10' to 20' high, minimum		110		16.70	4.09	20.79	24
015	Average		100		21.25	4.50	25.75	30
020	Maximum		80		26.50	5.60	32.10	37
15-001	GLASS BLOCK See division 4.3-60							
18-001	GLAZING VARIABLES							
050	For high rise glazing, from exterior, add per S.F. per story			Story		.07	.07	.08L
060	For glass replacement, add			S.F.		100%		
070	For gasket settings, add			L.F.	1.85		1.85	2.03M
080	For concrete reglet settings, add			S.F.	20%	25%		
090	For sloped glazing, add			"		25%		
200	Fabrication, polished edges, 1/4" thick			Inch	.55		.55	.60M
210	1/2" thick				.75		.75	.82M
250	Mitered edges, 1/4" thick				1.10		1.10	1.21M
260	1/2" thick				1.50		1.50	1.65M
21-001	INSULATING GLASS UNITS 2 lites 1/8" float, 1/2" thick	2 Glaz	95	S.F.	4.72	3.39	8.11	10.05
002	Clear		95		4.54	3.39	7.93	9.85
010	Tinted		95		5.85	3.39	9.24	11.30
011								
020	2 lites 3/16" float, for 5/8" thick unit, 15 to 30 S.F., clear	2 Glaz	90	S.F.	5.55	3.58	9.13	11.20
030	Tinted		90		6.80	3.58	10.38	12.60
040	1" thick, double glazed, 1/4" float, 30 to 70 S.F., clear		75		6.95	4.30	11.25	13.80
050	Tinted		75		8.20	4.30	12.50	15.15
060	1" thick double glazed, 1/4" float, 1/4" wire		75		11.25	4.30	15.55	18.50
070	1/4" float, 1/4" tempered		75		9.30	4.30	13.60	16.35
080	1/4" wire, 1/4" tempered		75		14.65	4.30	18.95	22
090	Both lites, 1/4" wire		75		15.95	4.30	20.25	24
200	Both lites, light & heat reflective		85		10.75	3.79	14.54	17.25
220								
250	Heat reflective, film inside, 1" thick unit, clear	2 Glaz	85	S.F.	10.20	3.79	13.99	16.65
260	Tinted		85		10.70	3.79	14.49	17.20
300	Film on weatherside, clear, 1/2" thick unit		95		7.40	3.39	10.79	13
310	5/8" thick unit		90		8.70	3.58	12.28	14.70
320	1" thick unit		85		9.85	3.79	13.64	16.25
24-001	LAMINATED GLASS Clear float, .03" vinyl, 1/4" thick	2 Glaz	90	S.F.	3.90	3.58	7.48	9.40
010	3/8" thick		78		7.95	4.13	12.08	14.65
020	.06" vinyl, 1/2" thick		65		9.60	4.96	14.56	17.65
100	5/16" thick		90		11.50	3.58	15.08	17.75
200	Bullet-resisting, 1-3/16" thick, to 15 S.F.		16		19.50	20	39.50	50
210	Over 15 S.F.		16		22.25	20	42.25	53
250	2" thick, to 15 S.F.		12		35	27	62	77
260	Over 15 S.F.		12		54	27	81	98
25-001	LEADED GLASS For X-ray, see division 13.1-60							
27-001	MIRRORS No frames, wall type, 1/4" plate glass, polished edge							
010	Up to 5 S.F.	2 Glaz	125	S.F.	5.20	2.58	7.78	9.40
020	Over 15 S.F.		160		5	2.02	7.02	8.40
050	Door type, 1/4" plate glass, up to 12 S.F.		160		4.70	2.02	6.72	8.05

For expanded coverage of these items see *Means' Interior Cost Data 1986*

8.8	Glass & Glazing		CREW	DAILY OUTPUT	UNIT	BARE COSTS			TOTAL INCL O&P
						MAT.	INST.	TOTAL	
27	100	Float glass, up to 10 S.F., 1/8" thick	2 Glaz	160	S.F.	3	2.02	5.02	6.20
	110	3/16" thick		150		4	2.15	6.15	7.45
	150	12" x 12" wall tiles, square edge, clear		195		1.25	1.65	2.90	3.73
	160	Veined		195		1.75	1.65	3.40	4.28
	200	1/4" thick, stock sizes, one way transparent		125		11	2.58	13.58	15.80
	201	Bathroom, unframed, laminated		160		6.50	2.02	8.52	10.05
30-001		OBSCURE GLASS 1/8" thick, minimum		140		3.60	2.30	5.90	7.25
	010	Maximum		125		4.55	2.58	7.13	8.70
	030	7/32" thick, minimum		120		4.75	2.69	7.44	9.05
	040	Maximum		105		5.25	3.07	8.32	10.15
33-001		PATTERNED GLASS Colored, 1/8" thick, minimum		140		3.65	2.30	5.95	7.30
	010	Maximum		125		4.75	2.58	7.33	8.90
	030	7/32" thick, minimum		120		4.95	2.69	7.64	9.30
	040	Maximum		105		5.40	3.07	8.47	10.30
36-001		FLOAT GLASS 3/16" thick, clear, plain		130		1.06	2.48	3.54	4.70
	010	Tinted		130		1.47	2.48	3.95	5.15
	020	Tempered, clear (103)		130		3.41	2.48	5.89	7.30
	030	Tempered, tinted		130		4.43	2.48	6.91	8.40
	060	1/4" thick, clear, plain		120		1.48	2.69	4.17	5.45
	070	Tinted		120		1.95	2.69	4.64	6
	080	Tempered, clear		120		3.43	2.69	6.12	7.60
	090	Tempered, tinted		120		4.48	2.69	7.17	8.75
	120	5/16" thick, clear, plain		100		2.95	3.22	6.17	7.85
	130	Tempered, clear		100		9.30	3.22	12.52	14.85
	160	3/8" thick, clear, plain		75		3.05	4.30	7.35	9.50
	170	Tinted		75		4.60	4.30	8.90	11.20
	180	Tempered, clear		75		9.95	4.30	14.25	17.10
	190	Tempered, tinted		75		11.80	4.30	16.10	19.10
	220	1/2" thick, clear, plain		55		4.65	5.85	10.50	13.50
	230	Tinted		55		7.15	5.85	13	16.25
	240	Tempered, clear		55		14.55	5.85	20.40	24
	250	Tempered, tinted		55		16.75	5.85	22.60	27
	280	5/8" thick, clear, plain		45		5.55	7.15	12.70	16.35
	290	Tempered, clear		45		17.50	7.15	24.65	29
	320	3/4" thick, clear, plain		35		6.60	9.20	15.80	20
	330	Tempered, clear		35		24	9.20	33.20	40
	360	1" thick, clear, plain		30		15.70	10.75	26.45	33
39-001		PLEXIGLASS ACRYLIC Clear, masked, MCM grade, 1/8" thick, cut sheets	2 Glaz	170	S.F.	1.74	1.90	3.64	4.62
	020	Full sheets		195		1.19	1.65	2.84	3.67
	050	1/4" thick, cut sheets		165		2.67	1.95	4.62	5.70
	060	Full sheets		185		2.15	1.74	3.89	4.85
	090	3/8" thick, cut sheets		155		4.30	2.08	6.38	7.70
	100	Full sheets		180		3.95	1.79	5.74	6.90
	130	1/2" thick, cut sheets		135		5.75	2.39	8.14	9.75
	140	Full sheets		150		5.25	2.15	7.40	8.85
	170	3/4" thick, cut sheets		115		13.25	2.80	16.05	18.60
	180	Full sheets		130		12.60	2.48	15.08	17.40
	210	1" thick, cut sheets		105		17.55	3.07	20.62	24
	220	Full sheets		125		17	2.58	19.58	22
	300	Colored, 1/8" thick, cut sheets		170		2.15	1.90	4.05	5.05
	320	Full sheets		195		1.75	1.65	3.40	4.28
	350	1/4" thick, cut sheets		165		2.95	1.95	4.90	6.05
	360	Full sheets		185		2.55	1.74	4.29	5.30
	400	Mirrors, untinted, cut sheets, 1/8" thick		185		2.65	1.74	4.39	5.40
	420	1/4" thick		180		3.95	1.79	5.74	6.90
42-001		POLYCARBONATE Clear, masked, cut sheets, 1/8" thick		170		2	1.90	3.90	4.91
	050	3/16" thick		165		3.15	1.95	5.10	6.25
	100	1/4" thick		155		3.68	2.08	5.76	7
	150	3/8" thick		150		6.85	2.15	9	10.60

For expanded coverage of these items see *Means' Interior Cost Data 1986*

8.8 Glass & Glazing

		CREW	DAILY OUTPUT	UNIT	BARE COSTS MAT.	BARE COSTS INST.	BARE COSTS TOTAL	TOTAL INCL O&P
45-001	**REFLECTIVE GLASS** 1/4" float with fused metallic oxide, tinted	2 Glaz	115	S.F.	5.20	2.80	8	9.70
050	1/4" float glass with reflective applied coating		115		3.10	2.80	5.90	7.40
200	Solar film on glass, not including glass, minimum		180		1	1.79	2.79	3.66
205	Maximum		225		1.85	1.43	3.28	4.08
48-001	**SANDBLASTED GLASS** Float glass, 1/8" thick		160		2.70	2.02	4.72	5.85
010	3/16" thick		130		3.20	2.48	5.68	7.05
050	Plate glass, 1/4" thick		120		3.40	2.69	6.09	7.55
060	3/8" thick		75		3.80	4.30	8.10	10.30
51-001	**SHEET GLASS** Gray, 1/8" thick		160		2.75	2.02	4.77	5.90
020	1/4" thick		130		3.40	2.48	5.88	7.30
54-001	**SPANDREL GLASS** 1/4" thick, standard colors, over 2000 S.F.		110		4	2.93	6.93	8.60
020	Under 2000 S.F.	↓	120	↓	5.30	2.69	7.99	9.65
030	For custom colors, add			Total	10%		10%	10%
050	For 3/8" thick, add			S.F.	3.80		3.80	4.18M
100	For double coated, 1/4" thick, add				1.30		1.30	1.43M
120	For insulation on panels, add				2.75		2.75	3.02M
200	Panels, insulated, with aluminum backed fiberglass, 1" thick	2 Glaz	120		6.65	2.69	9.34	11.15
210	2" thick	"	120		7.35	2.69	10.04	11.90
250	With galvanized steel backing, add				2.10		2.10	2.31M
300	Maximum size 72" x 168", for over 140", add				10%			
57-001	**VINYL GLASS** Steel mesh reinforced, stock sizes, .090" thick	2 Glaz	170		2.75	1.90	4.65	5.75
050	.120" thick		170		3.25	1.90	5.15	6.30
100	.250" thick	↓	155		5.10	2.08	7.18	8.60
150	For non-standard sizes, add				15%			
60-001	**WINDOW GLASS** Clear float, stops, putty bed, 1/8" thick	2 Glaz	480		2.05	.67	2.72	3.21
050	3/16" thick, clear		480		2.15	.67	2.82	3.32
060	Tinted		480		2.40	.67	3.07	3.60
070	Tempered	↓	480	↓	4.10	.67	4.77	5.45
63-001	**WINDOW WALL** See division 8.9							
66-001	**WIRE GLASS** 1/4" thick, rough obscure (chicken wire)	2 Glaz	135	S.F.	4.40	2.39	6.79	8.25
100	Polished wire, 1/4" thick, diamond, clear		135		4.95	2.39	7.34	8.85
150	Pinstripe, obscure	↓	135	↓	7.50	2.39	9.89	11.65

8.9 Window/Curtain Walls

		CREW	DAILY OUTPUT	UNIT	BARE COSTS MAT.	BARE COSTS INST.	BARE COSTS TOTAL	TOTAL INCL O&P
10-001	**CURTAIN WALLS** Aluminum, stock, including glazing, minimum	H-1	205	S.F.	13.90	3.27	17.17	20
005	Average, single glazed		195		18.55	3.43	21.98	26
015	Average, double glazed		180		32.50	3.72	36.22	41
020	Maximum	↓	160	↓	48	4.19	52.19	59
50-001	**TUBE FRAMING** For window walls and store fronts, aluminum, stock							
005	Plain tube frame, mill finish, 1-3/4" x 1-3/4"	2 Glaz	103	L.F.	3.95	3.13	7.08	8.80
010	1-3/4" x 3"		100		4.55	3.22	7.77	9.60
015	1-3/4" x 4"		98		5.40	3.29	8.69	10.65
020	1-3/4" x 4-1/2"		95		5.75	3.39	9.14	11.15
025	2" x 6"		89		7.55	3.62	11.17	13.45
030	3" x 3"		92		7.05	3.50	10.55	12.75
035	4" x 4"		87		7.75	3.71	11.46	13.80
040	4-1/2" x 4-1/2"		85		9.90	3.79	13.69	16.30
045	Glass bead		240		1.26	1.34	2.60	3.30
100	Flush tube frame, mill finish, 1/4" glass, 1-3/4" x 4", open header		80		3.95	4.03	7.98	10.10
105	Open sill		82		4.02	3.93	7.95	10.05
110	Closed back header		83		5.30	3.88	9.18	11.35
115	Closed back sill	↓	85	↓	5.65	3.79	9.44	11.65

For expanded coverage of these items see *Means' Interior Cost Data 1986*

8.9 Window/Curtain Walls

		CREW	DAILY OUTPUT	UNIT	BARE COSTS MAT.	BARE COSTS INST.	BARE COSTS TOTAL	TOTAL INCL O&P
120	Vertical mullion, one piece	2 Glaz	75	L.F.	5.45	4.30	9.75	12.15
125	Two piece		73		6.05	4.42	10.47	12.95
130	90° or 180° vertical corner post		75		9.80	4.30	14.10	16.90
140	1-3/4" x 4-1/2", open header		80		4.60	4.03	8.63	10.80
145	Open sill		82		4.75	3.93	8.68	10.85
150	Closed back header		83		5.35	3.88	9.23	11.45
155	Closed back sill		85		7.09	3.79	10.88	13.20
160	Vertical mullion, one piece		75		5.65	4.30	9.95	12.35
165	Two piece		73		7	4.42	11.42	14
170	90° or 180° vertical corner post		75		10.40	4.30	14.70	17.55
200	Flush tube frame, mill fin. for ins. glass, 2" x 4-1/2", open header		75		5.55	4.30	9.85	12.25
205	Open sill		77		5.75	4.19	9.94	12.30
210	Closed back header		78		7.50	4.13	11.63	14.15
215	Closed back sill		80		7.80	4.03	11.83	14.35
220	Vertical mullion, one piece		70		7.15	4.61	11.76	14.45
225	Two piece		68		8	4.74	12.74	15.55
230	90° or 180° vertical corner post		70		13.25	4.61	17.86	21
235								
500	Flush tube frame, mill fin., thermal brk., 2-1/4"x 4-1/2"	2 Glaz	74	L.F.	8.25	4.36	12.61	15.30
505	Open sill		75		8.30	4.30	12.60	15.25
510	Vertical mullion, one piece		69		9.40	4.67	14.07	17
515	Two piece		67		10.30	4.81	15.11	18.20
520	90° or 180° vertical corner post		69		17	4.67	21.67	25
525								
530	Door stop (snap in)	2 Glaz	380	L.F.	1.26	.85	2.11	2.60
535								
700	For joint other than 90°, add			Ea.	3.30		3.30	3.63M
705	Screw spline joint, add				2.75		2.75	3.02M
710	For joint other than 90°, add				6.60		6.60	7.25M
800	For bronze finish, add			L.F.	18%			
802	For black finish, add				27%			
805	For stainless steel, add				75%			
810	For monumental grade, add				50%			
815	For steel stiffener, add	2 Glaz	200		3.75	1.61	5.36	6.45
820	For 2 to 5 stories, add per story			Story		5%		
80-001	WINDOW WALLS Aluminum, stock, including glazing, minimum	H-2	160	S.F.	12	2.81	14.81	17.25
005	Average		140		19	3.21	22.21	26
010	Maximum		110		44	4.09	48.09	54
020								
050	For translucent sandwich wall systems, refer to division 7.4-36							
085	Cost of the above walls depends on material,							
086	finish, repetition, and size of units.							
087	The larger the opening, the lower the S.F. cost							
120	Double glazed acoustical window wall for airports,							
122	including 1" thick glass with 2" x 4-1/2" tube frame	H-2	40	S.F.	37.50	11.25	48.75	57

9.1 Lath & Plaster

		CREW	DAILY OUTPUT	UNIT	BARE COSTS MAT.	BARE COSTS INST.	BARE COSTS TOTAL	TOTAL INCL O&P
10-001	ACCESSORIES, PLASTER Casing bead, expanded flange, galvanized	1 Lath	2.70	C.L.F.	31	60	91	120
010	Zinc alloy	"	2.70		55	60	115	145
090	Channels, cold rolled, 16 ga., 3/4" deep, painted				16		16	17.60M
100	Galvanized				21		21	23M
110	1" deep, 16 ga., painted				21		21	23M
115	Galvanized				27		27	30M

For expanded coverage of these items see *Means' Interior Cost Data 1986*

9.1 Lath & Plaster

			CREW	DAILY OUTPUT	UNIT	BARE COSTS MAT.	BARE COSTS INST.	BARE COSTS TOTAL	TOTAL INCL O&P
10	120	1-1/2" deep, 16 ga., painted			C.L.F.	26		26	29M
	130	Galvanized				32		32	35M
	150	2" deep, 16 ga., painted				36		36	40M
	160	Galvanized				43		43	47M
	162	Corner bead, expanded bullnose, 3/4" radius, #10 galvanized	1 Lath	2.60		42	62	104	135
	164	Zinc alloy		2.70		75	60	135	165
	165	#1, galvanized		2.55		19	63	82	110
	166	Zinc alloy		2.70		39	60	99	125
	167	Expanded wing, 2-3/4" wide, galv. #1		2.65		19	61	80	105
	168	Zinc alloy		2.70		39	60	99	125
	170	Inside corner, (corner rite) 3" x 3", painted		2.60		9.50	62	71.50	98
	175	Strip-ex, 4" wide, painted		2.55		7.50	63	70.50	97
	180	Expansion joint, 3/4" grounds, limited expansion, galv., 1 piece		2.70		60	60	120	150
	190	Zinc alloy		2.70		100	60	160	195
	210	Extreme expansion, galvanized, 2 piece		2.60		110	62	172	210
	230	Zinc alloy		2.70		180	60	240	280
	250	Joist clips for lath, 2-1/2" flange		1.90	M	44	85	129	170
	260	4-1/2" flange		1.80	"	62	89	151	195
	280	Metal base, galvanized and painted, 2-1/2" high		2.40	C.L.F.	40	67	107	140
	290	Stud clips for gypsum lath, field clip		2.35	M	51	68	119	155
	310	Resilient		2.30		135	70	205	245
	320	Starter/finisher		2.20		50	73	123	160
	370	Tie wire, galv., 50 lb. units, 8 or 9 ga., hanks			Cwt.	80		80	88M
	380	Coils				65		65	72M
	400	18 ga., straightened hanks				110		110	120M
	410	Coils				95		95	105M
	500	Zee runner			C.L.F.	22		22	24M
15-001		FURRING Beams & columns, 3/4" galvanized channels,							
	003	12" O.C.	1 Lath	155	S.F.	.26	1.04	1.30	1.75
	005	16" O.C.		170		.21	.95	1.16	1.57
	007	24" O.C.		185		.16	.87	1.03	1.40
	010	Ceilings, on steel, 3/4" channels, galvanized, 12" O.C.		210		.26	.77	1.03	1.37
	030	16" O.C.		290		.21	.55	.76	1.01
	060	1-1/2" channels, galvanized, 12" O.C.		190		.40	.85	1.25	1.63
	070	16" O.C.		260		.32	.62	.94	1.22
	090	24" O.C.		390		.26	.41	.67	.87
	100	Walls, galvanized, 3/4" channels, 12" O.C.		235		.26	.68	.94	1.25
	120	16" O.C.		265		.21	.61	.82	1.09
	130	24" O.C.		350		.16	.46	.62	.82
	150	1-1/2" channels, galvanized, 12" O.C.,		210		.40	.77	1.17	1.52
	160	16" O.C.		240		.32	.67	.99	1.30
	180	24" O.C.		305		.26	.53	.79	1.03
	800	Suspended ceilings, including carriers							
	820	1-1/2" carriers, 24" O.C.							
	830	3/4" channels, 16" O.C.	1 Lath	165	S.F.	.47	.97	1.44	1.89
	832	24" OC		200		.42	.80	1.22	1.60
	840	1-1/2" channels, 16" O.C.		155		.58	1.04	1.62	2.10
	842	24" OC		190		.52	.85	1.37	1.77
	860	2" carriers, 24" OC							
	870	3/4" channels, 16" OC	1 Lath	155	S.F.	.51	1.04	1.55	2.02
	872	24" OC		190		.46	.85	1.31	1.70
	880	1-1/2" channels, 16" OC		145		.62	1.11	1.73	2.25
	882	24" OC		180		.56	.89	1.45	1.88
16-001		STUDDING, PLASTER For drywall studs, see division 9.2-20							
	003	For complete partitions, see division 9.2-31							
	400	LB studs, light gauge structural, painted, 18 ga., 2", 16" O.C.	1 Lath	435	S.F.	.30	.37	.67	.85
	410	24" O.C.		510		.24	.32	.56	.71
	420	2-1/2" studs, 16" O.C.		425		.32	.38	.70	.89
	424	24" O.C.		500		.26	.32	.58	.74

For expanded coverage of these items see *Means' Interior Cost Data 1986*

9.1 Lath & Plaster

		CREW	DAILY OUTPUT	UNIT	BARE COSTS MAT.	BARE COSTS INST.	BARE COSTS TOTAL	TOTAL INCL O&P
16 430	3-1/4" studs, 16" O.C. (24)	1 Lath	410	S.F.	.38	.39	.77	.97
435	24" O.C.		490		.30	.33	.63	.79
440	3-5/8" studs, 16" O.C.		400		.41	.40	.81	1.02
445	24" O.C.		480		.33	.34	.67	.84
450	4" studs, 16" O.C.		390		.44	.41	.85	1.07
455	24" O.C.		450		.35	.36	.71	.89
460	6" studs, 16" O.C.		360		.59	.45	1.04	1.28
465	24" O.C.		440		.47	.37	.84	1.03
17-001	**GAUGING PLASTER** 100 lb. bags, less than 1 ton			Bag	13.80		13.80	15.20M
010	Over 1 ton			"	13		13	14.30M
20-001	**GYPSUM LATH** Plain or perforated, nailed, 3/8" thick	1 Lath	85	S.Y.	2.10	1.89	3.99	4.98
010	1/2" thick, nailed		80		2.25	2.01	4.26	5.30
030	Clipped to steel studs, 3/8" thick (106)		75		2.30	2.14	4.44	5.55
040	1/2" thick		70		2.40	2.30	4.70	5.90
060	Firestop gypsum base, to steel studs, 1/2" thick		70		2.55	2.30	4.85	6.05
070	5/8" thick		65		2.70	2.47	5.17	6.45
090	Moisture resistant, 4' x 8' sheets, 1/2" thick		75		2.95	2.14	5.09	6.25
100	5/8" thick		70		3.25	2.30	5.55	6.80
120	Laminated, 1" thick, to steel studs		65		4.15	2.47	6.62	8.05
130	For foil facing, add to above				.25		.25	.27M
150	For ceiling installations, add	1 Lath	198			.81	.81	1.15L
160	For columns and beams, add	"	198			.81	.81	1.15L
21-001	**GYPSUM PLASTER** 80# bag, less than 1 ton			Bag	9		9	9.90M
010	Over 1 ton			"	8.35		8.35	9.20M
030	2 coats, no lath included, on walls (110)	J-1	105	S.Y.	2.52	7.40	9.92	13.25
040	On ceilings	"	92		2.52	8.45	10.97	14.70
060	2 coats on and incl. 3/8" gypsum lath on steel, on walls (106)	J-2	97		4.80	9.70	14.50	18.95
070	On ceilings	"	83		4.80	11.35	16.15	21
090	3 coats, no lath included, on walls	J-1	87		3.45	8.95	12.40	16.40
100	On ceilings	"	78		3.45	10	13.45	17.85
120	3 coats on and including painted metal lath, on wood studs	J-2	86		5.35	10.95	16.30	21
130	On ceilings	"	76.50		5.35	12.30	17.65	23
160	For irregular or curved surfaces, add				30%	30%		
180	For columns and beams, add				50%	50%		
22-001	**HIGH RISE** For lathing and plastering on high							
002	rise buildings, add to all prices			S.Y.		25%		
24-001	**KEENES CEMENT** In 100 lb. bags, less than 1 ton			Bag	16.45		16.45	18.10M
010	Over 1 ton			"	16		16	17.60M
030	Finish only, add to plaster prices, standard	J-1	215	S.Y.	1.60	3.63	5.23	6.85
040	High quality	"	144		1.60	5.40	7	9.40
25-001	**METAL LATH** Diamond, expanded, 2.5 lb. per S.Y., painted				1.40		1.40	1.54M
010	Galvanized, 2.5 lb. per S.Y.				1.79		1.79	1.96M
030	3.4 lb. per S.Y., painted				2		2	2.20M
040	Galvanized				2.25		2.25	2.47M
060	For paper backing, asphalt paper, add				.42		.42	.46M
070	30 lb. kraft, add				.30		.30	.33M
090	Flat rib, 1/8" high, 2.75 lb., painted				1.52		1.52	1.67M
100	Foil backed				2.30		2.30	2.53M
120	3.4 lb. per S.Y., painted				1.85		1.85	2.03M
130	Galvanized				2.40		2.40	2.64M
150	For paper backing, asphalt paper, add				.42		.42	.46M
160	30 lb. kraft, add				.30		.30	.33M
180	High rib, 3/8" high, 3.4 lb. per S.Y., painted				1.92		1.92	2.11M
190	Galvanized				2.70		2.70	2.97M
220	For 30 lb. kraft paper backing, add				.30		.30	.33M
240	High rib, 3/4" high, painted, .60 lb. per S.F.			S.F.	.38		.38	.41M
250	.75 lb. per S.F.			"	.46		.46	.50M

For expanded coverage of these items see *Means' Interior Cost Data 1986*

9.1 Lath & Plaster

			CREW	DAILY OUTPUT	UNIT	BARE COSTS MAT.	BARE COSTS INST.	BARE COSTS TOTAL	TOTAL INCL O&P
25	270	Stucco mesh, painted, 1.8 lb.			S.Y.	2		2	2.20M
	280	3.6 lb.				2.80		2.80	3.08M
	300	K-lath, perforated, absorbent paper, regular				1.86		1.86	2.04M
	310	Heavy duty				2.18		2.18	2.39M
	330	Waterproof, heavy duty, grade B backing				2.22		2.22	2.44M
	340	Fire resistant backing				2.44		2.44	2.68M
	360	2.5 lb. diamond painted, on wood framing, on walls	1 Lath	85		1.56	1.89	3.45	4.39
	370	On ceilings		75		1.56	2.14	3.70	4.74
	390	3.4 lb. diamond painted, on wood framing, on walls		80		2.16	2.01	4.17	5.20
	400	On ceilings		70		2.16	2.30	4.46	5.60
	420	3.4 lb. diamond painted, wired to steel framing, on walls		75		2.16	2.14	4.30	5.40
	430	On ceilings		60		2.16	2.68	4.84	6.15
	460	Cornices, wired to steel	↓	35	↓	2.53	4.59	7.12	9.25
	480	Screwed to steel studs, 2.5 lb.	1 Lath	80	S.Y.	1.57	2.01	3.58	4.56
	490	3.4 lb.		75		2.20	2.14	4.34	5.45
	510	Rib lath, painted, wired to steel, on walls, 2.75 lb.		75		2.09	2.14	4.23	5.30
	520	3.4 lb.		70		2.19	2.30	4.49	5.65
	540	4.0 lb.	↓	65		2.41	2.47	4.88	6.15
	550	For self-furring lath, add				.03		.03	.03M
	570	Suspended ceiling system, incl. 3.4 lb. diamond lath, painted	1 Lath	15		5.40	10.70	16.10	21
	580	Galvanized	"	15	↓	5.65	10.70	16.35	21
	600	Hollow metal stud partitions, 3.4 lb. painted lath both sides							
	601	Non-load bearing, 25 ga., 2-1/2" studs, 12" O.C.	1 Lath	20.30	S.Y.	6.25	7.90	14.15	18.05
	630	16" O.C.		21.10		5.80	7.60	13.40	17.15
	635	24" O.C., rib lath		22.70		5.55	7.10	12.65	16.10
	640	3-5/8" studs, 16" O.C.		19.50		6.05	8.25	14.30	18.30
	660	24" O.C., rib lath		20.40		5.80	7.90	13.70	17.50
	670	4" studs, 16" O.C.		20.40		6.60	7.90	14.50	18.40
	690	24" O.C., rib lath		21.60		6.15	7.45	13.60	17.25
	700	6" studs, 16" O.C.		19.50		7.15	8.25	15.40	19.50
	710	24" O.C., rib lath		21.10		6.70	7.60	14.30	18.10
	720	Load bearing partitions, 16" O.C., 2-1/2" studs, 16 ga.		20		12.30	8.05	20.35	25
	730	3-5/8" studs, 16 ga.		19.70		13.20	8.15	21.35	26
	750	4" studs, 16 ga.		19.50		13.55	8.25	21.80	27
	760	6" studs, 16 ga.	↓	18.70	↓	15.25	8.60	23.85	29
	775								
	776								
	780	Solid 2" thick partition on 3/4" cold rolled channel, 3.4#							
	790	diamond painted metal lath, 1 side, no plaster	1 Lath	21.60	S.Y.	3.45	7.45	10.90	14.30
30-001		PARTITION WALLS							
	030								
	040	Stud walls, 3.4 lb. metal lath, 3 coat gypsum plaster, 2 sides							
	060	2" x 4" wood studs, 16" O.C.	J-2	315	S.F.	1.25	2.98	4.23	5.60
	070	2-1/2" metal studs, 25 ga., 12" O.C.		325		1.21	2.89	4.10	5.40
	080	3-5/8" metal studs, 25 ga., 16" O.C.	↓	320	↓	1.19	2.94	4.13	5.45
	090	Gypsum lath, 2 coat vermiculite plaster, 2 sides							
	100	2" x 4" wood studs, 16" O.C.	J-2	355	S.F.	1.08	2.65	3.73	4.92
	120	2-1/2" metal studs, 25 ga., 12" O.C.		365		1.06	2.58	3.64	4.80
	130	3-5/8" metal studs, 25 ga., 16" O.C.	↓	360	↓	1.04	2.61	3.65	4.82
35-001		PERLITE OR VERMICULITE PLASTER Under 200 bags			Bag	5.15		5.15	5.65M
	010	Over 200 bags			"	5		5	5.50M
	030	2 coats, no lath included, on walls	J-1	92	S.Y.	2.46	8.45	10.91	14.65
	040	On ceilings	"	79		2.46	9.85	12.31	16.60
	060	2 coats, on and incl. 3/8" gypsum lath, on metal studs	J-2	84		4.76	11.20	15.96	21
	070	On ceilings	"	70		4.76	13.45	18.21	24
	090	3 coats, no lath included, on walls	J-1	74		3.77	10.55	14.32	19
	100	On ceilings	"	63		3.77	12.35	16.12	22
	120	3 coats, on and incl. painted metal lath, on metal studs	J-2	72		5.75	13.05	18.80	25
	130	On ceilings	"	61	↓	5.75	15.40	21.15	28

For expanded coverage of these items see *Means' Interior Cost Data 1986*

9.1 Lath & Plaster

		CREW	DAILY OUTPUT	UNIT	BARE COSTS MAT.	BARE COSTS INST.	BARE COSTS TOTAL	TOTAL INCL O&P
35 150	3 coats, on and incl. suspended metal lath ceiling	J-2	37	S.Y.	9.15	25	34.15	46
170	For irregular or curved surfaces, add to above				30%	30%		
180	For columns and beams, add to above				50%	50%		
190	For soffits, add to ceiling prices					40%		
40-001	**PORTLAND CEMENT PLASTER** For Refrigeration, see 13.1-57-240							
50-001 005	**SPRAYED** Mineral fiber or cementitious for fireproofing, not incl. tamping or canvas protection							
010	1" thick, on flat plate steel	G-2	3,000	S.F.	.30	.18	.48	.57
020	Flat decking		2,400		.30	.22	.52	.63
040	Beams		1,500		.30	.35	.65	.81
050	Corrugated or fluted decks		1,250		.30	.42	.72	.90
070	Columns, 1-1/8" thick		1,100		.36	.48	.84	1.05
080	2-3/16" thick		700		.60	.75	1.35	1.68
085	For tamping, add					10%		
090	For canvas protection, add	G-2	5,000		.03	.11	.14	.18
100	Acoustical sprayed, 1" thick, finished, straight work, minimum		520		.30	1.01	1.31	1.71
110	Maximum		200		.30	2.63	2.93	3.92
130	Difficult access, minimum		225		.30	2.34	2.64	3.52
140	Maximum		130		.30	4.04	4.34	5.85
150	Intumescent epoxy fireproofing on wire mesh, 3/16" thick							
155	1 hour rating, exterior use	G-2	136	S.F.	4.30	3.87	8.17	10
160	Magnesium oxychloride, 35# to 40# density, 1/4" thick		3,000		.70	.18	.88	1.01
165	1/2" thick		2,000		1.40	.26	1.66	1.90
170	60# to 70# density, 1/4" thick		3,000		.92	.18	1.10	1.25
175	1/2" thick		2,000		1.80	.26	2.06	2.34
200	Vermiculite cement, troweled or sprayed, 1/4" thick		3,000		.70	.18	.88	1.01
205	1/2" thick		2,000		1.40	.26	1.66	1.90
55-001	**STUCCO** 3 coats 1" thick, float finish, on frame construction	J-2	52	S.Y.	4.57	18.10	22.67	31
010	On masonry construction	J-1	55		1.50	14.15	15.65	22
030	For trowel finish, add	1 Plas	170			.94	.94	1.33L
040	For 3/4" thick, deduct	J-1	880		.55	.89	1.44	1.85
060	For coloring and special finish, add, minimum		685		.11	1.14	1.25	1.72
070	Maximum		200		.65	3.90	4.55	6.20
090 091	For soffits, add	J-2	155		1.55	6.05	7.60	10.25
100	Exterior plaster, with bonding agent, 1 coat, on walls	J-1	240	S.Y.	2.25	3.25	5.50	7.05
120	Ceilings		200		2.25	3.90	6.15	7.95
130	Beams		100		2.25	7.80	10.05	13.45
150	Columns		120		2.25	6.50	8.75	11.60
160	Mesh, painted, nailed to wood, 1.8 lb.	1 Lath	60		1.98	2.68	4.66	5.95
180	3.6 lb.		55		2.26	2.92	5.18	6.60
190	Wired to steel, painted, 1.8 lb.		53		2.17	3.03	5.20	6.65
210	3.6 lb.		50		2.50	3.22	5.72	7.30
220	Clinton cloth, on wood		60		3.40	2.68	6.08	7.50
240	On steel		53		3.45	3.03	6.48	8.10
56-001	**THIN COAT** Plaster, 1 coat veneer, not incl. lath	J-1	360		.63	2.17	2.80	3.74
57-001	**TILE OR TERRAZZO BASE** Scratch coat only	J-1	300	S.Y.	.37	2.60	2.97	4.07
050	Scratch and brown coat only		115		.85	6.80	7.65	10.50
60-001	**WOOD FIBER PLASTER** On walls, no furring, 2 coats		72		2.10	10.85	12.95	17.55
010	3 coats		57		2.40	13.65	16.05	22

For expanded coverage of these items see *Means' Interior Cost Data 1986*

9.2 Drywall

		CREW	DAILY OUTPUT	UNIT	BARE COSTS MAT.	BARE COSTS INST.	BARE COSTS TOTAL	TOTAL INCL O&P
02-001	ACCESSORIES, DRYWALL Casing bead, galvanized steel	1 Carp	2.90	C.L.F.	9.75	55	64.75	91
010	Vinyl		3		8.65	53	61.65	87
030	Corner bead, galvanized steel, 1" x 1"		2.90		8	55	63	89
040	1-1/4" x 1-1/4"		2.85		10.50	56	66.50	93
060	Vinyl corner bead		2.90		7	55	62	88
070	Door casing, vinyl, for 2" wall systems		2.50		22	64	86	115
090	Furring channel, galv. steel, 7/8" deep, standard		2.60		15.50	62	77.50	105
100	Resilient		2.55		15.50	63	78.50	110
110	J bead, galvanized steel, 1/2" wide		3		11.50	53	64.50	90
112	5/8" wide		2.95		12.50	54	66.50	92
114	L bead, galvanized		3		12.50	53	65.50	91
115	U bead, galvanized	↓	2.95	↓	12.50	54	66.50	92
116	Screws #6 x 1" A			M	11.40		11.40	12.55M
117	#6 x 1-5/8" A			"	17.25		17.25	18.95M
120	(108) Studs and runners for partitions, see also 9.1-16							
121								
150	Z bar, galvanized steel, 1-1/2" wide	1 Carp	2.60	C.L.F.	21	62	83	110
160	2" wide	"	2.55	"	23	63	86	115
07-001	DRYWALL Gypsum plasterboard, nailed or screwed to studs,							
010	unless otherwise noted							
015	3/8" thick, on walls, standard, no finish included	2 Carp	2,000	S.F.	.20	.16	.36	.45
020	On ceilings, standard, no finish included		1,400		.20	.23	.43	.55
025	On beams, columns, or soffits, no finish included	↓	750	↓	.20	.43	.63	.84
027								
030	1/2" thick, on walls, standard, no finish included	2 Carp	1,800	S.F.	.22	.18	.40	.50
035	Taped and finished		900		.24	.36	.60	.78
040	Fire resistant, no finish included		1,800		.24	.18	.42	.52
045	Taped and finished		900		.26	.36	.62	.80
050	Water resistant, no finish included		1,800		.30	.18	.48	.59
055	Taped and finished		900		.32	.36	.68	.87
060	Prefinished, vinyl, clipped to studs	↓	1,100	↓	.64	.29	.93	1.13
065								
100	On ceilings, standard, no finish included	2 Carp	1,200	S.F.	.22	.27	.49	.63
105	Taped and finished		800		.24	.40	.64	.84
110	Fire resistant, no finish included		1,200		.24	.27	.51	.65
115	Taped and finished		800		.26	.40	.66	.87
120	Water resistant, no finish included		1,200		.30	.27	.57	.72
125	Taped and finished		800		.32	.40	.72	.93
150	On beams, columns, or soffits, standard, no finish included		675		.22	.47	.69	.93
155	Taped and finished		475		.24	.67	.91	1.24
160	Fire resistant, no finish included		675		.24	.47	.71	.95
165	Taped and finished		475		.26	.67	.93	1.26
170	Water resistant, no finish included		675		.30	.47	.77	1.02
175	Taped and finished		475		.32	.67	.99	1.33
200	5/8" thick, on walls, standard, no finish included		1,700		.24	.19	.43	.54
205	Taped and finished		850		.26	.38	.64	.83
210	Fire resistant, no finish included		1,700		.25	.19	.44	.55
215	Taped and finished		850		.27	.38	.65	.84
220	Water resistant, no finish included		1,700		.33	.19	.52	.64
225	Taped and finished		850		.35	.38	.73	.93
230	Prefinished, vinyl, clipped to studs	↓	1,050	↓	.67	.30	.97	1.18
235								
300	On ceilings, standard, no finish included	2 Carp	1,100	S.F.	.24	.29	.53	.69
305	Taped and finished		750		.26	.43	.69	.90
310	Fire resistant, no finish included		1,100		.25	.29	.54	.70
315	Taped and finished		750		.27	.43	.70	.92
320	Water resistant, no finish included		1,100		.33	.29	.62	.78
325	Taped and finished		750		.35	.43	.78	1
350	On beams, columns, or soffits, standard, no finish included		650		.24	.49	.73	.98
355	Taped and finished	↓	450	↓	.26	.71	.97	1.32

For expanded coverage of these items see *Means' Interior Cost Data 1986*

9.2 Drywall

		Description	CREW	DAILY OUTPUT	UNIT	BARE COSTS MAT.	BARE COSTS INST.	BARE COSTS TOTAL	TOTAL INCL O&P
07	360	Fire resistant, no finish included	2 Carp	650	S.F.	.25	.49	.74	.99
	365	Taped and finished		450		.27	.71	.98	1.33
	370	Water resistant, no finish included		650		.33	.49	.82	1.08
	375	Taped and finished		450		.35	.71	1.06	1.42
	400	Fireproofing, beams or columns, 2 layers, 1/2" thick		330		.34	.97	1.31	1.78
	405	5/8" thick		300		.42	1.07	1.49	2.01
	410	3 layers, 1/2" thick		225		.51	1.42	1.93	2.62
	415	5/8" thick		210		.63	1.52	2.15	2.90
	460	Blueboard, 1/2" thick, standard		1,800		.22	.18	.40	.50
	465	Fireproof		1,800		.24	.18	.42	.52
	470	5/8" thick, fireproof		1,700		.25	.19	.44	.55
	505	For 1" thick coreboard on columns	↓	480		.35	.67	1.02	1.35
	510	For foil-backed board, add				.11		.11	.12M
	520	For high ceilings, over 8' high, add	2 Carp	3,060		.10	.10	.20	.26
	525	For prime coat (residential construction), add		2,400		.05	.13	.18	.25
	527	For textured spray, add		1,450		.09	.22	.31	.42
	530	For over 3 stories high, add per story	↓	6,100	↓	.05	.05	.10	.13
	540								
	550	For acoustical sealant, add per bead	1 Carp	500	L.F.	.02	.32	.34	.49
	555	Sealant, 1 quart tube			Ea.	5.95		5.95	6.55M
	560	Sound deadening board, 1/4" gypsum	2 Carp	1,800	S.F.	.15	.18	.33	.42
	565	1/2" wood fiber	"	1,800	"	.18	.18	.36	.46
20	001	**METAL STUDS, DRYWALL** Partitions, 10' high, with runners							
	005	See also Studding, division 9.1-16							
	200	Non-load bearing, galvanized, 25 ga. 1-5/8", 16" O.C.	1 Carp	420	S.F.	.17	.38	.55	.74
	210	24" O.C.		500		.14	.32	.46	.62
	220	2-1/2" wide, 16" O.C.		410		.20	.39	.59	.79
	225	24" O.C.		490		.16	.33	.49	.65
	230	3-5/8" wide, 16" O.C. (108)		400		.23	.40	.63	.83
	235	24" O.C.		480		.19	.33	.52	.69
	240	4" wide, 16" O.C.		390		.28	.41	.69	.90
	245	24" O.C.		450		.23	.36	.59	.77
	250	6" wide, 16" O.C.		360		.35	.44	.79	1.03
	255	24" O.C.		440		.29	.36	.65	.85
	260	20 ga studs, 1-5/8" wide, 16" O.C.		435		.33	.37	.70	.90
	265	24" O.C.		510		.28	.31	.59	.76
	270	2-1/2" wide, 16" O.C.		425		.35	.38	.73	.93
	275	24" O.C.		500		.29	.32	.61	.78
	280	3-5/8" wide, 16" O.C.		400		.41	.40	.81	1.03
	285	24" O.C.		480		.34	.33	.67	.86
	290	4" wide, 16" O.C.		390		.48	.41	.89	1.12
	295	24" O.C.		450		.40	.36	.76	.96
	300	6" wide, 16" O.C.		360		.58	.44	1.02	1.28
	305	24" O.C.		440		.48	.36	.84	1.05
	400	LB studs, light ga. structural, galv., 18 ga., 2-1/2"., 16" O.C.		425		.76	.38	1.14	1.38
	410	24" O.C.		500		.63	.32	.95	1.16
	420	3-5/8" wide, 16" O.C.		400		.84	.40	1.24	1.50
	425	24" O.C.		480		.70	.33	1.03	1.25
	430	4" wide, 16" O.C.		390		.88	.41	1.29	1.56
	435	24" O.C.		450		.73	.36	1.09	1.32
	440	6" wide, 16" O.C.		360		1.04	.44	1.48	1.79
	445	24" O.C.		440		.87	.36	1.23	1.48
	460	16 ga. studs, 2-1/2", 16" O.C.		400		.92	.40	1.32	1.59
	465	24" O.C.		480		.76	.33	1.09	1.32
	470	3-5/8" wide, 16" O.C.		390		1.02	.41	1.43	1.72
	475	24" O.C.		450		.85	.36	1.21	1.45
	480	4" wide, 16" O.C.		380		1.06	.42	1.48	1.78
	485	24" O.C.		440		.88	.36	1.24	1.49
	490	6" wide, 16" O.C.		340		1.25	.47	1.72	2.06
	495	24" O.C.	↓	415	↓	1.04	.39	1.43	1.70

For expanded coverage of these items see *Means' Interior Cost Data 1986*

9.2 Drywall

		Crew	Daily Output	Unit	Bare Costs Mat.	Bare Costs Inst.	Bare Costs Total	Total Incl O&P
31-001	**PARTITION WALL** Stud wall, 8' to 12' high. See also division 9.1-30							
002								
005	1/2", interior, gypsum drywall, standard, taped both sides							
050	Installed on and incl., 2" x 4" wood studs, 16" O.C.	2 Carp	310	S.F.	.76	1.03	1.79	2.33
100	Metal studs, NLB, 25 ga., 16" O.C., 3-5/8" wide		350		.71	.91	1.62	2.11
120	6" wide		330		.83	.97	1.80	2.32
140	Water resistant, on 2" x 4" wood studs, 16" O.C.		310		.92	1.03	1.95	2.51
160	Metal studs, NLB, 25 ga., 16" O.C., 3-5/8" wide		350		.87	.91	1.78	2.28
180	6" wide		330		.99	.97	1.96	2.49
200	Fire res.,2 layers,1-1/2 hr.,on 2" x 4" wood studs,16"O.C.		210		1.32	1.52	2.84	3.66
220	Metal studs, NLB, 25 ga., 16" O.C., 3-5/8" wide		250		1.27	1.28	2.55	3.25
240	6" wide		230		1.39	1.39	2.78	3.55
260	Fire & water res.,2 layers,1-1/2 hr., 2"x4" studs,16" O.C.		210		1.48	1.52	3	3.84
280	Metal studs, NLB, 25 ga., 16" O.C., 3-5/8" wide		250		1.43	1.28	2.71	3.43
300	6" wide		230		1.55	1.39	2.94	3.72
320	5/8", interior, gypsum drywall, standard, taped both sides							
340	Installed on and including 2" x 4" wood studs, 16" O.C.	2 Carp	300	S.F.	.80	1.07	1.87	2.43
360	24" O.C.		330		.73	.97	1.70	2.21
380	Metal studs, NLB, 25 ga., 16" O.C., 3-5/8" wide		340		.75	.94	1.69	2.19
400	6" wide		320		.87	1	1.87	2.41
420	24" O.C., 3-5/8" wide		360		.71	.89	1.60	2.07
440	6" wide		340		.81	.94	1.75	2.25
480	Water resistant, on 2" x 4" wood studs, 16" O.C.		300		.98	1.07	2.05	2.62
500	24" O.C.		330		.91	.97	1.88	2.41
520	Metal studs, NLB, 25 ga. 16" O.C., 3-5/8" wide		340		.93	.94	1.87	2.39
540	6" wide		320		1.05	1	2.05	2.60
560	24" O.C., 3-5/8" wide		360		.89	.89	1.78	2.27
580	6" wide		340		.99	.94	1.93	2.45
600	Fire res., 2 layers, 2 hr., on 2" x 4" wood studs, 16" O.C.		205		1.36	1.56	2.92	3.76
620	24" O.C.		235		1.29	1.36	2.65	3.39
640	Metal studs, NLB, 25 ga., 16" O.C., 3-5/8" wide		245		1.31	1.31	2.62	3.33
660	6" wide		225		1.43	1.42	2.85	3.63
680	24" O.C., 3-5/8" wide		265		1.27	1.21	2.48	3.15
700	6" wide		245		1.37	1.31	2.68	3.40
720	Fire & water res., 2 layers, 2 hr., 2" x 4" studs, 16" O.C.		205		1.52	1.56	3.08	3.93
740	24" O.C.		235		1.45	1.36	2.81	3.57
760	Metal studs, NLB, 25 ga., 16" O.C., 3-5/8" wide		245		1.47	1.31	2.78	3.51
780	6" wide		225		1.59	1.42	3.01	3.81
800	24" O.C., 3-5/8" wide		265		1.43	1.21	2.64	3.32
820	6" wide		245		1.53	1.31	2.84	3.58
860	1/2" blueboard, mesh tape both sides							
862	Installed on and including 2" x 4" wood studs, 16" O.C.	2 Carp	300	S.F.	.76	1.07	1.83	2.38
864	Metal studs, NLB, 25 ga., 16" O.C., 3-5/8" wide		340		.71	.94	1.65	2.14
866	6" wide		320		.83	1	1.83	2.36
868	For plaster partitions, see division 9.1-30							
875								
900	Exterior, 1/2" gypsum sheathing, 1/2" gypsum finished, interior,							
910	including foil faced insulation, metal studs, 20 ga.							
920	16" O.C., 3-5/8" wide	2 Carp	270	S.F.	1	1.19	2.19	2.82
940	6" wide	"	290	"	1.17	1.10	2.27	2.89
35-001	**CEILINGS** Gypsum drywall, fire rated, finished							
002								
010	Screwed to grid, channel or joists, 1/2" thick	2 Carp	770	S.F.	.26	.42	.68	.89
020	5/8" thick		750		.27	.43	.70	.92
030	Over 8' high, 1/2" thick		725		.26	.44	.70	.93
040	5/8" thick		685		.27	.47	.74	.97
050								
060	Grid suspension system, direct hung							
070	1-1/2" C.R.C., with 7/8" hi hat furring channel, 16" O.C.	2 Carp	600	S.F.	.42	.53	.95	1.23
080	24" O.C.	"	900	"	.34	.36	.70	.89

For expanded coverage of these items see *Means' Interior Cost Data 1986*

9.2 Drywall

			Crew	Daily Output	Unit	Mat.	Inst.	Total	Total Incl O&P
35	090	3-5/8" channel, 25 ga., with track, 16" O.C.	2 Carp	600	S.F.	.23	.53	.76	1.03
	100	24" O.C.	"	900	"	.19	.36	.55	.72
	45-001	**SHAFT WALL** Cavity type, 1" steel C-H studs, with 2 layers 5/8"							
	003	gypsum board 1 side, 2 hour fire rating	2 Carp	165	S.F.	1.44	1.94	3.38	4.39
	006	Laminated gypsum drywall, 2-1/2" solid or							
	010	3-3/4" core with steel H sections							
	030	24" wide units, to 10'-4" high	2 Carp	108	S.F.	1.60	2.96	4.56	6.05
	040	16" wide units, to 11'-7" high	"	92	"	1.95	3.48	5.43	7.20
	060	Solid 2" thick, steel edge gypsum in channels with							
	070	1/2" fire resistant gypsum							
	080	1 side, 2 hour fire rating	2 Carp	150	S.F.	1.25	2.13	3.38	4.47
	090	2 sides, 3 hour fire rating	"	135	"	1.59	2.37	3.96	5.20

9.3 Tile & Terrazzo

		Crew	Daily Output	Unit	Mat.	Inst.	Total	Total Incl O&P
05-001	**CERAMIC TILE** Base, using 1" x 1" tiles, 4" high, mud set	D-7	82	L.F.	3.15	3.45	6.60	8.30
010	Thin set	"	128		2.90	2.21	5.11	6.30
030	For 6" high base, 1" x 1" tiles, add				.29		.29	.31M
040	For 2" x 2" tiles, add to above				.16		.16	.17M
060	Cove base, 4-1/4" x 4-1/4" high, mud set	D-7	91		1.52	3.11	4.63	6.05
070	Thin set		128		1.42	2.21	3.63	4.66
090	6" x 4-1/4" high, mud set		100		1.51	2.83	4.34	5.65
100	Thin set		137		1.37	2.06	3.43	4.40
120	Sanitary cove base, 6" x 4-1/4" high, mud set		93		1.65	3.04	4.69	6.10
130	Thin set		124		1.61	2.28	3.89	4.97
150	6" x 6" high, mud set		84		1.74	3.37	5.11	6.65
160	Thin set		117		1.64	2.42	4.06	5.20
180	Bathroom accessories, average		82	Ea.	6.95	3.45	10.40	12.50
190	Bathtub, 5', recessed, modular wainscot, adhesive set, 6' high		4.30		140	66	206	245
210	7' high wainscot		4		150	71	221	265
220	8' high wainscot		3.80		165	74	239	285
240	Bullnose trim, 4-1/4" x 4-1/4", mud set		82	L.F.	1.10	3.45	4.55	6.05
250	Thin set		128		1.05	2.21	3.26	4.25
270	6" x 4-1/4" bullnose trim, mud set		84		1.10	3.37	4.47	5.95
280	Thin set		124		1.05	2.28	3.33	4.35
300	Floors, natural clay, random or uniform, thin set, color group 1		183	S.F.	1.60	1.55	3.15	3.93
310	Color group 2		183		1.72	1.55	3.27	4.06
330	Porcelain type, 1 color, color group 2, 1" x 1"		183		1.80	1.55	3.35	4.15
340	2" x 2" or 2" x 1", thin set		190		2	1.49	3.49	4.29
360	For random blend, 2 colors, add				.10		.10	.11M
370	4 colors, add				.20		.20	.22M
390	For color group 3, add				.15		.15	.16M
400	For abrasive non-slip tile, add				.42		.42	.46M
420	Conductive tile, 1" squares, black	D-7	109		3.86	2.59	6.45	7.90
422	4" x 8" or 4" x 4", 3/8" thick		120		4.06	2.36	6.42	7.75
424	Trim, bullnose, etc.		200	L.F.	1.47	1.41	2.88	3.60
430	Specialty tile, 3" x 6" x 1/2", decorator finish		183	S.F.	4	1.55	5.55	6.55
450	Add for epoxy grout, 1/16" joint, 1" x 1" tile		800		.61	.35	.96	1.17
460	2" x 2" tile		820		.47	.34	.81	1
480	Pregrouted sheets, walls, 4-1/4" x 4-1/4", 6" x 4-1/4"							
481	and 8-1/2" x 4-1/4", 4 S.F. sheets, silicone grout	D-7	240	S.F.	2	1.18	3.18	3.85
510	Floors, unglazed, 2 S.F. sheets,							
511	urethane adhesive	D-7	180	S.F.	3.30	1.57	4.87	5.85
540	Walls, interior, thin set, 4-1/4" x 4-1/4" tile		190		1.22	1.49	2.71	3.43
550	6" x 4-1/4" tile		190		1.55	1.49	3.04	3.79

For expanded coverage of these items see *Means' Interior Cost Data 1986*

9.3 Tile & Terrazzo

			CREW	DAILY OUTPUT	UNIT	BARE COSTS MAT.	BARE COSTS INST.	BARE COSTS TOTAL	TOTAL INCL O&P
05	570	8-1/2" x 4-1/4" tile	D-7	190	S.F.	1.94	1.49	3.43	4.22
	580	6" x 6" tile		200	"	1.70	1.41	3.11	3.85
	600	Decorated wall tile, 4-1/4" x 4-1/4", minimum		870	Ea.	.79	.33	1.12	1.32
	610	Maximum		580	"	9.50	.49	9.99	11.15
	630	Exterior walls, frostproof, mud set, 4-1/4" x 4-1/4"		102	S.F.	3.05	2.77	5.82	7.25
	640	1-3/8" x 1-3/8"		93		7.50	3.04	10.54	12.50
	660	Crystalline glazed, 4-1/4" x 4-1/4", mud set, plain		100		1.80	2.83	4.63	5.95
	670	4-1/4" x 4-1/4", scored tile		100		2	2.83	4.83	6.15
	690	1-3/8" squares		93		4.20	3.04	7.24	8.90
	700	For epoxy grout, 1/16" joints, 4-1/4" tile, add		800		.40	.35	.75	.94
	720	For tile set in dry mortar, add		1,735			.16	.16	.23L
	730	For tile set in portland cement mortar, add		290			.98	.98	1.37L
10-001		CERAMIC TILE PANELS Insulated, over 1000 S.F., 1-1/2" thick		220		6.60	1.29	7.89	9.05
010		2-1/2" thick		220		7.20	1.29	8.49	9.70
15-001		GLASS MOSAICS 3/4" tile on 12" sheets, color group 1 & 2, minimum		82		7.75	3.45	11.20	13.35
030		Maximum (latex set)		73		9.25	3.87	13.12	15.60
035		Color group 3		73		9.60	3.87	13.47	16
040		Color group 4		73		11	3.87	14.87	17.55
045		Color group 5		73		14	3.87	17.87	21
050		Color group 6		73		18	3.87	21.87	25
060		Color group 7		73		21	3.87	24.87	29
070		Color group 8, golds, silvers & specialties		64		30	4.42	34.42	39
20-001		MARBLE Thin gauge tile, 12" x 6", 9/32", White Carara		64		4.20	4.42	8.62	10.80
010		Filled Travertine		64		4.80	4.42	9.22	11.45
020		Synthetic tiles, 12" x 12" x 5/8", thin set, floors		64		5.25	4.42	9.67	11.95
030		On walls (see also division 4.4-30)		55		5.25	5.15	10.40	13
25-001		METAL TILE Cove base, standard colors, 4-1/4" square	1 Carp	150	L.F.	1	1.07	2.07	2.65
020		4-1/8" x 8-1/2"		200	"	1	.80	1.80	2.26
040		Wall tile, aluminum, 4-1/4" square, thin set, plain		80	S.F.	1.70	2	3.70	4.77
050		Epoxy enameled		75		1.90	2.13	4.03	5.20
070		Leather on aluminum, colors		65		16	2.46	18.46	21
080		Stainless steel		75		4.60	2.13	6.73	8.15
100		Suede on aluminum		65		16	2.46	18.46	21
110		For sizes other than 4-1/4" x 4-1/4", add				.13		.13	.14M
30-001		PLASTIC TILE Walls, 4-1/4" x 4-1/4", .050" thick	1 Carp	125		.90	1.28	2.18	2.84
010		.110" thick	"	120		1.10	1.33	2.43	3.14
35-001		QUARRY TILE Base, cove or sanitary, 2" or 5" high, mud set							
010		1/2" thick	D-7	110	L.F.	1.60	2.57	4.17	5.35
030		Bullnose trim, red, mud set, 6" x 6" x 1/2" thick		120		1.54	2.36	3.90	5
040		4" x 4" x 1/2" thick		110		1.35	2.57	3.92	5.10
060		4" x 8" x 1/2" thick, using 8" as edge		130		1.84	2.18	4.02	5.05
061									
070		Floors, mud set, 1000 S.F. lots, red, 4" x 4" x 1/2" thick	D-7	120	S.F.	2	2.36	4.36	5.50
090		6" x 6" x 1/2" thick		140		1.65	2.02	3.67	4.65
100		4" x 8" x 1/2" thick		130		1.71	2.18	3.89	4.93
130		For waxed coating, add				.25		.25	.27M
150		For colors other than green, add				.20		.20	.22M
160		For abrasive surface, add				.25		.25	.27M
180		Brown tile, imported, 6" x 6" x 7/8"	D-7	120		2.90	2.36	5.26	6.50
190		9" x 9" x 1-1/4"		110		4.50	2.57	7.07	8.55
210		For thin set mortar application, deduct		700			.40	.40	.57L
220		For epoxy grout & mortar, 6" x 6" x 1/2", add		350		.85	.81	1.66	2.07
270		Stair tread & riser, 6" x 6" x 3/4", plain		50		4	5.65	9.65	12.35
280		Abrasive		47		4.40	6	10.40	13.25
300		Wainscot, 6" x 6" x 1/2", thin set, red		105		1.42	2.69	4.11	5.35
310		Colors other than green		105		1.51	2.69	4.20	5.45
330		Window sill, 6" wide, 3/4" thick		90	L.F.	3.20	3.14	6.34	7.90
340		Corners		80	Ea.	2.20	3.54	5.74	7.40
37-001		SLATE TILE Vermont, 6" x 6" x 1/4" thick, thin set		180	S.F.	1.83	1.57	3.40	4.22
020		See also division 2.6-40-130							

9.3 Tile & Terrazzo

		CREW	DAILY OUTPUT	UNIT	BARE COSTS MAT.	BARE COSTS INST.	BARE COSTS TOTAL	TOTAL INCL O&P
40-001	**TERRAZZO, CAST IN PLACE** Cove base, 6" high	1 Mstz	23	L.F.	.88	6.90	7.78	10.65
010	Curb, 6" high and 6" wide	"	15		1.56	10.60	12.16	16.60
030	Divider strip for floors, 12 ga., 1-1/4" deep, zinc				.68		.68	.74M
040	Brass				1.83		1.83	2.01M
060	Solid 1/4" thick, 1-1/4" deep, zinc				2.33		2.33	2.56M
070	Brass				4.69		4.69	5.15M
090	Heavy top strip, galv. bottoms, 1/2" x 1-1/4", zinc				1.04		1.04	1.14M
100	Brass				3.05		3.05	3.35M
120	For thin set floors, 16 ga., 1/2" x 1/4", zinc				.35		.35	.38M
130	Brass				.67		.67	.73M
150	Floor, bonded to concrete, 1-3/4" thick, gray cement	J-3	130	S.F.	1.63	3.12	4.75	5.90
160	White cement		125		2.28	3.24	5.52	6.75
180	Not bonded, 3" total thickness, gray cement		100		2	4.05	6.05	7.50
190	White cement		95		2.65	4.27	6.92	8.50
210	For Venetian terrazzo, 1" topping, add				50%	50%		
220	For heavy duty abrasive terrazzo, add				50%	50%		
240	Bonded conductive floor for hospitals	J-3	110		1.80	3.68	5.48	6.80
250	Epoxy terrazzo, 1/4" thick, minimum		170		1.90	2.38	4.28	5.20
255	Average		130		2.25	3.12	5.37	6.55
260	Maximum		100		2.60	4.05	6.65	8.20
270	Monolithic terrazzo, 5/8" thick, incl. 3-1/2" base slab,							
271	10' panels, mesh and felt	J-3	230	S.F.	1.90	1.76	3.66	4.40
300	Stairs, cast in place, pan filled treads		55	L.F.	1.85	7.35	9.20	11.70
310	Treads and risers		20	"	2.90	20	22.90	30
330	Stair landings, add to floor prices		62	S.F.	.27	6.55	6.82	8.90
331								
340	Stair stringers and fascia	J-3	55	S.F.	1.88	7.35	9.23	11.75
360	For abrasive metal nosings on stairs, add		285	L.F.	2.50	1.42	3.92	4.62
370	For abrasive surface finish, add		600	S.F.	.43	.68	1.11	1.36
390	For flush abrasive strips, add		620	L.F.	.30	.65	.95	1.19
400	Wainscot, bonded, 1-1/2" thick		40	S.F.	1.80	10.15	11.95	15.30
420	Epoxy terrazzo, 1/4" thick		70	"	3.20	5.80	9	11.10
45-001	**TERRAZZO, PRECAST** Base, 6" high, straight	1 Mstz	120	L.F.	4.90	1.33	6.23	7.25
010	Cove		100		5.10	1.59	6.69	7.85
030	8" high base, straight		110		5.30	1.45	6.75	7.85
040	Cove		90		5.30	1.77	7.07	8.30
060	For white cement, add				.16		.16	.17M
070	For 16 ga. zinc toe strip, add				.40		.40	.44M
090	Curbs, 4" x 4" high	1 Mstz	55		10.10	2.89	12.99	15.15
100	8" x 8" high	"	45		12	3.54	15.54	18.15
120	Floor tiles, non-slip, 1" thick, 12" x 12"	D-1	60	S.F.	5.25	4.87	10.12	12.70
130	1-1/4" thick, 12" x 12"		60		6.20	4.87	11.07	13.75
150	16" x 16"		55		5.80	5.30	11.10	13.95
160	1-1/2" thick, 16" x 16"		50		6.60	5.85	12.45	15.60
180	For Venetian terrazzo, add				1.66		1.66	1.82M
190	For white cement, add				.30		.30	.33M
210	Floor tiles, 12" x 12", 3/16" thick, 1/4" to 1/2" chips	1 Tilf	130		3.50	1.22	4.72	5.55
220	1/4" to 1" chips	"	130		3.75	1.22	4.97	5.85
240	Stair treads, 1-1/2" thick, non-slip, diamond pattern	2 Mstz	95	L.F.	12.50	3.35	15.85	18.45
250	Line pattern		90		12.20	3.54	15.74	18.40
270	2" thick treads, straight		90		12.60	3.54	16.14	18.80
280	Curved		85		13.70	3.75	17.45	20
300	Stair risers, 1" thick, to 6" high, straight sections		160		5	1.99	6.99	8.30
310	Cove		150		6.20	2.12	8.32	9.80
330	Curved, 1" thick, to 6" high, vertical		135		7.20	2.36	9.56	11.25
340	Cove		130		9.05	2.45	11.50	13.40
360	Stair tread and riser, single piece, straight, minimum		65		18.70	4.90	23.60	27
370	Maximum		60		24	5.30	29.30	34
390	Curved tread and riser, minimum		60		31	5.30	36.30	42
400	Maximum		55		38	5.80	43.80	50

For expanded coverage of these items see *Means' Interior Cost Data 1986*

9.3 Tile & Terrazzo

			CREW	DAILY OUTPUT	UNIT	BARE COSTS MAT.	BARE COSTS INST.	BARE COSTS TOTAL	TOTAL INCL O&P
45	420	Stair stringers, notched, 1" thick	2 Mstz	70	L.F.	9.45	4.55	14	16.75
	430	2" thick		60	"	12.80	5.30	18.10	22
	450	Stair landings, structural, non-slip, 1-1/2" thick		105	S.F.	7.05	3.03	10.08	12
	460	3" thick	↓	95		9.60	3.35	12.95	15.25
	480	Wainscot, 12" x 12" x 1" tiles	1 Mstz	35		5.25	4.55	9.80	12.15
	490	16" x 16" x 1-1/2" tiles	"	30	↓	6.40	5.30	11.70	14.50
	50-001	**TERRA COTTA TILE** On walls, dry set, 1/2" thick							
	010	Square, hexagonal or lattice shapes, unglazed	1 Tilf	135	S.F.	3.70	1.17	4.87	5.70
	030	Glazed, plain colors		130		5	1.22	6.22	7.20
	040	Intense colors	↓	125		8	1.26	9.26	10.55

9.5 Acoustical Treatment

		CREW	DAILY OUTPUT	UNIT	BARE COSTS MAT.	BARE COSTS INST.	BARE COSTS TOTAL	TOTAL INCL O&P
05-001	**BARRIERS** Plenum, leaded vinyl, .48 lb. per S.F.	1 Carp	170	S.F.	1.35	.94	2.29	2.85
010	.87 lb. per S.F.		155		1.55	1.03	2.58	3.20
030	1.50 lb. per S.F.		140		2.40	1.14	3.54	4.30
040	3.0 lb. per S.F.		125		3.70	1.28	4.98	5.90
060	Aluminum foil, fiberglass reinf., parallel with joists		275		.56	.58	1.14	1.46
070	Perpendicular to joists		155		.56	1.03	1.59	2.11
090	Aluminum mesh, kraft paperbacked		275		.45	.58	1.03	1.34
095	Asbestos sheet, 1/8" thick		250		.90	.64	1.54	1.92
097	Fiberglass batts, kraft faced, 3/1/2" thick		1,400		.25	.11	.36	.44
098	6" thick		1,300		.35	.12	.47	.56
100	Sheet lead, 1 lb., 1/64" thick, perpendicular to joists		150		1.20	1.07	2.27	2.87
110	Vinyl foam reinforced, 1/8" thick, 1.0 lb. per S.F.	↓	150	↓	1.85	1.07	2.92	3.58
10-001	**CEILING TILE** Stapled, cemented or installed on suspension							
010	system, 12" x 12" or 12" x 24", not including furring							
060	Mineral fiber, plastic coated, 5/8" thick	1 Carp	410	S.F.	.43	.39	.82	1.04
070	3/4" thick		400		.49	.40	.89	1.12
090	Fire rated, 3/4" thick, plain faced		395		.49	.41	.90	1.13
100	Plastic coated face		390		.55	.41	.96	1.20
120	Aluminum faced, 5/8" thick, plain	↓	385	↓	1.60	.42	2.02	2.36
121								
150	Metal pan units, 24 ga. steel, not incl. pads, painted, 12" x 12"	1 Shee	350	S.F.	1.60	.52	2.12	2.52
160	12" x 36" or 12" x 24", 7% open area		340		1.20	.53	1.73	2.10
180	Aluminum, .025" thick, painted, 12" x 12"		350		2.90	.52	3.42	3.95
190	12" x 24"		360		1.50	.50	2	2.39
210	.032" thick, 12" x 12"		350		3.40	.52	3.92	4.50
220	12" x 24"		360		1.90	.50	2.40	2.83
240	Stainless steel, 12" x 24", 26 ga., solid		350		4.10	.52	4.62	5.25
250	5.2% open area	↓	340		4.50	.53	5.03	5.75
270	For 1" sound-absorbing pads, flameproof, wrapped, add				.30		.30	.33M
280	Unwrapped, add				.15		.15	.16M
370	Wall application of above, add	1 Carp	3,100			.05	.05	.07L
390	For ceiling primer, add				.08		.08	.08M
400	For ceiling cement, add			↓	.22		.22	.24M
15-001	**SUSPENDED ACOUSTIC CEILING BOARDS** Not including							
010	suspension system							
030	Fiberglass boards, film faced, 2' x 2' or 2' x 4', 5/8" thick	1 Carp	675	S.F.	.24	.24	.48	.61
040	3/4" thick		500		.60	.32	.92	1.12
050	3" thick, thermal, R11		450		.90	.36	1.26	1.51
060	Glass cloth faced fiberglass, 3/4" thick		500		.88	.32	1.20	1.43
070	1" thick		485		1	.33	1.33	1.58
082	1-1/2" thick, nubby face	↓	475	↓	1.20	.34	1.54	1.81

For expanded coverage of these items see *Means' Interior Cost Data 1986*

9.5 Acoustical Treatment

		CREW	DAILY OUTPUT	UNIT	BARE COSTS MAT.	BARE COSTS INST.	BARE COSTS TOTAL	TOTAL INCL O&P
15 090	Mineral fiber boards, 5/8" thick, aluminum faced, 24" x 24"	1 Carp	600	S.F.	.69	.27	.96	1.15
093	24" x 48"		650		.67	.25	.92	1.09
096	Standard face		675		.21	.24	.45	.57
100	Plastic coated face		400		.43	.40	.83	1.05
120	Mineral fiber, 2 hour rating, 5/8" thick		675		.28	.24	.52	.65
130	Mirror faced panels, 15/16" thick		500		8.70	.32	9.02	10.05
150	Air distributing ceilings, 5/8" thick, F.R.D. water felted board		400		.55	.40	.95	1.18
190	Eggcrate, acrylic, 1/2" x 1/2" x 1/2" cubes		500		1.68	.32	2	2.31
210	Polystyrene eggcrate, 3/8" x 3/8" x 1/2" cubes		510		.73	.31	1.04	1.26
220	1/2" x 1/2" x 1/2" cubes		500		1	.32	1.32	1.56
240	Luminous panels, prismatic, acrylic		400		.70	.40	1.10	1.35
250	Polystyrene		400		.40	.40	.80	1.02
270	Flat or ribbed, acrylic		400		1.12	.40	1.52	1.81
280	Polystyrene		400		.65	.40	1.05	1.29
300	Drop pan, white, acrylic		400		2.65	.40	3.05	3.49
310	Polystyrene		400		2.15	.40	2.55	2.94
360	Perforated aluminum sheets, .024" thick, corrugated, painted		490		1.24	.33	1.57	1.84
370	Plain		500		.94	.32	1.26	1.50
372	Mineral fiber, 24" x 24" or 48", reveal edge, painted, 5/8" thick		600		.34	.27	.61	.76
374	3/4" thick		575		.66	.28	.94	1.13
375	Wood fiber in cementitious binder, 2' x 2' or 4', painted, 1" thick		600		.65	.27	.92	1.10
376	2" thick		550		1.16	.29	1.45	1.70
377	2-1/2" thick		500		1.60	.32	1.92	2.22
378	3" thick		450		1.86	.36	2.22	2.56
390	Access panels, metal, 12" x 12"		20	Ea.	17	8	25	30
400	12" x 24"		15		24	10.65	34.65	42
410	18" x 18"		12		25	13.35	38.35	47
420	24" x 24"		10		35	16	51	62
430	24" x 36"		8		46	20	66	80
440	Stainless steel, 12" x 12"		20		66	8	74	84
450	18" x 18"		12		140	13.35	153.35	175
460	24" x 24"		10		175	16	191	215
470	Fire rated, with lock, 12" x 12"		20		84	8	92	105
480	18" x 18"		12		110	13.35	123.35	140
490	24" x 24"		10		140	16	156	175
510	36" x 48"		6		295	27	322	365
20 001	**SUSPENSION SYSTEMS** For boards and tile listed above							
002								
005	Class A suspension system, T bar, 2' x 4' grid	1 Carp	800	S.F.	.36	.20	.56	.69
030	2' x 2' grid		650		.38	.25	.63	.77
040	Concealed Z bar suspension system, 12" module		520		.46	.31	.77	.95
060	1-1/2" carrier channels, 4' O.C., add		470		.11	.34	.45	.61
070	Carrier channels for ceilings with							
090	recessed lighting fixtures, add	1 Carp	460	S.F.	.25	.35	.60	.78
104	Hanging wire, 12 ga., 4' long		35	C.S.F.	.06	4.57	4.63	6.70
108	8' long		39	"	.12	4.10	4.22	6.10
25-001	**SUSPENDED CEILINGS, COMPLETE** Including standard							
010	suspension system but not incl. 1-1/2" carrier channels							
030	Air distributing ceilings, incl. barriers, 2' x 2' board	1 Carp	300	S.F.	.72	.53	1.25	1.56
040	12" x 12" tile		240		.90	.67	1.57	1.96
060	Ceiling board system, 2' x 4', plain faced, supermarkets		500		.63	.32	.95	1.16
070	Offices		380		.68	.42	1.10	1.36
080	Wood fiber, cementitious binder, T bar susp. 2' x 2' x 1" board		345		1.10	.46	1.56	1.88
081	2' x 4' x 1" board		380		1.10	.42	1.52	1.82
090	Luminous panels, flat or ribbed, acrylic		255		1.53	.63	2.16	2.59
100	Polystyrene		255		1.10	.63	1.73	2.12
120	Metal pan with acoustic pad, steel		205		2.72	.78	3.50	4.12
130	Painted aluminum		205		3.90	.78	4.68	5.40
150	Aluminum, degreased finish		205		3.43	.78	4.21	4.90
160	Stainless steel		205		5.80	.78	6.58	7.50

For expanded coverage of these items see *Means' Interior Cost Data 1986*

9.5 Acoustical Treatment

			CREW	DAILY OUTPUT	UNIT	BARE COSTS MAT.	BARE COSTS INST.	BARE COSTS TOTAL	TOTAL INCL O&P
25	180	Tile, Z bar suspension, 5/8" mineral fiber tile	1 Carp	235	S.F.	1.01	.68	1.69	2.10
	190	3/4" mineral fiber tile	"	230	"	1.10	.70	1.80	2.22
	210	Reveal tile with drop, 2' x 2' grid							
	220	with colored suspension system	1 Carp	345	S.F.	1.30	.46	1.76	2.10
	240	For strip lighting, see division 16.6-10							
	250	For rooms under 500 S.F., add			S.F.		25%		
30-001		SOUND ABSORBING PANELS Perforated steel facing, painted with							
	010	fiberglass or mineral filler, no backs, 2-1/4" thick, modular							
	020	space units, ceiling or wall hung, white or colored	1 Carp	200	S.F.	3.95	.80	4.75	5.50
	030	Fiberboard sound deadening panels, 1/2" thick	"	600	"	.13	.27	.40	.53
	050	Fiberglass panels, 4' x 8' x 1" thick, with							
	060	glass cloth face for walls, cemented	1 Carp	155	S.F.	1.10	1.03	2.13	2.71
	070	1-1/2" thick, dacron covered, inner aluminum frame,							
	071	wall mounted	1 Carp	600	S.F.	5.85	.27	6.12	6.80
	090	Mineral fiberboard panels, fabric covered, 30"x 108",							
	100	3/4" thick, concealed spline, wall mounted	1 Carp	150	S.F.	3.95	1.07	5.02	5.90
40-001		SOUND ATTENUATION Blanket, 1" thick		925		.20	.17	.37	.47
	050	1-1/2" thick		920		.24	.17	.41	.52
	100	2" thick		915		.28	.17	.45	.56
	150	3" thick		910		.53	.18	.71	.84
	300	Thermal or acoustical batt above ceiling, 2" thick		2,100		.46	.08	.54	.62
	310	3" thick		2,000		.50	.08	.58	.67
	320	4" thick		1,900		.59	.08	.67	.77
	340	Urethane plastic foam, open cell, on wall, 2" thick	2 Carp	2,050		2	.16	2.16	2.43
	350	3" thick		1,550		2.60	.21	2.81	3.16
	360	4" thick		1,050		3.80	.30	4.10	4.62
	370	On ceiling, 2" thick		1,700		2	.19	2.19	2.47
	380	3" thick		1,300		2.60	.25	2.85	3.22
	390	4" thick		900		3.80	.36	4.16	4.70
	391								
	400	Nylon matting 0.4" thick, with carbon black spinerette							
	401	plus polyester fabric, on floor	J-4	4,000	S.F.	1.35	.07	1.42	1.58
	420	Fiberglass reinf. backer board underlayment, 7/16" thick, on floor	"	800	"	1.10	.35	1.45	1.71

9.6 Flooring

			CREW	DAILY OUTPUT	UNIT	BARE COSTS MAT.	BARE COSTS INST.	BARE COSTS TOTAL	TOTAL INCL O&P
05-001		CARPET Commercial grades, cemented							
	070	Acrylic, 26 oz., light to medium traffic	1 Tilf	57	S.Y.	11.65	2.77	14.42	16.70
	090	28 oz., medium traffic		57		14.35	2.77	17.12	19.65
	110	35 oz., medium to heavy traffic		57		16.15	2.77	18.92	22
	210	Nylon, non anti-static, 15 oz., light traffic		57		8	2.77	10.77	12.70
	280	Nylon, with anti-static, 17 oz., light to medium traffic		57		9	2.77	11.77	13.80
	290	20 oz., medium traffic		57		10.75	2.77	13.52	15.70
	300	22 oz., medium traffic		57		10.75	2.77	13.52	15.70
	310	24 oz., medium to heavy traffic		57		10.95	2.77	13.72	15.95
	320	26 oz., medium to heavy traffic		57		11.65	2.77	14.42	16.70
	330	28 oz., heavy traffic		57		14.35	2.77	17.12	19.65
	334	32 oz., heavy traffic		57		16.15	2.77	18.92	22
	337	42 oz., heavy traffic		49		19.10	3.22	22.32	26
	340	Needle bonded, 20 oz., no padding		57		9.85	2.77	12.62	14.70
	350	Polypropylene, 15 oz., light traffic		57		4.95	2.77	7.72	9.35
	365	22 oz., medium traffic		57		6.60	2.77	9.37	11.15
	366	24 oz., medium to heavy traffic		57		7.50	2.77	10.27	12.15
	367	26 oz., medium to heavy traffic		57		9	2.77	11.77	13.80

For expanded coverage of these items see *Means' Interior Cost Data 1986*

9.6 Flooring

		CREW	DAILY OUTPUT	UNIT	BARE COSTS MAT.	BARE COSTS INST.	BARE COSTS TOTAL	TOTAL INCL O&P
368	28 oz., heavy traffic	1 Tilf	57	S.Y.	10.50	2.77	13.27	15.45
370	32 oz., heavy traffic		57		13	2.77	15.77	18.20
373	42 oz., heavy traffic	↓	49	↓	15.50	3.22	18.72	22
375								
380	Scrim installed, nylon sponge back carpet, 20 oz.	1 Tilf	57	S.Y.	15	2.77	17.77	20
385	60 oz.		57	"	21	2.77	23.77	27
400	Tile, foam-backed, needle punch		570	S.F.	.70	.28	.98	1.16
410	Tufted loop or shag		570	"	1.25	.28	1.53	1.76
411	Wool, 30 oz., medium traffic		57	S.Y.	22.50	2.77	25.27	29
450	Wool, 36 oz., medium to heavy traffic		57		27	2.77	29.77	34
470	Sponge back, wool, 36 oz., medium to heavy traffic		57		24	2.77	26.77	30
490	42 oz., heavy traffic	↓	49		27	3.22	30.22	34
550	For stretched and edge fastened, deduct				.40		.40	.44M
800	For residential construction, add to carpet	1 Tilf	167		.65	.95	1.60	2.04
900	Padding, sponge rubber cushion, minimum		150		2	1.05	3.05	3.68
910	Maximum		75		4.50	2.11	6.61	7.90
920	Felt, 32 oz. to 56 oz., minimum		150		2.25	1.05	3.30	3.95
930	Maximum		75		3.50	2.11	5.61	6.80
940	Bonded urethane, 3/8" thick, minimum		150		2	1.05	3.05	3.68
950	Maximum		75		3	2.11	5.11	6.25
960	Prime urethane, 1/4" thick, minimum		150		1.75	1.05	2.80	3.40
970	Maximum	↓	75	↓	3	2.11	5.11	6.25
09-001	CARPET MAINTENANCE Steam clean, per cleaning, minimum	1 Clab	9,760	S.F.	.03	.01	.04	.05
050	Maximum	"	3,250		.04	.04	.08	.10
10-001	COMPOSITION FLOORING Acrylic, 1/4" thick	C-6	520		2.10	1.65	3.75	4.66
010	3/8" thick		450		2.45	1.91	4.36	5.40
030	Cupric oxychloride, on bond coat, minimum		480		2.70	1.79	4.49	5.50
040	Maximum		420		3.70	2.05	5.75	7
060	Epoxy, with colored quartz chips, broadcast, minimum		675		1.10	1.27	2.37	3.02
070	Maximum		490		1.80	1.75	3.55	4.47
090	Trowelled, minimum		560		1.40	1.54	2.94	3.72
100	Maximum	↓	480	↓	2.05	1.79	3.84	4.80
120	Heavy duty epoxy topping, 1/4" thick,							
130	500 to 1,000 S.F.	C-6	420	S.F.	2.50	2.05	4.55	5.65
150	1,000 to 2,000 S.F.		450		2.10	1.91	4.01	5
160	Over 10,000 S.F.		480		2.10	1.79	3.89	4.85
180	Epoxy terrazzo, 1/4" thick, chemical resistant, minimum		375		3.40	2.29	5.69	7
190	Maximum		280		6.50	3.07	9.57	11.50
210	Conductive, minimum		355		3.25	2.42	5.67	7
220	Maximum		270		6.25	3.18	9.43	11.40
240	Mastic, hot laid, 2 coat, 1-1/2" thick, standard, minimum		690		1.65	1.25	2.90	3.58
250	Maximum		520		2.10	1.65	3.75	4.66
270	Acidproof, minimum		605		2.15	1.42	3.57	4.38
280	Maximum		350		3.20	2.46	5.66	7
300	Neoprene, trowelled on, 1/4" thick, minimum		545		2.10	1.58	3.68	4.55
310	Maximum		430		4.50	2	6.50	7.80
315	Polyacrylate terrazzo, 1/4" thick, minimum		735		4.20	1.17	5.37	6.30
317	Maximum		480		7	1.79	8.79	10.25
320	3/8" thick, minimum		620		4.30	1.39	5.69	6.70
322	Maximum		480		7	1.79	8.79	10.25
330	Conductive terrazzo, 1/4" thick, minimum		450		3.35	1.91	5.26	6.40
333	Maximum		305		5.85	2.82	8.67	10.45
335	3/8" thick, minimum		365		3.40	2.36	5.76	7.10
337	Maximum		255		5.15	3.37	8.52	10.45
345	Granite, conductive, 1/4" thick, minimum		695		4.25	1.24	5.49	6.45
347	Maximum		420		5.10	2.05	7.15	8.50
350	3/8" thick, minimum		450		4.30	1.91	6.21	7.45
352	Maximum		380		5.20	2.26	7.46	8.95
360	Polyester, with colored quartz chips, 1/16" thick, minimum		1,065		1.25	.81	2.06	2.52
370	Maximum	↓	560	↓	1.85	1.54	3.39	4.22

For expanded coverage of these items see *Means' Interior Cost Data 1986*

9.6 Flooring		CREW	DAILY OUTPUT	UNIT	BARE COSTS			TOTAL INCL O&P
					MAT.	INST.	TOTAL	
10 390	1/8" thick, minimum	C-6	810	S.F.	1.60	1.06	2.66	3.27
400	Maximum		675		2.30	1.27	3.57	4.34
420	Polyester, heavy duty, compared to epoxy, add	↓	2,590	↓	.35	.33	.68	.86
421								
430	Polyurethane, with suspended vinyl chips, minimum	C-6	1,065	S.F.	3.75	.81	4.56	5.25
450	Maximum	"	860	"	5	1	6	6.90
15-001	**CONCRETE FLOORS** And toppings, see division 3.3-26							
17-001	**MASONRY FLOORS** See division 4.2-28							
20-001	**RESILIENT** Asphalt tile, on concrete, 1/8" thick							
005	Color group B	1 Tilf	400	S.F.	.60	.40	1	1.21
010	Color group C & D	"	400		.65	.40	1.05	1.27
030	For wood subfloor, add to above for felt underlayment				.07		.07	.07M
050	For less than 500 S.F., add	1 Tilf	750			.21	.21	.30L
060	For over 5000 S.F., deduct	"	1,600	↓		.10	.10	.14L
080	Base, cove, rubber or vinyl, .080" thick							
110	(113) Standard colors, 2-1/2" high	1 Tilf	315	L.F.	.35	.50	.85	1.09
115	4" high		315		.45	.50	.95	1.20
120	6" high		315		.55	.50	1.05	1.31
145	(113) 1/8" thick, standard colors, 2-1/2" high		315		.35	.50	.85	1.09
150	4" high		315		.50	.50	1	1.25
155	6" high		315	↓	.55	.50	1.05	1.31
160	Corners, 2-1/2" high		315	Ea.	.55	.50	1.05	1.31
163	4" high		315		.58	.50	1.08	1.34
166	6" high		315	↓	.70	.50	1.20	1.47
170	Conductive flooring, rubber tile, 1/8" thick		315	S.F.	2.60	.50	3.10	3.56
180	Homogeneous vinyl tile, 1/8" thick		315		3.50	.50	4	4.55
220	Cork tile, standard finish, 1/8" thick		315		1.10	.50	1.60	1.91
225	3/16" thick		315		1.25	.50	1.75	2.08
230	5/16" thick		315		1.40	.50	1.90	2.24
235	1/2" thick		315		1.75	.50	2.25	2.63
250	Urethane finish, 1/8" thick		315		1.50	.50	2	2.35
255	3/16" thick		315		1.55	.50	2.05	2.41
260	5/16" thick		315		1.75	.50	2.25	2.63
265	1/2" thick		315		2.10	.50	2.60	3.01
370	Polyethylene, in rolls, no base incl., landscape surfaces		275		1.49	.57	2.06	2.44
380	Nylon action surface, 1/8" thick		275		1.66	.57	2.23	2.63
390	1/4" thick		275		2.39	.57	2.96	3.43
400	3/8" thick		275		3.28	.57	3.85	4.41
410	Golf tee surface with foam back		235		3.36	.67	4.03	4.64
420	Practice putting, knitted nylon surface	↓	235	↓	2.73	.67	3.40	3.95
440	Polyurethane, thermoset, prefabricated in place, indoor							
450	3/8" thick for basketball, gyms, etc.	1 Tilf	100	S.F.	2.50	1.58	4.08	4.96
460	1/2" thick for professional sports		95		2.60	1.66	4.26	5.20
470	Outdoor, 1/4" thick, smooth, for tennis		100		2	1.58	3.58	4.41
480	Rough, for track, 3/8" thick		95		2.50	1.66	4.16	5.10
500	Poured in place, indoor, with finish, 1/4" thick		80		1.75	1.98	3.73	4.69
505	3/8" thick		65		2.15	2.43	4.58	5.75
510	1/2" thick		50		2.70	3.16	5.86	7.40
550	Polyvinyl chloride, sheet goods for gyms, 1/4" thick		80		2	1.98	3.98	4.97
560	3/8" thick		60		2.25	2.63	4.88	6.15
590	Rubber, sheet goods, 36" wide, 1/8" thick		210		1.60	.75	2.35	2.81
595	3/16" thick		210		2.30	.75	3.05	3.58
600	1/4" thick		200		2.55	.79	3.34	3.91
601	5/16" thick		185		2.75	.85	3.60	4.22
602	3/8" thick		170		3.30	.93	4.23	4.93
603	1/2" thick		160		4	.99	4.99	5.80
605	Tile, marbleized colors, 12" x 12", 1/8" thick		485		1.65	.33	1.98	2.27
610	3/16" thick	↓	485	↓	2.30	.33	2.63	2.99

For expanded coverage of these items see *Means' Interior Cost Data 1986*

9.6 Flooring

		CREW	DAILY OUTPUT	UNIT	BARE COSTS MAT.	BARE COSTS INST.	BARE COSTS TOTAL	TOTAL INCL O&P
615	1/4" thick	1 Tilf	485	S.F.	2.50	.33	2.83	3.21
630	Special tile, plain colors, 1/8" thick		485		2.40	.33	2.73	3.10
635	3/16" thick		485		3.25	.33	3.58	4.03
640	1/4" thick		485		4	.33	4.33	4.86
645	For golf course, skating rink, etc., 1/4" thick	↓	275	↓	4.30	.57	4.87	5.55
660								
670	Synthetic turf, 3/8" thick	1 Tilf	270	S.F.	2.25	.59	2.84	3.30
675	Interlocking 2' x 2' squares, 1/2" thick, not							
681	cemented, for playgrounds, minimum	1 Tilf	485	S.F.	2	.33	2.33	2.66
685	Maximum		400		3.70	.40	4.10	4.62
700	Vinyl composition tile, 12" x 12", 1/16" thick		315		.52	.50	1.02	1.27
705	Embossed		315		.66	.50	1.16	1.43
710	Marbleized		315		.66	.50	1.16	1.43
715	Plain		315		.72	.50	1.22	1.49
720	3/32" thick, embossed		315		.75	.50	1.25	1.53
725	Marbleized		315		.80	.50	1.30	1.58
730	Plain	↓	315	↓	.95	.50	1.45	1.75
731								
735	1/8" thick, marbleized	1 Tilf	315	S.F.	.95	.50	1.45	1.75
740	Plain		315		1	.50	1.50	1.80
750	Vinyl tile, 12" x 12", .050" thick, minimum		425		1	.37	1.37	1.62
755	Maximum		315		2.50	.50	3	3.45
760	1/8" thick, minimum		425		1.50	.37	1.87	2.17
765	Solid colors		390		2	.41	2.41	2.77
770	Marbleized or Travertine pattern		390		2.50	.41	2.91	3.32
775	Florentine pattern		390		3	.41	3.41	3.87
780	Maximum	↓	315	↓	7.25	.50	7.75	8.70
781								
800	Vinyl sheet goods, backed, .070" thick, minimum	1 Tilf	600	S.F.	.80	.26	1.06	1.25
805	Maximum		265		1.90	.60	2.50	2.93
810	.093" thick, minimum		600		1.05	.26	1.31	1.52
815	Maximum		265		2.35	.60	2.95	3.42
820	.125" thick, minimum		600		1.25	.26	1.51	1.74
825	Maximum		265		3	.60	3.60	4.14
830	.250" thick, minimum		600		1.70	.26	1.96	2.24
835	Maximum	↓	265	↓	4.35	.60	4.95	5.60
870	Adhesive cement, 1 gallon does 100 to 300 S.F.			Gal.	13.10		13.10	14.40M
880	Asphalt primer, 1 gallon per 300 S.F.				5		5	5.50M
890	Emulsion, 1 gallon per 140 S.F.				4.80		4.80	5.30M
895	Latex underlayment			↓	20		20	22M
22-001	**SLATE & STONE FLOORS** See division 2.6-40 & 4.4							
25-001	**STAIR TREADS AND RISERS** See index for materials other							
010	than rubber and vinyl							
030	Rubber, molded tread, 12" wide, 5/16" thick, black	1 Tilf	115	L.F.	4.20	1.37	5.57	6.55
040	Colors		115		5	1.37	6.37	7.45
060	1/4" thick, black		115		4.05	1.37	5.42	6.40
070	Colors		115		4.80	1.37	6.17	7.20
090	Grip strip safety tread, colors, 5/16" thick		115		6.25	1.37	7.62	8.80
100	3/16" thick		120	↓	5	1.32	6.32	7.35
120	Landings, smooth sheet rubber, 1/8" thick		275	S.F.	1.00	.57	2.17	2.57
130	3/16" thick		270	"	2.30	.59	2.89	3.35
150	Nosings, 1-1/2" deep, 3" wide, residential		140	L.F.	1.20	1.13	2.33	2.90
160	Commercial		140		1.45	1.13	2.58	3.18
180	Risers, 7" high, 1/8" thick, flat		175		1	.90	1.90	2.37
190	Coved		175		1.15	.90	2.05	2.53
210	Vinyl, molded tread, 12" wide, colors, 1/8" thick		115		2.10	1.37	3.47	4.24
220	1/4" thick		115	↓	3.40	1.37	4.77	5.65
230	Landing material, 1/8" thick		200	S.F.	1.37	.79	2.16	2.61
240	Riser, 7" high, 1/8" thick, coved	↓	175	L.F.	1.27	.90	2.17	2.66

For expanded coverage of these items see *Means' Interior Cost Data 1986*

9.6 Flooring

			CREW	DAILY OUTPUT	UNIT	BARE COSTS MAT.	BARE COSTS INST.	BARE COSTS TOTAL	TOTAL INCL O&P
25	245	Threshold, 5-1/2" wide	1 Tilf	100	L.F.	2.90	1.58	4.48	5.40
	250	Tread and riser combined, 1/8" thick	"	80	"	2.30	1.98	4.28	5.30
35-001		**TILE AND TERRAZZO** See division 9.3							
40-001		**WOOD** Fir, vertical grain, 1" x 4", not incl. finish, B & better	1 Carp	255	S.F.	1.70	.63	2.33	2.78
	010	C grade & better	"	255	"	1.57	.63	2.20	2.64
	050								
	060	Gym floor, in mastic, over 2 ply felt, #2 & better							
	070	25/32" thick maple, including finish	1 Carp	100	S.F.	2	1.60	3.60	4.52
	090	33/32" thick maple, incl. finish		98		2.65	1.63	4.28	5.30
	100	For 1/2" corkboard underlayment, add	↓	750	↓	.34	.21	.55	.68
	130	For #1 grade maple, add				.17		.17	.18M
	160	Maple flooring, over sleepers, #2 & better including							
	170	finish, 25/32" thick	1 Carp	85	S.F.	2.53	1.88	4.41	5.50
	190	33/32" thick	"	83		3.15	1.93	5.08	6.25
	200	For #1 grade, add				.17		.17	.18M
	220	For 3/4" subfloor, add	1 Carp	350		.48	.46	.94	1.19
	230	With two 1/2" subfloors, 25/32" thick	"	69	↓	3.30	2.32	5.62	7
	250	Maple, incl. finish, #2 & btr., 25/32" thick, on rubber							
	260	Sleepers, with two 1/2" subfloors	1 Carp	76	S.F.	3.36	2.11	5.47	6.75
	280	With steel spline, double connection to channels	"	73		3.52	2.19	5.71	7.05
	290	For 33/32" maple, add				.63		.63	.69M
	310	For #1 grade maple, add				.17		.17	.18M
	350	For termite proofing all of the above, add				.11		.11	.12M
	370	Portable hardwood, prefinished panels	1 Carp	83		3.80	1.93	5.73	6.95
	372	Insulated with polystyrene, add		165		2.10	.97	3.07	3.72
	375	Running tracks, Sitka spruce surface		62		5.45	2.58	8.03	9.75
	377	3/4" plywood surface	↓	100	↓	3.40	1.60	5	6.05
	380	See also resilient gym floors, division 9.6-20-440							
	381								
	400	Maple, strip, 25/32" x 2-1/4", not incl. finish, select	1 Carp	170	S.F.	1.72	.94	2.66	3.26
	410	#2 & better		170		1.57	.94	2.51	3.09
	430	33/32" x 3-1/4", not incl. finish, #1 grade		170		2.40	.94	3.34	4
	440	#2 & better	↓	170	↓	2.20	.94	3.14	3.78
	460	Oak, white or red, 25/32" x 2-1/4", not incl. finish							
	470	Clear quartered	1 Carp	170	S.F.	1.63	.94	2.57	3.16
	490	Clear/select, 2-1/4" wide		170		1.45	.94	2.39	2.96
	500	#1 common		185		1.15	.86	2.01	2.52
	520	Parquetry, standard, 5/16" thick, not incl. finish, oak, minimum		160		1	1	2	2.55
	530	Maximum		100		5.25	1.60	6.85	8.10
	550	Teak, minimum		160		2.45	1	3.45	4.14
	560	Maximum		100		5.20	1.60	6.80	8.05
	565	13/16" thick, select grade oak, minimum		160		8.60	1	9.60	10.90
	570	Maximum		100		10.85	1.60	12.45	14.25
	580	Custom parquetry, including finish, minimum		100		10.50	1.60	12.10	13.85
	590	Maximum		50		14.75	3.20	17.95	21
	610	Prefinished white oak, prime grade, 2-1/4" wide		170		1.68	.94	2.62	3.21
	620	3-1/4" wide		185		1.68	.86	2.54	3.10
	640	Ranch plank		145		2.85	1.10	3.95	4.73
	650	Hardwood blocks, 9" x 9", 25/32" thick		160		3.10	1	4.10	4.86
	670	Parquetry, 5/16" thick, oak, minimum		160		1	1	2	2.55
	680	Maximum		100		1.80	1.60	3.40	4.30
	700	Walnut or teak, parquetry, minimum		160		3.15	1	4.15	4.91
	710	Maximum	↓	100	↓	5.70	1.60	7.30	8.60
	720	Acrylic wood parquet blocks, 12" x 12" x 5/16",							
	721	irradiated, set in epoxy	1 Carp	160	S.F.	3.75	1	4.75	5.55
	740	Yellow pine, 3/4" x 3-1/8", T & G, C & better, not incl. finish	"	200	"	1.33	.80	2.13	2.62
	741								
	750	Refinish old floors, minimum	1 Carp	400	S.F.	.42	.40	.82	1.04
	760	Maximum	"	130	"	.57	1.23	1.80	2.41

For expanded coverage of these items see *Means' Interior Cost Data 1986*

9.6 Flooring

		CREW	DAILY OUTPUT	UNIT	BARE COSTS MAT.	BARE COSTS INST.	BARE COSTS TOTAL	TOTAL INCL O&P
40 780	Sanding and finishing, fill, shellac, wax	1 Carp	295	S.F.	.36	.54	.90	1.18
790	Subfloor and underlayment, see division 6.1-82 & 86							
45-001	**WOOD BLOCK FLOORING** End grain flooring, creosoted, 2" thick	1 Carp	295	S.F.	1.75	.54	2.29	2.71
040	Natural finish, 1" thick		275		2.10	.58	2.68	3.15
060	1-1/2" thick		255		2.75	.63	3.38	3.93
070	2" thick,	↓	240	↓	3	.67	3.67	4.27
50-001	**WOOD COMPOSITION** Gym floors							
010	2-1/4" x 6-7/8" x 3/8", on 2" grout setting bed	D-7	150	S.F.	3.40	1.89	5.29	6.40
020	Thin set, on concrete	"	250		2.90	1.13	4.03	4.78
030	Sanding and finishing, add	1 Carp	200	↓	.57	.80	1.37	1.79

9.8 Painting & Wall Covering

		CREW	DAILY OUTPUT	UNIT	BARE COSTS MAT.	BARE COSTS INST.	BARE COSTS TOTAL	TOTAL INCL O&P
05-001	**CORNER GUARDS** Rubber, 3" wide, standard	1 Pord	135	L.F.	2.20	1.14	3.34	4.05
010	Bullnose (see also division 5.4-14)		135		3.70	1.14	4.84	5.70
030	1/4" thick, 2-3/4" wide, rubber		135		2.50	1.14	3.64	4.38
040	Vinyl, 5/16" thick, 2-1/2" wide	↓	135	↓	2.10	1.14	3.24	3.94
16-001	**CABINETS AND CASEWORK**							
002	Labor cost includes protection of adjacent items not painted							
100	Primer coat, oil base, brushwork	1 Pord	400	S.F.	.03	.39	.42	.58
200	Paint, oil base, brushwork, 1 coat		380		.04	.41	.45	.62
300	Stain, brushwork, wipe off		360		.03	.43	.46	.64
400	Shellac, 1 coat, brushwork		380		.03	.41	.44	.61
450	Varnish, 3 coats, brushwork	↓	235		.09	.66	.75	1.03
500	For latex paint, deduct			↓	10%			
17-001	**DOORS AND WINDOWS**							
002	Labor cost includes protection of adjacent items not painted							
050	Flush door and frame, per side, oil base, primer coat, brushwork	1 Pord	14	Ea.	1.05	11	12.05	16.85
100	Paint, 1 coat		25		1.31	6.15	7.46	10.20
140	Stain, brushwork, wipe off		10		.98	15.40	16.38	23
160	Shellac, 1 coat, brushwork		12		.89	12.85	13.74	19.25
180	Varnish, 3 coats, brushwork	↓	5	↓	2.70	31	33.70	47
190								
200	Panel door and frame, per side, oil base, primer coat, brushwork	1 Pord	13	Ea.	1.15	11.85	13	18.15
220	Paint, 1 coat		12		1.41	12.85	14.26	19.85
260	Stain, brushwork, wipeoff		6		1.08	26	27.08	38
280	Shellac, 1 coat, brushwork		8		.98	19.25	20.23	29
300	Varnish, 3 coats, brushwork	↓	3	↓	2.95	51	53.95	76
440	Windows, including frame and trim, per side							
460	Colonial type, 2' x 3', oil base, primer coat, brushwork	1 Pord	24	Ea.	.85	6.40	7.25	10.10
580	Paint, 1 coat		22		1	7	8	11.05
620	3' x 5' opening, primer coat, brushwork		15		1.05	10.25	11.30	15.80
640	Paint, 1 coat		13		1.26	11.85	13.11	18.25
680	4' x 8' opening, primer coat, brushwork		12		1.26	12.85	14.11	19.65
700	Paint, 1 coat		10		1.51	15.40	16.91	24
800	Single lite type, 2' x 3', oil base, primer coat, brushwork		40		.80	3.85	4.65	6.35
820	Paint, 1 coat		37		.97	4.16	5.13	7
860	3' x 5' opening, primer coat, brushwork		27		1	5.70	6.70	9.25
880	Paint, 1 coat		25		1.23	6.15	7.38	10.15
920	4' x 8' opening, primer coat, brushwork		14		2.25	11	13.25	18.15
940	Paint, 1 coat	↓	13		2.45	11.85	14.30	19.60
980	For latex paint deduct			↓	10%			
18-001	**MISCELLANEOUS**							
002	Labor cost includes protection of adjacent items not painted							

For expanded coverage of these items see *Means' Interior Cost Data 1986*

9.8 Painting & Wall Covering

			CREW	DAILY OUTPUT	UNIT	BARE COSTS MAT.	BARE COSTS INST.	BARE COSTS TOTAL	TOTAL INCL O&P
18	070	Fence, chain link, per side, oil base, primer coat, brushwork	2 Pord	1,200	S.F.	.03	.26	.29	.40
	100	Spray		1,600		.03	.19	.22	.31
	120	Paint 1 coat, brushwork		1,150		.04	.27	.31	.43
	140	Spray		1,600		.04	.19	.23	.32
	160	Picket, wood, one side, primer coat, brushwork		1,600		.05	.19	.24	.33
	180	Spray		3,200		.05	.10	.15	.19
	200	Paint 1 coat, brushwork		1,600		.06	.19	.25	.34
	220	Spray		3,200		.06	.10	.16	.20
	240	Floors, concrete or wood, oil base, primer or sealer coat, brushwork		3,400		.03	.09	.12	.16
	245	Roller		3,800		.03	.08	.11	.15
	260	Spray		6,000		.03	.05	.08	.11
	265	Paint 1 coat, brushwork		3,200		.04	.10	.14	.18
	280	Roller		3,600		.04	.09	.13	.17
	285	Spray		6,000		.04	.05	.09	.12
	300	Stain, wood floor, brushwork		3,400		.03	.09	.12	.16
	320	Roller		3,800		.03	.08	.11	.15
	325	Spray		6,000		.03	.05	.08	.11
	340	Varnish, wood floor, brushwork		3,200		.04	.10	.14	.18
	345	Roller		3,400		.04	.09	.13	.17
	360	Spray		6,000		.04	.05	.09	.12
	365	For dust proofing or anti skid, see division 3.3-26							
	375								
	380	Grilles, per side, oil base, primer coat, brushwork	1 Pord	400	Ea.	.03	.39	.42	.58
	385	Spray		500		.03	.31	.34	.47
	392	Paint 2 coats, brushwork		200		.06	.77	.83	1.16
	394	Spray		250		.06	.62	.68	.94
	420	Gutters and downspouts, oil base, primer coat, brushwork	2 Pord	650	L.F.	.06	.47	.53	.74
	430	Paint 2 coats, brushwork		325		.13	.95	1.08	1.49
	500	Pipe, to 4" diameter, primer or sealer coat, oil base, brushwork		800		.05	.39	.44	.60
	510	Spray		1,100		.05	.28	.33	.45
	535	Paint 2 coats, brushwork		400		.11	.77	.88	1.22
	540	Spray		550		.11	.56	.67	.92
	630	To 16" diameter, primer or sealer coat, brushwork		192		.16	1.60	1.76	2.46
	635	Spray		240		.16	1.28	1.44	2
	650	Paint 2 coats, brushwork		100		.33	3.08	3.41	4.75
	655	Spray		130		.33	2.37	2.70	3.74
	700	Trim, wood, incl. puttying, under 6" wide							
	720	Primer coat, oil base, brushwork	1 Pord	900	L.F.	.02	.17	.19	.27
	725	Paint, 1 coat, brushwork		875		.03	.18	.21	.28
	745	3 coats		300		.08	.51	.59	.82
	747								
	750	Over 6" wide, primer coat, brushwork	1 Pord	600	L.F.	.03	.26	.29	.40
	755	Paint, 1 coat, brushwork		450		.04	.34	.38	.53
	765	3 coats		190		.11	.81	.92	1.28
	800	Cornice, simple design, primer coat, oil base, brushwork		275	S.F.	.03	.56	.59	.83
	825	Paint, 1 coat, brushwork		250		.04	.62	.66	.92
	835	Ornate design, primer coat		150		.03	1.03	1.06	1.50
	840	Paint, 1 coat		140		.04	1.10	1.14	1.61
	860	Balustrades, per side, primer coat, oil base, brushwork		300		.03	.51	.54	.76
	865	Paint, 1 coat		285		.04	.54	.58	.81
	890	Trusses and exposed wood frames, primer coat, oil base, brushwork		800		.03	.19	.22	.31
	895	Spray		1,200		.03	.13	.16	.22
	922	Paint 2 coats, brushwork		500		.07	.31	.38	.52
	924	Spray		600		.07	.26	.33	.44
	926	Stain, brushwork, wipe off		600		.03	.26	.29	.40
	928	Varnish, 3 coats, brushwork		275		.15	.56	.71	.96
	935	For latex paint, deduct				10%			
	19-001	**SIDING** Exterior							
	002	Labor cost includes protection of adjacent items not painted							

For expanded coverage of these items see *Means' Interior Cost Data 1986*

9.8 Painting & Wall Covering

		CREW	DAILY OUTPUT	UNIT	BARE COSTS MAT.	BARE COSTS INST.	BARE COSTS TOTAL	TOTAL INCL O&P
010	Steel siding, oil base, primer or sealer coat, brushwork	2 Pord	1,700	S.F.	.05	.18	.23	.31
050	Spray		3,200		.05	.10	.15	.19
080	Paint 2 coats, brushwork		1,350		.08	.23	.31	.41
100	Spray		2,600		.08	.12	.20	.26
120	Stucco, rough, oil base, paint 2 coats, brushwork		1,300		.10	.24	.34	.45
140	Roller		2,000		.10	.15	.25	.33
160	Spray	↓	2,600	↓	.10	.12	.22	.28
170								
180	Texture 1-11 or clapboard, oil base, primer coat, brushwork	2 Pord	2,600	S.F.	.06	.12	.18	.23
200	Spray		4,200		.06	.07	.13	.17
240	Paint 2 coats, brushwork		1,250		.09	.25	.34	.45
260	Spray		2,100		.09	.15	.24	.31
340	Stain 2 coats, brushwork		1,250		.12	.25	.37	.48
400	Spray		2,100		.12	.15	.27	.34
420	Wood shingles, oil base primer coat, brushwork		2,500		.07	.12	.19	.25
440	Spray		4,100		.07	.08	.15	.18
500	Paint 2 coats, brushwork		1,150		.10	.27	.37	.49
520	Spray		2,100		.10	.15	.25	.32
650	Stain 2 coats, brushwork		1,150		.13	.27	.40	.52
700	Spray	↓	2,100		.13	.15	.28	.35
800	For latex paint, deduct			↓	10%			
21-001	**WALL AND CEILINGS**							
002	Labor cost includes protection of adjacent items not painted							
010	Concrete, dry wall or plaster, oil base, primer or sealer coat							
020	Smooth finish, brushwork	1 Pord	1,900	S.F.	.04	.08	.12	.16
024	Roller		2,200		.04	.07	.11	.14
028	Spray		5,000		.04	.03	.07	.09
030	Sand finish, brushwork		1,700		.05	.09	.14	.18
034	Roller		2,100		.05	.07	.12	.16
038	Spray		3,750		.05	.04	.09	.11
080	Paint 2 coats, smooth finish, brushwork		975		.09	.16	.25	.32
084	Roller		1,125		.09	.14	.23	.29
088	Spray		2,250		.09	.07	.16	.20
090	Sand finish, brushwork		825		.11	.19	.30	.39
094	Roller		1,050		.11	.15	.26	.33
098	Spray	↓	2,250	↓	.11	.07	.18	.22
150								
160	Glaze coating, 5 coats, spray, clear	1 Pord	900	S.F.	.50	.17	.67	.79
164	Multicolor	"	900		.60	.17	.77	.90
170	For latex paint, deduct				10%			
180	For ceiling installations, add			↓		25%		
190								
200	Masonry or concrete block, oil base, primer or sealer coat							
210	Smooth finish, brushwork	1 Pord	1,725	S.F.	.06	.09	.15	.19
218	Spray		3,750		.06	.04	.10	.12
220	Sand finish, brushwork		1,400		.07	.11	.18	.23
228	Spray		3,750		.07	.04	.11	.14
280	Paint 2 coats, smooth finish, brushwork		825		.10	.19	.29	.38
288	Spray		2,250		.10	.07	.17	.21
290	Sand finish, brushwork		750		.11	.21	.32	.41
298	Spray		2,250		.11	.07	.18	.22
360	Glaze coating, 5 coats, spray, clear		900		.50	.17	.67	.79
362	Multicolor		900		.60	.17	.77	.90
400	Block filler, 1 coat, brushwork		1,350		.08	.11	.19	.25
410	Silicone, water repellent, 2 coats, spray	↓	900		.04	.17	.21	.29
412	For latex paint, deduct			↓	10%			
30-001	**VARNISH** 1 coat + sealer, on wood trim, no sanding included	1 Pord	900	S.F.	.07	.17	.24	.32
010	Hardwood floors, 2 coat, no sanding included	"	800	"	.07	.19	.26	.35

For expanded coverage of these items see *Means' Interior Cost Data 1986*

9.8 Painting & Wall Covering

		CREW	DAILY OUTPUT	UNIT	BARE COSTS MAT.	BARE COSTS INST.	BARE COSTS TOTAL	TOTAL INCL O&P
35-001	**WALL COATINGS** Acrylic glazed coatings, minimum	1 Pord	525	S.F.	.22	.29	.51	.66
010	Maximum		305		.43	.50	.93	1.19
030	Epoxy coatings, minimum		525		.25	.29	.54	.69
040	Maximum		170		.93	.91	1.84	2.31
060	Exposed aggregate, troweled on, 1/16" to 1/4", minimum		235		.59	.66	1.25	1.58
070	Maximum (epoxy or polyacrylate)		130		1.22	1.18	2.40	3.03
090	1/2" to 5/8" aggregate, minimum		130		1.05	1.18	2.23	2.84
100	Maximum		80		1.85	1.93	3.78	4.78
150	Exposed aggregate, sprayed on, 1/8" aggregate, minimum		295		.46	.52	.98	1.25
160	Maximum		145		.96	1.06	2.02	2.57
180	High build epoxy, 50 mil, minimum (see also 5.1-50-680)		390		.36	.39	.75	.96
190	Maximum		95		1.36	1.62	2.98	3.81
210	Laminated epoxy with fiberglass, minimum		295		.46	.52	.98	1.25
220	Maximum		145		.90	1.06	1.96	2.50
240	Sprayed perlite or vermiculite, 1/16" thick, minimum		2,935		.05	.05	.10	.13
250	Maximum		640		.34	.24	.58	.72
270	Vinyl plastic wall coating, minimum		735		.19	.21	.40	.51
280	Maximum		240		.68	.64	1.32	1.66
300	Urethane on smooth surface, 2 coat, minimum		1,135		.11	.14	.25	.31
310	Maximum		665		.24	.23	.47	.59
360	Ceramic-like glazed coating, cementitious, minimum		440		.28	.35	.63	.81
370	Maximum		345		.36	.45	.81	1.03
390	Resin base, minimum		640		.20	.24	.44	.56
400	Maximum	↓	330	↓	.33	.47	.80	1.03
40-001	**WALL COVERING**							
005	Aluminum foil	1 Pape	275	S.F.	.53	.57	1.10	1.39
010	Copper sheets, .025" thick, vinyl backing		240		2.90	.65	3.55	4.12
030	Phenolic backing		240		3.88	.65	4.53	5.20
060	Cork tiles, light or dark, 12" x 12" x 3/16"		240		1.25	.65	1.90	2.30
070	5/16" thick		235		1.35	.66	2.01	2.43
090	1/4" basketweave		240		2.90	.65	3.55	4.12
100	1/2" natural, non-directional pattern		240		2.75	.65	3.40	3.95
120	Granular surface, 12" x 36", 1/2" thick		385		.50	.41	.91	1.13
130	1" thick		370		.70	.42	1.12	1.37
150	Polyurethane coated, 12" x 12" x 3/16" thick		240		1.50	.65	2.15	2.58
160	5/16" thick		235		2	.66	2.66	3.15
180	Cork wallpaper, paperbacked, natural		480		.95	.33	1.28	1.51
190	Colors		480		1.05	.33	1.38	1.62
210	Flexible wood veneer, 1/32" thick, plain woods		100		1.20	1.56	2.76	3.54
220	Exotic woods	↓	95	↓	1.80	1.64	3.44	4.32
230								
240	Gypsum-based, fabric-backed, fire							
250	resistant for masonry walls, minimum	1 Pape	400	S.F.	.52	.39	.91	1.13
270	Maximum	"	300		.67	.52	1.19	1.48
275	Acrylic, modified, semi-rigid PVC, .028" thick	2 Carp	330		.65	.97	1.62	2.12
280	.040" thick	"	320		.85	1	1.85	2.38
300	Vinyl wall covering, fabric-backed, lightweight	1 Pape	640		.43	.24	.67	.82
330	Medium weight		480		.61	.33	.94	1.13
340	Heavy weight	↓	435	↓	.87	.36	1.23	1.47
360	Adhesive, 5 gal. lots			Gal.	6.50		6.50	7.15M
370	(114) Wallpaper at $8.00 per double roll, average workmanship	1 Pape	640	S.F.	.16	.24	.40	.52
390	Paper at $17 per double roll, average workmanship		535		.32	.29	.61	.77
400	Paper at $40 per double roll, quality workmanship	↓	435	↓	.70	.36	1.06	1.28
410	Linen wall covering, paper backed							
415	Flame treatment, minimum			S.F.	.38		.38	.41M
418	Maximum				.45		.45	.49M
420	Grass cloths with lining paper, minimum	1 Pape	400		.51	.39	.90	1.12
430	Maximum	"	350	↓	1.57	.45	2.02	2.36
45-001	**WALLGUARD** See division 5.4							
050	Neoprene with aluminum fastening strip, 1-1/2" x 2"	1 Carp	110	L.F.	3.40	1.45	4.85	5.85

For expanded coverage of these items see *Means' Interior Cost Data 1986*

9.8 Painting & Wall Covering

		CREW	DAILY OUTPUT	UNIT	BARE COSTS MAT.	BARE COSTS INST.	BARE COSTS TOTAL	TOTAL INCL O&P
100	Trolley rail, PVC, clipped to wall, 5" high	1 Carp	185	L.F.	2.75	.86	3.61	4.28
105	8" high		180	"	3.65	.89	4.54	5.30
120	Vinyl acrylic bed aligner and bumper, 37" long		10	Ea.	31	16	47	57
130	43" long	↓	9	"	34	17.80	51.80	63

10.1 Specialties

		CREW	DAILY OUTPUT	UNIT	BARE COSTS MAT.	BARE COSTS INST.	BARE COSTS TOTAL	TOTAL INCL O&P
02-001 016	**BATHROOM ACCESSORIES**							
020	Curtain rod, stainless steel, 5' long, 1" diameter	1 Carp	13	Ea.	13.75	12.30	26.05	33
030	1-1/4" diameter	"	13	"	17.85	12.30	30.15	37
050	Dispenser units, combined soap & towel dispensers,							
051	mirror and shelf, flush mounted	1 Carp	10	Ea.	205	16	221	250
060	Towel dispenser and waste receptacle,							
061	flush mounted	1 Carp	10	Ea.	295	16	311	350
080	Grab bar, straight, 1" diameter, stainless steel, 12" long		24		14.40	6.65	21.05	26
090	18" long		23		16.60	6.95	23.55	28
100	24" long		22		18.80	7.25	26.05	31
110	36" long		20		23	8	31	37
120	1-1/2" diameter, 18" long		23		26	6.95	32.95	39
130	36" long		20		31	8	39	46
150	Tub bar, 1" diameter, horizontal		14		47	11.45	58.45	68
160	Plus vertical arm		12		74	13.35	87.35	100
190	End tub bar, 1" diameter, 90° angle	↓	12		43	13.35	56.35	67
200	For 1-1/4" diameter bars, add				20%			
210	For 1-1/2" diameter bars, add			↓	30%			
226								
230	Hand dryer, surface mounted, electric, 110 volt	1 Carp	4	Ea.	285	40	325	370
240	220 volt		4		285	40	325	370
260	Hat and coat strip, stainless steel, 4 hook, 36" long		24		38	6.65	44.65	51
270	6 hook, 60" long		20		63	8	71	81
300	Mirror with stainless steel, 3/4" square frame, 18" x 24"		20		59	8	67	76
310	36" x 24"		15		105	10.65	115.65	130
320	48" x 24"		10		130	16	146	165
330	72" x 24"		6		190	27	217	250
350	Mirror with 5" stainless steel shelf, 3/4" sq. frame, 18" x 24"		20		75	8	83	94
360	36" x 24"		15		140	10.65	150.65	170
370	48" x 24"		10		190	16	206	230
380	72" x 24"		6		280	27	307	345
410	Mop holder strip, stainless steel, 6 holders, 60" long		20		67	8	75	85
420	Napkin/tampon dispenser, surface mounted		15		255	10.65	265.65	295
430	Robe hook, single, regular		36		11.20	4.44	15.64	18.75
440	Heavy duty, concealed mounting		36		14.50	4.44	18.94	22
460	Soap dispenser, chrome, surface mounted, liquid		20		52	8	60	69
470	Powder		20		45	8	53	61
500	Recessed stainless steel, liquid		10		56	16	72	85
510	Powder		10		86	16	102	120
530	Soap tank, stainless steel, 1 gallon		10		71	16	87	100
540	5 gallon		5		125	32	157	185
560	Shelf, stainless steel, 5" wide, 18 ga., 24" long		24		21	6.65	27.65	33
570	72" long		16		60	10	70	80
580	8" wide shelf, 18 ga., 24" long		22		32	7.25	39.25	46
590	72" long		14		81	11.45	92.45	105
600	Toilet seat cover dispenser, stainless steel, recessed		20		62	8	70	80
605	Surface mounted	↓	15	↓	27	10.65	37.65	45

10.1 Specialties

			CREW	DAILY OUTPUT	UNIT	BARE COSTS MAT.	BARE COSTS INST.	BARE COSTS TOTAL	TOTAL INCL O&P
02	610	Toilet tissue dispenser, surface mounted, stainless steel, single roll	1 Carp	30	Ea.	17.50	5.35	22.85	27
	620	Double roll		24		26.50	6.65	33.15	39
	640	Towel bar, stainless steel, 18" long		23		17.10	6.95	24.05	29
	650	30" long		21		19	7.60	26.60	32
	670	Towel dispenser, stainless steel, surface mounted		16		56	10	66	76
	680	Flush mounted, recessed		10		88	16	104	120
	700	Towel holder, hotel type, 2 guest size		20		9	8	17	21
	720	Towel shelf, stainless steel, 24" long, 8" wide		20		29	8	37	43
	740	Tumbler holder, tumbler only		30		10.10	5.35	15.45	18.85
	750	Soap, tumbler & toothbrush		30		17.50	5.35	22.85	27
	770	Wall urn ash receiver, recessed, 14" long		12		70	13.35	83.35	96
	780	Surface, 8" long		18		45	8.90	53.90	62
	800	Waste receptacles, stainless steel, with top, 13 gallon		10		135	16	151	170
	810	36 gallon		8		230	20	250	280
05-001		BULLETIN BOARD Cork sheets, unbacked, no frame, 1/8" thick	2 Carp	290	S.F.	1.95	1.10	3.05	3.74
	010	1/4" thick		290		3.05	1.10	4.15	4.95
	030	Burlap-faced cork, no frame, 1/8" thick,		290		2.45	1.10	3.55	4.29
	040	With 1/8" cork backing		290		3.45	1.10	4.55	5.40
	060	With 1/4" cork backing		290		4.40	1.10	5.50	6.45
	070	With 3/8" backboard		290		3.15	1.10	4.25	5.05
	090	1/16" vinyl cork, on 3/8" fiber board, no frame		280		3.20	1.14	4.34	5.20
	100	1/4" vinyl cork, on 7/16" backboard, no frame		280		2.80	1.14	3.94	4.74
	120	1/4" vinyl cork, on 1/4" hardboard, no frame		280		4.40	1.14	5.54	6.50
	130	No backing, no frame		280		3.10	1.14	4.24	5.05
	200	For map and display rail, economy, add		385	L.F.	1.20	.83	2.03	2.52
	210	Deluxe, add		350	"	2.15	.91	3.06	3.69
	212	Prefabricated, 1/4" cork, 4' x 4' with aluminum frame		14	Ea.	100	23	123	145
	214	Aluminum frame with glass door		13		465	25	490	545
	216	8' x 4' with aluminum frame		8		185	40	225	260
	218	Aluminum frame with glass door		7		870	46	916	1,025
	220	3' x 4' with wood frame		14		125	23	148	170
	221	Wood frame with glass door		13		410	25	435	485
	222	4' x 4' with wood frame		10		135	32	167	195
	223	6' x 4' with wood frame		8		185	40	225	260
	224	Prefabricated with vinyl covered cork, aluminum frame, 4' x 3'		12		110	27	137	160
	225	Vinyl on 1/8" vinyl cork plus 3/8" fiberboard, 4' x 3'		11		125	29	154	180
	226	Vinyl on 1/4" vinyl cork plus 1/4" hardboard, 8' x 4'		8		245	40	285	325
	227	Vinyl on 3/8" fiberboard, aluminum frame, 8' x 4'		7		270	46	316	365
	230	Prefabricated, sliding glass, enclosed, 3' x 4'		16		415	20	435	485
	240	5' x 3'		12		470	27	497	555
	250	5' x 4'		11		530	29	559	625
	260	6' x 4'		7		575	46	621	700
	280	For changeable letter type, add				10%			
	290	For lights, add per cabinet	1 Elec	13		180	13.80	193.80	220
	310	Horizontal sliding units with 2 sliders, 8' x 4'	2 Carp	3		940	105	1,045	1,200
	320	12' x 4'		2		1,175	160	1,335	1,525
	340	4 sliding units, 16' x 4'		2		2,475	160	2,635	2,950
	350	24' x 4'		1.50		2,875	215	3,090	3,475
07-001		CANOPIES Wall hung, aluminum, prefinished, 8' x 10'	K-2	1.30		795	515	1,310	1,600
	030	8' x 20'		1.10		1,375	610	1,985	2,375
	050	10' x 10'		1.30		935	515	1,450	1,750
	070	10' x 20'		1.10		1,575	610	2,185	2,600
	100	12' x 20'		1		1,775	670	2,445	2,900
	136	12' x 30'		.80		2,450	840	3,290	3,875
	170	12' x 40'		.60		3,200	1,125	4,325	5,100
	190	For free standing units, add				20%	10%		
	230	Aluminum entrance canopies, flat soffit							
	250	3'-6" x 4'-0", clear anodized	2 Carp	4	Ea.	325	80	405	475
	270	Bronze anodized		4		490	80	570	655
	300	Polyurethane painted		4		390	80	470	545

10.1 Specialties

		CREW	DAILY OUTPUT	UNIT	BARE COSTS MAT.	BARE COSTS INST.	BARE COSTS TOTAL	TOTAL INCL O&P
330	4'-6" x 10'-0", clear anodized	2 Carp	2	Ea.	625	160	785	920
350	Bronze anodized		2		1,000	160	1,160	1,325
370	Polyurethane painted	↓	2		760	160	920	1,075
400	Wall downspout, 10 L.F., clear anodized	1 Carp	7		35	23	58	72
430	Bronze anodized		7		70	23	93	110
450	Polyurethane painted	↓	7	↓	52	23	75	90
470	Canvas awnings, including canvas, frame & lettering							
500	Minimum	2 Carp	100	S.F.	5.50	3.20	8.70	10.70
530	Average		90		11	3.56	14.56	17.25
550	Maximum	↓	80	↓	16.50	4	20.50	24
700	Carport, baked vinyl finish, 20' x 10', no foundations, minimum	K-2	4	Car	900	170	1,070	1,225
725	Maximum		2	"	2,700	335	3,035	3,450
750	Walkway cover, to 12' wide, steel, vinyl finish, no fndtns., minimum		250	S.F.	3.25	2.68	5.93	7.40
775	Maximum	↓	200		7.85	3.35	11.20	13.40
10-001	**CHALKBOARD** Cement asbestos, no frame, economy	2 Carp	270		2.45	1.19	3.64	4.41
010	Deluxe		260		2.80	1.23	4.03	4.86
030	Hardboard, tempered, no frame, 1/4" thick		270		1.52	1.19	2.71	3.39
040	1/2" thick		260		2.78	1.23	4.01	4.84
060	Hardboard, not tempered, no frame, 1/4" thick		270		1.40	1.19	2.59	3.26
070	1/2" thick	↓	260	↓	2.10	1.23	3.33	4.09
080	Porcelain enamel, 24 ga. steel, 4' high, with aluminum							
090	trim, alum. foil backing, with core materials as follows:							
100	3/8" gypsum core	2 Carp	260	S.F.	5.30	1.23	6.53	7.60
110	1/4" hardboard core		270		5.25	1.19	6.44	7.50
120	7/16" hardboard core		255		5.40	1.25	6.65	7.75
130	3/8" honeycomb core		275		5.40	1.16	6.56	7.65
140	3/8" particleboard core		260		5.15	1.23	6.38	7.45
150	1/4" plywood core		270		5.40	1.19	6.59	7.65
155	3/8" plywood core	↓	260		5.10	1.23	6.33	7.40
160	For 20 ga. steel face, add				.79		.79	.86M
170	For 18 ga. steel face, add				1.42		1.42	1.56M
175	For liquid chalk marking on 24 ga. steel, add				.95		.95	1.04M
180	For aluminum sheet backing, add				1.20		1.20	1.32M
190	For steel backing, add				1.50		1.50	1.65M
210	Slate, 3/8" thick, frame not included, to 4' wide	2 Carp	240		7.55	1.33	8.88	10.25
220	To 4'-6" wide		230		8.30	1.39	9.69	11.15
240	Over 4'-6" to 5'-0" wide	↓	220		8.75	1.45	10.20	11.75
250	For over 3/8" thick, add per 1/8" increment				1.60		1.60	1.76M
260	Treated plastic on plywood, no frame, 1/4" thick	2 Carp	240		4.45	1.33	5.78	6.85
280	1/2" thick		230	↓	5.25	1.39	6.64	7.80
300	Frame for chalkboards, aluminum, chalk tray		290	L.F.	3.60	1.10	4.70	5.55
310	Trim		385		1.05	.83	1.88	2.36
330	Map and display rail, economy		385		1.40	.83	2.23	2.74
340	Deluxe		350	↓	2.20	.91	3.11	3.74
360	Factory fabricated, tempered hardbd. w/wood frame, 3'-6" high		265	S.F.	4.10	1.21	5.31	6.25
370	4' high		260		3.60	1.23	4.83	5.75
390	Aluminum frame, 3'-6" high		270		4.45	1.19	5.64	6.60
400	4' high		265		4.10	1.21	5.31	6.25
420	Porcelain steel, aluminum frame, 3' high		260		8.45	1.23	9.68	11.10
430	4' high		255	↓	6.35	1.25	7.60	8.80
450	Magnetic swing leaf panels, 36" x 24", 4 panels		3	Total	720	105	825	945
460	5 panels		3		830	105	935	1,075
470	6 panels		3		945	105	1,050	1,200
480	8 panels		3	↓	1,150	105	1,255	1,425
500	Vertical sliding with tempered hardboard, manual		150	S.F.	18.90	2.13	21.03	24
510	Electric		115		66	2.78	68.78	77
530	Horizontal sliding, 4' high, 10' long, 2 track		40		29	8	37	43
535	3 track		40		38	8	46	53
540	4 track	↓	40	↓	42	8	50	58
542								

10

10.1 Specialties

		Description	CREW	DAILY OUTPUT	UNIT	BARE COSTS MAT.	BARE COSTS INST.	BARE COSTS TOTAL	TOTAL INCL O&P
10	545	Horizontal sliding, 4' high, 20' long, 2 track	2 Carp	80	S.F.	22	4	26	30
	550	3 track		80		29	4	33	38
	555	4 track	↓	80		34	4	38	43
	570	For projection screens, add				4		4	4.40M
	590	For non-standard colors, add				10%			
	600	For installation on gypsum drywall, add			↓		20%		
	620	Portable, chalkboard 2 sides, 4' x 6', economy			Ea.	235		235	260M
	630	Deluxe				860		860	945M
	650	Chalkboard 1 side, tackboard other side, economy				250		250	275M
	660	Deluxe			↓	860		860	945M
12-001		**CHUTES** Linen or refuse, incl. sprinklers							
	002								
	005	Aluminized steel, 16 ga., 18" diameter	2 Shee	3.50	Floor	550	105	655	755
	010	24" diameter		3.20		625	115	740	855
	020	30" diameter		3		740	120	860	990
	030	36" diameter		2.80		890	130	1,020	1,175
	040	Galvanized steel, 16 ga., 18" diameter		3.50		510	105	615	715
	050	24" diameter		3.20		560	115	675	780
	060	30" diameter		3		670	120	790	915
	070	36" diameter		2.80		770	130	900	1,025
	080	Stainless steel, 18" diameter		3.50		745	105	850	970
	090	24" diameter		3.20		890	115	1,005	1,150
	100	30" diameter		3		965	120	1,085	1,250
	110	36" diameter		2.80	↓	1,150	130	1,280	1,450
	120	Linen bottom collector, aluminized steel		4	Ea.	495	91	586	675
	130	Stainless steel		4		545	91	636	730
	150	Refuse bottom hopper, aluminized steel, 18" diameter		3		425	120	545	645
	160	24" diameter		3		465	120	585	690
	180	36" diameter	↓	3	↓	530	120	650	760
	190								
	200	Mail chutes, aluminum & glass, 14-1/4" wide, 4-5/8" deep	2 Shee	4	Floor	525	91	616	710
	210	8-5/8" deep		3.80		630	96	726	835
	230	8-3/4" x 3-1/2", aluminum		5		445	73	518	595
	240	Bronze or stainless		4.50	↓	695	81	776	880
	260	Lobby collection boxes, aluminum		5	Ea.	795	73	868	980
	270	Bronze or stainless		4.50	"	1,100	81	1,181	1,325
	290	Package chutes, spiral type, minimum		4.50	Floor	1,445	81	1,526	1,700
	300	Maximum	↓	1.50	"	3,700	240	3,940	4,425
17-001		**CONTROL BOARDS** Magnetic, porcelain finish, 24" x 18", framed	2 Carp	8	Ea.	43	40	83	105
	010	36" x 24"		7.50		80	43	123	150
	020	48" x 36"		7		140	46	186	220
	030	72" x 48"		6		260	53	313	365
	040	96" x 48"	↓	5		335	64	399	460
	062	For custom gridding, add			↓	25%			
	100	Roll type, 49" x 74" case, with 2 rollers, 8 to 60 S.F. sleeves	2 Carp	125	S.F.	11.80	2.56	14.36	16.70
	120	49" x 95" case, with 4 rollers, 37 to 120 S.F. sleeves	"	130	"	10.25	2.46	12.71	14.85
	130	Motorized drive, add			Ea.	170		170	185M
20-001		**DIRECTORY BOARDS** Plastic, glass covered, 30" x 20"	2 Carp	3	Ea.	220	105	325	395
	010	36" x 48"		2		470	160	630	750
	030	Grooved cork, 30" x 20"		3		250	105	355	430
	040	36" x 48"		2		520	160	680	805
	060	Black felt, 30" x 20"		3		260	105	365	440
	070	36" x 48"		2		565	160	725	855
	090	Outdoor, weatherproof, black plastic, 36" x 24"		2		440	160	600	715
	100	36" x 36"		1.50		495	215	710	855
	120	Grooved cork, 36" x 24"		2		460	160	620	740
	130	36" x 36"		1.50		525	215	740	885
	150	Vinyl plastic, 36" x 24"		2		470	160	630	750
	160	36" x 36"	↓	1.50	↓	565	215	780	930

10.1 Specialties

		CREW	DAILY OUTPUT	UNIT	BARE COSTS MAT.	BARE COSTS INST.	BARE COSTS TOTAL	TOTAL INCL O&P
180	Indoor, economy, open face, 20" x 15"	2 Carp	7	Ea.	59	46	105	130
190	24" x 18"		7		63	46	109	135
200	30" x 20"		6		74	53	127	160
210	39" x 27"		6		91	53	144	175
240	Building directory boards, alum., black felt panels, 24" x 18"		4		220	80	300	360
250	39" x 22"		3.50		280	91	371	440
260	48" x 32"		3		365	105	470	555
270	36" x 48"		2.50		650	130	780	900
280	48" x 60"		2		780	160	940	1,100
290	48" x 72"	↓	1		960	320	1,280	1,525
310	For extruded bronze frame, add				50%			
320	For stainless steel frame, add				40%			
340	For illuminated directory, add				40%			
350	For recessed mounting, add			↓	10%			
22-001	**DISAPPEARING STAIRWAY** No trim included							
010	Custom grade, pine, 8'-6" ceiling, minimum	1 Carp	4	Ea.	54	40	94	115
015	Average		3.50		100	46	146	175
020	Maximum		3		160	53	213	255
050	Heavy duty, pivoted, 8'-6" ceiling pine		3		195	53	248	290
060	16'-0" ceiling		2		350	80	430	500
080	Economy folding, pine, 8'-6" ceiling		4		43	40	83	105
090	9'-6" ceiling	↓	4		56	40	96	120
100	Fire escape, galvanized steel, 8'-0" to 10'-4" ceiling	2 Carp	1		1,000	320	1,320	1,575
101	10'-6" to 13'-6" ceiling		1		1,250	320	1,570	1,850
110	Automatic electric, aluminum, floor to floor height, 8' to 9'		1		4,200	320	4,520	5,075
140	11' to 12'		.90		4,775	355	5,130	5,775
170	14' x 15'	↓	.70	↓	5,150	455	5,605	6,325
25-001	**DISPLAY CASES** Free standing, all glass							
002	Aluminum frame, 42" high x 36" x 12" deep	2 Carp	8	Ea.	580	40	620	695
010	70" high x 48" x 18" deep	"	6		1,075	53	1,128	1,250
050	For wood bases, add				10%			
060	For hardwood frames, deduct				35%			
070	For bronze, baked enamel finish, add			↓	20%			
200	Wall mounted, glass front, aluminum frame							
201	Non-illuminated, one section 3' x 4' x 1'-4"	2 Carp	5	Ea.	1,100	64	1,164	1,300
210	5' x 4' x 1'-4"		5		1,225	64	1,289	1,450
220	6' x 4' x 1'-4"		4		1,300	80	1,380	1,550
250	Two sections, 8' x 4' x 1'-4"		2		1,950	160	2,110	2,375
260	10' x 4' x 1'-4"		2		2,100	160	2,260	2,550
300	Three sections, 16' x 4' x 1'-4"	↓	1.50	↓	3,025	215	3,240	3,625
350	For fluorescent lights, add			Section	95		95	105M
400	Table exhibit cases, 2' wide, 3' high, 4' long, flat top	2 Carp	5	Ea.	700	64	764	865
410	3' wide, 3' high, 4' long, sloping top	"	3	"	1,125	105	1,230	1,400
27-001	**FIRE EXTINGUISHER** See division 15.4-20							
28-001	**FIRE HOSE EQUIPMENT** See division 15.4-30							
30-001	**FIREPLACE, PREFABRICATED** Free standing or wall hung							
010	with hood & screen, minimum	F-1	1.30	Ea.	350	130	480	570
015	Average		1		650	170	820	955
020	Maximum		.90	↓	2,000	185	2,185	2,475
050	For chimney heights over 8'-6", 7" diameter, add		33	V.L.F.	19	5.10	24.10	28
060	10" diameter, add		32		31	5.25	36.25	42
070	12" diameter, add		31	↓	46	5.45	51.45	58
080	14" diameter, add		30		65	5.60	70.60	80
100	Simulated brick chimney top, 4' high, 16" x 16"		10	Ea.	125	16.85	141.85	160
110	24" x 24"	↓	7		200	24	224	255
130	Woodburning stoves, cast iron, minimum	F-2	1.30		300	260	560	700
131	Average	"	1	↓	550	335	885	1,100

10.1 Specialties

			CREW	DAILY OUTPUT	UNIT	BARE COSTS MAT.	BARE COSTS INST.	BARE COSTS TOTAL	TOTAL INCL O&P
30	132	Maximum	F-2	.80	Ea.	1,300	420	1,720	2,025
	135	For gas log lighter, add			"	28		28	31M
	150	Simulated logs, gas fired, 40,000 BTU, 2' long, minimum	F-1	7	Set	125	24	149	170
	160	Maximum		6		300	28	328	370
	170	Electric, 1,500 BTU, 1'-6" long, minimum		7		60	24	84	100
	180	11,500 BTU, maximum		6		120	28	148	170
32-001		**FLAGPOLE** Not including base or foundation							
	010	Aluminum, tapered, ground set 20' high	K-1	2	Ea.	375	235	610	720
	020	25' high		1.70		800	275	1,075	1,250
	030	30' high		1.50		900	310	1,210	1,400
	040	35' high		1.40		1,300	335	1,635	1,875
	050	40' high		1.20		1,700	390	2,090	2,375
	060	50' high		1		2,800	470	3,270	3,700
	070	60' high		.90		4,600	520	5,120	5,750
	080	70' high		.80		9,000	585	9,585	10,700
	090	80' high		.70		10,500	670	11,170	12,400
	110	Counterbalanced, tapered, aluminum, 20' high		1.80		950	260	1,210	1,375
	120	30' high		1.50		1,400	310	1,710	1,950
	130	40' high		1.30		2,300	360	2,660	3,000
	140	50' high		1		4,100	470	4,570	5,125
	160	Aluminum, outrigger wall poles, including base							
	170	10' long, minimum	K-1	1.70	Ea.	700	275	975	1,125
	180	Maximum		1.40		1,000	335	1,335	1,550
	220	20' long outrigger, minimum		1.30		800	360	1,160	1,350
	225	Average		1.20		1,000	390	1,390	1,600
	240	Maximum		1		1,200	470	1,670	1,925
	270	Aluminum, vertical wall set, tapered, with base, 23' high		1.20		1,200	390	1,590	1,825
	280	28' high		1		1,475	470	1,945	2,225
	282	Aluminum, electronically operated, 30' high		1.40		2,050	335	2,385	2,700
	284	35' high		1.30		2,150	360	2,510	2,825
	286	40' high		1.10		2,250	425	2,675	3,025
	288	45' high		1		2,750	470	3,220	3,650
	290	50' high		.90		3,175	520	3,695	4,175
	292	60' high		.80		3,750	585	4,335	4,900
	300	Fiberglass, tapered, ground set, 25' high		2		650	235	885	1,025
	310	30' high		1.50		750	310	1,060	1,225
	320	35' high		1.40		975	335	1,310	1,500
	330	40' high		1.20		1,050	390	1,440	1,675
	340	50' high		1		2,800	470	3,270	3,700
	350	60' high		.90		3,800	520	4,320	4,850
	370	Outrigger poles with base, 12' long		1.30		360	360	720	870
	380	14' long		1		420	470	890	1,075
	400	Yardarms and rigging for poles, 6' total length		1.90		425	245	670	790
	410	12' total length		1.80		840	260	1,100	1,275
	430	Steel, sectional, lightweight, ground set, 20' high		1.30		250	360	610	745
	440	25' high		1.20		350	390	740	895
	450	30' high		1.10		475	425	900	1,075
	460	35' high		1		675	470	1,145	1,350
	470	40' high		.90		700	520	1,220	1,450
	480	50' high		.80		1,350	585	1,935	2,250
	500	Tapered, heavyweight steel, ground set, 35' high		.80		1,000	585	1,585	1,875
	510	50' high		.70		2,000	670	2,670	3,075
	520	60' high		.70		3,800	670	4,470	5,050
	530	75' high		.60		7,500	780	8,280	9,275
	550	For entasis tapered metal poles, add				50%			
	560	For bronze or stainless steel poles, add				125%			
	580	Bases, ornamental, minimum	1 Carp	6		27	27	54	68
	590	Average		4		85	40	125	150
	610	Maximum		2		1,100	80	1,180	1,325
	640	Wood poles, tapered, clear vertical grain fir with tilting							

10.1 Specialties

			CREW	DAILY OUTPUT	UNIT	BARE COSTS MAT.	BARE COSTS INST.	BARE COSTS TOTAL	TOTAL INCL O&P
32	641	base, not incl. foundation, 4" butt, 25' high	K-1	1.90	Ea.	310	245	555	665
	680	6" butt, 30' high	"	1.30	"	375	360	735	885
	730	Foundations for flagpoles, including							
	740	excavation and concrete, to 35 ft. high poles	C-1	10	Ea.	125	63	188	230
	760	40 ft. to 50 ft. high		3.50		405	180	585	705
	770	Over 60 ft. high	↓	2		640	315	955	1,150
35-001		**KEY CABINETS** Wall mounted, 50 key capacity	1 Carp	20		115	8	123	140
	010	1200 key capacity	"	10		1,000	16	1,016	1,125
	020	Drawer type, 600 key capacity	1 Clab	15		1,050	8.50	1,058.50	1,175
	030	2400 key capacity		20		2,375	6.35	2,381.35	2,625
	040	Tray type, 20 key capacity		50		46	2.54	48.54	54
	050	50 key capacity	↓	40	↓	90	3.18	93.18	105
37-001		**LOCKERS** Steel, baked enamel, 60" or 72", single tier, minimum	1 Shee	14	Opng.	54	12.95	66.95	78
	010	Maximum		12		99	15.15	114.15	130
	030	2 tier, 60" or 72" total height, minimum		26		37	7	44	51
	040	Maximum		20		60	9.10	69.10	79
	060	5 tier box lockers, minimum		30		17.10	6.05	23.15	28
	070	Maximum		24		23	7.55	30.55	36
	090	6 tier box lockers, minimum		36		16.60	5.05	21.65	26
	100	Maximum		30	↓	23	6.05	29.05	34
	120	Basket rack with 32 baskets, 9" x 13" x 8" basket		50	Basket	14.10	3.63	17.73	21
	130	24 baskets, 12" x 13" x 8" basket		50	"	17.35	3.63	20.98	24
	150	Athletic, wire mesh, no lock, 18" x 18" x 72"	↓	12	Ea.	75	15.15	90.15	105
	160	Overhead locker baskets on chains, 14" x 14" baskets	3 Shee	96	Basket	48	5.70	53.70	61
	180	Overhead locker framing system, add		600		7.65	.91	8.56	9.75
	190	Locking rail and bench units, add	↓	120	↓	11.25	4.54	15.79	19
	210	Locker bench, laminated maple, top only	1 Shee	100	L.F.	12.90	1.82	14.72	16.85
	220	Pedestals, steel pipe	"	25	Ea.	14.90	7.25	22.15	27
	240	Teacher and pupil wardrobes, enameled							
	250	22" x 15" x 61" high, minimum	1 Shee	10	Ea.	92	18.15	110.15	130
	255	Average		9		155	20	175	200
	270	Maximum		8		160	23	183	210
	300	Duplex lockers with 2 doors, 72" high, 15" x 15"		10		130	18.15	148.15	170
	310	15" x 21"	↓	10		145	18.15	163.15	185
	330	For built-in lock with 2 keys, add				2.80		2.80	3.08M
	340	For built-in combination lock, add				11.50		11.50	12.65M
	360	For hanger rods, add				1.25		1.25	1.37M
	370	For stainless steel lockers, add				100%			
40-001		**MAIL** Boxes, horizontal, key lock, 5"H x 6"W x 15"D, alum., rear loading	1 Carp	34		51	4.71	55.71	63
	010	Front loading		34		55	4.71	59.71	67
	020	Double, 5"H x 12"W x 15"D, rear loading		26		94	6.15	100.15	110
	030	Front loading		26		105	6.15	111.15	125
	050	Quadruple, 10"H x 12"W x 15"D, rear loading		20		165	8	173	195
	060	Front loading		20		175	8	183	205
	080	Vertical, front loading, 15"H x 5"W x 6"D, aluminum		34		44	4.71	48.71	55
	090	Bronze, duranodic finish		34		51	4.71	55.71	63
	100	Steel, enameled	↓	34	↓	48	4.71	52.71	60
	170	Alphabetical directories, 50 names	1 Carp	10	Ea.	120	16	136	155
	180	Letter collection box		6		550	27	577	645
	190	Letter slot, residential		20		28	8	36	42
	200	Post office type		8		110	20	130	150
	220	Post office counter window, with grille	↓	2	↓	325	80	405	475
	225	Key keeper, single key, aluminum	1 Carp	26	Ea.	49	6.15	55.15	63
	230	Steel, enameled		26		55	6.15	61.15	69
42-001		**MEDICINE CABINETS** With mirror, stock, 16" x 22", unlighted		14		43	11.45	54.45	64
	010	Lighted		6		84	27	111	130
	030	Sliding mirror doors, 34" x 21", unlighted	K-1	7		66	23	89	105
	040	Lighted	↓	5	↓	135	32	167	195

10.1 Specialties

			DAILY		BARE COSTS			TOTAL
		CREW	OUTPUT	UNIT	MAT.	INST.	TOTAL	INCL O&P
42 060	Center mirror, 2 end cabinets, unlighted, 48" long	1 Carp	7	Ea.	175	23	198	225
070	72" long	"	5		205	32	237	270
090	For lighting, 48" long, add	1 Elec	3.50		51	51	102	130
100	72" long, add	"	3		75	60	135	170
120	Hotel cabinets, stainless, with lower shelf, unlighted	1 Carp	10		145	16	161	185
130	Lighted	"	5	↓	205	32	237	270
45-001	**PARTITIONS, FOLDING ACCORDION** See also division 8.2-32							
010	Vinyl covered, over 150 S.F., frame not included							
030	Residential, 1.25 lb. per S.F., 8 ft. maximum height	2 Carp	300	S.F.	6.85	1.07	7.92	9.10
031								
040	Commercial, 1.75 lb. per S.F., 8 ft. maximum height	2 Carp	225	S.F.	7	1.42	8.42	9.75
060	2 lb. per S.F., 17 ft. maximum height		150		8.40	2.13	10.53	12.35
070	Industrial, 4 lb. per S.F., 27 ft. maximum height	↓	100	↓	11.35	3.20	14.55	17.10
071								
090	Acoustical, 3 lb. per S.F., 17 ft. maximum height	2 Carp	100	S.F.	8.30	3.20	11.50	13.75
120	5 lb. per S.F., 27 ft. maximum height		95		12	3.37	15.37	18.10
130	5.5 lb. per S.F., 17 ft. maximum height		90		14.60	3.56	18.16	21
140	Fire rated, 4.5 psf, 20 ft. maximum height		160		19.20	2	21.20	24
150	Vinyl clad wood or steel, electric operation, 5.0 psf		160		18.85	2	20.85	24
190	Wood, non-acoustic, birch or mahogany, to 10 ft. high	↓	300	↓	11	1.07	12.07	13.65
47-001	**PARTITIONS, FOLDING LEAF** Acoustic, wood							
010	Vinyl faced, to 18' high, 6 psf, minimum	2 Carp	60	S.F.	25	5.35	30.35	35
015	Average		45		26	7.10	33.10	39
020	Maximum		30		27	10.65	37.65	45
040	Formica or hardwood finish, minimum		60		25	5.35	30.35	35
050	Maximum		30		26	10.65	36.65	44
060	Wood, low acoustical type, 4.5 psf, to 12' high	↓	50	↓	15.25	6.40	21.65	26
061								
110	Steel, acoustical, 7.5 lb. per S.F., vinyl faced, minimum	2 Carp	60	S.F.	27	5.35	32.35	37
120	Maximum		30		30	10.65	40.65	48
170	Aluminum framed, acoustical, to 12' high, 5.5 psf, minimum		60		16.50	5.35	21.85	26
180	Maximum		30		19	10.65	29.65	36
200	6.5 lb. per S.F., minimum		60		18	5.35	23.35	28
210	Maximum	↓	30	↓	21	10.65	31.65	39
50-001	**PARTITIONS, HOSPITAL** Curtain track, box channel, ceiling mounted	1 Carp	135	L.F.	6.55	1.19	7.74	8.90
010	Suspended	"	100	"	9.30	1.60	10.90	12.55
030	Curtains, 8' to 9', nylon mesh tops, fire resistant							
031	Cotton, 2.85 lbs. per S.Y.	1 Carp	425	L.F.	7.60	.38	7.98	8.90
050	Anti-bacterial, thermoplastic		425		6.85	.38	7.23	8.10
070	Fiberglass, 7 oz. per S.Y.	↓	425	↓	10.40	.38	10.78	12
080	I.V. track systems							
082	I.V. track, 4'-0" x 7'-0" oval	1 Carp	135	L.F.	16.50	1.19	17.69	19.85
083	I.V. trolley		32	Ea.	32	5	37	42
084	I.V. pendent, (tree, 5 hook)	↓	32	"	225	5	230	255
52-001	**PARTITIONS, MOVABLE OFFICE** Demountable, add for doors							
010	Do not deduct door openings from total L.F.							
090	Asbestos cement, 1-3/4" thick, prefinished, low walls	2 Carp	80	L.F.	31	4	35	40
100	Full height	"	40		38	8	46	53
120	Economy grade, 1-3/4" thick, deduct				20%			
130	High quality, 4" thick, add				100%			
150	Gypsum, laminated 2-1/4" thick, 9' high, unpainted	2 Carp	40		11.40	8	19.40	24
160	Painted		40		12.60	8	20.60	25
180	3" thick, acoustical, unpainted		40		16	8	24	29
190	Painted	↓	40	↓	17.50	8	25.50	31
210	For architectural wood flitches, add to							
220	unpainted prices, minimum			S.F.	5.30		5.30	5.85M
225	Average				7.90		7.90	8.70M
230	Maximum			↓	10.60		10.60	11.65M
235	Vinyl clad drywall on 2-1/2" metal studs, to 9' high	2 Carp	60	L.F.	17.70	5.35	23.05	27
240	42" high, plus 10" glass	"	60	"	32	5.35	37.35	43

10.1 Specialties

			CREW	DAILY OUTPUT	UNIT	BARE COSTS MAT.	BARE COSTS INST.	BARE COSTS TOTAL	TOTAL INCL O&P
52	250	Vinyl clad gypsum with air space, 3" thick	2 Carp	60	L.F.	23	5.35	28.35	33
	251	Steel clad gypsum, as above	"	60	"	53	5.35	58.35	66
	255								
	260	Hardboard, vinyl faced, 7' high, 1-9/16" thick	2 Carp	60	L.F.	26	5.35	31.35	36
	280	2-1/4" thick, painted		50		22	6.40	28.40	33
	290	With 40" glass		40		37	8	45	52
	310	10' high, vinyl faced, 1-9/16" thick		40		25	8	33	39
	320	Enameled, 2-3/4" thick		30		32	10.65	42.65	51
	340	Metal, to 9'-6" high, enameled steel, no glass		40		66	8	74	84
	350	Steel frame, all glass		40		75	8	83	94
	370	Vinyl covered, no glass		40		76	8	84	95
	380	Steel frame with 52% glass		40		68	8	76	86
	400	Free standing, 4'-6" high, steel with glass		100		34	3.20	37.20	42
	410	Acoustical		100		48	3.20	51.20	57
	430	Low rails, 3'-3" high, enameled steel		100		33	3.20	36.20	41
	440	Vinyl covered		100		41	3.20	44.20	50
	460	Plywood, prefin., 1-3/4" thick, rotary cut veneers, minimum		80		35	4	39	44
	470	Maximum		80		84	4	88	98
	490	Sliced veneers, book matched, minimum		80		42	4	46	52
	500	Maximum		80		100	4	104	115
	520	Sliced veneers, random matched, minimum		80		35	4	39	44
	530	Maximum		80		100	4	104	115
	531	Trackless wall, cork finish, semi-acoustic, 1-5/8" thick, minimum		325	S.F.	12	.98	12.98	14.65
	532	Maximum		190		18	1.68	19.68	22
	533	Acoustic, 2" thick, minimum		305		12.80	1.05	13.85	15.60
	534	Maximum		225		20	1.42	21.42	24
	550	For acoustical partitions, add, minimum				.42		.42	.46M
	555	Maximum				.58		.58	.63M
	570	For doors, not incl. hardware, hollow metal door, add	2 Carp	4.30	Ea.	143	74	217	265
	580	Hardwood door, add		3.40		102	94	196	250
	600	Hardware for doors, not incl. closers, keyed		29		69	11.05	80.05	92
	610	Non-keyed		29		61	11.05	72.05	83
54-	001	**PARTITIONS, WORK STATIONS** Self contained, incl. top cabinet & desk top							
	002								
	020	Multi-station units, 1 to 3 person seating capacity							
	025	Minimum per person	1 Carp	6	Ea.	725	27	752	835
	026	Average per person		5		875	32	907	1,000
	027	Maximum per person		4		1,125	40	1,165	1,300
	028	4 to 6 person seating capacity							
	029	Minimum per person	1 Carp	4	Ea.	825	40	865	965
	030	Average per person		3		925	53	978	1,100
	032	Maximum per person		2		1,200	80	1,280	1,425
55-	001	**PARTITIONS, OPERABLE** Acoustic air wall, 1-5/8" thick, minimum	2 Carp	375	S.F.	13.35	.85	14.20	15.90
	010	Maximum		365		19.45	.88	20.33	23
	030	2-1/4" thick, minimum		360		14.80	.89	15.69	17.55
	040	Maximum		330		25.50	.97	26.47	29
	060	For track type, add to above			L.F.	55		55	61M
	065								
	070	Overhead track type, acoustical, 3" thick, 11 psf, minimum	2 Carp	350	S.F.	37	.91	37.91	42
	080	Maximum	"	300	"	43	1.07	44.07	49
56-	001	**PARTITIONS, PORTABLE** Divider walls, free standing, fiber core							
	002	Panels, 5' long, 4'-6" high							
	010	Burlap face, straight	2 Carp	160	L.F.	55	2	57	63
	020	Curved		158		65	2.03	67.03	74
	050	Carpeted face, straight		160		57	2	59	66
	060	Curved		158		70	2.03	72.03	80
	090	Plastic laminated face, straight		160		42	2	44	49
	100	Curved		158		55	2.03	57.03	63
	150	5 ft. high, burlap face, straight		150		59	2.13	61.13	68
	160	Curved		148		71	2.16	73.16	81

10.1 Specialties

			CREW	DAILY OUTPUT	UNIT	BARE COSTS			TOTAL INCL O&P
						MAT.	INST.	TOTAL	
56	180	Carpeted face, straight	2 Carp	150	L.F.	65	2.13	67.13	75
	190	Curved		148		77	2.16	79.16	88
	220	Plastic laminated face, straight		150		44	2.13	46.13	51
	230	Curved		148		64	2.16	66.16	74
	260	6 ft. high, burlap face, straight		125		54	2.56	56.56	63
	270	Curved		120		66	2.67	68.67	76
	300	Carpeted face, straight		125		70	2.56	72.56	81
	310	Curved		120		85	2.67	87.67	97
	340	Plastic laminated face, straight		125		49	2.56	51.56	58
	350	Curved		120		60	2.67	62.67	70
	400	Metal chalkboard, 6'-6" high, chalkboard, 1 side		125		56	2.56	58.56	65
	410	Metal chalkboard, 2 sides		120		65	2.67	67.67	75
	420	Hardboard chalkboard, 1 side		125		50	2.56	52.56	59
	430	Tackboard, both sides	↓	123	↓	52	2.60	54.60	61
57	001	**PARTITIONS, SHOWER** Economy, painted steel, steel							
	010	base, no door or plumbing included	2 Shee	5	Ea.	120	73	193	240
	030	Square, 32" x 32", stock, with receptor & door, fiberglass		4.50		265	81	346	410
	060	Galvanized and painted steel	↓	5	↓	180	73	253	305
	070	Shower stall, double wall, incl. receptor but not including							
	080	door or plumbing, enameled steel	2 Shee	5	Ea.	640	73	713	810
	110	Porcelain enameled steel		5		1,075	73	1,148	1,300
	120	Stainless steel	↓	5		1,100	73	1,173	1,325
	140	For double entry type, no doors, deduct				10%			
	150	Circular fiberglass, 36" diameter, no plumbing included	2 Shee	4		225	91	316	380
	170	One piece, 36" diameter, less door		4		255	91	346	415
	180	With door		3.50		410	105	515	605
	200	Curved shell shower, no door needed	↓	3		420	120	540	640
	230	For fiberglass seat, add to both above				53		53	58M
	240	Glass stalls, with doors, no receptors, chrome on brass	2 Shee	3		630	120	750	870
	270	Anodized aluminum	"	4		390	91	481	560
	290	Marble shower stall, stock design, with shower door	2 Marb	1.20		870	270	1,140	1,350
	300	With curtain		1.30		720	245	965	1,150
	320	Receptors, precast terrazzo, 32" x 32"		14		140	23	163	185
	330	48" x 34"		12		175	27	202	230
	350	Plastic, simulated terrazzo receptor, 32" x 32"		14		60	23	83	99
	360	32" x 48"		12		84	27	111	130
	380	Precast concrete, colors, 32" x 32"		14		115	23	138	160
	390	48" x 48"	↓	12		135	27	162	185
	410	Shower doors, economy plastic, 24" wide	1 Shee	9		63	20	83	99
	420	Tempered glass door, economy		8		94	23	117	135
	440	Folding, tempered glass, aluminum frame		6		140	30	170	200
	450	Sliding, tempered glass, 48" opening		6		100	30	130	155
	470	Deluxe, tempered glass, chrome on brass frame, minimum		5		265	36	301	345
	500	Maximum		5		490	36	526	590
	530	On anodized aluminum frame, minimum		5		74	36	110	135
	540	Maximum		5		265	36	301	345
	560	Tub enclosure, plastic panels, economy, sliding panel		4		65	45	110	140
	570	Folding panel		4		90	45	135	165
	590	Deluxe, tempered glass, anodized alum. frame, minimum		2		155	91	246	305
	620	Maximum		1.50		335	120	455	545
	650	On chrome-plated brass frame, minimum		2		245	91	336	400
	660	Maximum	↓	1.50	↓	515	120	635	745
60	001	**PARTITIONS, TOILET** For ceiling framing, see division 5.4-68							
	010	Cubicles, ceiling hung, marble	2 Marb	2	Ea.	480	160	640	755
	020	Painted metal	2 Carp	4		210	80	290	345
	030	Plastic laminate on particle board		4		310	80	390	455
	040	Porcelain enamel		4		635	80	715	815
	050	Stainless steel	↓	4		585	80	665	760
	060	For handicap units, add			↓	70		70	77M
	070								

10.1 Specialties

		CREW	DAILY OUTPUT	UNIT	BARE COSTS MAT.	BARE COSTS INST.	BARE COSTS TOTAL	TOTAL INCL O&P
080	Floor & ceiling anchored, marble	2 Marb	2.50	Ea.	475	130	605	705
100	Painted metal	2 Carp	5		210	64	274	325
110	Plastic laminate on particle board		5		370	64	434	500
120	Porcelain enamel		5		640	64	704	795
130	Stainless steel	↓	5		580	64	644	730
140	For handicap units, add				70		70	77M
160	Floor mounted, marble	2 Marb	3		410	105	515	605
170	Painted metal	2 Carp	7		185	46	231	270
180	Plastic laminate on particle board		7		175	46	221	260
190	Porcelain enamel		7		635	46	681	765
200	Stainless steel	↓	7		585	46	631	710
210	For handicap units, add				70		70	77M
220	For juvenile units, deduct			↓	20		20	22M
230								
240	Floor mounted, headrail braced, marble	2 Marb	3	Ea.	435	105	540	630
250	Painted metal	2 Carp	6		195	53	248	290
260	Plastic laminate on particle board		6		185	53	238	280
270	Porcelain enamel		6		630	53	683	770
280	Stainless steel	↓	6		570	53	623	705
290	For handicap units, add				70		70	77M
300	Wall hung partitions, painted metal	2 Carp	7		390	46	436	495
320	Porcelain enamel		7		550	46	596	670
330	Stainless steel	↓	7		550	46	596	670
340	For handicap units, add			↓	70		70	77M
400	Screens, entrance, floor mounted, 54" high							
410	Marble	D-1	35	L.F.	75	8.35	83.35	94
420	Painted metal	2 Carp	60		48	5.35	53.35	61
430	Plastic laminate on particle board		60		59	5.35	64.35	73
440	Porcelain enamel		60		115	5.35	120.35	135
450	Stainless steel	↓	60	↓	100	5.35	105.35	120
460	Urinal screen, 18" wide, ceiling braced, marble	D-1	6	Ea.	225	49	274	315
470	Painted metal	2 Carp	8		115	40	155	185
480	Plastic laminate on particle board		8		150	40	190	225
490	Porcelain enamel		8		220	40	260	300
500	Stainless steel	↓	8	↓	220	40	260	300
505								
510	Floor mounted, head rail braced							
520	Marble	D-1	6	Ea.	200	49	249	290
530	Painted metal	2 Carp	8		120	40	160	190
540	Plastic laminate on particle board		8		140	40	180	210
550	Porcelain enamel		8		220	40	260	300
560	Stainless steel	↓	8		220	40	260	300
570	Pilaster, flush, marble	D-1	9		240	32	272	310
580	Painted metal	2 Carp	10		65	32	97	120
590	Plastic laminate on particle board		10		80	32	112	135
600	Porcelain enamel		10		160	32	192	220
610	Stainless steel	↓	10	↓	160	32	192	220
615								
620	Post braced, marble	D-1	9	Ea.	235	32	267	305
630	Painted metal	2 Carp	10		65	32	97	120
640	Plastic laminate on particle board		10		80	32	112	135
650	Porcelain enamel		10		160	32	192	220
660	Stainless steel	↓	10	↓	160	32	192	220
670	Wall hung, bracket supported							
680	Painted metal	2 Carp	10	Ea.	96	32	128	150
690	Plastic laminate on particle board		10		91	32	123	145
700	Porcelain enamel		10		190	32	222	255
710	Stainless steel		10		175	32	207	240
740	Flange supported, painted metal		10		190	32	222	255
750	Plastic laminate on particle board	↓	10	↓	110	32	142	165

10.1 Specialties

			CREW	DAILY OUTPUT	UNIT	BARE COSTS MAT.	BARE COSTS INST.	BARE COSTS TOTAL	TOTAL INCL O&P
60	760	Porcelain enamel	2 Carp	10	Ea.	230	32	262	300
	770	Stainless steel		10		220	32	252	290
	780	Wedge type, painted metal		10		135	32	167	195
	800	Porcelain enamel		10		210	32	242	275
	810	Stainless steel	↓	10	↓	210	32	242	275
62-001		**PARTITIONS, WOVEN WIRE** For tool or stockroom enclosures							
	010	Channel frame, 1-1/2" diamond mesh, 10 ga. wire, painted							
	030	Wall panels, 4'-0" wide, 7' high	2 Carp	25	Ea.	52	12.80	64.80	76
	040	8' high		23		55	13.90	68.90	81
	060	10' high	↓	18		66	17.80	83.80	98
	070	For 5' wide panels, add				3%			
	090	Ceiling panels, 10' long, 2' wide	2 Carp	25		55	12.80	67.80	79
	100	4' wide		15		66	21	87	105
	120	Panel with service window & shelf, 5' long, 7' high		20		175	16	191	215
	130	10' high		15		195	21	216	245
	150	Sliding doors, 3' wide, 7' full height		6		250	53	303	350
	160	10' full height		5		285	64	349	405
	180	6' wide sliding door, 7' full height		5		325	64	389	450
	190	10' full height		4		380	80	460	535
	210	Swinging doors, 3' wide, 7' high, no transom		6		145	53	198	235
	220	7' high, 3' transom	↓	5		200	64	264	315
	250	For galvanizing, add to above				20%			
	300	For stainless steel or aluminum, add				400%			
65-001		**PARTS BINS** 1' deep, 3' wide, 6'-3" high, 31 bins	2 Clab	10		220	25	245	280
	010	48 shelf drawers & 42 bins		10		765	25	790	880
	030	144 bins		10		535	25	560	625
	040	96 shelf drawers		10		1,025	25	1,050	1,175
	060	108 drawers, 12" deep		10		1,350	25	1,375	1,525
	070	18" deep	↓	10	↓	1,650	25	1,675	1,850
67-001		**PROJECTION SCREENS** Wall or ceiling hung, glass beaded							
	010	Manually operated, economy	2 Carp	500	S.F.	3.15	.64	3.79	4.39
	030	Intermediate		450		5.75	.71	6.46	7.35
	040	Deluxe		400	↓	8.60	.80	9.40	10.60
	060	Electric operated, glass beaded, 50 S.F., economy		5	Ea.	500	64	564	645
	070	Deluxe		4		1,250	80	1,330	1,500
	090	100 S.F., economy		3		930	105	1,035	1,175
	100	Deluxe		2		1,625	160	1,785	2,025
	120	Heavy duty, electric operated, 200 S.F.		1.50		1,475	215	1,690	1,925
	130	400 S.F.	↓	1	↓	1,725	320	2,045	2,350
	150	Rigid acrylic in wall, for rear projection, 1/4" thick	2 Glaz	195	S.F.	33	1.65	34.65	39
	160	1/2" thick (maximum size 10' x 20')	"	130	"	54	2.48	56.48	63
70-001		**SCALES** Built-in floor scale, not incl. foundations							
	002								
	010	Dial type, 5 ton capacity, 8' x 6' platform	3 Carp	.50	Ea.	6,000	960	6,960	8,000
	030	9' x 7' platform		.40		6,500	1,200	7,700	8,900
	040	10 ton capacity, steel platform, 8' x 6' platform		.40		6,500	1,200	7,700	8,900
	060	9' x 7' platform	↓	.35	↓	7,500	1,375	8,875	10,200
	070	Truck scales, incl. steel weigh bridge,							
	080	not including foundations							
	090	Dial type, mech., 24' x 10' platform, 20 ton cap.	3 Carp	.30	Ea.	7,000	1,600	8,600	10,000
	100	30 ton capacity		.20		8,000	2,400	10,400	12,300
	110	50 ton capacity, 50' x 10' platform		.14		13,000	3,425	16,425	19,300
	120	70' x 10' platform		.12		16,000	4,000	20,000	23,400
	140	60 ton capacity, 60' x 10' platform		.13		15,000	3,700	18,700	21,900
	150	70' x 10' platform		.10		18,000	4,800	22,800	26,800
	155	Digital, electronic, 30 ton capacity, 12' x 10' platform		.20		12,500	2,400	14,900	17,200
	160	50 ton capacity, 40' x 10' platform		.14		17,000	3,425	20,425	23,700
	164	60 ton capacity, 60' x 10' platform		.13	↓	21,600	3,700	25,300	29,100
	168	75' x 10' platform	↓	.12		25,600	4,000	29,600	34,000

10.1 Specialties

		CREW	DAILY OUTPUT	UNIT	BARE COSTS MAT.	BARE COSTS INST.	BARE COSTS TOTAL	TOTAL INCL O&P
70 200	For standard automatic printing device, add			Ea.	1,200		1,200	1,325M
210	For remote reading electronic system, add				1,300		1,300	1,425M
230	Concrete foundation pits for above, 8' x 6' platform, 5 C.Y. required	C-1	.50		675	1,275	1,950	2,550
240	14' x 6' platform, 10 C.Y. required		.35		1,025	1,800	2,825	3,725
260	50' x 10' platform, 30 C.Y. required		.25		2,850	2,525	5,375	6,775
270	70' x 10' platform, 40 C.Y. required	▼	.15		3,425	4,225	7,650	9,825
275	Crane scales, dial, 1 ton capacity				1,900		1,900	2,100M
278	5 ton capacity				2,000		2,000	2,200M
280	Digital, 1 ton capacity				2,400		2,400	2,650M
285	25 ton capacity			▼	5,000		5,000	5,500M
290	Low profile electronic warehouse scale,							
300	not incl. printer, 4' x 6' platform, 6000 lb. capacity	2 Carp	.30	Ea.	4,500	1,075	5,575	6,500
330	5' x 7' platform, 10,000 lb. capacity		.25		5,100	1,275	6,375	7,475
340	20,000 lb. capacity	▼	.20		5,500	1,600	7,100	8,375
350	For printers, incl. time, date & numbering, add			▼	2,300		2,300	2,525M
351								
380	Portable, beam type, capacity 1000#, platform 18" x 27"			Ea.	510		510	560M
390	Dial type, capacity 2000#, platform 24" x 30"				3,200		3,200	3,525M
400	Digital type, capacity 1000#, platform 24" x 30"			▼	2,800		2,800	3,075M
405								
410	Portable contractor truck scales, 50 ton cap., 40' x 10' platform			Ea.	17,000		17,000	18,700M
420	60' x 10' platform			"	22,000		22,000	24,200M
72-001	**SECURITY GATES** For roll up type, see division 8.3-09 & 45							
030	Scissors type folding gate, ptd. 51" steel, single, 6' high, 4' wide	2 Sswk	2.20	Opng.	190	160	350	455
035	75" wide		2		230	175	405	525
040	99" wide		1.80		275	195	470	605
060	Double gate, 8' wide		1.80		355	195	550	695
065	10' wide		1.50		415	230	645	820
070	12' wide		1.20		440	290	730	940
075	14' wide	▼	1		495	345	840	1,100
090	Vertical member gates, add to above				50%			
100	For galvanized gates, add				20%			
120	For aluminum gates, add				125%			
130	For stainless steel gates, add				150%			
150	For bronze gates, add				250%			
160	For custom fabrication, add			▼	40%			
75-001	**SHELVING** Metal, industrial, cross-braced, 3' wide, 12" deep	1 Sswk	175	SF Shlf	2.95	.99	3.94	4.80
010	24" deep		330		2	.53	2.53	3.03
030	4' wide, 12" deep		185		2.60	.94	3.54	4.33
040	24" deep		380		1.80	.46	2.26	2.70
120	Enclosed sides, cross-braced back, 3' wide, 12" deep		175		3.90	.99	4.89	5.85
130	24" deep		290		2.65	.60	3.25	3.86
150	Fully enclosed, sides and back, 3' wide, 12" deep		150		4	1.16	5.16	6.20
160	24" deep		255		2.70	.68	3.38	4.04
180	4' wide, 12" deep		150		3.65	1.16	4.81	5.85
190	24" deep		290		2.45	.60	3.05	3.64
220	Wide span, 1600 lb. capacity per shelf, 7' wide, 36" deep		380		4.20	.46	4.66	5.35
240	48" deep		440		3.65	.39	4.04	4.63
260	8' wide, 36" deep		440		3.70	.39	4.09	4.69
280	48" deep	▼	520		3.20	.33	3.53	4.04
400	Pallet racks, steel frame 2500 lb. capacity, 6' long, 30" deep	2 Sswk	450		2.90	.77	3.67	4.40
420	36" deep		500		2.40	.69	3.09	3.73
440	42" deep	▼	520	▼	2.10	.67	2.77	3.36
450								
77-001	**SIGNS** Letters, individual, 2" high, cast aluminum	1 Carp	32	Ea.	9.80	5	14.80	18.05
010	Cast bronze		32		14	5	19	23
030	4" high, 5/8" deep, cast aluminum		24		14	6.65	20.65	25
040	Cast bronze		24		21	6.65	27.65	33
060	6" high, 1" deep, cast aluminum		20		18	8	26	31
070	Cast bronze	▼	20	▼	35	8	43	50

10.1 Specialties

		Description	CREW	DAILY OUTPUT	UNIT	BARE COSTS MAT.	BARE COSTS INST.	BARE COSTS TOTAL	TOTAL INCL O&P
77	090	12" high, 1-1/4" deep, cast aluminum	1 Carp	18	Ea.	32	8.90	40.90	48
	100	Cast bronze		18		70	8.90	78.90	90
	120	18" high, 1-1/4" deep, cast aluminum		12		91	13.35	104.35	120
	130	Cast bronze		12		115	13.35	128.35	145
	150	Fabricated aluminum, 12" high, 3" deep		18		49	8.90	57.90	67
	160	18" high, 3" deep		12		100	13.35	113.35	130
	180	Fabricated stainless steel, 6" high, 3" deep		20		46	8	54	62
	190	12" high, 3" deep		18		60	8.90	68.90	79
	210	18" high, 3" deep		12		100	13.35	113.35	130
	220	24" high, 4" deep		10		115	16	131	150
	240	Painted sheet steel, 12" high, 2" deep		18		35	8.90	43.90	51
	250	18" high, 3" deep		12		70	13.35	83.35	96
	270	Plastic, 6" high, 1" deep		20		7	8	15	19.30
	280	12" high, 2" deep		18		14	8.90	22.90	28
	300	Plastic face, alum. frame, 20" high, 15" wide		5		140	32	172	200
	330	36" high, 24" deep		4		195	40	235	270
	340	Stainless steel frame, 12" high, 6" deep		5		115	32	147	175
	370	24" high, 8" deep		4		250	40	290	335
	390	Plaques, 20" x 30", for up to 300 letters, cast aluminum	2 Carp	4		675	80	755	860
	400	Cast bronze		4		740	80	820	930
	420	30" x 40", cast aluminum		3		1,350	105	1,455	1,650
	430	Cast bronze		3		1,475	105	1,580	1,775
	450	36" x 48", for up to 600 letters, cast bronze		2		2,150	160	2,310	2,600
	460	For sculpture, bas-relief portrait, add				25%			
	480	Signs, cast aluminum street signs, 2-way	2 Carp	30		29	10.65	39.65	47
	490	4-way		30		57	10.65	67.65	78
	510	Acrylic exit signs, 15" x 6", surface mounted, minimum		30		9.80	10.65	20.45	26
	520	Maximum		20		18	16	34	43
	540	Bracket mounted, double face, minimum		30		19.50	10.65	30.15	37
	550	Maximum		20		39	16	55	66
	570	Plexiglass, exterior, illuminated, single face		100	S.F.	45	3.20	48.20	54
	580	Double face		75		58	4.27	62.27	70
	600	Interior, illuminated, single face		100		42	3.20	45.20	51
	610	Double face		75		50	4.27	54.27	61
	640	Painted plywood (MDO), over 4' x 8'		120		6.65	2.67	9.32	11.20
	660	Under 4' x 8'		100		7.50	3.20	10.70	12.90
	670	For metal edge moldings, add			L.F.	2.65		2.65	2.91M
80	001	**TELEPHONE ENCLOSURE** Desk-top type	2 Carp	8	Ea.	390	40	430	485
	002								
	030	Shelf type, wall hung, minimum	2 Carp	5	Ea.	400	64	464	535
	040	Maximum		5		1,900	64	1,964	2,175
	060	Booth type, painted steel, indoor or outdoor, minimum		1.50		1,225	215	1,440	1,650
	070	Maximum (stainless steel)		1.50		3,300	215	3,515	3,950
	130	Outdoor, acoustical, on post		3		940	105	1,045	1,200
	140	Phone carousel, pedestal mounted with dividers		.60		3,750	535	4,285	4,900
	190	Outdoor, drive-up type, wall mounted		4		520	80	600	690
	200	Post mounted, stainless steel posts		3		940	105	1,045	1,200
	220	Directory shelf, wall mounted, stainless steel							
	230	3 binders	2 Carp	8	Ea.	410	40	450	510
	250	4 binders		7		460	46	506	570
	260	7 binders		6		600	53	653	735
	280	Table type, stainless steel, 4 binders		8		660	40	700	785
	290	7 binders		7		805	46	851	950
82	001	**TURNSTILES** One way, 4 arm, 46" diameter, economy, manual		5		265	64	329	385
	010	Electric		1.20		1,000	265	1,265	1,475
	030	High security, 3 arm, 5'-5" diameter, 7' high, manual		1		2,575	320	2,895	3,300
	035	Electric		.60		3,025	535	3,560	4,100
	042	Three arm, 24" opening, light duty, manual		2		595	160	755	885
	045	Heavy duty		1.50		725	215	940	1,100

10.1 Specialties

		CREW	DAILY OUTPUT	UNIT	BARE COSTS MAT.	BARE COSTS INST.	BARE COSTS TOTAL	TOTAL INCL O&P
82 046	Manual, with registering & controls, light duty	2 Carp	2	Ea.	750	160	910	1,050
047	Heavy duty	↓	1.50		890	215	1,105	1,300
048	Electric, heavy duty	↓	1.10		1,000	290	1,290	1,525
049	For wall mounted, add				85		85	94M
050	For coin or token operating, add				250		250	275M
060	For portable models, add			↓	465		465	510M
120	One way gate with horizontal bars, 5'-5" diameter							
130	7 ft. high, recreation or transit type	2 Carp	.80	Ea.	2,350	400	2,750	3,175
140	For ticket box with visual panels, add				220		220	240M
150	For electronic counter, add			↓	120		120	130M

11.1 Architectural Equipment

		CREW	DAILY OUTPUT	UNIT	BARE COSTS MAT.	BARE COSTS INST.	BARE COSTS TOTAL	TOTAL INCL O&P
03-001	**APPLIANCES** Cooking range, 30" free standing, 1 oven, minimum	2 Clab	10	Ea.	300	25	325	365
005	Maximum		4		1,200	64	1,264	1,400
015	2 oven, minimum		10		680	25	705	785
020	Maximum	↓	4		1,500	64	1,564	1,750
035	Built-in, 30" wide, 1 oven, minimum	2 Carp	4		350	80	430	500
040	Maximum		2		800	160	960	1,100
050	2 oven, minimum		4		700	80	780	885
055	Maximum	↓	2		1,400	160	1,560	1,775
070	Free-standing, 21" wide range, 1 oven, minimum	2 Clab	10		300	25	325	365
075	Maximum	"	4		410	64	474	545
090	Counter top cook tops, 4 burner, standard, minimum	1 Elec	6		220	30	250	285
095	Maximum		3		430	60	490	560
105	As above, but with griddle, minimum		6		300	30	330	375
110	Maximum		3		520	60	580	660
125	Microwave oven, minimum		4		145	45	190	225
130	Maximum	↓	2		1,000	90	1,090	1,225
150	Combination range, refrigerator and sink, 30" wide, minimum	L-1	2		520	180	700	830
155	Maximum		1		1,180	360	1,540	1,825
157	60" wide, average		1.40		1,650	255	1,905	2,175
159	72" wide, average		1.20		1,870	300	2,170	2,500
160	Office model, 48" wide		2		1,470	180	1,650	1,875
162	Refrigerator and sink only	↓	2.40	↓	1,300	150	1,450	1,650
164	Combination range, refrigerator, sink, microwave							
166	oven and ice maker	L-1	.80	Ea.	3,175	450	3,625	4,150
175	Compactor, residential size, 4 to 1 compaction, minimum	1 Carp	5		340	32	372	420
180	Maximum	"	3		490	53	543	615
200	Deep freeze, 15 to 23 C.F., minimum	2 Clab	10		320	25	345	390
205	Maximum		5		660	51	711	800
220	30 C.F., minimum		8		650	32	682	760
225	Maximum	↓	3		750	85	835	950
245	Dehumidifier, portable, automatic, 15 pint				175		175	195M
255	30 pint				245		245	270M
275	Dishwasher, built-in, 2 cycles, minimum	L-1	4		270	90	360	425
280	Maximum		2		360	180	540	655
295	4 or more cycles, minimum		4		390	90	480	560
300	Maximum	↓	2		700	180	880	1,025
320	Dryer, automatic, minimum	L-2	3		250	95	345	415
325	Maximum	"	2		530	140	670	790
330	Garbage disposer, sink type, minimum	L-1	5		50	72	122	160
335	Maximum	"	3		190	120	310	380
355	Heater, electric, built-in, 1250 watt, ceiling type, minimum	1 Elec	4		40	45	85	110
360	Maximum	"	3	↓	100	60	160	195

11.1 Architectural Equipment

			CREW	DAILY OUTPUT	UNIT	BARE COSTS MAT.	BARE COSTS INST.	BARE COSTS TOTAL	TOTAL INCL O&P
03	370	Wall type, minimum	1 Elec	4	Ea.	38	45	83	105
	375	Maximum		3		75	60	135	170
	390	1500 watt wall type, with blower		4		75	45	120	145
	395	3000 watt	↓	3		130	60	190	230
	415	Hood for range, 2 speed, vented, 30" wide, minimum	L-3	5		40	68	108	145
	420	Maximum		3		230	115	345	415
	430	42" wide, minimum		5		155	68	223	270
	435	Maximum	↓	3		270	115	385	460
	450	For ventless hood, 2 speed, add				12		12	13.20M
	465	For vented 1 speed, deduct from maximum				25		25	28M
	485	Humidifier, portable, 7 gallons per day				80		80	88M
	500	15 gallons per day				160		160	175M
	520	Icemaker, automatic, 3 lb. per day	1 Plum	7		300	26	326	365
	535	35 lb. per day	"	2		725	90	815	930
	550	Refrigerator, no frost, 10 C.F. to 12 C.F. minimum	2 Clab	10		350	25	375	420
	560	Maximum		6		580	42	622	700
	575	14 C.F. to 16 C.F., minimum		9		500	28	528	590
	580	Maximum		5		700	51	751	845
	595	18 C.F. to 20 C.F., minimum		8		530	32	562	630
	600	Maximum		4		1,000	64	1,064	1,200
	615	21 C.F. to 29 C.F., minimum		7		850	36	886	990
	620	Maximum	↓	3		1,700	85	1,785	2,000
	640	Sump pump cellar drainer, 1/3 H.P., minimum	1 Plum	3		75	60	135	170
	645	Maximum		2		300	90	390	460
	665	Washing machine, automatic, minimum		3		300	60	360	415
	670	Maximum	↓	1		690	180	870	1,025
	690	Water heater, electric, glass lined, 30 gallon, minimum	L-1	5		125	72	197	240
	695	Maximum		3		260	120	380	460
	710	82 gallon, minimum		2		260	180	440	545
	715	Maximum	↓	1		440	360	800	1,000
	718	Water heater, gas, glass lined, 30 gallon, minimum	2 Plum	5		115	72	187	230
	722	Maximum		3		325	120	445	530
	726	50 gallon, minimum		2.50		185	145	330	410
	730	Maximum		1.50		450	240	690	845
	735	Water softener, automatic, to 30 grains per gallon		5		335	72	407	475
	740	To 75 grains per gallon	↓	4		655	90	745	850
	745	Vent kits for dryers	1 Carp	10	↓	12	16	28	36
06-001		AUTOMOTIVE Compressors, electric, 1-1/2 H.P., standard controls	L-4	1.50	Ea.	2,450	300	2,750	3,125
	055	Dual controls		1.50		2,650	300	2,950	3,350
	060	5 H.P., 115/230 volt, standard controls		1		3,250	450	3,700	4,225
	065	Dual controls		1		3,450	450	3,900	4,450
	100	Gasoline pumps, conventional, lighted, single		2.50		2,000	180	2,180	2,475
	101	Double		2		3,750	225	3,975	4,450
	220	Hoists, single post, 8000# capacity, swivel arms	↓	.40	↓	3,000	1,125	4,125	4,950
	221								
	240	Two posts, adjustable frames, 11,000# capacity	L-4	.25	Ea.	3,900	1,800	5,700	6,925
	250	24,000# capacity		.15		5,300	3,025	8,325	10,200
	270	7500# capacity, frame supports		.50		4,575	905	5,480	6,350
	280	Four post, roll on ramp		.50	↓	4,200	905	5,105	5,950
	300	Lube equipment, 3 reel type, with pumps, not including piping		.50	Set	5,200	905	6,105	7,050
	400	Spray painting booth, 26' long, complete	↓	.40	Ea.	9,200	1,125	10,325	11,800
09-001		BANK EQUIPMENT Alarm system, police	2 Elec	1.60		3,175	225	3,400	3,825
	010	With vault alarm	"	.40		12,600	895	13,495	15,100
	040	Bullet resistant teller window, 44" x 60"	1 Glaz	.60	↓	2,100	270	2,370	2,700
	050	48" x 60"	"	.60		2,525	270	2,795	3,150
	300	Counters for banks, frontal only	2 Carp	1	Station	1,250	320	1,570	1,850
	310	Complete with steel undercounter	"	.50	"	2,550	640	3,190	3,725
	460	Door and frame, bullet-resistant, with vision panel, minimum	2 Sswk	1.10	Ea.	1,700	315	2,015	2,375
	470	Maximum	"	1.10	"	2,500	315	2,815	3,250

11.1 Architectural Equipment

		Crew	Daily Output	Unit	Mat.	Inst.	Total	Total Incl O&P
09 480	Drive-up window, drawer & mike, not incl. glass, minimum	2 Sswk	1	Ea.	4,400	345	4,745	5,375
490	Maximum		.50		6,400	695	7,095	8,125
500	Night depository, with chest, minimum		1		7,000	345	7,345	8,250
510	Maximum		.50		10,000	695	10,695	12,100
520	Package receiver, painted		3.20		886	110	996	1,150
530	Stainless steel		3.20		1,500	110	1,610	1,825
540	Partitions, bullet-resistant, 1-3/16" glass, 8' high	2 Carp	10	L.F.	140	32	172	200
545	Acrylic	"	10	"	260	32	292	330
550	Pneumatic tube systems, 2 lane drive-up, complete	L-3	.25	Total	18,000	1,350	19,350	21,800
555	With T.V. viewer	"	.20	"	34,900	1,700	36,600	40,900
557	Safety deposit boxes, minimum	1 Sswk	44	Opng.	40	3.95	43.95	50
558	Maximum, 10" x 15" opening		19		85	9.15	94.15	110
559	Teller locker, average		15		1,100	11.55	1,111.55	1,225
560	Pass thru, bullet-resist. window, painted steel, 24" x 36"	2 Sswk	1.60	Ea.	1,100	215	1,315	1,550
570	48" x 48"		1.20		1,800	290	2,090	2,425
580	72" x 40"		.80		2,200	435	2,635	3,100
590	For stainless steel frames, add				20%			
600	Surveillance system, 16mm film camera, complete	2 Elec	1		3,400	360	3,760	4,250
601	For each additional camera, add				1,500		1,500	1,650M
610	Surveillance system, video camera, complete	2 Elec	1	Ea.	11,000	360	11,360	12,600
611	For each additional camera, add				700		700	770M
620	Twenty-four hour teller, single unit,							
630	automated deposit, cash and memo	L-3	.25	Ea.	33,000	1,350	34,350	38,300
700	Vault front, see division 8.3-69							
10-001	**BARBER EQUIPMENT** Chair, hydraulic, minimum	1 Carp	1.30	Ea.	400	125	525	620
005	Maximum		1		2,000	160	2,160	2,425
020	Console including mirrors, minimum		2		250	80	330	390
030	Maximum		.80		1,500	200	1,700	1,950
050	Sink, hair washing basin	1 Plum	1		315	180	495	605
100	Sterilizer, liquid solution for tools				80		80	88M
110	Total equipment, rule of thumb, per chair, minimum	L-8	1		1,500	410	1,910	2,250
115	Maximum	"	1		3,500	410	3,910	4,450
12-001	**CHURCH EQUIPMENT** Altar, wood, custom design, plain	1 Carp	1.40		1,150	115	1,265	1,425
005	Deluxe	"	.20		6,300	800	7,100	8,100
007	Granite or marble, average	2 Marb	.50		4,500	645	5,145	5,875
009	Deluxe	"	.20		10,000	1,600	11,600	13,300
010	Arks, prefabricated, plain	2 Carp	.80		2,000	400	2,400	2,775
013	Deluxe, maximum	"	.20		94,000	1,600	95,600	105,500
015	Baptistry, fiberglass, 3'-6" deep, x 13'-7" long,							
016	steps at both ends, incl. plumbing, minimum	L-8	1	Ea.	1,350	410	1,760	2,075
020	Maximum	"	.70		3,100	585	3,685	4,250
025	Add for filter, heater and lights				950		950	1,050M
030	Carillon, 4 octave (48 bells), with keyboard			System	230,000		230,000	253,000
032	2 octave (24 bells)				115,000		115,000	126,500
034	3 to 4 bell peal, minimum				13,750		13,750	15,100M
036	Maximum				36,500		36,500	40,200M
038	Cast bronze bell, average			Ea.	6,500		6,500	7,150M
045								
050	Confessional, wood, prefabricated, single, plain	1 Carp	.60	Ea.	2,750	265	3,015	3,400
055	Deluxe		.40		5,975	400	6,375	7,150
065	Double, plain		.40		5,675	400	6,075	6,825
070	Deluxe		.20		13,700	800	14,500	16,200
100	Lecterns, wood, plain		5		340	32	372	420
110	Deluxe		2		1,100	80	1,180	1,325
150	Pews, bench type, hardwood, minimum		20	L.F.	44	8	52	60
155	Maximum		15	"	71	10.65	81.65	94
200	Pulpits, hardwood, prefabricated, plain		2	Ea.	880	80	960	1,075
210	Deluxe		1.60	"	5,775	100	5,875	6,500
250	Railing, hardwood, average		25	L.F.	60	6.40	66.40	75
270	Safes, see division 11.1-43							

11.1 Architectural Equipment

		CREW	DAILY OUTPUT	UNIT	BARE COSTS MAT.	BARE COSTS INST.	BARE COSTS TOTAL	TOTAL INCL O&P
300	Seating, individual, oak, contour, laminated	1 Carp	21	Person	70	7.60	77.60	88
310	Cushion seat		21		80	7.60	87.60	99
320	Fully upholstered		21		85	7.60	92.60	105
330	Combination, self-rising	↓	21		110	7.60	117.60	130
340	For mahogany and maple, add				20%			
350	For black walnut and cherry, add			↓	40%			
400	Steeples, translucent fiberglass, 30" square, 15' high	F-3	2	Ea.	1,650	530	2,180	2,550
415	25' high		1.80		2,600	590	3,190	3,675
435	Painted fiberglass, 24" square, 14' high		2		1,700	530	2,230	2,600
450	28' high	↓	1.80		2,600	590	3,190	3,675
460	Aluminum, baked finish, 14' high, 16" square				1,000		1,000	1,100M
462	20' high, 3'-6" base				3,550		3,550	3,900M
464	35' high, 8' base				13,500		13,500	14,900M
466	60' high, 14' base				30,000		30,000	33,000M
468	152' high, custom			↓	300,000		300,000	330,000M
469								
470	Porcelain enamel steeples, custom, 40' high	F-3	.50	Ea.	9,750	2,125	11,875	13,600
480	60' high	"	.30	"	16,900	3,525	20,425	23,400
500	Wall cross, aluminum, extruded, 2" x 2"	1 Carp	34	L.F.	27	4.71	31.71	37
515	4" x 4"		29		39	5.50	44.50	51
530	Bronze, extruded, 1" x 2"		31		52	5.15	57.15	65
535	2-1/2" x 2-1/2"		34		80	4.71	84.71	95
545	Solid bar stock, 1/2" x 3"		29		105	5.50	110.50	125
560	Fiberglass, stock		34		20	4.71	24.71	29
570	Stainless steel, 4" deep, open back		29		85	5.50	90.50	100
580	Closed back	↓	29	↓	105	5.50	110.50	125
13-001	**CHECKOUT COUNTER** Supermarket conveyor, single belt	2 Clab	10	Ea.	1,800	25	1,825	2,025
010	Double belt		9		2,200	28	2,228	2,450
040	Power take-away		7		3,000	36	3,036	3,350
080	Warehouse or bulk type	↓	6	↓	4,500	42	4,542	5,000
090								
100	Scanning system, 2 lanes, w/registers, scanning guns and memory			System	12,000		12,000	13,200M
110	10 lanes, single processor, full scan				70,000		70,000	77,000M
115	With duplex/back-up processor			↓	125,000		125,000	137,500
200	Register, restaurant, minimum			Ea.	1,300		1,300	1,425M
210	Maximum				4,500		4,500	4,950M
215	Store, minimum				900		900	990M
220	Maximum			↓	3,500		3,500	3,850M
14-001	**CONTRACTOR EQUIPMENT** See division 1.5							
15-001	**DARKROOM EQUIPMENT** Developing tanks, 5" deep, 24" x 48"	Q-1	2	Ea.	860	160	1,020	1,175
005	48" x 52"		1.70		1,545	190	1,735	1,975
020	10" deep, 24" x 48"		1.70		970	190	1,160	1,350
025	24" x 108"	↓	1.50		1,750	215	1,965	2,250
050	Dryers, dehumidified filtered air, 36" x 25" x 68" high	L-7	6		3,100	89	3,189	3,550
055	48" x 25" x 68" high		5		3,300	105	3,405	3,775
200	Processors, automatic, color print, minimum		4		8,250	135	8,385	9,275
205	Maximum		1		56,600	535	57,135	63,000
230	Black and white print, minimum		4		6,180	135	6,315	7,000
235	Maximum		1		33,000	535	33,535	37,100
260	Manual processor, 16" x 20" maximum print size		4		4,650	135	4,785	5,300
265	20" x 24" maximum print size		1		5,800	535	6,335	7,150
300	Viewing lites, 20" x 24"		6		210	89	299	360
310	20" x 24" with color correction	↓	6		320	89	409	480
350	Washers, round, maximum sheet 11" x 14"	Q-1	2		1,080	160	1,240	1,425
355	Maximum sheet 20" x 24"		1		1,280	325	1,605	1,875
380	Square, maximum sheet 20" x 24"		1		1,170	325	1,495	1,750
390	Maximum sheet 50" x 56"	↓	.80	↓	1,850	405	2,255	2,625
450	Combination tank sink, tray sink, washers, with							
451	dry side tables, average	Q-1	1	Ea.	3,800	325	4,125	4,650

11.1 Architectural Equipment

		CREW	DAILY OUTPUT	UNIT	BARE COSTS MAT.	BARE COSTS INST.	BARE COSTS TOTAL	TOTAL INCL O&P
16-001	**DENTAL EQUIPMENT** Central suction system, minimum	1 Plum	1.20	Ea.	1,475	150	1,625	1,850
010	Maximum	"	.90		3,675	200	3,875	4,325
030	Air compressor, minimum	1 Skwk	.80		1,250	205	1,455	1,675
040	Maximum		.50		4,000	330	4,330	4,875
060	Chair, electric or hydraulic, minimum		.50		2,500	330	2,830	3,225
070	Maximum	↓	.25		5,925	655	6,580	7,475
080	Doctor's/assistant's stool, minimum				260		260	285M
085	Maximum				600		600	660M
100	Drill console with accessories, minimum	1 Skwk	.50		2,100	330	2,430	2,800
110	Maximum		.33		7,400	495	7,895	8,875
200	Light, floor or ceiling mounted, minimum		3.60		800	46	846	945
210	Maximum	↓	1.20		1,575	135	1,710	1,925
230	Sterilizers, steam portable, minimum				1,300		1,300	1,425M
235	Maximum				2,200		2,200	2,425M
260	Steam, institutional				8,425		8,425	9,275M
265	Dry heat, electric, portable, 3 trays				585		585	645M
270	Ultra-sonic cleaner, portable, minimum				300		300	330M
275	Maximum (institutional)				1,300		1,300	1,425M
300	X-ray unit, wall	1 Skwk	1.90		3,975	86	4,061	4,500
310	Panoramic unit		.60		15,150	275	15,425	17,100
350	Developers, X-ray, minimum		1		1,300	165	1,465	1,675
360	Maximum	↓	1	↓	3,800	165	3,965	4,425
19-001	**DETENTION EQUIPMENT** Bar front, rolling, 7/8" bars,							
050	4" O.C., 7' high, 5' wide, with hardware	E-4	2	Ea.	3,275	385	3,660	4,200
100	Doors & frames, 3' x 7', complete, with hardware, single plate		4		2,450	190	2,640	3,000
165	Double plate	↓	4	↓	3,075	190	3,265	3,675
200	Cells, prefab., 5' to 6' wide, 7' to 8' high, 7' to 8' deep,							
201	bar front, cot, not incl. plumbing	E-4	1.50	Ea.	4,450	510	4,960	5,675
250	Cot, bolted, single, painted steel		20		240	38	278	325
270	Stainless steel	↓	20		550	38	588	665
300	Toilet apparatus including wash basin, average	L-8	1.50		1,600	275	1,875	2,150
400	Visitor cubicle, vision panel, no intercom	E-4	2		1,775	385	2,160	2,550
22-001	**LOADING DOCK** Bumpers, rubber blocks 4-1/2" thick, 10" high, 14" long	1 Carp	26		40	6.15	46.15	53
020	24" long		22		49	7.25	56.25	64
030	36" long		17		55	9.40	64.40	74
050	12" high, 14" long		25		45	6.40	51.40	59
055	24" long		20		53	8	61	70
060	36" long		15		63	10.65	73.65	85
080	Rubber blocks 6" thick, 10" high, 14" long		22		54	7.25	61.25	70
085	24" long		18		63	8.90	71.90	82
090	36" long		13		76	12.30	88.30	100
091	20" high, 11" long		13		76	12.30	88.30	100
092	Extruded rubber bumpers, T section, 22" x 22" x 3" thick		41		58	3.90	61.90	69
094	Molded rubber bumpers, 24" x 12" x 3" thick	↓	20		40	8	48	56
100	Welded installation of above bumpers	E-14	8		2	31	33	47
110	For drilled anchors, add per anchor	1 Carp	36	↓	3.60	4.44	8.04	10.40
130	Steel bumpers, see division 5.4-12, 14 & 26							
135	Wood bumpers, see division 6.1-10							
220	Dock boards, heavy duty, 60" x 60", aluminum, 5000 lb. cap.			Ea.	880		880	970M
270	9000 lb. capacity				1,000		1,000	1,100M
320	15,000 lb. capacity			↓	1,200		1,200	1,325M
360	Door seal for door perimeter, 12" x 12", vinyl covered	1 Carp	26	L.F.	22	6.15	28.15	33
390	Folding gates, see division 10.1-72							
420	Platform lifter, 6' x 6', portable, 3000 lb. capacity			Ea.	4,800		4,800	5,275M
425	4000 lb. capacity				5,200		5,200	5,725M
440	Fixed, 6' x 8', 5000 lb. capacity	L-4	.70		5,400	645	6,045	6,875
450	Levelers, hinged for trucks, 10 ton capacity, 6' x 8'		1.90		2,700	240	2,940	3,325
465	7' x 8'		1.90		2,900	240	3,140	3,550
470	Hydraulic, 10 ton capacity, 6' x 8'		1.90		3,850	240	4,090	4,575
480	7' x 8'	↓	1.90	↓	4,050	240	4,290	4,800

11.1 Architectural Equipment

			CREW	DAILY OUTPUT	UNIT	BARE COSTS MAT.	BARE COSTS INST.	BARE COSTS TOTAL	TOTAL INCL O&P
22	500	Lights for loading docks, single arm, 24" long	1 Elec	3.80	Ea.	100	47	147	180
	570	Double arm, 60" long	"	3.80		140	47	187	220
	620	Shelters, fabric, for truck or train, scissor arms, minimum	1 Carp	1		750	160	910	1,050
	630	Maximum	"	.50		1,300	320	1,620	1,900
25-001		EQUIPMENT INSTALLATION Industrial equipment, minimum	E-2	12	Ton		165	165	225
	020	Maximum	"	2	"		990	990	1,350
27-001		HEALTH CLUB EQUIPMENT Abdominal rack, 2 board capacity			Ea.	450		450	495M
	005	Abdominal board, upholstered				460		460	505M
	020	Bicycle trainer, minimum				430		430	475M
	030	Deluxe, electric				1,800		1,800	1,975M
	040	Bar bell set, chrome plated steel, 25 lbs.				185		185	205M
	042	100 lbs.				400		400	440M
	045	200 lbs.				700		700	770M
	050	Weight plates, cast iron, per lb.			Lb.	3		3	3.30M
	052	Storage rack, 10 station			Ea.	750		750	825M
	060	Circuit training apparatus, 12 machines minimum	2 Clab	1.25	Set	20,000	205	20,205	22,300
	070	Average		1		28,200	255	28,455	31,400
	080	Maximum		.75		35,000	340	35,340	39,000
	082	Dumbell set, cast iron, with rack and 5 pair				545		545	600M
	090	Squat racks	2 Clab	5	Ea.	320	51	371	425
	120	Multi-station gym machine, 5 station				8,500		8,500	9,350M
	125	9 station				13,150		13,150	14,500M
	128	Rowing machine, hydraulic				1,225		1,225	1,350M
	129								
	130	Treadmill, manual			Ea.	600		600	660M
	132	Motorized				1,495		1,495	1,650M
	134	Electronic				3,200		3,200	3,525M
	136	Cardio-testing				5,250		5,250	5,775M
	140	Treatment/massage tables, minimum				305		305	335M
	142	Deluxe, with accessories				860		860	945M
	160	For saunas, see division 11.1-46							
	164	For steam baths, see division 11.1-58							
	168	For whirlpool baths, see division 15.2-04							
	172	For additional exercise equipment, see div. 11.1-49							
28-001		KITCHEN EQUIPMENT Bake oven, single deck	Q-1	8	Ea.	1,630	41	1,671	1,850
	030	Double deck		7		3,190	46	3,236	3,575
	060	Triple deck		6		4,760	54	4,814	5,325
	090	Electric convection, 40" x 45" x 57"	L-7	4		2,430	135	2,565	2,875
	130	Broiler, without oven, standard	Q-1	8		2,330	41	2,371	2,625
	155	Infra-red	L-7	4		3,270	135	3,405	3,800
	185	Coffee urns, twin 6 gallon urns				4,000		4,000	4,400M
	186								
	235	Cooler, reach-in, beverage, 6' long	Q-1	6	Ea.	2,200	54	2,254	2,500
	236								
	270	Dishwasher, commercial, rack type							
	272	10 to 12 racks per hour	Q-1	3.20	Ea.	2,000	100	2,100	2,350
	275	Semi-automatic 38 to 50 racks per hour	"	1.30		4,200	250	4,450	4,975
	280	Automatic 190 to 230 racks per hour	Q-2	.70		9,000	720	9,720	10,900
	282	235 to 275 racks per hour		.50		13,000	1,000	14,000	15,800
	284	8750 to 12,500 dishes per hour		.20		33,900	2,525	36,425	40,900
	300	Fast food equipment, total package, minimum	6 Skwk	.08		90,000	12,300	102,300	117,000
	310	Maximum	"	.07		130,000	14,100	144,100	163,500
	330	Food warmer, counter, 1.2 KW				480		480	530M
	355	1.65 KW				550		550	605M
	380	Food mixers, 20 quarts	L-7	7		2,600	77	2,677	2,975
	400	60 quarts	"	5		8,100	105	8,205	9,075
	430	Freezers, reach-in, 40 C.F.	Q-1	4		4,200	81	4,281	4,725
	450	70 C.F.		3		5,100	110	5,210	5,775
	475	Fryer, with submerger, single		7		2,650	46	2,696	2,975
	500	Double		5		3,225	65	3,290	3,650

11.1 Architectural Equipment

		Description	CREW	DAILY OUTPUT	UNIT	BARE COSTS MAT.	BARE COSTS INST.	BARE COSTS TOTAL	TOTAL INCL O&P
28	530	Griddle, 3' long	Q-1	7	Ea.	2,675	46	2,721	3,000
	555	4' long		6		3,250	54	3,304	3,650
	580	Ice cube maker, 50 pounds per day		6		1,375	54	1,429	1,600
	605	500 pounds per day		4		3,500	81	3,581	3,975
	635	Kettles, steam-jacketed, 20 gallons	L-7	7		3,900	77	3,977	4,400
	660	60 gallons	"	6		4,800	89	4,889	5,400
	690	Range, restaurant type, 6 burners and 1 oven, 36"	Q-1	7		1,775	46	1,821	2,025
	715	2 ovens, 60"		6		2,850	54	2,904	3,225
	745	Heavy duty, single 34" oven, open top		5		2,500	65	2,565	2,850
	770	Fry top		6		2,570	54	2,624	2,900
	795	Hood fire protection system, minimum		3		2,300	110	2,410	2,675
	805	Maximum		1		17,000	325	17,325	19,200
	830	Refrigerators, reach-in type, 44 C.F.		5		2,760	65	2,825	3,125
	855	With glass doors, 68 C.F.		4		4,100	81	4,181	4,625
	885	Steamer, electric 27 KW	L-7	7		5,300	77	5,377	5,950
	910	Electric, 10 KW or gas 100,000 BTU	"	5		2,800	105	2,905	3,225
	915	Toaster, conveyor type, 16-22 slices per minute				1,100		1,100	1,200M
	920	For deluxe models of above equipment, add				75%			
	940	Rule of thumb: Equipment cost based							
	941	on kitchen work area							
	942	Office buildings, minimum	L-7	77	S.F.	35	6.95	41.95	49
	945	Maximum		58		58	9.25	67.25	77
	955	Public eating facilities, minimum		77		46	6.95	52.95	61
	960	Maximum		46		73	11.65	84.65	97
	975	Hospitals, minimum		58		47	9.25	56.25	65
	980	Maximum		39		78	13.75	91.75	105
31	-001	LABORATORY EQUIPMENT Cabinets, base, door units, metal	2 Carp	18	L.F.	94	17.80	111.80	130
	030	Drawer units		18		180	17.80	197.80	225
	070	Tall storage cabinets, open, 7' high		20		110	16	126	145
	090	With glazed doors		20		210	16	226	255
	130	Wall cabinets, metal, 12-1/2" deep, open		20		59	16	75	88
	150	With doors		20		115	16	131	150
	155	Counter tops, not incl. base cabinets, acidproof, minimum		82	S.F.	14	3.90	17.90	21
	160	Maximum		70		31	4.57	35.57	41
	165	Stainless steel		82		42	3.90	45.90	52
	175								
	200	Fume hood, with countertop & base, not including HVAC							
	202	Simple, minimum	2 Carp	5.40	L.F.	470	59	529	605
	205	Complex, including fixtures		2.40		730	135	865	995
	210	Special, maximum		1.70		945	190	1,135	1,300
	220	Ductwork, minimum	2 Shee	1	Hood	605	365	970	1,200
	225	Maximum	"	.50	"	3,225	725	3,950	4,600
	230	Service fixtures, average			Ea.	31		31	34M
	250	For sink assembly with hot and cold water, add	1 Plum	1.40		340	130	470	560
	255	Glassware washer, distilled water rinse, minimum	L-1	1.80		2,375	200	2,575	2,900
	260	Maximum	"	1		14,000	360	14,360	15,900
	265	Glove box, fiberglass, bacteriological				3,325		3,325	3,650M
	270	Controlled atmosphere				3,475		3,475	3,825M
	275	Radioisotope				4,575		4,575	5,025M
	277								
	280	Sink, one piece plastic, flask wash, hose, free standing	1 Plum	1.60	Ea.	910	115	1,025	1,175
	285	Epoxy resin sink, 25" x 16" x 10"	"	2		205	90	295	355
	300	Utility table, acid resistant top with drawers	2 Carp	30	L.F.	71	10.65	81.65	94
	380	Titration unit, four 2000 ml. reservoirs			Ea.	3,650		3,650	4,025M
	420	Alternate pricing method: as percent of lab furniture							
	440	Installation, not incl. plumbing & duct work			% Furn.				22%
	480	Plumbing, final connections, simple system							10%
	500	Moderately complex system							15%
	520	Complex system							20%
	540	Electrical, simple system							10%

11.1 Architectural Equipment

			Crew	Daily Output	Unit	Bare Costs Mat.	Bare Costs Inst.	Bare Costs Total	Total Incl O&P
31	560	Moderately complex system			% Furn.				20%
	580	Complex system			"				35%
	600	Safety equipment, eye wash, hand held			Ea.	185		185	205M
	620	Deluge shower			"	110		110	120M
	630	Rule of thumb: lab furniture including installation & connection							
	632	High school			S.F.				21
	634	College							31
	636	Clinical, health care							26
	638	Industrial			↓				42
32-001		**LAUNDRY EQUIPMENT** Not incl. rough-in. Dryers, gas fired							
	050	Residential, 16 lb. capacity, average	1 Plum	3	Ea.	420	60	480	550
	100	Commercial, 30 lb. capacity, coin operated, single		3		1,600	60	1,660	1,850
	110	Double stacked		2		1,300	90	1,390	1,550
	150	Industrial, 30 lb. capacity		3		1,600	60	1,660	1,850
	160	50 lb. capacity		3		1,950	60	2,010	2,225
	200	Dry cleaners, electric, 20 lb. capacity		1		18,000	180	18,180	20,100
	205	25 lb. capacity		.80		20,000	225	20,225	22,300
	210	30 lb. capacity	↓	.20		25,000	900	25,900	28,800
	300	Extractors, industrial, 200 lb. capacity	L-6	.20		32,000	1,350	33,350	37,100
	350	Folders, blankets & sheets, minimum	1 Elec	.30		13,500	595	14,095	15,700
	370	King size with automatic stacker		.30		29,000	595	29,595	32,800
	380	For conveyor delivery, add		.30		4,500	595	5,095	5,800
	450	Ironers, institutional, 110", single roll	↓	.30		10,000	595	10,595	11,900
	470	Lint collector, ductwork not included, 8,000 to 10,000 C.F.M.	Q-10	.30		5,900	1,700	7,600	8,975
	500	Washers, residential, 4 cycle, average	1 Plum	3		560	60	620	705
	530	Commercial, coin operated, average	"	3		650	60	710	800
	600	Combination washer/extractor, 20 lb. capacity	L-6	2		2,200	135	2,335	2,625
	610	30 lb. capacity		.80		5,500	340	5,840	6,525
	620	50 lb. capacity		.80		6,500	340	6,840	7,625
	630	75 lb. capacity		.40		12,000	675	12,675	14,200
	635	125 lb. capacity	↓	.40	↓	19,000	675	19,675	21,900
34-001		**LIBRARY EQUIPMENT** Bookshelf, metal, 90" high, 10" shelf, double face	1 Carp	12	L.F.	60	13.35	73.35	85
	030	Single face	"	12		45	13.35	58.35	69
	060	For 8" shelving, subtract from above			↓	10%			
	070	For 12" shelving, add to above				10%			
	080	For 42" high with countertop, subtract from above			L.F.	30%			
	100								
	110	Magazine shelving, 90" high, 12" deep, single face, add			L.F.	57		57	63M
	140	Double face, add			"	105		105	115M
	250	Carrels, hardwood, 36" x 24", minimum	1 Carp	5	Ea.	425	32	457	515
	265	Maximum	"	4	↓	600	40	640	720
	270	Card catalog file, 60 trays, complete				3,100		3,100	3,400M
	272	Alternate method: each tray				50		50	55M
	310	Chairs, wood			↓	80		80	88M
	350	Charging desk, built-in, with counter, plastic laminated top	1 Carp	7	L.F.	300	23	323	365
	370	Reading table, laminated top, 60" x 36"			Ea.	460		460	505M
	375								
	380	Mobile compacted shelving, hand crank, 9'-0" high							
	382	Double face, including track, 3' section			Ea.	600		600	660M
	384	For electrical operation, add				25%			
37-001		**MEDICAL EQUIPMENT** Autopsy table, standard	1 Plum	1	Ea.	5,500	180	5,680	6,300
	020	Deluxe	"	.60	"	7,500	300	7,800	8,675
	030	Blood pressure unit, mercurial, wall				100		100	110M
	040	Diagnostic set, wall				500		500	550M
	070	Distiller, water, steam heated, 50 gal. capacity	1 Plum	1.40	Ea.	12,500	130	12,630	13,900
	071								
	075	Exam room furnishings, average per room			Ea.	3,000		3,000	3,300M
	180	Heat therapy unit, humidified, 26" x 78" x 28"			"	1,600		1,600	1,750M

11.1 Architectural Equipment

		Description	CREW	DAILY OUTPUT	UNIT	BARE COSTS MAT.	BARE COSTS INST.	BARE COSTS TOTAL	TOTAL INCL O&P
37	210	Hubbard tank with accessories, stainless steel,							
	211	125 GPM at 45 psi water pressure			Ea.	12,000		12,000	13,200M
	215	For electric overhead hoist, add			"	1,500		1,500	1,650M
	235								
	250	Incubators, minimum			Ea.	2,600		2,600	2,850M
	260	Maximum				9,000		9,000	9,900M
	290	K-Module for heat therapy, 20 oz. capacity, 75° to 110°F				220		220	240M
	291								
	320	Mortuary refrigerator, end operated, 2 capacity			Ea.	10,000		10,000	11,000M
	330	6 capacity				16,300		16,300	17,900M
	360	Paraffin bath, 126°F, auto controlled				1,200		1,200	1,325M
	390	Parallel bars for walking training, 12'-0"				500		500	550M
	420	Refrigerator, blood bank, 28.6 C.F. emergency signal				4,050		4,050	4,450M
	421	Reach-in, 16.9 C.F.				4,000		4,000	4,400M
	440	Scale, physician's, with height rod				225		225	250M
	450								
	460	Station, dietary, medium, with ice			Ea.	10,200		10,200	11,200M
	470	Medicine				4,600		4,600	5,050M
	500	Scrub, surgical, minimum				3,200		3,200	3,525M
	510	Maximum				4,300		4,300	4,725M
	560	Sterilizers, floor loading, 28" x 67" x 52", single door, steam				85,000		85,000	93,500M
	565	Double door, steam				130,000		130,000	143,000
	580	General purpose, 20" x 20" x 28", single door				6,000		6,000	6,600M
	600	Portable, counter top, steam, minimum				1,000		1,000	1,100M
	602	Maximum				3,000		3,000	3,300M
	605	Portable, counter top, gas				10,000		10,000	11,000M
	615	Automatic washer/sterilizer	1 Plum	2		21,500	90	21,590	23,800
	620	Steam generators, electric 10 KW to 180 KW							
	625	Minimum	1 Elec	3	Ea.	2,100	60	2,160	2,400
	630	Maximum	"	.70		18,000	255	18,255	20,200
	650	Surgery table, minor minimum	1 Sswk	.70		2,500	250	2,750	3,150
	652	Maximum		.50		7,000	345	7,345	8,250
	655	Major surgery table, minimum		.50		18,000	345	18,345	20,300
	657	Maximum		.50		60,000	345	60,345	66,500
	670	Surgical lights, single arm	2 Elec	.90		1,100	400	1,500	1,775
	675	Dual arm	"	.30		2,600	1,200	3,800	4,575
	678								
	700	Tables, physical therapy, walk off, electric	2 Carp	3	Ea.	1,600	105	1,705	1,925
	715	Standard, vinyl top with base cabinets, minimum		3		750	105	855	980
	720	Maximum		2		2,000	160	2,160	2,425
	750	Thermometer, electric, portable				270		270	295M
	810	Utensil washer-sanitizer	1 Plum	2		4,800	90	4,890	5,400
	840	Whirlpool bath, mobile, 18" x 24" x 60"				2,400		2,400	2,650M
	845	Fixed, incl. mixing valves	1 Plum	2		2,800	90	2,890	3,200
	870	X-ray, mobile, minimum				6,000		6,000	6,600M
	875	Maximum				40,000		40,000	44,000M
	890	Stationary, minimum				25,000		25,000	27,500M
	895	Maximum				80,000		80,000	88,000M
	915	Developing processors, minimum				4,000		4,000	4,400M
	920	Maximum				19,000		19,000	20,900M
	980	For hospital partitions, see division 10.1-50							
	990	For casework & furniture, see division 12.1-15 & 45							
38	001	**MOVIE EQUIPMENT** Changeover, minimum			Ea.	175		175	195M
	010	Maximum				350		350	385M
	040	Film transport, incl. platters and autowind, minimum				6,300		6,300	6,925M
	050	Maximum				8,000		8,000	8,800M
	080	Lamphouses, incl. rectifiers, xenon, 1000 watt	1 Elec	2		4,800	90	4,890	5,400
	090	1600 watt		2		5,800	90	5,890	6,500
	100	2000 watt		1.50		6,500	120	6,620	7,325
	110	4000 watt		1.50		9,200	120	9,320	10,300

11.1 Architectural Equipment

		Description	CREW	DAILY OUTPUT	UNIT	BARE COSTS MAT.	BARE COSTS INST.	BARE COSTS TOTAL	TOTAL INCL O&P
38	140	Lenses, anamorphic, minimum			Ea.	875		875	965M
	150	Maximum				2,000		2,000	2,200M
	180	Flat 35 mm, minimum				300		300	330M
	190	Maximum				1,000		1,000	1,100M
	220	Pedestals, for projectors, minimum				1,200		1,200	1,325M
	230	Console type, maximum				7,800		7,800	8,575M
	260	Projector mechanisms, 35 mm, minimum				5,300		5,300	5,825M
	270	Maximum				8,900		8,900	9,800M
	300	Projection screens, rigid, in wall, acrylic, 1/4" thick	2 Glaz	195	S.F.	30	1.65	31.65	35
	310	1/2" thick	"	130	"	35	2.48	37.48	42
	330	Electric operated, heavy duty, 400 S.F.	2 Carp	1	Ea.	1,850	320	2,170	2,500
	340	Also see division 10.1-67							
	370	Sound systems, incl. amplifier, single system, minimum	1 Elec	.90	Ea.	1,600	200	1,800	2,050
	380	Maximum		.40		1,900	450	2,350	2,725
	410	Dual system, minimum		.70		3,000	255	3,255	3,675
	420	Maximum		.40		3,600	450	4,050	4,600
	450	Sound heads, 35 mm, minimum				2,900		2,900	3,200M
	460	Maximum				3,200		3,200	3,525M
	490	Splicer, wet type, minimum				250		250	275M
	500	Maximum				400		400	440M
	530	Speakers, recessed behind screen, minimum	1 Elec	2		800	90	890	1,000
	540	Maximum	"	1		2,225	180	2,405	2,700
	570	Seating, painted steel, upholstered, minimum	2 Carp	35		80	9.15	89.15	100
	580	Maximum	"	28		140	11.45	151.45	170
	610	Rewind tables, minimum				900		900	990M
	620	Maximum				1,400		1,400	1,550M
	700	For automation, varying sophistication, minimum	1 Elec	1	System	850	180	1,030	1,200
	710	Maximum	2 Elec	.30	"	4,000	1,200	5,200	6,125
40-001		PARKING EQUIPMENT Traffic, detectors, magnetic		2.70	Ea.	355	135	490	580
	020	Single treadle		2.40		615	150	765	890
	050	Automatic gates, 8' arm, one way		1.10		2,350	325	2,675	3,050
	065	Two way		1.10		2,400	325	2,725	3,100
	100	Booth for attendant, minimum				2,600		2,600	2,850M
	105	Average				3,650		3,650	4,025M
	115	Maximum				5,200		5,200	5,725M
	140	Fee indicator, 1" display	2 Elec	4.10		1,150	87	1,237	1,400
	350	Ticket printer and dispenser, standard		1.40		4,050	255	4,305	4,825
	370	Rate computing		1.40		4,150	255	4,405	4,925
	400	Card control station, single period		4.10		415	87	502	580
	420	4 period		4.10		435	87	522	605
	450	Key station on pedestal		4.10		285	87	372	440
	475	Coin station, multiple coins		4.10		2,075	87	2,162	2,400
42-001		REFRIGERATED FOOD CASES Dairy, multi-deck, 12' long	Q-5	3		4,690	110	4,800	5,325
	010	For rear sliding doors, add				700		700	770M
	020	Delicatessen case, service deli, 12' long, single deck	Q-5	3.90		5,100	84	5,184	5,725
	030	Multi-deck, 18 S.F. shelf display		3		5,675	110	5,785	6,400
	040	Freezer, self-contained, chest-type, 30 C.F.		3.90		2,180	84	2,264	2,525
	050	Glass door, upright, 78 C.F.		3.30		4,250	99	4,349	4,825
	060	Frozen food, chest type, 12' long		3.30		3,800	99	3,899	4,325
	070	Glass door, reach-in, 5 door		3		6,800	110	6,910	7,650
	080	Island case, 12' long, single deck		3.30		4,100	99	4,199	4,650
	090	Multi-deck		3		8,600	110	8,710	9,625
	100	Meat case, 12' long, single deck		3.30		2,900	99	2,999	3,325
	105	Multi-deck		3.10		3,200	105	3,305	3,675
	110	Produce, 12' long, single deck		3.30		3,150	99	3,249	3,600
	120	Multi-deck		3.10		3,600	105	3,705	4,125
43-001		SAFE Office, 4 hr. rating, 30" x 18" x 18" inside				1,925		1,925	2,125M
	010	62" x 33" x 20"				4,300		4,300	4,725M
	020	1 hr. rating, 34" x 20" x 20"				1,500		1,500	1,650M
	030	62" x 33" x 20", double door				3,870		3,870	4,250M

11.1 Architectural Equipment

		CREW	DAILY OUTPUT	UNIT	BARE COSTS MAT.	BARE COSTS INST.	BARE COSTS TOTAL	TOTAL INCL O&P
43 040	Data, 4 hr. rating, 23-1/2" x 19-1/2" x 17" inside			Ea.	2,760		2,760	3,025M
050	66-3/4" x 36-3/4" x 17", double door				5,600		5,600	6,150M
060	1 hr. rating, 27" x 19" x 16"				2,100		2,100	2,300M
070	63" x 34" x 16"				4,700		4,700	5,175M
075	Diskette, 1 hr., 9" x 12" x 13" (inside)				1,500		1,500	1,650M
080	Money, "B" label, 9" x 14" x 14"				410		410	450M
090	Tool resistive, 24" x 24" x 20"				2,800		2,800	3,075M
100	Tool and torch resistive, 24" x 24" x 20"				7,000		7,000	7,700M
110	Jewelers, 23" x 20" x 18"				9,500		9,500	10,500M
120	63" x 25" x 18"				19,000		19,000	20,900M
130	For handling into building, add, minimum	A-2	8.50			52	52	73
140	Maximum	"	.78			565	565	795
46-001	SAUNA Prefabricated, incl. heater & controls, 7' high, 6' x 4'	L-7	2.20		2,400	245	2,645	3,000
040	6' x 5'		2		2,675	270	2,945	3,325
060	6' x 6'		1.80		2,975	300	3,275	3,700
080	6' x 9'		1.60		3,775	335	4,110	4,650
100	8' x 12'		1.10		5,225	490	5,715	6,450
120	8' x 8'		1.40		4,100	385	4,485	5,075
140	8' x 10'		1.20		4,650	445	5,095	5,775
160	10' x 12'		1		5,850	535	6,385	7,200
170	Door only, with tempered insulated glass window	2 Carp	3.40		150	94	244	300
180	Prehung, incl. jambs, pulls & hardware	"	12		190	27	217	250
250	Heaters only (incl. above), wall mounted, to 200 C.F.				355		355	390M
275	To 300 C.F.				545		545	600M
300	Floor standing, to 720 C.F., 10,000 watts	1 Elec	3		680	60	740	835
325	To 1,000 C.F., 12,500 watts	"	3		740	60	800	900
49-001	SCHOOL EQUIPMENT For exterior equipment see division 2.7							
020	For chairs & desks, see division 12.1-62							
030	For chalkboards & bulletin boards, see division 10.1-5 & 10							
040	(117) For lockers, see division 10.1-37							
100	Basketball backstops, wall mtd., 6' extended, fixed, minimum	L-2	1	Ea.	405	285	690	860
110	Maximum		1		1,050	285	1,335	1,575
120	Swing up, minimum		1		765	285	1,050	1,250
125	Maximum		1		1,300	285	1,585	1,850
130	Portable, manual, heavy duty, hydraulic		1.90		6,000	150	6,150	6,825
140	Ceiling suspended, stationary, minimum		.78		1,200	365	1,565	1,850
145	Fold up, with accessories, maximum		1		4,000	285	4,285	4,825
160	For electrically operated, add	1 Elec	1		450	180	630	755
200	Benches, folding, in wall, 14' table, 2 benches	L-4	2	Set	425	225	650	800
220								
300	Bleachers, telescoping, manual to 15 tier, minimum	F-5	65	Seat	30	10.35	40.35	48
310	Maximum		60		35	11.20	46.20	55
330	16 to 20 tier, minimum		60		32	11.20	43.20	51
340	Maximum		55		36	12.25	48.25	57
360	21 to 30 tier, minimum		50		34	13.45	47.45	57
370	Maximum		40		38	16.85	54.85	66
390	For integral power operation, add, minimum	2 Elec	300		9	1.19	10.19	11.60
400	Maximum	"	250		12	1.43	13.43	15.25
415	Exercise equipment, bicycle trainer			Ea.	430		430	475M
418	Chinning bar, adjustable, wall mounted	1 Carp	5		215	32	247	285
420	Exercise ladder, 16' x 1'-7", suspended	L-2	3		550	95	645	745
421	High bar, floor plate attached	1 Carp	4		840	40	880	980
424	Parallel bars, adjustable		4		1,750	40	1,790	1,975
427	Uneven parallel bars, adjustable		4		1,975	40	2,015	2,225
428	Wall mounted, adjustable	L-2	1.50	Set	575	190	765	910
430	Rope, ceiling mounted, 18' long	1 Carp	10	Ea.	160	16	176	200
433	Side horse, vaulting		5		550	32	582	650
436	Treadmill, motorized, deluxe, training type		5		3,200	32	3,232	3,575
439	Weight lifting multi-station, minimum	2 Clab	1		3,350	255	3,605	4,050
445	Maximum	"	.50		12,000	510	12,510	13,900

11.1 Architectural Equipment

		CREW	DAILY OUTPUT	UNIT	BARE COSTS MAT.	BARE COSTS INST.	BARE COSTS TOTAL	TOTAL INCL O&P
49 448	For additional exercise equipment, see div. 11.1-27							
449								
450	Gym divider curtain, mesh top, vinyl bottom, manual	L-4	500	S.F.	3.55	.90	4.45	5.25
470	Electric roll up	L-7	400		4.15	1.34	5.49	6.50
550	Gym mats, 2" thick, naugahyde covered				3.70		3.70	4.07M
560	Vinyl/nylon covered				3.60		3.60	3.96M
580	Wall pads, 1-1/2" thick				4.35		4.35	4.78M
600	Wrestling mats, 1" thick, heavy duty			↓	4.20		4.20	4.62M
700	Scoreboards, baseball, minimum	R-3	2.40	Ea.	1,800	225	2,025	2,300
720	Maximum		.05		16,000	10,700	26,700	32,400
730	Football, minimum		1.20		3,000	445	3,445	3,925
740	Maximum		.20		50,000	2,675	52,675	58,500
750	Basketball (one side), minimum		2.40		1,700	225	1,925	2,175
760	Maximum		.30		16,000	1,800	17,800	20,100
770	Hockey-basketball (four sides), minimum		.25		5,500	2,150	7,650	9,025
780	Maximum	↓	.15	↓	25,000	3,575	28,575	32,400
52-001	**STAGE EQUIPMENT** Control boards with dimmers & breakers							
005	Minimum	1 Elec	1	Ea.	2,000	180	2,180	2,450
010	Average		.50		7,000	360	7,360	8,225
015	Maximum	↓	.20	↓	30,000	895	30,895	34,300
050	Curtain track, straight, light duty	2 Carp	20	L.F.	9.55	16	25.55	34
060	Heavy duty		18		14	17.80	31.80	41
070	Curved sections		12	↓	46	27	73	89
080								
100	Curtains, velour, medium weight	2 Carp	600	S.F.	9.25	.53	9.78	10.95
110	Asbestos		50		10.25	6.40	16.65	21
115	Silica based yarn, fireproof	↓	50	↓	17.50	6.40	23.90	29
116								
150	Flooring, portable oak parquet, 3' x 3' sections			S.F.	6.75		6.75	7.40M
160	Cart to carry 180 S.F. of flooring			Ea.	150		150	165M
200	Lights, border, quartz, reflector, vented,							
210	colored or white	1 Elec	20	L.F.	115	8.95	123.95	140
250	Spotlight, follow spot, with transformer, 2100 watt	"	4	Ea.	1,100	45	1,145	1,275
260	For no transformer, deduct				400		400	440M
300	Stationary spot, fresnal quartz, 6" lens	1 Elec	4		95	45	140	170
310	8" lens		4		150	45	195	230
350	Ellipsoidal quartz, 1000W, 6" lens		4		200	45	245	285
360	12" lens		4		350	45	395	450
400	Strobe light, 1 to 15 flashes per second, quartz		3		450	60	510	580
450	Color wheel, portable, five hole, motorized	↓	4	↓	85	45	130	160
500	Stages, portable with steps, folding legs, stock, 8" high			SF Stage	6.60		6.60	7.25M
510	16" high				7.30		7.30	8.05M
520	32" high				7.70		7.70	8.45M
530	40" high				8.30		8.30	9.15M
600	Telescoping platforms, extruded alum., straight, minimum	4 Carp	157		19.85	4.08	23.93	28
610	Maximum		77		24	8.30	32.30	38
650	Pie-shaped, minimum		150		23	4.27	27.27	31
660	Maximum	↓	70		28	9.15	37.15	44
680	For 3/4" plywood covered deck, deduct			↓	2.20		2.20	2.42M
681								
700	Band risers, steel frame, plywood deck, minimum	4 Carp	275	SF Stage	16	2.33	18.33	21
710	Maximum	"	138	"	19	4.64	23.64	28
750	Chairs for above, self-storing, minimum	2 Carp	43	Ea.	39	7.45	46.45	54
760	Maximum	"	40	"	70	8	78	89
800	Rule of thumb: total stage equipment, minimum	4 Carp	100	SF Stage	60	6.40	66.40	75
810	Maximum	"	25	"	330	26	356	400
58-001	**STEAM BATH** Heater, timer & head, single, to 140 C.F.	1 Plum	1.20	Ea.	750	150	900	1,050
050	To 300 C.F.		1.10		850	165	1,015	1,175
100	Commercial size, to 800 C.F.		.90		1,395	200	1,595	1,825
150	To 2500 C.F.	↓	.80	↓	3,900	225	4,125	4,625

11.1 Architectural Equipment

		Crew	Daily Output	Unit	Bare Costs Mat.	Bare Costs Inst.	Bare Costs Total	Total Incl O&P
58 200	Multiple baths, motels, apartment, 2 baths	Q-1	1.30	Ea.	1,125	250	1,375	1,600
250	4 baths	"	.70		1,400	465	1,865	2,200
270	Conversion unit for residential tub, including door			↓	900		900	990M
61-001	**VACUUM CLEANING** Central, 3 inlet, residential	1 Skwk	.90	Total	600	180	780	925
020	Commercial		.70		1,200	235	1,435	1,650
040	5 inlet system		.50		690	330	1,020	1,225
060	7 inlet system		.40		780	410	1,190	1,450
080	9 inlet system	↓	.30	↓	870	545	1,415	1,750
62-001	**VOCATIONAL SHOP EQUIPMENT** Benches, work, wood, average	2 Carp	5	Ea.	345	64	409	470
010	Metal, average		5		220	64	284	335
040	Combination belt & disc sander, 6"		4		1,500	80	1,580	1,775
070	Drill press, floor mounted, 12", 1/2 H.P.	↓	4		535	80	615	705
080	Dust collector, not incl. ductwork, 6' diameter	1 Shee	1.10		2,250	165	2,415	2,725
100	Grinders, double wheel, 1/2 H.P.	2 Carp	5		250	64	314	370
130	Jointer, 4", 3/4 H.P.		4		1,325	80	1,405	1,575
160	Kilns, 16 C.F., to 2000°		4		1,975	80	2,055	2,300
190	Lathe, woodworking, 10", 1/2 H.P.		4		500	80	580	665
220	Planer, 13" x 6"		4		1,500	80	1,580	1,775
250	Potter's wheel, motorized		4		605	80	685	780
280	Saws, band, 14", 3/4 H.P.		4		860	80	940	1,050
310	Metal cutting band saw, 14"		4		1,725	80	1,805	2,025
340	Radial arm saw, 10", 2 H.P.		4		810	80	890	1,000
370	Scroll saw, 24"		4		1,200	80	1,280	1,425
400	Table saw, 10", 3 H.P.		4		1,525	80	1,605	1,800
430	Welder AC arc, 30 amp capacity	↓	4	↓	410	80	490	565
64-001	**WASTE HANDLING** Compactors, 115 volt, 250#/hr., chute fed	L-4	1	Ea.	6,200	450	6,650	7,475
010	Hand fed		2.40		4,150	190	4,340	4,850
030	Multi-bag, hand or chute fed, 230 volt, 600#/hr.	↓	1	↓	5,700	450	6,150	6,925
040								
050	Containerized, hand fed, 2 to 6 C.Y. containers, 250#/hr.	L-4	1	Ea.	6,000	450	6,450	7,250
055	For chute fed, add per floor		1		850	450	1,300	1,600
100	Heavy duty industrial compactor, 0.5 C.Y. capacity		1		4,500	450	4,950	5,600
105	1.0 C.Y. capacity		1		7,250	450	7,700	8,625
110	2.5 C.Y. capacity		.50		11,500	905	12,405	14,000
115	5.0 C.Y. capacity		.50		20,000	905	20,905	23,300
120	Combination shredder/compactor (5,000 lbs./hr.)	↓	.50	↓	25,000	905	25,905	28,800
135								
150	Crematory, not including building, 1 place	Q-3	.20	Ea.	42,000	3,450	45,450	51,000
175	2 place	"	.10		60,000	6,900	66,900	76,000
375	Incinerator, electric, 100 lb. per hr., minimum	L-9	.75		13,000	865	13,865	15,600
385	Maximum		.70		26,000	925	26,925	30,000
400	400 lb. per hr., minimum		.60		25,000	1,075	26,075	29,100
410	Maximum		.50		60,000	1,300	61,300	68,000
425	1000 lb. per hr., minimum		.25		68,000	2,600	70,600	78,500
435	Maximum	↓	.20		140,000	3,250	143,250	159,000
440	Gas, not incl. chimney, elec. or pipe, 50#/hr., minimum	Q-3	.80		14,500	860	15,360	17,200
442	Maximum		.70		19,000	985	19,985	22,300
444	200 lb. per hr., minimum (batch type)		.60		19,000	1,150	20,150	22,600
446	Maximum (with feeder)		.50		37,000	1,375	38,375	42,700
448	400 lb. per hr., minimum (batch type)		.30		22,500	2,300	24,800	28,100
450	Maximum (with feeder)		.25		42,500	2,750	45,250	50,500
452	800 lb. per hr., with feeder, minimum		.20		55,000	3,450	58,450	65,500
454	Maximum		.17		75,000	4,050	79,050	88,500
456	1200 lb. per hr., with feeder, minimum		.15		80,000	4,600	84,600	94,500
458	Maximum		.11		96,000	6,275	102,275	114,500
460	2000 lb. per hr., with feeder, minimum		.10		140,000	6,900	146,900	164,000
462	Maximum	↓	.05	↓	235,000	13,800	248,800	278,500

11.1 Architectural Equipment

		CREW	DAILY OUTPUT	UNIT	BARE COSTS MAT.	BARE COSTS INST.	BARE COSTS TOTAL	TOTAL INCL O&P
64 470	For heat recovery system, add, minimum	Q-3	.25	Ea.	50,000	2,750	52,750	59,000
471	Add, maximum		.11		160,000	6,275	166,275	185,000
472	For automatic ash conveyer, add	↓	.50	↓	20,000	1,375	21,375	24,000
473								
475	Large municipal incinerators, incl. stack, minimum	Q-3	.25	Ton/Day	12,800	2,750	15,550	18,100
485	Maximum	"	.10	"	34,000	6,900	40,900	47,400
550	Shredder, municipal use, 35 ton per hour			Ea.	190,000		190,000	209,000
560	60 ton per hour				405,000		405,000	445,500
575	Shredder & baler, 50 ton per day			↓	380,000		380,000	418,000
577								
580	Shredder, industrial, minimum			Ea.	15,000		15,000	16,500M
585	Maximum				80,000		80,000	88,000M
590	Baler, industrial, minimum			↓	6,000		6,000	6,600M
595	Maximum			↓	350,000		350,000	385,000
600	Transfer station compactor, with power unit							
605	and pedestal, not including pit, 50 ton per hour			Ea.	120,000		120,000	132,000
67-001	**WINE VAULT** Redwood, air conditioned, walk-in type							
002	6'-8" high, incl. racks, 2' x 4' for 156 bottles	2 Carp	2	Ea.	3,000	160	3,160	3,525
020	4' x 6' for 614 bottles		1.50		5,000	215	5,215	5,800
040	6' x 12' for 1940 bottles	↓	1	↓	9,500	320	9,820	10,900
060	Portable chillers, reach-in, 71" high x 27" wide x 23" deep							
065	One temperature, 235 bottles			Ea.	900		900	990M
070	Three temperature, 200 bottles				1,200		1,200	1,325M
075	31" x 23" x 23", 60 bottles			↓	600		600	660M

12.1 Furnishings

		CREW	DAILY OUTPUT	UNIT	BARE COSTS MAT.	BARE COSTS INST.	BARE COSTS TOTAL	TOTAL INCL O&P
06-001	**ASH/TRASH RECEIVERS**							
100	Ash urn, cylindrical metal							
102	8" diameter, 20" high	1 Clab	60	Ea.	84	2.12	86.12	95
106	10" diameter, 26" high	"	60	"	120	2.12	122.12	135
200	Combination ash/trash urn, metal							
202	8" diameter, 20" high	1 Clab	60	Ea.	73	2.12	75.12	83
205	10" diameter, 26" high	"	60	"	125	2.12	127.12	140
210								
400	Trash receptacle, metal							
402	8" diameter, 15" high	1 Clab	60	Ea.	46	2.12	48.12	54
404	10" diameter, 18" high		60		72	2.12	74.12	82
504	16" x 8" x 14" high	↓	60	↓	17	2.12	19.12	22
550	Trash receptacle, plastic, with lid							
552	35 gallon	1 Clab	60	Ea.	95	2.12	97.12	110
554	45 gallon	"	60	"	115	2.12	117.12	130
10-001	**BLINDS, INTERIOR** Solid colors							
002								
009	Horizontal, 1" aluminum slats, custom, minimum	1 Carp	590	S.F.	1.75	.27	2.02	2.32
010	Maximum		440		5	.36	5.36	6.05
025	2" aluminum slats, custom, minimum		590		1	.27	1.27	1.49
035	Maximum		440		4.50	.36	4.86	5.50
045	Stock, minimum		590		1	.27	1.27	1.49
050	Maximum		440		3.20	.36	3.56	4.05
060	2" steel slats, stock, minimum		590		.90	.27	1.17	1.38
063	Maximum		440		3	.36	3.36	3.83
075	Custom, minimum		590		1.20	.27	1.47	1.71
085	Maximum	↓	400	↓	4.70	.40	5.10	5.75

For expanded coverage of these items see *Means' Interior Cost Data 1986*

12.1 Furnishings

		Description	CREW	DAILY OUTPUT	UNIT	BARE COSTS MAT.	BARE COSTS INST.	BARE COSTS TOTAL	TOTAL INCL O&P
10	150	Vertical, 3" to 5" PVC or cloth strips, minimum	1 Carp	460	S.F.	3.80	.35	4.15	4.68
	160	Maximum		400		6.70	.40	7.10	7.95
	180	4" aluminum slats, minimum		460		2.25	.35	2.60	2.98
	190	Maximum		400		5.20	.40	5.60	6.30
	195	Mylar mirror-finish strips, to 8" wide, minimum		460		4.50	.35	4.85	5.45
	197	Maximum		400		11.50	.40	11.90	13.25
	300	Wood folding panels with movable louvers, 7" x 20" each		17	Pr.	14	9.40	23.40	29
	330	8" x 28" each		17		21	9.40	30.40	37
	345	9" x 36" each		17		29	9.40	38.40	46
	360	10" x 40" each		17		32	9.40	41.40	49
	400	Fixed louver type, stock units, 8" x 20" each		17		16	9.40	25.40	31
	415	10" x 28" each		17		25	9.40	34.40	41
	430	12" x 36" each		17		36	9.40	45.40	53
	445	18" x 40" each		17		48	9.40	57.40	66
	500	Insert panel type, stock, 7" x 20" each		17		10	9.40	19.40	25
	515	8" x 28" each		17		16	9.40	25.40	31
	530	9" x 36" each		17		21	9.40	30.40	37
	545	10" x 40" each		17		23	9.40	32.40	39
	560	Raised panel type, stock, 10" x 24" each		17		25	9.40	34.40	41
	565	12" x 26" each		17		27	9.40	36.40	43
	570	14" x 30" each		17		32	9.40	41.40	49
	575	16" x 36" each		17		38	9.40	47.40	55
	600	For custom built pine, add				20%			
	650	For custom built hardwood blinds, add				400%			
13-001		**BOOTHS**							
	100	Banquette, upholstered seat and back, custom							
	150	Straight, minimum	2 Carp	40	L.F.	72	8	80	91
	152	Maximum		36		120	8.90	128.90	145
	160	"L" or "U" shape, minimum		35		80	9.15	89.15	100
	162	Maximum		30		130	10.65	140.65	160
	180	Upholstered outside finished backs for							
	181	single booths and custom banquettes							
	182	Minimum	2 Carp	44	L.F.	9	7.25	16.25	20
	184	Maximum	"	40	"	24	8	32	38
	300	Fixed seating, one piece plastic chair and							
	301	plastic laminate table top							
	310	Two seat, 24" x 24" table, minimum	F-7	30	Ea.	195	19.70	214.70	245
	312	Maximum		26		270	23	293	330
	320	Four seat, 24" x 48" table, minimum		28		270	21	291	325
	322	Maximum		24		460	25	485	540
	500	Mount in floor, wood fiber core with							
	501	plastic laminate face, single booth							
	505	24" wide	F-7	30	Ea.	105	19.70	124.70	145
	510	48" wide	"	28	"	135	21	156	180
15-001		**CABINETS** For wood kitchen cabinets, see division 6.2-06							
	050	Hospital, base cabinets, laminated plastic	2 Carp	10	L.F.	114	32	146	170
	100	Stainless steel	"	10		151	32	183	210
	120	For all drawers, add				6.75		6.75	7.40M
	130	Cabinet base trim, 4" high, enameled steel	2 Carp	200		16	1.60	17.60	19.90
	140	Stainless steel		200		26	1.60	27.60	31
	145	Counter top, laminated plastic, no backsplash		40		18	8	26	31
	165	With backsplash		40		24	8	32	38
	180	For sink cutout, add		12.20	Ea.		26	26	38L
	185								
	190	Stainless steel counter top	2 Carp	40	L.F.	54	8	62	71
	200	For drop-in stainless 43" x 21" sink, add			Ea.	281		281	310M
	210	Nurses station, door type, laminated plastic	2 Carp	10	L.F.	140	32	172	200
	220	Enameled steel		10		120	32	152	180
	230	Stainless steel		10		161	32	193	225
	240	For drawer type, add				56		56	62M

For expanded coverage of these items see *Means' Interior Cost Data 1986*

12.1 Furnishings

			CREW	DAILY OUTPUT	UNIT	BARE COSTS MAT.	BARE COSTS INST.	BARE COSTS TOTAL	TOTAL INCL O&P
15	250	Wall cabinets, laminated plastic	2 Carp	15	L.F.	89	21	110	130
	260	Enameled steel		15		101	21	122	140
	270	Stainless steel	↓	15		156	21	177	205
	280	For glass doors, add				14		14	15.40M
	350	Kitchen, base cabinets, metal, minimum	2 Carp	30		29	10.65	39.65	47
	360	Maximum		25		76	12.80	88.80	100
	370	Wall cabinets, metal, minimum		30		29	10.65	39.65	47
	380	Maximum		25		68	12.80	80.80	93
	500	School, 24" deep, metal, 84" high units		15		146	21	167	190
	515	Counter height units		20		114	16	130	150
	545	Wood, custom fabricated, 32" high counter		20		96	16	112	130
	560	Add for counter top		56		11	5.70	16.70	20
	580	84" high wall units	↓	15	↓	104	21	125	145
	600	Laminated plastic finish is same price as wood							
19	001	**COAT RACKS & WARDROBES** Dormitory units, wood or metal							
	002	Stock units, 84" high, incl. door, minimum	1 Carp	10	L.F.	79	16	95	110
	005	Average		7		307	23	330	370
	010	Maximum		5		588	32	620	695
	050	Hospital type, 84" high, plastic faced wood		5		120	32	152	180
	065	Enameled steel		5		100	32	132	155
	080	Stainless steel	↓	5	↓	208	32	240	275
	140	School type, see division 10.1-37							
	150	Coat and hat rack, wall mounted, tubular steel, 1 shelf	1 Carp	70	L.F.	13	2.29	15.29	17.60
	165	3 shelves		50		32	3.20	35.20	40
	185	Floor mounted, double shelf, tubular steel, 1 face		50		33	3.20	36.20	41
	200	Double face	↓	50		45	3.20	48.20	54
	215	For triple shelf, add				30%			
	230	For custom construction, add				50%	50%		
	250	For aluminum, add				25%			
	265	For stainless steel, add			↓	200%			
28	001	**DRAPERY HARDWARE**							
	002								
	003	Standard traverse, per foot, minimum	1 Carp	59	L.F.	1.30	2.71	4.01	5.35
	010	Maximum		51	"	4.50	3.14	7.64	9.50
	400	Traverse rods, adjustable, 28" to 48"		22	Ea.	11.85	7.25	19.10	24
	402	48" to 84"		20		18.60	8	26.60	32
	404	66" to 120"		18		22	8.90	30.90	37
	406	84" to 156"		16		24	10	34	41
	408	120" to 220"		14		30	11.45	41.45	50
	410	228" to 312"	↓	13		42	12.30	54.30	64
	500	Stationary rods, first 2 feet			↓	5.33		5.33	5.85M
	502	Each additional foot, add			L.F.	1.97		1.97	2.16M
33	001	**FLOOR MATS** Recessed, in-laid black rubber, 3/8" thick, solid	1 Clab	155	S.F.	10.75	.82	11.57	13
	005	Perforated		155		11.50	.82	12.32	13.85
	010	1/2" thick, solid		155		14.30	.82	15.12	16.90
	015	Perforated		155		14.55	.82	15.37	17.20
	020	In colors, 3/8" thick, solid		155		14.90	.82	15.72	17.60
	025	Perforated		155		15.45	.82	16.27	18.20
	030	1/2" thick, solid		155		16.80	.82	17.62	19.65
	035	Perforated		155		19.15	.82	19.97	22
	050	Link mats, including nosings, aluminum, 3/8" thick		155		7.70	.82	8.52	9.65
	055	Black rubber with galvanized tie rods		155		8.60	.82	9.42	10.65
	060	Steel, galvanized, 3/8" thick		155		3.55	.82	4.37	5.10
	065	Vinyl, in colors	↓	155	↓	9.40	.82	10.22	11.55
	075	Add for nosings, rubber			L.F.	2.40		2.40	2.64M
	080								
	085	Recess frames for above mats, aluminum	1 Carp	100	L.F.	3.55	1.60	5.15	6.20
	087	Bronze	"	100	"	7.30	1.60	8.90	10.35
	090	Skate lock tile, 24" x 24" x 1/2" thick, rubber, black	1 Clab	125	S.F.	5.20	1.02	6.22	7.20
	095	Color	"	125	"	6.60	1.02	7.62	8.75

For expanded coverage of these items see *Means' Interior Cost Data 1986*

12.1 Furnishings

		CREW	DAILY OUTPUT	UNIT	BARE COSTS MAT.	BARE COSTS INST.	BARE COSTS TOTAL	TOTAL INCL O&P
100	12" x 24" border, black	1 Clab	75	L.F.	9.15	1.70	10.85	12.50
110	Color		75	"	10.90	1.70	12.60	14.45
115	12" x 12" outside corner, black		100	S.F.	4.16	1.27	5.43	6.40
120	Color		100		5.85	1.27	7.12	8.30
150	Duckboard, aluminum slats		155		11.50	.82	12.32	13.85
170	Hardwood strips on rubber base, to 54" wide		155		7.80	.82	8.62	9.75
180	Assembled with brass rods and vinyl spacers, to 48" wide		155		9.90	.82	10.72	12.10
185	Tire fabric, 3/4" thick		155		5.20	.82	6.02	6.90
190	Vinyl, 36" wide, in colors, hollow top & bottoms		155		2.75	.82	3.57	4.21
195	Solid top & bottom members	↓	155	↓	4.80	.82	5.62	6.45
38-001	**BANK FURNITURE** See division 12.1-53							
41-001	**CHURCH FURNITURE** See division 11.1-12							
42-001	**FURNITURE, DORMITORY**							
100	Chest, four drawer, minimum			Ea.	245		245	270M
102	Maximum			"	380		380	420M
105	Built-in, minimum	2 Carp	13	L.F.	70	25	95	115
115	Maximum		10		135	32	167	195
120	Desk top, built-in, laminated plastic, 24" deep, minimum		50		16	6.40	22.40	27
130	Maximum		40		45	8	53	61
145	30" deep, minimum		50		16	6.40	22.40	27
155	Maximum		40		65	8	73	83
175	Dressing unit, built-in, minimum		12		110	27	137	160
185	Maximum	↓	8	↓	325	40	365	415
400								
800	Rule of thumb: total cost for furniture, minimum			Student				1,800
805	Maximum			"				3,500
45-001	**FURNITURE, HOSPITAL** Beds, manual, minimum			Ea.	520		520	570M
010	Maximum				905		905	995M
030	Manual and electric beds, minimum				650		650	715M
040	Maximum				1,400		1,400	1,550M
060	All electric hospital beds, minimum				865		865	950M
070	Maximum				2,150		2,150	2,375M
090	Manual, nursing home beds, minimum				410		410	450M
100	Maximum				850		850	935M
102	Overbed table, laminated top, minimum				175		175	195M
104	Maximum			↓	435		435	480M
110	Patient wall systems, not incl. plumbing, minimum			Room	520		520	570M
120	Maximum			"	960		960	1,050M
200	Geriatric chairs, minimum			Ea.	190		190	210M
202	Maximum			"	355		355	390M
50-001	**FURNITURE, HOTEL** Standard quality, set, minimum			Room	1,625		1,625	1,800M
020	Maximum			"	5,000		5,000	5,500M
51-001	**FURNITURE, LIBRARY**							
010	Attendant desk, 36" x 62" x 29" high	1 Carp	16	Ea.	1,375	10	1,385	1,525
020	Book display, "A" frame display, both sides		16		1,550	10	1,560	1,725
022	Table with bulletin board		16		820	10	830	915
080	Card catalogue, 30 tray unit		16		1,650	10	1,660	1,825
084	60 tray unit	↓	16		2,800	10	2,810	3,100
088	72 tray unit	2 Carp	16	↓	2,900	20	2,920	3,225
090								
100	Carrels, single face, initial unit	1 Carp	16	Ea.	405	10	415	460
150	Double face, initial unit	2 Carp	16		600	20	620	690
400	Dictionary stand, stationary	1 Carp	16		460	10	470	520
402	Revolving		16		150	10	160	180
420	Exhibit case, table style, 60" x 28" x 36"		11		2,600	14.55	2,614.55	2,875
700	Tables, card catalog reference, 24" x 60" x 42"	↓	16	↓	520	10	530	585
53-001	**FURNITURE, OFFICE**							

For expanded coverage of these items see *Means' Interior Cost Data 1986*

12.1 Furnishings

					BARE COSTS			TOTAL
		CREW	DAILY OUTPUT	UNIT	MAT.	INST.	TOTAL	INCL O&P
53 002	Desks, 29" high, double pedestal, 30" x 60", metal, minimum			Ea.	205		205	225M
003	Maximum				560		560	615M
016	Wood, minimum				200		200	220M
018	Maximum				1,075		1,075	1,175M
060	Desks, single pedestal, 30" x 60", metal, minimum				185		185	205M
062	Maximum				475		475	525M
072	Desks, secretarial return, 18" x 42", metal, minimum				290		290	320M
074	Maximum				780		780	860M
080	Wood, minimum				310		310	340M
082	Maximum				1,375		1,375	1,525M
200	Chairs, office type, executive, minimum				155		155	170M
215	Maximum				985		985	1,075M
230	Secretarial, minimum				83		83	91M
232	Maximum			↓	285		285	315M
300								
600	Table, conference							
605	Boat, 96" x 42", minimum			Ea.	435		435	480M
615	Maximum				1,050		1,050	1,150M
672	Rectangle, 96" x 42", minimum				325		325	360M
674	Maximum			↓	965		965	1,050M
60-001	**FURNITURE, RESTAURANT** Bars, built-in, front bar	1 Carp	5	L.F.	130	32	162	190
020	Back bar	"	5	"	95	32	127	150
030	Booth seating see 12.1-13							
100								
200	Chair, bentwood side chair, metal, minimum			Ea.	48		48	53M
202	Maximum				60		60	66M
260	Upholstered seat & back, arms, minimum				135		135	150M
262	Maximum			↓	350		350	385M
62-001	**FURNITURE, SCHOOL**							
100	Chair, molded plastic,							
110	Integral tablet arm, minimum			Ea.	30		30	33M
115	Maximum				36		36	40M
200	Desk, single pedestal, top book compartment, minimum				33		33	36M
202	Maximum				35		35	39M
220	Flip top, minimum				29		29	32M
222	Maximum				31		31	34M
63-001	**IRONING CENTER** Including cabinet, board & light, minimum	1 Carp	2		222	80	302	360
010	Maximum	"	1.50	↓	390	105	495	585
67-001	**MULTI-MEDIA BOARDS** See division 10.1-10							
75-001	**PLANTERS**							
100	Fiberglass, hanging, 12" diameter, 7" high			Ea.	17		17	18.70M
150	Rectangular, 48" long, 16" high x 15" wide				198		198	220M
165	60" long, 30" high, 28" wide				390		390	430M
200	Round, 12" diameter, 13" high				53		53	58M
205	25" high				60		60	66M
500	Square, 10" side, 20" high				60		60	66M
510	14" side, 15" high				64		64	70M
600	Metal bowl, 32" diameter, 8" high, minimum				208		208	230M
605	Maximum			↓	229		229	250M
700								
875	Wood, fiberglass liner, square							
878	14" square, 15" high, minimum			Ea.	109		109	120M
880	Maximum				135		135	150M
940	Plastic cylinder, molded, 10" diameter, 10" high				4.70		4.70	5.15M
950	11" diameter, 11" high				13		13	14.30M
80-001	**POSTS** Portable for pedestrian traffic control, standard, minimum				50		50	55M
010	Maximum				88		88	97M
030	Deluxe posts, minimum				80		80	88M
040	Maximum			↓	220		220	240M

For expanded coverage of these items see *Means' Interior Cost Data 1986*

12.1 Furnishings

		CREW	DAILY OUTPUT	UNIT	BARE COSTS MAT.	BARE COSTS INST.	BARE COSTS TOTAL	TOTAL INCL O&P
060	Ropes for above posts, plastic covered, 1-1/2" diameter			L.F.	4.30		4.30	4.73M
070	Chain core			"	5.75		5.75	6.30M
81-001	**SEATING** Benches, see division 2.7-05, 10.1-37 & 11.1-49							
010	Bleachers, see division 2.7-07 & 11.1-49							
050	Classroom, movable chair & desk type, minimum			Set				65
060	Maximum			"				120
100	Lecture hall, pedestal type, minimum	2 Carp	35	Ea.	65	9.15	74.15	85
120	Maximum		20		205	16	221	250
200	Auditorium chair, all veneer construction		35		65	9.15	74.15	85
220	Veneer back, padded seat		35		81	9.15	90.15	100
235	Fully upholstered, spring seat	↓	35	↓	97	9.15	106.15	120
245	For tablet arms, add				22		22	24M
83-001	**SHADES** Basswood, roll-up, stain finish, 3/8" slats	1 Carp	300	S.F.	9.10	.53	9.63	10.80
020	7/8" slats		300		9.40	.53	9.93	11.10
030	Vertical side slide, stain finish, 3/8" slats		300		11.50	.53	12.03	13.40
040	7/8" slats		300		11	.53	11.53	12.85
050	For fire retardant finishes, add		300		15%			
060	For "B" rated finishes, add		300		20%			
090	Mylar, single layer, non-heat reflective		685		2.70	.23	2.93	3.31
100	Double layered, heat reflective		685		4.75	.23	4.98	5.55
110	Triple layered, heat reflective	↓	685	↓	5.50	.23	5.73	6.40
120	For metal roller instead of wood, add per			Shade	1.70		1.70	1.87M
130	Vinyl coated cotton, standard	1 Carp	685	S.F.	1	.23	1.23	1.44
140	Lightproof decorator shades		685		1.30	.23	1.53	1.77
150	Vinyl, lightweight, 4 gauge		685		.28	.23	.51	.65
160	Heavyweight, 6 gauge		685		.85	.23	1.08	1.27
170	Vinyl laminated fiberglass, 6 ga., translucent		685		1.35	.23	1.58	1.82
180	Lightproof		685		1.60	.23	1.83	2.10
300	Woven aluminum, 3/8" thick, lightproof and fireproof	↓	350	↓	2.25	.46	2.71	3.14
88-001	**TABLES, FOLDING** Laminated plastic tops							
100	Tubular steel legs with glides							
102	18" x 60", minimum			Ea.	87		87	96M
104	Maximum				136		136	150M
184	36" x 96", minimum				115		115	125M
186	Maximum				160		160	175M
200	Round, wood stained, plywood top, 60" diameter, minimum				185		185	205M
202	Maximum			↓	210		210	230M

For expanded coverage of these items see Means' *Interior Cost Data 1985*

13.1 Special Construction

		CREW	DAILY OUTPUT	UNIT	BARE COSTS MAT.	BARE COSTS INST.	BARE COSTS TOTAL	TOTAL INCL O&P
01-001	**ACOUSTICAL** Enclosure, 4" thick wall and ceiling panels							
002	8# per S.F., up to 12' span	3 Carp	72	S.F.Surf	12.50	6.65	19.15	23
030	Better quality panels, 10.5# per S.F.		64		16	7.50	23.50	28
040	Reverb-chamber, 4" thick, parallel walls		60		17.10	8	25.10	30
060	Skewed wall, parallel roof, 4" thick panels		55		17	8.75	25.75	31
070	Skewed walls, skewed roof, 4" layers, 4" air space		48		27	10	37	44
090	Sound-absorbing panels, painted metal, 2'-6" x 8', under 1000 S.F.		215		6.50	2.23	8.73	10.40
110	Over 2400 S.F.		240		6.50	2	8.50	10.05
120	Fabric faced	↓	240	↓	4.30	2	6.30	7.65
125								
150	Flexible transparent curtain, clear	3 Shee	215	S.F.Surf	4.35	2.53	6.88	8.50
160	50% foam	"	215	"	6.25	2.53	8.78	10.60

13.1 Special Construction

		CREW	DAILY OUTPUT	UNIT	BARE COSTS MAT.	BARE COSTS INST.	BARE COSTS TOTAL	TOTAL INCL O&P
01 170	75% foam	3 Shee	215	S.F.Surf	6.25	2.53	8.78	10.60
180	100% foam		215		6.25	2.53	8.78	10.60
220	Strip entrance, 2/3 overlap		135		5	4.04	9.04	11.40
225	Full overlap		115	↓	6	4.74	10.74	13.50
230	Add for suspension system			L.F.	9.25		9.25	10.15M
290								
310	Audio masking system, including speakers, amplification							
311	and signal generator							
320	Ceiling mounted, 5000 S.F.	2 Elec	2,400	S.F.	.80	.15	.95	1.09
330	10,000 S.F.		2,800		.63	.13	.76	.88
340	Plenum mounted, 5000 S.F.		3,800		.67	.09	.76	.87
350	10,000 S.F.	↓	4,400	↓	.44	.08	.52	.60
02-001	**AIR CURTAINS** Not incl. motor starters, transformers,							
005	door switches or temperature controls							
010	Shipping and receiving doors, unheated, minimal wind stoppage							
015	8' high, multiples of 3' wide	2 Shee	3.60	L.F.	170	100	270	335
016	Multiples of 5' wide		3.60		125	100	225	285
021	10' high, multiples of 4' wide		3.60		175	100	275	340
025	12' high, 3'-6" wide		3.30		170	110	280	350
026	12' wide		3.30		125	110	235	300
035	16' high, 3'-6" wide		3.30		180	110	290	360
036	12' wide	↓	3.30	↓	135	110	245	310
050	Maximum wind stoppage							
055	10' high, multiples of 4' wide	2 Shee	3.60	L.F.	350	100	450	530
065	14' high, multiples of 4' wide		3.25		450	110	560	660
075	20' high, multiples of 8' wide	↓	3	↓	785	120	905	1,050
110	Heated, maximum wind stoppage, steam heat							
115	10' high, multiples of 4' wide	2 Shee	3	L.F.	440	120	560	660
125	14' high, multiples of 4' wide		2.80		580	130	710	825
135	20' high, multiples of 8' wide	↓	2.60	↓	1,050	140	1,190	1,350
150	Customer entrance doors, unheated, minimal wind stoppage							
155	10' high, multiples of 3' wide	2 Shee	3.60	L.F.	165	100	265	330
156	Multiples of 5' wide		3.60		110	100	210	270
165	Maximum wind stoppage, 12' high, multiples of 4' wide	↓	3.30	↓	270	110	380	460
170	Heated, minimal wind stoppage, electric heat							
175	8' high, multiples of 3' wide	2 Shee	3.10	L.F.	345	115	460	550
176	Multiples of 5' wide	"	3.10	"	235	115	350	430
180								
185	10' high, multiples of 3' wide	2 Shee	3.10	L.F.	400	115	515	610
186	Multiples of 5' wide	"	3.10	"	275	115	390	475
195	Maximum wind stoppage, steam heat							
196	12' high, multiples of 4' wide	2 Shee	2.80	L.F.	370	130	500	595
200	Walk-in coolers and freezers, ambient air, minimal wind stoppage							
205	8' high, multiples of 3' wide	2 Shee	3.60	L.F.	175	100	275	340
206	Multiples of 5' wide		3.60		145	100	245	305
225	Maximum wind stoppage, 12' high, multiples of 3' wide		3.30		280	110	390	470
245	Conveyor openings or service windows, unheated, 5' high		4		115	91	206	260
246	Heated, electric, 5' high, 2'-6" wide	↓	3.30		190	110	300	370
270	Entrance with recirculating system, unheated, add			L.F.Wide	145		145	160M
280	Installation of each doorway	2 Shee	.10	Total		3,625	3,625	5,300L
05-001	**AIR SUPPORTED STRUCTURES**							
002	Site preparation, incl. anchor placement and utilities	B-11B	2,000	S.F.Flr.	.64	.56	1.20	1.37
003	For concrete curb, see division 3.3-14-170							
005	Warehouse, polyester/vinyl fabric, 24 oz., over 10 yr. life, welded							
006	(120) Seams, tension cables, primary & auxiliary inflation system,							
007	airlock, personnel doors and liner							
010	5000 S.F.	4 Clab	5,000	S.F.Flr.	7.85	.10	7.95	8.80
025	12,000 S.F.	"	6,000		5.30	.08	5.38	5.95
040	24,000 S.F.	8 Clab	12,000		3.91	.08	3.99	4.42
050	50,000 S.F.	"	12,500	↓	3.30	.08	3.38	3.75

13.1 Special Construction

		CREW	DAILY OUTPUT	UNIT	BARE COSTS MAT.	BARE COSTS INST.	BARE COSTS TOTAL	TOTAL INCL O&P
070	12 oz. reinforced vinyl fabric, 5 yr. life, sewn seams,							
071	accordian door, including liner							
075	3000 S.F.	4 Clab	3,000	S.F.Flr.	4.27	.17	4.44	4.94
080	12,000 S.F.	"	6,000		2.99	.08	3.07	3.41
085	24,000 S.F.	8 Clab	12,000		2.68	.08	2.76	3.07
095	Deduct for single layer				.42		.42	.46M
100	Add for welded seams				.54		.54	.59M
105	Add for double layer, welded seams included			↓	1.06		1.06	1.16M
125	Tedlar/vinyl fabric, 17 oz., with liner, over 10 yr. life,							
126	incl. overhead and personnel doors							
130	3000 S.F.	4 Clab	3,000	S.F.Flr.	7.35	.17	7.52	8.35
145	12,000 S.F.	"	6,000		4.61	.08	4.69	5.20
155	24,000 S.F.	8 Clab	12,000		3.61	.08	3.69	4.09
170	Deduct for single layer			↓	.74		.74	.81M
225	Greenhouse/shelter, woven polyethylene with liner, 2 yr. life,							
226	sewn seams, including doors							
230	3000 S.F.	4 Clab	3,000	S.F.Flr.	3.30	.17	3.47	3.88
235	12,000 S.F.	"	6,000		2.05	.08	2.13	2.38
245	24,000 S.F.	8 Clab	12,000		1.65	.08	1.73	1.94
255	Deduct for single layer			↓	.32		.32	.35M
260	Tennis/gymnasium, polyester/vinyl fabric, 24 oz., over 10 yr. life,							
261	including thermal liner, heat and lights							
265	7200 S.F.	4 Clab	6,000	S.F.Flr.	9.35	.08	9.43	10.40
275	13,000 S.F.	"	6,500		7.60	.08	7.68	8.45
285	Over 24,000 S.F.	8 Clab	12,000		5.80	.08	5.88	6.50
286	For low temperature conditions, add			↓	.32		.32	.35M
287	For average shipping charges, add			Total	1,225		1,225	1,350M
289								
290	Thermal liner, translucent reinforced vinyl			S.F.Flr.	.54		.54	.59M
295	Metalized mylar fabric and mesh, double liner			"	1.06		1.06	1.16M
305	Stadium/convention center, teflon coated fiberglass, heavy weight,							
306	over 20 yr. life, incl. thermal liner and heating system							
310	Minimum	9 Clab	26,000	S.F.Flr.	18.55	.04	18.59	20
311	Maximum	"	19,000	"	24	.06	24.06	26
340	Doors, air lock, 15' long, 10' x 10'	2 Carp	.80	Ea.	12,400	400	12,800	14,200
360	15' x 15'	"	.50		14,400	640	15,040	16,800
370	For each added 5' length, add				1,075		1,075	1,175M
390	Revolving personnel door, 6' diameter, 6'-6" high	2 Carp	.80	↓	6,500	400	6,900	7,725
420	Double wall, self supporting, shell only, minimum			S.F.Flr.				20
430	Maximum			"				38
07-001	**AIR SUPPORTED STORAGE TANK COVERS** Vinyl polyester							
010	scrim, double layer, with hardware, blower, standby & controls							
020	Round, 75' diameter	B-2	5,000	S.F.	3.02	.13	3.15	3.51
030	100' diameter		6,000		2.73	.11	2.84	3.16
040	150' diameter		6,000		2.15	.11	2.26	2.52
050	Rectangular, 20' x 20'		6,000		9.45	.11	9.56	10.55
060	30' x 40'		6,000		4.89	.11	5	5.55
070	50' x 60'	↓	6,000		3.35	.11	3.46	3.84
080	For single wall construction, deduct, minimum				.26		.26	.28M
090	Maximum				1.03		1.03	1.13M
100	For maximum resistance to atmosphere or cold, add			↓	.32		.32	.35M
110	For average shipping charges, add			Total	1,025		1,025	1,125M
10-001	**ANECHOIC CHAMBERS** Standard units, 7' ceiling heights							
010	Area for pricing is net inside dimensions							
030	200 cycles per second cutoff, 25 S.F. floor area			S.F.Flr.				775
040	50 S.F.							600
060	75 S.F.							540
070	100 S.F.							520
090	For 150 cycles per second cutoff, add				30%	30%		
100	For 100 cycles per second cutoff, add			↓	45%	45%		

13.1 Special Construction

		CREW	DAILY OUTPUT	UNIT	BARE COSTS MAT.	BARE COSTS INST.	BARE COSTS TOTAL	TOTAL INCL O&P
12-001	**AUDIOMETRIC ROOMS** Under 500 S.F. surface	4 Carp	180	S.F.Surf	31	3.56	34.56	39
010	Over 500 S.F. surface	"	200	"	29	3.20	32.20	37
15-001	**BOWLING ALLEYS** Including alley, pinsetter, scorer,							
002	counters and misc. supplies, minimum	4 Carp	.20	Lane	30,300	3,200	33,500	38,000
015	Average	↓	.19		32,000	3,375	35,375	40,100
030	Maximum		.18		35,000	3,550	38,550	43,700
060	For automatic scorer, add, minimum				9,500		9,500	10,500M
070	Maximum			↓	12,500		12,500	13,800M
17-001	**CHIMNEY** Foundations, add to all prices below, see division 3.3-14-050							
180	Metal, steel, guyed, 2' diameter, 1/4" thick shell	E-2	65	V.L.F.	72	31	103	120
190	5' diameter, 1/2" thick shell	"	30	"	535	66	601	680
195								
210	Poured concrete, brick lining, no foundation, 200' x 10' diameter			Ea.				715K
220	(122) 200' x 18' diameter							1,000K
240	250' x 9' diameter							930K
250	300' x 14' diameter							1,225K
270	400' x 18' diameter							2,160K
280	500' x 20' diameter							3,500K
300	Radial brick, foundations not incl., 75' x 3'-6" I.D.							130K
310	(122) 85' x 3'-6" I.D.							185K
330	85' x 5'-0" I.D.							185K
340	100' x 5'-0" I.D.							215K
360	125' x 5'-6" I.D.							235K
370	150' x 6'-6" I.D.							275K
390	160' x 6'-6" I.D.							360K
400	175' x 7'-0" I.D.			↓				390K
20-001	**COMFORT STATIONS** Prefab., stock, with doors, windows and fixtures							
010	Not incl. interior finish or electrical							
030	Mobile, on steel frame, minimum			S.F.	24		24	26M
035	Maximum				35		35	39M
040	Permanent, including concrete slab, minimum	B-12J	50		120	15	135	150
050	Maximum	"	43	↓	160	17.40	177.40	200
060	Alternate pricing method, mobile, minimum			Fixture	1,500		1,500	1,650M
065	Maximum				2,250		2,250	2,475M
070	Permanent, minimum	B-12J	.70		8,550	1,075	9,625	10,700
075	Maximum	"	.50	↓	14,500	1,500	16,000	17,800
22-001	**CONTROL TOWERS** Modular, 12' x 10', incl. instrumentation, minimum			Ea.				280,000
010	Maximum							375,000
050	With standard 40' tower, average			↓				600,000
075								
100	Temporary portable control towers, 8' x 12',							
101	complete with one position communications			Ea.				120,000
25-001	**DARKROOMS** Shell, complete except for door, 64 S.F., 8' high	2 Carp	128	S.F.Flr.	50	2.50	52.50	59
010	12' high		64		68	5	73	82
050	120 S.F. floor, 8' high		120		39	2.67	41.67	47
060	12' high		60		50	5.35	55.35	63
080	240 S.F. floor, 8' high		120		27.50	2.67	30.17	34
090	12' high		60	↓	35	5.35	40.35	46
120	Mini-cylindrical, revolving, unlined, 4' diameter		3.50	Ea.	3,150	91	3,241	3,600
140	5'-6" diameter	↓	2.50		4,850	130	4,980	5,525
160	Add for lead lining, inner cylinder, 1/32" thick				1,275		1,275	1,400M
170	1/16" thick				1,475		1,475	1,625M
180	Add for lead lining, inner and outer cylinder, 1/32" thick				2,350		2,350	2,575M
190	1/16" thick			↓	2,450		2,450	2,700M
200	For darkroom door, see division 8.3-12							
27-001	**DOMES** Revolving aluminum, electric drive,							
002	for astronomy observation, shell only, stock units							
060	10'-0" diameter, 800#, dome	2 Carp	.25	Ea.	9,400	1,275	10,675	12,200
070	Base	"	.67	"	3,250	480	3,730	4,275

13.1 Special Construction

		CREW	DAILY OUTPUT	UNIT	BARE COSTS MAT.	BARE COSTS INST.	BARE COSTS TOTAL	TOTAL INCL O&P
090	18'-0" diameter, 2,500#, dome	2 Carp	.17	Ea.	27,500	1,875	29,375	33,000
100	Base		.33		9,100	970	10,070	11,400
120	24'-0" diameter, 4,500#, dome		.08		50,850	4,000	54,850	61,500
130	Base	↓	.25	↓	16,400	1,275	17,675	19,900
150	Bulk storage, shell only, dual radius hemispherical arch, steel							
160	framing, corrugated steel covering, 150' diameter	E-2	550	S.F.Flr.	27	3.60	30.60	35
170	400' diameter	"	720		22	2.75	24.75	28
180	Wood framing, wood decking, to 400' diameter	F-4	400	↓	20	3.73	23.73	27
190	Radial framed wood (2"x6"), 1/2" thick							
200	plywood, asphalt shingles, 50' diameter	F-3	2,000	S.F.Flr.	25	.53	25.53	28
210	60' diameter		1,900		19	.56	19.56	22
220	72' diameter		1,800		17	.59	17.59	19.50
230	116' diameter		1,730		15	.61	15.61	17.35
240	150' diameter	↓	1,500		13	.71	13.71	15.25
30-001	**GARAGE COSTS** Public parking, average			Car				6,450
010	See also division 17.1-41							
030	Residential, prefab shell, stock, wood, single car, minimum	F-2	1	Total	1,175	335	1,510	1,775
035	Maximum		.67		1,750	505	2,255	2,650
040	Two car, minimum		.67		1,875	505	2,380	2,775
045	Maximum	↓	.50	↓	2,725	675	3,400	3,975
32-001	**GARDEN HOUSE** Prefab wood, no floors or foundations							
002								
010	48 to 200 S.F., minimum	F-2	200	S.F.Flr.	15	1.69	16.69	18.90
030	Maximum	"	48	"	25	7.05	32.05	38
33-001	**GEODESIC DOME** Shell only, interlocking plywood panels							
040	30' diameter	F-5	1.60	Ea.	5,950	420	6,370	7,150
050	35' diameter		1.14		7,150	590	7,740	8,725
060	39' diameter		1		8,400	675	9,075	10,200
070	45' diameter	↓	.90		8,500	750	9,250	10,400
080	60' diameter	F-3	.40	↓	37,500	2,650	40,150	44,900
110	Aluminum panel, stressed skin, with 1-1/2" insulation							
120	82' diameter	L-5	900	S.F.Flr.	16.80	1.79	18.59	21
130	232' diameter	"	1,300		15.50	1.24	16.74	18.85
140	For vinyl faced fiberglass insulation, add			↓	3.75		3.75	4.12M
160	Aluminum framed, plexiglass closure panels							
170	40' diameter	K-2	250	S.F.Flr.	54	2.68	56.68	63
180	200' diameter	L-5	1,000	"	36	1.61	37.61	42
190								
210	Aluminum framed, aluminum closure panels							
220	40' diameter	K-2	500	S.F.Flr.	17.30	1.34	18.64	21
230	320' diameter	C-17C	1,900		12	1.06	13.06	14.70
240	415' diameter	"	2,300	↓	10.35	.87	11.22	12.60
270	Aluminum framed, fiberglass sandwich panel closure							
280	6' diameter	F-2	150	S.F.Flr.	33	2.25	35.25	40
290	28' diameter	"	350	"	25	.96	25.96	29
35-001	**GRANDSTANDS** Permanent, municipal, closed deck, minimum			Seat				77
010	Maximum							390
030	Steel, minimum							62
040	Maximum							260
060	Aluminum, extruded, stock design, minimum							72
070	Maximum							125
090	Composite, steel, wood and plastic, stock design, minimum							41
100	Maximum			↓				88
37-001	**GREENHOUSE** Shell only, stock units, not incl. 2' stub walls, foundation,							
002	floors, heat or compartments							
030	Residential type, free standing, 8'-6" long x 7'-6" wide	2 Carp	59	S.F.Flr.	33	5.40	38.40	44
040	10'-6" wide		85		26	3.76	29.76	34
060	13'-6" wide		108		23	2.96	25.96	30
070	17'-0" wide	↓	160	↓	26	2	28	32

13.1 Special Construction

			CREW	DAILY OUTPUT	UNIT	BARE COSTS MAT.	BARE COSTS INST.	BARE COSTS TOTAL	TOTAL INCL O&P
37	090	Lean-to type, 3'-10" wide	2 Carp	34	S.F.Flr.	30	9.40	39.40	47
	100	6'-10" wide	"	58	"	23	5.50	28.50	33
	150	Commercial, custom, truss frame, incl. equip., plumbing, elec.,							
	151	benches and controls, under 2000 S.F., minimum			S.F.Flr.				29
	155	Maximum							36
	170	Over 5000 S.F., minimum							22
	175	Maximum							29
	200	Institutional, custom, rigid frame, including compartments and							
	201	multi-controls, under 500 S.F., minimum			S.F.Flr.				73
	205	Maximum							94
	215	Over 2000 S.F., minimum							37
	220	Maximum							62
	240	Concealed rigid frame, under 500 S.F., minimum							86
	245	Maximum							106
	255	Over 2000 S.F., minimum							65
	260	Maximum							75
	280	Lean-to type, under 500 S.F., minimum							76
	285	Maximum							115
	300	Over 2000 S.F., minimum							42
	305	Maximum							70
	360	For 1/4" clear plate glass, add			S.F.Surf	1.40		1.40	1.54M
	370	For 1/4" tempered glass, add			"	3.15		3.15	3.46M
	390	For cooling, add, minimum			S.F.Flr.	2.10		2.10	2.31M
	400	Maximum				5.20		5.20	5.70M
	420	For heaters, 13.6 MBH, add				4		4	4.40M
	430	60 MBH, add				1.50		1.50	1.65M
	450	For benches, 2' x 3'-6", add			S.F.Hor.	17.90		17.90	19.70M
	460	3' x 10', add			S.F.	9.70		9.70	10.65M
	480	For controls, add, minimum			Total	1,800		1,800	1,975M
	490	Maximum			"	10,700		10,700	11,800M
	510	For humidification equipment, add			M.C.F.	4.60		4.60	5.05M
	520	For vinyl shading, add			S.F.	.97		.97	1.06M
	600	Geodesic hemisphere, 1/8" plexiglass glazing							
	605	8' diameter	2 Carp	2	Ea.	2,025	160	2,185	2,450
	615	24' diameter		.35		10,400	915	11,315	12,800
	625	48' diameter		.20		28,200	1,600	29,800	33,300
	40-001	**HANGARS** Prefabricated steel T hangars, Galv. steel roof &							
	010	walls, incl. electric bi-folding doors, 4 or more units,							
	011	not including floors or foundations, minimum	E-2	1,275	S.F.Flr.	6.30	1.55	7.85	9.05
	013	Maximum		1,063		6.70	1.86	8.56	9.90
	090	With bottom rolling doors, minimum		1,386		5.70	1.43	7.13	8.20
	100	Maximum		966		6.70	2.05	8.75	10.15
	120	Alternate pricing method:							
	121								
	130	Galv. roof and walls, electric bi-folding doors, minimum	E-2	1.06	Plane	7,450	1,875	9,325	10,700
	150	Maximum		.91		8,375	2,175	10,550	12,200
	160	With bottom rolling doors, minimum		1.25		5,975	1,575	7,550	8,725
	180	Maximum		.97		7,125	2,050	9,175	10,600
	200	Circular type, prefab., steel frame, plastic skin, electric							
	201	door, including foundations, 80' diameter,							
	202	for up to 5 light planes, minimum	E-2	.50	Total	70,200	3,975	74,175	82,500
	220	Maximum	"	.25	"	78,000	7,925	85,925	96,500
	42-001	**ICE SKATING** Equipment incl. refrigeration and plumbing, not							
	002	including building or slab, 85' x 200' rink							
	030	55° system, 5 mos., 100 ton			Total				195,000
	070	90° system, 12 mos., 135 ton			"				205,000
	100	Dasher boards, polyethylene coated plywood, 3' acrylic							
	102	screen at sides, 5' acrylic ends, 85' x 200'	F-5	.09	Ea.	62,500	7,475	69,975	79,500
	110	Fiberglass & aluminum construction, same sides and ends	"	.08	"	80,000	8,425	88,425	100,000
	115								

13.1 Special Construction

		CREW	DAILY OUTPUT	UNIT	BARE COSTS MAT.	BARE COSTS INST.	BARE COSTS TOTAL	TOTAL INCL O&P
42 120	Subsoil heating system (recycled from compressor), 85' x 200'	Q-7	.30	Ea.	13,100	2,325	15,425	17,800
130	Subsoil insulation, 2 lb. polystyrene with vapor barrier, 85' x 200'	F-2	.16	"	18,250	2,100	20,350	23,100
45-001	**INTEGRATED CEILINGS** Lighting, ventilating & acoustical							
010	Luminaire, 5' x 5' modules, suspended, 50% lighted	L-3	90	S.F.	3.05	3.78	6.83	8.85
020	100% lighted	"	50		4.55	6.80	11.35	14.85
040	For ventilating capacity with perforations, add			↓				.26
060	For supply air diffuser, add			Ea.				58
065								
070	Dimensionaire, 2' x 4' board system, see also division 9.5-25	L-3	50	L.F.	7.10	6.80	13.90	17.70
090	Tile system	1 Carp	250	S.F.	1.85	.64	2.49	2.96
100	For air bar suspension, add, minimum							.53
110	Maximum			↓				.90
120	For vaulted coffer, including fixture, stock, add			Ea.	69		69	76M
130	Custom, add, minimum							120
140	Average							315
150	Maximum			↓				525
180	Radiant hot water system with finished acoustic ceiling,							
181	not including supply piping. Heating only (gross S.F.)							
210	Elementary schools, minimum			S.F.				4.45
220	Maximum							5.60
240	High schools and colleges, minimum							3.95
250	Maximum							5.60
270	Libraries, minimum							4
280	Maximum							5.30
300	Hospitals, minimum							5.45
310	Maximum							6.80
330	Office buildings, minimum							3.90
340	Maximum							5.15
360	For combined heating and cooling, add, minimum				30%	30%		
370	Maximum				40%	40%		
400	Radiant electric ceiling board, strapped between joists	1 Elec	250	↓	1.47	.72	2.19	2.65
410	2' x 4' heating panel for grid system, manila finish		25	Ea.	34	7.15	41.15	48
420	Textured epoxy finish		22		35	8.15	43.15	50
430	Vinyl finish		19		37	9.45	46.45	54
440	Hair cell, ABS plastic finish	↓	13	↓	60	13.80	73.80	86
445								
450	2' x 4' alternate blank panel, for use with above							
460	Manila finish	1 Elec	50	Ea.	8.10	3.58	11.68	14.05
470	Textured epoxy finish		45		17.90	3.98	21.88	25
480	Vinyl finish		40		19.85	4.48	24.33	28
490	Hair cell, ABS plastic finish	↓	25	↓	54	7.15	61.15	70
46-001	**KIOSKS** Round, 5' diameter, 8' high, 1/4" fiberglass wall			Ea.	2,850		2,850	3,125M
010	1" insulated double wall, fiberglass				3,250		3,250	3,575M
050	Rectangular, 5' x 9', 7'-6" high, 1/4" fiberglass wall				5,975		5,975	6,575M
060	1" insulated double wall, fiberglass			↓	6,800		6,800	7,475M
47-001	**MUSIC** Practice room, modular, perforated steel, under 500 S.F.	2 Carp	70	S.F.Surf	18.10	4.57	22.67	27
010	Over 500 S.F.	"	80	"	12.15	4	16.15	19.15
50-001	**PEDESTAL ACCESS FLOORS** Computer room application, metal							
002	Particle bd. or steel panels, no covering, under 6000 S.F.	2 Carp	1,000	S.F.	6	.32	6.32	7.05
030	Metal covered, over 6000 S.F.	↓	1,100		4.90	.29	5.19	5.80
040	Aluminum, 24" panels	↓	500		15	.64	15.64	17.45
060	For carpet covering, add				2.25		2.25	2.47M
070	For vinyl floor covering, add				.75		.75	.82M
090	For high pressure laminate covering, add				1.25		1.25	1.37M
091	For snap on stringer system, add	2 Carp	1,000	↓	.35	.32	.67	.85
095	Office applications, to 8" high, steel panels,							
096	no covering, over 6000 S.F.	2 Carp	400	S.F.	4.75	.80	5.55	6.40
100	Machine cutouts after initial installation	1 Carp	10	Ea.	5	16	21	29
105								

13.1 Special Construction

			CREW	DAILY OUTPUT	UNIT	BARE COSTS MAT.	BARE COSTS INST.	BARE COSTS TOTAL	TOTAL INCL O&P
50	110	Air conditioning grilles, 4" x 12"	1 Carp	17	Ea.	22	9.40	31.40	38
	115	6" x 18"	"	14	"	35	11.45	46.45	55
	120	Approach ramps, minimum	2 Carp	85	S.F.	9.60	3.76	13.36	16
	130	Maximum	"	60	"	18.20	5.35	23.55	28
	150	Handrail, 2 rail aluminum	1 Carp	15	L.F.	25	10.65	35.65	43
52-001		**PORTABLE BOOTHS** Prefab aluminum with doors, windows, lights,							
	010	wiring & insulation, 15 S.F. building, O.D., painted, minimum			S.F.	130		130	145M
	030	30 S.F. building, minimum				88		88	97M
	040	50 S.F. building, minimum				70		70	77M
	060	80 S.F. building, minimum				59		59	65M
	070	100 S.F. building, minimum			↓	56		56	62M
	090	Acoustical booth, 27 Db @ 1000 Hz, 15 S.F. floor			Ea.	1,650		1,650	1,825M
	100	7' x 7'-6", including light & ventilation				5,225		5,225	5,750M
	120	Ticket booth, galv. steel, not incl. foundations., 3' x 3'-6"				1,750		1,750	1,925M
	130	3'-6" x 10'-4"				3,200		3,200	3,525M
55-001		**RADIO TOWERS** Guyed, 50' high, 40 lb. section, wind load 30 psf	2 Sswk	.40		810	870	1,680	2,250
	010	Wind load 50 psf		.40		915	870	1,785	2,375
	030	200' high, 40 lb. section, wind load 30 psf		.20		4,850	1,725	6,575	8,050
	040	70 lb. section, wind load 50 psf		.20		6,300	1,725	8,025	9,650
	060	300' high, 70 lb. section, wind load 30 psf		.12		7,600	2,900	10,500	12,900
	070	90 lb. section, wind load 50 psf		.10		8,475	3,475	11,950	14,800
	080	400' high, 90 lb. section, 30 psf wind load	↓	.05	↓	11,800	6,950	18,750	23,900
	085								
	090	Self-supporting, 30 psf wind load, 60' high	2 Sswk	.80	Ea.	1,300	435	1,735	2,100
	100	120' high		.40		3,950	870	4,820	5,700
	120	200' high	↓	.20	↓	10,800	1,725	12,525	14,600
57-001		**REFRIGERATORS** Curbs, 12" high, 4" thick, concrete	2 Carp	58	L.F.	2.65	5.50	8.15	10.90
	100	Doors, see division 8.3-06							
	240	Finishes, 2 coat portland cement plaster, 1/2" thick	1 Plas	48	S.F.	.75	3.32	4.07	5.55
	250	For galvanized reinforcing mesh, add	1 Lath	335		.50	.48	.98	1.23
	270	3/16" thick latex cement	1 Plas	88		1.25	1.81	3.06	3.95
	290	For glass cloth reinforced ceilings, add	"	450		.30	.35	.65	.83
	310	Fiberglass panels, 1/8" thick	1 Carp	135		1.52	1.19	2.71	3.39
	320	Polystyrene, plastic finish ceiling, 1" thick		230		1.40	.70	2.10	2.55
	340	2" thick		230		1.63	.70	2.33	2.80
	350	4" thick	↓	205		1.75	.78	2.53	3.06
	380	Floors, concrete, 4" thick	1 Cefi	93		.73	1.65	2.38	3.13
	390	6" thick	"	85	↓	1.10	1.81	2.91	3.75
	400	Insulation, 1" to 6" thick, cork			B.F.	.57		.57	.62M
	410	Urethane				.50		.50	.55M
	430	Polystyrene, regular				.37		.37	.40M
	440	Bead board			↓	.27		.27	.29M
	460	Installation of above, add per layer	2 Carp	680	S.F.	.17	.47	.64	.87
	470	Wall and ceiling juncture		270	L.F.	.96	1.19	2.15	2.77
	490	Partitions, galvanized sandwich panels, 4" thick, stock	↓	220	S.F.	4.40	1.45	5.85	6.95
	500	Aluminum or fiberglass		220	"	4.70	1.45	6.15	7.30
	520	Prefab walk-in, 7'-6" high, aluminum, incl. door & floors,							
	521	not incl. partitions or refrigeration, 6' x 6' O.D. nominal	2 Carp	48	S.F.Flr.	76	6.65	82.65	93
	550	10' x 10' O.D. nominal		100		46	3.20	49.20	55
	570	12' x 14' O.D. nominal		134.40		37	2.38	39.38	44
	580	12' x 20' O.D. nominal	↓	160	↓	33	2	35	39
	581								
	600	For aluminum exterior and interior, add			S.F.Flr.	22%			
	610	For 8'-6" high, add				5%			
	630	Rule of thumb for complete units, not incl. doors, cooler	2 Carp	134.40		68	2.38	70.38	78
	640	Freezer		100	↓	82	3.20	85.20	95
	660	Shelving, plated or galvanized, steel wire type		360	S.F.Hor.	5	.89	5.89	6.80
	670	Slat shelf type	↓	375	"	7.25	.85	8.10	9.20

13.1 Special Construction

		Crew	Daily Output	Unit	Mat.	Inst.	Total	Total Incl O&P
57 690	For stainless steel shelving, add			S.F.Hor.	290%			
691								
700	Vapor barrier, on wood walls	2 Carp	1,815	S.F.	.07	.18	.25	.33
720	On masonry walls	"	1,360	"	.22	.24	.46	.58
750	For air curtain doors, see division 13.1-02							
59-001	**SHELTERS** Aluminum frame, acrylic glazing, 3' x 9' x 8' high	2 Sswk	2	Ea.	1,650	175	1,825	2,100
010	9' x 12' x 8' high	"	1	"	2,175	345	2,520	2,950
60-001	**SHIELDING LEAD**							
010	Laminated lead in wood doors, 1/16" thick			S.F.	20		20	22M
020	Lead lined door frame, not incl. steel frame							
021	or hardware, 1/16" thick	1 Lath	2.40	Ea.	130	67	197	240
030	Lead lath or sheets, 1/16" thick	2 Lath	135	S.F.	4.50	2.38	6.88	8.30
040	1/8" thick		120	"	8.50	2.68	11.18	13.15
060	Lead glass, 1/4" thick, 12" x 16"		13	Ea.	167	25	192	220
070	24" x 36"		8		750	40	790	880
080	36" x 60"		2		1,875	160	2,035	2,300
082								
085	Frame with 1/16" lead and voice passage, 36" x 60"	2 Lath	2	Ea.	545	160	705	825
087	24" x 36" frame		8	"	420	40	460	520
090	Lead gypsum board, 5/8" thick with 1/16" lead		160	S.F.	3.95	2.01	5.96	7.20
091	1/8" lead		140		7.35	2.30	9.65	11.35
093	1/32" lead		200		3.25	1.61	4.86	5.85
095	Lead headed nails (average 1 lb. per sheet)			Lb.	5		5	5.50M
100	Butt joints in 1/8" lead or thicker, lead strip, add	2 Lath	240	S.F.	.75	1.34	2.09	2.72
110								
120	X-ray protection, average radiography or fluoroscopy							
121	room, up to 300 S.F. floor, 1/16" lead, minimum	2 Lath	.25	Total	2,200	1,275	3,475	4,225
150	Maximum, 7'-0" walls	"	.15	"	3,500	2,150	5,650	6,875
160	Deep therapy X-ray room, 250 KV capacity,							
180	up to 300 S.F. floor, 1/4" lead, minimum	2 Lath	.08	Total	8,500	4,025	12,525	15,000
190	Maximum, 7'-0" walls	"	.06	"	11,500	5,350	16,850	20,200
62-001	**SHIELDING, RADIO FREQUENCY**							
002	Prefabricated or screen-type copper or steel, minimum	2 Carp	180	S.F.Surf	23	1.78	24.78	28
010	Average		155		25	2.06	27.06	30
015	Maximum		145		30	2.21	32.21	36
63-001	**SHOOTING RANGE** Incl. bullet traps, target provisions, controls,							
010	separators, ceiling system, etc. Not incl. structural shell							
020	Commercial	L-9	.88	Point	3,950	735	4,685	5,425
030	Law enforcement		.52		6,200	1,250	7,450	8,675
040	National Guard armories		.86		3,450	755	4,205	4,900
050	Reserve training centers		.86		2,250	755	3,005	3,600
060	Schools and colleges		.78		3,175	830	4,005	4,725
070	Major acadamies		.46		7,400	1,400	8,800	10,200
080	For acoustical treatment, add				8%	10%		
090	For lighting, add				20%	24%		
100	For plumbing, add				20%	20%		
110	For ventilating system, add, minimum				33%	35%		
120	Add, average				30%	30%		
130	Add, maximum				50%	50%		
65-001	**SILOS** Concrete stave industrial, not incl. foundations, conical or							
010	sloping bottoms, 12' diameter, 35' high	C-11	.30	Ea.	11,100	8,575	19,675	23,900
020	16' diameter, 45' high		.20		17,300	12,800	30,100	36,600
040	25' diameter, 75' high		.10		43,000	25,700	68,700	82,500
050	Steel, factory fab., 30,000 gallon cap., painted, minimum	L-5	1		10,200	1,600	11,800	13,500
070	Maximum		.50		15,900	3,225	19,125	22,100
080	Epoxy lined, minimum		1		16,400	1,600	18,000	20,400
100	Maximum		.50		20,500	3,225	23,725	27,200
67-001	**SPORT COURT** Floors, No. 2 & better maple, 25/32" thick			S.F.Flr.			5.25	6
010	Walls, laminated plastic bonded to galv. steel studs			S.F.Wall			6.85	8.05

13.1 Special Construction

		CREW	DAILY OUTPUT	UNIT	BARE COSTS MAT.	BARE COSTS INST.	BARE COSTS TOTAL	TOTAL INCL O&P
67	015	Laminated fiberglass surfacing, minimum			S.F.Wall			1.82	2
	018	Maximum			"			2.05	2.26
	030	Squash, regulation court in existing building, minimum			Court			1,200	13,800
	040	Maximum			"			22,500	25,800
	045	Rule of thumb for components:							
	047	Walls	3 Carp	.15	Court	9,650	3,200	12,850	15,300
	050	Floor	"	.25		3,900	1,925	5,825	7,075
	055	Lighting	2 Elec	.60		1,350	595	1,945	2,350
	060	Handball, racquetball court in existing building, minimum	C-1	.20		1,675	3,175	4,850	6,375
	080	Maximum	"	.10		22,300	6,325	28,625	33,600
	090	Rule of thumb for components: walls	3 Carp	.12		8,775	4,000	12,775	15,400
	100	Floor		.25		4,575	1,925	6,500	7,825
	110	Ceiling		.33		1,900	1,450	3,350	4,200
	120	Lighting	2 Elec	.60		1,375	595	1,970	2,375
70-001		SWIMMING POOL ENCLOSURE Translucent, free standing,							
	002	not including foundations, heat or light							
	020	Economy, minimum	2 Carp	200	S.F.Hor.	5.75	1.60	7.35	8.65
	030	Maximum		100		13.80	3.20	17	19.80
	040	Deluxe, minimum		100		19.85	3.20	23.05	26
	060	Maximum		70		205	4.57	209.57	230
	070	For motorized roof, 40% opening, solid roof, add				1.96		1.96	2.15M
	080	Skylight type roof, add				3.91		3.91	4.30M
	090	Air-inflated, including blowers and heaters, minimum							3.35
	100	Maximum							6.70
72-001		SWIMMING POOL EQUIPMENT Diving stand, stainless steel, 3 meter	2 Carp	.40	Ea.	2,525	800	3,325	3,925
	030	1 meter		2.70		1,700	120	1,820	2,050
	060	Diving boards, 16' long, aluminum		2.70		970	120	1,090	1,250
	070	Fiberglass		2.70		530	120	650	755
	090	Filter system, sand or diatomite type, incl. pump, 6000 gal./hr.	2 Plum	1.80	Total	740	200	940	1,100
	102	Add for chlorination system, 800 S.F. pool		3	Ea.	255	120	375	455
	104	5000 S.F. pool		3	"	535	120	655	760
	106								
	110	Gutter system, stainless steel, with grating, stock,							
	111	contains supply and drainage system	E-1	20	L.F.	120	29	149	175
	112	Integral gutter and 5' high wall system, stainless steel	"	10	"	175	58	233	280
	115								
	120	Ladders, heavy duty, stainless steel, 2 tread	2 Carp	7	Ea.	85	46	131	160
	150	4 tread		6		155	53	208	250
	180	Lifeguard chair, stainless steel, fixed		2.70		630	120	750	865
	190	Portable				980		980	1,075M
	210	Lights, underwater, 12 volt, with transformer, 300 watt	1 Elec	.40		175	450	625	835
	220	110 volt, 500 watt, standard		.40		60	450	510	710
	240	Low water cutoff type		.40		69	450	519	720
	280	Heaters, see division 15.5-86							
	300	Pool covers, reinforced vinyl			S.F.	.23		.23	.25M
	310	Vinyl water tube, minimum				.35		.35	.38M
	320	Maximum				.85		.85	.93M
	325	Sealed air bubble polyethylene solar blanket				.25		.25	.27M
	330	Slides, fiberglass, aluminum handrails & ladder, 6'-0", straight	2 Carp	1.60	Ea.	220	200	420	530
	332	7'-6", curved		3		325	105	430	510
	340	10'-6", curved		1		560	320	880	1,075
	342	12'-0", straight with platform		1.20		635	265	900	1,075
	450	Hydraulic lift, movable pool bottom, single ram							
	452	Under 1000 S.F. area	L-9	.03	Ea.	75,000	21,600	96,600	114,500
	460	Four ram lift, over 1000 S.F.	"	.02	"	90,000	32,400	122,400	147,000
	480								
	500	Removable access ramp, stainless steel	2 Clab	2	Ea.	6,000	125	6,125	6,775
	550	Removable stairs, stainless steel, collapsible	"	2	"	6,100	125	6,225	6,900
75-001		SWIMMING POOLS Residential in-ground, vinyl lined, concrete sides							
	002	Sides including equipment, sand bottom			S.F.Surf				16

13.1 Special Construction

		CREW	DAILY OUTPUT	UNIT	BARE COSTS MAT.	BARE COSTS INST.	BARE COSTS TOTAL	TOTAL INCL O&P
010	metal or polystyrene sides			S.F.Surf				12
020	(119) Add for vermiculite bottom			"	.65		.65	.71M
050	Gunite bottom and sides, white plaster finish							
060	350 S.F.			S.F.Surf				25
075	800 S.F.			"				16
080								
110	Motel, gunite with plaster finish, incl. medium							
115	capacity filtration & chlorination			S.F.Surf				22
120	Municipal, gunite with plaster finish, incl. high							
125	capacity filtration & chlorination			S.F.Surf				35
135	Add for formed gutters			L.F.	46		46	51M
136	Add for stainless steel gutters			"	125		125	140M
160	For water heating system, see division 15.5-86							
170	Filtration and deck equipment only, as % of			Total	20%	20%		
180	Deck equipment, rule of thumb, 20' x 40' pool			S.F.Pool				1.30
190	5000 S.F. pool			"				1.90
300	Painting pools, preparation + 3 coats, 20' x 40' pool, epoxy	2 Pord	.33	Total	525	935	1,460	1,900
310	Rubber base paint, 18 gallons	"	.33		395	935	1,330	1,775
350	42' x 82' pool, 75 gallons, epoxy paint	3 Pord	.14		2,425	3,300	5,725	7,375
360	Rubber base paint	"	.14	↓	1,700	3,300	5,000	6,575
77-001	**TANKS** Not incl. pipe or pumps, prestressed concrete, 250,000 gallons			Ea.				184K
010	500,000 gallons							294K
030	1,000,000 gallons							399K
040	2,000,000 gallons							625K
060	4,000,000 gallons							980K
070	6,000,000 gallons							1,300K
090	Steel, ground level, not incl. foundations, 100,000 gallons							84K
100	250,000 gallons							105K
120	500,000 gallons							184K
125	750,000 gallons							215K
130	1,000,000 gallons							278K
150	2,000,000 gallons							520K
160	4,000,000 gallons							835K
180	6,000,000 gallons							1,125K
185	8,000,000 gallons							1,500K
190	10,000,000 gallons			↓				2,000K
210	Steel standpipes, 100' to overflow, no foundations							
220	500,000 gallons			Ea.				245K
240	750,000 gallons							310K
250	1,000,000 gallons							368K
270	1,500,000 gallons							504K
280	2,000,000 gallons			↓				614K
300	Elevated water tanks, 100' to bottom capacity line							
301	50,000 gallons			Ea.				179K
330	100,000 gallons							225K
340	250,000 gallons							310K
360	500,000 gallons							500K
370	750,000 gallons							690K
390	1,000,000 gallons			↓				877K
391								
400	Fixed roof oil storage tanks, steel, (1 barrel=42 gallons)							
420	5,000 barrels			Ea.				69K
430	25,000 barrels							199K
450	55,000 barrels							336K
460	100,000 barrels							540K
480	150,000 barrels							760K
490	225,000 barrels			↓				1,145K
491								
510	Floating roof gasoline tanks, steel, 5,000 barrels			Ea.				120K
520	25,000 barrels			"				268K

13.1 Special Construction

		Description	CREW	DAILY OUTPUT	UNIT	BARE COSTS MAT.	BARE COSTS INST.	BARE COSTS TOTAL	TOTAL INCL O&P
77	540	55,000 barrels			Ea.				405K
	550	100,000 barrels							614K
	570	150,000 barrels							830K
	580	225,000 barrels							1,150K
	600	Wood tanks, ground level, 2" cypress, 3,000 gallons	C-1	.19		5,000	3,325	8,325	10,300
	610	2-1/2" cypress, 10,000 gallons		.12		11,200	5,275	16,475	19,900
	630	3" redwood or 3" fir, 20,000 gallons		.10		14,300	6,325	20,625	24,800
	640	30,000 gallons		.08		19,700	7,925	27,625	33,000
	660	45,000 gallons	↓	.07	↓	25,000	9,050	34,050	40,500
	661								
	670	Larger sizes, minimum			Gal.				.66
	690	Maximum			"				.81
	700	Vinyl coated fabric pillow tanks, freestanding, 5000 gallons	4 Clab	4	Ea.	1,550	125	1,675	1,900
	710	Supporting embankment not included, 25,000 gallons	6 Clab	2		5,450	380	5,830	6,550
	720	50,000 gallons	8 Clab	1.50		7,925	680	8,605	9,700
	730	100,000 gallons	9 Clab	.90		12,400	1,275	13,675	15,500
	740	150,000 gallons		.50		15,500	2,300	17,800	20,400
	750	200,000 gallons		.40		18,600	2,850	21,450	24,600
	760	250,000 gallons	↓	.30	↓	21,700	3,825	25,525	29,400
	78-001	**TENSION STRUCTURES** Rigid steel frame, vinyl coated polyester							
	010	fabric shell, 72' clear span, not incl. foundations or floors							
	020	4,800 S.F.	B-41	1,000	S.F.Flr.	7.05	.85	7.90	8.95
	030	12,000 S.F.		1,100		6.25	.77	7.02	7.95
	040	20,600 S.F.	↓	1,220		6.05	.70	6.75	7.65
	045	124' clear span, 36,900 S.F.	L-5	2,500		7	.64	7.64	8.65
	050	For roll-up door, 12' x 14', add	L-2	1	Ea.	5,850	285	6,135	6,850
	055								
	060	For side walls and doors, add, minimum			S.F.Flr.	10%			
	070	Add, maximum				25%			
	080	For sitework, simple foundation, etc., add, minimum						1.75	1.95
	090	Add, maximum			↓			2.75	3.05
	80-001	**THERAPEUTIC POOLS** See division 15.2-04							
	82-001	**VAULT FRONT** See division 8.3-69							

14.1 Conveying Systems

	Description	CREW	DAILY OUTPUT	UNIT	BARE COSTS MAT.	BARE COSTS INST.	BARE COSTS TOTAL	TOTAL INCL O&P
10-001	**CORRESPONDENCE LIFT** 1 floor 2 stop, 25 lb. capacity, electric	2 Elev	.20	Ea.	3,570	1,800	5,370	6,550
010	Hand, 5 lb. capacity		.20		1,050	1,800	2,850	3,800
15-001	**DUMBWAITERS** 2 stop, electric, minimum		.13		1,920	2,800	4,720	6,150
010	Maximum		.11		5,000	3,300	8,300	10,300
030	Hand, minimum		.23		620	1,575	2,195	2,975
040	Maximum		.19	↓	1,055	1,900	2,955	3,925
060	For each additional stop, electric, add		.54	Stop	750	670	1,420	1,800
070	Hand, add	↓	.60	"	545	605	1,150	1,475
20-001	**ELEVATORS** For multi-story buildings, housing project, minimum			% Total				2.50%
010	Maximum							4.50%
030	Office building, minimum							2.50%
040	Maximum			↓				8%
060	(125) Freight, 2 story, hydraulic, 4,000 lb. capacity, minimum	M-1	.09	Ea.	19,000	8,325	27,325	32,800
070	Maximum	"	.05	"	26,000	15,000	41,000	50,000

14.1 Conveying Systems

			CREW	DAILY OUTPUT	UNIT	BARE COSTS MAT.	BARE COSTS INST.	BARE COSTS TOTAL	TOTAL INCL O&P
20	090	10,000 lb. capacity, minimum	M-1	.07	Ea.	24,800	10,700	35,500	42,500
	100	Maximum		.05		34,500	15,000	49,500	59,500
	120	6 story hydraulic, 4,000 lb. capacity		.03		43,800	25,000	68,800	84,000
	130	10,000 lb. capacity		.03		50,500	25,000	75,500	91,000
	150	6 story geared electric 4,000 lb. capacity		.04		32,000	18,700	50,700	62,000
	160	10,000 lb. capacity		.03		55,400	25,000	80,400	96,500
	210	Passenger, self-service, to prices in rear							
	211	of section, add for all stainless steel cabs			Ea.	695		695	765M
	300	For glass enclosed panoramic cabs, add, minimum				7,250		7,250	7,975M
	310	Maximum				30,000		30,000	33,000M
	320	For glass enclosed shafts, see division 8.9							
	330								
	500	Passenger, pre-engineered, 5 story, hydraulic, 2,500 lb. capacity	M-1	.04	Ea.	45,400	18,700	64,100	76,500
	510	For less than 5 stops, deduct	"	.29	Stop	6,050	2,575	8,625	10,300
	520	For 4,000 lb. capacity, general purpose, add			Ea.	1,925		1,925	2,125M
	530	For hospital, add				9,625		9,625	10,600M
	540	10 story, geared traction, 200 FPM, 2,500 lb. capacity	M-1	.02		43,000	37,500	80,500	100,500
	550	For less than 10 stops, deduct		.34	Stop	1,875	2,200	4,075	5,200
	560	For 4,500 lb. capacity, general purpose		.02	Ea.	48,500	37,500	86,000	106,500
	570	For hospital		.02		52,500	37,500	90,000	111,000
	700	Residential, cab type, 1 floor, 2 stop, minimum	2 Elev	.20		5,200	1,800	7,000	8,350
	710	Maximum		.10		9,300	3,625	12,925	15,500
	720	2 floor, 3 stop, minimum		.12		6,300	3,025	9,325	11,300
	730	Maximum		.06		15,700	6,050	21,750	26,000
	770	Stair climber (chair lift), single seat, minimum		1		1,900	360	2,260	2,625
	780	Maximum		.20		2,500	1,800	4,300	5,375
	800	Wheelchair, porch lift, minimum		1		3,200	360	3,560	4,050
	850	Maximum		.50		6,375	725	7,100	8,075
	870	Stair lift, minimum		1		5,000	360	5,360	6,025
	890	Maximum		.20		8,000	1,800	9,800	11,400
25-001		ESCALATORS Per single unit, minimum	M-1	.06		54,000	12,500	66,500	77,000
	030	Maximum	"	.04		76,500	18,700	95,200	111,000
30-001		MATERIAL HANDLING Conveyers, gravity type 2" rollers, 3" O.C.							
	005	10' sections with supports, 50 lb./L.F. capacity, 15" wide			Ea.	275		275	305M
	010	27" wide				355		355	390M
	015	100 lb. capacity per L.F., 15" wide				285		285	315M
	020	27" wide				365		365	400M
	025	33" wide				400		400	440M
	035	Horizontal belt, center drive and takeup, 45 F.P.M.							
	040	16" belt, 26.5' length	M-2	.50	Ea.	2,200	700	2,900	3,400
	045	24" belt, 41.5' length		.40		3,100	875	3,975	4,625
	050	61.5' length		.30		4,100	1,175	5,275	6,150
	060	Inclined belt, 25° incline with horizontal loader and							
	062	end idler assembly, 34' length, 12" belt	M-2	.30	Ea.	3,100	1,175	4,275	5,050
	065	16" belt		.20		3,350	1,750	5,100	6,125
	070	24" belt		.15		3,900	2,325	6,225	7,550
	120	Conveyer, overhead, automatic powered							
	121	chain conveyer, 130 lb./L.F. capacity	M-2	17	L.F.	48	21	69	82
	150	Cranes, portable hydraulic, floor type, 2,000 lb. capacity			Ea.	1,575		1,575	1,725M
	160	4,000 lb. capacity				2,250		2,250	2,475M
	180	Movable gantry type, 12' to 15' range, 2,000 lb. capacity				1,825		1,825	2,000M
	190	6,000 lb. capacity				2,750		2,750	3,025M
	210	Hoists, electric overhead, chain, hook hung, 15' lift, 1 ton cap.				1,125		1,125	1,250M
	220	2 ton capacity				1,500		1,500	1,650M
	250	5 ton capacity				2,775		2,775	3,050M
	260	For hand-pushed trolley, add				15%			
	270	For geared trolley, add				30%			
	280	For motor trolley, add				60%			
	300	For lifts over 15 ft., 1 ton, add			L.F.	16		16	17.60M
	310	5 ton, add			"	28		28	31M

14.1 Conveying Systems

			CREW	DAILY OUTPUT	UNIT	BARE COSTS MAT.	BARE COSTS INST.	BARE COSTS TOTAL	TOTAL INCL O&P
30	330	Lifts, scissor type, portable, electric, 36" high, 1000 lb.			Ea.	2,150		2,150	2,375M
	340	42" high, 2000 lb.			"	2,450		2,450	2,700M
	360	Monorail, overhead, manual, channel type							
	370	125 lb. per L.F.	1 Mill	26	L.F.	6.60	6.40	13	16.30
	390	500 lb. per L.F.	"	21	"	9.90	7.90	17.80	22
	400	Trolleys for above, 2 wheel, 125 lb. capacity			Ea.	39		39	43M
	420	4 wheel, 500 lb. capacity				84		84	92M
	430	8 wheel, 2000 lb. capacity				200		200	220M
32	-001	MOTORIZED CAR Distribution systems, single track,							
	010	20 lb. per car capacity, material handling							
	020	Minimum	4 Mill	.19	Station	19,300	3,500	22,800	26,200
	030	Maximum	"	.15	"	27,000	4,425	31,425	36,000
	040	Larger system, incl. hospital transport, track type,							
	050	fully automated material handling system							
	060	Minimum	4 Mill	.05	Station	88,500	13,300	101,800	116,000
	070	Maximum	E-6	.05	"	520,000	75,500	595,500	680,000
35	-001	MOVING RAMPS AND WALKS Walk, 24" tread width, minimum	M-1	6.50	L.F.	310	115	425	505
	010	Maximum		4.43		450	170	620	735
	030	40" tread width walk, minimum		4.43		450	170	620	735
	040	Maximum		3.82		535	195	730	870
	060	Ramp, 12° incline, 32" tread width, minimum		5.27		380	140	520	620
	070	Maximum		3.82		535	195	730	870
	090	40" tread width, minimum		3.57		535	210	745	890
	100	Maximum		2.91		700	260	960	1,125
40	-001	PARCEL LIFT 20" x 20", 100 lb. capacity, electric, per floor	2 Mill	.25	Ea.	5,775	1,325	7,100	8,225
45	-001	PNEUMATIC TUBE SYSTEM Single tube, 2 stations,							
	002	100 ft. long, stock, economy,							
	010	3" diameter	2 Stpi	.12	Total	2,030	3,025	5,055	6,625
	030	4" diameter	"	.09	"	2,830	4,050	6,880	8,950
	040	Twin tube, two stations or more, conventional system							
	060	2-1/2" round	2 Stpi	62.50	L.F.	6.90	5.80	12.70	16
	070	3" round		46		7.90	7.90	15.80	20
	090	4" round		49.60		8.60	7.35	15.95	20
	100	4" x 7" oval		37.60		12.10	9.70	21.80	27
	105	Add for blower		2	System	1,980	180	2,160	2,450
	111	Plus for each round station, add		7.50	Ea.	175	49	224	265
	115	Plus for each oval station, add		7.50	"	175	49	224	265
	120	Alternate pricing method: base cost, minimum		.75	Total	1,700	485	2,185	2,575
	130	Maximum		.25	"	5,200	1,450	6,650	7,825
	150	Plus total system length, add, minimum		93.40	L.F.	4.60	3.90	8.50	10.70
	160	Maximum		37.60	"	12.10	9.70	21.80	27
	180	Completely automatic system, 4" round, 15 to 50 stations		.29	Station	9,900	1,250	11,150	12,700
	220	51 to 144 stations		.32		8,200	1,150	9,350	10,700
	240	6" round or 4" x 7" oval, 15 to 50 stations		.24		13,100	1,525	14,625	16,600
	280	51 to 144 stations		.23		11,500	1,575	13,075	14,900
50	-001	VERTICAL CONVEYER Automatic selective							
	010	central control, to 10 floors, base price	2 Mill	.04	Total	13,800	8,300	22,100	26,900
	020	For automatic service, any floor to any floor, add	"	1.15	Floor	1,545	290	1,835	2,100

15.1 Pipe & Fittings

		CREW	DAILY OUTPUT	UNIT	BARE COSTS MAT.	BARE COSTS INST.	BARE COSTS TOTAL	TOTAL INCL O&P
01-001	**AVERAGE** Square foot and percent of total							
010	job cost for plumbing, see division 17.1							
04-001	**BACKFLOW PREVENTER** Includes gate valves,							
002	and four test cocks, corrosion resistant, automatic operation							
400	Reduced pressure principle							
410	Threaded							
412	3/4" pipe size	1 Plum	16	Ea.	157	11.30	168.30	190
414	1" pipe size		14		189	12.90	201.90	225
415	1-1/4" pipe size		12		310	15.05	325.05	365
416	1-1/2" pipe size		10		341	18.05	359.05	400
418	2" pipe size	↓	7	↓	420	26	446	500
500	Flanged, bronze							
506	2-1/2" pipe size	Q-1	5	Ea.	1,260	65	1,325	1,475
508	3" pipe size		4.50		1,600	72	1,672	1,875
510	4" pipe size	↓	3		2,230	110	2,340	2,600
512	6" pipe size	Q-2	3	↓	4,700	170	4,870	5,425
560	Flanged, iron							
566	2-1/2" pipe size	Q-1	5	Ea.	1,150	65	1,215	1,350
568	3" pipe size		4.50		1,310	72	1,382	1,550
570	4" pipe size	↓	3		1,840	110	1,950	2,175
572	6" pipe size	Q-2	3		2,810	170	2,980	3,325
574	8" pipe size		2		6,140	255	6,395	7,125
576	10" pipe size	↓	1	↓	8,140	505	8,645	9,675
07-001	**CLEANOUTS**							
006	Floor type							
008	Round or square, scoriated nickel bronze top							
010	2" pipe size	1 Plum	10	Ea.	34.50	18.05	52.55	64
012	3" pipe size		8		36.75	23	59.75	73
014	4" pipe size	↓	6	↓	50.25	30	80.25	99
098	Round top, recessed for terrazzo							
100	2" pipe size	1 Plum	9	Ea.	62.50	20	82.50	98
108	3" pipe size		6		66.50	30	96.50	115
110	4" pipe size	↓	4		73.50	45	118.50	145
112	5" pipe size	Q-1	6	↓	118	54	172	210
10-001	**CLEANOUT TEE** Cast iron with countersunk plug							
020	2" pipe size	1 Plum	4	Ea.	13.75	45	58.75	80
022	3" pipe size		3.60		17.25	50	67.25	91
024	4" pipe size	↓	3.30		26.75	55	81.75	110
028	6" pipe size	Q-1	5	↓	76.25	65	141.25	180
050	For round smooth access cover, add				20%			
13-001	**CONNECTORS** Flexible, corrugated, 7/8" O.D., 1/2" I.D.							
005	Gas, seamless brass, steel fittings							
020	12" long	1 Plum	36	Ea.	4.60	5	9.60	12.30
022	18" long		36		5.70	5	10.70	13.50
024	24" long		34		6.80	5.30	12.10	15.15
028	36" long		32		8.05	5.65	13.70	17
034	60" long	↓	30	↓	12.20	6	18.20	22
200	Water, copper tubing, dielectric separators							
210	12" long	1 Plum	36	Ea.	4.34	5	9.34	12
226	24" long	"	34	"	6.45	5.30	11.75	14.75
16-001	**DRAINS**							
014	Cornice, C.I., 45° or 90° outlet							
018	1-1/2" & 2" pipe size	Q-1	14	Ea.	32.50	23	55.50	69
020	3" and 4" pipe size	"	12	↓	44	27	71	88
026	For galvanized body, add				8.55		8.55	9.40M
028	For polished bronze dome, add			↓	7.35		7.35	8.10M
040	Deck, auto park, C.I., 13" top							
044	3", 4", 5", and 6" pipe size	Q-1	8	Ea.	155	41	196	230

For expanded coverage of these items see *Means' Mechanical Cost Data 1986*

15.1 Pipe & Fittings

			CREW	DAILY OUTPUT	UNIT	BARE COSTS MAT.	BARE COSTS INST.	BARE COSTS TOTAL	TOTAL INCL O&P
16	048	For galvanized body, add			Ea.	69.50		69.50	76M
	200	Floor, medium duty, C.I., deep flange, 7" top							
	204	2" and 3" pipe size	Q-1	12	Ea.	26.75	27	53.75	69
	208	For galvanized body, add	↓		↓	11.45		11.45	12.60M
	212	For polished bronze top, add				14.75		14.75	16.20M
	240	Heavy duty, with sediment bucket, C.I., 12" loose grate							
	242	3", 4", 5", and 6" pipe size	Q-1	9	Ea.	91.50	36	127.50	155
	246	For polished bronze top, add			"	38.25		38.25	42M
	250	Heavy duty, cleanout & trap w/bucket, C.I., 15" top							
	254	2", 3", and 4" pipe size	Q-1	6	Ea.	663.75	54	717.75	810
	256	For galvanized body, add			↓	196		196	215M
	258	For polished bronze top, add				207		207	230M
	278	Shower, with strainer, uniform diam. trap, bronze top							
	280	1-1/2", 2" and 3" pipe size	Q-1	8	Ea.	56.25	41	97.25	120
	282	4" pipe size	"	7	↓	66.50	46	112.50	140
	284	For galvanized body, add				22.75		22.75	25M
	386	Roof, flat metal deck, C.I. body, 10" aluminum dome							
	389	3" pipe size	Q-1	14	Ea.	95.25	23	118.25	140
	392	6" pipe size	"	10	"	172	32	204	235
	398	Precast plank deck, C.I. body, aluminum dome							
	400	10" top, 2" pipe size	Q-1	14	Ea.	44	23	67	82
	410	10" top, 3" pipe size		13		52.50	25	77.50	94
	412	13" top, 4" pipe size		12		68.50	27	95.50	115
	414	13" top, 5" pipe size	↓	10		81	32	113	135
	422	For galvanized body, add			↓	37.25		37.25	41M
	462	Main, all aluminum, 12" low profile dome							
	464	2", 3" and 4" pipe size	Q-1	14	Ea.	135	23	158	180
	498	Scupper floor, oblique strainer, C.I.							
	500	6" x 7" top, 2", 3" and 4" pipe size	Q-1	16	Ea.	36.60	20	56.60	70
	510	8" x 12" top, 5" and 6" pipe size	"	14		71	23	94	110
	516	For galvanized body, add				40%			
	520	For polished bronze strainer, add			↓	100%			
	598	Trench, floor, heavy duty, modular, C.I., 12" x 12" top							
	600	2", 3", 4", 5", & 6" pipe size	Q-1	8	Ea.	119.25	41	160.25	190
	610	For polished bronze top, add			"	58		58	64M
	696	Backwater valve, in soil pipe, C.I. body,							
	698	bronze gate and automatic flapper valves							
	700	3" and 4" pipe size	Q-1	13	Ea.	237	25	262	295
	710	5" and 6" pipe size	"	13	"	381	25	406	455
	724	Bronze flapper valve, bolted cover							
	726	2" pipe size	Q-1	16	Ea.	79.50	20	99.50	115
	730	4" pipe size	"	13		153	25	178	205
	734	6" pipe size	Q-2	17	↓	220	30	250	285
	900	Sewer control system, valve unit, guardrail, leak detector							
	901	With fittings, monitor, 40' cable. No pit fabrication							
	902	4" pipe size	Q-2	.25	Ea.	4,350	2,025	6,375	7,700
	912	5" pipe size	↓	.22	↓	4,750	2,300	7,050	8,550
	914	6" pipe size		.20		5,650	2,525	8,175	9,875
	916	8" pipe size		.17		6,450	2,975	9,425	11,400
22	001	**FAUCETS/FITTINGS**							
	015	Bath, faucets, diverter spout combination, sweat	1 Plum	8	Ea.	37	23	60	73
	020	For integral stops, IPS unions, add				18		18	19.80M
	100	Kitchen sink faucets, top mount, cast spout	1 Plum	10		27	18.05	45.05	56
	110	For spray, add				8	10%		
	200	Laundry faucets, shelf type, I.P.S. or copper unions	1 Plum	12		25	15.05	40.05	49
	210	Lavatory faucet, centerset, without drain		10		19	18.05	37.05	47
	300	Service sink faucet, cast spout, pail hook, hose end		14		38	12.90	50.90	60
	400	Shower by-pass valve with union		18		29	10	39	46
	420	Shower thermostatic mixing valve, concealed	↓	8	↓	140	23	163	185

For expanded coverage of these items see *Means' Mechanical Cost Data 1986*

15.1 Pipe & Fittings

		CREW	DAILY OUTPUT	UNIT	BARE COSTS MAT.	BARE COSTS INST.	BARE COSTS TOTAL	TOTAL INCL O&P
22 500	Sillcock, compact, brass, I.P.S. or copper to hose	1 Plum	24	Ea.	4.07	7.50	11.57	15.35
28-001	**HYDRANTS**							
005	Wall type, moderate climate, bronze, encased							
020	3/4" IPS connection	1 Plum	16	Ea.	101	11.30	112.30	125
030	1" IPS connection	"	14		115	12.90	127.90	145
040	For 3/4" adapter type vacuum breaker, add				45.75		45.75	50M
050	For anti-siphon type, add			↓	55.25		55.25	61M
100	Non-freeze, bronze, exposed							
110	3/4" IPS connection, 4" to 9" thick wall	1 Plum	14	Ea.	71.50	12.90	84.40	97
112	10" to 14" thick wall		12		77.25	15.05	92.30	105
114	15" to 19" thick wall		12		85.75	15.05	100.80	115
116	20" to 24" thick wall	↓	10		92.50	18.05	110.55	130
120	For 1" IPS connection, add				13.40	10%		
124	For 3/4" adapter type vacuum breaker, add				8.55		8.55	9.40M
128	For anti-siphon type, add			↓	18.10		18.10	19.90M
200	Non-freeze bronze, encased							
210	3/4" IPS connection, 5" to 9" thick wall	1 Plum	14	Ea.	121	12.90	133.90	150
214	15" to 19" thick wall	"	12		135.25	15.05	150.30	170
228	For anti-siphon type, add			↓	55.25		55.25	61M
300	Ground box type, bronze frame, 3/4" IPS connection							
308	Non-freeze, all bronze, polished face, set flush							
310	2 feet depth of bury	1 Plum	8	Ea.	143	23	166	190
318	6 feet depth of bury		7		182	26	208	235
322	8 feet depth of bury	↓	5		201	36	237	275
340	For 1" IPS connection, add				20	10%		
355	For 2" connection, add				455	24%		
360	For tapped drain port in box, add			↓	11.45		11.45	12.60M
500	Moderate climate, all bronze, polished face							
502	and scoriated cover, set flush							
510	3/4" IPS connection	1 Plum	16	Ea.	113	11.30	124.30	140
512	1" IPS connection	"	14	↓	126	12.90	138.90	155
520	For tapped drain port in box, add			↓	11.45		11.45	12.60M
30-001	**PIPING** See also division 2.5 for sitework							
34-001	**PIPE, BRASS** Plain end,							
090	Field threaded, coupling & clevis hanger 10' OC							
092	Regular weight							
098	1/8" diameter	1 Plum	62	L.F.	1.60	2.91	4.51	5.95
112	1/2" diameter		48		3.87	3.76	7.63	9.70
114	3/4" diameter		46		5.20	3.92	9.12	11.40
116	1" diameter	↓	43		7.33	4.20	11.53	14.15
118	1-1/4" diameter	Q-1	72		10.90	4.51	15.41	18.50
120	1-1/2" diameter		65		12.72	5	17.72	21
122	2" diameter	↓	53	↓	16.55	6.15	22.70	27
37-001	**PIPE, CAST IRON** Soil, on hangers 5' O.C.							
002	Single hub, service wt., lead & oakum joints 10' O.C.							
212	2" diameter	Q-1	50	L.F.	2.57	6.50	9.07	12.20
214	3" diameter		48		3.67	6.75	10.42	13.80
216	4" diameter	↓	44		4.85	7.40	12.25	16
218	5" diameter	Q-2	62		6.35	8.15	14.50	18.75
220	6" diameter	"	59		9.27	8.55	17.82	23
222	8" diameter	Q-3	49		14.16	14.05	28.21	36
224	10" diameter		44		21.17	15.65	36.82	46
226	12" diameter	↓	40		26.80	17.25	44.05	54
232	For service weight, double hub, add				44%			
234	For extra heavy, single hub, add				47%	4%		
236	For extra heavy, double hub, add			↓	95%	4%		
240	Lead for caulking			Lb.	.90		.90	.99M

For expanded coverage of these items see *Means' Mechanical Cost Data 1986*

15.1 Pipe & Fittings

			CREW	DAILY OUTPUT	UNIT	BARE COSTS MAT.	BARE COSTS INST.	BARE COSTS TOTAL	TOTAL INCL O&P
37	242	Oakum for caulking			Lb.	1.15		1.15	1.26M
	400	No hub, couplings 10' O.C.							
	410	1-1/2" diameter	Q-1	65	L.F.	2.74	5	7.74	10.25
	412	2" diameter		61		2.81	5.30	8.11	10.80
	416	4" diameter	↓	53	↓	4.76	6.15	10.91	14.10
	40-001	**PIPE, COPPER** 50/50 solder joints,							
	002	Type K tubing, couplings & clevis hangers 10' O.C.							
	110	1/4" diameter	1 Plum	55	L.F.	.49	3.28	3.77	5.30
	120	1" diameter		34		1.78	5.30	7.08	9.65
	126	2" diameter	↓	22	↓	4.34	8.20	12.54	16.65
	199								
	200	Type L tubing, couplings & hangers 10' O.C.							
	210	1/4" diameter	1 Plum	59	L.F.	.43	3.06	3.49	4.89
	212	3/8" diameter		56		.57	3.22	3.79	5.30
	214	1/2" diameter		53		.69	3.40	4.09	5.70
	216	5/8" diameter		49		.96	3.68	4.64	6.40
	218	3/4" diameter		44		1	4.10	5.10	7.05
	220	1" diameter		36		1.44	5	6.44	8.85
	222	1-1/4" diameter		29		1.90	6.20	8.10	11.10
	224	1-1/2" diameter		27		2.40	6.70	9.10	12.30
	226	2" diameter	↓	23		3.61	7.85	11.46	15.30
	228	2-1/2" diameter	Q-1	33		5.23	9.85	15.08	20
	230	3" diameter		30		7.07	10.80	17.87	23
	232	3-1/2" diameter		25		9.83	13	22.83	30
	234	4" diameter		23		11.84	14.10	25.94	33
	236	5" diameter	↓	19		29.59	17.10	46.69	57
	238	6" diameter	Q-2	24		37.47	21	58.47	72
	240	8" diameter	"	21		93.85	24	117.85	140
	241	For other than full hard temper, add				25%			
	258	For 95/5 solder, add					6%		
	259	For silver solder, add			↓		15%		
	400	Type DWV tubing, couplings & hangers 10' O.C.							
	410	1-1/4" diameter	1 Plum	30	L.F.	1.54	6	7.54	10.40
	412	1-1/2" diameter		28		1.89	6.45	8.34	11.40
	414	2" diameter	↓	24		2.44	7.50	9.94	13.55
	416	3" diameter	Q-1	32		4.19	10.15	14.34	19.30
	418	4" diameter		24		7.20	13.55	20.75	27
	420	5" diameter	↓	20		20.49	16.25	36.74	46
	422	6" diameter	Q-2	25	↓	28.47	20	48.47	61
	43-001	**PIPE, CORROSION RESISTANT** No couplings or hangers							
	002	Iron alloy, drain, mechanical joint							
	100	1-1/2" diameter	Q-1	70	L.F.	17.85	4.64	22.49	26
	110	2" diameter		66		20.50	4.92	25.42	30
	112	3" diameter		60		29	5.40	34.40	40
	114	4" diameter	↓	52	↓	36.50	6.25	42.75	49
	298	Plastic, Epoxy, fiberglass filament wound							
	300	2" diameter	Q-1	62	L.F.	4.68	5.25	9.93	12.70
	310	3" diameter		51		6.60	6.35	12.95	16.45
	312	4" diameter		45		8.45	7.20	15.65	19.75
	314	6" diameter	↓	32	↓	12.35	10.15	22.50	28
	398	Polyester, fiberglass filament wound							
	400	2" diameter	Q-1	62	L.F.	9.45	5.25	14.70	17.95
	410	3" diameter		51		12.25	6.35	18.60	23
	412	4" diameter		45		14.50	7.20	21.70	26
	414	6" diameter	↓	32	↓	20.15	10.15	30.30	37
	498	Polypropylene, acid resistant, Schedule 40							
	500	1-1/2" diameter	Q-1	68	L.F.	1.46	4.78	6.24	8.50
	510	2" diameter		62		1.88	5.25	7.13	9.65
	512	3" diameter	↓	51	↓	3.34	6.35	9.69	12.90

15.1 Pipe & Fittings

			CREW	DAILY OUTPUT	UNIT	BARE COSTS MAT.	BARE COSTS INST.	BARE COSTS TOTAL	TOTAL INCL O&P
43	514	4" diameter	Q-1	45	L.F.	5.05	7.20	12.25	16
	598	Proxylene, fire retardent, Schedule 40							
	600	1-1/2" diameter	Q-1	68	L.F.	1.94	4.78	6.72	9.05
	610	2" diameter		62		2.60	5.25	7.85	10.45
	612	3" diameter		51		4.72	6.35	11.07	14.40
	614	4" diameter	↓	45	↓	6.70	7.20	13.90	17.80
46	001	PIPE, GLASS Borosilicate, couplings & supports 10' O.C.							
	002	Drainage							
	110	1-1/2" diameter	Q-1	47	L.F.	4.47	6.90	11.37	14.90
	112	2" diameter		39		5.98	8.35	14.33	18.60
	114	3" diameter		34		7.99	9.55	17.54	23
	116	4" diameter		25		13.71	13	26.71	34
	118	6" diameter	↓	21	↓	24.67	15.45	40.12	50
49	001	PIPE, PLASTIC See also division 15.1-43							
	002	Fiberglass reinforced, couplings 10' O.C., hangers 3 per 10'							
	020	High strength							
	024	2" diameter	Q-1	58	L.F.	6.19	5.60	11.79	14.90
	026	3" diameter		51		8.33	6.35	14.68	18.35
	028	4" diameter		47		10.57	6.90	17.47	22
	030	6" diameter	↓	38		16.40	8.55	24.95	30
	032	8" diameter	Q-2	48	↓	27.56	10.50	38.06	46
	180	PVC, couplings 10' O.C., hangers 3 per 10'							
	182	Schedule 40							
	186	1/2" diameter	1 Plum	54	L.F.	.67	3.34	4.01	5.55
	187	3/4" diameter		51		.73	3.54	4.27	5.90
	188	1" diameter		46		.90	3.92	4.82	6.65
	189	1-1/4" diameter		42		1.10	4.30	5.40	7.40
	190	1-1/2" diameter	↓	36		1.20	5	6.20	8.55
	191	2" diameter	Q-1	59		1.49	5.50	6.99	9.60
	192	2-1/2" diameter		56		2.10	5.80	7.90	10.70
	193	3" diameter		53		2.59	6.15	8.74	11.70
	194	4" diameter		48		3.60	6.75	10.35	13.75
	195	5" diameter		43		6.31	7.55	13.86	17.85
	196	6" diameter	↓	39	↓	6.57	8.35	14.92	19.25
	410	DWV type, schedule 40, couplings 10' O.C., hangers 3 per 10'							
	412	ABS							
	414	1-1/4" diameter	1 Plum	42	L.F.	.86	4.30	5.16	7.15
	415	1-1/2" diameter	"	36	↓	.89	5	5.89	8.20
	416	2" diameter	Q-1	59	↓	1.01	5.50	6.51	9.05
	440	PVC							
	441	1-1/4" diameter	1 Plum	42	L.F.	1.08	4.30	5.38	7.40
	442	1-1/2" diameter	"	36		1.07	5	6.07	8.40
	446	2" diameter	Q-1	59		1.27	5.50	6.77	9.35
	447	3" diameter		53		2.26	6.15	8.41	11.35
	448	4" diameter		48		3.12	6.75	9.87	13.20
	449	6" diameter	↓	39	↓	6.65	8.35	15	19.35
	536	CPVC, couplings 10' O.C., hangers 3 per 10'							
	538	Schedule 40							
	546	1/2" diameter	1 Plum	54	L.F.	1.27	3.34	4.61	6.25
	547	3/4" diameter		51		1.51	3.54	5.05	6.75
	548	1" diameter		46		1.91	3.92	5.83	7.75
	549	1-1/4" diameter		42		2.60	4.30	6.90	9.05
	550	1-1/2" diameter	↓	36		3.20	5	8.20	10.75
	551	2" diameter	Q-1	59		3.67	5.50	9.17	12
	552	2-1/2" diameter		56		5.81	5.80	11.61	14.80
	553	3" diameter	↓	53	↓	7.44	6.15	13.59	17.05
52	001	PIPE, STAINLESS STEEL							
	350	Threaded, couplings and hangers 10' O.C.							

For expanded coverage of these items see *Means' Mechanical Cost Data 1986*

15.1 Pipe & Fittings

			CREW	DAILY OUTPUT	UNIT	BARE COSTS MAT.	BARE COSTS INST.	BARE COSTS TOTAL	TOTAL INCL O&P
52	352	Schedule 40, type 304							
	354	1/4" diameter	1 Plum	54	L.F.	1.97	3.34	5.31	7
	355	3/8" diameter		53		2.39	3.40	5.79	7.55
	356	1/2" diameter		52		2.95	3.47	6.42	8.25
	358	1" diameter	↓	45		4.23	4.01	8.24	10.45
	361	2" diameter	Q-1	57		8.05	5.70	13.75	17.10
	364	4" diameter	Q-2	51		26.18	9.90	36.08	43
	374	For small quantities, add			↓	10%			
	425	Schedule 40, type 316							
	429	1/4" diameter	1 Plum	54	L.F.	2.46	3.34	5.80	7.55
	430	3/8" diameter		53		2.97	3.40	6.37	8.20
	431	1/2" diameter		52		3.77	3.47	7.24	9.15
	432	3/4" diameter		51		4.30	3.54	7.84	9.85
	433	1" diameter	↓	45		5.74	4.01	9.75	12.10
	436	2" diameter	Q-1	57		11.30	5.70	17	21
	439	4" diameter	Q-2	51		35.96	9.90	45.86	54
	449	For small quantities, add			↓	10%			
	55-001	PIPE, STEEL							
	002	Diameters thru 4" are Spec. A-120, 5" up are Spec. A-53							
	005	Schedule 40, threaded, with couplings, and clevis type							
	006	hangers sized for covering, 10' O.C.							
	054	Black, 1/4" diameter	1 Plum	66	L.F.	.62	2.73	3.35	4.63
	055 (129)	3/8" diameter		65		.68	2.78	3.46	4.76
	056	1/2" diameter		63		.84	2.86	3.70	5.05
	057	3/4" diameter		61		.96	2.96	3.92	5.35
	058	1" diameter	↓	53		1.39	3.40	4.79	6.45
	059	1-1/4" diameter	Q-1	89		1.73	3.65	5.38	7.20
	060	1-1/2" diameter		80		2.01	4.06	6.07	8.10
	061	2" diameter		64		2.76	5.05	7.81	10.35
	062	2-1/2" diameter		50		4.50	6.50	11	14.35
	063	3" diameter		43		5.81	7.55	13.36	17.30
	064	3-1/2" diameter		40		8.02	8.10	16.12	21
	065	4" diameter		36		9.07	9	18.07	23
	066	5" diameter	↓	26		21.09	12.50	33.59	41
	067	6" diameter	Q-2	31		24.04	16.30	40.34	50
	129	Galvanized, 1/4" diameter	1 Plum	66		.82	2.73	3.55	4.85
	130	3/8" diameter		65		.90	2.78	3.68	5
	131	1/2" diameter		63		.97	2.86	3.83	5.20
	132	3/4" diameter		61		1.10	2.96	4.06	5.50
	133	1" diameter	↓	53		1.59	3.40	4.99	6.65
	134	1-1/4" diameter	Q-1	89		2	3.65	5.65	7.50
15	135	1-1/2" diameter		80		2.31	4.06	6.37	8.40
	136	2" diameter		64		3.18	5.05	8.23	10.85
	137	2-1/2" diameter		50		5.24	6.50	11.74	15.15
	138	3" diameter		43		6.74	7.55	14.29	18.35
	139	3-1/2" diameter		40		9.47	8.10	17.57	22
	140	4" diameter		36		10.71	9	19.71	25
	141	5" diameter	↓	26		23.69	12.50	36.19	44
	142	6" diameter	Q-2	31		27.40	16.30	43.70	54
	143	8" diameter		27		41.74	18.70	60.44	73
	144	10" diameter		23		61.08	22	83.08	99
	145	12" diameter	↓	18	↓	75.42	28	103.42	125
	200	Welded, on yoke & roll hangers							
	201	sized for covering, 10' O.C.							
	204	Black, 1" diameter	Q-15	93	L.F.	1.50	3.69	5.19	6.90
	207	2" diameter		61		2.73	5.65	8.38	11.05
	209	3" diameter		43		4.72	8	12.72	16.60
	211	4" diameter		37		6.88	9.30	16.18	21
	212	5" diameter	↓	32	↓	17.18	10.75	27.93	34

For expanded coverage of these items see *Means' Mechanical Cost Data 1986*

15.1 Pipe & Fittings

			CREW	DAILY OUTPUT	UNIT	BARE COSTS MAT.	BARE COSTS INST.	BARE COSTS TOTAL	TOTAL INCL O&P
55	213	6" diameter	Q-16	36	L.F.	18.48	14.55	33.03	41
	214	8" diameter		29		24.50	18.05	42.55	53
	215	10" diameter		24		38.95	22	60.95	74
	216	12" diameter	↓	19	↓	43.20	28	71.20	87
61-001		**PIPE, GROOVED-JOINT STEEL FITTINGS & VALVES**							
	002	Pipe includes coupling & clevis type hanger 10' O.C.							
	100	Schedule 40, black							
	104	3/4" diameter	1 Plum	71	L.F.	1.30	2.54	3.84	5.10
	105	1" diameter		63		1.55	2.86	4.41	5.85
	106	1-1/4" diameter		58		1.97	3.11	5.08	6.65
	107	1-1/2" diameter		51		2.25	3.54	5.79	7.60
	108	2" diameter	↓	40		2.80	4.51	7.31	9.60
	109	2-1/2" diameter	Q-1	57		4.08	5.70	9.78	12.75
	110	3" diameter		50		5.10	6.50	11.60	15
	111	4" diameter		45		7.43	7.20	14.63	18.60
	112	5" diameter	↓	37		18.10	8.80	26.90	33
	113	6" diameter	Q-2	42	↓	19.57	12.05	31.62	39
	180	Galvanized							
	184	3/4" diameter	1 Plum	71	L.F.	1.48	2.54	4.02	5.30
	185	1" diameter		63		1.75	2.86	4.61	6.05
	186	1-1/4" diameter		58		2.23	3.11	5.34	6.95
	187	1-1/2" diameter		51		2.55	3.54	6.09	7.90
	188	2" diameter	↓	40		3.20	4.51	7.71	10.05
	189	2-1/2" diameter	Q-1	57		4.67	5.70	10.37	13.40
	190	3" diameter		50		5.82	6.50	12.32	15.80
	191	4" diameter		45		8.40	7.20	15.60	19.70
	192	5" diameter	↓	37		20.53	8.80	29.33	35
	193	6" diameter	Q-2	42	↓	23.41	12.05	35.46	43
	400	Elbow, 90° or 45°, black steel							
	403	3/4" diameter	1 Plum	50	Ea.	4.53	3.61	8.14	10.20
	404	1" diameter		50		4.53	3.61	8.14	10.20
	405	1-1/4" diameter		40		6	4.51	10.51	13.10
	406	1-1/2" diameter		33		6.40	5.45	11.85	14.95
	407	2" diameter	↓	25		6.40	7.20	13.60	17.45
	408	2-1/2" diameter	Q-1	40		8.75	8.10	16.85	21
	409	3" diameter		33		11.60	9.85	21.45	27
	410	4" diameter		25		17	13	30	37
	411	5" diameter	↓	20		41.25	16.25	57.50	69
	412	6" diameter	Q-2	25		48.50	20	68.50	83
	416	For galvanized elbows, add			↓	15%			
	469	Tee, black steel							
	470	3/4" diameter	1 Plum	38	Ea.	6.40	4.75	11.15	13.90
	474	1" diameter		33		6.40	5.45	11.85	14.95
	475	1-1/4" diameter		27		7.45	6.70	14.15	17.85
	476	1-1/2" diameter		22		8	8.20	16.20	21
	477	2" diameter	↓	17		9.95	10.60	20.55	26
	478	2-1/2" diameter	Q-1	27		13.50	12.05	25.55	32
	479	3" diameter		22		18.95	14.75	33.70	42
	480	4" diameter		17		29	19.10	48.10	60
	481	5" diameter	↓	13		68.25	25	93.25	110
	482	6" diameter	Q-2	17		78.75	30	108.75	130
	490	For galvanized tees, add			↓	15%			
67-001		**SHOCK ABSORBERS**							
	050	3/4" male I.P.S. For 1 to 11 fixtures	1 Plum	12	Ea.	28.50	15.05	43.55	53
	060	1" male I.P.S., For 12 to 32 fixtures		8		57.25	23	80.25	96
	070	For 33 to 60 fixtures		8		85.75	23	108.75	125
	080	For 61 to 113 fixtures		8		215	23	238	270
	090	For 114 to 154 fixtures		8		257	23	280	315
	100	For 155 to 330 fixtures	↓	4	↓	299	45	344	395

For expanded coverage of these items see *Means' Mechanical Cost Data 1986*

15.1 Pipe & Fittings

					Bare Costs			Total
		Crew	Daily Output	Unit	Mat.	Inst.	Total	Incl O&P
70-001	**SUPPORTS/CARRIERS** For plumbing fixtures							
050	Drinking fountain, wall mounted							
060	Plate type with studs, top back plate	1 Plum	7	Ea.	11.75	26	37.75	50
070	Top front and back plate		7		14.65	26	40.65	53
080	Top & bottom, front & back plates, w/bearing jacks	↓	7	↓	21.25	26	47.25	61
300	Lavatory, concealed arm							
305	Floor mounted, single							
310	High back fixture	1 Plum	6	Ea.	63	30	93	115
320	Flat slab fixture		6		73	30	103	125
322	Paraplegic	↓	6	↓	61	30	91	110
325	Floor mounted, back to back							
330	High back fixtures	1 Plum	5	Ea.	89.50	36	125.50	150
340	Flat slab fixtures		5		110	36	146	175
343	Paraplegic	↓	5	↓	75	36	111	135
350	Wall mounted, in stud or masonry							
360	High back fixture	1 Plum	6	Ea.	37.25	30	67.25	84
370	Flat slab fixture	"	6	"	47.25	30	77.25	95
460	Sink, floor mounted							
465	Exposed arm system							
470	Single heavy fixture	1 Plum	5	Ea.	99.50	36	135.50	160
475	Single heavy sink with slab		5		136	36	172	200
480	Back to back, standard fixtures		5		128	36	164	195
485	Back to back, heavy fixtures		5		147	36	183	215
490	Back to back, heavy sink with slab	↓	5	↓	147	36	183	215
495	Exposed offset arm system							
500	Single heavy deep fixture	1 Plum	5	Ea.	100	36	136	160
510	Plate type system							
520	With bearing jacks, single fixture	1 Plum	5	Ea.	73.25	36	109.25	135
530	With exposed arms, single heavy fixture		5		110	36	146	175
540	Wall mounted, exposed arms, single heavy fixture		5		52	36	88	110
600	Urinal, floor mounted, 2" or 3" coupling, blowout type		6		67	30	97	115
610	With fixture or hanger bolts, blowout or washout		6		47.75	30	77.75	96
620	With bearing plate		6		53.25	30	83.25	100
630	Wall mounted, plate type system	↓	6	↓	53.25	30	83.25	100
698	Water closet, siphon jet							
700	Horizontal, adjustable, caulk							
704	Single, 4" pipe size	1 Plum	6	Ea.	88.75	30	118.75	140
705	4" pipe size, paraplegic		6		88.75	30	118.75	140
706	5" pipe size		6		117	30	147	170
711	Double, 4" pipe size, paraplegic		5		166	36	202	235
712	5" pipe size	↓	5	↓	217	36	253	290
716	Horizontal, adjustable, extended, caulk							
718	Single, 4" pipe size	1 Plum	6	Ea.	115	30	145	170
720	5" pipe size		6		153	30	183	210
724	Double, 4" pipe size		5		200	36	236	270
726	5" pipe size	↓	5	↓	254	36	290	330
740	Vertical, adjustable, caulk or thread							
744	Single, 4" pipe size	1 Plum	6	Ea.	105	30	135	160
746	5" pipe size		6		133	30	163	190
748	6" pipe size		5		155	36	191	225
752	Double, 4" pipe size		5		182	36	218	250
754	5" pipe size		5		210	36	246	285
756	6" pipe size	↓	4	↓	232	45	277	320
760	Vertical, adjustable, extended, caulk							
762	Single, 4" pipe size	1 Plum	6	Ea.	153	30	183	210
764	5" pipe size		6		173	30	203	235
768	6" pipe size		5		270	36	306	350
772	Double, 4" pipe size		5		238	36	274	315
774	5" pipe size		5		290	36	326	370
776	6" pipe size	↓	4	↓	370	45	415	470

For expanded coverage of these items see *Means' Mechanical Cost Data 1986*

15.1 Pipe & Fittings

						BARE COSTS			TOTAL
			CREW	DAILY OUTPUT	UNIT	MAT.	INST.	TOTAL	INCL O&P
1	778	Water closet, blow out							
	780	Vertical offset, caulk or thread							
	782	Single, 4" pipe size	1 Plum	6	Ea.	90.50	30	120.50	145
	784	Double, 4" pipe size	"	5	"	155	36	191	225
	788	Vertical offset, extended, caulk							
	790	Single, 4" pipe size	1 Plum	6	Ea.	120	30	150	175
	792	Double, 4" pipe size	"	5	"	170	36	206	240
	796	Vertical, for floor mounted back-outlet							
	798	Single, 4" thread, 2" vent	1 Plum	6	Ea.	78	30	108	130
	800	Double, 4" thread, 2" vent	"	6	"	160	30	190	220
	804	Vertical, for floor mounted back-outlet, extended							
	806	Single, 4" caulk, 2" vent	1 Plum	6	Ea.	106	30	136	160
	808	Double, 4" caulk, 2" vent	"	6	"	148	30	178	205
	820	Water closet, residential							
	822	Vertical centerline, floor mount							
	824	Single, 3" caulk, 2" or 3" vent	1 Plum	6	Ea.	50	30	80	98
	826	4" caulk, 2" or 4" vent		6		68.50	30	98.50	120
	828	3" copper sweat, 3" vent		6		56	30	86	105
	830	4" copper sweat, 4" vent		6		100	30	130	155
	840	Vertical offset, floor mount							
	842	Single, 3" or 4" caulk, vent	1 Plum	4	Ea.	90.50	45	135.50	165
	844	3" or 4" copper sweat, vent		5		145	36	181	210
	846	Double, 3" or 4" caulk, vent		4		155	45	200	235
	848	3" or 4" copper sweat, vent		5		220	36	256	295
	900	Water cooler (electric), floor mounted							
	910	Plate type with bearing plate, single	1 Plum	6	Ea.	48.50	30	78.50	97
73-001		**TRAPS**							
	003	Cast iron, service weight							
	005	Long P trap, 2" pipe size							
	110	12" long	Q-1	16	Ea.	8.55	20	28.55	39
	114	18" long	"	16	"	8.90	20	28.90	39
	118	Running trap, single hub, with vent							
	208	3" pipe size, 3" vent	Q-1	14	Ea.	14.75	23	37.75	50
	212	4" pipe size, 4" vent	"	13		19.20	25	44.20	57
	230	For double hub, vent, add				10%	20%		
	300	P trap, 2" pipe size	Q-1	16		6.20	20	26.20	36
	304	3" pipe size	"	14		7.75	23	30.75	42
	335	Deep seal trap							
	340	1-1/4" pipe size	Q-1	14	Ea.	26.75	23	49.75	63
	341	1-1/2" pipe size		14		26.75	23	49.75	63
	342	2" pipe size		14		26.75	23	49.75	63
	344	3" pipe size		12		41	27	68	84
	380	Drum trap, 4" x 5", 1-1/2" tapping	Q-2	17		10.30	30	40.30	54
	382	2" tapping	"	17		10.30	30	40.30	54
	384	For galvanized, add				100%			
	470	Copper, drainage, drum trap							
	480	3" x 5" solid, 1-1/2" pipe size	1 Plum	16	Ea.	9.90	11.30	21.20	27
	484	3" x 6" swivel, 1-1/2" pipe size		16		13.90	11.30	25.20	32
	490	4" x 8" swivel, 1-1/2" pipe size		13		34	13.90	47.90	57
	492	2" pipe size		13		34	13.90	47.90	57
	510	P trap, standard pattern							
	520	1-1/4" pipe size	1 Plum	18	Ea.	8.25	10	18.25	24
	524	1-1/2" pipe size		17		7.15	10.60	17.75	23
	526	2" pipe size		15		14.45	12.05	26.50	33
	528	3" pipe size		11		29	16.40	45.40	56
	534	With cleanout and slip joint							
	536	1-1/4" pipe size	1 Plum	18	Ea.	7.90	10	17.90	23
	540	1-1/2" pipe size	"	17	"	6.70	10.60	17.30	23
76-001		**VACUUM BREAKERS** Hot or cold water							
	103	Anti-siphon, brass							

For expanded coverage of these items see *Means' Mechanical Cost Data 1986*

15.1 Pipe & Fittings

		CREW	DAILY OUTPUT	UNIT	BARE COSTS MAT.	BARE COSTS INST.	BARE COSTS TOTAL	TOTAL INCL O&P
76 104	1/4" size	1 Plum	24	Ea.	10.05	7.50	17.55	22
105	3/8" size		24		10.05	7.50	17.55	22
106	1/2" size		24		11.50	7.50	19	24
108	3/4" size		20		13.50	9	22.50	28
110	1" size		19		21	9.50	30.50	37
112	1-1/4" size		15		35	12.05	47.05	56
114	1-1/2" size		13		41	13.90	54.90	65
116	2" size	↓	11		64	16.40	80.40	94
130	For polished chrome, (1/4" thru 1"), add			↓	40%			
190	Vacuum relief, water service, bronze							
200	1/2" size	1 Plum	30	Ea.	9.15	6	15.15	18.75
79-001	**VALVES, BRASS**							
050	Gas stops, without checks							
053	1/2" size	1 Plum	24	Ea.	4.39	7.50	11.89	15.70
054	3/4" size		22		5.41	8.20	13.61	17.80
055	1" size		19		10.45	9.50	19.95	25
056	1-1/4" size	↓	15	↓	15.25	12.05	27.30	34
80-001	**VALVES, BRONZE**							
102	Angle, 150 lb., rising stem, threaded							
103	1/8" size	1 Plum	24	Ea.	17.35	7.50	24.85	30
104	1/4" size		24		17.35	7.50	24.85	30
105	3/8" size		24		22	7.50	29.50	35
106	1/2" size		22		22	8.20	30.20	36
107	3/4" size		20		26	9	35	42
108	1" size		19		37	9.50	46.50	54
110	1-1/2" size		13		65	13.90	78.90	92
111	2" size	↓	11	↓	100	16.40	116.40	135
138	Ball, 150 psi, threaded							
140	1/4" size	1 Plum	24	Ea.	3.54	7.50	11.04	14.75
143	3/8" size		24		3.54	7.50	11.04	14.75
145	1/2" size		22		3.54	8.20	11.74	15.75
146	3/4" size		20		5.55	9	14.55	19.15
147	1" size		19		7.15	9.50	16.65	22
148	1-1/4" size		15		12.45	12.05	24.50	31
149	1-1/2" size		13		15.80	13.90	29.70	37
150	2" size	↓	11	↓	19.80	16.40	36.20	45
175	Check, swing, class 150, regrinding disc, threaded							
180	1/8" size	1 Plum	24	Ea.	10.25	7.50	17.75	22
183	1/4" size		24		10.25	7.50	17.75	22
184	3/8" size		24		10.25	7.50	17.75	22
185	1/2" size		24		10.70	7.50	18.20	23
186	3/4" size		20		12.45	9	21.45	27
187	1" size		19		16.30	9.50	25.80	32
188	1-1/4" size		15		23	12.05	35.05	43
189	1-1/2" size		13		26	13.90	39.90	49
190	2" size	↓	11		40	16.40	56.40	68
191	2-1/2" size	Q-1	15		78	22	100	115
200	For 200 lb., add				20%	10%		
204	For 300 lb., add			↓	60%	15%		
285	Gate, NRS, soldered, 300 psi							
290	3/8" size	1 Plum	24	Ea.	10.45	7.50	17.95	22
292	1/2" size		24		10.45	7.50	17.95	22
294	3/4" size		20		13.55	9	22.55	28
295	1" size		19		16.30	9.50	25.80	32
296	1-1/4" size		15		22	12.05	34.05	42
297	1-1/2" size		13		28	13.90	41.90	51
298	2" size	↓	11		40	16.40	56.40	68
299	2-1/2" size	Q-1	15		82	22	104	120
300	3" size	"	13	↓	117	25	142	165

For expanded coverage of these items see *Means' Mechanical Cost Data 1986*

15.1 Pipe & Fittings

		CREW	DAILY OUTPUT	UNIT	BARE COSTS MAT.	BARE COSTS INST.	BARE COSTS TOTAL	TOTAL INCL O&P
385	Rising stem, soldered, 300 psi							
395	1" size	1 Plum	19	Ea.	16.30	9.50	25.80	32
398	2" size	"	11		40	16.40	56.40	68
400	3" size	Q-1	13	↓	115	25	140	165
425	Threaded, class 150							
430	1/8" size	1 Plum	24	Ea.	10.55	7.50	18.05	22
431	1/4" size		24		10.55	7.50	18.05	22
432	3/8" size		24		10.55	7.50	18.05	22
433	1/2" size		24		10.55	7.50	18.05	22
434	3/4" size		20		13.10	9	22.10	27
435	1" size		19		16.60	9.50	26.10	32
436	1-1/4" size		15		22	12.05	34.05	42
437	1-1/2" size		13		28	13.90	41.90	51
438	2" size	↓	11		39	16.40	55.40	67
439	2-1/2" size	Q-1	15		85	22	107	125
440	3" size	"	13		120	25	145	170
450	For 300 psi, threaded, add				100%	15%		
454	For chain operated type, add			↓	15%			
485	Globe, class 150, rising stem, threaded							
490	1/8" size	1 Plum	24	Ea.	13.85	7.50	21.35	26
492	1/4" size		24		13.85	7.50	21.35	26
494	3/8" size		24		14.30	7.50	21.80	27
495	1/2" size		24		14.65	7.50	22.15	27
496	3/4" size		20		18.40	9	27.40	33
497	1" size		19		30	9.50	39.50	47
498	1-1/4" size		15		43	12.05	55.05	65
499	1-1/2" size		13		53	13.90	66.90	78
500	2" size	↓	11		81	16.40	97.40	115
501	2-1/2" size	Q-1	15		155	22	177	200
502	3" size	"	13		215	25	240	275
512	For class 300, threaded, add			↓	150%	15%		
560	Relief, pressure & temperature, self-closing, ASME							
564	3/4" size	1 Plum	28	Ea.	30.75	6.45	37.20	43
565	1" size		24		43.60	7.50	51.10	59
566	1-1/4" size		20		89.25	9	98.25	110
567	1-1/2" size		18		171	10	181	205
568	2" size	↓	16	↓	186	11.30	197.30	220
595	Pressure, poppet type, threaded							
600	1/2" size	1 Plum	30	Ea.	8	6	14	17.50
604	3/4" size	"	28	"	8.60	6.45	15.05	18.75
640	Pressure, water, ASME, threaded							
644	3/4" size	1 Plum	28	Ea.	23.25	6.45	29.70	35
645	1" size		24		39.75	7.50	47.25	55
646	1-1/4" size		20		60	9	69	79
647	1-1/2" size		18		84	10	94	105
648	2" size	↓	16	↓	121	11.30	132.30	150
690	Reducing, water pressure							
692	300 psi to 25-75 psi, threaded or sweat							
694	1/2" size	1 Plum	24	Ea.	41	7.50	48.50	56
695	3/4" size		20		48.25	9	57.25	66
696	1" size		19		74.25	9.50	83.75	95
697	1-1/4" size		15		155	12.05	167.05	190
698	1-1/2" size	↓	13	↓	175	13.90	188.90	215
835	Tempering, water, sweat connections							
840	1/2" size	1 Plum	24	Ea.	18.55	7.50	26.05	31
844	3/4" size	"	20	"	21.50	9	30.50	37
865	Threaded connections							
870	1/2" size	1 Plum	24	Ea.	21.50	7.50	29	35
874	3/4" size		20		26.50	9	35.50	42
875	1" size	↓	19	↓	69.50	9.50	79	90

For expanded coverage of these items see *Means' Mechanical Cost Data 1986*

15.1 Pipe & Fittings

			CREW	DAILY OUTPUT	UNIT	BARE COSTS MAT.	BARE COSTS INST.	BARE COSTS TOTAL	TOTAL INCL O&P
80	876	1-1/4" size	1 Plum	15	Ea.	116	12.05	128.05	145
	877	1-1/2" size		13		126	13.90	139.90	160
	878	2" size	↓	11	↓	190	16.40	206.40	235
82-001		**VALVES, IRON BODY**							
	102	Butterfly, wafer type, lever actuator							
	103	2" size	1 Plum	8	Ea.	49.25	23	72.25	87
	104	2-1/2" size	Q-1	5		56.50	65	121.50	155
	105	3" size		4.50		58.75	72	130.75	170
	106	4" size	↓	3		74.50	110	184.50	240
	107	5" size	Q-2	3.40		94.75	150	244.75	320
	108	6" size	"	3	↓	120	170	290	375
	165	Gate, 125 lb., NRS, threaded							
	170	2" size	1 Plum	11	Ea.	119	16.40	135.40	155
	176	3" size	Q-1	13		157	25	182	210
	178	4" size	"	10	↓	200	32	232	265
	215	Flanged							
	220	2" size	1 Plum	5	Ea.	90	36	126	150
	224	2-1/2" size	Q-1	5		95	65	160	200
	226	3" size		4.50		110	72	182	225
	228	4" size	↓	3		150	110	260	320
	230	6" size	Q-2	3	↓	250	170	420	520
	355	OS&Y, flanged							
	360	2" size	1 Plum	5	Ea.	93	36	129	155
	366	3" size	Q-1	4.50		115	72	187	230
	368	4" size	"	3		160	110	270	335
	370	6" size	Q-2	3		255	170	425	525
	390	For 250 lb flanged, add			↓	120%	10%		
	435	Globe, OS&Y, class 125, threaded							
	440	2" size	1 Plum	11	Ea.	220	16.40	236.40	265
	444	2-1/2" size	Q-1	15		245	22	267	300
	445	3" size		13		290	25	315	355
	446	4" size	↓	10	↓	400	32	432	485
	545	Swing check, 125 lb., threaded							
	550	2" size	1 Plum	11	Ea.	80	16.40	96.40	110
	554	2-1/2" size	Q-1	15		96	22	118	135
	555	3" size		13		115	25	140	165
	556	4" size	↓	10	↓	175	32	207	240
	595	Flanged							
	600	2" size	1 Plum	5	Ea.	63	36	99	120
	604	2-1/2" size	Q-1	5		76	65	141	180
	605	3" size		4.50		85	72	157	200
	606	4" size	↓	3		130	110	240	300
	607	6" size	Q-2	3	↓	220	170	390	485
85-001		**VALVES, PLASTIC**							
	110	Angle, PVC, threaded							
	111	1/4" size	1 Plum	26	Ea.	12.10	6.95	19.05	23
	112	1/2" size		26		17.85	6.95	24.80	30
	113	3/4" size		25		24.50	7.20	31.70	37
	114	1" size	↓	23	↓	27.50	7.85	35.35	42
	115	Ball, PVC, socket or threaded, single union							
	120	1/4" size	1 Plum	26	Ea.	10	6.95	16.95	21
	122	3/8" size		26		10	6.95	16.95	21
	123	1/2" size		26		10	6.95	16.95	21
	124	3/4" size		25		11.50	7.20	18.70	23
	125	1" size		23		14.40	7.85	22.25	27
	126	1-1/4" size		21		18.55	8.60	27.15	33
	127	1-1/2" size		20		23.25	9	32.25	39
	128	2" size	↓	17	↓	33.50	10.60	44.10	52
	136	For PVC, flanged, add				45%	15%		

For expanded coverage of these items see *Means' Mechanical Cost Data 1986*

15.1 Pipe & Fittings

		CREW	DAILY OUTPUT	UNIT	BARE COSTS MAT.	BARE COSTS INST.	BARE COSTS TOTAL	TOTAL INCL O&P
140	For true union, socket or threaded, add			Ea.	40%	5%		
165	CPVC, socket or threaded, single union							
170	1/2" size	1 Plum	26	Ea.	14.90	6.95	21.85	26
172	3/4" size		25		16.60	7.20	23.80	29
173	1" size		23		20.75	7.85	28.60	34
175	1-1/4" size		21		28	8.60	36.60	43
176	1-1/2" size	↓	20		34.50	9	43.50	51
184	For CPVC, flanged, add				100%	15%		
188	For true union, socket or threaded, add			↓	50%	5%		
205	Polypropylene, threaded							
210	1/4" size	1 Plum	26	Ea.	14.80	6.95	21.75	26
212	3/8" size		26		14.80	6.95	21.75	26
213	1/2" size		26		14.80	6.95	21.75	26
214	3/4" size		25		17.35	7.20	24.55	30
215	1" size		23		21.75	7.85	29.60	35
216	1-1/4" size		21		28.80	8.60	37.40	44
217	1-1/2" size		20		36.45	9	45.45	53
218	2" size	↓	17	↓	49.40	10.60	60	70
485	Foot valve, PVC, socket or threaded							
490	1/2" size	1 Plum	34	Ea.	24.75	5.30	30.05	35
493	3/4" size		32		30.25	5.65	35.90	41
494	1" size		28		37	6.45	43.45	50
495	1-1/4" size		27		71.25	6.70	77.95	88
496	1-1/2" size	↓	26	↓	71.25	6.95	78.20	88
635	Y sediment strainer, PVC, socket or threaded							
640	1/2" size	1 Plum	26	Ea.	18.20	6.95	25.15	30
644	3/4" size		24		20.25	7.50	27.75	33
645	1" size		23		22.50	7.85	30.35	36
646	1-1/4" size		21		40.75	8.60	49.35	57
647	1-1/2" size	↓	20	↓	40.75	9	49.75	58
88-001	**VALVES, STEEL**							
080	Cast, angle, 150 lb., flanged							
083	2" size	1 Plum	8	Ea.	750	23	773	860
084	2-1/2" size	Q-1	5		780	65	845	950
085	3" size		4.50		780	72	852	960
086	4" size	↓	3		1,150	110	1,260	1,425
094	For 300 lb., flanged, add				30%	15%		
096	For 600 lb., flanged, add			↓	100%	20%		
135	Check valve, 150 lb., flanged							
140	2" size	1 Plum	8	Ea.	365	23	388	435
144	2-1/2" size	Q-1	5		455	65	520	595
145	3" size		4.50		455	72	527	605
146	4" size	↓	3		645	110	755	865
154	For 300 lb., flanged, add				50%	15%		
156	For 600 lb., flanged, add			↓	110%	20%		
195	Gate valve, 150 lb., flanged							
200	2" size	1 Plum	8	Ea.	380	23	403	450
204	2-1/2" size	Q-1	5		550	65	615	700
205	3" size		4.50		550	72	622	710
206	4" size	↓	3		645	110	755	865
207	6" size	Q-2	3	↓	1,050	170	1,220	1,400
365	Globe valve, 150 lb., flanged							
370	2" size	1 Plum	8	Ea.	505	23	528	590
374	2-1/2" size	Q-1	5		655	65	720	815
375	3" size		4.50		655	72	727	825
376	4" size	↓	3		950	110	1,060	1,200
377	6" size	Q-2	3	↓	1,450	170	1,620	1,850
515	Forged, angle, class 800, socket or threaded							
520	1/4" size	1 Plum	24	Ea.	77	7.50	84.50	96
522	3/8" size	"	24	"	77	7.50	84.50	96

For expanded coverage of these items see *Means' Mechanical Cost Data 1986*

15.1 Pipe & Fittings

		CREW	DAILY OUTPUT	UNIT	BARE COSTS MAT.	BARE COSTS INST.	BARE COSTS TOTAL	TOTAL INCL O&P
523	1/2" size	1 Plum	24	Ea.	77	7.50	84.50	96
524	3/4" size		20		91	9	100	115
525	1" size		19		100	9.50	109.50	125
528	2" size		11		375	16.40	391.40	435
565	Check valve, class 800, horizontal, socket or threaded							
570	1/4" size	1 Plum	24	Ea.	36	7.50	43.50	50
572	3/8" size		24		36	7.50	43.50	50
573	1/2" size		24		36	7.50	43.50	50
574	3/4" size		20		42	9	51	59
575	1" size		19		49	9.50	58.50	68
576	1-1/4" size		15		95	12.05	107.05	120
95-001	**VENT FLASHING**							
100	Aluminum with lead ring							
102	1-1/4" pipe	1 Plum	20	Ea.	3.80	9	12.80	17.20
103	1-1/2" pipe		20		3.80	9	12.80	17.20
104	2" pipe		18		3.99	10	13.99	18.90
105	3" pipe		17		4.88	10.60	15.48	21
106	4" pipe		16		5.30	11.30	16.60	22
135	Copper with lead ring							
140	1-1/4" pipe	1 Plum	20	Ea.	11.65	9	20.65	26
143	1-1/2" pipe		20		11.65	9	20.65	26
144	2" pipe		18		12.30	10	22.30	28
145	3" pipe		17		14.40	10.60	25	31
146	4" pipe		16		15.90	11.30	27.20	34
97-001	**WATER SUPPLY METERS**							
200	Domestic/commercial, bronze							
202	Threaded							
206	5/8" diameter, to 20 GPM	1 Plum	16	Ea.	38.50	11.30	49.80	59
208	3/4" diameter, to 30 GPM		14		73.50	12.90	86.40	99
210	1" diameter, to 50 GPM		12		90.60	15.05	105.65	120
230	Threaded/flanged							
234	1-1/2" diameter, to 100 GPM	1 Plum	8	Ea.	206	23	229	260
236	2" diameter, to 160 GPM	"	6	"	281	30	311	355
260	Flanged, compound							
264	3" diameter, 320 GPM	Q-1	3	Ea.	1,010	110	1,120	1,275
266	4" diameter, to 500 GPM		1.50		1,580	215	1,795	2,050
268	6" diameter, to 1,000 GPM		1		3,070	325	3,395	3,850
270	8" diameter, to 1,800 GPM		.80		5,810	405	6,215	6,975

15.2 Plumbing Fixtures

		CREW	DAILY OUTPUT	UNIT	BARE COSTS MAT.	BARE COSTS INST.	BARE COSTS TOTAL	TOTAL INCL O&P
01-001	**FIXTURES** Includes trim fittings unless otherwise noted							
008	For rough-in, supply, waste, and vent, see add for each type							
012	For electric water coolers, see Division 15.3-45							
016	For color, unless otherwise noted, add			Ea.	20%			
04-001	**BATHS**							
002								
010	Tubs, recessed porcelain enamel on cast iron, with trim							
014	42" x 37"	Q-1	5	Ea.	440	65	505	580
018	48" x 42"		4		700	81	781	885
022	72" x 36"		3		725	110	835	955
200	Enameled formed steel, 4'-6" long		5.80		175	56	231	275
220	5' long		5.50		165	59	224	265

For expanded coverage of these items see *Means' Mechanical Cost Data 1986*

15.2 Plumbing Fixtures

		CREW	DAILY OUTPUT	UNIT	BARE COSTS MAT.	INST.	TOTAL	TOTAL INCL O&P
400	Soaking, acrylic with pop-up drain, 40" x 40"	Q-1	5.50	Ea.	610	59	669	755
410	66" x 36" x 18-1/2" deep	"	5	"	1,170	65	1,235	1,375
600	Whirlpool, bath with vented overflow, molded fiberglass							
610	56" x 46" x 23"	Q-1	1	Ea.	2,150	325	2,475	2,825
960	Rough-in, supply, waste and vent, for all above tubs, add	"	1.73	"	68.79	190	258.79	345
12-001	**DENTAL FOUNTAIN** See also division 15.2-16.							
005	Stainless steel receptor	1 Plum	4	Ea.	145	45	190	225
010	Enameled steel receptor		4		110	45	155	185
960	For rough-in, supply and waste, add	↓	1.16	↓	38.40	155	193.40	265
16-001	**DRINKING FOUNTAIN** For connection to cold water supply							
100	Wall mounted, non-recessed							
140	Bronze, with no back	1 Plum	4	Ea.	520	45	565	635
180	Cast iron, enameled, for correctional institutions		4		220	45	265	305
200	Fiberglass, 12" back, single bubbler unit		4		235	45	280	325
204	Dual bubbler		3.20		320	56	376	435
240	Precast stone, no back	↓	4	↓	230	45	275	320
250								
270	Stainless steel, single bubbler, no back	1 Plum	4	Ea.	515	45	560	630
274	With back		4		725	45	770	865
278	Dual handle & wheelchair projection type		4		300	45	345	395
282	Dual level for handicapped type	↓	3.20	↓	565	56	621	705
330	Vitreous china							
334	7" back	1 Plum	4	Ea.	240	45	285	330
394	For vandal-resistant bottom plate, add				18		18	19.80M
396	For freeze-proof valve system, add	1 Plum	2		30	90	120	165
398	For rough-in, supply and waste, add	"	1.83	↓	24.76	99	123.76	170
399								
400	Wall mounted, semi-recessed							
420	Poly-marble, single bubbler	1 Plum	4	Ea.	285	45	330	380
460	Stainless steel, satin finish, single bubbler		4		325	45	370	425
490	Vitreous china, single bubbler		4		330	45	375	430
598	For rough-in, supply and waste, add	↓	1.83	↓	24.76	99	123.76	170
600	Wall mounted, fully recessed							
640	Poly-marble, single bubbler	1 Plum	4	Ea.	410	45	455	515
680	Stainless steel, single bubbler		4		350	45	395	450
756	For freeze-proof valve system, add		2		150	90	240	295
758	For rough-in, supply and waste, add	↓	1.83	↓	24.76	99	123.76	170
760	Floor mounted, pedestal type							
778	Wheelchair handicap unit	1 Plum	2	Ea.	595	90	685	785
840	Stainless steel, architectural style		2		640	90	730	835
860	Enameled iron, heavy duty service, 2 bubblers		2		560	90	650	745
866	4 bubblers		2		580	90	670	770
888	For freeze-proof valve system, add		2		160	90	250	305
890	For rough-in, supply and waste, add	↓	1.83	↓	24.76	99	123.76	170
910	Deck mounted							
950	Stainless steel, circular receptor	1 Plum	4	Ea.	180	45	225	265
976	White enameled steel, 14" x 9" receptor		4		107	45	152	185
986	White enameled cast iron, 24" x 16" receptor		3		120	60	180	220
998	For rough-in, supply and waste, add	↓	1.83	↓	24.76	99	123.76	170
20-001	**HOT WATER DISPENSERS**							
016	Commercial, 100 cup, 11.3 amp	1 Plum	14	Ea.	219	12.90	231.90	260
318	Household, 60 cup	"	14	"	121	12.90	133.90	150
24-001	**INDUSTRIAL SAFETY FIXTURES** Rough-in not included							
100	Eye wash fountain, aluminum bowl							
120	Wall mounted	Q-1	4	Ea.	190	81	271	325
140	Plastic bowl, pedestal mounted		4		110	81	191	240
160	Unmounted		4		70	81	151	195
180	Wall mounted	↓	4	↓	75	81	156	200

For expanded coverage of these items see *Means' Mechanical Cost Data 1986*

15.2 Plumbing Fixtures

			CREW	DAILY OUTPUT	UNIT	BARE COSTS MAT.	BARE COSTS INST.	BARE COSTS TOTAL	TOTAL INCL O&P
24	200	Stainless steel, pedestal mounted	Q-1	4	Ea.	150	81	231	280
	220	Unmounted		4		96	81	177	225
	240	Wall mounted	↓	4		99	81	180	225
	300	Eye wash, portable, self-contained			↓	225		225	250M
	400	Eye and face wash, combination fountain							
	420	Stainless steel, pedestal mounted	Q-1	4	Ea.	175	81	256	310
	440	Unmounted		4		120	81	201	250
	460	Wall Mounted		4		125	81	206	255
	500	Shower, single head, drench, ball valve, pull, freestanding		4		145	81	226	275
	520	Horizontal or vertical supply		4		125	81	206	255
	600	Multi-nozzle, eye/face wash combination	↓	4	↓	275	81	356	420
	620								
	640	Multi-nozzle, 12 spray, shower only	Q-1	4	Ea.	785	81	866	980
	660	For freeze-proof, add	"	6	"	250	54	304	355
28-001		**INTERCEPTORS**							
	002								
	015	Grease, cast iron, 4 GPM, 8 lb. fat capacity	1 Plum	4	Ea.	158	45	203	240
	020	7 GPM, 14 lb. fat capacity		4		219	45	264	305
	100	10 GPM, 20 lb. fat capacity		4		260	45	305	350
	104	15 GPM, 30 lb. fat capacity		4		380	45	425	485
	106	20 GPM, 40 lb. fat capacity	↓	3		465	60	525	600
	116	100 GPM, 200 lb. fat capacity	Q-1	2		1,715	160	1,875	2,125
	156	For chemical add-port, add				35		35	39M
	158	For seepage pan, add				5%			
	300	Hair, cast iron, 1-1/4" and 1-1/2" pipe connection	1 Plum	8		59	23	82	98
	310	For chrome-plated cast iron, add				44		44	48M
	400	Oil, fabricated steel, 10 GPM, 2" pipe size	1 Plum	4		335	45	380	435
	410	15 GPM, 2" or 3" pipe size		4		460	45	505	570
	412	20 GPM, 2" or 3" pipe size	↓	3		555	60	615	695
	422	100 GPM, 3" pipe size	Q-1	2		1,720	160	1,880	2,125
	600	Solids, precious metals recovery, C.I., 1-1/4" to 2" pipe	1 Plum	4		90	45	135	165
	610	Dental Lab., large, C.I., 1-1/2" to 2" pipe	"	3	↓	410	60	470	540
32-001		**LAVATORIES** With trim, white unless noted otherwise							
	050	Vanity top, porcelain enamel on cast iron							
	060	20" x 18"	Q-1	6.40	Ea.	105	51	156	190
	064	26" x 18" oval		6.40		135	51	186	220
	072	18" round	↓	6.40		97	51	148	180
	086	For color, add				25%			
	100	Cultured marble, 19" x 17", single bowl	Q-1	6.40		65	51	116	145
	104	25" x 19", single bowl		6.40		77	51	128	160
	132	73" x 22", double bowl	↓	4.80	↓	205	68	273	325
	158	For color, same price							
	190	Stainless steel, self-rimming, 25" x 22", single bowl, ledge	Q-1	6.40	Ea.	125	51	176	210
	196	17" x 22", single bowl		6.40		115	51	166	200
	260	Steel, enameled, 20" x 17", single bowl		5.80		60	56	116	145
	266	19" round		5.80		58	56	114	145
	290	Vitreous china, 20" x 16", single bowl		5.40		155	60	215	255
	296	20" x 17", single bowl		5.40		105	60	165	200
	302	19" round, single bowl		5.40		102	60	162	200
	320	22" x 13", single bowl	↓	5.40		102	60	162	200
	356	For color, add				20%			
	358	Rough-in, supply, waste and vent for all above lavatories	Q-1	1.96	↓	54.40	165	219.40	300
	400	Wall hung							
	404	Porcelain enamel on cast iron, 16" x 14", single bowl	Q-1	8	Ea.	187	41	228	265
	418	20" x 18", single bowl		8		121	41	162	190
	424	22" x 19", single bowl		8		189	41	230	265
	458	For color, add			↓	30%			
	550								
	600	Vitreous china, 18" x 15", single bowl with backsplash	Q-1	7	Ea.	120	46	166	200
	618	18" x 18", corner style	"	8	"	235	41	276	315

For expanded coverage of these items see *Means' Mechanical Cost Data 1986*

15.2 Plumbing Fixtures

		CREW	DAILY OUTPUT	UNIT	BARE COSTS MAT.	BARE COSTS INST.	BARE COSTS TOTAL	TOTAL INCL O&P
650	For color, add			Ea.	30%			
696	Rough-in, supply, waste and vent for above lavatories	Q-1	1.66	"	105.40	195	300.40	400
36-001	**LAUNDRY SINKS** With trim							
002	Porcelain enamel on cast iron, black iron frame							
005	24" x 20", single compartment	Q-1	6	Ea.	210	54	264	310
010	24" x 23", single compartment	"	6	"	235	54	289	335
200	Molded stone, on wall hanger or legs							
202	22" x 21", single compartment	Q-1	6	Ea.	90	54	144	175
210	45" x 21", double compartment	"	5	"	150	65	215	260
300	Plastic, on wall hanger or legs							
302	18" x 23", single compartment	Q-1	6.50	Ea.	52	50	102	130
330	40" x 24", double compartment		5.50		121	59	180	220
500	Stainless steel, counter top, 22" x 17" single compartment		6		140	54	194	230
510	19" x 22", single compartment		6		152	54	206	245
520	33" x 22", double compartment		5		156	65	221	265
960	Rough-in, supply, waste and vent, for all laundry sinks		1.84		58.30	175	233.30	320
40-001	**PUMPS** See also Division 15.4-40							
41-001	**PUMPS, CIRCULATING** Heated or chilled water application							
060	Bronze, sweat connections, 1/40 HP, in line							
064	3/4" size	Q-1	16	Ea.	71	20	91	105
100	Flange connection, 3/4" to 1-1/2" size							
104	1/12 HP	Q-1	6	Ea.	185	54	239	280
106	1/8 HP		6		290	54	344	395
110	1/3 HP		6		300	54	354	410
114	2" size, 1/6 HP		5		370	65	435	500
118	2-1/2" size, 1/4 HP		5		535	65	600	680
200	Cast iron, flange connection							
204	3/4" to 1-1/2" size, in line, 1/12 HP	Q-1	6	Ea.	125	54	179	215
210	1/3 HP		6		210	54	264	310
214	2" size, 1/6 HP		5		225	65	290	340
218	2-1/2" size, 1/4 HP		5		295	65	360	420
222	3" size, 1/4 HP		4		300	81	381	445
260	For non-ferrous impeller, add				3%			
43-001	**PUMPS, CONDENSATE RETURN SYSTEM**							
020	Simplex, motor, float switch, controls, cast iron receiver	Q-1	1	Ea.	895	325	1,220	1,450
100	Duplex, 2 pumps, motors, float switch,							
106	alternator asssembly, C.I. receiver	Q-1	.50	Ea.	2,420	650	3,070	3,600
45-001	**PUMPS, GENERAL UTILITY** With motor, mounted on base							
020	Multi-stage, horizontal split, for boiler feed applications							
030	Two stage, 3" discharge x 4" suction, 75 HP	Q-7	.30	Ea.	10,000	2,325	12,325	14,400
034	Four stage, 3" discharge x 4" suction, 150 HP	"	.18	"	17,000	3,875	20,875	24,300
200	Single stage							
206	End suction, 1"D. x 2"S., 3 HP	Q-1	.50	Ea.	2,500	650	3,150	3,700
210	1-1/2"D. x 3"S., 10 HP	"	.40		2,700	810	3,510	4,150
214	2"D. x 3"S., 15 HP	Q-2	.60		2,900	840	3,740	4,400
300	Double suction, 2"D. x 2-1/2"S., 10 HP	Q-1	.30		3,400	1,075	4,475	5,300
306	3"D. x 4"S., 15 HP	Q-2	.46		3,500	1,100	4,600	5,450
318	6"D. x 8"S., 60 HP	Q-3	.30		5,800	2,300	8,100	9,700
47-001	**PUMPS, GRINDER SYSTEM** Complete, includes check valve, tank,							
002	standard controls. Excavation not included							
026	Simplex, 11 GPM at 40 PSIG, 60 gal. tank			Ea.	1,750		1,750	1,925M
030	For manway, 26" I.D., 18" high, add				405		405	445M
034	26" I.D., 36" high, add				560		560	615M
038	43" I.D., 4' high, add				735		735	810M
49-001	**PUMPS, PRESSURE BOOSTER SYSTEM** Constant speed							
020	2 pump system with hydrocumulator							
030	100 GPM @ 50 psi	Q-2	.30	Ea.	9,975	1,675	11,650	13,400
040	300 GPM @ 100 psi	"	.26	"	13,900	1,950	15,850	18,100

For expanded coverage of these items see *Means' Mechanical Cost Data 1986*

15.2 Plumbing Fixtures

			CREW	DAILY OUTPUT	UNIT	BARE COSTS MAT.	BARE COSTS INST.	BARE COSTS TOTAL	TOTAL INCL O&P
49	100	3 pump system without hydrocumulator							
	110	300 GPM @ 100 psi	Q-2	.23	Ea.	14,400	2,200	16,600	19,000
	120	1000 GPM @ 100 psi	↓	.12	↓	27,300	4,200	31,500	36,100
	130	5000 GPM @ 100 psi		.08		46,700	6,325	53,025	60,500
51-001		PUMPS, PEDESTAL SUMP With float control							
	040	Molded base, 42 GPM at 15' head, 1/3 HP	1 Plum	5	Ea.	100	36	136	160
	080	Iron base, 42 GPM at 15' head, 1/3 HP	"	5	"	83	36	119	145
52-001		PUMPS, SEWAGE EJECTOR With operating and level controls							
	010	Simplex, bitumastic coated steel tank, cover, 10' head, 230 volt							
	060	Bronze pump							
	064	70 GPM, 1/3 HP	Q-1	3	Ea.	1,180	110	1,290	1,450
	068	143 GPM, 1/2 HP	"	2.50	"	1,230	130	1,360	1,550
	104	Cast iron pump							
	106	110 GPM, 1/2 HP	Q-1	2.50	Ea.	1,180	130	1,310	1,475
	108	173 GPM, 3/4 HP	"	2	"	2,050	160	2,210	2,500
53-001		PUMPS, SPRINKLER With check valve, steel base, 15' lift							
	010	37 GPM, 3/4 HP	1 Plum	4	Ea.	258	45	303	350
	014	56 GPM, 1-1/2 HP	"	2	↓	555	90	645	740
	018	68 GPM, 2 H.P.	Q-1	2		570	160	730	860
54-001		PUMPS, SUBMERSIBLE Dewatering							
	002	Sand & sludge, 20' head, starter & level control							
	005	Cast iron							
	010	2" discharge, 10 GPM	1 Plum	4	Ea.	340	45	385	440
	016	60 GPM	↓	3		470	60	530	605
	020	120 GPM		2		825	90	915	1,050
	100	160 GPM	Q-1	1.50		665	215	880	1,050
	110	3" discharge, 220 GPM	"	1.20	↓	1,660	270	1,930	2,225
	700	Sump pump, 10' head, automatic							
	710	Bronze, 22 GPM., 1/4 HP, 1-1/4" discharge	1 Plum	6	Ea.	200	30	230	265
	714	68 GPM, 1/2 HP, 1-1/4" or 1-1/2" dis.		6		325	30	355	400
	716	94 GPM, 1/2 HP, 1-1/4" or 1-1/2" dis.		5		440	36	476	535
	718	105 GPM, 1/2 HP, 2" or 3" discharge		4		420	45	465	525
	750	Cast iron, 23 GPM, 1/4 HP, 1-1/4" discharge		6		94	30	124	145
	754	35 GPM, 1/3 HP, 1-1/4" discharge		6		105	30	135	160
	756	68 GPM, 1/2 HP, 1-1/4" or 1-1/2" dis.	↓	5	↓	220	36	256	295
55-001		PUMPS, WELL Water system, with pressure control							
	002								
	100	Deep well, multi-stage jet, 42 gal. tank							
	104	110' lift, 40 lb. discharge, 5 GPM, 3/4 HP	1 Plum	.80	Ea.	455	225	680	825
	200	Shallow well, reciprocating, 25 gal. tank							
	204	25' lift, 5 GPM, 1/3 HP	1 Plum	2	Ea.	350	90	440	515
	300	Shallow well, single stage jet, 42 gal. tank,							
	304	15' lift, 40 lb. discharge, 16 GPM, 3/4 HP	1 Plum	2	Ea.	345	90	435	510
56-001		SHOWERS							
	150	Stall, with door and trim							
	300	Fiberglass, one piece, with 3 walls, 32" x 32" square	Q-1	2.40	Ea.	230	135	365	450
	310	36" x 36" square		2.40		265	135	400	485
	400	Polypropylene, with molded-stone floor, 30" x 30"	↓	2		245	160	405	505
	496	Rough-in, supply, waste and vent for above showers		1.71		39.30	190	229.30	320
	500	Built-in, head, arm, 4 GPM valve	1 Plum	4		42	45	87	110
	520	Head, arm, by-pass, integral stops, handles	"	3.60	↓	98	50	148	180
	540								
	600	Group, w/valve, rough-in and rigging not included							
	680	Column, 6 heads, no receptors, less partitions	Q-1	3	Ea.	1,575	110	1,685	1,900
	690	With enameled partitions		1		2,650	325	2,975	3,375
	760	5 heads, no receptors, less partitions		3		1,340	110	1,450	1,625
	770	With enameled partitions		1		2,180	325	2,505	2,875
	800	Corner, 2 heads, no receptors, less partitions		4		866	81	947	1,075
	810	With partitions	↓	2	↓	1,075	160	1,235	1,425

For expanded coverage of these items see *Means' Mechanical Cost Data 1986*

15.2 Plumbing Fixtures

		CREW	DAILY OUTPUT	UNIT	BARE COSTS MAT.	BARE COSTS INST.	BARE COSTS TOTAL	TOTAL INCL O&P
60-001	**SINKS** With faucets and drain							
200	Kitchen, counter top, P.E. on C.I., 24" x 21" single bowl	Q-1	3.20	Ea.	113	100	213	270
210	30" x 21" single bowl		3.20		134	100	234	295
220	32" x 21" double bowl		2.60		148	125	273	345
230	42" x 21" double bowl		2.60		255	125	380	460
260								
300	Stainless steel, self rimming, 19" x 18" single bowl	Q-1	3.20	Ea.	205	100	305	370
310	25" x 22" single bowl		3.20		228	100	328	400
400	Steel, enameled, with ledge, 24" x 21" single bowl		3.20		59	100	159	210
410	32" x 21" double bowl		2.60		67	125	192	255
496	For color sinks except stainless steel, add				10%			
498	For rough-in, supply, waste and vent, counter top sinks	Q-1	1.85		58.30	175	233.30	320
500	Kitchen, raised deck, P.E. on C.I.							
510	32" x 21", dual level, double bowl	Q-1	1.60	Ea.	205	205	410	520
570	For color, add				30%			
579	For rough-in, supply, waste & vent, sinks	Q-1	1.85		58.30	175	233.30	320
665	Service, floor, corner, P.E. on C.I., 28" x 28"		4		305	81	386	455
679	For rough-in, supply, waste & vent, floor service sinks		1.30		87.70	250	337.70	460
700	Service, wall, P.E. on C.I., roll rim, 22" x 18"		3		230	110	340	410
710	24" x 20"		3		255	110	365	435
860	Vitreous china, 22" x 20"		3		260	110	370	445
896	For stainless steel rim guard, front or side, add				18		18	19.80M
898	For rough-in, supply, waste & vent, wall service sinks	Q-1	1.30		145.22	250	395.22	520
68-001	**URINALS**							
002								
300	Wall hung, vitreous china, with hanger & self-closing valve	Q-1	3	Ea.	290	110	400	475
330	Rough-in, supply, waste & vent		1.99		53.01	165	218.01	295
500	Stall type, vitreous china, includes valve		2.50		360	130	490	585
510	3" seam cover, add		12		90	27	117	140
520	6" seam cover, add		12		125	27	152	175
698	Rough-in, supply, waste and vent		1.99		67.47	165	232.47	310
76-001	**WASH FOUNTAINS** Rigging not included							
190	Group, foot control							
200	Precast terrazzo, circular, 36" diam., 5 or 6 persons	Q-2	3	Ea.	880	170	1,050	1,200
210	54" diameter for 8 or 10 persons		2.50		1,025	200	1,225	1,425
240	Semi-circular, 36" diam. for 3 persons		3		800	170	970	1,125
250	54" diam. for 4 or 5 persons		2.50		985	200	1,185	1,375
270	Quarter circle (corner), 54" for 3 persons		3.50		1,045	145	1,190	1,350
285								
300	Stainless steel, circular, 36" diameter	Q-2	3.50	Ea.	1,050	145	1,195	1,375
310	54" diameter		2.80		1,360	180	1,540	1,750
340	Semi-circular, 36" diameter		3.50		905	145	1,050	1,200
350	54" diameter		2.80		1,180	180	1,360	1,550
570	Rough-in, supply, waste and vent for above wash fountains	Q-1	1.38		187.85	235	422.85	545
590								
620	Duo for small washrooms, stainless steel	Q-1	2	Ea.	485	160	645	770
650	Rough-in, supply, waste & vent for duo fountains	"	2.02	"	28.71	160	188.71	265
80-001	**WATER CLOSETS**							
015	Tank type, vitreous china, including seat, supply pipe with stop							
020	Wall hung, one piece	Q-1	5.30	Ea.	465	61	526	600
040	Two piece, close coupled		5.30		290	61	351	410
096	For rough-in, supply, waste, vent and carrier		2.24		121.66	145	266.66	345
100	Floor mounted, one piece		5.30		360	61	421	485
110	Two piece, close coupled, water saver		5.30		100	61	161	200
196	For color, add				30%			
198	For rough-in, supply, waste and vent	Q-1	1.94		83.25	165	248.25	335
200								
300	Bowl only, with flush valve, seat							
310	Wall hung	Q-1	5.80	Ea.	230	56	286	335

For expanded coverage of these items see *Means' Mechanical Cost Data 1986*

15.2 Plumbing Fixtures

			CREW	DAILY OUTPUT	UNIT	BARE COSTS MAT.	BARE COSTS INST.	BARE COSTS TOTAL	TOTAL INCL O&P
80	320	For rough-in, supply, waste and vent, single WC	Q-1	2.05	Ea.	127.75	160	287.75	370
	330	Floor mounted		5.80		210	56	266	310
	335	With wall outlet		5.80		275	56	331	385
	340	For rough-in, supply, waste and vent, single WC	↓	1.80	↓	89.34	180	269.34	360

15.3 Plumbing Appliances

			CREW	DAILY OUTPUT	UNIT	BARE COSTS MAT.	BARE COSTS INST.	BARE COSTS TOTAL	TOTAL INCL O&P
45-001		WATER COOLER							
	003	See line 15.3-45-980 for rough-in, waste & vent							
	004	for all water coolers							
	010	Wall mounted, non-recessed							
	014	4 GPH	Q-1	4	Ea.	285	81	366	430
	018	8.2 GPH	"	4		335	81	416	485
	060	For hot and cold water, add				123		123	135M
	064	For stainless steel cabinet, add				38		38	42M
	100	Dual height, 8.2 GPH	Q-1	3.80		455	85	540	625
	104	14.3 GPH		3.80		480	85	565	650
	108	16.1 GPH	↓	3.80		525	85	610	700
	124	For stainless steel cabinet, add				72		72	79M
	260	Wheelchair type, 8 GPH	Q-1	4	↓	865	81	946	1,075
	261								
	330	Semi-recessed, 8.1 GPH	Q-1	4	Ea.	440	81	521	600
	332	12 GPH	"	4	"	465	81	546	630
	460	Floor mounted, flush-to-wall							
	464	4 GPH	1 Plum	3	Ea.	300	60	360	415
	468	8.2 GPH		3		345	60	405	465
	472	14.3 GPH	↓	3		355	60	415	475
	496	For hot and cold water, add				110		110	120M
	498	For stainless steel cabinet, add				60		60	66M
	500	Dual height, 8.2 GPH	1 Plum	2		505	90	595	685
	504	14.3 GPH		2		510	90	600	690
	508	19.5 GPH	↓	2		520	90	610	700
	512	For stainless steel cabinet, add				60		60	66M
	980	For supply, waste & vent, all coolers	1 Plum	1.24	↓	24.76	145	169.76	240
50-001		WATER HEATERS							
	002	For solar, see Division 15.5-77							
	100	Residential, electric, glass lined tank, 10 gal., single element	1 Plum	2.30	Ea.	105	78	183	230
	104	20 gallon, single element		2.20		130	82	212	260
	106	30 gallon, double element		2.20		140	82	222	275
	108	40 gallon, double element		2		145	90	235	290
	118	120 gallon, double element	↓	1.40	↓	440	130	570	670
	119								
	200	Gas fired, glass lined tank, vent not incl., 20 gallon	1 Plum	2.10	Ea.	140	86	226	280
	204	30 gallon		2		150	90	240	295
	210	75 gallon		1.50		370	120	490	580
	212	100 gallon		1.30		635	140	775	900
	300	Oil fired, glass lined tank, vent not included, 30 gallon		2		700	90	790	900
	304	50 gallon		1.80		930	100	1,030	1,175
	306	70 gallon		1.50		1,035	120	1,155	1,300
	308	85 gallon	↓	1.40	↓	1,465	130	1,595	1,800
	400	Commercial, 100° rise, NOTE: for each size tank, a range of heaters							
	401	between the ones shown are available							
	402	Electric							
	410	5 gal., 3 KW, 12 GPH	1 Plum	2	Ea.	700	90	790	900

For expanded coverage of these items see *Means' Mechanical Cost Data 1986*

15.3 Plumbing Appliances

		CREW	DAILY OUTPUT	UNIT	BARE COSTS MAT.	INST.	TOTAL	TOTAL INCL O&P
412	10 gal., 6 KW, 25 GPH	1 Plum	2	Ea.	775	90	865	985
414	50 gal., 9 KW, 37 GPH		1.80		1,010	100	1,110	1,250
416	50 gal., 36 KW, 148 GPH		1.80		1,545	100	1,645	1,850
430	200 gal., 15 KW, 61 GPH	Q-1	1.70		4,955	190	5,145	5,725
432	200 gal., 120 KW, 490 GPH		1.70		7,250	190	7,440	8,250
446	400 gal., 30 KW, 123 GPH		1		6,840	325	7,165	8,000
540	Modulating step control, 2-5 steps	1 Elec	5.30		610	34	644	720
544	6-10 steps		3.20		790	56	846	950
546	11-15 steps		2.70		975	66	1,041	1,175
548	16-20 steps		1.60		1,150	110	1,260	1,425
550	21-25 steps		.30		1,325	595	1,920	2,325
552	26-30 steps		.26		1,510	690	2,200	2,650
600	Gas fired, flush jacket, std. controls, vent not incl.							
604	75 MBH input, 63 GPH	1 Plum	1.40	Ea.	660	130	790	910
606	96 MBH input, 81 GPH		1.40		810	130	940	1,075
608	120 MBH input, 101 GPH		1.20		850	150	1,000	1,150
618	200 MBH input, 192 GPH		.60		1,550	300	1,850	2,150
620	240 MBH input, 230 GPH		.50		1,900	360	2,260	2,600
634	1200 MBH input, 1150 GPH	Q-1	.40		7,500	810	8,310	9,425
636	1500 MBH input, 1440 GPH	Q-2	.60		9,275	840	10,115	11,400
690	For low water cutoff, add	1 Plum	8		120	23	143	165
696	For bronze body hot water circulator, add	"	4		500	45	545	615
800	Oil fired, flush jacket, std. controls, vent not incl.							
806	103 MBH gross output, 116 GPH	1 Plum	1.10	Ea.	1,150	165	1,315	1,500
808	122 MBH gross output, 141 GPH		1		1,175	180	1,355	1,550
810	137 MBH gross output, 161 GPH		.80		1,250	225	1,475	1,700
816	225 MBH gross output, 256 GPH	Q-1	.80		1,850	405	2,255	2,625
818	262 MBH gross output, 315 GPH		.70		2,075	465	2,540	2,950
828	735 MBH gross output, 880 GPH		.40		5,500	810	6,310	7,225
830	840 MBH gross output, 1000 GPH		.40		6,300	810	7,110	8,100
890	For low water cutoff, add	1 Plum	8		120	23	143	165
896	For bronze body hot water circulator, add	"	4		500	45	545	615

15.4 Fire Extinguishing Systems

		CREW	DAILY OUTPUT	UNIT	BARE COSTS MAT.	INST.	TOTAL	TOTAL INCL O&P
02-001	**AUTOMATIC FIRE SUPPRESSION SYSTEMS**							
004	For detectors and control stations, see division 16.8-15							
100	Dispersion nozzle, CO2, 3" x 5"	1 Plum	18	Ea.	55	10	65	75
110	Halon, 1-1/2"	"	14		65	12.90	77.90	90
200	Extinguisher, CO2 system, high pressure, 75 lb. cylinder	Q-1	6		550	54	604	685
240	Halon system, filled, with mounting bracket							
254	196 lb. container	Q-1	4	Ea.	2,850	81	2,931	3,250
300	Electro/mechanical release	L-1	4		160	90	250	305
340	Manual pull station	1 Plum	6		26	30	56	72
500								
600	Average halon system, minimum			C.F.				.55
602	Maximum			"				1.50
10-001	**FIRE EQUIPMENT CABINETS** Not equipped, 20 ga. steel box,							
004	recessed, D.S. glass in door, box size given							
100	Portable extinguisher, single, 8" x 12" x 27", alum. door & frame	Q-12	8	Ea.	72	42	114	140
300	Hose rack assy., 1-1/2" valve & 100' hose, 24" x 40" x 5-1/2"							
310	Aluminum door and frame	Q-12	6	Ea.	102	56	158	195
320	Steel door and frame		6		101	56	157	190
330	Stainless steel door and frame		6		228	56	284	330
400	Hose rack assy., 2-1/2" x 1-1/2" valve, 100' hose, 24" x 40" x 8"							

For expanded coverage of these items see *Means' Mechanical Cost Data 1986*

15.4 Fire Extinguishing Systems

			CREW	DAILY OUTPUT	UNIT	BARE COSTS MAT.	BARE COSTS INST.	BARE COSTS TOTAL	TOTAL INCL O&P
10	410	Aluminum door and frame	Q-12	6	Ea.	105	56	161	195
	420	Steel door and frame	↓	6	↓	104	56	160	195
	430	Stainless steel door and frame		6		230	56	286	335
	500	Hose rack assy., 2-1/2" x 1-1/2" valve, 100' hose							
	510	Aluminum door and frame	Q-12	5	Ea.	112	67	179	220
	520	Steel door and frame		5		113	67	180	220
	530	Stainless steel door and frame	↓	5	↓	244	67	311	365
	800	Valve cabinet for 2-1/2" F.D. angle valve, 18" x 18" x 8"							
	810	Aluminum door and frame	Q-12	12	Ea.	72	28	100	120
	820	Steel door and frame		12		66	28	94	115
	830	Stainless steel door and frame	↓	12	↓	131	28	159	185
20-001		**FIRE EXTINGUISHERS**							
	012	CO2, portable with swivel horn, 5 lb.			Ea.	61.25		61.25	67M
	014	With hose and "H" horn, 10 lb.				90		90	99M
	016	15 lb.				100		100	110M
	018	20 lb.			↓	135		135	150M
	020								
	036	Wheeled type, cart mounted, 50 lb.			Ea.	750		750	825M
	038	100 lb. (two 50 lb. cylinders)			"	1,200		1,200	1,325M
	100	Dry chemical, pressurized							
	104	Standard type, portable, painted, 2-1/2 lb.			Ea.	16.50		16.50	18.15M
	106	5 lb.				28		28	31M
	108	10 lb.				42		42	46M
	110	20 lb.			↓	65		65	72M
	120								
	130	Standard type, wheeled, 150 lb.			Ea.	1,250		1,250	1,375M
	136	350 lb.				2,050		2,050	2,250M
	200	ABC all purpose type, portable, 2-1/2 lb.				17.35		17.35	19.10M
	206	5 lb.				29		29	32M
	208	9-1/2 lb.				41		41	45M
	210	18 lb.				67		67	74M
	230	Wheeled, 45 lb.				905		905	995M
	236	150 lb.				1,250		1,250	1,375M
	238	315 lb.			↓	1,525		1,525	1,675M
	250								
	300	Dry chemical, outside cartridge to -65°F, painted, 9 lb.			Ea.	130		130	145M
	306	26 lb.				187		187	205M
	400	Halon, painted, wall bracket, 2-1/2 lb.				40		40	44M
	406	5 lb.				70		70	77M
	500	Pressurized water, 2-1/2 gallon, stainless steel				42.50		42.50	47M
	506	With anti-freeze				59.50		59.50	65M
15	940	Installation of extinguishers, 12 or more, on wood	1 Carp	30			5.35	5.35	7.75L
	942	On masonry or concrete	"	15	↓		10.65	10.65	15.45L
30-001		**FIRE HOSE AND EQUIPMENT**							
	020	Adapters, rough brass, straight hose threads							
	022	One piece, female to male, rocker lugs							
	024	1" x 1"			Ea.	8.25		8.25	9.05M
	210								
	220	Hose, less couplings							
	226	Synthetic jacket, lined, 300 lb. test, 1-1/2" diameter			L.F.	1.14		1.14	1.25M
	228	2-1/2" diameter				1.56		1.56	1.71M
	236	High strength, 500 lb. test, 1-1/2" diameter				1.32		1.32	1.45M
	238	2-1/2" diameter			↓	2.13		2.13	2.34M
	260	Hose rack, swinging, for 1-1/2" diameter hose,							
	262	Enameled steel, 50' & 75' lengths of hose	Q-12	20	Ea.	16.60	16.75	33.35	43
	264	100' and 125' lengths of hose	"	20	"	16.60	16.75	33.35	43
	375	Hydrants, wall, w/caps, single, flush, polished brass							
	380	2-1/2" x 2-1/2"	Q-12	5	Ea.	68.50	67	135.50	175
	384	2-1/2" x 3"	"	5	"	77	67	144	180

For expanded coverage of these items see *Means' Mechanical Cost Data 1986*

15.4 Fire Extinguishing Systems

			CREW	DAILY OUTPUT	UNIT	BARE COSTS MAT.	BARE COSTS INST.	BARE COSTS TOTAL	TOTAL INCL O&P
30	395	Double, flush, polished brass							
	400	2-1/2" x 2-1/2" x 4"	Q-12	5	Ea.	210	67	277	330
	404	2-1/2" x 2-1/2" x 6"	"	4.60		241	73	314	370
	420	For polished chrome, add				3%			
	435	Double, projecting, polished brass							
	440	2-1/2" x 2-1/2" x 4"	Q-12	5	Ea.	194	67	261	310
	444	2-1/2" x 2-1/2" x 5"		4.60		278	73	351	410
	445	2-1/2" x 2-1/2" x 6"		4.60		220	73	293	350
	446	Valve control, dbl. flush/projecting hydrant, cap &							
	447	chain, ext. rod & cplg., escutcheon, polished brass	Q-12	8	Ea.	108	42	150	180
	560	Nozzles, brass							
	562	Adjustable fog, 3/4" booster line			Ea.	61		61	67M
	563	1" booster line				61		61	67M
	564	1-1/2" leader line				61		61	67M
	566	2-1/2" direct connection				121		121	135M
	578	For chrome plated, add				6%			
	585	Electrical fire, adjustable fog, no shock							
	590	1-1/2"			Ea.	215		215	235M
	592	2-1/2"				278		278	305M
	598	For polished chrome, add				6%			
	620	Heavy duty, comb. adj. fog and str. stream, with handle							
	621	1" booster line			Ea.	173		173	190M
	714	Standpipe connections, wall, w/plugs & chains							
	716	Single, flush, brass, 2-1/2" x 2-1/2"	Q-12	5	Ea.	75.60	67	142.60	180
	718	2-1/2" x 3"	"	5		86.10	67	153.10	190
	724	For polished chrome, add				3%			
	728	Double, flush, polished brass							
	730	2-1/2" x 2-1/2" x 4"	Q-12	5	Ea.	231	67	298	350
	732	2-1/2" x 2-1/2" x 5"		4.60		300	73	373	435
	733	2-1/2" x 2-1/2" x 6"		4.60		268	73	341	400
	740	For polished chrome, add				3%			
	744	For sill cock combination, add				73.50		73.50	81M
	790	Three way, flush, polished brass							
	792	2-1/2" (3) x 4"	Q-12	4.80	Ea.	425	70	495	570
	793	2-1/2" (3) x 6"		4.80		440	70	510	585
	794	2-1/2" x 3-1/2" x 2-1/2" x 4"		4.80		595	70	665	755
	795	2-1/2" x 3-1/2" x 2-1/2" x 6"		4.80		535	70	605	690
	800	For polished chrome, add				6%			
	802	Three way, projecting, polished brass							
	804	2-1/2"(3) x 4"	Q-12	4.80	Ea.	320	70	390	455
40	001	**FIRE PUMPS** Including controller, fittings and relief valve							
	003	Diesel							
	005	500 GPM, 50 psi, 40 HP, 4" pump	Q-13	.64	Ea.	23,850	1,100	24,950	27,800
	020	750 GPM, 50 psi, 40 HP, 5" pump		.60		24,300	1,175	25,475	28,500
	040	1000 GPM, 100 psi, 99 HP, 5" pump		.56		28,350	1,275	29,625	33,000
	070	2000 GPM, 100 psi, 188 HP, 6" pump		.34		32,700	2,100	34,800	39,000
	095	3500 GPM, 100 psi, 320 HP, 10" pump		.24		41,650	2,950	44,600	50,000
	250								
	300	Electric							
	310	250 GPM, 40 psi, 15 HP, 3550 RPM, 3" pump	Q-13	.70	Ea.	9,220	1,025	10,245	11,600
	320	500 GPM, 50 psi, 30 HP, 3550 RPM, 4" pump		.68		9,620	1,050	10,670	12,100
	335	750 GPM, 50 psi, 50 HP, 1770 RPM, 5" pump		.64		10,000	1,100	11,100	12,600
	340	750 GPM, 100 psi, 100 HP, 1770 RPM, 5" pump		.58		12,900	1,225	14,125	16,000
	500	For jockey pump 1", 7-1/2 HP, add	Q-12	2		1,475	165	1,640	1,875
50	001	**FIRE VALVES**							
	002								
	008	Wheel handle, 300 lb., 1-1/2"	1 Spri	12	Ea.	21	15.50	36.50	46
	009	2-1/2"	"	7		48	27	75	91
	010	For polished brass, add				35%			
	011	For polished chrome, add				50%			

For expanded coverage of these items see *Means' Mechanical Cost Data 1986*

15.4 Fire Extinguishing Systems

		CREW	DAILY OUTPUT	UNIT	BARE COSTS MAT.	BARE COSTS INST.	BARE COSTS TOTAL	TOTAL INCL O&P
60-001	SPRINKLER SYSTEM COMPONENTS							
060	Accelerator	1 Spri	8	Ea.	220	23	243	275
260	Sprinkler heads, not including supply piping							
264	Dry, pendent or upright, 1/2" orifice, 3/4" or 1" NPT							
270	11" to 12-3/4" length	1 Spri	8	Ea.	22.50	23	45.50	59
271	13" to 14-3/4" length		7		24.50	27	51.50	66
272	15" to 16-3/4" length		7		24.50	27	51.50	66
273	17" to 18-3/4" length	↓	7		26	27	53	67
290	For recessed fitting, add				1.40		1.40	1.54M
360	Foam-water, pendent or upright, 1/2" NPT	1 Spri	8	↓	22.25	23	45.25	58
370	Standard spray, pendent or upright, brass, 135° to 286°F							
374	1/2" NPT, 1/2" orifice	1 Spri	10	Ea.	2.15	18.60	20.75	29
386	For wax and lead coating, add				1.65		1.65	1.81M
388	For wax coating, add				.65		.65	.71M
390	For lead coating, add				.95		.95	1.04M
392	For 360°F, same cost							
393	For 400°, add				4		4	4.40M
394	For 500°F, add			↓	4		4	4.40M
420	Sidewall, vertical, brass, 135° to 286°F							
422	1/2" NPT, 3/8" orifice	1 Spri	10	Ea.	3.18	18.60	21.78	31
423	1/2" NPT, 7/16" orifice	"	10	"	3.18	18.60	21.78	31
440								
450	Sidewall, horizontal, brass, 135° to 286°F							
452	1/2" NPT, 1/2" orifice	1 Spri	10	Ea.	2.80	18.60	21.40	30
454	For 360°F, same cost							
455	For 400°F, add				4		4	4.40M
456	For 500°F, add			↓	4		4	4.40M
480	Recessed pendent, brass, 135° to 286°F							
482	1/2" NPT, 3/8" orifice	1 Spri	8	Ea.	3.81	23	26.81	38
483	1/2" NPT, 7/16" orifice		8		3.81	23	26.81	38
484	1/2" NPT, 1/2" orifice	↓	8	↓	3.22	23	26.22	37
620	Valves							
650	Check, swing, C.I. body, brass fittings							
652	4" size	Q-12	3	Ea.	96	110	206	270
680	Check, wafer, butterfly type, C.I. body, bronze fittings							
682	4" size	Q-12	4	Ea.	185	84	269	325
700	Deluge, assembly, incl. trim, pressure							
702	operated relief, emergency release, gauges							
704	2" size	Q-12	2	Ea.	720	165	885	1,025
706	3" size	"	1.50	"	820	225	1,045	1,225

15.5 Heating

		CREW	DAILY OUTPUT	UNIT	BARE COSTS MAT.	BARE COSTS INST.	BARE COSTS TOTAL	TOTAL INCL O&P
02-001	AVERAGE Square foot and percent of total							
002	job cost, see division 17.1							
04-001	BOILERS, GENERAL Prices do not include flue piping, elec. wiring,							
002	gas or oil piping, boiler base, pad, or tankless unless noted							
010	Boiler horsepower: 10 KW = 34 lbs/steam/hr = 33,475 BTU/hr.							
015	To convert SFR to BTU rating: Hot water, 150 x SFR;							
016	Forced hot water, 180 x SFR; steam, 240 x SFR							
06-001	BOILERS, ELECTRIC, ASME Standard controls and trim							
100	Steam, 6 KW, 20.5 MBH	Q-19	1.20	Ea.	2,450	420	2,870	3,300
110	30 KW, 102 MBH		1.10		2,675	460	3,135	3,600
116	60 KW, 205 MBH	↓	1	↓	3,650	505	4,155	4,750

For expanded coverage of these items see *Means' Mechanical Cost Data 1986*

15.5 Heating

		CREW	DAILY OUTPUT	UNIT	BARE COSTS MAT.	BARE COSTS INST.	BARE COSTS TOTAL	TOTAL INCL O&P
122	120 KW, 409 MBH	Q-19	.75	Ea.	5,825	675	6,500	7,375
128	210 KW, 716 MBH	"	.55		7,625	920	8,545	9,725
138	510 KW, 1740 MBH	Q-21	.36		15,700	1,925	17,625	20,000
148	1080 KW, 3685 MBH		.25		22,900	2,750	25,650	29,200
160	2340 KW, 7984 MBH	↓	.16		47,700	4,300	52,000	58,500
200	Hot water, 12 KW, 41 MBH	Q-19	1.30		2,350	390	2,740	3,150
210	60 KW, 205 MBH		1.10		3,125	460	3,585	4,100
222	240 KW, 820 MBH	↓	.55		6,550	920	7,470	8,525
232	480 KW, 1636 MBH	Q-21	.46		11,100	1,500	12,600	14,400
250	1200 KW, 4095 MBH		.34		20,300	2,025	22,325	25,300
268	2400 KW, 8191 MBH		.25		34,600	2,750	37,350	42,000
282	3600 KW, 12,283 MBH	↓	.16	↓	44,600	4,300	48,900	55,500
08-001	**BOILERS, GAS FIRED** Natural or propane, standard controls							
100	Cast iron, with insulated jacket							
200	Steam, gross output, 81 MBH	Q-7	1.40	Ea.	895	495	1,390	1,700
208	203 MBH		.90		1,560	775	2,335	2,825
218	400 MBH		.60		2,680	1,150	3,830	4,625
224	816 MBH		.45		5,390	1,550	6,940	8,175
232	2000 MBH		.28		12,000	2,475	14,475	16,800
244	5032 MBH		.14		28,450	4,975	33,425	38,500
248	6100 MBH		.12		34,950	5,800	40,750	46,800
254	6970 MBH		.08		38,900	8,700	47,600	55,500
300	Hot water, gross output, 80 MBH		1.46		795	475	1,270	1,575
314	320 MBH		.80		2,080	870	2,950	3,550
326	1088 MBH		.50		6,720	1,400	8,120	9,400
338	3264 MBH		.30		18,500	2,325	20,825	23,700
348	6100 MBH		.12		36,800	5,800	42,600	48,900
354	6970 MBH	↓	.08	↓	40,750	8,700	49,450	57,500
400	Steel, insulating jacket							
450	Steam, not including burner, gross output							
550	1440 MBH	Q-6	.35	Ea.	7,575	1,450	9,025	10,400
558	3065 MBH	"	.18		10,800	2,825	13,625	16,000
564	7200 MBH	Q-7	.18		19,800	3,875	23,675	27,400
572	17,990 MBH	"	.10	↓	37,000	6,950	43,950	51,000
600	Hot water, including burner & one zone valve, gross output							
602	72 MBH	Q-6	2	Ea.	1,100	255	1,355	1,575
612	236 MBH		1.30		2,600	390	2,990	3,425
618	480 MBH		.70		4,450	730	5,180	5,950
624	960 MBH		.45		7,625	1,125	8,750	10,000
634	3000 MBH	↓	.15		21,300	3,400	24,700	28,300
700	For tankless water heater on smaller gas units, add			↓	10%			
10-001	**BOILERS, OIL FIRED** Standard controls, flame retention burner							
100	Cast iron, with insulated flush jacket							
200	Steam, gross output, 109 MBH	Q-7	1.20	Ea.	1,140	580	1,720	2,100
206	207 MBH		.90		1,640	775	2,415	2,925
218	940 MBH		.42		5,750	1,650	7,400	8,725
228	2920 MBH		.23		14,450	3,025	17,475	20,300
238	5520 MBH		.13		23,800	5,350	29,150	33,900
246	6970 MBH	↓	.08	↓	32,550	8,700	41,250	48,400
300	Hot water, same price as steam							
400	For tankless coil in smaller sizes, add			Ea.	15%			
500	Steel, insulated jacket, burner							
600	Steam, full water leg construction, gross output							
602	144 MBH	Q-6	1.60	Ea.	1,250	320	1,570	1,825
612	468 MBH		.80		2,350	635	2,985	3,500
618	1008 MBH		.45		4,250	1,125	5,375	6,300
628	2400 MBH	↓	.22	↓	8,150	2,325	10,475	12,300
640	Larger sizes are same as steel, gas fired.							
700	Hot water, gross output, 103 MBH	Q-6	1.90	Ea.	1,300	270	1,570	1,825

For expanded coverage of these items see *Means' Mechanical Cost Data 1986*

15.5 Heating

		CREW	DAILY OUTPUT	UNIT	BARE COSTS MAT.	BARE COSTS INST.	BARE COSTS TOTAL	TOTAL INCL O&P
10 712	420 MBH	Q-6	.80	Ea.	3,275	635	3,910	4,525
720	840 MBH		.50		6,425	1,025	7,450	8,550
726	1680 MBH		.33		11,700	1,550	13,250	15,100
732	3150 MBH	↓	.13		21,000	3,925	24,925	28,800
734	For tankless coil in steam or hot water, add			↓	7%			
12-001	**BOILERS, GAS/OIL** Combination with burners and controls							
100	Cast Iron with insulated jacket							
200	Steam, gross output, 720 MBH	Q-7	.40	Ea.	7,220	1,750	8,970	10,500
208	1600 MBH		.26		10,700	2,675	13,375	15,600
214	2700 MBH		.19		16,400	3,650	20,050	23,300
228	5520 MBH		.10		41,750	6,950	48,700	56,000
234	6390 MBH		.07		49,150	9,925	59,075	68,500
238	6970 MBH		.05		52,400	13,900	66,300	78,000
300	Hot water, gross output, 584 MBH		.54		7,900	1,300	9,200	10,600
306	1460 MBH		.45		19,500	1,550	21,050	23,700
316	4088 MBH		.22		54,500	3,150	57,650	64,500
330	13,500 MBH, 403.3 BHP	↓	.02	↓	83,600	34,800	118,400	142,000
400	Steel, insulated jacket, skid base, tubeless							
450	Steam, 150 psi gross output, 335 MBH, 10 BHP	Q-6	.65	Ea.	6,575	785	7,360	8,375
456	670 MBH, 20 BHP		.45		8,425	1,125	9,550	10,900
464	1339 MBH, 40 BHP		.28		13,400	1,825	15,225	17,400
472	2511 MBH, 75 BHP		.17		17,500	3,000	20,500	23,600
500	Hot water, gross output, 525 MBH		.60		6,100	850	6,950	7,950
508	1050 MBH		.38		9,750	1,350	11,100	12,700
514	2310 MBH		.20		18,200	2,550	20,750	23,700
518	3150 MBH	↓	.09	↓	24,100	5,650	29,750	34,700
27-001	**DUCT FURNACES** Includes burner, controls, stainless steel							
002	heat exchanger. Gas fired, electric ignition							
003	Indoor installation							
010	120 MBH output	Q-5	4	Ea.	770	82	852	965
013	200 MBH output		2.70		1,090	120	1,210	1,375
014	240 MBH output		2.30		1,140	140	1,280	1,450
018	320 MBH output	↓	1.60		1,420	205	1,625	1,850
030	For powered venter and adapter, add			↓	161		161	175M
050	For required flue pipe, see division 15.5-92							
100	Outdoor installation, with vent cap							
102	75 MBH output	Q-5	4	Ea.	885	82	967	1,100
106	120 MBH output		4		1,050	82	1,132	1,275
110	187 MBH output		3		1,400	110	1,510	1,700
114	300 MBH output		1.80		1,830	180	2,010	2,275
118	450 MBH output	↓	1.40		3,270	235	3,505	3,925
140	For powered venter, add			↓	10%			
29-001	**DUCT HEATERS** Electric, slip-in. Includes blast coil, controls, fused							
002	transformer, air flow switch, thermal cutouts, contactors							
010	120 volt, 1 phase, 0.5 KW, 8" wide x 8" high	Q-20	14	Ea.	210	30	240	275
012	1 KW, 10" wide x 10" high		12		215	35	250	285
014	2 KW, 14" wide x 14" high		11		226	38	264	305
016	5 KW, 16" wide x 16" high		10		252	42	294	340
200	208 or 240 volt, 3 phase, 1.5 KW, 8" wide x 8" high		12		241	35	276	315
204	10 KW, 30" wide x 30" high		8		355	52	407	465
212	50 KW, 60" wide x 30" high		6		925	69	994	1,125
216	100 KW, 72" wide x 30" high	↓	4	↓	1,710	105	1,815	2,025
350	To obtain BTU, multiply KW by 3413							
35-001	**FURNACES** Hot air heating, blowers, standard controls,							
002	not including gas, oil or flue piping.							
100	Electric, UL listed, heat staging, 240 volt							
102	30 MBH	Q-20	4	Ea.	430	105	535	625

For expanded coverage of these items see *Means' Mechanical Cost Data 1986*

15.5 Heating

		CREW	DAILY OUTPUT	UNIT	BARE COSTS MAT.	BARE COSTS INST.	BARE COSTS TOTAL	TOTAL INCL O&P
35 104	47 MBH	Q-20	4	Ea.	470	105	575	670
106	61 MBH		3.80		520	110	630	730
108	76 MBH		3.60		575	115	690	800
110	91 MBH	↓	3.40	↓	610	120	730	850
300	Gas, AGA certified, direct drive models							
302	42 MBH output	Q-9	4	Ea.	335	82	417	490
304	63 MBH output		3.80		350	86	436	510
306	79 MBH output		3.60		370	91	461	540
308	84 MBH output		3.40		400	96	496	580
310	105 MBH output	↓	3.20	↓	445	100	545	640
600	Oil, UL listed, atomizing gun type burner							
602	55 MBH output	Q-9	3.60	Ea.	625	91	716	820
603	84 MBH output		3.50		645	93	738	845
604	99 MBH output		3.40		700	96	796	910
606	125 MBH output		3.20		795	100	895	1,025
608	152 MBH output	↓	3	↓	1,000	110	1,110	1,250
40-001	**HEAT EXCHANGERS** 4 pass, 3/4" O.D. copper tubes,							
002	C.I. heads, C.I. tube sheet, steel shell							
010	Hot water 40°F to 180°F, by steam at 10 PSI							
012	8 GPM	Q-5	6	Ea.	455	55	510	580
014	10 GPM		5		700	66	766	865
016	40 GPM		4		1,100	82	1,182	1,325
018	64 GPM		2		1,625	165	1,790	2,025
020	96 GPM	↓	1		2,150	330	2,480	2,850
022	120 GPM	Q-6	1.50	↓	3,150	340	3,490	3,950
100	Hot water 40°F to 140°F, by water at 200°F							
102	7 GPM	Q-5	6	Ea.	545	55	600	680
104	16 GPM		5		790	66	856	965
106	34 GPM		4		1,225	82	1,307	1,475
110	74 GPM	↓	1.50	↓	2,250	220	2,470	2,800
42-001	**HEAT TRANSFER PACKAGES** Complete, controls,							
002	expansion tank, converter, air separator							
100	Hot water, 180°F enter, 200°F leaving, 15# steam							
101	One pump system, 28 GPM	Q-6	.75	Ea.	5,690	680	6,370	7,250
102	35 GPM		.70		6,410	730	7,140	8,100
104	55 GPM		.65		7,060	785	7,845	8,900
106	130 GPM		.55		9,160	925	10,085	11,400
108	255 GPM		.40		12,100	1,275	13,375	15,200
110	550 GPM	↓	.30	↓	16,250	1,700	17,950	20,300
45-001	**HEATING & VENTILATING UNITS** Classroom							
002	Includes filter, heating/cooling coils, standard controls							
008	750 CFM, 2 tons cooling	Q-6	2	Ea.	1,925	255	2,180	2,475
012	1250 CFM, 3 tons cooling		1.40		2,355	365	2,720	3,125
014	1500 CFM, 4 tons cooling	↓	.80		2,525	635	3,160	3,700
050	For electric heat, add				35%			
100	For no cooling, deduct			↓	25%	10%		
48-001	**HEAT RECOVERY PACKAGES**							
010	Air to air							
200	Kitchen exhaust, commercial, heat pipe exchanger							
204	Combined supply/exhaust air volume							
208	2.5 to 6.0 MCFM	Q-10	2.80	MCFM	2,950	180	3,130	3,500
212	6 to 16 MCFM		5		1,800	100	1,900	2,125
216	16 to 22 MCFM	↓	6	↓	1,400	85	1,485	1,675
795								
800	Ventilating systems, moderate heat, paper plates							
801	and fins exchanger							
802	Standard capacity							
806	23 CFM	1 Shee	16	Ea.	155	11.35	166.35	185

For expanded coverage of these items see *Means' Mechanical Cost Data 1986*

15.5 Heating

		CREW	DAILY OUTPUT	UNIT	BARE COSTS MAT.	BARE COSTS INST.	BARE COSTS TOTAL	TOTAL INCL O&P
814	71 CFM	1 Shee	12	Ea.	420	15.15	435.15	485
824	220 CFM	"	8		1,600	23	1,623	1,800
830	235 CFM	Q-9	9		280	36	316	360
850	735 CFM	"	4		740	82	822	935
880	For intermediate capacities, use multiples							
51-001	**HYDRONIC HEATING** Terminal units, not including pipe to main supply							
100	Radiation							
110	Panel, baseboard, C.I., including supports, no covers	Q-5	20	L.F.	18.40	16.40	34.80	44
115	Fin tube, wall hung, 14" slope top cover, with damper							
120	1-1/4" copper tube, 4-1/4" alum. fin	Q-5	16	L.F.	24.75	20	44.75	57
125	1-1/4" steel tube, 4-1/4" steel fin	"	16	"	23.10	20	43.10	55
150	Note: fin tube may also require corners, caps, etc.							
195								
199	Convector unit, floor recessed, flush, with trim							
200	for under large glass wall areas, no damper	Q-5	20	L.F.	20.55	16.40	36.95	46
210	For unit with damper, add			"	9		9	9.90M
300	Radiators, cast iron							
310	Free standing or wall hung, 6 tube, 25" high	Q-5	96	Section	16.60	3.41	20.01	23
320	4 tube, 19" high	"	96	"	11.55	3.41	14.96	17.65
325	Adj. brackets, 2 per wall radiator up to 30 sections	1 Stpi	32	Ea.	25	5.70	30.70	36
395	Unit heaters, propeller, 1 speed, 200° EWT							
400	Horizontal, 14.7 MBH	Q-5	12	Ea.	190	27	217	250
406	44.8 MBH		8		260	41	301	345
414	106.4 MBH		6		315	55	370	425
418	160.9 MBH		4		435	82	517	595
424	292.5 MBH		2		710	165	875	1,025
430	For vertical diffuser, add				76		76	84M
500	Vertical flow, 52.4 MBH	Q-5	11		260	30	290	330
502	71.4 MBH		8		285	41	326	375
508	140 MBH		4		415	82	497	575
514	296.7 MBH		2		830	165	995	1,150
516	408 MBH	Q-6	1.80		1,050	285	1,335	1,575
518	520 MBH	"	1.40		1,250	365	1,615	1,900
530	For capacity of unit heaters using 2 psi							
532	steam, multiply MBH by 1.43							
950	To convert SFR to BTU rating: Hot water, 150 x SFR							
951	Forced hot water, 180 x SFR; steam, 240 x SFR							
54-001	**HUMIDIFIERS**							
002								
003	Centrifugal atomizing							
005	5 lb. per hour	Q-5	12	Ea.	780	27	807	895
010	10 lb. per hour		10		1,005	33	1,038	1,150
012	24 lb. per hour		8		1,540	41	1,581	1,750
052	Steam, room or duct, filter, regulators, automatic controls, 220Volt							
054	10 lb. per hour	Q-5	6	Ea.	1,680	55	1,735	1,925
056	17 lb. per hour		5		1,740	66	1,806	2,000
058	30 lb. per hour		4		1,790	82	1,872	2,075
060	60 lb. per hour		4		2,350	82	2,432	2,700
062	90 lb. per hour		3		2,700	110	2,810	3,125
57-001	**INDUCED DRAFT FANS**							
100	Breeching installation							
180	Hot gas, 600°F, variable pitch pulley and motor							
186	8" diam. inlet, 1/4 H.P., 1phase, 1120 CFM	Q-9	4	Ea.	1,210	82	1,292	1,450
190	12" diam. inlet, 3/4 H.P., 3 phase, 2960 CFM		3		1,660	110	1,770	1,975
194	18" diam. inlet, 3 H.P., 3 phase, 9120 CFM		2		2,710	165	2,875	3,225
198	24" diam. inlet, 7-1/2 H.P., 3 phase, 17,760 CFM		.80		4,250	410	4,660	5,275
230	For multi-blade damper at fan inlet, add				20%			
62-001	**INFRA-RED UNIT**							

15.5 Heating

		CREW	DAILY OUTPUT	UNIT	BARE COSTS MAT.	BARE COSTS INST.	BARE COSTS TOTAL	TOTAL INCL O&P
002 003	Gas fired, unvented, electric ignition, 100% shutoff. Piping and wiring not included							
006	Input, 15 MBH	Q-5	7	Ea.	280	47	327	375
012	45 MBH		5		380	66	446	515
016	60 MBH		4		450	82	532	615
024	120 MBH	↓	2		685	165	850	990
100	For opt. screen, req'd. for hangar or garage install., add			↓	8%			
65-001	**INSULATION**							
010	Rule of thumb, as a percentage of total mechanical costs						10%	
100	Boiler, 1-1/2" calcium silicate, 1/2" cement finish	Q-14	50	S.F.	2.80	6.55	9.35	12.75
102	2" fiberglass	"	80	"	1.80	4.10	5.90	8
200	Breeching, 2" calcium silicate with 1/2" cement finish, no lath							
202	Rectangular	Q-14	50	S.F.	3.90	6.55	10.45	13.95
204	Round	"	40	"	3.90	8.20	12.10	16.35
300	Ductwork							
302	Blanket type, fiberglass, flexible							
303	Fire resistant liner, black coating one side							
305	1/2" thick, 2 lb. density	Q-14	380	S.F.	.25	.86	1.11	1.55
306	1" thick, 1-1/2 lb. density		350		.31	.94	1.25	1.72
307	1-1/2" thick, 1-1/2 lb. density		320		.45	1.02	1.47	2.01
308	2" thick, 1-1/2 lb. density	↓	300	↓	.60	1.09	1.69	2.27
314	FRK vapor barrier wrap, .75 lb. density							
316	1" thick	Q-14	350	S.F.	.18	.94	1.12	1.58
317	1-1/2" thick		320		.21	1.02	1.23	1.74
318	2" thick		300		.27	1.09	1.36	1.91
319	3" thick	↓	260	↓	.39	1.26	1.65	2.29
321	Vinyl jacket, same as FRK							
328	Unfaced, 1 lb. density							
331	1" thick	Q-14	360	S.F.	.22	.91	1.13	1.58
332	1-1/2" thick		330		.34	.99	1.33	1.84
333	2" thick		310		.45	1.06	1.51	2.05
349	Board type, fiberglass, 3 lb. density							
350	Fire resistant, black pigmented, 1 side							
352	1" thick	Q-14	150	S.F.	.59	2.18	2.77	3.87
354	1-1/2" thick		130		.70	2.52	3.22	4.49
356	2" thick	↓	120	↓	1	2.73	3.73	5.15
360	FRK vapor barrier							
362	1" thick	Q-14	150	S.F.	.97	2.18	3.15	4.29
363	1-1/2" thick		130		1.10	2.52	3.62	4.93
364	2" thick	↓	120	↓	1.35	2.73	4.08	5.50
368	No finish							
370	1" thick	Q-14	170	S.F.	.56	1.93	2.49	3.46
371	1-1/2" thick		140		.74	2.34	3.08	4.27
372	2" thick	↓	130	↓	.98	2.52	3.50	4.79
375	Finishes							
380	1/2" cement over 1" wire mesh, incl. corner bead	Q-14	116	S.F.	.89	2.82	3.71	5.15
382	Canvas, 8 oz. pasted on		246		1.50	1.33	2.83	3.61
390	Weatherproof, non-metallic, 2 lb. per S.F.	↓	100	↓	2.20	3.28	5.48	7.25
400	Pipe covering							
660	Fiberglass, with all service jacket							
684	1" wall, 1/2" iron pipe size	Q-14	240	L.F.	.86	1.37	2.23	2.96
687	1" iron pipe size		220		1.05	1.49	2.54	3.35
690	2" iron pipe size		200		1.40	1.64	3.04	3.96
692	3" iron pipe size		180		1.74	1.82	3.56	4.60
694	4" iron pipe size		150		2.27	2.18	4.45	5.70
732	2" wall, 1/2" iron pipe size		220		2.96	1.49	4.45	5.45
744	6" iron pipe size		100		5.92	3.28	9.20	11.35
746	8" iron pipe size		80		7.35	4.10	11.45	14.15
748	10" iron pipe size	↓	70	↓	8.82	4.68	13.50	16.60

For expanded coverage of these items see *Means' Mechanical Cost Data 1986*

15.5 Heating

			CREW	DAILY OUTPUT	UNIT	BARE COSTS MAT.	BARE COSTS INST.	BARE COSTS TOTAL	TOTAL INCL O&P
65	749	12" iron pipe size	Q-14	65	L.F.	9.76	5.05	14.81	18.15
	760	For fiberglass with standard canvas jacket, deduct			"	5%			
	766	For fittings, add 3 L.F. for each fitting							
	768	plus 4 L.F. for each flange of the fitting							
	770	For equipment or congested areas, add			L.F.		20%		
	772	Finishes, for .010" aluminum jacket, add	Q-14	120	S.F.	.26	2.73	2.99	4.31
	774	For .016" aluminum jacket, add		120		.37	2.73	3.10	4.43
	776	For .010" stainless steel, add	↓	100		.46	3.28	3.74	5.35
	778	For single layer of felt, add				10%	10%		
	780	For roofing paper, 45 lb. to 55 lb., add			↓	25%	10%		
	786	Rubber tubing, flexible closed cell foam							
	788	3/8" wall, 1/4" iron pipe size	1 Asbe	120	L.F.	.17	1.52	1.69	2.42
	791	1/2" iron pipe size		115		.22	1.58	1.80	2.58
	792	3/4" iron pipe size		115		.25	1.58	1.83	2.61
	793	1" iron pipe size		110		.28	1.65	1.93	2.75
	795	1-1/2" iron pipe size		110		.42	1.65	2.07	2.90
	796	2" iron pipe size		105		.52	1.73	2.25	3.13
	798	3" iron pipe size		100		.81	1.82	2.63	3.58
	810	1/2" wall, 1/4" iron pipe size		90		.26	2.02	2.28	3.27
	813	1/2" iron pipe size		89		.33	2.04	2.37	3.38
	814	3/4" iron pipe size		89		.36	2.04	2.40	3.41
	815	1" iron pipe size		88		.40	2.07	2.47	3.49
	817	1-1/2" iron pipe size		87		.56	2.09	2.65	3.70
	818	2" iron pipe size		86		.67	2.12	2.79	3.86
	820	3" iron pipe size		85		1.24	2.14	3.38	4.52
	830	3/4" wall, 1/4" iron pipe size		90		.39	2.02	2.41	3.41
	833	1/2" iron pipe size		89		.54	2.04	2.58	3.61
	834	3/4" iron pipe size		89		.65	2.04	2.69	3.73
	835	1" iron pipe size		88		.75	2.07	2.82	3.88
	837	1-1/2" iron pipe size		87		1.13	2.09	3.22	4.33
	838	2" iron pipe size		86		1.33	2.12	3.45	4.58
	840	3" iron pipe size	↓	85	↓	2.22	2.14	4.36	5.60
68-001		**MAKE-UP AIR UNIT**							
	002	Indoor suspension, natural/LP gas, direct fired,							
	003	standard control. For flue see Division 15.5-92.							
	004	70°F temperature rise, MBH is input							
	010	2000 CFM, 168 MBH	Q-6	3	Ea.	3,770	170	3,940	4,400
	016	6000 CFM, 502 MBH		1.50		4,750	340	5,090	5,725
	022	12,000 CFM, 1005 MBH	↓	1		5,910	510	6,420	7,250
	030	24,000 CFM, 2007 MBH	Q-7	1		7,710	695	8,405	9,475
	040	50,000 CFM, 4180 MBH	"	.80		10,600	870	11,470	12,900
	060	For discharge louver assembly, add				10%			
	070	For filters, add				20%			
	080	For air shut-off damper section, add				10%			
	090	For vertical unit, add			↓	10%			
77-001		**SOLAR ENERGY**							
	002	System/Package prices, not including connecting							
	003	pipe, insulation, or special heating/plumbing fixtures							
	050	Hot water, standard package, low temperature							
	054	2 collectors, circulator, fittings, no tank	Q-1	.50	Ea.	950	650	1,600	1,975
	058	2 collectors, circulator, fittings, 120 gal. tank		.40		1,630	810	2,440	2,975
	062	3 collectors, circulator, fittings, 120 gal. tank	↓	.40	↓	2,010	810	2,820	3,375
	070	Medium temperature package							
	074	2 collectors, circulator, fittings, 80 gal. tank	Q-1	.40	Ea.	1,750	810	2,560	3,100
	078	3 collectors, circulator, fittings, 120 gal. tank	"	.30	"	2,200	1,075	3,275	3,975
	090	Commercial/process							
	094	10 med. temp. collectors, fittings, 120 gal. tank	Q-2	.15	Ea.	6,100	3,375	9,475	11,600
	098	For each additional 120 gal. tank, add			"	650		650	715M

15.5 Heating

		CREW	DAILY OUTPUT	UNIT	BARE COSTS MAT.	INST.	TOTAL	TOTAL INCL O&P
77 130	Solar assist package, for space heating and domestic							
134	hot water, 10 collectors, fittings	Q-2	.13	Ea.	6,475	3,875	10,350	12,700
144	For heating, complete with heat pump, 9 collectors		.12		8,675	4,200	12,875	15,600
148	12 collectors	↓	.10	↓	10,100	5,050	15,150	18,400
230	Circulators, air							
231	Blowers							
232	30 to 100 S.F. system, 1/20 HP	Q-9	16	Ea.	43	20	63	77
240	Reversible fan, 20" diameter, 2 speed		18		150	18.15	168.15	190
252	Space & DHW system, less duct work	↓	.50	↓	2,000	655	2,655	3,150
270								
275	Circulators, liquid, 1/100 HP, 2 GPM	Q-1	14	Ea.	64	23	87	105
287	1/12 HP, 30 GPM	"	10	"	105	32	137	160
300	Collector panels, air with aluminum absorber plate							
301	Wall or roof mount							
304	Flat black, plastic glazing							
308	4' x 8'	Q-9	6	Ea.	350	54	404	465
320	Flush roof mount, 10' to 16' x 22" wide	"	96	L.F.	32	3.41	35.41	40
330	Collector panels, liquid with copper absorber plate							
333	Alum. frame, 3' x 8', 5/32" single glazing	Q-1	9.50	Ea.	285	34	319	365
336	Alum. frame, 3' x 8', 5/32" double glazing		9		315	36	351	400
339	Alum. frame, 4' x 8', 3/16" glazing	↓	6	↓	630	54	684	770
344	Flat black							
345	Alum. frame, 3' x 8', 5/32" single glazing	Q-1	9	Ea.	275	36	311	355
350	Alum. frame, 3' x 8', 5/32" double glazing		5.50		295	59	354	410
352	Alum. frame, 3' x 10', plastic glazing		10		185	32	217	250
354	Alum. frame, 3' x 8', 1/8" tempered glass		5		330	65	395	455
360	Liquid, full wetted, plastic, alum. frame, 3' x 10'		5		95	65	160	200
365	Collector panel mounting, flat roof or ground rack		7	↓	110	46	156	190
367	Roof clamps	↓	70	Set	3	4.64	7.64	10
370	Roof strap, teflon	1 Plum	205	L.F.	3.15	.88	4.03	4.74
390	Differential controller with two sensors							
393	Thermostat, hard wired	1 Plum	8	Ea.	59.50	23	82.50	98
410	Six station with digital read-out	"	3	"	230	60	290	340
430	Heat exchanger							
431	Fluid to air coil							
449	Horizontal, 110 MBH	Q-1	2	Ea.	189	160	349	445
458	Fluid to fluid package includes two circulating pumps							
459	expansion tank, check valve, relief valve							
460	controller, high temperature cutoff and sensors	Q-1	2.50	Ea.	585	130	715	830
462								
465	Heat transfer fluid							
470	Propylene glycol, inhibited anti-freeze	1 Plum	28	Gal.	5.65	6.45	12.10	15.55
825	Water storage tank with heat exchanger and electric element							
828	80 gal. with 2" x 1/2 lb. density insulation	1 Plum	1.60	Ea.	465	115	580	675
830	80 gal. with 2" x 2 lb. density insulation		1.60		500	115	615	715
835	120 gal. with 2" x 1/2 lb. density insulation		1.40		585	130	715	830
838	120 gal. with 2" x 2 lb. density insulation		1.40		610	130	740	855
840	120 gal. with 2" x 2 lb. density insul., 40 S.F. heat coil	↓	1.40	↓	740	130	870	1,000
80-001	**SPACE HEATERS** Cabinet, grilles, fan, controls, burner,							
002	thermostat, no piping. For flue see division 15.5-92							
100	Gas fired, floor mounted							
110	60 MBH output	Q-5	10	Ea.	365	33	398	450
114	100 MBH output		8		410	41	451	510
118	180 MBH output		6		555	55	610	690
200	Suspension mounted, propeller fan, 36 MBH output		8		330	41	371	420
204	60 MBH output		7		375	47	422	480
206	84 MBH output		6		425	55	480	545
210	120 MBH output		5		520	66	586	665
224	320 MBH output	↓	2		1,160	165	1,325	1,525
250	For powered venter and adapter, add			↓	164		164	180M

For expanded coverage of these items see *Means' Mechanical Cost Data 1986*

15.5 Heating

		CREW	DAILY OUTPUT	UNIT	BARE COSTS MAT.	BARE COSTS INST.	BARE COSTS TOTAL	TOTAL INCL O&P
80 500	Wall furnace, 14 MBH output	Q-5	6	Ea.	200	55	255	300
502	24 MBH output		5		250	66	316	370
504	49 MBH output		4		350	82	432	505
600	Oil fired, suspension mounted, 94 MBH output		4		980	82	1,062	1,200
604	140 MBH output		3		1,050	110	1,160	1,325
606	184 MBH output	↓	3	↓	1,125	110	1,235	1,400
86-001	**SWIMMING POOL HEATERS** Not including wiring, external							
002	piping, base or pad,							
006	Gas fired, gross output, 50 MBH	Q-6	3	Ea.	360	170	530	640
010	80 MBH		2		385	255	640	790
016	120 MBH		1.50		410	340	750	940
020	170 MBH		1		515	510	1,025	1,300
028	500 MBH		.40		1,650	1,275	2,925	3,650
034	890 MBH		.26		2,140	1,950	4,090	5,200
040	1440 MBH		.14		3,350	3,650	7,000	8,950
044	3000 MBH	↓	.09		5,820	5,650	11,470	14,600
150	Oil fired, deduct				3%			
200	Electric, 12 KW, 4800 gallon pool	Q-19	3		860	170	1,030	1,200
202	18 KW, 7200 gallon pool		2.80		885	180	1,065	1,225
204	24 KW, 9600 gallon pool		2.40		940	210	1,150	1,350
210	54 KW, 24,000 gallon pool		1.20		1,275	420	1,695	2,000
212	300 KW, 120,000 gallon pool		1		7,800	505	8,305	9,300
214	600 KW, 240,000 gallon pool		.80		12,900	635	13,535	15,100
216	1200 KW, 480,000 gallon pool	↓	.60	↓	20,300	845	21,145	23,500
89-001	**TANKS**							
002	Fiberglass, underground, U.L. listed, not including							
003	manway or hold-down strap							
014	2000 gallon capacity	Q-7	2	Ea.	2,400	350	2,750	3,150
016	4000 gallon capacity		1.30		3,140	535	3,675	4,225
018	6000 gallon capacity		1		3,970	695	4,665	5,375
022	10,000 gallon capacity		.60		5,290	1,150	6,440	7,500
026	15,000 gallon capacity		.40		8,680	1,750	10,430	12,100
028	20,000 gallon capacity	↓	.30		11,250	2,325	13,575	15,700
050	For manway, fittings and hold-downs, add				20%	15%		
200	Steel, liquid expansion, painted, 8 gallon capacity	Q-5	20		90	16.40	106.40	125
202	12 gallon capacity		18		150	18.20	168.20	190
204	15 gallon capacity		17		170	19.25	189.25	215
206	18 gallon capacity		16		190	20	210	240
208	24 gallon capacity		14		205	23	228	260
210	30 gallon capacity		12		225	27	252	285
212	40 gallon capacity		10		255	33	288	330
300	Steel ASME expansion, rubber diaphragm, 19 gallon capacity		17		540	19.25	559.25	620
302	31 gallon capacity		12		700	27	727	810
304	61 gallon capacity		9		1,075	36	1,111	1,225
308	119 gallon capacity		8		1,350	41	1,391	1,550
310	158 gallon capacity		7		1,950	47	1,997	2,225
314	317 gallon capacity		5		2,775	66	2,841	3,150
318	528 gallon capacity	↓	3	↓	4,575	110	4,685	5,200
400	Steel, storage, above ground, including supports, coating,							
402	fittings, not including mat, pumps or piping							
404	275 gallon capacity	Q-5	5	Ea.	155	66	221	265
406	550 gallon capacity	"	4		585	82	667	760
408	1000 gallon capacity	Q-7	4		815	175	990	1,150
410	1500 gallon capacity		3.70		1,225	190	1,415	1,625
412	2000 gallon capacity		3		1,250	230	1,480	1,700
414	5000 gallon capacity	↓	1	↓	2,950	695	3,645	4,250
500	Steel underground, coated, set in place, incl. hold-down bars.							
550	Excavation, pad, pumps and piping not included							
552	1000 gallon capacity, 7 gauge shell	Q-7	4	Ea.	725	175	900	1,050
554	5000 gallon capacity, 1/4" thick shell	"	1	"	2,625	695	3,320	3,900

For expanded coverage of these items see *Means' Mechanical Cost Data 1986*

15.5 Heating

		CREW	DAILY OUTPUT	UNIT	BARE COSTS MAT.	BARE COSTS INST.	BARE COSTS TOTAL	TOTAL INCL O&P
89 558	10,000 gallon capacity, 5/16" thick shell	Q-7	.60	Ea.	4,650	1,150	5,800	6,800
560	20,000 gallon capacity, 5/16" thick shell		.30		7,500	2,325	9,825	11,600
562	30,000 gallon capacity, 3/8" thick shell	↓	.20	↓	12,700	3,475	16,175	19,000
595								
600	For 20 year warranty galvanic protection,							
602	add to tank cost							
604	1000 gallon tank			Ea.	500		500	550M
610	10,000 gallon tank				1,625		1,625	1,800M
614	30,000 gallon tank			↓	3,575		3,575	3,925M
92-001	**VENT CHIMNEY** Prefab metal, U.L. listed							
002	Gas, double wall, galvanized steel							
008	3" diameter	Q-9	72	V.L.F.	2.10	4.54	6.64	8.95
010	4" diameter		68		2.56	4.81	7.37	9.85
012	5" diameter		64		3.03	5.10	8.13	10.80
014	6" diameter		60		3.54	5.45	8.99	11.85
016	7" diameter		56		4.82	5.85	10.67	13.85
018	8" diameter		52		5.36	6.30	11.66	15.10
020	10" diameter		48		11.35	6.80	18.15	22
022	12" diameter		44		15.15	7.45	22.60	28
026	16" diameter	↓	40		34.25	8.15	42.40	50
030	20" diameter	Q-10	36		50.50	14.10	64.60	76
048	38" diameter	"	24		153	21	174	200
300	All fuel, double wall, stainless steel, 6" diameter	Q-9	60		14.65	5.45	20.10	24
302	7" diameter		56		18.10	5.85	23.95	28
304	8" diameter		52		22	6.30	28.30	33
306	10" diameter		48		30.25	6.80	37.05	43
308	12" diameter		44		40.50	7.45	47.95	55
310	14" diameter	↓	42	↓	53	7.80	60.80	70
900	High temp. (2000°F), steel jacket, acid resist. refractory lining							
901	11 ga. galvanized jacket, U.L. listed							
902	Straight section, 48" long, 10" diameter	Q-10	13.30	V.L.F.	260	38	298	340
904	18" diameter		7.40		390	69	459	530
907	36" diameter	↓	2.70		800	190	990	1,150
909	48" diameter	Q-11	2.70	↓	1,300	255	1,555	1,800
912	Tee section, 10" diameter	Q-10	4.40	Ea.	430	115	545	640
914	18" diameter		2.40		650	210	860	1,025
917	36" diameter	↓	.80		1,975	635	2,610	3,100
919	48" diameter	Q-11	.90		3,175	770	3,945	4,625
922	Cleanout pier section, 10" diameter	Q-10	3.50		455	145	600	715
924	18" diameter		2		655	255	910	1,100
927	36" diameter	↓	.75		1,150	680	1,830	2,250
929	48" diameter	Q-11	.75		1,720	925	2,645	3,250
932	For drain, add				53%			
933	Elbow, 30° and 45°, 10" diameter	Q-10	6.60		250	77	327	390
935	18" diameter		3.70		445	135	580	690
938	36" diameter	↓	1.30		1,200	390	1,590	1,900
940	48" diameter	Q-11	1.35		2,500	515	3,015	3,500
943	For 60° and 90° elbow, add				112%			
944	End cap, 10" diameter	Q-10	26		515	19.55	534.55	595
946	18" diameter		15		760	34	794	885
949	36" diameter	↓	5		1,460	100	1,560	1,750
951	48" diameter	Q-11	5.30		2,210	130	2,340	2,625
954	Increaser (1 diameter), 10" diameter	Q-10	6.60		365	77	442	515
956	18" diameter		3.70		530	135	665	785
959	36" diameter	↓	1.30	↓	1,120	390	1,510	1,800
960	For expansion joints, add to straight section				7%			
961	For 1/4" hot rolled steel jacket, add				157%			
962	For 2950°F very high temperature, add				89%			
963	26 ga. aluminized jacket, straight section, 48" long							

For expanded coverage of these items see *Means' Mechanical Cost Data 1986*

15.5 Heating

			CREW	DAILY OUTPUT	UNIT	BARE COSTS MAT.	BARE COSTS INST.	BARE COSTS TOTAL	TOTAL INCL O&P
92	964	10" diameter	Q-10	15.30	V.L.F.	120	33	153	180
	966	18" diameter		8.50		208	60	268	315
	969	36" diameter		3.10	↓	475	165	640	760
	981	Draw band, 11 gauge, 10" diameter		32	Ea.	29.10	15.90	45	55
	982	12" diameter		30		31.20	16.95	48.15	59
	983	18" diameter		26		46.80	19.55	66.35	80
	986	36" diameter	↓	18		98.80	28	126.80	150
	988	48" diameter	Q-11	20	↓	140	35	175	205

15.6 HVAC Piping Specialties

			CREW	DAILY OUTPUT	UNIT	BARE COSTS MAT.	BARE COSTS INST.	BARE COSTS TOTAL	TOTAL INCL O&P
04-001		AUTOMATIC AIR VENT							
	002	Cast iron body, stainless steel internals							
	006	1/2" NPT inlet, 300 psi	1 Stpi	12	Ea.	39.75	15.15	54.90	66
	022	3/4" NPT inlet, 250 psi	"	10	↓	128	18.20	146.20	165
	034	1-1/2" NPT inlet, 250 psi	Q-5	12	↓	395	27	422	475
08-001		AIR CONTROL With strainer							
	004	2" diameter	Q-5	6	Ea.	290	55	345	400
	008	2-1/2" diameter		5		320	66	386	445
	010	3" diameter		4		495	82	577	665
	012	4" diameter	↓	3		710	110	820	940
	014	6" diameter	Q-6	3.40	↓	1,075	150	1,225	1,400
20-001		EXPANSION JOINTS Bellows type, neoprene cover, flanged spool							
	010	6" face to face, 1/2" diameter	1 Stpi	14	Ea.	171	13	184	205
	011	3/4" diameter		14		171	13	184	205
	012	1" diameter		13		171	14	185	210
	014	1-1/4" diameter		11		171	16.55	187.55	210
	016	1-1/2" diameter	↓	10.60		180	17.15	197.15	225
	018	2" diameter	Q-5	13.30		183	25	208	235
	020	3" diameter		11.40		204	29	233	265
	048	10" face to face, 2" diameter		13		234	25	259	295
	050	2-1/2" diameter		12		252	27	279	315
	052	3" diameter		11		259	30	289	330
	054	4" diameter		8		305	41	346	395
	056	5" diameter		7		340	47	387	440
	058	6" diameter	↓	6	↓	345	55	400	460
28-001		FLEXIBLE METAL HOSE Connectors, standard lengths							
	010	Bronze braided, bronze ends							
	012	3/8" diameter x 12"	1 Stpi	26	Ea.	14.95	7	21.95	27
	016	3/4" diameter x 12"		20		26	9.10	35.10	42
	018	1" diameter x 18"		19		43	9.60	52.60	61
	020	1-1/2" diameter x 18"		13		64	14	78	91
	022	2" diameter x 18"	↓	11	↓	90	16.55	106.55	125
32-001		HEATING CONTROL VALVES							
	010	Radiator supply, 1/2" diameter	1 Stpi	24	Ea.	56.75	7.60	64.35	73
	012	3/4" diameter		20		59.50	9.10	68.60	79
	014	1" diameter		19		71.60	9.60	81.20	93
	016	1-1/4" diameter	↓	15		79	12.15	91.15	105
	050	For low pressure steam, add				25%			
44-001		PRESSURE REGULATOR							
	300	Steam, high capacity, bronze body, stainless steel trim							
	302	Threaded, 1/2" diameter	1 Stpi	24	Ea.	242	7.60	249.60	275
	303	3/4" diameter	"	24	"	242	7.60	249.60	275

For expanded coverage of these items see *Means' Mechanical Cost Data 1986*

15.6 HVAC Piping Specialties

			CREW	DAILY OUTPUT	UNIT	BARE COSTS MAT.	BARE COSTS INST.	BARE COSTS TOTAL	TOTAL INCL O&P
44	304	1" diameter	1 Stpi	19	Ea.	270	9.60	279.60	310
	306	1-1/4" diameter		15		298	12.15	310.15	345
	308	1-1/2" diameter		13		340	14	354	395
	310	2" diameter	↓	11		420	16.55	436.55	485
	312	2-1/2" diameter	Q-5	12		525	27	552	615
	314	3" diameter	"	11	↓	595	30	625	700
	350	Flanged connection, iron body, 125 lb. W.S.P.							
	352	3" diameter	Q-5	11	Ea.	655	30	685	765
	354	4" diameter	"	5	"	825	66	891	1,000
55-001		**STEAM CONDENSATE METER**							
	005	250 lb. per hour	1 Stpi	16	Ea.	950	11.40	961.40	1,050
	010	500 lb. per hour		14		1,075	13	1,088	1,200
	012	750 lb. per hour		11		1,160	16.55	1,176.55	1,300
	014	1500 lb. per hour	↓	7	↓	1,340	26	1,366	1,500
60-001		**STEAM TRAP** Cast Iron							
	004	Inverted bucket							
	005	1/2" pipe size	1 Stpi	12	Ea.	52.75	15.15	67.90	80
	007	3/4" pipe size		10		88.75	18.20	106.95	125
	010	1" pipe size		9		138	20	158	180
	012	1-1/4" pipe size	↓	8	↓	208	23	231	260
	100	Float & thermostatic, 15 psig							
	101	3/4" pipe size	1 Stpi	16	Ea.	62	11.40	73.40	85
	102	1" pipe size		15		76.75	12.15	88.90	100
	104	1-1/2" pipe size		9		140	20	160	185
	106	2" pipe size	↓	6	↓	254	30	284	325
70-001		**STRAINERS, Y TYPE** Bronze body							
	005	Screwed, 150 lb., 1/4" pipe size	1 Stpi	24	Ea.	13.80	7.60	21.40	26
	007	3/8" pipe size		24		13.80	7.60	21.40	26
	010	1/2" pipe size		20		16.20	9.10	25.30	31
	014	1" pipe size		17		28.20	10.70	38.90	47
	016	1-1/2" pipe size		14		50.40	13	63.40	74
	018	2" pipe size	↓	13		82.20	14	96.20	110
	022	3" pipe size	Q-5	16		330	20	350	395
	024	4" pipe size	"	15		755	22	777	860
	050	For 300 lb. rating, add				15%			
	100	Flanged, 150 lb., 1-1/2" pipe size	1 Stpi	11		215	16.55	231.55	260
	102	2" pipe size	"	8		278	23	301	340
	104	3" pipe size	Q-5	4.50		530	73	603	690
	106	4" pipe size	"	3		800	110	910	1,050
	110	6" pipe size	Q-6	3		1,530	170	1,700	1,925
	150	For 300 lb. rating, add			↓	40%			
75-001		**STRAINERS, Y TYPE** Iron body							
	005	Screwed, 250 lb., 1/4" pipe size	1 Stpi	20	Ea.	6.60	9.10	15.70	20
	007	3/8" pipe size		20		6.60	9.10	15.70	20
	010	1/2" pipe size		20		6.60	9.10	15.70	20
	014	1" pipe size		16		10.20	11.40	21.60	28
	016	1-1/2" pipe size		12		17.10	15.15	32.25	41
	018	2" pipe size	↓	8		28.35	23	51.35	64
	022	3" pipe size	Q-5	11		64.25	30	94.25	115
	024	4" pipe size	"	5		225	66	291	340
	050	For galvanized body, add				50%			
	100	Flanged, 125 lb., 1-1/2" pipe size	1 Stpi	11		68.50	16.55	85.05	99
	102	2" pipe size	"	8		68.50	23	91.50	110
	104	3" pipe size	Q-5	4.50		80.50	73	153.50	195
	106	4" pipe size	"	3		156.75	110	266.75	330
	108	5" pipe size	Q-6	3.40		247	150	397	490
	110	6" pipe size	"	3	↓	298	170	468	575

For expanded coverage of these items see *Means' Mechanical Cost Data 1986*

15.6 HVAC Piping Specialties

			CREW	DAILY OUTPUT	UNIT	BARE COSTS MAT.	BARE COSTS INST.	BARE COSTS TOTAL	TOTAL INCL O&P
75	150	For 250 lb. rating, add			Ea.	20%			
	200	For galvanized body, add				50%			
	250	For steel body, add			↓	40%			
	85-001	**VENTURI FLOW** Measuring device							
	005	1/2" diameter	1 Stpi	24	Ea.	64	7.60	71.60	81
	012	1" diameter		19		72	9.60	81.60	93
	014	1-1/4" diameter		15		78.50	12.15	90.65	105
	016	1-1/2" diameter		13		86.50	14	100.50	115
	018	2" diameter	↓	11		105	16.55	121.55	140
	022	3" diameter	Q-5	14		176	23	199	225
	024	4" diameter	"	11		221	30	251	285
	028	6" diameter	Q-6	3.50		330	145	475	575
	050	For meter, add			↓	700		700	770M

15.7 Air Conditioning & Ventilating

		CREW	DAILY OUTPUT	UNIT	BARE COSTS MAT.	BARE COSTS INST.	BARE COSTS TOTAL	TOTAL INCL O&P
01-001	**AVERAGE** Square foot and percent of total							
010	job cost, see division 17.1							
04-001	**ABSORPTION COLD GENERATORS** Water chiller							
002	Steam or hot water, water cooled							
005	101 ton			Ea.	47,300	14,200	61,500	72,500
040	420 ton			"	104,700	31,400	136,100	160,500
07-001	**AIR CONDITIONING REQUIREMENTS** For various types of buildings (136)							
010	buildings							
10-001	**AIR FILTERS**							
005	Activated charcoal type, full flow			MCFM	450		450	495M
200	Electronic air cleaner, self-contained							
205	185 CFM	1 Shee	2.40	Ea.	205	76	281	335
215	500 CFM		2.30		570	79	649	740
220	1000 CFM		2.20		785	83	868	985
225	1200 CFM		2.10		1,160	86	1,246	1,400
230	2500 CFM	↓	2	↓	1,320	91	1,411	1,575
295	Mechanical media filtration units							
300	High efficiency type, with frame, non-supported			MCFM	35		35	39M
310	Supported type				47		47	52M
400	Medium efficiency, extended surface				4		4	4.40M
450	Permanent washable				25		25	28M
500	Renewable disposable roll			↓	70		70	77M
550	Throwaway glass or paper media ty			Ea.	1.90		1.90	2.09M
16-001	**BALANCING, AIR** (Subcontract including material & labor)							
090	Heating and ventilating equipment							
100	Centrifugal fans, utility sets			Ea.	159		159	175M
110	Heating and ventilating unit				239		239	265M
120	In-line fan				239		239	265M
130	Propeller and wall fan				45		45	50M
140	Roof exhaust fan				106		106	115M
200	Air conditioning equipment, central station				345		345	380M
210	Built-up low pressure unit				320		320	350M
220	Built-up high pressure unit				370		370	405M
250	Multi-zone A.C. and heating unit				239		239	265M
260	For each zone over one, add				53		53	58M
270	Package A.C. unit				133		133	145M
280	Rooftop heating and cooling unit			↓	185		185	205M

For expanded coverage of these items see *Means' Mechanical Cost Data 1986*

15.7 Air Conditioning & Ventilating

			CREW	DAILY OUTPUT	UNIT	BARE COSTS MAT.	BARE COSTS INST.	BARE COSTS TOTAL	TOTAL INCL O&P
16	300	Supply, return, exhaust, registers & diffusers, avg. height ceiling			Ea.	32		32	35M
	310	High ceiling				48		48	53M
	320	Floor height			↓	27		27	30M
19-001		**BALANCING, WATER** (Subcontract, including material & labor)							
	005	Air cooled condenser			Ea.	92		92	100M
	010	Cabinet unit heater				32		32	35M
	020	Chiller				224		224	245M
	030	Convector				26		26	29M
	050	Cooling tower				171		171	190M
	060	Fan coil unit, unit ventilator				47		47	52M
	070	Fin tube and radiant panels				53		53	58M
	080	Main and duct re-heat coils				49		49	54M
	100	Pumps				116		116	130M
	110	Unit heater			↓	37		37	41M
22-001		**CENTRAL STATION AIR-HANDLING UNIT, CHILL WATER**							
	100	Modular, capacity at 700 FPM face velocity							
	110	1300 CFM	Q-5	1.20	Ea.	1,320	275	1,595	1,850
	130	3200 CFM	"	.80		1,645	410	2,055	2,400
	150	8050 CFM	Q-6	.60		3,350	850	4,200	4,925
	170	18,410 CFM	↓	.33		6,290	1,550	7,840	9,150
	190	33,670 CFM	↓	.19		11,100	2,675	13,775	16,100
	210	63,000 CFM	Q-7	.13		23,050	5,350	28,400	33,100
	220	For hot water heating coil, add			↓	10%	5%		
25-001		**COILS, FLANGED**							
	050	Chilled water cooling, 6 rows, 24" x 48"	Q-5	2	Ea.	1,870	165	2,035	2,300
	100	Direct expansion cooling, 6 rows, 24" x 48"		2		2,040	165	2,205	2,475
	150	Hot water heating, 1 row, 24" x 48"		3		740	110	850	970
	200	Steam heating, 1 row, 24" x 48"	↓	3	↓	1,055	110	1,165	1,325
31-001		**COMPUTER ROOM UNITS**							
	100	Air cooled, includes remote condenser but not							
	102	interconnecting tubing or refrigerant							
	108	3 ton	Q-5	.50	Ea.	6,100	655	6,755	7,650
	112	5 ton		.45		6,820	730	7,550	8,550
	116	6 ton		.30		11,000	1,100	12,100	13,700
	120	8 ton		.27		12,100	1,225	13,325	15,100
	124	10 ton		.25		13,250	1,300	14,550	16,500
	128	15 ton	↓	.22		15,450	1,500	16,950	19,100
	132	20 ton	Q-6	.29		17,650	1,750	19,400	22,000
	136	23 ton	"	.28	↓	18,750	1,825	20,575	23,300
34-001		**CONDENSERS** Ratings are for 30°F TD, R-22							
	008	Air cooled, belt drive, propeller fan							
	010	20 ton	Q-5	1	Ea.	3,475	330	3,805	4,300
	018	30 ton	"	.60		3,675	545	4,220	4,825
	024	48 ton	Q-6	.48		5,370	1,050	6,420	7,450
	028	60 ton		.32		6,075	1,600	7,675	8,975
	032	72 ton		.29		7,160	1,750	8,910	10,400
	036	82 ton		.27		8,450	1,875	10,325	12,000
	038	90 ton	↓	.26	↓	9,150	1,950	11,100	12,900
	155	Air cooled, direct drive, propeller fan							
	160	1-1/2 ton	Q-5	3.60	Ea.	375	91	466	545
	162	2 ton		3.20		415	100	515	605
	164	5 ton		2		970	165	1,135	1,300
	166	10 ton		1.40		1,600	235	1,835	2,100
	169	16 ton		1.10		2,070	300	2,370	2,700
	172	26 ton	↓	.74	↓	2,900	445	3,345	3,825

For expanded coverage of these items see *Means' Mechanical Cost Data 1986*

15.7 Air Conditioning & Ventilating

			CREW	DAILY OUTPUT	UNIT	BARE COSTS MAT.	BARE COSTS INST.	BARE COSTS TOTAL	TOTAL INCL O&P
34	176	41 ton	Q-6	.54	Ea.	4,600	945	5,545	6,425
	180	63 ton	"	.32	"	6,400	1,600	8,000	9,350
	43-001	**CONTROL SYSTEMS, PNEUMATIC** (Subcontract includes material & labor)							
	010	Heating and Ventilating, split system							
	020	Mixed air control, economizer cycle, panel readout, tubing							
	022	Up to 10 tons			Ea.	2,500		2,500	2,750M
	024	For 10 to 20 tons, add				15%			
	026	For over 20 tons, add			↓	25%			
	030	Heating coil, hot water, 3 way valve,							
	032	freezestat, limit control on discharge readout			Ea.	1,890		1,890	2,075M
	050	Cooling coil, chilled water, room							
	052	thermostat, 3 way valve			Ea.	735		735	810M
	060	Cooling tower, fan cycle, damper control,							
	062	condenser, water readout in/out at panel			Ea.	1,900		1,900	2,100M
	100	Unit ventilator, day/night operation,							
	110	freezestat, ASHRAE, cycle 2			Ea.	1,845		1,845	2,025M
	200	Compensated hot water from boiler, valve control,							
	210	readout and reset at panel, up to 60 GPM			Ea.	2,535		2,535	2,800M
	212	For 120 GPM, add				7%			
	214	For 240 GPM, add				12%			
	300	Boiler room combustion air, damper, controls				895		895	985M
	350	Fan coil, heating and cooling valves, 4 pipe system				805		805	885M
	400	Pneumatic thermostat, controlling room radiator valve			↓	245		245	270M
	450	Air supply for pneumatic control system							
	460	Tank mounted duplex compressor, starter, alternator,							
	462	piping, dryer, PRV station and filter							
	463	1/2 HP			Ea.	5,820		5,820	6,400M
	466	1-1/2 HP				7,635		7,635	8,400M
	469	5 HP			↓	13,950		13,950	15,300M
	46-001	**CONTROL SYSTEMS, ELECTRONIC**							
	002	Pneumatic, costs, add to division 15.7-43			Ea.	15%			
	49-001	**COOLING TOWERS**							
	008	Draw thru, single flow							
	010	Belt drive, 60 tons	Q-6	16	Ton	58	32	90	110
	015	90 tons		20		53	25	78	95
	020	100 tons	↓	20	↓	50	25	75	92
	150	Induced air, double flow							
	180	Gear drive, 125 ton	Q-6	22	Ton	41.75	23	64.75	79
	190	150 ton		22		39.25	23	62.25	77
	200	300 ton		24		36.75	21	57.75	71
	210	600 ton		26		31	19.60	50.60	62
	215	840 ton		28		24.50	18.20	42.70	53
	220	Up to 1000 tons	↓	28	↓	23	18.20	41.20	52
	300	For higher capacities, use multiples							
	350	For pumps and piping, add	Q-6	38	Ton	29	13.40	42.40	51
	400	For absorption systems, add			"	75%	75%		
	450	For rigging, see division 1.5-20							
	52-001	**DIFFUSERS** Aluminum, opposed blade damper unless noted							
	010	Ceiling, linear, also for sidewall							
	050	Perforated, 24" x 24", panel size 6" x 6"	1 Shee	16	Ea.	48.75	11.35	60.10	70
	052	8" x 8"		15		50.75	12.10	62.85	74
	056	12" x 12"		12		53.25	15.15	68.40	81
	060	18" x 18"		10		63	18.15	81.15	96
	100	Rectangular, 1 to 4 way blow, 6" x 6"		16		29.75	11.35	41.10	49
	102	12" x 6"		15		38.50	12.10	50.60	60
	104	12" x 9"		14		45.50	12.95	58.45	69
	106	12" x 12"		12		52	15.15	67.15	79
	110	24" x 12"	↓	10	↓	85.75	18.15	103.90	120
	116	21" x 21"		8		122	23	145	165

15.7 Air Conditioning & Ventilating

			CREW	DAILY OUTPUT	UNIT	BARE COSTS MAT.	BARE COSTS INST.	BARE COSTS TOTAL	TOTAL INCL O&P
52	150	Round, butterfly damper, 6" diameter	1 Shee	18	Ea.	12.60	10.10	22.70	29
	152	8" diameter		16		13.90	11.35	25.25	32
	154	10" diameter		14		16.40	12.95	29.35	37
	156	12" diameter		12		21.25	15.15	36.40	45
	158	14" diameter		10		20.25	18.15	38.40	49
	200	T bar mounting, 24" x 24" lay-in frame, 6" x 6"		16		51.35	11.35	62.70	73
	202	9" x 9"		14		57.75	12.95	70.70	82
	204	12" x 12"		12		74.75	15.15	89.90	105
	206	15" x 15"		11		96.25	16.50	112.75	130
	208	18" x 18"		10		95.50	18.15	113.65	130
	600	For steel diffusers instead of aluminum, deduct				10%			
55-001		**GRILLES**							
	002	Aluminum							
	100	Air return, 6" x 6"	1 Shee	26	Ea.	6.45	7	13.45	17.30
	102	10" x 6"		24		6.95	7.55	14.50	18.70
	108	16" x 8"		22		9.90	8.25	18.15	23
	110	12" x 12"		22		9.90	8.25	18.15	23
	112	24" x 12"		18		18.10	10.10	28.20	35
	122	24" x 18"		16		23.50	11.35	34.85	42
	128	36" x 24"		14		50.50	12.95	63.45	75
	300	Filter grille with filter, 12" x 12"		24		16.65	7.55	24.20	29
	302	18" x 12"		20		23.35	9.10	32.45	39
	304	24" x 18"		18		40	10.10	50.10	59
	306	24" x 24"		16		54.25	11.35	65.60	76
	600	For steel grilles instead of aluminum in above, deduct				10%			
58-001		**REGISTERS**							
	098	Air supply							
	100	Ceiling/wall, O.B. damper, anodized aluminum							
	101	One or two way deflection, adj. curved face bars							
	102	8" x 4"	1 Shee	26	Ea.	23	7	30	36
	112	12" x 12"		18		38.50	10.10	48.60	57
	124	20" x 6"		18		44	10.10	54.10	63
	134	24" x 8"		13		55	13.95	68.95	81
	270	Above registers in steel instead of aluminum, deduct				10%			
	400	Floor, toe operated damper, enameled steel							
	402	4" x 6"	1 Shee	32	Ea.	17.30	5.70	23	27
	410	8" x 10"		22		10.90	8.25	19.15	24
	414	10" x 10"		20		27	9.10	36.10	43
	422	14" x 14"		16		23	11.35	34.35	42
	424	14" x 20"		15		76.75	12.10	88.85	100
	426	16" x 16"		15		81.25	12.10	93.35	105
	498	Air return							
	500	Ceiling or wall, fixed 45° face blades,							
	501	Adjustable O.B. damper, anodized aluminum							
	502	8" x 4"	1 Shee	26	Ea.	9.50	7	16.50	21
	506	10" x 6"		19		11.35	9.55	20.90	26
	528	24" x 24"		11		52	16.50	68.50	81
	530	36" x 24"		8		91	23	114	135
	600	Above registers in steel instead of aluminum, deduct				10%			
61-001		**DUCT ACCESSORIES**							
	002								
	005	Air extractors, 12" x 4"	1 Shee	24	Ea.	8.40	7.55	15.95	20
	010	8" x 6"		22		7.60	8.25	15.85	20
	020	20" x 8"		16		18.20	11.35	29.55	37
	028	24" x 12"		10		25.75	18.15	43.90	55
	100	Duct access door, insulated, 6" x 8"		14		11.05	12.95	24	31
	102	10" x 12"		11		14.30	16.50	30.80	40
	104	12" x 16"		10		20.75	18.15	38.90	49
	150	With insulated plexiglass window, 12" x 16"		10		42.25	18.15	60.40	73

For expanded coverage of these items see *Means' Mechanical Cost Data 1986*

15.7 Air Conditioning & Ventilating

			DAILY		BARE COSTS			TOTAL
		CREW	OUTPUT	UNIT	MAT.	INST.	TOTAL	INCL O&P
61 200	Fabrics for flexible connections, with metal edge	1 Shee	100	L.F.	1.06	1.82	2.88	3.82
210	Without metal edge		160	"	.59	1.14	1.73	2.31
300	Fire damper, curtain type, vertical, 8" x 4"		24	Ea.	21.50	7.55	29.05	35
302	12" x 4"		22		21.50	8.25	29.75	36
324	16" x 14"		18		23.50	10.10	33.60	41
340	24" x 20"		8		33.75	23	56.75	70
600	Multi-blade dampers, opposed blade, 12" x 12"		21		8	8.65	16.65	21
602	12" x 18"		18		11.10	10.10	21.20	27
608	24" x 24"	↓	8		26.25	23	49.25	62
618	48" x 36"	Q-9	7		80.75	47	127.75	155
700	Splitter damper assembly, self-locking, 1' rod	1 Shee	24		8.60	7.55	16.15	21
702	3' rod		22		11.25	8.25	19.50	24
704	4' rod		20		12.50	9.10	21.60	27
706	6' rod	↓	18	↓	15	10.10	25.10	31
800	Volume control, dampers							
810	8" x 8"	1 Shee	24	Ea.	14.75	7.55	22.30	27
814	16" x 10"		20		21	9.10	30.10	36
820	24" x 16"		11		40	16.50	56.50	68
826	30" x 18"	↓	7	↓	54	26	80	97
850								
900	Silencers, noise control for air flow, duct			MCFM	36		36	40M
910	Louvers, galvanized			"	50		50	55M
920	Plenums, measured by panel surface			S.F.	8		8	8.80M
930	Vent, air transfer			Ea.	135		135	150M
64-001	**DUCTWORK**							
002	Fabricated rectangular, includes fittings, joints, supports,							
003	-allowance for flexible connections, no insulation							
005	(137)							
010	Aluminum, alloy 3003-H14, under 300 lb.	Q-10	75	Lb.	190	6.80	196.80	220
016	Over 10,000 lb.		145		1.23	3.51	4.74	6.50
050	Galvanized steel, under 400 lb.		235		1.12	2.16	3.28	4.39
052	400 to 1000 lb.		255		.82	1.99	2.81	3.81
054	1000 to 2000 lb.		265		.64	1.92	2.56	3.51
056	2000 to 5000 lb.		275		.55	1.85	2.40	3.31
058	Over 10,000 lb.		300		.53	1.69	2.22	3.06
100	Stainless steel, type 304, under 400 lb.		165		2.59	3.08	5.67	7.35
106	Over 10,000 lb.	↓	235		1.09	2.16	3.25	4.36
110	For medium pressure ductwork, add					15%		
120	For high pressure ductwork, add			↓		40%		
130	Flexible, vinyl coated spring steel or aluminum,							
140	pressure to 10" (WG) UL-181							
150	Non-insulated, 3" diameter	Q-9	400	L.F.	.30	.82	1.12	1.52
154	5" diameter		320		.38	1.02	1.40	1.91
156	6" diameter		280		.41	1.17	1.58	2.16
158	7" diameter		240		.48	1.36	1.84	2.52
160	8" diameter		200		.60	1.63	2.23	3.05
164	10" diameter		160		.75	2.04	2.79	3.81
166	12" diameter		120		.94	2.72	3.66	5
190	Insulated, 4" diameter		340		.99	.96	1.95	2.49
192	5" diameter		300		1.15	1.09	2.24	2.86
194	6" diameter		260		1.39	1.26	2.65	3.37
196	7" diameter		220		1.51	1.49	3	3.83
198	8" diameter		180		1.68	1.82	3.50	4.50
202	10" diameter		140		2.03	2.33	4.36	5.65
204	12" diameter	↓	100	↓	2.40	3.27	5.67	7.40
300	Rigid fiberglass, round, .003" foil scrim jacket							
310	4" diameter	Q-9	310	L.F.	1.55	1.05	2.60	3.25
312	5" diameter		275		1.77	1.19	2.96	3.68
314	6" diameter		240		1.96	1.36	3.32	4.15
316	7" diameter	↓	220	↓	2.28	1.49	3.77	4.68

For expanded coverage of these items see *Means' Mechanical Cost Data 1986*

15.7 Air Conditioning & Ventilating

		CREW	DAILY OUTPUT	UNIT	BARE COSTS MAT.	BARE COSTS INST.	BARE COSTS TOTAL	TOTAL INCL O&P
318	8" diameter	Q-9	180	L.F.	2.60	1.82	4.42	5.50
322	10" diameter		140		3.18	2.33	5.51	6.90
324	12" diameter	↓	100		4.06	3.27	7.33	9.25
350	Rectangular, 1" thick, alum. faced, (FRK), std. weight	Q-10	350	S.F.Surf	.43	1.45	1.88	2.60
70-001	**FANS**							
002	Air conditioning and process air handling							
003	Axial flow, compact, low sound, 2.5" S.P.							
005	3800 CFM, 5 HP	Q-20	3.40	Ea.	2,110	120	2,230	2,500
008	6400 CFM, 5 HP		2.80		2,370	150	2,520	2,825
010	10,500 CFM, 7-1/2 HP		2.40		2,945	175	3,120	3,500
012	15,600 CFM, 10 HP	↓	1.60	↓	3,700	260	3,960	4,450
015								
020	In-line centrifugal, supply/exhaust booster,							
022	aluminum wheel/hub, disconnect switch, 1/4" S.P.							
024	500 CFM, 10" diameter connection	Q-20	3	Ea.	400	140	540	640
026	1380 CFM, 12" diameter connection		2		560	210	770	920
028	1520 CFM, 16" diameter connection		2		605	210	815	970
030	2560 CFM, 18" diameter connection		1		805	415	1,220	1,500
032	3480 CFM, 20" diameter connection		.80		970	520	1,490	1,825
150	Vane-axial, low pressure, 2000 CFM, 1/2 HP		3.60		1,300	115	1,415	1,600
152	4000 CFM, 1 HP		3.20		1,400	130	1,530	1,725
154	8000 CFM, 2 HP	↓	2.80	↓	1,700	150	1,850	2,075
250	Ceiling fan, right angle, extra quiet, 0.10" S.P.							
252	95 CFM	Q-20	20	Ea.	117	21	138	160
254	210 CFM		19		127	22	149	170
256	385 CFM		18		159	23	182	210
258	885 CFM		16		305	26	331	375
260	1650 CFM		13		405	32	437	490
262	2960 CFM	↓	11		545	38	583	655
264	For wall or roof cap, add	1 Shee	16		71	11.35	82.35	95
266	For straight thru fan, add				10%			
268	For speed control switch, add	1 Elec	16	↓	39	11.20	50.20	59
300	Paddle blade air circulator, 3 speed switch							
302	36", 4000 CFM high, 3000 CFM low	Q-20	6	Ea.	120	69	189	235
304	52", 7000 CFM high, 4000 CFM low	"	4		130	105	235	295
310	For antique white motor, add				7.50		7.50	8.25M
320	For brass plated motor, add				18.75		18.75	21M
330	For light adaptor kit, add			↓	15		15	16.50M
350	Centrifugal, airfoil, motor and drive, complete							
352	1000 CFM, 1/2 HP	Q-20	2.50	Ea.	1,000	165	1,165	1,350
354	2000 CFM, 1 HP		2		1,100	210	1,310	1,525
356	4000 CFM, 3 HP		1.80		1,500	230	1,730	1,975
358	8000 CFM, 7-1/2 HP		1.40		2,400	295	2,695	3,075
360	12,000 CFM, 10 HP	↓	1	↓	3,000	415	3,415	3,900
450	Corrosive fume resistant, plastic							
460	roof ventilators, centrifugal, V belt drive, motor							
462	1/4" S.P., 250 CFM, 1/4 HP	Q-20	6	Ea.	1,600	69	1,669	1,850
464	895 CFM, 1/3 HP		5		1,730	83	1,813	2,025
466	1630 CFM, 1/2 HP		4		2,050	105	2,155	2,400
468	2240 CFM, 1 HP	↓	3	↓	2,130	140	2,270	2,550
500	Utility set, centrifugal, V belt drive, motor							
502	1/4" S.P., 1900 CFM, 1/4 HP	Q-20	6	Ea.	2,700	69	2,769	3,075
504	2170 CFM, 1/3 HP		5		2,725	83	2,808	3,125
506	2680 CFM, 1/2 HP		4		2,725	105	2,830	3,150
508	3020 CFM, 3/4 HP		3		2,750	140	2,890	3,225
510	1/2" S.P., 3195 CFM, 1 HP		2		2,775	210	2,985	3,350
512	3610 CFM, 1-1/2 HP		1.60		2,800	260	3,060	3,450
514	4120 CFM, 2 HP	↓	1.40	↓	2,850	295	3,145	3,575
525	Vaneaxial, direct drive, 1/4" S.P.							
530	615 CFM, 1/4 HP	Q-20	3	Ea.	1,400	140	1,540	1,750

For expanded coverage of these items see *Means' Mechanical Cost Data 1986*

15.7 Air Conditioning & Ventilating

			CREW	DAILY OUTPUT	UNIT	BARE COSTS MAT.	BARE COSTS INST.	BARE COSTS TOTAL	TOTAL INCL O&P
70	532	1320 CFM, 1/4 HP	Q-20	2.60	Ea.	1,750	160	1,910	2,150
	550								
	600	Propeller exhaust, wall shutter, 1/4" S.P.							
	602	Direct drive, two speed							
	610	375 CFM, 1/10 HP	Q-20	10	Ea.	143	42	185	220
	612	730 CFM, 1/7 HP		9		175	46	221	260
	614	1000 CFM, 1/8 HP		8		238	52	290	340
	620	4720 CFM, 1 HP	↓	5	↓	460	83	543	625
	630	V-belt drive, 3 phase							
	632	6175 CFM, 3/4 HP	Q-20	5	Ea.	465	83	548	635
	634	7500 CFM, 3/4 HP		5		520	83	603	695
	636	10,100 CFM, 1 HP		4.50		665	93	758	865
	638	14,300 CFM, 1-1/2 HP	↓	4	↓	815	105	920	1,050
	660								
	665	Residential, bath exhaust, grille, back draft damper							
	666	50 CFM	Q-20	24	Ea.	28	17.35	45.35	56
	667	110 CFM		22		55	18.95	73.95	88
	668	Light combination, squirrel cage, 100 watt, 70 CFM	↓	24	↓	61	17.35	78.35	92
	670	Light/heater combination, ceiling mounted							
	671	70 CFM, 1450 watt	Q-20	24	Ea.	96	17.35	113.35	130
	680	Heater combination, recessed, 70 CFM		24		56	17.35	73.35	87
	682	With 2 infrared bulbs		23		71	18.10	89.10	105
	690	Kitchen exhaust, grille, complete, 160 CFM		22		47	18.95	65.95	79
	692	344 CFM	↓	18	↓	97	23	120	140
	700	Roof exhauster, centrifugal, aluminum housing, 12" galvanized							
	702	curb, bird screen, back draft damper, 1/4" S.P.							
	710	Direct drive, 420 CFM, 8" sq. damper	Q-20	7	Ea.	335	60	395	455
	712	675 CFM, 12" sq. damper		6		470	69	539	620
	714	770 CFM, 16" sq. damper		5		700	83	783	890
	716	1870 CFM, 20" sq. damper		4.20		970	99	1,069	1,200
	718	2150 CFM, 20" sq. damper		4		1,130	105	1,235	1,400
	720	V-belt drive, 1660 CFM, 12" sq. damper		6		560	69	629	715
	722	2830 CFM, 14" sq. damper		5		680	83	763	870
	724	4600 CFM, 20" sq. damper		4		895	105	1,000	1,125
	726	8750 CFM, 26" sq. damper		3		990	140	1,130	1,300
	728	12,500 CFM, 32" sq. damper		2		1,730	210	1,940	2,200
	730	21,600 CFM, 40" sq. damper	↓	1		2,885	415	3,300	3,775
	732	For 2 speed winding, add				15%			
	734	For explosion-proof motor, add				230		230	255M
	736	For belt drive, top discharge, add			↓	15%			
	750	Utility set, steel construction, pedestal, 1/4" S.P.							
	752	Direct drive, 150 CFM, 1/8 HP	Q-20	6.40	Ea.	380	65	445	515
	754	485 CFM, 1/6 HP		5.80		480	72	552	635
	756	1950 CFM, 1/2 HP		4.80		565	87	652	750
	758	2410 CFM, 3/4 HP		4.40		1,035	95	1,130	1,275
	760	3328 CFM, 1-1/2 HP	↓	3	↓	1,160	140	1,300	1,475
	768	V-belt drive, drive cover, 3 phase							
	770	800 CFM, 1/4 HP	Q-20	6	Ea.	330	69	399	465
	772	1300 CFM, 1/3 HP		5		370	83	453	530
	774	2000 CFM, 1 HP		4.60		390	91	481	560
	776	2900 CFM, 3/4 HP	↓	4.20	↓	600	99	699	805
	845								
	850	Wall exhausters, centrifugal, auto damper, 1/8" S.P.							
	852	Direct drive, 635 CFM, 1/20 HP	Q-20	14	Ea.	175	30	205	235
	854	845 CFM, 1/12 HP		13		201	32	233	270
	856	1005 CFM, 1/6 HP		12		291	35	326	370
	858	1220 CFM, 1/6 HP	↓	12	↓	325	35	360	410
	950	V-belt drive, 3 phase							
	952	2800 CFM, 1/4 HP	Q-20	9	Ea.	925	46	971	1,075
	954	3740 CFM, 1/2 HP	"	8	"	745	52	797	895

For expanded coverage of these items see *Means' Mechanical Cost Data 1986*

15.7 Air Conditioning & Ventilating

		CREW	DAILY OUTPUT	UNIT	BARE COSTS MAT.	BARE COSTS INST.	BARE COSTS TOTAL	TOTAL INCL O&P
73-001	**FAN COIL AIR CONDITIONING** Cabinet mounted, filters							
010	Chilled water, 1/2 ton cooling	Q-5	8	Ea.	575	41	616	690
012	1 ton cooling		6		655	55	710	800
018	3 ton cooling	↓	4		1,270	82	1,352	1,525
020	10 ton cooling	Q-6	2.80		1,660	180	1,840	2,100
022	15 ton cooling		1.50		2,200	340	2,540	2,900
024	20 ton cooling		.80		2,780	635	3,415	3,975
026	30 ton cooling	↓	.60		4,250	850	5,100	5,900
100	Direct expansion, air cooled condensing, 5 ton cooling	Q-5	3		540	110	650	750
104	10 ton cooling	Q-6	2.60		1,600	195	1,795	2,050
106	20 ton cooling		.70		3,000	730	3,730	4,350
110	40 ton cooling	↓	.45	↓	6,225	1,125	7,350	8,475
76-001	**HEAT PUMPS**							
100	Air to air, split system, not including curbs or pads							
102	2 ton cooling, 8.5 MBH heat @ 0°F	Q-5	1.20	Ea.	1,390	275	1,665	1,925
106	5 ton cooling, 27 MBH heat @ 0°F	↓	.50		3,030	655	3,685	4,275
108	7 ton cooling, 33 MBH heat @ 0°F	↓	.30		4,440	1,100	5,540	6,475
110	10 ton cooling, 50 MBH heat @ 0°F	Q-6	.38		6,240	1,350	7,590	8,800
112	15 ton cooling, 64 MBH heat @ 0°F	"	.26		9,460	1,950	11,410	13,200
118	40 ton cooling, 193 MBH heat @ 0°F	Q-7	.12		23,500	5,800	29,300	34,200
130	For supplementary electric heat coil, add			↓	9%			
150	Single package, not including curbs, pads, or plenums							
152	2 ton cooling, 6.5 MBH heat @ 0°F	Q-5	1.50	Ea.	1,375	220	1,595	1,825
158	4 ton cooling, 13 MBH heat @ 0°F	↓	.96		2,450	340	2,790	3,200
164	7 ton cooling, 35 MBH heat @ 0°F	↓	.40		4,260	820	5,080	5,875
166	15 ton cooling, 56 MBH heat @ 0°F	Q-6	.30		11,100	1,700	12,800	14,700
170	For supplementary electric heat coil, add			↓	5%			
200	Water source to air, single package							
210	1 ton cooling, 13 MBH heat @ 75°F	Q-5	2	Ea.	870	165	1,035	1,200
214	2 ton cooling, 19 MBH heat @ 75°F		1.70		980	195	1,175	1,350
222	5 ton cooling, 29 MBH heat @ 75°F	↓	.90		2,090	365	2,455	2,825
396	For supplementary heat coil, add			↓	10%			
400	For increase in capacity thru use							
402	of solar collector, size boiler at 60%							
79-001	**LOUVERS**							
002								
010	Aluminum, extruded, with screen, mill finish							
100	Brick vent, (see also division 4.1-80)							
110	Standard, 4" deep, 8" wide, 5" high	1 Shee	24	Ea.	19.80	7.55	27.35	33
120	Modular, 4" deep, 7-3/4" wide, 5" high		24		21	7.55	28.55	34
130	Speed brick, 4" deep, 11-5/8" wide, 3-7/8" high		24		21	7.55	28.55	34
140	Fuel oil brick, 4" deep, 8" wide, 5" high		24	↓	36.75	7.55	44.30	51
200	Cooling tower and mechanical equip., screens, light weight		40	S.F.	7.95	4.54	12.49	15.40
202	Standard weight		35		26.25	5.20	31.45	36
250	Dual combination, automatic, intake or exhaust		20		32.50	9.10	41.60	49
252	Manual operation		20		26.25	9.10	35.35	42
254	Electric or pneumatic operation		20	↓	26.25	9.10	35.35	42
256	Motor, for electric or pneumatic	↓	14	Ea.	121	12.95	133.95	150
300	Fixed blade, continuous line							
310	Mullion type, stormproof	1 Shee	28	S.F.	24.25	6.50	30.75	36
320	Stormproof		28		26.25	6.50	32.75	38
330	Vertical line	↓	28		32.50	6.50	39	45
350	For damper to use with above, add				50%	30%		
352	Motor, for damper, electric or pneumatic	1 Shee	14	Ea.	121	12.95	133.95	150
400	Operating, 45°, manual, electric or pneumatic		24	S.F.	26.25	7.55	33.80	40
410	Motor, for electric or pneumatic		14	Ea.	121	12.95	133.95	150
420	Penthouse, roof		56	S.F.	13.25	3.24	16.49	19.30
430	Walls		40		32.50	4.54	37.04	42
500	Thinline, under 4" thick, fixed blade	↓	40	↓	10.55	4.54	15.09	18.25
501	Finishes, applied by mfr. at additional cost, available in colors							

For expanded coverage of these items see *Means' Mechanical Cost Data 1986*

15.7 Air Conditioning & Ventilating

			CREW	DAILY OUTPUT	UNIT	BARE COSTS MAT.	BARE COSTS INST.	BARE COSTS TOTAL	TOTAL INCL O&P
79	502	Prime coat only, add			S.F.	1.40		1.40	1.54M
	504	Baked enamel finish coating, add				2.78		2.78	3.05M
	506	Anodized finish, add				3.34		3.34	3.67M
	508	Duranodic finish, add				6.30		6.30	6.95M
	510	Fluoropolymer finish coating, add				9.75		9.75	10.70M
	998	For small orders (under 10 pieces), add				20%			
82-001		**PACKAGED TERMINAL AIR CONDITIONER** Cabinet, wall sleeve,							
	010	louver, electric heat, thermostat, manual changeover, 208 V							
	020	6,000 BTUH cooling, 8800 BTU heat	Q-5	6	Ea.	685	55	740	830
	022	9,000 BTUH cooling, 13,900 BTU heat		5		700	66	766	865
	024	12,000 BTUH cooling, 13,900 BTU heat		4		755	82	837	950
	026	15,000 BTUH cooling, 13,900 BTU heat		3		795	110	905	1,025
	050	For hot water coil, increase heat by 10%, add				5%	10%		
	100	For steam, increase heat output by 30%, add				8%	10%		
85-001		**ROOF TOP AIR CONDITIONERS** Standard controls, curb							
	100	Single zone, electric cool, gas heat							
	110	2 ton cooling, 60 MBH heating	Q-5	1.50	Ea.	2,265	220	2,485	2,800
	112	4 ton cooling, 95 MBH heating		1.10		3,440	300	3,740	4,225
	114	5 ton cooling, 112 MBH heating		.56		3,375	585	3,960	4,550
	116	10 ton cooling, 200 MBH heating	Q-6	.46		6,485	1,100	7,585	8,725
	120	20 ton cooling, 360 MBH heating	Q-7	.32		12,200	2,175	14,375	16,600
	122	30 ton cooling, 540 MBH heating		.22		17,000	3,150	20,150	23,300
	126	50 ton cooling, 810 MBH heating		.13		28,300	5,350	33,650	38,900
	130	For hot water heat coil, deduct				6%			
	140	For steam heat coil, deduct				2%			
	150	For electric heat, deduct				3%	5%		
	158	With variable volume distribution, 20 ton	Q-7	.32		11,750	2,175	13,925	16,100
	160	30 ton		.14		16,950	4,975	21,925	25,800
	162	50 ton		.08		29,500	8,700	38,200	45,000
	200	Multizone, electric cool, gas heat, economizer							
	210	15 ton cooling, 360 MBH heating	Q-7	.22	Ea.	28,050	3,150	31,200	35,400
	212	20 ton cooling, 360 MBH heating		.21		29,575	3,300	32,875	37,300
	220	37 ton cooling, 540 MBH heating		.13		46,650	5,350	52,000	59,000
	222	70 ton cooling, 1500 MBH heating		.09		48,600	7,725	56,325	64,500
	224	80 ton cooling, 1500 MBH heating		.08		55,500	8,700	64,200	73,500
	226	90 ton cooling, 1500 MBH heating		.07		62,500	9,925	72,425	83,000
	228	105 ton cooling, 1500 MBH heating		.06		73,000	11,600	84,600	97,000
	240	For hot water heat coil, deduct				5%			
	250	For steam heat coil, deduct				2%			
	260	For electric heat, deduct				3%	5%		
88-001		**SELF-CONTAINED SINGLE PACKAGE**							
	010	Air cooled, for free blow or duct, not including remote condenser							
	020	3 ton cooling	Q-5	1	Ea.	2,800	330	3,130	3,550
	022	5 ton cooling	Q-6	1.20		3,540	425	3,965	4,500
	024	10 ton cooling	Q-7	1		6,275	695	6,970	7,900
	026	20 ton cooling		.90		10,300	775	11,075	12,400
	028	30 ton cooling		.80		13,600	870	14,470	16,200
	034	60 ton cooling	Q-8	.40		27,500	1,775	29,275	32,800
	040	For duct mounting, no price change							
	050	For hot water or steam heating coils, add			Ea.	10%	10%		
	100	Water cooled for free blow or duct, not including tower							
	110	3 ton cooling	Q-6	1	Ea.	2,190	510	2,700	3,150
	112	5 ton cooling	"	1		2,600	510	3,110	3,600
	114	10 ton cooling	Q-7	.90		5,290	775	6,065	6,925
	116	20 ton cooling		.80		9,650	870	10,520	11,900
	118	30 ton cooling		.70		12,200	995	13,195	14,900
96-001		**VENTILATORS** Base, damper & bird screen, CFM in 5 MPH wind							
	050	Rotary syphon, galvanized, 6" neck diameter, 185 CFM	Q-9	16	Ea.	33	20	53	66
	056	12" neck diameter, 310 CFM		10		45	33	78	97
	066	24" neck diameter, 1530 CFM		8		160	41	201	235

For expanded coverage of these items see *Means' Mechanical Cost Data 1986*

15.7 Air Conditioning & Ventilating

		Crew	Daily Output	Unit	Bare Costs Mat.	Bare Costs Inst.	Bare Costs Total	Total Incl O&P
070	36" neck diameter, 3800 CFM	Q-9	6	Ea.	345	54	399	460
072	42" neck diameter, 4500 CFM	"	4	"	565	82	647	740
128	Spinner ventilators, wind driven, galvanized							
130	4" neck diameter, 180 CFM	Q-9	20	Ea.	34	16.35	50.35	61
134	6" neck diameter, 250 CFM		16		38	20	58	72
140	12" neck diameter, 770 CFM		10		56	33	89	110
150	24" neck diameter, 3100 CFM		8		195	41	236	275
154	36" neck diameter, 5500 CFM	↓	6	↓	440	54	494	565
200	Stationary, gravity, syphon, galvanized							
216	6" neck diameter, 66 CFM	Q-9	16	Ea.	25	20	45	57
224	12" neck diameter, 160 CFM		10		39	33	72	91
234	24" neck diameter, 900 CFM		8		145	41	186	220
238	36" neck diameter, 2000 CFM	↓	6	↓	325	54	379	435
400								
420	Stationary mushroom, aluminum, 16" orifice diameter	Q-9	10	Ea.	260	33	293	335
422	24" orifice diameter		9		325	36	361	410
424	36" orifice diameter		8		530	41	571	645
426	50" orifice diameter	↓	7	↓	825	47	872	975
500	Relief vent							
550	Rectangular, aluminum, galvanized curb							
551	intake/exhaust, 0.05" SP							
560	500 CFM, 12" x 16"	Q-9	8	Ea.	350	41	391	445
564	1000 CFM, 12" x 24"		6.60		415	50	465	530
568	3000 CFM, 20" x 42"	↓	4	↓	710	82	792	900
98-001	**WATER CHILLERS**							
002	Centrifugal liquid chiller, water cooled							
003	not including water tower							
010	Open drive, 2000 ton			Ea.	378,000	121,000	499,000	590,500
045								
050	Reciprocating, air cooled, 20 ton cooling	Q-7	.35	Ea.	10,600	1,975	12,575	14,500
052	40 ton cooling		.24		19,200	2,900	22,100	25,300
054	65 ton cooling		.14		25,600	4,975	30,575	35,300
060	100 ton cooling	↓	.10	↓	38,500	6,950	45,450	52,500
068	Water cooled, single compressor, semi-hermetic							
070	2 to 5 ton cooling	Q-5	.44	Ea.	3,925	745	4,670	5,400
074	8 ton cooling	"	.27		6,100	1,225	7,325	8,475
076	10 ton cooling	Q-6	.36		7,050	1,425	8,475	9,800
080	20 ton cooling	Q-7	.38		10,300	1,825	12,125	14,000
082	30 ton cooling	"	.28	↓	12,600	2,475	15,075	17,500
098	Water cooled, multiple compressors, semi-hermetic							
100	15 ton cooling	Q-6	.30	Ea.	8,750	1,700	10,450	12,100
102	20 ton cooling	Q-7	.36		9,790	1,925	11,715	13,600
106	30 ton cooling		.27		13,200	2,575	15,775	18,200
110	50 ton cooling	↓	.21		16,400	3,300	19,700	22,800
116	100 ton cooling	Q-8	.12		29,000	5,950	34,950	40,500
118	120 ton cooling		.10		33,800	7,150	40,950	47,400
120	140 ton cooling	↓	.09	↓	38,700	7,925	46,625	54,000
145	Water cooled, dual compressors, direct drive							
150	80 ton cooling	Q-8	.10	Ea.	34,500	7,150	41,650	48,200
152	100 ton cooling		.09		38,000	7,925	45,925	53,000
154	120 ton cooling		.08		44,000	8,925	52,925	61,000
158	150 ton cooling		.07		53,500	10,200	63,700	73,500
162	200 ton cooling		.05		65,500	14,300	79,800	92,500
166	250 ton cooling	↓	.04	↓	74,500	17,900	92,400	107,500

For expanded coverage of these items see *Means' Mechanical Cost Data 1986*

16.0 Raceways

		CREW	DAILY OUTPUT	UNIT	BARE COSTS MAT.	BARE COSTS INST.	BARE COSTS TOTAL	TOTAL INCL O&P
10-001	**CABLE TRAY** Ladder type with fittings and supports, 4" deep							
016	Galvanized steel tray							
017	4" rung spacing, 6" wide	1 Elec	49	L.F.	4.35	3.66	8.01	10.05
020	12" wide		43		5.30	4.17	9.47	11.80
040	18" wide		41		6.15	4.37	10.52	13.05
060	24" wide		39		7.15	4.59	11.74	14.45
320	Aluminum tray, 4" deep, 6" rung spacing, 6" wide		67		6.50	2.67	9.17	11
322	12" wide		62		7.30	2.89	10.19	12.20
324	24" wide	↓	53	↓	9	3.38	12.38	14.75
20-001	**CONDUIT TO 15' HIGH** Includes 2 terminations, 2 elbows and							
002	10 beam clamps per 100 L.F.							
030	Aluminum, 1/2" diameter	1 Elec	100	L.F.	.50	1.79	2.29	3.13
050	(139) 3/4" diameter		90		.68	1.99	2.67	3.61
070	1" diameter		80		1	2.24	3.24	4.32
100	1-1/4" diameter		70		1.35	2.56	3.91	5.15
103	1-1/2" diameter		65		1.70	2.76	4.46	5.85
105	2" diameter		60		2.30	2.99	5.29	6.80
107	2-1/2" diameter		50		3.50	3.58	7.08	9
110	3" diameter		45		4.90	3.98	8.88	11.10
113	3-1/2" diameter		40		6.90	4.48	11.38	14.05
114	4" diameter		35		7.60	5.10	12.70	15.70
175	Rigid galvanized steel, 1/2" diameter		90		.54	1.99	2.53	3.46
177	3/4" diameter		80		.66	2.24	2.90	3.95
180	1" diameter		65		1	2.76	3.76	5.05
183	1-1/4" diameter		60		1.35	2.99	4.34	5.80
185	1-1/2" diameter		55		1.60	3.26	4.86	6.45
187	2" diameter		45		2.25	3.98	6.23	8.20
190	2-1/2" diameter		35		3.70	5.10	8.80	11.45
193	3" diameter		25		5.10	7.15	12.25	15.90
195	3-1/2" diameter		22		5.75	8.15	13.90	18.05
197	4" diameter		20		8	8.95	16.95	22
250	Steel, intermediate conduit (IMC), 1/2" diameter		100		.44	1.79	2.23	3.06
253	3/4" diameter		90		.54	1.99	2.53	3.46
255	1" diameter		70		.84	2.56	3.40	4.61
257	1-1/4" diameter		65		1.10	2.76	3.86	5.15
260	1-1/2" diameter		60		1.35	2.99	4.34	5.80
263	2" diameter		50		1.85	3.58	5.43	7.20
265	2-1/2" diameter		40		3.15	4.48	7.63	9.90
267	3" diameter		30		4.35	5.95	10.30	13.35
270	3-1/2" diameter		27		4.95	6.65	11.60	15
273	4" diameter		25		7.10	7.15	14.25	18.10
500	Electric metallic tubing (EMT), 1/2" diameter		170		.23	1.05	1.28	1.77
502	3/4" diameter		130		.33	1.38	1.71	2.35
504	1" diameter		115		.54	1.56	2.10	2.83
506	1-1/4" diameter		100		.84	1.79	2.63	3.50
508	1-1/2" diameter		90		1.06	1.99	3.05	4.03
510	2" diameter		80		1.40	2.24	3.64	4.76
512	2-1/2" diameter		60		2.97	2.99	5.96	7.55
514	3" diameter		50		3.70	3.58	7.28	9.20
516	3-1/2" diameter		45		4.67	3.98	8.65	10.85
518	4" diameter	↓	40	↓	5.50	4.48	9.98	12.50
990	Add to labor for higher elevated installation							
991	15' to 20' high, add					10%		
992	20' to 25' high, add					20%		
993	25' to 30' high, add					25%		
994	30' to 35' high, add					30%		
995	35' to 40' high, add					35%		
996	Over 40' high, add					40%		

For expanded coverage of these items see *Means' Electrical Cost Data 1986*

16.0 Raceways

		CREW	DAILY OUTPUT	UNIT	BARE COSTS MAT.	BARE COSTS INST.	BARE COSTS TOTAL	TOTAL INCL O&P
30-001	**CONDUIT IN CONCRETE SLAB** Including terminations,							
002	fittings and supports							
323	PVC, schedule 40, 1/2" diameter	1 Elec	270	L.F.	.16	.66	.82	1.13
325	3/4" diameter		230		.22	.78	1	1.36
327	1" diameter		200		.32	.90	1.22	1.64
330	1-1/4" diameter		170		.43	1.05	1.48	1.99
333	1-1/2" diameter		140		.48	1.28	1.76	2.37
335	2" diameter		120		.70	1.49	2.19	2.92
435	Rigid galvanized steel, 1/2" diameter		200		.51	.90	1.41	1.85
440	3/4" diameter		170		.61	1.05	1.66	2.19
445	1" diameter		130		.89	1.38	2.27	2.96
450	1-1/4" diameter		110		1.18	1.63	2.81	3.64
460	1-1/2" diameter		100		1.40	1.79	3.19	4.12
480	2" diameter		90		1.92	1.99	3.91	4.98
50-001	**CONDUIT IN TRENCH** Includes terminations and fittings							
002	Does not include excavation or backfill, see 2.3-19							
020	Rigid galvanized steel, 2" diameter	1 Elec	150	L.F.	1.90	1.19	3.09	3.81
040	2-1/2" diameter		100		3.10	1.79	4.89	6
060	3" diameter		80		4.25	2.24	6.49	7.90
080	3-1/2" diameter		70		5.60	2.56	8.16	9.85
100	4" diameter		50		6.60	3.58	10.18	12.40
120	5" diameter		40		14.30	4.48	18.78	22
140	6" diameter		30		20	5.95	25.95	31
70-001	**TRENCH DUCT** Steel with cover							
002	Standard adjustable, depths to 4"							
010	Straight, single compartment, 9" wide	1 Elec	20	L.F.	34	8.95	42.95	50
020	12" wide		16		40	11.20	51.20	60
040	18" wide		13		52	13.80	65.80	77
060	24" wide		11		69	16.30	85.30	99
080	30" wide		10		82	17.90	99.90	115
100	36" wide		8		100	22	122	140
120	Horizontal elbow, 9" wide		2.70	Ea.	130	66	196	240
140	12" wide		2.30		150	78	228	275
160	18" wide		2		190	90	280	340
180	24" wide		1.60		265	110	375	455
200	30" wide		1.30		360	140	500	595
220	36" wide		1.20		470	150	620	730
240	Vertical elbow, 9" wide		2.70		45	66	111	145
260	12" wide		2.30		49	78	127	165
280	18" wide		2		57	90	147	190
300	24" wide		1.60		66	110	176	235
320	30" wide		1.30		74	140	214	280
340	36" wide		1.20		82	150	232	305
360	Cross, 9" wide		2		205	90	295	355
380	12" wide		1.60		225	110	335	410
400	18" wide		1.30		265	140	405	490
420	24" wide		1.10		340	165	505	610
440	30" wide		1		445	180	625	745
460	36" wide		.90		555	200	755	895
480	End closure, 9" wide		7.20		12	25	37	49
500	12" wide		6		14	30	44	58
520	18" wide		5		24	36	60	78
540	24" wide		4		31	45	76	99
560	30" wide		3.30		37	54	91	120
580	36" wide		2.90		45	62	107	140
600	Tees, 9" wide		2		130	90	220	270
620	12" wide		1.80		150	100	250	310
640	18" wide		1.60		190	110	300	370
660	24" wide		1.50		265	120	385	465

For expanded coverage of these items see *Means' Electrical Cost Data 1986*

16.0 Raceways

			CREW	DAILY OUTPUT	UNIT	BARE COSTS MAT.	BARE COSTS INST.	BARE COSTS TOTAL	TOTAL INCL O&P
70	680	30" wide	1 Elec	1.30	Ea.	360	140	500	595
	700	36" wide		1		470	180	650	775
	720	Riser and cabinet connector, 9" wide		2.70		57	66	123	160
	740	12" wide		2.30		65	78	143	185
	760	18" wide		2		80	90	170	215
	780	24" wide		1.60		95	110	205	265
	800	30" wide		1.30		110	140	250	320
	820	36" wide		1		125	180	305	395
	840	Insert assembly, cell to conduit adapter, 1-1/4"	↓	16	↓	18	11.20	29.20	36
80-001		**UNDERFLOOR DUCT**							
	010	Duct, 1-3/8" x 3-1/8" blank, standard	1 Elec	80	L.F.	2.65	2.24	4.89	6.15
	020	1-3/8" x 7-1/4" blank, super duct		60		5.25	2.99	8.24	10.05
	040	7/8" or 1-3/8" insert type, 24" O.C., 1-3/8", x 3-1/8", std.		70		2.65	2.56	5.21	6.60
	060	1-3/8" x 7-1/4", super duct		50	↓	5.25	3.58	8.83	10.95
	080	Junction box, single duct, 1 level, 3-1/8"		4	Ea.	56	45	101	125
	100	Junction box, single duct, 1 level, 7-1/4"		2.70		95	66	161	200
	120	1 level, 2 duct, 3-1/8"		3.20		95	56	151	185
	140	Junction box, 1 level, 2 duct, 7-1/4"		2.30		260	78	338	400
	160	Triple duct, 3-1/8"		2.30		180	78	258	310
	180	Insert to conduit adapter, 3/4" & 1"	↓	32	↓	3.20	5.60	8.80	11.55
	190								
	200	Support, single cell	1 Elec	27	Ea.	4.05	6.65	10.70	14
	220	Super duct		16		4.25	11.20	15.45	21
	240	Double cell		16		4.25	11.20	15.45	21
	260	Triple cell		11		6.60	16.30	22.90	31
	280	Vertical elbow, standard duct		10		9	17.90	26.90	36
	300	Super duct		8		19	22	41	53
	320	Cabinet connector, standard duct		32		3.50	5.60	9.10	11.90
	340	Super duct		27		7.05	6.65	13.70	17.30
	360	Conduit adapter, 1" to 1-1/4"		32		4.90	5.60	10.50	13.45
	380	2" to 1-1/4"		27		6.40	6.65	13.05	16.60
	400	Outlet, low tension		8		15.15	22	37.15	49
	420	High tension, receptacle	↓	8	↓	18	22	40	52
90-001		**WIREMOLD RACEWAY**							
	009	Raceway, surface, metal, straight section							
	010	No. 500	1 Elec	100	L.F.	.45	1.79	2.24	3.07
	040	No. 1500, small pancake		90		.70	1.99	2.69	3.63
	060	No. 2000, base & cover		90		.75	1.99	2.74	3.69
	080	No. 3000, base & cover		75		1.45	2.39	3.84	5.05
	100	No. 4000, base & cover		65		2.70	2.76	5.46	6.95
	120	No. 6000, base & cover		50	↓	4.05	3.58	7.63	9.60
	240	Fittings, elbows, No. 500		40	Ea.	1	4.48	5.48	7.55
	280	Elbow cover, No. 2000		40		1.10	4.48	5.58	7.65
	300	Switch box, No. 500		16		2.65	11.20	13.85	19
	340	Telephone outlet, No. 1500		16		4.50	11.20	15.70	21
	360	Junction box, No. 1500	↓	16	↓	3.55	11.20	14.75	20
	380	Plugmold wired sections, No. 2000							
	400	1 circuit, 6 outlets, 3 ft. long	1 Elec	8	Ea.	13.20	22	35.20	47
	410	2 circuits, 8 outlets, 6 ft. long		5.30		21	34	55	72
	420	Tele-power poles, aluminum, 4 outlets	↓	2.70	↓	61	66	127	165
95-001		**WIREWAY**							
	010	Screw cover with fittings and supports, 2-1/2" x 2-1/2"	1 Elec	45	L.F.	4.70	3.98	8.68	10.90
	020	4" x 4"		40		4.85	4.48	9.33	11.80
	040	6" x 6"		30		9.85	5.95	15.80	19.40
	060	8" x 8"	↓	20	↓	16	8.95	24.95	30

For expanded coverage of these items see *Means' Electrical Cost Data 1986*

16.1 Conductors & Grounding

		CREW	DAILY OUTPUT	UNIT	BARE COSTS MAT.	BARE COSTS INST.	BARE COSTS TOTAL	TOTAL INCL O&P
10-001	WIRE							
002	600 volt, type THW, copper, solid, #14	1 Elec	13	C.L.F.	2.60	13.80	16.40	23
003	#12		11		3.70	16.30	20	28
004	#10		10		5.50	17.90	23.40	32
005	Stranded #14		13		3.10	13.80	16.90	23
010	#12		11		4.20	16.30	20.50	28
012	#10		10		6.50	17.90	24.40	33
014	#8		8		11.20	22	33.20	45
016	#6		6.50		14.50	28	42.50	56
018	#4		5.30		23	34	57	74
020	#3		5		29	36	65	83
022	#2		4.50		35	40	75	96
024	#1		4		46	45	91	115
026	1/0		3.30		54	54	108	135
028	2/0		2.90		65	62	127	160
030	3/0		2.50		79	72	151	190
035	4/0		2.20		98	81	179	225
040	250 MCM		2		115	90	205	255
042	300 MCM		1.90		150	94	244	300
045	350 MCM		1.80		160	100	260	320
048	400 MCM		1.70		200	105	305	370
049	500 MCM		1.60		220	110	330	405
053	Aluminum, stranded, #8		9		6.70	19.90	26.60	36
054	#6		8		7.60	22	29.60	41
056	#4		6.50		10.10	28	38.10	51
058	#2		5.30		13.40	34	47.40	63
060	#1		4.50		19.60	40	59.60	79
062	1/0		4		23	45	68	90
064	2/0		3.60		27	50	77	100
068	3/0		3.30		32	54	86	115
070	4/0		3.10		37	58	95	125
072	250 MCM		2.90		44	62	106	135
074	300 MCM		2.70		58	66	124	160
076	350 MCM		2.50		62	72	134	170
078	400 MCM		2.30		68	78	146	185
080	500 MCM		2		80	90	170	215
085	600 MCM		1.90		96	94	190	240
088	700 MCM		1.70		110	105	215	275
090	750 MCM		1.60		115	110	225	290
092	Type THWN-THHN, copper, solid, #14		13		2.40	13.80	16.20	22
094	#12		11		3.30	16.30	19.60	27
096	#10		10		5.50	17.90	23.40	32
100	Stranded, #14		13		2.85	13.80	16.65	23
120	#12		11		3.85	16.30	20.15	28
125	#10		10		6.35	17.90	24.25	33
130	#8		8		10.50	22	32.50	44
135	#6		6.50		15.30	28	43.30	56
140	#4	▼	5.30	▼	25	34	59	76
20-001	ARMORED CABLE							
005	600 volt, copper (BX), #14, 2 wire	1 Elec	2.40	C.L.F.	22	75	97	130
010	3 wire		2.20		27	81	108	145
015	#12, 2 wire		2.30		25	78	103	140
020	3 wire		2		35	90	125	165
025	#10, 2 wire		2		42	90	132	175
030	3 wire		1.60		54	110	164	220
035	#8, 3 wire		1.30		90	140	230	295
040	3 conductor with PVC jacket, in cable tray, #6		3.10		180	58	238	280
045	#4		2.70		230	66	296	350
050	#2		2.30		320	78	398	465
055	#1	▼	2	▼	430	90	520	600

For expanded coverage of these items see *Means' Electrical Cost Data 1986*

16.1 Conductors & Grounding

		CREW	DAILY OUTPUT	UNIT	BARE COSTS MAT.	BARE COSTS INST.	BARE COSTS TOTAL	TOTAL INCL O&P
20 060	1/0	1 Elec	1.80	C.L.F.	500	100	600	695
065	2/0		1.70		600	105	705	810
070	3/0		1.60		710	110	820	940
075	4/0		1.50		830	120	950	1,075
080	250 MCM		1.20		940	150	1,090	1,250
085	350 MCM		1.10		1,250	165	1,415	1,600
090	500 MCM	↓	1	↓	1,650	180	1,830	2,075
105	5 KV, copper, 3 conductor with PVC jacket,							
106	non-shielded, in cable tray, #4	1 Elec	190	L.F.	3.40	.94	4.34	5.10
110	#2		180		4.40	1	5.40	6.25
120	#1		150		5.60	1.19	6.79	7.90
140	1/0		145		6.50	1.24	7.74	8.95
160	2/0		130		7.55	1.38	8.93	10.30
200	4/0		120		10	1.49	11.49	13.15
210	250 MCM		110		14	1.63	15.63	17.75
215	350 MCM		105		17.20	1.71	18.91	21
220	500 MCM	↓	90	↓	21	1.99	22.99	26
240	15 KV, copper, 3 conductor with PVC jacket,							
250	grounded neutral, in cable tray, #2	1 Elec	150	L.F.	7.60	1.19	8.79	10.10
260	#1		140		8.10	1.28	9.38	10.75
280	1/0		130		9.90	1.38	11.28	12.85
290	2/0		110		12	1.63	13.63	15.55
300	4/0		95		13.80	1.89	15.69	17.90
310	250 MCM		90		15.50	1.99	17.49	19.90
315	350 MCM		80		18	2.24	20.24	23
320	500 MCM	↓	70	↓	24	2.56	26.56	30
340	15 KV, copper, 3 conductor with PVC jacket,							
345	ungrounded neutral, in cable tray, #2	1 Elec	130	L.F.	8	1.38	9.38	10.80
350	#1		115		8.80	1.56	10.36	11.90
360	1/0		100		10.20	1.79	11.99	13.80
370	2/0		95		12.40	1.89	14.29	16.35
380	4/0		80		15	2.24	17.24	19.70
400	250 MCM		70		17.50	2.56	20.06	23
405	350 MCM		65		24	2.76	26.76	30
410	500 MCM	↓	60	↓	29	2.99	31.99	36
30-001	**CONTROL CABLE**							
002	600 volt, copper, #14 THWN wire with PVC jacket, 2 wires	1 Elec	9	C.L.F.	11	19.90	30.90	41
010	4 wires		7		17.50	26	43.50	56
020	6 wires		6		28	30	58	74
030	8 wires		5.30		33	34	67	85
040	10 wires		4.80		40	37	77	98
050	12 wires		4.30		59	42	101	125
060	14 wires		3.80		64	47	111	140
070	16 wires		3.50		74	51	125	155
080	18 wires		3.30		84	54	138	170
090	20 wires		3		92	60	152	185
100	22 wires	↓	2.80	↓	97	64	161	200
60-001	**SHIELDED CABLE** Splicing & terminations not included							
005	Copper, CLP shielding, 5KV #4	1 Elec	2.20	C.L.F.	110	81	191	240
010	#2		2		125	90	215	265
020	#1		2		140	90	230	285
040	1/0		1.90		150	94	244	300
060	2/0		1.80		170	100	270	330
080	4/0		1.60		240	110	350	425
100	250 MCM		1.50		265	120	385	465
120	350 MCM		1.30		340	140	480	570
140	500 MCM		1.20		480	150	630	745
160	15 KV, ungrounded neutral, #1		2		170	90	260	315
180	1/0	↓	1.90	↓	205	94	299	360

16.1 Conductors & Grounding

		CREW	DAILY OUTPUT	UNIT	BARE COSTS MAT.	BARE COSTS INST.	BARE COSTS TOTAL	TOTAL INCL O&P
200	2/0	1 Elec	1.80	C.L.F.	250	100	350	420
220	4/0		1.60		300	110	410	490
240	250 MCM		1.50		340	120	460	545
260	350 MCM		1.30		400	140	540	640
280	500 MCM	↓	1.20	↓	525	150	675	790
70-001	**NON-METALLIC SHEATHED CABLE** 600 volt							
010	Copper with ground wire, (Romex)							
015	#14, 2 wire	1 Elec	2.70	C.L.F.	6.80	66	72.80	105
020	3 wire		2.40		12.80	75	87.80	120
025	#12, 2 wire		2.50		9.40	72	81.40	115
030	3 wire		2.20		18.50	81	99.50	135
035	#10, 2 wire		2.20		18	81	99	135
040	3 wire		1.80		28	100	128	175
045	#8, 3 wire		1.50		60	120	180	240
050	#6, 3 wire	↓	1.40	↓	84	130	214	275
055	SE type SER aluminum cable, 3 RHW and							
060	1 bare neutral, 3 #8 & 1 #8	1 Elec	1.60	C.L.F.	33	110	143	195
065	3 #6 & 1 #6		1.40		42	130	172	230
070	3 #4 & 1 #6		1.20		49	150	199	270
075	3 #2 & 1 #4		1.10		70	165	235	310
080	3 #1/0 & 1 #2		1		105	180	285	375
085	3 #2/0 & 1 #1		.90		120	200	320	420
090	3 #4/0 & 1 #2/0	↓	.80	↓	175	225	400	515
80-001	**GROUNDING**							
003	Rod, copper clad, 8' long, 1/2" diameter	1 Elec	5	Ea.	7.65	36	43.65	60
005	3/4" diameter		5.30		15	34	49	65
008	10' long, 1/2" diameter		4.80		9.40	37	46.40	64
010	3/4" diameter		4.40		18.50	41	59.50	79
013	15' long, 3/4" diameter		4	↓	50	45	95	120
040	Bare copper, #6 wire		10	C.L.F.	15.80	17.90	33.70	43
060	#2		5		38	36	74	93
080	3/0		3.30		77	54	131	165
100	4/0		2.85		95	63	158	195
120	250 MCM	↓	2.40	↓	110	75	185	230
160								
180	Water pipe ground clamps, heavy duty							
200	Bronze, 1/2" to 1" diameter	1 Elec	8	Ea.	4.65	22	26.65	37
210	1-1/4" to 2" diameter		8		6.70	22	28.70	40
220	2-1/2" to 3" diameter		6		20	30	50	65
280	Brazed connections, #6 wire		12		3.80	14.95	18.75	26
300	#2 wire		10		3.80	17.90	21.70	30
310	3/0 wire		8		3.80	22	25.80	36
320	4/0 wire		7		3.80	26	29.80	41
340	250 MCM wire		5		3.80	36	39.80	56
360	500 MCM wire	↓	4	↓	5.70	45	50.70	71

16.2 Boxes & Wiring Devices

		CREW	DAILY OUTPUT	UNIT	BARE COSTS MAT.	BARE COSTS INST.	BARE COSTS TOTAL	TOTAL INCL O&P
10-001	**PULL BOXES & CABINETS**							
010	Sheet metal, pull box, NEMA 1, type SC, 6"W x 6"H x 4"D	1 Elec	8	Ea.	5	22	27	38
020	8"W x 8"H x 4"D		8		7	22	29	40
030	10"W x 12"H x 6"D		5.30		12	34	46	62
040	16"W x 20"H x 8"D		4		46	45	91	115
050	20"W x 24"H x 8"D	↓	3.20	↓	52	56	108	140

For expanded coverage of these items see *Means' Electrical Cost Data 1986*

16.2 Boxes & Wiring Devices

			CREW	DAILY OUTPUT	UNIT	BARE COSTS MAT.	BARE COSTS INST.	BARE COSTS TOTAL	TOTAL INCL O&P
10	060	24"W x 36"H x 8"D	1 Elec	2.70	Ea.	75	66	141	180
	065	Hinged cabinets, type A, 6"W x 6"H x 4"D		8		5	22	27	38
	080	12"W x 16"H x 6"D		4.70		23	38	61	80
	100	20"W x 20"H x 6"D		3.60		39	50	89	115
	120	20"W x 20"H x 8"D		3.20		67	56	123	155
	140	24"W x 36"H x 8"D		2.70		120	66	186	225
	160	24"W x 42"H x 8"D	↓	2	↓	180	90	270	325
	210	NEMA 3R, raintight & weatherproof							
	215	6"L x 6"W x 6"D	1 Elec	10	Ea.	11	17.90	28.90	38
	220	8"L x 6"W x 6"D		8		11	22	33	44
	225	10"L x 6"W x 6"D		7		15	26	41	53
	230	12"L x 12"W x 6"D		5		20	36	56	74
	235	16"L x 16"W x 6"D		4.50		42	40	82	105
	240	20"L x 20"W x 6"D		4		54	45	99	125
	245	24"L x 18"W x 8"D		3		55	60	115	145
	250	24"L x 24"W x 10"D		2.50		130	72	202	245
	255	30"L x 24"W x 12"D		2		155	90	245	300
	260	36"L x 36"W x 12"D	↓	1.50	↓	265	120	385	465
	280	Cast iron, pull boxes for surface mounting							
	300	NEMA 4, watertight & dust tight							
	305	6"L x 6"W x 6"D	1 Elec	4	Ea.	67	45	112	140
	310	8"L x 6"W x 6"D		3.20		83	56	139	170
	315	10"L x 6"W x 6"D		2.50		110	72	182	225
	320	12"L x 12"W x 6"D		2		185	90	275	330
	325	16"L x 16"W x 6"D		1.30		375	140	515	610
	330	20"L x 20"W x 6"D		.80		645	225	870	1,025
	335	24"L x 18"W x 8"D		.70		680	255	935	1,125
	340	24"L x 24"W x 10"D		.50		1,050	360	1,410	1,675
	345	30"L x 24"W x 12"D		.40		1,425	450	1,875	2,200
	350	36"L x 36"W x 12"D	↓	.20	↓	2,550	895	3,445	4,100
	400	NEMA 7, explosionproof							
	405	6"L x 6"W x 6"D	1 Elec	2	Ea.	160	90	250	305
	410	8"L x 6"W x 6"D		1.80		210	100	310	375
	415	10"L x 6"W x 6"D		1.60		285	110	395	475
	420	12"L x 12"W x 6"D		1		470	180	650	775
	425	16"L x 14"W x 6"D		.60		780	300	1,080	1,275
	430	18"L x 18"W x 8"D		.50		1,150	360	1,510	1,775
	435	24"L x 18"W x 8"D		.40		1,550	450	2,000	2,350
	440	24"L x 24"W x 10"D		.30		2,000	595	2,595	3,050
	445	30"L x 24"W x 12"D	↓	.20	↓	3,050	895	3,945	4,650
	600	J.I.C. wiring boxes, NEMA 12, dust tight & drip tight							
	605	6"L x 8"W x 4"D	1 Elec	10	Ea.	20	17.90	37.90	48
	610	8"L x 10"W x 4"D		8		25	22	47	60
	615	12"L x 14"W x 6"D		5.30		40	34	74	93
	620	14"L x 16"W x 6"D		4.70		44	38	82	105
	625	16"L x 20"W x 6"D		4.40		100	41	141	170
	630	24"L x 30"W x 6"D		3.20		150	56	206	245
	635	24"L x 30"W x 8"D		2.90		155	62	217	260
	640	24"L x 36"W x 8"D		2.70		170	66	236	280
	645	24"L x 42"W x 8"D		2.30		190	78	268	320
	650	24"L x 48"W x 8"D	↓	2	↓	205	90	295	355
	700	Cabinets, current transformer							
	705	Single door, 24"H x 24"W x 10"D	1 Elec	1.60	Ea.	62	110	172	230
	710	30"H x 24"W x 10"D		1.30		76	140	216	280
	715	36"H x 24"W x 10"D		1.10		83	165	248	325
	720	30"H x 30"W x 10"D		1		115	180	295	385
	725	36"H x 30"W x 10"D		.90		135	200	335	435
	730	36"H x 36"W x 10"D		.80		150	225	375	485
	750	Double door, 48"H x 36"W x 10"D		.60		255	300	555	710
	755	24"H x 24"W x 12"D	↓	1	↓	125	180	305	395

For expanded coverage of these items see *Means' Electrical Cost Data 1986*

16.2 Boxes & Wiring Devices

		CREW	DAILY OUTPUT	UNIT	BARE COSTS MAT.	BARE COSTS INST.	BARE COSTS TOTAL	TOTAL INCL O&P
20-001	**OUTLET BOXES**							
002	Pressed steel, octagon, 4"	1 Elec	20	Ea.	.89	8.95	9.84	13.85
010	Extension		40		1.07	4.48	5.55	7.60
015	Square 4"		20		1.07	8.95	10.02	14.05
020	Extension		40		1.38	4.48	5.86	7.95
025	Covers, blank		64		.41	2.80	3.21	4.48
030	Plaster rings		64		.63	2.80	3.43	4.72
065	Switchbox		27		.93	6.65	7.58	10.55
110	Concrete, floor, 1 gang		5.30		34	34	68	86
200	Poke-thru fitting, fire rated, for 3-3/4" floor		6.80		49	26	75	92
204	For 7" floor		6.80		53	26	79	96
210	Pedestal, 15 amp, duplex receptacle & blank plate		24		17.50	7.45	24.95	30
212	Duplex receptacle and telephone plate		24		17.40	7.45	24.85	30
214	Pedestal, 20 amp, duplex receptacle & telephone		18		19.40	9.95	29.35	36
220	Abandonment plate		32		10.50	5.60	16.10	19.60
30-001	**WIRING DEVICES**							
020	Toggle switch, quiet type, single pole, 15 amp	1 Elec	40	Ea.	4.40	4.48	8.88	11.30
060	3 way, 15 amp		23		6.40	7.80	14.20	18.25
090	4 way, 15 amp		15		18.45	11.95	30.40	37
129								
165	Dimmer switch, 120 volt, incandescent, 600 watt, 1 pole	1 Elec	16	Ea.	6.50	11.20	17.70	23
246	Receptacle, duplex, 120 volt, grounded, 15 amp		40		3.85	4.48	8.33	10.70
247	20 amp		27		6.40	6.65	13.05	16.60
249	Dryer, 30 amp		15		4.10	11.95	16.05	22
250	Range, 50 amp		11		4.10	16.30	20.40	28
260	Wall plates, stainless steel, 1 gang		80		2.10	2.24	4.34	5.55
280	2 gang		53		4.35	3.38	7.73	9.65
320	Lampholder, keyless		26		1.50	6.90	8.40	11.55
340	Pullchain with receptacle		22		3.90	8.15	12.05	16
35-001	**LOW VOLTAGE SWITCHING**							
360	Relays, 120V or 277V standard	1 Elec	12	Ea.	15.75	14.95	30.70	39
380	Flush switch, standard		40		4.55	4.48	9.03	11.45
400	Interchangeable		40		8.35	4.48	12.83	15.65
410	Surface switch, standard		40		6.80	4.48	11.28	13.90
420	Transformer 115V to 25V		12		54	14.95	68.95	81
440	Master control, 9 circuit, manual		4		42	45	87	110
450	25 circuit, motorized		4		47	45	92	115
460	Rectifier, silicon		12		19.75	14.95	34.70	43
461								
480	Switchplates, 1 gang, 1, 2 or 3 switch, plastic	1 Elec	80	Ea.	1.35	2.24	3.59	4.71
500	Stainless steel		80		3.70	2.24	5.94	7.30
540	2 gang, 3 switch, stainless steel		53		7	3.38	10.38	12.55
550	4 switch, plastic		53		2.80	3.38	6.18	7.95
580	3 gang, 9 switch, stainless steel		32		27	5.60	32.60	38

16.3 Starters, Boards & Switches

		CREW	DAILY OUTPUT	UNIT	BARE COSTS MAT.	BARE COSTS INST.	BARE COSTS TOTAL	TOTAL INCL O&P
10-001	**CIRCUIT BREAKERS**							
010	Enclosed (NEMA 1), 600 volt, 3 pole, 30 amp	1 Elec	3.20	Ea.	190	56	246	290
020	60 amp		2.80		190	64	254	300
040	100 amp		2.30		230	78	308	365
060	225 amp		1.50		495	120	615	715
070	400 amp		.80		870	225	1,095	1,275

For expanded coverage of these items see *Means' Electrical Cost Data 1986*

16.3 Starters, Boards & Switches

			CREW	DAILY OUTPUT	UNIT	BARE COSTS MAT.	BARE COSTS INST.	BARE COSTS TOTAL	TOTAL INCL O&P
10	080	600 amp	1 Elec	.60	Ea.	1,425	300	1,725	2,000
	100	800 amp	"	.47	"	1,850	380	2,230	2,575
15-001		**CONTROL STATIONS**							
	005	NEMA 1, heavy duty, stop/start	1 Elec	8	Ea.	35	22	57	71
	010	Stop/start, pilot light		6.20		84	29	113	135
	020	Hand/off/automatic		6.20		35	29	64	80
	040	Stop/start/reverse	↓	5.30	↓	64	34	98	120
20-001		**FUSE CABINETS**							
	005	120/240 volts, 3 wire, 30 amp branches,							
	010	plug fuse not included							
	020	4 circuits	1 Elec	4	Ea.	20	45	65	86
	030	6 circuits		3.20		29	56	85	110
	040	8 circuits		2.70		39	66	105	140
	050	12 circuits	↓	2	↓	59	90	149	195
27-001		**MOTOR CONNECTIONS**							
	002	Flexible conduit and fittings, 1 HP motor	1 Elec	8	Ea.	3	22	25	36
	020	25 HP motor		2.70		10	66	76	105
	040	50 HP motor		2.20		28	81	109	150
	060	100 HP motor	↓	1.50	↓	115	120	235	300
30-001		**MOTOR STARTERS & CONTROLS**							
	005	Magnetic, FVNR, with enclosure and heaters, 480 volt							
	010	5 HP, size 0	1 Elec	2.30	Ea.	120	78	198	245
	020	10 HP, size 1		1.60		130	110	240	305
	030 (138)	25 HP, size 2		1.10		250	165	415	510
	040	50 HP, size 3		.90		405	200	605	730
	050	100 HP, size 4		.60		900	300	1,200	1,425
	060	200 HP, size 5		.45		2,100	400	2,500	2,875
	070	Combination, with motor circuit protectors, 5 HP, size 0		1.80		385	100	485	565
	080	10 HP, size 1		1.30		395	140	535	635
	090	25 HP, size 2		1		555	180	735	870
	100	50 HP, size 3		.66		805	270	1,075	1,275
	120	100 HP, size 4	↓	.40	↓	1,750	450	2,200	2,575
	121								
	140	Combination, with fused switch, 5 HP, size 0	1 Elec	1.80	Ea.	305	100	405	480
	160	10 HP, size 1		1.30		320	140	460	550
	180	25 HP, size 2		1		495	180	675	800
	200	50 HP, size 3		.66		830	270	1,100	1,300
	220	100 HP, size 4	↓	.40	↓	1,550	450	2,000	2,350
50-001		**PANELBOARDS** (Including breakers)							
	005	NQOB, w/20 amp 1 pole bolt-on circuit breakers							
	010	3 wire, 120/240 volts, 100 amp main lugs							
	015	10 circuits	1 Elec	1	Ea.	265	180	445	550
	020	14 circuits		.88		310	205	515	635
	025	18 circuits		.75		365	240	605	745
	030	20 circuits		.65		390	275	665	825
	035	225 amp main lugs, 24 circuits		.60		445	300	745	920
	040	30 circuits		.45		525	400	925	1,150
	045	36 circuits		.40		600	450	1,050	1,300
	050	38 circuits		.36		625	500	1,125	1,400
	055	42 circuits		.33		675	545	1,220	1,525
	060	4 wire, 120/208 volts, 100 amp main lugs, 12 circuits		1		310	180	490	600
	065	16 circuits		.75		360	240	600	740
	070	20 circuits		.65		410	275	685	845
	075	24 circuits		.60		465	300	765	940
	080	30 circuits		.53		535	340	875	1,075
	085	225 amp main lugs, 32 circuits	↓	.45	↓	580	400	980	1,200

For expanded coverage of these items see *Means' Electrical Cost Data 1986*

16.3 Starters, Boards & Switches

		CREW	DAILY OUTPUT	UNIT	BARE COSTS MAT.	BARE COSTS INST.	BARE COSTS TOTAL	TOTAL INCL O&P
50 090	34 circuits	1 Elec	.42	Ea.	605	425	1,030	1,275
095	36 circuits		.40		630	450	1,080	1,325
100	42 circuits	↓	.34	↓	705	525	1,230	1,525
110								
120	NEHB,w/20 amp, 1 pole bolt-on circuit breakers							
125	4 wire, 277/480 volts, 100 amp main lugs, 12 circuits	1 Elec	.88	Ea.	600	205	805	955
130	20 circuits		.60		875	300	1,175	1,400
135	225 amp main lugs, 24 circuits		.45		1,050	400	1,450	1,725
140	30 circuits		.40		1,250	450	1,700	2,025
145	36 circuits	↓	.36	↓	1,450	500	1,950	2,300
160	NQOB panel,w/20 amp, 1 pole, circuit breakers							
165	3 wire, 120/240 volt with main circuit breaker							
170	100 amp main, 12 circuits	1 Elec	.80	Ea.	390	225	615	750
175	20 circuits		.60		495	300	795	975
180	225 amp main, 30 circuits		.34		945	525	1,470	1,800
185	42 circuits		.26		1,100	690	1,790	2,200
190	400 amp main, 30 circuits		.27		1,400	665	2,065	2,500
195	42 circuits	↓	.25	↓	1,550	715	2,265	2,725
197								
200	4 wire, 120/208 volts with main circuit breaker							
205	100 amp main, 24 circuits	1 Elec	.47	Ea.	605	380	985	1,225
210	30 circuits		.40		675	450	1,125	1,375
220	225 amp main, 32 circuits		.36		1,100	500	1,600	1,925
225	42 circuits		.28		1,225	640	1,865	2,275
230	400 amp main, 42 circuits		.24		1,725	745	2,470	2,975
235	600 amp main, 42 circuits	↓	.20	↓	1,925	895	2,820	3,400
240	NEHB,with/20 amp, 1 pole circuit breaker							
245	4 wire, 227/480 volts with main circuit breaker							
250	100 amp main, 24 circuits	1 Elec	.42	Ea.	1,200	425	1,625	1,925
255	30 circuits		.38		1,400	470	1,870	2,225
260	225 amp main, 30 circuits		.36		1,850	500	2,350	2,750
265	42 circuits	↓	.28	↓	2,200	640	2,840	3,350
53-001	**PANELBOARD CIRCUIT BREAKERS**							
005	Bolt-on, 10,000 amp I.C., 120 volt, 1 pole							
010	15 to 50 amp	1 Elec	10	Ea.	7.25	17.90	25.15	34
020	60 amp		8		7.25	22	29.25	40
030	70 amp	↓	8	↓	14.10	22	36.10	48
035	240 volt, 2 pole							
040	15 to 50 amp	1 Elec	8	Ea.	16	22	38	50
050	60 amp		7.50		16	24	40	52
060	80 to 100 amp		5		43	36	79	99
070	3 pole, bolt-on, 15 to 60 amp		6.20		52	29	81	99
080	70 amp		5		67	36	103	125
090	80 to 100 amp		3.60		76	50	126	155
100	22,000 amp I.C., 240 volt, 2 pole, 70 to 225 amp		2.70		215	66	281	330
110	3 pole, 70 to 225 amp		2.30		330	78	408	475
120	14,000 amp I.C., 277 volts, 1 pole, 15 to 30 amp	↓	8	↓	27	22	49	62
125								
130	22,000 amp I.C., 480 volts, 2 pole, 70 to 225 amp	1 Elec	2.70	Ea.	365	66	431	495
140	3 pole, 70 to 225 amp	"	2.30	"	460	78	538	620
55-001	**SAFETY SWITCHES**							
010	General duty, 240 volt, 3 pole, fused, 30 amp	1 Elec	3.20	Ea.	35	56	91	120
020	60 amp		2.30		61	78	139	180
030	100 amp		1.90		105	94	199	250
040	200 amp		1.30		230	140	370	450
050	400 amp		.90		500	200	700	835
060	600 amp	↓	.60	↓	1,000	300	1,300	1,525
290	240 volt, 3 pole, fused							
291	30 amp	1 Elec	3.20	Ea.	58	56	114	145
300	60 amp	"	2.30	"	100	78	178	220

For expanded coverage of these items see *Means' Electrical Cost Data 1986*

16.3 Starters, Boards & Switches

			CREW	DAILY OUTPUT	UNIT	BARE COSTS MAT.	BARE COSTS INST.	BARE COSTS TOTAL	TOTAL INCL O&P
55	330	100 amp	1 Elec	1.90	Ea.	155	94	249	305
	350	200 amp		1.30		270	140	410	495
	370	400 amp		.90		625	200	825	975
	390	600 amp	↓	.60	↓	1,050	300	1,350	1,575
60-001		**SWITCHBOARDS** Aluminum bus bars incoming section,							
	010	not including CT's or PT's							
	020	No main disconnect, includes CT compartment							
	030	120/208 volt, 4 wire, 600 amp	1 Elec	.50	Ea.	1,725	360	2,085	2,425
	040	800 amp		.44		1,900	405	2,305	2,675
	050	1000 amp		.40		2,225	450	2,675	3,100
	060	1200 amp		.36		2,350	500	2,850	3,300
	070	1600 amp		.33		2,575	545	3,120	3,625
	080	2000 amp		.31		2,750	580	3,330	3,850
	100	3000 amp	↓	.28	↓	3,525	640	4,165	4,800
	200	Fused switch & CT compartment							
	210	120/208 volt, 4 wire, 400 amp	1 Elec	.56	Ea.	2,450	320	2,770	3,150
	220	600 amp		.47		2,850	380	3,230	3,675
	230	800 amp		.42		3,825	425	4,250	4,825
	240	1200 amp	↓	.34	↓	4,800	525	5,325	6,050
	290	Pressure switch & CT compartment							
	300	120/208 volt, 4 wire, 800 amp	1 Elec	.40	Ea.	5,475	450	5,925	6,675
	310	1200 amp		.33		6,225	545	6,770	7,625
	320	1600 amp		.31		6,925	580	7,505	8,450
	330	2000 amp	↓	.28	↓	8,100	640	8,740	9,825
	440	Circuit breaker, molded case & CT compartment							
	460	3 pole, 4 wire, 600 amp	1 Elec	.47	Ea.	4,050	380	4,430	5,000
	480	800 amp		.42		4,275	425	4,700	5,325
	500	1200 amp	↓	.34	↓	5,975	525	6,500	7,325
65-001		**DISTRIBUTION SECTION**							
	010	Aluminum bus bars, not including breakers							
	020	120/208 or 277/480 volt, 4 wire, 600 amp	1 Elec	.50	Ea.	1,300	360	1,660	1,950
	030	800 amp		.44		1,375	405	1,780	2,100
	040	1000 amp		.40		1,500	450	1,950	2,300
	050	1200 amp		.36		1,600	500	2,100	2,475
	060	1600 amp		.33		1,825	545	2,370	2,800
	070	2000 amp		.31		1,925	580	2,505	2,950
	080	2500 amp		.30		2,325	595	2,920	3,425
	090	3000 amp		.28		2,650	640	3,290	3,825
	095	4000 amp	↓	.26	↓	3,400	690	4,090	4,725
70-001		**FEEDER SECTION** Group mounted devices							
	003	Circuit breakers							
	016	FA frame, 15 to 60 amp, 240 volt, 1 pole	1 Elec	8	Ea.	36	22	58	72
	028	FA frame, 70 to 100 amp, 240 volt, 1 pole		7		47	26	73	89
	042	KA frame, 70 to 225 amp		3.20		445	56	501	570
	043	LA frame, 125 to 400 amp		2.30		790	78	868	980
	046	MA frame, 450 to 600 amp		1.60		1,300	110	1,410	1,600
	047	700 to 800 amp		1.30		1,700	140	1,840	2,075
	048	1000 amp		1		2,150	180	2,330	2,625
	049	PA frame, 1200 amp		.80		3,675	225	3,900	4,375
	050	Branch circuit, fusible switch, 600 volt, double 30/30 amp		4		235	45	280	325
	055	60/60 amp		3.20		235	56	291	340
	060	100/100 amp	↓	2.70	↓	375	66	441	510
	062								
	065	Single, 30 amp	1 Elec	5.30	Ea.	120	34	154	180
	070	60 amp		4.70		120	38	158	185
	075	100 amp		4		185	45	230	270
	080	200 amp		2.70		430	66	496	570
	085	400 amp		2.30		985	78	1,063	1,200
	090	600 amp		1.80		1,200	100	1,300	1,475

For expanded coverage of these items see *Means' Electrical Cost Data 1986*

16.3 Starters, Boards & Switches

		CREW	DAILY OUTPUT	UNIT	BARE COSTS MAT.	BARE COSTS INST.	BARE COSTS TOTAL	TOTAL INCL O&P
70 095	800 amp	1 Elec	1.30	Ea.	1,975	140	2,115	2,375
100	1200 amp	"	.80	"	2,500	225	2,725	3,075
80-001	**SWITCHBOARD INSTRUMENTS** 3 phase, 4 wire							
010	AC indicating, ammeter & switch	1 Elec	8	Ea.	950	22	972	1,075
020	Voltmeter & switch		8		950	22	972	1,075
030	Wattmeter		8		1,525	22	1,547	1,700
040	AC recording, ammeter		4		4,375	45	4,420	4,875
050	Voltmeter		4		4,375	45	4,420	4,875
060	Ground fault protection, zero sequence		2.70		3,075	66	3,141	3,475
070	Ground return path		2.70		3,075	66	3,141	3,475
080	3 current transformers, 5 to 800 amp		2		740	90	830	945
090	1000 to 1500 amp		1.30		1,375	140	1,515	1,700
120	2000 to 4000 amp	↓	1	↓	1,825	180	2,005	2,275
121								
130	Fused potential transformer, maximum 600 volt	1 Elec	8	Ea.	485	22	507	565
135								
140	Transition section between switchboard and transformer							
150	or motor control center, 4 wire, alum. bus, 600 amp	1 Elec	.57	Ea.	1,200	315	1,515	1,775
160	800 amp		.50		1,425	360	1,785	2,075
170	1000 amp		.44		1,475	405	1,880	2,200
180	1200 amp		.40		1,675	450	2,125	2,475
190	1600 amp		.36		1,875	500	2,375	2,775
200	2000 amp		.33		2,100	545	2,645	3,100
210	2500 amp		.31		2,300	580	2,880	3,350
220	3000 amp		.28		2,525	640	3,165	3,700
240	Weatherproof construction, per vertical section	↓	.88	↓	1,975	205	2,180	2,475

16.4 Transformers & Bus Duct

		CREW	DAILY OUTPUT	UNIT	BARE COSTS MAT.	BARE COSTS INST.	BARE COSTS TOTAL	TOTAL INCL O&P
01 001	**OIL FILLED TRANSFORMER** Pad mounted, Primary delta or Y,							
005	5 KV or 15 KV, with taps, 277/480 secondary, 3 phase							
010	150 KVA	R-3	.65	Ea.	5,225	825	6,050	6,900
020	300 KVA		.45		7,825	1,200	9,025	10,300
030	500 KVA		.40		9,750	1,350	11,100	12,600
040	750 KVA		.38		12,700	1,400	14,100	15,900
050	1000 KVA		.26		14,900	2,075	16,975	19,200
060	1500 KVA		.23		19,000	2,325	21,325	24,100
070	2000 KVA		.20		23,600	2,675	26,275	29,700
080	3750 KVA	↓	.16	↓	32,400	3,350	35,750	40,300
10-001	**DRY TYPE TRANSFORMER**							
005	Single phase, 240/480 volt primary 120/240 volt secondary							
010	1 KVA	1 Elec	2	Ea.	100	90	190	240
030	2 KVA		1.60		145	110	255	320
050	3 KVA		1.40		185	130	315	390
070	5 KVA		1.20		255	150	405	495
090	7.5 KVA		1.10		360	165	525	630
110	10 KVA	↓	.80		440	225	665	805
130	15 KVA	2 Elec	1.20		595	300	895	1,075
150	25 KVA		1		775	360	1,135	1,375
170	37.5 KVA		.80		1,025	450	1,475	1,775
190	50 KVA		.70		1,250	510	1,760	2,100
210	75 KVA	↓	.65	↓	1,700	550	2,250	2,675
230	3 phase, 240/480 volt primary, 120/208 volt secondary							
231	3 KVA	1 Elec	1	Ea.	330	180	510	620
270	6 KVA	"	.80	"	380	225	605	740

For expanded coverage of these items see *Means' Electrical Cost Data 1986*

16.4 Transformers & Bus Duct

			CREW	DAILY OUTPUT	UNIT	BARE COSTS MAT.	BARE COSTS INST.	BARE COSTS TOTAL	TOTAL INCL O&P
10	290	9 KVA	1 Elec	.70	Ea.	505	255	760	925
	310	15 KVA	2 Elec	1.10		760	325	1,085	1,300
	330	30 KVA		.90		1,025	400	1,425	1,700
	350	45 KVA		.80		1,225	450	1,675	2,000
	370	75 KVA	↓	.70		1,850	510	2,360	2,775
	390	112.5 KVA	R-3	.90		2,450	595	3,045	3,525
	410	150 KVA		.85		3,200	630	3,830	4,400
	430	225 KVA		.65		4,275	825	5,100	5,850
	450	300 KVA		.55		5,450	975	6,425	7,350
	470	500 KVA		.45		8,650	1,200	9,850	11,200
	480	750 KVA	↓	.35	↓	14,000	1,525	15,525	17,500
	30-001	**COPPER BUS DUCT** Plug-in, indoor							
	005	3 pole 4 wire, bus duct, straight section, 225 amp	1 Elec	20	L.F.	43	8.95	51.95	60
	100	400 amp		16		71	11.20	82.20	94
	150	600 amp		13		80	13.80	93.80	110
	240	800 amp		10		120	17.90	137.90	160
	245	1000 amp		9		140	19.90	159.90	185
	250	1350 amp		8		185	22	207	235
	251	1600 amp		6		225	30	255	290
	252	2000 amp		5		275	36	311	355
	255	Feeder, 600 amp		14		80	12.80	92.80	105
	260	800 amp		11		120	16.30	136.30	155
	270	1000 amp		10		140	17.90	157.90	180
	280	1350 amp		9		185	19.90	204.90	230
	290	1600 amp		7		225	26	251	285
	300	2000 amp		6	↓	275	30	305	345
	310	Elbows, 225 amp		2	Ea.	380	90	470	545
	320	400 amp		1.80		465	100	565	655
	330	600 amp		1.60		490	110	600	700
	340	800 amp		1.40		490	130	620	725
	350	1000 amp		1.30		525	140	665	775
	360	1350 amp		1.20		720	150	870	1,000
	370	1600 amp		1.10		910	165	1,075	1,225
	380	2000 amp		.90		1,025	200	1,225	1,425
	400	End box, 225 amp		17		61	10.55	71.55	82
	410	400 amp		16		61	11.20	72.20	83
	420	600 amp		14		61	12.80	73.80	86
	430	800 amp		13		61	13.80	74.80	87
	440	1000 amp		12		61	14.95	75.95	89
	450	1350 amp		11		61	16.30	77.30	91
	460	1600 amp		10		61	17.90	78.90	93
	470	2000 amp		9		81	19.90	100.90	120
	480	Cable tap box, end, 225 amp		1.60		300	110	410	490
	500	400 amp		1.30		540	140	680	790
	510	600 amp		1.10		695	165	860	1,000
	520	800 amp		1		735	180	915	1,075
	530	1000 amp		.80		770	225	995	1,175
	540	1350 amp		.70		830	255	1,085	1,275
	550	1600 amp		.60		885	300	1,185	1,400
	560	2000 amp		.50		980	360	1,340	1,600
	570	Switchboard stub, 225 amp		2.70		180	66	246	295
	580	400 amp		2.30		230	78	308	365
	590	600 amp		2		280	90	370	435
	600	800 amp		1.60		345	110	455	540
	610	1000 amp		1.50		405	120	525	615
	620	1350 amp		1.30		485	140	625	730
	630	1600 amp		1.20		570	150	720	840
	640	2000 amp		1		680	180	860	1,000
	649	Tee fittings, 225 amp	↓	1.20	↓	490	150	640	755

16.4 Transformers & Bus Duct

		CREW	DAILY OUTPUT	UNIT	BARE COSTS MAT.	BARE COSTS INST.	BARE COSTS TOTAL	TOTAL INCL O&P
30 650	400 amp	1 Elec	1	Ea.	615	180	795	935
660	600 amp		.90		655	200	855	1,000
670	800 amp		.80		820	225	1,045	1,225
680	1350 amp		.60		1,250	300	1,550	1,800
700	1600 amp		.50		1,425	360	1,785	2,075
710	2000 amp		.40		2,875	450	3,325	3,800
720	Plug-in switches, 600 volt, 3 pole, 30 amp		4		130	45	175	205
730	60 amp		3.60		140	50	190	225
740	100 amp		2.70		200	66	266	315
750	200 amp		1.60		355	110	465	550
760	400 amp		.70		925	255	1,180	1,375
770	600 amp		.45		1,325	400	1,725	2,025
780	800 amp		.33		2,325	545	2,870	3,350
790	1200 amp		.25		4,350	715	5,065	5,825
800	Plug-in circuit breakers, molded case, 15 to 50 amp		4.40		260	41	301	345
810	70 to 100 amp		3.10		290	58	348	400
820	150 to 225 amp		1.70		650	105	755	865
830	250 to 400 amp		.70		1,350	255	1,605	1,850
840	500 to 600 amp		.50		1,925	360	2,285	2,625
850	700 to 800 amp		.32		2,325	560	2,885	3,375
860	900 to 1000 amp		.28		2,750	640	3,390	3,950
870	1200 amp		.22		4,350	815	5,165	5,950

16.5 Power Systems & Capacitors

		CREW	DAILY OUTPUT	UNIT	BARE COSTS MAT.	BARE COSTS INST.	BARE COSTS TOTAL	TOTAL INCL O&P
05-001	**CAPACITORS** Indoor							
002	240 volts, single & 3 phase, 0.5 KVAR	1 Elec	2.70	Ea.	165	66	231	275
010	1.0 KVAR		2.70		200	66	266	315
015	2.5 KVAR		2		260	90	350	415
020	5.0 KVAR		1.80		290	100	390	460
025	7.5 KVAR		1.60		320	110	430	515
030	10 KVAR		1.50		390	120	510	600
035	15 KVAR		1.30		535	140	675	785
040	20 KVAR		1.10		665	165	830	965
045	25 KVAR		1		780	180	960	1,125
100	480 volts, single & 3 phrase KVAR		2.70		145	66	211	255
105	2 KVAR		2.70		165	66	231	275
110	5 KVAR		2		230	90	320	380
115	7.5 KVAR		2		245	90	335	400
120	10 KVAR		2		275	90	365	430
125	15 KVAR		2		340	90	430	505
130	20 KVAR		1.60		375	110	485	575
135	30 KVAR		1.50		470	120	590	690
140	40 KVAR		1.20		615	150	765	890
145	50 KVAR		1.10		700	165	865	1,000
10-001	**GENERATOR SET**							
002	Gas or gasoline operated, includes battery,							
005	charger, muffler & transfer switch							
020	3 phase, 4 wire, 277/480 volt, 7.5 KW	R-3	.83	Ea.	4,975	645	5,620	6,375
030	10 KW		.71		6,700	755	7,455	8,425
040	15 KW		.63		7,550	850	8,400	9,475
050	30 KW		.55		11,300	975	12,275	13,800
060	70 KW		.40		19,300	1,350	20,650	23,100
070	85 KW		.33		22,500	1,625	24,125	27,000

For expanded coverage of these items see *Means' Electrical Cost Data 1986*

16.5 Power Systems & Capacitors

			CREW	DAILY OUTPUT	UNIT	BARE COSTS MAT.	INST.	TOTAL	TOTAL INCL O&P
10	080	115 KW	R-3	.28	Ea.	38,500	1,925	40,425	45,000
	090	170 KW	"	.25	"	71,500	2,150	73,650	81,500
	200	Diesel engine, including battery, charger,							
	201	muffler, transfer switch & fuel tank, 30 KW	R-3	.55	Ea.	13,900	975	14,875	16,600
	210	50 KW		.42		17,000	1,275	18,275	20,500
	220	75 KW		.35		22,100	1,525	23,625	26,400
	230	100 KW		.31		24,500	1,725	26,225	29,300
	240	125 KW		.29		26,000	1,850	27,850	31,200
	250	150 KW		.26		30,000	2,075	32,075	35,900
	260	175 KW		.25		31,000	2,150	33,150	37,100
	270	200 KW		.24		32,000	2,225	34,225	38,300
	280	250 KW		.23		35,000	2,325	37,325	41,700
	290	300 KW		.22		42,000	2,450	44,450	49,600
	300	350 KW		.20		45,500	2,675	48,175	54,000
	310	400 KW		.19		55,000	2,825	57,825	64,500
	320	500 KW	↓	.18	↓	63,000	2,975	65,975	73,500

16.6 Lighting

			CREW	DAILY OUTPUT	UNIT	BARE COSTS MAT.	INST.	TOTAL	TOTAL INCL O&P
05-001		AVERAGE Square foot and percent of total							
	010	job cost, see division 17.1							
10-001		INTERIOR LIGHTING FIXTURES Including lamps, mounting							
	003	hardware and connections							
	010	Fluorescent, C.W. lamps, ceiling, recess mounted in grid, RS							
	020	Acrylic lens, 1'W x 4'L, two 40 watt	1 Elec	5.70	Ea.	35	31	66	84
	021	1'W x 4'L, three 40 watt		5.40		58	33	91	110
	030	2'W x 2'L, two U40 watt		5.70		47	31	78	97
	040	2'W x 4'L, two 40 watt		5.30		38	34	72	90
	050	2'W x 4'L, three 40 watt		5		48	36	84	105
	060	2'W x 4'L, four 40 watt		4.70		49	38	87	110
	070	4'W x 4'L, four 40 watt		3.20		125	56	181	220
	080	4'W x 4'L, six 40 watt		3.10		140	58	198	235
	090	4'W x 4'L, eight 40 watt	↓	2.90	↓	155	62	217	260
	100	Surface mounted, RS							
	103	Acrylic lens with hinged & latched door frame							
	110	1'W x 4'L, two 40 watt	1 Elec	7	Ea.	38	26	64	79
	111	1'W x 4'L, three 40 watt		6.70		59	27	86	105
	120	2'W x 2'L, two U40 watt		7		60	26	86	105
	130	2'W x 4'L, two 40 watt		6.20		49	29	78	95
	140	2'W x 4'L, three 40 watt		5.70		63	31	94	115
	150	2'W x 4'L, four 40 watt		5.30		64	34	98	120
	160	4'W x 4'L, four 40 watt		3.60		125	50	175	210
	170	4'W x 4'L, six 40 watt		3.30		145	54	199	240
	180	4'W x 4'L, eight 40 watt		3.10		165	58	223	265
	190	2'W x 8'L, four 40 watt		3.20		98	56	154	190
	200	2'W x 8'L, eight 40 watt	↓	3.10	↓	125	58	183	220
	201								
	210	Strip fixture, surface mounted							
	220	4' long, one 40 watt RS	1 Elec	8.50	Ea.	18	21	39	50
	230	4' long, two 40 watt RS		8		19	22	41	53
	240	4' long, one 40 watt, SL		8		34	22	56	70
	260	8' long, one 75 watt, SL		6.70		32	27	59	74
	270	8' long, two 75 watt, SL		6.20		37	29	66	82
	280	4' long, two 60 watt, HO		6.70		51	27	78	95
	290	8' long, two 110 watt, HO	↓	5.30	↓	53	34	87	105

For expanded coverage of these items see *Means' Electrical Cost Data 1986*

16.6 Lighting

			DAILY		BARE COSTS			TOTAL
		CREW	OUTPUT	UNIT	MAT.	INST.	TOTAL	INCL O&P
10	300	Pendent mounted, industrial, white porcelain enamel							
	310	4' long, two 40 watt, RS	1 Elec	5.70	Ea.	47	31	78	97
	320	4' long, two 60 watt, HO		5		75	36	111	135
	330	8' long, two 75 watt, SL		4.40		75	41	116	140
	340	8' long, two 110 watt, HO		4		88	45	133	160
	347	Troffer, air handling, 2'W x 4'L with four 40 watt, RS		4		105	45	150	180
	348	2'W x 2'L with two U40 watt RS		5.50		88	33	121	145
	349	Air connector insulated, 5" diameter		20		34	8.95	42.95	50
	350	6" diameter		20		34	8.95	42.95	50
	358	Mercury vapor, integral ballast, ceiling, recess mounted,							
	359	prismatic glass lens, floating door							
	360	2'W x 2'L, 250 watt DX lamp	1 Elec	3.20	Ea.	225	56	281	330
	370	2'W x 2'L, 400 watt DX lamp		2.90		235	62	297	345
	380	Surface mtd., prismatic lens, 2'W x 2'L, 250 watt DX lamp		2.70		210	66	276	325
	390	2'W x 2'L, 400 watt DX lamp		2.40		225	75	300	355
	400	High bay, aluminum reflector							
	403	Single unit, 400 watt DX lamp	1 Elec	2.30	Ea.	200	78	278	330
	410	Single unit, 1000 watt DX lamp		2		360	90	450	525
	420	Twin unit, two 400 watt DX lamps		1.60		400	110	510	600
	445	Incandescent, ceiling, recess mtd., round alzak reflector, prewired							
	447	100 watt	1 Elec	8	Ea.	38	22	60	74
	448	150 watt		8		39	22	61	75
	450	300 watt		6.70		42	27	69	85
	460	Square glass lens with metal trim, prewired							
	463	100 watt	1 Elec	6.70	Ea.	21	27	48	62
	470	200 watt		6.70		26	27	53	67
	480	300 watt		5.70		41	31	72	90
	490	Ceiling/wall, surface mounted, metal cylinder, 75 watt		10		20	17.90	37.90	48
	492	150 watt		10		31	17.90	48.90	60
	520	Ceiling, surface mounted, opal glass drum							
	530	8", one 60 watt	1 Elec	10	Ea.	21	17.90	38.90	49
	540	10", two 60 watt lamps		8		26	22	48	61
	550	12", four 60 watt lamps		6.70		55	27	82	99
	601	Vapor tight, incandescent, ceiling mounted, 200 watt		6.20		30	29	59	75
	610	Fluorescent, surface mounted, 2 lamps, 4'L, RS, 40 watt		3.20		54	56	110	140
	620	Mercury vapor with ballast, 175 watt		3.20		200	56	256	300
	625								
	630	Explosionproof							
	651	Incandescent, ceiling mounted, 200 watts	1 Elec	4	Ea.	200	45	245	285
	660	Fluorescent, RS, 4' long, ceiling mounted, two 40 watt		2.70		1,275	66	1,341	1,500
	670	Mercury vapor with ballast, surface mounted, 175 watt		2.70		475	66	541	620
	685	Vandalproof, surface mounted, fluorescent, two 40 watt		3.20		95	56	151	185
	686	Incandescent, one 150 watt		8		34	22	56	70
25-001		**EXIT AND EMERGENCY LIGHTING**							
	002								
	008	Exit light, ceiling or wall mount, incandescent, single face	1 Elec	8	Ea.	36	22	58	72
	010	Double face	"	6.70	"	36	27	63	78
	030	Emergency light units, battery operated							
	035	Twin sealed beam light, 25 watt, 6 volt each							
	050	Lead battery operated	1 Elec	4	Ea.	195	45	240	280
	070	Nickel cadmium battery operated		4		330	45	375	425
	090	Self-contained fluorescent lamp pack		10		150	17.90	167.90	190
50-001		**EXTERIOR FIXTURES** With lamps							
	020	Wall mounted, incandescent, 100 watt	1 Elec	8	Ea.	22	22	44	56
	040	Quartz, 500 watt		5.30		69	34	103	125
	060	Mercury vapor, 100 watt		5.30		190	34	224	260
	080	Wall pack, mercury vapor, 175 watt		4		195	45	240	280
	100	250 Watt		4		220	45	265	305

For expanded coverage of these items see *Means' Electrical Cost Data 1986*

16.6 Lighting

			CREW	DAILY OUTPUT	UNIT	BARE COSTS MAT.	INST.	TOTAL	TOTAL INCL O&P
50	110	Low pressure sodium, 35 watt	1 Elec	4	Ea.	155	45	200	235
	115	55 watt	"	4	"	235	45	280	325
	120	Floodlights with ballast and lamp,							
	140	pole mounted, pole not included							
	150	Mercury vapor, 250 watt	1 Elec	2.40	Ea.	230	75	305	360
	160	400 watt		2.20		275	81	356	420
	180	1000 watt		2		405	90	495	575
	195	Metal halide, 175 watt		2.70		240	66	306	360
	200	400 watt		2.20		340	81	421	490
	220	1000 watt		2		500	90	590	680
	234	High pressure sodium, 70 watt		2.70		210	66	276	325
	240	400 watt		2.20		325	81	406	475
	260	1000 watt		2		635	90	725	825
	265	Roadway area luminaire, low pressure sodium, 135 watt		2		430	90	520	600
	270	180 watt	↓	2	↓	465	90	555	640
	271								
	272	Mercury vapor, 400 watt	1 Elec	2.20	Ea.	485	81	566	650
	273	1000 watt		2		550	90	640	735
	275	Metal halide, 400 watt		2.20		550	81	631	720
	276	1000 watt		2		600	90	690	790
	278	High pressure sodium, 400 watt		2.20		640	81	721	820
	279	1000 watt	↓	2	↓	750	90	840	955
	280	Light poles, anchor base,							
	282	not including concrete bases							
	284	Aluminum pole, 8' high	1 Elec	4	Ea.	200	45	245	285
	300	20' high	R-3	2.90		485	185	670	790
	320	30' high	"	2.60	↓	945	205	1,150	1,325
	340	35' high	R-3	2.30	Ea.	1,325	235	1,560	1,775
	360	40' high	"	2		1,475	270	1,745	2,000
	380	Bracket arms, 1 arm	1 Elec	8		48	22	70	85
	400	2 arms		8		75	22	97	115
	420	3 arms		5.30		100	34	134	160
	440	4 arms		5.30		130	34	164	190
	450	Steel pole, galvanized, 8' high	↓	3.80		340	47	387	440
	460	20' high	R-3	2.60		740	205	945	1,100
	480	30' high		2.30		1,025	235	1,260	1,450
	500	35' high		2.20		1,125	245	1,370	1,575
	520	40' high	↓	1.70		1,250	315	1,565	1,800
	540	Bracket arms, 1 arm	1 Elec	8		48	22	70	85
	560	2 arms		8		110	22	132	155
	580	3 arms		5.30		130	34	164	190
	600	4 arms	↓	5.30	↓	165	34	199	230
75-001		**LAMPS**							
	002								
	008	Fluorescent, rapid start, cool white, 2' long, 20 watt	1 Elec	1	C	320	180	500	610
	010	4' long, 40 watt		.90		180	200	380	485
	020	Slimline, 4' long, 40 watt		.90		485	200	685	820
	030	8' long, 75 watt		.80		495	225	720	865
	040	High output, 4' long, 60 watt		.90		595	200	795	940
	050	8' long, 110 watt		.80		555	225	780	935
	060	Mercury vapor, mogul base, deluxe white, 100 watt		.30		1,875	595	2,470	2,925
	070	250 watt		.30		2,475	595	3,070	3,575
	080	400 watt		.30		1,975	595	2,570	3,025
	090	1000 watt		.20		4,575	895	5,470	6,325
	100	Metal halide, mogul base, 175 watt		.30		3,400	595	3,995	4,600
	120	400 watt		.30		3,900	595	4,495	5,150
	130	1000 watt	↓	.20	↓	9,000	895	9,895	11,200
	132								

For expanded coverage of these items see *Means' Electrical Cost Data 1986*

16.6 Lighting

		CREW	DAILY OUTPUT	UNIT	BARE COSTS MAT.	INST.	TOTAL	TOTAL INCL O&P
135	Sodium high pressure, 70W	1 Elec	.30	C	4,400	595	4,995	5,700
138	250 watt		.30		5,025	595	5,620	6,375
140	400 watt		.30		5,350	595	5,945	6,750
145	1000 watt		.20		11,500	895	12,395	13,900
300	Guards, fluorescent lamp, 4' long		1		300	180	480	590
320	8' long	↓	.90	↓	480	200	680	815

16.7 Lighting Utilities

		CREW	DAILY OUTPUT	UNIT	BARE COSTS MAT.	INST.	TOTAL	TOTAL INCL O&P
01-001	**ELECTRIC & TELEPHONE SITEWORK** Not including excavation, backfill							
020	and cast in place concrete							
040	Hand holes, precast concrete with concrete cover							
060	2' x 2' x 3' deep	R-3	2.40	Ea.	230	225	455	560
080	3' x 3' x 3' deep		1.90		315	280	595	735
100	4' x 4' x 4' deep	↓	1.40	↓	680	385	1,065	1,275
120	Manholes, precast, with iron racks, pulling irons, C.I. frame							
140	and cover, 4' x 6' x 7' deep	R-3	1.20	Ea.	1,150	445	1,595	1,875
160	6' x 8' x 7' deep		1		1,475	535	2,010	2,375
180	6' x 10' x 7' deep		.80		1,675	670	2,345	2,775
200	Poles, wood, creosoted, see also division 16.6-50, 20' high		3.10		80	175	255	325
240	25' high		2.90		90	185	275	355
260	30' high		2.60		100	205	305	395
280	35' high		2.40		155	225	380	480
300	40' high		2.30		175	235	410	515
320	45' high	↓	1.70	↓	215	315	530	675
340	Cross arms with hardware & insulators							
360	4' long	1 Elec	2.50	Ea.	22	72	94	125
380	5' long		2.40		24	75	99	135
400	6' long	↓	2.20	↓	30	81	111	150
420	Underground duct, banks ready for concrete fill, min. of 1-1/2"							
440	between ducts. For wire & cable see division 16.1							
458	PVC, type EB, 1 @ 2" diameter	1 Elec	240	L.F.	.29	.75	1.04	1.39
460	2 @ 2" diameter		120		.58	1.49	2.07	2.79
480	4 @ 2" diameter		60		1.16	2.99	4.15	5.55
500	2 @ 3" diameter		100		.78	1.79	2.57	3.43
520	4 @ 3" diameter		50		1.56	3.58	5.14	6.85
540	2 @ 4" diameter		80		1.24	2.24	3.48	4.59
560	4 @ 4" diameter		40		2.48	4.48	6.96	9.15
580	6 @ 4" diameter		27		3.72	6.65	10.37	13.65
620	Rigid galvanized steel, 2 @ 2" diameter		90		3.80	1.99	5.79	7.05
640	4 @ 2" diameter		45		7.60	3.98	11.58	14.10
680	2 @ 3" diameter		50		8.50	3.58	12.08	14.50
700	4 @ 3" diameter		25		17	7.15	24.15	29
720	2 @ 4" diameter		35		13.20	5.10	18.30	22
740	4 @ 4" diameter		17		26	10.55	36.55	44
760	6 @ 4" diameter	↓	11	↓	40	16.30	56.30	67
90-001	**TRANSITE DUCT** (Asbestos cement)							
008	Type 1, 3" diameter	1 Elec	130	L.F.	.67	1.38	2.05	2.72
010	3-1/2" diameter		110		.72	1.63	2.35	3.13
020	4" diameter	↓	100	↓	.82	1.79	2.61	3.48
050	Fittings, bends, 45°, 3" diameter		8	Ea.	22	22	44	56
060	3-1/2" diameter		7		24	26	50	63
070	4" diameter		6.50		24	28	52	66
100	90°, 3" diameter	↓	8	↓	40	22	62	76

For expanded coverage of these items see *Means' Electrical Cost Data 1986*

16.7 Lighting Utilities

		CREW	DAILY OUTPUT	UNIT	BARE COSTS MAT.	BARE COSTS INST.	BARE COSTS TOTAL	TOTAL INCL O&P
90	110	3-1/2" diameter	1 Elec	7	Ea.	41	26	67	82
	115	4" diameter		6.50		41	28	69	85
	150	Adapters to metal conduit, 3" diameter		20		3	8.95	11.95	16.20
	160	3-1/2" diameter		16		3	11.20	14.20	19.40
	170	4" diameter		16		3	11.20	14.20	19.40
	200	End bells, 3" diameter		11		5	16.30	21.30	29
	210	3-1/2" diameter		10		5.50	17.90	23.40	32
	220	4" diameter		10		5.50	17.90	23.40	32
	250	Plastic spacers, 3" diameter		100		.35	1.79	2.14	2.96
	260	3-1/2" diameter		100		.40	1.79	2.19	3.02
	270	4" diameter	↓	100	↓	.40	1.79	2.19	3.02

16.8 Special Systems

		CREW	DAILY OUTPUT	UNIT	BARE COSTS MAT.	BARE COSTS INST.	BARE COSTS TOTAL	TOTAL INCL O&P
01-001	CLOCKS							
008	12" diameter, single face	1 Elec	8	Ea.	42	22	64	78
010	Double face	"	6.20	"	85	29	114	135
05-001	CLOCK SYSTEMS							
002								
010	Time system components, master controller	1 Elec	.33	Ea.	1,050	545	1,595	1,925
020	Program bell		8		31	22	53	66
040	Combination clock & speaker		3.20		105	56	161	195
060	Frequency generator		2		4,000	90	4,090	4,525
080	Job time automatic stamp recorder, minimum		4		305	45	350	400
100	Maximum	↓	4		425	45	470	530
120	Time stamp for correspondence, hand operated				245		245	270M
140	Fully automatic				360		360	395M
160	Master time clock system, clocks & bells, 20 room	4 Elec	.20		2,550	3,575	6,125	7,950
180	50 room	"	.08		6,250	8,950	15,200	19,800
200	Time clock, 100 cards in & out, 1 color	1 Elec	3.20		670	56	726	820
220	2 colors		3.20		720	56	776	875
240	With 3 circuit program device, minimum		2		275	90	365	430
260	Maximum		2		370	90	460	535
280	Metal rack for 25 cards	↓	7	↓	18	26	44	57
290								
300	Watchman's tour station	1 Elec	8	Ea.	27	22	49	62
320	Annunciator with zone indication		1		150	180	330	425
340	Time clock with tape	↓	1	↓	360	180	540	655
15-001	DETECTION SYSTEMS							
002								
010	Burglar alarm, battery operated, mechanical trigger	1 Elec	4	Ea.	125	45	170	200
020	Electrical trigger		4		155	45	200	235
040	For outside key control, add		8		21	22	43	55
060	For remote signaling circuitry, add		8		37	22	59	73
080	Card reader, flush type, standard		2.70		540	66	606	690
100	Multi-code		2.70		675	66	741	840
120	Door switches, hinge switch		5.30		38	34	72	90
140	Magnetic switch		5.30		56	34	90	110
160	Exit control locks, horn alarm		4		210	45	255	295
180	Flashing light alarm		4		255	45	300	345
200	Indicating panels, 1 channel		2.70		245	66	311	365
220	10 channel	↓	1.60	↓	840	110	950	1,075

For expanded coverage of these items see *Means' Electrical Cost Data 1986*

16.8 Special Systems

		CREW	DAILY OUTPUT	UNIT	BARE COSTS MAT.	BARE COSTS INST.	BARE COSTS TOTAL	TOTAL INCL O&P
240	20 channel	1 Elec	1	Ea.	1,625	180	1,805	2,050
260	40 channel		.57		3,000	315	3,315	3,750
280	Ultrasonic motion detector, 12 volt		2.30		190	78	268	320
300	Infrared photoelectric detector		2.30		190	78	268	320
360	Fire, sprinkler & standpipe alarm, control panel, 4 zone		2		660	90	750	855
380	8 zone		1		950	180	1,130	1,300
400	12 zone		.66		1,325	270	1,595	1,850
402	Alarm device		8		85	22	107	125
405	Actuating device	▼	8	▼	225	22	247	280
415								
420	Battery and rack	1 Elec	4	Ea.	530	45	575	645
440	Automatic charger		8		340	22	362	405
460	Signal bell		8		32	22	54	67
480	Trouble buzzer or manual station		8		21	22	43	55
500	Detector, rate of rise		8		12	22	34	45
520	Smoke detector, ceiling type		6.20		64	29	93	110
540	Duct type		3.20		185	56	241	285
560	Light and horn		5.30		75	34	109	130
580	Fire alarm horn	▼	6.70	▼	21	27	48	62
590								
600	Door holder, electro-magnetic	1 Elec	4	Ea.	48	45	93	115
620	Combination holder and closer		3.20		320	56	376	435
640	Code transmitter		4		530	45	575	645
660	Drill switch		8		32	22	54	67
680	Master box		2.70		1,150	66	1,216	1,350
700	Break glass station		8		27	22	49	62
780	Remote annunciator, 8 zone lamp		1.80		160	100	260	320
800	12 zone lamp		1.30		210	140	350	430
820	16 zone lamp	▼	1.10	▼	265	165	430	525
25-001	**DOCTORS IN-OUT REGISTER**							
005	Register, 200 names	4 Elec	.64	Ea.	7,150	1,125	8,275	9,475
010	Combination control and recall, 200 names	"	.64		9,350	1,125	10,475	11,900
020	Recording register	1 Elec	.50		3,625	360	3,985	4,500
030	Transformers	"	4		120	45	165	195
040	Pocket pages			▼	550		550	605M
30-001	**DOORBELL SYSTEM** Incl. transformer, button & signal							
010	6" bell	1 Elec	4	Ea.	37	45	82	105
020	Buzzer	"	4	"	32	45	77	100
33-001	**ELECTRIC HEATING**							
020	Snow melting for paved surface embedded mat heaters & controls	1 Elec	130	S.F.	4.30	1.38	5.68	6.70
040	Cable heating, radiant heat plaster, no controls, in South		130		4.60	1.38	5.98	7.05
060	In North		90		4.60	1.99	6.59	7.90
080	Cable on 1/2" board, not incl. controls, tract housing		90		4.10	1.99	6.09	7.35
100	Custom housing		80	▼	4.60	2.24	6.84	8.30
110	Rule of thumb: Baseboard units, including control		4.40	KW	44	41	85	105
120	Duct heaters, including controls		5.30	"	38	34	72	90
130	Baseboard heaters, 2' long, 375 watt		8	Ea.	28	22	50	63
140	3' long, 500 watt		8		35	22	57	71
160	4' long, 750 watt		6.70		43	27	70	86
180	5' long, 935 watt		5.70		57	31	88	110
200	6' long, 1125 watt		5		64	36	100	120
240	8' long, 1500 watt	▼	4	▼	86	45	131	160
295	Wall heaters with fan, 120 to 277 volt							
297	surface mounted, residential, 750 watt	1 Elec	7	Ea.	45	26	71	86
300	1500 watt		5		72	36	108	130
304	2250 watt		4		115	45	160	190
307	4000 watt	▼	3.50	▼	135	51	186	220
350								

For expanded coverage of these items see *Means' Electrical Cost Data 1986*

16.8 Special Systems

			CREW	DAILY OUTPUT	UNIT	BARE COSTS MAT.	BARE COSTS INST.	BARE COSTS TOTAL	TOTAL INCL O&P
33	360	Thermostats, integral	1 Elec	16	Ea.	15	11.20	26.20	33
	380	Line voltage, 1 pole	"	8	"	14	22	36	48
35-001		**NURSE CALL SYSTEMS**							
	010	Single bedside call station	1 Elec	8	Ea.	130	22	152	175
	020	Ceiling speaker station		8		34	22	56	70
	040	Emergency call station		8		49	22	71	86
	060	Pillow speaker		8		115	22	137	160
	080	Double bedside call station		4		230	45	275	315
	100	Duty station		4		100	45	145	175
	120	Standard call button		8		39	22	61	75
	140	Lights, corridor, dome or zone indicator		8		35	22	57	71
	160	Master control station for 20 stations	2 Elec	.65	Total	2,725	550	3,275	3,800
38-001		**PUBLIC ADDRESS SYSTEM**							
	002								
	010	Conventional, office	1 Elec	5.33	Speaker	66	34	100	120
	020	Industrial	"	2.70	"	130	66	196	240
	040	Explosionproof system is 3 times cost of central control							
	060	Installation costs run about 120% of material cost							
40-001		**SOUND SYSTEM**							
	002								
	010	Components, outlet, projector	1 Elec	8	Ea.	18	22	40	52
	020	Microphone		4		14	45	59	80
	040	Speakers, ceiling or wall		8		44	22	66	81
	060	Trumpets		4		100	45	145	175
	080	Privacy switch		8		39	22	61	75
	100	Monitor panel		4		180	45	225	260
	120	Antenna, AM/FM		4		88	45	133	160
	140	Volume control		8		30	22	52	65
	160	Amplifier, 250 watts		1		830	180	1,010	1,175
	180	Cabinets		1		330	180	510	620
	200	Intercom, 25 station capacity, master station		1		950	180	1,130	1,300
	220	Remote station		8		68	22	90	105
	240	Intercom outlets		8		37	22	59	73
	260	Handset		4		140	45	185	220
	280	Emergency call system, 12 zones, annunciator		1.30		440	140	580	680
	300	Bell		5.30		39	34	73	92
	320	Light or relay		8		20	22	42	54
	340	Transformer		4		100	45	145	175
	360	House telephone, talking station		1.60		205	110	315	385
	380	Press to talk, release to listen		5.30		48	34	82	100
	400	System-on button				21		21	23M
	420	Door release	1 Elec	4		47	45	92	115
	440	Combination speaker and microphone		8		82	22	104	120
	460	Termination box		3.20		26	56	82	110
	480	Amplifier or power supply		5.30		330	34	364	410
	500	Vestibule door unit		16	Name	58	11.20	69.20	80
	520	Strip cabinet		27	Ea.	115	6.65	121.65	135
	540	Directory		16	"	32	11.20	43.20	51
50-001		**T.V. SYSTEMS**							
	002								
	010	Master TV antenna system							
	020	VHF reception & distribution, 12 outlets	1 Elec	6	Outlet	60	30	90	110
	040	30 outlets		10		56	17.90	73.90	87
	060	100 outlets		13		50	13.80	63.80	75
	080	VHF & UHF reception & distribution, 12 outlets		6		60	30	90	110
	100	30 outlets		10		56	17.90	73.90	87
	120	100 outlets		13		50	13.80	63.80	75
	140	School and deluxe systems, 12 outlets		2.40		130	75	205	250
	160	30 outlets		4		115	45	160	190
	180	80 outlets		5.30		105	34	139	165

For expanded coverage of these items see *Means' Electrical Cost Data 1986*

16.8 Special Systems

		CREW	DAILY OUTPUT	UNIT	BARE COSTS MAT.	BARE COSTS INST.	BARE COSTS TOTAL	TOTAL INCL O&P
50 200	Closed circuit, surveillance, one station (camera & monitor)	1 Elec	1.30	Total	500	140	640	750
220	For additional camera stations, add		2.70	Ea.	315	66	381	440
240	Industrial quality, one station (camera & monitor)		1.30	Total	1,250	140	1,390	1,575
260	For additional camera stations, add		2.70	Ea.	900	66	966	1,075
261	For low light, add		2.70		720	66	786	885
262	For very low light, add		2.70		5,500	66	5,566	6,150
280	For weatherproof camera station, add		1.30		585	140	725	840
300	For pan and tilt, add		1.30		1,500	140	1,640	1,850
320	For zoom lens - remote control, add, minimum		2		1,400	90	1,490	1,675
340	Maximum		2		5,050	90	5,140	5,675
341	For automatic iris for low light, add	↓	2	↓	1,200	90	1,290	1,450
360	Educational T.V. studio, basic 3 camera system, black & white,							
380	electrical & electronic equip. only, minimum	4 Elec	.80	Total	7,200	895	8,095	9,200
400	Maximum (full console)		.28		30,500	2,550	33,050	37,200
410	As above, but color system, minimum		.28		40,500	2,550	43,050	48,200
412	Maximum	↓	.12	↓	176,500	5,975	182,475	202,500
420	For film chain, black & white, add	1 Elec	1	Ea.	8,200	180	8,380	9,275
425	Color, add		.25		10,000	715	10,715	12,000
440	For video tape recorders, add, minimum	↓	1		2,250	180	2,430	2,725
460	Maximum	4 Elec	.40	↓	14,400	1,800	16,200	18,400
81-001	**WIRING, RESIDENTIAL** 20' average runs							
002								
018	Air conditioning receptacle							
020	Using non-metallic sheathed cable	1 Elec	10	Ea.	6	17.90	23.90	32
024	BX cable		8.30		10	22	32	42
026	EMT conduit	↓	6.70	↓	10	27	37	49
039	Disposal wiring							
040	Using non-metallic sheathed cable	1 Elec	9	Ea.	5	19.90	24.90	34
044	BX cable		7.50		8	24	32	43
046	EMT conduit	↓	6	↓	8	30	38	52
058	Dryer circuits							
060	Using non-metallic sheathed cable	1 Elec	5.50	Ea.	13	33	46	61
062	BX cable		4.60		20	39	59	78
064	EMT conduit	↓	3.70	↓	15	48	63	86
078	Duplex receptacles							
080	Using non-metallic sheathed cable	1 Elec	13	Ea.	5.50	13.80	19.30	26
084	BX cable		10.80		9	16.60	25.60	34
086	EMT conduit	↓	8.70	↓	9	21	30	40
098	Exhaust fan wiring							
100	Using non-metallic sheathed cable	1 Elec	10	Ea.	5	17.90	22.90	31
102	BX cable		8.30		8	22	30	40
106	EMT conduit	↓	6.70	↓	8	27	35	47
118	Fire alarm smoke detector & horn							
120	Using non-metallic sheathed cable	1 Elec	10	Ea.	36	17.90	53.90	65
122	BX cable		8.30		39	22	61	74
124	EMT conduit		6.70		39	27	66	81
140	Front doorbell	↓	5.50	↓	15	33	48	63
145								
158	Furnace circuit and switch							
160	Using non-metallic sheathed cable	1 Elec	6	Ea.	7.10	30	37.10	51
162	BX cable		5		11	36	47	64
164	EMT conduit	↓	4	↓	11	45	56	77
170	Ground fault receptacle							
172	Using non-metallic sheathed cable	1 Elec	8	Ea.	36	22	58	72
174	BX cable		6.60		41	27	68	84
176	EMT conduit	↓	5.40	↓	41	33	74	93
179	Heater circuits							
180	Using non-metallic sheathed cable	1 Elec	8	Ea.	5	22	27	38
182	BX cable		6.60		8	27	35	48
184	EMT conduit	↓	5.40	↓	8	33	41	57

For expanded coverage of these items see *Means' Electrical Cost Data 1986*

16.8 Special Systems

		Special Systems	CREW	DAILY OUTPUT	UNIT	BARE COSTS MAT.	BARE COSTS INST.	BARE COSTS TOTAL	TOTAL INCL O&P
81	198	Intercom, 8 stations							
	200	Using non-metallic sound cable	1 Elec	.70	Total	260	255	515	655
	202	BX cable		.58		315	310	625	790
	204	EMT conduit	↓	.47	↓	380	380	760	965
	220	Light fixtures, average		16	Ea.	20	11.20	31.20	38
	225								
	238	Lighting wiring							
	240	Using non-metallic sheathed cable	1 Elec	16	Ea.	5	11.20	16.20	22
	242	BX cable		13.30		9	13.45	22.45	29
	244	EMT conduit	↓	10.70	↓	9	16.75	25.75	34
	258	Range circuits							
	260	Using non-metallic sheathed cable	1 Elec	4	Ea.	30	45	75	97
	262	BX cable		3.30		42	54	96	125
	264	EMT conduit		2.70		26	66	92	125
	280	Service and panel, 100 amp		1.20		220	150	370	455
	300	200 amp	↓	.90	↓	420	200	620	750
	318	Switch, single pole							
	320	Using non-metallic sheathed cable	1 Elec	16	Ea.	5.40	11.20	16.60	22
	322	BX cable		13.30		9	13.45	22.45	29
	324	EMT conduit	↓	10.70	↓	9	16.75	25.75	34
	339	Switch, 3-way							
	340	Using non-metallic sheathed cable	1 Elec	12	Ea.	7	14.95	21.95	29
	342	BX cable		10		12	17.90	29.90	39
	343	EMT conduit	↓	8	↓	12	22	34	45
	358	Water heater circuit							
	360	Using non-metallic sheathed cable	1 Elec	5	Ea.	7	36	43	59
	362	BX cable		4.20		12	43	55	75
	364	EMT conduit	↓	3.40	↓	12	53	65	89
	378	Weatherproof receptacle with ground fault breaker at panel							
	380	Using non-metallic sheathed cable	1 Elec	6	Ea.	51	30	81	99
	382	BX cable		5		55	36	91	110
	384	EMT conduit	↓	4	↓	55	45	100	125

For expanded coverage of these items see Means' Electrical Cost Data 1986

17.1 S.F., C.F. and % of Total Costs

			UNIT	UNIT COSTS 1/4	UNIT COSTS MEDIAN	UNIT COSTS 3/4	% OF TOTAL 1/4	% OF TOTAL MEDIAN	% OF TOTAL 3/4
01-001		APARTMENTS Low Rise (1 to 3 story)	S.F.	30.85	38.90	51.65			
002		Total project cost	C.F.	2.77	3.57	4.49			
010	(141)	Sitework	S.F.	2.01	3.49	5.85	5.80%	10.60%	13.70%
050		Masonry		.53	1.46	2.39	1.20%	4%	6%
150	(142)	Finishes		3.26	3.89	5.25	8.80%	10.50%	12.80%
180		Equipment		1.05	1.58	2.32	2.90%	4.20%	6.80%
272		Plumbing		2.42	3.17	4.05	6.80%	9%	10.20%
277		Heating, ventilating, air conditioning		1.54	1.87	2.80	4.20%	5.80%	7.60%
290		Electrical		1.79	2.38	3.33	5.20%	6.70%	8.80%
310		Total: Mechanical & Electrical	↓	5.35	6.55	8.45	15.90%	18.30%	22.40%
900		Per apartment unit, total cost	Apt.	25,100	36,000	54,000			
950		Total: Mechanical & Electrical	"	4,450	6,475	8,975			
02-001		APARTMENTS Mid Rise (4 to 7 story)	S.F.	41.35	49	60.90			
002		Total project costs	C.F.	3.22	4.50	5.30			
010		Sitework	S.F.	2.58	3.40	6.45	5.70%	6.70%	10.10%
050		Masonry		2.91	3.87	5.20	5.90%	7.60%	10.70%
150		Finishes		5.20	6.40	8.15	10.30%	11.60%	16.10%
180		Equipment		1.37	1.93	2.42	2.80%	3.50%	4.50%
250		Conveying equipment		.99	1.19	1.48	2.10%	2.20%	2.60%
272		Plumbing	↓	2.42	3.05	4.19	6.20%	7.40%	9%

For expanded coverage of these items see Means' Square Foot Costs Data 1986

17.1 S.F., C.F. and % of Total Costs

			UNIT	UNIT COSTS			% OF TOTAL		
				1/4	MEDIAN	3/4	1/4	MEDIAN	3/4
02	290	Electrical	S.F.	2.75	3.69	4.53	6.30%	7%	8.90%
	310	Total: Mechanical & Electrical	"	7.60	9.40	11.40	17.10%	19.70%	21.50%
	900	Per apartment unit, total cost	Apt.	38,000	49,400	66,500			
	950	Total: Mechanical & Electrical	"	9,100	10,600	15,600			
03-001		**APARTMENTS** High Rise (8 to 24 story)	S.F.	46.70	54.35	62.75			
	002	Total project costs	C.F.	3.78	5	6.25			
	010	Sitework	S.F.	1.43	2.74	3.61	2.50%	4.70%	6.10%
	050	Masonry		3.65	5.55	6.25	5.30%	9.60%	11%
	150	Finishes		5.20	6.50	7.70	10.40%	11.70%	13.50%
	180	Equipment		1.50	1.83	2.36	2.70%	3.30%	4.20%
	250	Conveying equipment		1.06	1.56	2.23	2%	2.70%	3.30%
	272	Plumbing		2.99	4.07	5.10	6.70%	9.10%	10.40%
	290	Electrical		3.40	4.24	5.50	7%	7.70%	9.10%
	310	Total: Mechanical & Electrical	↓	9.65	11.80	14.50	19.70%	22.40%	24.50%
	900	Per apartment unit, total cost	Apt.	42,700	50,500	55,500			
	950	Total: Mechanical & Electrical	"	10,400	12,000	12,700			
04-001		**AUDITORIUMS**	S.F.	46.05	65.35	84.35			
	002	Total project costs	C.F.	2.89	4.23	5.85			
	272	Plumbing	S.F.	3.01	4	5.20	5.70%	6.80%	8.40%
	277	Heating, ventilating, air conditioning		6.25	15.15	17.60	6.90%	16%	19.80%
	290	Electrical		3.72	5.30	6.95	6.80%	8.80%	10.90%
	310	Total: Mechanical & Electrical		7.80	9.95	16	14.40%	17.80%	23.30%
05-001		**AUTOMOTIVE SALES**	↓	32.65	41.35	53.85			
	002	Total project costs	C.F.	2.44	2.84	3.69			
	272	Plumbing	S.F.	1.63	2.29	3.11	2.80%	5.90%	6.40%
	277	Heating, ventilating, air conditioning		2.38	3.93	5.65	6.30%	10.20%	10.80%
	290	Electrical		2.79	4.91	6.55	7.30%	9.90%	12.40%
	310	Total: Mechanical & Electrical		5.90	11.35	13.20	15.40%	19.20%	30.30%
06-001		**BANKS**	↓	72.05	90.25	118.45			
	002	Total project costs	C.F.	5.10	6.85	9.05			
	010	Sitework	S.F.	6.55	13	19.15	5.30%	12.90%	17.10%
	050	Masonry		2.70	5.55	9.85	2.50%	5.50%	8.30%
	150	Finishes		5.15	7.60	10.52	5.10%	7.10%	8.90%
	180	Equipment		2.16	6.60	14.55	2.40%	8.50%	14.90%
	272	Plumbing		2.36	3.37	4.97	2.90%	4%	5%
	277	Heating, ventilating, air conditioning		4.55	6.20	8.85	5.20%	7.40%	8.70%
	290	Electrical		7	9.15	12.25	8.40%	10.30%	12.30%
	310	Total: Mechanical & Electrical	↓	11.65	16.15	23	14.20%	18.10%	23.50%
	350	See also division 11.1-09							
13-001		**CHURCHES**	S.F.	48.15	59.75	75.05			
	002	Total project costs	C.F.	3.04	3.80	4.95			
	180	Equipment	S.F.	.60	1.53	3.03	1.10%	2.50%	5%
	272	Plumbing		1.91	2.75	4	3.60%	4.90%	6.30%
	277	Heating, ventilating, air conditioning		4.46	5.85	8.30	7.70%	10%	12.10%
	290	Electrical		3.96	5.35	7	7.30%	8.80%	10.90%
	310	Total: Mechanical & Electrical	↓	8.60	12.35	16.30	16.10%	21.80%	26.50%
	350	See also division 11.1-12							
15-001		**CLUBS, COUNTRY**	S.F.	48.30	58	75.90			
	002	Total project costs	C.F.	3.99	5	6.85			
	272	Plumbing	S.F.	2.96	4.14	7.50	5.40%	8.90%	10%
	277	Heating, ventilating, air conditioning		2.85	6.15	9.15	6.70%	10%	10.70%
	290	Electrical		3.45	6.10	7.55	7.80%	9.70%	11%
	310	Total: Mechanical & Electrical		9.35	14.25	20.60	17.20%	24.20%	30.90%
17-001		**CLUBS, SOCIAL** Fraternal	↓	41.50	57.10	74.60			
	002	Total project costs	C.F.	2.54	3.81	4.70			
	272	Plumbing	S.F.	1.94	3.08	3.52	4.90%	6.70%	7.80%
	277	Heating, ventilating, air conditioning		3.68	5.45	6.17	8.20%	10.90%	14.40%
	290	Electrical		2.91	5.01	5.66	6.70%	9.50%	11.40%
	310	Total: Mechanical & Electrical	↓	9.01	11.99	16.35	18.50%	28.70%	33.10%

For expanded coverage of these items see *Means' Square Foot Costs Data 1986*

17.1 S.F., C.F. and % of Total Costs

		UNIT	UNIT COSTS 1/4	UNIT COSTS MEDIAN	UNIT COSTS 3/4	% OF TOTAL 1/4	% OF TOTAL MEDIAN	% OF TOTAL 3/4
18-001	**CLUBS, Y.M.C.A.**	S.F.	47.60	61.30	73.80			
002	Total project costs	C.F.	2.35	3.91	4.69			
272	Plumbing	S.F.	3.48	5.25	7.20	6%	8.40%	11%
290	Electrical		3.36	4.79	6.20	6%	8.20%	10.10%
310	Total: Mechanical & Electrical	↓	9.70	15	22.75	17.20%	28.50%	32.10%
19-001	**COLLEGES** Classrooms & Administration	S.F.	60.20	79.10	102.20			
002	Total project costs	C.F.	4.17	5.93	8.80			
050	Masonry	S.F.	6.15	7.55	10.55	4%	7.40%	16.10%
180	Equipment		1.45	2.43	3.40	2.60%	4%	6.20%
272	Plumbing		3.13	4.23	6.73	4%	6.40%	8%
277	Heating, ventilating, air conditioning		6.30	10.05	14.85	8.70%	12.20%	14.60%
290	Electrical		4.57	7.45	11.25	7.80%	9.80%	11.50%
310	Total: Mechanical & Electrical		13.15	21.95	29.80	18.70%	28%	35.10%
21-001	**COLLEGES** Science, Engineering, Laboratories	↓	67.95	92.45	112.30			
002	Total project costs	C.F.	5.10	7	8.15			
180	Equipment	S.F.	3.22	12.05	14.40	3.80%	10.60%	15%
272	Plumbing		3.36	5.05	7.03	5.90%	6.80%	8%
277	Heating, ventilating, air conditioning		5.30	10.80	12.75	8.20%	14.40%	19.10%
290	Electrical		6.10	8.75	12.95	8.10%	9.60%	11.40%
310	Total: Mechanical & Electrical	↓	19.20	27.70	45.50	25.40%	34.90%	38.70%
350	See also division 11.1-49							
23-001	**COLLEGES** Student Unions	S.F.	61.25	85.20	100.65			
002	Total project costs	C.F.	3.46	4.37	5.75			
180	Equipment	S.F.	4.27	5.55	8.65	4.20%	6.70%	9.70%
272	Plumbing		3.77	4.92	6.25	4.20%	5.20%	8.60%
277	Heating, ventilating, air conditioning		9.25	10.80	15.65	10.90%	14.10%	17.40%
290	Electrical		4.91	7.30	9.80	7.90%	9.70%	10.70%
310	Total: Mechanical & Electrical	↓	15.95	22.90	27.45	22.90%	26.20%	31.70
25-001	**COMMUNITY CENTERS**	S.F.	50.50	61.85	78.45			
002	Total project costs	C.F.	3.09	4.43	5.70			
180	Equipment	S.F.	1.27	1.94	3.06	2%	3.10%	5.40%
272	Plumbing		2.65	4.25	5.75	5.40%	7%	9.10%
277	Heating, ventilating, air conditioning		4.21	6	8	7.60%	10.20%	12.90%
290	Electrical		4.22	5.40	7.70	7.50%	9.10%	10.80%
310	Total: Mechanical & Electrical	↓	10.25	15.15	21.70	19.20%	26%	30.70%
28-001	**COURT HOUSES**	S.F.	67.15	82.60	101.70			
002	Total project costs	C.F.	5.45	6.15	7.15			
050	Masonry	S.F.	4.15	9.65	9.80	4.40%	5.40%	8.70%
114	Roofing		.53	1.03	1.51	.70%	.90%	1.20%
272	Plumbing		3.47	4.72	5.55	3.90%	6.30%	7.40%
277	Heating, ventilating, air conditioning		9.20	10.25	15.05	10.30%	11.90%	14.80%
290	Electrical		6.35	7.90	12.75	8.40%	9.80%	12%
310	Total: Mechanical & Electrical		14.70	21	30.20	20.10%	25.70%	28.70%
30-001	**DEPARTMENT STORES**	↓	27.35	36.50	41.65			
002	Total project costs	C.F.	1.23	1.80	2.39			
272	Plumbing	S.F.	.89	1.24	1.61	2.80%	3.90%	5.30%
277	Heating, ventilating, air conditioning		2.84	3.78	5.70	8.30%	11.80%	14.80%
290	Electrical		3.13	4.24	5.20	10.40%	11.90%	13.60%
310	Total: Mechanical & Electrical		5.90	10.35	11.85	22.10%	26.70%	32.30%
31-001	**DORMITORIES** Low Rise (1 to 3 story)	↓	44.10	58.85	75.90			
002	Total project costs	C.F.	3.29	5.45	6.95			
272	Plumbing	S.F.	3.07	3.85	5.20	8%	8.90%	9.60%
277	Heating, ventilating, air conditioning		3.25	3.74	5.20	4.60%	7.60%	9.90%
290	Electrical		3.15	4.46	5.45	6.40%	8.70%	9.50%
310	Total: Mechanical & Electrical	↓	7.15	11.45	16.40	18.20%	21.80%	28.40%
900	Per bed, total cost	Bed	9,275	14,800	26,500			

For expanded coverage of these items see *Means' Square Foot Costs Data 1986*

17.1 S.F., C.F. and % of Total Costs

		UNIT	UNIT COSTS 1/4	UNIT COSTS MEDIAN	UNIT COSTS 3/4	% OF TOTAL 1/4	% OF TOTAL MEDIAN	% OF TOTAL 3/4
32-001	**DORMITORIES** Mid Rise (4 to 8 story)	S.F.	65.25	75.90	91.45			
002	Total project costs	C.F.	5.67	6.92	7.60			
272	Plumbing	S.F.	5.05	5.10	7.40	6.40%	6.60%	10.30%
290	Electrical		4.53	7.30	8.90	8.20%	9.40%	10.10%
310	Total: Mechanical & Electrical	↓	13.45	16.80	20.95	17.70%	22.90%	25.70%
900	Per bed, total cost	Bed	8,875	15,300	18,750			
34-001	**FACTORIES**	S.F.	21.55	31.20	52.05			
002	Total project costs	C.F.	1.68	2.13	3.25			
010	Sitework	S.F.	2.33	4.92	5.50	6.80%	8.60%	9%
272	Plumbing		2.41	2.83	4.25	4.60%	4.80%	10.40%
277	Heating, ventilating, air conditioning		2.21	3.57	3.98	3.20%	5%	6.90%
290	Electrical		3.14	7.09	7.91	8%	11.30%	15.80%
310	Total: Mechanical & Electrical	↓	8.45	12.95	19.40	16.50%	25.90%	33%
36-001	**FIRE STATIONS**	S.F.	47.25	63.55	77.60			
002	Total project costs	C.F.	3.02	4.06	5.10			
050	Masonry	S.F.	7.80	12.75	17.65	10.80%	16.10%	19.20%
114	Roofing		2.13	3.21	4.45	1.80%	3.30%	4.90%
135	Glass & glazing		.54	.75	1.62	.50%	1%	1.40%
157	Floor covering		.31	.38	.55	.20%	.30%	.80%
158	Painting		1.38	1.61	1.81	1.40%	1.60%	2.10%
180	Equipment		.86	1.41	3.38	1.10%	2.50%	4.40%
272	Plumbing		3.06	4.78	6.50	5.90%	7.30%	9.50%
277	Heating, ventilating, air conditioning		2.62	4.22	6.70	4.80%	7.30%	9.20%
290	Electrical		3.48	5.95	8.55	7.10%	9.70%	12.10%
310	Total: Mechanical & Electrical		8.95	14.10	19.40	17.50%	22.60%	27.60%
37-001	**FRATERNITY HOUSES** And Sorority Houses	↓	47.20	56.35	62.85			
002	Total project costs	C.F.	4.51	5.50	6.10			
272	Plumbing	S.F.	3.56	4.16	5.55	5.90%	8%	10.80%
290	Electrical		3.11	4.10	7.45	6.50%	8.80%	10.40%
310	Total: Mechanical & Electrical	↓	8.80	12.50	15.10	14.60%	20.70%	24.20%
38-001	**FUNERAL HOMES**	S.F.	44.90	56.45	82.75			
002	Total project costs	C.F.	3.17	4.56	5.40			
272	Plumbing	S.F.	1.78	2.47	2.70	4.10%	4.40%	4.70%
277	Heating, ventilating, air conditioning		3.96	4	4.81	7%	9.20%	10.40%
290	Electrical		2.97	3.69	5.75	6.20%	7.70%	11%
310	Total: Mechanical & Electrical		8.10	10.70	12.40	18.80%	20.80%	27.20%
39-001	**GARAGES, COMMERCIAL**	↓	26.90	42.60	56.55			
002	Total project costs	C.F.	1.71	2.50	3.57			
180	Equipment	S.F.	1.13	3.41	5.55	2.70%	6.30%	8.40%
272	Plumbing		1.74	2.74	5.50	4.90%	7.30%	11%
273	Heating & ventilating		2.81	3.91	4.40	7%	11.20%	11.30%
290	Electrical		2.41	3.97	5.55	7.10%	9%	11.40%
310	Total: Mechanical & Electrical	↓	5.60	9.95	14.35	15.70%	21.90%	27.80%
40-001	**GARAGES, MUNICIPAL**	S.F.	28.95	45.25	63.20			
002	Total project costs	C.F.	2.07	2.80	3.76			
050	Masonry	S.F.	3.88	5.20	8	7%	11.80%	19%
114	Roofing		2.25	3.09	5.25	6.50%	7.30%	10.10%
272	Plumbing		1.80	3.40	5.55	4.10%	6.90%	8.60%
273	Heating & ventilating		1.97	2.81	5.80	6%	7.90%	11.30%
290	Electrical		2.53	3.95	5.40	6.30%	8%	10.10%
310	Total: Mechanical & Electrical		5.80	11.55	17.05	15.50%	24.10%	31.50%
41-001	**GARAGES, PARKING**	↓	15.55	19.45	32.85			
002	Total project costs	C.F.	1.34	1.77	2.92			
272	Plumbing	S.F.	.28	.55	.79	2.10%	2.80%	3.80%
290	Electrical		.65	.95	1.54	4.20%	5.20%	6.30%
310	Total: Mechanical & Electrical	↓	.97	1.53	1.95	6.80%	8.30%	9.40%
320								

For expanded coverage of these items see *Means' Square Foot Costs Data 1986*

17.1 S.F., C.F. and % of Total Costs

			UNIT	UNIT COSTS 1/4	UNIT COSTS MEDIAN	UNIT COSTS 3/4	% OF TOTAL 1/4	% OF TOTAL MEDIAN	% OF TOTAL 3/4
41	900	Per car, total cost	Car	4,975	6,800	9,550			
	950	Total: Mechanical & Electrical	"	355	535	660			
43-001		**GYMNASIUMS**	S.F.	40.90	55	70.45			
	002	Total project costs	C.F.	2.07	2.66	3.62			
	180	Equipment	S.F.	.94	1.76	3.09	2%	3.20%	6.70%
	272	Plumbing		2.52	3.43	4.34	4.80%	7.20%	8.50%
	277	Heating, ventilating, air conditioning		2.68	4.64	7.60	7.40%	9.70%	14%
	290	Electrical		3.44	4.26	6.25	6.20%	9%	10.70%
	310	Total: Mechanical & Electrical	↓	6.50	11.55	15	16.60%	21.80%	27%
	350	See also division 11.1-49							
46-001		**HOSPITALS**	S.F.	90.30	111	147			
	002	Total project costs	C.F.	6.65	7.95	11.10			
	112	Roofing	S.F.	.66	1.64	2.67	.50%	1.20%	2.90%
	132	Finish hardware		.89	.97	1.17	.60%	1%	1.20%
	154	Floor covering		.62	1.13	2.84	.50%	1.10%	1.60%
	180	Equipment		2.09	4.12	6.10	1.50%	3.80%	5.30%
	272	Plumbing		7.95	10.15	13.75	7.50%	9.10%	10.70%
	277	Heating, ventilating, air conditioning		8.65	14.40	19.80	8.40%	13%	16.60%
	290	Electrical		9.30	12.65	19.20	10.30%	12.30%	15.20%
	310	Total: Mechanical & Electrical	↓	27.30	37.20	55.05	26.90%	37.70%	40.30%
	900	Per bed or person, total cost	Bed	27,100	41,800	62,700			
	990	See also division 11.1-37							
48-001		**HOUSING** For the Elderly	S.F.	43.70	54.90	69.05			
	002	Total project costs	C.F.	3.08	4.28	5.60			
	010	Sitework	S.F.	3.19	4.74	6.70	6.10%	8.20%	12%
	050	Masonry		1.32	4.76	7.45	2%	6.50%	10.70%
	073	Miscellaneous metals		.91	1.72	2.40	1.20%	2.10%	2.40%
	112	Roofing		.85	1.50	2.60	1.20%	2.10%	3%
	114	Dampproofing		.25	.34	.79	.20%	.40%	.70%
	134	Windows		.60	1.01	2.18	1.10%	1.50%	2.40%
	135	Glass & glazing		.14	.45	.81	.20%	.40%	.90%
	153	Drywall		2.61	3.42	5.50	3.70%	4.10%	4.60%
	154	Floor covering		.71	1.04	1.51	.90%	1.40%	1.90%
	157	Tile & marble		.33	.49	.73	.50%	.60%	.80%
	158	Painting		1.41	1.98	3.08	2%	2.60%	3.10%
	180	Equipment		1.02	1.41	2.24	1.80%	3.20%	4.40%
	251	Conveying systems		1.04	1.41	1.87	1.80%	2.30%	2.80%
	272	Plumbing		3.32	4.60	6.85	8.30%	9.70%	10.90%
	273	Heating, ventilating, air conditioning		1.49	2.30	3.24	3.20%	5.60%	7.10%
	290	Electrical		3.20	4.51	6.35	7.50%	9%	10.60%
	291	Electrical incl. electric heat		3.69	6.95	8.25	9.60%	11%	13.30%
	310	Total: Mechanical & Electrical	↓	8.05	11.30	14.65	18.40%	21.90%	24.60%
	900	Per rental unit, total cost	Unit	38,800	45,900	50,500			
	950	Total: Mechanical & Electrical	"	7,700	9,650	11,400			
50-001		**HOUSING** Public (low-rise)	S.F.	33.15	46.10	62.95			
	002	Total project costs	C.F.	2.79	3.63	4.55			
	010	Sitework	S.F.	4.79	6.50	10.35	9%	11.70%	16.40%
	180	Equipment		.95	1.69	2.56	2.20%	3.20%	4.20%
	272	Plumbing		2.41	3.33	4.27	7.10%	9%	11.50%
	273	Heating, ventilating, air conditioning		1.28	2.41	2.73	4.40%	6%	6.40%
	290	Electrical		2.11	3.05	4.28	4.90%	6.50%	8.20%
	310	Total: Mechanical & Electrical	↓	6.45	9.30	12.15	15.70%	19.20%	22.10%
	900	Per apartment, total cost	Apt.	36,800	41,700	52,100			
	950	Total: Mechanical & Electrical	"	6,175	8,475	10,600			
51-001		**ICE SKATING RINKS**	S.F.	31.40	43.90	72.05			
	002	Total project costs	C.F.	1.78	2.23	2.63			
	272	Plumbing	S.F.	.94	1.39	2.13	3.10%	3.20%	4.60%
	290	Electrical		2.48	3.25	4.51	5.70%	7%	10.10%
	310	Total: Mechanical & Electrical	↓	4.49	6.37	9.59	12.40%	16.40%	25.90%

For expanded coverage of these items see *Means' Square Foot Costs Data 1986*

17.1 S.F., C.F. and % of Total Costs

		UNIT	UNIT COSTS			% OF TOTAL		
			1/4	MEDIAN	3/4	1/4	MEDIAN	3/4
52-001	JAILS	S.F.	101	113	132			
002	Total project costs	C.F.	7.45	9.25	11.60			
180	Equipment	S.F.	3.87	10.25	17.15	3.80%	8.90%	14.80%
272	Plumbing		5.90	10.15	12.25	7%	8.30%	12%
277	Heating, ventilating, air conditioning		5.90	10.70	15.50	6.30%	9.40%	12.10%
290	Electrical		8.60	11.35	14.35	7.80%	9.80%	12.40%
310	Total: Mechanical & Electrical		22.15	31.80	41.25	23.20%	29.60%	35.30%
53-001	LIBRARIES	S.F.	56.80	69.55	86.75			
002	Total project costs	C.F.	3.94	4.77	5.95			
050	Masonry	S.F.	3.14	5.70	9.40	4%	6.50%	9.40%
114	Roofing		2.05	2.59	3.70	1.90%	3.30%	3.50%
158	Painting		1.08	1.54	3.12	.90%	1.90%	3%
180	Equipment		.69	1.75	3.42	1.40%	2.70%	4.80%
272	Plumbing		2.35	3.24	4.32	3.60%	4.50%	5.80%
277	Heating, ventilating, air conditioning		5.75	8.15	10.60	8.70%	11%	13.20%
290	Electrical		5.80	7.10	9.45	8.40%	10.90%	12.10%
310	Total: Mechanical & Electrical		12.25	17.15	24.25	19.40%	25.50%	29.60%
55-001	MEDICAL CLINICS		54.40	67.05	83.20			
002	Total project costs	C.F.	4.03	5.31	7.05			
180	Equipment	S.F.	1.44	2.81	4.64	1.90%	4.30%	6.80%
272	Plumbing		3.75	5.15	7.10	6.10%	8.40%	10.20%
277	Heating, ventilating, air conditioning		4.56	5.95	8.65	6.70%	9%	11.70%
290	Electrical		5.15	6.60	8.60	8.10%	9.90%	12%
310	Total: Mechanical & Electrical		11.85	15.20	21.30	19%	24.20%	30.10%
350	See also division 11.1-37							
57-001	MEDICAL OFFICES	S.F.	50.25	63.30	76.20			
002	Total project costs	C.F.	3.85	5.10	6.80			
180	Equipment	S.F.	1.74	3.38	4.78	3.40%	5.90%	7.10%
272	Plumbing		3.08	4.57	6.30	5.70%	6.90%	9.40%
277	Heating, ventilating, air conditioning		3.65	5.40	6.90	6.50%	8%	10.40%
290	Electrical		4.26	6.15	8	7.60%	9.80%	11.70%
310	Total: Mechanical & Electrical		9.80	13.75	18.25	17.20%	22.40%	27.40%
59-001	MOTELS	S.F.	47.85	49.70	72.80			
002	Total project costs	C.F.	2.94	5.55	6.75			
272	Plumbing	S.F.	2.22	4.03	6.85	3.80%	8.90%	12.60%
277	Heating, ventilating, air conditioning		1.75	3.01	3.21	4.10%	6.20%	8.20%
290	Electrical		2.62	4.75	7.10	4.70%	9.50%	10.90%
310	Total: Mechanical & Electrical		8.35	14.05	16.40	17.30%	27.70%	33.30%
500								
900	Per rental unit, total cost	Unit	13,500	22,700	32,300			
950	Total: Mechanical & Electrical	"	3,750	5,025	5,525			
60-001	NURSING HOMES	S.F.	51.55	66.80	82.40			
002	Total project costs	C.F.	4.09	5.25	7			
180	Equipment	S.F.	1.61	2.09	3.27	2%	3.60%	6%
272	Plumbing		4.48	5.45	8.15	8.30%	10.30%	14.10%
277	Heating, ventilating, air conditioning		4.80	6.95	8.15	10.60%	11.70%	11.80%
290	Electrical		5.15	6.50	8	9.70%	11%	12.50%
310	Total: Mechanical & Electrical		11.75	16.05	23.15	22%	28.10%	33.20%
320								
900	Per bed or person, total cost	Bed	20,100	25,700	32,200			
61-001	OFFICES Low-Rise (1 to 4 story)	S.F.	41.25	53.35	70.15			
002	Total project costs	C.F.	3	4.26	5.70			
010	Sitework	S.F.	3.04	5	7.85	5.30%	9.20%	13.40%
050	Masonry		1.67	3.28	6.26	3%	5.90%	8.50%
180	Equipment		.58	.99	2.64	1.40%	1.70%	4.40%
272	Plumbing		1.58	2.38	3.41	3.60%	4.50%	6.10%

For expanded coverage of these items see *Means' Square Foot Costs Data 1986*

17.1 S.F., C.F. and % of Total Costs

			UNIT	UNIT COSTS 1/4	UNIT COSTS MEDIAN	UNIT COSTS 3/4	% OF TOTAL 1/4	% OF TOTAL MEDIAN	% OF TOTAL 3/4
61	277	Heating, ventilating, air conditioning	S.F.	3.35	4.75	7	7.20%	10.40%	11.90%
	290	Electrical		3.50	4.84	6.65	7.50%	9.60%	11.10%
	310	Total: Mechanical & Electrical	↓	7	10.60	15.55	14.50%	20.40%	27%
62-001	OFFICES Mid-Rise (5 to 10 story)		S.F.	46.70	57.15	77.45			
	002	Total project costs	C.F.	3.19	4.12	5.95			
	272	Plumbing	S.F.	1.40	2.14	3.07	2.80%	3.60%	4.50%
	277	Heating, ventilating, air conditioning		3.46	4.96	7.90	7.60%	9.30%	11%
	290	Electrical		2.85	4.23	7.15	6.50%	8.20%	10%
	310	Total: Mechanical & Electrical		7.50	10.50	18.05	14.50%	21.80%	25.50%
63-001	OFFICES High-Rise (11 to 20 story)		↓	51.10	70.60	89.70			
	002	Total project costs	C.F.	3.41	5	7.15			
	290	Electrical	S.F.	2.75	4.07	6.15	6.20%	8%	10.50%
	310	Total: Mechanical & Electrical		9.35	12.75	23.35	17.20%	21.70%	29.40%
64-001	POLICE STATIONS		↓	67.45	88.45	113			
	002	Total project costs	C.F.	5.05	6.55	8.05			
	050	Masonry	S.F.	7.25	11.25	13.85	6.80%	10.50%	13.20%
	114	Roofing		2.17	2.33	5.20	2%	2.20%	4.20%
	135	Glass & glazing		.63	.83	.89	.60%	.80%	.80%
	157	Floor covering		.31	.62	.69	.30%	.70%	.70%
	158	Painting		1.12	1.34	1.86	1.10%	1.50%	1.90%
	180	Equipment		1.08	5.65	9.45	2%	6%	13.30%
	272	Plumbing		3.88	5.50	9.05	5.70%	6.80%	10.60%
	277	Heating, ventilating, air conditioning		5.50	7.45	10.60	7%	10.50%	12%
	290	Electrical		6.90	10.75	13.80	9.30%	11.60%	14.50%
	310	Total: Mechanical & Electrical		18.40	23.15	30	22.20%	27.50%	33%
65-001	POST OFFICES		↓	55.45	66.80	87.45			
	002	Total project costs	C.F.	3.07	3.83	4.58			
	272	Plumbing	S.F.	2.39	3.06	3.83	4.10%	5.30%	5.60%
	277	Heating, ventilating, air conditioning		3.50	4.61	7.05	6.70%	8%	9.80%
	290	Electrical		4.39	6.30	7.45	7.50%	9.70%	11.10%
	310	Total: Mechanical & Electrical		9.65	13.20	18.20	16.50%	22.20%	26.60%
66-001	POWER PLANTS		↓	265	440	570			
	002	Total project costs	C.F.	9	15.70	22.10			
	290	Electrical	S.F.	19.35	38.85	71.30	9.50%	12.70%	17.20%
	810	Total: Mechanical & Electrical		44.10	72.45	265	16.30%	32.50%	47.70%
67-001	RELIGIOUS EDUCATION		↓	42	49.45	61.10			
	002	Total project costs	C.F.	2.44	3.40	4.60			
	272	Plumbing	S.F.	1.79	2.64	3.78	4.10%	5.10%	7.10%
	277	Heating, ventilating, air conditioning		4.08	4.85	6.35	8.10%	9.90%	11.20%
	290	Electrical		3.15	4.21	5.40	6.90%	8.50%	10%
	310	Total: Mechanical & Electrical		6.90	9.45	13.95	14.70%	19.70%	24.30%
69-001	RESEARCH Laboratories and facilities		↓	58.45	86.95	134			
	002	Total project costs	C.F.	3.51	6.60	10.20			
	180	Equipment	S.F.	1.43	4.78	9.60	.90%	4.70%	9%
	272	Plumbing		6.10	8.40	11.75	5.20%	8.30%	10.80%
	277	Heating, ventilating, air conditioning		5.95	19.35	22.85	7.20%	16.50%	17.70%
	290	Electrical		7.05	10.45	22.10	9.20%	12.40%	16.20%
	310	Total: Mechanical & Electrical	↓	14.85	30.10	59.05	20.60%	30.60%	43.20%
70-001	RESTAURANTS		S.F.	59.65	78.40	100			
	002	Total project costs	C.F.	5.25	6.80	8.60			
	180	Equipment	S.F.	3.68	9.85	15.40	6%	13.90%	16.80%
	272	Plumbing		4.95	6.25	8.50	6%	8.20%	9.30%
	277	Heating, ventilating, air conditioning		6.85	9.30	11.85	9.60%	12.30%	13.30%
	290	Electrical		6.40	8.20	10.90	8.30%	10.40%	12%
	310	Total: Mechanical & Electrical	↓	14.20	19.30	25.75	18.30%	23.20%	32.20%
500									
900	Per seat unit, total cost		Seat	1,925	2,800	3,675			
950	Total: Mechanical & Electrical		"	430	565	880			

For expanded coverage of these items see *Means' Square Foot Costs Data 1986*

17.1 S.F., C.F. and % of Total Costs

		UNIT	UNIT COSTS			% OF TOTAL		
			1/4	MEDIAN	3/4	1/4	MEDIAN	3/4
72-001	**RETAIL STORES**	S.F.	27.74	38.30	50.20			
002	Total project costs	C.F.	1.95	2.74	3.82			
272	Plumbing	S.F.	1.08	1.78	3.12	3.20%	4.50%	7%
277	Heating, ventilating, air conditioning		3.32	3.17	4.91	6.80%	8.70%	10.20%
290	Electrical		2.65	3.59	5.25	7.40%	10.10%	12.30%
310	Total: Mechanical & Electrical		4.71	7.15	10.40	14.80%	18.30%	24.30%
74-001	**SCHOOLS** Elementary		47.20	58.15	69.45			
002	Total project costs	C.F.	3.17	3.98	5.05			
050	Masonry	S.F.	5.55	8.30	10.20	8.50%	10.60%	13.20%
112	Water & dampproofing		.35	.37	.49	.30%	.40%	.60%
114	Roofing		1.44	.15	2.19	1.90%	2%	2.20%
134	Windows		.65	1.80	1.87	.50%	1.90%	2.50%
135	Glass & glazing		.20	.52	.90	.20%	.60%	1.20%
157	Floor covering		.22	.47	1.14	.10%	.60%	1.30%
158	Painting		1.28	1.38	1.79	1.50%	1.60%	2.20%
180	Equipment		1.39	2.70	4.73	2.50%	4.20%	9.60%
272	Plumbing		2.76	3.98	5.30	5.60%	7.10%	9.20%
273	Heating & ventilating		4.08	6.60	9.05	8.10%	10.80%	15.10%
290	Electrical		4.30	5.50	7.15	8.40%	10%	11.80%
310	Total: Mechanical & Electrical		9.85	14.30	18.45	19.80%	26%	32%
900	Per pupil, total cost	Pupil	4,000	6,425	8,750			
950	Total: Mechanical & Electrical	"	1,250	1,825	2,975			
76-001	**SCHOOLS** Junior High & Middle	S.F.	49.65	57.95	69.70			
002	Total project costs	C.F.	3.11	3.81	4.53			
050	Masonry	S.F.	6.65	7.60	8.20	9.10%	12.20%	14.20%
180	Equipment		1.64	2.90	4.55	2.60%	5.10%	8%
272	Plumbing		2.90	3.50	4.55	5.40%	6.90%	8.10%
277	Heating, ventilating, air conditioning		3.56	6.80	9.60	8.70%	11.50%	17.40%
290	Electrical		4.77	5.70	6.75	8%	9.60%	10.60%
310	Total: Mechanical & Electrical		10.20	14.20	20.05	19.20%	27%	32.20%
900	Per pupil, total cost	Pupil	4,450	6,625	8,700			
78-001	**SCHOOLS** Senior High	S.F.	48.90	57	77.40			
002	Total project costs	C.F.	3.03	3.73	4.60			
180	Equipment	S.F.	1.33	3.24	4.65	2.30%	3.70%	7.70%
272	Plumbing		2.59	4.36	6.95	5%	6.50%	8%
277	Heating, ventilating, air conditioning		5.95	6.65	9.60	8.90%	11.50%	14.20%
290	Electrical		4.95	6.30	8.75	8.30%	10.10%	12.30%
310	Total: Mechanical & Electrical		10	15.85	20.35	16.80%	25.60%	30%
900	Per pupil, total cost	Pupil	5,050	8,150	10,600			
80-001	**SCHOOLS** Vocational	S.F.	40.85	55.05	73.85			
002	Total project costs	C.F.	2.57	3.55	4.82			
050	Masonry	S.F.	2.40	5.95	9.15	4.60%	9.60%	10.90%
114	Roofing		.70	1.20	2.28	.80%	1.20%	2.20%
180	Equipment		1.02	1.77	3.48	1.90%	3.20%	4.60%
272	Plumbing		2.74	4.03	5.80	5.40%	7%	8.50%
277	Heating, ventilating, air conditioning		5.20	7.15	11.90	9.30%	12.50%	17%
290	Electrical		4.44	5.95	8.95	9.50%	11.80%	13.90%
310	Total: Mechanical & Electrical		10.20	14.80	22.45	21.10%	29.70%	36%
900	Per pupil, total cost	Pupil	2,500	15,100	22,300			
950	Total: Mechanical & Electrical	"	845	2,050	5,250			
83-001	**SPORTS ARENAS**	S.F.	35.85	45.15	57.60			
002	Total project costs	C.F.	1.95	3.55	4.50			
272	Plumbing	S.F.	1.83	3.15	5.60	4.30%	6.30%	8.50%
277	Heating, ventilating, air conditioning		3.81	5.30	6.90%	5.80%	10.20%	13.50%
290	Electrical		2.96	4.96	6.10	7.10%	9.70%	12.20%
310	Total: Mechanical & Electrical		6.55	11.65	14.95	13.40%	22.50%	30.80%
85-001	**SUPERMARKETS**		32.80	37.75	43.45			
002	Total project costs	C.F.	1.82	2.15	2.76			

For expanded coverage of these items see *Means' Square Foot Costs Data 1986*

327

17.1 S.F., C.F. and % of Total Costs

				UNIT COSTS			% OF TOTAL		
			UNIT	1/4	MEDIAN	3/4	1/4	MEDIAN	3/4
85	272	Plumbing	S.F.	1.85	2.36	2.74	5%	6%	6.90%
	277	Heating, ventilating, air conditioning		2.25	3.25	3.78	8.50%	8.60%	9.50%
	290	Electrical		3.86	4.71	5.60	10.40%	12.50%	13.40%
	310	Total: Mechanical & Electrical		6.10	8.80	11	17.80%	21.70%	27.60%
86-001		SWIMMING POOLS		50.35	65.15	91.85			
	002	Total project costs	C.F.	4.39	5.10	5.95			
	272	Plumbing	S.F.	3.53	5.80	8.10	4.60%	9.60%	12.40%
	290	Electrical		3.83	5.40	8.15	6.50%	7.60%	8%
	310	Total: Mechanical & Electrical		8.85	15.95	31.30	17.50%	24.90%	31%
87-001		TELEPHONE EXCHANGES	S.F.	72.80	99.65	132			
	002	Total project costs	C.F.	4.41	6.55	9.45			
	272	Plumbing	S.F.	2.52	4.33	6.40	3.50%	5.70%	6.60%
	277	Heating, ventilating, air conditioning		6.10	14.05	17.50	11.70%	16%	18.40%
	290	Electrical		6.65	11.40	20.75	10.70%	14%	17.80%
	310	Total: Mechanical & Electrical		13.85	20.30	40.05	19.80%	27%	34.70%
89-001		TERMINALS Bus		33.15	43.90	60.20			
	002	Total project costs	C.F.	1.63	2.84	3.63			
	272	Plumbing	S.F.	1.10	2.48	3.81	2.30%	7.20%	8.80%
	290	Electrical		.77	2.28	5.50	3.70%	7.50%	10.50%
	310	Total: Mechanical & Electrical		1.74	4.19	8.50	8.30%	16.40%	19.60%
91-001		THEATERS	S.F.	41.95	53.25	81.55			
	002	Total project costs	C.F.	2.19	3.06	4.53			
	272	Plumbing	S.F.	1.39	1.62	4.87	2.90%	4.60%	6.10%
	277	Heating, ventilating, air conditioning		3.88	5.20	6	7.30%	11.60%	13.30%
	290	Electrical		3.93	5.30	8.20	8%	9.90%	11.80%
	310	Total: Mechanical & Electrical		8.60	11	20.15	16%	24.90%	27.40%
94-001		TOWN HALLS City Halls & Municipal Buildings		50.70	63.70	83.40			
	002	Total project costs	C.F.	3.42	5.05	6.45			
	272	Plumbing	S.F.	1.78	3.48	6.10	4.20%	5.90%	7.90%
	277	Heating, ventilating, air conditioning		3.78	7.50	8.60	7%	9%	13.20%
	290	Electrical		4.02	6.10	8.50	7.90%	9.40%	11.60%
	310	Total: Mechanical & Electrical		8	13.05	20.70	15.50%	20.80%	29.90%
97-001		WAREHOUSES And Storage Buildings		18.25	25.40	39			
	002	Total project costs	C.F.	.95	1.54	2.50			
	010	Sitework	S.F.	1.93	3.78	5.70	6.20%	12.50%	18.90%
	050	Masonry		1.16	2.88	5.75	5%	9.10%	14.20%
	180	Equipment		.30	.61	2.39	.70%	2.40%	5.60%
	272	Plumbing		.62	1.06	2.13	2.90%	4.70%	6.50%
	273	Heating & ventilating		.71	1.84	2.69	2.40%	5%	8.80%
	290	Electrical		1.13	2.08	3.60	5.10%	7.50%	10.10%
	310	Total: Mechanical & Electrical		2.09	3.55	8.10	9.60%	15%	21.40%
99-001		WAREHOUSE & OFFICES Combination	S.F.	21.65	27.75	40			
	002	Total project costs	C.F.	1.15	1.68	2.48			
	180	Equipment	S.F.	.28	.71	1.20	1%	2.30%	2.80%
	272	Plumbing		.87	1.49	2.30	3.60%	4.60%	6.20%
	277	Heating, ventilating, air conditioning		1.40	2.18	3.04	5%	5.60%	9.50%
	290	Electrical		1.52	2.24	3.52	5.90%	7.90%	9.90%
	310	Total: Mechanical & Electrical		2.88	4.90	7.55	11.50%	16%	21.30%

For expanded coverage of these items see *Means' Square Foot Costs Data 1986*

18.1 Repair & Remodeling

		CREW	DAILY OUTPUT	UNIT	BARE COSTS EQUIP.	LABOR	TOTAL	TOTAL INCL O&P
08-001	CONTINGENCIES See division 1.1-11 (143)							
12-001	**CUTOUTS** Openings to 5 S.F., interior walls, not incl. re-framing							
601								
610	Drywall to 5/8" thick	F-1	24	Ea.	.36	6.65	7.01	10.05
620	Paneling to 3/4" thick		20		.43	8	8.43	12.05
630	Plaster on gypsum lath		20		.43	8	8.43	12.05
634	On wire lath	↓	14	↓	.61	11.45	12.06	17.25
700	Wood frame, openings to 5 S.F., not including re-framing							
710								
720	Floors, incl. subfloor, underlayment, wood or resilient flooring							
721	up to 2" thick	F-1	5	Ea.	1.72	32	33.72	48
730	Roofs, sheathing to 1" thick		6		1.43	27	28.43	40
740	Walls, sheathing to 1" thick	↓	7	↓	1.23	23	24.23	34
15-001	**DEMOLITION** See division 2.1 and 2.2							

					BARE COSTS MAT.	INST.	TOTAL	
20-001	**FACTORS** To adjust figures in other sections of this							
002	book for repair and remodeling projects:							
050	Cut & patch to match existing construction, add, minimum (143)			Costs	2%	3%		
055	Maximum				5%	9%		
080	Dust protection, add, minimum				1%	2%		
085	Maximum				4%	11%		
110	Equipment usage curtailment, add, minimum				1%	1%		
115	Maximum				3%	10%		
140	Material handling & storage limitation, add, minimum				1%	1%		
145	Maximum				6%	7%		
170	Protection of existing work, add, minimum				2%	2%		
175	Maximum				5%	7%		
200	Shift work requirements, add, minimum					5%		
205	Maximum					30%		
230	Temporary shoring and bracing, add, minimum				2%	5%		
235	Maximum			↓	5%	12%		
25-001	**CLEANING MASONRY** No staging included							
020	Chemical cleaning, brush and wash, average	D-1	1,000	S.F.	.04	.29	.33	.46
040	High pressure, water only, average	B-9	3,000			.27	.27	.37
080	Water and chemical, average	"	1,250	↓	.04	.64	.68	.93
120	Sand blasting, see division 4.2-44							
140	Steam cleaning, see division 4.2-52							
150	Washing brick, see division 4.2-68							
30-001	**CAULKING MASONRY** No staging included							
002	Re-caulk only, 3/8" deep x 5/8" wide							
005	joints, 82 L.F. per 12 tubes, oil base	1 Bric	225	L.F.	.23	.73	.96	1.29
010	Butyl		205		.49	.80	1.29	1.68
020	Polysulfide		200		.38	.82	1.20	1.59
030	Silicone rubber	↓	195	↓	.81	.84	1.65	2.09
040	See also division 7.1-20							
35-001	**NEEDLE MASONRY** Including shoring							
040	Block, concrete, 8" thick	B-9	7.10	Ea.	33	115	148	195
080	Brick, 4" thick with 8" backup block		5.70		33	140	173	230
100	Brick solid, 8" thick	↓	6.20	↓	33	130	163	215
37-001	**PAINTING** See division 9.8							
40-001	**PATCHING CONCRETE** See division 3.3-37							
45-001	**POINTING MASONRY**							
002								

For expanded coverage of these items see Means' *Repair & Remodeling Cost Data 1986*

18.1 Repair & Remodeling

			CREW	DAILY OUTPUT	UNIT	BARE COSTS MAT.	BARE COSTS INST.	BARE COSTS TOTAL	TOTAL INCL O&P
45	030	Cut and repoint brick, common bond, hard mortar	1 Bric	80	S.F.	.15	2.05	2.20	3.08
	060	Soft old mortar		100	"	.15	1.64	1.79	2.50
	070	Stone work, hard mortar		140	L.F.	.23	1.17	1.40	1.92
	072	Soft old mortar		160	"	.23	1.03	1.26	1.71
	100	Repoint, mask and grout method common bond	↓	90	S.F.	.24	1.82	2.06	2.86
	190								
	200	Scrub coat, sand grout on walls, minimum	1 Bric	120	S.F.	.25	1.37	1.62	2.22
	202	Maximum	"	100	"	.28	1.64	1.92	2.64
50-001		**TOOTHING MASONRY** See division 2.2-08							
55-001		**INTERIOR CLEANUP**							
	010	Window sill repair, to 1.25 S.F. area	1 Pord	32	Ea.	.58	4.81	5.39	7.50
	050	Floor, clean and touchup		1,600	S.F.		.10	.10	.14L
	100	Wax and polish		2,400		.06	.06	.12	.16
	150	Wallpaper removal, 3 layer, minimum		400		.05	.39	.44	.60
	200	Maximum		560		.05	.28	.33	.45
	250	Wash walls and ceilings	↓	2,000	↓	.01	.08	.09	.12
60-001		**REMOVE EXISTING LEAD PAINT**							
	002	Refinish with 2 coats of paint							
	005	Baseboard, to 6" wide	1 Pord	190	L.F.	.13	.81	.94	1.30
	010	Balustrades, one side		150	S.F.	.28	1.03	1.31	1.77
	030	Cabinets, simple design		85		.13	1.81	1.94	2.72
	100	Ornate design		40		.15	3.85	4	5.65
	150	Cornice, simple design	↓	65	↓	.13	2.37	2.50	3.52
	170								
	200	Doors, one side, flush	1 Pord	125	S.F.	.13	1.23	1.36	1.90
	250	Four panel		95		.13	1.62	1.75	2.45
	300	Fence, picket, one side		80		.14	1.93	2.07	2.90
	350	Grilles, one side, simple design		95	↓	.15	1.62	1.77	2.47
	400	Pipes, to 4" diameter		200	L.F.	.13	.77	.90	1.24
	450	To 8" diameter		90	"	.25	1.71	1.96	2.71
	500	Siding	↓	170	S.F.	.13	.91	1.04	1.43
	550	Windows, one side only, double hung							
	600	1/1 light, 24" x 48" high	1 Pord	12	Ea.	1.01	12.85	13.86	19.40
	650	36" x 72" high		8		2.30	19.25	21.55	30
	700	Colonial window, 6/6 light, 24" x 48" high		7		1.30	22	23.30	33
	750	36" x 72" high		4		2.55	39	41.55	58
	800	12/12 light, 24" x 48" high	↓	5	↓	1.90	31	32.90	46
62-001		**RUBBISH HANDLING** See division 2.1-43							
65-001		**SAW CUTTING** See division 2.1-44 or 4.2-46							
70-001		**SCRAPE AND SEAL AFTER FIRE DAMAGE**							
	003								
	005	Boards, floor, 1" x 4"	1 Pord	145	L.F.	.07	1.06	1.13	1.59
	010	1" x 6"		110		.08	1.40	1.48	2.08
	015	Wall, 1" x 4"		116		.07	1.33	1.40	1.97
	020	1" x 6"		88		.08	1.75	1.83	2.58
	030	Framing, 2" x 4"		110		.08	1.40	1.48	2.08
	100	2" x 6"		90		.11	1.71	1.82	2.56
	150	2" x 8"		70		.12	2.20	2.32	3.27
	200	Heavy framing 3" x 4"		110		.08	1.40	1.48	2.08
	250	4" x 6"		90		.12	1.71	1.83	2.57
	300	4" x 10"		50	↓	.16	3.08	3.24	4.56
	350	For sealing only, minimum		889	S.F.	.08	.17	.25	.33
	400	Maximum	↓	470	"	.08	.33	.41	.55

For expanded coverage of these items see Means' *Repair & Remodeling Cost Data 1986*.

18.1 Repair & Remodeling

		CREW	DAILY OUTPUT	UNIT	BARE COSTS MAT.	BARE COSTS INST.	BARE COSTS TOTAL	TOTAL INCL O&P
450	For sand blasting, see division 4.2-44							
75-001	**REMOVE AND RESET DOORS**							
030	Combination, storm and screen	F-2	14	Ea.		24	24	34
160	Entrance door, flush birch, solid core	"	10	"		34	34	48
200								
400	Interior passage door, solid core	F-2	13	Ea.		26	26	37
460	Hollow core	"	14	"		24	24	34
80-001	**RETROFIT GLASS**							
002	Additional plate glass in existing metal sash							
010	Minimum	1 Glaz	50	S.F.	1	3.22	4.22	5.70
020	Maximum	"	25	"	24	6.45	30.45	36
85-001	**ROOF REPAIR**							
002								
010	Clay, mission tile, extensive replacement	1 Rots	1	Sq.	480	150	630	760
015	Individual pieces		19	Ea.	2.95	8	10.95	15.45
020	Slate, Pennsylvania tile, extensive replacement		1	Sq.	590	150	740	880
025	Individual pieces		19	Ea.	4.25	8	12.25	16.90
030	Wood shingles, 16" no. 1 red cedar, extensive replacement		1.50	Sq.	92	100	192	255
035	Individual pieces		40	Ea.	.61	3.79	4.40	6.45
87-001	**TORCH CUTTING** See division 2.1-60							
90-001	**WINDOW GLASS** Replace broken window lites							
010	1/8" float glass (9 S.F. maximum size)	1 Glaz	48	S.F.	1.25	3.36	4.61	6.15
020	3/16" float glass (16 S.F. maximum size)	"	48	"	1.50	3.36	4.86	6.45
95-001	**WOOD FRAMING**							
010	Sister, joist or rafter, 2" x 4"	F-2	940	L.F.	.25	.36	.61	.79
020	2" x 6"		800		.37	.42	.79	1.01
030	2" x 8"		640		.56	.53	1.09	1.37

For expanded coverage of these items see Means' *Repair & Remodeling Cost Data 1986*

CIRCLE REFERENCE NUMBERS

Historical Cost Indexes (Div. 1.1-16)

The table below lists both the Means City Cost Index based on Jan. 1, 1975 = 100 as well as the computed value of an index based on January 1, 1986 costs. Since the Jan. 1, 1986 figure is estimated, space is left to write in the actual index figures as they become available thru either the quarterly "Means Construction Cost Indexes" or as printed in the "Engineering News-Record". To compute the actual index based on Jan. 1, 1986 = 100, divide the Quarterly City Cost Index for a particular year by the actual Jan. 1, 1986 Quarterly City Cost Index. Space has been left to advance the index figures as the year progresses.

Year	"Quarterly City Cost Index" Jan. 1, 1975 = 100		Current Index Based on Jan. 1, 1986 = 100		Year	"Quarterly City Cost Index" Jan. 1, 1975 = 100	Current Index Based on Jan. 1, 1986 = 100		Year	"Quarterly City Cost Index" Jan. 1, 1975 = 100	Current Index Based on Jan. 1, 1986 = 100	
	Est.	Actual	Est.	Actual		Actual	Est.	Actual		Actual	Est.	Actual
Oct. 1986					July 1973	86.3	44.9		July 1957	42.2	22.0	
July 1986					1972	79.7	41.5		1956	40.4	21.0	
April 1986					1971	73.5	38.3		1955	38.1	19.8	
Jan. 1986	192.0		100.0	100.0	1970	65.8	34.3		1954	36.7	19.1	
July 1985		189.1	98.5		1969	61.6	32.1		1953	36.2	18.9	
1984		187.6	97.7		1968	56.9	29.6		1952	35.3	18.4	
1983		183.5	95.6		1967	53.9	28.1		1951	34.4	17.9	
1982		174.3	90.8		1966	51.9	27.0		1950	31.4	16.4	
1981		160.2	83.4		1965	49.7	25.9		1949	30.4	15.8	
1980		144.0	75.0		1964	48.6	25.3		1948	30.4	15.8	
1979		132.3	68.9		1963	47.3	24.6		1947	27.6	14.4	
1978		122.4	63.8		1962	46.2	24.1		1946	23.2	12.1	
1977		113.3	59.0		1961	45.4	23.6		1945	20.2	10.5	
1976		107.3	55.9		1960	45.0	23.4		1944	19.3	10.1	
1975		102.6	53.4		1959	44.2	23.0		1943	18.6	9.7	
1974		94.7	49.3		1958	43.0	22.4		1942	18.0	9.4	

City Cost Indexes (Div. 1.1-06)

Tabulated on the following pages are average construction cost indexes for 162 major U.S. and Canadian cities. Index figures for both material and installation are based on the 30 major city average of 100 and represent the cost relationship as of July 1, 1985. The index for each division is computed from representative material and labor quantities for that division. The weighted average for each city is a weighted total of the components listed above it, but does not include relative productivity between trades or cities.

The material index for the weighted average includes about 100 basic construction materials with appropriate quantities of each material to represent typical "average" building construction projects.

The installation index for the weighted average includes the contribution of about 30 construction trades with their representative man-days in proportion to the material items installed. Also included in the installation costs are the representative equipment costs for those items requiring equipment.

Since each division of the book contains many different items, any particular item multiplied by the particular city index may give incorrect results. However, when all the book costs for a particular division are summarized and then factored, the result should be very close to the actual costs for that particular division for that city.

If a project has a preponderance of materials from any particular division (say structural steel), then the weighted average index should be adjusted in proportion to the value of the factor for that division.

19.1 CITY COST INDEXES

		ALABAMA											ALASKA			ARIZONA		
DIVISION		BIRMINGHAM			HUNTSVILLE			MOBILE			MONTGOMERY			ANCHORAGE			PHOENIX		
		MAT.	INST.	TOTAL	MAT.	INST.	TOTAL	MAT.	INST.	TOTAL	MAT.	INST.	TOTAL	MAT.	INST.	TOTAL	MAT.	INST.	TOTAL
2	SITE WORK	96.7	90.8	94.1	115.6	92.1	104.9	118.4	87.0	104.2	88.2	89.9	88.9	154.7	129.1	143.1	88.2	96.6	92.0
3.1	FORMWORK	90.4	72.2	76.2	92.4	73.7	77.8	97.0	74.6	79.5	102.2	72.9	79.3	114.1	145.8	138.8	109.1	92.7	96.3
3.2	REINFORCING	94.6	76.8	87.0	95.8	73.5	86.4	83.0	73.1	78.8	83.0	76.8	80.3	117.8	138.8	126.7	113.3	100.3	107.8
3.3	CAST IN PLACE CONC.	89.4	91.8	90.9	102.0	93.5	96.8	100.1	93.1	95.7	101.3	92.9	96.1	226.0	112.5	155.4	109.1	93.7	99.5
3	CONCRETE	90.7	82.8	85.6	98.7	84.0	89.2	95.7	84.1	88.2	97.5	83.7	88.5	179.7	127.8	146.2	110.0	93.9	99.6
4	MASONRY	79.9	75.0	76.1	87.3	73.6	76.7	92.4	79.7	82.7	86.1	72.3	75.5	139.4	148.4	146.3	93.0	89.7	90.5
5	METALS	96.0	82.4	91.1	100.5	80.3	93.2	93.8	80.6	89.0	94.7	82.3	90.2	116.7	129.8	121.4	98.9	97.7	98.5
6	WOOD & PLASTICS	91.5	73.4	81.4	104.6	73.9	87.4	92.0	77.4	83.8	101.2	73.2	85.5	117.8	145.2	133.2	100.8	92.2	96.0
7	MOISTURE PROTECTION	84.5	68.8	79.6	92.3	70.0	85.3	87.1	70.8	82.0	88.3	69.3	82.3	102.5	146.8	116.4	92.4	87.7	90.9
8	DOORS, WINDOWS, GLASS	91.2	73.8	82.1	101.6	73.6	86.9	99.2	75.5	86.8	98.5	73.1	85.2	129.6	145.4	137.9	103.4	88.6	95.6
9.1	LATH & PLASTER	96.2	69.7	76.0	86.4	73.9	76.8	92.0	81.4	83.9	108.5	75.0	82.9	120.4	146.0	139.9	90.6	92.2	91.8
9.2	DRYWALL	100.5	72.9	87.4	110.4	73.5	92.9	92.4	78.1	85.6	100.6	73.7	87.9	122.0	146.9	133.8	90.2	91.2	90.7
9.5	ACOUSTICAL WORK	98.8	73.0	84.7	101.2	73.6	86.1	94.2	76.9	84.7	94.2	72.3	82.2	125.5	146.9	137.2	103.5	91.3	96.9
9.6	FLOORING	111.9	78.1	102.9	94.0	73.6	88.6	113.9	81.9	105.4	99.9	72.3	92.6	117.2	146.9	125.1	92.8	92.8	92.8
9.8	PAINTING	107.0	72.2	79.0	110.6	73.5	80.8	121.5	77.8	86.4	119.8	72.2	81.6	123.1	146.8	142.2	94.9	89.0	90.2
9	FINISHES	103.5	72.8	86.9	105.4	73.6	88.2	100.3	78.4	88.5	102.1	73.1	86.4	121.3	146.8	135.1	92.3	90.6	91.4
10-14	TOTAL DIV. 10-14	100.0	76.0	92.7	100.0	73.6	92.0	100.0	76.1	92.7	100.0	72.3	91.6	100.0	146.9	114.1	100.0	99.5	99.8
15	MECHANICAL	96.4	77.3	86.5	99.7	76.5	87.7	97.5	77.4	87.1	99.2	73.3	85.8	107.7	146.6	127.9	98.8	83.8	91.0
16	ELECTRICAL	94.0	77.3	82.1	91.9	77.1	81.4	89.7	78.2	81.5	90.7	72.2	77.5	110.8	146.8	136.5	105.1	89.4	93.9
1-16	WEIGHTED AVERAGE	94.3	78.0	85.4	99.7	77.8	87.7	97.3	79.6	87.6	96.1	76.2	85.2	125.3	141.3	134.1	99.2	90.6	94.5

		ARIZONA			ARKANSAS						CALIFORNIA								
DIVISION		TUCSON			FORT SMITH			LITTLE ROCK			ANAHEIM			BAKERSFIELD			FRESNO		
		MAT.	INST.	TOTAL	MAT.	INST.	TOTAL	MAT.	INST.	TOTAL	MAT.	INST.	TOTAL	MAT.	INST.	TOTAL	MAT.	INST.	TOTAL
2	SITE WORK	106.6	96.9	102.2	96.7	93.0	95.0	103.3	95.6	99.8	101.1	112.6	106.3	93.2	113.6	102.4	91.7	120.5	104.7
3.1	FORMWORK	102.1	92.5	94.6	102.6	72.8	79.4	95.6	75.3	79.8	94.9	130.7	122.8	113.7	130.7	127.0	99.6	122.1	117.1
3.2	REINFORCING	95.1	100.3	97.3	124.6	72.3	102.4	117.8	74.8	99.6	99.4	128.5	111.7	96.1	128.5	109.8	106.5	128.5	115.8
3.3	CAST IN PLACE CONC.	105.7	98.3	101.1	90.6	92.8	92.0	98.5	93.8	95.6	109.5	109.3	109.4	103.4	109.5	107.2	93.1	107.2	101.9
3	CONCRETE	102.7	96.2	98.5	100.5	83.2	89.3	102.2	84.9	91.0	104.3	119.3	114.0	103.9	119.5	114.0	97.4	114.9	108.7
4	MASONRY	90.2	89.7	89.8	93.6	72.4	77.3	93.7	74.9	79.2	106.5	125.2	120.9	99.6	117.3	113.2	111.1	119.6	117.7
5	METALS	91.0	91.6	91.0	97.0	79.7	90.7	106.6	81.5	97.5	99.4	121.5	107.4	99.8	121.8	107.7	95.2	121.7	104.8
6	WOOD & PLASTICS	107.3	91.6	98.5	106.7	73.3	88.0	94.3	75.7	83.9	95.7	128.0	113.8	96.4	128.0	114.1	96.9	118.6	109.1
7	MOISTURE PROTECTION	106.8	76.8	97.3	84.7	72.4	80.8	84.2	74.9	81.3	108.2	131.4	115.5	84.7	116.8	94.8	107.7	110.5	108.6
8	DOORS, WINDOWS, GLASS	88.7	88.6	88.7	93.5	72.4	82.4	95.9	74.9	84.8	94.0	127.2	111.4	100.6	124.3	113.0	101.6	121.3	111.9
9.1	LATH & PLASTER	107.6	89.9	94.1	92.9	73.1	77.8	98.5	75.5	81.0	97.1	131.7	123.5	96.8	115.6	111.2	101.8	123.9	118.6
9.2	DRYWALL	87.1	91.2	89.1	95.2	72.4	84.4	114.8	74.9	95.8	97.5	129.0	112.4	98.3	123.9	110.5	98.7	121.5	109.5
9.5	ACOUSTICAL WORK	114.9	91.3	102.0	84.5	72.4	77.9	84.5	74.9	79.2	82.2	129.0	107.8	89.7	129.0	111.2	97.5	119.7	109.6
9.6	FLOORING	109.8	92.8	105.3	89.1	72.4	84.7	88.3	74.9	84.7	117.5	125.8	119.6	112.0	118.9	113.8	88.3	119.7	96.6
9.8	PAINTING	96.5	88.7	90.3	111.0	72.3	80.0	104.8	65.9	73.5	108.3	125.3	122.0	120.2	116.5	117.3	108.1	105.7	106.2
9	FINISHES	95.9	90.4	92.9	94.5	72.4	82.6	105.0	71.7	87.0	101.9	127.7	115.8	102.9	120.9	112.6	97.3	115.8	107.3
10-14	TOTAL DIV. 10-14	100.0	98.8	99.6	100.0	72.4	91.6	100.0	74.9	92.4	100.0	125.7	107.7	100.0	122.8	106.9	100.0	143.9	113.2
15	MECHANICAL	98.8	83.8	91.0	97.5	72.5	84.5	97.0	73.1	84.6	96.7	125.7	111.8	94.9	116.6	106.1	92.4	125.2	109.4
16	ELECTRICAL	102.3	88.7	92.6	99.2	75.5	82.3	93.4	76.2	81.1	98.7	125.3	117.7	106.1	116.5	113.5	109.7	119.6	116.8
1-16	WEIGHTED AVERAGE	98.9	90.7	94.4	96.7	76.4	85.5	98.7	77.8	87.2	100.6	123.9	113.4	98.5	118.8	109.6	98.6	120.3	110.5

		CALIFORNIA																	
DIVISION		LOS ANGELES			OXNARD			RIVERSIDE			SACRAMENTO			SAN DIEGO			SAN FRANCISCO		
		MAT.	INST.	TOTAL	MAT.	INST.	TOTAL	MAT.	INST.	TOTAL	MAT.	INST.	TOTAL	MAT.	INST.	TOTAL	MAT.	INST.	TOTAL
2	SITE WORK	95.2	114.9	104.1	98.0	104.3	100.9	95.2	111.4	102.5	83.1	105.8	93.4	98.5	108.3	102.9	103.1	113.9	108.0
3.1	FORMWORK	112.8	131.1	127.1	90.0	131.0	122.0	102.5	130.7	124.5	100.9	127.1	121.4	105.8	128.5	123.5	104.9	134.6	128.1
3.2	REINFORCING	62.7	128.5	90.6	99.4	128.5	111.7	124.6	128.5	126.2	99.4	128.5	111.7	145.3	128.5	138.2	123.5	128.5	125.6
3.3	CAST IN PLACE CONC.	92.4	112.3	104.8	102.5	110.0	107.2	102.5	109.6	107.0	116.1	106.4	110.1	103.8	104.1	104.0	101.6	116.3	110.7
3	CONCRETE	90.0	121.1	110.1	99.3	119.8	112.6	107.3	119.5	115.2	109.3	116.4	113.9	113.3	115.8	114.9	107.1	124.5	118.3
4	MASONRY	110.4	125.2	121.8	85.5	125.4	119.1	102.9	124.8	119.7	101.4	125.9	120.3	109.1	113.3	112.4	130.9	138.0	136.3
5	METALS	101.7	122.5	109.2	105.6	121.8	111.5	99.5	121.6	107.5	111.4	121.7	115.1	99.2	120.5	106.9	104.4	124.8	111.8
6	WOOD & PLASTICS	100.6	129.1	116.6	92.6	128.5	112.8	98.5	128.0	115.0	78.4	125.0	104.6	93.9	125.3	111.5	95.0	134.2	117.0
7	MOISTURE PROTECTION	103.9	131.8	112.6	89.8	131.4	102.9	90.3	127.5	102.0	85.3	122.0	96.9	94.6	119.6	102.5	101.0	129.2	109.9
8	DOORS, WINDOWS, GLASS	101.6	127.2	115.1	103.1	127.2	115.8	103.7	127.2	116.1	92.3	126.5	110.2	107.5	120.6	114.4	113.2	133.5	123.8
9.1	LATH & PLASTER	96.3	131.7	123.3	97.5	126.6	119.7	97.5	125.7	119.0	98.8	123.1	117.4	103.2	114.8	112.0	101.7	144.7	134.5
9.2	DRYWALL	90.0	129.0	108.5	98.8	127.4	112.4	94.8	129.0	111.0	97.2	123.6	109.7	106.0	124.1	114.6	81.7	136.0	107.4
9.5	ACOUSTICAL WORK	98.4	129.0	115.1	88.4	129.0	110.6	88.4	129.0	110.6	86.6	126.0	108.1	100.5	126.3	114.6	100.5	135.8	119.8
9.6	FLOORING	96.6	125.5	104.2	95.2	125.5	103.2	95.2	125.5	103.2	85.7	126.0	96.3	98.0	117.5	103.2	106.1	138.1	114.6
9.8	PAINTING	85.6	126.9	118.8	92.3	125.1	118.7	100.5	124.5	119.8	112.4	126.0	123.3	92.1	128.5	121.3	100.6	140.5	132.6
9	FINISHES	91.9	128.2	111.5	96.5	126.6	112.7	95.0	127.0	112.3	95.3	124.8	111.2	102.3	124.8	114.4	91.1	138.2	116.5
10-14	TOTAL DIV. 10-14	100.0	126.1	107.8	100.0	125.7	107.7	100.0	125.5	107.7	100.0	146.0	113.9	100.0	124.2	107.3	100.0	150.5	115.2
15	MECHANICAL	97.4	126.2	112.3	98.7	125.7	112.8	96.5	124.7	112.1	98.2	126.4	112.8	102.9	125.0	114.4	100.8	160.0	131.5
16	ELECTRICAL	101.8	133.2	124.2	98.7	125.1	117.5	98.2	124.5	117.0	109.7	125.9	121.3	106.1	113.8	111.6	110.7	152.7	140.6
1-16	WEIGHTED AVERAGE	98.1	125.6	113.2	98.8	123.6	112.4	99.0	123.8	112.6	99.2	123.6	112.5	103.4	118.5	111.7	104.0	139.3	123.3

19.1 CITY COST INDEXES

		CALIFORNIA									COLORADO						CONNECTICUT		
	DIVISION	SANTA BARBARA			STOCKTON			VALLEJO			COLORADO SPRINGS			DENVER			BRIDGEPORT		
		MAT.	INST.	TOTAL	MAT.	INST.	TOTAL	MAT.	INST.	TOTAL	MAT.	INST.	TOTAL	MAT.	INST.	TOTAL	MAT.	INST.	TOTAL
2	SITE WORK	121.3	112.4	117.3	117.0	113.0	115.2	103.9	112.9	108.0	95.9	97.9	96.8	103.4	104.9	104.1	117.7	100.6	110.0
3.1	FORMWORK	105.0	130.7	125.1	96.6	123.9	117.9	103.6	131.9	125.7	93.9	90.3	91.1	93.8	87.1	88.6	111.4	101.7	103.8
3.2	REINFORCING	99.4	128.5	111.7	83.5	128.5	102.5	99.4	128.5	111.7	96.1	91.2	94.0	112.1	91.2	103.2	112.8	101.9	108.2
3.3	CAST IN PLACE CONC.	125.8	121.6	123.1	103.1	106.8	105.4	103.1	106.4	105.2	114.0	97.5	103.7	126.4	94.3	106.5	103.0	101.7	102.2
3	CONCRETE	115.8	125.7	122.2	97.5	115.4	109.0	102.3	118.3	112.7	106.0	94.1	98.3	116.7	91.2	100.2	106.9	101.7	103.5
4	MASONRY	115.7	123.6	121.8	110.0	106.1	107.0	107.5	132.1	126.4	104.0	90.0	93.2	106.1	94.3	97.0	104.1	104.4	104.3
5	METALS	96.6	125.5	107.0	91.9	121.4	102.6	89.0	121.6	100.8	91.7	94.4	92.7	95.0	93.6	94.5	92.2	101.4	95.5
6	WOOD & PLASTICS	108.3	128.2	119.5	86.0	121.2	105.8	99.4	131.0	117.1	86.1	90.3	88.4	90.1	86.7	88.2	107.1	101.2	103.8
7	MOISTURE PROTECTION	89.5	131.0	102.6	89.0	114.5	97.0	87.1	125.1	99.1	85.4	90.1	86.9	117.0	85.8	107.2	100.1	103.4	101.1
8	DOORS, WINDOWS, GLASS	104.5	127.2	116.4	96.0	123.2	110.3	99.7	131.9	116.6	98.9	92.2	95.4	91.6	90.7	91.1	103.0	106.7	104.9
9.1	LATH & PLASTER	104.0	119.3	115.7	99.9	122.1	116.8	99.9	126.9	120.5	101.7	90.2	93.0	92.5	102.4	100.1	105.9	105.2	105.4
9.2	DRYWALL	122.5	129.1	125.6	107.2	122.4	114.4	107.9	127.3	117.1	93.1	90.0	91.6	88.6	92.8	90.6	113.0	102.0	107.8
9.5	ACOUSTICAL WORK	88.4	129.0	110.6	86.6	122.0	105.9	89.6	132.2	112.9	95.4	90.0	92.4	96.4	86.6	91.0	106.8	102.0	104.2
9.6	FLOORING	101.5	124.3	107.6	84.8	113.8	92.5	84.1	137.4	98.2	107.9	90.0	103.2	99.5	105.7	101.1	85.3	105.1	90.6
9.8	PAINTING	121.6	123.2	122.9	103.2	105.9	105.4	106.0	132.1	127.0	117.5	90.0	95.4	104.6	103.2	103.5	121.5	102.0	105.8
9	FINISHES	114.5	126.1	120.8	99.9	116.0	108.6	100.7	130.1	116.5	99.3	90.0	94.2	93.4	97.4	95.6	106.9	102.4	104.5
10-14	TOTAL DIV. 10-14	100.0	125.0	107.5	100.0	144.7	113.5	100.0	148.1	114.5	100.0	90.0	96.9	100.0	92.7	97.8	100.0	104.1	101.2
15	MECHANICAL	98.7	125.3	112.5	97.1	119.3	108.6	95.8	132.0	114.6	98.0	91.3	94.5	97.2	95.2	96.2	104.1	102.3	103.2
16	ELECTRICAL	97.9	123.2	116.0	100.3	121.9	115.7	110.0	132.1	125.8	97.9	89.9	92.2	95.2	91.4	92.5	102.8	101.9	102.2
1-16	WEIGHTED AVERAGE	104.8	124.7	115.7	98.6	117.6	109.0	98.8	128.1	114.8	98.0	91.8	94.6	101.6	93.5	97.2	103.6	102.7	103.1

		CONNECTICUT												DELAWARE			D.C.		
	DIVISION	HARTFORD			NEW HAVEN			STAMFORD			WATERBURY			WILMINGTON			WASHINGTON		
		MAT.	INST.	TOTAL	MAT.	INST.	TOTAL	MAT.	INST.	TOTAL	MAT.	INST.	TOTAL	MAT.	INST.	TOTAL	MAT.	INST.	TOTAL
2	SITE WORK	97.4	101.3	99.2	113.6	98.7	106.9	121.3	101.1	112.2	106.2	100.2	103.5	111.7	102.2	107.4	91.8	92.2	92.0
3.1	FORMWORK	100.3	104.1	103.3	101.3	102.0	101.8	102.7	102.8	102.8	93.3	101.0	99.3	96.5	103.9	102.3	106.3	82.8	88.0
3.2	REINFORCING	115.3	104.8	110.9	112.8	101.9	108.2	130.8	102.9	119.0	115.3	100.7	109.1	112.8	102.8	108.6	101.7	83.0	93.8
3.3	CAST IN PLACE CONC.	98.6	102.0	100.7	95.3	100.5	98.5	118.0	101.7	107.9	115.1	101.7	106.8	99.2	114.4	108.7	101.8	89.1	93.9
3	CONCRETE	102.6	103.1	102.9	100.3	101.2	100.9	117.7	102.2	107.7	110.8	101.4	104.7	101.7	109.3	106.6	102.7	86.1	91.9
4	MASONRY	97.3	102.0	100.9	120.2	101.9	106.1	120.1	103.3	107.2	107.0	101.8	103.0	100.1	94.0	95.4	90.6	104.6	101.4
5	METALS	93.0	103.5	96.8	85.5	101.4	91.3	86.1	102.2	91.9	88.5	100.5	92.8	86.0	107.1	93.7	102.9	86.8	97.0
6	WOOD & PLASTICS	113.5	104.0	108.2	115.2	101.9	107.7	110.8	102.8	106.4	113.0	100.7	106.1	102.8	104.0	103.5	105.3	84.0	93.4
7	MOISTURE PROTECTION	101.3	105.4	102.6	87.8	102.0	92.3	87.5	104.2	92.8	88.1	102.7	92.7	88.8	116.2	97.4	106.4	88.9	100.9
8	DOORS, WINDOWS, GLASS	93.1	108.7	101.2	99.6	106.7	103.3	95.3	107.3	101.6	89.0	105.9	97.9	87.2	106.1	97.1	98.8	88.2	93.3
9.1	LATH & PLASTER	113.9	106.6	108.3	121.6	101.9	106.5	98.6	105.4	103.8	114.0	104.8	107.0	94.2	95.4	95.1	99.2	103.2	102.2
9.2	DRYWALL	108.3	104.9	106.7	113.0	102.0	107.8	117.9	103.0	110.8	113.5	100.8	107.4	102.4	104.0	103.1	121.4	83.7	103.5
9.5	ACOUSTICAL WORK	106.8	104.9	105.7	106.8	102.0	104.2	106.2	103.0	104.4	87.8	100.8	94.9	90.9	104.2	98.1	107.6	83.8	94.6
9.6	FLOORING	93.8	105.6	96.9	98.2	105.1	100.0	95.4	106.5	98.4	102.2	104.9	102.9	93.7	97.6	94.7	91.5	92.2	91.7
9.8	PAINTING	107.3	104.8	105.3	121.4	95.2	100.3	121.4	103.0	106.6	114.5	100.7	103.5	100.7	95.2	96.2	102.8	93.7	95.5
9	FINISHES	104.9	105.0	105.0	110.2	99.8	104.6	111.8	103.4	107.2	109.0	101.3	104.8	99.1	99.9	99.6	111.1	89.0	99.2
10-14	TOTAL DIV. 10-14	100.0	105.0	101.5	100.0	104.1	101.2	100.0	104.4	101.3	100.0	103.7	101.1	100.0	107.7	102.3	100.0	86.0	95.7
15	MECHANICAL	102.0	102.1	102.0	102.3	101.4	101.9	101.8	106.8	104.4	101.8	97.3	99.1	100.3	103.8	102.1	101.5	84.5	92.7
16	ELECTRICAL	98.6	104.8	103.0	92.9	101.9	99.4	92.9	102.9	100.1	91.6	100.7	98.1	105.8	102.8	103.7	99.0	90.3	92.8
1-16	WEIGHTED AVERAGE	100.1	103.6	102.0	101.0	101.7	101.4	103.7	103.8	103.8	100.4	100.9	100.7	98.4	103.8	101.4	101.5	89.7	95.1

		FLORIDA															GEORGIA		
	DIVISION	FT LAUDERDALE			JACKSONVILLE			MIAMI			ORLANDO			TAMPA			ATLANTA		
		MAT.	INST.	TOTAL	MAT.	INST.	TOTAL	MAT.	INST.	TOTAL	MAT.	INST.	TOTAL	MAT.	INST.	TOTAL	MAT.	INST.	TOTAL
2	SITE WORK	104.6	93.1	99.4	114.5	83.9	100.6	93.8	89.7	92.0	93.6	90.3	92.1	106.4	93.6	100.6	101.9	91.5	97.2
3.1	FORMWORK	98.9	89.3	91.4	95.9	73.7	78.6	100.2	90.6	92.7	95.2	75.0	79.5	91.3	80.9	83.2	85.6	72.3	75.2
3.2	REINFORCING	100.2	88.8	95.4	87.4	76.0	82.6	100.2	90.3	96.0	100.2	74.5	89.4	100.2	80.5	91.9	89.6	69.6	81.1
3.3	CAST IN PLACE CONC.	91.6	99.5	96.5	97.1	92.1	94.0	88.8	104.1	98.3	94.5	92.4	93.2	99.3	112.4	107.4	86.1	95.1	91.7
3	CONCRETE	95.0	94.5	94.7	94.7	83.5	87.5	93.6	97.6	96.2	95.9	84.1	88.2	97.9	97.3	97.5	86.8	83.9	84.9
4	MASONRY	97.0	93.6	94.4	89.5	76.1	79.2	91.1	90.4	90.5	88.1	63.1	68.9	95.1	80.6	83.9	88.7	74.7	77.9
5	METALS	87.3	91.9	89.0	95.1	82.9	90.7	86.9	95.4	89.9	86.9	81.8	85.0	98.3	92.0	96.0	111.0	78.6	99.3
6	WOOD & PLASTICS	106.5	89.7	97.1	102.0	76.6	87.8	107.2	90.7	97.9	102.0	75.4	87.1	101.5	81.2	90.1	89.8	75.3	81.7
7	MOISTURE PROTECTION	87.9	90.6	88.8	87.9	74.4	83.7	86.4	86.4	86.4	87.3	68.9	81.5	104.8	68.6	93.4	98.6	69.4	89.4
8	DOORS, WINDOWS, GLASS	87.1	90.0	88.6	90.2	73.5	81.5	93.8	93.0	93.4	88.8	74.6	81.3	97.1	80.6	88.4	92.2	82.2	86.9
9.1	LATH & PLASTER	100.9	94.4	95.9	103.0	70.1	77.9	103.8	91.3	94.2	102.3	69.0	76.9	100.9	81.6	86.1	112.0	78.4	86.4
9.2	DRYWALL	103.1	88.9	96.4	107.1	76.1	92.4	102.7	90.4	96.9	103.1	74.4	89.6	96.9	80.6	89.2	113.2	77.0	96.0
9.5	ACOUSTICAL WORK	91.8	88.9	90.2	91.8	76.1	83.2	100.9	90.4	95.1	91.8	74.6	82.4	93.6	80.6	86.5	91.3	74.1	81.9
9.6	FLOORING	104.0	95.8	101.8	104.0	76.1	96.6	105.2	90.4	101.3	102.7	66.7	93.2	99.8	80.6	94.7	103.2	80.6	97.3
9.8	PAINTING	113.1	88.8	93.6	102.6	76.0	81.3	111.8	90.3	94.6	94.8	74.5	78.5	108.6	80.5	86.1	90.4	97.8	96.3
9	FINISHES	103.4	89.7	96.0	104.6	75.7	89.0	104.1	90.4	96.7	101.3	73.7	86.4	98.6	80.6	88.9	106.8	84.4	94.7
10-14	TOTAL DIV. 10-14	100.0	93.7	98.1	100.0	76.1	92.7	100.0	94.2	98.2	100.0	74.6	92.3	100.0	80.6	94.1	100.0	74.6	92.3
15	MECHANICAL	101.2	88.9	94.8	100.4	81.1	90.4	97.8	94.3	96.0	97.1	79.7	88.1	97.4	81.5	89.1	102.8	78.3	90.1
16	ELECTRICAL	98.3	87.6	90.7	99.2	76.0	82.7	99.8	98.3	98.7	92.8	74.5	79.8	92.8	83.6	86.2	96.5	81.1	85.5
1-16	WEIGHTED AVERAGE	97.1	91.4	93.9	98.1	79.0	87.6	95.6	94.0	94.8	94.6	76.7	84.8	98.8	85.5	91.5	98.9	80.3	88.7

19.1 CITY COST INDEXES

DIVISION		GEORGIA									HAWAII			IDAHO			ILLINOIS		
		COLUMBUS			MACON			SAVANNAH			HONOLULU			BOISE			CHICAGO		
		MAT.	INST.	TOTAL	MAT.	INST.	TOTAL	MAT.	INST.	TOTAL	MAT.	INST.	TOTAL	MAT.	INST.	TOTAL	MAT.	INST.	TOTAL
2	SITE WORK	118.6	87.7	104.6	111.8	89.8	101.8	114.2	87.6	102.1	127.9	106.3	118.1	89.1	97.2	92.7	100.4	103.1	101.6
3.1	FORMWORK	96.3	66.1	72.7	90.3	70.5	74.8	92.8	69.4	74.6	121.6	113.9	115.6	100.6	95.5	96.6	81.5	104.8	99.7
3.2	REINFORCING	97.2	69.6	85.5	86.3	69.6	79.2	107.2	70.4	91.6	117.8	103.7	111.9	117.8	87.7	105.1	105.3	109.3	107.0
3.3	CAST IN PLACE CONC.	108.4	91.0	97.6	102.5	104.0	103.4	95.0	104.9	101.2	113.8	103.3	107.3	96.7	99.6	98.5	101.0	99.2	99.9
3	CONCRETE	103.5	79.4	87.9	96.5	87.9	90.9	97.2	88.0	91.3	116.3	107.5	110.6	102.1	97.0	98.8	98.0	102.3	100.8
4	MASONRY	91.1	51.7	60.9	82.6	70.0	72.9	90.6	70.5	75.1	125.1	117.6	119.3	100.2	88.1	90.9	92.6	100.8	98.9
5	METALS	96.0	77.8	89.4	90.8	82.1	87.7	100.8	83.2	94.4	110.7	104.2	108.3	93.8	92.0	93.1	89.9	105.5	95.5
6	WOOD & PLASTICS	101.6	66.5	81.9	92.2	71.0	80.3	101.8	71.5	84.8	121.4	115.2	117.9	92.2	94.2	93.3	94.4	103.9	99.7
7	MOISTURE PROTECTION	92.6	66.5	84.4	92.6	68.2	84.9	90.8	70.8	84.5	107.2	115.1	109.7	106.1	90.6	101.2	95.5	108.9	99.7
8	DOORS, WINDOWS, GLASS	90.0	66.1	77.5	93.9	69.9	81.3	91.7	70.5	80.6	115.0	113.4	114.2	105.7	86.4	95.6	110.7	101.7	106.0
9.1	LATH & PLASTER	103.4	58.0	68.7	103.0	70.4	78.1	97.7	71.3	77.5	116.7	123.3	121.8	114.2	90.6	96.2	106.7	100.7	102.1
9.2	DRYWALL	91.9	65.3	79.3	91.9	70.0	81.5	111.9	70.4	92.3	151.2	117.9	135.4	94.6	91.5	93.1	97.3	104.2	100.6
9.5	ACOUSTICAL WORK	102.5	65.3	82.1	102.5	70.0	84.7	102.7	70.5	85.1	120.0	115.6	117.6	114.0	94.0	103.1	97.4	104.4	101.2
9.6	FLOORING	82.3	55.2	75.2	88.7	70.0	83.7	96.7	70.5	89.8	134.3	121.5	130.9	93.6	88.9	92.4	99.5	99.8	99.6
9.8	PAINTING	117.1	65.2	75.4	116.6	70.0	79.1	116.9	70.4	79.6	124.5	115.5	117.3	106.2	87.7	91.4	91.5	88.2	88.9
9	FINISHES	93.3	64.1	77.5	94.7	70.0	81.4	107.9	70.5	87.7	141.4	117.4	128.5	97.5	90.2	93.5	97.4	98.1	97.8
10-14	TOTAL DIV. 10-14	100.0	69.8	90.9	100.0	71.4	91.3	100.0	72.5	91.6	100.0	114.2	104.3	100.0	83.4	95.0	100.0	103.9	101.2
15	MECHANICAL	98.6	68.5	83.0	98.9	72.0	85.0	98.6	71.8	84.7	111.9	112.9	112.4	96.6	88.1	92.2	97.5	95.9	96.7
16	ELECTRICAL	97.0	64.8	74.0	107.5	69.0	80.1	100.8	70.4	79.2	105.2	107.6	106.9	95.8	87.7	90.1	93.5	96.3	95.5
1-16	WEIGHTED AVERAGE	98.5	68.6	82.1	97.1	75.5	85.3	99.4	75.9	86.5	115.1	111.7	113.3	98.5	90.6	94.2	97.5	100.2	99.0

DIVISION		ILLINOIS									INDIANA								
		PEORIA			ROCKFORD			SPRINGFIELD			EVANSVILLE			FORT WAYNE			GARY		
		MAT.	INST.	TOTAL	MAT.	INST.	TOTAL	MAT.	INST.	TOTAL	MAT.	INST.	TOTAL	MAT.	INST.	TOTAL	MAT.	INST.	TOTAL
2	SITE WORK	110.7	98.6	105.2	108.4	100.5	104.8	105.7	97.8	102.1	100.1	99.1	99.6	93.9	97.4	95.5	107.5	97.2	102.8
3.1	FORMWORK	106.1	95.5	97.8	113.7	100.9	103.7	95.7	93.6	94.0	97.1	94.6	95.1	109.0	92.1	95.8	103.7	100.0	100.8
3.2	REINFORCING	112.8	95.4	105.4	113.0	101.1	108.0	92.8	93.4	93.1	93.9	94.3	94.1	93.9	104.3	98.3	93.9	100.5	96.7
3.3	CAST IN PLACE CONC.	86.9	82.7	84.3	97.6	100.4	99.4	101.9	98.5	99.8	113.9	96.0	102.7	91.9	97.6	95.5	97.4	112.6	106.9
3	CONCRETE	96.4	88.8	91.5	104.3	100.7	101.9	98.6	96.1	97.0	106.1	95.3	99.1	95.8	96.1	96.0	97.9	106.6	103.5
4	MASONRY	101.1	95.5	96.8	90.0	101.1	98.6	104.3	93.5	96.0	86.6	107.3	102.5	93.4	91.7	92.1	93.4	100.5	98.9
5	METALS	88.0	90.5	88.9	108.4	100.8	105.7	93.3	95.2	94.0	94.1	95.0	94.4	100.2	101.6	100.7	89.9	105.0	95.3
6	WOOD & PLASTICS	113.6	95.6	103.5	106.8	100.7	103.4	103.2	93.7	97.9	93.0	94.5	93.9	102.5	91.9	96.6	112.6	100.5	105.8
7	MOISTURE PROTECTION	91.8	101.6	94.9	107.1	105.7	106.6	99.0	93.5	97.2	87.8	95.2	90.1	96.3	91.7	94.9	93.6	100.4	95.7
8	DOORS, WINDOWS, GLASS	95.0	95.5	95.2	101.4	101.2	101.3	106.1	93.5	99.5	107.9	94.4	100.8	98.7	91.5	94.9	99.4	100.6	100.0
9.1	LATH & PLASTER	103.8	95.7	97.6	105.1	100.9	101.9	103.5	93.4	95.8	101.2	95.3	96.7	101.9	91.8	94.2	117.4	101.2	105.0
9.2	DRYWALL	110.2	95.5	103.2	94.6	101.2	97.7	116.0	93.5	105.3	113.4	94.4	104.4	100.6	91.7	96.4	100.3	100.6	100.4
9.5	ACOUSTICAL WORK	85.0	95.5	90.7	84.9	101.2	93.8	84.9	93.5	89.6	105.2	94.4	99.3	86.1	91.7	89.1	91.3	100.6	96.4
9.6	FLOORING	104.3	95.5	101.9	123.7	101.2	117.7	115.1	93.5	109.4	96.7	94.4	96.1	98.2	91.7	96.5	99.7	100.6	99.9
9.8	PAINTING	115.9	95.4	99.5	97.2	101.1	100.4	98.3	93.4	94.4	103.7	94.3	96.2	95.1	91.6	92.3	92.6	100.5	99.0
9	FINISHES	107.3	95.5	100.9	100.9	101.1	101.0	111.3	93.4	101.6	107.7	94.4	100.5	98.4	91.7	94.8	99.1	100.6	99.9
10-14	TOTAL DIV. 10-14	100.0	92.2	97.6	100.0	101.2	100.3	100.0	92.7	97.8	100.0	94.4	98.3	100.0	91.7	97.4	100.0	88.9	96.6
15	MECHANICAL	96.1	95.8	95.9	102.0	99.9	100.9	97.1	92.1	94.5	98.2	95.6	96.8	98.7	91.7	95.1	97.7	100.5	99.2
16	ELECTRICAL	93.8	95.4	95.0	92.5	101.1	98.7	94.5	93.4	93.8	94.0	94.3	94.3	96.6	91.6	93.1	100.8	100.5	100.6
1-16	WEIGHTED AVERAGE	97.9	94.0	95.8	102.2	100.9	101.5	100.3	94.0	96.9	99.1	96.0	97.9	97.7	93.3	95.3	98.0	101.4	99.9

DIVISION		INDIANA									IOWA						KANSAS		
		INDIANAPOLIS			SOUTH BEND			TERRE HAUTE			DAVENPORT			DES MOINES			TOPEKA		
		MAT.	INST.	TOTAL	MAT.	INST.	TOTAL	MAT.	INST.	TOTAL	MAT.	INST.	TOTAL	MAT.	INST.	TOTAL	MAT.	INST.	TOTAL
2	SITE WORK	102.4	94.3	98.7	97.8	99.0	98.3	88.7	88.5	88.6	104.1	90.0	97.7	99.9	93.4	96.9	106.8	87.4	98.0
3.1	FORMWORK	108.5	97.1	99.6	99.8	92.8	94.3	103.7	94.0	96.1	94.1	91.9	92.4	102.3	81.4	86.0	92.6	83.6	85.6
3.2	REINFORCING	112.8	102.4	108.4	106.0	92.4	100.3	93.9	93.8	93.9	97.8	90.6	94.8	101.9	82.0	93.5	87.7	98.4	92.2
3.3	CAST IN PLACE CONC.	88.3	95.6	92.8	90.2	104.0	98.8	93.0	96.9	95.4	91.8	97.5	95.3	115.5	94.3	102.3	92.0	99.0	96.4
3	CONCRETE	97.7	96.8	97.1	95.6	98.6	97.5	95.4	95.5	95.4	93.6	94.7	94.3	109.9	88.2	95.8	91.3	93.0	92.3
4	MASONRY	101.6	94.9	96.4	92.2	92.5	92.4	85.4	94.6	92.5	113.1	90.7	95.9	99.5	84.5	88.0	98.2	79.4	83.7
5	METALS	99.2	99.9	99.4	101.6	96.6	99.8	91.8	95.4	93.1	89.2	93.1	90.6	88.9	86.4	88.0	101.9	98.1	100.5
6	WOOD & PLASTICS	100.6	98.5	99.4	106.0	93.0	98.7	108.5	94.1	100.4	104.9	93.6	98.5	104.5	79.8	90.7	94.0	83.4	88.1
7	MOISTURE PROTECTION	92.9	95.8	93.8	89.8	92.5	90.7	92.9	91.6	92.5	90.5	90.7	90.6	87.1	81.2	85.3	96.7	82.9	92.3
8	DOORS, WINDOWS, GLASS	110.4	99.7	104.8	105.9	92.5	98.8	99.9	93.9	96.7	92.8	90.7	91.7	89.7	80.5	84.9	99.3	84.4	91.5
9.1	LATH & PLASTER	97.0	93.3	94.2	105.7	93.1	96.1	107.7	94.7	97.8	100.0	88.0	90.8	106.0	81.7	87.5	87.2	83.3	84.2
9.2	DRYWALL	89.0	97.7	93.2	101.0	92.5	97.0	100.9	93.9	97.5	102.6	90.7	97.0	115.7	80.0	98.8	91.4	82.9	87.4
9.5	ACOUSTICAL WORK	101.5	97.9	99.5	89.4	92.5	91.1	92.7	93.9	93.3	92.0	90.7	91.3	105.6	79.1	91.1	94.8	82.9	88.3
9.6	FLOORING	103.9	99.0	102.6	103.1	92.5	100.3	107.3	97.5	104.7	107.3	90.7	102.9	92.5	80.8	89.4	107.1	79.6	99.8
9.8	PAINTING	95.5	92.2	92.9	97.6	92.4	93.5	120.5	93.8	99.1	103.3	90.6	93.1	104.3	81.2	85.8	89.8	82.8	84.2
9	FINISHES	94.2	95.6	95.0	100.3	92.5	96.1	103.8	94.2	98.6	102.9	90.5	96.2	108.3	80.5	93.3	95.0	82.7	88.3
10-14	TOTAL DIV. 10-14	100.0	98.5	99.5	100.0	92.5	97.7	100.0	97.3	99.2	100.0	78.6	93.5	100.0	81.3	94.3	100.0	91.4	97.4
15	MECHANICAL	100.6	96.2	98.3	98.1	92.5	95.2	98.1	94.2	96.1	100.0	90.7	95.2	96.6	88.5	92.4	99.5	84.5	91.7
16	ELECTRICAL	98.3	96.1	96.8	97.1	92.4	93.8	108.4	94.8	98.7	97.2	90.6	92.5	98.5	81.2	86.2	97.8	81.1	85.9
1-16	WEIGHTED AVERAGE	99.4	96.5	97.8	98.1	94.3	96.0	97.2	94.5	95.7	98.0	91.1	94.2	98.8	85.2	91.3	98.1	85.9	91.4

19.1 CITY COST INDEXES

DIVISION		KANSAS WICHITA			KENTUCKY LEXINGTON			KENTUCKY LOUISVILLE			LOUISIANA BATON ROUGE			LOUISIANA NEW ORLEANS			LOUISIANA SHREVEPORT		
		MAT.	INST.	TOTAL	MAT.	INST.	TOTAL	MAT.	INST.	TOTAL	MAT.	INST.	TOTAL	MAT.	INST.	TOTAL	MAT.	INST.	TOTAL
2	SITE WORK	110.9	93.4	102.9	89.4	91.7	90.4	103.5	94.0	99.2	94.9	79.1	87.7	105.8	90.1	98.7	102.6	83.2	93.8
3.1	FORMWORK	93.4	68.6	74.0	97.7	81.4	85.0	90.7	82.5	84.3	111.3	76.1	83.8	95.0	87.8	89.4	94.8	73.9	78.5
3.2	REINFORCING	87.7	78.1	83.7	101.4	81.0	92.8	117.8	82.2	102.8	100.5	80.6	92.1	96.8	80.3	89.8	100.5	73.2	89.0
3.3	CAST IN PLACE CONC.	94.6	102.1	99.3	84.0	95.0	90.9	78.0	104.5	94.5	97.4	88.1	91.6	99.6	91.0	94.3	86.6	93.2	90.7
3	CONCRETE	92.8	86.9	89.0	90.6	88.5	89.2	89.3	93.9	92.3	100.9	82.8	89.2	98.1	88.8	92.1	91.3	83.9	86.5
4	MASONRY	94.9	77.7	81.7	88.0	81.1	82.7	90.5	82.3	84.2	103.5	80.5	85.8	105.1	83.2	88.3	112.4	73.3	82.3
5	METALS	98.4	87.2	94.4	83.9	86.1	84.7	80.2	89.9	83.7	105.1	82.4	96.9	95.3	85.1	91.6	92.6	80.3	88.2
6	WOOD & PLASTICS	93.7	70.2	80.5	105.8	81.7	92.3	101.4	82.9	91.0	103.8	78.4	89.5	96.5	90.5	93.1	87.8	74.7	80.4
7	MOISTURE PROTECTION	97.1	76.0	90.4	84.6	81.8	83.5	84.7	82.3	83.9	111.7	79.5	101.6	112.4	78.5	101.7	86.9	73.3	82.6
8	DOORS, WINDOWS, GLASS	93.7	74.3	83.5	103.8	81.1	91.9	99.7	82.3	90.6	90.0	79.3	84.4	99.5	85.4	92.1	103.6	73.3	87.7
9.1	LATH & PLASTER	94.8	78.5	82.4	88.0	81.6	83.1	95.9	83.4	86.3	88.1	80.8	82.5	89.6	79.9	82.2	103.4	74.5	81.3
9.2	DRYWALL	90.9	73.0	82.4	92.1	81.1	86.9	90.3	82.3	86.5	100.2	77.5	89.4	87.6	84.7	86.2	90.9	73.3	82.5
9.5	ACOUSTICAL WORK	86.7	69.5	77.3	86.2	81.1	83.4	91.3	82.3	86.4	80.9	77.6	79.1	105.5	89.9	97.0	101.5	73.3	86.0
9.6	FLOORING	97.4	76.5	91.9	104.4	81.1	98.2	113.4	82.3	105.1	94.0	80.1	90.3	93.4	79.3	89.6	98.1	73.3	91.5
9.8	PAINTING	104.5	78.1	83.3	120.5	70.4	80.3	110.9	82.2	87.9	115.7	80.6	87.5	110.4	80.5	86.4	107.3	60.4	69.6
9	FINISHES	93.5	75.1	83.6	97.2	77.4	86.5	97.8	82.3	89.4	98.6	79.0	88.0	92.7	83.0	87.4	95.3	68.8	81.0
10-14	TOTAL DIV. 10-14	100.0	79.3	93.7	100.0	81.1	94.2	100.0	82.3	94.6	100.0	80.7	94.1	100.0	82.0	94.5	100.0	73.3	91.9
15	MECHANICAL	97.0	78.2	87.3	100.7	81.1	90.6	101.5	85.4	93.2	96.6	81.2	88.6	99.3	86.9	92.9	95.9	74.6	84.9
16	ELECTRICAL	102.4	78.1	85.1	102.4	81.0	87.2	100.2	82.2	87.4	97.3	80.6	85.4	97.0	87.6	90.3	94.6	73.2	79.4
1-16	WEIGHTED AVERAGE	97.4	80.4	88.1	95.0	83.0	88.4	95.3	86.2	90.3	99.8	80.9	89.4	99.7	86.1	92.2	96.2	76.2	85.2

DIVISION		MAINE LEWISTON			MAINE PORTLAND			MARYLAND BALTIMORE			MASSACHUSETTS BOSTON			MASSACHUSETTS LAWRENCE			MASSACHUSETTS LOWELL		
		MAT.	INST.	TOTAL	MAT.	INST.	TOTAL	MAT.	INST.	TOTAL	MAT.	INST.	TOTAL	MAT.	INST.	TOTAL	MAT.	INST.	TOTAL
2	SITE WORK	94.8	92.1	93.6	97.9	102.7	100.1	99.8	85.1	93.2	109.8	99.7	105.2	104.2	100.6	102.5	99.9	99.2	99.6
3.1	FORMWORK	97.1	78.1	82.3	97.1	78.1	82.3	112.7	83.0	89.5	105.7	107.0	106.7	108.4	106.6	107.0	101.9	109.5	107.8
3.2	REINFORCING	117.8	87.3	104.9	117.8	87.3	104.9	108.0	91.7	101.1	113.8	107.0	110.9	98.3	107.0	102.0	116.3	107.0	112.3
3.3	CAST IN PLACE CONC.	92.9	94.3	93.8	92.9	94.7	94.0	120.6	90.5	101.9	115.4	117.4	116.6	111.7	103.7	106.7	108.5	103.3	105.3
3	CONCRETE	99.2	87.4	91.6	99.2	87.6	91.7	116.2	87.7	97.8	113.1	112.4	112.6	108.1	105.1	106.2	108.9	106.0	107.0
4	MASONRY	92.3	77.6	81.0	92.5	77.6	81.1	81.4	71.8	74.0	108.6	111.3	110.7	103.4	111.3	109.5	108.0	111.5	110.7
5	METALS	94.4	89.4	92.6	107.7	89.4	101.1	97.5	91.3	95.3	110.8	109.9	110.5	96.5	105.4	99.7	94.0	105.4	98.1
6	WOOD & PLASTICS	107.6	77.0	90.4	104.4	77.0	89.0	102.6	85.9	93.3	104.8	105.9	105.4	112.6	105.0	108.3	110.2	108.6	109.3
7	MOISTURE PROTECTION	88.1	77.8	84.9	86.6	77.8	83.8	92.7	82.1	89.4	108.9	117.9	111.7	98.6	117.8	104.6	99.1	117.7	105.0
8	DOORS, WINDOWS, GLASS	107.6	79.3	92.7	95.7	79.3	87.1	96.8	91.6	94.1	103.6	107.8	105.8	107.6	107.3	107.4	102.0	109.4	105.9
9.1	LATH & PLASTER	107.5	88.8	93.2	108.4	89.2	93.7	110.9	89.2	94.3	131.0	112.4	116.8	99.2	111.6	108.7	97.0	109.4	106.5
9.2	DRYWALL	105.9	89.5	98.1	105.9	89.5	98.1	112.7	85.0	99.6	134.2	106.8	121.3	105.7	106.8	106.3	106.7	109.1	107.8
9.5	ACOUSTICAL WORK	95.7	77.3	85.6	95.7	77.3	85.6	99.4	85.2	91.6	86.2	106.1	97.2	106.8	106.2	106.5	106.8	110.0	108.5
9.6	FLOORING	105.3	77.6	98.0	102.9	77.6	96.2	103.4	92.8	100.6	118.5	110.5	116.4	98.4	110.5	101.6	104.3	111.2	106.1
9.8	PAINTING	98.2	52.1	61.2	102.1	52.1	61.9	95.4	85.2	87.2	111.4	126.6	123.6	101.3	108.5	107.1	96.6	110.0	107.3
9	FINISHES	104.2	74.5	88.2	104.1	74.5	88.1	107.8	85.9	96.0	124.5	114.3	119.0	103.6	107.9	105.9	104.9	109.6	107.5
10-14	TOTAL DIV. 10-14	100.0	78.1	93.4	100.0	78.1	93.4	100.0	92.6	97.7	100.0	107.1	102.1	100.0	105.8	101.7	100.0	106.3	101.9
15	MECHANICAL	96.7	78.6	87.3	96.7	78.6	87.3	99.4	93.1	96.1	104.3	109.9	107.2	98.9	105.5	102.3	98.6	107.0	103.0
16	ELECTRICAL	93.6	77.6	82.2	97.0	77.6	83.2	101.0	98.1	98.9	95.5	116.8	110.7	100.3	108.5	106.2	100.3	109.9	107.2
1-16	WEIGHTED AVERAGE	97.7	80.9	88.5	98.4	81.4	89.1	101.1	87.8	93.8	108.0	111.4	109.9	102.0	107.1	104.8	101.6	108.1	105.2

DIVISION		MASSACHUSETTS SPRINGFIELD			MASSACHUSETTS WORCESTER			MICHIGAN ANN ARBOR			MICHIGAN DETROIT			MICHIGAN FLINT			MICHIGAN GRAND RAPIDS		
		MAT.	INST.	TOTAL	MAT.	INST.	TOTAL	MAT.	INST.	TOTAL	MAT.	INST.	TOTAL	MAT.	INST.	TOTAL	MAT.	INST.	TOTAL
2	SITE WORK	92.9	100.8	96.5	102.5	101.4	102.0	94.9	102.2	98.2	92.0	102.9	96.9	94.4	99.3	96.6	78.3	90.0	83.6
3.1	FORMWORK	102.7	96.5	97.9	103.9	106.2	105.7	100.8	96.6	97.6	95.4	106.7	104.2	102.2	88.2	91.3	109.2	86.2	91.2
3.2	REINFORCING	98.3	96.4	97.5	98.3	107.0	102.0	109.2	102.6	106.4	89.6	102.6	95.1	109.2	102.6	106.4	117.8	81.9	102.6
3.3	CAST IN PLACE CONC.	99.5	105.5	103.3	106.4	103.6	104.6	89.1	102.9	97.7	98.2	110.5	105.9	103.9	98.6	100.6	96.4	94.9	95.5
3	CONCRETE	99.9	101.2	100.7	104.1	104.9	104.6	95.9	100.5	98.8	95.7	108.3	103.9	104.7	94.9	98.3	103.7	90.3	95.1
4	MASONRY	110.5	96.4	99.7	100.9	108.6	106.8	96.5	106.0	103.8	98.8	107.7	105.7	105.4	94.7	97.2	92.7	82.2	84.7
5	METALS	96.5	99.8	97.7	95.5	105.4	99.1	92.0	103.5	96.1	106.0	110.8	107.7	91.5	97.1	93.5	104.9	89.0	99.2
6	WOOD & PLASTICS	97.4	96.5	96.9	114.2	104.5	108.8	118.4	94.4	105.0	92.1	106.7	100.3	111.1	86.5	97.3	115.6	87.1	99.6
7	MOISTURE PROTECTION	97.6	96.5	97.2	100.4	107.8	102.8	90.3	103.7	94.5	95.1	109.6	99.7	84.7	94.0	87.6	112.1	81.9	102.6
8	DOORS, WINDOWS, GLASS	99.1	96.5	97.7	110.3	107.6	108.9	104.9	98.9	101.7	104.0	101.3	102.6	108.4	91.1	99.3	104.8	84.0	93.9
9.1	LATH & PLASTER	96.4	97.0	96.8	93.8	110.1	106.2	87.1	108.4	103.4	98.6	104.7	103.2	90.2	95.5	94.2	91.4	81.7	84.0
9.2	DRYWALL	104.5	96.5	100.7	106.8	105.8	106.3	88.1	99.6	93.5	89.4	107.0	97.7	111.7	89.7	101.3	114.8	84.7	100.6
9.5	ACOUSTICAL WORK	117.3	96.5	105.9	110.0	105.8	107.7	80.9	94.2	88.2	99.4	106.2	103.4	98.0	86.0	91.4	105.2	86.6	95.1
9.6	FLOORING	97.4	96.5	97.2	115.3	112.9	114.7	89.4	107.7	94.3	108.8	107.0	108.5	89.6	93.5	90.7	102.4	82.8	97.2
9.8	PAINTING	105.5	96.4	98.2	112.8	105.7	107.1	97.1	107.2	103.6	111.8	103.2	104.9	92.1	95.2	94.6	105.5	82.0	86.4
9	FINISHES	103.8	96.5	99.9	109.3	106.5	107.8	89.7	101.3	96.0	97.1	105.5	101.6	103.1	92.0	97.1	109.7	83.6	95.6
10-14	TOTAL DIV. 10-14	100.0	95.0	98.5	100.0	105.8	101.7	100.0	98.6	99.5	100.0	100.3	100.1	100.0	94.3	98.2	100.0	79.5	93.8
15	MECHANICAL	97.6	96.5	97.1	101.3	104.8	103.1	99.8	103.4	101.7	103.8	102.1	102.9	99.2	95.9	97.5	98.2	82.6	90.1
16	ELECTRICAL	94.6	94.0	94.2	96.3	105.7	103.0	98.5	102.6	101.4	100.6	103.9	102.9	98.5	95.2	96.2	97.2	81.9	86.3
1-16	WEIGHTED AVERAGE	99.0	97.5	98.2	102.2	105.8	104.2	96.9	102.2	99.8	99.6	105.4	102.8	99.4	94.7	96.8	101.1	84.8	92.2

19.1 CITY COST INDEXES

	DIVISION	MICHIGAN									MINNESOTA						MISSISSIPPI		
		KALAMAZOO			LANSING			SAGINAW			DULUTH			MINNEAPOLIS			JACKSON		
		MAT.	INST.	TOTAL	MAT.	INST.	TOTAL	MAT.	INST.	TOTAL	MAT.	INST.	TOTAL	MAT.	INST.	TOTAL	MAT.	INST.	TOTAL
2	SITE WORK	92.8	94.0	93.3	111.7	95.3	104.3	110.1	94.6	103.1	119.6	92.1	107.1	122.3	103.9	113.9	98.0	84.8	92.0
3.1	FORMWORK	103.4	86.5	90.2	106.2	90.7	94.1	96.8	87.2	89.3	110.4	93.0	96.8	91.7	101.2	99.1	94.2	69.4	74.9
3.2	REINFORCING	117.8	86.5	104.6	117.8	102.6	111.4	109.2	102.6	106.4	105.4	92.8	100.1	75.0	97.0	84.4	104.0	68.7	89.0
3.3	CAST IN PLACE CONC.	98.2	96.7	97.3	95.2	85.1	88.9	91.4	97.9	95.4	95.9	98.9	97.7	95.3	97.4	96.6	97.1	91.7	93.7
3	CONCRETE	103.6	91.8	96.0	102.4	88.8	93.6	96.4	94.1	94.9	100.9	96.1	97.8	90.1	98.8	95.8	98.0	80.9	87.0
4	MASONRY	95.7	86.5	88.7	105.0	85.0	89.7	96.4	90.0	91.5	108.1	92.9	96.4	98.5	98.6	98.5	100.3	68.8	76.1
5	METALS	101.8	91.7	98.2	110.8	91.0	103.6	100.5	94.0	98.2	105.6	95.2	101.8	95.3	96.6	95.8	86.1	77.0	82.8
6	WOOD & PLASTICS	96.6	86.0	90.7	114.0	91.8	101.6	106.6	86.5	95.3	100.8	93.1	96.5	99.8	99.9	99.8	93.1	70.1	80.2
7	MOISTURE PROTECTION	89.3	87.1	88.6	111.5	89.7	104.7	93.2	89.7	92.1	94.7	92.9	94.1	90.2	103.9	94.5	83.3	68.8	78.7
8	DOORS, WINDOWS, GLASS	95.7	86.6	90.9	110.2	91.6	100.4	99.5	88.4	93.7	95.2	89.4	92.1	105.5	98.6	101.9	99.0	68.8	83.1
9.1	LATH & PLASTER	89.2	86.6	87.2	90.4	84.3	85.7	100.0	91.2	93.3	106.0	93.2	96.2	91.2	101.7	99.3	97.3	69.7	76.3
9.2	DRYWALL	107.6	86.6	94.3	116.4	91.6	104.6	94.4	87.7	91.2	107.1	92.9	100.3	99.0	100.1	99.5	114.6	68.8	92.9
9.5	ACOUSTICAL WORK	100.4	86.6	92.8	100.4	91.6	95.6	105.5	86.0	94.9	97.1	92.9	94.8	97.4	100.3	99.0	89.1	68.8	78.0
9.6	FLOORING	97.9	86.6	94.9	103.5	91.6	100.4	102.4	89.4	98.9	93.4	92.9	93.2	92.1	99.6	94.1	99.7	68.8	91.5
9.8	PAINTING	110.3	86.5	91.2	111.7	91.5	95.5	121.4	90.2	96.4	119.9	97.8	102.2	108.0	99.2	100.9	115.2	68.7	77.9
9	FINISHES	104.7	86.6	94.9	111.2	91.1	100.3	99.9	88.8	93.9	104.4	94.6	99.1	98.0	99.9	99.0	108.9	68.8	87.2
10-14	TOTAL DIV. 10-14	100.0	88.2	96.4	100.0	95.7	98.7	100.0	92.6	97.7	100.0	87.8	96.3	100.0	100.5	100.1	100.0	70.4	91.0
15	MECHANICAL	98.0	87.4	92.5	99.8	93.8	96.7	99.8	90.3	94.9	99.4	96.0	97.6	101.4	99.9	100.6	99.7	68.9	83.8
16	ELECTRICAL	93.3	86.5	88.5	99.6	91.5	93.9	98.5	90.2	92.6	97.2	92.8	94.1	92.8	105.0	101.5	94.8	68.7	76.2
1-16	WEIGHTED AVERAGE	98.5	88.5	93.0	105.2	90.6	97.2	99.4	91.2	94.9	101.8	93.9	97.5	98.7	100.2	99.5	97.3	72.6	83.8

	DIVISION	MISSOURI						MONTANA						NEBRASKA					
		KANSAS CITY			ST LOUIS			BILLINGS			GREAT FALLS			LINCOLN			OMAHA		
		MAT.	INST.	TOTAL	MAT.	INST.	TOTAL	MAT.	INST.	TOTAL	MAT.	INST.	TOTAL	MAT.	INST.	TOTAL	MAT.	INST.	TOTAL
2	SITE WORK	92.2	104.4	97.7	81.3	95.6	87.8	95.9	96.1	96.0	90.8	95.3	92.9	100.2	89.8	95.5	109.9	97.3	104.2
3.1	FORMWORK	100.2	97.4	98.1	94.3	107.3	104.4	93.9	84.5	86.6	112.4	85.4	91.4	104.5	67.0	75.2	93.9	79.1	82.3
3.2	REINFORCING	78.7	98.4	87.0	82.3	93.4	87.0	112.6	84.2	100.6	112.8	85.1	101.1	107.5	73.8	93.2	96.8	79.9	89.7
3.3	CAST IN PLACE CONC.	100.4	95.1	97.1	86.9	101.8	96.2	103.5	95.9	98.7	115.2	96.1	103.3	104.8	93.3	97.6	93.8	94.6	94.3
3	CONCRETE	95.6	96.3	96.1	87.4	103.2	97.6	103.5	90.4	95.1	114.1	91.0	99.1	105.4	81.3	89.8	94.5	87.2	89.8
4	MASONRY	105.3	95.5	97.8	90.0	104.3	101.0	107.4	84.3	89.6	117.7	85.2	92.7	98.9	73.3	79.2	104.1	79.9	85.5
5	METALS	105.1	97.6	102.4	102.1	96.3	100.0	95.9	88.4	93.2	89.0	89.1	89.0	86.0	80.7	84.0	108.9	85.2	100.3
6	WOOD & PLASTICS	111.2	96.6	103.0	84.8	105.3	96.3	88.9	84.8	86.6	99.2	85.7	91.6	102.8	64.7	81.4	100.5	79.4	88.7
7	MOISTURE PROTECTION	99.6	101.1	100.0	93.2	104.9	96.9	100.4	84.3	95.2	102.5	85.2	97.0	88.3	68.3	82.0	106.0	79.7	97.8
8	DOORS, WINDOWS, GLASS	93.1	98.9	96.1	101.5	111.0	106.5	90.3	78.1	83.9	102.3	85.2	93.3	99.3	69.2	83.5	112.4	79.4	95.1
9.1	LATH & PLASTER	103.7	92.8	95.4	113.3	103.4	105.7	104.3	84.7	89.3	109.4	85.6	91.2	96.9	74.6	79.8	91.3	80.5	83.1
9.2	DRYWALL	99.3	96.1	97.7	93.7	105.4	99.2	103.2	84.3	94.2	105.8	85.1	96.0	90.1	67.6	79.4	92.1	79.2	86.0
9.5	ACOUSTICAL WORK	108.6	96.2	101.8	101.5	105.5	103.7	100.9	84.3	91.8	101.6	85.2	92.6	90.0	63.4	75.5	95.0	78.7	86.1
9.6	FLOORING	96.8	104.5	98.9	101.5	98.7	100.8	97.4	84.3	93.9	106.3	85.2	100.7	110.7	71.9	100.5	103.0	77.9	96.4
9.8	PAINTING	76.9	95.0	91.5	98.5	101.1	100.6	121.0	84.2	91.5	115.9	85.1	91.2	101.4	71.0	76.9	88.9	71.0	74.5
9	FINISHES	97.3	96.1	96.7	97.0	103.3	100.4	103.5	84.3	93.1	106.7	85.2	95.1	96.1	69.2	81.5	94.5	76.2	84.6
10-14	TOTAL DIV. 10-14	100.0	90.3	97.0	100.0	99.1	99.7	100.0	91.3	97.3	100.0	91.6	97.4	100.0	82.1	94.6	100.0	84.1	95.2
15	MECHANICAL	102.3	94.5	98.2	97.8	104.8	101.4	99.5	96.3	97.8	99.2	87.3	93.0	100.0	74.0	86.5	100.2	80.5	90.0
16	ELECTRICAL	104.7	94.3	97.3	103.5	113.3	110.5	98.0	84.2	88.2	97.4	85.1	88.7	95.6	73.8	80.1	91.4	79.9	83.2
1-16	WEIGHTED AVERAGE	99.8	96.0	97.7	95.4	104.6	100.5	99.6	88.5	93.5	102.0	87.7	94.2	97.9	75.7	85.7	101.1	82.4	90.8

	DIVISION	NEVADA						NEW HAMPSHIRE						NEW JERSEY					
		LAS VEGAS			RENO			MANCHESTER			NASHUA			JERSEY CITY			NEWARK		
		MAT.	INST.	TOTAL	MAT.	INST.	TOTAL	MAT.	INST.	TOTAL	MAT.	INST.	TOTAL	MAT.	INST.	TOTAL	MAT.	INST.	TOTAL
2	SITE WORK	86.9	108.1	96.5	87.5	105.1	95.5	89.0	90.9	89.9	95.5	95.4	95.5	109.6	105.3	107.6	104.1	104.4	104.2
3.1	FORMWORK	103.2	116.5	113.6	99.2	114.4	111.1	102.9	87.7	91.1	101.6	88.2	91.2	103.0	117.3	114.2	112.0	117.5	116.3
3.2	REINFORCING	118.5	128.5	122.7	77.4	128.5	99.1	117.8	87.5	105.0	117.8	87.5	105.0	109.2	113.0	110.9	109.2	113.0	110.9
3.3	CAST IN PLACE CONC.	99.0	110.1	105.9	109.5	109.0	109.2	93.3	86.4	89.0	100.6	97.0	98.4	99.1	103.4	101.7	105.8	99.4	101.8
3	CONCRETE	104.1	114.2	110.7	100.4	112.8	108.4	100.6	87.0	91.8	104.6	92.8	96.9	102.1	109.7	107.0	107.8	107.6	107.7
4	MASONRY	106.3	116.4	114.1	121.2	94.7	100.8	101.4	87.6	90.8	104.9	87.6	91.6	104.6	113.4	111.3	108.5	109.4	109.2
5	METALS	111.2	121.6	114.9	98.2	121.2	106.5	101.1	87.2	96.0	88.4	90.8	89.3	101.4	109.6	104.4	98.6	108.1	102.1
6	WOOD & PLASTICS	88.1	115.9	103.7	87.6	114.4	102.6	105.0	88.0	95.5	110.3	89.1	98.4	111.3	118.2	115.2	109.2	118.4	114.3
7	MOISTURE PROTECTION	88.5	120.1	98.4	103.0	117.5	107.6	96.4	87.0	93.6	102.3	87.6	97.6	118.1	113.9	116.8	105.0	113.2	107.8
8	DOORS, WINDOWS, GLASS	95.3	118.7	107.6	100.5	116.8	109.1	102.0	88.5	95.0	97.2	88.5	92.7	99.5	120.1	110.3	101.4	118.6	110.4
9.1	LATH & PLASTER	86.7	116.1	109.1	101.3	109.9	107.9	110.8	88.5	93.8	110.7	88.4	93.6	98.7	112.7	109.4	99.8	112.7	109.6
9.2	DRYWALL	92.5	116.5	103.9	99.3	121.4	109.8	113.4	87.6	101.1	114.6	87.6	101.8	108.1	116.5	112.1	109.6	116.6	113.1
9.5	ACOUSTICAL WORK	111.1	116.5	114.0	102.5	114.9	109.3	107.5	87.6	96.6	107.5	87.6	96.6	101.8	118.8	111.1	101.8	119.0	111.2
9.6	FLOORING	118.6	116.5	118.0	99.7	93.9	98.1	83.6	87.6	84.7	88.5	87.6	88.3	106.1	114.1	108.2	98.3	114.1	102.5
9.8	PAINTING	105.3	116.4	114.3	115.1	113.3	113.7	126.9	87.5	95.3	105.7	87.5	91.1	92.7	98.6	97.5	89.7	101.3	99.0
9	FINISHES	101.1	116.4	109.4	101.3	115.4	108.9	107.4	87.6	96.7	107.1	87.6	96.6	105.4	110.0	107.9	104.3	111.0	108.0
10-14	TOTAL DIV. 10-14	100.0	116.5	104.9	100.0	134.8	110.5	100.0	96.0	98.8	100.0	96.0	98.8	100.0	111.1	103.3	100.0	111.1	103.3
15	MECHANICAL	100.6	116.4	108.8	103.8	108.4	105.9	97.3	88.8	93.0	98.7	88.9	93.6	100.3	112.8	106.8	97.3	113.1	105.5
16	ELECTRICAL	102.3	116.4	112.4	107.7	118.5	115.4	101.1	87.5	91.4	97.1	87.5	90.3	95.7	120.2	113.2	96.5	113.0	108.3
1-16	WEIGHTED AVERAGE	100.1	116.0	108.8	101.5	111.7	107.1	99.7	88.2	93.4	100.0	89.9	94.4	103.2	112.9	108.5	102.0	110.9	106.9

19.1 CITY COST INDEXES

	DIVISION	NEW JERSEY						NEW MEXICO			NEW YORK								
		PATERSON			TRENTON			ALBUQUERQUE			ALBANY			BINGHAMTON			BUFFALO		
		MAT.	INST.	TOTAL	MAT.	INST.	TOTAL	MAT.	INST.	TOTAL	MAT.	INST.	TOTAL	MAT.	INST.	TOTAL	MAT.	INST.	TOTAL
2	SITE WORK	112.9	105.3	109.5	103.8	105.5	104.5	108.3	92.3	101.1	99.8	97.3	98.7	89.2	89.4	89.3	99.5	98.5	99.0
3.1	FORMWORK	100.5	117.8	114.0	113.7	117.4	116.6	114.8	77.7	85.8	106.2	93.1	96.0	101.2	89.1	91.7	122.0	111.9	114.2
3.2	REINFORCING	109.0	114.6	111.4	109.2	112.6	110.7	117.8	79.2	101.5	80.2	92.8	85.5	80.2	85.2	82.3	96.8	108.8	101.9
3.3	CAST IN PLACE CONC.	102.4	105.1	104.1	89.3	104.4	98.7	102.1	102.2	102.1	77.9	98.7	90.8	91.3	100.4	97.0	108.2	99.8	102.9
3	CONCRETE	103.5	110.9	108.3	98.6	110.2	106.1	108.1	90.6	96.8	84.1	96.0	91.8	90.8	94.7	93.3	108.5	105.3	106.4
4	MASONRY	107.6	115.5	113.7	103.4	109.2	107.8	101.2	79.3	84.3	86.1	92.9	91.3	98.6	88.8	91.1	99.9	108.0	106.1
5	METALS	96.5	110.8	101.6	98.5	109.3	102.4	107.7	87.9	100.6	97.5	95.1	96.6	99.6	89.9	96.1	105.2	105.0	105.2
6	WOOD & PLASTICS	116.9	118.4	117.7	119.9	118.9	119.3	100.0	80.0	88.7	96.4	93.1	94.6	103.4	86.0	93.6	111.1	111.4	111.3
7	MOISTURE PROTECTION	114.5	116.5	115.1	103.0	113.9	106.4	97.4	78.6	91.5	105.9	92.9	101.8	97.5	87.8	94.4	100.4	107.2	102.6
8	DOORS, WINDOWS, GLASS	99.4	116.6	108.4	105.7	120.1	113.2	100.2	79.3	89.2	104.9	92.9	98.6	100.1	85.2	92.4	95.0	107.7	101.7
9.1	LATH & PLASTER	99.7	117.1	113.0	115.5	112.8	113.4	119.3	79.8	89.2	106.0	94.3	97.0	109.0	90.0	94.5	104.9	102.3	102.9
9.2	DRYWALL	113.1	117.3	115.1	109.5	116.1	112.6	85.6	79.3	82.6	105.8	92.9	99.7	113.6	85.3	100.2	113.6	112.0	112.8
9.5	ACOUSTICAL WORK	101.8	119.0	111.2	98.2	118.4	109.3	93.1	79.3	85.5	112.3	92.9	101.7	111.7	85.5	97.4	115.7	112.2	113.8
9.6	FLOORING	91.4	119.6	98.9	101.0	113.9	104.4	104.2	79.3	97.6	85.8	92.9	87.7	98.8	91.2	96.8	104.6	111.0	106.3
9.8	PAINTING	99.8	114.6	111.7	96.4	98.6	98.2	110.4	79.2	85.4	117.0	92.8	97.6	103.5	85.2	88.8	116.2	102.3	105.1
9	FINISHES	105.6	116.6	111.6	105.5	109.8	107.8	93.7	79.3	85.9	102.9	92.9	97.5	108.9	86.0	96.6	111.8	108.0	109.7
10-14	TOTAL DIV. 10-14	100.0	111.6	103.5	100.0	111.0	103.3	100.0	79.3	93.7	100.0	89.2	96.7	100.0	83.9	95.1	100.0	103.4	101.0
15	MECHANICAL	100.3	112.5	106.6	100.2	112.0	106.3	100.5	79.3	89.5	96.0	90.5	93.1	100.7	85.3	92.7	96.8	99.1	98.0
16	ELECTRICAL	95.7	114.6	109.2	93.9	115.3	109.2	94.1	79.2	83.5	91.6	93.6	93.0	91.3	85.2	87.0	97.9	106.2	103.8
1-16	WEIGHTED AVERAGE	103.1	113.3	108.7	101.2	111.6	106.9	101.4	82.7	91.1	96.4	93.3	94.7	98.2	88.3	92.8	102.0	104.9	103.5

	DIVISION	NEW YORK															NORTH CAROLINA		
		NEW YORK			ROCHESTER			SYRACUSE			UTICA			YONKERS			CHARLOTTE		
		MAT.	INST.	TOTAL	MAT.	INST.	TOTAL	MAT.	INST.	TOTAL	MAT.	INST.	TOTAL	MAT.	INST.	TOTAL	MAT.	INST.	TOTAL
2	SITE WORK	119.4	115.2	117.5	104.6	93.9	99.8	93.5	94.5	94.0	113.2	94.1	104.5	119.9	113.5	117.0	112.2	81.9	98.5
3.1	FORMWORK	111.7	135.2	130.0	97.9	98.6	98.4	101.4	95.6	96.9	103.8	80.9	86.0	105.3	126.5	121.8	101.5	63.3	71.7
3.2	REINFORCING	101.7	157.9	125.5	106.5	98.7	103.2	106.5	95.5	101.9	106.5	90.1	99.6	80.2	119.7	96.9	87.8	65.7	78.5
3.3	CAST IN PLACE CONC.	153.0	104.8	123.0	126.9	96.1	107.7	103.2	77.9	87.4	84.3	97.3	92.4	116.6	105.7	109.9	110.9	96.3	101.8
3	CONCRETE	133.4	121.3	125.6	116.6	97.3	104.1	103.6	86.4	92.4	93.1	90.3	91.3	106.3	115.1	112.0	104.0	80.7	88.9
4	MASONRY	107.8	130.2	125.0	100.0	98.7	99.0	100.3	95.6	96.7	98.2	89.5	91.5	120.2	120.3	120.3	88.6	65.8	71.0
5	METALS	103.4	137.6	115.8	101.8	97.6	100.3	103.6	89.2	98.4	104.3	92.7	100.1	97.9	114.5	103.9	100.7	77.2	92.2
6	WOOD & PLASTICS	111.1	135.8	125.0	97.6	98.4	98.0	107.0	95.7	100.7	112.5	80.1	94.3	103.1	127.7	116.9	104.3	66.9	83.3
7	MOISTURE PROTECTION	108.8	133.7	116.7	97.4	98.7	97.8	96.7	95.6	96.3	97.6	88.3	94.7	105.8	129.5	113.2	88.5	51.1	76.7
8	DOORS, WINDOWS, GLASS	96.3	139.6	119.0	97.4	98.8	98.1	98.5	88.6	93.3	104.5	80.5	91.9	105.9	128.8	117.9	96.6	65.8	80.4
9.1	LATH & PLASTER	87.4	120.0	112.3	107.5	98.7	100.9	106.1	95.5	98.0	109.3	86.2	91.7	93.9	119.9	113.7	100.1	55.5	66.1
9.2	DRYWALL	118.1	134.9	126.0	94.1	98.8	96.3	108.7	95.6	102.5	110.3	79.3	95.6	97.5	125.2	110.6	88.1	65.8	77.5
9.5	ACOUSTICAL WORK	102.5	137.5	121.6	117.0	98.8	107.0	100.1	95.6	97.6	117.0	79.4	96.5	117.0	128.9	123.5	93.5	65.8	78.3
9.6	FLOORING	97.6	129.2	105.9	93.5	98.8	94.9	86.0	95.6	88.5	87.5	88.1	87.6	103.2	121.5	108.0	90.1	65.8	83.6
9.8	PAINTING	115.6	123.2	121.7	97.9	98.4	98.3	103.2	95.5	97.1	108.6	90.1	93.8	107.8	119.7	117.4	96.8	66.4	72.4
9	FINISHES	111.2	129.7	121.2	96.5	98.6	97.6	102.2	95.5	98.6	105.4	84.1	93.9	101.3	123.0	113.0	90.1	65.4	76.7
10-14	TOTAL DIV. 10-14	100.0	115.3	104.6	100.0	100.5	100.1	100.0	95.7	98.7	100.0	91.9	97.5	100.0	112.9	103.9	100.0	68.1	90.3
15	MECHANICAL	99.2	129.8	115.0	97.2	98.8	98.0	101.2	95.6	98.3	99.6	90.2	94.7	96.5	119.7	108.5	97.4	66.6	81.5
16	ELECTRICAL	95.7	130.8	120.7	97.5	98.7	98.4	97.9	95.5	96.2	94.9	90.1	91.5	103.4	119.7	115.0	97.2	65.7	74.8
1-16	WEIGHTED AVERAGE	108.1	128.0	119.0	101.4	98.2	99.7	100.6	92.8	96.3	100.6	89.1	94.3	103.5	119.2	112.1	98.1	70.0	82.7

	DIVISION	NORTH CAROLINA						OHIO											
		GREENSBORO			RALEIGH			AKRON			CANTON			CINCINNATI			CLEVELAND		
		MAT.	INST.	TOTAL	MAT.	INST.	TOTAL	MAT.	INST.	TOTAL	MAT.	INST.	TOTAL	MAT.	INST.	TOTAL	MAT.	INST.	TOTAL
2	SITE WORK	87.8	86.3	87.1	96.2	89.6	93.2	113.1	104.1	109.0	101.7	100.9	101.4	93.4	103.0	97.8	120.2	110.8	116.0
3.1	FORMWORK	91.0	64.0	69.9	94.9	63.6	70.5	101.3	110.0	108.9	99.7	102.4	101.8	99.0	104.1	103.0	120.4	127.3	125.8
3.2	REINFORCING	82.3	66.6	75.7	93.9	66.2	82.2	98.3	114.3	105.1	98.3	102.4	100.1	104.1	96.8	101.0	87.4	114.3	98.8
3.3	CAST IN PLACE CONC.	99.6	94.2	96.2	107.4	99.0	102.2	89.8	102.9	97.9	89.8	100.6	96.5	79.4	96.1	89.8	90.6	114.4	105.4
3	CONCRETE	94.1	80.0	85.0	101.9	82.3	89.2	94.0	107.1	102.5	93.7	101.5	98.7	88.8	99.3	95.6	95.9	119.4	111.1
4	MASONRY	101.1	66.7	74.6	87.8	66.3	71.3	91.9	111.1	106.7	105.9	102.6	103.3	73.9	94.7	89.8	94.6	127.2	119.6
5	METALS	91.6	77.0	86.3	91.5	78.4	86.8	99.4	110.2	103.3	99.4	101.8	100.3	98.6	96.8	98.0	103.5	112.3	106.7
6	WOOD & PLASTICS	91.5	67.8	78.2	94.8	67.4	79.4	104.0	110.8	107.8	102.6	102.4	102.5	107.9	102.7	105.0	142.9	124.1	132.4
7	MOISTURE PROTECTION	86.5	51.4	75.5	87.0	51.3	75.7	99.0	111.1	102.8	99.0	102.5	100.1	95.6	106.9	99.2	108.3	128.2	114.5
8	DOORS, WINDOWS, GLASS	91.6	65.5	77.9	85.9	65.2	75.0	101.7	113.3	107.8	89.2	102.5	96.1	98.2	99.0	98.6	94.5	119.3	107.5
9.1	LATH & PLASTER	98.0	68.0	75.1	106.8	67.6	76.9	115.9	111.2	112.3	112.2	102.7	104.9	102.2	99.4	100.2	105.4	123.2	119.0
9.2	DRYWALL	85.9	66.7	76.8	95.5	66.2	81.6	112.7	111.1	112.0	110.8	102.5	106.9	98.2	101.7	99.9	100.2	124.6	111.8
9.5	ACOUSTICAL WORK	98.7	66.7	81.2	110.1	66.3	86.1	83.6	111.1	98.7	102.9	102.5	102.6	96.4	102.9	100.0	98.4	124.8	112.8
9.6	FLOORING	90.3	66.7	84.1	102.9	66.3	93.2	82.3	111.2	89.9	113.9	102.5	110.9	93.2	96.2	94.0	86.0	123.2	95.9
9.8	PAINTING	90.9	66.4	71.2	90.3	66.4	71.1	107.9	111.1	110.5	101.6	102.4	102.3	108.3	94.3	97.1	106.1	123.5	120.1
9	FINISHES	88.7	66.7	76.8	98.1	66.4	81.0	103.1	111.2	107.4	110.0	102.5	105.9	98.1	98.7	98.4	97.6	124.0	111.9
10-14	TOTAL DIV. 10-14	100.0	70.5	91.1	100.0	71.1	91.2	100.0	112.0	103.3	100.0	102.5	100.7	100.0	100.4	100.1	100.0	113.4	104.0
15	MECHANICAL	95.9	66.8	80.8	97.3	66.4	81.3	99.7	111.1	105.6	99.4	102.5	101.0	99.9	96.4	98.1	101.8	109.8	105.9
16	ELECTRICAL	96.8	66.6	75.3	98.6	66.2	75.5	93.7	111.1	106.1	92.5	102.4	99.6	100.4	89.3	92.5	96.1	113.3	108.4
1-16	WEIGHTED AVERAGE	93.8	70.6	81.0	95.8	71.1	82.2	99.4	110.0	105.2	99.2	102.2	100.8	95.9	97.2	96.6	101.6	117.9	110.5

19.1 CITY COST INDEXES

		OHIO															OKLAHOMA		
	DIVISION	COLUMBUS			DAYTON			LORAIN			TOLEDO			YOUNGSTOWN			OKLAHOMA CITY		
		MAT.	INST.	TOTAL	MAT.	INST.	TOTAL	MAT.	INST.	TOTAL	MAT.	INST.	TOTAL	MAT.	INST.	TOTAL	MAT.	INST.	TOTAL
2	SITE WORK	88.4	105.2	96.0	87.5	99.8	93.1	100.9	104.5	102.5	120.3	102.3	112.1	94.3	101.3	97.5	119.0	93.4	107.4
3.1	FORMWORK	106.9	94.5	97.2	107.7	97.8	99.9	100.7	111.0	108.8	106.4	105.5	105.7	98.0	103.6	102.4	106.6	85.8	90.3
3.2	REINFORCING	116.2	99.1	108.9	98.3	97.3	97.9	98.3	114.3	105.1	98.3	105.0	101.2	98.3	103.6	100.6	94.4	85.4	90.6
3.3	CAST IN PLACE CONC.	93.3	97.2	95.7	106.1	99.9	102.2	95.6	102.9	100.1	92.9	101.5	98.3	82.6	100.9	94.0	100.7	96.0	97.8
3	CONCRETE	101.1	96.3	98.0	104.7	98.8	100.9	97.2	107.1	103.6	96.8	103.4	101.1	89.1	102.2	97.6	100.5	91.1	94.4
4	MASONRY	88.5	97.2	95.2	84.7	97.3	94.4	99.5	111.1	108.5	107.5	101.7	103.0	94.3	103.6	101.5	98.0	85.5	88.4
5	METALS	101.5	99.4	100.7	101.4	98.6	100.4	99.2	110.3	103.2	99.2	103.5	100.8	98.8	102.7	100.2	94.7	89.4	92.8
6	WOOD & PLASTICS	107.8	93.2	99.6	108.5	97.3	102.2	99.6	110.8	105.9	119.2	105.2	111.3	105.4	103.5	104.4	115.2	86.0	98.8
7	MOISTURE PROTECTION	98.6	99.3	98.8	100.6	97.5	99.6	97.7	122.2	105.4	91.2	105.2	95.6	101.3	103.6	102.0	94.6	85.5	91.7
8	DOORS, WINDOWS, GLASS	98.0	92.0	94.8	108.7	97.4	102.7	97.1	113.3	105.6	100.8	105.1	103.0	98.3	103.7	101.1	103.9	83.3	93.1
9.1	LATH & PLASTER	91.4	91.6	91.6	110.8	97.7	100.8	110.4	111.2	111.0	114.9	104.8	107.2	115.5	103.0	106.0	99.7	87.1	90.1
9.2	DRYWALL	101.8	93.2	97.7	112.6	97.4	105.4	107.8	111.2	109.4	109.7	105.1	107.5	96.4	103.7	99.8	108.6	85.5	97.6
9.5	ACOUSTICAL WORK	101.5	93.3	97.0	88.0	97.4	93.1	92.7	111.1	102.8	116.3	105.1	110.1	96.5	103.7	100.4	88.0	85.5	86.6
9.6	FLOORING	99.4	90.3	97.0	113.4	97.4	109.1	118.6	111.2	116.6	98.7	105.1	100.4	87.8	103.7	92.0	106.4	85.5	100.9
9.8	PAINTING	110.0	96.4	99.1	84.3	97.3	94.8	103.5	123.5	119.6	115.2	97.8	101.2	74.7	103.6	98.0	105.7	85.4	89.4
9	FINISHES	101.8	94.0	97.6	107.9	97.4	102.2	108.7	115.5	112.4	108.4	102.5	105.2	92.7	103.6	98.6	106.0	85.6	94.9
10-14	TOTAL DIV. 10-14	100.0	95.8	98.7	100.0	100.0	100.0	100.0	110.2	103.1	100.0	106.2	101.8	100.0	103.7	101.1	100.0	86.8	96.0
15	MECHANICAL	102.2	100.9	101.5	99.3	97.3	98.3	99.2	111.0	105.3	99.2	104.7	102.0	100.1	103.6	101.9	94.4	85.6	89.8
16	ELECTRICAL	93.6	94.8	94.4	97.4	97.3	97.4	96.4	111.1	106.9	98.3	106.4	104.1	96.4	103.6	101.6	102.0	85.4	90.2
1-16	WEIGHTED AVERAGE	99.1	97.1	98.0	100.4	98.0	99.1	99.6	110.7	105.7	101.4	103.9	102.8	96.8	103.2	100.3	100.3	87.2	93.1

		OKLAHOMA			OREGON						PENNSYLVANIA								
	DIVISION	TULSA			EUGENE			PORTLAND			ALLENTOWN			ERIE			HARRISBURG		
		MAT.	INST.	TOTAL	MAT.	INST.	TOTAL	MAT.	INST.	TOTAL	MAT.	INST.	TOTAL	MAT.	INST.	TOTAL	MAT.	INST.	TOTAL
2	SITE WORK	87.7	98.1	92.4	92.7	106.1	98.8	106.4	103.1	104.9	122.8	101.3	113.1	116.3	98.6	108.3	94.5	99.1	96.6
3.1	FORMWORK	114.1	84.7	91.1	109.7	100.2	102.3	111.1	103.1	104.9	100.5	103.1	102.5	82.6	96.0	93.1	103.3	88.3	91.6
3.2	REINFORCING	91.5	84.3	88.5	101.4	97.7	99.8	101.4	97.7	99.8	114.7	102.1	109.4	117.6	96.3	108.6	114.7	102.8	109.7
3.3	CAST IN PLACE CONC.	88.6	100.7	96.1	100.9	125.1	116.0	117.0	101.8	107.5	103.3	100.7	101.7	83.5	98.9	93.1	104.6	102.8	103.5
3	CONCRETE	94.4	93.0	93.5	102.8	112.9	109.4	112.3	101.9	105.6	105.3	101.8	103.0	90.8	97.5	95.2	106.6	97.1	100.5
4	MASONRY	94.2	84.4	86.6	112.4	100.2	103.1	124.2	102.2	107.3	84.2	100.9	97.0	103.3	95.8	97.6	81.0	89.3	87.3
5	METALS	105.5	90.1	99.9	103.8	107.7	105.2	107.3	99.0	104.3	105.2	102.8	104.3	101.3	97.6	100.0	97.0	105.6	100.2
6	WOOD & PLASTICS	116.9	84.9	98.9	91.9	100.2	96.6	89.1	100.2	96.3	103.7	105.1	104.5	89.5	95.9	93.1	109.6	89.5	98.3
7	MOISTURE PROTECTION	96.8	84.4	92.9	86.4	100.3	90.8	86.4	98.4	90.2	96.9	117.7	103.4	97.4	96.0	97.0	78.7	88.5	81.8
8	DOORS, WINDOWS, GLASS	101.3	84.4	92.4	102.8	99.9	101.3	108.5	100.1	104.1	103.1	103.6	103.3	92.7	96.0	94.4	104.7	91.8	97.9
9.1	LATH & PLASTER	99.9	85.4	88.9	113.2	100.2	103.3	118.0	102.2	105.9	99.6	102.2	101.6	110.2	95.6	99.1	96.7	83.4	86.5
9.2	DRYWALL	111.7	84.4	98.8	111.7	100.3	106.3	109.9	102.1	106.2	107.0	104.1	105.6	93.9	95.9	94.8	107.3	89.3	98.7
9.5	ACOUSTICAL WORK	88.0	84.4	86.0	111.6	100.3	105.4	111.6	102.1	106.4	101.8	105.4	103.7	101.1	95.9	98.2	99.8	89.3	94.1
9.6	FLOORING	110.5	84.4	103.5	123.5	100.3	117.3	103.6	102.2	103.3	97.7	98.3	97.9	103.4	95.9	101.4	97.6	89.3	95.4
9.8	PAINTING	79.2	84.3	83.3	119.9	100.2	104.1	97.4	102.2	101.3	104.5	102.1	102.6	88.7	95.8	94.5	91.4	89.2	89.7
9	FINISHES	106.0	84.4	94.3	115.2	100.2	107.1	107.5	102.2	104.6	104.1	103.0	103.5	96.5	95.8	96.1	102.7	88.9	95.2
10-14	TOTAL DIV. 10-14	100.0	84.4	95.2	100.0	100.3	100.0	100.0	102.3	100.6	100.0	98.0	99.4	100.0	97.0	99.1	100.0	90.5	97.1
15	MECHANICAL	96.9	84.5	90.5	98.9	101.3	100.1	98.3	103.1	100.8	100.7	101.1	100.9	100.6	96.3	98.2	101.5	89.6	95.3
16	ELECTRICAL	99.7	84.3	88.8	101.9	100.2	100.7	101.9	102.2	102.1	97.7	102.1	100.9	89.5	95.8	94.0	92.1	89.2	90.1
1-16	WEIGHTED AVERAGE	98.7	87.1	92.3	101.3	103.8	102.7	104.0	102.0	102.9	102.3	102.2	102.2	98.2	96.6	97.3	98.2	92.3	95.0

		PENNSYLVANIA												RHODE ISLAND			SOUTH CAROLINA		
	DIVISION	PHILADELPHIA			PITTSBURGH			READING			SCRANTON			PROVIDENCE			CHARLESTON		
		MAT.	INST.	TOTAL	MAT.	INST.	TOTAL	MAT.	INST.	TOTAL	MAT.	INST.	TOTAL	MAT.	INST.	TOTAL	MAT.	INST.	TOTAL
2	SITE WORK	102.3	106.8	104.3	120.4	103.2	112.6	84.0	99.9	91.2	99.0	107.3	102.7	80.9	100.8	89.9	117.8	87.9	104.2
3.1	FORMWORK	92.6	112.1	107.8	103.4	106.8	106.0	100.8	92.3	94.2	98.1	94.2	95.0	104.6	101.9	102.5	95.5	66.4	72.8
3.2	REINFORCING	82.5	115.7	96.6	82.3	107.9	93.1	114.7	91.9	105.1	114.7	92.4	105.3	117.8	100.3	110.4	96.0	65.7	83.2
3.3	CAST IN PLACE CONC.	93.9	105.8	101.3	107.8	99.5	102.7	98.2	98.2	98.2	92.6	124.3	112.3	87.7	101.6	96.4	100.9	91.1	94.8
3	CONCRETE	91.1	109.1	102.7	101.3	103.1	102.5	102.3	95.4	97.8	98.6	109.7	105.8	98.7	101.6	100.3	98.7	79.2	86.1
4	MASONRY	89.2	105.5	101.7	99.1	111.1	108.3	88.3	92.0	91.1	103.4	92.5	95.0	99.3	102.7	101.9	90.6	65.8	71.5
5	METALS	104.0	103.5	103.8	100.2	104.9	101.9	96.7	96.6	96.6	98.7	106.1	101.3	111.7	99.8	107.4	99.2	75.4	90.6
6	WOOD & PLASTICS	90.3	112.1	102.5	111.0	105.6	108.0	100.0	92.1	95.6	101.5	92.6	96.5	90.9	100.3	96.2	98.1	66.9	80.6
7	MOISTURE PROTECTION	102.2	120.4	107.9	96.0	110.7	100.6	87.7	111.6	95.2	98.8	92.9	96.9	106.4	103.8	105.6	91.1	65.1	83.0
8	DOORS, WINDOWS, GLASS	93.1	107.5	100.7	100.7	106.2	103.6	104.4	92.0	97.9	93.4	92.5	92.9	111.1	97.3	103.9	101.7	65.8	82.8
9.1	LATH & PLASTER	98.7	113.1	109.7	106.7	97.5	99.7	95.2	92.7	93.3	101.0	93.1	95.0	106.5	101.0	102.3	94.0	66.7	73.2
9.2	DRYWALL	115.7	112.8	114.3	113.1	103.5	108.6	109.3	92.0	101.1	104.5	92.5	98.8	95.8	100.4	97.9	108.0	65.8	88.0
9.5	ACOUSTICAL WORK	97.4	113.0	105.9	104.5	105.9	105.3	99.8	92.0	95.5	100.8	92.5	96.2	104.7	100.4	102.3	96.1	65.8	79.5
9.6	FLOORING	96.6	101.4	97.8	112.8	103.8	110.4	103.6	92.0	100.5	101.2	92.5	98.9	89.1	100.4	92.1	89.6	64.8	83.3
9.8	PAINTING	121.7	104.1	107.6	85.3	102.7	99.3	89.7	92.0	91.5	109.4	92.4	95.8	105.5	100.3	101.3	112.0	64.6	73.9
9	FINISHES	110.1	109.0	109.5	109.4	103.1	106.0	105.0	92.0	98.0	103.9	92.5	97.7	96.2	100.4	98.4	102.9	65.4	82.7
10-14	TOTAL DIV. 10-14	100.0	108.9	102.7	100.0	103.3	101.0	100.0	94.7	98.4	100.0	91.5	97.4	100.0	100.9	100.2	100.0	67.7	90.2
15	MECHANICAL	97.3	109.1	103.4	98.7	99.6	99.2	99.5	92.1	95.6	99.5	92.6	95.9	100.0	100.2	100.1	98.6	64.5	80.9
16	ELECTRICAL	98.5	109.5	106.4	95.3	93.0	93.6	93.6	89.0	90.3	97.3	92.4	93.8	98.4	100.3	99.8	97.5	65.7	74.9
1-16	WEIGHTED AVERAGE	98.6	108.6	104.1	101.8	102.9	102.4	97.7	93.6	95.4	99.4	97.5	98.3	100.1	100.9	100.5	99.6	69.9	83.3

19.1 CITY COST INDEXES

	DIVISION	SOUTH CAROLINA COLUMBIA			SOUTH DAKOTA SIOUX FALLS			TENNESSEE										
								CHATTANOOGA			KNOXVILLE			MEMPHIS			NASHVILLE		
		MAT.	INST.	TOTAL	MAT.	INST.	TOTAL	MAT.	INST.	TOTAL	MAT.	INST.	TOTAL	MAT.	INST.	TOTAL	MAT.	INST.	TOTAL
2	SITE WORK	108.6	93.9	102.0	90.1	85.8	88.2	94.9	90.4	92.8	99.0	97.4	98.3	86.7	91.9	89.1	84.6	87.5	85.9
3.1	FORMWORK	82.4	67.3	70.6	95.6	71.5	76.8	97.4	77.2	81.6	98.6	73.5	79.0	88.7	82.3	83.7	88.6	64.9	70.1
3.2	REINFORCING	91.5	66.2	80.8	105.0	70.9	90.6	98.3	76.7	89.2	98.3	73.1	87.6	96.8	69.1	85.1	96.8	66.9	84.2
3.3	CAST IN PLACE CONC.	81.3	97.5	91.4	98.8	82.9	88.9	81.9	93.9	89.4	86.0	92.6	90.1	92.9	91.2	91.8	85.5	84.9	85.1
3	CONCRETE	83.8	82.9	83.2	99.5	77.4	85.2	88.6	85.9	86.8	91.2	83.4	86.2	92.9	85.8	88.3	88.6	75.5	80.1
4	MASONRY	86.0	66.3	70.9	105.0	71.0	78.8	77.8	76.8	77.0	78.5	73.2	74.4	87.5	79.7	81.5	92.6	73.0	77.5
5	METALS	105.1	77.9	95.3	107.5	75.3	95.8	91.9	82.9	88.7	103.7	80.2	95.2	90.7	76.8	85.8	97.9	73.5	89.1
6	WOOD & PLASTICS	85.7	67.9	75.7	97.6	72.1	83.3	105.3	77.6	89.7	106.9	73.7	88.3	85.4	86.3	85.9	107.8	66.7	84.7
7	MOISTURE PROTECTION	102.5	65.6	90.7	98.3	70.8	89.7	103.6	76.8	95.1	87.8	73.2	83.2	96.7	77.6	90.7	102.4	64.3	90.4
8	DOORS, WINDOWS, GLASS	103.2	66.5	84.0	104.1	71.0	86.7	104.1	72.5	87.5	108.3	73.2	89.9	95.8	83.0	89.1	95.1	69.9	81.9
9.1	LATH & PLASTER	104.8	67.1	76.1	110.3	71.8	80.9	94.1	77.4	81.4	92.0	73.4	77.8	94.4	82.6	85.4	90.0	66.7	72.2
9.2	DRYWALL	83.5	66.5	75.4	105.9	71.0	89.3	109.4	76.8	93.9	107.5	73.2	91.2	96.4	90.8	93.7	103.5	63.7	84.7
9.5	ACOUSTICAL WORK	94.9	66.8	79.5	102.7	71.0	85.3	98.7	76.8	86.7	91.0	73.2	81.2	99.4	85.8	92.0	100.5	65.2	81.2
9.6	FLOORING	95.4	66.3	87.7	111.8	71.0	101.0	108.7	76.8	100.2	100.8	73.2	93.5	102.2	74.4	94.8	129.9	74.2	115.2
9.8	PAINTING	121.4	64.6	75.8	113.1	71.0	79.2	91.0	76.7	79.6	94.4	73.1	77.3	89.2	90.8	90.5	93.9	65.9	71.4
9	FINISHES	91.4	65.9	77.6	107.8	71.0	87.9	106.2	76.8	90.3	103.0	73.2	86.9	97.2	88.7	92.6	108.0	65.5	85.1
10-14	TOTAL DIV. 10-14	100.0	67.8	90.2	100.0	80.7	94.1	100.0	76.8	92.9	100.0	73.2	91.9	100.0	83.7	95.0	100.0	71.3	91.3
15	MECHANICAL	100.0	65.1	81.9	99.5	70.9	84.7	99.4	76.9	87.7	100.3	73.3	86.3	99.9	89.3	94.4	96.7	76.0	85.9
16	ELECTRICAL	103.5	66.2	77.0	95.9	70.9	78.1	93.7	76.7	81.6	95.3	73.1	79.5	109.7	90.3	95.9	100.0	55.0	67.9
1-16	WEIGHTED AVERAGE	97.4	71.4	83.2	100.8	73.6	85.9	96.8	79.4	87.3	97.6	76.8	86.2	96.1	85.8	90.5	96.9	71.1	82.8

	DIVISION	TEXAS																	
		AMARILLO			AUSTIN			BEAUMONT			CORPUS CHRISTI			DALLAS			EL PASO		
		MAT.	INST.	TOTAL	MAT.	INST.	TOTAL	MAT.	INST.	TOTAL	MAT.	INST.	TOTAL	MAT.	INST.	TOTAL	MAT.	INST.	TOTAL
2	SITE WORK	114.3	90.0	103.3	85.1	88.5	86.6	123.4	97.7	111.8	110.0	88.8	100.4	115.4	90.0	103.9	104.6	101.7	103.3
3.1	FORMWORK	89.3	80.3	82.3	99.0	85.4	88.4	86.9	93.9	92.4	96.5	73.7	78.7	81.0	79.2	79.6	96.8	67.8	74.2
3.2	REINFORCING	113.6	80.2	99.5	97.2	85.1	92.1	93.9	93.5	93.8	104.5	73.2	91.3	94.4	85.7	90.7	117.8	65.7	95.8
3.3	CAST IN PLACE CONC.	114.3	94.6	102.1	96.8	96.1	96.4	111.5	98.3	103.3	106.3	93.0	98.0	99.5	96.9	97.9	92.3	84.2	87.3
3	CONCRETE	109.1	87.8	95.3	97.3	91.0	93.2	102.7	96.2	98.5	103.9	83.7	90.9	94.7	89.0	91.0	98.8	76.2	84.2
4	MASONRY	104.8	80.2	85.9	102.3	85.2	89.1	109.8	93.6	97.3	105.5	73.3	80.7	103.1	93.5	95.8	94.2	65.8	72.4
5	METALS	98.3	87.2	94.3	91.2	89.1	90.4	98.2	95.2	97.1	98.2	80.3	91.7	91.9	90.7	91.4	104.6	72.2	92.9
6	WOOD & PLASTICS	93.8	79.8	85.9	96.5	85.7	90.4	94.9	94.4	94.6	101.7	74.2	86.2	95.0	82.6	88.0	91.6	66.9	77.8
7	MOISTURE PROTECTION	93.9	80.2	89.6	97.6	85.5	93.8	96.1	94.2	95.5	97.2	66.3	87.5	103.5	80.4	96.3	98.7	66.2	88.5
8	DOORS, WINDOWS, GLASS	93.0	80.3	86.3	97.3	85.2	90.9	103.4	93.6	98.2	101.5	73.3	86.7	101.4	86.4	93.6	100.4	65.8	82.3
9.1	LATH & PLASTER	92.8	80.8	83.6	80.8	85.6	84.4	86.0	90.2	89.2	92.0	74.0	78.3	90.6	93.3	92.7	90.8	66.2	72.0
9.2	DRYWALL	85.0	80.3	82.7	84.9	85.2	85.0	90.1	93.6	91.8	87.1	73.3	80.5	78.7	81.3	79.9	87.3	65.8	77.1
9.5	ACOUSTICAL WORK	92.7	80.3	85.9	95.3	85.2	89.8	95.3	93.6	94.4	91.4	73.3	81.5	95.4	81.4	87.7	91.7	65.8	77.5
9.6	FLOORING	104.1	80.3	97.8	102.9	85.2	98.2	102.9	93.6	100.5	94.6	73.3	88.9	87.5	87.6	87.5	104.3	65.8	94.1
9.8	PAINTING	104.1	69.0	75.9	108.7	85.1	89.8	104.9	93.5	95.8	97.0	73.2	77.9	101.5	95.4	96.6	89.5	65.7	70.4
9	FINISHES	92.0	76.4	83.6	92.2	85.2	88.4	94.9	93.3	94.0	90.2	73.3	81.1	84.6	87.4	86.1	91.8	65.8	77.8
10-14	TOTAL DIV. 10-14	100.0	83.4	95.0	100.0	85.2	95.5	100.0	95.1	98.5	100.0	88.4	96.5	100.0	90.9	97.2	100.0	65.8	89.6
15	MECHANICAL	99.4	80.8	89.7	99.2	85.2	91.9	99.2	94.6	96.8	98.7	69.9	83.8	99.1	78.9	88.6	101.8	65.9	83.2
16	ELECTRICAL	103.1	80.2	86.8	101.7	85.1	89.9	105.0	96.4	98.8	99.6	68.1	77.1	97.8	90.7	92.8	93.4	65.7	73.7
1-16	WEIGHTED AVERAGE	100.6	82.5	90.6	96.7	86.8	91.2	101.6	95.0	98.0	100.0	75.8	86.7	98.0	87.6	92.3	99.2	69.9	83.1

	DIVISION	TEXAS												UTAH			VERMONT		
		FORT WORTH			HOUSTON			LUBBOCK			SAN ANTONIO			SALT LAKE CITY			BURLINGTON		
		MAT.	INST.	TOTAL	MAT.	INST.	TOTAL	MAT.	INST.	TOTAL	MAT.	INST.	TOTAL	MAT.	INST.	TOTAL	MAT.	INST.	TOTAL
2	SITE WORK	111.2	91.6	102.3	117.4	93.8	106.7	118.6	98.9	109.7	81.5	92.7	86.6	81.4	92.1	86.2	93.3	92.0	92.7
3.1	FORMWORK	94.0	79.2	82.5	98.1	80.3	84.2	100.1	78.3	83.1	95.3	75.9	80.2	104.0	81.7	86.6	106.1	79.8	85.6
3.2	REINFORCING	101.1	85.7	94.6	113.3	87.6	102.4	106.7	77.7	94.4	101.7	75.1	90.4	133.5	89.0	114.7	111.6	79.6	98.0
3.3	CAST IN PLACE CONC.	109.3	93.3	99.3	111.3	90.8	98.5	99.5	94.4	96.3	75.8	106.8	95.1	87.9	96.0	92.9	101.3	103.2	102.5
3	CONCRETE	104.4	87.1	93.2	109.1	86.4	94.4	101.2	86.7	91.8	85.4	91.9	89.6	101.2	89.8	93.8	104.5	92.0	96.4
4	MASONRY	110.7	85.8	91.6	117.0	88.0	94.7	111.3	77.8	85.5	96.9	76.5	81.2	90.2	88.6	89.0	108.7	79.7	86.4
5	METALS	97.3	90.6	94.9	93.5	90.1	92.3	89.2	85.6	87.9	92.3	87.3	90.5	106.7	92.1	101.4	104.1	88.0	98.3
6	WOOD & PLASTICS	99.1	82.7	89.9	104.4	81.3	91.4	91.0	78.5	84.0	89.6	76.8	83.6	83.6	81.1	82.2	119.1	79.8	97.1
7	MOISTURE PROTECTION	95.5	83.5	91.8	89.9	79.4	86.6	93.5	78.3	88.7	84.7	65.4	78.6	90.1	87.6	89.3	87.5	79.7	85.0
8	DOORS, WINDOWS, GLASS	100.0	86.4	92.9	104.4	82.8	93.1	99.7	77.8	88.2	102.2	78.8	89.9	100.9	85.4	92.8	106.3	79.7	92.4
9.1	LATH & PLASTER	101.7	86.4	90.0	92.3	87.4	88.6	92.3	78.4	81.7	95.3	77.2	81.5	117.9	88.7	95.6	91.5	80.7	83.3
9.2	DRYWALL	102.0	81.3	92.2	104.5	82.9	94.3	86.1	77.8	82.1	87.2	77.5	82.6	92.1	84.2	88.3	108.5	79.6	94.8
9.5	ACOUSTICAL WORK	92.7	81.4	86.5	96.4	80.7	87.8	92.7	77.8	84.5	97.4	78.3	87.0	99.9	80.8	89.5	100.2	79.7	88.9
9.6	FLOORING	104.5	85.2	99.4	94.8	87.2	92.8	106.9	77.8	99.2	95.4	79.0	91.0	100.6	87.5	97.1	109.7	79.7	101.7
9.8	PAINTING	99.9	92.7	94.1	100.7	87.7	90.2	102.7	77.7	82.6	93.4	67.7	72.8	97.4	89.0	90.7	110.3	79.6	85.7
9	FINISHES	101.6	85.9	93.1	101.0	85.0	92.4	93.2	77.8	84.9	90.6	74.2	81.8	95.8	86.1	90.5	107.9	79.7	92.7
10-14	TOTAL DIV. 10-14	100.0	90.2	97.0	100.0	94.1	98.2	100.0	87.5	96.2	100.0	82.3	94.6	100.0	90.6	97.1	100.0	88.3	96.4
15	MECHANICAL	100.4	76.1	87.8	100.4	85.2	92.5	100.0	78.7	88.9	99.9	87.5	93.4	104.6	87.4	95.7	101.0	80.7	90.5
16	ELECTRICAL	97.9	86.0	89.5	103.7	84.9	90.3	95.4	77.7	84.3	108.3	77.8	86.6	101.5	89.0	92.6	100.4	79.6	85.6
1-16	WEIGHTED AVERAGE	101.4	84.9	92.4	102.7	88.1	94.7	99.6	81.6	89.7	94.2	83.1	88.1	98.7	88.5	93.1	101.8	83.8	92.0

19.1 CITY COST INDEXES

	DIVISION	VIRGINIA												WASHINGTON					
		NEWPORT NEWS			NORFOLK			RICHMOND			ROANOKE			SEATTLE			SPOKANE		
		MAT.	INST.	TOTAL	MAT.	INST.	TOTAL	MAT.	INST.	TOTAL	MAT.	INST.	TOTAL	MAT.	INST.	TOTAL	MAT.	INST.	TOTAL
2	SITE WORK	117.4	86.4	103.3	107.6	84.9	97.3	80.2	88.8	84.1	97.7	86.7	92.7	103.1	100.7	102.0	104.9	101.3	103.3
3.1	FORMWORK	93.8	65.9	72.1	98.1	66.2	73.2	91.0	68.4	73.4	96.3	65.3	72.1	82.5	103.8	99.1	97.7	102.1	101.1
3.2	REINFORCING	101.0	67.1	86.6	101.0	67.6	86.8	97.0	74.6	87.5	105.9	64.6	88.4	108.9	97.7	104.2	113.0	97.7	106.5
3.3	CAST IN PLACE CONC.	115.6	88.4	98.7	113.4	90.7	99.3	109.7	90.9	98.0	111.8	89.4	97.9	94.4	111.7	105.2	108.5	100.4	103.5
3	CONCRETE	108.0	77.7	88.4	107.6	79.1	89.2	103.2	80.7	88.6	107.4	77.8	88.3	95.2	107.4	103.1	107.3	100.8	103.1
4	MASONRY	102.0	67.2	75.2	102.1	67.7	75.6	100.8	70.3	77.3	100.5	53.3	64.3	132.8	107.8	113.6	113.5	102.6	105.2
5	METALS	87.0	75.3	82.7	86.9	76.1	83.0	96.3	80.8	90.7	97.0	74.2	88.7	106.6	102.4	105.1	103.1	98.8	101.6
6	WOOD & PLASTICS	104.0	68.2	83.9	103.8	68.6	84.1	103.4	70.9	85.2	102.7	65.7	82.0	75.7	100.9	89.8	101.0	101.9	101.5
7	MOISTURE PROTECTION	85.3	66.6	79.4	86.5	67.1	80.4	109.1	69.5	96.7	85.7	63.0	78.6	112.9	112.0	112.7	96.3	102.6	98.3
8	DOORS, WINDOWS, GLASS	91.1	67.7	78.5	90.2	67.7	78.4	90.6	71.1	80.4	95.3	64.7	79.2	94.6	101.8	98.4	101.0	98.3	99.5
9.1	LATH & PLASTER	113.8	68.5	79.3	98.3	69.4	76.2	105.5	71.5	79.5	107.4	58.7	70.2	98.7	104.2	102.9	115.5	101.8	105.0
9.2	DRYWALL	87.5	67.2	77.8	87.5	67.7	78.1	97.8	70.3	84.7	91.8	64.6	79.0	79.6	101.6	90.0	96.1	102.7	99.2
9.5	ACOUSTICAL WORK	98.5	67.2	81.4	95.6	67.7	80.3	105.2	70.3	86.1	98.0	64.7	79.8	98.4	100.7	99.6	105.8	102.7	104.1
9.6	FLOORING	99.4	67.2	90.8	94.7	67.7	87.5	87.9	70.3	83.2	103.4	56.6	91.0	90.9	105.7	94.8	105.5	102.7	104.8
9.8	PAINTING	91.1	62.1	67.8	101.7	67.6	74.3	96.6	70.2	75.4	90.6	64.6	69.8	83.6	102.4	98.7	101.3	102.6	102.4
9	FINISHES	92.0	65.5	77.7	91.4	67.8	78.6	96.2	70.3	82.2	95.2	63.7	78.2	84.5	102.2	94.1	100.0	102.6	101.4
10-14	TOTAL DIV. 10-14	100.0	67.2	90.0	100.0	67.7	90.2	100.0	73.9	92.1	100.0	69.0	90.6	100.0	111.4	103.4	100.0	102.9	100.8
15	MECHANICAL	103.2	67.4	84.7	103.0	67.8	84.7	101.9	70.4	85.6	103.7	62.1	82.2	99.3	107.9	103.7	103.7	103.1	103.4
16	ELECTRICAL	105.4	67.1	78.1	105.4	67.6	78.5	102.5	70.2	79.5	101.5	64.6	75.2	106.3	97.0	99.8	103.3	102.6	102.8
1-16	WEIGHTED AVERAGE	99.7	70.5	83.7	99.0	71.3	83.8	99.3	74.0	85.4	99.7	66.9	81.7	100.8	105.1	103.2	103.1	101.8	102.4

	DIVISION	WASHINGTON			WEST VIRGINIA						WISCONSIN						WYOMING		
		TACOMA			CHARLESTON			HUNTINGTON			MADISON			MILWAUKEE			CHEYENNE		
		MAT.	INST.	TOTAL	MAT.	INST.	TOTAL	MAT.	INST.	TOTAL	MAT.	INST.	TOTAL	MAT.	INST.	TOTAL	MAT.	INST.	TOTAL
2	SITE WORK	102.0	98.2	100.3	115.3	100.7	108.7	120.4	104.2	113.0	80.8	100.0	89.5	78.1	97.8	87.0	106.5	94.6	101.1
3.1	FORMWORK	104.5	103.5	103.8	107.5	99.9	101.6	90.9	99.9	97.9	99.2	86.6	89.4	105.1	98.9	100.2	99.1	85.6	88.6
3.2	REINFORCING	100.7	97.7	99.4	120.9	96.9	110.7	116.4	102.1	110.4	106.3	86.2	97.8	105.3	104.0	104.7	103.4	85.3	95.7
3.3	CAST IN PLACE CONC.	100.7	103.1	102.2	117.9	99.3	106.4	114.5	100.3	105.7	100.2	103.0	102.0	86.1	95.2	91.7	98.4	96.2	97.0
3	CONCRETE	101.5	102.8	102.3	116.5	99.3	105.4	110.2	100.3	103.8	101.4	95.1	97.3	94.1	97.4	96.2	99.6	91.1	94.1
4	MASONRY	120.5	107.2	110.3	96.7	99.6	99.0	100.6	96.1	97.1	99.9	86.3	89.4	104.8	98.2	99.7	110.0	85.4	91.1
5	METALS	99.3	98.9	99.2	112.2	97.5	106.9	102.7	100.8	102.0	96.9	93.7	95.7	94.8	100.7	96.9	87.7	89.2	88.3
6	WOOD & PLASTICS	101.0	100.6	100.8	117.5	101.1	108.3	102.4	99.3	100.7	99.5	86.7	92.3	96.2	97.0	96.7	96.3	85.9	90.4
7	MOISTURE PROTECTION	110.9	111.7	111.2	89.3	98.4	92.1	88.7	100.0	92.3	87.3	86.3	87.0	104.3	97.7	102.2	89.0	84.6	87.6
8	DOORS, WINDOWS, GLASS	104.1	101.8	102.9	107.5	93.1	99.9	113.5	93.8	103.2	98.3	86.3	92.0	95.7	97.3	96.5	109.4	85.5	96.9
9.1	LATH & PLASTER	108.1	103.4	104.5	118.7	94.8	100.5	116.0	99.5	103.4	105.4	86.3	90.8	103.0	94.6	96.6	107.7	85.8	91.0
9.2	DRYWALL	99.5	101.6	100.5	100.6	98.4	99.5	98.1	98.4	98.2	109.2	86.3	98.3	94.5	97.2	95.8	95.4	85.4	90.6
9.5	ACOUSTICAL WORK	105.8	100.7	103.0	109.8	100.6	104.8	111.6	100.1	105.3	105.5	86.3	96.8	99.4	97.3	98.3	96.3	85.4	90.3
9.6	FLOORING	97.7	106.3	100.0	85.1	102.9	89.8	81.4	99.0	86.0	96.9	86.3	94.1	101.9	91.2	99.1	93.2	85.4	91.1
9.8	PAINTING	110.1	102.4	103.9	123.9	84.6	92.3	114.7	90.7	95.4	99.1	86.2	88.8	102.5	91.2	93.5	100.4	85.3	88.3
9	FINISHES	100.8	102.2	101.6	100.6	93.8	96.9	97.5	95.9	96.6	105.3	86.3	95.0	97.6	94.5	95.9	95.8	85.4	90.1
10-14	TOTAL DIV. 10-14	100.0	111.4	103.4	100.0	96.0	98.8	100.0	96.6	98.9	100.0	86.3	95.8	100.0	98.3	99.5	100.0	90.2	97.0
15	MECHANICAL	102.1	107.5	104.9	101.8	92.6	97.1	104.5	96.4	100.3	102.2	86.7	94.1	99.7	93.1	96.3	103.4	85.2	94.0
16	ELECTRICAL	103.3	107.6	106.4	94.7	95.0	94.9	97.7	95.8	96.4	88.2	86.2	86.8	93.7	89.4	90.6	97.6	85.3	88.9
1-16	WEIGHTED AVERAGE	103.2	105.0	104.2	104.3	96.6	100.1	103.6	97.6	100.3	97.8	89.2	93.1	96.8	95.7	96.2	99.5	87.4	92.9

	DIVISION	CANADA																	
		EDMONTON			MONTREAL			QUEBEC			TORONTO			VANCOUVER			WINNIPEG		
		MAT.	INST.	TOTAL	MAT.	INST.	TOTAL	MAT.	INST.	TOTAL	MAT.	INST.	TOTAL	MAT.	INST.	TOTAL	MAT.	INST.	TOTAL
2	SITE WORK	103.3	102.9	103.1	91.3	95.9	93.3	90.8	71.1	81.9	112.4	103.9	108.6	112.6	107.7	110.4	105.4	103.5	104.5
3.1	FORMWORK	110.6	105.4	106.5	108.8	92.7	96.3	103.6	92.7	95.1	108.2	115.2	113.6	99.7	122.5	117.5	98.4	94.9	95.7
3.2	REINFORCING	119.8	105.6	113.8	119.5	79.7	102.7	117.8	79.7	101.7	80.2	100.0	88.6	100.2	112.1	105.3	118.0	88.4	105.4
3.3	CAST IN PLACE CONC.	119.3	100.8	107.8	107.8	97.5	101.4	146.0	73.0	100.6	164.4	103.7	126.6	117.4	107.3	111.1	108.7	105.6	106.7
3	CONCRETE	117.6	103.0	108.2	110.6	94.1	99.9	131.3	81.3	98.9	134.6	107.8	117.3	110.0	113.7	112.4	108.7	99.9	103.0
4	MASONRY	110.6	105.6	106.8	116.3	92.4	98.0	106.9	92.4	95.8	121.1	110.1	112.7	123.6	117.6	119.0	126.4	95.3	102.5
5	METALS	102.2	104.1	102.9	89.3	92.0	90.3	87.1	83.2	85.7	103.9	103.6	103.8	104.9	108.6	106.3	108.5	98.1	104.7
6	WOOD & PLASTICS	94.0	105.6	100.5	109.4	93.1	100.3	106.1	93.1	98.8	104.2	113.7	109.5	93.1	119.3	107.8	94.4	94.8	94.6
7	MOISTURE PROTECTION	99.4	105.8	101.4	92.3	93.2	92.5	91.7	93.2	92.2	97.2	108.1	100.7	97.0	123.3	112.1	104.0	94.7	101.1
8	DOORS, WINDOWS, GLASS	101.6	103.8	102.8	102.1	91.5	96.5	102.8	91.7	97.0	96.4	107.9	102.4	108.6	116.7	112.9	111.3	94.4	102.4
9.1	LATH & PLASTER	109.2	105.1	106.0	99.2	91.3	93.2	95.1	91.5	92.4	100.6	101.1	101.0	111.3	118.5	116.8	111.7	93.5	97.8
9.2	DRYWALL	108.3	105.7	107.0	111.4	92.0	102.2	114.4	92.2	103.9	107.1	106.7	106.9	112.1	119.7	115.7	105.3	94.4	100.1
9.5	ACOUSTICAL WORK	80.9	105.7	94.4	102.6	92.9	97.3	102.6	92.9	97.3	102.6	114.8	109.3	84.0	120.0	103.7	101.8	94.4	97.7
9.6	FLOORING	99.3	105.7	101.0	92.6	95.4	93.3	89.1	95.4	90.8	96.3	104.4	98.4	99.5	119.6	104.9	109.9	93.2	105.5
9.8	PAINTING	112.9	95.6	99.0	124.4	89.1	96.0	134.6	89.1	98.0	117.9	105.4	107.8	121.4	121.5	121.5	115.9	94.3	98.6
9	FINISHES	104.6	102.1	103.2	107.5	91.2	98.7	109.3	91.4	99.6	105.2	106.4	105.9	107.9	120.3	114.6	107.3	94.2	100.2
10-14	TOTAL DIV. 10-14	100.0	105.7	101.7	100.0	94.5	98.3	100.0	94.7	98.4	100.0	104.1	101.2	100.0	110.1	103.0	100.0	97.6	99.2
15	MECHANICAL	99.8	105.8	102.9	101.4	86.0	93.4	100.4	86.0	92.9	103.7	103.6	103.7	98.1	109.5	104.0	99.0	98.9	98.9
16	ELECTRICAL	107.3	104.4	105.2	103.3	86.7	91.5	105.1	86.8	92.0	103.6	104.1	104.0	99.8	110.9	107.7	110.7	100.6	103.5
1-16	WEIGHTED AVERAGE	104.3	104.3	104.3	101.8	91.0	95.9	104.1	86.8	94.6	108.4	106.4	107.3	105.3	113.7	109.9	105.5	97.6	101.3

CIRCLE REFERENCE NUMBERS

① Performance Bond (Div. 1.1-34)

The table below shows the performance bond rate for a job scheduled to be completed in about 24 months. The rates are "preferred" rates which are offered to contractors that the bonding company considers financially sound and capable of doing the work.

Contractors should prequalify through a bonding company agency before submitting a bid on a contract which requires a bond.

Contract Amount	Building Construction Projects	Highways & Bridges — New Construction	Highways & Bridges — Highway Resurfacing
First $ 500,000 bid	$9.00 per M	$7.20 per M	$4.50 per M
Next 2,000,000 bid	$ 4,500 plus $5.85 per M	$ 3,600 plus $4.50 per M	$ 2,250 plus $4.20 per M
Next 2,500,000 bid	16,200 plus 4.90 per M	12,600 plus 3.85 per M	10,650 plus 3.60 per M
Next 2,500,000 bid	28,450 plus 4.20 per M	22,225 plus 3.40 per M	19,650 plus 3.15 per M
Over 7,500,000 bid	38,950 plus 4.10 per M	30,725 plus 3.10 per M	27,525 plus 2.80 per M

② Builders Risk Insurance (Div. 1.1-18)

Builder's Risk Insurance is insurance on a building during construction. Premiums are paid by the owner or the contractor. Blasting, collapse & underground insurance would raise total insurance costs above those listed. Floater policy for materials delivered to the job runs $1.00 to $1.50 per $100 value. Contractor equipment insurance runs $.50 to $2.50 per $100 value.

Tabulated below are New England Builder's Risk insurance rates in dollars per $100 value for $1000 deductible. For $25,000 deductible, rates can be reduced 13% to 34%. On contracts over $1,000,000, rates may be lower than those tabulated. High liability limits may be difficult to obtain. Policies are written annually for the total completed value in place. For "all risk" insurance (excluding flood, earthquake and certain other perils) add $.022 to total rates below.

Coverage	Frame Construction (Class 1) Range	Frame Construction (Class 1) Average	Brick Construction (Class 4) Range	Brick Construction (Class 4) Average	Fire Resistive (Class 6) Range	Fire Resistive (Class 6) Average
Fire Insurance	$.165 to .378	$.287	$.100 to .177	$.170	$.059 to .168	$.134
Extended Coverage	.077 to .111	.098	.042 to .112	.095	.029 to .075	.063
Vandalism	.008 to .011	.010	.008 to .011	.010	.008 to .011	.010
Total Annual Rate	$.250 to .500	$.395	$.150 to .300	$.275	$.096 to .254	$.207

③ Overtime (Div. 1.1-32)

One way to improve the completion date of a project or eliminate negative float from a schedule, is to compress activity duration times. This can be achieved by increasing the crew size or working overtime with the proposed crew.

To determine the costs of working overtime to compress activity duration times, consider the following examples. Below is an overtime efficiency and cost chart based on a 5, 6, or 7 day week with an eight through twelve hour day. Payroll percentage increases for time and one half and double time are shown for the various working days.

Days per Week	Hours per Day	Production Efficiency 1 Week	2 Weeks	3 Weeks	4 Weeks	Average 4 Weeks	Payroll Cost Factors @ 1 1/2 Times	@ 2 Times
5	8	100%	100%	100%	100%	100%	100%	100%
5	9	100	100	95	90	96.25	105.6	111.1
5	10	100	95	90	85	91.25	110.0	120.0
5	11	95	90	75	65	81.25	113.6	127.3
5	12	90	85	70	60	76.25	116.7	133.3
6	8	100	100	95	90	96.25	108.3	116.7
6	9	100	95	90	85	92.50	113.0	125.9
6	10	95	90	85	80	87.50	116.7	133.3
6	11	95	85	70	65	78.75	119.7	139.4
6	12	90	80	65	60	73.75	122.2	144.4
7	8	100	95	85	75	88.75	114.3	128.6
7	9	95	90	80	70	83.75	118.3	136.5
7	10	90	85	75	65	78.75	121.4	142.9
7	11	85	80	65	60	72.50	124.0	148.1
7	12	85	75	60	55	68.75	126.2	152.4

CIRCLE REFERENCE NUMBERS

④ General Contractor's Overhead (Div. 1.1)

The table below shows a contractor's overhead as a percentage of direct cost in two ways. The figures on the right are for the overhead, markup based on both material and labor. The figures on the left are based on the entire overhead applied only to the labor. This figure would be used if the owner supplied the materials or if a contract is for labor only.

Items of General Contractor's Indirect Costs	% of Direct Costs	
	As a Markup of Labor Only	As a Markup of Both Material and Labor
Field Supervision	6.0%	2.4%
Main Office Expense (see details below)	9.2	7.7
Tools and Minor Equipment	1.0	0.4
Workers' Compensation & Employers' Liability. See ⑦	9.3	3.7
Field Office, Sheds, Photos, Etc.	2.0	0.8
Performance and Payment Bond, .5% to .9%. See ①	0.7	0.7
Unemployment Tax See ⑤ (Combined Federal and State)	5.5	2.2
Social Security and Medicare (7.05% of first $40,000)	7.0	2.7
Sales Tax — add if applicable 48/80 x % as markup of total direct costs including both material and labor. See ⑧		
Sub Total	40.7%	20.6%
*Builder's Risk Insurance ranges from .1% to .5%. See ②	0.4	0.4
*Public Liability Insurance	0.8	0.8
Grand Total	41.9%	21.8%

Paid by Owner or Contractor

Main Office Expense

A General Contractor's main office expense consists of many items not detailed in the front portion of the book. The percentage of main office expense declines with increased annual volume of the contractor. Typical main office expense ranges from 2% to 20% with the median about 7.2% of total volume. This equals about 7.7% of direct costs.

The following are approximate percentages for different items usually included in a General Contractor's main office overhead. With different accounting procedures, these percentages may vary.

Item	Typical Range	Average
Managers', clerical and estimators' salaries	40% to 55%	48%
Profit sharing, pension and bonus plans	2 to 20	12
Insurance	5 to 8	6
Estimating and project management (not including salaries)	5 to 9	7
Legal, accounting and data processing	0.5 to 5	3
Automobile and light truck expense	2 to 8	5
Depreciation of overhead capital expenditures	2 to 6	4
Maintenance of office equipment	0.1 to 1.5	1
Office rental	3 to 5	4
Utilities including phone and light	1 to 3	2
Miscellaneous	5 to 15	8
Total		100%

Unemployment Taxes and Social Security Taxes (Div. 1.1-56)

Mass. State Unemployment tax ranges from 1.5% to 5.7% plus an experience rating assessment the following year, on the first $7,000 of wages. Federal Unemployment tax is 3.5% of the first $7,000 of wages. This is reduced by a credit for payment to the state. The minimum Federal Unemployment tax is .8% after all credits.

Combined rates in Mass. thus vary from 2.3% to 6.5% of the first $7,000 of wages. Combined average U.S. rate is about 5.5% of the first $7,000. Contractors with permanent workers will pay less since the average annual wages for skilled workers is $20.50 x 2,000 hours or about $41,000 per year. The average combined rate for U.S. would thus be 5.5% x 7,000 ÷ 41,000 = .94% of total wages for permanent employees.

Rates not only vary from state to state but also with the experience rating of the contractor.

Social Security (FICA) for 1986 is estimated at time of publication to be 7.05% of wages up to $40,000.

CIRCLE REFERENCE NUMBERS

⑥ Subcontractor's Overhead & Profit (Div. 1.1)

Listed below in the last two columns are **average** billing rates for Subcontractor's labor.

The Base Rate is for the building construction industry and includes the usual negotiated fringe benefits. These wage rates with trends are listed on the inside back cover. Workers' Compensation is a national average of states which have established rates in each trade. Average Fixed Overhead is a total of average rates for U.S. and State Unemployment, 5.5%; Social Security (FICA), 7.05%; Builders' Risk, 0.38%; and Public Liability, 0.82%. These are analyzed in ② and ⑤. All the rates except Social Security vary from state to state as well as from company to company.

The Subcontractor's Overhead presumes annual billing between 0.5 and 1.5 million dollars. A Subcontractor with lower annual billing usually has a higher overhead percentage.

Subcontractor's Overhead varies greatly within each trade. Some controlling factors are Annual Volume, Type Job, Size Job, Location, Local Economic Conditions, Engineering and Logistical Support Staff and Equipment Requirements. These should be examined carefully for each job.

Abbr.	Trade	Base Rate Incl. Fringes Hourly	Base Rate Incl. Fringes Daily	Workers' Comp. Ins.	Average Fixed Overhead	Subs Overhead	Subs Profit	Subs Total Overhead & Profit %	Subs Total Overhead & Profit Amount	Rate with Subs O & P Hourly	Rate with Subs O & P Daily
Skwk	Skilled Workers Average (35 trades)	$20.50	$164.00	9.3%	13.8%	12.8%	10%	45.9%	$ 9.40	$29.90	$239.20
	Helpers Average (5 trades)	15.55	124.40	9.8		13.0		46.6	7.25	22.80	182.40
	Foremen Average, Inside (50¢ over trade)	21.00	168.00	9.3		12.8		45.9	9.65	30.65	245.20
	Foremen Average, Outside ($2.00 over trade)	22.50	180.00	9.3		12.8		45.9	10.35	32.85	262.80
Clab	Common Building Laborers	15.90	127.20	10.1		11.0		44.9	7.15	23.05	184.40
Asbe	Asbestos Workers	22.75	182.00	7.7		16.0		47.5	10.80	33.55	268.40
Boil	Boilermakers	22.75	182.00	6.6		16.0		46.4	10.55	33.30	266.40
Bric	Bricklayers	20.50	164.00	7.6		11.0		42.4	8.70	29.20	233.60
Brhe	Bricklayer Helpers	16.00	128.00	7.6		11.0		42.4	6.80	22.80	182.40
Carp	Carpenters	20.00	160.00	10.1		11.0		44.9	9.00	29.00	232.00
Cefi	Cement Finishers	19.20	153.60	5.9		11.0		40.7	7.80	27.00	216.00
Elec	Electricians	22.40	179.20	4.0		16.0		43.8	9.80	32.20	257.60
Elev	Elevator Constructors	22.65	181.20	5.5		16.0		45.3	10.25	32.90	263.20
Eqhv	Equipment Operators, Crane or Shovel	21.05	168.40	7.2		14.0		45.0	9.45	30.50	244.00
Eqmd	Equipment Operators, Medium Equipment	20.60	164.80	7.2		14.0		45.0	9.25	29.85	238.80
Eqlt	Equipment Operators, Light Equipment	19.45	155.60	7.2		14.0		45.0	8.75	28.20	225.60
Eqol	Equipment Operators, Oilers	17.50	140.00	7.2		14.0		45.0	7.90	25.40	203.20
Eqmm	Equipment Operators, Master Mechanics	21.80	174.40	7.2		14.0		45.0	9.80	31.60	252.80
Glaz	Glaziers	20.15	161.20	7.9		11.0		42.7	8.60	28.75	230.00
Lath	Lathers	20.10	160.80	6.3		11.0		41.1	8.25	28.35	226.80
Marb	Marble Setters	20.10	160.80	7.6		11.0		42.4	8.50	28.60	228.80
Mill	Millwrights	20.75	166.00	6.6		11.0		41.4	8.60	29.35	234.80
Mstz	Mosaic and Terrazzo Workers	19.90	159.20	5.4		11.0		40.2	8.00	27.90	223.20
Pord	Painters, Ordinary	19.25	154.00	7.7		11.0		42.5	8.20	27.45	219.60
Psst	Painters, Structural Steel	20.00	160.00	27.0		11.0		61.8	12.35	32.35	258.80
Pape	Paper Hangers	19.50	156.00	7.7		11.0		42.5	8.30	27.80	222.40
Pile	Pile Drivers	20.10	160.80	17.0		16.0		56.8	11.40	31.50	252.00
Plas	Plasterers	19.90	159.20	7.7		11.0		42.5	8.45	28.35	226.80
Plah	Plasterer Helpers	16.50	132.00	7.7		11.0		42.5	7.00	23.50	188.00
Plum	Plumbers	22.55	180.40	4.8		16.0		44.6	10.05	32.60	260.80
Rodm	Rodmen (Reinforcing)	21.75	174.00	16.8		14.0		54.6	11.90	33.65	269.20
Rofc	Roofers, Composition	18.80	150.40	18.2		11.0		53.0	9.95	28.75	230.00
Rots	Roofers, Tile & Slate	18.95	151.60	18.2		11.0		53.0	10.05	29.00	232.00
Rohe	Roofer Helpers (Composition)	13.75	110.00	18.2		11.0		53.0	7.30	21.05	168.40
Shee	Sheet Metal Workers	22.70	181.60	6.3		16.0		46.1	10.45	33.15	265.20
Spri	Sprinkler Installers	23.25	186.00	5.5		16.0		45.3	10.55	33.80	270.40
Stpi	Steamfitters or Pipefitters	22.75	182.00	4.8		16.0		44.6	10.15	32.90	263.20
Ston	Stone Masons	20.30	162.40	7.6		11.0		42.4	8.60	28.90	231.20
Sswk	Structural Steel Workers	21.70	173.60	19.3		14.0		57.1	12.40	34.10	272.80
Tilf	Tile Layers (Floor)	19.75	158.00	5.4		11.0		40.2	7.95	27.70	221.60
Tilh	Tile Layer Helpers	15.60	124.80	5.4		11.0		40.2	6.30	21.90	175.20
Trlt	Truck Drivers, Light	16.35	130.80	8.6		11.0		43.4	7.10	23.45	187.60
Trhv	Truck Drivers, Heavy	16.60	132.80	8.6		11.0		43.4	7.20	23.80	190.40
Sswl	Welders, Structural Steel	21.70	173.60	19.3		14.0		57.1	12.40	34.10	272.80
Wrck	*Wrecking	15.90	127.20	20.7		11.0		55.5	8.80	24.70	197.60

*Not included in Averages.

CIRCLE REFERENCE NUMBERS

⑦ Workers' Compensation (Div. 1.1)

The table below tabulates the national averages for Workers' Compensation insurance rates by trade and type of building. The average "Insurance Rate" is multiplied by the "% of Building Cost" for each trade. This produces the "Workers' Compensation Cost" by % of total labor cost, to be added for each trade by building type to determine the weighted average Workers' Compensation rate for the building types analyzed.

Trade	Insurance Rate (% of Labor Cost)		% of Building Cost			Workers' Compensation Cost		
	Range	Average	Office Bldgs.	Schools & Apts.	Mfg.	Office Bldgs.	Schools & Apts.	Mfg.
Excavation, Grading, etc.	2.1% to 27.8%	7.2%	4.8%	4.9%	4.5%	.35%	.35%	.32%
Piles & Foundations	4.0 to 47.6	17.0	7.1	5.2	8.7	1.21	.88	1.48
Concrete	2.3 to 27.1	9.2	5.0	14.8	3.7	.46	1.36	.34
Masonry	1.6 to 18.7	7.6	6.9	7.5	1.9	.52	.57	.14
Structural Steel	3.1 to 42.9	19.3	10.7	3.9	17.6	2.07	.75	3.40
Miscellaneous & Ornamental Metals	1.3 to 15.7	6.8	2.8	4.0	3.6	.19	.27	.24
Carpentry & Millwork	2.9 to 54.1	10.1	3.7	4.0	0.5	.37	.40	.05
Metal or Composition Siding	2.7 to 23.3	7.8	2.3	0.3	4.3	.18	.02	.34
Roofing	3.7 to 52.7	18.2	2.3	2.6	3.1	.42	.47	.56
Doors & Hardware	2.0 to 15.9	5.9	0.9	1.4	0.4	.05	.08	.02
Sash & Glazing	3.0 to 33.3	7.9	3.5	4.0	1.0	.28	.32	.08
Lath & Plaster	2.3 to 22.9	7.7	3.3	6.9	0.8	.25	.53	.06
Tile, Marble & Floors	1.1 to 18.7	5.4	2.6	3.0	0.5	.14	.16	.03
Acoustical Ceilings	1.6 to 15.4	6.3	2.4	0.2	0.3	.15	.01	.02
Painting	2.6 to 20.0	7.7	1.5	1.6	1.6	.12	.12	.12
Interior Partitions	2.9 to 54.1	10.1	3.9	4.3	4.4	.39	.43	.44
Miscellaneous Items	1.2 to 92.4	9.6	5.2	3.7	9.7	.50	.36	.93
Elevators	1.4 to 15.6	5.5	2.1	1.1	2.2	.12	.06	.12
Sprinklers	1.4 to 16.8	5.5	0.5	—	2.0	.03	—	.11
Plumbing	1.3 to 12.5	4.8	4.9	7.2	5.2	.24	.35	.25
Heat., Vent., Air Conditioning	1.8 to 14.5	6.3	13.5	11.0	12.9	.85	.69	.81
Electrical	1.0 to 13.1	4.0	10.1	8.4	11.1	.40	.34	.44
Total	1.0% to 92.4%	—	100.0%	100.0%	100.0%	9.29%	8.52%	10.30%
Overall Weighted Average							9.37%	

The table below lists the weighted average Workers' Compensation base rate for each state with a factor comparing this with the national average of 9.3%.

State	Weighted Average	Factor	State	Weighted Average	Factor	State	Weighted Average	Factor
Alabama	6.4%	69	Kentucky	5.7%	61	North Dakota	6.9%	74
Alaska	12.4	133	Louisiana	6.4	69	Ohio	7.0	75
Arizona	10.5	113	Maine	14.3	154	Oklahoma	8.1	87
Arkansas	6.4	69	Maryland	14.7	158	Oregon	21.5	231
California	11.5	124	Massachusetts	12.5	134	Pennsylvania	10.2	110
Colorado	10.9	117	Michigan	11.3	122	Rhode Island	12.2	131
Connecticut	15.0	161	Minnesota	13.6	146	South Carolina	7.4	80
Delaware	9.7	104	Mississippi	5.4	58	South Dakota	6.1	66
District of Columbia	16.5	177	Missouri	5.2	56	Tennessee	4.8	52
Florida	11.2	120	Montana	12.0	129	Texas	6.0	65
Georgia	5.4	58	Nebraska	6.1	66	Utah	5.3	57
Hawaii	25.1	270	Nevada	10.6	114	Vermont	6.1	66
Idaho	8.1	87	New Hampshire	13.3	143	Virginia	7.3	78
Illinois	11.4	123	New Jersey	6.6	71	Washington	6.8	73
Indiana	2.5	27	New Mexico	14.4	155	West Virginia	8.7	94
Iowa	6.3	68	New York	8.9	96	Wisconsin	6.5	70
Kansas	6.3	68	North Carolina	4.7	51	Wyoming	5.4	58
Weighted Average for U.S. is 9.3% of payroll = 100								

Rates in the following table are the base or manual costs per $100 of payroll for Workers' Compensation in each state. Rates are usually applied to straight time wages only and not to premium time wages and bonuses.

The weighted average skilled worker rate for 35 trades is 9.3%. For bidding purposes, apply the full value of Workers' Compensation directly to total labor costs, or if labor is 32%, materials 48% and overhead and profit 20% of total cost, carry 32/80 x 9.3% = 3.7% of cost (before overhead and profit) into overhead. Rates vary not only from state to state but also with the experience rating of the contractor.

Rates are the most current available at the time of publication.

CIRCLE REFERENCE NUMBERS

⑦ Worker's Compensation (cont.)

STATE	CARPENTRY — 3 stories or less	CARPENTRY — interior cab. work	CARPENTRY — general	CONCRETE WORK—NOC	CONCRETE WORK — flat (flr., sdwk.)	ELECTRICAL WIRING — inside	EXCAVATION — earth NOC	EXCAVATION — rock	GLAZIERS	INSULATION WORK	LATHING	MASONRY	PAINTING & DECORATING	PILE DRIVING	PLASTERING	PLUMBING	ROOFING	SHEET METAL WORK (HVAC)	STEEL ERECTION — door & sash	STEEL ERECTION — inter. ornam.	STEEL ERECTION — structure	STEEL ERECTION — NOC	TILE WORK — (interior ceramic)	WATERPROOFING	WRECKING
	5651	5437	5403	5213	5221	5190	6217	6217	5462	5479	5443	5022	5474	6003	5480	5183	5551	5538	5102	5102	5040	5057	5348	9014	5701
AL	6.32	3.56	6.49	6.53	3.67	3.00	6.05	6.05	5.61	4.85	3.51	4.27	6.46	16.85	5.66	2.61	8.74	5.68	3.87	3.87	8.60	11.69	4.33	3.18	11.69
AK	12.75	6.91	10.68	10.22	7.68	7.32	7.04	7.04	9.62	10.60	8.45	3.45	9.92	19.71	10.75	7.43	19.85	7.43	12.52	12.52	34.45	19.77	6.11	3.81	34.45
AZ	8.39	7.90	19.10	8.03	6.73	4.78	6.52	6.52	11.14	8.44	7.46	11.09	7.39	18.57	16.39	4.48	19.01	6.75	11.03	11.03	8.86	12.98	6.45	4.32	19.10
AR	6.31	3.50	6.71	6.33	3.71	2.51	5.21	5.21	4.78	4.78	4.27	3.61	4.69	14.97	4.61	2.47	10.77	4.38	4.20	4.20	20.68	7.73	2.93	2.66	20.68
CA	11.39	5.44	11.46	6.08	6.08	4.54	6.38	6.38	13.00	13.81	8.01	8.62	10.61	22.68	11.70	6.29	27.70	7.24	8.71	8.71	19.30	14.97	6.48	10.61	19.83
CO	7.79	5.90	10.41	8.64	6.56	3.93	8.99	8.99	9.11	8.87	7.03	12.63	7.23	31.08	12.42	5.16	25.99	6.63	6.85	6.85	15.40	18.09	3.99	3.92	18.09
CT	12.79	10.36	14.44	14.56	11.45	4.05	8.72	8.72	11.37	14.56	9.30	16.89	10.88	19.75	9.29	6.75	31.69	9.46	7.59	7.59	42.88	37.61	8.09	3.27	42.88
DE	9.39	9.39	9.39	8.44	5.76	3.28	7.51	7.51	9.36	9.39	6.96	6.04	10.20	10.90	6.96	3.35	25.31	8.00	9.19	9.19	17.99	9.19	6.43	6.04	24.81
DC	8.82	7.14	12.71	20.40	8.80	10.71	12.97	12.97	11.41	11.21	8.56	17.57	9.86	30.30	10.12	12.48	22.50	9.20	15.72	15.72	40.45	32.71	18.65	4.13	40.45
FL	10.58	5.58	12.58	13.37	8.66	5.55	9.73	9.73	9.58	8.89	8.22	9.62	8.78	17.74	9.44	6.86	24.62	8.10	7.50	7.50	18.28	17.36	5.08	5.73	18.28
GA	4.43	3.10	6.70	4.75	3.14	2.53	5.58	5.58	6.52	4.33	2.50	3.67	3.63	9.63	4.33	3.44	7.70	4.68	3.63	3.63	9.96	11.50	2.35	2.24	11.50
HI	15.42	15.87	54.13	27.08	13.77	13.13	27.75	27.75	33.28	16.18	15.39	18.68	16.60	44.30	22.92	11.57	44.35	11.43	11.46	11.46	37.11	42.12	13.90	11.62	37.11
ID	9.19	5.16	9.86	5.94	4.34	3.79	6.71	6.71	6.77	7.23	5.40	7.35	7.05	14.45	7.94	3.55	14.22	5.55	5.77	5.77	13.89	13.88	4.56	4.13	13.89
IL	8.76	5.99	9.61	14.06	5.83	4.76	6.40	6.40	8.51	7.19	6.08	9.34	8.96	18.88	6.14	7.12	21.60	7.26	9.23	9.23	29.05	33.16	5.81	2.92	33.16
IN	2.73	1.98	2.86	2.32	1.61	1.00	2.09	2.09	3.00	2.14	1.55	1.59	2.58	3.95	2.28	1.26	3.71	1.82	1.33	1.33	3.12	7.14	1.12	1.16	3.12
IA	3.51	3.43	6.90	7.09	3.54	3.65	4.27	4.27	3.86	4.77	3.73	4.54	6.82	9.15	4.13	4.83	9.07	4.59	5.37	5.37	14.64	16.97	2.61	2.46	14.64
KS	4.82	4.20	5.40	4.02	4.15	1.93	3.17	3.17	5.64	8.25	4.53	4.87	6.75	16.77	4.13	2.85	12.72	3.85	3.86	3.86	7.97	17.39	3.11	1.87	17.39
KY	6.94	3.02	6.87	3.80	3.75	2.55	4.27	4.27	4.58	5.04	4.34	4.60	5.99	9.80	4.12	2.72	11.54	3.94	4.14	4.14	10.81	7.33	3.36	4.09	10.81
LA	8.06	4.35	7.60	4.99	3.64	2.47	5.34	5.34	5.24	4.26	3.08	3.35	6.07	17.46	5.05	3.87	10.71	4.42	4.36	4.36	10.96	8.65	4.26	3.15	10.96
ME	6.59	7.64	19.90	14.58	7.71	5.73	11.59	11.59	11.18	10.36	7.54	10.04	11.33	30.32	10.32	7.39	27.20	9.41	9.07	9.07	36.50	29.85	7.24	4.96	36.50
MD	10.09	8.83	10.31	22.62	7.82	7.25	10.20	10.20	13.13	12.45	8.24	9.36	11.23	26.39	9.00	9.48	26.49	11.38	11.42	11.42	37.95	28.06	8.37	4.24	37.95
MA	8.52	4.57	18.12	12.64	5.57	3.50	5.32	5.32	8.46	6.20	6.03	10.41	11.00	14.09	7.20	5.25	52.69	7.33	7.92	7.92	30.32	28.70	5.65	5.05	26.38
MI	7.54	6.27	10.24	14.50	8.88	3.52	9.97	9.97	11.79	9.40	9.08	9.60	9.71	17.68	10.09	5.01	19.17	7.52	6.74	6.74	17.29	19.78	10.12	NA	17.29
MN	10.16	10.16	20.31	13.72	8.15	4.15	8.36	8.36	9.07	8.83	9.18	10.01	10.02	23.58	9.18	8.10	28.30	8.07	11.37	11.37	27.75	30.08	6.83	5.71	27.75
MS	4.90	3.31	6.08	3.97	3.36	3.12	4.87	4.87	4.25	3.36	3.69	2.98	3.16	17.09	5.12	1.61	7.40	4.11	3.94	3.94	10.64	7.11	3.84	2.37	10.64
MO	3.67	3.73	3.83	4.79	2.91	1.75	3.47	3.47	4.62	5.47	3.44	4.09	5.81	13.56	3.85	2.49	8.22	3.49	4.00	4.00	8.19	11.36	2.40	3.05	11.36
MT	9.65	6.44	13.37	5.89	5.90	4.33	9.15	9.15	7.62	10.66	6.66	12.75	10.38	27.11	9.09	7.00	25.10	7.32	7.16	7.16	32.17	20.61	5.29	5.70	32.17
NE	5.80	3.40	5.70	5.30	3.98	2.54	5.12	5.12	5.54	4.54	4.67	4.84	4.15	12.41	6.35	3.16	12.29	3.89	3.27	3.27	8.65	13.99	3.03	4.59	13.99
NV	10.82	10.82	10.82	8.51	8.51	5.48	8.51	8.51	7.34	10.82	10.82	10.24	6.92	8.51	6.92	6.74	17.31	6.74	10.82	10.82	17.11	17.11	7.34	8.51	17.11
NH	11.45	5.85	11.18	15.59	7.44	5.03	15.17	15.17	8.77	10.49	9.66	8.86	10.76	26.95	12.67	6.85	36.18	6.82	8.09	8.09	24.11	19.47	6.43	4.28	24.11
NJ	5.48	4.60	5.48	5.81	4.52	2.43	5.11	5.11	4.21	5.60	6.34	6.94	8.10	8.05	6.34	2.81	13.39	3.82	8.02	8.02	14.52	6.88	2.70	2.91	19.97
NM	11.33	10.26	12.73	17.13	9.18	4.79	7.69	7.69	12.36	13.76	8.22	11.01	8.17	47.64	13.16	8.25	28.26	9.27	12.17	12.17	19.90	21.01	8.75	5.28	19.90
NY	6.63	4.03	8.89	12.97	7.04	3.92	8.40	8.40	9.92	4.73	7.20	8.65	7.28	14.54	12.46	5.39	NA	7.00	4.19	4.19	18.60	16.60	5.70	3.33	26.03
NC	3.36	2.56	5.20	3.22	2.45	2.76	3.40	3.40	3.55	3.85	2.92	2.51	4.08	13.25	3.30	2.45	8.24	3.48	2.88	2.88	16.93	5.36	2.22	1.15	16.93
ND	5.93	5.93	5.93	5.25	5.25	2.20	5.42	5.42	8.16	3.95	3.08	3.10	4.92	13.65	3.08	6.22	8.20	6.22	5.93	5.93	13.65	13.65	2.90	8.20	NA
OH	4.77	4.77	4.77	4.81	4.81	1.98	4.81	4.81	5.38	5.01	5.01	4.40	5.38	4.81	5.01	3.26	10.95	8.26	NA	NA	18.31	18.31	2.32	10.95	NA
OK	7.54	4.73	7.21	6.36	4.30	2.95	5.98	5.98	6.15	6.69	4.78	4.36	4.54	16.30	7.28	4.39	19.81	5.51	5.70	5.70	23.36	12.95	3.59	3.70	23.36
OR	23.29	11.36	20.02	20.24	12.97	6.02	14.97	14.97	13.68	22.40	15.14	16.70	20.00	39.32	16.21	8.92	47.47	14.54	15.61	15.61	38.51	39.13	17.39	13.57	38.51
PA	8.64	7.65	8.64	11.55	5.70	4.25	7.95	7.95	9.67	8.64	8.49	8.84	11.84	11.24	8.49	5.01	17.57	8.21	8.87	8.87	24.68	8.87	6.33	8.84	42.10
RI	8.14	5.37	9.28	10.02	8.89	5.14	7.64	7.64	9.93	7.82	5.97	8.37	9.18	24.41	8.30	3.74	24.12	6.21	7.33	7.33	40.04	38.15	4.77	4.11	40.04
SC	8.04	5.35	8.10	5.13	3.82	6.19	5.21	5.21	4.96	7.19	4.38	6.63	8.68	12.04	5.85	2.55	15.79	7.24	3.77	3.77	15.49	13.35	3.89	2.45	15.49
SD	5.04	3.70	6.91	5.66	3.96	3.20	6.57	6.57	4.85	5.21	3.85	3.85	5.00	11.20	4.69	3.39	13.55	3.83	4.03	4.03	11.08	10.81	3.27	1.94	11.08
TN	4.03	3.81	6.52	4.00	3.16	2.30	4.17	4.17	4.51	4.56	3.71	4.09	4.65	8.59	3.58	2.65	7.22	4.16	3.62	3.62	9.45	6.59	2.43	1.99	9.45
TX	5.44	3.90	5.44	5.13	3.57	2.94	3.66	3.66	4.69	5.97	2.39	3.72	3.78	11.43	7.02	2.97	14.31	5.68	3.35	3.35	16.95	6.59	2.88	3.54	14.11
UT	NA	NA	4.97	6.15	3.10	1.72	3.74	3.74	4.22	5.47	4.42	4.97	5.12	8.67	4.94	2.65	14.99	2.93	3.64	3.64	NA	8.17	2.64	2.44	8.17
VT	4.15	3.14	5.32	6.92	3.25	3.38	5.24	5.24	5.02	5.07	3.81	5.30	4.24	12.40	4.75	2.69	11.71	4.41	4.64	4.64	13.07	10.62	3.13	2.51	13.07
VA	4.86	5.63	7.62	7.75	3.70	3.49	4.89	4.89	4.67	7.85	8.94	5.91	6.25	10.65	4.54	3.57	13.76	6.92	4.37	4.37	14.34	16.97	3.86	2.35	14.34
WA	6.66	6.66	6.66	5.38	4.61	2.31	6.25	6.25	6.99	5.35	6.66	7.43	6.17	12.85	6.76	2.44	6.38	2.83	6.17	6.17	8.58	8.58	5.16	6.97	7.64
WV	9.49	9.49	9.49	11.46	11.46	3.89	9.02	9.02	4.98	4.98	12.16	6.82	12.16	7.62	12.16	3.91	10.98	4.98	6.67	6.67	9.10	6.67	6.82	2.80	9.10
WI	4.83	3.29	8.06	6.57	3.68	3.32	3.92	3.92	5.38	5.88	4.64	5.06	4.38	10.98	5.08	3.34	10.38	5.09	4.07	4.07	18.57	13.59	3.16	3.50	22.81
WY	5.00	5.00	5.00	5.00	5.00	5.00	5.00	5.00	5.00	5.00	5.00	5.00	5.00	5.00	5.00	5.00	5.00	5.00	5.00	5.00	5.00	5.00	5.00	5.00	5.00
AVG.	7.80	5.90	10.11	9.12	5.83	4.04	7.16	7.16	7.89	7.69	6.36	7.62	7.73	17.04	7.76	4.85	18.28	6.31	6.80	6.80	19.30	16.74	5.39	4.54	20.71

346

CIRCLE REFERENCE NUMBERS

⑦ Workers' Compensation (cont.) (Canada in Canadian dollars)

PROVINCE		Alberta	British Columbia	Manitoba	Ontario	New Brunswick	Newfndld & Labrador	Northwest Territories	Nova Scotia	Prince Edward Island	Quebec	Saskatchewan	Yukon
CARPENTRY—3 stories or less	Rate	5.00	3.50	3.06	3.47	3.80	4.50	4.75	2.00	4.25	7.83	4.50	3.00
	Code	8-04	060412	401	062-08	403	4-3	4-41	700	401	42251	B01-11	4
CARPENTRY—interior cab. work	Rate	5.00	3.50	3.06	3.47	3.80	4.50	4.75	2.00	4.25	7.83	4.50	3.00
	Code	8-04	060412	401	062-08	403	4-3	4-41	700	401	42251	B01-11	4
CARPENTRY—general	Rate	5.00	3.50	3.06	3.47	3.80	4.50	4.75	2.00	4.25	7.83	4.50	3.00
	Code	8-04	060412	401	062-08	403	4-3	4-41	700	401	42251	B01-11	4
CONCRETE WORK—NOC	Rate	5.50	6.02	3.06	7.15	3.80	4.50	4.75	2.00	4.25	10.75	4.50	4.50
	Code	8-01	070604	401	744-09	403	4-3	4-41	700	401	42221	B01-22	3
CONCRETE WORK—flat (flr., sidewalk)	Rate	4.50	6.02	3.06	7.15	3.80	4.50	4.75	2.00	4.25	10.75	4.50	4.50
	Code	6-01	070604	401	744-09	403	4-3	4-41	700	401	42221	B01-22	3
ELECTRICAL Wiring—inside	Rate	2.60	2.69	1.68	3.95	3.60	2.25	2.75	1.00	2.00	4.58	4.50	3.00
	Code	6-06	071100	402	864-07	404	4-2	4-46	701	402	42612	B01-03	4
EXCAVATION—earth NOC	Rate	5.55	5.35	4.74	9.75	3.80	4.50	5.00	2.00	4.25	9.21	3.50	6.00
	Code	6-07	072607	407	753-13	403	4-3	4-43	700	401	42141	B01-06	2
EXCAVATION—rock	Rate	5.55	5.35	4.74	9.75	3.80	4.50	5.00	2.00	4.25	9.21	3.50	6.00
	Code	6-07	072607	407	753-13	403	4-3	4-43	700	401	42141	B01-06	2
GLAZIERS	Rate	4.75	3.33	3.06	8.31	3.60	2.25	4.75	2.00	2.00	10.75	4.50	3.00
	Code	6-03	060236	401	873-11	404	4-2	4-41	700	402	42321	B01-20	4
INSULATION WORK	Rate	4.75	5.18	3.06	8.31	3.60	2.25	4.75	2.00	4.25	5.52	4.50	4.50
	Code	6-03	070504	401	873-11	404	4-2	4-41	700	401	42561	B01-17	3
LATHING	Rate	4.75	5.18	3.06	7.57	3.60	2.25	4.75	2.00	2.00	6.60	4.50	4.50
	Code	6-03	070500	401	854-12	404	4-2	4-41	700	402	42711	B01-17	3
MASONRY	Rate	4.85	6.18	3.06	7.57	3.80	4.50	4.75	2.00	4.25	7.83	4.50	4.50
	Code	6-04	070602	401	854-12	403	4-3	4-41	700	401	42311	B01-18	3
PAINTING & DECORATING	Rate	4.75	5.18	3.06	8.31	3.60	2.25	3.25	2.00	2.00	6.60	4.50	4.50
	Code	6-03	070501	401	873-11	404	4-2	4-49	700	402	42751	B01-02	3
PILE DRIVING	Rate	4.50	8.85	4.74	10.01	3.80	4.50	5.00	2.00	4.25	14.29	4.50	6.00
	Code	6-01	072502	407	836-13	404	4-3	4-43	700	401	42211	B01-38	2
PLASTERING	Rate	4.75	5.18	3.06	7.57	3.60	2.25	4.75	2.00	2.00	6.60	4.50	4.50
	Code	6-03	070502	401	854-12	404	4-2	4-41	700	402	42711	B01-17	3
PLUMBING	Rate	3.25	4.34	1.68	3.95	3.60	2.25	2.75	1.35	2.00	6.60	4.50	3.00
	Code	6-02	070712	402	864-07	404	4-2	4-46	703	402	42412	B01-01	4
ROOFING	Rate	9.00	5.87	6.12	7.57	3.60	4.50	4.75	2.00	4.25	10.75	4.50	4.50
	Code	6-05	070600	404	854-12	404	4-3	4-41	700	401	42351	B01-28	3
SHEET METAL WORK (HVAC)	Rate	3.25	4.54	6.12	3.95	3.60	4.50	4.75	2.00	2.00	7.83	4.50	4.50
	Code	6-02	070714	404	864-07	404	4-3	4-41	700	402	42183	B01-04	3
STEEL ERECTION—door & sash	Rate	4.25	8.85	8.64	8.03	3.80	4.50	4.75	2.00	4.25	7.83	4.50	6.00
	Code	8-03	072509	405	827-09	403	4-3	4-41	700	401	42231	B01-32	2
STEEL ERECTION—inter., ornam.	Rate	4.25	8.85	8.64	8.03	3.80	4.50	4.75	2.00	4.25	7.83	4.50	6.00
	Code	8-03	072509	405	827-09	403	4-3	4-41	700	401	42231	B01-32	2
STEEL ERECTION—structure	Rate	4.25	8.85	8.64	17.81	3.80	4.50	6.00	4.35	4.25	7.83	4.50	6.00
	Code	8-03	072509	405	827-14	403	4-3	4-44	532	401	42231	B01-37	2
STEEL ERECTION—NOC	Rate	4.25	8.85	8.64	8.03	3.80	4.50	4.75	4.35	4.25	7.83	4.50	6.00
	Code	8-03	072509	405	827-09	403	4-3	4-41	532	401	42231	B01-32	2
TILE WORK—inter. (ceramic)	Rate	4.75	5.18	3.06	3.95	3.60	2.25	3.25	2.00	2.00	6.60	4.50	4.50
	Code	6-03	070506	401	864-07	404	4-2	4-49	700	402	42781	B01-16	3
WATERPROOFING	Rate	3.25	3.30	4.74	8.31	3.60	4.50	4.75	2.00	2.00	7.83	4.50	3.00
	Code	6-02	060237	407	873-11	404	4-3	4-41	700	402	42391	B01-18	4
WRECKING	Rate	10.35	5.87	3.06	22.36	3.80	4.50	5.00	2.00	4.25	18.38	4.50	6.00
	Code	6-08	070600	401	859-15	403	4-3	4-43	700	401	42211	B01-39	2

CIRCLE REFERENCE NUMBERS

⑧ Sales Tax (Div. 1.1-56)

State sales tax on materials is tabulated below (5 states have no sales tax). Many states allow local jurisdictions, such as a county or city, to levy additional sales tax.

Some projects may be sales tax exempt, particularly those constructed with public funds.

State	Tax	State	Tax	State	Tax	State	Tax
Alabama	4%	Illinois	5%	Montana	0%	Rhode Island	6%
Alaska	0	Indiana	5	Nebraska	3.5	South Carolina	5
Arizona	5	Iowa	4	Nevada	5.75	South Dakota	4
Arkansas	4	Kansas	3	New Hampshire	0	Tennessee	5.5
California	6	Kentucky	5	New Jersey	6	Texas	4
Colorado	3	Louisiana	4	New Mexico	3.75	Utah	5.5
Connecticut	7.5	Maine	5	New York	4	Vermont	4
Delaware	0	Maryland	5	North Carolina	3	Virginia	4
District of Columbia	6	Massachusetts	5	North Dakota	4	Washington	7.9
Florida	5	Michigan	4	Ohio	5.5	West Virginia	5
Georgia	3	Minnesota	6	Oklahoma	3	Wisconsin	5
Hawaii	4	Mississippi	6	Oregon	0	Wyoming	3
Idaho	4	Missouri	6.125	Pennsylvania	6	Average	4.28%

⑨ Construction Time Requirements (Div. 1.1-10)

Table below is average construction time in months for different types of building projects. Also shown is the construction time in months for different size projects. Design time runs 25% to 40% of construction time.

Type Building	Construction Time	Project Value	Construction Time
Industrial Buildings	13 Months	Under $1,000,000	10 Months
Commercial Buildings	15 Months	Up to $3,000,000	15 Months
Research & Development	17 Months	Up to $15,000,000	21 Months
Institutional Buildings	18 Months	Over $15,000,000	28 Months

⑩ Architectural Fees (Div. 1.1-02)

Tabulated below are typical percentage fees, below which adequate service cannot be expected. Fees may vary from those listed due to economic conditions.

Rates can be interpolated horizontally and vertically. Various portions of the same project requiring different rates should be adjusted proportionately. For alterations, add 50% to the fee for the first $500,000 of project cost and add 25% to the fee for project cost over $500,000.

Architectural fees tabulated below include Engineering Fees.

Building Type	Total Project Size in Thousands of Dollars						
	100	250	500	1,000	2,500	5,000	10,000
Factories, garages, warehouses repetitive housing	9.0%	8.0%	7.0%	6.2%	5.6%	5.3%	4.9%
Apartments, banks, schools, libraries, offices, municipal buildings	11.7	10.8	8.5	7.3	6.7	6.4	6.0
Churches, hospitals, homes, laboratories, museums, research	14.0	12.8	11.9	10.9	9.5	8.5	7.8
Memorials, monumental work, decorative furnishings	—	16.0	14.5	13.1	11.3	10.0	9.0

CIRCLE REFERENCE NUMBERS

⑪ Engineering Fees (Div. 1.1-15)

Typical **Structural Engineering Fees** based on type of construction and total project size. These fees are included in Architectural Fees.

Type of Construction	Total Project Size			
	To $250,000	$250,000-$500,000	$1,000,000	$5,000,000 & over
Industrial buildings, factories & warehouses	Technical payroll times 2.0 to 2.5	1.60%	1.25%	1.00%
Hotels, apartments, offices, dormitories, hospitals, public buildings, food stores	↓	2.00%	1.70%	1.20%
Museums, banks, churches and cathedrals		2.00%	1.75%	1.25%
Thin shells, prestressed concrete, earthquake resistive		2.00%	1.75%	1.50%
Parking ramps, auditoriums, stadiums, convention halls, hangars & boiler houses		2.50%	2.00%	1.75%
Special buildings, major alterations, underpinning & future expansion		Add to above 0.5%	Add to above 0.5%	Add to above 0.5%

For complex reinforced concrete or unusually complicated structures, add 20% to 50%.

Typical **Mechanical and Electrical Engineering Fees** based on the size of the subcontract. These fees are included in Architectural Fees.

Type of Construction	Subcontract Size							
	$25,000	$50,000	$100,000	$225,000	$350,000	$500,000	$750,000	$1,000,000
Simple structures	6.4%	5.7%	4.8%	4.5%	4.4%	4.3%	4.2%	4.1%
Intermediate structures	8.0	7.3	6.5	5.6	5.1	5.0	4.9	4.8
Complex structures	12.0	9.0	9.0	8.0	7.5	7.5	7.0	7.0

For renovations, add 15% to 25% to applicable fee.

⑫ Heavy Lifting (Div. 1.5-20-600)

Hydraulic Climbing Jacks

The use of hydraulic heavy lift systems is an alternative to conventional type crane equipment. The lifting, lowering, pushing, or pulling mechanism is a hydraulic climbing jack moving on a square steel jackrod from 1-5/8" to 4" square, or a steel cable. The jackrod or cable can be vertical or horizontal, stationary or movable, depending on the individual application. When the jackrod is stationary, the climbing jack will climb the rod and push or pull the load along with itself. When the climbing jack is stationary, the jackrod is movable with the load attached to the end and the climbing jack will lift or lower the jackrod with the attached load. The heavy lift system is normally operated by a single control lever located at the hydraulic pump.

The system is flexible in that one or more climbing jacks can be applied wherever a load support point is required, and the rate of lift synchronized.

Economic benefits have been demonstrated on projects such as: erection of ground assembled roofs and floors, complete bridge spans, girders and trusses, towers, chimney liners and steel vessels, storage tanks, and heavy machinery. Other uses are raising and lowering offshore work platforms, caissons, tunnel sections and pipelines.

CIRCLE REFERENCE NUMBERS

⑬ Contractor Equipment (Div. 1.5)

Rental Rates shown in the front of the book pertain to late model high quality machines in excellent working condition, rented from equipment dealers. Rental rates from contractors may be substantially lower than the rental rates from equipment dealers depending upon economic conditions. For older, less productive machines, reduce rates by a maximum of 12%. Any overtime must be added to the base rates. For shift work, rates are lower. Usual rule of thumb is 150% of one shift rate for two shifts; 200% for three shifts.

Daily Operated rates indicated by (*) include operator (and oiler if required), fuel, maintenance and repair. For periods of less than one week, operated equipment is usually more economical to rent than renting bare equipment and hiring an operator.

Equipment moving and mobilization costs must be added to rental rates where applicable. A large crane, for instance, may take two days to erect and two days to dismantle.

Rental rates vary throughout the country with larger cities generally having lower rates. Lease plans for new equipment are available for periods in excess of six months with a percentage of payments applying toward purchase. Recently long-term rental costs for the larger cranes has not increased due to intense competition between manufacturers. During a recession, rental rates do not rise as fast as dealer costs, due to market conditions.

Monthly rental rates vary from 2% to 5% of the cost of the equipment depending on the anticipated life of the equipment and its wearing parts. Weekly rates are about 1/3 the monthly rates and daily rental rates about 1/3 the weekly rate.

The hourly operating costs for each piece of equipment includes costs to the user such as fuel, oil, lubrication, normal expendables for the equipment, and a percentage of mechanic's wages chargeable to maintenance. Hourly operating costs do not include the operator's wages and do not apply to fully operated equipment designated by an*.

The daily cost for equipment used in the standard crews (foreword) is figured by dividing the weekly rate by five, then adding eight times the hourly operating cost to give the total daily equipment cost, not including the operator. This figure is in the right hand column of Division 1.5 under Crew Equip. Cost.

Pile Driving rates shown for pile hammer and extractor do not include leads, crane, boiler or compressor. Vibratory pile driving requires an added field specialist at $275 per day during set-up and pile driving operation for the electric model. The hydraulic model requires a field specialist for set-up only. Up to 125 reuses of sheet piling are possible using vibratory drivers. For normal conditions, crane capacity for hammer type and size are as follows.

Crane Capacity	Hammer Type and Size		
	Air or Steam	Diesel	Vibratory
25 ton	to 8,750 ft.-lb.		70 H.P.
40 ton	15,000 ft.-lb.	to 32,000 ft.-lb.	170 H.P.
60 ton	25,000 ft.-lb.		300 H.P.
100 ton		112,000 ft.-lb.	

Cranes should be specified for the job by size, building and site characteristics, availability, performance characteristics, and duration of time required.

Backhoes & Shovels rent for about the same as equivalent size cranes but maintenance and operating expense is higher. Crane operators rate must be adjusted for high boom heights as follows: for 150' boom add 50¢ per hour; over 185', add $1.00 per hour; over 210', add $1.25 per hour; over 250', add $1.90 per hour and over 295', add $2.65 per hour.

Tower Cranes

Capacity in Kip-Feet	Typical Jib Length in Feet	Speed at Maximum Reach and Load	Purchase Price (New)		Monthly Rental, to 6 mo.	
			Crane & 80' Mast	Mast Sections	Crane & 80' Mast	Mast Sections
725	100	350 FPM	$171,000	$430/L.F.	$ 4,580	$12/L.F.
900	100	500	210,000	500	5,020	13
*1100	130	1000	285,000	570	7,530	16
1450	150	1000	400,000	790	10,680	19
2150	200	1000	505,000	965	13,660	23
3000	200	1000	715,000	1,045	17,600	28

*Most widely used.

Tower Cranes of the climbing or static type have jibs from 50' to 200' and capacities at maximum reach range from 4,000 to 14,000 pounds. Lifting capacities increase up to maximum load as the hook radius decreases.

Typical rental rates, based on purchase price are about 2% to 3% per month.

Erection and dismantling run between $9,500 and $63,000. Climbing operation takes three men three hours per 20' climb. Crane dead time is about five hours per 40' climb. If crane is bolted to side of the building add cost of ties and extra mast sections. Mast sections cost $330 to $930 per vertical foot or can be rented at 2% to 3% of purchase price per month. Contractors using climbers claim savings of $1.40 per C.Y. of concrete placed, plus 8¢ per S.F. of formwork. Climbing cranes have from 80' to 180' of mast while static cranes have 80' to 800' of mast.

Truck Cranes can be converted to tower cranes by using tower attachments. Mast heights over 400' have been used. See Division 1.5-20 for rental rates of high boom cranes.

A single 100' high material **Hoist and Tower** can be erected and dismantled for about $11,500; a double 100' high hoist and tower for about $16,600. Erection costs for additional heights are $75 and $95 per vertical foot respectively up to 150' and $80 to $135 per vertical foot over 150' high. A 40' high portable Buck hoist costs about $4,100 to erect and dismantle. Additional heights run $65 per vertical foot to 80' and $87 per vertical foot for the next 100'. Most material hoists do not meet local code requirements for carrying personnel.

A 150' high **Personnel Hoist** requires about 500 to 800 man-hours to erect and dismantle with costs ranging from $10,300 to $22,500. Budget erection cost is $115 per vertical foot for all trades. Local code requirements or labor scarcity requiring overtime can add up to 50% to any of the above erection costs.

Earthmoving Equipment: The selection of earthmoving equipment depends upon the type and quantity of material, moisture content, haul distance, haul road, time available, and equipment available. Short haul cut and fill operations may require dozers only, while another operation may require excavators, a fleet of trucks, and spreading and compaction equipment. Stockpiled material and granular material is easily excavated with front end loaders. Scrapers are most economically used with hauls between 300' and 1-1/2 miles if adequate haul roads can be maintained. Shovels are often used for blasted rock and any material where a vertical face of 8' or more can be excavated. Special conditions may dictate the use of draglines, clamshells, or backhoes. Spreading and compaction equipment must be matched to the soil characteristics, the compaction required and the rate the fill is being supplied.

CIRCLE REFERENCE NUMBERS

⑭ Steel Tubular Scaffolding (Div. 1.1-44)

On new construction, tubular scaffolding is efficient up to 60' high or five stories. Above this it is usually better to use a hung scaffolding if construction permits.

In repairing or cleaning the front of an existing building the cost of tubular scaffolding per S.F. of building front increases as the height increases above the first tier. The first tier cost is relatively high due to leveling and alignment. Swing scaffolding operations may interfere with tenants. In this case the tubular is more practical at all heights.

The minimum efficient crew for erection is three men. For heights over 50', a four-man crew is more efficient. Use two or more on top and two at the bottom for handing up or hoisting. Four men can erect and dismantle about nine frames per hour up to five stories. From five to eight stories they will average six frames per hour. With 7' horizontal spacing this will run about 300 S.F. and 200 S.F. of wall surface, respectively. Time for placing planks must be added to the above. On heights above 50', five planks can be placed per man-hour.

The cost per 1,000 S.F. of building front in the table below was developed by pricing the materials required for a typical tubular scaffolding system eleven frames long and two frames high. Planks were figured five wide for standing plus two wide for materials.

Frames are 2', 4' and 5' wide and usually spaced 7' O.C. horizontally. Sidewalk frames are 6' wide. Rental rates will be lower for jobs over three months duration.

For jobs under twenty-five frames, figure rental at $6.00 per frame. For jobs over one hundred frames, rental can go as low as $2.25 per frame. These figures do not include accessories which are listed separately below. Large quantities for long periods can reduce rental rates by 20%.

Item	Unit	Purchase, Each		Monthly Rent, Each		Per 1,000 S.F. of Building Front	
		Regular	Heavy Duty	Regular	Heavy Duty	No. of Frames	Rental per Mo.
5' Wide Frames, 3' High	Ea.	$45	$52	$3.10	$3.40	—	—
*5'-0" High		56	57	3.10	3.40	—	—
*6'-6" High		71	76	3.10	3.40	24	$ 74.40
2' & 4' Wide, 5' High		—	68	—	3.40	—	—
6'-0" High		—	79	—	3.40	—	—
6' Wide Frame, 7'-6" High		100	110	7.25	9.25	—	—
Sidewalk Bracket, 20"		17	19	.94	1.10	12	11.28
Guardrail Post		13	18	.53	.63	12	6.36
Guardrail, 7' section		5	6	.47	.52	11	5.17
Cross Braces		15	16	.47	.58	44	20.68
Screw Jacks & Plates		20	28	1.30	1.40	24	31.20
8" Casters		70	—	5.70	6.00	—	—
16' Plank, 2" x 10"		19	—	3.60	3.85	35	126.00
8' Plank, 2" x 10"		10	—	1.85	2.10	7	12.95
1' to 6' Extension Tube		—	65	—	2.10	—	—
Shoring Stringers, steel, 10' to 12' long	L.F.	—	6	—	.35	—	—
Aluminum, 12' to 16' long		—	12	—	.58	—	—
Aluminum joists with nailers, 10' to 22' long		—	7.60	—	.35	—	—
Flying Truss System, Aluminum	S.F.C.A.	—	8.00	—	.45	—	—
						Total	$288.04
						2 Use/Mo.	$144.02

*Most commonly used

Scaffolding is often used as falsework over 15' high during construction of cast-in-place concrete beams and slabs. Two ft. wide scaffolding is generally used for heavy beam construction. The span between frames depends upon the load to be carried with a maximum span of 5'.

Heavy duty scaffolding with a capacity of 10,000#/leg can be spaced up to 10' O.C. depending upon form support design and loading.

Scaffolding used as horizontal shoring requires less than half the material required with conventional shoring.

On new construction, erection is done by carpenters.

Rolling towers supporting horizontal shores can reduce labor and speed the job. For manintenance work, catwalks with spans up to 70' can be supported by the rolling towers.

⑮ Concrete Pipe (Div. 2.5-27)

Prices given are for inside 20 mile delivery zone. Add $1.30 per ton of pipe for each additional 10 miles. Minimum truckload is 10 tons. The non-reinforced pipe listed in the front of the book is designation ASTM C14-59 extra strength. The reinforced pipe listed is ASTM C76-65T class 3, no gaskets. The installation cost given includes placing the pipe, shaping bottom of the trench and backfilling and tamping to the top of the pipe only.

CIRCLE REFERENCE NUMBERS

⑯ Excavating (Div. 2.3)

Bulk excavating on building jobs today is done entirely by machine. Hand work is used only to square out corners, for leveling off, cutting out trenches and footings below the general excavating grade and for minor miscellaneous cuts.

Classifying the soils and rock to be excavated is an exact science and goes beyond the scope of an estimating manual. Some basic factors to consider; (1) type of material; (2) weight of material in place or Bank Cubic Yard; (3) % of swell or expansion in the volume when excavated; (4) moisture content; and (5) loadability or the ease of filling bucket or dipper and the size of bucket or dipper.

A typical example using the above basics would be to find the excavation cost for a 1-1/2 C.Y. Hydraulic Excavator (backhoe) excavating 15' deep in sand and gravel with 18% swell.

$$\text{Swell Factor} = \frac{100}{100 + \% \text{ swell}} = \frac{100}{100 + 18} = .85$$

Swell Factor converts the net volume of B.C.Y. or pay yards from the loose volume (L.C.Y.) which is actually handled.

Or bucket heaped capacity x swell factor = 1.5 x .85 = 1.28 B.C.Y.

Hauler units matched up with excavator bucket size and haul distance using 10 wheel truck with 12 C.Y. water level (struck capacity)

Number of cycles required to fill truck = 1.5 C.Y. bucket x 8 cycles = 12 loose C.Y.
= 1.28 C.Y. x 8 = 10.2 B.C.Y.

Truck haul cycle:
- Load truck 8 passes = 4 minutes
- Haul distance 1 mile = 9 minutes
- Dump time = 2 minutes
- Return 1 mile = 7 minutes
- Spot under machine = 1 minute

23 minute cycle

Fleet Haul Production per day in B.C.Y.

$$4 \text{ trucks} \times \frac{50 \text{ min. hr.}}{23 \text{ min. haul cycle}} \times 8 \text{ hrs.} \times 10.2 \text{ B.C.Y.}$$

= 4 x 2.2 x 8 x 10.2 = 718 B.C.Y./day

The following figures on machine excavating apply to basement pits, large footings and other excavation for buildings. On general open field excavation this performance can be bettered by 33% to 50%. The figures presume loading directly into trucks.

Allowance has been made for idle time due to truck delay and moving about on the job. Figures include no water problems or slowdown due to sheeting and bracing operations. They also do not include truck spotters or clean-up labor.

Excavation Cost with a Two Mile Round Trip Distance for a 1-1/2 C.Y. Hydraulic Excavator Backhoe 15' Deep	Daily Cost	Sub Total	Unit Price
Equipment operator @ $21.05 per hr. x 8 hrs.	$ 168.40	$ 308.40 Labor	$ .43 Labor
Oiler @ $17.50 per hr. x 8 hrs.	140.00		
Equip. rental 1-1/2 C.Y. Hyd. Backhoe @ $2,200/wk. ÷ 5 days	440.00		
Operating expenses, fuel oil, lube @ $17.45 per hr. x 8 hrs.	139.60		
4 - 12 C.Y. Tandem trucks & drivers @ $51.55 per hr. x 8 hrs.	1,649.60	2,229.20 Equip.	3.09 Equip.
Production rate taken from ⑰ sand & gravel with a conversion factor of .75 x 120 x 8 hr. = 720 C.Y. per day	$2,537.60	$2,537.60 Total	$3.52 Total

Description	1-1/2 C.Y. Hyd. Backhoe 15' deep	1-1/2 C.Y. Power Shovel 7' deep	1-1/2 C.Y. Dragline 7' deep	2-1/2 C.Y. Trackloader Stockpile
Operator (& Oiler, if required)	$ 308.40	$ 308.40	$ 308.40	$ 164.80
Equipment rental (weekly basis ÷ 5)	440.00	452.00	381.80	300.00
Operating expenses	139.60	152.95	133.05	143.20
20 C.Y. Trailer dump truck w/drivers @ $56.05 per hr.	3 ea. 1,345.20	4 ea. 1,793.60	3 ea. 1,345.20	4 ea. 1,793.60
Total Cost per day	$2,233.20	$2,706.95	$2,168.45	$2,401.60
Daily Production in C.Y., bank measure	720	960	640	1000
Cost per C.Y.	$3.10	$2.82	$3.39	$2.40

Add for mobilizing and demobilizing $85 to $255 respectively, spread over total yardage. When equipment is rented for more than three days there is usually no mobilizing cost.

On larger jobs outside of downtown areas, scrapers can move earth economically providing a dump site is available near the excavation. Excavating within sheeting bracing or cofferdam bracing is usually done with a clamshell and production is low.

Cost can run between $3.50 to $13.00 per C.Y., since the clamshell may have to be guided by hand when the crane operator cannot see what he is excavating. When excavating and backfilling on a site that is enclosed with a wellpoint system, add 10% to 50% to cost because of restricted access. When estimating normal earth excavation it is well to figure side slopes at 1 to 1.

CIRCLE REFERENCE NUMBERS

(17) Excavating Equipment (Div. 2.3-16)

The table below lists THEORETICAL hourly production in C.Y. bank measure for the various type and size equipment in use. Figures assume 50 minute hours, 83% job efficiency, 100% operator efficiency, 90° swing and properly sized hauling units, but do not include consideration of hard digging and loading conditions. Actual production costs in the front of the book average about 50% of the theoretical values listed here.

Equipment	Soil Type	B.C.Y Weight	% Swell	1 C.Y.	1½ C.Y.	2 C.Y.	2½ C.Y.	3 C.Y.	3½ C.Y.	4 C.Y.
Hydraulic Excavator "Backhoe" 15' deep cut	Moist loam, sandy clay	3400 lb.	40%	85	125	175	220	275	330	380
	Sand and gravel	3100	18	80	120	160	205	260	310	365
	Common earth	2800	30	70	105	150	190	240	280	330
	Clay, hard, dense	3000	33	65	100	130	170	210	255	300
Power Shovel Optimum Depth (ft.)	Moist loam, sandy clay	3400	40	170 (6.0)	245 (7.0)	295 (7.8)	335 (8.4)	385 (8.8)	435 (9.1)	475 (9.4)
	Sand and gravel	3100	18	165 (6.0)	225 (7.0)	275 (7.8)	325 (8.4)	375 (8.8)	420 (9.1)	460 (9.4)
	Common earth	2800	30	145 (7.8)	200 (9.2)	250 (10.2)	295 (11.2)	335 (12.1)	375 (13.0)	425 (13.8)
	Clay, hard, dense	3000	33	120 (9.0)	175 (10.7)	220 (12.2)	255 (13.3)	300 (14.2)	335 (15.1)	375 (16.0)
Drag line Optimum Depth (ft.)	Moist loam, sandy clay	3400	40	130 (6.6)	180 (7.4)	220 (8.0)	250 (8.5)	290 (9.0)	325 (9.5)	385 (10.0)
	Sand and gravel	3100	18	130 (6.6)	175 (7.4)	210 (8.0)	245 (8.5)	280 (9.0)	315 (9.5)	375 (10.0)
	Common earth	2800	30	110 (8.0)	160 (9.0)	190 (9.9)	220 (10.5)	250 (11.0)	280 (11.5)	310 (12.0)
	Clay, hard, dense	3000	33	90 (9.3)	130 (10.7)	160 (11.8)	190 (12.3)	225 (12.8)	250 (13.3)	280 (12.0)

				Wheel Loaders				Track Loaders		
Equipment	Soil Type	B.C.Y Weight	% Swell	3 C.Y.	4 C.Y.	6 C.Y.	8 C.Y.	2¼ C.Y.	3 C.Y.	4 C.Y.
Loading Tractors	Moist loam, sandy clay	3400	40	260	340	510	690	135	180	250
	Sand and gravel	3100	18	245	320	480	650	130	170	235
	Common earth	2800	30	230	300	460	620	120	155	220
	Clay, hard, dense	3000	33	200	270	415	560	110	145	200
	Rock, well blasted	4000	50	180	245	380	520	100	130	180

(18) Compacting Backfill (Div. 2.3-08)

When quantities of 1000 C.Y. or more are involved, a soil analysis should be made before bids are taken. See division 1.1-60 for testing prices. Field checks of moisture content, material and density run about $75 each. Geotechnical engineering analysis runs $275 to $475 per day.

Figures below are based on compaction of 27 C.Y. per hour to a 95% Modified Proctor Density with one 24" vibratory plate 5000# compactor and Crew A-1.

Example: Compact 1000 cubic yards of backfilled granular material around a building foundation using a 21" wide x 24" vibratory plate 5000 lb. centrifugal force compactor in 8" lifts. Operator moves at 50 F.P.M. working a 50 minute hour developing 95% modified Proctor Density after 4 complete passes.

Production Rate:
$$\frac{1.75' \text{ plate width} \times 50 \text{ F.P.M.} \times 50 \text{ min.} \times .67' \text{ lift}}{27 \text{ C.F. per C.Y.}} = 108.5 \text{ C.Y.}$$

Production Rate for 4 Passes:
$$\frac{108.5 \text{ C.Y.}}{4 \text{ passes}} = 27 \text{ C.Y. per hr. @ 95\% compaction}$$

Labor = 1000 ÷ 27 C.Y. per hr. = 37 hrs. x $15.90 per hr. =   Labor Cost $588.30
Equipment rental for 24" vibratory plate compactor = $170 ÷ 40 x 37 =   Equipment Rental 157.25
Operating Cost, fuel, oil, lube, etc. = $.73 per hr. x 37 hrs. =   Equipment Operation 27.00
Total Cost $772.55

$$\text{Compaction Cost} = \frac{\$772.55}{1000 \text{ C.Y.}} = 77¢ \text{ per C.Y.}$$

Furnish backfill material, delivered to the job	$6.00
Adjust moisture content	.10
Placement of soil using dozer and some hand labor	.72
Actual compaction using 24" vibratory plate compactor	.77
Total per C.Y.	$7.59

When compacting over 1000 C.Y. of well graded gravel a fully loaded 10 wheel truck can be used if there is enough room for the truck to maneuver. Also a 15 ton or heavier dozer can be used at high speed with four complete passes. The cost of compaction to a 95% Modified Proctor Density for these methods runs $.60 to $1.65 per C.Y.

CIRCLE REFERENCE NUMBERS

 Caissons (Div. 2.4-05)

The three principal types of caissons are (1) Belled Caisson which, except for shallow depths and poor soil conditions, are generally recommended. They provide more bearing than shaft area. Because of its conical shape, no horizontal reinforcement of the bell is required.

SOIL CONDITIONS FOR BELLING		
Good	Requires Handwork	Not Recommended
Clay	Hard Shale	Silt
Sandy Clay	Limestone	Sand
Silty Clay	Sandstone	Gravel
Clayey Silt	Weathered Mica	Igneous Rock
Hard-pan		
Soft Shale		
Decomposed Rock		

(2) Straight Shaft Caissons are used where relatively light loads are to be supported by a caisson that rests on high value bearing strata. While the shaft is larger in diameter than for belled types this is more than offset by the saving in time and labor. (3) Keyed Caissons are used when extremely heavy loads are to be carried. A keyed or socketed caisson transfers its load into rock by a combination of end-bearing and shear reinforcing of the shaft. The most economical shaft often consists of a steel casing, a steel wide flange core and concrete. Allowable compressive stresses of 0.225 f'c for concrete, 16,000 psi for the wide flange core, and 9,000 psi for the steel casing are commonly used. The usual range of shaft diameter is from 18" to 84". The number of sizes specified for any one project should be limited due to the problems of casing and auger storage. When hand work is to be performed shaft diameters should not be less than 32". When inspection of borings is required a minimum shaft diameter of 30" is recommended. Concrete caissons are intended to be poured against earth excavation so permanent forms which add to cost should not be used if the excavation is clean and the earth sufficiently impervious to prevent excessive loss of concrete.

⑳ Wood Sheet Piling (Div. 2.3-38-390)

Wood sheet piling may be used for depths to 20' where there is no ground water. If moderate ground water is encountered Tongue & Groove sheeting will help to keep it out. When considerable ground water is present, steel sheeting must be used.

For estimating purposes on trench excavation, sizes are as follows:

Depth	Sheeting	Wales	Braces	B.F. per S.F.
To 8'	3 x 12's	6 x 8's, 2 line	6 x 8's, @ 10'	4.0 @ 8'
8' x 12'	3 x 12's	10 x 10's, 2 line	10 x 10's, @ 9'	5.0 average
12' to 20'	3 x 12's	12 x 12's, 3 line	12 x 12's, @ 8'	7.0 average

Sheeting to be 2' longer than depth of wall. A four man crew with air compressor and sheeting driver can place, drive and brace 55 S.F. per hour at 8' depth; 45 S.F. at 12' and 40 S.F. at 16'. For normal soils, piling can be pulled in 1/3 the time to install. Pulling difficulty increases with the time in the ground.

Production can be increased by high pressure jetting. Figures below assume 50% of lumber is salvaged and includes pulling costs. Some jurisdictions require an equipment operator in addition to Crew B-31.

Daily Cost, Crew B-31		8' Depth, 440 S.F. Installed		16' Depth, 320 S.F. Installed	
		To Drive	To Pull	To Drive	To Pull
1 - Foreman @ $17.90 per hour	$143.20	$ 143.20	$ 47.75	$ 143.20	$ 47.75
3 - Laborers @ $15.90 per hour	381.60	381.60	127.20	381.60	127.20
1 - Carpenter @ $20.00 per hour	160.00	160.00	53.35	160.00	53.35
1 - Air Compressor & Sheeting Driver	123.05	123.05	41.00	123.05	41.00
Lumber 50% cost @ $390 per M.B.F.	—	1.76M 343.20	—	2.24M 436.80	—
Total in Place	$807.85	$1,151.05	$269.30	$1,244.65	$269.30
Total per S.F.		$ 2.62	$ .61	$ 3.89	$ .84
Grand Total per S.F.		$ 3.23		$ 4.73	

Sheeting Left in Place

Daily Cost	8' Depth 440 S.F./Day		10' Depth 400 S.F./Day		12' Depth 360 S.F./Day		16' Depth 320 S.F./Day		18' Depth 305 S.F./Day		20' Depth 280 S.F./Day	
Crew B-31		$ 807.85		$ 807.85		$ 807.85		$ 807.85		$ 807.85		$ 807.85
Lumber @ $390 M.B.F.	1.76M	686.40	1.8M	702.00	1.8M	702.00	2.24M	873.60	2.1M	819.00	1.9M	741.00
Total in Place		$1,494.25		$1,509.85		$1,509.85		$1,681.45		$1,626.85		$1,548.85
Total per S.F.		$ 3.40		$ 3.77		$ 4.19		$ 5.25		$ 5.33		$ 5.53
Total per M.B.F. (Labor & Material)		$ 850.00		$ 840.00		$ 840.00		$ 750.00		$ 775.00		$ 815.00

When trench sheeting is left in place the cross brace and nalers are salvaged. Assumed reuse 10 times.

CIRCLE REFERENCE NUMBERS

㉑ Steel Sheet Piling (Div. 2.3-38)

Limiting weights are 22 to 38#/S.F. of wall surface with 27#/S.F. average for usual types and sizes. (Weights of piles themselves are from 30.7#/L.F. to 57#/L.F. but they are 15" to 21" wide.) Lightweight sections 12" to 28" wide from 3 ga. to 12 ga. thick are also available for shallow excavations. Piles may be driven two at a time with an impact vibratory hammer (use vibratory to pull) hung from a crane without leads. A reasonable estimate of the life of a sheet pile is 10 uses with up to 125 uses possible if a vibratory hammer is used. Typical cost of new plain piling at the mill is $570 per ton for shallow sections; $580 per ton for Z sections. Fabricated corners, tees and crosses cost about three times as much per ton as plain material. Used piling costs from $370 to $585 per ton depending on location and market conditions. Sheet piling and H piles can be rented for about 30% of the delivered mill price or $160 per ton for the first month and $15 per month thereafter. Cleaning and trimming after driving runs about $27 per pile. Vibratory drivers are faster in wet granular soils and are excellent for pile extraction. Pulling difficulty increases with the time in the ground and may cost more than driving. It is often economical to abandon the sheet piling, especially if it can be used as the outer wall form. Allow about 1/3 additional length or more, for toeing into ground. Add bracing, waler and strut costs. These costs increase with depth and hydrostatic head. Waler costs can range up to $500 per ton of sheeting.

Cost of Sheet Piling & Production Rate by Ton & S.F.						
Depth of Excavation	15' Depth		20' Depth		25' Depth	
Description of Pile	22 psf = 90.9 S.F./ton		27 psf = 74 S.F./ton		38 psf = 52.6 S.F./ton	
Type of operation: Left in place or removed	Drive & Left	Drive & Extract*	Drive & Left	Drive & Extract*	Drive & Left	Drive & Extract*
Labor & Equipment to drive 1 ton	$256.00	$384.00	$214.00	$320.00	$146.00	$218.00
Piling, rented or bought including accessories @ $625/ton with 75% salvage value	625.00	156.00	625.00	156.00	625.00	156.00
Cost per ton, total	$881.00	$540.00	$839.00	$476.00	$771.00	$374.00
Production rate, tons per day	10.81	7.22	12.95	8.65	19.00	12.75
Cost per S.F. in place	$9.69	$5.94	$11.34	$6.43	$14.66	$7.11
Production Rate, S.F. per day including 33% toe-in	983.00	656.00	960.00	640.00	1,000.00	670.00

Crew B-40	Bare Cost		Incl. Subs O & P		
	Hr.	Daily	Hr.	Daily	
1 Foreman	$22.10	$176.80	$34.65	$277.20	
4 Pile Drivers	20.10		31.50	1,008.00	
2 Equip. Operators (crane)	21.05	336.80	30.50	488.00	
1 Oiler	17.50	140.00	25.10	200.80	
One-40 ton crane & accessories		483.40		531.75	
Vibratory hammer & generator		993.00		1,092.30	
64 M.H. Daily Total		$2,773.20		$3,598.05	

Production Rate Formula per ton for 15' Deep Excavation plus 33% toe-in

$2,773.20 per day ÷ 10.81 tons = $257 per ton for installation

For S.F. Production

$$\frac{2000\#}{22\#/S.F.} = 90.9 \text{ S.F.} \times 10.81 \text{ tons} = 983 \text{ S.F. per day}$$

*For driving & extracting two mobilizations & demobilizations will be necessary.

㉒ Wood Bearing Piles (Div. 2.4-25)

Untreated Southern Yellow Pine is most generally used for building piles cut off below low water line. Drive untreated piles with bark on. All piles cut off above low water line should be treated with 12# creosote per C.Y. or concrete encased at $15 to $25 per L.F. Creosoting costs about 75¢ per C.F. of wood treated. For heavier retention add 25¢ to 30¢ per C.F.

Item	Quantities	Total Cost	Each Pile	L.F.
50' long treated piles, 12" butt, 9" tip	200 each	$59,500.00	$297.50	$5.95
Crew B-19 @ $2290.70 per day	13 days	29,779.10	148.90	2.98
Mobilization & Demobilization, crew B-19	3 days	6,872.10	34.35	.69
Transportation equipment one way	Allow	750.00	3.75	.08
Totals		$96,901.20	$484.50	$9.70

The above figures are based on driving 800 L.F. daily which can be considered average. Time is included for moving rig, cutoff & ordinary delays. A general observation is that the cost of a pile in place complete, is about two times the cost of pile only. See also equipment rental division 1.5-05 and ⑬ for equipment capacities.

CIRCLE REFERENCE NUMBERS

㉓ Wellpoints (Div. 2.3-55)

A single stage wellpoint system is usually limited to dewatering to an average 15′ depth below normal ground water level. Multi-stage systems are employed for greater depth with the pumping equipment installed only at the lowest header level. Ejectors, with unlimited lift capacity, can be economical when two or more stages of wellpoints can be replaced or when horizontal clearance is restricted, such as in deep trenches or tunneling projects, and where low water flows are expected.

Wellpoints are usually spaced on 2-1/2′ to 10′ centers along a header pipe. Wellpoint spacing, header size, and pump size are all determined by the expected flow, as dictated by soil conditions.

In almost all points encountered in wellpoint dewatering, the wellpoints may be jetted into place. Cemented soils and stiff clays may require sand wicks about 12″ in diameter around each wellpoint to increase efficiency and eliminate weeping into the excavation. These sand wicks require 1/2 to 3 C.Y. of washed filter sand and are installed by using a 12″ diameter steel casing and hole puncher jetted into the ground 2′ deeper than the wellpoint. Rock may require predrilled holes.

Labor required for the complete installation and removal of a single stage wellpoint system is in the range of 3/4 to 2 man-hours per linear foot of header, depending upon jetting conditions, wellpoint spacing, etc.

Continuous pumping is necessary except in some free draining soil where temporary flooding is permissible (as in trenches which are backfilled after each day's work). Good practice requires provision of a stand-by pump during the continuous pumping operation.

Systems for continuous trenching below the water table should be three to four times the length of expected daily progress to insure uninterrupted digging and header pipe size should not be changed during the job.

For pervious free draining soils, deep wells may be economical because of lower installation and maintenance costs. Daily production ranges between two to three wells per day for 25′ to 40′ depths to one well per day for depths over 50′.

Detailed analysis and estimating for any dewatering problem is available at no cost from wellpoint manufacturers. Major firms will quote "sufficient equipment" quotes or their affiliates offer lump sum proposals to cover complete dewatering responsibility.

	Description for 200 ft. System with 8″ Header	Quantities	1st Month Unit $	1st Month Total	Thereafter Unit $	Thereafter Total
Equipment	Wellpoints 25′ long, 2″ diameter @ 5′ O.C.	40 Each	$ 25.75	$ 1,030	$ 18.38	$ 735
	Header pipe, 8″ diameter	200 L.F.	4.74	948	2.34	468
	Discharge pipe, 8″ diameter	100 L.F.	3.62	362	1.50	150
	8″ valves	3 Each	100.00	300	60.00	180
	Combination Jetting & Wellpoint pump (standby)	1 Each	2,150.00	2,150	1,520.00	1,520
	Operating 8″ diameter pump	1 Each	2,420.00	2,420	1,440.00	1,440
	Transportation to and from yard	Allow		525		
	Fuel 30 days x 60 G.P.D.	1800 Gallons	1.20	2,160	1.20	2,160
	Grease & oil 30 days x 2 G.P.D.	60 Gallons	5.75	345	5.75	330
	Material and Equipment Sub Total			$10,240		$6,983
Labor & Material	Technician to help set up	5 Man-days	$ 345.00	$ 1,725		
	Install & remove system	300 Man-hours	15.90	4,770		
	4 Operators regular pay @ 40 hr./wk. x 4.3 wk.	688 hr. (6 hr. shifts)	19.45	13,382	$ 19.45	$13,382
	4 Operators overtime @ 2 hr./wk. x 4.3 wk.	34 hr. (6 hr. shifts)	38.90	1,323	38.90	1,323
	Sand for points	40 C.Y.	8.00	320		
	Labor Sub Total			$21,520		$14,705
	Total Cost			$31,760		$21,688
	Monthly Cost per Header L.F.			$ 159		$ 108

㉔ Pressure Injected Footings (Div. 2.4-40)

Pressure Injected Footings are end bearing foundation units consisting of expanded bulbous bases formed by high energy blows, with concrete shafts either cased or uncased.

They are practical only in granular, rock or hardpan soils. Bearing is achieved at a predetermined depth so that depth is usually less than for friction piles. High load capacity reduces number of units required and cap quantities.

Mobilization and demobilization costs assume maximum distance of 50 miles.

	100 Uncased Pressure Injected Footings 25′ Long	Total	Each	L.F.
Installation	Crew B-44 @ $2,165.95 per day. Drive 8 per day = 13 days driving time	$28,157	$281.55	$11.26
	Mobilization & Demobilization (2 days for crew B-44)	4,332	43.30	1.73
	Transportation equipment, one way	750	7.50	.30
Material	Shafts: 17″ diameter x 25′ deep = 155 C.Y. Pressure bulbs: 100 ea. @ .75 C.Y. = 75 C.Y. } 265 C.Y. @ $65/C.Y. Waste: 15% of above = 35 C.Y.	17,225	172.25	6.89
	Total Cost in Place	$50,464	$504.60	$20.18

CIRCLE REFERENCE NUMBERS

㉕ Vibroflotation Soil Compaction (Div.2.3-08)

Vibroflotation is a proprietary system of compacting sandy soils in place to increase relative density to about 70%. Typical bearing capacities attained will be 6000 psf for saturated sand and 12,000 psf for dry sand. Usual range is 4000 to 8000 psf capacity. Costs in the front of the book are for a vertical foot of compacted cylinder 6' to 10' in diameter.

Vibro replacement is a proprietary system of improving cohesive soils in place to increase bearing capacity. Most silts and clays above or below the water table can be strengthened by installation of stone columns.

The process consists of radial displacement of the soil by vibration. The created hole is then backfilled in stages with coarse granular fill which is thoroughly compacted and displaced into the surrounding soil in the form of a column.

The total project cost would depend on the number and depth of the compacted cylinders. The installing company guarantees relative soil density of the sand cylinders after compaction and the bearing capacity of the soil after the replacement process.

Detailed estimating information is available from the installer at no cost.

㉖ Bituminous Paving (Div. 2.6-10)

City	Bituminous Asphalt per Ton	3" Thick Roads & Parking Areas (6.13 S.Y./ton)				2" Thick Sidewalks (9.2 S.Y./ton)			
		Cost Per Square Yard			Ton	Cost Per Square Yard			Ton
		Material	Installation	Total	Total	Material	Installation	Total	Total
Atlanta	$24.25	$3.96	$0.96	$4.92	$30.15	$2.64	$0.89	$3.53	$32.50
Baltimore	25.75	4.20	1.04	5.24	32.10	2.80	1.01	3.81	35.05
Boston	27.00	4.40	1.35	5.75	35.25	2.94	1.46	4.40	40.50
Buffalo	31.75	5.20	1.36	6.56	40.20	3.45	1.51	4.96	45.65
Chicago	28.00	4.57	1.33	5.90	36.15	3.04	1.44	4.48	41.20
Cincinnati	27.00	4.40	1.34	5.74	35.20	2.94	1.47	4.41	40.60
Cleveland	29.50	4.81	1.56	6.37	39.05	3.21	1.82	5.03	46.30
Columbus	24.00	3.92	1.26	5.18	31.75	2.61	1.35	3.96	36.45
Dallas	29.50	4.81	1.02	5.83	35.75	3.21	0.99	4.20	38.65
Denver	26.50	4.32	1.18	5.50	33.70	2.88	1.24	4.12	37.90
Detroit	25.75	4.20	1.33	5.53	33.90	2.80	1.43	4.23	38.90
Houston	29.50	4.81	1.09	5.90	36.15	3.21	1.08	4.29	39.45
Indianapolis	23.75	3.87	1.19	5.06	31.00	2.58	1.22	3.80	34.95
Kansas City	23.75	3.87	1.27	5.14	31.50	2.58	1.38	3.96	36.45
Los Angeles	23.00	3.75	1.59	5.34	32.75	2.50	1.85	4.35	40.00
Memphis	25.50	4.16	1.00	5.16	31.65	2.77	0.96	3.73	34.30
Milwaukee	27.00	4.40	1.30	5.70	34.95	2.94	1.41	4.35	40.00
Minneapolis	22.00	3.59	1.30	4.89	30.00	2.39	1.41	3.80	34.95
Nashville	24.50	4.00	0.93	4.93	30.20	2.66	0.86	3.52	32.40
New Orleans	25.75	4.20	1.10	5.30	32.50	2.80	1.08	3.88	35.70
New York City	33.00	5.40	1.47	6.87	42.10	3.59	1.67	5.26	48.40
Philadelphia	32.75	5.35	1.36	6.71	41.15	3.56	1.49	5.05	46.45
Phoenix	24.75	4.04	1.22	5.26	32.25	2.69	1.30	3.99	36.70
Pittsburgh	31.50	5.15	1.35	6.50	39.85	3.43	1.48	4.91	45.15
St. Louis	25.50	4.16	1.39	5.55	34.00	2.77	1.54	4.31	39.65
San Antonio	29.50	4.81	0.99	5.80	35.55	3.21	0.93	4.14	38.10
San Diego	24.00	3.92	1.59	5.51	33.75	2.61	1.85	4.46	41.05
San Francisco	32.00	5.20	1.55	6.75	41.40	3.48	1.77	5.25	48.30
Seattle	29.50	4.81	1.39	6.20	38.00	3.21	1.57	4.78	43.95
Washington, D.C.	32.00	5.20	1.08	6.28	38.50	3.48	1.08	4.56	41.95
Average	$27.30	$4.45	$1.26	$5.71	$35.00	$2.97	$1.35	$4.32	$39.70

Assumed density is 145 lb. per C.F.

Table below shows quantities and bare costs for 1000 S.Y. of Bituminous Paving.

Item	Roads and Parking Areas, 3" Thick		Sidewalks, 2" Thick	
	Quantities	Cost	Quantities	Cost
Bituminous asphalt	163 tons @ $27.30 per ton	$4450.00	109 tons @ $27.30 per ton	$2976.00
Installation using	Crew B-25 @ $2080.00 for 0.6 days	1248.00	Crew B-37 @ $892.60 for 1.5 days	1339.00
Total per 1000 S.Y.		$5698.00		$4315.00
Total per S.Y.		$ 5.70		$ 4.32
Total per Ton		$ 34.95		$ 39.75

CIRCLE REFERENCE NUMBERS

㉗ Single Track R.R. Siding (Div. 2.9-15)

Description and Bare Costs	Bare Costs	Incl. Subs O&P
100 lb. rail @ $570 per ton, prime grade (Relayer rail cost $230 per ton)	$19.00	$20.90
Spikes, plates, bolts (36 plates @ $4.50 ea., 2 splices @ $30.00 ea., plus 8 bolts @ $2.00 per 33')	7.20	7.90
6" x 8" x 8'-6" treated timber ties 22" O.C. at $16.00 ea. (C.L. lots)	8.75	9.65
Ballast 1/2 C.Y. (crushed stone @ $9.00 per ton, delivered using 2/3 ton per L.F.)	6.00	6.60
Labor 1 hr. @ $15.90 per hour	15.90	23.05
Total per L.F.	$56.85	$68.10

Add $3.10 per L.F. for curves over 14°.

Rail lighter than 100 lbs. per yard or special chemical or shape requirements cause price of rail to vary upwards from mill base price.

Light rail sections cost from $65 to $120 per ton more than heavy sections. Typical turnout quantities are 3700 B.F. timber, 5 tons of rail and 50 tons of ballast. A prefabricated #8 turnout with new 100 lb. rail costs $17,900.

㉘ Single Track, Steel Ties, Concrete Bed (Div. 2.9-15)

Description and Bare Costs	Bare Costs	Incl. Subs O&P
100 lb. rail @ $570 per ton	$19.00	$ 20.90
Fasteners and plates	7.20	7.90
6" WF beam @ 30" O.C., 6'-6" long, 15.5 or 25#/L.F.	18.10	19.90
Concrete 9' wide x 10" thick incl. reinforcing and forms	15.90	17.50
Labor 2.7 hrs. @ $15.90 per hour	42.95	62.25
Curves not included. Total per L.F.	$103.15	$128.45

㉙ Seeding Div. 2.8-45)

The type of grass is determined by light, shade & moisture content of soil plus intended use. Fertilizer should be disked 4" before seeding. For steep slopes disk five tons of mulch per acre and lay two tons of hay or straw on surface after seeding. Surface mulch can be staked, lightly disked or tar sprayed. Material for mulch can be wood chips, peat moss, partially rotted hay or straw, wood fibers and sprayed emulsions. Hemp seed blankets with fertilizer are also available. For spring planting, watering is necessary. Late fall planting may have to be reseeded in the spring. Aerial seeding can be used on areas larger than 100,000 S.F.

㉚ Cost of Trees (Deciduous) (Div. 2.8-65)

Tree Diameter	Normal Height	Catalog List Price of Tree	Guying Material	Equipment Charge	Installation Labor	Total
2 to 3 inch	14 feet	$ 80	$ 10	$ 25	$ 65	$ 180
3 to 4 inch	16 feet	155	25	44	116	340
4 to 5 inch	18 feet	225	40	52	138	455
6 to 7 inch	22 feet	600	70	69	181	920
8 to 9 inch	26 feet	1200	85	90	235	1610

Installation Time & Cost for Planting Trees, Bare Costs

Ball Size Diam. x Depth	Soil in Ball	Weight of Ball	Hole Diam. Req'd.	Hole Excavation	Amount of Soil Displ.	Topsoil Handled	Time Required in Man-Hours					Cost	
							Dig & Lace	Handle Ball	Dig Hole	Plant & Prune	Water & Guy	Total M.H.	Crew	Total per Tree
Inches	C.F.	Lbs.	Feet	C.F.	C.F.	C.F.								
12x12	.7	56	2	4	3	11	.25	.17	.33	.25	.07	1.1	1 Clab	$ 17
18x16	2	160	2.5	8	6	21	.50	.33	.47	.35	.08	1.7	2 Clab	27
24x18	4	320	3	13	9	38	1.00	.67	1.08	.82	.20	3.8	3 Clab	60
30x21	7.5	600	4	27	19.5	76	.82	.71	.79	1.22	.26	3.8		$ 90
36x24	12.5	980	4.5	38	25.5	114	1.08	.95	1.11	1.32	.30	4.76		115
42x27	19	1520	5.5	64	45	185	1.90	1.27	1.87	1.43	.34	6.8	B-6	160
48x30	28	2040	6	85	57	254	2.41	1.60	2.06	1.55	.39	8.0	@	190
54x33	38.5	3060	7	127	88.5	370	2.86	1.90	2.39	1.76	.45	9.4	$23.58	220
60x36	52	4160	7.5	159	107	474	3.26	2.17	2.73	2.00	.51	10.7	per	250
66x39	68	5440	8	196	128	596	3.61	2.41	3.07	2.26	.58	11.9	man-	280
72x42	87	7160	9	267	180	785	3.90	2.60	3.71	2.78	.70	13.7	hour	325

CIRCLE REFERENCE NUMBERS

(31) Formwork Labor Hours (Div. 3.1)

Item	Unit	Hours Required			Total Hours 1 Use	Multiple Use		
		Fabricate	Erect & Strip	Clean & Move		2 Use	3 Use	4 Use
Beam and Girder, interior beams, 12" wide	100 S.F.	6.4	8.3	1.3	16.0	13.3	12.4	12.0
Hung from steel beams		5.8	7.7	1.3	14.8	12.4	11.6	11.2
Beam sides only, 36" high		5.8	7.2	1.3	14.3	11.9	11.1	10.7
Beam bottoms only, 24" wide		6.6	13.0	1.3	20.9	18.1	17.2	16.7
Box out for openings		9.9	10.0	1.1	21.0	16.6	15.1	14.3
Buttress forms, to 8' high		6.0	6.5	1.2	13.7	11.2	10.4	10.0
Centering, steel, 3/4" rib lath			1.0		1.0			
3/8" rib lath or slab form			0.9		0.9			
Chamfer strip or keyway	100 L.F.		1.5		1.5	1.5	1.5	1.5
Columns, fiber tube 8" diameter			20.6		20.6			
12"			21.3		21.3			
16"			22.9		22.9			
20"			23.7		23.7			
24"			24.6		24.6			
30"			25.6		25.6			
Round Steel, 12" diameter			22.0		22.0	22.0	22.0	22.0
16"			25.6		25.6	25.6	25.6	25.6
20"			30.5		30.5	30.5	30.5	30.5
24"			37.7		37.7	37.7	37.7	37.7
Plywood 8" x 8"	100 S.F.	7.0	11.0	1.2	19.2	16.2	15.2	14.7
12" x 12"		6.0	10.5	1.2	17.7	15.2	14.4	14.0
16" x 16"		5.9	10.0	1.2	17.1	14.7	13.8	13.4
24" x 24"		5.8	9.8	1.2	16.8	14.4	13.6	13.2
Steel framed plywood 8" x 8"			10.0	1.0	11.0	11.0	11.0	11.0
12" x 12"			9.3	1.0	10.3	10.3	10.3	10.3
16" x 16"			8.5	1.0	9.5	9.5	9.5	9.5
24" x 24"			7.8	1.0	8.8	8.8	8.8	8.8
Drop head forms, plywood		9.0	12.5	1.5	23.0	19.0	17.7	17.0
Coping forms		8.5	15.0	1.5	25.0	21.3	20.0	19.4
Culvert, box			14.5	4.3	18.8	18.8	18.8	18.8
Curb forms, 6" to 12" high, on grade		5.0	8.5	1.2	14.7	12.7	12.1	11.7
On elevated slabs		6.0	10.8	1.2	18.0	15.5	14.7	14.3
Edge forms to 6" high, on grade	100 L.F.	2.0	3.5	0.6	6.1	5.6	5.4	5.3
7" to 12" high	100 S.F.	2.5	5.0	1.0	8.5	7.8	7.5	7.4
Equipment foundations		10.0	18.0	2.0	30.0	25.5	24.0	23.3
Flat slabs, including drops		3.5	6.0	1.2	10.7	9.5	9.0	8.8
Hung from steel		3.0	5.5	1.2	9.7	8.7	8.4	8.2
Closed deck for domes		3.0	5.8	1.2	10.0	9.0	8.7	8.5
Open deck for pans		2.2	5.3	1.0	8.5	7.9	7.7	7.6
Footings, continuous, 12" high		3.5	3.5	1.5	8.5	7.3	6.8	6.6
Spread, 12" high		4.7	4.2	1.6	10.5	8.7	8.0	7.7
Pile caps, square or rectangular		4.5	5.0	1.5	11.0	9.3	8.7	8.4
Grade beams, 24" deep		2.5	5.3	1.2	9.0	8.3	8.0	7.9
Lintel or Sill forms		8.0	17.0	2.0	27.0	23.5	22.3	21.8
Spandrel beams, 12" wide		9.0	11.2	1.3	21.5	17.5	16.2	15.5
Stairs			25.0	4.0	29.0	29.0	29.0	29.0
Trench forms in floor		4.5	14.0	1.5	20.0	18.3	17.7	17.4
Walls, Plywood, at grade, to 8' high		5.0	6.5	1.5	13.0	11.0	9.7	9.5
8' to 16'		7.5	8.0	1.5	17.0	13.8	12.7	12.1
16' to 20'		9.0	10.0	1.5	20.5	16.5	15.2	14.5
Foundation walls, to 8' high		4.5	6.5	1.0	12.0	10.3	9.7	9.4
8' to 16' high		5.5	7.5	1.0	14.0	11.8	11.0	10.6
Retaining wall to 12' high, battered		6.0	8.5	1.5	16.0	13.5	12.7	12.3
Radial walls to 12' high, smooth		8.0	9.5	2.0	19.5	16.0	14.8	14.3
But in 2' chords		7.0	8.0	1.5	16.5	13.5	12.5	12.0
Prefabricated modular, to 8' high		—	4.3	1.0	5.3	5.3	5.3	5.3
Steel, to 8' high		1.0	5.8	1.2	8.0	8.0	8.0	8.0
8' to 16' high		1.5	7.6	1.5	10.5	10.3	10.2	10.2
Steel framed plywood to 8' high		2.0	4.8	1.2	8.0	7.5	7.3	7.2
8' to 16' high		3.0	6.3	1.2	10.5	9.5	9.2	9.0

㉜ Forms for Reinforced Concrete (Div. 3.1)

Design Economy

Avoid many sizes in proportioning beams and columns.

From story to story avoid changing column dimensions. Gain strength by adding steel or using a richer mix. If a change in size of column is necessary vary one dimension only to minimize form alterations. Keep beams and columns the same width.

From floor to floor in a multi-story building vary beam depth not width as that will leave slab panel form unchanged. It is cheaper to vary the strength of a beam from floor to floor by means of steel area than by 2" changes in either width or depth.

Cost Factors

Material includes the cost of lumber, cost of rent or metal pans or forms if used, nails, form ties, form oil, bolts and accessories.

Labor includes the cost of carpenters to make up, erect, remove and repair, plus common labor to clean and move. Having carpenters remove forms minimizes repairs.

Improper alignment and condition of forms will increase finishing cost. When forms are heavily oiled, concrete surfaces must be neutralized before finishing. Special curing compounds will cause spillages to spall off in first frost. Gang forming methods will reduce costs on large projects.

Materials Used

Boards are seldom used unless their architectural finish is required. Generally, steel, fiberglass and plywood are used for contact surfaces. Labor on plywood is 10% less than with boards. The plywood is backed up with 2 x 4's at 12" to 32" O.C. Walers are generally 2 - 2 x 4's. Column forms are held together with steel yokes or bands. Shoring is with adjustable shoring or scaffolding for high ceilings. See also ㉝ ㉞ (82) (86).

Reuse

Floor and column forms can be reused four or possibly five times without excessive repair. Remember to allow for 10% waste on each reuse.

When modular sized wall forms are made, up to twenty uses can be expected with exterior plyform.

When forms are reused, the cost to erect, strip, clean and move will not be affected. 10% replacement of lumber should be included and about one hour of carpenter time for repairs on each reuse per 100 S.F.

The reuse cost for certain accessory items normally rented on a monthly basis will be lower than the cost for the first use.

After fifth use, new material required plus time needed for repair prevent form cost from dropping further and it may go up. Much depends on care in stripping, the number of special bays, changes in beam or column sizes and other factors.

1. Costs for multiple use of formwork may be developed as follows:

2 Uses

$$\frac{\text{1st Use + Reuse}}{2} = \text{avg. cost/2 uses}$$

3 Uses

$$\frac{\text{1st Use + 2 Reuse}}{3} = \text{avg. cost/3 uses}$$

4 Uses

$$\frac{\text{1st Use + 3 Reuse}}{4} = \text{avg. cost/4 uses}$$

㉝ Forms In Place (Div. 3.1)

This section assumes that all cuts are made with power saws, that adjustable shores are employed and that maximum use is made of commercial form ties and accessories. Bare costs are used in the table below.

BEAM AND GIRDER, INTERIOR, 12" Wide (Line 20-200)	First Use			Reuse		
	Quantities	Material	Installation	Quantities	Material	Installation
5/8" exterior plyform at $615 per M.S.F.	115 S.F.	$ 70.75		11.5 S.F.	$ 7.10	
Lumber at $315 per M.B.F.	200 B.F.	63.00		20.0 B.F.	6.30	
Accessories, incl. adjustable shores	Allow	20.25		Allow	20.25	
Make up, crew C-2 at $20.37 per man-hour	6.4 M.H.		$130.35	1.0 M.H.		$ 20.35
Erect and strip	8.3 M.H.		169.05	8.3 M.H.		169.05
Clean and move	1.3 M.H.		26.50	1.3 M.H.		26.50
Total per 100 S.F.C.A.	16.0 M.H.	$154.00	$325.90	10.6 M.H.	$33.65	$215.90

For structural steel frame with beams encased, subtract 1.2 man-hours, and 50 M.B.F. lumber or about $37 per 100 S.F.C.A. for the first use and $24 for each reuse.

BOX CULVERT, 5' to 8' Square or Rectangular (Line 30-001)	First Use			Reuse		
	Quantities	Material	Installation	Quantities	Material	Installation
3/4" exterior plyform at $670 per M.S.F.	110 S.F.	$ 73.70		11.0 S.F.	$ 7.35	
Lumber at $315 per M.B.F.	170 B.F.	53.55		17.0 B.F.	5.35	
Accessories	Allow	12.75		Allow	15.30	
Build in place, crew C-1 at $19.78 per man-hour	14.5 M.H.		$286.80	14.5 M.H.		$286.80
Strip and salvage	4.3 M.H.		85.05	4.3 M.H.		85.05
Total per 100 S.F.C.A.	18.8 M.H.	$140.00	$371.85	18.8 M.H.	$28.00	$371.85

CIRCLE REFERENCE NUMBERS

(33) Forms in Place (cont.)

COLUMNS, 24" x 24" (Line 25-650)	First Use			Reuse		
	Quantities	Material	Installation	Quantities	Material	Installation
5/8" exterior plyform @ $615 per M.S.F.	120 S.F.	$ 73.80		12.0 S.F.	$ 7.40	
Lumber @ $315 per M.B.F.	125 B.F.	39.40		12.5 B.F.	3.95	
Clamps, chamfer strips and accessories	Allow	35.80		Allow	5.65	
Make up, crew C-1 at $19.78 per man-hour	5.8 M.H.		114.70	1.0 M.H.		19.80
Erect and strip	9.8 M.H.		193.85	9.8 M.H.		193.85
Clean and move	1.2 M.H.		23.75	1.2 M.H.		23.75
Total per 100 S.F.C.A.	16.8 M.H.	$149.00	$332.30	12.0 M.H.	$17.00	$237.40

FLAT SLAB (Line 35-200)	First Use			Reuse		
	Quantities	Material	Installation	Quantities	Material	Installation
5/8" exterior plyform @ $615 per M.S.F.	115 S.F.	$ 70.75		11.5 S.F.	$ 7.10	
Lumber @ $315 per M.B.F.	210 B.F.	66.15		21 B.F.	6.60	
Accessories, incl. adjustable shores	Allow	22.10		Allow	22.10	
Make up, crew C-2 at $20.37 per man-hour	3.5 M.H.		$ 71.30	1.0 M.H.		$ 20.35
Erect and strip	6.0 M.H.		122.20	6.0 M.H.		122.20
Clean and move	1.2 M.H.		24.45	1.2 M.H.		24.45
Total per 100 S.F.C.A.	10.7 M.H.	$159.00	$217.95	8.2 M.H.	$35.80	$167.00

Drop panels included but column caps figure with columns.

FOOTINGS, SPREAD Line (45-500)	First Use			Reuse		
	Quantities	Material	Installation	Quantities	Material	Installation
Lumber @ $315 per M.B.F.	260 B.F.	$ 81.90		26 B.F.	$ 8.20	
Accessories	Allow	6.10		Allow	3.80	
Make up, crew C-1 at $19.78 per man-hour	4.7 M.H.		$ 92.95	1.0 M.H.		$ 19.80
Erect and strip	4.2 M.H.		83.05	4.2 M.H.		83.05
Clean and move	1.6 M.H.		31.65	1.6 M.H.		31.65
Total per 100 S.F.C.A.	10.5 M.H.	$ 88.00	$207.65	6.8 M.H.	$12.00	$134.50

FOUNDATION WALL, 8' High (Line 65-200)	First Use			Reuse		
	Quantities	Material	Installation	Quantities	Material	Installation
5/8" exterior plyform @ $615 per M.S.F.	110 S.F.	$ 67.65		11.0 S.F.	$ 6.75	
Lumber @ $315 per M.B.F.	140 B.F.	44.10		14.0 B.F.	4.40	
Accessories	Allow	15.25		Allow	13.85	
Make up, crew C-2 at $20.37 per man-hour	5.0 M.H.		$101.85	1.0 M.H.		$ 20.35
Erect and strip	6.5 M.H.		132.40	6.5 M.H.		132.40
Clean and move	1.5 M.H.		30.55	1.5 M.H.		30.55
Total per 100 S.F.C.A.	13.0 M.H.	$127.00	$264.80	9.0 M.H.	$25.00	$183.30

PILE CAPS, Square or Rectangular (Line 45-300)	First Use			Reuse		
	Quantities	Material	Installation	Quantities	Material	Installation
5/8" exterior plyform @ $615 per M.S.F.	110 S.F.	$ 67.65		11.0 S.F.	$ 6.75	
Lumber @ $315 per M.B.F.	160 B.F.	44.40		16.0 B.F.	4.45	
Accessories	Allow	11.95		Allow	10.80	
Make up, crew C-1 at $19.78 per man-hour	4.5 M.H.		$ 89.00	1.0 M.H.		$ 19.80
Erect and strip	5.0 M.H.		98.90	5.0 M.H.		98.90
Clean and move	1.5 M.H.		29.65	1.5 M.H.		29.65
Total per 100 S.F.C.A.	11.0 M.H.	$124.00	$217.55	7.5 M.H.	$22.00	$148.35

STAIRS, Average Run (Inclined Length x Width) (Line 60-001)	First Use			Reuse		
	Quantities	Material	Installation	Quantities	Material	Installation
5/8" exterior plyform @ $615 per M.S.F.	110 S.F.	$ 67.65		11.0 S.F.	$ 6.75	
Lumber @ $315 per M.B.F.	425 B.F.	133.90		42.5 B.F.	13.40	
Accessories	Allow	16.45		Allow	15.85	
Build in place, crew C-2 at $20.37 per man-hour	25.0 M.H.		$509.25	25.0 M.H.		$509.25
Strip and salvage	4.0 M.H.		81.50	4.0 M.H.		81.50
Total per 100 S.F.	29.0 M.H.	$218.00	$590.75	29.0 M.H.	$36.00	$590.75

CIRCLE REFERENCE NUMBERS

㉞ Wall Form Materials (Div. 3.1-65)

Aluminum Forms

Approximate weight is 3 lbs. per S.F.C.A. Standard widths are available from 4" to 36" with 36" most common. Standard lengths of 2', 4', 6' to 8' are available. Forms are lightweight and fewer ties are needed with the wider widths. The form face is either smooth or textured.

Cost of bare forms per S.F.C.A. with different facing surface is listed below. Typical material cost including usual accessories but not including form ties is $12.10 per S.F.C.A.

Forms may also be rented.

Finish	3' x 8'	2' x 8'	12" x 8'	6" x 8'	3' x 4'	2' x 4'	12" x 4'	6" x 4'
Smooth Aluminum (.096" Face sheet)	$8.40	$10.45	$14.45	$23.15	$ 9.45	$12.10	$16.30	$25.75
Textured Brick Aluminum	9.80	12.20	16.85	27.00	11.00	11.50	18.95	30.00

Metal Framed Plywood Forms

Manufacturers claim over 75 reuses of plywood and over 300 reuses of steel frames. Sale price for steel framed forms is $7.05 per S.F. for 2' x 8' to $9.25 per S.F. for 2' x 3'. Narrower forms range between $8.30 per S.F. to $13.15 per S.F. for 1' x 3'. Many specials such as corners, fillers, pilasters, etc. are available. Monthly rental is generally about 15% of purchase price for first month and 9% per month thereafter with 100% of rental applied to purchase for first two months and decreasing percentages thereafter. Aluminum framed forms cost 25% to 30% more than steel framed.

Rule of thumb purchase price including corners, specials, etc.; steel framed $9.20 per S.F.; aluminum framed $12.65 per S.F.

Reconditioned steel framed forms are rented for 80¢ per S.F per month. Aluminum framed forms are rented for $1.80 per S.F. for the first month and $1.10 per S.F. per month thereafter.

After the first month, extra days may be prorated from the monthly charge. Rental rates do not include ties, accessories, cleaning, loss of hardware or freight in and out. Approximate weight is 5 lbs. per S.F. for steel; 3 lbs. per S.F. for aluminum.

Forms can be rented with option to buy.

Plywood Forms, Job Fabricated

There are two types of plywood used for concrete forms.

1. Exterior plyform which is completely waterproof. This is face oiled to facilitate stripping. Ten reuses can be expected with this type with 25 reuses possible.
2. An overlaid type consists of a resin fiber fused to exterior plyform. No oiling is required except to facilitate cleaning. This is available in both high density (HDO) and medium density overlaid (MDO). Using HDO, 50 reuses can be expected with 200 possible.

Plyform is available in 5/8" and 3/4" thickness. High density overlaid is available in 3/8", 1/2", 5/8" and 3/4" thickness.

5/8" thick is sufficient for most building forms, while 3/4" is best on heavy construction.

For prices on plywood and framing lumber see ㊓ and ㊏.

Plywood Forms, Modular, Prefabricated

There are many plywood forming systems without frames. Most of these are manufactured from 1-1/8" (HDO) plywood and have some hardware attached. These are used principally for foundation walls 8' or less high. With care and maintenance, 100 reuses can be attained with decreasing quality of surface finish. Sale price of 2' x 8' panels is $4.00 with 4" x 8' fillers costing 16.30 per S.F. Typical forms and accessories for 1000 S.F. of form area cost $5,425 not including form ties.

Steel Forms

Approximate weight is 6-1/2 lbs. per S.F.C.A. including accessories. Standard widths are available from 2" to 24", with 24" most common. Standard lengths are from 2' to 8', with 4' the most common. Forms are easily ganged into modular units.

Sale price for typical hand set job is $11.00 per S.F.C.A. including all usual accessories, specials, corners, etc., but not including ties or freight. Cost of bare forms runs from $8.20 for wide forms to over $20 per S.F.C.A. for narrow widths and/or short lengths. Forms are usually leased for 8.7% of the purchase price per month prorated daily over 30 days. Standard 6000 lb. wall ties for 12" walls cost $50 per hundred and 24" long ties cost $71 per hundred.

Rental may be applied to sale price and usually rental forms are bought. With careful handling and cleaning 200 to 400 reuses are possible.

Straight wall gang forms up to 12' x 20' or 8' x 30' can be fabricated. These crane handled forms usually cost from $18 to $23 per S.F.C.A. or can be leased for 8.7% per month. Straight wall gang forms utilizing 40,000 lb. and 60,000 lb. ties with sizes from 24' x 10' to 24' x 24' cost from $20 to $25 per S.F. including all accessories and taper ties. Rental rates on these gang forms run from 30¢ to 45¢ per week per S.F.C.A.

Individual job analysis is available from the manufacturer at no charge.

㉟ Concrete For Entire Building (Div. 3.3-14-010)

These figures give an average cost per C.Y. for all concrete in a typical multi-storied reinforced concrete building. It is assumed that ready mixed concrete is used and that forms will be used four times. Rubbing is not included; 5% concrete waste is included. When forms and reinforcing quantity ratios are available compute the total C.Y. costs from the unit prices in front of the book. See ㉝ for handling equipment.

Item	Material	Labor	Hoist & Equipment	Total
Concrete in place	$51.60	$15.60	$ 2.80	$ 61.60
Forms	20.55	83.90	—	104.45
Reinforcing steel	25.40	14.60	.75	40.75
Total per C.Y.	$97.55	$114.10	$3.55	$206.80

CIRCLE REFERENCE NUMBERS

㊱ Floor Pans and Domes (Div. 3.1-35)

For 8' to 15' Ceiling Heights Using Crew C-2 at $20.37 per Man-hour	20" Pans				19" Domes			
	(Line 350)	1 Use	(Line 365)	4 Use	(Line 400)	1 Use	(Line 415)	4 Use
Rent 20" pans or 19" domes			$ 60.00		$ 19.50		$ 60.00		$ 19.50
Adjustable shores and accessories			32.00		17.00		32.00		17.00
110 S.F. 3/4" plyform at $670 per M.S.F.			73.70		23.95		73.70		23.95
210 B.F. supporting joists, girts and braces at $315 per MBF			66.15		21.50		66.15		21.50
Labor handle, place, strip, oil and clean pans and domes	1.5 hr.	30.55	1.0 hr.	20.35	1.8 hr.	36.65	1.2 hr.	24.45
Make up, erect, remove wood decking	10.0 hr.	203.70	8.5 hr.	173.15	10.0 hr.	203.70	8.5 hr.	173.15
Total per 100 S.F. of floor area		$466.10		$275.45		$472.20		$279.55

The figures above are for closed deck forming. For open deck forming deduct $20 per 100 S.F. for four uses. For pan rental, figure 60¢ per S.F. of total area for one use; 33¢ for two uses; 23¢ for three uses; and 20¢ for four uses, which is about the most that can be expected for any one project. The purchase price of fiberglass long pans is $4.25 per S.F.C.A. or an average $8.55 per S.F. of floor area.*

For two-way grid system, 19" steel domes are leased for 20¢ to 60¢ and 30" fiberglass domes are leased for 45¢ to $1.90 per S.F. of floor area. The purchase price of custom size fiberglass forms runs from $5.20 to $13.00 per S.F.C.A.*

For slab height from 15' to 20' add $38 per 100 S.F. and for 20' to 35' add $53 per 100 S.F., both for four uses.

*It is necessary to divide the purchase cost by the number of expected uses to determine the cost per use.

㊲ Slipforms (Div. 3.1-85)

The slipform method of forming may be used for forming circular silo and multi-celled storage bin type structures over 30' high, and building core shear walls over eight stories high. The shear walls usually enclose elevator shafts, stairwells, mechanical spaces, and toilet rooms. Reuse of the form on duplicate structures will reduce the height necessary to spread the cost of building the form. Slipform systems can be used to cast chimneys, towers, piers, dams, underground shafts or other structures capable of being extruded.

Slipforms are usually 4' high and are raised semi-continuously by jacks climbing on rods which are embedded in the concrete. The jacks are powered by a hydraulic, pneumatic, or electric source and are available in 3, 6, and 22 ton capacities. Interior work decks and exterior scaffolds must be provided for placing inserts, embedded items, reinforcing steel and concrete. Scaffolds below the form for finishers may be required. The interior work decks are often used as roof slab forms on silos and bin work. Form raising rates will range from 6" to 20" per hour for silos; 6" to 30" per hour for buildings and 6" to 48" per hour for shaft work. Reinforcing bars and stressing strands are usually hoisted by crane or gin pole and the concrete material can be hoisted by crane, winch powered skip or pumps. The slipform system is operated on a continuous 24 hour day when a monolithic structure is desired. For least cost, the system is operated only during normal working hours.

The cost of the form, including equipment setting, will range from $7.00 to $15.00 per square foot of actual form built. The cost of work decks and scaffolds will range from $2.00 to $5.00 per square foot of total work deck area. Slipform equipment rental will range from 15¢ to 30¢ per square foot of surface formed. Operating costs of slipform systems will range from 5¢ to 20¢ per S.F. of surface formed.

Placing concrete will range from .5 to 1.25 man hours per cubic yard. Finishing (if required) and curing costs will range from 3¢ to 20¢ per square foot of surface. Bucks, blockouts, keyways, weldplates, etc. are extra. Recent in place total costs have ranged from $225 to $750 per cubic yard.

㊳ Reinforcing Extras (Div. 3.2)

There are two "base prices" used, the price for bars shipped directly from the mill and the base price for the bars shipped from the steel fabricator's shops. The extras for both methods of pricing are listed below and are listed in dollars per ton for A615-76 steel.

Extras	Item	From Mill	From Fabricator	Extras	Item	From Mill	From Fabricator
Base Price		$250	$365				$500 minimum
Size Extras	#3	$100	$100	Quantity Extras	Under 20 tons	$3	25
	#4	40	40		20-49 tons	—	15
	#5	30	30		50-99 tons	—	10
	#6	25	25		100-300 tons	—	5
	#7 to 11	20	20		Over 300 tons	—	—
	#14	40	40	Bending	Light #2 thru #11	N.A.	80
	#18	40	40		Heavy #4 up		35
Grade Extras	Grade 40	—	—	Detailing	Under 50 tons		40
	Grade 60	$ 10	$ 14		50 to 150 tons		30
	Grade 75	NA	NA		150 to 500 tons		25
					Over 500 tons		20

Listing only runs $3 per ton.

CIRCLE REFERENCE NUMBERS

(39) Reinforcing Key Prices (Div. 3.2)

Cost given below are for over 50 tons delivered and bent from fabricators in metropolitan area. Over 500 tons, subtract $10 per ton. Figures are for domestic A615, grade 60. Economic conditions can alter the prices substantially.

City	Bare Costs					Costs Incl. Subs O & P		
	Material		Installation		Total per Ton in Place	Material Incl. 10% Mark-up	Installation Incl. 52.5% Mark-up	Total per Ton
	50 Tons Fabricated	Allow for Accessories	Unload, Sort and Pile	Place & Tie				
Atlanta	$450	$45	$12	$210	$ 717	$545	$342	$ 887
Baltimore	440		15	277	777	534	450	984
Boston	520		16	323	904	622	523	1,145
Buffalo	470		17	328	860	567	531	1,098
Chicago	465		17	330	857	567	535	1,102
Cincinnati	460		15	292	812	556	473	1,029
Cleveland	500		17	345	907	600	558	1,158
Columbus	485		16	299	845	589	486	1,075
Dallas	460		14	259	778	556	421	977
Denver	460		15	275	795	556	446	1,002
Detroit	480		16	310	851	578	503	1,081
Houston	485		14	264	808	589	428	1,017
Indianapolis	450		16	309	820	545	501	1,046
Kansas City	470		16	297	828	567	482	1,049
Los Angeles	460		19	388	912	556	473	1,029
Memphis	435		12	209	701	528	340	868
Milwaukee	430		16	314	805	523	509	1,032
Minneapolis	420		15	293	773	512	474	986
Nashville	410		12	203	670	500	331	831
New Orleans	420		14	243	722	512	396	908
New York (17 hrs.)*	510*		22	580	1,157	611	928	1,539
Philadelphia	460		17	349	871	556	564	1,120
Phoenix	450		16	315	826	545	510	1,055
Pittsburgh	435		17	326	823	528	523	1,051
St. Louis	430		16	298	789	523	484	1,007
San Antonio	440		13	227	725	534	369	903
San Diego	500		19	388	952	600	627	1,227
San Francisco	485		19	388	937	589	627	1,216
Seattle	465		16	295	821	561	479	1,040
Washington, D.C.	460		14	250	769	556	407	963
Average per ton	$460	$45	$16	$306	$ 827	$557	$491	$1,048

*Bent and cut on the job.
Place and tie costs do not include crane charge.

Base price of A615, Grade 40 at Eastern mills is $250 per ton.

Bare Material Costs	
Item	Cost per Ton
Base price at mill	$250
Grade 60 extra	10
Size and length extras	35
Freight from mill	25
Warehouse handling & storage	25
Shearing and shop bending	55
Drafting, detailing and listing	40
Trucking to jobsite	20
Reinforcing Delivered	$460
Accessories Delivered	45
Total Material Delivered	$505

Total Installed Costs		
Description	Bare Costs	Cost Incl. Subs O&P
Unload, sort & pile, C-5 crew	$ 16 (41.9%)	$ 23
Place & tie, 14 hours at $21.75 per hour	305** (54.7%)	472
Total Installation Cost	$321	$ 495
Bare Material Cost	$505	
Material Cost + 10%		$ 556
Total Installed Cost	$826	$1,051

** Note: the $2 bare cost and the $4 Subs O & P cost difference is due to New York labor figured at 17 hours instead of 14 for the other 29 cities.

Engineering or design fees are not included above.
Material delivered for small jobs can range up to $700 per ton with erection going as high as $700 per ton if job consists of small bars.
Restrictive union practices can increase field labor costs above those listed above.

CIRCLE REFERENCE NUMBERS

40 Prestressed Concrete, post-tensioned (Div. 3.2-03)

In post-tensioned concrete the steel tendons are tensioned after the concrete has reached about 3/4 of its ultimate strength. The cableways are grouted after tensioning to provide bond between the steel and concrete. If bond is to be prevented, the tendons are coated with a corrosion-preventative grease and wrapped with waterproof paper or plastic. Bonded tendons are usually used when ultimate strength (beams & girders) is a controlling factor.

High strength concrete is used to fully utilize the steel, thereby reducing the size and weight of the member. A plasticizing agent may be added to reduce water content. Maximum size aggregate ranges from 1/2" to 1-1/2" depending on the spacing of the tendons.

The type of steel commonly used are bars and strands. Job conditions determine which is best suited. Bars are best for vertical prestresses since they are easy to support. The trend is for steel manufacturers to supply a finished package, cut to length, which reduces field preparation to a minimum.

Bars vary from 3/4" to 1-3/8" diameter. Table below gives time in man-hours per tendon and the labor cost per pound for placing, tensioning and grouting (if required) a 75' beam. Tendons used in buildings are not usually grouted; tendons for bridges usually are grouted. For strands the table indicates the man-hours and labor cost per pound for typical prestressed units 100' long. Simple span beams usually require one end stressing regardless of lengths. Continuous beams are usually stressed from two ends. Long slabs are poured from the center outward and stressed in 75' increments after the initial 150' center pour. Prices below do not include subcontractor's overhead and profit. Prices in the front of the book do not include subcontractor's overhead and profit, except in the column headed Total Incl. O & P.

Man Hours per Tendon and Labor Costs per Pound of Prestress Steel						
Length	100' Beam		75' Beam		100' Slab	
Type Steel	Strand		Bars		Strand	
Diameter	0.5"		3/4"	1-3/8"	0.5"	0.6"
Number	12	24	1	1	1	1
Force in Kips	298	595	42	143	25	35
Preparation & Placing Cables	6.6	10.0	0.7	2.3	0.8	0.8
Stressing Cables	2.4	3.0	0.6	1.3	0.4	0.4
Grouting, if required	3.0	3.5	0.5	1.0		
Total Man Hours	12.0	16.5	1.8	4.6	1.2	1.2
Prestressing Steel Weights	640#	1280#	115#	380#	53#	74#
Labor cost per lb. Bonded	52¢ to 60¢	40¢ to 47¢	40¢ to 45¢	25¢ to 30¢		
Non-bonded			35¢ to 40¢	30¢ to 35¢	65¢ to 78¢	65¢ to 78¢

Flat Slab construction — 4000 psi concrete with span to depth ratio between 36 and 44. Two way post-tensioned steel averages 1.0 lb. per S.F. for 24' to 28' bays (usually strand) and additional reinforcing steel averages .5 lb. per S.F.

Pan and Joist construction — 4000 psi concrete with span to depth ratio 28 to 30. Post-tensioned steel averages .8 lb. per S.F. and reinforcing steel about 1.0 lb. per S.F. Placing and stressing averages 40 hours per ton of total material.

Beam construction — 4000 to 5000 psi concrete. Steel weights vary greatly. Tendon material cost runs from $.80 to $1.30 per lb. for grouted and $.70 to $.85 per lb. for non-grouted strand. Bar costs are from $.70 to $.95 per lb.

From the chart above placing and stressing cost for grouted tendons runs from $.40 to $.60 per lb; non-grouted placing costs run from $.30 to $.70 per lb. Labor cost per pound goes down as the size and length of the tendon increases. The primary economic consideration is the cost per kip for the member.

Post-tensioning becomes feasible for beams and girders over 30' long; for continuous two-way slabs over 20' clear; also in transferring upper building loads over longer spans at lower levels. Post-tension suppliers will provide engineering services at no cost to the user. Substantial economies are possible by using post-tensioned Lift Slabs. See 48 for Lift Slabs.

41 Materials for One C.Y. of Concrete (Div. 3.3)

This is an approximate method of figuring quantities of cement, sand and gravel for a field mix with waste allowance included.

With gravel as coarse aggregate for barrels of cement required, divide 10 by total mix; that is, for 1:2:4 mix, 10 divided by 7 = 1-3/7 barrels.

For tons of sand multiply barrels of cement by parts of sand and then by 0.2; that is, for the 1:2:4 mix, as above, 1-3/7 x 2 x .2 = .57 tons. Tons of gravel are in the same ratio to tons of sand as parts in mix, or 4/2 x .57 = 1.14 tons.

If course aggregate is crushed stone, use 10-1/2 instead of 10 as given for gravel.

- 1 bag cement = 94#
- 4 bags = 1 barrel
- 1 C.Y. sand or gravel = 2700#
- 1 ton = 20 C.F.
- 1 C.Y. crushed stone = 2575#
- 1 ton = 21 C.F.

Average carload of cement is 692 bags; of sand or gravel is 56 tons.

Do not stack stored cement over 10 bags high.

CIRCLE REFERENCE NUMBERS

㊷ Concrete Material Net Prices (Div. 3.3)

Costs below are C.Y. of concrete delivered; per ton of bulk cement; per bag cement delivered T.L.L.; per ton for stone and sand aggregates loaded at plant (no trucking included) and per 4 C.F. bag for perlite or vermiculite aggregate delivered T.L.L.

City	Ready Mix Concrete Regular Weight		Cement T.L. Lots		Aggregates per Ton			Vermiculite or Perlite 4 C.F. Bag
	3000 psi	5000 psi	Bulk per Ton	Bags per Bag	Crushed Stone 1-1/2"	3/4"	Sand	
Atlanta	$42.15	$48.70	$64.75	$4.70	$7.50	$7.50	$6.90	$4.75
Baltimore	50.20	56.10	63.00	5.00	5.80	5.80	5.20	4.70
Boston	51.00	57.15	64.50	5.20	7.00	7.00	6.25	4.85
Buffalo	54.00	60.50	67.00	5.00	6.00	6.25	5.15	5.20
Chicago	49.00	54.90	65.50	5.30	6.00	6.50	5.00	4.65
Cincinnati	45.20	51.25	61.50	4.90	5.25	5.25	5.00	5.30
Cleveland	46.25	52.45	63.50	5.20	8.00	8.50	7.20	5.00
Columbus	48.50	55.10	70.50	5.50	6.20	6.20	5.20	5.25
Dallas	50.85	57.00	63.00	4.80	8.20	9.00	6.40	4.75
Denver	59.35	67.70	79.50	5.20	7.50	7.60	6.10	5.00
Detroit	46.50	52.10	62.50	4.75	5.60	6.00	5.00	5.10
Houston	50.50	54.90	65.00	5.20	8.20	8.50	7.40	4.60
Indianapolis	49.75	56.05	66.00	5.30	7.00	7.20	6.00	5.20
Kansas City	48.35	53.70	73.00	5.50	8.20	8.20	6.25	5.50
Los Angeles	48.15	55.20	73.20	5.35	7.90	8.90	6.50	6.50
Memphis	46.50	52.25	63.50	5.25	7.20	7.20	5.50	5.30
Milwaukee	47.10	53.05	68.50	5.15	7.50	7.50	5.40	5.40
Minneapolis	46.20	52.95	73.40	5.00	8.30	8.30	6.00	5.75
Nashville	46.20	51.75	65.20	5.00	7.50	7.50	6.20	5.20
New Orleans	49.25	55.00	61.00	4.80	8.10	8.50	7.15	5.35
New York City	67.50	73.70	65.50	5.50	10.00	10.00	8.50	4.80
Philadelphia	43.00	50.95	66.50	4.80	8.75	8.75	8.00	4.65
Phoenix	45.10	52.30	76.50	5.75	7.00	7.00	6.00	5.10
Pittsburgh	48.50	54.70	66.50	4.75	10.50	11.50	8.60	5.15
St. Louis	44.90	50.35	59.50	4.50	7.80	7.80	6.50	5.50
San Antonio	44.50	50.30	63.50	5.00	7.20	7.20	5.50	4.80
San Diego	47.20	54.20	68.50	5.50	9.00	9.20	8.00	4.90
San Francisco	48.50	55.65	70.50	5.50	9.25	9.50	9.00	4.75
Seattle	46.50	53.90	64.50	5.00	8.10	8.10	8.00	5.75
Washington, D.C.	49.50	55.35	63.50	5.35	9.40	9.70	8.65	5.30
Average	$48.65	$55.00	$66.65	$5.15	$7.65	$7.85	$6.55	$5.15

㊸ Ready Mix Material Prices (Div. 3.3)

Table below lists national average prices per C.Y. of concrete. Prices in the key cities for different strengths can be closely estimated by factoring against the 3000 psi or 5000 psi price from ㊷ above or from Div. 19, City Cost Indexes, cast-in-place concrete material factor.

Design Mix		Heavy Weight		Light Weight		Admixtures and Special Items	
Strength in psi	Nominal Mix	Bags per C.Y.	Regular	High Early	110# per C.F.	All Light Weight	Add to Each C.Y. of Concrete for the Following Items:	
2,000	1:3:5	4.5	$45.55	$49.35	$59.30	$69.30	Calcium chloride, 1%	$1.10 per C.Y.
2,500	1:2½:4½	5	47.25	51.50	61.00	71.00	" " 2%	2.20 per C.Y.
3,000	1:2:4	5.5	48.65	53.55	62.65	72.65	Lampblack	2.85 per lb.
3,500	1:2:3½	6	50.55	55.65	64.30	74.30	Water reducing agent	.30 per bag
3,750	1:2:3¼	6.3	51.50	56.85	65.70	75.70	Set retarder	.30
4,000	1:2:3	6.5	52.20	57.70	66.40	76.40	High early cement	.85
4,500	1:2:2½	7	54.15	60.10	68.35	78.35	White cement	9.75
5,000	1:1½:2½	7.5	55.00	61.50	69.30	79.30	Pump aid	1.60 per C.Y.
	1:1:2	8	56.75	63.55	Perlite 1:6= $67.00		Winter concrete	2.75 per C.Y.
	1:4 topping	7.5	55.20	61.55				
	1:3 topping	8.5	58.45	65.65				
	1:2 topping	11	66.55	75.85				

CIRCLE REFERENCE NUMBERS

 Bend, Place and Tie Reinforcing (Div. 3.2-04)

See ㊳ for shop bending costs. Field bending costs are usually higher.

Placing and tieing by rodmen for footings and slabs runs from nine hrs. per ton for heavy bars to fifteen hrs. per ton for light bars. For beams, columns and walls, production runs from eight hrs. per ton for heavy bars to twenty hrs. per ton for light bars. Overall average for typical reinforced concrete buildings is about fourteen hrs. per ton. These production figures include placing of accessories and usual inserts. Equipment handling is necessary for the larger size bars so that installation costs for the very heavy bars will not decrease proportionately.

For tie wire, figure 5 lbs. 16 ga. black annealed wire at 60¢ per lb. for each ton of bars. Use of commercial accessories improves placing accuracy and the cost of $45 per ton of bars is offset by reduced labor time.

Installation costs for splicing reinforcing bars includes allowance for equipment to hold the bars in place while splicing as well as necessary scaffolding for iron workers.

 Field Mix Concrete (Div. 3.3-10)

Presently most building jobs are built with ready mix concrete except for isolated locations and some larger jobs requiring over 10,000 C.Y. where land is readily available for setting up a temporary batch plant.

The most economical mix is a controlled mix using local aggregate proportioned by trial to give the required strength with the least cost of material.

Costs tabulated below are based on a job that requires 10,000 C.Y. of concrete to be placed within an eight month period.

28 day Strength	(Line 002)	3000 psi	(Line 001)	2250 psi
Mix Proportions	1:2½:3		1:2¾:4	
Cement, trucked in bulk @ $3.33 per Cwt.	6-1/4 bags @ 94 lb./bag	$19.56	5 bags @ 94 lb./bag	$15.65
Sand @ $6.55 per ton	.68 tons	4.45	.7 tons	4.60
Stone @ $7.65 per ton, maximum size aggregate 1-1/2"	.9 tons	6.89	1.02 tons	7.80
Set up plant, foundations, piping, etc.		1.20		1.20
Plant, trucks and equipment rental		5.45		5.45
Labor to mix and transport to forms		8.90		8.90
Supervision		1.00		1.00
Cost per C.Y. FOB Forms		$47.45		$44.60

See ㊷ for material prices in the major cities.

㊻ **Placing Ready Mixed Concrete** (Div. 3.3-38)

For ground pours allow for 5% waste when figuring quantities.

Prices in the front of the book assume normal deliveries. If deliveries are made before 8 A.M. or after 5 P.M. or on Saturdays add $10 per C.Y. Large volume discounts are not included in prices in front of the book.

For the lower floors without truck access, concrete may be wheeled in rubber tired buggies, conveyer handled, crane handled or pumped. Pumping is economical if there is top steel. Conveyers are more efficient for thick slabs. Concrete pump with an operator can be rented from $600 per day for up to 25 C.Y. to $800 per day for a 400 C.Y. pour. Figures include travel time if done at straight time. Pumping lightweight concrete costs an extra $1.80 per C.Y.

At higher floors the rubber tired buggies may be hoisted by a hoisting tower and then wheeled to location. Placement by a conveyer is limited to three floors and is best for high volume pours. Pumped concrete is best when building has no crane access. Concrete may be pumped directly as high as thirty-six stories using special pumping techniques. Normal maximum height is about fifteen stories.

Best pumping aggregate is screened and graded bank gravel rather than crushed stone.

Pumping downward is more difficult than pumping upwards. Hoizontal distance from pump to pour may increase preparation time prior to pour. Placing by cranes, either mobile, climbing or tower types continues as the most efficient method for high rise concrete buildings.

Cost per C.Y. for Wheeled Concrete, dumped only (Add to appropriate placing cost)								
	10 C.F. Walking Cart				18 C.F. Riding Cart			
Item	Hourly Cost	Wheeled up to 50 ft.	Ea. Added 100 feet	Ea. Added 5 Minutes	Hourly Cost	Wheeled up to 50 ft.	Ea. Added 100 feet	Ea. Added 5 Minutes
Building laborer	$15.90	$3.97	$5.08	$3.58	$15.90	$1.59	$ .79	$1.99
Foreman, outside	22.50	5.62	7.19	5.07	22.50	2.25	1.12	2.82
Concrete cart	4.55	1.13	1.46	1.02	7.20	.71	.35	.89
Total Cost/C.Y.	—	$10.72	$13.73	$9.67	—	$4.55	$2.26	$5.70
Hourly production		4 C.Y.				10 C.Y.		

CIRCLE REFERENCE NUMBERS

㊼ Proportionate Quantities (Div. 3.3)

The tables below show both quantities per S.F. of floor areas as well as form and reinforcing quantities per C.Y. Unusual structural requirements would increase the ratios below. High strength reinforcing would reduce the steel weights. Figures are for 3000 psi concrete and 60,000 psi reinforcing unless specified otherwise.

Type of Construction	Live Load	Span	Per S.F. of Floor Area				Per C.Y. of Concrete		
			Concrete	Forms	Reinf.	Pans	Forms	Reinf.	Pans
Flat Plate	50 psf	15 Ft.	.46 C.F.	1.06 S.F.	1.71 lb.		62 S.F.	101 lb.	
		20	.63	1.02	2.4		44	104	
		25	.79	1.02	3.03		35	104	
	100	15	.46	1.04	2.14		61	126	
		20	.71	1.02	2.72		39	104	
		25	.83	1.01	3.47		33	113	
Flat Plate (waffle construction) 20" domes	50	20	.43	1.0	2.1	.84 S.F.	63	135	53 S.F.
		25	.52	1.0	2.9	.89	52	150	46
		30	.64	1.0	3.7	.87	42	155	37
	100	20	.51	1.0	2.3	.84	53	125	45
		25	.64	1.0	3.2	.83	42	135	35
		30	.76	1.0	4.4	.81	36	160	29
Waffle Construction 30" domes	50	25	.69	1.06	1.83	.68	42	72	40
		30	.74	1.06	2.39	.69	39	87	39
		35	.86	1.05	2.71	.69	33	85	39
		40	.78	1.0	4.8	.68	35	165	40
Flat Slab (two way with drop panels)	50	20	.62	1.03	2.34		45	102	
		25	.77	1.03	2.99		36	105	
		30	.95	1.03	4.09		29	116	
	100	20	.64	1.03	2.83		43	119	
		25	.79	1.03	3.88		35	133	
		30	.96	1.03	4.66		29	131	
	200	20	.73	1.03	3.03		38	112	
		25	.86	1.03	4.23		32	133	
		30	1.06	1.03	5.3		26	135	
One Way Joists 20" pans	50	15	.36	1.04	1.4	.93	78	105	70
		20	.42	1.05	1.8	.94	67	120	60
		25	.47	1.05	2.6	.94	60	150	54
	100	15	.38	1.07	1.9	.93	77	140	66
		20	.44	1.08	2.4	.94	67	150	58
		25	.52	1.07	3.5	.94	55	185	49
One Way Joists 8" x 16" filler blocks	50	15	.34	1.06	1.8	.81 Ea.	84	145	64 Ea.
		20	.40	1.08	2.2	.82	73	145	55
		25	.46	1.07	3.2	.83	63	190	49
	100	15	.39	1.07	1.9	.81	74	130	56
		20	.46	1.09	2.8	.82	64	160	48
		25	.53	1.10	3.6	.83	56	190	42
One Way Beam & Slab	50	15	.42	1.30	1.73		84	111	
		20	.51	1.28	2.61		68	138	
		25	.64	1.25	2.78		53	117	
	100	15	.42	1.30	1.9		84	122	
		20	.54	1.35	2.69		68	154	
		25	.69	1.37	3.93		54	154	
	200	15	.44	1.31	2.24		80	137	
		20	.58	1.40	3.30		65	163	
		25	.69	1.42	4.89		53	183	
Two Way Beam & Slab	100	15	.47	1.20	2.26		69	130	
		20	.63	1.29	3.06		55	131	
		25	.83	1.33	3.79		43	123	
	200	15	.49	1.25	2.70		41	149	
		20	.66	1.32	4.04		54	165	
		25	.88	1.32	6.08		41	187	

CIRCLE REFERENCE NUMBERS

㊼ Proportionate Quantities (cont.)

4000 psi Concrete and 60,000 psi Reinforcing — Form and Reinforcing Quantities per C.Y.

Item	Size	Forms	Reinforcing	Minimum	Maximum
Columns (square tied)	10" x 10"	130 S.F.C.A.	#5 to #11	220 lbs.	875 lbs.
	12" x 12"	108	#6 to #14	200	955
	14" x 14"	92	#7 to #14	190	900
	16" x 16"	81	#6 to #14	187	1082
	18" x 18"	72	#6 to #14	170	906
	20" x 20"	65	#7 to #18	150	1080
	22" x 22"	59	#8 to #18	153	902
	24" x 24"	54	#8 to #18	164	884
	26" x 26"	50	#9 to #18	169	994
	28" x 28"	46	#9 & #18	147	864
	30" x 30"	43	#10 to #18	146	983
	32" x 32"	40	#10 to #18	175	866
	34" x 34"	38	#10 to #18	157	772
	36" x 36"	36	#10 to #18	175	852
	38" x 38"	34	#10 to #18	158	765
	40" x 40"	32	#10 to #18	143	692

Item	Size	Forms	Spirals	Reinforcing	Minimum	Maximum
Columns (spirally reinforced)	12" diameter	34.5 L.F.	190 lb.	#4 to #11	165 lb.	1505 lb.
		34.5	190	#14 & #18	—	1100
	14"	25	170	#4 to #11	150	970
		25	170	#14 & #18	800	1000
	16"	19	160	#4 to #11	160	950
		19	160	#14 & #18	605	1080
	18"	15	150	#4 to #11	160	915
		15	150	#14 & #18	480	1075
	20"	12	130	#4 to #11	155	865
		12	130	#14 & #18	385	1020
	22"	10	125	#4 to #11	165	775
		10	125	#14 & #18	320	995
	24"	9	120	#4 to #11	195	800
		9	120	#14 & #18	290	1150
	26"	7.3	100	#4 to #11	200	729
		7.3	100	#14 & #18	235	1035
	28"	6.3	95	#4 to #11	175	700
		6.3	95	#14 & #18	200	1075
	30"	5.5	90	#4 to #11	180	670
		5.5	90	#14 & #18	175	1015
	32"	4.8	85	#4 to #11	185	615
		4.8	85	#14 & #18	155	955
	34"	4.3	80	#4 to #11	180	600
		4.3	80	#14 & #18	170	855
	36"	3.8	75	#4 to #11	165	570
		3.8	75	#14 & #18	155	865
	40"	3.0	70	#4 to #11	165	500
		3.0	70	#14 & #18	145	765

CIRCLE REFERENCE NUMBERS

㊼ Proportionate Quantities (cont.)

3000 psi Concrete and 60,000 psi Reinforcing — Form and Reinforcing Quantities per C.Y.

Item	Type	Loading	Height	C.Y./L.F.	Forms/C.Y.	Reinf./C.Y.
Retaining Walls	Cantilever	Level Backfill	4 Ft.	0.2 C.Y.	49 S.F.	35 lb.
			8	0.5	42	45
			12	0.8	35	70
			16	1.1	32	85
			20	1.6	28	105
		Highway Surcharge	4	0.3	41	35
			8	0.5	36	55
			12	0.8	33	90
			16	1.2	30	120
			20	1.7	27	155
		Railroad Surcharge	4	0.4	28	45
			8	0.8	25	65
			12	1.3	22	90
			16	1.9	20	100
			20	2.6	18	120
	Gravity, with vertical face	Level Backfill	4	0.4	37	None
			7	0.6	27	↓
			10	1.2	20	
		Sloping Surcharge	4	0.3	31	
			7	0.8	21	
			10	1.6	15	↓

		Live Load in Kips per Linear Foot							
	Span	Under 1 Kip		2 to 3 Kips		4 to 5 Kips		6 to 7 Kips	
		Forms	Reinf.	Forms	Reinf.	Forms	Reinf.	Forms	Reinf.
Beams	10 Ft.	—	—	90 S.F.	170#	85 S.F.	175#	75 S.F.	185#
	16	130 S.F.	165#	85	180	75	180	65	225
	20	110	170	75	185	62	200	51	200
	26	90	170	65	215	62	215	—	—
	30	85	175	60	200	—	—	—	—

Item	Size	Type	Forms per C.Y.	Reinforcing per C.Y.
Spread Footings	Under 1 C.Y.	1,000 psf soil	24 S.F.	44 lb.
		5,000	24	42
		10,000	24	52
	1 C.Y. to 5 C.Y.	1,000	14	49
		5,000	14	50
		10,000	14	50
	Over 5 C.Y.	1,000	9	54
		5,000	9	52
		10,000	9	56
Pile Caps (30 ton Concrete Piles)	Under 5 C.Y.	shallow caps	20	65
		medium	20	50
		deep	20	40
	5 C.Y. to 10 C.Y.	shallow	14	55
		medium	15	45
		deep	15	40
	10 C.Y. to 20 C.Y.	shallow	11	60
		medium	11	45
		deep	12	35
	Over 20 C.Y.	shallow	9	60
		medium	9	45
		deep	10	40

(47) Proportionate Quantities (cont.)

3000 psi Concrete and 60,000 psi Reinforcing — Form and Reinforcing Quantities per C.Y.

Item	Size	Pile Spacing	50 T Pile	100 T Pile	50 T Pile	100 T Pile
Pile Caps (Steel H Piles)	Under 5 C.Y.	24″ O.C.	24 S.F.	24 S.F.	75 lb.	90 lb.
		30″	25	25	80	100
		36″	24	24	80	110
	5 C.Y. to 10 C.Y.	24″	15	15	80	110
		30″	15	15	85	110
		36″	15	15	75	90
	Over 10 C.Y.	24″	13	13	85	90
		30″	11	11	85	95
		36″	10	10	85	90

Item		8″ Thick		10″ Thick		12″ Thick		15″ Thick	
	Height	Forms	Reinf.	Forms	Reinf.	Forms	Reinf.	Forms	Reinf.
Basement Walls	7 Ft.	81 S.F.	44 lb.	65 S.F.	45 lb.	54 S.F.	44 lb.	41 S.F.	43 lb.
	8		44		45		44		43
	9		46		45		44		43
	10		57		45		44		43
	12		83		50		52		43
	14		116		65		64		51
	16				86		90		65
	18						106		70

(48) Lift Slabs (Div. 3.3-14-520)

The cost advantage of the lift slab method is due to placing all concrete, reinforcing steel, inserts and electrical conduit at ground level and in reduction of formwork. Minimum economical project size is about 30,000 S.F. Slabs may be tilted for parking garage ramps.

It is now used in all types of buildings and has gone up to 22 stories high in apartment buildings. Current trend is to use post-tensioned flat plate slabs with spans from 22′ to 35′. Cylindrical void forms are used when deep slabs are required. One pound of prestressing steel is about equal to seven pounds of conventional reinforcing.

To be considered cured for stressing and lifting, a slab must have attained 75% of design strength. Seven days are usually sufficient with four to five days possible if high early strength cement is used. Slabs can be stacked using two coats of a non-bonding agent to insure that slabs do not stick to each other. Lifting is done by companies specializing in this work. Lift rate is 5′ to 15′ per hour with an average of 10′ per hour. Total areas up to 33,000 S.F. have been lifted at one time. 24 to 36 jacking columns are common. Most economical bay sizes are 24′ to 28′ with four to fourteen stories most efficient. Continuous design reduces reinforcing steel cost. Use of post-tensioned slabs allow larger bay sizes. Supplementary reinforcing, post-tensioned tendons, accessories and field labor cost about $1.80 per pound of tendon.

Table below shows the usual range and average S.F. cost of a typical lift slab project that has been designed to take advantage of lift slab techniques. Figures include Subs Overhead and Profit. Add 15% for Northeast.

Typical Cost per S.F. for Lift Slabs Project Incl. Subs O & P

Item	Description	Typical Range	Average	Sub Total
Concrete Slabs	Edge and bulkhead forms	$ .05 to $ .30	$ .20	$3.98
	Reinforcing and post-tensioning steel	1.25 to 2.00	1.75	
	Reinforcing and accessories	.06 to .12	.08	
	Concrete cast at ground level	1.25 to 2.00	1.60	
	Float or trowel finish	.25 to .50	.35	
Lifting Slabs	Separating and curing compound	.03 to .06	.04	1.94
	Lifting collars	.20 to .40	.30	
	Lifting and welding	1.00 to 2.20	1.60	
Columns	Fabricated columns & accessories	.55 to 1.50	1.00	1.27
	Set initial stage	.05 to .15	.10	
	Column splicing	.10 to .25	.17	
Miscellaneous	Grout plates, patching, bolts, etc.	.04 to .15	.10	.10
Total in place per S.F.		$4.79 to $9.63	$7.29	$7.29

CIRCLE REFERENCE NUMBERS

㊾ Average C.Y. of Concrete (Div. 3.3-14)
Rubbing and floor finish not included — 4 uses of forms assumed, 5 story building.

Item	Line Number	Description	Strength in psi	Cost for 1 C.Y. of Concrete				
				Ready Mix	Place	Forms	Reinforcing	Total
Beams, 5 kip/L.F.	030	10' span	3000	$48.90	$43.40	$222.85	$139.20	$454.35
	035	25'		48.90	42.80	203.45	79.90	375.05
Beam & Slab, 1 way	270	15'	3000	48.90	47.85	221.45	46.35	364.55
100 psf L.L.	275	25'		48.90	38.10	149.20	60.45	296.65
Beam & Slab, 2 way	290	15'	3000	48.90	57.35	181.50	49.40	337.15
100 psf L.L.	295	25'		48.90	48.10	115.20	46.75	258.95
Columns, square	082	16" x 16"	4000	52.20	43.05	257.60	260.30	613.15
tied	092	24" x 24"		52.20	32.15	170.65	214.30	469.30
Columns, Tied	122	16" diameter	4000	52.20	33.50	176.15	279.75	541.60
reinforced	142	24" diameter		52.20	11.85	126.65	248.85	451.90
Flat Plate,	210	15' span	4000	52.20	53.70	133.30	47.85	287.05
100 psf L.L.	215	25'		52.20	40.50	73.00	42.55	208.25
Flat Slab,	190	20'	4000	52.20	39.10	108.75	45.60	245.65
100 psf L.L.	195	30'		52.20	34.50	72.55	49.75	209.00
Grade Wall,	420	8" thick	3000	48.90	13.40*	197.65	16.75	276.70
8' high	430	15" thick		48.90	8.50*	100.05	20.55	178.00
Metal Pan Joists,	250	15' span	4000	52.20	44.85	156.75	17.85	271.65
100 psf L.L.	255	25' span		52.20	29.30	174.50	30.75	286.75
Pile Caps	590	under 5 C.Y.	3000	51.35	10.75*	33.45	16.75	112.30
	595	over 10 C.Y.		51.35	5.75*	17.45	20.55	95.40
Slab on Grade	465	4" thick	3500	53.10	8.85*	11.20	14.25	87.40
	470	6" thick		53.10	5.95*	7.30	9.55	75.90
Spread Footings	380	under 1 C.Y.	3000	51.35	17.45*	44.40	16.75	129.95
	385	over 5 C.Y.		51.35	8.50*	16.65	19.75	96.25
Strip Footings	390	9" x 18" plain	3000	51.35	8.15*	56.85	—	116.35
	395	12" x 36" reinforced		51.35	10.15*	28.45	15.20	105.15
Waffle 30" Domes	230	20' span	4000	52.20	33.40	123.60	44.95	254.15
100 psf L.L.	235	30' span		52.20	27.40	98.60	45.90	224.10

*Direct chuting assumed. For equipment handling add 40% to 50% to placing costs.

When form and reinforcing quantity ratios are available compute the C.Y. costs directly from the unit prices in the front of the book. See also ㊾ for relative quantities per C.Y. for additional items. Higher strength concrete would change the ratios below.

The tables below show typical examples of how the total costs per C.Y. have been derived. Ready mix concrete prices are U.S. average and equipment handling is assumed.

Beams 5 kip/L.F.	(Line 030)	10' Span			(Line 035)	25' Span		
	Material		Installation		Material		Installation	
3000 psi concrete	1 C.Y.	$48.90	1 C.Y.	$43.40	1 C.Y.	$48.90	1 C.Y.	$42.80
Formwork, 4 uses	69 S.F. @ 64¢	44.15	69 S.F. @ $2.59	178.70	63 S.F. @ 64¢	40.30	63 S.F. @ $2.59	163.15
Reinforcing steel	290# @ 26¢	75.40	290# @ .22¢	63.80	170# @ 26¢	44.20	170# @ .21¢	35.70
Total per C.Y.		$168.45		$285.90		$133.40		$241.65

Beam & Slab, One Way	(Line 270)	100 psf, 15' Span			(Line 275)	100 psf, 25' Span		
	Material		Installation		Material		Installation	
3000 psi concrete	1 C.Y.	$48.90	1 C.Y.	$47.85	1 C.Y.	$48.90	1 C.Y.	$38.10
Formwork, 4 uses	64 S.F. @ 71¢	45.45	64 S.F. @ $2.75	176.00	40 S.F. @ 76¢	30.40	40 S.F. @ $2.97	118.80
Reinforcing steel	122# @ 26¢	31.70	122# @ 12¢	14.65	159# @ 26¢	41.35	159# @ 12¢	19.10
Total per C.Y.		$126.05		$238.50		$120.65		$176.00

Beam & Slab, Two Way	(Line 290)	100 psf, 15' Span			(Line 295)	100 psf, 25' Span		
	Material		Installation		Material		Installation	
3000 psi concrete	1 C.Y.	$48.90	1 C.Y.	$57.35	1 C.Y.	$48.90	1 C.Y.	$48.10
Formwork, 4 uses	58 S.F. @ 64¢	37.10	58 S.F. @ $2.49	144.40	32 S.F. @ 73¢	23.35	32 S.F. @ $2.87	91.85
Reinforcing steel	130# @ 26¢	33.80	130# @ 12¢	15.60	123# @ 26¢	32.00	123# @ 12¢	14.75
Total per C.Y.		$119.80		$217.35		$104.25		$154.70

CIRCLE REFERENCE NUMBERS

(49) Average C.Y. of Concrete (cont.)

Columns, Square Square Tied	(Line 082)	16" Square		(Line 092)	24" Square	
	Material		Installation	Material		Installation
4000 psi concrete	1 C.Y.	$52.20	1 C.Y. $43.05	1 C.Y.	$52.20	1 C.Y. $32.15
Formwork, 4 uses	81 S.F. @ 49¢	39.70	81 S.F. @ $2.69 217.90	54 S.F. @ 50¢	27.00	54 S.F. @ $2.66 143.65
Reinforcing steel, avg.	634# @ 25¢	158.85	634# @ 16¢ 101.45	524# @ 25¢	131.20	524# @ 16¢ 83.10
Total per C.Y.		$250.75	$362.40		$210.40	$258.90

Note: Reinforcing of 16" and 24" square columns can vary from 144 lb. to 972 lb. per C.Y.

Columns, Round Tied Reinforced	(Line 122)	16" Diameter		(Line 142)	24" Diameter	
	Material		Installation	Material		Installation
4000 psi concrete	1 C.Y.	$52.20	1 C.Y. $33.50	1 C.Y.	$52.20	1 C.Y. $24.20
Formwork, fiber forms	19 L.F. @ $4.75	90.25	19 L.F. @ $4.52 85.90	9 L.F. @ $9.20	82.80	9 L.F. @ $4.87 43.85
Reinforcing steel, avg.	606# @ 25¢	151.50	600# @ 16¢ 96.00	600# @ 25¢	150.00	600# @ 15¢ 90.00
Ties	70# @ 23¢	16.10	70# @ 23¢ 16.15	18# @ 26¢	4.70	18# @ 23¢ 4.15
Total per C.Y.		$310.05	$231.55		$289.70	$162.20

Note: Reinforcing of 16" and 24" diameter columns can vary from 160 lb. to 1150 lb. per C.Y.

Flat Plate	(Line 210)	100 psf, 15' Span		(Line 215)	100 psf, 25' Span	
	Material		Installation	Material		Installation
4000 psi concrete	1 C.Y.	$52.20	1 C.Y. $53.70	1 C.Y.	$52.20	1 C.Y. $40.50
Formwork, 4 uses	59 S.F. @ 48¢	28.30	59 S.F. @ $1.78 105.00	32 S.F. @ 49¢	15.70	32 S.F. @ $1.79 57.30
Reinforcing steel	126# @ 26¢	32.75	126# @ 12¢ 15.10	112# @ 26¢	29.10	112# @ 12¢ 13.45
Total per C.Y.		$113.25	$173.80		$97.00	$111.25

Flat Slab	(Line 190)	100 psf, 20' Span		(Line 195)	100 psf, 30' Span	
	Material		Installation	Material		Installation
4000 psi concrete	1 C.Y.	$52.20	1 C.Y. $39.10	1 C.Y.	$52.20	1 C.Y. $34.50
Formwork, 4 uses	42 S.F. @ 68¢	28.55	42 S.F. @ $1.91 80.20	28 S.F. @ 68¢	19.05	28 S.F. @ $1.91 53.50
Reinforcing steel	120# @ 26¢	31.20	120# @ 12¢ 14.40	131# @ 26¢	34.05	131# @ 12¢ 15.70
Total per C.Y.		$111.95	$133.70		$105.30	$103.70

Grade Wall, 8' High	(Line 420)	8" Thick		(Line 430)	15" Thick	
	Material		Installation	Material		Installation
3000 psi concrete	1 C.Y.	$48.90	1 C.Y. $13.40	1 C.Y.	$48.90	1 C.Y. $8.50
Formwork, 4 uses	81 S.F. @ 50¢	40.50	81 S.F. @ $1.94 157.15	41 S.F. @ 50¢	20.50	41 S.F. @ $1.94 79.55
Reinforcing steel	44# @ 26¢	11.45	44# @ 12¢ 5.30	44# @ 26¢	14.05	44# @ 12¢ 6.50
Total per C.Y.		$100.85	$175.85		$83.45	$94.55

Metal Pan Joists, 30" One Way	(Line 250)	100 psf, 15' Span		(Line 255)	100 psf, 25' Span	
	Material		Installation	Material		Installation
4000 psi concrete	1 C.Y.	$52.20	1 C.Y. $44.85	1 C.Y.	$52.20	1 C.Y. $29.30
Formwork, 4 uses	49 S.F. @ 73¢	35.75	49 S.F. @ $2.03 99.45	40 S.F. @ $1.08	43.20	40 S.F. @ $2.83 113.20
Reinforcing steel	47# @ 26¢	12.20	47# @ 12¢ 5.65	81# @ 26¢	21.05	81# @ 12¢ 9.70
30" metal pans	44 S.F. @ 23¢	10.10	44 S.F. @ 26¢ 11.45	37 S.F. @ 23¢	8.50	37 S.F. @ 26¢ 9.60
Total per C.Y.		$110.25	$161.40		$124.95	$161.80

Pile Caps	(Line 590)	Under 5 C.Y.		(Line 595)	Over 10 C.Y.	
	Material		Installation	Material		Installation
3000 psi concrete	1.05 C.Y.	$51.35	1.05 C.Y. $10.75	1.05 C.Y.	$51.35	1.05 C.Y. $5.75
Formwork, 4 uses	16 S.F. @ 44¢	7.05	16 S.F. @ $1.65 26.40	8.5 S.F. @ 44¢	3.75	8.5 S.F. @ $1.65 14.00
Reinforcing steel	44# @ 26¢	11.45	44# @ 12¢ 5.30	54# @ 26¢	14.05	54# @ 12¢ 6.50
Total per C.Y.		$69.85	$42.25		$69.15	$26.25

Slab on Grade	(Line 465)	4" Thick		(Line 470)	6" Thick	
	Material		Installation	Material		Installation
3500 psi concrete	1.05 C.Y.	$53.10	1.05 C.Y. $8.85	1.05 C.Y.	$53.10	1.05 C.Y. $5.95
Formwork, 4 uses	9.2 L.F. @ 16¢	1.45	9.2 L.F. @ $1.06 9.75	6 L.F. @ 16¢	.95	6 L.F. @ $1.06 6.35
W.W. Fabric	81 S.F. @ 7.65¢	6.20	81 S.F. @ 9.95¢ 8.05	54 S.F. @ 7.65¢	4.15	54 S.F. @ 9.95¢ 5.40
Total per C.Y.		$60.75	$26.65		$58.20	$17.70

CIRCLE REFERENCE NUMBERS

㊾ Average C.Y. of Concrete (cont.)

Spread Footings	(Line 380)	Under 1 C.Y.			(Line 385)	Over 5 C.Y.		
	Material		Installation		Material		Installation	
3000 psi concrete	1.05 C.Y.	$51.35	1.05 C.Y.	$17.45	1.05 C.Y.	$51.35	1.05 C.Y.	$8.50
Formwork, 4 uses	24 S.F. @ 32¢	7.70	24 S.F. @ $1.53	36.70	9 S.F. @ 32¢	2.90	9 S.F. @ $1.53	13.75
Reinforcing steel	44# @ 26¢	11.45	44# @ 12¢	5.30	52# @ 26¢	13.50	52# @ 12¢	6.25
Total per C.Y.		$70.50		$59.45		$67.75		$28.50

Strip Footings	(Line 390)	9' x 18' Plain			(Line 395)	12" x 36" Reinforced		
	Material		Installation		Material		Installation	
3000 psi concrete	1.05 C.Y.	$51.35	1.05 C.Y.	$8.15	1.05 C.Y.	$51.35	1.05 C.Y.	$10.15
Formwork, 4 uses	36 S.F. @ 27¢	9.70	36 S.F. @ $1.31	47.15	18 S.F. @ 27¢	4.85	18 S.F. @ $1.31	23.60
Reinforcing steel	—	—	—	—	40# @ 26¢	10.40	40# @ 12¢	4.80
Total per C.Y.		$61.05		$55.30		$66.60		$38.55

Waffle, 30" Domes	(Line 230)	100 psf, 20' Span			(Line 235)	100 psf, 30' Span		
	Material		Installation		Material		Installation	
4000 psi concrete	1 C.Y.	$52.20	1 C.Y.	$33.40	1 C.Y.	$52.20	1 C.Y.	$27.40
Formwork, 4 uses	39 S.F. @ 74¢	28.85	39 S.F. @ $1.93	75.25	31 S.F. @ 74¢	22.95	31 S.F. @ $1.94	60.15
Reinforcing steel & mesh	107# @ 28¢	29.95	107# @ 14¢	15.00	102# @ 30¢	30.60	102# @ 15¢	15.30
Metal Domes	39 S.F. @ 23¢	8.95	39 S.F. @ 27¢	10.55	31 S.F. @ 23¢	7.15	31 S.F. @ 27¢	8.35
Total per C.Y.		$119.95		$134.20		$112.90		$111.20

㊿ Granolithic Finish and Base (Div. 3.3)

Description	Granolithic Topping, Mix 1:1:1-1/2				Granolithic Base	
	(Line 26-080) 1" Topping		(Line 26-095) 2" Topping		(Line 14-020) 1" x 5" Straight	
Cement @ $5.15 per bag	3.6 bags	$18.55	7.2 bags	$37.10	1.7 bags	$8.75
Sand @ $9.15 per C.Y.	3.6 C.F.	1.20	7.2 C.F.	2.45	1.7 C.F.	.60
Peastone @ $11.50 per C.Y.	5.4 C.F.	2.30	10.8 C.F.	4.65	1.85 C.F.	.80
Clean, mix, place forms and finish. Crew C-10 @ $61.40/hour	1.39 hrs.	85.35	1.60 hrs.	98.25	4.57 hrs.	280.60
Total	100 S.F.	$107.40	100 S.F.	$142.45	100 L.F.	$290.75

㊶ Lightweight Concrete (Div. 3.5-25)

Vermiculite or Perlite comes in bags of 4 C.F. under various trade names. Weight is about 8 lbs. per C.F. For insulating roof fill use 1:6 mix. For structural deck use 1:4 mix over gypsum boards, steeltex, steel centering, etc. supported by closely spaced joists or bulb tees. For structural slabs use 1:3:2 vermiculite sand concrete over steeltex, metal lath, steel centering, etc. on joists spaced 2'-0" O.C. for maximum L.L. of 80#/S.F. Use same mix for slab base fill over steel flooring or regular reinforced concrete slab when tile, terrazzo or other finish is to be laid over.

For slabs on grade use 1:3:2 mix when tile, etc. finish is to be laid over. If radiant heating units are installed use a 1:6 mix for a base. After coils are in place, cover with a regular granolithic finish (mix 1:3:2) to a minimum depth of 1-1/2" over top of units.

Reinforce all slabs with 6 x 6 or 10 x 10 welded wire mesh.

Vermiculite concrete can be purchased ready mixed but the following breakdown is included for field mix. Prices given below are for a one story building and assume 50 C.Y. or more. For less than 50 C.Y. add 10%. For over one story add $2.25 per C.Y. Screed finish cost is included below. Ready mix 1:6 costs $67.00 per C.Y. delivered in truckload lots.

See ㊸ for prices of ready mix lightweight concrete.

Quantities per C.Y., Field Mix	(Line 010) 1:6 Mix Insulating Roof Fill		1:3:2 Mix Lightweight Structural Concrete	
Portland cement @ $5.15 per bag	5.0 bags	$25.75	6.2 bags	$31.95
Vermiculite or Perlite @ $5.15 per bag	7.5 bags	38.60		
treated type @ $6.25 per bag			4.7 bags	29.40
Sand @ $9.15 per C.Y.			12.5 C.F.	4.25
Plant and water		5.10		5.10
Labor, machine mix, hoist and place, Crew C-8 at $188.50/hr.	.16 Hrs.	30.15	.13 Hrs.	24.50
Total in place, per C.Y.		$99.60		$95.20

CIRCLE REFERENCE NUMBERS

(52) Precast Concrete Wall Panels (Div. 3.4-65)

Panels are either solid or insulated with plain, colored or textured finishes. Transportation is an important cost factor. Prices below are based on delivery within 50 miles of a plant including fabricators' overhead and profit. Engineering data is available from fabricators to assist with construction details. Usual minimum job size for economical use of panels is about 5000 S.F. Small jobs can double the prices below. For large, highly repetitive jobs, deduct up to 15% from the prices below.

Panel Cost and Maximum Size Base price for panels based on 50 S.F. or larger panels is as follows:

Thickness	Cost per S.F.	Maximum Size	Thickness	Cost per S.F.	Maximum Size
3"	$6.10	50 S.F.	6"	$8.90	300 S.F.
4"	7.00	150 S.F.	7"	9.00	300 S.F.
5"	8.75	200 S.F.	8"	9.10	300 S.F.

The above prices are for gray smooth form or board finish one side. Add $1.25 per S.F. for white facing; $7.00 per C.F. for solid white for full thickness. For fluted surface (not broken) add 40¢ to 75¢ per S.F. For broken rib finish add 85¢ to $1.25 to fluted surface. Add 35¢ to $1.00 per S.F. for exposed local aggregate and $1.00 to $3.00 per S.F. for special facing aggregates. Sandblasting runs 45¢ to 70¢ per S.F., and bush hammering runs $1.60 to $2.70 per S.F. Loose hardware (not included above) runs 80¢ to $1.25 per S.F.

2" thick panels cost about the same as 3" thick panels and maximum panel size is less. For building panels faced with granite, marble or stone, add the material prices from division 4.4 to the plain panel price above. There is a growing trend toward aggregate facings and broken rib finish rather than plain gray concrete panels.

Composite Panels Add to above panel prices for core insulation, mesh and shear ties per S.F.

Type	1" Thick	1-1/2" Thick	2" Thick
Fiberglass	$1.10	$1.60	$2.00
Polystyrene (E.P.S. Board)	.60	.85	1.05

Erection

Table shows cost ranges for erection including Subcontractor's O & P using a six man crew, crane, operator & oiler.

Panel Size L x H	Area per Panel	Low Rise — Hyd. 55 Ton Crane @ $2960 Daily Cost				High Rise — Daily Production Range		55 Ton Crane $2960 Daily Cost		90 Ton Crane $3250 Daily Cost		150 Ton Crane $3515 Daily Cost	
		Daily Production Range		Erection Cost									
		Pieces	Area	Per Piece	Per S.F.	Pieces	Area	Piece Cost	S.F. Cost	Piece Cost	S.F. Cost	Piece Cost	S.F. Cost
4' x 4'	16 S.F.	10	160 S.F.	$296	$18.50	9	144 S.F.	$329	$20.55	$361	$22.55	$391	$24.40
		20	320	148	9.25	18	288	164	10.30	181	11.30	195	12.20
4' x 8'	32	10	320	296	9.25	9	288	329	10.30	361	11.30	391	12.20
		19	608	156	4.85	17	544	174	5.45	191	6.00	207	6.45
8' x 8'	64	9	576	329	5.15	8	512	370	5.80	406	6.35	439	6.85
		18	1152	164	2.55	16	1024	185	2.90	203	3.20	220	3.45
10' x 10'	100	8	800	370	3.70	7	700	423	4.20	464	4.65	502	5.00
		15	1500	197	2.00	14	1400	211	2.10	232	2.30	251	2.50
15' x 10'	150	7	1050	423	2.80	6	900	—	—	542	3.60	586	3.90
		12	1800	247	1.65	11	1650	—	—	295	1.95	320	2.15
20' x 10'	200	6	1200	493	2.45	5	1000	—	—	650	3.25	703	3.50
		8	1600	370	1.85	7	1400	—	—	464	2.30	502	2.50
30' x 10'	300	5	1500	592	2.00	4	1200	—	—	813	2.70	879	2.90
		7	2100	423	1.40	7	2100	—	—	464	1.55	502	1.70

Total Cost in Place for Low Rise Construction Including Subcontractor's O & P

Description	Gray				White Face				Exposed Aggregate			
	4' x 8'		20' x 10'		4' x 8'		20' x 10'		4' x 8'		20' x 10'	
Panel, steel form, broomed finish	4"	$7.70	6"	$9.80	4"	$9.10	6"	$11.15	4"	$8.60	6"	$10.65
Caulking, grout, etc. (runs about $1.75 per L.F.)		.65		.25		.65		.25		.65		.25
Erect, plumb, align		7.05		2.15		7.05		2.15		7.05		2.15
Total in place per S.F.		$15.40		$12.20		$16.80		13.55		$16.30		$13.05

No allowance has been made for supporting steel framework. On one story buildings, panels may rest on grade beams and require only wind bracing and fasteners. On multi-story buildings panels can span from column to column and floor to floor. Plastic designed steel framed structures may have large deflections which slow down erection and raise costs.

Large panels are more economical than small panels on a S.F. basis. When figuring areas include all protrusions, returns, etc. Overhangs can triple erection costs. Panels over 45' have been produced. Larger flat units should be prestressed. Vacuum lifting of smooth finish panels eliminates inserts and can speed erection.

CIRCLE REFERENCE NUMBERS

⑬ Tilt Up Concrete Panels (Div. 3.4-60)

The advantage of tilt up construction is in the low cost of forms and the placing of concrete and reinforcing. Panels up to 75' high and 5-1/2" thick have been tilted using strongbacks. Tilt up has been used for one to five story buildings and is well suited for warehouses, stores, offices, schools and residences.

The panels are cast in forms on the floor slab. Most jobs use 5-1/2" thick solid reinforced concrete panels. Sandwich panels with a layer of insulating materials are also used. Where dampness is a factor, lightweight aggregate is used at an added cost of 35¢ per square foot. Optimum panel size is 300 to 500 S.F.

Slabs are usually poured with 3000 psi concrete which permits tilting seven days after pouring. Slabs may be stacked on top of each other and are separated from each other by either two coats of bond breaker or a film of polyethylene. Use of high early strength cement allows tilting two days after a pour. Tilting up is done with a roller outrigger crane with a capacity of at least 1-1/2 times the weight of the panel at the required reach. Exterior precast columns can be set at the same time as the panels; interior precast columns can be set first and the panels clipped directly to them. The use of cast-in-place concrete columns is diminishing due to shrinkage problems. Structural steel columns are sometimes used if crane rails are planned. Panels can be clipped to the columns or lowered between the flanges. Steel channels with anchors may be used as edge forms for the slab. When the panels are lifted the channels form an integral steel column to take structural loads. Roof loads can be carried directly by the panels for wall heights to 14'. For soft ground requiring mats for cranes, add 100% to costs below.

Below are typical costs per S.F. for panels of 300 to 500 S.F., 20' high.

Item	5½" Thick			7½" Thick		
	Material	Installation	Total	Material	Installation	Total
Prepare pouring surface	$ .01	$ .03	$ .04	$ .01	$ .03	$ .04
Erect, strip side forms	.04	.25	.29	.04	.24	.28
Place concrete, 3000 psi	.83	.13	.96	1.12	.18	1.30
Steel trowel finish & curing	.02	.36	.38	.02	.36	.38
Reinforcing, inserts & misc. items	.79	.43	1.22	1.22	.72	1.94
Panel erection and aligning	—	.28	.28	—	.23	.23
Total before columns	$ 1.69	$ 1.48	$ 3.17	$ 2.41	$ 1.76	$ 4.17
Precast concrete columns, add	.83	.90	1.73	.88	.95	1.83
Total per S.F. of wall	$ 2.52	$ 2.38	$ 4.90	$ 3.29	$ 2.71	$ 6.00
Panels only, per C.Y.	$106.10	$91.65	$197.75	$103.30	$75.45	$178.75

Requirements of local building codes may be a limiting factor and should be checked. Building floor slabs should be poured first and should be a minimum of 5" thick with 100% compaction of soil or 6" thick with less than 100% compaction.

Setting times as fast as nine minutes per panel have been observed, but a safer expectation would be four panels per hour with a crane and a four man setting crew. If crane erects from inside building, some provision must be made to get crane out after walls are erected. Good yarding procedure is important to minimize delays. Equalizing three point lifting beams and self-releasing pick-up hooks speed erection. If panels must be carried to their final location, setting time per panel will be increased and erection costs may fall in the erection cost range of architectural precast wall panels. Placing panels into slots formed in continuous footers will speed erection.

Reinforcing should be with #5 bars with vertical bars on the bottom. If surface is to be sandblasted, stainless steel chairs should be used to prevent rust staining.

Use of a broom finish is popular since the unavoidable surface blemishes are concealed. Many West Coast jobs have used exposed aggregate panels which cost an additional 35¢ to $3.00 per S.F. depending on the aggregate and construction method employed. Eastern prices for exposed aggregate tend to be considerably higher. Panel connections run $1.65 to $5.00 per L.F. Precast columns run from three to five times the C.Y. price of the panels only.

⑭ Integral Floor Finish (Div. 3.3-26)

Mix 1:1:2	(Line 045) 1/2" Thick		(Line 060) 1" Thick	
Cement at $5.15 per bag	1.7 bag	$ 8.75	3.4 bag	$17.50
Sand at $9.15 per C.Y.	0.8 C.F.	.25	1.6 C.F.	.55
Gravel at $11.50 per C.Y.	3.2 C.F.	1.40	6.4 C.F.	2.75
Mix, place and finish using Crew C-10 @ $61.40 per hr.	.84 hrs.	51.60	1.07 hrs.	65.70
Total for 100 S.F.		$62.00		$86.50

CIRCLE REFERENCE NUMBERS

(55) Prestressed Precast Concrete Structural Units (Div. 3.4)

See also (40) for post-tensioned prestressed concrete.

Type	Location	Depth	Span in Ft.	Live Load Lb. per S.F.	Cost per S.F. Incl. Subs O & P	
					Delivered	Erected
Double Tee	Floor	28" to 34"	60 to 80	50 to 80	$4.50 to 7.75	$5.00 to 8.25
	Roof	12" to 24"	30 to 50	40	$4.70 to 5.00	$5.40 to 5.50
	Wall	Width 8'	Up to 55' high	Wind	$4.50 to $7.00	$5.50 to $8.50
Multiple Tee	Roof	8" to 12"	15 to 40	40	$3.00	$3.45
	Floor	8" to 12"	15 to 30	100	$3.10	$3.60
Plank	Roof or Floor	4"	Roof 13 / Floor 12	40 for Roof	$2.75	$3.45
		6"	22 / 18		2.85	3.45
		8"	26 / 25		3.15	3.60
		10"	33 / 29	100 for Floor	3.40	3.80
		12"	42 / 32		3.85	4.20
Single Tee	Roof	28"	40	40	$5.25*	$6.40*
		32"	80		6.60*	7.20*
		36"	100		9.35*	9.80*
		48"	120		9.45*	10.30*
AASHO Girder	Bridges	Type 4	100	Highway	$90/L.F.	$110/L.F.
		5	110		115	135
		6	125		140	160
Box Beam	Bridges	15"	40 to 100	Highway	$90/L.F.	$110/L.F.
		27"			125	145
		33"			145	165

*Costs are for 10' wide members; for 8' wide members add $.40 per S.F.

The costs above are based on a project of 10,000 to 20,000 S.F. with a haul distance of 25 to 50 miles.

The majority of precast projects today utilize double tees rather than single tees because of speed and ease of installation. As a result casting beds at manufacturing plants are normally formed for double tees. Single tee projects will therefore require an initial set up charge of approximately $6000 to be spread over the individual single tee costs. The prices above for single tees includes this cost based on a project size of 15,000 square feet.

For floors, a 2" to 3" topping is field cast over the shapes. For roofs, insulating concrete or rigid insulation is placed over the shapes.

Hauling costs (incl. above) run from 35¢ per S.F. for short haul to 60¢ for 50 mile haul. Member lengths up to 40' are standard haul, 40' to 60' require special permits and lengths over 60' must be escorted. Over width and/or over length can add up to 100% on hauling costs.

Multiple tee erection runs between 50¢ to $1.50 per S.F. and single tee erection runs between $1.00 to $1.50 per S.F. Large heavy members may require two cranes for lifting which would increase these erection costs by about 45%. Eight man erection crew and crane rent for $2960 to $3515 per day depending on location and size of crane. The crew can install 12 to 20 double tees, or 45 to 70 quad tees or planks per day.

The cost of supporting beams must be added to the above costs. Simple support beams run from $40 to $90 per L.F. delivered. Inverted tee beams run from $70 to $220 per L.F. delivered. Standard sized columns run $20 to $120 per L.F. delivered.

Grouting of connections must be added to the above. Typical costs are about $18 per connection but can go as high as $40 per connection.

Grouting planks run between 35¢ to 45¢ per S.F.

Single story buildings, including double tee roof members, supporting columns and girders, but no foundations, cost from $9.00 to $15.00 per S.F. of floor area. Parking garages run from $11.00 to $14.00 per S.F. above the foundations. Optimum parking garage design runs .02 C.Y. per S.F with overall costs for concrete in place running between $575 to $700 per C.Y.

Several system buildings utilizing precast members are available. Heights can go to 22 stories for apartment buildings with costs ranging from $11.50 to $22.50 per S.F. depending on the system, its components, its location and the degree of interior finish supplied. Optimum design ratio is 3 S.F. of surface to 1 S.F. of floor area.

CIRCLE REFERENCE NUMBERS

⑤⑥ Cement Mortar (material only) (Div. 4.1-50)

Type N — 1:1:6 mix by volume. Use everywhere above grade except as noted below.

Type M — 1:¼:3 mix by volume, or 1 part cement, 1/4 (10% by wt.) lime, 3 parts sand. Use for heavy loads and where earthquakes or hurricanes may occur. Also for reinforced brick, sewers, manholes and everywhere below grade.

Type N — 1:3 mix using conventional masonry cement which saves handling two separate bagged materials.

Cost and Mix Proportions of Various Types of Mortar and Grout

Components	Type Mortar and Mix Proportions by Volume										
	M		S		N		O		K	PM	PL
	1:1:6	1:¼:3	½:1:4	1:½:4	1:3	1:1:6	1:3	1:2:9	1:3:12	1:1:6	1:½:4
Portland cement @ $5.15 per bag	$5.15	$5.15	$2.60	$5.15	—	$5.15	—	$5.15	$5.15	$5.15	$5.15
Masonry cement @ $4.75 per bag	4.75	—	4.75	—	$4.75	—	$4.75	—	—	4.75	—
Lime @ $3.90 per 50 lb. bag	—	1.00	—	1.95	—	3.90	—	7.80	11.70	—	1.95
Masonry sand @ $11.20 per C.Y.	2.50	1.25	1.65	1.65	1.25	2.50	1.25	3.75	5.00	2.50	1.65
Mixing machine incl. fuel*	1.10	.55	.75	.75	.55	1.10	.55	1.60	2.15	1.10	.75
Total for Materials	$13.50	$7.95	$9.75	$9.50	$6.55	$12.65	$6.55	$18.30	$24.00	$13.50	$9.50
Total C.F.	6	3	4	4	3	6	3	9	12	6	4
Approximate Cost per C.F.	$2.25	$2.65	$2.45	$2.40	$2.20	$2.10	$2.20	$2.05	$2.00	$2.25	$2.40

* Based on a daily rental, 10 C.F. mixer capacity, 30 mixes per day, see division 1.5

Mix Proportions by Volume, Compressive Strength and Cost of Mortar & Grout

Where Used	Mortar Type	Allowable Proportions by Volume				Compressive Strength @28 days	Cost Per Cubic Foot
		Portland Cement	Masonry Cement	Hydrated Lime	Masonry Sand		
Plain Masonry	M	1	1	—	6	2500 psi	$2.25
		1	—	1/4	3		2.65
	S	1/2	1	—	4	1800 psi	2.45
		1	—	1/4 to 1/2	4		2.40
	N	—	1	—	3	750 psi	2.20
		1	—	1/2 to 1-1/4	6		2.10
	O	—	1	—	3		2.20
		1	—	1-1/4 to 2-1/2	9		2.05
	K		1	2-1/2 to 4	12	75 psi	2.00
Reinforced Masonry	PM	1	1	—	6	2500 psi	2.25
	PL	1	—	1/4 to 1/2	4	2500 psi	2.40

Note: The total aggregate should be between 2.25 to 3 times the sum of the cement and lime used.

The labor cost to mix is included in the labor cost on brickwork, etc.

Machine mixing is usually specified on jobs of any size. There is a large price saving over hand mixing and mortar is more uniform.

There are two types of mortar color used. Prices in Section 4.1-15 are for the inert additive type with about 100 lbs. per M brick as the typical quantity required. These colors are also available in smaller batch size bags (1 lb. to 15 lb.) which can be placed directly into the mixer without measuring. The other type is premixed and replaces the masonry cement with ranges in price from $9 to $18 per 70 lb. bag. Dark green color has the highest cost.

⑤⑦ Miscellaneous Mortar (material only) (Div. 4.1-50)

Quantities	Glass Block Mortar		Gypsum Cement Mortar	
White waterproofed portland cement at $13.00 per bag	7 bags	$91.00		
Gypsum cement at $8.75 per 80 lb. bag			11.25 bags	$98.45
Lime at $3.90 per 50 lb. bag	280 lbs.	21.85		
Sand at $11.20 per C.Y.	1 C.Y.	11.20	1 C.Y.	11.20
Mixing machine and fuel		4.80		4.80
Total per C.Y.		$128.85		$114.45
Approximate Total per C.F.		$4.80		$4.25

CIRCLE REFERENCE NUMBERS

(58) Masonry Reinforcing (Div. 4.1-35)

Reinforcing prevents wall cracks where there is earth movement due to freezing and thawing. Reinforcing strips come in 10' and 12' lengths. Field labor runs between 2.7 to 5.3 hours per 1000 L.F. for wall thicknesses up to 12". Table below is cost per thousand L.F., material only, all galvanized. For hot dip galvanizing, add 65%.

Size	Type	Wall Thickness						
		4"	6"	8"	10"	12"	14"	16"
9 ga. sides, 9 ga. ties	Regular truss	$120	$125	$130	$135	$155	$175	$180
3/16" sides, 9 ga. ties		165	170	175	180	185	195	215
3/16" sides, 3/16" ties		185	185	215	225	235	245	260
9 ga. sides, 9 ga. ties	Cavity truss	125	130	135	145	160	180	185
3/16" sides, 9 ga. ties		170	175	180	185	190	205	220
8 ga. sides, 9 ga. ties	Ladder	95	100	100	110	115	120	125
9 ga. sides, 9 ga. ties	Cavity Ladder	110	115	125	130	135	145	150
3/16" sides, 9 ga. ties		165	175	180	190	190	195	210

(59) Brick, Block & Mortar Quantities (Div. 4)

Running Bond						For Other Bonds Standard Size add to S.F. Quantities in Table to left		
Number of brick per S.F. of wall - single wythe with 1/2" joints				C.F. of mortar per M bricks, waste included				
Type Brick	Nominal Size (incl. mortar) L H W	Modular Coursing	Number of Brick per S.F.	1 wythe	2 wythe	Bond Type	Description	Factor
Standard	8 x 2-2/3 x 4	3C=8"	6.75	13.0	16.5	Common	full header every fifth course	+20%
Economy	8 x 4 x 4	1C=4"	4.50	14.6	19.6		full header every sixth course	+16.7%
Engineer	8 x 3-1/5 x 4	5C=16"	5.63	13.6	17.6	English	full header every second course	+50%
Fire	9 x 2-1/2 x 4-1/2	2C=5"	6.40	550# fireclay	—	Flemish	alternate headers every course	+33.3%
Jumbo	12 x 4 x 6 or 8	1C=4"	3.00	34.0	41.4		every sixth course	+5.6%
Norman	12 x 2-2/3 x 4	3C=8"	4.50	17.8	22.8	Header = W x H exposed		+100%
Norwegian	12 x 3-1/5 x 4	5C=16"	3.75	18.5	24.4	Rowlock = H x W exposed		+100%
Roman	12 x 2 x 4	2C=4"	6.00	17.0	20.7	Rowlock stretcher = L x W exposed		+33.3%
SCR	12 x 2-2/3 x 6	3C=8"	4.50	26.7	31.7	Soldier = H x L exposed		—
Utility	12 x 4 x 4	1C=4"	3.00	19.4	26.8	Sailor = W x L exposed		-33.3%

Concrete Blocks Nominal Size	Approximate weight per S.F.		Blocks per 100 S.F.	Mortar per M block	
	Standard	Lightweight		Partitions	Back up
2" x 8" x 16"	20 PSF	15 PSF	113	16 C.F.	36 C.F.
4"	30	20		31	51
6"	42	30		46	66
8"	55	38		62	82
10"	70	47		77	97
12"	85	55		92	112

(60) Economy in Bricklaying (Div. 4)

Have adequate supervision. Be sure bricklayers are always supplied with materials so there is no waiting. Place best bricklayers at corners and openings.

Use only screened sand for mortar, otherwise men will waste time picking out pebbles. Use seamless metal tubs for mortar as they do not leak or catch trowel. Locate stack and mortar for easy wheeling.

Have brick delivered for stacking. This makes for faster handling, reduces chipping and breakage, and requires less storage space. Many dealers will deliver select common in 2' x 3' x 4' pallets or face brick packaged. This affords quick handling with a crane or forklift and easy tonging in units of ten, which reduces waste.

Use wider bricks for one wythe wall construction. Keep scaffolding away from wall to allow mortar to fall clear and not stain wall.

On large jobs develop specialized crews for each type of masonry unit.

Consider designing for prefabricated panel construction on high rise projects.

Avoid excessive corners or openings. Each opening adds about 50% to labor cost for area of opening.

Bolting stone panels and using window frames as stops reduces labor costs and speeds up erection.

CIRCLE REFERENCE NUMBERS

(61) Common and Face Brick Prices (Div. 4.2)

Prices are truckload lots on Common, carloads on Face Brick. Prices are per M brick.

City	Material		Mortar & Scaffold Rental	Installation				Total			
	Brick per M Delivered			Common in 8" Wall		Face Brick, 4" Veneer		Common in 8" Wall		Face Brick, 4" Veneer	
	Common	Face		Bare Costs	Incl. Subs O&P	Bare Costs	Incl. Subs O&P	Bare Costs	Incl. Subs O&P	Bare Costs	Incl. Subs O&P
Atlanta	$140	$170	$45.00 for 8" Wall	$306	$438	$387	$553	$495	$646	$611	$800
Baltimore	155	210		321	459	407	582	526	684	672	1,266
Boston	240	270		487	696	613	877	779	1,017	940	1,237
Buffalo	240	260		479	685	605	865	771	1,006	922	1,213
Chicago	160	230		449	642	566	809	659	873	852	1,123
Cincinnati	160	185	$19 for 4" Veneer	439	628	554	792	649	859	794	1,056
Cleveland	170	225		582	832	731	1,045	802	1,074	1,012	1,354
Columbus	155	210		431	616	542	775	636	841	807	1,067
Dallas	230	250		364	495	459	656	646	805	766	993
Denver	180	200		409	585	515	736	639	838	770	1,017
Detroit	210	220		467	668	589	842	728	955	865	1,145
Houston	170	195		366	523	461	659	586	765	711	934
Indianapolis	165	220		411	588	519	742	626	824	795	1,045
Kansas City	180	210		428	689	539	771	658	942	804	1,063
Los Angeles	195	235		578	827	729	1,042	824	1,097	1,020	1,362
Memphis	185	200		331	473	419	599	567	732	674	880
Milwaukee	180	235		440	629	554	792	670	882	845	1,112
Minneapolis	185	235		443	633	559	799	679	892	850	1,119
Nashville	155	165		299	428	377	539	504	653	596	780
New Orleans	145	170		356	509	451	645	550	723	675	890
New York City	200	250		563	805	712	1,018	814	1,081	1,019	1,355
Philadelphia	180	230		471	674	595	851	701	927	881	1,165
Phoenix	170	220		399	571	504	721	619	813	780	1,024
Pittsburgh	195	240		487	696	613	877	733	966	909	1,203
St. Louis	175	215		476	681	599	857	701	929	869	1,155
San Antonio	175	225		316	452	400	572	541	700	681	881
San Diego	230	265		529	756	669	957	811	1,066	991	1,311
San Francisco	280	310		592	847	746	1,067	925	1,214	1,114	1,472
Seattle	185	215		490	701	615	879	726	960	885	1,177
Washington, D.C.	175	210		416	595	523	748	641	843	788	1,040
Average	$185	$223	$45+$49	$438	$627	$552	$789	$674	$887	$830	$1,098

In figuring above installation costs, a D-2 Crew (with a daily output of 1.5M) was used for the 4" veneer. A D-3 Crew (with a daily output of 1.8M) was used for the 8" solid wall.

In figuring the total cost including overhead and profit, an allowance of 10% was added to the sum of the cost of the brick and mortar and scaffold. Also, 3% breakage was included for both the bare costs and the costs with overhead and profit. If bricks are delivered palletized with 280 to 300 per pallet, or packaged allow only 1-1/2% for breakage. Then add $8 per M to cost of brick and deduct two hours helper time. The net result is a savings of $15 to $30 per M in place. Packaged or palletized delivery is practical when a job is big enough to have a crane or other equipment available to handle a package of brick. This is so on all industrial work but not always true on small commercial buildings.

There are many types of red face brick. The prices above are for the most usual type used in commercial, apartment house or industrial construction. If it is possible to obtain the price of the actual brick to be used, it should be done and substituted in the table. The use of buff and gray face is increasing, and there is a continuing trend to the Norman, Roman, Jumbo and SCR brick.

See (39) for brick quantities per S.F. and mortar quantities per M brick. (Average prices for the various sizes are listed in the front of the book.)

The use of common red clay brick for backup has decreased to near the vanishing point. Concrete block is the most usual backup material with occasional use of sand lime or cement brick. Sand lime cost about $15 per M less than red clay and cement brick are about $5 per M less than red clay. These figures may be substituted in the common brick breakdown for the cost of these items in place, as labor is about the same. Occasionally common brick is being used in solid walls for strength and as a fire stop.

Several recent jobs have brick panels built on the ground floor and then crane erected to the upper floors. This allows the work to be done under cover and without scaffolding.

CIRCLE REFERENCE NUMBERS

(62) Brick in Place (Div. 4.2-56 & 60)

Table below is for common bond with 1/2″ concave joints and includes 3% waste for brick and 25% waste for mortar. Crew costs are bare costs.

Item	8″ Common Brick Wall 8″ x 2-2/3″ x 4″		Select Common Face 8″ x 2-2/3″ x 4″		Red Face Brick 8″ x 2-2/3″ x 4″	
1030 brick delivered	$185 per M	$190.55	$205 per M	$211.15	$225 per M	$231.75
Type N mortar @ $2.10 per C.F.	17 C.F.	35.70	13 C.F.	27.30	13 C.F.	27.30
Scaffold rental		10.65		21.30		21.30
Installation incl. scaffolding using indicated crew	Crew D-3 @ 0.556 Days	438.15	Crew D-2 @ 0.645 Days	534.05	Crew D-2 @ 0.667 Days	552.30
Total per M in place		$675.05		$793.80		$832.65
Total per S.F. of wall	13.5 bricks/S.F.	$ 9.10	6.75 bricks/S.F.	$ 5.35	6.75 bricks/S.F.	$ 5.60

4

(63) Brick Veneer in Place (Div. 4.2-56)

Table below is for common bond with 1/2″ concave joints and includes 3% waste.

Item	Buff Face Brick 8″ x 2-2/3″ x 4″		Red Norman Face Brick 12″ x 2-2/3″ x 4″		Buff Roman Face Brick 12″ x 2″ x 4″	
1030 brick delivered	$260 per M	$267.80	$395 per M	$ 406.85	$470 per M	$ 484.10
Type N mortar @ $2.10 per C.F.	13 C.F.	27.30	18 C.F.	37.80	17 C.F.	35.70
Scaffold rental		21.30		32.00		24.00
Wall ties, galvanized	40 each	3.60	56 each	5.05	42 each	3.80
Installation incl. scaffolding & ties, using indicated crew	Crew D-2 @ 0.667 days	552.30	Crew D-2 @ 0.690 days	571.30	Crew D-2 @ 0.667 days	552.30
Total per M in place		$872.30		$1,053.00		$1,099.85
Total per S.F. of wall	6.75 brick/S.F.	$ 5.90	4.50 brick/S.F.	$ 4.75	6.00 brick/S.F.	$ 6.60

(64) Reinforced Brick in Walls (Div. 4.2-56)

Table below is for common bond with 1/2″ concave joints and includes 3% waste. Standard 8″ x 2-2/3″ x 4″ bricks (select).

Item	8″ to 9″ Thick Wall		16″ to 17″ Thick Wall	
1030 brick (straight hard)	$205 per M	$211.15	$205 per M	$211.15
Type PL mortar @ $2.40 per C.F.	17 C.F.	40.80	19 C.F.	45.60
Scaffold rental		10.65		5.35
Reinforcing bars delivered	50 lb. @ 30¢ per lb.	15.00	50 lb. @ 30¢ per lb.	15.00
Installation incl. scaffolding & reinforcing using indicated crew	Crew D-3 @ 0.571 Days	450.00	Crew D-3 @ 0.513 Days	404.25
Total per M in place		$727.60		$681.35
Total per S.F. of wall	8″ wall, 13.5/S.F.	$ 9.85	16″ wall, 27.0/S.F.	$ 18.40

(65) Brick Chimneys (Div. 4.2-06)

Quantities	16″ x 16″		20″ x 20″		20″ x 24″		20″ x 32″	
Brick at $205 per M	28 brick	$ 5.75	37 brick	$ 7.60	42 brick	$ 8.60	51 brick	$10.45
Type M mortar at $2.25 per C.F.	.5 C.F.	1.15	.6 C.F.	1.35	1.0 C.F.	2.25	1.3 C.F.	2.95
Flue tile (square)	8″ x 8″	2.60	12″ x 12″	5.00	2 @ 8″ x 12″	7.50	2 @ 12″ x 12″	10.00
Scaffold rental		.80		1.00		1.10		1.25
Scaffolding erection 1 Carp.	.010 day	1.60	.013 day	2.10	.015 day	2.40	.017 day	2.70
Install tile & brick, crew D-1	.055 day	16.05	.073 day	21.30	.083 day	24.25	.10 day	29.20
Total per L.F. high		$27.95		$38.35		$46.10		$56.55

Labor costs are bare costs and do not include subcontractor's O&P.

Only use 2/3 of the above figures if the chimney butts against a straight wall. Labor for chimney brick using D-1 crew is 15.8 hours per thousand brick or about $577 per thousand brick. An 8″ x 12″ flue takes 33 brick and two 8″ x 8″ flues take 37 brick.

CIRCLE REFERENCE NUMBERS

⑥⑥ Cleaning Face Brick (Div. 4.2-68)

On smooth brick a man can clean 70 S.F. an hour; on rough brick 50 S.F. per hour. Use one gallon muriatic acid to 20 gallons of water for 1000 S.F. Do not use acid solution until wall is at least seven days old, but a mild soap solution may be used after two days. Commercial cleaners cost from $3 to $14 per gallon.

Time has been allowed for clean-up in brick prices.

⑥⑦ Glass Block (Div. 4.3-60)

Cost of blocks each, all blocks 4" thick, zone 1 contractor prices.

Nominal Size (Including Mortar)	Truckload or Carload				Less Truckload or Carload			
	Regular	Thinline	Chiaro & Intaglio	Solar Reflective	Regular	Thinline	Chiaro & Intaglio	Solar Reflective
6" x 6"	$ 3.40	$2.85	—	—	$ 3.80	$3.00	—	—
8" x 8"	4.60	3.65	$10.50	$ 6.10	5.70	4.80	$13.00	$ 7.65
Solid 8" x 8"	15.50	—	—	—	17.80	—	—	—
4" x 12"	—	—	—	—	—	—	—	—
4" x 8"	2.90	2.85	5.70	—	3.65	2.95	6.70	—
12" x 12"	11.00	—	—	14.00	12.00	—	—	15.75

Size	Per 100 S.F.		Per 1000 Block				
	No. of Block	Mortar 1/4" Joint	Asphalt Emulsion	Caulk	Expansion Joint	Panel Anchors	Wall Mesh
6" x 6"	410 ea.	5.0 C.F.	.17 gal.	1.5 gal.	80 L.F.	20 ea.	500 L.F.
8" x 8"	230	3.6	.33	2.8	140	36	670
12" x 12"	102	2.3	.67	6.0	312	80	1000
Approximate quantity per 100 S.F.			.07 gal.	.6 gal.	32 L.F.	9 ea.	51, 68, 102 L.F.

Accessories required for installation — Wall ties: run galvanized double steel mesh full length of joint at $.29 per L.F. — fiberglass expansion joints at sides and top at $.52 per L.F. — caulking compound one gallon at $6.90 does 180 L.F. — oakum one lb. at $1.15 per lb. does 30 L.F. — asphalt emulsion one gallon does 600 L.F. at $3.40 per gallon in one gallon lots. If blocks are not set in wall chase use 2'-0" long wall anchors at $.48 each at 2'-0" O.C.

Quantities	Cost per 100 S.F. (Truckload or Carload price)					
	6" x 6" Block		8" x 8" Block		12" x 12" Block	
Block, delivered	410 @ $3.40	$1,394.00	230 @ $4.60	$ 1,058.00	102 @ $11.00	$1,122.00
Mortar @ $4.80 per C.F.	5.0 C.F.	24.00	3.6 C.F.	17.30	2.3 C.F.	11.05
Ties, expansion joints, etc.		41.50		41.50		41.50
Installation with crew D-3	.690 days	543.70	.465 days	336.40	.417 days	328.60
Total per 100 S.F.		$2,003.20		$1,483.20		$1,503.15

⑥⑧ Ashlar Veneer (Div. 4.4-05)

Quantities	Low Price Stone		High Price Stone	
4" Stone, random ashlar, 2.5 tons average	$140 per ton	$350.00	$400 per ton	$1,000.00
Type N mortar at $2.10 per C.F.	20 C.F.	42.00	20 C.F.	42.00
50 - 1" x 1/4" x 6" galv. stone anchors, at $.86 each		43.00		43.00
Scaffold rental		14.40		14.40
Installation incl. scaffold using crew D-8 @ $748.00 per day	.714 days	534.10	.833 days	623.10
Total per 100 S.F.		$983.50		$1,722.50

Stone coverage varies from 30 S.F. to 50 S.F. per ton. Mortar quantities include 1" backing bed.

CIRCLE REFERENCE NUMBERS

⑥⑨ Concrete Block (Div. 4.3)

8" x 16" block, sand aggregate blocks with 3/8 joints for partitions, tooled joint one side, not including Subs O & P.

City	Material				Mortar & Scaffold	Installation		Total Per 100 S.F.	
	Per Block, Delivered		113 Block, Delivered			4" Thick	8" Thick	4" Thick	8" Thick
	4" Thick	8" Thick	4" Thick	8" Thick					
Atlanta	$.55	$.85	$62.15	$ 96.05	$21.75 for 4"	$130.25	$144.20	$214	$269
Baltimore	.45	.65	50.85	73.45	$29.10 for 8"	135.75	152.90	208	255
Boston	.48	.82	54.25	92.65		208.75	235.15	285	357
Buffalo	.45	.74	50.85	83.60		202.50	228.10	275	341
Chicago	.46	.70	52.00	79.10		191.50	215.70	265	324
Cincinnati	.49	.59	55.40	66.65		187.25	210.90	264	307
Cleveland	.57	.65	64.40	73.45		249.25	280.75	335	383
Columbus	.40	.60	45.20	67.80		184.75	208.10	252	305
Dallas	.50	.89	56.50	100.55		155.50	175.15	234	305
Denver	.60	.90	67.80	101.70		175.25	197.40	265	328
Detroit	.60	.70	67.80	79.10		199.50	224.70	289	333
Houston	.51	.85	57.65	96.05		156.75	176.55	236	302
Indianapolis	.55	.79	62.15	89.25		175.00	197.10	259	315
Kansas City	.60	.90	67.80	101.70		183.00	206.15	273	337
Los Angeles	.54	.89	61.00	100.55		247.50	278.75	330	408
Memphis	.49	.65	55.40	73.45		140.25	158.00	217	261
Milwaukee	.60	.75	67.80	84.75		188.25	212.05	278	316
Minneapolis	.46	.75	52.00	84.75		189.50	213.45	263	327
Nashville	.60	.70	67.80	79.10		128.25	144.45	218	253
New Orleans	.60	.89	67.80	100.55		151.25	170.35	241	300
New York City	.68	1.00	76.85	113.00		239.50	269.75	338	412
Philadelphia	.49	.68	55.40	76.85		200.50	225.85	278	332
Phoenix	.58	.68	65.55	76.85		170.75	192.35	258	298
Pittsburgh	.48	.74	54.25	83.60		208.00	234.30	284	347
St. Louis	.54	.73	61.00	82.50		203.50	229.20	286	341
San Antonio	.55	.80	62.15	90.40		134.50	151.50	218	271
San Diego	.54	.85	61.00	96.05		225.75	254.30	309	379
San Francisco	.58	.98	65.55	110.75		252.75	284.70	340	425
Seattle	.68	.95	76.85	107.35		210.50	237.10	309	425
Washington D.C.	.50	.68	56.50	76.85		178.75	201.35	257	307
Average	$.53	$.78	$60.70	$ 87.95	$21.75 & 29.10	$186.80	$210.35	$269	$328.75

Cost for 100 S.F. of 8" x 16" Concrete Block Partitions to Four Stories High, Tooled Joint One Side							
8" x 16" Sand Aggregate	4" Thick Block		8" Thick Block		12" Thick Block		
113 block delivered	$.53 ea.	$ 59.90	$.78 ea.	$ 88.15	$1.25 ea.	$141.25	
Type N mortar at $2.10 per C.F.	3.5 C.F.	7.35	7.0 C.F.	14.70	10.4 C.F.	21.85	
Scaffold rental		14.40		14.40		14.40	
Installation crew	D-3 .233 days	183.60	D-3 .267 days	210.40	D-6 .294 days	269.30	
Total per 100 S.F.		$265.25		$327.65		$446.80	
Add for filling cores solid	6.7 C.F.	$ 78.00	25.8 C.F.	$140.00	12.2 C.F.	$181.00	

Cost for 100 S.F. of 8" x 16" Concrete Block Backup with Trowel Cut Joints						
8" x 16" Sand Aggregate	4" Thick Block		8" Thick Block		12" Thick Block	
113 block delivered	$.53 Ea.	$ 59.90	$.78 Ea.	$ 88.15	$1.25 Ea.	$141.25
Type N mortar at $2.10 per C.F.	5.8 C.F.	12.20	9.3 C.F.	19.55	12.7 C.F.	26.65
Scaffold rental		14.40		14.40		14.40
Installation crew	D-3 .225 days	177.30	D-3 .250 days	197.00	D-6 .270 days	247.30
Total per 100 S.F.		$263.80		$319.10		$429.60

Special block: corner, jamb and head block are same price as ordinary block of same size. Tabulated on the next page are National average prices per block. Labor on specials is about the same as equal sized regular block. Bond beam and 16" high lintel blocks cost 30% more than regular units of equal size.

Lintel blocks are 8" long and 8" or 16" high. Costs in individual cities may be factored from the table above.

Use of motorized mortar spreader box will speed construction of continuous walls.

383

CIRCLE REFERENCE NUMBERS

(69) Concrete Block (cont.)

Size in Inches	Type	Net Area Strength in PSI, Material Cost per Block									Lightweight	
		2000	2500	3000	3500	4000	4500	5000	5500	6000	Std.	Prem.
2 x 8 x 16	Solid	$ .40	$ .42	$ .44	$ .46	$ .48	$ .50	$ .52	$ .57	$ .61	$ .54	$ .73
3 x 8 x 16	Solid	.74	.76	.78	.80	.82	.84	.86	.90	.94	.93	1.34
4 x 8 x 16	Hollow	.53	.56	.60	.62	.66	.72	.75	.79	.83	.66	.89
	75% Solid	.69	.71	.74	.78	.82	.88	.92	.96	1.00	.84	1.19
	Solid	.72	.74	.77	.79	.83	.86	.89	.95	1.02	.92	1.29
6 x 8 x 16	Hollow	.62	.65	.71	.73	.77	.83	.88	.93	.98	.84	1.14
	75% Solid	.82	.85	.90	.92	.96	1.04	1.12	1.20	1.25	1.14	1.29
	Solid	.92	.95	1.01	1.02	1.06	1.14	1.21	1.30	1.36	1.22	1.39
8 x 8 x 16	Hollow	.78	.81	1.01	1.06	1.11	1.23	1.28	1.33	1.43	1.04	1.36
	75% Solid	.83	1.03	1.15	1.21	1.29	1.37	1.46	1.54	1.62	1.39	1.68
	Solid	1.18	1.25	1.35	1.42	1.45	1.53	1.61	1.67	1.77	1.54	1.84
	Bond Beam	1.04	—	—	—	—	—	—	—	—	1.29	1.44
	Sash	.94	—	—	—	—	—	—	—	—	1.14	1.39
	Interlocking Grout	.90	—	1.08	—	1.16	—	—	—	—	—	—
10 x 8 x 16	Hollow	1.24	1.26	1.38	1.42	1.50	1.61	1.72	1.74	1.86	1.54	2.00
	75% Solid	1.66	1.73	1.82	1.86	1.96	2.07	2.19	2.31	2.43	2.04	2.96
	Solid	1.86	1.94	2.14	2.19	2.31	2.43	2.49	2.55	2.71	2.54	3.24
12 x 8 x 16	Hollow	1.24	1.26	1.38	1.42	1.50	1.61	1.72	1.74	1.86	1.54	2.00
	75% Solid	1.66	1.73	1.82	1.86	1.96	2.07	2.19	2.31	2.43	2.04	2.96
	Solid	1.86	1.96	2.14	2.19	2.31	2.43	2.47	2.59	2.71	2.54	3.24
	Bond Beam	1.44	—	—	—	—	—	—	—	—	1.74	2.29
	Sash	1.24	—	—	—	—	—	—	—	—	1.59	2.00
	Fire Rated	1.64	—	—	—	—	—	—	—	—	2.16	2.88
	Interlocking Grout	1.38	—	1.65	—	1.79	—	—	—	—	—	—
16 x 8 x 16	Interlocking Grout	1.74	—	2.14	—	2.30	—	—	—	—	—	—

(70) Interlocking Grout Block Walls (Div. 4.3-37)

	Cost per 100 S.F. of 8" x 16" Interlocking Grout Block, "Open End" Type					
8" x 16" Sand Aggregate	8" Thick Block		12" Thick Block		16" Thick Block	
113 block delivered	$ .90 ea.	$ 101.70	$1.38 ea.	$155.95	$1.74 ea.	$196.60
Type N mortar @ $2.10 per C.F.	7.0 C.F.	14.70	10.4 C.F.	21.85	13.8 C.F.	29.00
Minimum reinf. req. @ $.30 per lb.						
8" block #4 @ 24" horiz., 1 face	33.4 lb.	10.00				
32" vert., 1 face	40.0 lb	12.00				
12" to 16" block 8" horiz., 1 face			99.7 lb.	29.90	99.7 lb.	29.90
16" vert., 1 face			50.2 lb.	15.05	50.2 lb.	15.05
Grout type PM @ $2.25 C.F	25.8 C.F.	58.05	42.2 C.F.	94.95	56.1 C.F.	126.20
Scaffolding		14.40		14.40		14.40
Installation Crew D-4 @ $730.40 per day	.408 days	298.00	.455 days	332.35	.555 days	405.40
Total per 100 S.F.		$508.85		$664.45		$816.55
1 S.F.		5.09		6.65		8.17

CIRCLE REFERENCE NUMBERS

(71) Structural Steel (Div. 5.1)

1 to 2 Story Building	Bare Costs for 100 Ton Job with Common Bolts					Including Subs O&P		
City	Fabricated and Delivered	Unloading and Sorting	Erection Equipment	Field Erection Labor	Total per Ton in Place	Material	Installation	Total
Atlanta	$775	$18	$23	$117	$ 934	$ 852	$250	$1,103
Baltimore	775	25	25	163	989	852	337	1,190
Boston	851	28	27	179	1,086	936	369	1,305
Buffalo	775	28	27	182	1,014	852	375	1,228
Chicago	802	28	28	183	1,043	882	378	1,260
Cincinnati	759	24	26	158	969	834	329	1,164
Cleveland	861	30	28	191	1,111	947	392	1,340
Columbus	920	26	26	166	1,140	1,012	344	1,357
Dallas	775	22	24	144	967	852	301	1,154
Denver	888	24	27	153	1,093	977	321	1,298
Detroit	995	28	27	180	1,232	1,095	372	1,467
Houston	775	23	25	147	971	852	308	1,161
Indianapolis	775	27	25	171	999	852	351	1,204
Kansas City	861	26	27	165	1,079	947	343	1,290
Los Angeles	920	33	31	214	1,201	1,012	440	1,453
Memphis	775	18	23	117	934	852	249	1,102
Milwaukee	802	28	27	179	1,037	882	369	1,252
Minneapolis	748	25	26	163	963	822	337	1,160
Nashville	930	17	22	113	1,084	1,024	241	1,265
New Orleans	802	21	24	135	984	882	285	1,168
New York City	770	41	30	263	1,105	847	527	1,374
Philadelphia	726	26	27	170	951	799	352	1,151
Phoenix	775	27	25	175	1,004	852	359	1,212
Pittsburgh	796	28	28	180	1,034	876	373	1,250
St. Louis	775	26	27	165	994	852	344	1,197
San Antonio	877	20	23	126	1,047	965	267	1,233
San Diego	920	33	31	214	1,201	1,012	440	1,453
San Francisco	812	33	31	214	1,092	893	439	1,333
Seattle	861	25	28	164	1,079	947	343	1,290
Washington, D.C.	861	22	24	143	1,051	947	299	1,246
U.S. Average	$825	$26	$26	$168	$1,046	$907	$348	$1,255

Note: For high strength bolts, add $35 per ton to the base cost figures and $48 per ton to the O & P figures.

Adjustments to the Base Price Above	Bare Costs	Incl. Subs O&P
For 3 to 6 story building with high strength bolting, add per ton	$59	$85
For 7 to 15 stories with high strength bolting, add per ton	75	106
For over 15 stories with high strength bolting, add per ton	89	127
For field welded connections, add to these prices per ton	23—32	32—46
For multi-story masonry wall bearing construction, add to erection costs	30%	30%

General Average per Ton in Place for A 36 Steel (High Strength Steel Base Price May Be Substituted)				
Item	Description	Bare Costs Itemized	Bare Costs Summary	Incl. Subs O&P
Material	Base price	$400		
	Extras and delivery to shop (with common bolts)	60		
	Drafting	45	$ 825	$ 907
	Shop fabrication & warehouse rehandling	265		
	Shop coat paint	28		
	Trucking to job site	27		
Installation	Unload and shake out, 1.2 hours at $22.10	27		
	Erect and plumb, 5.6 hours at $22.10	124	$ 221	$ 348
	Field bolt, 2.0 hours at $22.10	44		
	Crane and minor erection equipment (incl. operator & oiler) 25 tons per day	26		
	Total per ton in place	$1,046	$1,046	$1,255

Note: Due to the influx of foreign steel and area economic conditions, material and fabrication costs for structural steel are negotiable.

CIRCLE REFERENCE NUMBERS

(72) Steel Estimating Quantities (Div. 5.1)

One estimate on erection is that a crane can handle 35 to 60 pieces per day. Say average is 45. With usual sizes of beams, girders, and columns, this would amount to about 25 tons. Allowing $650 per day rental for the crane including fuel and operator, this would amount to about $26 per ton for crane charge. The type of connection greatly affects the speed of erection. Moment connections for continuous design slow down production and increase erection costs.

Short open web bar joists can be set at the rate of 75 to 80 per day, with 50 per day being the average for setting long span joists.

After main members are calculated, add the following for usual allowances: base plates 2% to 3%; column splices 4% to 5%; and miscellaneous details 4% to 5%, for a total of 10% to 13% in addition to main members.

Ratio of column to beam tonnage varies depending on type of steels used, typical spans, story heights and live loads.

It is more economical to keep the column size constant and to vary the strength of the column by using high strength steels. This also saves floor space. Buildings have recently gone as high as ten stories with 8" high strength columns. For light columns under W8X31 lb. sections, concrete filled steel columns are economical.

High strength steels may be used in columns and beams to save floor space and to meet head room requirements. High strength steels in some sizes sometimes require long lead times.

Round, square and rectangular columns, both plain and concrete filled, are readily available and save floor area, but are higher in cost per pound than rolled columns. For high unbraced columns, tube columns may be less expensive.

Below are average minimum figures for the weights of the structural steel frame for different types of buildings using A36 steel, rolled shapes and simple joints. For economy in domes, rise to span ratio = 0.13. Open web joist framing systems will reduce weights by 10% to 40%. Composite design can reduce figures up to 25% but additional concrete floor slab thickness may be required. Continuous design can reduce the weights up to 20%. There are many building codes with different live load requirements and different structural requirements, such as hurricane and earthquake loadings which can alter the figures.

*See (121) for Domes and Thin Shell Structures.

Structural Steel Weights per S.F. of Floor Area										
Type of Building	No. of Stories	Avg. Spans	L.L. #/S.F.	Lbs. Per S.F.	Type of Building	No. of Stories	Avg. Spans	L.L. #/S.F.	Lbs. Per S.F.
Steel Frame Mfg.	1	20'x20' 30'x30' 40'x40'	30	6 10 14	Apartments	2-8 9-25	20'x20'	40	8 14
Parking garage	4	Various	80	8.5	Office	to 10 20 30 over 50	Various	80	10 18 26 35
Domes (Schwedler)*	1	200' 300'	30	10 15					

(73) High Strength Steels (Div. 5.1)

The mill price of these steels is higher than A36 carbon steel but their proper use can achieve overall savings thru total reduced weights. For columns with L/r over 100, A36 steel is best; under 100, high strength steels are economical. For heavy columns high strength steels are economical when cover plates are eliminated. There is no economy using high strength steels for clip angles or supports or for beams where deflection governs. Thinner members are more economical than thick.

Below is a table of the high strength steels available and their base price with specification extras at Eastern mills. A36 is included for comparison purposes. See also (71) for typical in place costs of A36 steel. The per ton erection and fabricating costs of the high strength steels will be higher than for A36 since the same number of pieces, but less weight, will be installed. See (76) for extras for Jumbo 14" A36 columns.

Mill Base Prices and Range of Size and Specification Extras per Ton for Structural Shapes, Eastern Mills											
Type Steel	Yield ksi	Web Thickness	Size Extra	Spec. Extra	Base + Size & Spec. Extra	Type Steel	Yield ksi	Web Thickness	Size Extra	Spec. Extra	Base + Size & Spec. Extra
A36 Carbon	36	Jumbo 14" Cols.	$75 to 85	$33	$586 to 596	A572 Alloy	42	Jumbo 14" Cols.	$75 to 85	$84	$637 to 647
A441 Manganese Vanadium	50 to 42	Thru 1-7/8"	25 to 80	66	569 to 624		50	Jumbo 14" Cols.	75 to 85	93	646 to 656
A242, Type 2 Corrosion Resistant	50	Most hvy. 14" Cols.	30	141	649	A572 Low Alloy Colum- bium Vana- dium Steels	42	Thru 1 7/8"	25 to 80	50	553 to 608
Mayari R-50	50	Jumbo 14" Cols.	75 to 85	141	694 to 704		50		25 to 80	56	559 to 614
Mayari R-60	60	Jumbo 14" Cols.	75 to 85	148	701 to 711		60 A6 Gr. 1 & 2		25 to 80	72	575 to 630

CIRCLE REFERENCE NUMBERS

(74) Pre Engineered Steel Buildings (Div. 5.1-35)

These buildings are manufactured by many companies and normally erected by franchised dealers throughout the U.S. The four basic types are: Rigid Frames, Truss type, Post and Beam and the Sloped Beam type. Most popular roof slope is low pitch of 1" in 12". The minimum economical area of these buildings is about 3000 S.F. of floor area. Bay sizes are usually 20' to 24' but can go as high as 30' with heavier girts and purlins. Eave heights are usually 12' to 24' with 18' to 20' most typical. Pre-engineered buildings become increasingly economical with higher eave heights.

Prices shown here are for the building shell only and do not include floors, foundations, interior finishes or utilities. Erection of the frame and insulated roof runs $.95 to $1.20 per S.F. Insulated side wall installation runs about $.95 per S.F. of skin. Typical erection cost including both siding and roofing depends on the building shape and runs $1.15 to $2.15 for one in twelve roof slope and $1.20 to $3.25 per S.F. of floor for four in twelve roof slope. Site, weather, labor source, shape and size of project will determine the erection cost of each job. Prices are for the New England area and include erector's overhead and profit.

Table below is based on 30 psf roof load, 20 psf wind load and no unusual structural requirements. Costs assume at least three bays of 24' each. Material costs include the structural frame, 26 ga. colored steel roofing, 26 ga. colored steel siding, fasteners, closures and flashing but no allowance for doors, windows, gutters or skylights. Very large projects would generally cost less than the prices listed below. Typical budget figures for above material delivered to the job runs $1025 to $1200 per ton. Fasteners and flashings (included below) run $.40 to $.55 per S.F.

Material Costs per S.F. of Floor Area Above the Foundations

Type of Building	Total Width in Feet	Eave Height				
		10 Ft.	14 Ft.	16 Ft.	20 Ft.	24 Ft.
Rigid Frame Clear Span	30-40	$3.70	$4.45	$4.90	$5.80	$6.45
	50-100	3.65	4.35	4.75	5.55	6.00
	110	—	4.45	4.70	5.45	5.95
	120	—	4.65	4.90	5.45	6.00
	130	—	4.75	4.95	5.70	6.15
Tapered Beam Clear Span	30	4.20	5.00	5.25	6.00	—
	40	3.75	4.50	4.70	5.75	—
	50-80	3.65	4.35	4.60	5.50	—
Post & Beam 1 Post at Center	80	—	4.00	4.35	4.20	4.45
	100	—	4.35	4.45	4.45	4.55
	120	—	4.45	4.50	4.60	4.70
Post & Beam 2 Posts at 1/3 Points	120	—	3.95	4.20	4.40	4.55
	150	—	4.15	4.40	4.55	4.70
	180	—	4.40	4.50	4.70	4.85
Post & Beam 3 Posts at 1/4 points	160	—	3.90	4.10	4.30	4.45
	200	—	4.10	4.30	4.50	4.60
	240	—	4.35	4.50	4.60	4.75

Typical accessory items are listed in the front of the book. All normal interior work, floors, foundations, utilities and sitework should be figured the same as usual.

The table below indicates typical total building shell costs in the New England area. These costs include allowance for erection, normal doors, windows, gutters and erector's overhead and profit. Figures do not include foundations, floors, interior finishes, electrical, mechanical or installed equipment.

Total Cost per S.F. Above the Foundations, 16' Eave Height

Project Size: Rigid Frame 30' to 60' Spans 1 in 12 Roof Slope	Basic Building Using 26 ga. Galvanized Roof & Siding S.F. Floor Area	R20 Field Insulation S.F. Floor Area	Add to Basic Building Price	
			Exterior Finish	S.F. of Skin
4,000 S.F.	$7.05	$2.05	Sandwich wall	$2.20 to $4.65
10,000 S.F.	6.40	1.65	Vinyl clad steel	.25
20,000 S.F.	5.80	1.45	Corrugated fiberglass	.41
			5 year paint	.16
			15 year paint	.28

(75) Lightgage Joists (Div. 5.2-20)

Material price, delivered in truckload lots is 62¢ per pound for galvanized; 50¢ per pound for painted. Weights run from 4.3# per L.F. to 7# per L.F. depending on gauge, depth and flange width.

Connections are made with self tapping screws. Prices in front of book include allowance for usual clip angles, welding plates, anchor bolts, etc.

CIRCLE REFERENCE NUMBERS

⑦⑥ Structural Steel Extras (Div. 5.1)

Principal Extras in Dollars Per Ton

Item quantity — using 5 tons per size as base price. Under 5 tons to 3 tons inclusive add $5 per ton. Under 3 tons to 2 tons inclusive add $10 per ton. Under 2 tons to 1 ton inclusive add $15 per ton. Under 1 ton to 1/2 ton add $50 per ton. Under 1/2 ton add $100 per ton.

Lightest beams in each size add $45 for 24"; $45 for 14"; $70 for 8"; and $80 for 6" depth.

Equal leg angles add $55 for 8" x 8"; $50 for 6" x 6"; $54 for 5" x 5" x 5/16"; $50 for other gauges of 5" x 5"; $31 for 4" x 4" x 1/4"; $28 for 4" x 4" x 3/4"; $46 for 3" x 3" x 3/16"; $42 for 3" x 3" x 1/2".

Special 14" column sections add $75 for 455 lb. to $85 for 730 lb. in A36 steel.

Standard beams add $45 for 24" to 10"; $47 for 8"; $55 for 6"; and $125 for 3" depth.

Standard channels $40 for 15"; $40 for 12"; $60 for 8"; $60 for 6"; $80 for 4"; and $100 for 3" depth.

Unequal leg angles add $60 for 8" x 6"; $60 for 8" x 4"; $50 for 6" x 4"; $38 for 5" x 3" x 1/4" to $35 for 5" x 3" x 1/2"; $41 for 4" x 3" x 1/4" to $37 for 4" x 3" x 1/2".

Most common WF sections add $25 to a maximum of $80 per ton.

*Cambering $20 to $40 per ton, varies with wt./ft. and length.

Galvanizing under 1 ton $350; over 20 tons $250 per ton. For color coating of galvanizing add 40% to prices.

Government Specifications, medium grade $24, high tensile $77 per ton.

High strength steels see ⑦③.

*Length 10' to 20', $25; 20' to 30', $10; 30' to 40', $5; 60' to 65', $4; 40' to 50', $3; 65' to 80', $21; 80' to 90', $24; 90' to 100', $26.

Milling, one or two ends, 10' to 25' inclusive, members weighing 10 thru 50 lbs., $80; 51 lbs. thru 200 lbs., $48; 201 lbs. thru 426 lbs., $36; over 426 lbs., $24. Members over 25' weighing 10 thru 50 lbs., $70; 51 lbs. thru 200 lbs., $43; 201 lbs. thru 426 lbs., $31; over 426 lbs., $24.

Special testing runs $5 to $7 per ton. Handling and loading runs from $2 under 10,000 lbs.; $3 under 6000 lbs.; $4 under 4000 lbs.

*Splitting beams to produce T's $35 for heavy beams to $60 for light members.

Note: * Subject to mill tolerance.

⑦⑦ High Strength Bolts (Div. 5.1-50-520)

In factory buildings, common bolts are used in secondary connections.

Allow 20 field bolts per ton of steel for a 6 story office building, apartment house or light industrial building. For 6 to 12 stories allow 18 bolts and above 12 stories, 25 bolts. On power stations 20 to 25 bolts per ton are needed.

Cost per bolt, nut and washer combination, material only. Under 5000 lbs. A325 and A490.

Length	1½"		2"		2½"		3"		3½"		4"		5"		6"		7"		8"	
ASTM Designation	A325	A490	A325	A490	A325	A490	A325	A490	A325	A490	A325	A490	A325	A490	A325	A490	A325	A490	A325	A490
5/8" diameter	$.49	$.64	$.51	$.67	$.55	$.73	$.60	$.79	$.67	$.84	$.72	$.90	$.81	$1.02	—	—	—	—	—	—
3/4"	.66	.86	.70	.91	.75	.99	.81	1.07	.86	1.14	.96	1.22	1.08	1.37	1.20	1.51	1.48	1.88	1.64	—
7/8"	—	—	1.12	1.48	1.16	1.53	1.24	1.64	1.32	1.75	1.40	1.86	1.56	2.07	1.72	2.29	2.00	2.58	2.21	$2.81
1"	—	—	—	—	1.70	2.25	1.81	2.40	1.91	2.54	2.02	2.69	2.23	2.98	2.45	3.26	2.66	3.55	3.06	3.96
1-1/8"	—	—	—	—	—	3.17	3.20	4.33	3.33	4.51	3.47	4.70	3.73	5.07	4.00	5.44	4.27	5.81	4.77	6.18
1-1/4"	—	—	—	—	—	—	3.80	5.30	3.91	5.30	4.08	5.53	4.41	6.00	4.74	6.45	5.07	6.91	5.41	7.37

⑦⑧ Subpurlins (Div. 5.2-10)

Table is based on subpurlins 32-3/4" O.C. with simple spans, 40 psf, L.L., material only, 10,000 lb. to 30,000 lb. lots.

	Bulb Tees, Painted			Truss Tees, Painted						
Type	Wt. per L.F.	Cost per L.F.	Max. Span	Size	Wt. per L.F.	Cost per L.F.	Max. Span	Size	Wt. per L.F.	Cost per L.F.	Max. Span
112	1.44#	$ .42	5'-6"	2"	1.1 #	$.53	5'-9"	2-1/2"	1.39#	$ .69	8'-7"
158	1.63	.47	6'-5"		1.27	.62	7'-3"		1.85	.96	10'-0"
168	1.85	.52	7'-8"		1.33	.65	7'-7"	3"	1.14	.58	7'-3"
178	2.13	.60	8'-9"		1.78	.91	8'-9"		1.88	1.01	9'-2"
218	3.06	.89	10'-2"	2-1/2"	1.12	.56	6'-7"	3-1/2"	1.17	.61	9'-7"
228	3.69	1.16	12'-1"		1.34	.67	8'-3"		1.9	1.04	11'-2"

CIRCLE REFERENCE NUMBERS

 Coating Structural Steel (Div. 5.1-50)

See for galvanizing and shop coat painting.

For field coats use red oxide rust inhibitive paint at $13.00 a gallon or if steel remains exposed use aluminum paint at $13.50 a gallon.

On field welded jobs, shop coat is necessarily omitted. All painting must be done in the field and usually consists of two coats. Cleaning with wire brushes runs $.50 per S.F.

Sandblasting on the ground runs $.60 to $1.10; above grade $.75 to $1.35 and to sandblast previously painted steel above grade runs $1.10 to $1.45 per S.F. White metal sandblasts run an additional $.40 per S.F.

Table below shows paint coverage and daily production for field painting.

Type Construction	Surface Area per Ton	Coat	One Gallon Covers		In 8 hrs. Man Covers		Average per Ton Spray	
			Brush	Spray	Brush	Spray	Gallons	Man-hours
Light Structural	300 S.F. to 500 S.F.	1st	500 S.F.	455 S.F.	640 S.F.	2000 S.F.	0.9 gals.	1.6 M.H.
		2nd	450	410	800	2400	1.0	1.3
		3rd	450	410	960	3200	1.0	1.0
Medium	150 S.F. to 300 S.F.	All	400	365	1600	3200	0.6	0.6
Heavy Structural	50 S.F. to 100 S.F.	1st	400	365	1920	4000	0.2	0.2
		2nd	400	365	2000	4000	0.2	0.2
		3rd	400	365	2000	4000	0.2	0.2
Weighted Average	225 S.F.	All	400	365	1350	3000	0.6	0.6

⑧⓪ Welded Structural Steel (Div. 5.1-65)

Usual weight reductions with welded design run 10% to 20% compared with bolted or riveted connections. This amounts to about the same total cost compared with bolted structures since field welded runs up to $90 per ton higher than bolts.

For normal spans of 18' to 24' figures 6 to 7 connections per ton. Cost of typical patented connector averages about $2 per set or about $12 to $14 per ton of steel.

Trusses — For welded trusses add 4% to weight of main members for connections. Up to 15% less steel can be expected in a welded compared to one that is shop bolted. Cost of erection is the same whether shop bolted or welded.

General — Usual electrodes for structural steel welding are E6010, E6011, E60T and E70T. Usual buildings vary between 2# to 8# of weld rod per ton of steel. Buildings utilizing continuous design require about three times as much welding as conventional welded structures. In estimating field erection by welding, it is best to use the average linear feet of weld per ton to arrive at the welding cost per ton. The type, size and position of the weld will have a direct bearing on the cost per linear foot. A typical field welder will deposit 1.8# to 2# of weld rod per hour manually. Using semiautomatic methods can increase production by as much as 50% to 75%. Below is the cost per hour for manual welding.

Welded Structural Steel in Field			
Item	No Operating Engr.	1/2 Operating Engr.	1 Operating Engr.
3 lb. weld rod $.66 per lb.	$ 2.00	$ 2.00	$ 2.00
Equipment (for welding only) 1/40 x $110	2.75	2.75	2.75
Operating cost at $4.35 per hour	4.35	4.35	4.35
Welder 1 hr. @ $21.70 per hour	21.70	21.70	21.70
Operating engineer @ $19.45 per hour	—	9.75	19.45
Total per Welder Hour (Bare Costs)	$30.80	$40.55	$50.25

The cost per ton of structural steel will vary from $25 to $150 per ton depending on Union requirements, design and inspection required.

CIRCLE REFERENCE NUMBERS

(81) Aluminum Floor Grating (Div. 5.4-36)

The table below lists the material weights and costs in dollars per S.F. for aluminum grating, alloy 6063.

Bearing Bar Size in Inches	Bearing Bars 1-3/16" O.C.								For Close Mesh Add to 4" O.C. Chart	
	Cross Bars 4" O.C.				Cross Bars 2" O.C.					
	Wt.	1 to 75 S.F.	76 to 300 S.F.	301 to 1000 S.F.	Wt.	1 to 75 S.F.	76 to 300 S.F.	301 to 1000 S.F.	Total Wt.	Add to Cost
1 x 1/8	2.0#	$ 7.95	$ 6.35	$ 5.75	2.1#	$ 9.10	$ 7.30	$ 6.65	3.1#	75%
1-1/4 x 1/8	2.4	9.35	7.50	6.80	2.5	10.70	8.65	7.75	3.8	75%
1-1/4 x 3/16	3.3	12.80	10.30	9.20	3.5	14.70	16.60	10.60	5.1	60%
1-1/2 x 1/8	2.9	10.70	8.65	7.75	3.0	12.30	9.90	8.90	4.6	75%
1-3/4 x 3/16	4.6	17.00	13.60	12.25	4.8	19.35	15.65	14.00	7.2	60%
2 x 3/16	5.3	19.35	15.35	13.80	5.5	22.10	17.65	15.85	8.1	60%
2-1/4 x 3/16	5.9	21.25	17.00	15.30	6.1	24.00	17.95	17.45	9.1	60%
For widths 18" or less add to the above costs:			135%				135%			

(82) Steel Floor Grating and Treads (Div. 5.4-41)

The table below lists the material weights and cost in dollars per S.F. for welded steel grating, material prices only. Steel cross bars are 4" O.C. for upper tables.

Bearing Bar Size	Steel Bearing Bars, 1-3/16" O.C.							Steel Bearing Bars, 15/16" O.C.						
	Wt.	1-75 S.F.		76-300 S.F.		301-1000 S.F.		Wt.	1-75 S.F.		76-300 S.F.		301-1000 S.F.	
		Paint	Galv.	Paint	Galv.	Paint	Galv.		Paint	Galv.	Paint	Galv.	Paint	Galv.
3/4 x 1/8	4.1#	$ 4.80	$ 5.30	$3.65	$ 4.20	$3.80	$ 3.95	5.0	$ 6.20	$ 6.70	$ 4.80	$ 5.35	$ 4.30	$ 4.95
1 x 1/8	5.2	5.55	6.30	4.30	4.95	3.85	4.55	6.4	7.15	8.00	5.55	6.30	5.05	5.85
1-1/4 x 1/8	6.3	6.30	7.00	4.80	5.55	4.25	5.05	7.9	8.30	9.30	6.30	7.30	5.75	6.70
1-1/4 x 3/16	9.1	7.65	8.75	5.85	6.80	5.25	6.20	11.5	10.25	11.95	7.65	9.10	7.00	8.35
1-1/2 x 1/8	7.4	6.90	7.65	5.30	6.20	4.80	5.65	9.3	9.30	10.35	7.20	8.25	6.40	7.90
1-3/4 x 3/16	12.5	9.85	11.30	7.40	8.85	6.70	8.10	15.8	13.10	14.80	9.85	11.50	8.85	10.55
2 x 3/16	14.1	10.80	12.45	8.10	9.50	7.35	9.10	18.0	13.85	15.75	10.55	12.35	9.50	11.40
2-1/4 x 3/16	15.7	11.60	12.70	8.80	10.40	7.90	9.70	20.0	15.20	17.30	11.50	13.60	10.35	12.45

Bearing Bar Size	As Above, but Cross Bars 2" O.C.							As Above, but Cross Bars 2" O.C.						
	Wt.	1-75 S.F.		76-300 S.F.		301-1000 S.F.		Wt.	1-75 S.F.		76-300 S.F.		301-1000 S.F.	
		Paint	Galv.	Paint	Galv.	Paint	Galv.		Paint	Galv.	Paint	Galv.	Paint	Galv.
3/4 x 1/8	4.8#	$ 6.05	$ 6.55	$ 4.65	$ 5.25	$4.25	$ 4.80	5.7	$ 7.75	$ 8.60	$ 6.10	$ 6.70	$ 5.55	$ 6.20
1 x 1/8	5.9	6.70	7.45	5.15	5.75	4.80	5.45	7.1	8.85	9.70	7.05	7.65	6.20	7.05
1-1/4 x 1/8	7.0	7.45	8.40	5.75	6.50	5.45	6.05	8.6	9.85	10.80	7.65	8.55	6.90	7.80
1-1/4 x 3/16	9.8	8.85	10.00	6.80	7.90	6.15	7.30	12.2	11.50	13.00	8.85	10.15	8.00	9.25
1-1/2 x 1/8	8.1	8.10	9.10	6.25	7.15	5.70	6.60	10.0	10.75	11.85	8.35	9.30	7.45	8.45
1-3/4 x 3/16	13.2	11.00	12.45	8.45	9.90	7.55	9.10	16.5	14.15	16.00	10.85	12.60	9.80	11.60
2 x 3/16	14.8	11.95	14.00	9.10	10.75	8.20	9.80	18.7	15.55	17.40	11.70	13.65	10.65	12.55
2-1/4 x 3/16	16.4	12.90	14.70	9.80	11.60	8.80	10.55	20.7	16.65	19.45	12.75	14.90	11.50	13.65

Prices below are for 3'-0" long stair treads in dollars per tread.

Bearing Bar Size	6" Wide Tread				9" Wide Tread				12" Wide Tread			
	Plain		Serrated		Plain		Serrated		Plain		Serrated	
	Paint	Galv.	Paint	Galv.	Paint	Galv.	Paint	Galv.	Paint	Galv.	Paint	Galv.
3/4 x 1/8	$10.90	$12.15	$11.25	$12.65	$13.00	$14.90	$13.70	$15.55	$13.95	$16.40	$14.95	$17.30
1 x 1/8	11.55	14.00	12.00	14.50	14.15	16.55	14.85	17.25	15.35	18.10	16.30	20.00
1-1/4 x 1/8	12.75	14.65	13.25	15.25	15.85	18.85	16.50	19.50	17.60	21.25	18.55	22.25
1-1/4 x 3/16	13.45	15.90	14.40	16.80	17.05	20.75	17.90	21.50	19.35	24.75	20.25	25.00

CIRCLE REFERENCE NUMBERS

⑧³ Plywood (Div. 6.1)

There are two types of plywood used in construction: interior, which is moisture resistant but not waterproofed, and exterior, which is waterproofed.

The grade of the exterior surface of the plywood sheets is designated by the first letter: A, for smooth surface with patches allowed; B, for solid surface with patches and plugs allowed; C, which may be surface plugged or may have knot holes up to 1" wide; and D, which is used only for interior type plywood and may have knot holes up to 2-1/2" wide. "Structural Grade" is specifically designed for engineered applications such as box beams. All CC & DD grades have roof and floor spans marked on them.

Underlayment grade plywood runs from 1/4" to 1-1/4" thick. Thicknesses 5/8" and over have optional tongue and groove joints which eliminates the need for blocking the edges. Underlayment 19/32" and over may be referred to as Sturd-i-Floor.

The price of plywood can fluctuate widely due to geographic and economic conditions. When one or two local prices are known, the relative prices for other types and sizes may be found by direct factoring of the prices in the table below.

Typical uses for various plywood grades are as follows:
AA-AD Interior — cupboards, shelving, paneling, furniture
B-B Plyform — concrete form plywood
CDX — wall and roof sheathing
Structural — box beams, girders, stressed skin panels
AA-AC Exterior — fences, signs, siding, soffits, etc.
Underlayment — base for resilient floor coverings
Overlaid HDO — high density for concrete forms & highway signs
Overlaid MDO — medium density for painting, siding, soffits & signs
303 siding — exterior siding, textured, striated, embossed, etc.

Grade	National Average Price in Lots of 10 MSF, per MSF-January 1986					
	Type	4'x8'	Type	4'x8'	4'x10'	
Sanded Grade	1/4" Interior AD	$ 345	1/4" Exterior AC	$375	$ 405	
	3/8"	455	3/8"	480	520	
	1/2"	525	1/2"	540	560	
	5/8"	640	5/8"	650	690	
	3/4"	725	3/4"	720	760	
	1"	980	1"	985	1,015	
	1-1/4"	1,160	Exterior AA, add	90	90	
	Interior AA, add	85	Exterior AB, add	70	70	
			CD Structural 1	Underlayment		
Un-sanded Grade 4'x8' Sheets	5/16" CDX	$ 255	5/16", 4'x8' sheets	$285	3/8", 4'x8' sheets	$ 350
	3/8"	290	3/8"	330	1/2'	400
	1/2"	370	1/2"	405	5/8"	500
	5/8"	405	5/8"	445	3/4	565
	3/4"	455	3/4"	520	1-1/8" 2-4-1	765
	3/4" T&G	485				
Form Plywood	5/8" Exterior, oiled BB, plyform	$615	5/8" HDO (overlay 2 sides)		$1,350	
	3/4" Exterior, oiled BB, plyform	670	3/4" HDO (overlay 2 sides)		1,525	
	Overlay 2 Sides MDO		Overlay 1 Side MDO			
Overlaid 4'x8' Sheets	3/8" thick	$ 700	3/8" thick		$ 565	
	1/2"	870	1/2"		730	
	5/8"	985	5/8"		840	
	3/4"	1,100	3/4"		910	
303 Siding	Fir, rough sawn, natural finish, 3/8" thick	$ 370	Texture 1-11 5/8" thick, Fir		$ 560	
	Redwood	1,120	Redwood		1,350	
	Cedar	960	Cedar		1,300	
	Southern Yellow Pine	360	Southern Yellow		515	
Wafer Board	1/4" sheathing	$ 200	5/8" underlayment		$ 400	
	7/16" sheathing	250	3/4" underlayment		500	

For 2 MSF to 10 MSF, add 10%. For less than 2 MSF, add 15%.

CIRCLE REFERENCE NUMBERS

(84) Wood Stair, Residential (Div. 6.2-76)

One Flight with 8'-6" Story Height, 3'-6" Wide Oak Treads Open One Side, Built in Place				
Item	Quantity	Unit Cost	Bare Costs	Costs Incl. Subs O & P
Treads 10-1/2" x 1-1/16" thick	11 Ea.	$ 16.25	$ 178.75	$ 196.35
Landing tread nosing	1 Ea.	3.00	3.00	3.30
Risers 3/4" thick	12 Ea.	10.50	126.00	138.60
Single end starting step (range $100 to $155)	1 Ea.	100.00	100.00	110.00
Balusters (range $3.70 to $9.60)	22 Ea.	4.75	104.50	114.95
Newels, starting & landing (range $32 to $155)	2 Ea.	32.00	64.00	70.40
Rail starter (range $34 to $85)	1 Ea.	45.00	45.00	49.50
Handrail (range $3.90 to $7.90)	26 L.F.	4.25	110.50	121.55
Cove trim	50 L.F.	.43	21.50	23.65
Rough stringers three - 2 x 12's, 14' long	84 B.F.	.40	33.60	36.95
Carpenters installation: Bare Cost	36 Hrs.	$ 20.00	$ 720.00	
Cost incl. Subs O & P		29.00		$1,044.00
	Total per Flight		$1,506.85	$1,909.25

Add for rail return on second floor and for varnishing or other finish. Adjoining walls or landings must be figured separately.

(85) Wood Roof Trusses (Div. 6.1-66)

Material prices for trusses below are based on a lumber price of $310 per MBF. Since the cost of lumber fluctuates, add or subtract 5% for every $40 differential in the current price of the lumber used in the manufacture of the truss. All prices are given delivered to the job site (50 mile radius).

Loading figures represent live load, an additional load of 10 psf on the top chord and 10 psf on the bottom chord is included in the truss design, spacing is 24" O.C.

Span in Feet	Cost per Truss for Different Live Loads and Roof Pitches					
	Flat	4 in 12 Pitch		5 in 12 Pitch		8 in 12 Pitch
	40 psf	30 psf	40 psf	30 psf	40 psf	30 psf
20	$ 48.75	$36.25	$37.75	$39.50	$39.50	$42.00
22	54.50	37.50	39.50	42.25	42.25	44.00
24	61.00	43.25	44.00	46.00	46.00	47.50
26	68.75	44.25	46.75	50.50	51.00	52.25
28	77.00	48.25	55.50	53.55	59.00	64.00
30	85.75	56.25	60.50	62.25	62.75	68.50
32	91.00	56.25	65.75	67.75	67.75	70.25
34	99.50	58.75	68.75	62.25	63.25	77.75
36	108.00	63.75	76.75	64.31	67.25	81.50
38	115.50	68.75	89.75	67.75	83.75	103.00
40	124.00	76.50	94.50	76.25	98.75	114.75

CIRCLE REFERENCE NUMBERS

(86) Thirty City Lumber Prices (Jan. 1st, 1986) (Div. 6.1)

Prices for boards are for #2 or better or sterling, whichever is in best supply. Dimension lumber is "Standard or Better" either Southern Yellow Pine (S.Y.P.), Spruce-Pine-Fir (S.P.F.), Hem-Fir (H.F.) or Douglas Fir (D.F.). The species of lumber used in a geographic area is listed by city. Rough Sawn lumber is Douglas Fir, Hem-Fir, or a variety of hardwood, sheathing or lagging grade. Plyform is 3/4" BB oil sealed fir or S.Y.P. whichever prevails locally, 5/8" CDX is S.Y.P. or Fir.

For 10 MBF lots add 5%; for retail add 10% to prices.

These are prices at the time of publication and should be checked against the current market price. Relative differences between cities will stay approximately constant.

City	Species	Carload Lots per M.B.F.					Rough Sawn Lumber			Carload Lots per M.S.F.	
		S4S									
		Dimensions			Boards		3"x12"	6"x12"	12"x12"	3/4 Ext. Plyform	5/8" Thick CDX
		2"x4"	2"x6"	2"x10"	1"x6"	1"x12"					
Atlanta	S.Y.P.	$295	$295	$340	$725	$855	$425	$455	$455	$615	$345
Baltimore	S.P.F.	290	305	340	710	900	385	410	410	640	510
Boston	S.P.F.	285	295	345	730	925	455	490	480	690	450
Buffalo	S.P.F.	325	340	385	735	930	445	480	470	680	450
Chicago	S.P.F.	340	355	400	735	905	435	470	460	680	425
Cincinnati	S.Y.P.	315	325	370	730	880	435	475	470	640	345
Cleveland	S.P.F.	295	310	350	720	910	450	485	480	690	450
Columbus	S.P.F.	310	325	365	715	895	445	485	480	690	440
Dallas	S.Y.P.	335	340	380	775	905	445	475	465	615	330
Denver	H.F.	305	300	320	705	890	460	490	485	730	440
Detroit	S.P.F.	290	305	355	715	895	465	495	480	690	445
Houston	S.Y.P.	335	340	380	770	910	450	480	460	610	330
Indianapolis	S.P.F.	305	320	365	735	925	450	475	465	690	440
Kansas City	D.F.	365	355	420	705	885	450	490	480	740	425
Los Angeles	D.F.	320	310	380	685	845	430	460	455	690	400
Memphis	S.Y.P.	325	330	370	720	865	455	485	480	615	350
Milwaukee	S.P.F.	280	295	340	715	885	440	465	460	685	440
Minneapolis	S.P.F.	275	290	335	720	905	425	465	450	670	430
Nashville	S.Y.P.	335	340	380	730	855	465	495	485	620	360
New Orleans	S.Y.P.	305	305	350	740	875	435	470	460	620	330
New York City	H.F.	335	325	350	655	925	485	520	510	700	445
Philadelphia	H.F.	320	310	340	735	900	405	445	450	690	445
Phoenix	S.Y.P.	355	360	400	775	910	465	495	485	640	340
Pittsburgh	S.P.F.	295	310	355	730	895	485	520	515	695	440
St. Louis	S.Y.P.	335	340	380	725	865	425	445	450	740	350
San Antonio	S.Y.P.	345	350	390	760	945	455	485	485	620	330
San Diego	D.F.	315	305	370	685	855	450	475	470	695	400
San Francisco	D.F.	310	300	365	665	845	425	465	460	680	405
Seattle	D.F.	275	265	330	640	815	410	445	440	665	380
Washington, DC	H.F.	315	305	330	715	905	370	410	405	645	440
Average		$315	$320	$365	$720	$890	$440	$475	$465	$670	$405

To convert square feet of surface to board feet, 4% waste included

S4S Size	Multiply S.F. by	T & G size	Multiply S.F. by	Flooring Size	Multiply S.F. by
1 x 4	1.18	1 x 4	1.27	25/32" x 2-1/4"	1.37
1 x 6	1.13	1 x 6	1.18	25/32" x 3-1/4"	1.29
1 x 8	1.11	1 x 8	1.14	15/32" x 1-1/2"	1.54
1 x 10	1.09	2 x 6	2.36	1" x 3"	1.28
				1" x 4"	1.24

CIRCLE REFERENCE NUMBERS

(87) Lumber Product Material Prices (Div. 6.1, 6.2, 7.3 & 9.6)

The price of forest products fluctuates widely from location to location and from season to season depending upon economic conditions. The table below indicates National Average material prices in effect Jan. 1, 1986. The table shows relative differences between various sizes, grades and species. These percentage differentials remain fairly constant even though lumber prices in general may change significantly during the year.

Availability of certain items depends upon geographic location and must be checked prior to firm price bidding.

For less than carload lots, add 5% to prices below.

For retail, add 10% to prices below.

National Average Price in Carload Lots

Dimension Lumber, S4S, #2 & Better, KD — Heavy Timbers, Fir

	Species	2"x4"	2"x6"	2"x8"	2"x10"	2"x12"		
Framing Lumber per MBF	Douglas Fir	$330	$320	$320	$390	$395	3"x4" thru 3"x12"	$505
	Spruce	295	300	305	345	360	4"x4" thru 4"x12"	500
	Southern Yellow Pine	365	360	355	385	430	6"x6" thru 6"x12"	500
	Hem-Fir	345	345	350	370	390	8"x8" thru 8"x12"	510
	Redwood	870	860	870	910	870	10"x10" and 10"x12"	520

S4S "D" Quality or Clear, KD — S4S #2 & Better or Sterling, KD

	Species	1"x4"	1"x6"	1"x8"	1"x10"	1"x12"	Species	1"x4"	1"x6"	1"x8"	1"x10"	1"x12"
Boards per MBF	Sugar Pine	$850	$1,000	$900	$980	$1,200	Sugar Pine	—	$670	$650	$675	$840
	Idaho Pine	1,050	1,070	1,100	1,200	1,300	Idaho Pine	$1,000	1,000	990	1,020	1,040
	Engleman Spruce	775	875	950	940	1,575	Engleman Spruce	620	695	650	695	825
	Southern Yellow Pine	835	960	865	870	950	Southern Yellow Pine	570	625	640	625	705
	Ponderosa Pine	900	900	1,000	1,025	1,025	Ponderosa Pine	695	690	685	710	805
	Redwood	1,900	1,980	1,980	1,980	2,100						

Flooring per MSF	1"x4" Vertical grain, Fir "B" & better	$1,650	2-1/4"x25/32" Maple, select	$1,640
	2-1/4"x25/32", Oak, clear	1,750	#2 & better	1,500
	Select	1,580	2-1/4"x33/32" Maple, #2 & better	1,960
	#1 common	1,250	3-1/4"x33/32" Maple, #2 & better	2,100
	Oak, prefinished, standard & better	1,800	Parquet, unfinished, 5/16", minimum	950
	Standard	1,720	Maximum	5,000
Siding per MBF	Clapboard, Cedar, beveled		Rough sawn, Cedar, tongue & groove, 1" x 4"	$1,210
	1/2"x6" thru 1/2"x8", clear	$1,300	1"x12" Board & batten, #3 & better	860
	"A" grade	920	Factory stained	890
	"B" grade	900	4" battens, per M.L.F.	210
	3/4"x10" "clear"	1,125	Factory stained	220
	"A" grade	1,100	Cedar channel siding 1x8, #3 & better	$980
	Redwood, beveled		Factory stained	1,020
	1/2"x6" thru 1/2"x8", vertical grain, clear	930	White Pine siding, T&G, rough sawn	$360
	3/4"x10" vertical grain, clear	1,360	Factory stained	400
Shingles per CSF	Red Cedar		White Cedar shingles	
	5X—16" long #1 regular	$65	16" long, extra grade	$60
	#2	48	Clear, 1st grade	48
	18" long perfections #1	67		
	#2	36	Fire retardant Red Cedar shakes	
	Resquared & Rebutted #1	53	5X — 16" long	$150
	#2	48	18" long perfections	153
	Handsplit shakes, resawn		Handsplit & resawn	
	24" long, 1/2" to 3/4"	77	24" long, 1/2" to 3/4"	170
	18" long, 1/2" to 3/4"	63	3/4" to 5/4"	210

CIRCLE REFERENCE NUMBERS

(88) Roof Slate (Div. 7.3-35)

16", 18" and 20" are standard lengths and slate usually comes in random widths. For standard 3/16" thickness use 1-1/4" copper nails. Allow for 3% breakage.

Boston Area Prices	Unfading Vermont Colored	Weathering Sea Green	Buckingham, Virginia Black, Clear
Slate delivered (incl. punching)	$570.00	$580.00	$580.00
# 30 Felt, copper nails	20.00	20.00	20.00
Slate roofer 4.6 hrs. @ $18.95 per hr.	87.20	87.20	87.20
Total Bare Cost per Square	$677.20	$687.20	$687.20

(89) 1/2" Pargeting (rough dampproofing plaster) (Div. 7.1-25)

1:2-1/2 Mix, 4.5 C.F. Covers 100 S.F., Waste Included	Regular Portland Cement		Waterproofed Portland Cement	
1.7 lbs. integral waterproofing admixture			72¢ per lb.	$ 1.20
1.7 bags portland cement	$5.15 per bag	$ 8.75	$5.15 per bag	8.75
4.25 C.F. sand at $11.20 per C.Y.		1.75		1.75
Labor mix, apply, crew D-1 at $292.00 per day	.231 Days	67.45	.231 Days	67.45
Total Bare Cost per 100 S.F.		$77.95		$79.15

(90) Metal and Fiberglass Sandwich Panels (Div. 7.2-80 & 7.4-06)

Aluminum facing panels may be field erected with corrugated sheets or perforated acoustical type on the inside face and either corrugated, V-beam or ribbed exterior sheets. The most usual gauges are .032" for the exterior and .024" for the interior sheets. The V-beam type is available in .032", .040" and .050". The insulation is generally 1" thick. Individual sheets can be as large as 30' x 2'.

The cost of the particular sandwich can be figured by adding the appropriate siding costs (Div. 7.4-06) to the insulation costs (Div. 7.2-80). Then add 15% per S.F. for multi-story construction and 25% per S.F. additional for jobs less than 2000 S.F. Add 2% to 4% for the usual flashings. These figures do not include any supporting framework or flashings.

(91) Mineral Fiber Roofing Shingles (Div. 7.3-10)

Quantities per Square	Strip Shingles 14" x 30" x 5/32"		Shakes 9.35" x 16" x 1/4"	
Colored Shingles	325 lb.	$105.00	500 lb.	$155.00
Felt, galvanized nails, special pieces		10.00		10.00
Carpenter at $20.00 per hour	2.0 hrs.	40.00	3.64 hrs.	72.80
Total Bare Cost per Square		155.00		237.80

(92) Front Door, Residential (Div. 8.2-32)

Figures below do not include Subs O & P.

Size 3'-0" x 6'-8" x 1-3/4"	Item Location	Deluxe Colonial Design Pine		Modern Flush Solid Core Birch	
Door, glazed with small panels	8.2-32		$185.00		$140.00
Frame and sill, stock unit	8.2-12		50.00		50.00
Entrance and interior trim (Colonial, range $100 to $1,000)	8.2-12		175.00		100.00
Hardware (with brass cylinder set)	8.7-40+32		92.00		92.00
Install door, carpenter @ $20.00 per hour		1.0 hr.	20.00	1.0 hr.	20.00
Install hardware		1.0 hr.	20.00	1.0 hr.	20.00
Install entrance, frame and trim		3.0 hrs.	60.00	1.5 hrs.	30.00
Complete in place			$602.00		$452.00

CIRCLE REFERENCE NUMBERS

⑬ Roof Decks — Comparisons (Div. 7.4-54)

Poured gypsum is cheaper than concrete plank but should not be used where interior humidity is excessive as in laundries, paper mills, swimming pools, etc. A 2" thickness of gypsum is standard. This can be poured on insulating board, 1" fiberglass, acoustical board or various combinations.

Precast lightweight concrete plank is not affected by moisture or temperature but costs more than poured gypsum. With metal edging it is suitable for spans up to 8'. Plank can be used on a sloped roof. They make a good nailing base for slate and are permanent, requiring no upkeep.

Steel roof deck is lighter, hence lighter joists, columns and varying insulation characteristics.

Wood fiber planks cost more than gypsum but have good acoustical and insulation characteristics.

See ⑱ for sub-purlin prices.

⑭ Built-Up Roofing (Div. 7.4-15)

Asphalt is available in kegs of 100 lbs. each; coal tar pitch in 600 lb. kegs. Prepared roofing felts are available in a wide range of sizes, weights & characteristics. However, the most commonly used are #15 (432 S.F. per roll, 13 lbs. per square) and #30 (216 S.F. per roll, 27 lbs. per square).

Inter-ply bitumen varies from 20 lbs. per sq. (asphalt) to 30 lbs. per sq. (coal tar) per ply, ± 25%. Flood coat bitumen also varies from 60 lbs. per sq. (asphalt) to 75 lbs. per sq. (coal tar), ± 25%. Expendable equipment (mops, brooms, screeds, etc.) runs about 16% of the bitumen cost. For new, inexperienced crews this factor may be much higher.

Rigid insulation boards are typically applied in two layers. The first is mechanically attached to nailable decks or spot or solid mopped to non-nailable decks; the second layer is then spot or solid mopped to the first layer. Membrane application follows the insulation, except in protected membrane roofs where the membrane goes down first and the insulation on top, followed with ballast (stone or concrete pavers). Insulation and related labor costs are NOT included in Div. 7.4-15, they can be found in Div. 7.2-50.

Prices shown are bare costs only.

4-Ply Organic Felt on Insulated Deck	Asphalt		Coal Tar	
4 ply #15 organic felt (1/4 roll per ply)	$3.00 per ply	$12.00	$4.75 per ply	$ 19.00
175 lbs. hot asphalt/245 lbs. hot tar	$290 per ton	25.40	$500 per ton	61.25
500 lbs. gravel (1/4" to 1/2")	$8.00 per ton	2.00	$8.00 per ton	2.00
Installation, crew G-1, at $1089.35 per day	.045 days	49.00	.048 days	52.30
Total Bare Cost per Square in-place		$88.40		$134.55

⑮ Hollow Metal Doors (Div. 8.1-21 & 24)

Table below lists material prices only, not including hardware or labor.

Door Thickness and Size		Full Flush Doors, 18 Ga.				Flush Fire Doors			
		Hollow Core		Composite Core		Hollow Core "B"		Composite Core "A"	
		Plain	Glazed	Plain	Glazed	20 Ga.	18 Ga.	16 Ga.	18 Ga.
1-3/4"	3'-0" x 6'-8"	$150	$175	$165	$220	$165	$195	—	$215
	3'-0" x 7'-0"	160	220	175	235	170	200	$270	235
	3'-6" x 7'-0"	180	240	215	250	—	220	290	250
	3'-0" x 8'-0"	200	260	235	280	—	245	315	295
	4'-0" x 8'-0"	225	280	255	305	—	265	—	305
1-3/8"	2'-0" x 6'-8"	108*	145*	—	—	140	—	—	—
	2'-6" x 6'-8"	114*	150*	—	—	155	—	—	—
	3'-0" x 7'-0"	130*	165*	—	—	165	—	—	—

*Indicates 20 gauge doors.

⑯ Welded Metal Door Frames for Masonry (Div. 8.1-10)

Table below lists material prices only.

Frame for Opening Size	16 Ga. Frames, FM Label			18 Ga. Frames			Fire Door Frames, UL Label	
	3" Deep	4-3/4" Deep	8-3/4" Deep	3" Deep	4-3/4" Deep	6-3/4" Deep	4-3/4" Deep	6-3/4" Deep
2'-0" x 6'-8"	$76	$78	$ 86	$67	$69	$75	$ 84	$ 90
2'-6" x 6'-8"	76	78	86	67	69	75	85	90
3'-0" x 7'-0"	79	80	92	70	70	81	85	96
3'-6" x 7'-0"	80	80	92	70	70	81	85	96
4'-0" x 7'-0"	80	80	94	70	70	82	85	97
6'-0" x 7'-0"	88	90	102	77	79	89	94	104
8'-0" x 8'-0"	98	98	112	86	86	98	101	113

For KD Masonry Frames, deduct $15.00.   For KD Drywall Frames, deduct $10.00.   For Galvanized Frames, add 10%.

CIRCLE REFERENCE NUMBERS

(97) Wood Doors (Div. 8.2)

Table below lists price per door only, not including frame, hardware or labor. For pre-hung exterior door units up to 3' x 7', add $125 per door for wood frame and hardware for types not listed under pre-hung. Pricing is for ten or more doors.

Door Thickness and Size		Flush Type Doors				Panel Type Doors, Plain				Pre-hung	
		Hollow Core		Solid Particle Core			Pine		Fir	Pine Panel	Flush Birch Solid Core
		Lauan	Birch	Lauan	Birch	Pine	Plain	Glazed	Plain	Glazed		
1-3/4"	2'-6" x 6'-8"	$27	$35	$37	$44*	$175	$130	$140	$ 95	$100	$220	$134
	3'-0" x 6'-8"	29	39	41	47*	185	148	161	100	105	230	140
	3'-0" x 7'-0"	33	43	45	51*	195	168	176	107	118	256	147
	3'-6" x 7'-0"	—	—	—	68*	—	—	—	—	—	—	—
	4'-0" x 7'-0"	—	—	—	74*	—	—	—	—	—	—	—
1-3/8"	2'-0" x 6'-8"	16	23	—	38	—	85	—	79	—	132	124
	2'-6" x 6'-8"	18	26	—	44	—	90	—	85	—	143	124
	3'-0" x 6'-8"	20	29	—	47	—	102	—	97	—	152	129
	3'-0" x 7'-0"	24	34	—	51	—	—	—	115	—	167	—

*Add to the above for the following birch face door types:

Solid wood core, add $30
3/4 hour label door, add 60%
1 hour label door, add 70%

1-1/2 hour label door, add 100%
8' high door, add 30%
Acoustical door, add 220%

Static shielded door, add 500%
Lead lined door, add 450%
Vinyl laminated door, add 100%

(98) Wood Interior Door, Residential (Div. 8.2-32)

Figures below do not include Subs Overhead & Profit.

Item	Item Location	6 Panel Pine	Flush Birch Hollow Core	Louvered Pine
Door 3'-0" x 6'-8" x 1-3/8" thick	8.2-32	$105.00	$ 29.00	$ 78.00
Frame 1-3/8" x 4-5/8", stock pine	8.2-12	20.00	20.00	20.00
Trim interior	6.2-37	19.00	19.00	19.00
Hardware incl. hinge & lockset	8.7-40	36.00	36.00	36.00
Install door, 2 Carp. @ $20.00 per hr./each	.9 hrs.	36.00	36.00	36.00
Install hardware, 1 Carp.	.7 hrs.	14.00	14.00	14.00
Install frame and trim, 1 Carp.	2.0 hrs.	40.00	40.00	40.00
Total in Place		$270.00	$194.00	$243.00

(99) Tin Clad Fire Door (Div. 8.3-66)

6' x 7' Opening, A Label	Double Sliding Door		Double Swing Door	
Doors, 3 Ply	($9.00 to $10.00/S.F.)	$ 400.00	($9.00 to $10.00/S.F.)	$ 400.00
Frame, 9" channel (15 lb. per L.F.), installed	325 lb.	610.00	325 lb.	610.00
Track, hangers, hardware	13.75 L.F.	680.00	Total	620.00
2 Carpenters @ $20.00 per hour	10 hrs. total	400.00	10 hrs. total	400.00
Complete in place		$2,090.00		$2,030.00

(100) Steel Sash (Div. 8.5-60 & 70)

Ironworker crew will erect 25 S.F. or 1.3 sash unit per hour, whichever is less.

Mechanic will point 30 L.F. per hour.
Painter will paint 90 S.F. per coat per hour.
Glazier production depends on light size.
Allow 1 lb. special steel sash putty per 16" x 20" light.

CIRCLE REFERENCE NUMBERS

(101) Hinges (Div. 8.7-33)

All closer equipped doors should have ball bearing hinges. Lead lined or extremely heavy doors require special strength hinges.

Usually 1-1/2 pair of hinges are used per door up to 7'-6" openings. Table below shows typical hinge requirements.

Use Frequency	Type Hinge Required	Type of Opening	Type of Structure
High	Heavy weight ball bearing	Entrances Toilet Rooms	Banks, Office buildings, Schools, Stores & Theaters Office buildings and Schools
Average	Standard weight ball bearing	Entrances Corridors Toilet Rooms	Dwellings Office buildings and Schools Stores
Low	Plain bearing	Interior	Dwellings

Door Thickness	Weight of Doors in Pounds per Square Foot				
	White Pine	Oak	Hollow Core	Solid Core	Hollow Metal
1-3/8"	3 psf	6 psf	1-1/2 psf	3-1/2 — 4 psf	6-1/2 psf
1-3/4"	3-1/2	7	2	4-1/2 — 5-1/4	6-1/2
2-1/4"	4-1/2	9	—	5-1/2 — 6-3/4	6-1/2

(102) D. H. Wood Window (ready hung) (Div. 8.6-40)

Ponderosa pine sash, glazed and exterior primed.

Description	2'-0" x 3'-0"			3'-0" x 4'-0"		
	Plain Glazed		Insulating Glass	Plain Glazed		Insulating Glass
Window and frame, 2 lights		$ 68.00	$100.00		$ 91.00	$145.00
Removable grilles		22.00	22.00		24.00	24.00
Aluminum storm/screen		32.00	32.00		45.00	45.00
Interior trim set		11.75	11.75		13.00	13.00
Carpenter @ $20.00 per hr.	2.5 hr.	50.00	2.5 hr. 50.00	3.2 hr.	64.00	3.2 hr. 64.00
Complete in place		$183.75	$215.75		$237.00	$291.00

Mullions for above windows are about $22 each. Vinyl clad double-hung windows run about $10.00 per square foot of glass.

Aluminum clad double-hung windows run about $13.00 per square foot of glass.

(103) Glazing Labor (Div. 8.8)

Glass sizes are estimated by the "united inch" (height + width). Table below shows the number of lights glazed in an eight hour period by the crew size indicated, for glass up to 1/4" thick. Square or nearly square lights are more economical on a S.F. basis. Long slender lights will have a high S.F. installation cost.

For insulated glass reduce production by 33%. For 1/2" plate glass reduce production by 50%. Production time for glazing with two glazers per day averages; 1/4" plate glass 120 S.F.; 1/2" plate glass 55 S.F.; 1/2" insulated glass 95 S.F.; insulated glass 75 S.F.

Glazing Method	United Inches per Light							
	40"	60"	80"	100"	135"	165"	200"	240"
Number of Men in Crew	1	1	1	1	2	3	3	4
Industrial sash, putty	60	45	24	15	18	—	—	—
With stops, putty bed	50	36	21	12	16	8	4	3
Wood stops, rubber	40	27	15	9	11	6	3	2
Metal stops, rubber	30	24	14	9	9	6	3	2
Structural glass	10	7	4	3	—	—	—	—
Corrugated glass	12	9	7	4	4	4	3	—
Store fronts	16	15	13	11	7	6	4	4
Skylights, puttyglass	60	36	21	12	16	—	—	—
Thiokol set	15	15	11	9	9	6	3	2
Vinyl set, snap on	18	18	13	12	12	7	5	4
Maximum area per light	2.8 S.F.	6.3 S.F.	11.1 S.F.	17.4 S.F.	31.6 S.F.	47 S.F.	69 S.F.	100 S.F.
Daily Crew Cost	$161.20	$161.20	$161.20	$161.20	$332.40	$483.60	$483.60	$644.80

CIRCLE REFERENCE NUMBERS

(104) Commercial Window Glass (Div. 8.8-60)

Description	1/2" Insulated Glass		3/8" Float Glass	
Glass per S.F.	$4.62	R Value	$2.83	R Value
Allow for breakage 10%	.46	1.68	.28	.93
Glazing compound 1/2 lb. per S.F. at 89¢ per lb.	.46	Average	.46	Average
Labor, 2 glaziers @ $20.15 per hour each (75 S.F./day)	4.29		4.29	
Total in place, per S.F.	$9.83		$7.86	

(105) Window Walls (Div. 8.9-80)

The table below shows the S.F. costs for 1-3/4" x 4-1/2" clear anodized tubular aluminum framing, flush glazed with fixed 1/4" clear polished float glass for jobs over 250 S.F. and includes overhead & profit.

Description	Total Window Wall Height								
	3'-0"			6'-0"			10'-0"		
Mullion Spacing	3'	5'	7'	3'	5'	7'	3'	5'	7'
Mullions only	$24.30	$19.45	$17.00	$18.75	$14.25	—	$13.90	—	—
1 Intermediate horizontal member	—	25.25	22.30	22.25	17.65	$15.50	18.20	$14.35	—
2	—	—	—	25.65	20.40	16.05	19.70	16.10	$14.20
3	—	—	—	28.76	22.95	20.30	22.30	17.55	15.50

Add to the above erected costs for the following:
5/8" insulating glass, plus 2" x 4-1/2" clear tube frame, add $8.20 per S.F.
1" insulating glass, plus 2" x 4-1/2" clear tube frame, add $14.25 per S.F.

Bronze anodized aluminum tubing, add 15% to 20%.
For screw applied square stops, add 8%.
For operating sash, add cost of each sash to total job.

(106) Gypsum Lath (Div. 9.1-20)

Item	Nailed to Wood Studs			Clipped to Steel Studs		
	Quantities and Bare Cost		Incl. O & P	Quantities and Bare Cost		Incl. O & P
105 S.Y. 3/8" gypsum lath	$2.10 per S.Y.	$220.50	$242.55	$2.10 per S.Y.	$220.50	$242.55
Fasteners, nails & clips	8 lb. @ 87¢/lb.	6.95	7.65	600 @ 5¢ ea.	30.00	33.00
Lather at $20.10 and $28.35 per hr.	9.4 hours	188.95	266.50	10.7 hours	215.05	303.35
Total per 100 S.Y. in place		$416.40	$516.70		$465.55	$578.90

Regular lath comes in 16" x 32" and 16" x 48" sheets. Firestop gypsum base comes in 4' x 8' sheets. For nailing use 1-1/8" No. 13 ga. flathead blued nails.

(107) Metal Lath (Div. 9.1-25)

Painted Diamond Expanded	Nailed to Wood Studs			Screwed to Steel Studs		
	Quantity and Bare Cost		Incl. O & P	Quantity and Bare Cost		Incl. O & P
105 S.Y. 3.4 lb. lath	$2.00 per S.Y.	$210.00	$231.00	$2.00 per S.Y.	$210.00	$231.00
Fasteners, corner beads, etc.		10.80	11.90		11.55	12.70
Lather at $20.10 and $28.35 per hr.	10.0 hrs	201.00	283.50	10.7 hrs.	215.05	303.35
Total per 100 S.Y. in place		$421.80	$526.40		$436.60	$547.05

Material prices of other types of lath in Div. 9.1 may be substituted in the table above to arrive at their costs in place. For nailing on studs use a 1" roofing nail with 7/16" head. For nailing on ceiling use 1-1/2" No. 11 ga. barbed roofers nail with 7/16" head for holding power.

CIRCLE REFERENCE NUMBERS

(108) Studs, Joists and Track (Div. 9.1-16 & 9.2-20)

Material prices per 1000 L.F., for galvanized studs, joists and tracks. Panhead, framing screws, 7/16" long are $14.00 per thousand; 1-5/8" long are $17.25 per thousand.

Non-load bearing, 20 ga. stud and track are primarily used for curtain wall. (C.W.)

Size	Non-Load Bearing				Load Bearing 1-5/8" Flange—Light Gauge Structural					
	25 Ga.		20 Ga. (C.W.)		18 Ga.		16 Ga.		14 Ga.	
	Stud	Track	Stud	Track	Stud	Track	Stud	Track	Stud	Track
1-5/8"	$145	$135	$280	$270						
2-1/2"	165	160	295	290	$640	$575	$775	$710		
3-5/8"	195	190	345	330	710	640	865	795	$1,050	$980
4"	230	225	400	385	745	675	890	825	1,080	1,015
6"	295	290	485	470	880	810	1,050	980	1,280	1,215
8"					1,150	1,080	1,350	1,280	1,550	1,485

Size	Non-Load Bearing				Load Bearing-Extra Wide Flange - 2" Flange							
	25 Ga.		22 Ga.		18 Ga.		16 Ga.		14 Ga.		12 Ga.	
	C-H Stud	J-Track	C-H Stud	J-Track	Joist	Track	Joist	Track	Joist	Track	Joist	Track
2-1/2"	$455	$380	$660	$495								
4"	595	485	865	650								
6"					$1,100	$1,080	$1,330	$1,300	$1,650	$1,610	$2,140	$2,100
8"					1,130	1,100	1,370	1,340	1,670	1,630	2,820	2,780
10"							1,560	1,520	1,920	1,880	3,070	3,020
12"							2,340	2,290	2,830	2,750	3,700	3,640

(109) Vermiculite or Perlite Plaster (Div. 9.1-35)

Proportions: Over lath, scratch coat and brown coat 100# gypsum plaster to 2 C.F. aggregate; over masonry, 100# gypsum plaster to 3 C.F. aggregate.

Quantities for 100 S.Y.	2 Coat, 5/8" Thick			3 Coat, 3/4" Thick		
	Quantities	Bare Cost	Incl. O & P	Quantities	Bare Cost	Incl. O & P
Gypsum plaster @ $9.00 per 80 lb. bag	1250 lb.	$140.65	$154.70	2250 lb.	$253.15	$278.45
Vermiculite or perlite @ $5.15 per bag	7.8 bags	40.15	44.20	11.3 bags	58.20	64.00
Finish hydrated lime @ $6.15 per 50 lb. bag	340 lb.	41.80	46.00	340 lb.	41.80	46.00
Gauging plaster @ $13.80 per 100 lb. bag	170 lb.	23.45	25.80	170 lb.	23.45	25.80
J-1 crew @ $779.40 & $1,098.00 per day	.85 days	662.50	933.30	1.00 days	779.40	1,098.00
Cleaning, staging, handling, patching		89.15	117.65		115.40	149.60
Total per 100 S.Y. in place		$997.70	$1,321.65		$1,271.40	$1,661.85

(110) Gypsum Plaster (Div. 9.1-21)

Quantities for 100 S.Y.	2 Coat, 5/8" Thick		3 Coat, 3/4" Thick		
	Base	Finish	Scratch	Brown	Finish
	1:3 Mix	2:1 Mix	1:2 Mix	1:3 Mix	2:1 Mix
Gypsum plaster	1300 lb.		1350 lb.	650 lb.	
Sand	2.6 C.Y.		1.85 C.Y.	1.35 C.Y.	
Finish hydrated lime		340 lb.			340 lb.
Gauging plaster		170 lb.			170 lb.

Total, in Place for 100 S.Y. on Walls	2 Coat, 5/8" Grounds			3 Coat, 3/4" Grounds		
	Quantities	Bare Cost	Incl. O & P	Quantities	Bare Cost	Incl. O & P
Gypsum plaster @ $9.00 per 80 lb. bag	1300 lb.	$146.25	$160.90	2000 lb.	$225.00	$247.50
Finish hydrated lime @ $6.15 per 50 lb. bag	340 lb.	41.80	46.00	340 lb.	41.80	46.00
Gauging plaster @ $13.80 per 100 lb. bag	170 lb.	23.45	25.80	170 lb.	23.45	25.80
Sand @ $11.20 per C.Y.	2.6 C.Y.	29.10	32.00	3.2 C.Y.	35.85	39.45
J-1 crew @ $779.40 & $1,098.00 per day	.87 days	678.10	955.25	1.05 days	818.35	1,152.90
Cleaning, staging, patching after trades (labor & mat'l.)		92.40	114.85		109.75	143.15
Total per 100 S.Y. in place		$1,011.10	$1,334.80		$1,254.20	$1,654.80

CIRCLE REFERENCE NUMBERS

⑪ Stucco (Div. 9.1-55)

Quantities for 100 S.Y. 3 Coats, 1" Thick	On Wood Frame			On Masonry		
	Quantities	Bare Cost	Incl. O & P	Quantities	Bare Cost	Incl. O & P
Portland cement at $5.15 per bag	29 bags	$ 149.35	$ 164.30	21 bags	$ 108.15	$ 118.95
Sand at $11.20 per C.Y.	3.2 C.Y.	35.85	39.45	2.4 C.Y.	26.90	29.60
Hydrated lime at $6.15 per 50 lb. bag	180 lb.	22.15	24.35	120 lb.	14.75	16.25
Painted stucco mesh at $2.26 per S.Y.	105 S.Y.	237.30	261.05	—	—	—
Furring nails and jute fiber		12.50	13.75		—	—
Mix and install with crew indicated	J-2 @ 1.35 days	1,269.25	1,788.45	J-1 @ 1.30 days	1,013.20	1,427.35
Cleaning, staging, handling, patching		180.15	241.65		117.90	166.50
Total per 100 S.Y. in place		$1,906.55	$2,533.00		$1,280.90	$1,758.65

⑫ Terrazzo Floor (Div. 9.3-40)

5,000 S.F. 5/8" Terrazzo Topping Quantities for 100 S.F.	Bonded to Concrete 1-1/8" Bed, 1:4 Mix			Not Bonded 2-1/8" Bed and 1/4" Sand		
	Quantities	Bare Cost	Incl. O & P	Quantities	Bare Cost	Incl. O & P
Portland cement at $5.15 per bag	6 bags	$ 30.90	$ 34.00	8 bags	$ 41.20	$ 45.30
Sand at $11.20 per C.Y.	10 C.F.	4.15	4.55	20 C.F.	8.30	9.15
Divider strips for 4' square panels, zinc, 12 ga.	55 L.F.	37.40	41.15	55 L.F.	37.40	41.15
Terrazzo fill, domestic aggregates	600 lb.	100.00	110.00	600 lb.	100.00	110.00
15 lb. tarred felt		—	—		4.50	4.95
Mesh 2 x 2 #16 galvanized		—	—		11.00	12.10
Crew J-3 @ $405.20 & $532.25 per day	.77 days	312.00	409.85	1.00 days	405.20	532.25
Total per 100 S.F. in place	Bonded	$484.45	$599.55	Not Bonded	$607.60	$754.90

2' x 2' panels required 1 L.F. divider strip per S.F.; 6' x 6' panels need .33 L.F. per S.F.

⑬ Resilient Flooring and Base (Div. 9.6-20)

Description	12" x 12" x 3/22" V.C. Tile			4" x 1/8" Vinyl Base		
	Quantities	Bare Cost	Incl. O & P	Quantities	Bare Cost	Incl. O & P
Vinyl composition, tile, standard line	100 S.F.	$ 66.00	$ 72.60	100 L.F.	$41.25	$ 45.40
Vinyl cement at $13.10 per gallon	0.8 gallon	10.50	11.55	0.5 gallon	6.55	7.20
Prime and wax		5.30	5.85		—	—
Tile layer at $19.75 and $27.70 per hour	1.5 hrs.	29.65	41.55	2.7 hrs.	53.35	74.80
Total per 100 S.F. in place		$111.45	$131.55	100 L.F.	$101.15	$127.40

⑭ Wall Covering (Div. 9.8-40)

Quantities for 100 S.F.	Medium Price Paper			Expensive Paper		
	Quantities	Bare Cost	Incl. O & P	Quantities	Bare Cost	Incl. O & P
Paper at $17.00 and $40.00 per double roll	1.6 dbl. rolls	$27.20	$29.90	1.6 dbl. rolls	$ 64.00	$ 70.40
Acrylic sizing at $10.00 per gal.	0.2 gallon	2.00	2.20	0.2 gallon	2.00	2.20
Wheat paste at $1.50 per 2 lb. box	1 lb.	.75	.85	1 lb.	.75	.85
Apply sizing at $19.50 and $27.80 per hour	0.3 hour	5.85	8.35	0.3 hour	5.85	8.35
Apply paper at $19.50 and $27.80 per hour	1.2 hours	23.40	33.35	1.5 hours	29.25	41.70
Total including waste allowance per 100 S.F.		$59.20	$74.65		$101.85	$123.50

This is equivalent to about $47.00 and $77.00 per roll complete in place. Most wallpapers now come in double rolls only. To remove old paper allow 1.3 hours per 100 S.F.

CIRCLE REFERENCE NUMBERS

(115) Paint (Div. 9.8)

Material prices per gallon in 5 gallon lots, up to 25 gallons. For 100 gallons, deduct 10%.

Exterior, Alkyd (oil base)	
Flat	$15.00
Gloss	16.25
Primer	14.80

Exterior, Latex (water base)	
Acrylic stain	12.90
Gloss enamel	16.50
Flat	13.60
Primer	13.50
Semi-gloss	15.75

Interior, Alkyd (oil base)	
Enamel undercoater	12.25
Flat	12.00
Gloss	16.35
Primer sealer	12.00
Semi-gloss	15.40

Interior, Latex (water base)	
Enamel undercoater	9.50
Flat	10.50
Floor and deck	13.65
Gloss	16.25
Primer sealer	9.50
Semi-gloss	13.50

Masonry, Exterior	
Alkali resistant primer	$16.00
Block filler, epoxy	14.30
Block filler, latex	8.00
Latex, flat or semi-gloss	12.30

Masonry, Interior	
Alkali resistant primer	16.00
Block filler, epoxy	14.30
Block filler, latex	8.00
Floor, alkyd	13.80
Floor, latex	12.00
Latex, flat acrylic	11.00
Latex, flat emulsion	12.50
Latex, sealer	9.20
Latex, semi-gloss	13.50

Varnish and Stain	
Alkyd clear	15.00
Polyurethane, clear	15.25
Primer sealer	12.00
Semi-transparent stain	12.50
Solid color stain	13.00

Metal Coatings	
Galvanized	16.95
High heat	51.30
Machinery enamel, alkyd	16.75
Normal heat	14.50
Rust inhibitor ferrous metal	$13.00
Zinc chromate	18.50

Heavy Duty Coatings	
Acrylic urethane	43.50
Chlorinated rubber	22.00
Coal tar epoxy	22.00
Metal pretreatment (polyvinyl butyral)	22.50
Polyamide epoxy finish	22.50
Polyamide epoxy primer	23.50
Silicone alkyd	30.00
2 component solvent based acrylic epoxy	28.00
2 component solvent based polyester epoxy	28.00
Vinyl	19.00
Zinc rich primer	32.50

Special Coatings/Miscellaneous	
Aluminum	13.50
Creosote	7.00
Dry fall out, flat	8.50
Fire retardant, intumescent	22.00
Linseed oil	8.25
Shellac	12.50
Swimming pool, epoxy or urethane base	35.00
Swimming pool, rubber base	22.00
Texture paint	8.50
Turpentine	8.80
Water repellent 5% silicone	9.50

(116) Painting (Div. 9.8)

Item	Coat	One Gallon Covers			In 8 Hrs. Man Covers			Man Hours per 100 S.F.		
		Brush	Roller	Spray	Brush	Roller	Spray	Brush	Roller	Spray
Paint wood siding	prime	275 S.F.	250 S.F.	325 S.F.	1150 S.F.	1400 S.F.	4000 S.F.	.695	.571	.200
	others	300	275	325	1600	2200	4000	.500	.364	.200
Paint exterior trim	prime	450	—	—	650	—	—	1.230	—	—
	1st	525	—	—	700	—	—	1.143	—	—
	2nd	575	—	—	750	—	—	1.067	—	—
Paint shingle siding	prime	300	285	335	1050	1700	2800	.763	.470	.286
	others	400	375	425	1200	2000	3200	.667	.400	.250
Stain shingle siding	1st	200	190	220	1200	1400	3200	.667	.571	.250
	2nd	300	275	325	1300	1700	4000	.615	.471	.200
Paint brick masonry	prime	200	150	175	850	1700	4000	.941	.471	.200
	1st	300	250	320	1200	2200	4400	.364	.364	.182
	2nd	375	340	400	1300	2400	4400	.615	.333	.182
Paint interior plaster or drywall	prime	450	425	550	1600	2500	4000	.500	.320	.200
	others	500	475	550	1400	3000	4000	.571	.267	.200
Paint interior doors and windows	prime	450	—	—	1300	—	—	.333	—	—
	1st	475	—	—	1150	—	—	.696	—	—
	2nd	500	—	—	1000	—	—	.800	—	—

CIRCLE REFERENCE NUMBERS

(117) Steel Lockers (Div. 10.1-37)

No. of Tiers	Size in Inches W x D	Price per Opening, Material Only, Based on 100 Openings, Not Including Locks				Add for the Following	
		60" Total Height		72" Total Height		Sloping Tops	Closed Base
		1 Locker Wide	3 Lockers Wide	1 Locker Wide	3 Lockers Wide		
Single Tier High	12 x 12	$64.75	$54.25	$71.80	$59.70	$2.45	$2.50
	12 x 15	68.00	56.70	74.50	62.05	2.50	2.50
	12 x 18	72.10	60.05	78.85	65.75	3.00	2.50
	12 x 21	75.60	63.65	81.20	68.05	3.05	2.50
	15 x 15	72.45	59.50	80.15	66.70	3.00	2.65
	15 x 18	76.80	63.75	86.15	71.80	3.10	2.65
	15 x 21	79.80	66.85	87.45	73.85	3.65	2.65
	18 x 18			93.10	76.95	3.70	3.05
	18 x 21			94.90	78.50	4.10	3.05
	18 x 24			99.25	85.55	4.90	3.05
Two Tier High	12 x 12	43.50	37.15	46.50	39.30	1.20	1.25
	12 x 15	43.95	42.70	49.45	41.25	1.25	1.25
	12 x 18	49.25	44.40	50.00	43.05	1.50	1.25
	15 x 15			52.15	44.45	1.50	1.30
	15 x 18			60.00	44.90	1.55	1.30
Five Tier High	12 x 12	19.80	17.10			.49	.50
	12 x 15	21.00	17.50			.50	.50
	15 x 15	22.70	19.75			.60	.53
Six Tier High	12 x 12			19.20	16.60	.41	.42
	12 x 15			20.35	17.00	.42	.42
	12 x 18			22.05	19.15	.50	.42
	15 x 15			20.75	17.30	.50	.44
	15 x 18			22.50	19.50	.52	.44

For over 500 openings, deduct 10%.

(118) Public Garage Costs (Div. 13.1-30 & 17.1-41)

Cars	Per Car	Levels	Per Car	Levels	Per Car	Levels	Per S.F.
200 to 400	$6,150	3	$6,475	6	$6,350	Surface	$17.00
400 to 600	6,300	4	6,475	7	6,600	2	17.40
600 to 800	6,575	5	6,350	8 to 12	6,875	3 or 4	21.30
800 to 1200	6,775					5 or 6	21.00
Average	$6,475	Average per Car (310 to 320 S.F. per car)			$6,475	7 or 8	22.05

The above costs refer to open wall construction. For enclosed structures, costs must be added for sprinkler and ventilation systems.

(119) Swimming Pools (Div. 13.1-75)

Pool prices given per square foot of surface area include pool structure, filter and chlorination equipment where required, pumps, related piping, diving boards, ladders, maintenance kit, skimmer and vacuum system. Decks and electrical service to equipment are not included.

Residential in-ground pool construction can be divided into two categories: vinyl lined and gunite. Vinyl lined pool walls are constructed of different materials including wood, concrete, plastic or metal. The bottom is often graded with sand over which the vinyl liner is installed. Costs are generally in the $10 to $17 per S.F. range. Vermiculite or soil cement bottoms may be substituted for an added cost of $.70 per S.F. surface.

Gunite pool construction is used both in residential and municipal installations. These structures are steel reinforced for strength and finished with a white cement limestone plaster. Residential costs run from $16 to $26 per S.F. surface. Municipal costs vary from $25 to $45 because plumbing codes require more expensive materials, chlorination equipment and higher filtration rates.

Municipal pools greater than 1,600 S.F. require gutter systems to control waves. This gutter may be formed into the concrete wall. Often a vinyl, stainless steel gutter or gutter and wall system is specified, which will raise the pool cost an additional $90 to $150 per L.F. of gutter installed and up to $265 per L.F. if a gutter and wall system is installed.

Competition pools usually require tile bottoms and sides with contrasting lane striping. Add $12 per S.F. of wall or bottom to be tiled.

CIRCLE REFERENCE NUMBERS

(120) Air Supported Structures (Div. 13.1-05)

Air supported structures are made from fabrics that can be classified into two groups; temporary and permanent. Temporary fabrics include nylon, woven polyethylene, vinyl film, and vinyl coated dacron. These have lifespans that range from two to ten plus years, and costs that vary accordingly from $2 to $10 per S.F. floor area. This is a package cost for a fabric shell, tension cables, primary and back-up inflation systems and doors. The lower cost structures are used for construction shelters, bulk storage and pond covers. The more expensive are used for recreational structures and warehouses.

Permanent fabrics are teflon coated fiberglass. The package cost runs $20 to $26 per S.F. floor area. The life of this structure is twenty plus years. The high cost limits its application to architectural designed structures which call for a clear span covered area, such as stadiums and convention centers. Both temporary and permanent structures are available in translucent fabrics which eliminates the need for daytime lighting.

Areas to be covered vary from 10,000 S.F. to 10 acres. Height restrictions range from a maximum of 1/2 of width to a minimum of 1/6 of the width. Erection of even the largest of the temporary structures requires no more than a week ($.04 to $.17 per S.F. floor area).

Centrifugal fans provide the inflation necessary to support the structure during application of live loads. Air locks are usually used at large entrances to prevent loss of static pressure. Some manufacturers employ propeller fans which generate sufficient airflow (30,000 CFM) to eliminate the need for air locks. These fans may also be automatically controlled to resist high wind conditions, regulate humidity (air changes), and provide cooling and heat.

Insulation can be provided with the addition of a second or even third interior liner, creating a dead air space with an "R" value of four to nine. Some structures allow for the liner to be collapsed into the outer shell to enable the internal heat to melt accumulated snow. For cooling or air conditioning, the exterior face of the liner can be aluminized to reflect the sun's heat. Costs for the liner run about $.60 to $1.75 per S.F. floor area.

(121) Dome Structures (Div. 13.1-27)

Steel — The four types in order of increasing cost and weight per S.F. are: Lamella, Schwedler, Arch and Geodesic. For maximum economy, diameter should equal spherical radius. Most common diameters are in the 200' to 300' range. Lamella domes weigh about 5 P.S.F. of floor area less than Schwedler domes. Schwedler dome weight in lb. per S.F. approaches 0.046 times the diameter. Domes below 125' diameter weigh .07 times diameter and the cost per ton of steel is higher. Recent costs per ton erected range from $2,650 to $5,575 per ton of steel. See (72) for estimating weights.

Wood — Small domes are of sawn lumber, larger ones are laminated. In larger sizes, triaxial and triangular cost about the same; radial domes cost more. Radial domes are economical in the 60' to 70' diameter range. Most economical range of all types is 80' to 200' diameters. Diameters can run over 400'. Costs above foundations run $23 per S.F. for under 50' diameter. At 77' diameter, radial cost $15 and triaxial cost $17 per S.F. Cost is $10.25 for 120'; $14.50 for 200'; $18.75 for 300' and $24.00 for 400' diameter. Prices include 2" decking and tension tie ring in place.

Plywood — See division 6.1-74 for stressed skin and folded plates. Stock prefab geodesic domes are available with diameters from 23' to 60'. Shells cost between $5.50 to $14.00 per S.F. of floor with erection running between $.48 and $2.20 per S.F. Canopies without sides run about $1.00 per S.F. less.

Fiberglass — Aluminum framed translucent sandwich panels with spans from 5' to 45' are commercially available. Material cost is a minimum of $29 per S.F. for 28' diameters up to $40 per S.F. for 6' diameter units. Erection runs between $2.00 to $4.00 per S.F.

Aluminum — Stressed skin aluminum panels form geodesic domes with spans ranging from 80' to 230'. Costs per S.F. of floor vary from $18 to $21 total in-place costs. For vinyl faced fiberglass insulation with an R rating of 20, add $4.00 per S.F. of floor. An aluminum tubular space truss, triangulated or nontriangulated, with aluminum or clear acrylic closure panels can be used for clear spans of 30' to 300'. The aluminum panel costs range from $12 to $21 per S.F. floor and the acrylic panel from $40 to $65 per S.F. floor total in-place costs.

(122) Industrial Chimneys (Div. 13.1-17)

Foundation requirements in C.Y. of concrete for various sized chimneys.

Size Chimney	2 Ton Soil	3 Ton Soil	Size Chimney	2 Ton Soil	3 Ton Soil	Size Chimney	2 Ton Soil	3 Ton Soil
75' x 3'-0"	13 C.Y.	11 C.Y.	160' x 6'-6"	86 C.Y.	76 C.Y.	300' x 10'-0"	325 C.Y.	245 C.Y.
85' x 5'-6"	19	16	175' x 7'-0"	108	95	350' x 12'-0"	422	320
100' x 5'-0"	24	20	200' x 6'-0"	125	105	400' x 14'-0"	520	400
125' x 5'-6"	43	36	250' x 8'-0"	230	175	500' x 18'-0"	725	575

CIRCLE REFERENCE NUMBERS

(123) Elevator Selective Costs (Div. 14.1-20)

	Passenger		Freight		Hospital	
A. Base Unit	Hydraulic	Electric	Hydraulic	Electric	Hydraulic	Electric
Capacity	1500 Lb.	2000 Lb.	2000 Lb.	4000 Lb.	3500 Lb.	3500 Lb.
Speed	50 F.P.M.	100 F.P.M.	25 F.P.M.	50 F.P.M.	50 F.P.M.	100 F.P.M.
#Stops/Travel Ft.	2/10	4/40	2/10	4/40	2/10	4/40
Push Button Oper.	Yes	Yes	Yes	Yes	Yes	Yes
Telephone Box & Wire	"	"	"	"	"	"
Emergency Lighting	"	"	No	No	"	"
Cab	Painted Steel	Painted Steel	Painted Steel	Painted Steel	S.S. Wainscot, Baked	Enamel Above
Cove Lighting	Yes	Yes	No	No	Yes	Yes
Floor	V.A.T.	V.A.T.	Wood w/Safety Treads	Wood w/Safety Treads	V.A.T.	V.A.T.
Doors, & Speedside Slide	Yes	Yes	No	No	Yes	Yes
Gates, Manual	No	No	Yes	Yes	No	No
Signals, Lighted Buttons	Car Only	Car Only	In Use Light	In Use Light	Car and Hall	Car and Hall
O.H. Geared Machine	N.A.	Yes	N.A.	Yes	N.A.	Yes
Variable Voltage Contr.	"	"	"	"	"	"
Emergency Alarm	"	"	"	"	"	"
Class "A" Loading	"	N.A.	Yes	"	"	N.A.
Base Cost	$37,500	$49,000	$29,900	$47,500	$41,700	$58,500
B. Capacity Adjustment						
2,000 Lb.	$ 2,350					
2,500	4,500	$ 2,350	$ 2,150			
3,000	5,000	2,800	3,500			
3,500	6,200	4,000	4,000			
4,000	6,600	4,600	5,200		$ 4,200	$ 2,925
4,500	7,800	5,700	5,600		4,850	3,350
5,000	9,000	7,000	6,800	$ 2,350	7,000	4,600
6,000			8,000	2,800		
7,000			9,600	4,000		
8,000			12,800	12,000		
10,000			15,000	15,000		
12,000			19,250	19,250		
16,000			22,500	22,500		
20,000			25,000	27,600		
C. Travel Over Base	$ 535 V.L.F.	$ 185 V.L.F.	$ 640 V.L.F.	$ 185 V.L.F.	$ 640 V.L.F.	$ 185 V.L.F.
D. Additional Stops	$ 5,025 Ea.	$ 5,025 Ea.	$ 4,300 Ea.	$ 4,300 Ea.	$ 5,200 Ea.	$ 5,025 Ea.
E. Speed Adjustment						
50 F.P.M.			$ 535			
75	535		750	$ 1,875	$ 590	
100	940		1,500	2,975	965	
125	1,300		2,375	4,100	1,400	
150	3,575		3,200	5,100	2,450	
175				6,000	3,000	
Geared 4 Flrs. Min. 200		$ 4,800		6,800		$ 5,800
250		6,950		7,800		7,950
300		8,550		10,200		9,550
350		9,600		11,700		10,600
400		10,700		12,800		11,700
Gearless 10 Flrs. Min. 500		25,700		16,400		26,700
600		28,900		18,000		29,900
700		29,900		21,000		30,900
800		37,500		23,500		38,500
1,000		Spec. Applic.		Spec. Applic.		Spec. Applic.
1,200		"		"		"
F. Other Than Class "A" Loading						
"B"			$ 1,225	$ 1,225		
"C-1"			2,150	2,150		
"C-2"			1,350	1,350		
"C-3"			2,780	2,780		

CIRCLE REFERENCE NUMBERS

(123) Elevator Selective Costs (cont.)

	Passenger	Freight	Hospital
G. Options			
1. Controls			
Automatic, 2 car group	$ 3,200		$ 3,200
3 car group	5,500		5,500
4 car group	6,950		6,950
5 car group	8,550		8,550
6 car group	12,800		12,800
Emergency, fireman service	2,000		2,000
Intercom service	1,500		1,500
Selective collective, single car	2,350		2,350
Duplex car	3,850		3,850
2. Doors			
Center opening, 1 speed	$ 750		$ 750
2 speed	800		800
Rear opening-opposite front	—		5,350
Side opening, 2 speed	800		800
Freight, bi-parting	—	$3,000	—
Power operated door and gate	—	6,900	—
3. Emergency power switching, automatic	$ 2,125		$ 2,125
Manual	1,075		1,075
4. Finishes based on 3500# cab			
Ceilings, acrylic panel	$ 225		—
Aluminum egg crate	275		$ 215
Doors, stainless steel	325		215
Floors, carpet, class "A"	80		—
Epoxy	215		215
Quarry tile	125		215
Slate	140		
Steel plate	—	$ 535	—
Textured rubber	65		54
Walls, plastic laminate	300		215
Stainless steel	765		535
Return at door	430		430
Steel plate, 1/4" x 4' high, 14 ga. above	—	1,175	—
Entrance, doors, baked enamel	250		180
Stainless steel	430		430
Frames, baked enamel	250		180
Stainless steel	535		535
5. Maintenance contract - 12 months	$ 2,600	$1,650	$ 2,850
6. Signal devices			
Hall lantern, each	$ 375	$ 375	$ 375
Position indicator, car or lobby	270	270	270
Add for over three each	65	65	65
7. Specialties			
High speed, heavy duty door opener	$ 450		$ 450
Variable voltage, O.H. gearless machine	26,700 - 53,400		26,700 - 53,400
Basement installed geared machine	6,000	$6,000	6,000

CIRCLE REFERENCE NUMBERS

 Elevator Cost Development (Div. 14.1-20)

Requirement:

One freight elevator, five story hydraulic, 4,000 lb. capacity, 12' floor to floor, speed 75 F.P.M., power doors, class "B" loading and maintenance contract.

Description	Total Cost
A. Base Elevator, (Hydraulic Freight)	$ 29,900
B. Capacity Adjustment (4,000 lb.)	5,200
C. Excess Travel Over Base (5 x 12' = 60' - (2 @ 10' for Base Unit = 20') = 40' x $640 per V.L.F.	25,600
D. Stops Over Base (5-2 for Base Unit = 3 x $4,300)	12,900
E. Speed Adjustment (75 F.P.M.)	750
F. Alternate Loading Class (Class "B")	1,225
G. Options:	
2. Power door and gate, 5 doors @ $6,900 ea., less 2 manual from Base Unit @ $3,000 ea.	28,500
5. Maintenance Contract (12 Months)	1,650
Total Cost	$105,725

 Passenger Elevators (Div. 14.1-20)

Electric elevators are used generally but hydraulic elevators can be used for lifts up to 70' and where large capacities are required. Hydraulic speeds are limited to 150 F.P.M. but cars are self leveling at the stops. On low rises, hydraulic installation runs about 15% less than standard electric types but on higher rises this installation cost advantage is reduced. Maintenance of hydraulic elevators is about the same as electric type but underground portion is not included in the maintenance contract.

In standard electric elevators there are two basic control systems; rheostatic system for speeds up to 150 F.P.M. and variable voltage for speeds over 150 F.P.M. The two types of drives are geared for low speeds and gearless for 450 F.P.M. and over. As a rule of thumb, each added 100 F.P.M. adds about 20% to the total cost.

The tables on the preceding pages illustrate typical installed costs of the various types of elevators available.

 Freight Elevators (Div. 14.1-20-060)

Capacities run from 1,500 lbs. to over 100,000 lbs. with 3,000 lbs. to 10,000 lbs. most common. Travel speeds are generally lower and control less intricate than on passenger elevators. Figures in division 14.1-20 are for hydraulic and geared elevators.

 Escalators (Div. 14.1-25)

Moving stairs can be used for buildings where 600 or more people are to be carried to the second floor or beyond. Freight cannot be carried on escalators and at least one elevator must be available for this function. Carrying capacity is 5,000 to 8,000 people per hour. Power requirement is 2 to 3 KW per hour and incline angle is 30°.

Table below indicates approximate installed costs per unit. Installation outside major cities would increase the costs below, due to mechanics additional travel expense.

Story Height	Metal Balustrade		Glass Paneled Balustrade	
	32" Wide	48" Wide	32" Wide	48" Wide
10'	$72,000	$79,500	$81,500	$ 89,000
15'	76,500	86,500	86,500	94,500
20'	81,000	91,500	92,500	100,000
25'	85,500	98,000	98,500	105,000

 Moving Ramps and Walks (Div. 14.1-35)

These are a specialized form of belt conveyor 2' to 9' wide with capacities of 3,600 to 18,000 persons per hour. Maximum speed is 140 F.P.M. and normal incline is 0° to 15°. With high use by wheeled vehicles, a maximuim of 12° is recommended.

Local codes will determine the maximum angle. For a 40" width with a 10' to 23' story height, moving ramps cost $27,000 to $75,000 per unit. Low incline angle involves considerable lost floor space.

CIRCLE REFERENCE NUMBERS

(129) Pipe Material Costs and Considerations (Div. 2.5 & 15.1)

1. Malleable fittings should be used for gas service.
2. Malleable fittings are used where there are stresses/strains due to expansion and vibration.
3. Cast fittings may be broken as an aid to disassembling of heating lines frozen by long use, temperature and minerals.
4. Cast iron pipe is extensively used for underground and submerged service.
5. Type M (light wall) copper tubing is available in hard temper only and is used for nonpressure and less severe applications than K and L.
6. Type L (medium wall) copper tubing, available hard or soft for interior service.
7. Type K (heavy wall) copper tubing, available in hard or soft temper for use where conditions are severe. For underground and interior service.
8. Hard drawn tubing requires fewer hangers or supports but should not be bent. Silver brazed fittings are recommended, however 50/50 and 95/5 are occasionally used.

(130) Plumbing Fixtures (Div. 15.2)

Total labor hours to install fixtures.

Item	Rough-In	Set	Total Hours	Item	Rough-In	Set	Total Hours
Bath tub	5	5	10	Shower head only	2	1	3
Bath tub and shower, cast iron	6	6	12	Shower drain	3	1	4
Fire hose reel and cabinet	4	2	6	Shower stall, slate		15	15
Floor drain to 4 inch diameter	3	1	4	Slop sink	5	3	8
Grease trap, single, cast iron	5	3	8	Test 6 fixtures			14
Kitchen gas range		4	4	Urinal, wall	6	2	8
Kitchen sink, single	4	4	8	Urinal, pedestal or floor	6	4	10
Kitchen sink, double	6	6	12	Water closet and tank	4	3	7
Laundry tubs	4	2	6	Water closet and tank, wall hung	5	3	8
Lavatory wall hung	5	3	8	Water heater, 45 gals. gas, automatic	5	2	7
Lavatory pedestal	5	3	8	Water heaters, 65 gals. gas, automatic	5	2	7
Shower and stall	6	4	10	Water heaters, electric, plumbing only	4	2	6

Fixture prices in front of book are based on the cost per fixture set in place. The rough-in cost, which must be added for each fixture, includes carrier, if required, some supply, waste and vent pipe connecting fittings and stops. The lengths of rough-in pipe are nominal runs which would connect to the larger runs and stacks. The supply runs and DWV runs and stacks must be accounted for in separate entries. In the eastern half of the United States it is common for the plumber to carry these to a point 5 feet outside the building.

(131) Water Cooler Application (Div. 15.3)

Type of Service	Requirement
Office, School or Hospital	12 persons per gallon per hour
Office, Lobby or Department Store	4 or 5 gallons per hour per fountain
Light manufacturing	7 persons per gallon per hour
Heavy manufacturing	5 persons per gallon per hour
Hot heavy manufacturing	4 persons per gallon per hour
Hotel	.08 gallons per hour per room
Theatre	1 gallon per hour per 100 seats

(132) Heating (Boston latitude) (Div. 15.5)

$$\text{Approximate S.F. radiation} = \frac{\text{S.F. sash}}{2} + \frac{\text{S.F. wall + roof} - \text{sash}}{20} + \frac{\text{C.F. building}}{200}$$

CIRCLE REFERENCE NUMBERS

⑬ Quick Heat Loss Approximation (Div. 15.5)

1. Calculate the cubical volume of the room or building.
2. Select the appropriate factor from Table 1.
3. If the building has a bad north and west exposure, multiply the heat loss factor by 1.1
4. If the outside design temperature is other than 0°F, multiply the factor from Table 1 by the factor from Table 2.
5. Multiply the cubic volume by the factor selected from Table 1. This will give the estimated BTUH heat loss.
6. The heat loss is the heat which must be made up to maintain inside temperature.

Table 1	Factor for Determining Heat Loss for Various Type Buildings			Table 2	
Type Buildings	Conditions	Qualifications	Loss Factor	Outside Design Temperature Correction Factor	
Factories & Industrial plants at 70° F	One Story	Skylights in Roof	6.2	Outside Design Temp.	Correction Factor
	Multiple Story	Two Story	4.6		
		Three Story	4.3		
Warehouses etc. at 60° F	All Walls Exposed	Skylights in Roof	5.5		
		No Skylights in Roof	5.1		
	One Warm Common Wall	Skylights in Roof	5.0	30° F	0.57
		No Skylights in Roof	4.9		
		Heated Space Above	3.4	20	0.72
	Warm Common Walls On Both Long Sides	Skylights in Roof	4.7		
		No Skylights in Roof	4.4	10	0.86
General Office areas at 70° F	All Walls Exposed	Flat Roof	6.9		
		Heated Space Above	5.2	0	1.00
	One Long Warm Common Wall	Flat Roof	6.3		
		Heated Space Above	4.7	-10	1.14
	Warm Common Walls On Both Long Sides	Flat Roof	5.8		
		Heated Space Above	4.1	-20	1.28

⑭ Solar Heating (basic requirements) (Div. 15.5-77)

Face collectors as close to due south as practical. Locate collectors so they are not shaded from sun's rays. Incline collectors at a slope of latitude minus 5° for domestic hot water and latitude plus 15° for space heating. Insulate piping and storage tank well to minimize heat losses. Size domestic water heating storage tanks to hold 20 gallons water per user, minimum, 30 gallons per user preferable. For domestic water heating an optimum collector size is approximately 3/4 square foot of area per gallon of water storage. For space heating of residences and small commercial applications the collector is commonly sized between 30% and 50% of the internal floor area. For space heating of large commercial applications, collector areas less than 30% of the internal floor area can still provide significant heating reductions.

The price of the collector varies from $15 to $34 per square foot of collector depending on the installation and includes all but the terminal pump, controls, thermostat and storage tank. See Divisions 15.5-77 and 15.5-89 for tanks. Size the tank 1-1/2 gallons capacity per square foot of collector area.

A supplementary heat source is recommended for Northern states for December through February. The solar energy transmission per square foot of collector surface varies greatly with the material used.

Initial cost, heat transmittance and useful life are obviously interrelated.

⑮ Recommended Ventilation Air Changes (Div. 15.7)

Table below lists range of time in minutes per change for various types of facilities.

Assembly Halls 2-10	Dance Halls 2-10	Laundries 1-3
Auditoriums 2-10	Dining Rooms 3-10	Markets 2-10
Bakeries 2-3	Dry Cleaners 1-5	Offices 2-10
Banks 3-10	Factories 2-5	Pool Rooms 2-5
Bars 2-5	Garages 2-10	Recreation Rooms 2-10
Beauty Parlors 2-5	Generator Rooms 2-5	Sales Rooms 2-10
Boiler Rooms 1-5	Gymnasiums 2-10	Theaters 2-8
Bowling Alleys 2-10	Kitchens-Hospitals 2-5	Toilets 2-5
Churches 5-10	Kitchens-Restaurant 1-3	Transformer Rooms 1-5

CFM air required for changes = Volume of room in cubic feet ÷ Minutes per change.

CIRCLE REFERENCE NUMBERS

(136) Air Conditioning Requirements (Div. 15.7)

BTU's per hour per S.F. of floor area and S.F. per ton of air conditioning.

Type Building	BTU per S.F.	S.F. per Ton	Type Building	BTU per S.F.	S.F. per Ton	Type Building	BTU per S.F.	S.F. per Ton
Apartments, Individual	26	450	Dormitory, Rooms	40	300	Libraries	50	240
Corridors	22	550	Corridors	30	400	Low Rise Office, Exterior	38	320
Auditoriums & Theaters	666	18*	Dress Shops	43	280	Interior	33	360
Banks	50	240	Drug Stores	80	150	Medical Centers	28	425
Barber Shops	48	250	Factories	40	300	Motels	28	425
Bars & Taverns	133	90	High Rise Office—Ext. Rms.	46	263	Office (small suite)	43	280
Beauty Parlors	66	180	Interior Rooms	37	325	Post Office, Individual Office	42	285
Bowling Alleys	68	175	Hospitals, core	43	280	Central Area	46	260
Churches	600	20*	Perimeter	46	260	Residences	20	600
Cocktail Lounges	68	175	Hotel, Guest Rooms	44	275	Restaurants	60	200
Computer Rooms	141	85	Public Spaces	55	220	Schools & Colleges	46	260
Dental Offices	52	230	Corridors	30	400	Shoe Stores	55	220
Dept. Stores, Basement	34	350	Industrial Plants, Offices	38	320	Shop'g. Ctrs., Super Markets	34	350
Main Floor	40	300	General Offices	34	350	Retail Stores	48	250
Upper Floor	30	400	Plant Areas	40	300	Specialty	60	200

*Persons per ton

12,000 BTU = 1 ton of air conditioning

(137) Ductwork (Div. 15.7-64)

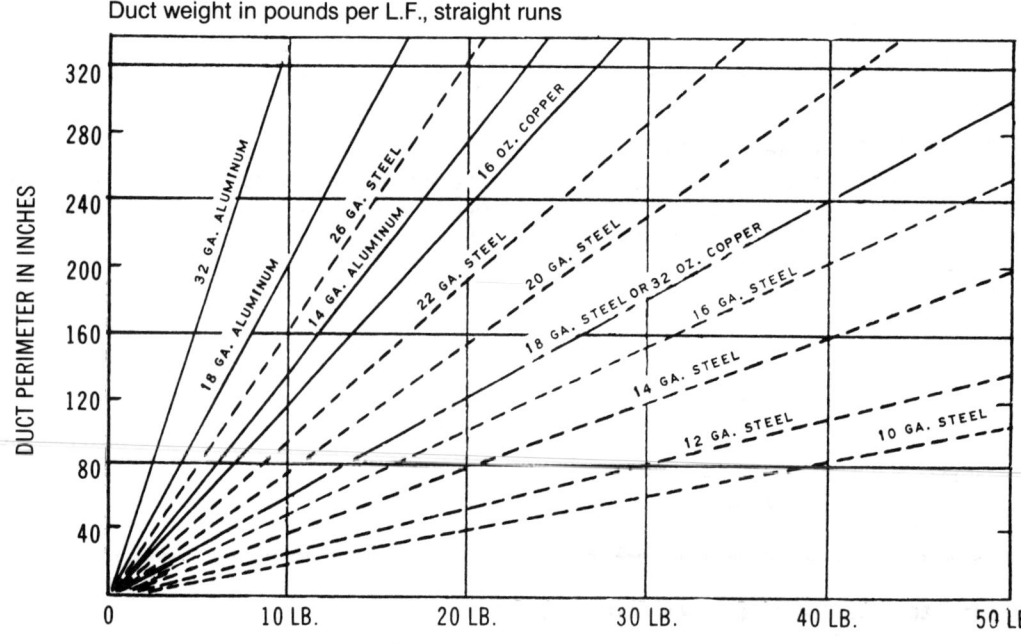

Duct weight in pounds per L.F., straight runs

Add to the above for fittings; 90° elbow is 3 L.F.; 45° elbow is 2.5 L.F.; offset is 4 L.F.; transition offset is 6 L.F.; square-to-round transition is 4 L.F.; 90° reducing elbow is 5 L.F. For bracing and waste, add 20% to aluminum and copper, 15% to steel.

CIRCLE REFERENCE NUMBERS

(138) Motors (Div. 16.3)

230/460 Volt A.C., three phase 60 cycle ball bearing squirrel cage induction motors, NEMA design B Standard Line, including installation and Subs O&P.

Horse Power	Synchronous Speed in RPM	Drip Proof, Class B Insulation 1.15 Service Factor			Totally Enclosed, Class B Insulation, 1.0 Service Factor		
		No Starter	Manual Starter	Magnetic Starter	No Starter	Manual Starter	Magnetic Starter
1	1,200	$ 215	$375	$ 400	$ 220	$380	$ 405
	1,800	185	345	370	185	345	370
2	1,200	245	405	490	255	415	500
	1,800	215	375	460	205	365	450
	3,600	220	380	465	215	375	460
3	1,200	310	470	555	310	470	555
	1,800	230	390	480	255	410	495
	3,600	245	405	490	245	405	490
5	1,200	405	635	710	430	660	735
	1,800	255	480	555	275	505	580
	3,600	270	500	580	300	530	605
7.5	1,800	340	570	850	325	555	835
10		405	635	915	385	615	895
15		475		985	545		1,055
20		600		1,330	655		1,385
25		715		1,440	815		1,540
30		815		1,540	910		1,640
40		1,025		2,450	1,200		2,625
50		1,225		2,665	1,500		2,940
60		1,725		4,580	2,375		5,240
75		2,050		4,925	2,875		5,750
100		2,650		5,525	3,850		6,715
125		3,475		10,625	5,375		12,500
150		5,350		12,450	6,525		13,650
200		6,875		14,050	9,325		16,450

CIRCLE REFERENCE NUMBERS

(139) Conductors in Conduit (Div. 16.0)

Table below lists maximum number of conductors for various sized conduit using THW, TW or THWN insulations.

Copper Wire Size	1/2" TW	1/2" THW	1/2" THWN	3/4" TW	3/4" THW	3/4" THWN	1" TW	1" THW	1" THWN	1-1/4" TW	1-1/4" THW	1-1/4" THWN	1-1/2" TW	1-1/2" THW	1-1/2" THWN	2" TW	2" THW	2" THWN	2-1/2" TW	2-1/2" THW	2-1/2" THWN	3" THW	3" THWN	3-1/2" THW	3-1/2" THWN	4" THW	4" THWN
#14	9	6	13	15	10	24	25	16	39	44	29	69	60	40	94	99	65	154	142	93		143		192			
#12	7	4	10	12	8	18	19	13	29	35	24	51	47	32	70	78	53	114	111	76	164	117		157			
#10	5	4	6	9	6	11	15	11	18	26	19	32	36	26	44	60	43	73	85	61	104	95	160	127		163	
#8	2	1	3	4	3	5	7	5	9	12	10	16	17	13	22	28	22	36	40	32	51	49	79	66	106	85	136
#6		1	1	2	2	4	4	4	6	7	7	11	10	10	15	16	16	26	23	23	37	36	57	48	76	62	98
#4		1	1	1	1	2	3	3	4	5	5	7	7	7	9	12	12	16	17	17	22	27	35	36	47	47	60
#3		1	1	1	1	1	2	2	3	4	4	6	6	6	8	10	10	13	15	15	19	23	29	31	39	40	51
#2		1	1	1	1	1	2	2	3	4	4	5	5	5	7	9	9	11	13	13	16	20	25	27	33	34	43
#1				1	1	1	1	1	1	3	3	3	4	4	5	6	6	8	9	9	12	14	18	19	25	25	32
1/0				1	1	1	1	1	1	2	2	3	3	3	4	5	5	7	8	8	10	12	15	16	21	21	27
2/0				1	1	1	1	1	1	1	1	2	3	3	3	5	5	6	7	7	8	10	13	14	17	18	22
3/0					1	1	1	1	1	1	1	1	2	2	3	4	4	5	6	6	7	9	11	12	14	15	18
4/0						1	1	1	1	1	1	1	1	1	2	3	3	4	5	5	6	7	9	10	12	13	15
250MCM							1	1	1	1	1	1	1	1	1	2	2	3	4	4	4	6	7	8	10	10	12
300							1	1	1	1	1	1	1	1	1	2	2	3	3	3	4	5	6	7	8	9	11
350									1	1	1	1	1	1	1	1	1	2	3	3	3	4	5	6	7	8	9
400										1	1	1	1	1	1	1	1	1	2	2	3	4	5	5	6	7	8
500										1	1	1	1	1	1	1	1	1	1	1	2	3	4	4	5	6	7
600												1	1	1	1	1	1	1	1	1	1	3	3	4	4	5	5
700													1	1	1	1	1	1	1	1	1	2	3	3	4	4	5
750													1	1	1	1	1	1	1	1	1	2	2	3	3	4	4

(140) Cable Cost Comparisons (Div. 16.1)

Table below lists material prices per L.F. for copper conductor cables. Aluminum wiring generally requires larger conductor sizes than copper wiring and is subject to very close tolerance in torque tightening of connections. The cost of aluminum wire of equivalent ampacity is about 45% of the material cost of copper. Size of conduit must allow for the increased size of the aluminum conductors.

600V Wire Capacity Aluminum THW	600V Wire Capacity Copper THW	Size	TW	THW	THWN/THHN	Bare Copper	5KV CLP Shielded	15KV CLP Shielded
	15 amp	#14	$.023	$ .026	$ .024	$ .026		
	20	#12	.033	.037	.033	.036		
	30	#10	.051	.056	.055	.056		
40 amp	45	#8	.103	.113	.105	.096		
50	65	#6		.150	.154	.157		
65	85	#4		.23	.25	.26		
90	115	#2		.35	.44	.39	$1.25	$1.50
100	130	#1		.46	.58	.45	1.40	1.70
120	150	1/0		.54	.70	.53	1.50	2.05
135	175	2/0		.65	.80	.64	1.70	2.50
155	200	3/0		.79	.98	.78	—	—
180	230	4/0		.98	1.20	.96	2.40	3.00
205	255	250 MCM		1.15	1.35	1.10	2.65	3.40
230	285	300		1.50	1.65	1.45	—	—
250	310	350		1.60	1.85	1.55	3.40	4.00
310	380	400		2.00	2.25	1.90	—	—
		500		2.20	2.60	2.15	4.80	5.25
		600		3.25		2.70		
		750		4.00		3.40		
		1,000		5.75		4.90		

CIRCLE REFERENCE NUMBERS

(141) Square Foot Project Size Modifier (Div. 17.1)

One factor that affects the S.F. cost of a particular building is the size. In general, for buildings built to the same specifications in the same locality, the larger building will have the lower S.F. Cost. This is due mainly to the decreasing contribution of the exterior walls plus the economy of scale usually achievable in larger buildings. The Area Conversion Scale shown below will give a factor to convert costs for the typical size building to an adjusted cost for the particular project.

The Square Foot Base Size lists the median costs, most typical project size in our accumulated data and the range in size of the projects.

The Size Factor for your project is determined by dividing your project area in S.F. by the typical project size for the particular Building Type. With this factor, enter the Area Conversion Scale at the appropriate Size Factor and determine the appropriate cost multiplier for your building size.

EXAMPLE: Determine the cost per S.F. for a 100,000 S.F. Mid-rise apartment building.

$$\frac{\text{Proposed building area} = 100,000 \text{ S.F.}}{\text{Typical size from below} = 50,000 \text{ S.F.}} = 2.00$$

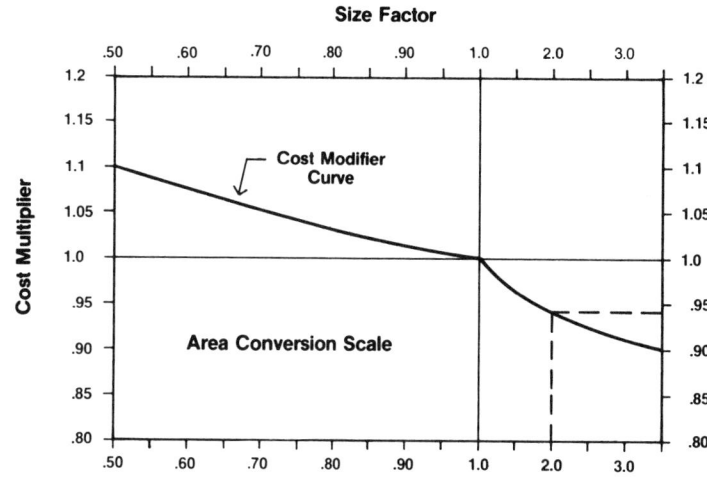

Enter Area Conversion scale at 2.0, intersect curve, read horizontally the appropriate cost multiplier of 0.94. Size adjusted cost becomes 0.94 x $49.00 = $46.06 based on national average costs.

Building Type	Median Cost Per S.F.	Typical Size Gross S.F.	Typical Range Gross S.F.	Building Type	Median Cost Per S.F.	Typical Size Gross S.F.	Typical Range Gross S.F.
Apartments, Low Rise	$ 38.90	21,000	9,700 - 37,200	Jails	$113.00	13,700	7,500 - 28,000
Apartments, Mid Rise	49.00	50,000	32,000 - 100,000	Libraries	69.55	12,000	7,000 - 31,000
Apartments, High Rise	54.35	310,000	100,000 - 650,000	Medical Clinics	67.05	7,200	4,200 - 15,700
Auditoriums	65.35	25,000	7,600 - 39,000	Medical Offices	63.30	6,000	4,000 - 15,000
Auto Sales	41.35	20,000	10,800 - 28,600	Motels	49.70	27,000	15,800 - 51,000
Banks	90.25	4,200	2,500 - 7,500	Nursing Homes	66.80	23,000	15,000 - 37,000
Churches	59.75	9,000	5,300 - 13,200	Offices, Low Rise	53.35	8,600	4,700 - 19,000
Clubs, Country	58.00	6,500	4,500 - 15,000	Offices, Mid Rise	57.15	52,000	31,300 - 83,100
Clubs, Social	57.10	10,000	6,000 - 13,500	Offices, High Rise	70.60	260,000	151,000 - 468,000
Clubs, YMCA	61.30	28,300	12,800 - 39,400	Police Stations	88.45	10,500	4,000 - 19,000
Colleges (Class)	79.10	50,000	23,500 - 98,500	Post Offices	66.80	12,400	6,800 - 30,000
Colleges (Science Lab)	92.45	45,600	16,600 - 80,000	Power Plants	440.00	7,500	1,000 - 20,000
College (Student Union)	85.20	33,400	16,000 - 85,000	Religious Education	49.45	9,000	6,000 - 12,000
Community Center	61.85	9,400	5,300 - 16,700	Research	86.95	19,000	6,300 - 45,000
Court Houses	82.60	32,400	17,800 - 106,000	Restaurants	78.40	4,400	2,800 - 6,000
Dept. Stores	36.50	90,000	44,000 - 122,000	Retail Stores	38.30	7,200	4,000 - 17,600
Dormitories, Low Rise	58.85	24,500	13,400 - 40,000	Schools, Elementary	58.15	41,000	24,500 - 55,000
Dormitories, Mid Rise	75.90	55,600	36,100 - 90,000	Schools, Jr. High	57.95	92,000	52,000 - 119,000
Factories	31.20	26,400	12,900 - 50,000	Schools, Sr. High	57.00	101,000	50,500 - 175,000
Fire Stations	63.55	5,800	4,000 - 8,700	Schools, Vocational	55.05	37,000	20,500 - 82,000
Fraternity Houses	56.35	12,500	8,200 - 14,800	Sports Arenas	45.15	15,000	5,000 - 40,000
Funeral Homes	56.45	7,800	4,500 - 11,000	Supermarkets	37.75	20,000	12,000 - 30,000
Garages, Commercial	42.60	9,300	5,000 - 13,600	Swimming Pools	65.15	13,000	7,800 - 22,000
Garages, Municipal	45.25	8,300	4,500 - 12,600	Telephone Exchange	99.65	4,500	1,200 - 10,600
Garages, Parking	19.45	163,000	76,400 - 225,300	Terminals, Bus	43.90	11,400	6,300 - 16,500
Gymnasiums	55.00	19,200	11,600 - 41,000	Theaters	53.25	10,500	8,800 - 17,500
Hospitals	111.00	55,000	27,200 - 125,000	Town Halls	63.70	10,800	4,800 - 23,400
House (Elderly)	54.90	37,000	21,000 - 66,000	Warehouses	25.40	25,000	8,000 - 72,000
Housing (Public)	46.10	36,000	14,400 - 74,400	Warehouse & Office	27.75	25,000	8,000 - 72,000
Ice Rinks	43.90	29,000	27,200 - 33,600				

CIRCLE REFERENCE NUMBERS

(142) Square Foot and Cubic Foot Building Costs (Div. 17.1)

The cost figures in division 17.1 were derived from more than 10,500 projects contained in the Means Data Bank of Construction Costs and include the contractor's overhead and profit, but do not include architectural fees or land costs. The figures have been adjusted to January 1, 1986. New projects are added to our files each year and projects over ten years old are discarded. For this reason, certain costs may not show a uniform annual progression. In no case are all subdivisions of a project listed.

These projects were located throughout the U.S. and reflect tremendous differences in S.F. and C.F. costs. This is due to both differences in labor and material costs, plus differences in the owner's requirements. For instance, a bank in a large city would have different features than one in a rural area. This is true of all different types of buildings analyzed. As a general rule, the projects on the low side did not include any site work or equipment, but the projects on the high side may include both equipment and site work. The median figures do not generally include site work.

None of the figures "go with" any others. All individual cost items were computed and tabulated separately. Thus the sum of the median figures for Plumbing, HVAC and Electrical will not normally total up to the total Mechanical and Electrical costs arrived at by separate analysis and tabulation of the projects.

Each building was analyzed as to total and component costs and percentages. The figures were arranged in ascending order with the results tabulated as shown. The 1/4 column shows that 25% of the projects had lower costs, 75% higher. The 3/4 column shows that 75% of the projects had lower costs, 25% had higher. The median column shows that 50% of the projects had lower costs, 50% had higher.

There are two times when square foot costs are useful. The first is in the conceptual stage when no details are available. Then square foot costs make a useful starting point. The second is after the bids are in and the costs can be worked back into their appropriate units for information purposes. As soon as details become available in the project design, the square foot approach should be discontinued and the project priced as to its particular components. When more precision is required or for estimating the replacement cost of specific buildings, the "Means Square Foot Costs 1986" should be used.

In using the figures in division 17.1, it is recommended that the median column be used for preliminary figures if no additional information is available. The median figures, when multiplied by the total city construction cost index figures (see division 19.1) and then multiplied by the project size modifier on the preceding page, should present a fairly accurate base figure, which would then have to be adjusted in view of the estimator's experience, local economic conditions, code requirements and the owner's particular requirements. There is no need to factor the percentage figures as these should remain constant from city to city. All tabulations mentioning air conditioning had at least partial air conditioning.

The editors of this book would greatly appreciate receiving cost figures on one or more of your recent projects which would then be included in the averages for next year. All cost figures received will be kept confidential except that they will be averaged with other similar projects to arrive at S.F. and C.F. cost figures for next year's book. See the last two pages of the book for details and the discount available for submitting one or more of your projects.

(143) Repair and Remodeling (Div. 18.1)

Cost figures in BUILDING CONSTRUCTION COST DATA are based on new construction utilizing the most cost-effective combination of labor, equipment and material with the work scheduled in proper sequence to allow the various trades to accomplish their work in an efficient manner.

The costs for repair and remodeling work must be modified due to the following factors that may be present in any given repair and remodeling project:

1. Equipment usage curtailment due to the physical limitations of the project, with only hand-operated equipment being used.
2. Increased requirement for shoring and bracing to hold up the building while structural changes are being made and to allow for temporary storage of construction materials on above-grade floors.
3. Material handling becomes more costly due to having to move within the confines of an enclosed building. For multi-story construction, low capacity elevators and stairwells may be the only access to the upper floors.
4. Large amount of cutting and patching and attempting to match the existing construction is required. It is often more economical to remove entire walls rather than create many new door and window openings. This sort of trade-off has to be carefully analyzed.
5. Cost of protection of completed work is increased since the usual sequence of construction usually can not be accomplished.
6. Economies of scale usually associated with new construction may not be present. If small quantities of components must be custom fabricated due to job requirements, unit costs will naturally increase. Also, if only small work areas are available at a given time, job scheduling between trades becomes difficult and subcontractor quotations may reflect the excessive start-up and shut-down phases of the job.
7. Work may have to be done on other than normal shifts and may have to be done around an existing production facility which has to stay in production during the course of the repair and remodeling.
8. Dust and noise protection of adjoining non-construction areas can involve substantial special protection and alter usual construction methods.
9. Job may be delayed due to unexpected conditions discovered during demolition or removal. These delays ultimately increase construction costs.
10. Piping and ductwork runs may not be as simple as for new construction. Wiring may have to be snaked through walls and floors.
11. Matching "existing construction" may be impossible because materials may no longer be manufactured. Substitutions may be expensive.
12. Weather protection of existing structure requires additional temporary structures to protect building at opening.
13. On small projects, because of local conditions, it may be necessary to pay a tradesman for a minimum of four hours for a task that is completed in one hour.

All of the above areas can contribute to increased costs for a repair and remodeling project. Each of the above factors should be considered in the planning, bidding and construction stage in order to minimize the increased costs associated with repair and remodeling jobs.

Abbreviations

A	Area Square Feet; Ampere	Calc	Calculated	D.H.	Double Hung
ABS	Acrylonitrile Butadiene Styrene; Asbestos Bonded Steel	Cap.	Capacity	DHW	Domestic Hot Water
		Carp.	Carpenter	Diag.	Diagonal
A.C.	Alternating Current; Air Conditioning; Asbestos Cement	C.B.	Circuit Breaker	Diam.	Diameter
		C.C.F.	Hundred Cubic Feet	Distrib.	Distribution
		cd	Candela	Dk.	Deck
A.C.I.	American Concrete Institute	cd/sf	Candela per Square Foot	D.L.	Dead Load; Diesel
Addit.	Additional	CD	Grade of Plywood Face & Back	Do.	Ditto
Adj.	Adjustable	CDX	Plywood, grade C&D, exterior glue	Dp.	Depth
af	Audio-frequency	Cefi.	Cement Finisher	D.P.S.T.	Double Pole, Single Throw
A.G.A.	American Gas Association	Cem.	Cement	Dr.	Driver
Agg.	Aggregate	CF	Hundred Feet	Drink.	Drinking
A.H.	Ampere Hours	C.F.	Cubic Feet	D.S.	Double Strength
A hr	Ampere-hour	CFM	Cubic Feet per Minute	D.S.A.	Double Strength A Grade
A.I.A.	American Institute of Architects	c.g.	Center of Gravity	D.S.B.	Double Strength B Grade
AIC	Ampere Interrupting Capacity	CHW	Commercial Hot Water	Dty.	Duty
Allow.	Allowance	C.I.	Cast Iron	DWV	Drain Waste Vent
alt.	Altitude	C.I.P.	Cast in Place	DX	Deluxe White, Direct Expansion
Alum.	Aluminum	Circ.	Circuit	dyn	Dyne
a.m.	ante meridiem	C.L.	Carload Lot	e	Eccentricity
Amp.	Ampere	Clab.	Common Laborer	E	Equipment Only; East
Approx.	Approximate	C.L.F.	Hundred Linear Feet	Ea.	Each
Apt.	Apartment	CLF	Current Limiting Fuse	Econ.	Economy
Asb.	Asbestos	CLP	Cross Linked Polyethylene	EDP	Electronic Data Processing
A.S.B.C.	American Standard Building Code	cm	Centimeter	E.D.R.	Equiv. Direct Radiation
Asbe.	Asbestos Worker	CMP	Corr. Metal Pipe	Eq.	Equation
A.S.H.R.A.E.	American Society of Heating, Refrig. & AC Engineers	C.M.U.	Concrete Masonry Unit	Elec.	Electrician; Electrical
		Col.	Column	Elev.	Elevator; Elevating
A.S.M.E.	American Society of Mechanical Engineers	CO_2	Carbon Dioxide	EMT	Electrical Metallic Conduit; Thin Wall Conduit
		Comb.	Combination		
A.S.T.M.	American Society for Testing and Materials	Compr.	Compressor	Eng.	Engine
		Conc.	Concrete	EPDM	Ethylene Propylene Diene Monomer
Attchmt.	Attachment	Cont.	Continuous; Continued		
Avg.	Average			Eqhv.	Equip. Oper., heavy
Bbl.	Barrel	Corr.	Corrugated	Eqlt.	Equip. Oper., light
B.&B.	Grade B and Better; Balled & Burlapped	Cos	Cosine	Eqmd.	Equip. Oper., medium
		Cot	Cotangent	Eqmm.	Equip. Oper., Master Mechanic
B.&S.	Bell and Spigot	Cov.	Cover	Eqol.	Equip. Oper., oilers
B.&W.	Black and White	CPA	Control Point Adjustment	Equip.	Equipment
b.c.c.	Body-centered Cubic	Cplg.	Coupling	ERW	Electric Resistance Welded
B.F.	Board Feet	C.P.M.	Critical Path Method	Est.	Estimated
Bg. Cem.	Bag of Cement	CPVC	Chlorinated Polyvinyl Chloride	esu	Electrostatic Units
BHP	Brake Horse Power	C. Pr.	Hundred Pair	E.W.	Each Way
B.I.	Black Iron	CRC	Cold Rolled Channel	EWT	Entering Water Temperature
Bit.; Bitum	Bituminous	Creos.	Creosote	Excav.	Excavation
		Crpt.	Carpet & Linoleum Layer	Exp.	Expansion
Bk.	Backed	CRT	Cathode-ray Tube	Ext.	Exterior
Bkrs.	Breakers	CS	Carbon Steel	Extru.	Extrusion
Bldg.	Building	Csc	Cosecant	f.	Fiber stress
Blk.	Block	C.S.F.	Hundred Square Feet	F	Fahrenheit; Female; Fill
Bm.	Beam	C.S.I.	Construction Specification Institute	Fab.	Fabricated
Boil.	Boilermaker			FBGS	Fiberglass
B.P.M.	Blows per Minute	C.T.	Current Transformer	F.C.	Footcandles
BR	Bedroom	CTS	Copper Tube Size	f.c.c.	Face-centered Cubic
Brg.	Bearing	Cu	Cubic	f'c.	Compressive Stress in Concrete; Extreme Compressive Stress
Brhe.	Bricklayer Helper	Cu. Ft.	Cubic Foot		
Bric.	Bricklayer	cw	Continuous Wave	F.E.	Front End
Brk.	Brick	C.W.	Cool White	FEP	Fluorinated Ethylene Propylene (Teflon)
Brng.	Bearing	Cwt.	100 Pounds		
Brs.	Brass	C.W.X.	Cool White Deluxe	F.G.	Flat Grain
Brz.	Bronze	C.Y.	Cubic Yard (27 cubic feet)	F.H.A.	Federal Housing Administration
Bsn.	Basin	C.Y./Hr.	Cubic Yard per Hour	Fig.	Figure
Btr.	Better	Cyl.	Cylinder	Fin.	Finished
BTU	British Thermal Unit	d	Penny (nail size)	Fixt.	Fixture
BTUH	BTU per Hour	D	Deep; Depth; Discharge	Fl. Oz.	Fluid Ounces
BX	Interlocked Armored Cable	Dis.; Disch.	Discharge	Flr.	Floor
c	Conductivity			F.M.	Frequency Modulation; Factory Mutual
C	Hundred; Centigrade	Db.	Decibel		
		Dbl.	Double	Fmg.	Framing
C/C	Center to Center	DC	Direct Current	Fndtn.	Foundation
Cab.	Cabinet	Demob.	Demobilization	Fori.	Foreman, inside
Cair.	Air Tool Laborer	d.f.u.	Drainage Fixture Units	Foro.	Foreman, outside

Abbreviations

Fount.	Fountain	I.P.S.	Iron Pipe Size	M.C.M.	Thousand Circular Mils
FPM	Feet per Minute	I.P.T.	Iron Pipe Threaded	M.C.P.	Motor Circuit Protector
FPT	Female Pipe Thread	J	Joule	MD	Medium Duty
Fr.	Frame	J.I.C.	Joint Industrial Council	M.D.O.	Medium Density Overlaid
F.R.	Fire Rating	K.	Thousand; Thousand Pounds	Med.	Medium
FRK	Foil Reinforced Kraft	K.D.A.T.	Kiln Dried After Treatment	MF	Thousand Feet
FRP	Fiberglass Reinforced Plastic	kg	Kilogram	M.F.B.M.	Thousand Feet Board Measure
FS	Forged Steel	kG	Kilogauss	Mfg.	Manufacturing
FSC	Cast Body; Cast Switch Box	kgf	Kilogram force	Mfrs.	Manufacturers
Ft.	Foot; Feet	kHz	Kilohertz	mg	Milligram
Ftng.	Fitting	Kip.	1000 Pounds	MGD	Million Gallons per Day
Ftg.	Footing	KJ	Kiljoule	MGPH	Thousand Gallons per Hour
Ft. Lb.	Foot Pound	K.L.	Effective Length Factor	MH	Manhole; Metal Halide; Man Hour
Furn.	Furniture	Km	Kilometer	MHz	Megahertz
FVNR	Full Voltage Non Reversing	K.L.F.	Kips per Linear Foot	Mi.	Mile
FXM	Female by Male	K.S.F.	Kips per Square Foot	MI	Malleable Iron; Mineral Insulated
Fy.	Minimum Yield Stress of Steel	K.S.I.	Kips per Square Inch	mm	Millimeter
g	Gram	K.V.	Kilo Volt	Mill.	Millwright
G	Gauss	K.V.A.	Kilo Volt Ampere	Min.	Minimum
Ga.	Gauge	K.V.A.R.	Kilovar (Reactance)	Misc.	Miscellaneous
Gal.	Gallon	KW	Kilo Watt	ml	Milliliter
Gal./Min.	Gallon Per Minute	KWh	Kilowatt-hour	M.L.F.	Thousand Linear Feet
Galv.	Galvanized	L	Labor Only; Length; Long	Mo.	Month
Gen.	General	Lab.	Labor	Mobil.	Mobilization
Glaz.	Glazier	lat	Latitude	Mog.	Mogul Base
GPD	Gallons per Day	Lath.	Lather	MPH	Miles per Hour
GPH	Gallons per Hour	Lav.	Lavatory	MPT	Male Pipe Thread
GPM	Gallons per Minute	lb.; #	Pound	MRT	Mile Round Trip
GR	Grade	L.B.	Load Bearing; L Conduit Body	ms	millisecond
Gran.	Granular	L. & E.	Labor & Equipment	M.S.F.	Thousand Square Feet
Grnd.	Ground	lb./hr.	Pounds per Hour	Mstz.	Mosaic & Terrazzo Worker
H	High; High Strength Bar Joist;	lb./L.F.	Pounds per Linear Foot	M.S.Y.	Thousand Square Yards
	Henry	lbf/sq in.	Pound-force per Square Inch	Mtd.	Mounted
H.C.	High Capacity	L.C.L.	Less than Carload Lot	Mthe.	Mosaic & Terrazzo Helper
H.D.	Heavy Duty; High Density	Ld.	Load	Mtng.	Mounting
H.D.O.	High Density Overlaid	L.F.	Linear Foot	Mult.	Multi; Multiply
Hdr.	Header	Lg.	Long; Length; Large	MV	Megavolt
Hdwe.	Hardware	L. & H.	Light and Heat	MW	Megawatt
Help.	Helper average	L.H.	Long Span High Strength Bar Joist	MXM	Male by Male
HEPA	High Efficiency Particulate Air Filter	L.J.	Long Span Standard Strength	MYD	Thousand yards
Hg	Mercury		Bar Joist	N	Natural; North
H.O.	High Output	L.L.	Live Load	nA	nanoampere
Horiz.	Horizontal	L.L.D.	Lamp Lumen Depreciation	NA	Not Available; Not Applicable
H.P.	Horsepower; High Pressure	lm	Lumen	N.B.C.	National Building Code
H.P.F.	High Power Factor	lm/sf	Lumen per Square Foot	NC	Normally Closed
Hr.	Hour	lm/W	Lumen Per Watt	N.E.M.A.	National Electrical
Hrs./Day	Hours Per Day	L.O.A.	Length Over All		Manufacturers Association
HSC	High Short Circuit	log	Logarithm	NEHB	Bolted Circuit Breaker to 600V.
Ht.	Height	L.P.	Liquefied Petroleum;	N.L.B.	Non-Load-Bearing
Htg.	Heating		Low Pressure	nm	nanometer
Htrs.	Heaters	L.P.F.	Low Power Factor	No.	Number
HVAC	Heating, Ventilating &	Lt.	Light	NO	Normally Open
	Air Conditioning	Lt. Ga.	Light Gauge	N.O.C.	Not Otherwise Classified
Hvy.	Heavy	L.T.L.	Less than Truckload Lot	Nose.	Nosing
HW	Hot Water	Lt. Wt.	Lightweight	N.P.T.	National Pipe Thread
Hyd.;		L.V.	Low Voltage	NQOB	Bolted Circuit Breaker to 240V.
Hydr.	Hydraulic	M	Thousand; Material; Male;	N.R.C.	Noise Reduction Coefficient
Hz.	Hertz (cycles)		Medium Wall Copper	N.R.S.	Non Rising Stem
I.	Moment of Inertia	m/hr	Manhour	ns	nanosecond
I.C.	Interrupting Capacity	mA	Milliampere	nW	nanowatt
ID	Inside Diameter	Mach.	Machine	OB	Opposing Blade
I.D.	Inside Dimension;	Mag. Str.	Magnetic Starter	OC	On Center
	Identification	Maint.	Maintenance	OD	Outside Diameter
I.F.	Inside Frosted	Marb.	Marble Setter	O.D.	Outside Dimension
I.M.C.	Intermediate Metal Conduit	Mat.	Material	ODS	Overhead Distribution System
In.	Inch	Mat'l.	Material	O & P	Overhead and Profit
Incan.	Incandescent	Max.	Maximum	Oper.	Operator
Incl.	Included; Including	MBF	Thousand Board Feet	Opng.	Opening
Int.	Interior	MBH	Thousand BTU's per hr.	Orna.	Ornamental
Inst.	Installation	M.C.F.	Thousand Cubic Feet	O.S.&Y.	Outside Stem and Yoke
Insul.	Insulation	M.C.F.M.	Thousand Cubic Feet	Ovhd	Overhead
I.P.	Iron Pipe		per Minute	Oz.	Ounce

Abbreviations

P.	Pole; Applied Load; Projection	S.	Suction; Single Entrance; South	T.S.	Trigger Start
p.	Page			Tr.	Trade
Pape.	Paperhanger	Scaf.	Scaffold	Transf.	Transformer
PAR	Weatherproof Reflector	Sch.; Sched.	Schedule	Trhv.	Truck Driver, Heavy
Pc.	Piece	S.C.R.	Modular Brick	Trlr.	Trailer
P.C.	Portland Cement; Power Connector	S.D.R.	Standard Dimension Ratio	Trlt.	Truck Driver, Light
P.C.F.	Pounds per Cubic Foot	S.E.	Surfaced Edge	TV	Television
P.E.	Professional Engineer; Porcelain Enamel; Polyethylene; Plain End	S.E.R.; S.E.U.	Service Entrance Cable	T.W.	Thermoplastic Water Resistant Wire
		S.F.	Square Foot	UCI	Uniform Construction Index
Perf.	Perforated	S.F.C.A.	Square Foot Contact Area	UF	Underground Feeder
Ph.	Phase	S.F.G.	Square Foot of Ground	U.H.F.	Ultra High Frequency
P.I.	Pressure Injected	S.F. Hor.	Square Foot Horizontal	U.L.	Underwriters Laboratory
Pile.	Pile Driver	S.F.R.	Square Feet of Radiation	Unfin.	Unfinished
Pkg.	Package	S.F.Shlf.	Square Foot of Shelf	URD	Underground Residential Distribution
Pl.	Plate	S4S	Surface 4 Sides	V	Volt
Plah.	Plasterer Helper	Shee.	Sheet Metal Worker	VA	Volt/amp
Plas.	Plasterer	Sin	Sine	V.A.T.	Vinyl Asbestos Tile
Pluh.	Plumbers Helper	Skwk.	Skilled Worker	VAV	Variable Air Volume
Plum.	Plumber	SL	Saran Lined	Vent.	Ventilating
Ply.	Plywood	S.L.	Slimline	Vert.	Vertical
p.m.	Post Meridiem	Sldr.	Solder	V.G.	Vertical Grain
Pord.	Painter, Ordinary	S.N.	Solid Neutral	V.H.F.	Very High Frequency
pp	Pages	S.P.	Static Pressure; Single Pole; Self-Propelled	VHO	Very High Output
PP; PPL	Polypropylene			Vib.	Vibrating
P.P.M.	Parts Per Million	Spri.	Sprinkler Installer	V.L.F.	Vertical Linear Foot
Pr.	Pair	Sq.	Square; 100 square feet	Vol.	Volume
Prefab.	Prefabricated	S.P.D.T.	Single Pole, Double Throw	W	Wire; Watt; Wide; West
Prefin.	Prefinished	S.P.S.T.	Single Pole, Single Throw	w/	With
Prop.	Propelled	SPT	Standard Pipe Thread	W.C.	Water Column; Water Closet
PSF; psf	Pounds per Square Foot	Sq. Hd.	Square Head	W.F.	Wide Flange
PSI; psi	Pounds per Square Inch	S.S.	Single Strength; Stainless Steel	W.G.	Water Gauge
PSIG	Pounds per Square Inch Gauge	S.S.B.	Single Strength B Grade	Wldg.	Welding
PSP	Plastic Sewer Pipe	Sswk.	Structural Steel Worker	Wrck.	Wrecker
Pspr.	Painter, Spray	Sswl.	Structural Steel Welder	W.S.P.	Water, Steam, Petroleum
Psst.	Painter, Structural Steel	St.; Stl.	Steel	WT, Wt.	Weight
P.T.	Potential Transformer	S.T.C.	Sound Transmission Coefficient	WWF	Welded Wire Fabric
P. & T.	Pressure & Temperature	Std.	Standard	XFMR	Transformer
Ptd.	Painted	STP	Standard Temperature & Pressure	XHD	Extra Heavy Duty
Ptns.	Partitions	Stpi.	Steamfitter; Pipefitter	Y	Wye
Pu	Ultimate Load	Str.	Strength; Starter; Straight	yd	Yard
PVC	Polyvinyl Chloride	Strd.	Stranded	yr	Year
Pvmt.	Pavement	Struct.	Structural	Δ	Delta
Pwr.	Power	Sty.	Story	%	Percent
Q	Quantity Heat Flow	Subj.	Subject	~	Approximately
Quan.; Qty.	Quantity	Subs.	Subcontractors	Ø	Phase
Q.C.	Quick Coupling	Surf.	Surface	@	At
r	Radius of Gyration	Sw.	Switch	#	Pound; Number
R	Resistance	Swbd	Switchboard	<	Less Than
R.C.P.	Reinforced Concrete Pipe	S.Y.	Square Yard	>	Greater Than
Rect.	Rectangle	Syn.	Synthetic		
Reg.	Regular	Sys.	System		
Reinf.	Reinforced	t.	Thickness		
Req'd.	Required	T	Temperature; Ton		
Resi	Residential	Tan	Tangent		
Rgh.	Rough	T.C.	Terra Cotta		
R.H.W.	Rubber, Heat & Water Resistant; Residential Hot Water	T.D.	Temperature Difference		
		TFE	Tetrafluoroethylene (Teflon)		
rms	Root Mean Square	T. & G.	Tongue & Groove; Tar & Gravel		
Rnd.	Round				
Rodm.	Rodman	Th.; Thk.	Thick		
Rofc.	Roofer, Composition	Thn.	Thin		
Rofp.	Roofer, Precast	Thrded	Threaded		
Rohe	Roofer Helpers (Composition)	Tilf.	Tile Layer, Floor		
Rots.	Roofer, Tile & Slate	Tilh.	Tile Layer Helper		
R.O.W.	Right of Way	THW	Insulated Strand Wire		
RPM	Revolutions per Minute	THWN; THHN	Nylon Jacketed Wire		
R.R.	Direct Burial Feeder Conduit				
R.S.	Rapid Start	T.L.	Truckload		
RT	Round Trip	Tot.	Total		

417

INDEX

A

Abandon catch basin	19
Abbreviations	415-417
ABC extinguisher	274
Abrasive aggregate	73, 375
floor	77, 401
floor tile	194
nosing	108, 110
stair tread	79
terrazzo	195, 401
tile	193
tread	194
ABS DWV pipe	257
Absorption chiller	288
cold generator	288
testing	7
Accelerator set	74
sprinkler system	276
Access door basement	80
door ceiling	156
door duct	291
door floor	168
door roof	156
floor	245
panel	197
Accessory bathroom	193, 207
drainage	152
drywall	190
duct	291
formwork	59, 61
galvanized reinforcing	69
plaster	185
reinforcing	69
Accordion door	165, 214
Acid proof floor	86, 199
resistant pipe	256
Acoustical barrier	196
batt	198
block	92
booth	246
ceiling	196, 245
door	167, 214, 397
enclosure	239, 246
folding partition	214
glass	182
metal deck	104
panel	214-215
partition	214-215
phone booth	220
room	241, 245
sealant	135, 191
sprayed	189
underlayment	198
wall tile	196
wallboard	191
window wall	185
Acrylic carpet	198
caulking	136
ceiling	197
dome closure	404
floor	199
latex	73
sign	220
wall coating	206
wallcovering	206
wood block	202
Adapter fire-hose	274
Adhesive butyl	136
cement	201
floor	201
neoprene	137
PVC	137
roof	144
wallpaper	206, 401
Adjustable astragal	177
Adjustment factor	1, 329
Admixture	82
cement	73, 77
concrete	73, 366
masonry	82, 378
water reducing	74, 366
Adobe brick	84
Aeration plant	46
Aerial lift	11
photography	3
seeding	56
survey	5
Aggregate abrasive	73
concrete	72, 366
exposed	76
lightweight	72
marble	73
masonry	84, 378
panel	146
plaster	72, 400
roof	73, 146, 396
spreader	9
stone	56
testing	7
Air balancing	288
cleaner electronic	288
compressor	11
compressor dental	225
compressor mobilization	36
conditioner removal	23
conditioning	410
conditioning wine vault	234
control	286
curtain	240
distributing ceiling	197
entraining agent	73
extractor	291
handling fan	293
handling troffer	313
hose	12
lock	241
make-up unit	282
register	291
spade	12
supported building	6, 240, 404
supported enclosure	248, 404
supported storage tank cover	241
tool	11
tube system	252
vent roof	140
wall	215
Air-compressor control-system	290
Air-conditioner cooling & heating	296
gas heat	296
packaged terminal	296
receptacle wiring	319
rooftop	296
self-contained	296
thru-wall	296
Air-conditioning central	288
central station	289
computer	289
direct expansion	295
fan	293
fan-coil	295
ventilating	288, 290, 292-294, 296-297
Air-cooled belt drive condenser	289
Air-filter	288
roll type	288
Air-return grille	291
Air-supply pneumatic control	290
register	291
Air-vent automatic	286
Airfoil fan centrifugal	293
Airplane hangar door	168
hanger	244
Alarm burglar	316
exit control	316
sprinkler	317
standpipe	317
valve sprinkler	276
All fuel vent-chimney	285
Alley bowling	242
Altar	223
Alteration fee	1
Aluminum astragal	176
bench	51
blind	234
cable-tray	298
ceiling	196
ceiling tile	196
coat rack	236
column	98, 125
column base	106
conduit	298, 412
coping	85
cross	224
curtain wall	184
diffuser perforated	290
directory board	211
dome	243, 404
door	160, 168
door frame	160
downspout	150
duct	410
ductwork	292
expansion joint	112, 152
fence	51, 111
flagpole	212
flashing	144, 152, 266
foil	135, 138, 196, 306
framing	98, 399
grandstand	243
grating	107, 390
gravel stop	154
grille	291
gutter	154
handrail	246
ladder	108
louver	295
manhole cover	109
mansard	155
miscellaneous	106, 390
nail	115
operating louver	295
pipe	42-43
push pull plate	181
rail	109
register	291
reglet	155
rivet	115
roof	143
salvage	25
sash	172
screw	144
service entrance cable	303
shade	239
sheet metal	150, 410
shelter	247
shingle	141
shore	67
siding	143, 395
sign	219
stair tread	79, 110
steeple	224
storefront	171
storm door	160
structural	98
tile	141, 194
transom	160
trench cover	111
tube frame	184
tubular	99
wall	147
weatherstripping	176
weld rod	103
window	172, 399
window demolition	26
wire	111, 301
Ammeter	309
Analysis petrographic	7
sieve	7-8
Anchor bolt	59, 82, 114
buck	84
bumper	225
expansion	113
hollow wall	113
joist	116
machinery	114
masonry	59, 84
nailing	113
screw	69, 113
stone	84
wall	112
Anechoic chamber	241
Angle corner	106
framing	99
valve	262, 264-265
Antenna system	318
Anti-siphon water	262
Apartment call system	318
S.F. & C.F.	320, 413
Appliance	221
plumbing	272-273
receptacle	319
residential	221, 228
wiring	319
Application water cooler	408
Approach ramp	246
Apron wood	127
Arch laminated	132
radial	132
Architectural equipment	221
fee	1, 348
panel	147
Area wall	106
Areaway grating	107
Arena sport S.F. & C.F.	327
Ark	223
Armored cable	301-302, 412
Arrow	49
Artificial turf	55
Asbestos barrier	196
base sheet	145
cement partition	214
cement pipe	41, 43-44
control	20
encapsulation	21
felt	143, 145-146
formboard	81
removal	20
shingle	141, 395
waterproofing	137
Asbestos-cement duct	315
Ash conveyor	234
receiver	208, 234
urn	234
Ashlar stone	96
veneer	95, 382
Asphalt binder	47
block	47
block floor	47
coating	135
curb	48
cutting	19
distributor	12
expansion joint	61
felt	144, 146, 396
flashing	152
flood coat	144
floor tile	200
lined pipe	41
panels	144
paper	135
pavement	47, 55, 357
paver	13
primer	135, 201
roof	396
sheathing	123
shingle	142
sidewalk	49, 357
testing	7
Aspirator dental	225
Astragal adjustable	177
aluminum	176
magnetic	176-177
molding	127
overlapping	177
rubber	176-177
split	177
steel	176
Astronomy observation dome	242
Athletic equipment	51, 231
locker	213
pole	53
Atomizing humidifier	280
Attendant booth	230
Attic stair	211
Audio masking	240
Audiometric room	242
Auditorium chair	239
S.F. & C.F.	321, 413
Auger	9
boring	8
hole	16
Auto park drain	253
Automatic air-vent	286
fire suppression	273
gate	230
opener	177
operator	177
teller	223
washing machine	222
Automotive equipment	222
lift	222
sales S.F. & C.F.	321
service S.F. & C.F.	413
Autopsy equipment	228
Average S.F. cost	413
Awning	209
window	172, 174
Axial flow fan	293

B

Back splash	235
Back-up block	89
Backer board	137
rod	61, 135
Backfill	26, 30, 353
trench	30
Backflow preventer	253
Backhoe	28, 353
excavation	29, 353
rental	9, 353
Backsplash counter top	125
Backstop baseball	51
basketball	51, 231
electric	231
Back-up control sewer	254
Backwater valve	254
Baffle roof	140
Baked enamel doors	159
enamel frame	158
Balanced door	171
Balancing air	288
water	289
Balcony fire escape	107
rail	109
Baler	234
Ball valve	262, 264-265
wrecking	14
Ballast railroad	58, 358
roof	149
Balled & burlapped tree	57, 358
Baluster	131
stair	392
Balustrades painting	204
Band molding	127
riser	232
Banding grating	108
iron	59
Bank equipment	222-223
S.F. & C.F.	321, 413
window	222
Bankrun gravel	26, 72
Banquette booth	235
Baptistry	223
Bar bell	226
chair	69
front	225
grab	207
joist	105, 387-388
joist painting	105
panic	180-181
parallel	229
prestressed	70
reinforcing	69, 362
restaurant	238
tie	69
touch	181
towel	208
tub	207
zee	190

418

INDEX

Barbed wire 111
- wire fence 51

Barber equipment 223
Barrel 12
Barricade 6, 12, 49
Barrier acoustical 196
- impact 49
- median 49
- moisture 135
- parking 49, 230
- plenum 196
- waterstop 61

Base cabinet 124, 133
- ceramic tile 193
- column 98, 102, 106, 116
- concrete 74, 374
- course 47
- cove 200
- flagpole 212
- light 67
- masonry 93
- metal 186, 194
- plate column 102
- quarry tile 194
- resilient 200
- road 47
- sheet 145-146
- sign 67
- sink 124
- stone 96
- terrazzo 195, 401
- vanity 132
- wood 127

Baseball backstop 51
- scoreboard 232

Baseboard demolition 24
- heater electric 317
- heating 280

Basement stair 80
Basket locker 213
Basketball backstop 51, 231
- scoreboard 232

Batch trial 7
Bath 266
- exhaust fan 294
- faucet 254
- heater & fan 294
- paraffin 229
- soaking 267
- steam 232
- tub removal 24
- whirlpool 229, 267

Bathroom accessory 193, 207
- fixture 266, 268, 270-271

Bathtub 266, 408
- bar 207
- enclosure 216

Batt acoustical 198
- insulation 141
- thermal 198

Battered brick 88
Battery light 313
Bay window 174
Bead board 130 140
- board insulation 130, 240
- casing 190
- corner 186, 190
- parting 128

Beam bolster 69
- bond 83, 90, 379
- bondcrete 189
- bottom 62
- box 123
- castellated 102
- ceiling 124
- concrete 74-75, 78, 366-368, 372
- drywall 190
- fireproofing 189
- formwork 61, 360
- grade 65, 78
- hanger 60, 116
- laminated 132
- mantel 126
- plaster 187, 189, 400
- precast 80, 377
- reinforcing 70, 362
- side 62
- soldier 33
- spandrel 61
- steel 102, 386, 388
- tee 80, 377
- test 7
- wide flange 101, 102
- wood 117, 120

Bearing pad 103
Bed molding 127
- plant 56

Beech tread 131
Bell & spigot pipe 255
- bar 226
- system 317

Bellow expansion joint pipe 286
Bells 223
Belt material handling 251
Bench aluminum 51
- fiberglass 51
- folding 231
- greenhouse 244
- locker 213
- park 51
- planter 53
- players 51
- wood 51
- work 233

Bentonite 34, 135
Berm paving 48
Bevel siding 130
Beveled glass 182
Bi-passing closet doors 165
Bicycle rack 53
- trainer 226, 231

Bifolding door 160, 165
Bin part 218
Binder asphalt 47
Birch door 165-166, 395, 397
- molding 128
- paneling 129
- stair 131
- wood frame 161

Bituminous block 47
- coating 135
- concrete 48, 357
- curb 48
- expansion joint 61
- fiber pipe 41, 43
- lined pipe 41
- pavement 47, 55, 357
- paver 13
- paving 47
- sidewalk 49, 357

Blackboard 209, 216
Blank leaf door 164
Blanket curing 76
- insulation 140, 281
- sound attenuation 198

Blasting 28
- cap 28
- mat 28

Bleacher 243
- outdoor 51
- telescoping 231

Blind exterior 129
- venetian 234
- window 234

Block acoustical 92
- asphalt 47
- asphalt floor 47
- back-up 89
- bituminous 47
- cap 93
- chimney 93
- concrete 89-92, 379
- concrete bond beam 90
- concrete foundation 91
- corner 93
- decorative concrete 90-91
- filler 205
- floor 202
- floor wood 203
- glass 93, 382
- glazed 93
- grooved 90
- ground face 90
- high strength 91
- insulation 94
- interlocking 91
- lead lined 92
- lightweight 379
- lintel 92
- manhole 39
- masonry 379
- partition 92
- patio 97
- pilasters column 93
- planter 53
- profile 91
- reflective 93
- scored 91
- sill 93
- slotted 92
- slump 90
- solar screen 93
- stud 92
- wall grout 384

Blocking 117
- wood 117, 121

Blockout slab 65
Blood pressure unit 228
Bloodbank refrigeration 229
Blower pneumatic tube 252
Blown-in cellulose 137
- fiberglass 138
- insulation 137

Blueboard 191
- partition 192

Bluestone 95
- sidewalk 50
- sill 89
- step 50

Board 394
- & batten fence 53
- & batten siding 130, 394
- backer 137
- bead 139
- bulletin 208
- ceiling 196-197
- control 210, 232
- cork 208
- directory 210
- dock 225
- drain 135
- insulation 140
- paneling 128
- ridge 119
- sheathing 122
- valance 125
- verge 127
- wood 393

Boat dock 58
Boiler demolition 23
- electric 276-277
- electric steam 276
- feed pump 269
- gas-fired 277
- gas/oil combination 278
- general 276
- hot-water 277-278
- insulation 281
- large control-system 290
- mobilization 36
- oil-fired 277
- steam 277-278

Bolster beam 69
- slab 69

Bolt 69, 112
- anchor 59, 82, 114
- dead 179
- door 177
- expansion 113
- eye 69
- flush 177
- roof 58
- steel 116
- structural steel 102, 388
- toggle 113
- wedge 114

Bolt-on circuit breaker 307
Bond beam 83, 90, 379
- performance 3, 342
- roof 146

Bondcrete 189
Bonding agent 73
- surface 94

Bookcase 129, 133
Bookshelf 228
Boom lift 11
Booster fan 293
Boot pile 38
Booth acoustical 246
- attendant 230
- banquette 235
- fixed 235
- mounted 235
- painting 222
- parking 230
- portable 246
- telephone 220
- ticket 246

Border light 232
Boring 16
- auger 8
- horizontal 32

Borosilicate pipe 257
Borrow 26, 48
Bottle storage 234
Bottom beam 62
Boulder excavation 28
Bow window 174
Bowling alley 242
Bowstring truss 132
Box beam 123
- distribution 46
- electrical 304
- explosionproof 304
- locker 213
- mail 213
- out 65
- out opening 64
- pull 303
- safety deposit 223
- stair 131
- termination 318
- vent 84

Boxes & wiring devices 303-304
Brace cross 99
- formwork 60

Bracing let-in 117
Braided bronze hose 286
Branch circuit fusible switch 308
Brass hinge 178-179
- pipe 255
- salvage 25
- screw 117
- valve 262

Brazed wire to grounding 303
Break glass station 317
Breaker circuit 305
- vacuum 262

Breeching draft inducer 280
- insulation 281

Brick 85-86, 88-89, 380-381
- adobe 84
- anchor 60, 84
- battered 88
- cart 12
- catch basin 39
- chimney 242, 381, 404
- chimney simulated 211
- demolition 22-23
- edging 56
- face 85
- floor 49
- forklift 12
- in place 381
- masonry 379, 381
- molding 127
- needle 329
- paving 48
- reinforced 89
- removal 19
- repoint 330
- sawing 87
- shelf 65
- sidewalk 49
- sill 89
- simulated 87
- soldier 381
- stair 88
- step 50
- testing 8
- veneer 88, 379
- veneer demolition 24
- vent 295
- vent box 84
- wall 89, 381
- wash 89

Bridge 58
- concrete 58
- pedestrian 58
- sidewalk 4

Bridging 105, 117
Broiler 226
Bronze body strainer 287
- cross 224
- expansion joint 112
- letter 219
- plaque 220
- push-pull plate 181
- stair tread 110
- valve 262

Broom cabinet 125
- finish 48, 76

Brown coat 189
Brownstone 97
Brush clearing 16
- cutter 9

Bubbler 267
- drinking 267

Buck anchor 84
- rough 118

Bucket concrete 11
- crane 9

Buff face brick 379
Buffalo roadway box 47
Buggy concrete 11, 79
Builder risk insurance 1, 342
Building air supported 6, 240
- demolition 17
- directory 210-211, 318
- greenhouse 243
- gutting 18
- hangar 244
- hardware 176
- model 3
- move 19
- paper 135
- permit 3
- portable 6
- pre-engineered 387
- prefabricated 243
- sprinkler 276
- steel 100
- temporary 6
- tension 250

Built-in range 221
- shower 270

Built-up roof 144-145, 396
Bulb tee 82, 104, 388
Bulk storage dome 243, 404
Bulkhead door 167
- formwork 64-65, 359

Bull winch clamp 59
Bulldozer 10, 26, 28, 31
- mobilization 32

Bullet-resistant window 223
Bulletin board 208
Bulletproof glass 182
Bullnose block 91
- trim 193

Bumper car 58, 106

INDEX

corner	106
dock	117, 225
door	107, 177
parking	49
rail	106
railroad	58
wall	178, 206
Burglar alarm indicating-panel	316
Burlap curing	76
rubbing	77
Bus terminal S.F. & C.F.	328
Bus-duct cable tap box	310
copper	310
fitting	310
plug-in	310-311
switch	311
tap box	310
Bush hammer concrete	77, 375
Butt splice	71
weld	71
Butterfly valve	45, 264
Buttress formwork	66
Butyl adhesive	136
caulking	136
expansion joint	152
flashing	153
waterproofing	136

C

Cabinet base	133
broom	125
casework	133
current transformer	304
demolition	24
door	133
electrical	303
electrical hinged	304
fire equipment	273
fuse	306
hardware	134
hinge	134
hose rack	273
hotel	214
key	213
kitchen	124, 236
laboratory	227
medicine	213
metal	236
oven	125
painting	203
school	236
shower	216
stain	203
storage	227
strip	318
transformer	304
valve	274
varnish	203
Cable armored	301-302
control	302
copper	302
copper shielded	302
electric	301-302, 412
guardrail	49
heating	317
jack	15
prestressed	70, 365
PVC jacket	302
sheathed nonmetallic	303
sheathed romex	303
shielded	302
tap box bus-duct	310
wire rope	111
Cable-tray aluminum	298
ladder type	298
Cafe door	163
Caisson	35, 354
bell	36, 354
displacement	39, 354
foundation	35, 354
Calcium aluminate cement	73
chloride	48, 73
Call system apartment	318
system emergency	318
system nurse	318
Camera TV closed circuit	319
Canopy	208
entrance	100, 208
framing	98
wood	120
Cant roof	119, 146
Cantilever retaining wall	54
Canvas awning	209
Cap blasting	28
block	93
column	98
pile	38, 64, 76, 78
Capacitor indoor	311
Capital column	63
Car bumper	49, 58, 106
motorized	252
Carbon black	73
dioxide extinguisher	273-274
Carborundum	77
Card catalog file	228

station	230
Carillon	223
Carousel compactor	233
telephone	220
Carpentry finish	124
rough	391-394
Carpet	198
computer room	245
floor	198
maintenance	199
padding	199
removal	22
tile	199
Carport	209
Carrel	228
Carrier ceiling	186
channel	197
fixture	260
Cart brick	12
concrete	11, 79, 367
mounted extinguisher	274
Carving stone	96
Case display	211
exhibit	211
refrigerated	230
work	133
Cased boring	16
Casement window	172-174
Casework cabinet	133
demolition	24
ground	121
painting	203
varnish	203
Cash register	224
Casing bead	190
door	190
metal	185, 396
wood	127, 395, 397
Cast in place concrete	72, 74-75, 365-368, 371-372, 374
in place pile	36
in place terrazzo	195
iron bench	51
iron casting	106
iron damper	82
iron manhole cover	40, 109
iron miscellaneous	106
iron pull box	304
iron stair	110
iron stair tread	79, 110
iron weld rod	103
steel valve	265
stone floor	86
trim lock	180
Cast-iron drain	253
pipe	255
radiator	280
trap	261
Castellated beam	102
Casting construction	106
Catch basin	39
basin frame and cover	40
basin masonry	39
basin removal	19
door	134
Catwalk	4, 6
Caulking	136
masonry	329
sealant	135
Cavity wall	83
wall anchor	60
wall brick	88
wall insulation	138
wall reinforcing	83
Cedar closet	129
fence	53
paneling	128
post	125
roof deck	133
roof plank	122
shingle	143
siding	130
stair	131
Ceiling access panel	197
acoustical	196
beam	197
board	196-197
bondcrete	189
carrier	186
demolition	21
diffuser	290
dome	156
drill	112
drywall	190, 192
eggcrate	197
expansion joint	113
fan	293
framing	118
furring	121, 186
hatch	156
heater	221
insert	60
insulation	137
integrated	245
lath	187-188, 399

luminous	197, 314
molding	127
painting	205, 402
plaster	187-188, 399
polystyrene	246
radiant	245
register	291
stair	211
stressed skin	101, 123
suspended	186, 192, 197, 245
suspension system	197
tile	198
wash	330
woven wire	218
Cell prison	225
Cellar door	167
Cellular concrete	75
deck concrete	80, 377
fill	82
metal deck	104
Cellulose blown-in	137
insulation	138
Cement adhesive	201
admixture	73, 366
color	73, 82, 378
flashing	135
grout	31, 77, 83, 378
gunite	77
gypsum	82
keene	187
lined pipe	41
masonry	73, 82, 378
masonry unit	90-92
mortar	194
parging	136
plaster	246, 401
portland	73, 366, 378
testing	7
vermiculite	189
Cementitious deck	80-81, 377
plank	82
waterproofing	136
Center bulb waterstop	68
Central air conditioning	410
air-conditioning	288
station air handling unit	289
Centrifugal airfoil fan	293
fan	293
humidifier	280
liquid chiller	297
pump	13
Ceramic coating	206
tile	193-194
tile counter top	126
tile demolition	22
tile floor	193
Certification welding	8
Chain core rope	239
hoist	15, 251
hoist door	168
link fence	6, 111
link fence removal	19
saw	13
trencher	10
Chair	239
bar	69
barber	223
dental	225
high	69
hydraulic	225
joist	69
library	228
life guard	248
locker bench	213
molding	127
movie	230, 239
portable	232
rail demolition	24
reinforcing	69
seating	239
subgrade	70
Chalk tray	209
Chalkboard	209
electric	209
frame	209
metal	216
portable	210
Chamber anechoic	241
echo	239
Chamfer strip	59
Channel carrier	197
frame	107
framing	99
furring	186, 190
pipe	44
siding	130, 394
slab concrete	81, 377
slotted	99
steel	185-186
Charge dump	18
Charging desk	228
Check floor	178
swing valve	262, 264
valve	265-266, 276
Checkered plate	106, 111

plate nosing	108
Checkout scanning	224
supermarket	224
Chemical cleaning	329
dry extinguisher	274
grout	31
toilet	13
water-closet	46
Chest	237
Chilled water coil	289
Chiller absorption	288
portable	234
water	288, 297
Chimney	242
accessory	211
block	93
brick	85, 381, 404
concrete	242, 404
demolition	23
flue	87, 381
foundation	74, 404
metal	211, 285
radial brick	404
screen	82
simulated brick	211
vent	285
China cabinet	124
Chipper brush	9
Chipping hammer	12
Chlorination system	248
Church equipment	223
pew	223
S.F. & C.F.	321, 413
Chute	210
linen	210
mail	210
package	210
refuse	210
rubbish	19
Cinema equipment	229
Circuit breaker bolt-on	307
breaker plug-in	311
electric dryer	319
wiring heater	319
wiring water-heater	320
Circuit-breaker	305
feeder section	308
Circular saw	13
Circulating pump	269
Circulator air solar-energy	283
solar-energy-system	283
City hall S.F. & C.F.	328
Clamp column	59
rod	61
water pipe ground	303
Clamshell	27-28, 353
bucket	9, 353
Clapboard painting	205
wood	394
Classroom heating & cooling	279
seating	239
Clay fire	86
pipe	43-44
roofing tile	142
tennis court	55
tile coping	85
tile roof repair	331
Clean floor	330
steam	88
Cleaner steam	13
Cleaning brick	89
masonry	85, 87, 329, 382
up	1
Cleanout door	82
pipe	253
tee	253
Cleanup interior	330
Clear & grub	16
pine	394
Clearing brush	16
Climbing crane	14
jack	15
Clinic S.F. & C.F.	325
Clinton cloth	189
Clip flange	69
insulation	138
joist	186
plywood	116
sleeper	70
stud	186
tie	69
Clock	316
& bell system	316
system	316
Closed circuit TV	223
deck grandstand	243
Closer concealed	178
door	178
electronic	178
floor	178
holder	178
Closet	213
cedar	129
door	160, 165
pole	127

420

INDEX

rod ... 129
Closure trash ... 55
Cloth clinton ... 189
  hardware ... 111
Clothes dryer commercial ... 228
  dryer residential ... 221
Club country S.F. & C.F. ... 321
  S.F. & C.F. ... 321, 413
Coal tar pitch ... 145
Coat brown ... 189
  glaze ... 205
  hook ... 207
  rack ... 133, 236
  scratch ... 189
  scrub ... 330
Coated reinforcing ... 70
Coating bituminous ... 135
  ceramic ... 206
  epoxy ... 77, 206
  flood ... 147
  metallic ... 137
  roof ... 135
  rubber ... 137
  silicone ... 137
  wall ... 206
  waterproofing ... 135
Coffee maker ... 207
  urn ... 226
Cofferdam excavation ... 29
Cohesion testing ... 7
Coil cooling ... 289
  flanged ... 289
Coin operated gate ... 221
  station ... 230
Cold generator absorption ... 288
  roofing ... 145
  storage door ... 167
Collar friction ... 59
Collector solar ... 409
  solar-energy ... 282
  solar-energy-system ... 283
College S.F. & C.F. ... 322, 413
Colonial door ... 165-166
  wood frame ... 161
Color concrete ... 48, 74
  floor ... 77
  wheel ... 232
Column ... 125
  aluminum ... 98
  base ... 98, 106, 116
  base plate ... 102
  block pilasters ... 93
  bondcrete ... 189
  brick ... 85
  cap ... 98
  capital ... 63
  clamp ... 59
  concrete ... 74, 78, 80-81, 365-368, 372
  demolition ... 18, 23
  drywall ... 190
  fireproof ... 99
  fireproofing ... 189
  form ... 59
  formwork ... 62, 360
  lally ... 98
  laminated wood ... 133
  lath ... 187
  pipe ... 98
  plaster ... 187, 189
  precast ... 80
  reinforcing ... 71, 363
  removal ... 23
  residential ... 98
  splice ... 71
  steel ... 98, 386, 388
  stone ... 96
  wood ... 119-120, 125
Combination storm door ... 165-166
Comfort station ... 242
Command dog ... 8
Commercial door ... 158-159, 164, 168, 396-397
  folding partition ... 214
  greenhouse ... 244
  refrigeration ... 246
  water-heater ... 272-273
Common brick ... 85, 380-381
  nail ... 115
Communicating lockset ... 179-180
Communication system ... 318
Community center
  S.F. & C.F. ... 322, 413
Compact fill ... 31
Compaction ... 27, 353
  soil ... 26, 353, 357
  test proctor ... 8, 353
Compactor ... 233
  earth ... 9, 353
  residential ... 221
  waste ... 233
Compensation worker ... 1
Component sound system ... 318
  sprinkler system ... 276
Composite door ... 159

joist ... 105
  metal deck ... 105
  railing ... 109
Composition floor ... 199, 203
  flooring removal ... 22
Compressive strength ... 7
Compressor air ... 11
  mobilization air ... 36
  tank mounted ... 290
Computer air-conditioning ... 289
  floor ... 245
  room carpet ... 245
Concealed closer ... 178
Concrete admixture ... 73, 366
  aggregate ... 72, 366-367
  base ... 74, 374
  beam ... 74-75, 80, 372
  bituminous ... 48
  block ... 89-92, 379
  block bond beam ... 90
  block decorative ... 90-91
  block demolition ... 21
  block foundation ... 91
  block insulation ... 138
  block planter ... 53
  bridge ... 58
  bucket ... 11
  buggy ... 79, 367
  bush hammer ... 77
  caisson ... 35
  cart ... 11, 79
  cast in place ... 72, 74, 365-367, 376
  catch basin ... 39
  cellular ... 75, 81
  cellular deck ... 80
  channel slab ... 81, 377
  chimney ... 242, 404
  color ... 48, 74
  column ... 74, 81, 365-368, 372
  conveyer ... 11
  coping ... 85
  cribbing ... 54
  curb ... 48, 75
  curing ... 48, 73, 76
  curtain wall ... 81, 375-377
  cutout ... 21
  cutting ... 19-20
  cylinder ... 7
  deck ... 377
  demolition ... 17-18, 21
  drill ... 17
  encasement ... 72, 74
  facing panel ... 81, 375-377
  field mix ... 74
  filled column ... 74
  finish ... 48, 76-77, 376
  float ... 11
  floor ... 73, 365-368, 372, 374
  footing ... 75, 78, 365-368, 372
  formwork ... 59, 61, 359-360, 362-363, 368, 372
  foundation ... 75, 365-368, 372, 404
  furring ... 120
  gravel ... 73
  grout ... 83
  hand hole ... 315
  hardener ... 73, 77
  insulation ... 76
  joist ... 75, 80
  lightweight ... 74-76, 81, 366, 374
  lintel ... 83
  manhole ... 39, 315
  mix design ... 7
  mixer ... 11
  nail ... 115
  painting ... 205
  panel ... 81
  patching ... 76
  paver ... 13
  paving ... 48
  perlite ... 72, 374
  pier ... 75
  pile ... 36
  pipe ... 41, 351
  pipe removal ... 20
  placing ... 78, 367
  plank ... 81
  planter ... 53
  pool ... 403
  post ... 49
  post-tensioned ... 70, 79, 365
  poured ... 72
  precast ... 80, 375, 377
  prestressed ... 70, 80, 365, 377
  protection ... 79
  pump ... 11, 78, 366-367
  ready mix ... 74, 366
  reinforcing ... 69-70, 363-364
  removal ... 20
  retaining wall ... 54, 365-368, 372
  roof ... 80, 377
  roof deck ... 80, 377
  sand ... 72
  sandblasting ... 77
  saw ... 11

sealer ... 74
  septic tank ... 45
  shingle ... 142
  sidewalk ... 49
  siding ... 81
  sill ... 89
  silo ... 247
  slab ... 75
  spreader ... 13
  stair ... 76, 80
  stair tread ... 79
  substructure ... 74
  superstructure ... 74
  tank ... 249
  testing ... 7
  tie ... 58
  topping ... 76
  trowel ... 11
  truck ... 11
  utility vault ... 46
  vibrator ... 11
  wall ... 75
  wheeling ... 79
Condensate pump ... 269
  steam meter ... 287
Condenser air-cooled belt drive ... 289
Conductive floor ... 193, 195, 199-200
Conductor & grounding ... 301-303
  wire ... 301
Conduit & fitting flexible ... 306
  aluminum ... 298
  electric ... 412
  electrical ... 298
  high-installation ... 298
  in-slab ... 299
  in-slab PVC ... 299
  in-trench electrical ... 299
  in-trench steel ... 299
  intermediate steel ... 298
  raceway ... 298
  rigid in-slab ... 299
  rigid steel ... 298
Cone traffic ... 6
Confessional ... 223
Connection motor ... 306
  standpipe ... 275
Connector flexible ... 253
  joist ... 116
  shear ... 116
  stud ... 115
  timber ... 116
  water copper tubing ... 253
Consolidation test ... 8
Construction cost index ... 1, 333-341
  management ... 1
  shelter ... 404
  time ... 1, 348
Contingencies ... 1, 414
Continuous hinge ... 179
Contractor equipment ... 9, 350
  overhead ... 1, 4, 342-351
  pump ... 13
  scale ... 219
Control & motor-starter ... 306
  air ... 286
  board ... 210, 232
  cable ... 302
  hand/off/automatic ... 306
  joint ... 82
  radiator supply ... 286
  station ... 306
  system electronic ... 290
  tower ... 242
  valve heating ... 286
Control-station stop/start ... 306
Control-system air-compressor ... 290
  boiler large ... 290
  cooling ... 290
  split-system ... 290
  ventilator ... 290
Controller solar-energy-system ... 283
  time system ... 316
Convector cover ... 125
  heating unit ... 280
Convent S.F. & C.F. ... 413
Conveying system ... 250, 405, 407
Conveyor ... 11
  ash ... 234
  door ... 240
  material handling ... 251
  vertical ... 252
Cooking equipment ... 221, 226
  range ... 221
Cooler ... 246
  beverage ... 226
  door ... 240
  water ... 272
Cooling & heating
  air-conditioner ... 296
  coil ... 289
  control-system ... 290
  tower ... 290
  tower louver ... 295
Coping ... 85, 96
  clay tile ... 85

removal ... 24
  terra cotta ... 85, 94
Copper bus-duct ... 310
  cable ... 301-302
  downspout ... 151
  drum trap ... 261
  DWV tubing ... 256
  flashing ... 152, 266
  gravel stop ... 154
  gutter ... 154
  pipe ... 44, 256
  reglet ... 155
  rivet ... 115
  roof ... 150
  salvage ... 25
  sheet metal ... 410
  shielded cable ... 302
  tubing ... 44, 408
  wall covering ... 206
  wire ... 301
Corbel ... 88
  formwork ... 66
Core drill ... 7, 11, 17
  testing ... 7
Coreboard ... 191
Cork board ... 208
  bulletin board ... 208
  expansion joint ... 61
  floor ... 200
  insulation ... 246
  tile ... 200
  wall tile ... 206
Corner base cabinet ... 124
  bead ... 186, 190
  block ... 93
  guard ... 106, 203
  post ... 52
  protection ... 106
  wall cabinet ... 125
Cornice drain ... 253
  molding ... 127
  painting ... 204
  stone ... 96
Correspondence lift ... 250
Corrosion-resistant fan ... 293
  pipe ... 256
  plastic pipe ... 256
Corrugated metal pipe ... 42
  pipe ... 43
  roof tile ... 142
  siding ... 143, 147-149
Cost control ... 5
  mark up ... 3
  plumbing average ... 253
Cot prison ... 225
Counter bank ... 222
  checkout ... 224
  door ... 167
  flashing ... 155
  hospital ... 235
  top ... 125-126, 134, 227
  top sink ... 271
  window ... 213
Countertop demolition ... 24
Country club S.F. & C.F. ... 321, 413
Course base ... 47
  sand ... 48
Court air supported ... 241
  fence tennis ... 52
  handball ... 51, 248
  paddle tennis ... 55
  racquetball ... 248
  squash ... 51, 247
  tennis ... 51, 55
Courthouse S.F. & C.F. ... 322, 413
Cove base ... 200
  base ceramic tile ... 193
  base terrazzo ... 195
  molding ... 127
  scotia molding ... 127
Cover convector ... 125
  ground ... 56
  manhole ... 40
  pool ... 248
  stadium ... 241
  stair tread ... 7
  tank ... 241
  trench ... 111
  walkway ... 209
  weather protection ... 8
Covering decorative ... 107
  wall ... 206
CPE roof ... 148
CPVC pipe ... 257
  valve ... 265
Crane ... 14, 354-355, 357
  bucket ... 9
  climbing ... 14
  construction ... 350
  crawler ... 14
  hydraulic ... 14
  material handling ... 14, 354
  mobilization ... 36
  rail ... 107
  scale ... 219

421

INDEX

tower 14
Crawler crane 354-355, 357
Crematory 233
Creosote 402
   floor 203
   lumber 121
Crew 1
   cost 1
   survey 5
Cribbing concrete 54
Critical path scheduling . 5
Cross brace 99
   wall 224
Cross-arm pole 315
Crossing pedestrian ... 50
Crown molding 127
Crushed stone sidewalk . 50
Cubic foot building cost . 320, 392-393
Cubicle 214
   curtain 214
   detention 225
   shower 270
   toilet 216
Culvert formwork 63, 360
   pipe 41-42
Cupola 126
Cupric oxychloride floor . 199
Curb and gutter monolithic . 48
   asphalt 48
   bituminous 48
   concrete 48, 75
   edging 48, 107
   form 363
   formwork 64-65
   granite 48, 96
   highway 96
   inlet 48
   precast 48
   removal 20
   roof 119
   sealing 49
   terrazzo 195, 401
Curing blanket 76
   concrete 48, 73, 76
   paper 135
Current transformer ... 309
   transformer cabinet . 304
Curtain air 240
   cubicle 214
   divider 232
   rod 207
   sound 239
   stage 232
   track 232
   type fire-damper .. 292
   wall 395
   wall concrete 81, 375, 377
   wall glass 184, 399
   wall metal 147, 395
Curved stair 132
Cut & patch 329
   & point 330
   stone trim 97
Cutoff pile 38
Cutout 21
   counter 126
   interior 329
Cutter brush 9
Cutting asphalt 19
   block 126
   concrete 19-20
   masonry 19
   oxygen lance 20
   saw 19
   torch 13, 20
Cylinder concrete 7
   lockset 179
   recore 179-180

D

Dacron coated fabric structure . 404
Dairy case 230
Damper 292
   fireplace 82
   foundation vent ... 84
   multi-blade 292
Dampproofing 135
Darby finish 76
Darkroom 242
   door 167
   equipment 224
   revolving 242
Dasher hockey 244
Data safe 231
Day gate 171
Dead bolt 179
Deadbolt lock 179
Deadlocking latch 179
Deck cementitious 80-81, 377
   concrete cellular . 80
   drain 253
   edge form 105
   formwork 360
   gypsum 81, 396
   metal 104

roof 105, 122-123
steel 104
traffic 146
wood 122, 133
Decorative block 90
   covering 107
Deep freeze 221
   seal trap 261
   therapy room 247
   well pump 270
Dehumidifier 221
Delicatessen case 230
Deluge valve assembly sprinkler . 276
Demolition 23-25
   baseboard 24
   boiler 23
   brick 22-23
   building 17
   cabinet 24
   casework 24
   ceiling 21
   chimney 23
   column 23
   concrete 21
   countertop 24
   door 22
   drywall 21
   ductwork 23
   explosive 18
   exterior 17
   fireplace 24
   flooring 22
   framing 23
   girder 18
   glass 26
   granite 24
   gutter 25
   hammer 9
   house 17
   HVAC 23
   interior 414
   joist 23
   lath 21
   masonry 21, 23, 25
   millwork 24
   paneling 24
   partition 25
   plaster 21, 25
   plumbing 24
   plywood 21, 25
   post 23
   rafter 23
   railing 24
   roofing 25
   siding 25
   terrazzo 22
   tile 21
   trim 24
   wall 25
   window 26
   wood 21
Dental equipment 225
   fountain 267
   fountain rough-in . 267
   metal interceptor . 268
Department store S.F. & C.F. . 322, 413
Depository night 223
Derail 58
Derrick 15
   guyed 15
   stiffleg 15
Desk 237, 239
   top 237
Detection system 316
Detector infra-red 317
   motion 317
   smoke 317
   temperature rise .. 317
   ultrasonic motion . 317
   wiring smoke 319
Detention cubicle 225
   equipment 225
Developing tank 224
Device exit 180
   wiring 305
Devices & boxes wiring . 304
Dewater 15, 34, 356
   pump 270
   pumping 27, 35, 356
Diamond lath 187, 399
Diaphragm pump 13
Diesel fire pump 275
   hammer 9
   operated generator . 312
Diffuser ceiling 290
   opposed blade damper . 290
   perforated aluminum . 290
   rectangular 290
   steel 291
   T-bar mount 291
Dimension lumber 393
Dimensionaire ceiling . 245
Dimmer switch 305
Direct expansion air-conditioning . 295

Directional sign 50
Directory board 210
   building 210-211, 318
   shelf 220
Disappearing stair 211
Discharge hose 13
Dishwasher 221, 226
Diskette safe 231
Dispenser hot-water ... 267
   napkin 207
   soap 207
   toilet tissue 207
   towel 207-208
Dispersion nozzle 273
Displacement caisson .. 39, 354
Display case 211
Disposal 18
   field 46
   waste 233
   wiring 319
Disposer garbage 221
Distiller medical 228
   water 229
Distribution box 46
   section electric .. 308
Distributor asphalt ... 12
Ditching 27, 34
Divider strip terrazzo . 195
Diving board 248
Dock board 225
   boat 58
   bumper 117, 225
   leveler 225
   loading 225
   shelter 226
   truck 225
Doctor register 317
Dog command 8
Dome 404
   bulk storage 404
   ceiling 156
   drain 254
   fiberglass 64, 404
   formwork 363
   geodesic 243-244
   metal 64, 404
   observation 242
   plywood 404
   roof 157, 404
   steel 101, 404
   structures 404
   wood 123, 404
Door accordion 165, 214
   acoustical 167, 214
   air curtain 240
   air lock 241
   aluminum 160
   balanced 159
   bifolding 160
   blind 130
   bolt 177
   bulkhead 167
   bulletresistant ... 222
   bumper 107, 177
   cabinet 133
   cafe 163
   casing 190, 395-397
   catch 134
   ceiling 156
   cleanout 82
   closer 178
   closet 160
   commercial 158-159, 164, 396-397
   composite 159
   conveyor 240
   cooler 240
   counter 167
   darkroom 167
   decorative covering . 107
   demolition 22
   duct access 291
   dutch 165
   dutch oven 83
   entrance 160, 165, 171-172
   fire 159, 163-164, 170, 317
   flexible 170
   floor 168
   flush 162
   folding 214
   frame 100, 107, 128, 158, 161, 396
   frame grout 83
   frame lead lined .. 247
   frame metal 396
   frame wood 395, 397
   french 166
   glass 168, 171
   handle 134
   hangar 168, 244
   hanger 244
   hardware 176, 395, 397-398
   industrial 169, 397
   jalousie 168
   kennel 168
   kick plate 179
   knob 180

   labeled 159, 163-164
   metal 100, 169, 396-397
   mirror 182
   molding 128
   moulded 164
   opener 169-170, 177
   operator 177
   overhead 168-169
   painting 203
   panel 164
   paneled 159
   partition 215
   passage 166
   pre-hung 166, 395, 397
   prefinished 162
   prison 225
   protection 107
   pull 181
   refrigerator 167
   release 318
   removal 22
   remove reset 331
   residential 160-161, 164, 397
   revolving 167-168, 171, 241
   rolling 169
   roof 156
   sauna 231
   seal 225
   shower 216
   sill 128, 161, 181
   sliding 170
   special 167
   stain 203
   steel 158-159, 396-397
   stop 178, 185
   storm 160
   swing 170
   switch burglar alarm . 316
   telescoping 170
   threshold 161
   varnish 203
   vault 170
   vertical lift 171
   weatherstrip 181
   weatherstrip garage . 181
   wire partition 218
   wood 161-162, 164, 395, 397
Door-bell system 317
   wiring 319
Dormer gable 120
Dormitory S.F. & C.F. . 322, 413
   wardrobe 236
Double acting door 168
   hung window 173-175
   tee beam 80
   weight hinge 179
Douglas fir 394
Dovetail anchor 59
Dowel & cap 71
   support 59, 64-65
Downspout 150-152, 209
   aluminum 150
   copper 151
   elbow 151
   steel 151
   strainer 151
Dozer 10, 26, 28, 31
   mobilization 32
Draft inducer breeching . 280
Dragline 27-28, 353
   bucket 9
Drain 253-254
   board 135
   cast-iron 253
   deck 253
   dome 254
   floor 254
   main 254
   pipe 27
   roof 158, 254
   scupper 254
   shower 254
   trap 254
   trench 254
Drainage accessory 152
   mat 237
   pipe 256-257
   site 39, 41
   trap 261
Drapery hardware 236
Drawer 134
   kitchen 124
   track 134
   type cabinet 235
   wood 134
Dredge 27
   hydraulic 28
Dressing unit 237
Drill auger 8
   ceiling 112
   concrete 17
   console dental 224
   core 7, 11, 17
   drywall 112
   earth 16, 35

422

INDEX

hammer ... 12
horizontal ... 32
plaster ... 112
rig ... 16
rig mobilization ... 36
rock ... 16, 28
shop ... 233
steel ... 12
wagon ... 12
wall ... 112
wood ... 116
Drilled pier ... 36
Drinking bubbler ... 267
fountain ... 267, 272
fountain deck rough-in ... 267
fountain floor rough-in ... 267
fountain handicap ... 267
fountain support ... 260
fountain wall rough-in ... 267
Drip edge ... 151
Drive pin ... 115
Drive-up window ... 223
Driver sheeting ... 12
stud ... 115
Driveway removal ... 19
Drop pan ceiling ... 197
panel ... 63
Drum trap ... 261
trap copper ... 261
Dry chemical extinguisher ... 274
kiln ... 122
pipe sprinkler head ... 276
transformer ... 309
wall leaded ... 247
Drycleaner ... 228
Dryer circuit electric ... 319
clothes ... 221
commercial clothes ... 228
darkroom ... 224
hand ... 207
residential ... 228
vent ... 222
Drywall ... 190
accessory ... 190
ceiling ... 192
cutout ... 22, 329
demolition ... 21, 25
drill ... 112
frame ... 158
gypsum ... 190
nail ... 115
painting ... 205, 402
partition ... 192
prefinished ... 190
removal ... 25
screw ... 190
Duck tarpaulin ... 5
Duckboard ... 237
Duct access door ... 291
accessory ... 291
asbestos-cement ... 315
connection fabric flexible ... 292
electric ... 310
electrical ... 299
electrical underfloor ... 300
fiberglass ... 292
fitting transite ... 315
fitting trench ... 299
fitting underfloor ... 300
flexible insulated ... 292
flexible noninsulated ... 292
furnace ... 278
grille ... 246
heater ... 278
humidifier ... 280
insulation ... 281
mechanical ... 410
silencer ... 292
steel trench ... 299
trench ... 299
underground ... 315
utility ... 315
Ductile iron fitting ... 45
iron pipe ... 44
Ductwork ... 292
aluminum ... 292
demolition ... 23
fabricated ... 292
galvanized ... 292
heater ... 292
laboratory ... 227
rectangular ... 292
rigid ... 292
vinyl-coated flexible ... 292
Dumbbell ... 226
waterstop ... 68
Dumbwaiter ... 250
Dump charge ... 18
truck ... 10, 31
Dumpster ... 19
Duplex locker ... 213
pump ... 269
receptacle ... 305, 319
Dust collector shop ... 233
partition ... 19

protection ... 21, 329
Dustproofing ... 73, 77
Dutch door ... 165
oven door ... 83
DWV pipe abs ... 257
PVC pipe ... 257
tubing copper ... 256
Dynamite ... 28

E

Earth compactor ... 9, 353
drill ... 16, 35
rolling ... 27, 353
vibration ... 353
vibrator ... 26-27, 30
Earthwork ... 28, 353
equipment ... 9, 26, 350
equipment rental ... 9
haul ... 48
Eave overhang ... 100
Echo chamber ... 239
Econo brick ... 86
Edge drip ... 151
form ... 105
formwork ... 64-65
Edging curb ... 48, 107
landscape ... 56
Education religious S.F. & C.F. ... 326
Educational facility S.F. & C.F. ... 327
TV studio ... 319
Efflorescence testing ... 8
Eggcrate ceiling ... 197
Ejector pump ... 270
Elastomeric roof ... 146
waterproofing ... 136
Elbow downspout ... 151
pipe ... 259
Elderly housing S.F. & C.F. ... 324
Electric ... 412
appliance ... 221
backstop ... 231
baseboard heater ... 317
bed ... 237
boiler ... 276-277
cable ... 301-302
capacitor ... 311
chalkboard ... 209
curing ... 76
duct ... 310
elevator ... 405-406
feeder ... 310
fire pump ... 275
fixture ... 312-313
furnace ... 278
generator ... 12
generator set ... 311
heat ... 278, 317
heater ... 221
hinge ... 179
incinerator ... 233
lamp ... 314
log ... 212
metallic tubing ... 298
meter ... 309
panelboard ... 306
pool heater ... 284
release lock ... 177
service ... 305
stair ... 211
switch ... 305, 307
trench ... 105
unit heater ... 317
utilities ... 315-316
water-heater ... 272
wire ... 301
Electrical box ... 304
cabinet ... 303
conduit ... 298
duct ... 299
engineering fee ... 349
fee ... 1
laboratory ... 227
pole ... 315
sitework ... 315
tubing ... 298
Electronic air cleaner ... 288
closer ... 178
control system ... 290
Elementary school S.F. & C.F. ... 327
Elevated conduit ... 298
floor ... 75
slab ... 75, 78, 366, 371
slab formwork ... 63, 363
slab reinforcing ... 71, 363-364
water tank ... 249
Elevating scraper ... 29
Elevator ... 250, 405-407
building ... 405-407
construction ... 15, 350, 405, 407
fee ... 1
passenger ... 405-407
shaft wall ... 193
Elliptical pipe ... 42
Embankment fill ... 48, 353
Embossed print door ... 164

Emergency call system ... 318
lighting ... 313
Employer liability ... 1, 345-347
EMT ... 298
Emulsion adhesive ... 201
pavement ... 49
paving ... 47
sprayer ... 12
Encapsulation asbestos ... 21
Encasement concrete ... 72, 74, 365-367, 372
pile ... 38
plaster ... 187, 400
Enclosure acoustical ... 239
bathtub ... 216
swimming pool ... 248
telephone ... 220
tub ... 216
Energy solar ... 409
Engineer brick ... 86
Engineering fee ... 1, 349
Engleman spruce ... 394
English bond ... 88
Entrance canopy ... 100, 208
door ... 160, 165-166, 171-172
frame ... 161
lock ... 179
screen ... 217
sliding ... 171
strip ... 240
EPDM roof ... 148
Epoxy coated reinforcing ... 70
coating ... 77, 206
dustproofing ... 73
facing panel ... 146
fiberglass wound pipe ... 256
floor ... 195, 199
grating ... 108
grout ... 78, 193-194
lined pipe ... 42
lined silo ... 247
painting ... 402
panel ... 146
resin ... 73
terrazzo ... 195, 199
wall coating ... 206
Equipment athletic ... 51
cinema ... 229
contractor ... 350
earthwork ... 9, 26, 353
fire ... 274
formwork ... 64
foundation ... 64
general ... 2
health club ... 226
hospital ... 228
installation ... 226
insurance ... 1
laundry ... 228
medical ... 228
mobilization ... 32, 36, 39
playground ... 51, 53
rental ... 14
rental earthwork ... 353
shop ... 233
stage ... 232
Erosion control ... 56
Escalator ... 251, 407
Escape fire ... 107
Estimate electrical heating ... 317
Excavation ... 28, 31, 40, 353
backhoe ... 29, 353
cofferdam ... 29
hand ... 26, 29-30
heavy equipment ... 29, 353
machine ... 26
quarry ... 28
rock ... 28
septic tank ... 46
structural ... 29
trench ... 29-30
Exchanger heat ... 279, 283
Exercise equipment ... 231
ladder ... 231
rope ... 231
weight ... 226
Exhaust fan ... 409
hood ... 222, 227
vent ... 296
Exhauster roof ... 294
Exhibit case ... 211
Exit control alarm ... 316
device ... 180
light ... 313
Expanded aluminum grating ... 107
steel grating ... 108
Expansion anchor ... 113
bolt ... 113
joint ... 61, 113, 152, 186
joint cover ... 113
joint masonry ... 82
joint roof ... 152
shield ... 113
tank ... 284
Expansion-joint ... 286

Expense office ... 2, 343
Explosionproof box ... 304
fixture ... 313
lighting ... 313
mercuryvapor fixtur ... 313
PA system ... 318
Explosive ... 28
demolition ... 18
Exposed aggregate ... 50, 76
aggregate coating ... 206
Exterior blind ... 129
door frame ... 161
fixture mercury-vapor ... 313
light fixture ... 313
lighting ... 314
molding ... 127
plaster ... 189
pre-hung door ... 166
residential door ... 165
sprinkler ... 54
tile ... 194
wood frame ... 161
Extinguisher fire ... 274
installation ... 274
Extra work ... 2
Extractor ... 10
air ... 291
industrial ... 228
Extrusion aluminum ... 98
Eye bolt ... 69
wash fountain ... 267
wash portable ... 268

F

Fabric filter ... 135
flashing ... 153
shell structure ... 404
stabilization ... 47
stile ... 177
structure ... 240
waterproofing ... 137
wire ... 111
Fabricated ductwork ... 292
Face brick ... 85, 380
wash fountain ... 268
Faceted glass ... 182
Facing panel ... 97
panel concrete ... 81, 366, 372, 375
panel metal ... 395
stone ... 96
Factor cost adjustment ... 332-341, 413
Factors repair remodel ... 329
Factory S.F. & C.F. ... 323, 413
Fan ... 293, 409
air handling ... 293
air-conditioning ... 293
axial flow ... 293
bath exhaust ... 294
booster ... 293
ceiling ... 293
centrifugal ... 293
corrosion-resistant ... 293
inline ... 293
induced draft ... 280
kitchen exhaust ... 294
low sound ... 293
propeller exhaust ... 294
roof ... 294
utility set ... 293-294
vane-axial ... 293
ventilator ... 293
wall exhaust ... 294
Fan-coil air-conditioning ... 295
Farm type siding ... 143
Fascia board demolition ... 23
metal ... 152
wood ... 119, 127
Fast food equipment ... 226
Fastener anchor bolt ... 82
metal ... 113
steel ... 114
structural ... 102
timber ... 115
wood ... 117
Faucet & fitting ... 254
bath ... 254
Fee architectural ... 1
consulting ... 348-349
engineering ... 1
Feeder electric ... 310
section ... 308
section branch circuit ... 308
section circuit-breaker ... 308
section frame ... 308
Felt ... 146, 396
asbestos ... 143, 146
asphalt ... 144, 146
carpet ... 199
fiberglass ... 145
tarred ... 146
underlayment ... 200
waterproofing ... 137
Fence board & batten ... 53
chain link ... 6, 51, 111
helical topping ... 111

423

INDEX

metal	51
painting	204
plywood	6
removal	19
residential	52
security	52
snow	52
steel	52
temporary	6
tennis court	52
wire	6, 111
Fertilizer	56, 358
Fiber pipe bituminous	41
tube formwork	62
Fiberboard roof deck	138
Fiberglass area wall cap	106
bench	51
blown-in	138
ceiling board	196
coated fabric structure	404
cross	224
cupola	126
dome	64, 404
door	168
duct	292
felt	145
flagpole	212
formboard	81
formwork	62, 363
insulation	100, 140-141, 281
panel	8, 101, 147, 198
pipe-covering	281
planter	53
receptacle	55
reinforced ceiling	246
roof deck	139
septic tank	45
shade	239
shower stall	216
stair tread	110
steeple	224
tank	284
wall lamination	206
waterproofing	137
wool	138
Field disposal	46
engineer	2
mix concrete	74, 365, 367-368
office	6
personnel	2
seeding	56
Fieldstone	97
Fill	26, 30, 353
cellular	82
embankment	48, 353
floor	75, 353
gravel	26, 31, 40, 47, 72, 353
Filled column concrete	98
Filler block	205
strip	177
Fillet weld	104
Film equipment	224, 229
Filter air	288
fabric	135
grille	291
mechanical media	288
stone	32
swimming pool	248
Filtration equipment	249
Fine grade	31, 49, 56
Finish	374
carpentry	124
concrete	48, 76-77, 374, 376
float	77
floor	205
grading	31
granolithic	76
hardware	176, 398
integral concrete	76
keene cement	187
lime	83
nail	115
refrigeration	246
steel trowel	76
Finisher floor	11
Finishing floor	203
Fir column	125
floor	202
molding	127
roof deck	133
roof plank	122
Fire alarm	317
brick	86
call pullbox	317
clay	86
damage repair	330
door	159, 163-164, 170, 317, 396-397
door frame	158
equipment	274
equipment cabinet	273
escape	107
escape disappearing	211
extinguisher	274

extinguishing system	273-274, 276
horn	317
hose	274
hydrant	45, 274
protection kitchen	227
pump	275
rated closer	178
rated tile	94
resistant drywall	190-191
resistant wall	192
retardant lumber	121
retardant plywood	121
retardent pipe	257
signal bell	317
station S.F. & C.F.	323, 413
suppression automatic	273
Fire-damper curtain type	292
Fire-hose adapter	274
nozzle	275
storage cabinet	273
valve	275
Fireplace accessory	82
box	86
damper	82
demolition	24
form	83
free standing	211
mantel	126
masonry	86
prefabricated	211
Fireproof column	99
shingle	394
Fireproofing plaster	189
plastic	189
sprayed	189
Firestop gypsum	187
wood	118
Fitting bus-duct	310
poke-thru	305
tee	45
transite duct	315
underfloor duct	300
vent-chimney	285
wye	45
Fittings waterpipe	45
Fixed blade louver	295
booth	235
roof tank	249
Fixture bathroom	266, 268, 270-271
carrier	260
electric	312
explosionproof	313
fluorescent	312
incandescent	313
incandescent vaportight	313
interior light	312-313
mercury-vapor	313
plumbing	266-271, 408
removal	24
sodium-high-pressure	314
sodium-low-pressure	314
support handicap	260
support/carrier	260
vandalproof	313
wiring	320
Flagging	50, 87
slate	50
Flagpole	212
base	212
foundation	213
Flange clip	69
Flanged coil	289
valve	265
Flared end pipe	41
Flashing	152-154
aluminum	144, 266
asphalt	152
butyl	153
cement	135
copper	152, 266
counter	155
fabric	153
membrane	146
metal	100
PVC	153
stainless	153
valley	141
vent	266
Flat plate	63
slab	63
Flatbed truck	10
Flemish bond	88
Flexible conduit & fitting	306
connector	253
door	170
duct connection fabric	292
ductwork vinyl-coated	292
insulated duct	292
metal hose	286
Float concrete	11
finish	76-77
Floater equipment	1
Floating dock	58
pin	178

roof tank	249
Flood coating	147
Floodback sewer control	254
Floodlight	12-13
pole mounted	314
Floor abrasive	77
acid proof	86
adhesive	201
asphalt block	47
brick	49, 86
bumper	178
carpet	198
ceramic tile	193
check	178
clean	330
cleaning	1
cleanout	253
closer	178
color	77
composition	199
concrete	73, 372
conductive	193, 195, 199-200
cutout	329
deck steel	104
demolition	17
door	168
drain	254
elevated	75
epoxy	195, 199
expansion joint	113
fill	75
finish	205, 376
finisher	11
flagging	50
framing removal	22
grating	107-108, 390
hatch	168
marble	96
mastic	199
mat	236
metal deck	104
nail	115
neoprene	199
painting	204
patching concrete	77-78
pedestal	245
plank	108, 120
plywood	123
polyacrylate	199
polyester	199
polyethylene	200
polyurethane	200
portable	232
quarry tile	194
refrigeration	246
register	291
rubber	200
scale	218
scupper	254
slab formwork	63
sleeper	119
stain	204
stone	49
stressed skin	123
subfloor	123
terrazzo	195, 401
tile	195
tile removal	22
tile terrazzo	195
topping	77
underlayment	124
varnish	204
vinyl	201
wood	202
Flooring conversion table	393
demolition	22
slate	97
Flow meter	288
Flue chimney	87
chimney metal	285
lining	85, 87, 381
prefab metal	285
screen	82
Fluid heat transfer	283
Fluorescent fixture	312
lamp	314
Fluoroscopy room	247
Flush bolt	177
door	162, 165
tube framing	184
wood door	161
Flying truss	67
Foam glass roof deck	139
insulation	138-140
phenolic	139
pipe-covering	282
urethane	136
Foamglass insulation	140
Foil aluminum	135, 138, 196, 206
back insulation	141
barrier	196
Folded plate roof	123
Folder laundry	228
Folding accordion door	165

accordion partition	214
bench	231
door	214
door shower	216
partition	214
table	239
window	172
Food mixer	226
service equipment	226
warmer	226
Foot valve	265
Football goalpost	53
scoreboard	232
Footing concrete	75, 78, 368
demolition	17
formwork	64, 360
pressure injected	39, 356
reinforcing	71
spread	60, 65
Forged valve	265
Forklift brick	12
Form column	59
edge	105
fireplace	83
liner	66
patch	60
plywood	393
release	74
slab	105
wedge	61
work	59, 359-360
Formbloc	90
Formboard asbestos	81
roof	81
Formwork	62-63, 66, 360, 362
accessory	59, 61
anchor	59
concrete	61, 359-360, 362-363, 368, 372
culvert	63
equipment	64
foundation	65
in place	61
insert	60
labor hours	359
liner	66
oil	60
plywood	66, 391
Fossil-fuel boiler	278
Foundation caisson	35, 354
chimney	74
concrete	75, 363, 365-368, 372, 404
concrete block	91
equipment	64
flagpole	213
formwork	75, 65
mat	75, 78
pile	36, 354-355
scale pit	219
underpin	34
vent	84
wall	91
Fountain	53
dental	267
drinking	267, 272
eye wash	267
face wash	268
wash	271
Frame chalkboard	209
door	100, 107, 128, 158, 161, 396
drywall	158
entrance	161
grating	108
labeled	158
metal	158
roof	99
steel	158
welded	158
window	176
wood	133, 161
Framing	393
aluminum	98, 399
anchor	116
canopy	98
ceiling	118
demolition	23
laminated	132, 404
lightgage	99
opening	100
partition	191
roof truss	122
steel	102, 386, 388-389, 400
timber	120
tube	184
wall	120
window	100
window wall	184
wood	118-120, 391
Fraternal club S.F. & C.F.	321
Fraternity house S.F. & C.F.	323, 413
Freeze deep	221
Freezer	226, 230, 246
door	240
Freight elevator	250, 407

424

INDEX

French door . . . 166
Friction collar . . . 59
pile . . . 36, 38
Front end loader . . . 28
shovel . . . 10
vault . . . 170
Fryer . . . 226
Fuel storage tank . . . 249
tank . . . 284
Full vision door . . . 160
vision glass . . . 182
Fume hood . . . 227
Funeral home S.F. & C.F. . . . 323, 413
Furnace duct . . . 278
electric . . . 278
gas . . . 278
gas-fired . . . 279
hot-air . . . 278
oil-fired . . . 279
wall . . . 284
wiring . . . 319
Furniture hospital . . . 237
hotel . . . 237
library . . . 237
office . . . 238
restaurant . . . 238
school . . . 238
Furring anchor . . . 60
ceiling . . . 121, 186
channel . . . 190
metal . . . 186
tile . . . 94
wall . . . 120, 186
Fuse cabinet . . . 306
Fusible link closer . . . 178
switch branch circuit . . . 308

G

Gabion . . . 32, 54
Gable dormer . . . 120
Galvanic tank protection . . . 285
Galvanized ductwork . . . 292
reinforcing . . . 70
roof . . . 149
structural steel . . . 102, 388
wire gabion . . . 54
Galvanizing . . . 102
Gang wall formwork . . . 66
Gantry crane . . . 251
Garage . . . 243, 403
door . . . 168-169
door weatherstrip . . . 181
residential . . . 243
S.F. & C.F. . . . 323, 413
Garbage disposer . . . 221
Garden house . . . 243
Gas connector . . . 253
fired space heater . . . 283
furnace . . . 278
heat air-conditioner . . . 296
incinerator . . . 233
log . . . 212
operated generator . . . 311
pipe . . . 40
service and distribution . . . 40
station formwork . . . 67
station tank . . . 284
stop valve . . . 262
vent . . . 285
water-heater . . . 273
Gas-fired boiler . . . 277
furnace . . . 279
infra-red heater . . . 281
pool heater . . . 284
space heater . . . 284
Gasket neoprene . . . 136
pipe . . . 41
Gasoline operated generator . . . 311
pump . . . 222
tank . . . 249
Gate automatic . . . 230
day . . . 171
fence . . . 52
operator . . . 52
security . . . 219
valve . . . 44, 264-265
valve soldered . . . 262
Gauging plaster . . . 187, 400
General equipment . . . 2
Generator construction . . . 12
diesel operated . . . 312
emergency . . . 311
gas operated . . . 311
gasoline operated . . . 311
set . . . 311
steam . . . 229, 276
Geodesic dome . . . 243-244, 404
Girder demolition . . . 18
formwork . . . 61-62
reinforcing . . . 70
wood . . . 117, 120
Glass . . . 183-184, 398-399
acoustical . . . 182
and glazing . . . 182, 398
bead . . . 184, 399
bead molding . . . 128
beveled . . . 182
block . . . 93, 382
block mortar . . . 83, 382
block skylight . . . 157
bulletproof . . . 182
curtain wall . . . 184
demolition . . . 26
door . . . 160, 168, 171
door astragal . . . 177
door shower . . . 216
faceted . . . 182
full vision . . . 182
heat reflective . . . 182
insulating . . . 182
insulation fiber . . . 140
laminated . . . 182
lead . . . 247
lined water heater . . . 222
mirror . . . 182, 207
mosaic . . . 194
obscure . . . 183
patterned . . . 183
pipe . . . 257
plate . . . 183
reflective . . . 184
replace . . . 331
retrofit . . . 331
safety . . . 182
sandblasted . . . 184
sheet . . . 184, 398
shower stall . . . 216
spandrel . . . 184
tempered . . . 183
tile . . . 183
tinted . . . 183
vinyl . . . 184
window . . . 184
window wall . . . 185
wire . . . 184
Glassware washer . . . 227
Glasweld . . . 148
Glaze coat . . . 205
Glazed block . . . 93
brick . . . 86, 88
ceramic tile . . . 193
wall coating . . . 206
Glazing . . . 182
glass . . . 398-399
polycarbonate . . . 183
Globe valve . . . 263-265
Glove box . . . 227
Glued laminated . . . 132
Goalpost . . . 53
Golf shelter . . . 53
tee surface . . . 200
Gore line . . . 49
Grab bar . . . 207
Gradall . . . 9, 29
Grade beam . . . 65, 78
fine . . . 31, 49, 56
wall . . . 75
Grader motorized . . . 9
Grading . . . 32, 56
site . . . 31
Grandstand . . . 51, 243
Granite . . . 95
building . . . 95
conductive floor . . . 199
curb . . . 48, 96
demolition . . . 24
paver . . . 96
paving block . . . 50
sidewalk . . . 50
Granolithic finish . . . 76, 374
topping . . . 374
Grass . . . 56
cloth wallpaper . . . 206
seed . . . 56, 358
sprinkler . . . 54
Grating . . . 107
area wall . . . 106
area way . . . 106
floor . . . 107-108, 390
frame . . . 108
sewer . . . 40
solar screen . . . 110
steel . . . 111
Gravel bankrun . . . 26, 353
concrete . . . 73, 366
fill . . . 26, 31, 40, 47, 72, 353
pack well . . . 47
roof . . . 72, 396
stop . . . 154
surfacing . . . 48, 353
Gravity retaining wall . . . 54
Grease interceptor . . . 268, 408
Greenhouse . . . 243-244
air supported . . . 241
cooling . . . 244
Grid spike . . . 116
Griddle . . . 227
Grille air-return . . . 291
aluminum . . . 291
convector cover . . . 125
decorative wood . . . 126
duct . . . 246
filter . . . 291
painting . . . 204
plastic . . . 291
roll up . . . 170
window . . . 175
Grinder pump system . . . 269
shop . . . 233
terrazzo . . . 11
Grooved block . . . 90
Grooved-joint pipe . . . 259
pipe-fitting . . . 259
Ground . . . 121
box hydrant . . . 255
clamp water pipe . . . 303
cover . . . 56
face block . . . 90
fault protection . . . 309
fault receptacle wiring . . . 319
rod . . . 303
slab . . . 75
socket . . . 53
Grounding . . . 303
& conductor . . . 301-303
wire brazed . . . 303
Group shower . . . 270
wash fountain . . . 271
Grout . . . 83
block wall . . . 384
cavity wall . . . 83
cement . . . 77, 83
column base . . . 77
concrete block . . . 83
door frame . . . 83
epoxy . . . 78, 193-194
pressure . . . 31
prestressed concrete . . . 70, 365
soil . . . 31
tile . . . 193
wall . . . 83
Guard corner . . . 106, 203
gutter . . . 154
lamp . . . 315
rail . . . 49
rail removal . . . 19
service . . . 8
snow . . . 158
window . . . 111
Guardrail temporary . . . 6
Guidance inertial . . . 5
Guide rail . . . 49
Gunite . . . 77
pool . . . 403
reinforcing . . . 72
Gutter . . . 100, 154
aluminum . . . 154
copper . . . 154
demolition . . . 25
guard . . . 154
monolithic . . . 48
painting . . . 204
road . . . 48
stainless . . . 154
steel . . . 154
strainer . . . 154
swimming pool . . . 248, 403
vinyl . . . 154
wood . . . 154
Gutting building . . . 18
Guyed chimney . . . 242
derrick . . . 15
tower . . . 246
tree . . . 57, 358
Gym equipment . . . 226, 231
floor . . . 200, 202
floor expansion joint . . . 113
floor underlayment . . . 202
locker . . . 213
mat . . . 232
squash court . . . 248
Gymnasium divider curtain . . . 232
S.F. & C.F. . . . 324, 413
Gypsum block demolition . . . 21
board . . . 190
board leaded . . . 247
cement . . . 82
deck . . . 81, 396
drywall . . . 190
fabric wallcovering . . . 206
firestop . . . 187
laminated . . . 193
lath . . . 187, 399
mortar . . . 83-84, 378
partition . . . 192, 214
plaster . . . 187, 400
poured . . . 82
roof . . . 81
roof plank demolition . . . 25
shaft wall . . . 193
sheathing . . . 122
weatherproof . . . 122

H

Hair interceptor . . . 268
Half round molding . . . 127
Halon fire extinguisher . . . 273
Hammer bush . . . 77
chipping . . . 12
demolition . . . 9
drill . . . 12
hydraulic . . . 12
jack . . . 28
pile . . . 9
pile mobilization . . . 36
Hand carved door . . . 163
clearing . . . 16
dryer . . . 207
excavation . . . 26, 29-30, 40
hole . . . 46
hole concrete . . . 315
split shake . . . 143
Hand/off/automatic control . . . 306
Handball court . . . 51, 248
Handicap drinking fountain . . . 267
fixture support . . . 260
opener . . . 177
water cooler . . . 272
Handicapped lever . . . 180
Handle door . . . 134
Handling reinforcing . . . 71
waste . . . 233
Handrail . . . 109
aluminum . . . 246
wood . . . 127, 129, 392
Handsplit shingle . . . 394
Hangar . . . 244
door . . . 168
Hanger beam . . . 60
formwork . . . 59-60
joist . . . 116
Hanging wire . . . 60, 197
Hardboard cabinet . . . 124
chalk board . . . 209
overhead door . . . 169
partition . . . 215
siding . . . 130
soffit . . . 131
tempered . . . 128
underlayment . . . 124
Hardener concrete . . . 73, 77
Hardware cabinet . . . 134
cloth . . . 111
door . . . 176, 215, 395-397
drapery . . . 236
finish . . . 176, 398
panic . . . 180
rough . . . 122
window . . . 176
Hardwood carrel . . . 228
floor stage . . . 232
grille . . . 126
Hat and coat strip . . . 207
rack . . . 200
Hatch ceiling . . . 156, 197
floor . . . 168
roof . . . 156-157
smoke . . . 157
Haul earthwork . . . 48
Hauling . . . 377
trailer . . . 32
truck . . . 32
Head sprinkler . . . 276
Header pipe . . . 15
wood . . . 120
Health club equipment . . . 226
Hearth . . . 86
Heat . . . 409
electric . . . 278, 317
exchanger . . . 279, 283
exchanger solar-energy-system . . . 283
greenhouse . . . 244
loss approximation . . . 409
pump . . . 295
radiant . . . 245, 317
recovery . . . 234
reflective glass . . . 182
temporary . . . 6, 79
therapy . . . 229
transfer fluid . . . 283
transfer package . . . 279
Heat-recovery air to air . . . 279
ventilation . . . 279
Heated doorway . . . 240
Heater & fan bath . . . 294
circuit wiring . . . 319
contractor . . . 12
duct . . . 278
ductwork . . . 292
electric . . . 221
electric baseboard . . . 317
electric residential . . . 317
electric wall . . . 317
floor mounted space . . . 283
infra-red . . . 280
sauna . . . 231
swimming pool . . . 284

INDEX

terminal 281
unit 279
warm air 278
water 222, 272-273
Heating 276-279, 281-285
& cooling classroom 279
baseboard 280
cable 317
estimate electrical 317
hot-air 278-279
hot-water 277, 280
hydronic 277, 280
insulation 281
subsoil 245
unit convector 280
Heavy construction 58
equipment excavation 353
framing 120
Hedge 56
Hem-fir 394
Hex nut 112
Hexagonal face block 91
High bay lighting 313
build coating 206, 402
chair 69
early strength cement 73
intensity discharge lighting 313-314
rib lath 187
rise glazing 182
rise plaster 187
school S.F. & C.F. 327, 413
strength block 90-91, 379
strength concrete 74, 366
strength steel 102
temperature vent-chimney 285
High-installation conduit 298
Highway curb 96
paver 50
sign 50
Hinge 398
brass 178-179
cabinet 134
continuous 179
electric 179
hospital 179
paumelle 179
prison 179
residential 178
school 179
security 179
special 179
steel 178
Hip rafter 119
Historical cost index 1, 332
labor index 332
material index 332
Hockey dasher 244
scoreboard 232
Hoist and tower 15
automotive 222
contractor 15, 350
electric 251
lift equipment 14, 349
personnel 15
Holder closer 178
Holding tank 46
Hole drill 112
Hollow core door 162, 166
metal 158
metal door 158-159, 396
metal frame 396
metal stud partition 188
wall anchor 113
Honed block 91
Hood exhaust 227
fire protection 227
fume 227
range 222
Hook coat 207
robe 207
Hopper refuse 210
Horizontal auger 9
boring 32
drill 32
shore 67
Horn fire 317
Hose air 12
braided bronze 286
discharge 13
equipment 274
fire 274
metal flexible 286
nozzle 275
rack 274
rack cabinet 273
suction 12
valve cabinet 274
water 12
Hospital cabinet 235
door hardware 176
elevator 251, 405-406
equipment 228
furniture 237
hinge 179
kitchen equipment 227
partition 214
S.F. & C.F. 324, 413
tip pin 178
Hot-air furnace 278
heating 278-279
Hot-water boiler 277-278
dispenser 267
heating 277, 280
Hot-water/steam exchange 279
Hot-water/water exchange 279
Hotel cabinet 214
furniture 237
lockset 179-180
House demolition 17
garden 243
telephone 318
Housing for the elderly S.F. & C.F. 324, 413
project S.F. & C.F. 413
public S.F. & C.F. 324
S.F. & C.F. 324, 413
Hubbard tank 229
Humidification equipment 244
Humidifier 222
centrifugal 280
duct 280
room 280
HVAC demolition 23
piping specialties 286-287
Hydrant fire 45, 274
ground box 255
removal 19
water 255
Hydrated lime 83
Hydraulic chair 225
crane 14, 251, 355
dredge 28
elevator 405-407
hammer 12
jack 15
lift 349
seeding 56
Hydrocumulator 269
Hydronic heating 277, 280
Hypalon caulking 135-136
neoprene roofing 146

I

I.V. track system 214
Ice cube maker 227
skating equipment 244
skating rink S.F. & C.F. 324, 413
Icemaker 222
Idaho pine 394
Illuminated sign 220
Impact barrier 49
wrench 12
Improvement site 51
In-line fan 293
In-slab conduit 299
Incandescent fixture 313
Incinerator electric 233
gas 233
municipal 234
waste 233
Inclined ladder 110
ramp 252
Increaser vent-chimney 285
Incubator 229
Index construction cost 1, 332-341
Indicating-panel burglar alarm 316
Induced draft fan 280
Industrial address system 318
chimney 404
door 168
equipment installation 226
folding partition 214
lighting 313
safety fixture 267
window 173
Inertial guidance 5
Inflation system 404
Infra-red broiler 226
detector 317
heater 280
heater gas-fired 281
Inlet curb 48
Insecticide 34
Insert formwork 60
slab 69
Inspection technician 8
Installation extinguisher 274
Instrument switchboard 309
Insulating concrete 366
glass 182
glass spandrel 184
Insulation 137-140, 281-282
batt 141
blanket 281
blown-in 137
boiler 281
breeching 281
building 138
cavity wall 138
cellulose 138
concrete 76, 81, 366
duct 281
fiberglass 100, 281
foam 138-139
glass 398-399
insert 94
masonry 138
perlite 139
pipe 282
polystyrene 147
refrigeration 246
roof 82, 100
roof deck 138
sandwich panel 144
sandwich wall 375, 395
screw 138
sheathing 140
shingle 141
sprayed 139
subsoil 245
vapor barrier 135
wall 94, 140
Insurance 1, 342
builder risk 1, 342
equipment 1
public liability 1, 342
Intake-exhaust louver 295
Intake/exhaust vent 297
Integral waterproofing 74, 79
Integrated ceiling 245
siding 147
Interceptor grease 268
metal recovery 268
oil 268
Intercom 318
Interior cleanup 330
cutout 329
demolition door 22
door 396-397
door frame 161
light fixture 312-313
pre-hung door 166
residential door 165
wood frame 161
Interlocking block 91
block wall 384
Intermediate conduit 298
Invert manhole 40
Inverted bucket steam trap 287
tee beam 80
Iron alloy mechanical joint pipe 256
banding 59
body valve 264
compound waterproofing 137
Ironer laundry 228
Ironing center 238
Ironspot brick 87
Island formwork 67

J

Jack cable 15
hammer 28
hydraulic 15
mud 11
post 99
roof 119
screw 34
Jackhammer 11
Jacking 32, 349
Jail equipment 225
S.F. & C.F. 325, 413
Jalousie 173
door 168
Jet water system 270
Jetting pump 15
Jeweler safe 231
Job condition 2
Jockey pump fire 275
Joint control 82
expansion 113, 152, 186
reinforcing 83
roof 152
Jointer shop 233
Joist anchor 116
chair 69
clip 186
composite 105
concrete 75, 80
connector 116
demolition 23
hanger 116
lightgage 104, 387
open web 105, 388
sister 331
steel 105
wood 118, 123
Jumbo brick 86, 88
Jute mesh 56

K

K-lath 188
Kalamein door 168, 396
Keene cement 187
Kennel door 168
fence 52
Kettle 227
Key cabinet 213
keeper 213
station 230
Keyless lock 180
Keyway 64
Kick plate 108-109
plate door 179
Kiln dried lumber 121
dry 122
vocational 233
Kingsize brick 86
Kiosk 245
Kitchen appliance 221
cabinet 124, 236
equipment 226
equipment fee 1
exhaust fan 294
heat-recovery 279
sink 271
sink faucet 254
Knob door 180
Kraft paper 135

L

Labeled door 159, 163-164
frame 158
Labor index 2, 333-341
rates 344
Laboratory equipment 227
research S.F. & C.F. 326
sink 227
table 227
Ladder 13
building 108, 110
exercise 231
fire escape 107
rolling 4
swimming pool 248
towel 208
type cable-tray 298
Lag screw 114
Lagging 33-34
Lally column 98
Laminated beam 132
counter top 126
counter top plastic 126
epoxy & fiberglass 206
framing 132
glass 182
gypsum 193
gypsum lath 187
lead 247
roof deck 133
timber 404
wood 132
Lamp fluorescent 314
guard 315
mercury-vapor 314
metal-halide 314
post 108
Lampholder 305
Lamphouse 229
Landing newel 131
stair 110, 196, 392
Landscape 56-57, 358
edging 56
surface 200
Landscaping fee 1
Laser 13
Latch deadlocking 179
set 179-180
Latchset 180
Latex acrylic 73, 402
caulking 136
underlayment 201
Lath 399
cutout 329
demolition 21
gypsum 187, 399
lead 247
metal 187-188, 399
rib 187
wire 185, 399
Lathe shop 233
Lattice molding 127
Lauan door 166
Laundry equipment 228
faucet 254
tray 269
Lava stone 96
Lavatory 408
faucet 254
support 260
vanity top 268
Lawn planting 358
seeding 56, 358

INDEX

Lazy susan 124
Leaching field chamber 46
  pit 46
Lead 108
  barrier 196
  coated downspout 151
  coated flashing 153
  flashing 153
  glass 247
  gypsum board 247
  lath 247
  lined block 92
  lined darkroom 242
  lined door frame 247
  paint removal 330
  pile 9, 354-355
  roof 155
  salvage 25
  screw anchor 113
  shielding 247
  wool 61
Leaded vinyl 196
Lean-to type greenhouse 244
Leather tile 194
Lectern 223
Lecture hall seating 239
Ledger 119
  wood 119
Lens movie 230
Let-in bracing 117
Letter sign 219
  slot 213
Leveler truck 225
Lever handicapped 180
Lexan 183
Liability employer 1, 345-347
Library chair 228
  equipment 228
  furniture 237
  S.F. & C.F. 325, 413
  shelf 228
Life guard chair 248
Lift aerial 11
  automotive 222
  commercial 250
  correspondence 250
  parcel 252
  porch 251
  scissor 252
  slab 69, 76, 371
Lifter platform 225
Light base 67
  border 232
  dental 225
  exit 313
  fixture exterior 313
  fixture interior 312-313
  fixture wiring 320
  gauge joist 387
  loading dock 226
  nurse call 318
  pole 314-315
  pole aluminum 314
  pole steel 314
  portable 12
  shield 49
  strobe 232
  temporary 6
  tower 13
  traffic 50
  underwater 248
Lightgage framing 99
  joist 104
Lighting 312-314
  ceiling 245
  darkroom 224
  emergency 313
  explosionproof 313
  exterior 314
  high intensity discharge 313-314
  incandescent 313
  industrial 313
  mercury-vapor 313
  outdoor 12
  roadway 314
  stage 232
  strip 312
  surgical 229
  wiring 320
Lightweight aggregate 72
  block 92, 379
  concrete 74-75, 366, 374
  concrete plank 81
  natural stone 96
Lime finish 83
  hydrated 83
Limestone 56, 96
  coping 85
Line gore 49
Linen chute 210
  collector 210
  wallcovering 206
Liner formwork 66
  pool 403
Lining flue 87
  pipe 41
Link mat 236
Linseed oil 402
Lint collector 228
Lintel 83, 96, 99, 109
  block 92
  formwork 67
Liquid chiller centrifugal 297
Load bearing stud 191
  test 39
  test pile 38
  test soil 353
Loader front end 28
  tractor 10
Loading dock 225
  dock light 226
Loam 27, 56, 358
Lobby collection box 210
Lock deadbolt 179
  electric release 180
  entrance 179
  keyless 180
  tile 236
  time 170
  tubular 179
Locker 213, 403
Lockset communicating 179-180
  cylinder 179
  hotel 179-180
  mortise 180
Log electric 212
  gas 212
Louver 155, 295
  aluminum 295
  aluminum operating 295
  coating 295
  cooling tower 295
  fixed blade 295
  intake-exhaust 295
  mullion type 295
  redwood 126
  silencer 292
  ventilation 126
  wood 126
Louvered door 166
Low-voltage silicon rectifier 305
  switching 305
  switchplate 305
  transformer 305
Lube equipment 222
Lumber 117, 393-394
  core paneling 129
  creosote 121
  kiln dried 121
  treatment 121
Luminaire ceiling 245
Luminous ceiling 197, 314
  panel 197

M

Macadam penstration 47
Machine excavation 26
  screw 114
  welding 14
Machinery anchor 114
Magazine shelving 228
Magnesium dock board 225
  oxychloride 189
Magnetic astragal 176-177
  motor-starter 306
  particle testing 8
Mahogany door 163
Mail box 213
  box call system 318
  chute 210
  slot 213
Main drain 254
  office expense 2, 342-351
Maintenance carpet 199
Make-up air unit 282
Mall front 171
Management construction 1
  fee 1
Manager project 2
Manhole 39
  concrete 315
  cover 40, 109
  electric service 315
  invert 40
  removal 19
  step 40
Mansard aluminum 155
Mantel beam 126
  fireplace 126
Map rail 208-209
Maple counter top 126
  floor 202
  wall 248
Marble 96
  aggregate 73
  coping 85
  counter top 126
  floor 96
  screen 217
  shower stall 216
  sill 89
  soffit 97
  stair 97
  synthetic 194
  tile 194
  toilet partition 217
Marina small boat 58
Mark up cost 2
Mark-up cost 344
Mask and grout 330
Masonry aggregate 84
  anchor 59, 84
  base 93
  block 379
  brick 85-86, 88, 381
  catch basin 39
  caulking 329
  cement 73, 82
  cleaning 85, 87, 329
  color 82
  cutting 19
  demolition 21, 23, 25
  expansion joint 82
  fireplace 86
  flashing 152
  furring 120
  insulation 138
  manhole 39
  mortar 84, 378
  nail 115
  needle 329
  point 85
  pointing 330
  reinforcing 83, 379
  removal 19, 23
  restoration 329, 382
  saw 13
  sidewalk 49
  sill 89
  step 50
  testing 8
  toothing 21
  wall 55, 90
  wall tie 84
  waterproofing 84
Massage table 226
Master clock system 316
Mastic floor 199
Mat blasting 28
  floor 236
  foundation 65, 75, 78
  gym 232
Material handling 251
  hoist 15, 350
  index 3, 333-341
Meat case 230
Mechanical cost 253
  engineering fee 349
  equipment demolition 23
  fee 1
  items 408-410
  media filter 288
  splice 72
Median barrier 49
  precast 49
  strip 57
Medical clinic S.F. & C.F. 325, 413
  equipment 228
  office S.F. & C.F. 325
  X-ray 229
Medicine cabinet 213
Melting snow 317
Membrane curing 76
  flashing 146
  roof 148
  waterproofing 137
Mercury-vapor exterior fixture 313
  fixture 313
  fixture vaportight 313
  lamp 314
Mercury—vapor
  fixture explosionproof 313
Mesh fence 51
  partition 218
  reinforcing 72, 77
  steel 77
  stucco 188
  wire 72, 111
Metal base 186, 194
  bookshelf 228
  building 100
  butt frame 158
  cabinet 236
  chalkboard 209, 216
  chimney 211, 285
  deck 104-105
  deck demolition 25
  dome 64, 404
  door 100, 160, 169, 396
  door frame 100, 160, 397
  door residential 160
  ductwork 292
  facing panel 147, 395
  fascia 152
  fastener 113
  fence 51
  flashing 100
  flexible hose 286
  flue chimney 285
  frame 158
  furring 186
  gutter 100
  hollow 158
  lath 187-188, 399
  locker 213
  miscellaneous 106, 390
  molding 186
  ornamental 106
  overhead door 169
  pan 63, 363
  pan ceiling 196-197
  panel 144
  partition 215
  pipe 42, 44, 408
  protected 148
  railing 109
  recovery interceptor 268
  roof 143, 149
  sash 173
  screen 173, 217
  sheet 150, 410
  shelf 219
  shingle 142
  siding 143-144, 395
  sign 220
  soffit 155
  stair 110
  stair tread 79
  structural steel 386, 388-389
  stud 186, 188, 191
  stud demolition 25
  stud partitions 191
  threshold 181
  tile 194
  toilet partition 217
  trash receptacle 234
  truss 102
  tube frame 184
  window 100, 172-173, 399
Metal-halide lamp 314
Metallic coating 137
  foil 138
  hardener 73, 77
  waterproofing 74, 137
Meter electric 309
  steam condensate 287
  venturi flow 288
  water supply 266
  water supply domestic 266
Microphone 318
Microwave oven 221
Middle school S.F. & C.F. 327
Mill construction 120
Millwork 133
  demolition 24
Mineral fiber ceiling 196-197
  fiber formboard 81
  fiber insulation 141
  roof 145, 396
  shingle 141
  siding 147-148, 396
  wool blown-in 138
Mirror 207
  ceiling boards 197
  door 182
  glass 182
  plexiglass 183
  wall 183
Miscellaneous aluminum 106
  metal 106, 390
Mixer concrete 11
  food 226
  mortar 11, 13
  plaster 13
  transit 11
Mobile shelving 228
  X-ray 229
Mobilization 32
  equipment 27, 32, 36, 39
Model building 3
Modification to cost 2
Modulus of elasticity 7
Moil point 12
Moisture barrier 135
  content test 8
  testing 8
Molding 127
  base 127
  brick 127
  ceiling 127
  exterior 127
  hardboard 128
  metal 186
  wood 131
Monel rivet 115
  roof 155

INDEX

Money safe	231
Monkey ladder	53
Monolithic curb and gutter	48
finish	48, 76
gutter	48
terrazzo	195
Monorail	252
Monument survey	5
Mop holder strip	207
roof	146
sink	271
Mortar	82, 378
and masonry accessory	82
cement	194
gypsum	84
masonry	84
mixer	11, 13
point	84
portland cement	84
sand	84
testing	8
thinset	194
Mortise lockset	180
Mortuary refrigeration	229
Mosaic glass	194
Motel S.F. & C.F.	325, 413
Motion detector	317
Motor connection	306
electric	411
generator	12
Motor-starter	306
& control	306
magnetic	306
w/circuit protector	306
w/fused switch	306
Motorized car	252
roof	248
Moulded door	164
Mounted booth	235
Mounting board plywood	122
Movable louver blind	235
office partition	214
Move building	19
Movie equipment	229
lens	230
projector	230
S.F. & C.F.	328
screen	218
Moving ramp	252
shrub	56
stairs & walks	407
tree	56
walk	252
Mud jack	11
sill	119
Mulch	56, 358
Mullion type louver	295
vertical	185
Multi plate pipe	43
Multi-blade damper	292
Multizone air-conditioner	
rooftop	296
Municipal building S.F. & C.F.	413
incinerator	234
Muntin window	175
Mushroom ventilator stationary	297
Music room	245
Mylar tarpaulin	6

N

Nail	115
lead head	247
stake	60
Nailable concrete plank	81
Nailer wood	118
Napkin dispenser	207
Needle masonry	329
Neoprene adhesive	137
expansion joint	61, 152
flashing	153
floor	199
gasket	136
roof	146, 149
vibration pad	103
waterproofing	137
Net safety	4
tennis	55
Newel	131
wood	392
Newspaper rack	228
Night depository	223
No-hub pipe	256
Non metallic hardener	74
Non-destructive testing	8
Non-removable pin	178
Non-shrink grout	77
Norman face brick	379
Norwegian brick	86
Nosing checkered plate	108
mat	236
rubber	201
stair	79, 110, 195, 201
Nozzle dispersion	273
fire-hose	275
fog	275
Nurse call light	318
call system	318
speaker station	318
station cabinet	235
Nursing home bed	237
home S.F. & C.F.	325, 413
Nut hex	112
Nylon carpet	198

O

Oak door frame	161
floor	202
molding	128
paneling	129
stair tread	131
threshold	161
Oakum	256, 382
Obscure glass	183
Observation dome	242
well	47
Off highway truck	11
Office expense	2, 343
field	6
floor	245
furniture	238
medical S.F. & C.F.	325
partition	214
S.F. & C.F.	326, 328, 413
safe	230
trailer	6
Oil filled transformer	309
formwork	60
heater temporary	12
interceptor	268
storage tank	249
tank	284
water-heater	272-273
Oil-fired boiler	277
furnace	279
space heater	284
Olive knuckle hinge	179
Omitted work	3
One piece astragal	176
Open web joist	105
Opener automatic	
door	169-170, 177
handicap	177
Opening framing	100
Operable partition	215
Operating cost	9
cost equipment	9, 350
Operator automatic	177
gate	52
Ornamental metal	106, 390
rail	109
Outdoor bleacher	51
lighting	12
Outlet-box steel	305
Outrigger wall pole	212
Oven	221, 226
cabinet	125
microwave	221
Overbed table	237
Overhaul	19, 32
Overhead & profit	3, 342-351
contractor	1, 4
conveyor	251
door	168-170
hoist	251
support	109, 111
Overlapping astragal	177
Overlay face door	162-164
pavement	48
Overpass	58
Overtime	3, 348
Oxygen lance cutting	20

P

P trap	261
P&T relief valve	263
Package chute	210
receiver	223
Packaged terminal	
air-conditioner	296
Pad vibration	103
Padding carpet	199
Paddle blade air circulator	293
tennis court	55
Paint lead removal	330
Painting balustrades	204
bar joist	105
booth	222
building	402
bulb tee	104
cabinet	203
casework	203
ceiling	205
clapboard	205
concrete	205
concrete floor	204
cornice	204
door	203
drywall	191, 205, 402
fence	204
floor	204
grille	204
gutter	204
pavement	49
pipe	204
plaster	205
reflective	49
siding	205
sprayer	13
stair tread	110
steel	103
steel siding	205
structural steel	388
stucco	205
swimming pool	249, 402
tennis court	55
texture 1-11	205
thermoplastic	49
trim	204
truss	204
wall	205
window	203
wood floor	204
wood shingle	205
Pallet rack	219
Pan form stair	110
formwork	363
metal	63
shower	153
slab	75
stair	79
Panel access	197
acoustical	198
aggregate	146
architectural	147
brick wall	89
ceramic tile	194
collecting	283
concrete	375
dome closure	404
door	164-165
epoxy	146
facing	97
fiberglass	147
fire	317
luminous	197
metal	144, 395
metal facing	147
polyester	146
polystyrene	146
precast	81, 375, 377
sandwich	147
sound	239
spandrel	184
wall	76, 80, 87, 96
woven wire	218
Panelboard circuit breaker	307
electric	306
w/circuit-breaker	306
Paneled door	159
pine door	166
Paneling	128, 391
board	128
cutout	22, 329
demolition	24
plywood	129
wood	128
Panelized shingle	143
Panel asphalt	144
Panic bar	180-181
device door hardware	176
Panoramic elevator	251
Paper building	135
curing	76
sheathing	135
tubing	69
Paperhanging	206, 401
Paperholder	208
Paraffin bath	229
Parallel bar	229, 231
Parcel lift	252
Pargeting	395
Parging cement	136
Park bench	51
Parking barrier	49, 230
barrier precast	49
booth	230
bumper	49
equipment	230
garage	243
garage S.F. & C.F.	323, 413
site	357
stall painting	49
Parquet floor	202
Part bin	218
Particle board siding	130
board underlayment	124
core door	162
Parting bead	128
Partition	215
acoustical	214-215
block	92
blueboard	192
bulletproof	223
demolition	25
door	215
drywall	192
dust	19
folding	214
folding leaf	214
framing	191
gypsum	192, 214
hospital	214
masonry	379
mesh	218
metal	215
office	214
operable	215
plaster	188, 399-401
plywood	215
portable	215-216
refrigeration	246
sheetrock	192
shower	97, 216
solid	193
steel	214
stud	188
thin plaster	192
tile	94
toilet	97, 216-217
wall	188, 192
wire	218
wood frame	121
work stations	215
woven wire	218
Passage door	166
Passenger elevator	251, 405
Patch core hole	7
form	60
roof	135
Patching asphalt	48
concrete	78
concrete floor	77
utility trench	48
Patient nurse call	318
Patio	50, 87
block	97, 150
door	168
Patterned glass	183
Paumelle hinge	179
Pavement asphalt	357
emulsion	49
overlay	48
painting	49
removal	19
sealer	55
Paver bituminous	13
concrete	13
floor	87
highway	50
Paving	357
berm	48
block granite	50
brick	48
concrete	48
slate	97
stone	47
Peastone	56, 396
Peat moss	56
Pedestal floor	245
type seating	239
Pedestrian bridge	58
crossing	50
traffic control	238
Pegboard	128
Pencil rod	61
Penetration macadam	47
testing	7
Penthouse roof louver	295
Perforated ceiling	197
pipe	43
siding	144
Performance bond	3
Perimeter insulation	138
Perlite concrete	72, 81, 366, 374
insulation	139-140
plaster	188, 400
sprayed	206
Permit building	3
Personnel field	2
hoist	15, 350
Petrographic analysis	7
Pew church	223
Phenolic foam	139
Phone booth	220
Photography	3
aerial	3
time lapse	3
Physician's scale	229
Pib roof	149
Pickup truck	14
Picture window	172-173, 175
Pier brick	85
concrete	75, 365-368, 372
formwork	360
Pilaster formwork	67
toilet partition	217

INDEX

  wood column 125
Pile 36
  bearing 355
  boot 38, 355
  cap 38, 64, 76, 78, 355
  cast in place 36
  concrete 36
  cutoff 38, 354-355
  driver 9, 350, 354-355
  driving mobilization 36
  encasement 38
  extractor 354-355
  foundation 36, 355
  H 33
  hammer 9
  high strength 33
  lightweight 33
  load test 38
  mobilization hammer 36
  pipe 36-37
  point 37, 355
  precast 37
  sheet 33, 354-355
  splice 37-38, 355
  steel 37
  step tapered 37
  testing 38
  wood 38, 355
  wood sheet 33
Pillow tank 250
Pin floating 178
  non-removable 178
Pine 394
  column 125
  door 165
  door frame 161
  fireplace mantel 126
  floor 202
  molding 127
  roof deck 133
  shelving 129
  siding 130
  stair 131
  stair tread 131
Pipe 39, 42-44, 408
  & fittings 253-261, 263-265
  acid resistant 256
  asbestos cement 41
  bedding trench 40
  bellow expansion joint 286
  brass 255
  bumper 106
  cast-iron 255
  cleanout 253
  column 98
  concrete 41, 351
  copper 44, 256
  corrosion-resistant 256
  corrugated 43
  CPVC 257
  culvert 41
  drain 27
  drainage 256
  drainage and sewage 41
  DWV ABS 257
  DWV PVC 257
  epoxy fiberglass wound 256
  fire retardant 257
  gas 40
  glass 257
  grooved-joint 259
  header 15
  insulation 282
  iron alloy mechanical joint 256
  lining 41
  no-hub 256
  painting 204
  pile 36-37
  plastic 257
  polyethylene 40
  polypropylene 256
  proxylene 257
  quick coupling 15
  rail 109
  reinforced concrete 41
  relay 27
  removal 20
  sewer 41
  shock absorber 259
  single hub 255
  soil 255-256
  specialties 408
  stainless-steel 257-258
  steel 40, 98, 258-259
  support 109
  transite 41
  water distribution 44
  weld joint 258
Pipe-covering 281-282
  fiberglass 281
  foam 282
Pipe-fitting grooved-joint 259
Piping elevated installation 255
  hydrant 44
  removal 25
  specialties HVAC 286-287
Pit cover 106
  excavation 29
  leaching 46
  scale 219
  sump 27
  test 20
Pitch coal tar 145
  emulsion tar 49
  roof 137
Pivoted window 173
Placing concrete 78, 367
Plain tube framing 184
Planer shop 233
Plank concrete 81, 377
  floor 108, 120
  grating 108
  roof 122
  sheathing 122
  wood fiber 82
Plant aeration 46
  bed 56
  power S.F. & C.F. 326
Planter 53, 238
  bench 53
Planting 56
Plaque 220
Plaster 400
  accessory 185
  aggregate 400
  ceiling 188, 400
  cement 246
  cutout 22, 329
  demolition 21, 25
  drill 112
  expansion joint 186
  gauging 187, 400
  ground 121
  gypsum 187, 400
  mixer 13
  painting 205
  partition 188, 400
  partition thin 192
  perlite 188, 400
  portland 401
  stud 186
  thincoat 189
  vermiculite 188
  wall 188
  white cement 403
  wood fiber 189
Plasterboard 190
Plastic convector cover 125
  faced hardboard 128
  fireproofing 189
  grating 108
  grille 291
  laminate door 162, 164
  laminated counter top 126
  pipe 43, 257
  pipe corrosion-resistant 256
  roof ventilator 293
  sign 220
  skylight 160
  sleeve 00
  tubing 40, 43, 45
  valve 264
  wall tile 194
Plate checkered 106, 111
  flat 63
  glass 183, 398
  glass greenhouse 244
  push pull 181
  shear 116
  wall switch 305
  wood 120
Platform checkered plate 106
  grating 108
  lifter 225
  telescoping 232
  tennis 55
  trailer 14
Plating zinc 115
Players bench 51
Playground equipment 51, 53
  surface 201
Plenum barrier 196
  silencer 292
Plexiglass 183
  dome closure 404
  mirror 183
  sign 220
Plinth 65
Plug wall 84
Plug-in bus-duct 310-311
  circuit breaker 311
  switch 311
Plugmold raceway 300
Plumbing 253, 408
  appliance 272-273
  average cost 253
  demolition 24
  fixture 266-271
  laboratory 227
Plyform 393
Plywood clip 116
  demolition 21, 25
  dome 404
  fence 6
  floor 123
  formwork 61, 63, 66, 359-360, 362-363, 391
  joist 123
  mounting board 122
  paneling 129, 391
  partition 215
  sheathing 122, 391
  shelving 129
  sidewalk 7
  siding 130, 391
  sign 220
  soffit 131
  stressed skin 123
  subfloor 123, 391
  treatment 121
  underlayment 124, 391
Pneumatic control-system 290
  tube 223
  tube system 252
Pocket door frame 161
Point masonry 85
  moil 12
  mortar 84
  pile 37-38, 355
Pointing masonry 330, 378
Poisoning soil 34
Poke-thru fitting 305
Pole aluminum light 314
  athletic 53
  closet 127
  cross-arm 315
  electrical 315
  light 314
  portable decorative 238
  steel light 314
  tele-power 300
  utility 314-315
Police station S.F. & C.F. 326, 413
Polyacrylate floor 199
Polycarbonate glazing 183
Polyester fabric structure 404
  floor 199
  panel 146
Polyethylene expansion joint 61
  fabric structure 404
  film 135
  floor 200
  pipe 40
  pool cover 248
  tarpaulin 5
  waterproofing 137
Polypropylene carpet 198
  pipe 256
  valve 265
Polystyrene blind 130
  ceiling 197, 246
  ceiling panel 197
  insulation 138-140, 147, 246
  panel 146
Polysulfide caulking 136
Polyurethane caulking 136
  expansion joint 61
  floor 200
  roof 146
  varnish 206, 402
Polyvinyl chloride pipe 45
  chloride roof 149
  soffit 155
  tarpaulin 5
Ponderosa pine 394
Pool accessory 248
  concrete 403
  cover 248
  gunite 403
  heater electric 284
  heater gas-fired 284
  liner 403
  residential 403
  swimming 248, 403
  swimming S.F. & C.F. 328, 413
  therapeutic 250
Porcelain enamel doors 159
  enamel frame 158
  enamel shingle 142
  facing panel 147
  tile 193
Porch lift 251
  molding 127
Porous wall pipe 43
Portable air compressor 11
  booth 246
  building 6
  chalkboard 210
  chiller 234
  eye wash 268
  fire extinguisher 273
  floor 202
  heater 12
  light 12
  partition 215-216
  post 238
  scale 219
  stage 232
Portland cement 73, 365-368, 374
  cement mortar 84
  cement plaster 189, 401
Post 238
  athletic 53
  cap 116
  cedar 125
  concrete 49
  corner 52
  demolition 23
  fence 52
  hydrant 45
  jack 99
  lamp 108
  office S.F. & C.F. 326
  portable 238
  recreational 53
  tennis 55
Post-tensioned concrete 70, 79, 377
Postal specialty 213
Potential transformer fused 309
Potters wheel 233
Poured concrete 72, 362-368, 371-372, 374, 376
  gypsum 82, 396
  insulation 138
Power capacitors 311
  plant S.F. & C.F. 326
  shovel 353
  systems & capacitors 312
  temporary 6
  wiring 305
Pre-engineered building 100, 387
Pre-hung door 166, 395, 397
Preblast survey 28
Precast bridge 58
  catch basin 39
  column 80
  concrete 80, 377
  coping 85
  curb 48
  lintel 83
  manhole 39
  median 49
  panel 81, 377
  parking barrier 49
  pile 37
  receptor 216
  roof 81
  septic tank 45
  sidewalk 50
  sill 89
  stair 80
  tee 80
  terrazzo 195
Prefabricated building 243
  comfort station 242
Prefinished door 162
  drywall 190
  floor 202
Premolded expansion joint 61
Pressure booster system 269
  grout 31
  grouting cement 31
  injected footing 39, 356
  pipe 44-45
  regulator oil 286
  regulator steam 286-287
  relief valve 263
  valve relief 263
Pressurized fire extinguisher 274
Prestressed cable 365
  concrete 70, 79-80, 377
  pile 37
Prime coat 47
Primer asphalt 135, 201
  steel 103
Prison door 225
  equipment 225
  fence 52
  hinge 179
  S.F. & C.F. 325
  toilet 225
Process air handling fan 293
Proctor compaction test 8
Produce case 230
Production efficiency 348
Profile block 90-91
Project manager 2
  overhead 3
  size modifier 413
Projected window 172-173
Projection screen 210, 218, 230
Projector movie 230
Propeller exhaust fan 294
  fan 410
Property line survey 5

INDEX

Proportionate quantities	
concrete	368
Protected metal	148
Protection corner	106
door	107
dust	21
of work	414
slope	56
stile	177
termite	34
winter	8
work	329
Proxylene pipe	257
Public address system	318
garage	403
housing S.F. & C.F.	324
Pull box	303
box explosionproof	304
door	134
Pulpit church	223
Pump	269-270
circulating	269
concrete	11, 78, 367
contractor	27
diaphragm	13
duplex	269
fire	275
gasoline	222
heat	295
jetting	15
operator	34
rental	13
shallow well	270
sprinkler	270
submersible	13, 270
sump	222, 270
trash	13
water	13, 15, 47, 269
wellpoint	15, 356
Pumping	27
dewater	27, 35, 356
station	46
Purlin roof	120
sub	104
Push button lock	179
pull plate	181
Putlog	4
Putting surface	200
PVC adhesive	137
blind	235
chamfer strip	59
conduit in-slab	299
control joint	82
DWV pipe	257
expansion joint	61
flashing	153
gasket	136
gravel stop	154
pipe	43, 45, 257
sheet	137, 200
siding	130, 150
subdrain	44
underground duct	315
valve	264-265
waterstop	68
Pyrex pipe	257

Q

Quarry excavation	28
tile	194
Quarter round molding	128

R

Raceway	298-299
conduit	298
plugmold	300
wiremold	300
wireway	300
Rack bicycle	53
coat	133
hose	274
pallet	219
Racquetball court	248
Radial arch	132
brick chimney	242, 404
rib dome	404
wall	66
Radiant ceiling	245
heat	245, 317
Radiator cast-iron	280
supply control	286
thermostat control-system	290
Radio frequency shielding	247
tower	246
Radiography test	8
Rafter anchor	116
demolition	23
hip	119
sister	331
tie	119
valley	119
wood	119
Rail balcony	109
bumper	106
crane	107
dock shelter	226
guard	49
map	208
ornamental	109
pipe	109
trolley	58, 207
wall	109
Railing	109
church	223
demolition	24
metal	109
temporary	6
wood	127, 129, 131, 392
Railroad bolt	116, 358
bumper	58
siding	58, 358
switch	358
tie	358
tie step	50, 358
track	58, 358
track removal	20
work	58, 358
Rain curtain	53
Raised floor	245
Ramp approach	246
moving	252
temporary	6
Ranch plank floor	202
Range cooking	221
hood	222
receptacle	305
restaurant	227
shooting	247
wiring	320
Ratio water cement	8
Re-point masonry	378
Reading table	228
Ready mix concrete	74, 365-368, 372, 374
Rebutted shingle	394
Receivers ash	234
trash	234
Receptacle appliance	319
duplex	305, 319
fiberglass	55
range	305
trash	55
waste	208
wiring weatherproof	320
Receptor shower	216, 270, 408
Recessed mat	236
Reciprocating water chiller	297
Recirculating chemical toilet	46
Recorder videotape	319
Recore cylinder	179-180
Recreational post	53
Rectangular diffuser	290
ductwork	292
structural tubing	99
Rectifier low-voltage silicon	305
Red face brick	381
Redwood	394
cupola	126
louver	126
paneling	128
siding	130
tank	250
wine vault	234
Refinish floor	202
Reflective block	93
glass	184
insulation	138
painting	49
sign	50
Refrigerated case	230
Refrigeration	246
bloodbank	229
commercial	227, 230
equipment	244
floor	246
insulation	246
mortuary	229
panel fiberglass	246
partition	222
residential	222
walk-in	246
Refrigerator door	167
Refuse chute	210
hopper	210
Register air-supply	291
cash	224
doctor	317
in-out	317
return	291
steel	291
wall	291
Reglet	60, 155
Regulator oil pressure	286
steam pressure	286-287
Reinforced brick	89, 381
Reinforcing	72
accessory	69
bar	362-364
chair	69
coated	70
concrete	69-70, 364
galvanized	70
masonry	83, 379
mesh	77
spacer	70
steel	48
testing	8
Relay pipe	27
railroad track	58
Release door	318
form	74
Relief pressure valve	263
valve P&T	263
valve pressure	263
vent ventilator	297
water vacuum	262
Religious education	
S.F. & C.F.	326, 413
Remodeling factor	414
Removal asbestos	20
catch basin	19
chain link fence	19
concrete	20
curb	20
driveway	19
fence	19
guard rail	19
hydrant	19
interior	414
masonry	19
pavement	19
piping	25
railroad track	20
shingle	25
sidewalk	20
stone	19
stump	16
tree	16
wallpaper	330
window	26
Remove lead paint	330
reset door	331
Rendering	3
Renewal seat	51
Rental equipment	14
equipment rate	4, 9
generator	12
Repair & remodeling	414
remodel factors	329
roof	331
sill	330
Repellent water	205
Replace glass	331
Replacement sash	176
Resaturant roof	135
Research building S.F. & C.F.	413
laboratory S.F. & C.F.	326
Reshoring	67
Residential appliance	221, 228
closet door	160
column	98
door	160-161, 164, 395
dryer	228
elevator	251
fence	52
folding partition	214
garage	243
greenhouse	243
heater electric	317
hinge	178
lock	179
overhead door	169
pool	403
rail	109
refrigeration	222
stair	132, 392
storm door	160
washer	228
water-heater	228
wiring	319-320
wood window	398
Resilient floor	200, 401
pavement	54
tile	200
Resin epoxy	73
Resquared shingle	143, 394
Restaurant furniture	238
range	227
S.F. & C.F.	326, 413
Restoration masonry	329, 414
Resurfacing railroad	58
wall	109
Retail store S.F. & C.F.	327, 413
Retaining wall	54-55, 76, 368
wall formwork	66
Retrofit glass	331
Return register	291
Reveal tile ceiling	198
Revolving darkroom	242
dome	242
door	167-168, 171, 241
Rewind table	230
Rib lath	187, 399
Ribbed siding	144
waterstop	68
Ridge board	119
cap	141, 143
flashing	100
roll	149
shingle	142
vent	155
Rig drill	16
Rigid conduit in-trench	299
in-slab conduit	299
insulation	138
Ring split	116
toothed	116
Rip rap	32
Ripper	31
Riser rubber	201
stair	195
terrazzo	195
wood	392
wood stair	131
Rivet	115
tool	115
Road	47
& walk	48, 357
gutter	48
scraper	10
sign	50
temporary	7
Roadway box buffalo	47
lighting	314
Robe hook	207
Rock drill	16, 28
excavation	28
Rod backer	61, 135
clamp	61
closet	129
curtain	207
ground	303
pencil	61
threaded	61
tie	33, 99
weld	103
Roll ridge	149
roof	144
type air-filter	288
up grille	170
Roller compaction	26
earth	10
sheepsfoot	26
Rolling door	169, 171
earth	27, 353
grille support	110
ladder	4
roof	157
service door	170
tower	4
Roman brick	88
face brick	379
Romex copper	303
Roof adhesive	144
aggregate	73, 396
air supported	404
aluminum	143, 396
asphalt	396
baffle	140
ballast	149
beam	132
bolt	58
bond	146
built-up	145, 396
cant	119, 146
clay tile	142
coating	135
concrete	80, 377
copper	150
CPE	148
CSPE	148
curb	119
cutout	329
deck	105, 122-123, 377
deck concrete	80
deck formboard	81
deck gypsum	396
deck insulation	138
deck laminated	133
deck truss	392
deck wood	133
dome	157, 404
drain	158, 254
elastomeric	146
EPDM	148
exhauster	294
expansion joint	113, 152
fiberglass	147
folded plate	123
formboard	81
frame	99
framing removal	22
gravel	72, 396
gypsum	81
hatch	156
insulation	82, 100

430

INDEX

jack . . . 119
joint . . . 152
lead . . . 155
members concrete . . . 80
metal . . . 143, 396
mineral . . . 145
modified bitumen . . . 149
monel . . . 155
mop . . . 146
nail . . . 115
neoprene . . . 149
patch . . . 135
PIB . . . 149
pitch . . . 137
plywood . . . 123
polyvinyl chloride . . . 149
precast . . . 81, 377
purlin . . . 120
reinforced PVC . . . 149
repair . . . 331
resaturant . . . 135
roll . . . 144
rolling . . . 157
sheathing . . . 122
sheet metal . . . 150
shingle . . . 141, 395-396
single-ply . . . 148
skylight . . . 156
slate . . . 142
stainless steel . . . 155
stressed skin . . . 123, 404
tile . . . 142, 395
truss . . . 102, 120, 122, 132
vent . . . 101, 156
ventilator . . . 140
ventilator plastic . . . 293
walkway . . . 150
wood . . . 123
zinc . . . 156
Roofing cold . . . 145
demolition . . . 25
Rooftop air-conditioner . . . 296
multizone air-conditioner . . . 296
Room acoustical . . . 245
audiometric . . . 242
humidifier . . . 280
Rope decorative . . . 239
exercise . . . 231
wire . . . 111
Rosewood door . . . 163
Rosin paper . . . 135
Rotary hammer drill . . . 12
Rough buck . . . 118
carpentry . . . 391
grade . . . 31
hardware . . . 122
sawn lumber . . . 393
stone wall . . . 97
Rough-in dental fountain . . . 267
drinking fountain deck . . . 267
drinking fountain floor . . . 267
drinking fountain wall . . . 267
sink counter top . . . 271
sink raised deck . . . 271
sink service floor . . . 271
sink service wall . . . 271
tub . . . 267
Round diffuser . . . 291
rail fence . . . 53
table . . . 239
Rowing machine . . . 226
Rowlock course . . . 88
Rubber astragal . . . 176 177
base . . . 200
coating . . . 137
control joint . . . 82
corner guard . . . 203
dock bumper . . . 225
expansion joint . . . 61
floor . . . 200
floor tile . . . 200
mat . . . 236
nosing . . . 201
pipe insulation . . . 282
sheet . . . 200
stair . . . 201
threshold . . . 181
tired roller . . . 10
waterproofing . . . 136
waterstop . . . 68
Rubberized expansion joint . . . 61
pavement . . . 54
Rubbing wall . . . 77
Rubbish chute . . . 19
handling . . . 19
trucking . . . 19
Runner zee . . . 186
Running track . . . 54, 202
trap . . . 261
Rupture testing . . . 8
Rustication strip . . . 66

S

Safe data . . . 231
diskette . . . 231
office . . . 230
Safety deposit box . . . 223
equipment laboratory . . . 228
fixture . . . 267
fixture industrial . . . 267
glass . . . 182
glass nosing . . . 111
net . . . 4
nosing . . . 201
shower . . . 268
switch . . . 307
Sailor course . . . 88
Salamander . . . 13
Sales tax . . . 6, 348
Salt treatment lumber . . . 121
Sand . . . 84
blast steel . . . 103
blasting . . . 87
concrete . . . 72, 366
course . . . 48
fill . . . 27
grout . . . 330
masonry . . . 378
pool bottom . . . 403
screened . . . 84
Sandblasted glass . . . 184
Sandblasting concrete . . . 77
equipment . . . 13
wall . . . 87
Sanding floor . . . 203
Sandstone . . . 97
flagging . . . 50
Sandwich panel . . . 147, 150
panel aluminum . . . 144
panel skylight . . . 157
wall concrete . . . 375
wall metal . . . 395
wall panel . . . 150
Sanitary cove base . . . 193
Sash . . . 100, 399
aluminum . . . 172
metal . . . 173
replacement . . . 176
steel . . . 173
wood . . . 175
Sauna . . . 231
door . . . 231
Saw . . . 13
chain . . . 13
circular . . . 13
concrete . . . 11
cutting . . . 19
masonry . . . 13
shop . . . 233
table . . . 233
Sawing brick . . . 87
Scaffold . . . 351
Scaffolding specialties . . . 4-5
stair . . . 5
tubular . . . 4
Scale . . . 218
contractor . . . 219
crane . . . 219
floor . . . 218
physician's . . . 229
pit . . . 219
portable . . . 219
truck . . . 218
warehouse . . . 219
Scanning checkout . . . 224
Scheduling . . . 5
board . . . 210
critical path . . . 5
School . . . 327
cabinet . . . 236
crossing . . . 51
door hardware . . . 176
elementary S.F. & C.F. . . . 327, 413
equipment . . . 231
furniture . . . 238
hinge . . . 179
jr. high & middle S.F.& C.F . . . 413
S.F. & C.F. . . . 327, 413
TV . . . 319
wardrobe . . . 213
Scissor gate . . . 219
lift . . . 252
Scoreboard baseball . . . 232
Scored block . . . 91
Scrape and seal . . . 330
Scraper . . . 10, 29, 31
elevating . . . 29
mobilization . . . 32
self propelled . . . 29, 31-32
Scratch coat . . . 189
Screed . . . 65, 69
Screen chimney . . . 82
entrance . . . 217
fence . . . 53
metal . . . 173
molding . . . 127
projection . . . 210, 218, 230
security . . . 111, 173
squirrel and bird . . . 83
urinal . . . 217
window . . . 173, 175-176
wood . . . 176
Screened sand . . . 84
Screw aluminum . . . 144
anchor . . . 69, 113
brass . . . 117
drywall . . . 190
insulation . . . 138
jack . . . 34
lag . . . 114
machine . . . 114
sheet metal . . . 115
steel . . . 117
wood . . . 117
Scrub coat . . . 330
station . . . 229
Scupper . . . 248
drain . . . 254
Seal pavement . . . 49
security . . . 176
Sealant . . . 135
acoustical . . . 191
caulking . . . 135
tape . . . 136
Sealcoat . . . 49
Sealer concrete . . . 74
pavement . . . 55
Sealing after fire . . . 330
curb . . . 49
Seamless floor . . . 199
Seat renewal . . . 51
Seating . . . 239
church . . . 224
movie . . . 230
Secondary treatment plant . . . 46
Section steel . . . 99
Sectional door . . . 168-169
overhead door . . . 169
Security fence . . . 52
gate . . . 219
hinge . . . 179
sash . . . 173
screen . . . 111, 173
seal . . . 176
vault door . . . 171
Sediment bucket drain . . . 254
strainer Y valve . . . 265
See-saw . . . 53
Seeding . . . 56, 358
Select brick . . . 381
Self propelled crane . . . 15
propelled scraper . . . 32
supporting tower . . . 246
Self-closing relief valve . . . 263
Self-contained air-conditioner . . . 296
Sentry dog . . . 8
Septic tank . . . 45
Service door . . . 160
electric . . . 305
entrance cable aluminum . . . 303
sink . . . 271
sink faucet . . . 254
station equipment . . . 222
wiring . . . 320
Set accelerator . . . 74
Sewage holding tank . . . 46
municipal waste water . . . 46
pump . . . 270
pumping station . . . 46
treatment plant . . . 46
Sewer backup control . . . 254
control floodback . . . 254
control system . . . 254
grating . . . 40
pipe . . . 41
Shade . . . 239
Shaft wall . . . 193
Shake aluminum . . . 141
shingle . . . 394
wood . . . 143
Shear connector . . . 116
plate . . . 116
test . . . 8
wall . . . 122
Sheathed nonmetallic cable . . . 303
Sheathing . . . 122, 391
asphalt . . . 123
gypsum . . . 122
insulation . . . 140
paper . . . 135
plywood . . . 122
roof . . . 122
Sheepsfoot roller . . . 10, 26-27
Sheet base . . . 145-146
floor . . . 199
glass . . . 184, 398-399
metal . . . 150, 410
metal screw . . . 115, 144
pile . . . 33, 354-355
Sheeting . . . 33
driver . . . 12
open . . . 34
tie back . . . 34
wale . . . 33
wood . . . 27, 33
Sheetrock . . . 190
partition . . . 192
removal . . . 25
Shelf bathroom . . . 207
bin . . . 218
brick . . . 65
directory . . . 220
library . . . 228
metal . . . 219
refrigeration . . . 246
Shell fabric structure . . . 404
Shellac door . . . 203
Shelter aluminum . . . 247
construction . . . 404
dock . . . 226
golf . . . 53
Shelving mobile . . . 228
steel . . . 219
wood . . . 129
Shield expansion . . . 113
light . . . 49
termite . . . 156
Shielded cable . . . 302
copper cable . . . 302
Shielding lead . . . 247
radio frequency . . . 247
Shift work . . . 3, 348
work requirement . . . 329
Shim washer . . . 60
Shingle . . . 141, 394-395
asbestos . . . 141
asphalt . . . 142
concrete . . . 142
metal . . . 142
panelized . . . 143
removal . . . 25
repair . . . 331
roof . . . 141
wood . . . 143
Shipping door . . . 240
Shock absorber pipe . . . 259
absorbing door . . . 170
Shooting range . . . 247
Shop drill . . . 233
equipment . . . 233
Shoring . . . 34, 67
and bracing . . . 329
frame system . . . 68
horizontal . . . 67
installation . . . 67
temporary . . . 33
vertical . . . 68
Shovel . . . 29
front . . . 10
Shower arm . . . 270
built-in . . . 270
by-pass valve . . . 254
door . . . 216
drain . . . 254
group . . . 270
pan . . . 153
partition . . . 97, 216
receptor . . . 216, 270
safety . . . 268
stall . . . 216, 270
Shredder . . . 234
compactor . . . 233
Shrinkage test . . . 7
Shrub . . . 56-57, 358
moving . . . 56
Shutter . . . 235
Siamese . . . 275
Side beam . . . 62
Sidelight . . . 161
Sidewalk . . . 49, 87
asphalt . . . 49, 357
bridge . . . 4
removal . . . 20
temporary . . . 7
Siding . . . 394
aluminum . . . 143, 395
concrete . . . 81, 375
corrugated . . . 148
demolition . . . 25
fiberglass . . . 147
hardboard . . . 130
integrated . . . 147
metal . . . 143, 395
mineral . . . 147-148, 396
nail . . . 115
painting . . . 205
plywood . . . 391
railroad . . . 358
redwood . . . 130
removal . . . 25
steel . . . 149
vinyl . . . 150
wood . . . 130

INDEX

Sieve analysis	7-8
Sign	7, 219
base	67
letter	219
reflective	50
road	50
stop	50
street	220
Signal bell fire	317
traffic	50
Silencer duct	292
louver	292
plenum	292
Silicon carbide abrasive	77
carbide aggregate	73
Silicone caulking	136
coating	137
Sill	96
block	93
door	128, 161, 181
formwork	67
masonry	89
mud	119
quarry tile	194
repair	330
stone	95, 98
window	176
wood	119
Sillcock	255
Silo	68, 247, 363
Simulated brick	87
stone	97
Single hub pipe	255
hung window	172
tee beam	80
zone rooftop unit	296
Single-ply roof	148
Sink	408
barber	223
base	124
counter top	271
counter top rough-in	271
darkroom	224
laboratory	227
laundry	269
lavatory	268
raised deck rough-in	271
removal	24
service	271
service floor rough-in	271
service wall rough-in	271
support	260
Site drainage	39, 41
grading	31
improvement	51, 53
parking	357
plan	3
removal	19
utility	40
Sitework electrical	315
Size modifier	413
Skating mat	236
rink S.F. & C.F.	324, 413
Skirtboard	131
Skylight	101, 144, 156
removal	25
roof	156
Skyroof	157
Slab blockout	65
bolster	69
concrete	75, 79, 365-368, 371-372, 374
cutout	21
demolition	18
elevated	75, 78
flat	63
form	105
ground	75
insert	69
lift	69, 76, 371
on grade	65
on grade removal	20
pan	75
precast	80, 365
reinforcing	71, 363-364
support	4
waffle	75
Slate	97-98
chalkboard	209, 216
flagging	50
paving	97
removal	25
roof	142, 395
roof repair	331
shingle	142
sidewalk	50
sill	89
stair	97
tile	194
Sleeper clip	70
wood	119
Sleeve anchor bolt	59
plastic	60
Slide gate	52
playground	53
swimming pool	248
Sliding door	170, 397
door shower	216
entrance	171
glass board	208
glass door	168
mirror	213
panel door	171
window	172, 175
Slimline fluorescent tube	314
Slipform	68, 363
Slop sink	269
Slope protection	32, 56
Slot anchor	59
letter	213
Slotted block	92
channel	99
insert	60
Slump block	90
Slurry trench	34
Small tool	5
Smoke detector	317
detector wiring	319
hatch	157
vent	157
vent-chimney	285
Smokestack	242
Snap tie	60, 61
Snow fence	52
guard	158
melting	317
Soaking bath	267
Soap dispenser	207
holder	208
tank	207
Soccer goalpost	53
Social club S.F. & C.F.	321
security tax	343, 345-347
Socket ground	53
Sod tennis court	55
Sodding	56, 358
Sodium-high-pressure fixture	314
Sodium-low-pressure fixture	314
Soffit drywall	190
marble	97
metal	155
plaster	189
plywood	131, 391
steel	100
stucco	189
vent	155
vinyl	150
wood	120, 131
Softener water	222
Soil cement pool bottom	403
compaction	26, 353
compactor	9, 353
grout	31
pipe	255-256, 408
poisoning	34
samples	8
tamping	26, 353
testing	7-8, 353
Solar energy	409
film glass	184
roof drain	158
screen	109
screen block	93
Solar-energy	282-283
circulator air	283
Solar-energy-system circulator	283
collector	283
controller	283
heat exchanger	283
Soldier beam	33
brick	381
course	88
Solid core door	162
partition	193
wood door	164
Sorority house S.F. & C.F.	323
Sound attenuation	198
curtain	239
movie	230
panel	239
proof enclosure	242
system component	318
Space frame	101, 123
heater floor mounted	283
heater rental	12
truss	404
truss aluminum	404
Spacer reinforcing	70
Spade air	12
Spandrel beam	61
glass	184
Spanish roof tile	142
Speaker movie	230
sound system	318
station nurse	318
Special construction	239, 403-404
door	167
hinge	179
systems	316, 318-320
Specialties	207, 403
piping HVAC	286-287
scaffolding	4-5
Specific gravity	8
gravity testing	7
Spike grid	116
railroad	116, 358
Spinner ventilator	297
Spiral reinforcing	71, 364, 372
stair	110, 132
Spire church	224
Splice butt	71
column	71
mechanical	72
pile	37-38, 355
reinforcing	71
Splicer movie	230
Split astragal	177
rail fence	53
rib block	90
ring	116
Split-system control-system	290
Splitter damper assembly	292
Sport arena S.F. & C.F.	327
Spotlight	232
Spotter	32
Sprayed acoustical	189
coating	135
fireproofing	189
insulation	139-140
paint	402
Sprayer emulsion	12
Spread footing	60, 65, 75, 360
Spreader concrete	13
Spring bolt astragal	176
bronze weatherstripping	177
hinge	179
Sprinkler alarm	317
head	276
pump	270
system	54
system accelerator	276
system component	276
Spruce	394
Square foot	
building cost	320, 413-414
structural tubing	99
Squash court	51, 247
Stabilization fabric	47
Stacked bond block	91
bond brick	88
Stadium	51, 243
cover	241
Stage curtain	232
equipment	232
lighting	232
portable	232
Staging	351
Stain cabinet	203
door	203
floor	204
truss	204
Staining lumber	132
siding	130
Stainless ceiling tile	196
duct	292
flashing	153
gutter	154
reglet	155
screen	217
sign	220
steel bolt	112, 114
steel corner guard	107
steel cot	225
steel cross	224
steel downspout	151
steel grating	108
steel gravel stop	154
steel hinge	178-179
steel pipe	257-258
steel rivet	115
steel roof	155
steel shelf	207
steel storefront	171
steel weld rod	104
tile	194
Stainless-steel pipe	257-258
Stair	131
basement	80, 131
brick	88
ceiling	211
climber	251
concrete	76, 80
disappearing	211
electric	211
finish	77
fire escape	107
formwork	65
grating	107
installation	392
landing	196
marble	97
metal	110
nosing	79, 110, 195, 201
pan	79
pan form	110
precast	80
railroad tie	50
removal	23
residential	132
riser	195
riser vinyl	201
rubber	201
scaffolding	5
slate	97
spiral	132
stone	50
stringer	196
stringer wood	118
temporary protection	7
terrazzo	195
tread	79, 95, 110, 201
tread aluminum	110
tread formwork	360
tread grating	390
tread quarry tile	194
tread terrazzo	195
tread wood	131
wood	131-132
Stairlift wheelchair	251
Stairway door hardware	176
Stake nail	60
subgrade	70
Stall shower	216, 270
toilet	217
urinal	271
Stamp time	316
Standard brick	86
Standpipe alarm	317
connection	275
steel	249
Starter wall	64
Starters boards & switches	307-308
Starting newel	131
Station control	306
hospital	229
transfer	234
weather	8
Stationary ventilator	297
ventilator mushroom	297
Steam bath	232
bath, residential	233
boiler	277-278
boiler electric	276
clean	88
cleaner	13
condensate meter	287
generator	276
humidifier	280
pressure valve	286
regulator pressure	286
trap	287
Steam-jacketed kettle	227
Steamer	227
Stearate coating	137
Steel astragal	176
beam	386, 388
bin wall	54
blind	234
blocking	117
bolt	112, 116
bridge	58
bridging	117
building	100, 387
cast valve	265
channel	185
column	388
conduit in-slab	299
conduit in-trench	299
conduit intermediate	298
conduit rigid	298
corner guard	106
deck	104
demolition	18
diffuser	291
dome	101, 243, 404
door	158-159, 167, 169-170, 396-397
downspout	151
drill	12
edging	56
fastener	114
fence	51-52
flagpole	212
floor grating	108
formwork	362
frame	158
framed formwork	63, 67
furring	186
grating	111
gravel stop	154
gutter	154
hinge	178
joist	105
ladder	108
lath	187
lintel	83, 109

432

INDEX

locker	403
mesh	77
mesh box	32
metal deck	104
painting	103
partition	214
pile	37
pipe	40, 42, 98, 258-259, 408
pipe downspout	151
pipe removal	20
pool wall	403
primer	103
protective coating	103
railing	109
register	291
reinforcing	48
roof	149
roofing & siding	395
salvage	25
sash	173
screw	117
section	99
sheet pile	355
shelving	219
shingle	142
shore	67
siding	149
sign	220
sill	89
silo	247
stair	110
stair tread	110
stressed	101
structural	98, 102, 386-390
stud	187, 191
stud wall	400
surface cleaning	103
tank	249, 284
testing	7
tile	143
tower	246
tube	98
underground duct	315
valve	265
weld rod	103
welding	103
window	100, 173, 399
window demolition	26
wire	111
wire mesh	72
wire prestressed	377
wire tie	362
Steeple	224
aluminum	224
tip pin	178
Step	50
manhole	40
masonry	50
stone	96
tapered pile	37
Sterilizer barber	223
dental	225
medical	229
Sterling pine	304
Stiffleg derrick	15
Stile fabric	177
protection	177
Stone aggregate	56, 72
anchor	60, 84
ashlar	96
base	96
cast	87
column	96
fill	27
filter	32
floor	49, 96
ground cover	56
paver	96
paving	47
removal	19
repoint	330
sill	95, 98
simulated	97
stair	50
step	96
stool	98
threshold	89
tread	95
trim cut	97
veneer	96-97
wall	55, 95, 382
Stonework	382
Stool cap	128
doctor	225
stone	98
window	89, 97
Stop door	128
gravel	154
sign	50
wheel	58
Stop/start control-station	306
Storage bldg	404
bottle	234
building S.F. & C.F.	328, 413

cabinet	227
dome	243
dome bulk	243
door cold	167
tank	249-250, 284
tank cover	241
trailer	14
van	7
Store retail S.F. & C.F.	327, 413
S.F. & C.F.	322
Storefront	171, 399
aluminum	171
Storm door	160
window	101, 173
Stove	227
wood burning	211
Strainer bronze body	287
downspout	151
gutter	154
roof	151
wire	151
Y type	287
Y type iron body	287
Strand prestressed	70
Stranded wire	301
Strap tie	116
Street sign	220
Streetlight	314
Strength compressive	7
Stressed skin	101, 123, 404
skin roof	123
steel	101
Stringer stair	196
stair terrazzo	196
wood	392
Strip cabinet	318
chamfer	59
entrance	240
filler	177
floor	202
footing	75
lighting	312
shingle	142, 395
soil	56
Striping pool lane	403
Strobe light	232
Structural engineering fee	349
excavation	29, 353
face tile	94
fastener	102
fee	1
steel	98-99, 101-102, 386-390
steel coating	388
tile	94
tubing	99
welding	102
Structure air supported	404
fabric	240
tension	250
Stub switchboard	310
Stucco	189, 401
mesh	188
painting	205
Stud block	92
clip	186
connector	115
demolition	23
driver	115
metal	186, 188, 191, 400
partition	121, 188
plaster	186
steel	187, 191
wall	120-121, 188
welded	117
wood	120, 188, 393
Stump removal	16
Sub purlin	82, 104
Subcontractor O & P	3, 342-351
Subdrain PVC	44
Subdrainage pipe	43
Subfloor	123, 391
plywood	123
Subgrade chair	70
stake	70
Submersible pump	13, 47, 270
Subpurlin	388
Subsoil heating	245
Substructure concrete	74
Subsurface exploration	8
Suction hose	12
Sugar pine	394
Sump hole construction	27
pit	27
pump	222, 270
Superintendent	2
Supermarket checkout	224
S.F. & C.F.	328, 413
scanning	224
Superstructure concrete	74, 345-348, 372
Support dowel	64-65
drinking fountain	260
lavatory	260
overhead	111

pipe	109
slab	4
Support/carrier fixture	260
Surface bonding	94
landscape	200
playground	201
Surfacing gravel	48
Surgery equipment	229
lighting	229
table	229
Surveillance system	223
system TV	319
Survey aerial	5
crew	5
monument	5
preblast	28
property line	5
stake	7
topographic	5
Suspended ceiling	118, 186, 192, 197, 245
Suspension system ceiling	197
Swell testing	7
Swimming pool	248, 403
pool blanket	248
pool enclosure	248
pool equipment	248
pool gutter	403
pool heater	284
pool hydraulic lift	248
pool painting	249
pool ramp	248
pool S.F. & C.F.	328, 413
Swing	54
check valve	262
clear hinge	179
door	170, 172
Swing-up overhead door	169
Switch bus-duct	311
dimmer	305
electric	305, 307
general duty	307
plug-in	311
railroad	58, 358
safety	307
timber	58
toggle	305
wiring	320
Switchboard electric	306
instrument	309
stub	310
transition section	309
w/bus bar	308
w/CT compartment	308
Switchbox	305
Switching low-voltage	305
Switchplate low-voltage	305
Synthetic marble	194
turf	201
System antenna	318
clock	316
control	290
fire extinguishing	273-274, 276
grinder pump	269
sewer control	254
sprinkler	54, 276
TV surveillance	319
UHF	318
VHF	318

T

T&G board	393
siding	394
T-bar mount diffuser	291
Table folding	231, 239
laboratory	227
massage	226
overbed	237
physical therapy	229
reading	228
rewind	230
round	239
saw	233
surgery	229
top	126, 134
Tack board	210
Tamper	12, 30
Tamping	353
soil	26, 353
Tandem roller	10
Tank cover	241
darkroom	224
developing	224
expansion	284
fiberglass	284
gasoline	249
holding	46
hubbard	229
oil & gas	284
pillow	250
protection galvanic	285
septic	45
soap	207
steel	249, 284

storage	249-250, 284
water	249
water storage solar	283
Tap box bus-duct	310
Tape sealant	136
Tar pitch emulsion	49
roof	137
Target range	247
Tarpaulin	5, 8
duck	5
mylar	6
polyethylene	5
polyvinyl	5
Tarred felt	146
Tax	6, 343-347
sales	6, 348
social security	6, 345-347
unemployment	6
Teak floor	202
molding	128
paneling	129
Technician inspection	8
Tee beam	80, 377
bulb	82, 104
cleanout	253
fitting	45
pipe	259
precast	80
Tele-power pole	300
Telephone booth	220
enclosure	220
exchange S.F. & C.F.	328, 413
house	318
manhole	315
pole	315
Telescoping bleacher	231
door	170
platform	232
Television equipment	318
Teller automatic	223
window	223
Temperature relief valve	263
rise detector	317
Tempered glass	183
glass greenhouse	244
hardboard	128
Tempering valve	263
Temporary building	6
construction	6
facility	6
guardrail	6
heat	6, 79
oil heater	12
road painting	49
shoring	33
toilet	13
Tennis court	51, 55
court air supported	241
court fence	52, 111
court surface	200
Tensile test	7-8
Tension structure	250
Terminal air-conditioner packaged	296
heater	281
S.F. & C.F.	328, 413
Termination box	318
Termite protection	34
shield	156
Terne coated flashing	154
Terra cotta	94-95
cotta coping	85
cotta demolition	21, 26
cotta pipe	43
cotta tile	196
probe	27
Terrace	50
Terrazzo	195, 401
base	195
demolition	22
epoxy	195, 199
floor	195, 401
precast	195
receptor	216
wainscot	195-196
Test beams	7
load	39
load pile	38
pit	20
well	47
Testing	7, 353
pile	38, 355
soil	353
sulfate soundness	7
Texture 1-11 painting	205
Theater S.F. & C.F.	328, 413
Therapeutic pool	250
Thermal batt	198
Thermometer	229
Thermoplastic painting	49
Thermostat control-system radiator	290
integral	318
Thin plaster partition	192

433

INDEX

set ceramic tile 193
shell construction 102, 123
Thincoat plaster 189
Thinset mortar 194
Threaded insert 60
rod 61
Threshold 181
door 161
stone 89, 97
vinyl 202
wood 128
Thru-wall air-conditioner 296
Ticket booth 246
dispenser 230
Tie back sheeting 34
bar 69
clip 69
concrete 58
masonry 84, 379
rafter 119
railroad 358
rod 33, 99
snap 60
strap 116
wall 60
wire 70, 186, 362
wood 58
Tier locker 213
Tile 200
abrasive 193
aluminum 141
carpet 199
ceiling 196, 198, 245
ceramic 193-194
concrete 142
cork wall 206
demolition 21
exterior 194
floor 195
flue 87
furring 94
glass 183, 194
grout 193
marble 194
metal 194
partition 94
plastic 194
pool 403
porcelain 193
quarry 194
resilient 200
reveal 198
roof 142, 395
slate 194
steel 143
structural 94
terra cotta 196
vinyl asbestos 201
wall 194
Tilt up construction 81, 376
Timber connector 116
fastener 115
framing 120, 391
guardrail 49
laminated 132, 404
roof deck 122
switch 58
Time lapse photography 3
lock 170
requirement 348
stamp 316
system controller 316
Timekeeper 2
Tin clad door 170
Tinted glass 183
Toaster 227
Toe plate 108-109
Toggle bolt 113
switch 305
Toilet 46
accessory 207
bowl 271
chemical 13
door hardware 176
partition 97, 216-217
partition removal 26
partition support 111
prison 225
stall 217
temporary 13
tissue dispenser 207-208
trailer 14
Tool air 11
rivet 115
small 5
Toothed ring 116
Toothing masonry 21
Top counter 125-126, 134
dressing 56
table 126, 134
Topographic survey 5
Topping concrete 76
epoxy 199
floor 77
Topsoil 27, 56, 358
Torch cutting 13, 20
Touch bar 181
Tour station watchman 316
Towel bar 208
dispenser 207-208
Tower control 242
cooling 290
crane 14
hoist 15
light 13
radio 246
rolling 4
Town hall S.F. & C.F. 328, 413
Track curtain 232
drawer 134
railroad 58, 358
running 202
surface 200
traverse 236
work 58
Tractor 10, 26, 28, 30-31
loader 10
loading 333
mobilization 32
truck 14
Traffic cone 6
control pedestrian 238
deck 146
detector 230
line 49
signal 50
Trailer hauling 32
office 6
platform 14
storage 14
toilet 14
truck 10, 14
Trainer bicycle 226
Transfer station 234
Transformer 309
& bus duct 309, 311
cabinet 304
current 309
dry type 309
fused potential 309
low-voltage 305
oil filled 309
Transit 14
mixer 11
Transite duct 315
Transition section switchboard . . . 309
Transom aluminum 160
lite frame 158
Trap cast-iron 261
drain 254
drainage 261
grease 268
inverted bucket steam 287
rock surface 77
steam 287
Trash closure 55
pump 13
receivers 234
receptacle 55
Traverse 236
track 236
Travertine 97
Tray cable 298
laundry 269
Tread abrasive 194
cast iron 110
rubber 201
stair 79, 95, 110
stone 97
vinyl 201
wood 131
Treadmill 226, 231
Treatment plant secondary 46
plywood 121
wood 121, 402
Tree 56-57, 358
guyed 57, 358
moving 56
removal 16
Trench backfill 30
box 14
cover 106, 111
drain 254
duct 299
duct fitting 299
duct steel 299
electric 105
excavation 29-30, 40, 353
fill 40
formwork 65
slurry 34
Trencher 10
chain 10
Trenching 27-28
Trial batch 7
Trim demolition 24
exterior 127
metal 186
painting 204
tile 193
wood 127
Triple weight hinge 179
Troffer air handling 313
Trolley rail 58, 207
Trowel concrete 11
finish 76
Troweled coating 135
Truck concrete 11
dock 225
dump 10
flatbed 10
hauling 32
leveler 225
loading 29
mounted crane 14
off highway 11
pickup 14
rental 10
scale 218
tractor 14
trailer 10, 14
Trucking 32, 377
rubbish 19
Truss bowstring 132
flying 67
metal 102
painting 204
plate 116
roof 120, 122, 132
space 404
stain 204
varnish 204
wood 392
Tub bar 207
bath 266
enclosure 216
rough-in 267
soaking 267
Tube framing 184
framing aluminum 404
pneumatic 223
steel 98
Tubing copper 44, 256, 408
electric metallic 298
electrical 298
metallic 408
paper 69
plastic 40, 43, 45, 408
structural 99, 399
Tubular aluminum 99
fence 52
lock 179
scaffolding 4
steel joist 123
Tumbler holder 208
Turf artificial 55
synthetic 201
Turned column 125
Turnout 58
railroad 358
Turnstile 220
TV closed circuit 318
closed circuit camera 319
Two piece astragal 177

U

UHF system 318
Ultra-sonic cleaner 225
Ultrasonic motion detector 317
test 8
testing 8
Under eave vent 155
Underdrain 39, 43
Underfloor duct electrical 300
duct fitting 300
Underground duct 315
tank 284
Underlayment 124, 391
acoustical 198
felt 200
gym floor 202
latex 201
Underpin foundation 34
Underwater light 248
Undisturbed soil 8
Unemployment tax 6, 343-347
Unit heater 279
heater electric 317
United inch 399
Unload reinforcing 71
Upstanding beam form 62
Urethane foam 136, 198
insulation 140, 246
roof deck 139
sprayed insulation 140
wall coating 206
Urinal 408
removal 24
screen 217
stall 271
support 260
Urn ash 234
Utensil washer medical 229
Utilities electric 315-316
Utility brick 86
duct 315
pole 314-315
pump 269
set fan 293-294
site 40
sitework 315
trench 30
vault 46

V

Vacuum breaker 262
cleaning 233
relief water 262
Valance board 125
Valley flashing 141
gutter 100
rafter 119
Valve 45, 262-265
angle 262, 264-265
assembly sprinkler deluge 276
backwater 254
ball 262, 264-265
bronze 262
butterfly 264
cabinet 274
check 265-266, 276
check swing 264
CPVC 265
fire-hose 275
flanged 265
foot 265
forged 265
gas stop 262
gate 44, 264-265
globe 263-265
hot-water radiator 286
iron body 264
plastic 264
polypropylene 265
pressure relief 263
PVC 264
shower by-pass 254
soldered gate 262
sprinkler alarm 276
steam pressure 286
steel 265
steel cast 265
swing check 262
tempering 263
water pressure 263
Y sediment strainer 265
Van storage 7
Vandalproof fixture 313
Vane-axial fan 293
Vanity base 132
top lavatory 268
Vapor barrier 135, 247
barrier sheathing 123
Vaportight fixture incandescent . . . 313
mercury-vapor fixture 313
Varnish 402
cabinet 203
casework 203
door 203
floor 204-205
polyurethane 206
truss 204
Vault door 170
front 250
utility 46
wine 234
Vaulting side horse 231
Veneer ashlar 95
brick 88, 379, 381
core paneling 129
granite 95
removal 24
stone 96-97, 382
Venetian blind 234
terrazzo 195
Vent box 84
brick 295
chimney 285
dryer 222
exhaust 296
flashing 266
foundation 84
intake/exhaust 297
metal chimney 285
metal deck 104
ridge 155
ridge strip 155
roof 101, 156
silencer 292
smoke 157
soffit 155
Vent-chimney 285
all fuel 285
fitting 285
high temperature 285
increaser 285

INDEX

Ventilating 409
air-conditioning 288, 290, 292-294, 296-297
Ventilation heat-recovery 279
louver 126
Ventilator control-system 290
fan 293
masonry 84
mushroom stationary 297
relief vent 297
roof 140
spinner 297
Venturi flow meter 288
Verge board 127
Vermiculite cement 189
insulation 138
plaster 188
pool bottom 403
Vertical blind 235
conveyor 252
lift door 171
shore 67
VHF system 318
Vibration pad 103
Vibrator concrete 11
earth 10, 26-27, 30, 353
plate 26, 353
Vibratory equipment 9-10
Vibroflotation 27, 357
Videotape recorder 319
Vinyl asbestos tile 201
blind 130
coated fabric structure 404
composition floor 201
corner guard 203
downspout 151
faced wallboard 190
floor 201
glass 184
gutter 154
lead barrier 196
leaded 196
mat 236
plastic waterproofing 137
pool liner 403
roof 146
siding 150
soffit 150
stair riser 201
stair tread 201
threshold 202
tread 201
wall coating 206
wall covering 401
wallpaper 206
Vitrified clay pipe 43-44
Vocational kiln 233
school S.F. & C.F. 327, 413
shop equipment 233
Void formwork 64
Voltmeter 309
Volume control damper 292

W

Waffle slab 76
Wagon drill 12
Wainscot ceramic tile 193
molding 128
quarry tile 194
terrazzo 195-196
Wale sheeting 33
Walk 50
moving 252
Walk-in refrigeration 246
Walkway cover 209
roof 150
Wall aluminum 143
anchor 112-113
area 106
brick 89, 381
bumper 178, 206
cabinet 124, 227, 236
canopy 208
cavity 83
ceramic tile 193
coating 206
concrete 75, 79, 365-368, 372
covering 206, 401
cross 224
cutout 21-22, 329
demolition 17, 25
drill 112
drywall 190
exhaust fan 294
expansion joint 113
finish 77
flagpole 212
formwork 65, 359-360, 362
foundation 91
framing 120
framing removal 22
furnace 284
furring 120, 186
grout 83, 378
guard 106
heater 222
heater electric 317
hydrant 45, 255, 274
insulation 94, 100, 138, 140
lath 187
masonry 55, 90
mat 232
metal sandwich 395
mirror 183
painting 205
panel 76, 80, 87, 96
panel brick 89, 379, 381
panel ceramic tile 194
panel sheathing 391
panel woven wire 218
paneling 129
partition 188, 192
patching concrete 78
plaster 187-189
plug 84
rail 109
register 291
reinforcing 71, 363-364
resurfacing 109
retaining 54-55, 76, 368
rubbing 77
sandblasting 87
shaft 193
shear 122
sheathing 122
siding 130
starter 64
stone 95
stucco 189
stud 120-121, 187-188, 191-192
switch plate 305
tie 60
tile 194
tile cork 206
urn ash receiver 208
window 184-185
Wallcovering acrylic 206
Wallguard 206
Wallpaper 206
removal 330
Walls wash 330
Walnut door frame 161
floor 202
Wardrobe 213, 236
dormitory 236
school 213
wood 133
Warehouse 240
S.F. & C.F. 328, 413
scale 219
Warm air heater 278
Wash bowl 268
brick 89, 382
fountain 271
walls 330
Washer 58, 116
commercial 228
darkroom 224
residential 222, 228
shim 60
Washing machine automatic 222
Waste compactor 233
disposal 233
incinerator 233
receptacle 208
Wastewater treatment system 47
Watchdog 8
Watchman service 8
tour station 316
Water anti-siphon 262
balancing 289
cement ratio 8
chiller 288, 297
closet removal 24
coil chilled 289
cooler application 408
cooler handicap 272
copper tubing connector 253
dispenser hot 267
distiller 229
distribution system 44
effect testing 7
fountain removal 25
heater 222, 408
heater removal 25
hose 12
hydrant 255
pipe ground clamp 303
pressure relief valve 263
pressure valve 263
pump 13, 15, 47, 222, 269, 356
pump fire 275
pumping 27, 356
repellent 205
softener 222
storage solar tank 283
supply domestic meter 266
supply meter 266
tank 249
tank elevated 249
tempering valve 263
trailer 14
vacuum relief 262
well 47
Water-closet 271
chemical 46
support 260-261
Water-cooler 272
support 261
Water-heater 273
circuit wiring 320
commercial 272-273
electric 272
gas 273
oil 272-273
Waterpipe fittings 45
Waterproofing 74, 135-137
coating 135
elastomeric 136
integral 79
masonry 84
membrane 137
Waterstop 68
fitting 68
rubber 68
Watertank sprayer 14
Watt meter 309
Weather station 8
strip window 181
Weatherproof receptacle wiring . . . 320
Weatherstrip 181
door 181
Weatherstripping 170
aluminum 176
Weathervane 111
Wedge bolt 114
form 61
insert 60
Weight exercise 226
lifting multi station 231
Weld butt 71
fillet 104
joint pipe 258
rod 103
rod aluminum 103
Welded frame 158
shear connector 116
stud 117
wire mesh 72
Welder arc 233
Welding 104
certification 8
machine 14
steel 103
structural 102, 389
Well 35, 47
gravel pack 47
pump 356
pump deep 270
pump shallow 270
Wellpoint 34, 356
equipment 15, 356
header pipe 15, 356
pump 15, 356
Wheel chair lift 251
color 232
guard 106
potters 233
stop 58
Wheelbarrow 14
Wheelchair stairlift 251
Whirler playground 54
Whirlpool bath 229, 267
Wide flange beam 101-102
throw hinge 179
Window 172, 398-399
aluminum 172
bank 222
blind 129, 234
bullet-resistant 223
casement 174
casing 127
counter 213
demolition 26
double hung 174-175
drive-up 223
frame 176
framing 100
glass 184
glass replacement 331
grille 175
guard 111
hardware 176
metal 100, 172
muntin 175
painting 203
picture 175
removal 26
screen 173, 175-176
sill 89
sill marble 96
sill quarry tile 194
sliding 175
steel 100, 173, 399
stool 89, 97
storm 173
teller 223
trim set 128
wall 185, 399
wall framing 99, 184, 399
weather strip 181
wood 174-176
Wine vault 234
Winter protection 8, 79
Wire 111
aluminum 301
copper 301
electric 301, 412
fence 6, 52, 111
glass 184
ground 303
hanging 60, 197
lath 185, 399
mesh 72, 111, 189
partition 218
prestressed concrete 70, 365
reinforcing 72
rope 111
strainer 151
THW 301
THWN-THHN 301
tie 70, 186, 362
to grounding brazed 303
window guard 111
Wiremold raceway 300
Wireway raceway 300
Wiring appliance 319
device 304-305
devices & boxes 304
disposal 319
furnace 319
light fixture 320
lighting 320
power 305
range 320
residential 319-320
service 320
switch 320
Wood 394
base 127
beam 117
bench 51
blind 129, 235
block floor 203
block floor demolition 22
blocking 117, 121
bridge 58
cabinet 236
canopy 120
casing 127
chamfer strip 59
column 119-120, 125
composition floor 203
conversion table 393
cupola 126
deck 122, 133
demolition 21
dome 123, 243, 404
door 161-162, 164, 395, 397
door frame 395, 397
drawer 134
fascia 119, 127
fastener 117
fiber ceiling 197
fiber formboard 81
fiber insulation 138
fiber plank 82
fiber plaster 189
fiber sheathing 122
fiber soffit 131
fiber subfloor 123
fiber underlayment 124
firestop 118
flagpole 212
floor 202, 394
floor demolition 22
folding partition 214
formwork 59, 359-360, 362-363
frame 133, 161
framing 391, 393
framing demolition 18
furring 120
girder 117
glue laminated 404
gutter 154
handrail 127
header 120
joist 118, 123
laminated 132
ledger 119
louver 126
molding 131
nailer 118
overhead door 169
panel door 164
paneling 128, 391

INDEX

partition 121
pile 38, 355
planter 53
plate 120
pole 315
rafter 119
railing 129, 131
roof 123
roof deck 122
roof deck demolition 25
roof truss 122
sash 175
screen 176
screw 117
shade 239
shake 143
sheathing 122, 391
sheathing cutting 19
sheeting 27, 33, 354
shelving 129
shingle 143
shingle repair 331
sidewalk 50
siding 130
siding demolition 25
sill 119
sleeper 119
soffit 120, 131
stair 131-132
stair stringer 118
storm door 166
stud 120, 188
subfloor 123
tank 250
threshold 128
tie 58
timber 394
tread 131
treatment 121, 402
trim 127
truss 392
veneer wallpaper 206
wardrobe 133
window 174-176, 398
window demolition 26
Wool carpet 199
  fiberglass 138
  lead 61
Work extra 2
  protection 329
  station partition 215
Worker compensation 1
Workers' compensation
  Canada 345-347
  compensation USA 345-347
Woven wire partition 218
Wrecking 17
  ball 14
Wrench impact 12
Wrestling mat 232
Wrought trim lock 180
Wye fitting 45

X

X-ray dental 225
  medical 229
  mobile 229
  protection 247

Y

Y sediment strainer valve 265
  type iron body strainer 287
Yardarm 212
Yellow pine 394
  pine floor 202
YMCA club S.F. & C.F. 322

Z

Z bar suspension 197
Zee bar 190
  runner 186
Zinc divider strip 195
  flashing 154
  plating 115
  roof 156
  terrazzo strip 195, 401
  weatherstrip 181

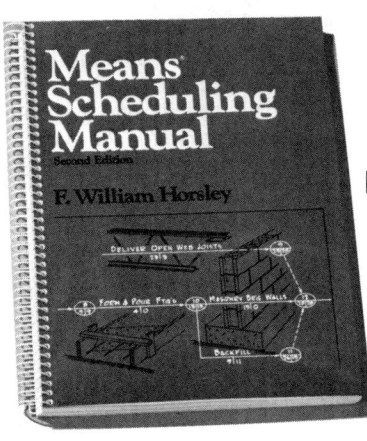

Means Scheduling Manual

By F. William Horsley

2nd edition
Over 200 pages, illustrated

Tightly written in one easy-to-use spiral-bound volume, here are today's most advanced methods for scheduling and managing building construction projects in the intermediate size range. This is a masterfully prepared construction management tool equal to today's — and tomorrow's — burgeoning scheduling problems.

Means Scheduling Manual is filled with the kind of sound practical advice and guidance you've always wanted in a scheduling book. This is a scheduling aid that's truly one step ahead of any other you've ever seen.

Book No. 67152 ... $32.95/copy   ISBN 0-911950-36-2

Labor Rates for the Construction Industry 1986

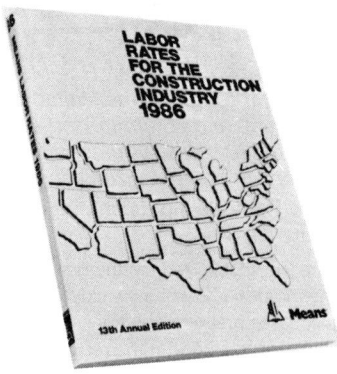

13th Annual Edition
Over 325 pages

- detailed wage rates by trade for over 300 U.S. and Canadian cities
- forty-six construction trades listed by local union number in each city

CITY
- base hourly wage rates plus fringe benefit package costs gathered from reliable sources
- dependable estimates for the trade wage rates not reported at press time

STATE
- effective dates for newly negotiated union contracts for both 1986 and 1987

NATIONAL
- factors for comparing each trade rate by city, state, and national averages
- historical 1984-1985 wage rates also included for comparison purposes
- each city chart is alphabetically arranged with handy visual flip tabs for quick reference

Book No. 60126 ... $45.95/copy   ISBN 0-87629-010-1

Means Construction Cost Indexes 1986

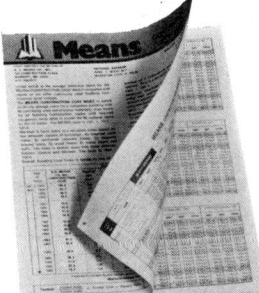

$49.95 per year

(Individual and back issues available at $17.95 per copy)

The index service providing updated cost adjustment factors.

Whether updating construction cost estimates based on information in the Means cost manuals, or those from other sources, the construction cost index service is the efficient way to assure 90-day cost accuracy.

Published quarterly (January, April, July, October), this handy report provides cost adjustment factors for the preparation of more precise estimates no matter how late in the year. It's also the ideal method for making continuous cost revisions on on-going projects as the year progresses.

The report comes in four unique sections:
- Breakdowns for 209 major cities
- National averages for 30 key cities
- Five large city averages
- Historical construction cost indexes

Book No. 60146 ... $49.95/year

Means Historical Cost Indexes 1986

$16.95 per copy

(20-page booklet)

Your latest estimating resource for historical cost information.

This sixth annual edition of *Means Historical Cost Indexes* is designed especially for professionals with estimating activities involving construction costs not only in various *cities* but in various *years*. The booklet indexes historical construction costs for 209 U.S. and Canadian cities from 1940 through 1986.

These historical indexes are the result of careful monitoring of specific material quantities, man-hours, building trades, equipment rental, and installation. This extensive coverage can be depended upon for functional and reliable assistance in all of your historical cost estimating activities. *Means Historical Cost Indexes* includes detailed explanations of uses, with complete examples.

Book No. 60136 ... $16.95/copy   ISBN 0-87629-011-X

1986—Means Seminar Information—1986

During the upcoming year, R.S. Means offers you a series of 2-day seminars oriented to a wide range of construction-related topics. Conducted throughout the United States, scheduled seminars thoroughly cover the subjects of cost estimating, scheduling, management, and computerized systems for construction. R.S. Means is proud to present these seminars, each specifically designed to give you the expertise you need to be successful in your profession. All seminars include comprehensive workbooks plus the latest innovations in computerized estimating and scheduling.

Repair & Remodeling Estimating
(two days) #104

Here are the specialized techniques you need to approach repair and remodeling projects with new skill and confidence. This seminar sorts out and discusses practical solutions to the difficult and varied cost estimating problems associated with the reuse and conversion of existing structures. A wealth of eye-opening procedures is presented, including information about computerized cost estimating as applied to repair and remodeling projects. Based on optimum use of Means cost manual *Repair & Remodeling Cost Data*. A "must" for new estimators.

Unit Price Estimating
(two days) #100

For those with estimating experience, this course shows you how today's advanced estimating techniques and cost information sources (such as computerized estimating systems and cost data banks) can be used to develop unit price estimates for projects of any size. You'll get down-to-earth know-how and easy-to-apply directions on organizing information effectively, using plans efficiently, ensuring error-free results, and maximizing your use of cost data resources, including *Building Construction Cost Data*.

Square Foot Cost Estimating
(two days) #106

For estimators with experience, here's the expert guidance you've asked for. We'll help you develop the practical know-how for producing rapid, reliable square foot cost estimates using only minimal design details. You'll perfect your skills for "before plans" estimating, replacement cost estimating, and assessing buildings of all types. In square foot costing methods, participants receive clear-cut explanations and suggestions for dealing with several types of construction projects at various stages of completion. Centered around techniques used in *Means Assemblies Cost Data, Means Square Foot Costs,* and *Square Foot Estimating*.

Scheduling & Project Management
(two days) #102

This seminar helps you successfully establish project priorities, develop realistic schedules, and apply today's advanced management techniques to your construction projects. Included are the "network" approach, Precedence, and Critical Path Methods. Special emphasis is placed on cost interfaced with various monitoring setups such as computer-based control systems. Through this seminar you'll perfect your scheduling and management skills, ensuring completion of your projects *on time* and *within budget*. Includes hands-on application of *Means Scheduling Manual* and *Building Construction Cost Data*.

Electrical/Mechanical Estimating
(two days) #114

This seminar is tailored to fit the needs of Electrical/Mechanical estimators, general contractors, sub-contractors and facilities managers seeking to prepare more comprehensive estimates of Electrical/Mechanical additions and changes. Featured are: unit price estimating, assemblies and special takeoff and pricing techniques. Highlights include:

ELECTRICAL	MECHANICAL
Raceways	Plumbing
Conductors & Grounding	Sprinkler & Fire Extinguishing
Boxes & Wiring Devices	Heating & Ventilating
Starters, Boards & Switches	Air Conditioning
Power Systems & Capacitors	Piping
Lighting	
Special Systems	

In combination with the use of *Means Electrical Cost Data* and *Means Mechanical Cost Data*, this seminar will ensure more accurate and complete Electrical/Mechanical estimates for both unit price and preliminary estimating procedures.

Registration Fees
- One 2-day seminar, $565 per person
- One 2-day seminar, two or more people registering at the same time, $515 per person
- Two 2-day seminars, $915 per person
- Two 2-day seminars, two or more people registering at the same time, $830 per person

Full payment must be received at least 15 days prior to seminar dates in order to confirm attendance (U.S. Funds).

How to Register
Complete the registration form and return with your full fee (or minimum deposit of $220/person for one 2-day seminar, or minimum deposit of $360/person for two 2-day seminars) to:

Seminar Division
R.S. Means Company, Inc.
100 Construction Plaza
P.O. Box 800
Kingston, MA 02364-0800

Class sizes are limited so please return your registration form as soon as possible.

Refunds
Advance payments are fully refundable up to 15 days prior to the seminar dates. After that time, a 2-day seminar cancellation is subject to a service charge of $160/person and two 2-day seminar cancellations $270/person.

Tax Deduction
A federal income tax deduction is allowed for expenses (including registration fees, travel, meals and lodging) for your attendance at a Means Seminar (see Treas. Reg. 1.162-5).

AACE Approved Courses
The R.S. Means Construction Estimating and Management Seminars described and offered to you here have each been approved for 14 hours (1.4 CEU) of credit by the American Association of Cost Engineers (AACE, Inc.) Certification Board toward meeting the continuing education requirements for re-certification as a Certified Cost Engineer/Certified Cost Consultant.

Coffee/Cola Breaks
Your registration includes the cost of a coffee break in the morning and an afternoon cola break. These informal segments will allow you to discuss topics of mutual interest with other members of the seminar.

Late Telephone Registrations
If you wish to register for any seminar with less than two weeks remaining before the seminar date, please call our Kingston office (617) 747-1270 and ask for the Seminar Registrar. Arrangements may be made if space is available.

Daily Course Schedule
The first day of each seminar session begins at 8:30 A.M. and ends at 4:30 P.M. The second day is 8:00 - 4:00. Participants are urged to bring a hand-held calculator since many actual problems will be worked out in each session.

Tranportation/ Hotel Accommodations
For your convenience you can obtain assistance in making transportation and hotel arrangements by calling Waterfront Travel in Boston, MA. at 1-800-343-6524, and notifying them of your planned attendance at the Means Seminar of your choice. Airline reservations, hotel accommodations, and car rental arrangements will be made for you to fit your schedule. We suggest that you contact Waterfront Travel as soon as you return your registration form in order to reserve the lodging of your choice. Reservations should be received by Waterfront Travel Service one month prior to the seminar date.

In-House Presentation
All R.S. Means seminars are available for in-house presentation and can be tailored to meet individual client/company needs. If you are interested in having a program conducted at your location, just contact Robert Gair, Director of Sales, R.S. Means Co., Inc., 150 Construction Plaza, Kingston, MA 02364. Telephone: (617) 747-1270.

1986—Means Seminar Schedule—1986

SPRING SEMINARS

BEVERLY HILLS, CA
Ramada Hotel
1150 S. Beverly Drive
Beverly Hills, CA 90035
(213) 553-6561

- 10463BH Repair & Remodeling .... March 3 & 4
- 10663BH Square Foot Costs .... March 3 & 4
- 10063BH Unit Price Est. .... March 5 & 6
- 10263BH Sched. & Proj. Mgmt. ...... March 5 & 6
- 11463BH Elect. & Mech. Est. .... March 5 & 6

SEATTLE, WA
Hyatt Seattle
17001 Pacific Highway
So. Seattle, WA 98188
(206) 244-6000

- 10463ST Repair & Remodeling ..... March 10 & 11
- 10663ST Square Foot Costs .... March 10 & 11
- 10063ST Unit Price Est. .... March 12 & 13
- 10263ST Sched. & Proj. Mgmt. .... March 12 & 13
- 11463ST Elect. & Mech. Est. ..... March 12 & 13

TAMPA, FL
Holiday Inn
4732 North Dale Mabry Highway
Tampa, FL 33614
(813) 877-6061

- 10463TA Repair & Remodeling ..... March 24 & 25
- 10663TA Square Foot Costs .... March 24 & 25
- 10063TA Unit Price Est. .... March 26 & 27
- 10263TA Sched. & Proj. Mgmt. .... March 26 & 27
- 11463TA Elect. & Mech. Est. ..... March 26 & 27

DALLAS, TX
Amfac Hotel & Resort
P.O. Box 619025
Dallas/Ft. Worth Airport
Dallas, TX 75261
(214) 453-8400

- 10464DA Repair & Remodeling .... April 7 & 8
- 10664DA Square Foot Costs .... April 7 & 8
- 10064DA Unit Price Est. .... April 9 & 10
- 10264DA Sched & Proj. Mgmt .... April 9 & 10
- 11464DA Elect. & Mech. Est. .... April 9 & 10

PITTSBURGH, PA
The Sheraton Hotel At Station Sq.
Smithfield and Carson St.
Pittsburgh, PA 15219
(412) 261-2000

- 10464PI Repair & Remodeling ...... April 28 & 29
- 10664PI Square Foot Costs .... April 28 & 29
- 10064PI Unit Price Est. .... April 30 & May 1
- 10264PI Sched. & Proj. Mgmt. ..... April 30 & May 1
- 11464PI Elect. & Mech. Est. ...... April 30 & May 1

NEW ORLEANS, LA
International Hotel
300 Canal Street
New Orleans, LA 70140
(504) 581-1300

- 10465NO Repair & Remodeling .... May 5 & 6
- 10665NO Square Foot Costs .... May 5 & 6
- 10065NO Unit Price Est. .... May 7 & 8
- 10265NO Sched. & Proj. Mgmt. .... May 7 & 8
- 11465NO Elect. & Mech. Est. .... May 7 & 8

WASHINGTON, DC
Quality Inn Pentagon City
300 Army Navy Drive
Arlington, VA 22202
(800) 848-7000

- 10465AR Repair & Remodeling ...... May 19 & 20
- 10665AR Square Foot Costs .... May 19 & 20
- 10065AR Unit Price Est. .... May 21 & 22
- 10265AR Sched. & Proj. Mgmt. ..... May 21 & 22
- 11465AR Elect. & Mech. Est. ...... May 21 & 22

DENVER, CO
Holiday Inn
1475 S. Colorado Blvd.
Denver, CO 80222
(303) 757-7731

- 10466DE Repair & Remodeling .... June 2 & 3
- 10666DE Square Foot Costs .... June 2 & 3
- 10066DE Unit Price Est. .... June 4 & 5
- 10266DE Sched. & Proj. Mgmt. .... June 4 & 5
- 11466DE Elect. & Mech. Est. .... June 4 & 5

PLYMOUTH, MA
Sheraton Regal Inn
180 Water Street
Plymouth, MA 02360
(617) 747-4900

- 10466PL Repair & Remodeling ...... June 16 & 17
- 10666PL Square Foot Costs .... June 16 & 17
- 10066PL Unit Price Est. .... June 18 & 19
- 10266PL Sched. & Proj. Mgmt. ..... June 18 & 19
- 11466PL Elect. & Mech. Est. ...... June 18 & 19

NEW YORK, NY
Westchester Marriott Hotel
670 White Plains Road
Tarrytown, NY 10591
(914) 631-2200

- 10466TT Repair & Remodeling ...... June 23 & 24
- 10666TT Square Foot Costs .... June 23 & 24
- 10066TT Unit Price Est. .... June 25 & 26
- 10266TT Sched. & Proj. Mgmt. ..... June 25 & 26
- 11466TT Elect. & Mech. Est. ...... June 25 & 26

FALL SEMINARS

CAPE COD, MA
Dunfey Hyannis
West End Circle
Hyannis, MA 02601
(617) 775-7775

- 10469HY Repair & Remodeling .... Sept. 8 & 9
- 10669HY Square Foot Costs .... Sept. 8 & 9
- 10069HY Unit Price Est. .... Sept. 10 & 11
- 10269HY Sched. & Proj. Mgmt. .... Sept. 10 & 11
- 11469HY Elect. & Mech. Est. .... Sept. 10 & 11

ST. LOUIS, MO
Holiday Inn of St. Peters/
St. Charles
4221 South Outer Road
P.O. Box 309
St. Peters, MO 63376
(314) 928-1500

- 10469SP Repair & Remodeling ...... Sept. 22 & 23
- 10669SP Square Foot Costs .... Sept. 22 & 23
- 10069SP Unit Price Est. .... Sept. 24 & 25
- 10269SP Sched. & Proj. Mgmt. ..... Sept. 24 & 25
- 11469SP Elect. & Mech. Est. ...... Sept. 24 & 25

PHILADELPHIA, PA
Philadelphia Marriott
City Line Ave. & Monument Rd.
Philadelphia, PA 19131
(215) 667-0200

- 10469PH Repair & Remodeling ...... Sept. 29 & 30
- 10669PH Square Foot Costs .... Sept. 29 & 30
- 10069PH Unit Price Est. .... Oct. 1 & 2
- 10269PH Sched. & Proj. Mgmt ...... Oct. 1 & 2
- 11469PH Elect. & Mech. Est. ...... Oct. 1 & 2

SAN FRANCISCO, CA
San Franciscan Hotel
1231 Market Street
San Francisco, CA 94103
(415) 626-8000

- 1046ASF Repair & Remodeling ...... Oct. 20 & 21
- 1066ASF Square Foot Costs .... Oct. 20 & 21
- 1006ASF Unit Price Est. .... Oct. 22 & 23
- 1026ASF Sched. & Proj. Mgmt. ..... Oct. 22 & 23
- 1146ASF Elect. & Mech. Est. ...... Oct. 22 & 23

CHICAGO, IL
Ramada O'Hare Hotel
6600 North Mannheim Road
Rosemont, IL 60018
(312) 827-5131

- 1046BRO Repair & Remodeling ..... Nov. 10 & 11
- 1066BRO Square Foot Costs .... Nov. 10 & 11
- 1006BRO Unit Price Est. .... Nov. 12 & 13
- 1026BRO Sched. & Proj. Mgmt ..... Nov. 12 & 13
- 1146BRO Elect. & Mech. Est. ..... Nov. 12 & 13

ATLANTA, GA
Ramada Inn Central
1630 Peachtree St. NW
Atlanta, GA 30367
(800) 241-5601

- 1046BAT Repair & Remodeling ..... Nov. 17 & 18
- 1066BAT Square Foot Costs .... Nov. 17 & 18
- 1006BAT Unit Price Est. .... Nov. 19 & 20
- 1026BAT Sched. & Proj. Mgmt. .... Nov. 19 & 20
- 1146BAT Elect. & Mech. Est. ..... Nov. 19 & 20

SAN ANTONIO, TX
St. Anthony Inter-Continental
300 East Travis
P.O. Box 2411
San Antonio, TX 78298
(512) 227-4392

- 1046CSA Repair & Remodeling .... Dec. 1 & 2
- 1066CSA Square Foot Costs .... Dec. 1 & 2
- 1006CSA Unit Price Est. .... Dec. 3 & 4
- 1026CSA Sched. & Proj. Mgmt. ...... Dec. 3 & 4
- 1146CSA Elect. & Mech. Est. .... Dec. 3 & 4

Registration Form

Please register the following people for the Means Construction Seminars as shown here. Full payment or deposit is enclosed, and we understand we must make our own hotel reservations if overnight stays are necessary.

- ☐ Full payment of $ _____ enclosed.
- ☐ Deposit of $ _____ enclosed.
  Balance due $ _____
  (U.S. Funds)

Firm Name _____
Address _____
City/State/Zip _____
Telephone Number _____

☐ Charge our registration(s) to:
___ American Express ___ Visa ___ MasterCard
Account No. _____ Exp. Date _____

CARDHOLDER'S SIGNATURE

NAME OF REGISTRANT(S) (To appear on certificate of completion)	SEMINAR NAME AND NUMBER	CITY	DATES

Please mail check to: **R.S. MEANS CO., INC., 150 Construction Plaza, P.O. Box 800, Kingston, MA 02364-0800**

GALAXY
Automated Quantity Takeoff and Pricing System

Modern technology can save you money by cutting takeoff time at least 50%

NEW features include paging (material checklist), earthwork calculator and perimeter/area/volume enhancements.

Quantity takeoffs don't have to be difficult - not when you can take advantage of modern technology. GALAXY Automated Takeoff and Pricing System lets you determine the detailed cost information you need with the aid of a computer.

You can reduce the time it takes to do an estimate by as much as 50 percent, and more for many projects, thanks to GALAXY's electronic digitizer table. Working directly from a set of plans, an estimator easily feeds the information into the computer, which then makes the proper calculations to determine length, height, width, volume, area or numeric count — all in a fraction of the time it normally would take.

Using a stylus (electronic pen), specific points are selected from the plans and "read" into the computer. The estimator chooses the items in convenient order and is able to quickly determine the quantities as desired. Designed for use with the ASTRO General Estimating Program, GALAXY provides you with access to your own files of known costs as well as any of the R.S. Means standard cost information files of up-to-date unit price information.

Best of all, GALAXY is easy to operate. Any construction professional can learn to run this system after attending the training workshop. It's all part of The Total Solution by R.S. Means Company, Inc., the people you depend on for your construction cost data.

GALAXY
Main Benefits
- Saves valuable time during takeoffs
- Easy to operate
- Increases productivity
- Reduces need for additional manpower
- Streamlines the estimating process
- Produces detailed estimates
- Makes changes easily
- Customizes estimates to specific requirements

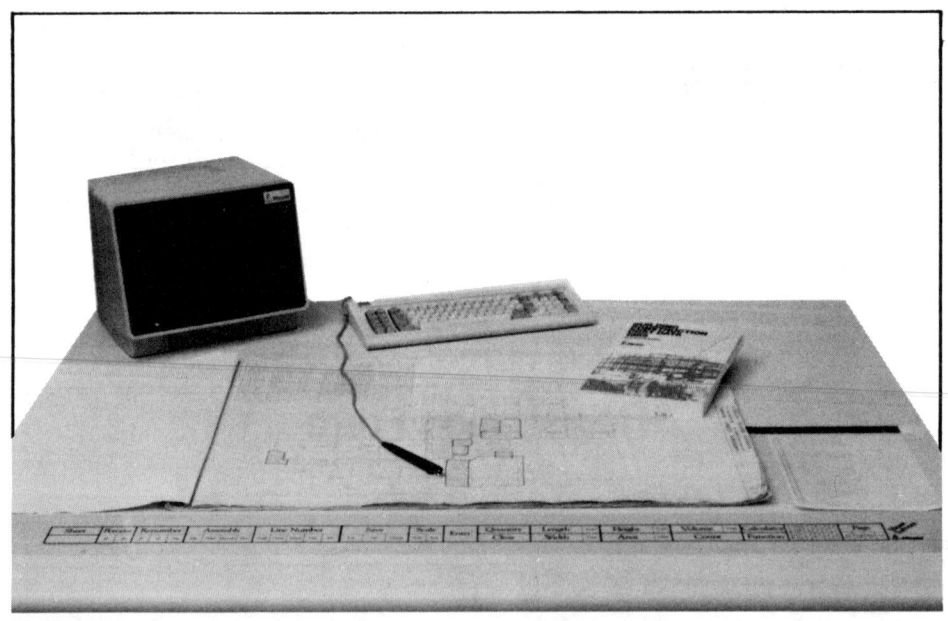

"Takeoffs" right from drawings.

Detail of Digitizer Menu

Clear	Keyboard	Renumber			Line No.			Save			Scale	
	Sheet	#1	#2	Tag	Cost	User	Job	Item	Sub	Change	Arch. ☐	Eng. ☐

Take Quantity	Length	☐	Height	☐	Volume	☐	Holding	Calculator
	Width	☐	Area	☐	Count		Function	

STAR
General Estimating Program

Helps develop fast and accurate estimates and manage vital cost data

Now you can use the speed and dependability of a computer to simplify your estimating process. STAR General Estimating Program is easy to use, reduces errors, provides more accurate estimates and enables you to increase productivity and improve profit margins.

Using known construction costs, which reflect your marketplace, STAR helps you to quickly develop your data file, which is organized in the UCI-16 division format. This file then gives you, the estimator, access to necessary information, such as item description, crew make-up, daily productivity, unit of measure and unit prices for material, labor, equipment and total.

In addition STAR can also access any of the Means standard cost files.

With STAR, you can store and update estimates with ease and make modifications with the touch of a button. Changes are no problem because the computer does all the work.

More importantly, STAR enables you to save time — and money! An estimator can now evaluate more jobs, which means more work, increased productivity and higher profits. It's all part of The Solution by R.S. Means Company, Inc., the people you can depend on for your construction cost data.

Main Benefits

- Develop your own cost file -quickly
- Provides fast, accurate estimates
- Easy to operate
- Provides up-to-date unit prices
- Saves hours of tedious calculating
- Two methods — daily output and man-hour analysis
- Helps in scheduling and planning
- Evaluated overhead and profit

VEGA
A scheduling service to help you plan projects

Scheduling in the construction industry can be a major headache, especially when you consider all the elements that must go into a project checklist. Resources, labor, material and equipment can all be a source of concern if the planning groundwork hasn't been properly laid.

The *VEGA Scheduling Program* addresses those concerns. It is designed to create a realistic schedule early in the conceptual stage of a project and later on during the bidding and negotiating phase. VEGA can provide you with a checklist of the different tasks involved in the project and can adjust the target date.

The input for the *VEGA Scheduling Program* is the quantity, crew, productivity and cost information from your ASTRO/GALAXY estimate.

The VEGA Program produces a bar chart on a printer (plotter is not needed) showing tasks in the i-j mode.

This useful program enables you to develop scheduling for your specific needs. You can determine duration times and cost data for each selected activity. The VEGA printout of the schedule includes: all activities, the critical path identified, the cash flow, and early start or late finish.

The *VEGA Scheduling Program* is an absolute must for the construction professional who has to stay on top of his scheduling situation. It's all part of the Total Solution by R.S. Means Company, Inc., the people you depend on for your construction cost data.

Main Benefits

- Creates schedules faster
- Uses data from the ASTRO/GALAXY program for input
- Tracks activities throughout the schedule
- Produces a cashflow chart
- Provides printed schedules
- Makes changes easily for adjustment to schedule

UNIT PRICE COST FILES

Specific construction cost files to help you create detailed estimates

Rapidly changing costs in the construction industry are no longer a problem thanks to a series of specific building information files from R.S. Means. The Unit Price Cost files provide you with up-to-date cost data.

These cost data files, which can be purchased together or separately, contain thousands of thoroughly-researched unit prices for labor, materials and installation. They can be used in conjunction with the STAR General Estimating Program and the GALAXY Automated Quantity Takeoff and Pricing System to determine construction costs in specific building applications. These important Unit Price Cost files are annually updated and are offered on a subscription basis.

UNIT PRICE COST FILES

- Building Construction
- Repair and Remodeling
- Concrete and Masonry
- Residential/Light Commercial
- Mechanical
- Electrical
- Site Work
- Interiors

Residential/Light Commercial Cost Data 1986

Square Foot • Systems Costs • Unit Costs

Here's complete cost estimating assistance for contractors, builders, and designers in the light construction market

$41.95 per copy — 5th annual edition, over 425 pages, illustrated

The *Means Residential/Light Commercial Cost Data* manual is exclusively designed for the residential/light commercial construction professional needing the ultimate in cost estimating reliability. This tool makes all similar references obsolete! It's simple and fast . . . It compares costs . . . points out money-saving opportunities . . . helps you stay within budgets.

For rapid "ball park" estimates, the guide includes a unique square foot cost section with sample residential and commercial buildings. This novel price format enables you to quickly move through excavation, foundation, framing . . . right on to the complete square foot estimate more efficiently and accurately than ever before.

You'll also find a comprehensive 100-page unit cost section in the Uniform Construction Index arrangement for estimates requiring the unit cost approach. This section provides you with the in-depth backup necessary for detailed component cost estimates.

Examples, explanations and work sheets will guide you through the estimating procedure, making sure you've included every cost component. With *Residential/Light Commercial Cost Data 1986*, you'll have at hand the invaluable extra "know-how" for making the most out of your residential *and* light commercial estimating opportunities!

ISBN: 0-87629-007-1   Book No. 60086 . . . . $41.95/copy

Concrete & Masonry Cost Data 1986

The first all-inclusive estimating manual devoted to the broad range of today's advanced concrete and masonry methods and materials

$40.95 per copy — 4th annual edition, over 350 pages, illustrated

Here's your all-inclusive price guide for estimating concrete and masonry construction.

Up-to-the minute costs for every application of concrete masonry construction — from intricate formwork to custom brickwork — are provided in exhaustive detail. The popular Means unit and systems direct you to the right costs instantly.

The mammoth unit cost section provides detailed crew, man-hour, materials, labor cost, equipment, and profit data for thousands of components.

The easy-to-use systems costs actually "picture" each system with three-dimensional illustrations. The systems cost groupings are quickly "adjustable" by using alternate price selections on the same, or adjoining pages.

Engineering data, city cost adjustment factors, and other data give you full design references, together with the ability to compensate for any cost variable.

ISBN: 0-87629-009-8   Book No. 60116 . . . . $40.95/copy

Means Assemblies Cost Data 1986

(Formerly *Means Systems Costs*)

The Means illustrated assemblies cost manual . . . preferred by more construction professionals for its superb organization and price reliability

This indispensable tool provides you with thousands of detailed assemblies costs, accompanied by hundreds of drawings, explanations, component breakdowns and tables — all within *one* easy-to-use format.

You'll find separate design aids to assist you in the conceptual phase of estimating any job. Tables of unit prices allow you to cost-out virtually any combination of building systems in logical, step-by-step sequence, customizing your project as more information becomes available.

Includes the UNIFORMAT numbering system as a guideline for consistent estimating and reporting. It enables you to price and compare various framing and envelope assemblies with ease.

Use the manual's complete examples, suggestions, and standardized format to compare past and future jobs and build your own cost data file.

$44.95 per copy — 11th annual edition, over 525 pages, illustrated

Book No. 60066 . . . . $44.95/copy   ISBN 0-87629-005-5

Bidding for the General Contractor

By Paul J. Cook

The techniques of successful bidding and how to apply them in your own construction business . . .

Here for the first time is a concise guide for making competitive bids in the one to twenty million dollar project range. It sheds new light on bidding procedures and techniques . . . gives you a way to see and compare your approaches with those of other successful bidders. You'll have in-depth discussion and illustrations covering every step of the bid management process. It's a book designed for quick reference, you'll use it over and over again.

$37.95 per copy — 1st edition, over 225 pages, with graphics

Book No. 67180 . . . . $37.95/copy   ISBN 0-911950-77-X

Square Foot Estimating

by F. William Horsley
and Billy Joe Cox

A new generation of techniques for conceptual and design-stage cost planning.

Square Foot Estimating gives you the expertise to transform conceptual and design-stage estimating into a powerful tool for new business.

You'll see how to maximize the use of your time and effort to achieve new heights of creativity in designs, materials, and construction approaches . . . all within given budget limitations!

This fast-moving new book begins with a clear look at four estimating techniques used in design-stage costing. Each is evaluated by degree of accuracy in relation to time and cost.

It spotlights the use of daring new departures from traditional thinking about square foot costs. It shows how one method — systems estimating — achieves remarkable reliability, typically in a day's work.

The editors demonstrate how building systems can be tried in an infinite number of combinations to quickly determine the direction to take for budgets, codes, loads, insulation, fire-proofing, acoustics, energy use, and special requirements.

First Edition — over 300 pages, illustrated

Hardcover Book No. 67145 .... $36.95/copy ISBN 0-911950-61-3
Softcover Book No. 67146 .... $31.95/copy ISBN 0-911950-62-1

Estimating for the General Contractor

by Paul J. Cook

The first book prepared specifically to help contractors and estimators in medium-sized companies estimate more efficiently and accurately

Light on theory, heavy on practical estimating methods and ideas, here's powerful help for the contractor/estimator who wants to evaluate and polish every facet of their estimating procedure.

No matter how much estimating experience you've had, no matter what your particular construction specialty, there's something of importance for you in this extraordinary volume.

Prepared by one of the industry's most successful estimating consultants, this guide is designed to give your estimating a refreshing breath of new ideas . . . insights . . . better techniques. It's a book you'll use over and over again, providing the "extra" dimension you need to make your estimates more reliable and usable than ever before.

First Edition — over 225 pages, illustrated

Hardcover Book No. 67160 .... $35.95/copy ISBN 0-911950-48-6
Softcover Book No. 67170 .... $30.95/copy ISBN 0-911950-49-4

Means Site Work Cost Data 1986

You don't need to waste time hunting for site work cost facts — they're all here...

- complete unit and systems costs
- crew and man-hour data
- adjustable factors for regional prices
- expert guidance

$41.95 per copy — 5th annual edition, over 400 pages, illustrated

Completely updated for 1986!

Means Site Work Cost Data 1986 provides timely assistance for the vast majority of site work estimating problems you'll encounter.

Its objective is to give you a full range of unit costs, illustrated systems prices, and the extensive backup and adjustment information you need for the **unusual as well as routine** estimate.

It helps you compensate for the fact that you don't have the time to develop or have convenient access to price data for handling modern site work challenges. Challenges like site explorations, roads, utilities, bridges, landscaping and the dozens of other specialized costs associated with site work.

Full-scope unit costs

Comprehensive site work unit prices for all major site work classifications showing:

crew — makeup of the typical crew needed to perform the function

units — the measure on which productivity, material, and labor are based

daily output — the number of units completed in a full working day

man-hours — the number of hours needed to complete a single unit of measure

material — the unit cost of the material or product used

labor — the unit cost of the crew specified

equipment — the unit cost for the equipment specified

total costs — the sum of the previous columns with — and without — overhead and profit added in

For fast, but accurate estimates, use site work systems costs

Complete systems costs for over 50 site work categories enable you to cut estimating time, yet retain accuracy.

cost per unit in which categories are most commonly expressed — for both materials and labor/equipment needed to install

quantity — measurement of the components comprising each system

description — variations within a particular division by size, number, or material

illustration — cut-away drawings of system components to aid in identification with helpful commentary

Extensive crew tables for labor costs

To enable you to analyze the work crews specified in the unit cost section, Standard Crew Tables are provided. They include base hourly and daily costs for each grouping of workers and associated equipment with subcontractors overhead and profit information.

Adjustable for regional price differences

These pages provide City Cost Indexes for 162 major U.S. and Canadian geographical areas. The cost adjustment factors enable you to allow for materials and installation costs which are higher or lower than the national average. They can be applied to both unit and systems site work costs.

Expert guidance and assistance

Where necessary or helpful, all unit price data is referenced by the "circle numbers" which provide detailed notes further explaining and pointing out potential pitfalls in making your estimates.

Book No. 60076 . . . . $41.95/copy

ISBN 0-87629-006-3

Means Graphic Construction Standards

... the way to vastly expand your construction planning expertise

Means Graphic Construction Standards bridges the gap between rough construction concepts and actual construction methods.

With illustrations of unit assemblies, systems and components, you can see quickly which construction methods work best to meet design, budget and time objectives.

You benefit from incisive discussion of each assembly... features, uses, advantages, limitations.

Accompanying charts provide man-hour per unit in place by component and entire assemblies. The man-hour data enables you to compare assemblies by the time and cost factors in minutes.

Each graphic is labeled and sized with unit and component terms to ensure you're completely familiar with what each segment of the assembly is called.

Create construction concepts/designs using new insight and approaches with this expansive reference.

$87.95 per copy
First Edition
1,000 illustrations
500 pages, hardcover

(Available March 1986)

ISBN 0-911950-79-6

Book No. 67210 . . . . $87.95/copy

Means Man-Hour Standards

The "professional's choice" for uncompromised trade and labor productivity data

Means Man-Hour Standards is the working encyclopedia of the building industry. This massive achievement is by far the most comprehensive and easy-to-use reference of its type available.

Arranged in the U.C.I. format, every division contains the labor requirements for specific construction tasks. The "quick-find" indexes guide you directly to each division.

You'll find every bit of labor information you may need, and because it's Means, you know it's reliable.

With *Means Man-Hour Standards*, you simply turn to the page containing the construction activity in which you're interested, and there in one place is the data you need. No shuffling back and forth. Never before has obtaining and using man-hour information been so complete and easy to do.

$89.95 per copy
First Edition
over 575 pages,
hardcover

ISBN 0-911950-60-5

Book No. 67140 . . . . $89.95/copy

Here's the new guide that shows you — start to finish — how to make expert commercial rehab estimates.

Estimating and Analysis for Commercial Renovation

By Edward B. Wetherill

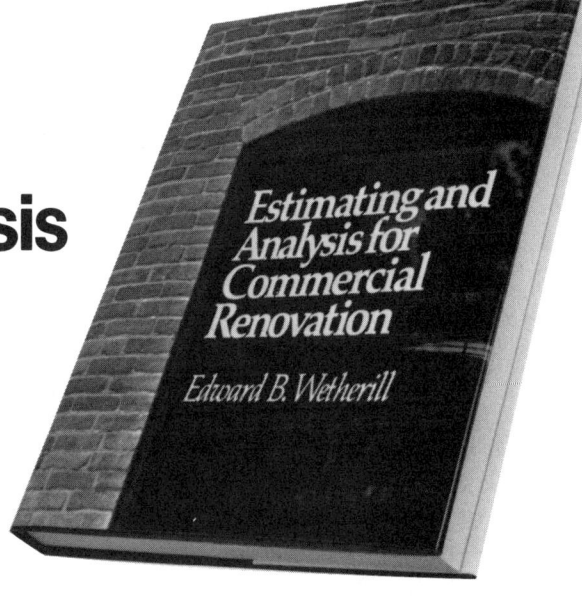

1st Edition Over 300 pages illustrated, hardcover

Why this book will be so important to you

Today, one of the biggest challenges contractors and estimators undertake is commercial renovation.

Even the best set of plans cannot forecast the complications of selective demolition, sizing, fitting, matching, and rebuilding. Every part of the renovation process poses its own set of problems to be solved.

Estimating and Analysis for Commercial Renovation prepares the estimator for these special challenges. Thanks to the author's extensive experience, every part of estimating renovation is explored in complete detail.

This is accomplished by focusing on four key areas: choosing the right estimating approach, site visitation and evaluation, quantity take-off and pricing, and practice through examples.

With time-tested pointers, step-by-step procedures, and real-life examples, it provides the help you need to make the most out of your rehab estimating opportunities . . . whether you're just starting out or trying to build upon previous success.

Leading-edge techniques solve your toughest remodeling estimating problems

This fact-filled guide is literally packed with fresh, practical know-how for your toughest rehab estimating . . .

- detailed methods for inspection and evaluation of existing structures and site locations
- how to choose the right estimating technique for the job at hand
- invaluable estimating "do's and don'ts" you won't find anywhere else
- dozens of helpful estimating aids . . . forms . . . charts . . . pricing sheets . . . checklists
- complete sample estimates that demonstrate the author's ideas in real situations

Although designed primarily for contractors and architects with experience, the concepts in *Estimating and Analysis for Commercial Renovation* are useful to anyone who wants to enhance their understanding of renovation estimating.

Book No. 67200 . . . . $39.95/copy

ISBN 0-911950-81-8

Means Mechanical Cost Data 1986

$41.95 per copy — 9th annual edition, over 425 pages, illustrated

Your all-inclusive source of mechanical unit and systems costs . . . featuring many new line items, more labor data, more expert help for your estimating.

What's new for 1986

Separated from our electrical data, *Means Mechanical Cost Data 1986* offers you the benefits of greatly-expanded unit and systems costs and estimating guidance . . . devoted *exclusively* to plumbing, HVAC, fire protection, and all related preparation and finish construction.

This new manual will be even more valuable and important to your work than ever before, providing you with *more* unit cost line items, *more* systems prices, *more* illustrations of parts and materials, and *more* general help covering every aspect of your cost needs.

Whether you need this cost assistance for complete competitive bids, or for help in specific areas of mechanical construction, you'll find all the cost facts you want quickly.

So many ways to benefit from this storehouse of mechanical cost information

- up-to-date unit cost entries, more illustrations of parts and materials for easy selection
- new "systems" cost groupings for faster, more reliable estimates
- new man-hour labor and productivity data for better comparisons
- more engineering sizing guides, notes, specifications
- updated mechanical square foot costs for 100 typical buildings and installations
- newest products, materials and installation methods included
- extensive energy use and conservation data, heat reclaiming equipment, retrofit and replacement materials cost information
- fixtures for the handicapped
- regional cost adjustment factors for labor and materials
- helpful diagrams, notes, explanations

Book No. 60026 . . . . $41.95/copy

ISBN 0-87629-001-2

Means Electrical Cost Data 1986

$41.95 per copy — 9th annual edition over 350 pages, illustrated

Completely updated with 100% focus on electrical estimating cost solutions...

latest electrical unit and systems costs

quick-find illustrations and indexing

time-saving design and specification guidance

With today's unpredictable costs for electrical construction, you need a fast, convenient way to keep up with it all.

That's what *Means Electrical Cost Data 1986* does for you. It gives you nearly 18,000 unit and systems costs covering every kind of electrical — or related — construction question you're likely to have.

For 1986 there are completely updated price entries, new estimating models, and a man-hour format which helps you in job scheduling as well as estimating.

For every phase of the electrical estimating process, you'll have sharp, to-the-point answers. You'll have clear specifications, accurate materials and labor costs, and complete guidance in the dozens of notes, formulas and explanations.

For you, the busy professional, it's the fast, efficient way to estimate electrical work.

In fact, it will more than pay for itself the first time you use it for an electrical estimate.

- provides you with more electrical estimating data for better application to your work
- 10,000 fully current unit costs with additional three dimensional drawings of parts for faster identification
- man hour format allows you to more easily interface with material deliveries and project work scheduling
- detailed estimating procedural models, many with complete diagrams, give you basic electrical frameworks to adapt to your projects
- design reference tables covering every possible code, requirement, or fact you might need
- includes demolition and remodeling, square foot costs, geographical materials and labor cost factors, crew data, and electrical "systems" costs

Book No. 60036 . . . . $41.95/copy   **ISBN: 0-87629-002-0**

Repair & Remodeling Cost Data 1986

Accuracy, efficiency, and thoroughness for estimating rebuilding jobs

$42.95 per copy — 7th annual edition, over 450 pages, illustrated

For contractors, subcontractors, builders, estimators, engineers, architects, governmental and insurance agencies, corporate facilities managers, firms of all sizes . . .
Reliable unit and systems cost guidance for repair and remodeling estimating problems

Everything you need to make better rebuilding estimates

- comprehensive unit costs under UCI groupings with crew, man-hours, unit, bare materials, labor, equipment and totals with adjustable overhead and profit columns
- illustrated systems prices for most widely-used remodeling methods with two additional sections for popular alternative systems and custom installations . . . presented in UNIFORMAT numbering sytem
- comprehensive labor cost information . . . union rates, crew output, productivity, every needed variation to calculate this vital cost area
- city cost modifiers for 162 U.S. and Canadian metro regions to factor in costs for your locality
- costs for the latest materials, methods, and equipment used in renovation work
- detailed diagrams, illustrations, explanations for ease of understanding and use . . . even a visual table of contents
- adjustable costs for contractor advantages such as owned equipment, low-cost materials purchases, low overhead, minimum daily charges, etc.

The *Means Repair & Remodeling Cost Data* book is the undisputed choice of anyone who must make accurate estimates for renovation projects. It's the most versatile, easy-to-use preliminary cost and bidding reference available from any source.

With over 450 pages containing 10,000 unit and building systems prices, this exciting guidebook handles almost any repair and remodeling challenge. With it you can accurately assemble rapid early-stage estimates as well as comprehensive competitive bids.

Most important of all, the Means repair and remodeling manual is designed to admirably fulfill varied user needs — whether a small-scale builder seeking to protect profits on remodeling work — right up to the most discriminating user needing prices for custom installations.

It's easy to use every section, provides thorough, easy-to-understand explanations to achieve the desired level of precision. Its superb layout of helpful diagrams, notes, and tips allows you to put the information to work for you quickly.

ISBN 0-87629-003-9   **Book No. 60046 . . . . $42.95/copy**

Means Square Foot Costs 1986

$42.95 per copy — 7th annual edition, over 425 pages, photographs, illustrations

Residential • Commercial
Industrial • Institutional

Here's what you'll find packed into the 1986 edition:

- over 6,000 costs for typical buildings and attachments . . . residential, commercial, industrial, institutional
- reliable costs for use in 1986
- four classes of residential quality: economy, average, custom, luxury
- over 350 illustrations of unit-in-place components and costs, fully detailed for systems style estimates
- worked-out examples and clear instructions
- cost adjustment factors for all construction-related variables
- over 6,500 unit-in-place costs illustrated for detailed budgets
- factors for local cost differences by Zip Code
- depreciation guidelines
- samples of convenient work sheets for preparing the estimate
- nearly 150 photographs of typical building types and qualities

Designed for effective use at all levels of building cost budgeting, estimating and valuation . . . the planning guide and replacement cost resource for architects, design/build contractors, real estate developers, corporate planners, appraisers and assessors.

Useful for anyone who needs rapid budget cost estimates in the office, with a client, or in the field, this manual provides clear descriptions, photographs, and illustrations of hundreds of residential, commercial, industrial, and institutional buildings. Corresponding costs by square foot are conveniently located on the same page or adjoining pages, minimizing effort for "on-the-spot" estimates.

Convenient tables provide calculation directions and costs for attached structures and special kinds of additions or adjustments.

For those preparing more detailed, itemized estimates, costs are broken down into "systems style" component specifications and unit costs. This fully illustrated unit-in-place section contains over 6,500 costs simplifying construction component identification and pricing.

You can use this new manual to achieve any degree of cost precision you might need.

Book No. 60056 . . . . $42.95

ISBN 0-87629-004-7

NEW!

1st Edition | over 275 pages illustrated

Means Home Improvement Cost Guide

How to plan and price out home improvements quickly, easily

Here for the first time is a reference which offers frank discussion and cost facts for homeowners planning improvements to their house.

Prepared by the nation's foremost construction cost consulting firm, it's a guide which delves deeply into dozens of typical home projects explaining what you should know before you attempt the work yourself or hire a contractor.

It will help you budget better, understand the work better, develop the most desirable plans to achieve your goals.

Book No. 67220 . . . . $24.95/copy   **ISBN 0-911950-80-X**

3rd Annual Edition | over 400 pages illustrated

Commercial and Industrial Construction

Partitions — Ceilings — Finishes — Floors — Furnishings

Interior Cost Data 1986

. . . the unit and systems cost tool every designer and estimator has been waiting for

Here are the up-to-date cost facts for solving the complex estimating problems of modern interior construction.

Due to constantly changing styles and new design directions, thousands of new materials and hardware items for interior construction are introduced each year. Few designers and estimators can keep up with this avalanche of price options for finish work.

But now, you can keep up with — and stay ahead of — the infinite varieties of specialized materials and installation expenses involved in commercial and industrial interior construction . . . thanks to *Means Interior Cost Data 1986.*

Book No. 60096 . . . . $40.95/copy   **ISBN 0-87629-008-X**

ANNOUNCING

Means Illustrated Construction Dictionary

Never be in doubt about a construction term or concept

Here's the on-the-job reference work even the most experienced professionals can turn to for immediate answers about construction terms.

It's the dictionary prepared especially for front-line professionals like you, who can't afford time-wasting hunts for precise definitions to construction terminology.

Whatever your specialty — architect, contractor, engineer, estimator — you'll welcome this practical construction dictionary written in everyday language.

Its no nonsense guidance, illustrations, abbreviations and easy-to-use format will serve you for years to come.

The Means Dictionary is packed with thousands of up-to-date explanations of construction terminology. It covers every area of construction from design right on through everyday "buzz words" used by tradesmen. It's "alive" with entries covering every conceivable new technique, product.

It includes illustrations of design and material use, definitions of abbreviations . . . even provides help on comparative costs of various construction techniques.

1st Edition — over 500 pages, hardcover

Construction answer book — *The Means Illustrated Construction Dictionary*

- more than 12,000 definitions of terms and abbreviations commonly used in construction work, thousands entirely unique to the industry
- covers every area of construction . . . architecture, design, materials, contracting, specifications, testing, trades, new products, techniques . . . even slang and lingo
- written in simple, everyday American language and filled with illustrations
- includes contemporary terms . . . computers, electronics, CPM, HVAC, you name it
- handy size, readable typeface, durable hard-cover construction

Book No. 67190 . . . . $59.95/copy

ISBN 0-911950-82-6

Open Shop Building Construction Cost Data 1986

— for open shop labor estimating

$44.95 per copy — 2nd annual edition, over 450 pages

You depend on BCCD for union labor... now you can have the same comprehensive prices for open shop work

One flip through your copy of *Open Shop Building Construction Cost Data 1986* will convince you of its value to your estimating process.

Here at last are over 20,000 reliable unit cost entries based on open shop skilled and trade labor costs. No longer will you need to use time-consuming, inefficient ways to get these prices. The Open Shop BCCD gives you every price you need in the familiar R.S. Means layout.

The labor cost data is broken down into man-hours, crew, bare labor, equipment, and overhead and profit.

You can break out labor cost facts in any way you need for substitutions, comparisons, and adjustments.

Where do the prices come from?

Prices quoted in *Open Shop Building Construction Cost Data 1986* are developed from averages of the 30 largest urban areas of the U.S. and Canada.

All price data is carefully prepared from the *actual experience* of hundreds of contractors and suppliers in the last 12 month period.

The labor cost data is based on an average for the U.S. and Canada and applies to skilled workers and those licensed in installation trades.

The City Cost Index provides adjustment factors for materials and labor price differences from region to region. 162 cities are included in the Index. Other adjustment data is furnished where desirable.

Powerful help for open shop estimating

- eliminates time-consuming hunts for open shop labor cost information
- makes it easier for you to bid on open shop projects
- allows you to make dependable comparisons of open shop prices
- enables you to substitute costs, make adjustments faster
- helps you prepare realistic work schedules
- gives you access to national average costs you wouldn't have otherwise
- covers every possible open shop unit cost fact

Partial Contents

General requirements	Specialties
Site work	Equipment
Concrete	Furnishings
Masonry	Special construction
Metals	Conveying Systems
Wood and plastics	Mechanical and electrical
Moisture protection	Square and cubic foot
Doors, windows, glass	Repair and remodeling
Finishes	City cost indexes

Book No. 60156 .... $44.95/copy

ISBN 0-87629-026-8

MEANS PROJECT COST REPORT

Discount Products Available　　　　For U.S. Customers Only　　　　Strictly confidential

Receive $16.00 discount per product for each report you submit.

By filling out and returning the Project Description, you can receive a discount of $16.00 off any one of the Means products advertised in the preceding pages. The cost information required includes all items marked (*) except those where no costs occurred. The sum of all major items should equal the Total Project Cost.

PROJECT DESCRIPTION

No remodeling projects, please

Type building _____　　Owner _____
Location _____　　Architect _____
Capacity _____　　General Contractor _____
Frame _____　　Bid Date _____
Exterior _____　　Typical Bay Size _____
Basement: full ☐　part ☐　none ☐　crawl ☐　　Labor Force: ____ % Union ____ % non-union
Height in Stories _____
Total Floor Area _____　　Project Description (Circle one number in each column)
Ground Floor Area _____　　1. Economy　　　　　1. Square
Volume in C.F. _____　　2. Average　　　　　2. Rectangular
% Air Conditioned _____ Tons _____　　3. Good　　　　　　3. Irregular
Comments:　　　　　　　　　　　　　　　　　4. Luxury　　　　　　4. Very irregular

	TOTAL PROJECT COST	$				$
A	★ **GENERAL CONDITIONS**	$	**J**	★ **FINISHES**		$
B	★ **SITE WORK**	$	JL	Lath & Plaster (S.Y.)		
BS	Site Clearing & Improvement	$	JD	Drywall (S.F.)		
BE	Excavation		JM	Tile & Marble (S.F.)		
BF	Caissons & Piling		JT	Terrazzo (S.F.)		
BU	Site Utilities		JA	Acoustical Treatment (S.F.)		
BP	Roads & Walks (Exterior Paving)		JF	Flooring (S.F.)		
			JP	Painting & Wall Covering (S.F.)		
C	★ **CONCRETE**	$	**K**	★ **SPECIALTIES**		$
C	Cast in Place (C.Y.)		KF	Partitions		
CP	Precast		KL	Lockers		
D	★ **MASONRY**	$	**L**	★ **EQUIPMENT**		$
DB	Brick		LK	Kitchen		
DC	Block		LS	School		
DT	Tile		**M**	★ **FURNISHINGS**		$
DS	Stone		MC	Carpet		
			MS	Seating		
E	★ **METALS**	$	**N**	★ **SPECIAL CONSTRUCTION**		$
ES	Structural Steel (tons)		NA	Acoustical		
EM	Misc. & Ornamental Metals		NB	Prefab. Bldgs.		
F	★ **WOOD & PLASTICS**	$	**P**	★ **CONVEYING SYSTEMS**		$
FR	Rough Carpentry (MFBM)		PE	Elevators		
FF	Finish Carpentry		PS	Escalators		
G	★ **MOISTURE PROTECTION**	$	**Q**	★ **MECHANICAL**		$
GW	Waterproofing-Dampproofing		QP	Plumbing (fixtures)		
GN	Insulation		QS	Fire Extinguishing Systems		
GR	Roofing & Metal Work		QB	Heating, Ventilating & A.C.		
H	★ **DOORS, WINDOWS & GLASS**	$	**R**	★ **ELECTRICAL**		$
HD	Doors (Ea.)		RL	Lighting		
HW	Windows (S.F.)		RP	Power		
HH	Finish Hardware		RA	Alarms		
HG	Glass & Glazing (S.F.)		RG	Special Systems		
			S	MECHANICAL/ ELECTRICAL COMBINED		$

Please specify the Means product you wish to receive. Complete the address information as requested. Return this form with your check (product cost less $16.00) to:

R. S. Means Company, Inc.
Square Foot Costs Department
100 Construction Plaza
P.O. Box 800
Kingston, MA 02364

(Mass. residents please add 5% sales tax)

Product Name _____
Product No. _____
Your Name _____
Title _____
Company _____
☐ Company Street Address _____
☐ Home
City, State, Zip _____

R.S. Means Co., Inc. — 1987 ORDER FORM

150 Construction Plaza
P.O. Box 800 Kingston, MA 02364-0800
(617) 747-1270

Enclosed is our check for the following publications:

QTY.	BOOK NO.	TITLE	UNIT PRICE	TOTAL
	60017	Building Construction Cost Data 1987	$ 44.95	$
	60027	Means Mechanical Cost Data 1987	46.95	
	60037	Means Electrical Cost Data 1987	46.95	
	60047	Repair & Remodeling Cost Data 1987	47.95	
	60057	Means Square Foot Costs 1987	45.95	
	60067	Means Assemblies Cost Data 1987	49.95	
	60077	Means Site Work Cost Data 1987	46.95	
	60087	Residential/Light Commercial Cost Data 1987	46.95	
	60097	Interior Cost Data 1987	45.95	
	60117	Concrete & Masonry Cost Data 1987	45.95	
	60127	Labor Rates for the Construction Industry 1987	52.95	
	60137	Means Historical Cost Indexes 1987	26.95	
	60147	Means Construction Cost Indexes *(Yearly Subscription)*	59.95	
	67140	Means Man-Hour Standards *(Hardcover only)*	89.95	
	67152	Means Scheduling Manual *(Softcover only)*	32.95	
	67145	Square Foot Estimating *(Hardcover)*	36.95	
	67146	Square Foot Estimating *(Softcover)*	31.95	
	67160	Estimating for the General Contractor *(Hardcover)*	35.95	
	67170	Estimating for the General Contractor *(Softcover)*	30.95	
	60157	Open Shop Building Construction Cost Data 1987	48.95	
	67180	Bidding for the General Contractor	37.95	
	67210	Means Graphic Construction Standards	87.95	
	67200	Estimating & Analysis for Commercial Renovation	39.95	
	67190	Means Illustrated Construction Dictionary	59.95	
	67220	Means Home Improvement Cost Guide	24.95	
			TOTAL	$

Note: Prices subject to change without notice. Mass. residents add 5% state sales tax.
Postage and Handling extra when billed.

Canadian customers ONLY — Please order books from:
Southam Communications, Ltd.
1450 Don Mills Road
Don Mills, Ontario, Canada, M3B-2X7
(Canadian Prices are higher)

SEND MY ORDER TO:
(Please print)
Name _____
Company _____
☐ Company
☐ Home Address _____
City/State/Zip _____
(Please Indicate If Company ☐ or Home ☐ Address)

R.S. Means Co., Inc. — 1986 ORDER FORM

150 Construction Plaza
P.O. Box 800 Kingston, MA 02364-0800
(617) 747-1270

Enclosed is our check for the following publications:

QTY.	BOOK NO.	TITLE	UNIT PRICE	TOTAL
	60016	Building Construction Cost Data 1986	$ 39.95	$
	60026	Means Mechanical Cost Data 1986	41.95	
	60036	Means Electrical Cost Data 1986	41.95	
	60046	Repair & Remodeling Cost Data 1986	42.95	
	60056	Means Square Foot Costs 1986	42.95	
	60066	Means Assemblies Cost Data 1986	44.95	
	60076	Means Site Work Cost Data 1986	41.95	
	60086	Residential/Light Commercial Cost Data 1986	41.95	
	60096	Interior Cost Data 1986	40.95	
	60116	Concrete & Masonry Cost Data 1986	40.95	
	60126	Labor Rates for the Construction Industry 1986	45.95	
	60136	Means Historical Cost Indexes 1986	16.95	
	60146	Means Construction Cost Indexes *(Yearly Subscription)*	49.95	
	67140	Means Man-Hour Standards *(Hardcover only)*	89.95	
	67152	Means Scheduling Manual *(Softcover only)*	32.95	
	67145	Square Foot Estimating *(Hardcover)*	36.95	
	67146	Square Foot Estimating *(Softcover)*	31.95	
	67160	Estimating for the General Contractor *(Hardcover)*	35.95	
	67170	Estimating for the General Contractor *(Softcover)*	30.95	
	60156	Open Shop Building Construction Cost Data 1986	44.95	
	67180	Bidding for the General Contractor	37.95	
	67210	Means Graphic Construction Standards	87.95	
	67200	Estimating & Analysis for Commercial Renovation	39.95	
	67190	Means Illustrated Construction Dictionary	59.95	
	67220	Means Home Improvement Cost Guide	24.95	
			TOTAL	$

Note: Prices subject to change without notice. Mass. residents add 5% state sales tax.
Postage and Handling extra when billed.

Canadian customers ONLY — Please order books from:
Southam Communications, Ltd.
1450 Don Mills Road
Don Mills, Ontario, Canada, M3B-2X7
(Canadian Prices are higher)

SEND MY ORDER TO:
(Please print)
Name _____
Company _____
☐ Company
☐ Home Address _____
City/State/Zip _____
(Please Indicate If Company ☐ or Home ☐ Address)